Lecture Notes in Artificial Intelligence 5781

Edited by R. Goebel, J. Siekmann, and W. Wahlster

Subseries of Lecture Notes in Computer Science

Wray Buntine Marko Grobelnik
Dunja Mladenić John Shawe-Taylor (Eds.)

Machine Learning and Knowledge Discovery in Databases

European Conference, ECML PKDD 2009
Bled, Slovenia, September 7-11, 2009
Proceedings, Part I

Springer

Series Editors

Randy Goebel, University of Alberta, Edmonton, Canada
Jörg Siekmann, University of Saarland, Saarbrücken, Germany
Wolfgang Wahlster, DFKI and University of Saarland, Saarbrücken, Germany

Volume Editors

Wray Buntine
NICTA
Locked Bag 8001, Canberra, 2601, Australia
and
Helsinki Institute of IT
Finland
E-mail: wray.buntine@nicta.com.au

Marko Grobelnik
Dunja Mladenić
Jožef Stefan Institute
Jamova 39, 1000 Ljubljana, Slovenia
E-mail: {marko.grobelnik,dunja.mladenic}@ijs.si

John Shawe-Taylor
University College London
Gower St., London, WC1E 6BT, UK
E-mail: jst@cs.ucl.ac.uk

Library of Congress Control Number: 2009933615

CR Subject Classification (1998): I.2, H.2.8, H.3, G.3, H.5, G.2, I.7

LNCS Sublibrary: SL 7 – Artificial Intelligence

ISSN	0302-9743
ISBN-10	3-642-04179-5 Springer Berlin Heidelberg New York
ISBN-13	978-3-642-04179-2 Springer Berlin Heidelberg New York

springer.com

Printed in Germany

Typesetting: Camera-ready by author, data conversion by Scientific Publishing Services, Chennai, India
Printed on acid-free paper   SPIN: 12753031   06/3180   5 4 3 2 1 0

Preface

The year 2008 was the first year that the previously separate European Conferences on Machine Learning (ECML) and the Conference on Principles and Practice of Knowledge Discovery in Databases (PKDD) were merged into a unified event. This is a natural evolution after eight consecutive years of their being collocated after the first joint conference in Freiburg in 2001. The European Conference on Machine Learning (ECML) traces its origins to 1986, when the first European Working Session on Learning was held in Orsay, France followed by the second European Working Session on Learning held in Bled, the location of this year's ECML PKDD 2009 conference. The European Conference on Principles and Practice of Knowledge Discovery in Databases (PKDD) was first held in 1997 in Trondheim, Norway. Over the years, the ECML/PKDD series has evolved into one of the largest and most selective international conferences in machine learning and data mining, the only one that provides a common forum for the two closely related fields. In 2009, ECML PKDD conference was held during September 7–11 in Bled, Slovenia.

The conference used a hierarchical reviewing process. We nominated 26 Area Chairs, each of them responsible for one sub-field or several closely related research topics. Suitable areas were selected on the basis of the submission statistics for ECML PKDD 2008 and from last year's International Conference on Machine Learning (ICML 2008) and International Conference on Knowledge Discovery and Data Mining (KDD 2008) to ensure a proper load balance among the Area Chairs. A joint Program Committee (PC) was nominated consisting of some 300 renowned researchers, mostly proposed by the Area Chairs. In order to make best use of the reviewing capabilities we initially only requested that two reviews be sought. However, in the event of an inconsistency between the two assessments a third review was requested. Papers receiving two very positive reviews were considered for inclusion in the two special issues of *Machine Learning* and *Data Mining and Knowledge Discovery* appearing in time for the conference. A further review was also sought for these papers in order to assess their suitability to appear in journal form. Aleksander Kolcz was the Best Papers Chair responsible for overseeing the selection of papers for these special issues.

ECML PKDD 2009 received 679 abstract submissions resulting in a final total of 422 papers that were submitted and not withdrawn during the reviewing process. Based on the reviews, and on discussions among the reviewers, the Area Chairs provided a recommendation for each paper with a ranking of the borderline papers. The three Program Chairs made the final program decisions after merging the opinions of the 26 Area Chairs.

All accepted papers were of equal status with an oral presentation, poster presentation and 16 pages in the proceedings, with the exception of those accepted for the special issues of journals that were only allocated a single page abstract in the proceedings. We have selected a total of 106 papers of which 14 were be equally divided between the two special issues. The acceptance rate for all papers is therefore 25%, in line with the high-quality standards of the conference series. It is inevitable with such a low acceptance rate that some good papers were rejected and we hope that authors of these papers were not discouraged by their disappointment. We are, however, confident that the accepted papers are of a high quality, making a very exciting and stimulating conference. In addition to research papers, 15 demo papers were accepted, each having 4 pages in the proceedings and demo of the system during the poster session. In addition to the paper and poster/demo sessions, ECML PKDD 2009 also featured five invited talks, ten workshops, six tutorials, and the ECML PKDD discovery challenge and industrial track. The selection of Invited Speakers covered a broad range from theoretical to leading application-orientated research. Together they made a very strong addition to the conference program. We are grateful to Shai Ben-David (University of Waterloo, Canada), Nello Cristianini (University of Bristol, UK), Mark Greaves (Vulcan Inc.), Rosie Jones (Yahoo! Research), Ralf Steinberger (European Commission - Joint Research Centre) for their participation in ECML PKDD 2009. The abstracts of their presentations are included in this volume.

This year we continued to promote an Industrial Track chaired by Marko Grobelnik (Jožef Stefan Institute, Slovenia) and Nataša Milić-Frayling (Microsoft Research, Cambridge, UK) consisting of selected talks with a strong industrial component presenting research from the area covered by the ECML PKDD conference. We have also included a Demonstration Track chaired by Alejandro Jaimes Larrarte, providing a venue for exciting exemplars of applications of novel technologies.

As in recent years, the conference proceedings were available on-line to conference participants during the conference. We are grateful to Springer for accommodating this access channel for the proceedings.

As in previous years we will continue with the recently established tradition of videorecording the event, ensuring an enduring record of the event made accessible at http://videolectures.net/. Mitja Jermol is the Video Chair overseeing this aspect of the organization.

This year's Discovery Challenge was coordinated by Andreas Hotho together with Folke Eisterlehner and Robert Jäschke. It involved three tasks in the area of tag recommendation.

We are all indebted to the Area Chairs, Program Committee members and external reviewers for their commitment and hard work that resulted in a rich but selective scientific program for ECML PKDD 2009. We are particularly grateful to those reviewers who helped with additional reviews at a very short notice to assist us in a small number of difficult decisions. We further thank the Workshop and Tutorial Chairs Ravid Ghani and Cédric Archambeau for selecting and coordinating the ten workshops and six tutorials that accompany the conference; the workshop organizers, tutorial presenters, and the organizers of the discovery challenge, the Industrial and Demonstration Tracks; the Video Chair;

the Publicity Chair David Hardoon; and Richard van de Stadt and CyberChair-PRO for highly competent and flexible support when confronted by novel features in our handling of the papers. Special thanks are due to the Local Chair, Tina Anžič, for the many hours spent ensuring the success of the conference. Finally, we are grateful to the Steering Committee and the ECML PKDD community that entrusted us with the organization of the ECML PKDD 2009.

Most of all, however, we would like to thank all the authors who trusted us with their submissions, thereby contributing to the main yearly European-focussed international event in the life of our expanding research community.

June 2009

Dunja Mladenić
Wray Buntine
Marko Grobelnik
John Shawe-Taylor

Organization

General Chair

Dunja Mladenić	Jožef Stefan Institute, Slovenia

Program Chairs

Wray Buntine	Helsinki Institute of IT, Finland; NICTA, Australia
Marko Grobelnik	Jožef Stefan Institute, Slovenia
John Shawe-Taylor	University College London, UK

Local Chair

Tina Anžic	Jožef Stefan Institute, Slovenia

Tutorial Chair

Cédric Archambeau	University College London, UK

Workshop Chair

Rayid Ghani	Accenture Technology Labs, USA

Discovery Challenge Chairs

Robert Jäschke	University of Kassel, Germany
Andreas Hotho	University of Kassel, Germany
Folke Eisterlehner	University of Kassel, Germany

Industrial Track Chairs

Marko Grobelnik	Jožef Stefan Institute, Slovenia
Nataša Milić-Frayling	Microsoft Research Cambridge, UK

Demo Chair

Alejandro Jaimes Larrarte	Telefonica Research, Spain

Best Paper Chair

Aleksander Kolcz	Microsoft Live Labs, USA

Publicity Chair

David Hardoon	University College London, UK

Video Chair

Mitja Jermol	Jožef Stefan Institute, Slovenia

Steering Committee

Walter Daelemans
Bart Goethals
Katharina Morik
Johannes Fürnkranz
Joost N. Kok
Stan Matwin
Dunja Mladenić
Tobias Scheffer
Andrzej Skowron
Myra Spiliopoulou

Area Chairs

Francis Bach
Hendrik Blockeel
Francesco Bonchi
Pavel Brazdil
Toon Calders
Nitesh Chawla
Walter Daelemans
Tijl De Bie
Johannes Fürnkranz
Thomas Gärtner
João Gama
Bart Goethals
Eamonn Keogh
Joost Kok
Alek Kolcz
Jure Leskovec
Stan Matwin
Taneli Mielikainen
Claire Nedellec
Martin Scholz
David Silver
Steffen Staab
Gerd Stumme
Luis Torgo
Michael Witbrock
Stefan Wrobel

Program Committee

Ameen Abu-Hanna
Osman Abul
Lada Adamic
Abdullah Al Mueen
Enrique Alfonseca
Erick Alphonse
Carlos Alzate
Massih-Reza Amini
Gennady Andrienko
Annalisa Appice
Hiroki Arimura
Andrew Arnold
Sitaram Asur
Martin Atzmueller
Nathalie Aussenac-Gilles
Paulo Azevedo
Lars Backstrom
Tony Bagnall
Roberto Basili
Vladimir Batagelj
Ron Bekkerman
Marc Bellemare
Paul Bennett
Bettina Berendt
Tanya Berger-Wolf
Michael Berthold
Sourangshu Bhattacharya
Concha Bielza
Misha Bilenko
Stephan Bloehdorn
Christian Bockermann
Mario Boley
Christian Borgelt
Karsten Borgwardt
Henrik Bostrom
Guillaume Bouchard
Jean-François Boulicaut
Janez Brank
Ulf Brefeld
Bjorn Bringmann
Paul Buitelaar
Rui Camacho
Stephane Canu
Olivier Cappe
Andre Carvalho
Carlos Castillo
Ciro Cattuto
Vineet Chaoji
Sanjay Chawla
David Cieslak
Philipp Cimiano
Lucchese Claudio
Vincent Claveau
Fabrice Colas
Antoine Cornuejols
Christophe Costa Florencio
Fabrizio Costa
Bruno Cremilleux
Padraig Cunningham
Alfredo Cuzzocrea
Florence D'Alche-Buc
Claudia d'Amato
Gautam Das
Kamalika Das
Jesse Davis
Alneu de Andrade Lopes
Jeroen de Bruin
Marco de Gemmis
Jeroen De Knijf
Thomas Degris-Dard
Jose del Campo-Avila
Krzysztof Dembczynski
Laura Dietz
Carlos Diuk
Kurt Driessens
Pierre Dupont
Jennifer Dy
Saso Dzeroski
Charles Elkan
Tapio Elomaa
Damien Ernst
Floriana Esposito
Fazel Famili
Nicola Fanizzi
Amir-massoud Farahmand
Ad Feelders

Xiaoli Fern
Daan Fierens
Ilias Flaounas
George Forman
Blaz Fortuna
Eibe Frank
Jordan Frank
Mohamed Gaber
Dragan Gamberger
Gemma Garriga
Gilles Gasso
Eric Gaussier
Ricard Gavalda
Floris Geerts
Peter Geibel
Lise Getoor
Olivier Gevaert
Rayid Ghani
Fosca Gianotti
Melanie Gnasa
Henrik Grosskreutz
Amit Gruber
Vincent Guigue
Robert Gwadera
Larry Hall
Zaid Harchaoui
Hannes Heikinheimo
Iris Hendrickx
Mark Herbster
Tom Heskes
Melanie Hilario
Alexander Hinneburg
Susanne Hoche
Frank Höppner
Geoff Holmes
Tamas Horvath
Bettina Hoser
Veronique Hoste
Andreas Hotho
Eyke Hüllermeier
Inaki Inza
Mariya Ishteva
Robert Jaeschke
Longin Jan Latecki
Nathalie Japkowicz
Szymon Jaroszewicz
Thorsten Joachims
Alipio Jorge
Felix Jungermann
Matti Kaariainen
Alexandros Kalousis
Murat Kantarcioglu
Samuel Kaski
Philip Kegelmeyer
Kristian Kersting
Svetlana Kiritchenko
Igor Kononenko
Anna Koop
Walter Kosters
Wojciech Kotlowski
Stefan Kramer
Andreas Krause
Yuval Krymolowski
Miroslav Kubat
Ravi Kumar
James Kwok
Bill Lampos
Niels Landwehr
Mark Last
Nada Lavrac
Lihong Li
Jessica Lin
Charles Ling
Huan Liu
XiaoHui Liu
Eneldo Loza Mencia
Peter Lucas
Elliot Ludvig
Yiming Ma
Sofus Macskassy
Michael Madden
Donato Malerba
Bradley Malin
Lluis Marquez
Michael May
Prem Melville
Rosa Meo
Pauli Miettinen
Roser Morante
Fabian Morchen

Katharina Morik
Flavia Moser
Fabien Moutarde
Klaus-Robert Müller
Ion Muslea
Amedeo Napoli
Olfa Nasraoui
Vivi Nastase
James Neufeld
Alexandru Niculescu-Mizil
Siegfried Nijssen
Joakim Nivre
Blaž Novak
Ann Nowe
Alexandros Ntoulas
Andreas Nuernberger
Guillaume Obozinski
Arlindo Oliveira
Martijn van Otterlo
Gerhard Paass
Cosmin Paduraru
Georgios Paliouras
Themis Palpanas
Junfeng Pan
Rong Pan
Andrea Passerini
Mykola Pechenizkiy
Dmitry Pechyony
Dino Pedreschi
Kristiaan Pelckmans
Jose-Maria Pena
Ruggero Pensa
Raffaele Perego
Bernhard Pfahringer
Christian Plagemann
Barnabas Poczos
Doina Precup
Philippe Preux
Kai Puolamaki
Peter van der Putten
Sampo Pyysalo
Predrag Radivojac
Troy Raeder Davood Rafiei
Chedy Raissi
Shyam Rajaram
Alain Rakotomamonjy
Liva Ralaivola
Jan Ramon
Chotirat Ratanamahatana
Chandan Reddy
Ehud Reiter
Elisa Ricci
Martin Riedmiller
Celine Robardet
Marko Robnik-Sikonja
Pedro Rodrigues
Teemu Roos
Fabrice Rossi
Volker Roth
Celine Rouveirol
Ulrich Rueckert
Stefan Rüping
Yvan Saeys
Lorenza Saitta
Scott Sanner
Vitor Santos Costa
Craig Saunders
Yucel Saygin
Lars Schmidt-Thieme
Jouni Seppanen
Shashi Shekhar
Jin Shieh
Stefan Siersdorfer
Tomi Silander
Ricardo Silva
Ozgur Simsek
Ajit Singh
Sergej Sizov
Carlos Soares
Maarten van Someren
Yang Song
Elaine Sousa
Myra Spiliopoulou
Karsten Steinhaeuser
David Stern
Jan Struyf
Jiang Su
Masashi Sugiyama
Johan Suykens
Vojtech Svatek

Sandor Szedmak
Nikolaj Tatti
Evimaria Terzi
Gerald Tesauro
Hanghang Tong
Volker Tresp
Koji Tsuda
Ville Tuulos
Rasmus Ulslev Pedersen
Dries Van Dyck
Stijn Vanderlooy
Sergei Vassilvitskii
Cor Veenman
Paola Velardi
Shankar Vembu
Celine Vens
Jean-Philippe Vert
Ricardo Vilalta
Michalis Vlachos
Christel Vrain
Jilles Vreeken
Christian Walder
Xiaoyue Wang
Markus Weimer
David Wingate
Michael Wurst
Dragomir Yankov
Lexiang Ye
Jie Yin
François Yvon
Menno van Zaanen
Bianca Zadrozny
Osmar Zaiane
Mikhail Zaslavskiy
Gerson Zaverucha
Filip Zelezny
Justin Zhan
Bin Zhang
Zhi-Hua Zhou
Qiang Zhu
Xiaojin Zhu
Albrecht Zimmermann

Additional Reviewers

Dima Alberg
Anelia Angelova
Mohammad Aziz
Michele Berlingerio
Marenglen Biba
Alexander Binder
Zoran Bosnić
Christos Boutsidis
Fabian Buchwald
Markus Bundschus
Wray Buntine
Lijuan Cai
Michelangelo Ceci
Weiwei Cheng
Joaquim Costa
Dave DeBarr
Marcos Domingues
Jun Du
Charles Elkan
Daan Fierens
Nuno A. Fonseca
Dmitriy Fradkin
Thomas Gabel
Zeno Gantner
Robby Goetschalckx
Habiba
Niina Haiminen
Katja Hansen
Andreas Hapfelmeier
Jingrui He
Raymond Heatherly
Thibault Helleputte
James Henderson
Yi Huang
Ali Inan
Tsuyoshi Ide
Szymon Jaroszewicz
Takafumi Kanamori
Alexandros Karatzoglou
Hisashi Kashima
Tsuyoshi Kato
Maarten Keijzer

Evan Kirshenbaum
Arto Klami
Thoralf Klein
Marius Kloft
Arne Koopman
Da Kuang
Mayank Lahiri
Tobias Lang
Lieven De Lathauwer
Gayle Leen
Jens Lehmann
Guy Lever
Biao Li
Nan Li
Shuyan Li
Yu-Feng Li
Yuan Li
Grigorios Loukides
Jose A. Lozano
Juan Luo
Hamid Reza Maei
Michael Mampaey
Alain-Pierre Manine
Leandro Marinho
Eneldo Loza Mencia
João Mendes-Moreira
Olana Missura
Matteo Mordacchini
Marianne Mueller
Eileen A. Ni
Martijn van Otterlo
Aline Paes
Indranil Palit
Sang-Hyeun Park
Aritz Perez
Claudia Perlich
Georgios Petasis
Dimitrios Pierrakos
Fabio Pinelli
Cristiano Pitangui
Troy Raeder
Alain Rakotomamonjy
Steffen Rendle
Achim Rettinger
François Rioult
Jan Rupnik
Jarkko Salojärvi
Leander Schietgat
Jana Schmidt
Armin Shmilovici
David Silver
Karsten Steinhäuser
Erik Štrumbelj
Jan Struyf
Ilija Subašić
Taiji Suzuki
Takashi Takenouchi
Nicola Tonellotto
Grigorios Tsoumakas
Mikalai Tsytsarau
Stijn Vanderlooy
Guy Van den Broeck
Sicco Verwer
Ricardo Vilalta
Dimitrios Vogiatzis
Petar Vračar
Chris Watkins
Joerg Wicker
Eric Wiewiora
Fuxiao Xin
Xingwei Yan
Xingwei Yang
Monika Zakova
De-Chuan Zhan
Indre Zliobaite

Sponsors

We wish to express our gratitude to the sponsors of ECML PKDD 2009 for their essential contribution to the conference. We wish to thank Jožef Stefan Institute, Slovenia, for providing financial and organizational means for the conference; the European Office of Aerospace Research and Development, a detachment of U.S. Air Force Office of Scientific Research (EOARD) for generous financial support; Pascal European Network of Excellence (PASCAL2) for sponsoring the Invited Speaker program and the videorecording of the conference; Slovenia Research Agency (ARRS); Google for supporting a Poster Reception; Microsoft Research Ltd., Cambridge, UK for supporting the Industrial track; Yahoo! Research, Quintelligence, Hewlett-Packard Labs, and ACTIVE European Integrated project for their financial support; the *Machine Learning* Journal for supporting the Student Best Paper Award; the *Data Mining and Knowledge Discovery* Journal for supporting the Student Best Paper Award; Nokia for sponsoring the Discovery Challenge Awards and the Best Demo Award.

Table of Contents – Part I

Invited Talks (Abstracts)

Machine Learning Journal Abstracts

Data Mining and Knowledge Discovery Journal Abstracts

Regular Papers

Table of Contents – Part II

Regular Papers

Demo Papers

Theory-Practice Interplay in Machine Learning – Emerging Theoretical Challenges

Shai Ben-David

University of Waterloo, Canada

Abstract. Theoretical analysis has played a major role in some of the most prominent practical successes of statistical machine learning. However, mainstream machine learning theory assumes some strong simplifying assumptions which are often unrealistic. In the past decade, the practice of machine learning has led to the development of various heuristic paradigms that answer the needs of a vastly growing range of applications. Many useful such paradigms fall beyond the scope of the currently available analysis. Will theory play a similar pivotal role in the newly emerging sub areas of machine learning?

In this talk, I will survey some such application-motivated theoretical challenges. In particular, I will discuss recent developments in the theoretical analysis of semi-supervised learning, multi-task learning, "learning to learn", privacy-preserving learning and more.

W. Buntine et al. (Eds.): ECML PKDD 2009, Part I, LNAI 5781, p. 1, 2009.

Are We There Yet?

Nello Cristianini

University of Bristol, UK

Abstract. Statistical approaches to Artificial Intelligence are behind most success stories of the field in the past decade. The idea of generating non-trivial behaviour by analysing vast amounts of data has enabled recommendation systems, search engines, spam filters, optical character recognition, machine translation and speech recognition. As we celebrate the spectacular achievements of this line of research, we need to assess its full potential, its limitations and its position within the larger scheme of things. What are the next steps to take towards machine intelligence?

W. Buntine et al. (Eds.): ECML PKDD 2009, Part I, LNAI 5781, p. 2, 2009.

The Growing Semantic Web

Mark Greaves

Vulcan Inc., USA

Abstract. From its beginnings in 2004, the data available on the web in Semantic Web formats has typically been both eclectic and relatively small, and closely linked the interests of particular researchers. In the past year, however, the quantity and scope of data published on the public semantic web has exploded, and the size of the semantic web is now measured in the billions of assertions. It is a significant and growing resource for applications which depend on web-based resources for some or all of their knowledge. With this massive increase in quantity and scope come many opportunities, as well as the usual issues of scale on the web: inconsistency, mapping problems, incompleteness and data variability. This talk will cover the history and current state of the Semantic Web and the Linked Data Cloud, describe some of the uses to which web-based semantic data is currently put, and discuss prospects for the ECML/PKDD community to leverage this growing web of data.

W. Buntine et al. (Eds.): ECML PKDD 2009, Part I, LNAI 5781, p. 3, 2009.

Privacy in Web Search Query Log Mining

Rosie Jones

Yahoo!, Inc, USA

Abstract. Web search engines have changed our lives - enabling instant access to information about subjects that are both deeply important to us, as well as passing whims. The search engines that provide answers to our search queries also log those queries, in order to improve their algorithms. Academic research on search queries has shown that they can provide valuable information on diverse topics including word and phrase similarity, topical seasonality and may even have potential for sociology, as well as providing a barometer of the popularity of many subjects. At the same time, individuals are rightly concerned about what the consequences of accidental leaking or deliberate sharing of this information may mean for their privacy. In this talk I will cover the applications which have benefited from mining query logs, the risks that privacy can be breached by sharing query logs, and current algorithms for mining logs in a way to prevent privacy breaches.

W. Buntine et al. (Eds.): ECML PKDD 2009, Part I, LNAI 5781, p. 4, 2009.

Highly Multilingual News Analysis Applications

Ralf Steinberger

European Commission - Joint Research Centre, Italy

Abstract. The publicly accessible Europe Media Monitor (EMM) family of applications (http://press.jrc.it/overview.html) gather and analyse an average of 80,000 to 100,000 online news articles per day in up to 43 languages. Through the extraction of meta-information in these articles, they provide an aggregated view of the news; they allow to monitor trends and to navigate the news over time and even across languages. EMM-NewsExplorer additionally collects historical information about persons and organisations from the multilingual news, generates co-occurrence and quotation-based social networks, and more. All EMM applications were entirely developed at, and are being maintained by, the European Commission's Joint Research Centre (JRC) in Ispra, Italy.

The applications make combined use of a variety of text analysis tools, including clustering, multi-label document classification, named entity recognition, name variant matching across languages and writing systems, topic detection and tracking, event scenario template filling, and more. Due to the high number of languages covered, linguistics-poor methods were used for the development of these text mining components. See the site http://langtech.jrc.it/ for technical details and a list of publications.

The speaker will give an overview of the various applications and will then explain the workings of selected text analysis components.

W. Buntine et al. (Eds.): ECML PKDD 2009, Part I, LNAI 5781, p. 5, 2009.

Combining Instance-Based Learning and Logistic Regression for Multilabel Classification*

Weiwei Cheng and Eyke Hüllermeier

Department of Mathematics and Computer Science
University of Marburg, Germany
{cheng,eyke}@mathematik.uni-marburg.de

Abstract. Multilabel classification is an extension of conventional classification in which a single instance can be associated with multiple labels. Recent research has shown that, just like for conventional classification, instance-based learning algorithms relying on the nearest neighbor estimation principle can be used quite successfully in this context. However, since hitherto existing algorithms do not take correlations and interdependencies between labels into account, their potential has not yet been fully exploited. In this paper, we propose a new approach to multilabel classification, which is based on a framework that unifies instance-based learning and logistic regression, comprising both methods as special cases. This approach allows one to capture interdependencies between labels and, moreover, to combine model-based and similarity-based inference for multilabel classification. As will be shown by experimental studies, our approach is able to improve predictive accuracy in terms of several evaluation criteria for multilabel prediction.

Reference

1. Cheng, W., Hüllermeier, E.: Combining Instance-Based Learning and Logistic Regression for Multilabel Classification. Machine Learning (2009) DOI: 10.1007/s10994-009-5127-5

* This is an extended abstract of an article published in the machine learning journal [1].

W. Buntine et al. (Eds.): ECML PKDD 2009, Part I, LNAI 5781, p. 6, 2009.

On Structured Output Training: Hard Cases and an Efficient Alternative*

Thomas Gärtner and Shankar Vembu

Fraunhofer IAIS, Schloß Birlinghoven, 53754 Sankt Augustin, Germany
{thomas.gaertner,shankar.vembu}@iais.fraunhofer.de

State-of-the-art structured prediction algorithms can be applied using off-the-shelf tools by implementing a joint kernel for inputs and outputs, and an algorithm for inference. The kernel is used for mapping the data to an appropriate feature space, while the inference algorithm is used for successively adding violated constraints to the optimisation problem. While this approach leads to efficient learning algorithms for many important real world problems, there are also many cases in which successively adding violated constraints is infeasible. As a simple yet relevant problem, we consider the prediction of routes (cyclic permutations) over a given set of points of interest. Solving this problem has many potential applications. For car drivers, prediction of individual routes can be used for intelligent car sharing applications or help optimise a hybrid vehicle's charge/discharge schedule. We show that state-of-the-art structured prediction algorithms cannot guarantee polynomial runtime for this output set of cyclic permutations.

Despite these hardness results, we show that efficient formulae for 'super-structure' counting can be derived and propose an alternative structured output training algorithm based on these counting formulae. By deriving 'super-structure' counting formulae for various combinatorial structures, we show that our approach subsumes many machine learning problems including multi-class, multi-label and hierarchical classification. Furthermore, our approach can be used for training complex combinatorial output sets for which the assumptions made in the literature do not hold. We empirically compare our algorithm with state-of-the-art general and special purpose algorithms on different structures. For multi-label and hierarchical classification, inference is trivial and our experiments demonstrate that our approach is competitive or better than the state-of-the-art. For route prediction, inference is hard and we focus on the training part of the algorithms, i.e., on synthetic data we compare the policy estimated by our approach to the policy estimated by SVM-Struct using approximate inference.

Reference

1. Gärtner, T., Vembu, S.: On Structured Output Training: Hard Cases and an Efficient Alternative. Machine Learning (2009) DOI: 10.1007/s10994-009-5129-3

* This is an extended abstract of an article published in the machine learning journal [1].

W. Buntine et al. (Eds.): ECML PKDD 2009, Part I, LNAI 5781, p. 7, 2009.

Sparse Kernel SVMs via Cutting-Plane Training*

Thorsten Joachims and Chun-Nam John Yu

Cornell University, Dept. of Computer Science, Ithaca, NY 14853 USA
{tj,cnyu}@cs.cornell.edu

While Support Vector Machines (SVMs) with kernels offer great flexibility and prediction performance on many application problems, their practical use is often hindered by the following two problems. Both problems can be traced back to the number of Support Vectors (SVs), which is known to generally grow linearly with the data set size [1]. First, training is slower than other methods and linear SVMs, where recent advances in training algorithms vastly improved training time. Second, since the prediction rule takes the form $h(x) = \text{sign}\left[\sum_{i=1}^{\#SV} \alpha_i K(x_i, x)\right]$ it is too expensive to evaluate in many applications when the number of SVs is large.

This paper tackles these two problems by generalizing the notion of Support Vector to arbitrary points in input space, not just training vectors. Unlike Wu et al. [2], who explore making the location of the points part of a large non-convex optimization problem, we propose an algorithm that iteratively constructs the set of basis vectors from a cutting-plane model. This makes our algorithm, called *Cutting-Plane Subspace Pursuit* (CPSP), efficient and modular. We analyze the training efficiency and the solution quality of the CPSP algorithm both theoretically and empirically. We find that its classification rules can be orders of magnitude sparser than the conventional support-vector representation while providing comparable prediction accuracy. The sparsity of the CPSP representation not only makes predictions substantially more efficient, it also allows the user to control training time. Especially for large datasets with sparse feature vectors (e.g. text classification), the CPSP methods is substantially faster than methods that only consider basis vectors from the training set.

Acknowledgments. This work was funded in part under NSF award IIS-0713483.

References

1. Steinwart, I.: Sparseness of support vector machines. JMLR 4, 1071–1105 (2003)
2. Wu, M., Schölkopf, B., Bakir, G.H.: A direct method for building sparse kernel learning algorithms. JMLR 7, 603–624 (2006)
3. Joachims, T., Yu, C.-N.J.: Sparse Kernel SVMs via Cutting-Plane Training. Machine Learning (2009), DOI:10.1007/s10994-009-5126-6

* This is an extended abstract of an article published in the machine learning journal [3].

W. Buntine et al. (Eds.): ECML PKDD 2009, Part I, LNAI 5781, p. 8, 2009.

Hybrid Least-Squares Algorithms for Approximate Policy Evaluation*

Jeff Johns, Marek Petrik, and Sridhar Mahadevan

Department of Computer Science
University of Massachusetts Amherst
{johns,petrik,mahadeva}@cs.umass.edu

Abstract. The goal of approximate policy evaluation is to "best" represent a target value function according to a specific criterion. Different algorithms offer different choices of the optimization criterion. Two popular least-squares algorithms for performing this task are the *Bellman residual* method, which minimizes the Bellman residual, and the *fixed point* method, which minimizes the *projection* of the Bellman residual. When used within policy iteration, the fixed point algorithm tends to ultimately find better performing policies whereas the Bellman residual algorithm exhibits more stable behavior between rounds of policy iteration. We propose two *hybrid least-squares algorithms* to try to combine the advantages of these algorithms. We provide an analytical and geometric interpretation of hybrid algorithms and demonstrate their utility on a simple problem. Experimental results on both small and large domains suggest hybrid algorithms may find solutions that lead to better policies when performing policy iteration.

Reference

1. Johns, J., Petrik, M., Mahadevan, S.: Hybrid Least-Squares Algorithms for Approximate Policy Evaluation. Machine Learning (2009) DOI: 10.1007/s10994-009-5128-4

* This is an extended abstract of an article published in the machine learning journal [1].

W. Buntine et al. (Eds.): ECML PKDD 2009, Part I, LNAI 5781, p. 9, 2009.

A Self-training Approach to Cost Sensitive Uncertainty Sampling*

Alexander Liu, Goo Jun, and Joydeep Ghosh

Department of Electrical and Computer Engineering
The University of Texas at Austin
Austin, Texas 78712
U.S.A.
{ayliu,gjun}@mail.utexas.edu, ghosh@ece.utexas.edu

Abstract. Uncertainty sampling is an effective method for performing active learning that is computationally efficient compared to other active learning methods such as loss-reduction methods. However, unlike loss-reduction methods, uncertainty sampling cannot minimize total misclassification costs when errors incur different costs. This paper introduces a method for performing cost-sensitive uncertainty sampling that makes use of self-training. We show that, even when misclassification costs are equal, this self-training approach results in faster reduction of loss as a function of number of points labeled and more reliable posterior probability estimates as compared to standard uncertainty sampling. We also show why other more naive methods of modifying uncertainty sampling to minimize total misclassification costs will not always work well.

Reference

1. Liu, A., Jun, G., Ghosh, J.: A Self-training Approach to Cost Sensitive Uncertainty Sampling. Machine Learning (2009) DOI: 10.1007/s10994-009-5131-9

* This is an extended abstract of an article published in the machine learning journal [1].

W. Buntine et al. (Eds.): ECML PKDD 2009, Part I, LNAI 5781, p. 10, 2009.

Learning Multi-linear Representations of Distributions for Efficient Inference*

Dan Roth and Rajhans Samdani

Department of Computer Science
University of Illinois at Urbana-Champaign
{danr,rsamdan2}@illinois.edu

Abstract. We examine the class of multi-linear polynomial representations (MLR) for expressing probability distributions over discrete variables. Recently, MLR have been considered as intermediate representations that facilitate inference in distributions represented as graphical models.

We show that MLR is an expressive representation of discrete distributions and can be used to concisely represent classes of distributions which have exponential size in other commonly used representations, while supporting probabilistic inference in time linear in the size of the representation.

Our key contribution is presenting techniques for learning bounded-size distributions represented using MLR, which support efficient probabilistic inference. We propose algorithms for exact and approximate learning for MLR and, through a comparison with Bayes Net representations, demonstrate experimentally that MLR representations provide faster inference without sacrificing inference accuracy.

Keywords: Learning Distributions, Multi-linear Polynomials, Probabilistic Inference, Graphical Models.

Acknowledgments. The authors wish to thank Daniel Lowd for his help with the Gibbs sampling program. This work is partly supported by an ONR Award on "Guiding Learning and Decision Making in the Presence of Multiple Forms of Information" and by the Siebel Scholars Foundation.

Reference

1. Roth, D., Samdani, R.: Learning Multi-linear Representations of Distributions for Efficient Inference. Machine Learning (2009) DOI: 10.1007/s10994-009-5130-X

* This is an extended abstract of an article published in the machine learning journal [1].

W. Buntine et al. (Eds.): ECML PKDD 2009, Part I, LNAI 5781, p. 11, 2009.

Cost-Sensitive Learning Based on Bregman Divergences*

Raúl Santos-Rodríguez[1], Alicia Guerrero-Curieses[2], Rocío Alaiz-Rodríguez[3], and Jesús Cid-Sueiro[1]

[1] Department of Signal Theory and Communications, Universidad Carlos III de Madrid, Leganés (Madrid), Spain
{rsrodriguez,jcid}@tsc.uc3m.es

[2] Department of Signal Theory and Communications, Universidad Rey Juan Carlos, Fuenlabrada (Madrid), Spain
alicia.guerrero@urjc.es

[3] Department of Electrical and Electronic Engineering, Universidad de León, León, Spain
rocio.alaiz@unileon.es

Abstract. This paper analyzes the application of a particular class of Bregman divergences to design cost-sensitive classifiers for multiclass problems. We show that these divergence measures can be used to estimate posterior probabilities with maximal accuracy for the probability values that are close to the decision boundaries. Asymptotically, the proposed divergence measures provide classifiers minimizing the sum of decision costs in non-separable problems, and maximizing a margin in separable MAP problems.

Keywords: Cost sensitive learning, Bregman divergence, posterior class probabilities, maximum margin.

Reference

1. Santos-Rodríguez, R., Guerrero-Curieses, A., Alaiz-Rodríguez, R., Cid-Sueiro, J.: Cost-sensitive learning based on Bregman divergences. Machine Learning (2009) DOI: 10.1007/s10994-009-5132-8

* This is an extended abstract of an article published in the machine learning journal [1].

W. Buntine et al. (Eds.): ECML PKDD 2009, Part I, LNAI 5781, p. 12, 2009.

RTG: A Recursive Realistic Graph Generator Using Random Typing

Leman Akoglu and Christos Faloutsos

Carnegie Mellon University
School of Computer Science
{lakoglu,christos}@cs.cmu.edu

Abstract. We propose a new, recursive model to generate realistic graphs, evolving over time. Our model has the following properties: it is (a) flexible, capable of generating the cross product of weighted/unweighted, directed/undirected, uni/bipartite graphs; (b) realistic, giving graphs that obey *eleven* static and dynamic laws that real graphs follow (we formally prove that for several of the (power) laws and we estimate their exponents as a function of the model parameters); (c) parsimonious, requiring only *four* parameters. (d) fast, being linear on the number of edges; (e) simple, intuitively leading to the generation of macroscopic patterns. We empirically show that our model mimics two real-world graphs very well: Blognet (unipartite, undirected, unweighted) with 27K nodes and 125K edges; and Committee-to-Candidate campaign donations (bipartite, directed, weighted) with 23K nodes and 880K edges. We also show how to handle time so that edge/weight additions are bursty and self-similar.

1 Introduction

Study of complex graphs such as computer and biological networks, the link structure of the WWW, the topology of the Internet, and recently with the widespread use of the Internet, large social networks, has been a vital research area. Many fascinating properties have been discovered, such as small and shrinking diameter [2,20], power-laws [5,11,16,24,22,28,29,20], and community structures [12,13,27]. As a result of such interesting patterns being discovered, and for many other reasons which we will discuss next, how to find a model that would produce synthetic but realistic graphs is a natural question to ask. There are several applications and advantages of modeling real-world graphs:

- *Simulation studies*: if we want to run tests for, say a spam detection algorithm, and want to observe how the algorithm behaves on graphs with different sizes and structural properties, we can use graph generators to produce such graphs by changing the parameters. This is also true when it is difficult to collect any kind of real data.
- *Sampling/Extrapolation*: we can generate a smaller graph for example for visualization purposes or in case the original graph is too big to run tests

W. Buntine et al. (Eds.): ECML PKDD 2009, Part I, LNAI 5781, pp. 13–28, 2009.

on it; or conversely to generate a larger graph for instance to make future prediction and answer what-if questions.
- *Summarization/Compression*: model parameters can be used to summarize and compress a given graph as well as to measure similarity to other graphs.
- *Motivation to understand pattern generating processes*: graph generators give intuition and shed light upon what kind of processes can (or cannot) yield the emergence of certain patterns. Moreover, modeling addresses the question of what patterns real networks exhibit that needs to be matched and provides motivation to figure out such properties.

Graph generator models are surveyed in [4]. Ideally, we would like a graph generator that is:

1. *simple*: it would be easy to understand and it would intuitively lead to the emergence of macroscopic patterns.
2. *realistic*: it would produce graphs that obey all the discovered "laws" of real-world graphs with appropriate values.
3. *parsimonious*: it would require only a few number of parameters.
4. *flexible*: it would be able to generate the cross product of weighted/unweighted, directed/undirected and unipartite/bipartite graphs.
5. *fast*: the generation process would ideally take linear time with respect to the number of edges in the output graph.

In this paper we propose RTG, for *Random Typing Generator*. Our model uses a process of 'random typing', to generate source and destination node identifiers, and it meets all the above requirements. In fact, we show that it can generate graphs that obey all *eleven* patterns that real graphs typically exhibit.

Next, we provide a survey on related work. Section 3 describes our RTG generator in detail. Section 4 provides experimental results and discussion. We conclude in Section 5. Appendix gives proofs showing some of the power-laws that the model generates.

2 Related Work

Graph patterns: Many interesting patterns that real graphs obey have been found, which we give a detailed list of in the next section. Ideally, a generator should be able to produce all of such properties.

Graph generators: The vast majority of earlier graph generators have focused on modeling a small number of common properties, but fail to mimic others. Such models include the *Erdos & Renyi* model [8], the *preferential attachment* model [3] and numerous more, like the *'small-world'*, *'winners don't take all'*, *'forest fire'* and *'butterfly'* models [31,26,20,22]. See [4] for a recent survey and discussion. In general, these methods are limited in trying to model some static graph property while neglecting others as well as dynamic properties or cannot be generalized to produce *weighted* graphs.

Random dot product graphs [17,32] assign each vertex a random vector in some d-dimensional space and an edge is put between two vertexes with probability equal to the dot product of the endpoints. This model does not generate weighted graphs and by definition only produces undirected graphs. It also seems to require the computation of the dot product for each pair of nodes which takes *quadratic* time.

A different family of models is utility-based, where agents try to optimize a predefined utility function and the network structure takes shape from their collective strategic behavior [10,9,18]. This class of models, however, is usually hard to analyze.

Kronecker graph generators [19] and their tensor followups [1] are successful in the sense that they match several of the properties of real graphs and they have proved useful for generating self-similar properties of graphs. However, they have two disadvantages: The first is that they generate multinomial/lognormal distributions for their degree and eigenvalue distribution, instead of a power-law one. The second disadvantage is that it is not easy to grow the graph incrementally: They have a fixed, predetermined number of nodes (say, N^k, where N is the number of nodes of the generator graph, and k is the number of iterations); where adding more edges than expected does *not* create additional nodes. In contrast, in our model, nodes emerge naturally.

3 Proposed Model

We first give a concise list of the *static* and *dynamic* 'laws' that real graphs obey, which a graph generator should be able to match.

L01. *Power-law degree distribution*: the degree distribution should follow a power-law in the form of $f(d) \propto d^{\gamma}$, with the exponent $\gamma < 0$ [5,11,16,24]

L02. *Densification Power Law (DPL)*: the number of nodes N and the number of edges E should follow a power-law in the form of $E(t) \propto N(t)^{\alpha}$, with $\alpha > 1$, over time [20].

L03. *Weight Power Law (WPL)*: the total weight of the edges W and the number of edges E should follow a power-law in the form of $W(t) \propto E(t)^{\beta}$, with $\beta > 1$, over time [22].

L04. *Snapshot Power Law (SPL)*: the total weight of the edges W_n attached to each node and the number of such edges, that is, the degree d_n should follow a power-law in the form of $W_n \propto d_n^{\theta}$, with $\theta > 1$ [22].

L05. *Triangle Power Law (TPL)*: the number of triangles Δ and the number of nodes that participate in Δ number of triangles should follow a power-law in the form of $f(\Delta) \propto \Delta^{\sigma}$, with $\sigma < 0$ [29].

L06. *Eigenvalue Power Law (EPL)*: the eigenvalues of the adjacency matrix of the graph should be power-law distributed [28].

L07. *Principal Eigenvalue Power Law (λ_1PL)*: the largest eigenvalue λ_1 of the adjacency matrix of the graph and the number of edges E should follow a power-law in the form of $\lambda_1(t) \propto E(t)^{\delta}$, with $\delta < 0.5$, over time [1].

L08. *small and shrinking diameter*: the (effective) diameter of the graph should be small [2] with a possible spike at the 'gelling point' [22]. It should also shrink over time [20].

L09. *constant size secondary and tertiary connected components*: while the 'giant connected component' keeps growing, the secondary and tertiary connected components tend to remain constant in size with small oscillations [22].

L10. *community structure*: the graph should exhibit a modular structure, with nodes forming groups, and possibly groups within groups [12,13,27].

L11. *bursty/self-similar edge/weight additions*: Edge (weight) additions to the graph over time should be self-similar and bursty rather than uniform with possible spikes [7,14,15,22].

Zipf introduced probably the earliest power law [33], stating that, in many natural languages, the rank r and the frequency f_r of vocabulary words follow a power-law $f_r \propto 1/r$. Mandelbrot [21] argued that Zipf's law is the result of optimizing the average amount of information per unit transmission cost. Miller [23] showed that a random process also leads to Zipf-like power laws. He suggested the following experiment: "A monkey types randomly on a keyboard with k characters and a space bar. A space is hit with probability q; all other characters are hit with equal probability, $\frac{(1-q)}{k}$. A space is used to separate words". The distribution of the resulting words of this random typing process follow a power-law. Conrad and Mitzenmacher [6] showed that this relation still holds when the keys are hit with *unequal* probability.

Our model generalizes the above model of natural human behavior, using 'random typing'. We build our model RTG (*Random Typing Generator*) in *three* steps, incrementally. In the next two steps, we introduce the base version of the proposed model to give an insight. However, as will become clear, it has *two* shortcomings. In particular, the base model does not capture (1) homophily, the tendency to associate and bond with similar others- people tend to be acquainted with others similar in age, class, geographical area, etc. and (2) community structure, the existence of groups of nodes that are more densely connected internally than with the rest of the graph.

3.1 RTG-IE: RTG with Independent Equiprobable Keys

As in Miller's experimental setting, we propose each unique word typed by the monkey to represent a node in the output graph (one can think of each unique word as the label of the corresponding node). To form links between nodes, we mark the sequence of words as 'source' and 'destination', alternatingly. That is, we divide the sequence of words into groups of two and link the first node to the second node in each pair. If two nodes are already linked, the weight of the edge is simply increased by 1. Therefore, if W words are typed, the total weight of the output graph is $W/2$. See Figure 1 for an example illustration. Intuitively, random typing introduces new nodes to the graph as more words are typed, because the possibility of generating longer words increases with increasing number of words typed.

Due to its simple structure, this model is very easy to implement and is indeed mathematically tractable. If W words are typed on a keyboard with k keys and

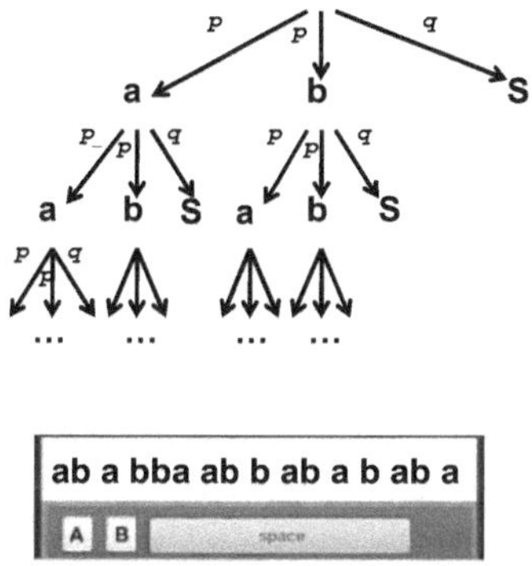

Time	Source	Destination	Weight
T1	ab	a	1
T2	bba	ab	1
T3	b	ab	1
T4	a	b	1
T5	ab	a	1

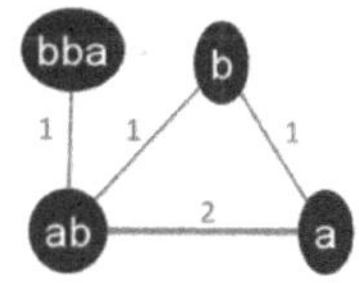

Fig. 1. Illustration of the RTG-IE. Upper left: how words are (recursively) generated on a keyboard with two equiprobable keys, 'a' and 'b', and a space bar; lower left: a keyboard is used to randomly type words, separated by the space character; upper right: how words are organized in pairs to create source and destination nodes in the graph over time; lower right: the output graph; each node label corresponds to a unique word, while labels on edges denote weights

a space bar, the probability p of hitting a key being the same for all keys and the probability of hitting the space bar being denoted as q=$(1 - kp)$:

Lemma 1. *The expected number of nodes N in the output graph G of the RTG-IE model is*

$$N \propto W^{-log_p k}.$$

Proof: In the Appendix. □

Lemma 2. *The expected number of edges E in the output graph G of the RTG-IE model is*

$$E \approx W^{-log_p k} * (1 + c' logW), \quad for \ c' = \frac{q^{-log_p k}}{-logp} > 0.$$

Proof: In the Appendix. □

Lemma 3. *The in(out)-degree d_n of a node in the output graph G of the RTG-IE model is power law related to its total in(out)-weight W_n, that is,*

$$W_n \propto d_n^{-log_k p}$$

with expected exponent $-log_k p > 1$.

Proof: In the Appendix. □

Even though most of the properties listed at the beginning of this section are matched, there are two problems with this model: (1) the degree distribution follows a power-law only for small degrees and then shows multinomial characteristics (See Figure 2), and (2) it does not generate homophily and community structure, because it is possible for every node to get connected to every other node, rather than to 'similar' nodes in the graph.

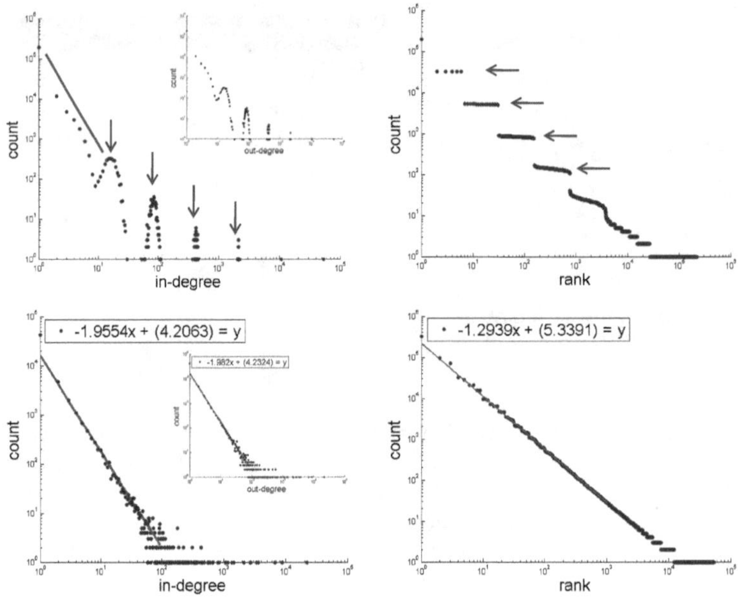

Fig. 2. Top row: Results of RTG-IE ($k = 5$, $p = 0.16$, $W = 1M$). The problem with this model is that in(out)-degrees form multinomial clusters (left). This is because nodes with labels of the same length are expected to have the same degree. This can be observed on the rank-frequency plot (right) where we see many words with the same frequency. Notice the 'staircase effect'. Bottom row: Results of RTG-IU ($k = 5$, $p = [0.03, 0.05, 0.1, 0.22, 0.30]$, $W = 1M$). Unequal probabilities introduce smoothing on the frequency of words that are of the same length (right). As a result, the degree distribution follows a power-law with expected heavy tails (left).

3.2 RTG-IU: RTG with Independent Un-Equiprobable Keys

We can spread the degrees so that nodes with the same-length but otherwise distinct labels would have different degrees by making keys have *unequal* probabilities. This procedure introduces smoothing in the distribution of degrees, which remedies the first problem introduced by the RTG-IE model. In addition, thanks to [6], we are still guaranteed to obtain the desired power-law characteristics as before. See Figure 2.

3.3 RTG: Random Typing Graphs

What the previous model fails to capture is the homophily and community structure. In a real network, we would expect nodes to get connected to similar nodes (homophily), and form groups and possibly groups within groups (modular structure). In our model, for example on a keyboard with two keys 'a' and 'b', we would like nodes with many 'a's in their labels to be connected to similar nodes, as opposed to nodes labeled with many 'b's. However, in both RTG-IE and

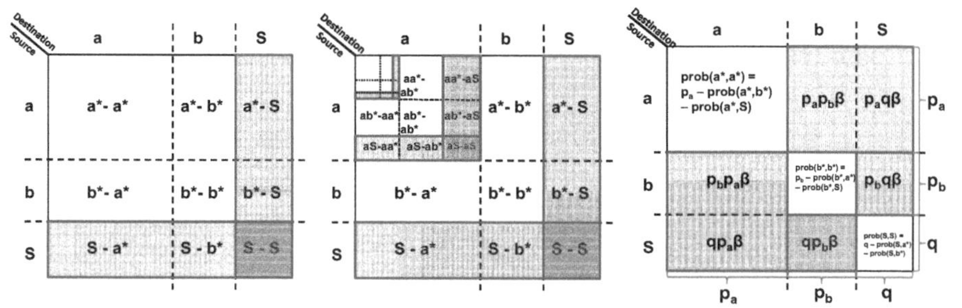

(a) first level (b) recursion (c) communities

Fig. 3. The **RTG** model: random typing on a 2-d keyboard, generating edges (source-destination pairs). See Algorithm 1. (a) an example 2-d keyboard (nine keys), hitting a key generates the row(column) character for source(destination), shaded keys terminate source and/or destination words. (b) illustrates recursive nature. (c) the imbalance factor β favors diagonal keys and leads to homophily.

RTG-IU it is possible for every node to connect to every other node. In fact, this yields a tightly connected core of nodes with rather short labels.

Our proposal to fix this is to envision a two-dimensional keyboard that generates source and destination labels in one shot, as shown in Figure 3. The previous model generates a word for source, and, completely independently, another word for destination. In the example with two keys, we can envision this process as picking one of the nine keys in Figure 3(a), using the independence assumption: the probability for each key is the product of the probability of the corresponding row times the probability of the corresponding column: p_l for letter l, and q for space ('S'). After a key is selected, its row character is appended to the source label, and the column character to the destination label. This process repeats recursively as in Figure 3(b), until the space character is hit on the first dimension in which case the source label is terminated and also on the second dimension in which case the destination label is terminated.

In order to model homophily and communities, rather than assigning cross-product probabilities to keys on the 2-d keyboard, we introduce an imbalance factor β, which will decrease the chance of a-to-b edges, and increase the chance for a-to-a and b-to-b edges, as shown in Figure 3(c). Thus, for the example that we have, the formulas for the probabilities of the nine keys become:

$$\begin{aligned}
prob(a,b) = prob(b,a) = p_a p_b \beta \ , \ prob(a,a) = p_a - (prob(a,b) + prob(a,S)),\\
prob(S,a) = prob(a,S) = q p_a \beta \ , \ prob(b,b) = p_b - (prob(b,a) + prob(b,S)),\\
prob(S,b) = prob(b,S) = q p_b \beta \ , \ prob(S,S) = q - (prob(S,a) + prob(S,b)).
\end{aligned}$$

By boosting the probabilities of the diagonal keys and down-rating the probabilities of the off-diagonal keys, we are guaranteed that nodes with similar labels will have higher chance to get connected. The pseudo-code of generating edges as described above is shown in Algorithm 1.

Next, before showing the experimental results of RTG, we take a detour to describe how we handle time so that edge/weight additions are bursty and self-similar. We also discuss the generalizations of the model in order to produce all types of uni/bipartite, (un)weighted, and (un)directed graphs.

Algorithm 1. RTG

```
Input: k, q, W, β
Output: edge-list L for output graph 𝒢
Initialize (k + 1)-by-(k + 1) matrix P with cross-product probabilities
// in order to ensure homophily and community structure
Multiply off-diagonal probabilities by β, 0 < β < 1
Boost diagonal probabilities s.t. sum of row(column) probabilities remain the same.
Initialize edge list L
for 1 to W do
   L1, L2 ← SelectNodeLabels(P)
   Append L1, L2 to L
end for

function SelectNodeLabels (P) : L1, L2
Initialize L1 and L2 to empty string
while not terminated L1 and not terminated L2 do
   Draw i,j with probability P(i, j)
   if i ≤ k, j ≤ k then
      Append character 'i' to L1 and 'j' to L2 if not terminated
   else if i ≤ k, j=k + 1 then
      Append character 'i' to L1 if not terminated
      Terminate L2
   else if i=k + 1, j ≤ k then
      Append character 'j' to L2 if not terminated
      Terminate L1
   else
      Terminate L1 and L2
   end if
end while
Return L1 and L2
end function
```

3.4 Burstiness and Self-Similarity

Most real-world traffic as well as edge/weight additions to real-world graphs have been found to be self-similar and bursty [7,14,15,22]. Therefore, in this section we give a brief overview of how to aggregate time so that edge and weight additions, that is ΔE and ΔW, are bursty and self-similar.

Notice that when we link two nodes at each step, we add 1 to the total weight W. So, if every step is represented as a single time-tick, the weight additions are uniform. However, to generate bursty traffic, we need to have a *bias factor* $b > 0.5$, such that b-fraction of the additions happen in one half and the remaining in the other half. We will use the b-model [30], which generates such self-similar and bursty traffic. Specifically, starting with a uniform interval, we will recursively subdivide weight additions to each half, quarter, and so on, according to the bias b. To create randomness, at each step we will randomly swap the order of fractions b and $(1-b)$.

Among many methods that measure self-similarity we use the entropy plot [30], which plots the entropy $H(r)$ versus the resolution r. The resolution is the scale, that is, at resolution r, we divide our time interval into 2^r equal sub-intervals, compute ΔE in each sub-interval $k(k = 1 \ldots 2^r)$, normalize into fractions $p_k = \frac{\Delta E}{E}$, and compute the Shannon entropy $H(r)$ of the sequence p_k. If the plot $H(r)$ is linear, the corresponding time sequence is said to be *self-similar*, and the slope of the plot is defined as the fractal dimension f_d of the time sequence. Notice that a uniform Δ distribution yields f_d=1; a lower value of f_d corresponds to a more bursty time sequence, with a single burst having the lowest f_d=0: the fractal dimension of a point.

3.5 Generalizations

We can easily generalize RTG to model all type of graphs. To generate undirected graphs, we can simply assume edges from source to destination to be undirected as the formation of source and destination labels is the same and symmetric. For unweighted graphs, we can simply ignore *duplicate* edges, that is, edges that connect already linked nodes. Finally, for bipartite graphs, we can use *two* different sets of keys such that on the 2-d keyboard, source dimension contains keys from the first set, and the destination dimension from the other set. This assures source and destination labels to be completely different, as desired.

4 Experimental Results

The question we wish to answer here is how RTG is able to model real-world graphs. The datasets we used are:
Blognet: a social network of blogs based on citations (undirected, unipartite and unweighted with N=27,726; E=126,227; over 80 time ticks).
Com2Cand: the U.S. electoral campaign donations network from organizations to candidates (directed, bipartite and weighted with N=23,191; E=877,721; and W=4,383,105,580 over 29 time ticks). Weights on edges indicate donated dollar amounts.

In Figures 4 and 5, we show the related patterns for *Blognet* and *Com2Cand* as well as synthetic results, respectively. In order to model these networks, we ran experiments for different parameter values k, q, W, and β. Here, we show the closest results that RTG generated, though fitting the parameters is a challenging future direction. We observe that RTG is able to match the *long* wish-list of static and dynamic properties we presented earlier for the two real graphs.

In order to evaluate community structure, we use the modularity measure in [25]. Figure 6(left) shows that modularity increases with smaller imbalance factor β. Without any imbalance, β=1, modularity is as low as 0.35, which indicates that no significant modularity exists. In Figure 6(right), we also show

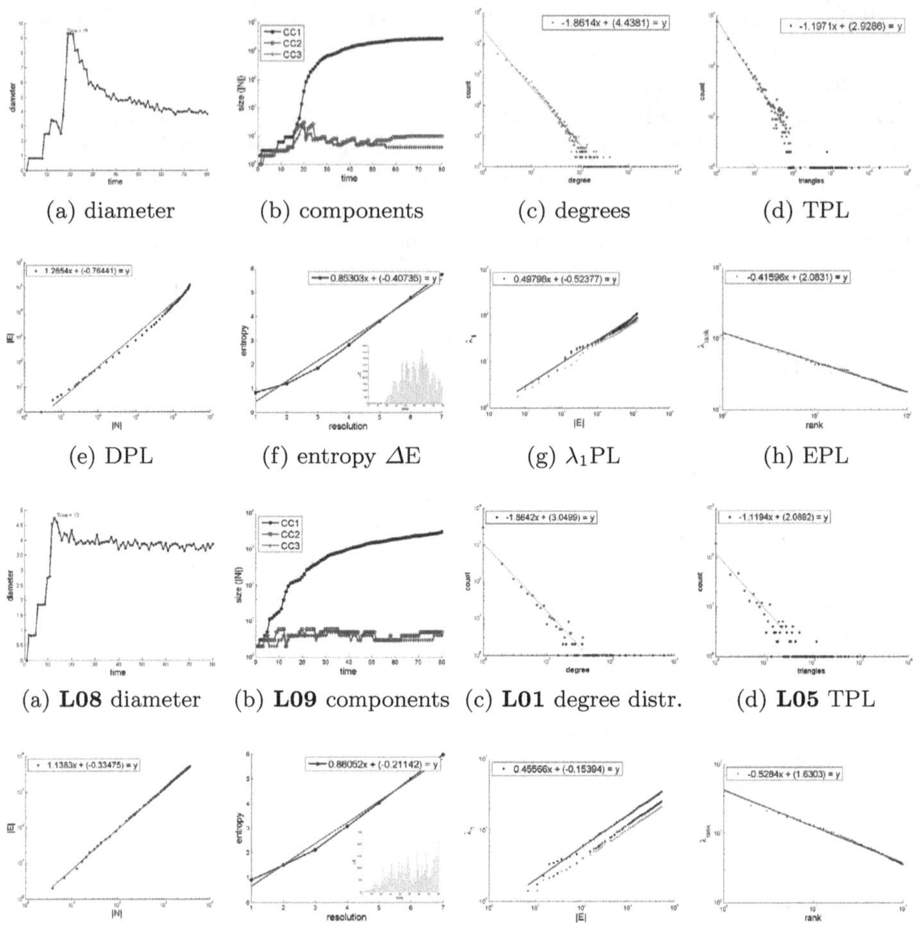

(a) diameter (b) components (c) degrees (d) TPL

(e) DPL (f) entropy ΔE (g) λ_1PL (h) EPL

(a) **L08** diameter (b) **L09** components (c) **L01** degree distr. (d) **L05** TPL

(e) **L02** DPL (f) **L11** entropy ΔE (g) **L07** λ_1PL (h) **L06** EPL

Fig. 4. Top two rows: properties of *Blognet*: (a) small and shrinking diameter; (b) largest 3 connected components; (c) degree distribution; (d) triangles Δ vs number of nodes with Δ triangles; (e) densification; (f) bursty edge additions; (g) largest 3 eigenvalues wrt E; (h) rank spectrum of the adjacency matrix. Bottom two rows: results of RTG. Notice the similar qualitative behavior for all *eight* laws.

the running time of RTG wrt the number of duplicate edges (that is, number of iterations W). Notice the *linear* growth with increasing W.

5 Conclusion

We have designed a generator that meets all the five desirable properties in the introduction. Particularly, our model is

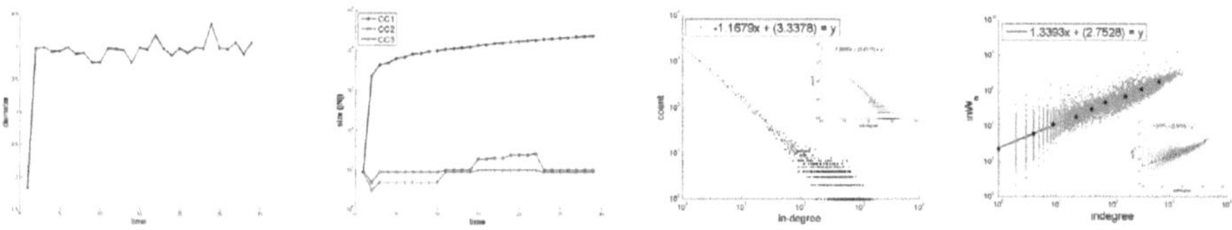

(a) diameter (b) components (c) degree distr. (d) SPL

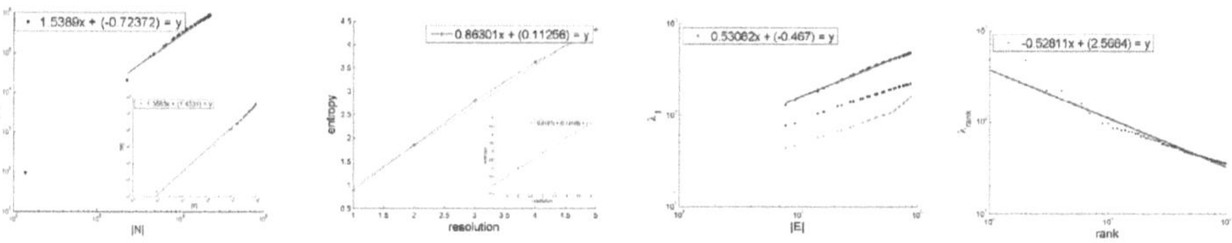

(e) D(W)PL (f) entropy ΔW(E) (g) λ_1PL (h) EPL

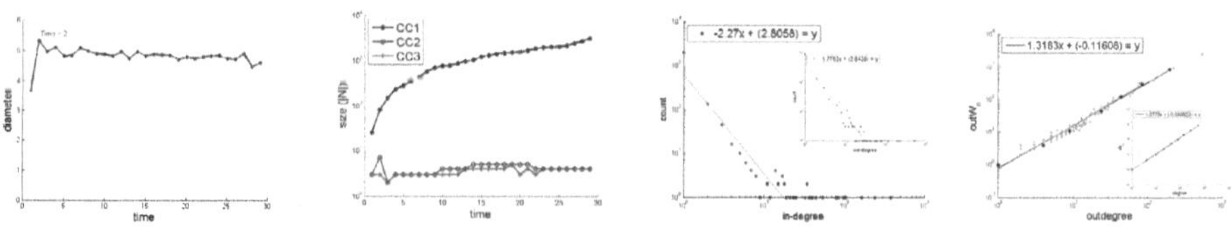

(a) **L08** diameter (b) **L09** components (c) **L01** degree distr. (d) **L04** SPL

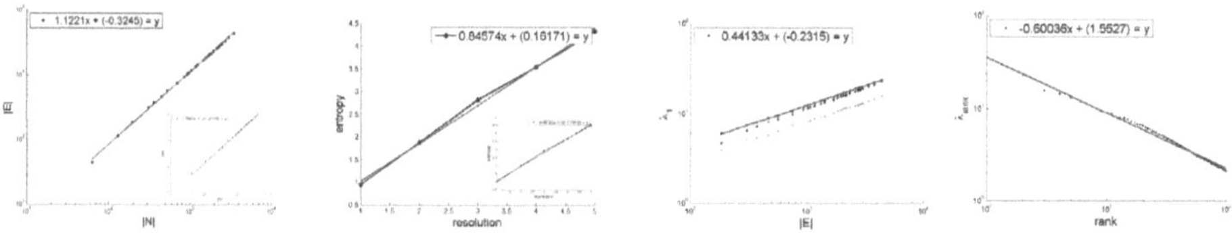

(e) **L02, L03** D(W)PL (f) **L11** entropy ΔW(E) (g) **L07** λ_1PL (h) **L06** EPL

Fig. 5. Top two rows: properties of *Com2Cand*; as opposed to *Blognet*, *Com2Cand* is *weighted*. So, different from above we show: (d) node weight vs in(inset: out)degree; (e) total weight vs number of edges(inset); (f) bursty weight additions(inset); Bottom two rows: results of RTG. Notice the similar qualitative behavior for all *nine* laws.

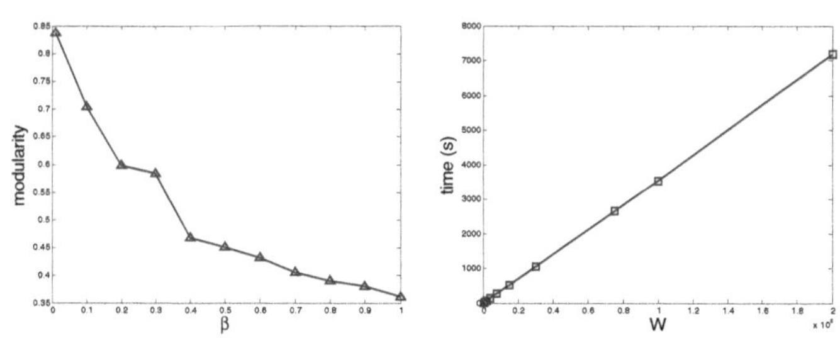

Fig. 6. Left: modularity score vs. imbalance factor β, modularity increases with decreasing β. For β=1, the score is very low indicating no significant modularity. Right: computation time vs. W, time grows *linearly* with increasing number of iterations W.

1. simple and intuitive, yet it generates the emergent, macroscopic patterns that we see in real graphs.
2. realistic, generating graphs that obey all *eleven* properties that real graphs obey - no other generator has been shown to achieve that.
3. parsimonious, requiring only a handful of parameters.
4. flexible, capable of generating weighted/unweighted, directed/undirected, and unipartite/bipartite graphs, and any combination of the above.
5. fast, being linear on the number of iterations (on a par with the number of duplicate edges in the output graph).

Moreover, we showed how well RTG can mimic some large, real graphs. We have also proven that an early version of RTG generates several of the desired (power) laws, formulated in terms of model parameters.

Acknowledgments. This material is based upon work supported by the National Science Foundation under Grants No. IIS-0705359, IIS-0705215 and IIS-0808661, also by the iCAST project sponsored by the National Science Council, Taiwan, under Grants No. NSC97-2745-P-001-001 and by the Ministry of Economic Affairs, Taiwan, under Grants No. 97-EC-17-A-02-R7-0823 and under the auspices of the U.S. Department of Energy by University of California Lawrence Livermore National Laboratory under contract DE-AC52-07NA27344 (LLNL-CONF-404625), subcontracts B579447, B580840. Any opinions, findings, and conclusions or recommendations expressed in this material are those of the author(s) and do not necessarily reflect the views of any of the funding parties.

References

1. Akoglu, L., McGlohon, M., Faloutsos, C.: Rtm: Laws and a recursive generator for weighted time-evolving graphs. In: ICDM (2008)
2. Albert, R., Jeong, H., Barabasi, A.-L.: Diameter of the World Wide Web. Nature 401, 130–131 (1999)
3. Barabasi, A.L., Albert, R.: Emergence of scaling in random networks. Science 286(5439), 509–512 (1999)
4. Chakrabarti, D., Faloutsos, C.: Graph mining: Laws, generators, and algorithms. ACM Comput. Surv. 38(1) (2006)
5. Chakrabarti, D., Zhan, Y., Faloutsos, C.: R-MAT: A recursive model for graph mining. In: SIAM Int. Conf. on Data Mining (April 2004)
6. Conrad, B., Mitzenmacher, M.: Power laws for monkeys typing randomly: the case of unequal probabilities. IEEE Transactions on Information Theory 50(7), 1403–1414 (2004)
7. Crovella, M., Bestavros, A.: Self-similarity in world wide web traffic, evidence and possible causes. Sigmetrics, 160–169 (1996)
8. Erdos, P., Renyi, A.: On the evolution of random graphs. Publ. Math. Inst. Hungary. Acad. Sci. 5, 17–61 (1960)
9. Even-Bar, E., Kearns, M., Suri, S.: A network formation game for bipartite exchange economies. In: SODA (2007)
10. Fabrikant, A., Luthra, A., Maneva, E.N., Papadimitriou, C.H., Shenker, S.: On a network creation game. In: PODC (2003)

11. Faloutsos, M., Faloutsos, P., Faloutsos, C.: On power-law relationships of the internet topology. In: SIGCOMM, August-September 1999, pp. 251–262 (1999)
12. Flake, G.W., Lawrence, S., Giles, C.L., Coetzee, F.M.: Self-organization and identification of web communities. IEEE Computer 35, 66–71 (2002)
13. Girvan, M., Newman, M.E.J.: Community structure in social and biological networks. PNAS 99, 7821 (2002)
14. Gomez, M.E., Santonja, V.: Self-similarity in i/o workload: Analysis and modeling. In: WWC (1998)
15. Gribble, S.D., Manku, G.S., Roselli, D., Brewer, E.A., Gibson, T.J., Miller, E.L.: Self-similarity in file systems. In: SIGMETRICS 1998 (1998)
16. Kleinberg, J.M., Kumar, R., Raghavan, P., Rajagopalan, S., Tomkins, A.S.: The Web as a graph: Measurements, models and methods. In: Asano, T., Imai, H., Lee, D.T., Nakano, S.-i., Tokuyama, T. (eds.) COCOON 1999. LNCS, vol. 1627, pp. 1–17. Springer, Heidelberg (1999)
17. Scheinerman, E., Kraetzl, M., Nickel, C.: Random dot product graphs: a model for social networks (Preliminary Manuscript) (2005)
18. Laoutaris, N., Poplawski, L.J., Rajaraman, R., Sundaram, R., Teng, S.-H.: Bounded budget connection (bbc) games or how to make friends and influence people, on a budget. In: PODC (2008)
19. Leskovec, J., Chakrabarti, D., Kleinberg, J.M., Faloutsos, C.: Realistic, mathematically tractable graph generation and evolution, using kronecker multiplication. In: Jorge, A.M., Torgo, L., Brazdil, P.B., Camacho, R., Gama, J. (eds.) PKDD 2005. LNCS (LNAI), vol. 3721, pp. 133–145. Springer, Heidelberg (2005)
20. Leskovec, J., Kleinberg, J., Faloutsos, C.: Graphs over time: densification laws, shrinking diameters and possible explanations. In: ACM SIGKDD (2005)
21. Mandelbrot, B.: An informational theory of the statistical structure of language. Communication Theory (1953)
22. McGlohon, M., Akoglu, L., Faloutsos, C.: Weighted graphs and disconnected components: Patterns and a generator. In: ACM SIGKDD, Las Vegas (August 2008)
23. Miller, G.A.: Some effects of intermittent silence. American Journal of Psychology 70, 311–314 (1957)
24. Newman, M.E.J.: Power laws, Pareto distributions and Zipf's law (December 2004)
25. Newman, M.E.J., Girvan, M.: Finding and evaluating community structure in networks. Physical Review E 69, 026113 (2004)
26. Pennock, D.M., Flake, G.W., Lawrence, S., Glover, E.J., Giles, C.L.: Winners don't take all: Characterizing the competition for links on the web. Proceedings of the National Academy of Sciences, 5207–5211 (2002)
27. Schwartz, M.F., Wood, D.C.M.: Discovering shared interests among people using graph analysis of global electronic mail traffic. Communications of the ACM 36, 78–89 (1992)
28. Siganos, G., Faloutsos, M., Faloutsos, P., Faloutsos, C.: Power laws and the AS-level internet topology (2003)
29. Tsourakakis, C.E.: Fast counting of triangles in large real networks without counting: Algorithms and laws. In: ICDM (2008)
30. Wang, M., Madhyastha, T., Chan, N.H., Papadimitriou, S., Faloutsos, C.: Data mining meets performance evaluation: Fast algorithms for modeling bursty traffic. In: ICDE, pp. 507–516 (2002)
31. Watts, D.J., Strogatz, S.H.: Collective dynamics of 'small-world' networks. Nature 393(6684), 440–442 (1998)

32. Young, S.J., Scheinerman, E.R.: Random dot product graph models for social networks. In: Bonato, A., Chung, F.R.K. (eds.) WAW 2007. LNCS, vol. 4863, pp. 138–149. Springer, Heidelberg (2007)
33. Zipf, G.K.: Selective Studies and the Principle of Relative Frequency in Language. Harvard University Press (1932)

Appendix

Consider the following setting: W words are typed on a keyboard with k keys and a space bar, the probability of hitting a key p being the same for all keys and probability of hitting the space bar being denoted as q=$(1-kp)$, in the output graph G of the RTG-IE model:

Lemma 1. *The expected number of nodes N is*

$$N \propto W^{-log_p k}.$$

Proof. Given the number of words W, we want to find the expected number of nodes N that the RTG-IE graph consists of. This question can be reformulated as follows: "Given W words typed by a monkey on a keyboard with k keys and a space bar, what is the size of the vocabulary V?" The number of unique words V is basically equal to the number of nodes N in the output graph.

Let w denote a single word generated by the defined random process. Then, w can recursively be written as follows: "$w : c_i w|S$", where c_i is the character that corresponds to key i, $1 \leq i \leq k$, and S is the space character. So, V as a function of model parameters can be formulated as:

$$\begin{aligned} V(W) &= V(c_1, Wp) + V(c_2, Wp) + \ldots + V(c_k, Wp) + V(S) \\ &= k * V(Wp) + V(S) = \ k * V(Wp) + \begin{cases} 1, 1-(1-q)^W \\ 0, (1-q)^W \end{cases} \end{aligned}$$

where q denotes the probability of hitting the space bar, i.e. $q = 1 - kp$. Given the fact that W is often large, and $(1-q) < 1$, it is almost always the case that w=S is generated; but since this adds only a constant factor, we can ignore it in the rest of the computation. That is,

$$V(W) \approx k * V(Wp) = k * (k * V(Wp^2)) = k^n * V(1)$$

where $n = log_p(1/W) = -log_p W$. By definition, when W=1, that is, in case only one word is generated, the vocabulary size is 1, i.e. V(1)=1. Therefore,

$$V(W) = N \propto k^n = k^{-log_p W} = W^{-log_p k}.$$

□

The above proof shown using recursion is in agreement with the early result of Miller [23], who showed that in the monkey-typing experiment with k equiprobable keys (with probability p) and a space bar (with probability q), the rank-frequency distribution of words follow a power law. In particular,

$$f(r) \propto r^{-1+log_k(1-q)-1} = r^{log_k p}.$$

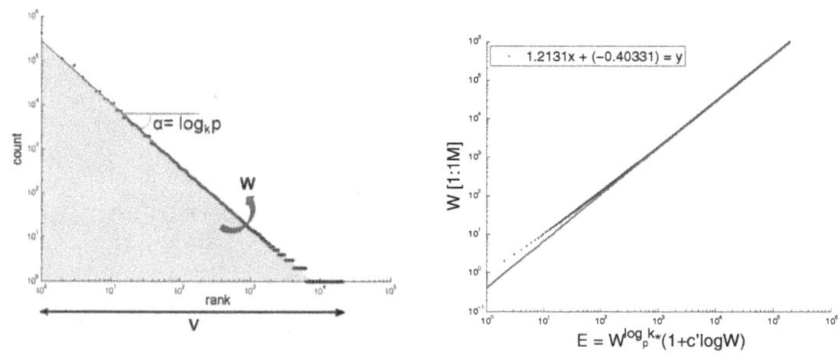

Fig. 7. (a) Rank vs count of vocabulary words typed randomly on a keyboard with k equiprobable keys (with probability p) and a space bar (with probability q), follow a power law with exponent $\alpha = log_k p$. Approximately, the area under the curve gives the total number of words typed. (b) The relationship between number of edges E and total weight W behaves like a power-law (k=2, p=0.4).

In this case, the number of ranks corresponds to the number of unique words, that is, the vocabulary size V. And, the sum of the counts of occurrences of all words in the vocabulary should give W, the number of words typed. The total count can be approximated by the area under the curve on the rank-count plot. See Figure 7(a). Next, we give a second proof of Lemma 1 using Miller's result.

Proof. Let $\alpha = log_k p$ and $C(r)$ denote the number of times that the word with rank r is typed. Then, $C(r) = cr^\alpha$, where $C(r)_{min} = C(V) = cV^\alpha$ and the constant $c = C(V)V^{-\alpha}$. Then we can write W as

$$W = C(V)V^{-\alpha}\left(\sum_{r=1}^{V} r^\alpha\right) \approx C(V)V^{-\alpha}\left(\int_{r=1}^{V} r^\alpha dr\right) = C(V)V^{-\alpha}\left(\frac{r^{\alpha+1}}{\alpha+1}\Big|_{r=1}^{V}\right)$$

$$= C(V)V^{-\alpha}\left(\frac{1}{-\alpha-1} - \frac{1}{(-\alpha-1)V^{-\alpha-1}}\right) \approx c'V^{-\alpha}.$$

where $c' = \frac{C(V)}{-\alpha-1}$, where $\alpha < -1$ and $C(V)$ is very small (usually 1). Therefore,

$$V = N \propto W^{-\frac{1}{\alpha}} = W^{-log_p k}.$$

□

Lemma 2. *The expected number of edges E is*

$$E \approx W^{-log_p k} * (1 + c' logW), \quad for\ c' = \frac{q^{-log_p k}}{-logp} > 0.$$

Proof. Given the number of words W, we want to find the expected number of edges E that the RTG-IE graph consists of. The number of edges E is the same as the unique number of *pairs* of words. We can think of a pair of words as a single word e, the generation of which is stopped after the *second* hit to the space bar. So, e always contains a single space character. Recursively, "$e : c_i e|Sw$", where "$w : c_i w|S$". So, E can be formulated as:

$$E(W) = k * E(Wp) + V(Wq) \tag{1}$$

$$V(Wq) = \quad k * V(Wqp) + \begin{cases} 1, 1-(1-q)^{Wq} \\ 0, (1-q)^{Wq} \end{cases} \tag{2}$$

From Lemma 1, Equ.(2) can be approximately written as $V(Wq) = (Wq)^{-log_p k}$. Then, Equ.(1) becomes $E(W) = k * E(Wp) + cW^{\alpha}$, where $c = q^{-log_p k}$ and $\alpha = -log_p k$. Given that E(W=1)=1, we can solve the recursion as follows:

$$\begin{aligned} E(W) &\approx k * (k * E(Wp^2) + c(Wp)^{\alpha}) + cW^{\alpha} \\ &= k * (k * (k * V(Wp^3) + c(Wp^2)^{\alpha}) + c(Wp)^{\alpha}) + cW^{\alpha} \\ &= k^n * V(1) + k^{n-1} * c(Wp^{n-1}) + k^{n-2} * c(Wp^{n-2})^{\alpha} + \ldots + cW^{\alpha} \\ &= k^n * V(1) + cW^{\alpha}((kp^{\alpha})^{n-1} + (kp^{\alpha})^{n-2} + \ldots + 1) \end{aligned}$$

where $n = log_p(1/W) = -log_p W$. Since $kp^{\alpha} = kp^{-log_p k} = 1$,

$$E(W) \approx k^n * V(1) + n * cW^{\alpha} = k^{-log_p W} + c\frac{-log\frac{1}{W}}{-logp}W^{-log_p k} = W^{-log_p k}(1 + c' logW)$$

where $c' = \frac{c}{-logp} = \frac{q^{-log_p k}}{-logp} > 0$.

□

The above function of E in terms of W and other model parameters looks like a power-law for a wide range of W. See Figure 7(b).

Lemma 3. *The in/out-degree d_n of a node is power law related to its total in/out-weight W_n, that is,*

$$W_n \propto d_n^{-log_k p}$$

with expected exponent $-log_k p > 1$.

Proof. We will show that $W_n \propto d_n^{-log_k p}$ for out-edges, and a similar argument holds for in-edges. Given that the experiment is repeated W times, let W_n denote the number of times a unique word is typed as a source. Each such unique word corresponds to a node in the final graph and W_n is basically its out-weight, since the node appears as a source node. Then, the out-degree d_n of a node is simply the number of unique words typed as a destination. From Lemma 1,

$$W_n \propto d_n^{-log_k p}, \quad for \; -log_k p > 1.$$

□

Taxonomy-Driven Lumping for Sequence Mining*

Francesco Bonchi, Carlos Castillo, Debora Donato, and Aristides Gionis

Yahoo! Research
Diagonal 177, Barcelona, 080018, Spain
{bonchi,chato,debora,gionis}@yahoo-inc.com

In many application domains, events are naturally organized in a hierarchy. Whether events describe human activities, system failures, coordinates in a trajectory, or biomedical phenomena, there is often a taxonomy that should be taken into consideration. A taxonomy allow us to represent the information at a more general description level, if we choose carefully the most suitable level of granularity.

Given a taxonomy of events and a dataset of sequences of these events, we study the problem of finding efficient and effective ways to produce a compact representation of the sequences. This can be valuable by itself, or can be used to help solving other problems, such as clustering.

We model sequences with Markov models whose states correspond to nodes in the provided taxonomy, and each state represents the events in the subtree under the corresponding node. By lumping observed events to states that correspond to internal nodes in the taxonomy, we allow more compact models that are easier to understand and visualize, at the expense of a decrease in the data likelihood.

We formally define and characterize our problem, and we propose a scalable search method for finding a good trade-off between two conflicting goals: maximizing the data likelihood, and minimizing the model complexity. We implement these ideas in TAXOMO, a taxonomy-driven modeler.

TAXOMO receives a database of sequences of symbols, and a taxonomy over those symbols. An initial Markov model is created for the sequences without considering the taxonomy, and then refine it iteratively by merging states driven by the taxonomy. The likelihood of the data given a new model generated by this merging procedure, can be computed directly from the likelihood of the data given the model before the merging. This yields a fast model evaluation method that can explore many configurations in a short time. We also implement efficient strategies that guide the search process. We apply TAXOMO in two different domains, query-log mining and mining of moving-object trajectories. The empirical evaluation confirms the feasibility and usefulness of our approach.

This is an extended abstract of an article published in the Data Mining and Knowledge Discovery journal [1].

Reference

1. Bonchi, F., Castillo, C., Donato, D., Gionia, A.: Taxonomy-driven lumping for sequence mining. Data Mining and Knowledge Discovery (2009) DOI: 10.1007/s10618-009-0141-6

* This is an extended abstract of an article published in the Data Mining and Knowledge Discovery Journal [1].

W. Buntine et al. (Eds.): ECML PKDD 2009, Part I, LNAI 5781, p. 29, 2009.

On Subgroup Discovery in Numerical Domains*

Henrik Grosskreutz and Stefan Rüping

Fraunhofer IAIS, Schloss Birlinghoven, Sankt Augustin, Germany
{henrik.grosskreutz,stefan.rueping}@iais.fraunhofer.de

Abstract. Subgroup discovery is a Knowledge Discovery task that aims at finding subgroups of a population with high generality and distributional unusualness. While several subgroup discovery algorithms have been presented in the past, they focus on databases with nominal attributes or make use of discretization to get rid of the numerical attributes. In this paper, we illustrate why the replacement of numerical attributes by nominal attributes can result in suboptimal results. Thereafter, we present a new subgroup discovery algorithm that prunes large parts of the search space by exploiting bounds between related numerical subgroup descriptions. The same algorithm can also be applied to ordinal attributes. In an experimental section, we show that the use of our new pruning scheme results in a huge performance gain when more that just a few split-points are considered for the numerical attributes.

Reference

1. Grosskreutz, H., Rüping, S.: On Subgroup Discovery in Numerical Domains. Data Mining and Knowledge Discovery (2009) DOI: 10.1007/s10618-009-0136-3

* This is an extended abstract of an article published in the Data Mining and Knowledge Discovery journal [1].

W. Buntine et al. (Eds.): ECML PKDD 2009, Part I, LNAI 5781, p. 30, 2009.

Harnessing the Strengths of Anytime Algorithms for Constant Data Streams*

Philipp Kranen and Thomas Seidl

Data Manangement and Exploration Department
RWTH Aachen University, Germany
{kranen,seidl}@cs.rwth-aachen.de

Abstract. Anytime algorithms have been proposed for many different applications e.g. in data mining. Their strengths are the ability to first provide a result after a very short initialization and second to improve their result with additional time. Therefore, anytime algorithms have so far been used when the available processing time varies, e.g. on varying data streams. In this paper we propose to employ anytime algorithms on constant data streams, i.e. for tasks with constant time allowance. We introduce two approaches that harness the strengths of anytime algorithms on constant data streams and thereby improve the over all quality of the result with respect to the corresponding budget algorithm. We derive formulas for the expected performance gain and demonstrate the effectiveness of our novel approaches using existing anytime algorithms on benchmark data sets.

The goal that was set and reached in this paper is to improve the quality of the result over that of traditional budget approaches, which are used in an abundance of stream mining applications. Using anytime classification as an example application we show for SVM, Bayes and nearest neighbor classifiers that both our novel approaches improve the classification accuracy for slow and fast data streams. The results confirm our general theoretic models and show the effectiveness of our approaches. The simple yet effective idea can be employed for any anytime algorithm along with a quality measure and motivates further research in e.g. classification confidence measures or anytime algorithms.

Acknowledgments

This work has been supported by the UMIC Research Centre, RWTH Aachen University.

Reference

1. Kranen, P., Seidl, T.: Harnessing the Strengths of Anytime Algorithms for Constant Data Streams. Data Mining and Knowledge Discovery (2009) DOI: 10.1007/s10618-009-0139-0

* This is an extended abstract of an article published in the Data Mining and Knowledge Discovery journal [1].

W. Buntine et al. (Eds.): ECML PKDD 2009, Part I, LNAI 5781, p. 31, 2009.

Identifying the Components*

Matthijs van Leeuwen, Jilles Vreeken, and Arno Siebes

Department of Computer Science
Universiteit Utrecht
{mleeuwen,jillesv,arno}@cs.uu.nl

Most, if not all, databases are mixtures of samples from different distributions. In many cases, however, nothing is known about the source components of these mixtures. Therefore, many methods that induce models regard a database as sampled from a single data distribution. Models that do take into account that databases actually are sampled from mixtures of distributions are often superior to those that do not, independent of whether this is modelled explicitly or implicitly.

Transaction databases are no different with regard to data distribution. For the prototypical example, supermarket basket analysis, one also expects a mixture of different buying behaviours. Households of retired people buy different collections of items than households with young children, although overlap may exist. By extracting both the groups of people and their corresponding buying patterns, a company can learn a lot about its customers.

But, what does "different buying behaviour" mean? It certainly does not mean that the different groups should buy completely different sets of items. Also, it does not mean that these groups cannot have common frequent item sets. Rather, it means that the *characteristics* of the sampled distributions are different. This may seem like a play of words, but it is not. Sampled distributions of transaction data can be characterised precisely through the use of a pattern-based compressor.

We introduce two MDL-based algorithms that follow orthogonal approaches to identify the components in a transaction database. The first follows a model-based approach, while the second is data-driven. Both are parameter-free: the number of components and the components themselves are chosen such that the combined complexity of data and model is minimised. Further, neither prior knowledge on the distributions nor a distance metric on the data is required.

Experiments show that highly characteristic components are identified. Both algorithms are evaluated on basis of total compressed sizes and component purity, but we also look at (dis)similarities between the components and their characteristic patterns. The results show that both our orthogonal methods identify the components of the database. Visual inspection confirms that characteristic decompositions are identified.

Reference

[1] van Leeuwen, M., Vreeken, J., Siebes, A.: Identifying the Components. Data Mining and Knowledge Discovery (2009) DOI: 10.1007/s10618-009-0137-2

* This is an extended abstract of an article published in the Data Mining and Knowledge Discovery Journal [1].

W. Buntine et al. (Eds.): ECML PKDD 2009, Part I, LNAI 5781, p. 32, 2009.

Two-Way Analysis of High-Dimensional Collinear Data*

Ilkka Huopaniemi[1,**], Tommi Suvitaival[1], Janne Nikkilä[1,2], Matej Orešič[3], and Samuel Kaski[1]

[1] Department of Information and Computer Science, Helsinki University of Technology, P.O. Box 5400, FI-02015 TKK, Finland
{ilkka.huopaniemi,tommi.suvitaival,janne.nikkila,samuel.kaski}@tkk.fi
[2] Department of Basic Veterinary Sciences, (Division of Microbiology and Epidemiology), Faculty of Veterinary Medicine, University of Helsinki, P.O. Box 66, University of Helsinki FIN-00014 Finland
[3] VTT Technical Research Centre of Finland, P.O. Box 1000, FIN-02044 VTT, Espoo, Finland
matej.oresic@vtt.fi
http://www.cis.hut.fi/projects/mi/

Abstract. We present a Bayesian model for two-way ANOVA-type analysis of high-dimensional, small sample-size datasets with highly correlated groups of variables. Modern cellular measurement methods are a main application area; typically the task is differential analysis between diseased and healthy samples, complicated by additional covariates requiring a multi-way analysis. The main complication is the combination of high dimensionality and low sample size, which renders classical multivariate techniques useless. We introduce a hierarchical model which does dimensionality reduction by assuming that the input variables come in similarly-behaving groups, and performs an ANOVA-type decomposition for the set of reduced-dimensional latent variables. We apply the methods to study lipidomic profiles of a recent large-cohort human diabetes study.

Keywords: ANOVA, factor analysis, hierarchical model, metabolomics, multi-way analysis, small sample-size.

Reference

1. Huopaniemi, I., Suvitaival, T., Nikkilä, J., Orešič, M., Kaski, S.: Two-Way Analysis of High-Dimensional Collinear Data. Data Mining and Knowledge Discovery (2009) DOI: 10.1007/s10618-009-0137-2

---

* This is an extended abstract of an article published in the Data Mining and Knowledge Discovery journal [1].

** The project was funded by Tekes MASI program. I.H., T.S and S.K belong to the Adaptive Informatics Research Centre and Helsinki Institute for Information Technology. I.H. is funded by the Graduate School of Computer Science and Engineering. S.K is partially supported by EU FP7 NoE PASCAL2, ICT 216886.

W. Buntine et al. (Eds.): ECML PKDD 2009, Part I, LNAI 5781, p. 33, 2009.

A Fast Ensemble Pruning Algorithm Based on Pattern Mining Process*

Qiang-Li Zhao, Yan-Huang Jiang, and Ming Xu

School of Computer Science, National University of Defense Technology, Changsha, Hunan Province, PR. China
zhao-qiangli@163.com, yhjiang@nudt.edu.cn

Abstract. Ensemble pruning deals with the reduction of base classifiers prior to combination in order to improve generalization and prediction efficiency. Existing ensemble pruning algorithms require much pruning time. This paper presents a fast pruning approach: PMEP (Pattern Mining based Ensemble Pruning). In this algorithm, the prediction results of all base classifiers are organized as a transaction database, and FP-Tree structure is used to compact the prediction results. Then a greedy pattern mining method is explored to find the ensemble of size *k*. After obtaining the ensembles of all possible sizes, the one with the best accuracy is outputted. Compared with Bagging, GASEN, and Forward Selection, experimental results show that PMEP achieves the best prediction accuracy and keeps the size of the final ensemble small, more importantly, its pruning time is much less than other ensemble pruning algorithms.

Keywords: PMEP (Pattern Mining based Ensemble Pruning), FP-Tree, Bagging, back-propagation neural network.

Reference

1. Zhao, Q.-L., Jiang, Y.-H., Xu, M.: A Fast Ensemble Pruning Algorithm based on Pattern Mining Process. Data Mining and Knowledge Discovery (2009) DOI: 10.1007/s10618-009-0138-1

---

* This is an extended abstract of an article published in the Data Mining and Knowledge Discovery journal [1].

W. Buntine et al. (Eds.): ECML PKDD 2009, Part I, LNAI 5781, p. 34, 2009.

Evaluation Measures for Multi-class Subgroup Discovery

Tarek Abudawood and Peter Flach

Department of Computer Science, University of Bristol, United Kingdom
dawood@cs.bris.ac.uk,
Peter.Flach@bristol.ac.uk
www.cs.bristol.ac.uk/~dawood/
www.cs.bristol.ac.uk/~flach/

Abstract. Subgroup discovery aims at finding subsets of a population whose class distribution is significantly different from the overall distribution. It has previously predominantly been investigated in a two-class context. This paper investigates multi-class subgroup discovery methods. We consider six evaluation measures for multi-class subgroups, four of them new, and study their theoretical properties. We extend the two-class subgroup discovery algorithm CN2-SD to incorporate the new evaluation measures and a new weighting scheme inspired by AdaBoost. We demonstrate the usefulness of multi-class subgroup discovery experimentally, using discovered subgroups as features for a decision tree learner. Not only is the number of leaves of the decision tree reduced with a factor between 8 and 16 on average, but significant improvements in accuracy and AUC are achieved with particular evaluation measures and settings. Similar performance improvements can be observed when using naive Bayes.

1 Introduction

Rule induction is a common form of machine learning and data mining often used in classification and association rule learning. Classification rule learning is a predictive task aimed at constructing a set of rules, based on training examples and their observed features, to predict the class of unseen future examples. Association rule learning, on the other hand, is a form of descriptive induction aimed at the discovery of individual rules that express interesting patterns in data.

In classification rule learning a target concept is pre-defined and so the search heuristic is usually some form of accuracy. On the other hand, in descriptive rule learning no target concept is given and the heuristic function evaluates measures of interestingness and unusualness in the data, e.g. support and confidence. Subgroup discovery can be seen as being halfway between predictive and descriptive rule learning, as there is a target concept but the goal of subgroup discovery is not necessarily to achieve high accuracy. Rather, the target concept helps us to achieve a trade-off between accuracy and interestingness. In [1] this trade-off was achieved using weighted relative accuracy, but their CN2-SD algorithm is restricted to two classes. In this paper we extend the approach to more than two classes, and perform an extensive study of different measures for multi-class subgroup discovery. To the best of our knowledge, multi-class subgroup

W. Buntine et al. (Eds.): ECML PKDD 2009, Part I, LNAI 5781, pp. 35–50, 2009.

discovery was previously only studied by Klösgen [2, 3], who proposed two of the multi-class measures we study in this paper but did not compare them experimentally.

Our goal in this paper is to investigate methods that allow multi-class subgroup discovery. Our main contribution is a systematic study of possible heuristics for multi-class subgroup discovery, including theoretical analysis and experimental evaluation. We show that a careful choice of heuristic and learning setting results in the discovery of significant subgroups, in a reasonable amount of time, that have high predictive power and can be used to build classification models that are an order of magnitude smaller.

The paper is organized as follows. In Section 2 we introduce the general framework of rule learning for subgroup discovery and describe CN2-SD as well as improvements leading to our CN2-MSD rule learner. Section 3 introduces and analyses six multi-class subgroup evaluation measures. An empirical evaluation of multi-class subgroup discovery for feature construction over 20 UCI data sets is presented in Section 4. Section 5 considers related work, and we discuss possible future work and conclude the paper in Section 6.

2 Rule Learning for Subgroup Discovery

Let T be a training set of examples labelled by n classes $C_1,\ldots,C_n$. We denote the total number of examples in the training set by E and the number of examples belonging to class C_i by E_i. The number of examples in T covered by a subgroup b is denoted by e, and the number of examples belonging to C_i and covered by b is denoted by e_i. This notation is summarised in Table 1; all evaluation measures considered in this paper are based on numbers in such a contingency table.

A *heuristic* is defined as an n-dimensional function $\mathbb{R}^n \to \mathbb{R}$ such that $h(e_1,...,e_n)$ represents the quality of subgroup b, where e_i are as in Table 1. We will often abbreviate this to $h(b)$ if no confusion can arise. We will use the terms heuristic and evaluation measure interchangeably in the rest of the paper. Following [4], we will use the following definition for comparing the different heuristics.

Definition 1 (Compatibility and Antagonism [4]). *Two heuristic functions h_1 and h_2 are* compatible *iff for all subgroups b_1, b_2: $h_1(b_1) > h_1(b_2) \iff h_2(b_1) > h_2(b_2)$. h_1 and h_2 are* antagonistic *iff for all subgroups b_1, b_2: $h_1(b_1) > h_1(b_2) \iff h_2(b_1) < h_2(b_2)$. h_1 and h_2 are* equivalent *($h_1 \sim h_2$) if they are either compatible or antagonistic.*

Table 1. Notational conventions for the frequencies in a multi-class contingency table tabulating examples covered or not covered by a given subgroup b

	Subgroup b	Complement $\bar{b}$	
Class 1	e_1	$E_1 - e_1$	E_1
...	...	...	...
Class n	e_n	$E_n - e_n$	E_n
	e	$E - e$	E

The intuition is that equivalent heuristics order the search space in the same way, even if they differ in numerical value. For instance, $h(x) \sim \frac{E}{n}h(x)$, since E and n are constants for a given data set. However, $h(x) \not\sim \frac{E}{E-e}h(x)$, since e depends on the subgroup.

CN2 is a rule induction algorithm that can be applied to propositional data sets for inducing classification rules [5, 6]. A propositional data set can be thought of as a single database table with rows representing examples and columns representing features. CN2 consists of two main algorithms: the search algorithm that performs beam search[1] in order to find a single rule and the covering algorithm that repeatedly executes the search for rules until all examples are covered. The search algorithm searches the rule space top-down, evaluating the quality of rules using precision (the relative frequency of positives among the examples covered). A rule takes the form of (*head* $\leftarrow$ *body*) where the body is a conjunction of body literals and the head is a single head literal indicating a class. A literal takes the form of (*feature Op value*) where *Op* can be one of the following three operators: $>$, $<$ or $=$.

CN2 can apply a significance test to each rule being learned. If there is a regularity unlikely to have occurred by chance, then the rule is regarded as significant. Furthermore, a minimum evaluation threshold can be used as a stopping criterion. When a rule achieves an evaluation value lower than this threshold, the search is pruned.

CN2 can induce rules in two forms: ordered list of rules and unordered set of rules. In the former, the search algorithm finds the best rule in the current set of training examples. The rule predicts the class with the highest frequency among the examples it covers. All examples covered by the newly induced rule are removed before starting another search iteration. The rule search is repeated until all the examples are covered. In the unordered setting, the main algorithm is iterated for each class in turn. For each induced rule, only covered examples belonging to the class being learned are removed.

CN2-SD is an extension of CN2 particularly geared towards subgroup discovery [1]. In subgroup discovery, one wants to find independent rules that may overlap. To that end, CN2-SD implements two major changes compared to CN2: replacing precision as a search heuristic with *weighted relative accuracy* or WRAcc (see Definition 2 in the next section), and the use of a weighted covering algorithm. While in the original covering algorithm examples are removed once they are covered by a rule, in the weighted covering algorithm examples are never removed but their weight is decreased according to one of two schemes: additive or multiplicative. Let $w_t(x)$ denote the weight of example x after being covered by t rules: in the additive method we have $w_t(x) = \frac{1}{1+t}$ while in the multiplicative method $w_t(x) = \gamma^t, 0 < \gamma < 1$.

The use of a weighting scheme, particularly the multiplicative weights, is related to the use of weights in AdaBoost [7], where an example x at time t is re-weighted according to $w_t(x) = w_{t-1}(x)e^{-\alpha_t p}$; here, $w_{t-1}(x)$ is the current weight, $\alpha_t = \frac{1}{2}\ln\frac{1-\varepsilon_t}{\varepsilon_t}$ with $\varepsilon_t < 0.5$ representing the error of the current hypothesis, and p is either 1 or -1 reflecting a correct or an incorrect prediction respectively. To obtain an update rule that is independent of the current hypothesis we set $p = 1$ and $\varepsilon_t = E_{min}$, the size of the

[1] In beam search, the k most promising candidate hypotheses are considered instead of all possible candidate hypotheses for efficiency reasons.

minority class, which gives $w_t(x) = w_{t-1}(x)\sqrt{\frac{E_{min}}{E-E_{min}}}$. We therefore extended CN2-SD with an option to set the γ parameter for multiplicative weights to $\sqrt{\frac{E_{min}}{E-E_{min}}}$.

Another improvement in *CN2-MSD*, our version of CN2-SD for more than two classes, is the use of inequality with nominal values. Like CN2, CN2-SD uses only equality when constructing a new literal for a nominal feature, e.g. $feature = val$, whereas CN2-MSD also considers literals of the form $feature \neq val$. Experiments suggest that the use of inequality slightly improves AUC as well as the accuracy.

3 Heuristics for Multi-class Subgroup Discovery

In this section we investigate possible heuristics for multi-class subgroup discovery. We start with considering multi-class versions of weighted relative accuracy in Section 3.1, followed by a consideration of other measures that are inherently multi-class in Section 3.2. In Section 3.3 we consider the question whether a subgroup or its complement is the more interesting one.

3.1 Multi-class Versions of Weighted Relative Accuracy

Weighted relative accuracy was defined in a binary classification context [8]: here, we adapt it such that a given class C_i is taken as positive and all other classes together as the negative class. The idea is that the rule *accuracy* (which is actually the precision $\frac{e_i}{e}$) should be taken *relative* to the accuracy obtained by always guessing C_i ($\frac{E_i}{E}$), and *weighted* by the rule's coverage ($\frac{e}{E}$).

Definition 2 (Weighted Relative Accuracy [8]). *The weighted relative accuracy of subgroup b for class C_i is defined as* $WRAcc_i(b) = \frac{e}{E}\left(\frac{e_i}{e} - \frac{E_i}{E}\right) = \frac{e_i}{E} - \frac{e}{E}\frac{E_i}{E}$.

Previous research [9, 10, 11, 12] has investigated methods of multi-class classification. The methods usually decompose a multi-class problem into several binary problems either by considering all pairwise combinations of classes (one-vs-one) or considering each class against the union of the other classes (one-vs-rest). One model is trained for each binary problem and, according to the chosen method, the classification for an unseen example is determined based on a competition of these binary models. It has been shown by [12, 10] that it is computationally expensive to perform the classification in such a manner because it requires $O(n)$ comparisons for each example to be classified. We are interested in obtaining a single final model, thus avoiding the computational costs of multiple binary comparisons during the classification phase.

The first idea might be to simply average $WRAcc_i$ over all classes. However, this will fail due to the following simple result.

Lemma 1. $\Sigma_{i=1}^{n} WRAcc_i(b) = 0$.

Proof. $\Sigma_{i=1}^{n} WRAcc_i(b) = \Sigma_{i=1}^{n} \frac{e}{E}\left(\frac{e_i}{e} - \frac{E_i}{E}\right) = \frac{e}{E}\left[\Sigma_{i=1}^{n}\frac{e_i}{e} - \Sigma_{i=1}^{n}\frac{E_i}{E}\right] = \frac{e}{E}[1-1] = 0.$

This justifies the use of the absolute value in the following definition.

Definition 3 (Multi-class WRAcc). *The (one-vs-rest) multi-class weighted relative accuracy is defined as* $MWRAcc(b) = \frac{1}{n}\sum_{i=1}^{n}|WRAcc_i(b)|$.

Clearly, in a two-class setting, $MWRAcc(b) = |WRAcc_i(b)|$ for $i = 1,2$.

Rather than taking an unweighted average, we may want to take a weighted average using the class prior.

Definition 4 (Weighted Multi-class WRAcc). *The (one-vs-rest) weighted multi-class weighted relative accuracy is defined as* $WMWRAcc(b) = \sum_{i=1}^{n}\frac{E_i}{E}|WRAcc_i(b)|$.

We can also define a one-vs-one version.

Definition 5 (One-vs-One Multi-class WRAcc). *The one-vs-one multi-class weighted relative accuracy is defined as follows:*

$$MWRAcc^{1vs1}(b) = \frac{1}{n(n-1)}\sum_{i=1}^{n}\sum_{j=1;j\neq i}^{n}|WRAcc_{ij}(b)|$$

where $WRAcc_{ij}(b) = \frac{e_i+e_j}{E_i+E_j}\left(\frac{e_i}{e_i+e_j} - \frac{E_i}{E_i+E_j}\right)$.

Note that, while $WRAcc_i(\bar{b}) = -WRAcc_i(b)$, for the multi-class versions defined here we have $MWRAcc(\bar{b}) = MWRAcc(b)$ and similar for the other measures.

We can further understand the similarities and differences between these heuristics by reducing them to simpler equivalent forms.

Theorem 1. $MWRAcc \sim \sum_{i=1}^{n}|e_iE - eE_i|$.

Proof. $MWRAcc = \frac{1}{n}\sum_{i=1}^{n}|\frac{e}{E}\left(\frac{e_i}{e} - \frac{E_i}{E}\right)| = \frac{1}{nE^2}\sum_{i=1}^{n}|e_iE - eE_i| \sim \sum_{i=1}^{n}|e_iE - eE_i|$.

Theorem 2. $WMWRAcc \sim \sum_{i=1}^{n}E_i|e_iE - eE_i|$.

Proof. $WMWRAcc = \sum_{i=1}^{n}\frac{E_i}{E}|\frac{e}{E}\left(\frac{e_i}{e} - \frac{E_i}{E}\right)| = \frac{1}{E^3}\sum_{i=1}^{n}E_i|e_iE - eE_i| \sim \sum_{i=1}^{n}E_i|e_iE - eE_i|$.

Theorem 3. $MWRAcc^{1vs1} \sim \sum_{i=1}^{n}\sum_{j=i+1}^{n}\frac{|e_iE_j - e_jE_i|}{(E_i+E_j)^2}$.

Proof. Notice that $WRAcc_{ij}(b) = -WRAcc_{ji}(b)$. Then,

$$\begin{aligned} MWRAcc^{1vs1} &= \frac{1}{n(n-1)}\sum_{i=1}^{n}\sum_{j=1;j\neq i}^{n}|WRAcc_{ij}| = \frac{2}{n(n-1)}\sum_{i=1}^{n}\sum_{j=i+1}^{n}|WRAcc_{ij}| \\ &\sim \sum_{i=1}^{n}\sum_{j=i+1}^{n}\frac{e_i+e_j}{E_i+E_j}\left|\frac{e_i}{e_i+e_j} - \frac{E_i}{E_i+E_j}\right| \\ &= \sum_{i=1}^{n}\sum_{j=i+1}^{n}\frac{|e_i(E_i+E_j) - E_i(e_i+e_j)|}{(E_i+E_j)^2} = \sum_{i=1}^{n}\sum_{j=i+1}^{n}\frac{|e_iE_j - e_jE_i|}{(E_i+E_j)^2} \end{aligned}$$

Thus, the key term in each of these heuristics is of the form $|e_iE - eE_i|$ for one-vs-rest measures and $|e_iE_j - e_jE_i|$ for one-vs-one measures, possibly with weights depending on the size of the classes.

3.2 Other Multi-class Subgroup Evaluation Measures

In this section we consider other ways of evaluating the frequencies observed in an n-by-2 contingency table.

One possibility is to consider two random variables: B is a binary variable indicating whether or not a random example is in the subgroup, and L is an n-ary variable indicating which class applies to that example. Given a contingency table, the marginal and joint entropies of these random variables are then as follows:

$$H(L) = -\sum_{i=1}^{n} \frac{E_i}{E} \log \frac{E_i}{E}$$

$$H(B) = -\frac{e}{E} \log \frac{e}{E} - \frac{E-e}{E} \log \frac{E-e}{E}$$

$$H(L,B) = -\sum_{i=1}^{n} \left(\frac{e_i}{E} \log \frac{e_i}{E} + \frac{E_i - e_i}{E} \log \frac{E_i - e_i}{E} \right)$$

We can then define their mutual information in the usual way: $MI(L,B) = H(L) + H(B) - H(L,B)$. Mutual information tells us how much knowledge about one variable would be increased by knowing the value of the other. The higher the mutual information, the more interesting the distribution of classes amongst the subgroup and its complement is.

Definition 6 (Mutual Information). *The mutual information score of a subgroup b is defined as follows:*

$$MI(b) = \sum_{i=1}^{n} \left(\frac{e_i}{E} \log \frac{e_i}{E} + \frac{E_i - e_i}{E} \log \frac{E_i - e_i}{E} \right) - \sum_{i=1}^{n} \frac{E_i}{E} \log \frac{E_i}{E} - \frac{e}{E} \log \frac{e}{E} - \frac{E-e}{E} \log \frac{E-e}{E}$$

Alternatively, we can use the *Chi-squared* test to decide whether L and B are statistically independent. Assuming that the data in Table 1 represents the observed frequencies, the expected values under the null hypothesis of independence of columns and rows can be calculated from the marginal frequencies. E.g., the expected value of e_i is $\frac{eE_i}{E}$. The Chi-squared statistic is then the sum of the squared differences between observed and expected frequencies divided by the expected frequencies.

Definition 7 (Chi-Squared). *The Chi-squared score of a subgroup b is defined as follows:*

$$Chi^2(b) = \sum_{i=1}^{n} \left(\frac{[e_i - \frac{E_i e}{E}]^2}{\frac{E_i e}{E}} + \frac{[(E_i - e_i) - \frac{E_i(E-e)}{E}]^2}{\frac{E_i(E-e)}{E}} \right)$$

Chi-squared has a wide range of uses including feature selection [13].

Finally, we consider a decision tree splitting criterion based on the Gini index. Consider calculating the utility of a binary split in a multi-class context. A splitting criterion calculates the decrease in impurity when going from parent to children. The impurity of the children is calculated as the weighted average of their individual impurities, using the relative frequency of examples covered by the child as weight. The Gini index calculates the impurity of a node as $\frac{1}{n}\sum_{i=1}^{n} p_i\,(1-p_i)$, where p_i is the relative frequency of examples of class C_i.

Definition 8 (Gini-split). *The Gini-split score of a subgroup b is defined as follows:*

$$GS(b) = \frac{1}{n}\sum_{i=1}^{n}\frac{E_i\,(E-E_i)}{E^2} - \frac{1}{n}\sum_{i=1}^{n}\frac{e}{E}\frac{e_i\,(e-e_i)}{e^2}$$
$$-\frac{1}{n}\sum_{i=1}^{n}\frac{E-e}{E}\frac{(E_i-e_i)\,((E-E_i)-(e-e_i))}{(E-e)^2}$$

Theorem 4. $Chi^2 = \sum_{i=1}^{n}\frac{[e_iE-eE_i]^2}{eE_i(E-e)}$.

Proof. $Chi^2 = \sum_{i=1}^{n}\frac{[e_i-\frac{eE_i}{E}]^2}{\frac{eE_i}{E}} + \frac{[(E_i-e_i)-\frac{E_i(E-e)}{E}]^2}{\frac{E_i(E-e)}{E}} = \frac{1}{E}\sum_{i=1}^{n}\frac{[e_iE-eE_i]^2}{eE_i} + \frac{[E_iE-e_iE-E_iE+eE_i]^2}{E_i(E-e)}$
$= \frac{1}{E}\sum_{i=1}^{n}[e_iE-eE_i]^2\left(\frac{1}{eE_i}+\frac{1}{E_i(E-e)}\right) = \frac{1}{E}\sum_{i=1}^{n}[e_iE-eE_i]^2\frac{E_i(E-e+e)}{eE_i^2(E-e)} = \sum_{i=1}^{n}\frac{[e_iE-eE_i]^2}{eE_i(E-e)}$.

Theorem 5. $GS \sim \sum_{i=1}^{n}\frac{[eE_i-e_iE]^2}{e(E-e)}$.

Proof. $GS = \frac{1}{n}\sum_{i=1}^{n}\frac{E_i(E-E_i)}{E^2} - \frac{e}{E}\frac{e_i(e-e_i)}{e^2} - \frac{E-e}{E}\frac{(E_i-e_i)[(E-E_i)-(e-e_i)]}{(E-e)^2} = \frac{1}{nE}\sum_{i=1}^{n}\frac{E_i(E-E_i)}{E}$
$-\frac{e_i(e-e_i)}{e} - \frac{(E_i-e_i)[(E-E_i)-(e-e_i)]}{E-e} = \frac{1}{nE^2}\sum_{i=1}^{n}\frac{[eE_i-e_iE]^2}{e(E-e)} \sim \sum_{i=1}^{n}\frac{[eE_i-e_iE]^2}{e(E-e)}$.

If the i-th term in Chi^2 is x_i, the i-th term in GS is equivalent to E_ix_i. We can therefore consider GS to be a class-weighted version of Chi^2. This shows that, in general, $Chi^2 \not\sim GS$, thereby negatively answering an open question in [4]. They proved the equivalence in the binary case, and conjectured that this would extend to more than two classes.

Theorem 6 (Equivalence of binary Chi-squared and Gini-split [4]). *For $n=2$ classes, $Chi^2 \sim GS$.*

Proof. First of all, $e_1E - eE_1 = e_1\,(E_1+E_2) - (e_1+e_2)\,E_1 = e_1E_2 - e_2E_1$. By symmetry, $e_2E - eE_2 = e_2E_1 - e_1E_2$. Therefore, $[e_1E-eE_1]^2 = [e_2E-eE_2]^2 = [e_1E_2-e_2E_1]^2$. It follows that $Chi^2 = \left(\frac{1}{E_1}+\frac{1}{E_2}\right)\frac{[e_1E_2-e_2E_1]^2}{e(E-e)} = \frac{E}{E_1E_2}\frac{[e_1E_2-e_2E_1]^2}{e(E-e)}$. On the other hand, $GS = \frac{1}{2E^2}\frac{2[e_1E_2-e_2E_1]^2}{e(E-e)} = \frac{E_1E_2}{E^3}Chi^2$.

It is interesting to note that our proof of Theorem 6 is much more succinct than the one given in [4]. It appears the multi-class notation is beneficial here.

Measures similar to Chi^2 and GS were used previously in the Explora system [2,3]. Explora is an interactive knowledge discovery system that incorporates several search strategies, refinement methods and evaluation functions. If we distinguish Klösgen's definitions by a subscript *Kl*, we can show the equivalence between our Chi^2 and GS measures and Klösgen's as follows:

$$GS_{Kl}(b) = \frac{\frac{e}{E}}{\frac{E-e}{E}} \sum_{i=1}^{n} \left(\frac{e_i}{e} - \frac{E_i}{E}\right)^2 = \frac{e}{E-e} \sum_{i=1}^{n} \left(\frac{e_i^2}{e^2} + \frac{E_i^2}{E^2} - \frac{2e_iE_i}{eE}\right)$$

$$= \frac{e}{E^2(E-e)} \sum_{i=1}^{n} \frac{e^2E_i^2 + e_i^2E^2 - 2e_ieE_i}{e^2} = \frac{1}{E^2} \sum_{i=1}^{n} \frac{[eE_i - e_iE]^2}{e(E-e)} = \frac{1}{n} GS(b)$$

$$Chi^2_{Kl}(b) = \frac{1}{E^2} \sum_{i=1}^{n} \frac{\frac{[eE_i - e_iE]^2}{e(E-e)}}{\frac{E_i}{E}} = \frac{1}{E} \sum_{i=1}^{n} \frac{[eE_i - e_iE]^2}{eE_i(E-e)} = \frac{1}{E} Chi^2(b)$$

Although our definitions of Chi^2 and GS are not equal to Klösgen's definitions they are indeed equivalent, hence $Chi^2 \sim Chi^2_{Kl}$ and $GS \sim GS_{Kl}$.

3.3 Sign of a Subgroup

In two-class subgroup discovery a subgroup correlates positively with one class if and only if its complement correlates negatively with the other class. This can easily be established by, e.g., the sign of $WRAcc_i(b)$; we generally restrict attention to the subgroup that has positive weighted relative accuracy for the designated positive class. For the multi-class case we propose a criterion for determining whether a subgroup or its complement is the more interesting one based on conditional entropy. We have $H(L|B=b) = \sum_{i=1}^{n} \frac{e_i}{e} \log \frac{e_i}{e}$, the entropy of the left column of the contingency table; and similarly $H(L|B=\bar{b}) = \sum_{i=1}^{n} \frac{E_i - e_i}{E-e} \log \frac{E_i - e_i}{E-e}$. The sign of the subgroup is then $\text{sign}(H(L|B=\bar{b}) - H(L|B=b))$. If the amount of uncertainty remaining about L assuming $B=b$ is smaller than when assuming $B=\bar{b}$, then the sign of $H(L|B=\bar{b}) - H(L|B=b)$ is positive, favouring b over $\bar{b}$.

4 Empirical Evaluation

We performed extensive experiments to test the behaviour of the six subgroup evaluation measures defined in the previous section, as well as the usefulness of multi-class subgroup discovery for feature generation in a classification context. We used 20 UCI data sets [14], which are listed in Table 2. Numerical attributes with more than 100 distinct values have been discretised. Our implementation of CN2-MSD is an adaptation of the CN2-SD implementation provided by the authors of [1], which was implemented in Java as part of the Weka data mining workbench [15]. Six different settings have been applied as follows:

- setting 0: unweighted covering;
- setting 1: multiplicative weights, $\gamma = .25$;
- setting 2: multiplicative weights, $\gamma = .5$;

Table 2. UCI data sets used for the experiments. CMC stands for Contraceptive Method Choice while WPBC stands for Wisconsin Prognostic Breast Cancer.

Dataset Name	#exs.	# attrs.	# cls.	Dist.
1 Abalone	4176	9	3	1527, 1342 and 1307
2 Balance-scale	624	5	3	288, 288 and 48
3 Car	1727	7	4	1209, 384, 69 and 65
4 CMC	1472	10	3	628, 511 and 333
5 Contact-lenses	24	5	3	15, 5 and 4
6 Credit	589	16	2	383 and 306
7 Dermatology	365	35	6	112,72, 60, 52, 49 and 20
8 Glass	213	11	6	76, 69, 29, 17, 13 and 9
9 Haberman	305	4	2	224 and 81
10 Hayes-roth	131	5	3	51, 50 and 30
11 House-votes	434	17	2	267 and 167
12 Ionosphere	350	34	2	224 and 126
13 Iris	150	5	3	50, 50 and 50
14 Labor	57	17	2	37 and 20
15 Mushroom	8123	23	2	4208 and 3915
16 Pima-indians	767	9	2	500 and 267
17 Soybean	683	36	19	92, 2 × 91, 88, 2 × 44, 9 × 20, 16, 15, 14 and 8
18 Tic-Tac-Toe	957	10	2	625 and 332
19 WPBC	197	34	2	150 and 47
20 Zoo	100	18	7	40, 20, 13, 10, 8, 5 and 4

Table 3. Average number of subgroups found on 20 UCI data sets using different heuristics and settings

Setting	$MWRAcc^{1vs1}$	$MWRAcc$	$WMWRAcc$	MI	Chi^2	GS	average
0	5.70	6.25	5.95	6.90	6.80	7.15	6.46
1	9.20	10.05	9.50	14.55	13.55	14.25	11.85
2	11.50	12.15	12.30	20.30	20.65	20.70	16.27
3	16.65	16.40	17.00	26.20	27.90	28.05	22.03
4	14.50	15.30	15.65	30.10	31.00	31.10	22.94
5	10.40	9.60	9.40	15.30	15.55	17.00	12.88
Average	11.33	11.63	11.63	18.89	19.24	19.71	

- setting 3: multiplicative weights, $\gamma = .75$;
- setting 4: additive weights; and
- setting 5: multiplicative weights, $\gamma = \sqrt{\frac{E_{min}}{E-E_{min}}}$.

The significance and minimum evaluation threshold parameters were fixed to 0.95 and 0.01, respectively.

The first result concerns the number of subgroups found (as reported in Table 3). It is clear that the WRAcc-based methods find significantly fewer subgroups than the other three. Weighted covering clearly helps in finding more subgroups. Additive weights and multiplicative weights with large γ (i.e., slow decay of the weights) result in the most subgroups. Setting $\gamma = \sqrt{\frac{E_{min}}{E-E_{min}}}$ gives performance similar to small to medium fixed γ.

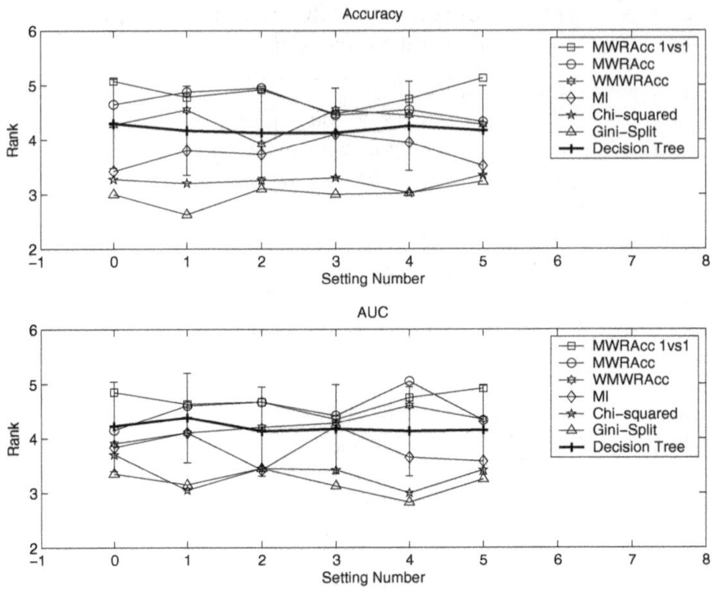

Fig. 1. Accuracies and AUCs of the multi-class subgroup discovery methods compared with the J48 control learner (bold line). For each setting, performances outside the vertical bar are significantly better (bottom) or worse (top) than the control.

In order to evaluate the quality and utility of the induced subgroups, we use them as features for a decision tree learner. We use the Weka implementation of C4.5 which is called J48, with default parameters. 10-fold cross-validated accuracy and AUC are recorded on each data set, for J48 run directly on the original data set (labelled J48 in our result tables) and J48 run using subgroups as features, where the subgroups are learned using CN2-MSD with one of the six evaluation heuristics (labelled with that heuristic in the table), and for each of the six settings.

For space reasons we only report averages over all 20 data sets in Tables 4 and 5 (all subsequent tables can be found in the Appendix). Average accuracies and AUCs have limited meaning because the values are not necessarily commensurate across data sets, and so we also report the average rank (1 is best, 7 is worst) of a method across all data sets. We use the Friedman test on these average ranks ($p = 0.10$) with Bonferroni-Dunn post-hoc test to check significance against J48 as a control learner. The Friedman test records wins and losses in the form of ranks, but ignores the magnitude of these wins and losses, which is considered more appropriate when comparing multiple classifiers on multiple data sets; see [16] for more details. A graphical illustration of the post-hoc test results is given in Fig. 1. For each setting the corresponding critical difference diagram is shown vertically. So, for instance, in setting 4 both *MI* and *GS* perform significantly better than J48, and in setting 5 $MWRAcc^{1vs1}$ performs significantly worse than J48.

Tables 6-8 show the results for number of leaves, number of nodes in the tree and the model construction times. As can be seen, the average tree has roughly between 8 and 16 times fewer leaves when trained with subgroups as features, compared with standard

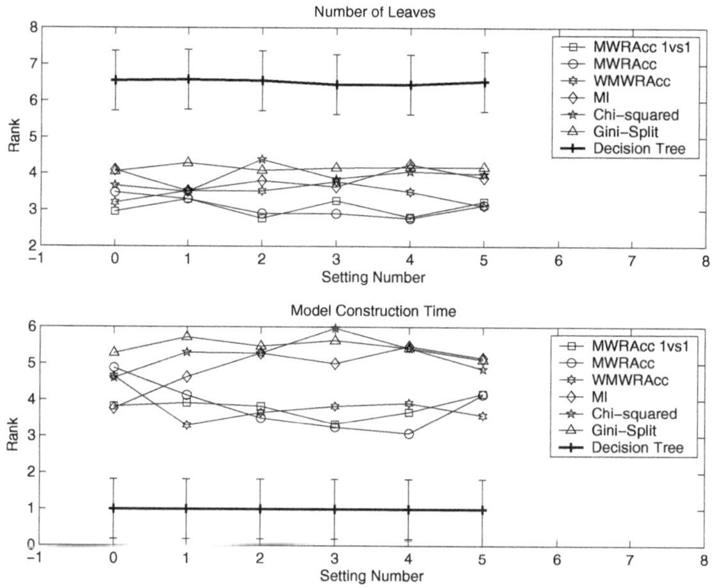

Fig. 2. Number of leaves and model construction times of the multi-class subgroup discovery methods compared with the J48 control learner (bold line)

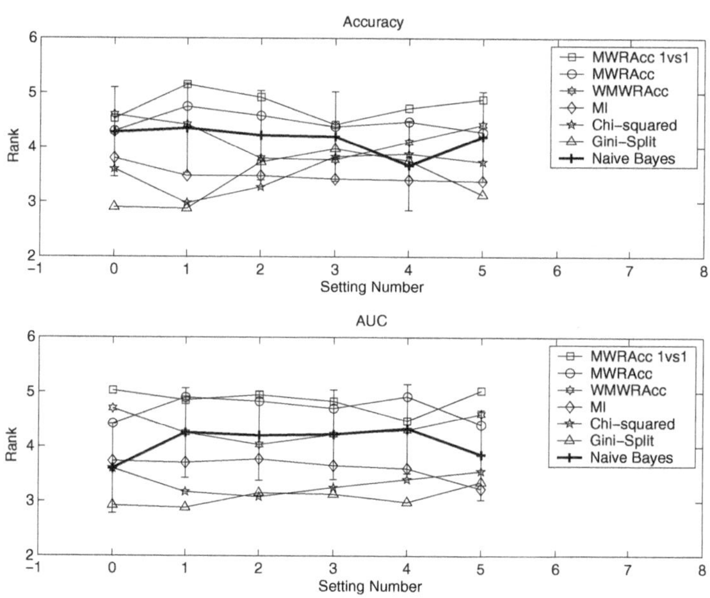

Fig. 3. Accuracies and AUCs of the multi-class subgroup discovery methods compared with the naive Bayes control learner (bold line)

J48, and roughly between 6 and 12 times fewer nodes. This comes, of course, at the expense of considerable additional execution time. Fig. 2 shows this graphically (the ranks in Tables 6 and 7 are equal and therefore the two graphs for number of leaves and number of nodes are identical).

Finally, we report accuracy and AUC results using naive Bayes as the learner (Fig. 3). The results are similar as for J48, although differences between evaluation measures are less pronounced. Tables 9-11 show the average accuracies, AUCs and model construction times of the multi-class subgroup discovery methods compared with naive Bayes as a control classifier.

Our conclusions from these experiments are that the multi-class subgroup evaluation measures considered in this paper can be divided into two groups. The WRAcc-based measures result on average in much fewer subgroups (although the model construction time is not significantly reduced). If the subgroups are used for prediction, the additional subgroups learned by *MI* and particularly Chi^2 and *GS* result in additional predictive power compared to the base learner. A consistently good performer is *GS* in combination with multiplicative weights, $\gamma = 0.25$ (setting 1) or additive weights (setting 4). Given that setting 4 is an order of magnitude slower, we finally settle on setting 1 as our recommendation. This combination achieves significantly higher predictive power than the base learners, building trees that on average have 11 times fewer leaves.

5 Related Work

As discussed earlier, CN2-MSD is an upgrade of the CN2-SD rule learner [1] which learns two-class subgroups and was in turn based on the CN2 inductive rule learner [5, 6]. More details about CN2 and CN2-SD were given in Section 2. Our three multi-class Weighted Relative Accuracy versions are multi-class adaptations of the two-class Weighted Relative Accuracy that originated in [8] and was incorporated in CN2-SD.

Mutual information is used to measure dependencies between random variables. Mutual information is appropriate for assessing the information content of features in complex classification tasks. It has been used for feature selection [17, 18]. The difference between the usage of MI in [17, 18] and CN2-MSD is that CN2-MSD uses MI to evaluate a combination of one or more features, rather than comparing single features. MI as used in [18] also differs from our approach in that it selects features that maximise the MI with respect to a single class but not all the classes. In addition, CN2-MSD only assesses the constructed feature and its complement while this is not the case in the other two approaches due to having different goals.

Chi-squared has been used widely as a statistical significance test in various contexts including classification and rule learning. Chi-squared was found useful for feature selection [13]. The original CN2-SD uses a Chi-squared significance test to filter a nominated subgroup in case it yields a value lower than a certain minimum value.[2] CN2-MSD implements the Chi-squared as a heuristic function similarly to the Explora system [2].

Gini-split is well-known in decision tree learning as a splitting criterion based on the Gini index that evaluates the decrease in impurity when going from parent to

[2] The minimum significance value is set by the user.

children. Gini-split is traditionally used in the divide and conquer (DAQ) paradigm where the learning algorithm divides the entire training set recursively according to a new selected literal. An example can not be covered by more than one rule or leaf in the final model. CN2-MSD, however, uses Gini-split in the separate and conquer (SAC) paradigm as it implements a sequential covering algorithm (and its weighted version). In SAC, single or multiple rules are learned iteratively and the examples covered by the learned rules are removed from the training set before proceeding to the next iteration until no examples are left or a stopping criterion met. In covering algorithms, a subset of the training examples (or weighted training examples) is used to construct rules iteratively. An example could be covered by more than one rule or leaf in the final model. Induction systems that implement SAC are less conservative in their learning model than the ones that implement DAC due to their ability to explore a larger search space [19].

Explora is an interactive knowledge discovery system for databases [2, 3]. Explora was designed to support analysts in finding new knowledge about a domain. It can be used for predictive learning as well as descriptive learning. Explora is a complex system that incorporates various search strategies (exhaustive or heuristic), refinement methods and evaluation functions for binary and multi-class prediction. Explora can also work with relational data. It can be viewed as a generic pattern discovery system in relational data mining.

Another system related to CN2-MSD is called PRIM [20]. PRIM finds subregions of the instance space within which the value of the (continuous) output variable is considerably larger or smaller than its average value over the entire space. PRIM searches for these subregions by top-down specialisation followed by a bottom-up generalisation on the induced subgroups. Further criteria are needed to ensure capturing subgroups of reasonable size. One of the advantages of using WRAcc is that it captures the the generality of the subgroup $\frac{e}{E}$ and its relative accuracy $\frac{e_i}{e} - \frac{E_i}{E}$ in a single score. While CN2 uses a greedy refinement strategy on partial candidate hypotheses, PRIM examines all possible solutions which makes it unsuitable for large datasets.

The work of [4] provided us with the basis for the theoretical analysis of multi-class subgroup evaluation measures. Two-class Weighted Relative Accuracy, Chi-squared and Gini-split were discussed and analysed in [4] and their ROC isometrics visually and analytically compared. Unfortunately, a full n-class ROC analysis requires $n(n-1)/2$ dimensions, and so it is hard – if not impossible – to visualise multi-class evaluation measures through their ROC isometrics.

6 Conclusions

In this paper we upgraded existing approaches for two-class subgroup discovery to handle more than two classes. We defined six multi-class subgroup evaluation measures and investigated their properties theoretically and experimentally. While multi-class subgroup discovery is an interesting task in its own right, located between predictive and descriptive learning, we have also shown that – if the additional computational cost can be justified – the learned subgroups lead to additional predictive power. In effect, the decision trees learned branch on more complex multivariate conditions, and the naive Bayes classifier relaxes its assumptions of statistical independence.

In future work we plan to study whether subgroup discovery can be exploited in other predictive tasks, such as probability estimation and regression. We also aim to investigate multi-class subgroup discovery in a relational context. Furthermore, we will focus on reducing the computational overhead of subgroup discovery.

References

1. Lavrač, N., Kavšek, B., Flach, P., Todorovski, L.: Subgroup Discovery with CN2-SD. Journal of Machine Learning Research 5, 153–188 (2004)
2. Klösgen, W.: Explora: A multipattern and multistrategy discovery assistant. In: Fayyad, U.M., Piatetsky-Shapiro, G., Smyth, P., Uthurusamy, R. (eds.) Advances in Knowledge Discovery and Data Mining, pp. 249–271. MIT Press, Cambridge (2004)
3. Klösgen, W.: Subgroup discovery. In: Klösgen, W., Zytkow, J.M. (eds.) Handbook of Data Mining and Knowledge Discovery, pp. 354–361. Oxford University Press, Oxford (2002)
4. Fürnkranz, J., Flach, P.: ROC 'n' rule learning: Towards a better understanding of covering algorithms. Machine Learning 58, 39–77 (2005)
5. Clark, P., Niblett, T.: The CN2 Induction Algorithm. Machine Learning 3, 261–283 (1989)
6. Clark, P., Boswell, R.: Rule induction with CN2: Some recent improvements. In: Kodratoff, Y. (ed.) EWSL 1991. LNCS (LNAI), vol. 482, pp. 151–163. Springer, Heidelberg (1991)
7. Schapire, R.E.: The boosting approach to machine learning: An overview. In: Nonlinear Estimation and Classification. Lecture Notes in Statistics. Springer, Heidelberg (2003)
8. Lavrač, N., Flach, P., Zupan, B.: Rule evaluation measures: A unifying view. In: Džeroski, S., Flach, P.A. (eds.) ILP 1999. LNCS (LNAI), vol. 1634, pp. 174–185. Springer, Heidelberg (1999)
9. Friedman, J.H.: Another approach to polychotomous classification. Technical report, Stanford University, Department of Statistics (1996)
10. Kijsirikul, B., Ussivakul, N., Meknavin, S.: Adaptive directed acyclic graphs for multiclass classification. In: Ishizuka, M., Sattar, A. (eds.) PRICAI 2002. LNCS (LNAI), vol. 2417, pp. 158–168. Springer, Heidelberg (2002)
11. Platt, J.C., Cristianini, N.: Large margin DAGs for multiclass classification. In: Advance in Neural Information Processing Systems, vol. 12. MIT Press, Cambridge (2000)
12. Hsu, C.W., Lin, C.J.: A comparison of methods for multiclass support vector machines. Neural Networks 13, 415–425 (2002)
13. Jin, X., Xu, A., Bie, R., Guo, P.: Machine Learning Techniques and Chi-Square Feature Selection for Cancer Classification Using SAGE Gene Expression Profiles. In: Li, J., Yang, Q., Tan, A.-H. (eds.) BioDM 2006. LNCS (LNBI), vol. 3916, pp. 106–115. Springer, Heidelberg (2006)
14. Newman, D., Hettich, S., Blake, C., Merz, C.: UCI repository of machine learning databases (1998)
15. Witten, I.H., Frank, E.: Data Mining: Practical Machine Learning Tools and Techniques, 2nd edn. Morgan Kaufmann, San Francisco (2005)
16. Demšar, J.: Statistical comparisons of classifiers over multiple data sets. Journal of Machine Learning Research 7, 1–30 (2006)
17. Battiti, R.: Using mutual information for selecting features in supervised neural net learning. Transactions on Neural Networks 5(4) (1994)
18. Fleuret, F.: Fast binary feature selection with conditional mutual information. Journal of Machine Learning Research 5, 1531–1555 (2004)
19. Bostrom, H.: Covering vs. divide-and-conquer for top-down induction of logic programs. In: Proceedings of the Fourteenth International Joint Conference on Artificial Intelligence, pp. 1194–1200. Morgan Kaufmann, San Francisco (1995)

20. Friedman, J.H., Fisher, N.I.: Bump hunting in high-dimensional data. Statistics and Computing 9, 123–143 (1999)

Appendix: Detailed Experimental Results

Table 4. Accuracies (ranks in brackets) of a decision tree learner using subgroups as features, averaged over 20 UCI data sets

Setting	$MWRAcc^{1vs1}$	$MWRAcc$	$WMWRAcc$	MI	Chi^2	GS	$J48$
0	75.11 (5.08)	76.95 (4.65)	77.90 (4.28)	77.40 (3.42)	78.24 (3.27)	79.87 (3.00)	81.21 (4.30)
1	76.90 (4.78)	78.40 (4.88)	79.95 (4.55)	78.14 (3.80)	80.47 (3.20)	81.85 (2.63)	81.21 (4.17)
2	76.75 (4.92)	78.80 (4.95)	80.52 (3.92)	80.19 (3.73)	81.37 (3.25)	81.34 (3.10)	81.21 (4.13)
3	77.85 (4.47)	78.95 (4.45)	80.31 (4.55)	80.03 (4.10)	81.08 (3.30)	81.64 (3.00)	81.21 (4.13)
4	76.63 (4.75)	78.83 (4.55)	80.34 (4.45)	80.33 (3.95)	81.33 (3.02)	81.48 (3.02)	81.21 (4.25)
5	75.28 (5.13)	77.70 (4.33)	79.05 (4.28)	78.94 (3.52)	79.42 (3.35)	79.96 (3.23)	81.21 (4.17)

Table 5. AUCs (ranks in brackets) of a decision tree learner using subgroups as features, averaged over 20 UCI data sets

Setting	$MWRAcc^{1vs1}$	$MWRAcc$	$WMWRAcc$	MI	Chi^2	GS	$J48$
0	0.83 (4.85)	0.85 (4.15)	0.86 (3.90)	0.86 (3.83)	0.86 (3.70)	0.87 (3.35)	0.84 (4.22)
1	0.86 (4.63)	0.86 (4.60)	0.87 (4.10)	0.87 (4.10)	0.88 (3.05)	0.88 (3.15)	0.84 (4.38)
2	0.85 (4.67)	0.86 (4.67)	0.86 (4.20)	0.88 (3.42)	0.88 (3.45)	0.88 (3.45)	0.84 (4.13)
3	0.86 (4.35)	0.86 (4.42)	0.86 (4.28)	0.86 (4.22)	0.88 (3.42)	0.88 (3.13)	0.84 (4.17)
4	0.85 (4.75)	0.85 (5.05)	0.86 (4.60)	0.87 (3.65)	0.88 (3.00)	0.88 (2.83)	0.84 (4.13)
5	0.85 (4.92)	0.85 (4.33)	0.86 (4.35)	0.87 (3.58)	0.87 (3.42)	0.87 (3.25)	0.84 (4.15)

Table 6. Numbers of leaves (ranks in brackets) of a decision tree learner using subgroups as features, averaged over 20 UCI data sets

Setting	$MWRAcc^{1vs1}$	$MWRAcc$	$WMWRAcc$	MI	Chi^2	GS	$J48$
0	7.55 (2.95)	8.30 (3.48)	8.05 (3.20)	8.25 (4.10)	9.45 (3.67)	9.40 (4.05)	122.05 (6.55)
1	8.65 (3.30)	10.00 (3.30)	9.15 (3.52)	9.85 (3.52)	9.25 (3.50)	11.00 (4.28)	122.05 (6.58)
2	8.45 (2.77)	9.50 (2.90)	10.45 (3.52)	9.20 (3.80)	11.30 (4.38)	11.35 (4.08)	122.05 (6.55)
3	11.80 (3.25)	11.00 (2.90)	13.05 (3.77)	11.70 (3.63)	11.95 (3.85)	13.00 (4.15)	122.05 (6.45)
4	8.80 (2.80)	10.90 (2.77)	12.40 (3.50)	13.90 (4.25)	12.00 (4.05)	12.75 (4.17)	122.05 (6.45)
5	10.45 (3.23)	8.90 (3.13)	7.70 (3.10)	11.60 (3.88)	11.60 (3.98)	14.55 (4.17)	122.05 (6.53)

Table 7. Tree sizes (ranks in brackets) of a decision tree learner using subgroups as features, averaged over 20 UCI data sets

Setting	$MWRAcc^{1vs1}$	$MWRAcc$	$WMWRAcc$	MI	Chi^2	GS	$J48$
0	14.10 (2.95)	15.60 (3.48)	15.10 (3.20)	15.50 (4.10)	17.90 (3.67)	17.80 (4.05)	162.05 (6.55)
1	16.30 (3.30)	19.00 (3.30)	17.30 (3.52)	18.70 (3.52)	17.50 (3.50)	21.00 (4.28)	162.05 (6.58)
2	15.90 (2.77)	18.00 (2.90)	19.90 (3.52)	17.40 (3.80)	21.60 (4.38)	21.70 (4.08)	162.05 (6.55)
3	22.60 (3.25)	21.00 (2.90)	25.10 (3.77)	22.40 (3.63)	22.90 (3.85)	25.00 (4.15)	162.05 (6.45)
4	16.60 (2.80)	20.80 (2.77)	23.80 (3.50)	26.80 (4.25)	23.00 (4.05)	24.50 (4.17)	162.05 (6.45)
5	19.90 (3.23)	16.80 (3.13)	14.40 (3.10)	22.20 (3.88)	22.20 (4.03)	28.10 (4.22)	162.05 (6.42)

Table 8. Model construction times (ranks in brackets) of a decision tree learner using subgroups as features, averaged over 20 UCI data sets

Setting	$MWRAcc^{1vs1}$	$MWRAcc$	$WMWRAcc$	MI	Chi^2	GS	$J48$
0	5.93 (3.83)	6.40 (4.88)	6.61 (4.65)	6.33 (3.77)	6.67 (4.60)	6.46 (5.28)	0.23 (1.00)
1	21.73 (3.92)	22.61 (4.13)	20.22 (3.30)	27.80 (4.63)	32.40 (5.30)	33.57 (5.72)	0.23 (1.00)
2	43.00 (3.83)	40.36 (3.50)	42.80 (3.65)	51.36 (5.28)	45.68 (5.28)	41.39 (5.47)	0.23 (1.00)
3	90.32 (3.33)	83.43 (3.25)	89.39 (3.83)	106.30 (5.00)	105.01 (5.97)	104.17 (5.63)	0.23 (1.00)
4	232.41 (3.67)	238.21 (3.08)	232.53 (3.92)	147.71 (5.47)	158.16 (5.42)	167.92 (5.42)	0.23 (1.00)
5	100.99 (4.17)	111.30 (4.15)	105.34 (3.58)	99.43 (5.15)	67.40 (4.85)	70.05 (5.10)	0.23 (1.00)

Table 9. Accuracies (ranks in brackets) of a naive Bayes learner using subgroups as features, averaged over 20 UCI data sets

Setting	$MWRAcc^{1vs1}$	$MWRAcc$	$WMWRAcc$	MI	Chi^2	GS	NB
0	73.76 (4.53)	75.08 (4.30)	75.42 (4.60)	76.31 (3.80)	76.78 (3.60)	78.82 (2.90)	79.86 (4.28)
1	75.74 (5.15)	77.49 (4.75)	79.32 (4.42)	78.39 (3.48)	80.04 (2.98)	81.44 (2.88)	79.86 (4.35)
2	75.39 (4.92)	77.73 (4.58)	79.79 (3.80)	79.37 (3.48)	80.51 (3.27)	79.83 (3.73)	79.86 (4.22)
3	75.63 (4.42)	77.20 (4.38)	79.58 (3.77)	78.85 (3.42)	79.22 (3.83)	79.39 (3.98)	79.86 (4.20)
4	74.98 (4.72)	77.02 (4.47)	79.18 (4.10)	78.24 (3.40)	77.66 (3.88)	78.37 (3.75)	79.86 (3.67)
5	74.99 (4.88)	77.14 (4.28)	78.25 (4.42)	78.47 (3.38)	78.03 (3.73)	79.27 (3.13)	79.86 (4.20)

Table 10. AUCs (ranks in brackets) of a naive Bayes learner using subgroups as features, averaged over 20 UCI data sets

Setting	$MWRAcc^{1vs1}$	$MWRAcc$	$WMWRAcc$	MI	Chi^2	GS	NB
0	0.83 (5.03)	0.86 (4.42)	0.86 (4.70)	0.87 (3.73)	0.87 (3.60)	0.88 (2.92)	0.88 (3.60)
1	0.87 (4.85)	0.88 (4.90)	0.89 (4.25)	0.89 (3.70)	0.90 (3.17)	0.90 (2.88)	0.88 (4.25)
2	0.87 (4.95)	0.88 (4.83)	0.89 (4.03)	0.90 (3.77)	0.91 (3.08)	0.90 (3.15)	0.88 (4.20)
3	0.88 (4.83)	0.88 (4.70)	0.89 (4.22)	0.89 (3.65)	0.90 (3.25)	0.90 (3.13)	0.88 (4.22)
4	0.87 (4.47)	0.88 (4.92)	0.89 (4.30)	0.90 (3.60)	0.90 (3.40)	0.90 (2.98)	0.88 (4.33)
5	0.86 (5.03)	0.87 (4.40)	0.87 (4.60)	0.89 (3.23)	0.89 (3.55)	0.89 (3.35)	0.88 (3.85)

Table 11. Model construction times (ranks in brackets) of a naive Bayes learner using subgroups as features, averaged over 20 UCI data sets

Setting	$MWRAcc^{1vs1}$	$MWRAcc$	$WMWRAcc$	MI	Chi^2	GS	NB
0	6.13 (3.92)	6.38 (4.53)	6.53 (3.95)	6.82 (4.45)	6.97 (5.13)	6.27 (5.03)	0.05 (1.00)
1	21.34 (3.85)	21.17 (4.03)	21.62 (4.03)	27.14 (4.33)	30.06 (5.25)	32.01 (5.53)	0.05 (1.00)
2	40.80 (4.08)	41.21 (3.50)	42.54 (3.50)	50.33 (5.25)	44.47 (5.17)	42.89 (5.50)	0.05 (1.00)
3	86.51 (3.42)	84.38 (3.63)	89.06 (3.35)	107.75 (5.00)	104.10 (5.70)	113.92 (5.90)	0.05 (1.00)
4	218.62 (3.35)	228.58 (3.60)	242.22 (4.13)	135.36 (5.22)	163.41 (5.28)	160.14 (5.42)	0.05 (1.00)
5	96.66 (3.90)	97.59 (3.92)	99.19 (4.22)	94.14 (4.92)	69.53 (5.13)	70.79 (4.90)	0.05 (1.00)

Empirical Study of Relational Learning Algorithms in the Phase Transition Framework

Erick Alphonse and Aomar Osmani

LIPN-CNRS UMR 7030, Université Paris 13, France
erick.alphonse@lipn.univ-paris13.fr,
aomar.osmani@lipn.univ-paris13.fr

Abstract. Relational Learning (RL) has aroused interest to fill the gap between efficient attribute-value learners and growing applications stored in multi-relational databases. However, current systems use general-purpose problem solvers that do not scale-up well. This is in contrast with the past decade of success in combinatorics communities where studies of random problems, in the phase transition framework, allowed to evaluate and develop better specialised algorithms able to solve real-world applications up to millions of variables. A number of studies have been proposed in RL, like the analysis of the phase transition of a NP-complete sub-problem, the subsumption test, but none has directly studied the phase transition of RL. As RL, in general, is $\Sigma_2 - hard$, we propose a first random problem generator, which exhibits the phase transition of its decision version, beyond NP. We study the learning cost of several learners on inherently easy and hard instances, and conclude on expected benefits of this new benchmarking tool for RL.

1 Introduction

Even though the expressiveness of (supervised) Relational Learning (RL), also known as Inductive Logic Programming (ILP), is attractive for many modern applications[1], such as life sciences, environmental sciences, engineering, natural language processing or arts (see also [1]), RL has to face the well-known trade-off between expressivity and efficiency.

From the efficiency perspective, one of the major obstacles is search efficiency, and several authors have acknowledged that a step forward would be in the design of novel search techniques (e.g. [2,1]). RL, as a sub-domain of symbolic learning, has been cast more than 25 years ago as search into a state space[3]: given a hypothesis space defined a priori, identified by its representation language, find a hypothesis consistent with the learning data. This seminal paper, relating symbolic learning to search in a state space, has enabled machine learning to integrate techniques from problem solving, operational research and combinatorics: greedy search in FOIL, beam search in ICL, breadth-first search in Aleph, 'A' search in PROGOL, IDA (Iterative-Deepening A) search in MIO

[1] http://www-ai.ijs.si/~ilpnet2/apps/index.html

W. Buntine et al. (Eds.): ECML PKDD 2009, Part I, LNAI 5781, pp. 51–66, 2009.

to name a few systems[2]. Besides very few exceptions, all systems are rooted in the generate-and-test paradigm [4] and therefore rely on *general-purpose* search strategies.

This approach has to be contrasted with the important progress, made during the past few years, in the performance of *specialised* SAT or CSP solvers, which can deal with problems' sizes that are orders of magnitude larger than those research communities would expect to solve only a decade ago. This progress has been driven by studies of randomly generated problem instances, in the phase transition framework, where hard to solve instances can be reliably generated, independently from the solver used. This framework relies on the conjecture that a phase transition in the probability of solubility of NP-complete problems can be exhibited along so-called order parameters and that the phase transition region contains the most difficult problem instances, those on which the computational complexity shows an exponential increase in the problem size [5,6,7]. This tool allows to design benchmark datasets to point out and filter "bad" algorithms, and to promote "good" algorithms which are close to the "easy-hard-easy" complexity resolution pattern, standard nowadays [8,9,10]. Since then, it has been strongly developed in many combinatorics domains and has changed the way search algorithms are empirically evaluated. This has lead to new designs of search algorithms, from incomplete to complete solvers and from deterministic to randomised solvers (see e.g. [11]).

We think that RL, as a combinatorics field, must follow the same path and study learning as a decision problem in the phase transition (PT) framework in order to re-new and improve its algorithmic approach to answer the new challenges of modern applications. As the consistency problem is at the core of the Statistical Learning Theory, notably studied in the PAC framework (see [12,13] for details), the PT framework will be able to go beyond the worst-case complexity analysis and allow to study the behaviour of learning algorithms in the "average" or "typical" complexity case. Most importantly, it is also at the core of learners' optimisation algorithms, typical of real-world systems: optimisation procedures are obtained by adding branch-and-bound techniques, or treated as subsequent decision problems[3]. This *a fortiori* is true in RL where almost all noise-resistant learners are relaxation of this problem [15], therefore studying this problem will benefit search strategies for learning.

Machine Learning has known several developments in the study of phase transition phenomena since the early 90s. The first works were in neural networks [16,17,18,19] and studied the phase transition of the generalization error, where the number of examples are shown as order parameters. Other works studied the impact of the phase transition of the solubility probability of the NP-complete subsumption test, a sub-problem of Relational Learning, on the generalisation

[2] Relevant information on these systems can be found at `http://www-ai.ijs.si/~ilpnet2/systems`

[3] This latter strategy, for instance, is used to solve pseudo-boolean constraints, an optimisation version of SAT: one of the best pseudo-boolean solver, Minisat+ is based on the SAT solver Minisat [14].

error [20] and heuristic search [21]. However, as surprising as it may seem, the phase transition framework strongly developped in combinatorics has almost not been imported to the realm of symbolic learning. To the best of our knowledge, the only work along this line has been done by Ruckert et al. [22] who exhibited the phase transition of the probability of solubility of a *learning problem*, the k-term DNF consistency problem, a well-known NP-complete problem [13], showing that the number of variables, the number of positive and negative examples were order parameters.

RL is arguably harder than attribute-value learning, like k-term DNF learning, which has been formalised by Gottlob et al. [23] who showed that the simple bounded ILP consistency problem, which will be discussed in later, is Σ_2-complete. This is one class higher in the polynomial hierarchy than NP-complete (or Σ_1-complete) problems. Some authors [24,25], have conjectured that a phase transition could be exhibited further up the polynomial hierarchy and therefore that this framework could be useful to other PSPACE problems. This was supported by results on planning and QBF-2 (Quantified Boolean Formulas with two alternating quantifiers, see also [26]).

In this paper, we show that this also holds true for the bounded ILP consistency problem and we present two main results. First, we exhibit the phase transition of the probability of solubility of the bounded ILP consistency problem, with the number of positive and negative examples as order parameters. Second, we exploit this framework as a benchmarking tool to evaluate for the first time learning algorithms on inherently hard and inherently easy problems, from a complexity point of view. We evaluate classical complete relational learners found in the learning systems Aleph [27], Progol [28] and Propal [29]: informed search algorithms such as Best-First Top-down Generate-and-Test (BESTF-TGT), A-search Top-down Generate-and-Test (A-TGT), and non-informed ones, such as Breadth-First Top-down Generate-and-Test and Data-Driven (BF-TGT and BF-TDD), Depth-First Top-Down Generate-and-Test (DF-TGT). We also use a lgg-based learner: Depth-First Bottom-up Data-Driven (DF-BDD).

The expected benefit is the same as for other combinatorics domains: (1) pointing out particularly bad algorithms; (2) importing and adapting the best search strategies developed in other domains; (3) developing a unified framework for the empirical evaluation of learners, not only based on real-world applications (as this is always necessary), but also on problems with controlled complexity; (4) understanding scaling-up problems acknowledged in Relational Learning (see e.g. [2]) by studying their behaviour on inherent hard instances in the phase transition; and last (5) finding ways to generate hard instances to understand the problem's complexity and to provide challenging benchmark datasets [30,31].

The paper is organised in the following manner. Section 2 presents background information about relational learning and search strategies, as well as main results on the phase transition framework in combinatorics and the "easy-hard-easy" pattern. Section 3 describes the random problem instance generator, which has been already proposed to study the bounded ILP consistency problem in [21], although they did not study its PT as they did not investigate the relevant parameters for this work. Next, section 4 shows that RL exhibits a phase transition of the probability of solubility and therefore can be used to reliably assess inherent complexity of learning problems. This is used in section 5 to evaluate and discuss the behaviour of popular complete learners on inherently hard and inherently easy problems. Finally, section 6 discusses the expected benefit of the development of the phase transition framework in RL and conclude on future works.

2 Background

2.1 Relational Learning

In machine learning, we are given a learning set $E = E^+ \cup E^-$, with positive and negative examples of the unknown target concept, drawn from an example language $\mathcal{L}_e$, a hypothesis language $\mathcal{L}_h$, a generality relation named subsumption test $\geq$ which relates $\mathcal{L}_e$ and $\mathcal{L}_h$ and partially order the hypotheses. After [3], (symbolic) learning is defined as search in $\mathcal{L}_h$. The problem, known as the consistency problem, is to find a hypothesis $h \in \mathcal{L}_h$ such that h is consistent with the data. A given hypothesis h is consistent iff it is complete: $\forall e^+ \in E^+, h \geq e^+$ and correct: $\forall e^- \in E^-, h \not\geq e^-$.

In this article, we study a typical instance of this problem in relational learning, known as the ILP consistency problem for function-free Horn clauses [23]: given $\mathcal{L}_e$, a language of function-free ground Horn clauses, $\mathcal{L}_h$, a language of (non-recursive) function-free Horn clauses and an integer l polynomial in $|E^+ \cup E^-|$, does there exist $h \in \mathcal{L}_h$ with no more than l literals such that h logically implies each element in E^+ and none element in E^-. In such hypothesis space, the logical implication is equivalent to θ-subsumption[4] which is NP-complete and therefore decidable [32]. This result implies that relational learning is higher than attribute-value learning in the polynomial hierarchy. [23] proved that this problem is Σ_2^P-complete (or equivalently NP^{NP}): the search is NP-complete and it is guided by the subsumption test which is NP-complete.

A central idea in symbolic learning is the use of a generality partial order between hypotheses to guide the resolution of the consistency problem [3]. Mitchell refines the search strategy into the generate-and-test (GT) and data-driven (DD) strategies. Virtually all GT algorithms are top-down, as it appeared early that a bottom-up approach would start with a too specific hypothesis to be efficiently guided by a heuristic function (see [33] for details). In this paradigm, the top-down refinement operator, noted ρ, is only based on the structure of the hypothesis space, independently of the learning data: Let $h \in \mathcal{L}_h : \rho(h) = \{h' \in \mathcal{L}_h | h \geq h'\}$.

Therefore, generate-and-test algorithms have to deal with many refinements that are not relevant with respect to the discrimination task. They only rely on the evaluation function to prune the irrelevant branches. On the contrary, the top-down DD (TDD) strategy searches the space of hypotheses that are more

[4] Let C, D two clauses. C θ-subsumes D, noted $C \geq_\theta D$ iff there exists a substitution θ such that $C\theta \subseteq D$.

general than or equal to a given positive example, named the seed example, and uses negative examples to prune irrelevant branches in the refinement graph. Formally, the TDD refinement operator is defined as a binary operator: Let $h \in \mathcal{L}_h, e^- \in E^- : \rho(h, e^-) = \{h' \in \mathcal{L}_h | h \geq h' \text{ and } h' \not\geq e^-\}$.

As opposed to a TGT approach, a TDD approach can therefore compensate for a poor evaluation function by using the learning data [29]. Moreover, some TDD strategies make the most of the negative instances in order to select informative negative examples, e.g. *near-misses* according to Winston. Dually to the TDD strategy, the Bottom-up Data Driven (BDD) strategy relies on positive examples to guide its generalisation step. Its BDD refinement operator, noted δ is given as follows: Let $h \in \mathcal{L}_h, e^+ \in E^+ : \delta(h, e^+) = \{h' \in \mathcal{L}_h | h \leq h' \text{ and } h' \geq e^+\}$.

This strategy has been first formalised by [34], who made the link between generalisation in learning and lowest-upper bound (lub) in lattice theory. Such an operator, also known as least-general generalisation (lgg) or most-specific generalisation (msg), has known several theoretic developments [35,12,36] but has been seldom used in learning systems, whose the best-known system is probably GOLEM [37].

2.2 Phase Transition and "Easy-Hard-Easy" Pattern

Phase transition is a term originally used in physics to describe the changes of state of matter [38]. Even though originally referring to gas, liquid, or solid, within the framework of thermodynamics, it is used, by extension, to describe an abrupt and suddenly change in one of the parameters describing the state (in thermodynamic sense) of an arbitrary system. Thus, the transition from ferromagnetic to paramagnetic state, the emergence of super-fluidity, changing the type of crystal (with broken symmetry), or denaturation transition of DNA are characterised as phase transitions. A well-known example in everyday life is water, which is boiling (at normal pressure) at 100 ^{o}C. At the transition point there is a coexistence of liquid water and vapor (a "first order" phase transition). When plotting the density as a function of the temperature, a jump at the transition temperature can be observed. In this example, using physics terminology, density corresponds to the order parameter, whereas temperature corresponds to the external parameter of the phase transition. Notice that, in computer sciences, the term order parameter is used instead of external parameter, and we will keep this terminology in the following.

The phase transition of the probability of NP-complete decision problems, and beyond, certainly is the most studied phase transition framework in Computer Sciences as it has important consequences in practice on the average search complexity [5,39,7,8,9,40,41,25,42,43]. The so-called order parameters allow to wander from an under-constrained region, named the "yes" region, where this is almost surely a solution, to an over-constrained region, named the "no" region, where there is almost surely no solution. In between, the phase transition region contains the most difficult problem instances, those on which the computational complexity shows an exponential increase in the problem size, independently of the solver used [5,6,41,7]. The under-constrained problems from the "yes" region

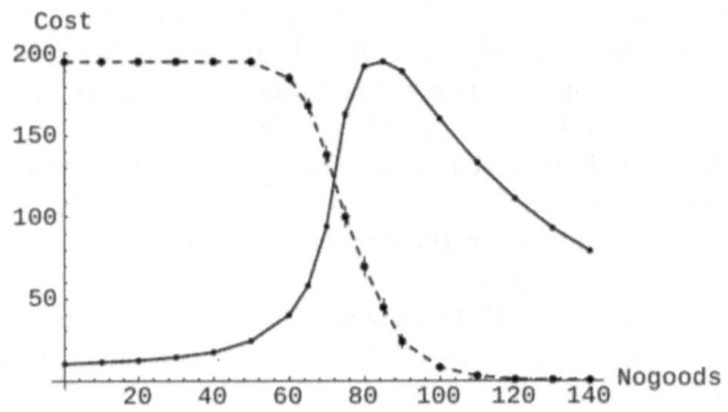

Fig. 1. Typical "easy-hard-easy" pattern [10]. Solid line shows median solution cost for dynamic backtracking and dashed line shows probability of a solution as a function of assignments to a pair of variables in random CSP that are considered to be inconsistent(number of no-goods).

appear to be easily soluble, as there are many solutions. This is the same for over-constrained problems from the "no" region as it is easy to prove that they are insoluble. These findings have been corroborated on several problems, with different types of algorithms, and it is considered that the problem instances appearing in the phase transition are inherently hard, independently of the algorithms used. In the "yes" and "no" regions, the easy ones, the complexity appears to be very dependent of the algorithm. There are, in these regions, some problems exceptionally hard, whose complexity dominates the complexity of instance problems in the phase transition region for certain types of algorithms [9,41,40]. In other words, a "good" algorithm when studied along the three different regions has to exhibit an average complexity following the so-called "easy-hard-easy" pattern. Such a typical pattern is shown in figure 1, from [10]. The search cost varies as a function of a given order parameter (refereed as nogoods) for a class of problems, independent of particular search algorithms.

The interest of the framework is two-fold as it gives a way to empirically assess the efficiency of algorithms on classes of problems whose inherent complexity is controlled by order parameters, and as finding ways to generate hard instances for a problem is important to understand the complexity of the problem [43,30]. We present in the next section the model RLPG for the bounded ILP consistency problem that we will use to exhibit the PT.

3 Random Generator for Relational Learning Problem

A learning problem instance in this model is denoted $RLPG(k, n, \alpha, N, Pos, Neg)$. The parameters k, n, α, N are related to the definition of the hypothesis and example spaces. Pos and Neg are the number of positive and negative examples respectively. The first four parameters are defined in order to ensure that a subsumption test between a hypothesis and an example during search encode a valid CSP problem following the model RB for random CSP [43]. We recall their meaning and

Table 1. Example of a random learning problem generated with RLPG, with no solution

$\bot$	$p0(A) \leftarrow p1(A,B,C), p2(A,B,D), p3(A,C,D)$
$+$	$p0(e1) \leftarrow p1(e1,b,c), p2(e1,c,d), p3(e1,e,f)$
$-$	$p0(e2) \leftarrow p1(e2,c,f), p2(e2,d,e), p3(e2,d,c)$

Table 2. Example of a random learning problem generated with RLPG, with a solution

$\bot$	$p0(A) \leftarrow p1(A,B,C), p2(A,B,D), p3(A,C,D)$
$+$	$p0(e1) \leftarrow p1(e1,b,c), p2(e1,d,e), p3(e1,e,e)$
$-$	$p0(e2) \leftarrow p1(e2,b,b), p2(e2,e,e), p3(e2,e,c)$

focus on the last two parameters, which were not studied before and which will be shown to be order parameters of the phase transition of the ILP consistency problem.

$k \geq 2$ denotes the arity of each predicate present in the learning language, $n \geq 2$ the number of variables in the hypothesis space, α the domain size for all variables as being equal to n^α, and finally, N the number of literals in the examples built on a given predicate symbol. Given k and n, the size of the bottom clause of the hypothesis space $\mathcal{L}_h$ is $\binom{n}{k}$, and encodes the largest constraint network of the underlying CSP model. Each constraint between variables is encoded by a literal built on a unique predicate symbol. $\mathcal{L}_h$ is then defined as the power set of the bottom clause, which is isomorphic to a boolean lattice. Its size is $2^{\binom{n}{k}}$.

Learning examples are randomly drawn, independently and identically distributed, given k, n, α and N. Their size is $N.\binom{n}{k}$. Each example defines N literals for each predicate symbol. The N tuples of constants used to define those literals are drawn uniformly and without replacement from the possible set of $\binom{n^\alpha}{k}$ tuples.

As an illustration, table 1 shows a random $RLPG(2, 3, \alpha, 1, 1, 1)$ problem, with α such that $n^\alpha = 5$. The first line shows the bottom-most element of the hypothesis space, which encodes all binary constraints between 3 variables. The next two lines show the positive and the negative example, respectively, allowing only one matching of a given predicate symbol (as $N = 1$). The search space is of size 2^3 and consists of all hypotheses built with the same head as the bottom clause, and with a subset of its body as body. In such a space, it is easy to see that there is no solution, given that no hypothesis subsumes the positive example without subsuming the negative example. Whereas the problem illustrated in table 2 accepts the following clause as solution: $p0(A) \leftarrow p2(A,B,D), p3(A,C,D)$.

4 Number of Positive and Negative Examples as Order Parameters

In this section, we study the effect of the number of positive and negative examples on the solubility probability of the ILP consistency problem. If we refer to the previous section, $RLPG$ is parametrised with 6 parameters but we only

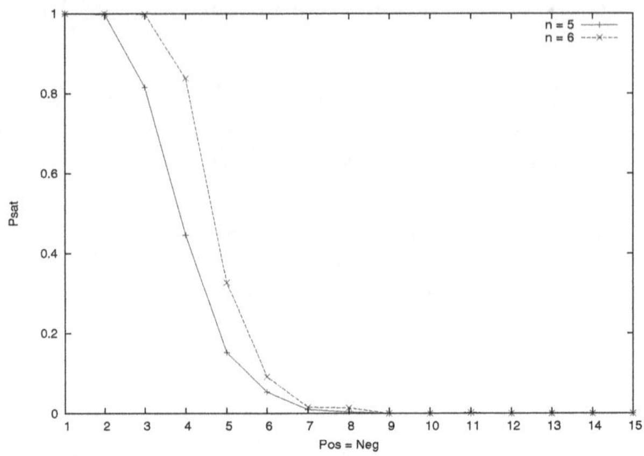

Fig. 2. Probability of satisfiability according to the number of learning examples (= Pos = Neg), with $n = 5$ and $n = 6$

study the last two, Pos and Neg, as the effect of the other parameters have already been studied in [21] for constant number of positive and negative examples. Here, we focus on few settings for these parameters, with $k = 2$, $n = 5$ and $n = 6$, to study different problem sizes, $\alpha = 1.4$ and $N = 10$. The choice of these parameters ensures that we do not generate trivially insoluble problems (see [25] for details), but also various experiments, not shown here, indicated that they were representative of the phase transition behaviour of the ILP consistency problem. In all experiments below, statistics were computed from a sample of 500 learning problems, solved with a complete learner.

We start by varying both Pos and Neg. Figure 2 shows the solubility probability of the ILP consistency problem when $Pos = Neg$ are varied from 1 to 15, for $n = 5$ and $n = 6$. As we can see, when the number of examples is small, there is almost surely a consistent hypothesis, and when the number is large, it is almost surely impossible to find a consistent hypothesis. The cross-over point, where the probability of solubility is about 0.5, is around 4 for $n = 5$ and 5 with $n = 6$. It is not surprising that it increases with bigger problems. For $n = 5$, the hypothesis space size is 2^{10} and 2^{15} for $n = 6$. We could not conduct experiments for larger values of n as the hypothesis space grows too fast in $RLPG$. For instance, $n = 7$ sets a hypothesis space of size 2^{21}, which cannot be handled by our complete solver. In the future, it would be interesting to modify $RLPG$ to specify the size of the bottom clause and then draw the number of variables accordingly.

We study now the phase transition along the number of positive examples, for constant values of Neg. Figures 3 and 4 show the phase transition when Pos varies from 1 to 25, for $n = 5$ and $n = 6$ respectively. With no positive examples, the bottom element of the search space is solution, but as Pos increases, complete hypotheses get more general and eventually subsume a negative example. The transition becomes sharper as Neg increases, which is not surprising as the subset of correct hypotheses shrinks as Neg increases. The second order parameter is

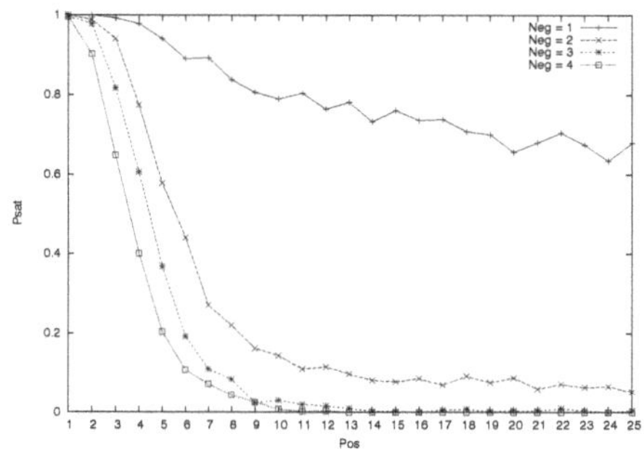

Fig. 3. Probability of satisfiability according to the number of positive examples with $n = 5$, for $Neg = 1, 2, 3, 4$

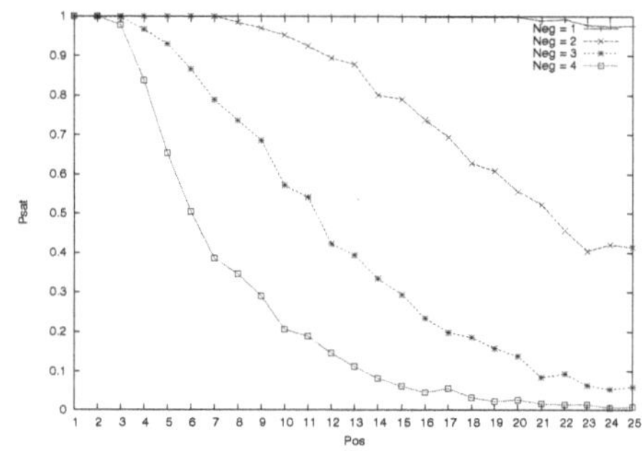

Fig. 4. Probability of satisfiability according to the number of positive examples with $n = 6$, for $Neg = 1, 2, 3, 4$

the number of negative examples Neg but it is not shown here because of the space requirements. The plots essentially exhibit the same profile.

5 Evaluating Complete Learners

We evaluate complete learners representative of the search strategies described in section 2.1. As non-informed searches, we use the Breadth-First TGT search (BF-TGT) and the Depth-First TGT search (DF-TGT). As informed searches, we use the A TGT search (A-TGT) and the Best-First TGT search (BESTF-TGT). Informed search makes use of an evaluation function to minimise, whose general form is $f = g + h$. g is defined as the cost from the start to the current hypothesis and h as an estimation of the distance from the current hypothesis to the goal. We define A-TGT according to the Progol system: g is defined as the length of the current hypothesis and h as the difference between the number of negative examples and the number of positive examples. In our context, as all positive examples must be subsumed, it simplifies to the number of negative examples. BESTF-TGT is not biased towards shorter hypotheses and defines $g = 0$. We refer to [27,28] for details about their implementations.

The next learning strategy we study is the one used in the TDD learner Propal. This is an incomplete learner as it performs a beam search guided by the Laplace function. So we set Propal with a beam of unlimited size, which basically turns down to a non-informed Breadth-First search (BF-TDD). The only difference is that when the solution is reached at a level of the search, it will be the first picked up at the next level. Note also that, as an incomplete learner, it does not have an optimal refinement operator, like the other learners, and may evaluate the same hypothesis several times.

The last learning strategy is Depth-First BDD (DF-BDD), based on Plotkin's lgg operator, and we refer to [36] for implementation details. Briefly, starting from the bottom element, the algorithm generalises the current hypothesis to subsume each positive example in turn, until it outputs a consistent hypothesis, or until it proves that no correct hypothesis subsumes all positive examples. Note that as the hypothesis space is not a lattice, which is the case here under θ-subsumption as the hypothesis space is finite, the lgg operator outputs all possible generalisations on subsequent backtracks.

We evaluate complete RL learners on random problem instances whose inherent complexity is controlled by the order parameter of the PT. We plot their search cost as a function of the order parameter to compare their complexity pattern to the standard "easy-hard-easy" pattern, as it is an indication of search efficiency (see section 2.2). As the consistency problem in RL is Σ_2^p-complete (see section 2.1), the search cost measurement has to take into account both the cost of the exploration of the hypothesis space and the cost of the consistency check. We propose to measure both the number of backtracks of the subsumption procedure and the time in milliseconds needed to solve a learning problem. The former measure is relevant for GT approaches, as the cost of the refinement operator is negligible compared to the subsumption cost, and it reflects the number of evaluated hypotheses. This is also the case for DF-BDD, as the lgg operator uses the subsumption test to find the common generalisations of two given clauses. However, it is not appropriate for BF-TDD which is based on the Propal system. Propal delegates the computation of refinements to a Weighted CSP solver [29] whose cost does not translate into backtracks of the subsumption test. It would be interesting to propose a relevant cost measure for all RL learners, independently on the implementation but we leave it for future research. Thus, we use the resolution time as cost measure for this strategy, and although it does not allow a direct comparison with other approaches, it is still relevant to study its expected cost pattern.

All experiments are done using instances from $RLPG(k, n, \alpha, N, Pos, Neg)$, with $k = 2$, $n = 5$ and $n = 6$ to study different problem sizes, $\alpha = 1.4$ and $N = 10$. Additional experiments using different parameter values (not shown here) have been conducted and result in similar findings. In the following figures every plot is averaged over 500 randomly drawn learning problems.

Figure 5 shows the results obtained with A-TGT, for $n = 6$. We can see that the easy problems resolution from the "yes" region follows the standard pattern. The superposition of the solubility probability plot shows the PT region. The cost sharply increases as soon as the solubility probability is no longer 1 (when both the number of positive and negative examples are greater than 3). This increase stops when the probability gets close to 0. However, the plot does not reach a maximum right after the PT. This is indicative of a bad search algorithm, as the backtracking cost keeps increasing, as the number of examples increases, in the region theoretically easy, dominating then the cost in the PT.

We are going to see that this behaviour is typical of the top-down approaches: interestingly, in the "no" region, extra examples do not help enough pruning the hypothesis space to compensate the increase in subsumption cost of those extra examples.

For various percentiles, figure 6 shows that BF-TGT, as a non-informed search strategy, is costly very early in the "yes" region. However, after the PT, A-TGT

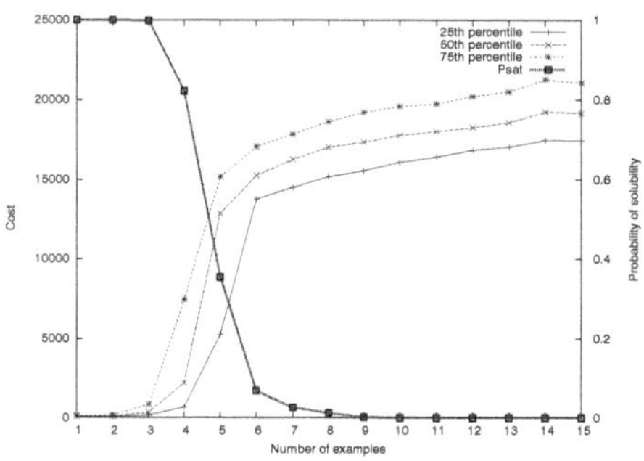

Fig. 5. Backtracking cost using A-TGT strategy for various percentiles and probability of solubility, for $n = 6$

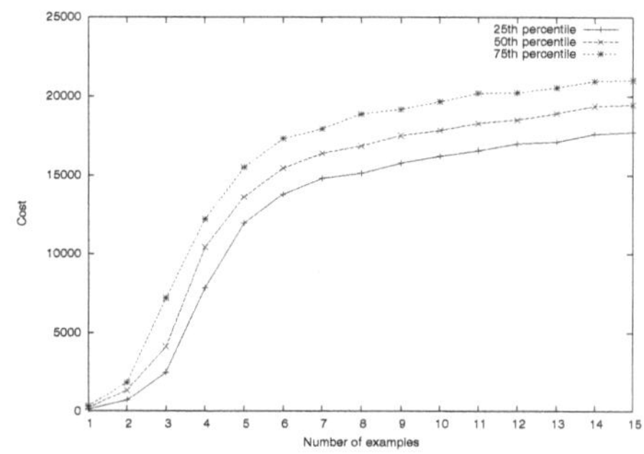

Fig. 6. Backtracking cost using BF-TGT strategy for various percentiles, for $n = 6$

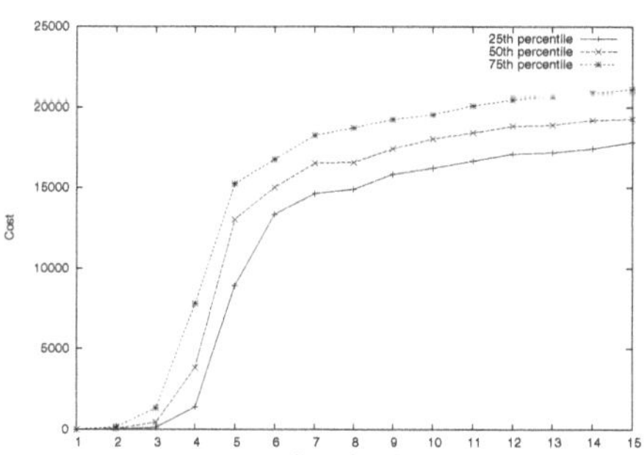

Fig. 7. Backtracking cost using DF-TGT strategy for various percentiles, for $n = 6$

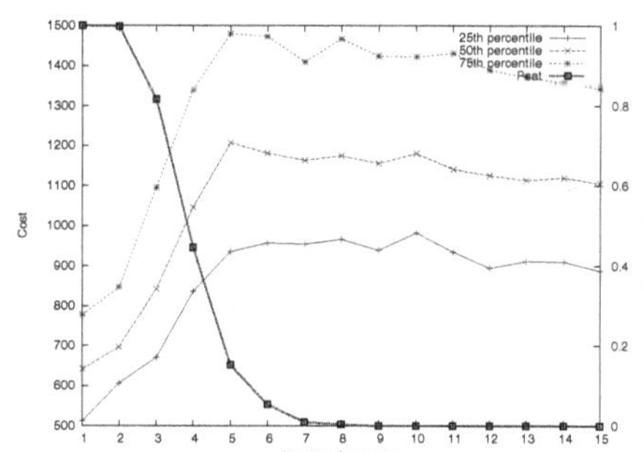

Fig. 8. Backtracking cost using DF-BDD strategy for various percentiles for $n = 5$

and BF-TGT are about equivalent: they cannot cope with an increasing number of examples, and the cost in the "no" region dominates the cost in the PT region.

In the "yes" region, DF-TGT behaves better than BF-TGT. This is particularly true in the "yes" region where there are a lot of solutions. In that case, to keep specialising a complete hypothesis leads almost surely to a consistent hypothesis. DF-TGT is as good as A-TGT in this region, but when they get closer to the PT, A-TGT performs better. Its heuristic function prioritizes hypotheses which discriminate negative examples the most and this seems to lead to consistent hypotheses faster. In the "no" region, we see again that DF-TGT degenerates as A-TGT and BF-TGT.

In figure 8, we show results for the data-driven search, DF-BDD, first on problems of size $n = 5$. It gets close to the standard pattern for the "yes' region problems. We note however that for higher percentiles (e.g. the median) the algorithm has a non negligible cost even for 1 positive and 1 negative example. Moreover, the superposition of the solubility plot shows the cross-over point of the PT between 4 and 5 examples and that the complexity peak is slightly shifted to the right with respect to this point, which indicates that DF-BDD's

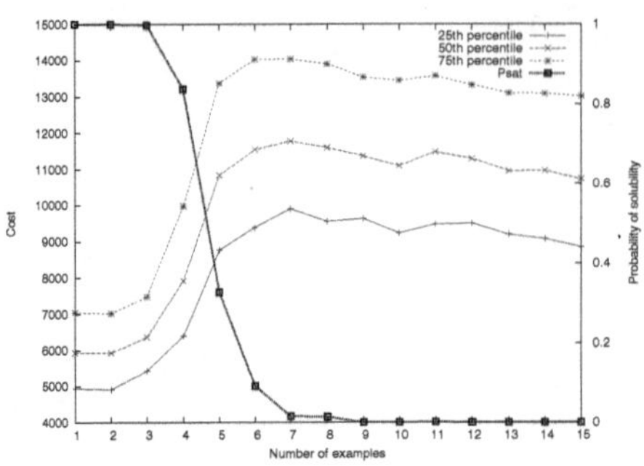

Fig. 9. Backtracking cost using DF-BDD strategy for various percentiles, for $n = 6$

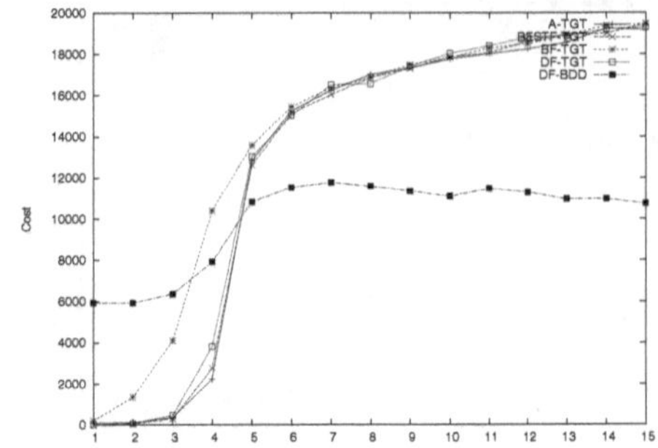

Fig. 10. Median (50th percentile) backtracking cost for A-TGT, DF-TGT, BF-TGT, BESTF-TGT and DF-BDD strategies, for $n = 6$

cost pattern is close to the "easy-hard-easy" pattern. Also, we see that for all percentiles, the cost slowly decreases after the PT. We can say that this algorithm is a good search algorithm, although some improvements can be done.

Figure 9 shows results for the same algorithm, but on larger problems, with $n = 6$. The cross-over point is now around 5, and DF-BDD's behaviour gets closer to the standard pattern in the "yes" region. Although there is a minimum cost (6000 backtracks as median cost), certainly due to the naive implementation of the lgg refinement operator, this cost does not vary much in the "yes" region. Among all tested algorithms, it is the only one exhibiting the "easy-hard-easy" pattern.

Figure 10 summarises the backtracking cost of the search algorithms discussed above, with the addition of BESTF-TGT. The results are clear: all GT approaches are interesting for problems with many solutions but are particularly bad when there are few or no solutions. Moreover, either informed or non-informed search strategies, they all have the same profile in this latter case, which is an interesting point to detail in the future. Conversely, DF-BDD, although penalised in the "yes" region, is more efficient in those problems with few or no solutions, with a decrease in cost as the number of examples increases.

The complexity analysis limited to the number of backtracks of the subsumption test is not enough for this study because it does not take into account the cost of the refinement operators for all approaches, such as for BF-TDD (see above). We then complete it by plotting the resolution time of BF-TDD in figure 11. Although the search cost cannot be directly compared, we see that it behaves similarly to the other top-down approaches. In the "yes" region, the TDD operator cannot compensate the breadth-first search with its smaller branching factor, and therefore behaves like BF-TGT. After the exponential increase in cost on the inherent hard instances, the cost keeps increasing as the number of examples grows in the "no" region. The penalty here is that the number of calls to the Weighted CSP solver to compute a near-miss is proportional to the number of negative examples. This is clearly too costly and the trade-off between

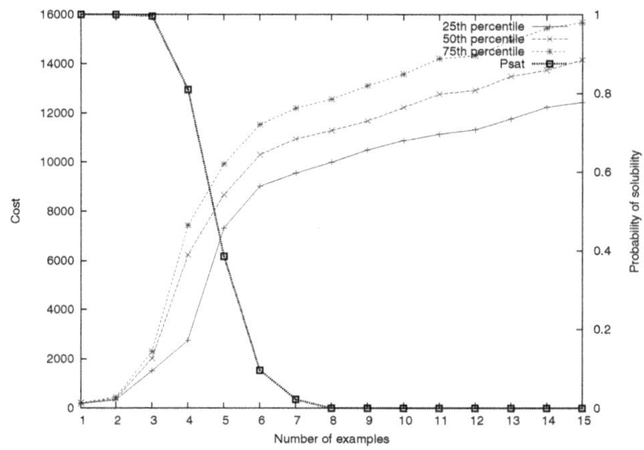

Fig. 11. Time cost using BF-TDD strategy for various percentiles, for $n = 6$

the quality of the near-miss and the reduction of the search space has to be evaluated.

6 Conclusion

Although Relational Learning has been cast, more than 25 years ago, as search, it has known very few developments from the search strategy point of view and most learners rely on general-purpose solvers. This is a strong limitation to its applicability on many modern applications, as it prevents RL to scale-up well. On the other hand, important progress has been made in other combinatorics communities, such as SAT and CSP, in the development of efficient specialised solvers, through the study of random NP-complete problem generators in the phase transition framework. RL has a higher complexity, being Σ_2-hard in the general case. However, we argue that this framework will benefit RL, based on the conjecture that the phase transition can be exhibited further up the polynomial hierarchy. We show that this conjecture holds true with the bounded ILP consistency problem, a Σ_2-complete problem, representative of RL problems. We propose a first simple random generator that exhibits a phase transition in the problem's solubility, with the number of positive and negative examples as order parameters. We used this framework to generate benchmark datasets with controlled complexity, based on conjectures linking the probability of problem solubility with inherent problem hardness. First, this study shows that all well-known top-down relational algorithms, rooted either in the generate-and-test or the data-driven paradigm, are bad as they fail to exhibit the standard "easy-hard-easy" pattern. Their complexity tend to increase with the number of examples, although the extra examples do not change the solubility of the problem, and therefore they exhibit an "easy-hard-hard" pattern. This has to be contrasted with DF-BDD, a lgg-based learner, which does not perform as well on the easy problems in the "yes" region, but well on the easy problems of the "no" region, as well as in the phase transition compared to the other algorithms.

This study shows that search strategies standard in RL lag behind what is considered state of the art in other combinatorics communities. It is clear that this study does not take into account all the dimensions of learning problems: optimization instead of consistency, presence of noise, etc. However, the first idea is to understand the complexity landscape of learning problems and to define order parameters to control this complexity. The most important advantage of the proposed approach to evaluate algorithm complexity is that contrary to results obtained directly on real-world applications, which hardly transpose when the size of the problems change of scale, the phenomena observed with few variables are the same as those observed with thousands of variables. We hope that it will enable RL and ILP to import and/or develop better search algorithms, to eventually benefit to better scaling relational learners. For instance, we plan to investigate lgg-based learning algorithms, which have been seldom used in learning systems but seem to be efficient solvers.

References

1. Domingos, P.: Prospects and challenges for multi-relational data mining. SIGKDD Explorations 5(1), 80–83 (2003)
2. Page, D., Srinivasan, A.: Ilp: a short look back and a longer look forward. J. Mach. Learn. Res. 4, 415–430 (2003)
3. Mitchell, T.M.: Generalization as search. Artificial Intelligence 18, 203–226 (1982)
4. Newell, A., Simon, H.A.: Human Problem Solving. Prentice-Hall, Englewood Cliffs (1972)
5. Cheeseman, P., Kanefsky, B., Taylor, W.: Where the really hard problems are. In: Proc. of the 12th International Joint Conference on Artificial Intelligence, pp. 331–340. Morgan Kaufmann, San Francisco (1991)
6. Mitchell, D., Selman, B., Levesque, H.: Hard and easy distribution of SAT problems. In: Proceedings of the Tenth National Conference on Artificial Intelligence (AAAI 1992), pp. 440–446 (1992)
7. Monasson, R., Zecchina, R., Kirkpatrick, S., Selman, B., Troyansky, L.: Determining computational complexity from characteristic 'phase transitions. Nature 400, 133–137 (1999)
8. Smith, B.M., Dyer, M.E.: Locating the phase transition in binary constraint satisfaction problems. Artificial Intelligence 81(1-2), 155–181 (1996)
9. Hogg, T., Williams, C.: The hardest constraint problems: A double phase transition. Artificial Intelligence 69(1–2), 359–377 (1994)
10. Mammen, D.L., Hogg, T.: A new look at the easy-hard-easy pattern of combinatorial search difficulty. Journal of Artificial Intelligence Research 7, 47–66 (1997)
11. Gomes, C., Heny Kautz, A.S., Selman, B.: Satisfiability solvers. In: Handbook of Knowledge Representation (2007)
12. Haussler, D.: Learning conjunctive concepts in structural domains. Machine Learning 4(1), 7–40 (1989)
13. Kearns, M.J., Vazirani, U.V.: An Introduction to Computational Learning Theory. The MIT Press, Cambridge (1994)
14. En, N., Srensson, N.: Translating pseudo-boolean constraints into SAT. Journal on Satisfiability, Boolean Modeling and Computation 2, 1–26 (2006)

15. Fürnkranz, J.: Pruning algorithms for rule learning. Mach. Learn. 27(2), 139–172 (1997)
16. Shim, G.M., Choi, M.Y., Kim, D.: Phase transitions in a dynamic model of neural networks. Physics Review A 43, 1079–1089 (1991)
17. Nagashino, H., Kelso, J.A.: Phase transitions in oscillatory neural networks. In: Society of Photo-Optical Instrumentation Engineers (SPIE) Conference Series, vol. 1710, pp. 279–287 (1992)
18. Schottky, B.: Phase transitions in the generalization behaviour of multilayer neural networks. Journal of Physics A Mathematical General 28, 4515–4531 (1995)
19. Biehl, M., Ahr, M., Schlsser, E.: Statistical physics of learning: Phase transitions in multilayered neural networks. Advances in Solid State Physics 40/2000, 819–826 (2000)
20. Botta, M., Giordana, A., Saitta, L., Sebag, M.: Relational learning as search in a critical region. Journal of Machine Learning Research 4, 431–463 (2003)
21. Alphonse, E., Osmani, A.: A model to study phase transition and plateaus in relational learning. In: Proc. of Conf. on Inductive Logic Programming, pp. 6–23 (2008)
22. Rückert, U., Kramer, S., Raedt, L.D.: Phase transitions and stochastic local search in k-term DNF learning. In: Elomaa, T., Mannila, H., Toivonen, H. (eds.) ECML 2002. LNCS (LNAI), vol. 2430, pp. 405–417. Springer, Heidelberg (2002)
23. Gottlob, G., Leone, N., Scarcello, F.: On the complexity of some inductive logic programming problems. In: Džeroski, S., Lavrač, N. (eds.) ILP 1997. LNCS, vol. 1297, pp. 17–32. Springer, Heidelberg (1997)
24. Bylander, T.: A probabilistic analysis of propositional strips planning. Artificial Intelligence 81(1-2), 241–271 (1996)
25. Gent, I.P., Walsh, T.: Beyond NP: the QSAT phase transition. In: AAAI 1999/IAAI 1999: Proceedings of the sixteenth national conference on Artificial intelligence and the eleventh Innovative applications of artificial intelligence conference, pp. 648–653 (1999)
26. Chen, H., Interian, Y.: A model for generating random quantified boolean formulas. In: Kaelbling, L.P., Saffiotti, A. (eds.) Proceedings of the Nineteenth International Joint Conference on Artificial Intelligence, pp. 66–71. Professional Book Center (2005)
27. Srinivasan, A.: A learning engine for proposing hypotheses (Aleph) (1999), `http://web.comlab.ox.ac.uk/oucl/research/areas/machlearn/Aleph`
28. Muggleton, S.: Inverse entailment and PROGOL. New Generation Computing 13, 245–286 (1995)
29. Alphonse, É., Rouveirol, C.: Extension of the top-down data-driven strategy to ILP. In: Muggleton, S.H., Otero, R., Tamaddoni-Nezhad, A. (eds.) ILP 2006. LNCS (LNAI), vol. 4455, pp. 49–63. Springer, Heidelberg (2007)
30. Cook, S.A., Mitchell, D.G.: Finding hard instances of the satisfiability problem: A survey. In: DIMACS Series in Discrete Mathematics and Theoretical Computer Science, pp. 1–17. American Mathematical Society (1997)
31. Xu, K., Li, W.: Many hard examples in exact phase transitions. Theor. Comput. Sci. 355(3), 291–302 (2006)
32. Gottlob, G.: Subsumption and implication. Information Processing Letters 24(2), 109–111 (1987)
33. Fürnkranz, J.: A pathology of bottom-up hill-climbing in inductive rule learning. In: Cesa-Bianchi, N., Numao, M., Reischuk, R. (eds.) ALT 2002. LNCS (LNAI), vol. 2533, pp. 263–277. Springer, Heidelberg (2002)

34. Plotkin, G.: A note on inductive generalization. In: Machine Intelligence, pp. 153–163. Edinburgh University Press (1970)
35. Valiant, L.G.: A theory of the learnable. In: ACM Symposium on Theory of Computing (STOC 1984), Baltimore, USA, pp. 436–445. ACM Press, New York (1984)
36. Kietz, J.U.: A comparative study of structural most specific generalisations used in machine learning. In: Proc. Third Workshop on ILP, pp. 149–164 (1993)
37. Muggleton, S., Feng, C.: Efficient induction of logic programs. In: Proc. of the 1st Conference on Algorithmic Learning Theory, Ohmsma, Tokyo, Japan, pp. 368–381 (1990)
38. Paskal, Y.I.: The meaning of the terms phase and phase transition. Russian Physics Journal 31(8), 664–666 (1988)
39. Selman, B., Levesque, H.J., Mitchell, D.: A new method for solving hard satisfiability problems. In: Proc. of the Tenth National Conference on Artificial Intelligence, Menlo Park, California, pp. 440–446 (1992)
40. Gent, I.P., Walsh, T.: Easy problems are sometimes hard. Artificial Intelligence 70(1–2), 335–345 (1994)
41. Davenport, A.: A comparison of complete and incomplete algorithms in the easy and hard regions. In: Workshop on Studying and Solving Really Hard Problems, CP 1995, pp. 43–51 (1995)
42. Smith, B.M.: Constructing an asymptotic phase transition in random binary constraint satisfaction problems. Theoretical Computer Science 265(1–2), 265–283 (2001)
43. Xu, K., Boussemart, F., Hemery, F., Lecoutre, C.: Random constraint satisfaction: Easy generation of hard (satisfiable) instances. Artif. Intell. 171(8-9), 514–534 (2007)

Topic Significance Ranking of LDA Generative Models

Loulwah AlSumait[1], Daniel Barbará[1], James Gentle[2], and Carlotta Domeniconi[1]

[1] Department of Computer Science, George Mason University, Fairfax VA 22030, USA
[2] Department of Computational and Data Sciences, George Mason University, Fairfax VA 22030, USA

Abstract. Topic models, like Latent Dirichlet Allocation (LDA), have been recently used to automatically generate text corpora topics, and to subdivide the corpus words among those topics. However, not all the estimated topics are of equal importance or correspond to genuine themes of the domain. Some of the topics can be a collection of irrelevant words, or represent insignificant themes. Current approaches to topic modeling perform manual examination to find meaningful topics. This paper presents the first automated unsupervised analysis of LDA models to identify junk topics from legitimate ones, and to rank the topic significance. Basically, the distance between a topic distribution and three definitions of "junk distribution" is computed using a variety of measures, from which an expressive figure of the topic significance is implemented using 4-phase Weighted Combination approach. Our experiments on synthetic and benchmark datasets show the effectiveness of the proposed approach in ranking the topic significance.

1 Introduction

Probabilistic Topic Modeling (PTM) is an emerging Bayesian approach to summarize data, such as text, in terms of (a small set of) latent variables that correspond (ideally) to the underlying themes or topics. It is a statistical generative model that represents documents as a mixture of probabilistic topics and topics as a mixture of words. Among the variety of topic models proposed, Latent Dirichlet Allocation (LDA) [4] is a truly generative model that is capable of generalizing the topic distributions so that it can be used to generate unseen documents as well. The completeness of the generative process for documents is achieved by considering Dirichlet priors on the document distributions over topics and on the topic distributions over words.

The setting of the number of latent variables K is extremely critical and directly effects the quality of the model and the interpretability of the estimated topics. Models with very few topics would result in broad topic definitions that could be a mixture of two or more distributions. On the other hand, models with too many topics are expected to have very specific descriptions that are

W. Buntine et al. (Eds.): ECML PKDD 2009, Part I, LNAI 5781, pp. 67–82, 2009.

Table 1. The Reuters: examples of topics estimated by LDA

Class	Top Words
coffee	export, coffee , quota , product , market , price , Brazil
ship	ship , gulf , attack , Iran , American , oil , tanker , water
oil/crude	oil , price, barrel , crude (0.045), increase, product , petroleum , energy
Na	was , report , official, any , did , said , ask , told , made , comment , time
Na	two , on , three , five, six, four , month, seven, eight

uninterpretable [8]. Since the actual number of underlying topics is unknown and there is no definite and efficient approach to accurately estimate it, the inferred topics of PTM does not always represent meaningful themes. For example, Table 1 lists five topics discovered by LDA when run on the Reuters-21578 dataset with K set to 50. It can be seen that the first three topics correspond to legitimate classes of the data. However, the last two topics are collections of insignificant words that are meaningless to the thematic structure of Reuters corpus.

Although LDA is heavily investigated and cited in the literature, none of the research provided an automatic analysis of the discovered topics to validate their importance and genuineness. Almost all the previous work manually examines the output to identify genuine topics in order to justify their work. Some work [10] computed the average distance of word distributions between all pairs of topics to measure how distinct they are. However, this figure evaluates the model in general and not the individual topics. In addition, the distance of a topic from the others does not provide any insight on the significance of the semantic content of the topic. Other approaches have used the probability of the topic as an indication of its importance [6,1]. However, as will be seen later, some meaningless topics that consist of common words across documents with different content can have a high probability.

This paper introduces a novel approach to automatically rank the LDA topics based on their semantic importance and, eventually, identify junk and insignificant topics. The idea is to measure the amount of insignificance that an inferred topic carries in its distribution by measuring how "different" the topic distribution is from a "junk" distribution. In this work, three definitions of Junk and Insignificant (J/I) topics are introduced. To quantify the difference between an estimated topic and a J/I distribution, a number of distance measures are used. Based on a Weighted Combination of multi-criteria decision analysis, this paper introduces a novel unsupervised quantification of the topic significance. Our experiments on synthetic and benchmark datasets show the effectiveness of the proposed topic significance ranking, and its ability to identify junk and insignificant topics. To the best of our knowledge, this is the first attempt to evaluate and rank topic significance of PTM models.

The rest of this paper is organized as follows. An overview of the problem definition and notations including a brief description of the Latent Dirichlet Allocation (LDA) topic model is given in Section 2. Section 3 introduces three definitions of J/I distributions and lists three distance measures by which the difference of the estimated topics of PTM from the J/I topics is computed. Then, the proposed Topic Significance Ranking (TSR) approach is defined in Section 4 followed by the experimental results that we obtained from applying the TSR on simulated and real data. Our final conclusions and future work are discussed in Section 6.

2 Problem Definition

LDA is a hierarchical Bayesian network that represents the generative model of a corpus of documents [4]. LDA assumes the standard bag-of-words representation, where each document d is represented as a vector of counts with W components, where W is the size of the dictionary. The documents of the corpus are modeled as mixtures over K topics each of which is a multinomial distribution over the dictionary of words. Each topic, $\boldsymbol{\phi}^{(k)}$, is drawn from a Dirichlet with parameter β, while each document, $\boldsymbol{\theta}^{(d)}$, is sampled from a Dirichlet with parameter α. For each word token i in document d, a topic assignment z_i is sampled from $\boldsymbol{\theta}^{(d)}$ which is introduced to represent the responsibility of a particular topic in using that word in the document. Then, the specific word x_i is drawn from $\boldsymbol{\phi}^{(z_i)}$. An exact estimation of $\boldsymbol{\phi}^{(k)}$ and $\boldsymbol{\theta}^{(d)}$ is found to be intractable [4], thus approximations such as Gibbs sampling [6] and variational inference [4] are used.

To identify genuine themes from the LDA estimated topics, the following learning setting is considered. Given a dataset of D documents with a total of N token words and W unique terms, a topic model $\mathcal{T}$ is generated from fitting its parameters, $\boldsymbol{\phi}$ and $\boldsymbol{\theta}$, to the dataset assuming that the number of topics is set to K. The matrix $\boldsymbol{\theta}$ is a $D \times K$ parameter matrix in which each row $\boldsymbol{\theta}^{(d)}$ is the multinomial distribution of document d. The matrix $\boldsymbol{\phi}$ consists of $W \times K$ parameters in which each column $\boldsymbol{\phi}^{(k)}$ represents the multinomial distribution of topic j [1]. Thus, the parameters in $\boldsymbol{\phi}$ and $\boldsymbol{\theta}$ indicate the relative importance of words in topics (i.e. $\phi_{w,k} = p(w|k)$) and the relative importance of topics in documents (i.e. $\theta_{d,k} = p(k|d)$), respectively.

In practice, a topic model $\mathcal{T}$ includes different sets of "*Junk and Insignificant*" (J/I) topics. A junk topic is an "uninterpretable topic that picks out idiosyncratic word combinations" [8]. An insignificant topic is a topic that consists of general words, known as "background words", which are commonly used in general or across a broad range of documents within each corpus/domain [5]. For domain experts and text miners, the content of these topics is low in significance and often meaningless.

To identify J/I topics, the approach is to define a decision criterion $\mathcal{C}$ as the distance D of the topic from a common J/I topic description Ω. If the distance is large, then this would provide a fair indication of the topic significance. However, if the distance of a topic to the J/I distribution is small, then the topic is more likely to be irrelevant to the domain structure.

[1] The notation $\boldsymbol{\phi}^{(k)}$ ($\boldsymbol{\theta}^{(d)}$) is used to indicate the topic (document) distribution. To refer to a particular probability value, this is noted by $\phi_{w,k}$ ($\theta_{d,k}$).

3 Junk/Insignificance Based Decision Criteria

This section introduces the J/I topic definitions and the distance measures that are used to evaluate the significance of a topic distribution.

3.1 Junk/Insignificance Topic Definitions

Uniform Distribution Over Words (W-Uniform). Aligning with the Zipf law for words [7], a genuine topic is expected to be modeled by a distribution that is skewed toward a small set of words, called "*salient words*", out of the total dictionary. A topic distribution under which a large number of terms are highly probable is more likely to be insignificant or "junk".

To illustrate this, the number of salient terms of topics estimated by LDA on the 20-Newsgroups dataset is computed. These words are defined to be the ones that have the highest conditional probability under a topic k. For each estimated topic, the number of salient words for which the total conditional probability is equal to some percentage, X, of the topic probability is counted. Then, these counts are averaged over all the topics. Table 2 lists the average and percentage of salient words for X ranged from 60% to 100%. The values are reported for experiments done with K set to 40 components.

It can be seen that most of the topic density corresponds to less than 3% of the total vocabulary. In fact, when $X = 100\%$, the average value was biased toward a set of extreme topics that have nonzero probability for the whole dictionary. When these topics were excluded from the average, the percentage of words dropped from 52% to 3.9%. This value is given in the table between parenthesis. Such topics are more likely to be junk topics that are irrelevant to the domain.

Under this frame, an extreme version of a junk topic will take the form of a uniform distribution over the dictionary. This topic, which is named *W-Uniform*, is the first junk definition in this paper. Formally, W-Uniform is a junk topic, $\boldsymbol{\Omega}^{\mathcal{U}}$, in which all the terms of the dictionary are equally probable

$$P(w_i|\boldsymbol{\Omega}^{\mathcal{U}}) = \frac{1}{W}, \forall i \in \{1, 2, \ldots, W\} \tag{1}$$

The degree of "*uniformity*", $\mathcal{U}$, of an estimated topic, $\phi^{(k)}$, can be quantified by computing its distance from the W-Uniform junk distribution, $\boldsymbol{\Omega}^{\mathcal{U}}$. The computed distance will provide a reasonable figure of the topic significance. The

Table 2. 20-Newsgroups: average and percentage of terms and documents that hold X percent of the topic density estimated by LDA

$D \times W$	X	Average # of terms	%age of dictionary	Average # of docs	%age of total docs
11269 × 53795	60%	227.25	0.42%	544.25	4.82%
	70%	342.7	0.64%	856.2	7.6%
	80%	521.35	0.97%	1382.9	12.27%
	90%	846.3	1.6%	2442.95	21.7%
	100%	28026.5 (2085.1)	52% (3.9%)	8538.8 (5202)	75.8% (46.1%)

larger the distance is, i.e. the farther a topic description is from the uniform distribution over the dictionary, the higher its significance is, and vise versa. Other definitions of junk topics are given next.

The Vacuous Semantic Distribution (W-Vacuous). The empirical distribution (the total word frequencies of the whole sample) is a convex combination of the probability distributions of the underlying themes that reveals no significant information if taken as a whole. A distribution of a real topic is expected to have a unique characteristic rather than a mixture model. Thus, the closer the topic distribution is to the empirical distribution of the sample, the less its significance is expected to be.

So, the second junk topic, namely the vacuous semantic distribution (W-Vacuous), is defined to be the empirical distribution of the sample set. It is equivalent to the marginal distribution of words over the latent variables. The probability of each term w_i under the W-Vacuous ($\boldsymbol{\Omega}^{\mathcal{V}}$) topic is given by

$$p(w_i|\boldsymbol{\Omega}^{\mathcal{V}}) = \sum_{k=1}^{K} p(w_i|k)p(k) \tag{2}$$

where $p(w_i|k) = \phi_{i,k}$, by definition, and $p(k)$ is the probability of the topics under the PTM which can be computed from

$$p(k) = \frac{\sum_{d=1}^{D} \theta_{d,k}}{N} \tag{3}$$

In order to detect junk topics, the "*vacuousness*" decision criterion of a topic, $\mathcal{V}$, is measured by computing the distance between the estimated distribution and the W-Vacuous. Lower $\mathcal{V}$ distances correspond to distributions with probability mixture models that represent insignificant topics.

The Background Distribution (D-BGround). The previous two definitions of junk topics are characterized by their distribution over words. However, investigating the distribution of topics over documents would identify another class of insignificant topics. In real datasets, well defined topics are usually covered in a subset (not all) of the documents. If a topic is estimated to be responsible of generating words in a wide range of documents, or all documents in the extreme case, then it is far from having a definite and authentic identity. Such topics are most likely to be constructed of the background terms, which are irrelevant to the domain structure.

Table 2 also provides the average and percentage of documents in which X percent of the topic density appears. In general, topics are inclined to appear heavily in a small subset of documents. Yet, nearly half of the topics are estimated to appear in a much larger fraction of documents, and in the extreme, in the whole the dataset. Examples of such topics are given in Table 3, in addition to examples of "normal" topics that appear in fewer documents.

To show reasonable significance for consideration, a topic is required to be far (enough) from being a "*background topic*", which can be defined as a topic that

Table 3. 20-Newsgroup: examples of background and legitimate topics

TopicID (Class)	Top Words
9(NA)	edu writes article cs apr cc michael andrew bitnet colorado cmu ohio acs cwru au
36(NA)	university information research national april center washington san california dr
4(space)	space nasa gov earth launch moon orbit satellite shuttle henry lunar flight mission
5(Crypt)	encryption government clipper chip technology key law phone security escrow

has a nonzero weight in all the documents. In the extreme case, the background topic (D-BGround) is found equally probable in all the documents. Formally, under the D-BGround topic, $\boldsymbol{\Omega}^{\mathcal{B}}$, the probability of each document d_m is given by

$$p(d_m|\boldsymbol{\Omega}^{\mathcal{B}}) = \frac{1}{D}, m \in \{1, 2, \ldots, D\}. \tag{4}$$

The distance between a topic and the D-BGround topic would determine how much "*background*" does it carry and, ultimately, grade the significance of the topic. Thus, given a topic k, defined as a distribution over documents

$$\boldsymbol{\vartheta}^{(k)} = (\theta_{1,k} \ldots \theta_{d,k} \ldots \theta_{D,k}), \tag{5}$$

then the background, $\mathcal{B}$, of a topic is measured by computing the distance of the topic distribution over documents from the D-BGround.

3.2 Distance Measures

Kullback-Leibler (KL) Divergence. The KL-Divergence D_{KL} is a distance measure that is defined based on the KL-divergence (or relative entropy) [2]. Thus, using D_{KL}, the distance of the topic distribution over words $\boldsymbol{\phi}^{(k)}$ from W-Uniform $\boldsymbol{\Omega}^{\mathcal{U}}$ and W-Vacuous $\boldsymbol{\Omega}^{\mathcal{V}}$, and the distance of the topic distribution over documents $\boldsymbol{\vartheta}^{(k)}$ from D-BGround $\boldsymbol{\Omega}^{\mathcal{B}}$ can be computed as follows

$$\mathcal{U}_k^{\mathrm{KL}} = D_{\mathrm{KL}}(\boldsymbol{\phi}^{(k)}, \boldsymbol{\Omega}^{\mathcal{U}}) \tag{6}$$
$$\mathcal{V}_k^{\mathrm{KL}} = D_{\mathrm{KL}}(\boldsymbol{\phi}^{(k)}, \boldsymbol{\Omega}^{\mathcal{V}}) \tag{7}$$
$$\mathcal{B}_k^{\mathrm{KL}} = D_{\mathrm{KL}}(\boldsymbol{\vartheta}^{(k)}, \boldsymbol{\Omega}^{\mathcal{B}}) \tag{8}$$

Cosine Dissimilarity. The cosine dissimilarity D_{COS} is a distance measure that is constructed based on the cosine similarity [9]. Similar to the D_{KL} in Equations (6), (7), and (8), the cosine distance is used to measure the uniformity ($\mathcal{U}_k^{\mathrm{COS}}$), vacuousness ($\mathcal{V}_k^{\mathrm{COS}}$), and background ($\mathcal{B}_k^{\mathrm{COS}}$) of topic k based on the cosine angle between the inferred topic vector and the W-Uniform, W-Vacuous, and D-BGround vectors, respectively. A cosine distance of value 0 (1) corresponds to completely related (unrelated) topics.

Correlation Coefficient. The correlation coefficient distance measure D_{COR} is a dissimilarity measure that is based on the correlation coefficient statistic [9]. The uniformity ($\mathcal{U}_k^{\mathrm{COR}}$), vacuousness ($\mathcal{V}_k^{\mathrm{COR}}$), and background ($\mathcal{B}_k^{\mathrm{COR}}$) of topic

k under the correlation-based distance is computed by measuring the correlation between the topic distribution and the W-Uniform, W-Vacuous, and D-BGround vectors, respectively. The distance is bounded by the closed interval $[0, 2]$, where independent and negatively related topics will result in distances greater than or equal to one. This fits with the definition of our problem since semantic relatedness between topics is evinced by positive correlations only.

4 Topic Significance Ranking

Due to the uncertainty that surround the data and the statistical modeling, it is very appealing to have an expressive quantitative measure of the topic significance that can assist in discriminating genuine topics from J/I ones.

In this paper, three different categories of topic significance criteria are defined each of which is quantified by a variety of distance measures. The objective is to construct a qualitatively representative figure of the topic significance by combining the information from these "*multi-criteria measures*" to form a single index of evaluation based on a "*Weighted Linear Combination*" (WLC) decision strategy [3].

WLC is a simple technique that is widely used in the area of multi-criteria decision analysis [3]. The simplest form of WLC evaluates each topic by the following formula

$$A_k = \sum_{m=1}^{N_m} \Psi_m \mathcal{S}_{m,k} \tag{9}$$

where N_m is the number of different measures to be combined, and Ψ_m is the weight of the measure in the total score, and $\mathcal{S}_{m,k}$ is the score of the kth topic with respect to the mth measure. Because of the different scales upon which these criteria are measured, it is necessary that the measures be standardized before combination.

In this work, a 4-phase weighted combination approach is introduced. The idea is to use the computed measurements to construct both the scores $\mathcal{S}_{m,k}$ and weights Ψ_m of the different criteria. To do so, two "*standardization procedures*" are performed in the first phase to transfer each distance measure from its true value into two standardized scores, one is a relative score of the distances and the other is a weight value between 0 and 1. Then, the standardized measurements of each topic within each J/I definition are combined into a single figure during the intra-criterion phase. In the third phase, two different techniques of "*Weighted Combination*" (WC) are performed to combine the J/I scores to construct a weight and a total score for each topic from which the final rank of the topic significance is computed. The following subsections describe each phase of the TSR in further details.

4.1 Standardization Procedure

Given the distance measures m, where $m \in \{KL, COR, COS\}$, under each J/I definition criterion $\mathcal{C}$, where $\mathcal{C} \in \{\mathcal{U}, \mathcal{V}, \mathcal{B}\}$, and for each topic k, the first phase is concerned with linearly transforming each distance value into a standardized score, denoted $\acute{\mathcal{C}}_k^m$, that maintains the relative order of distance magnitude with respect to the other topics instead of the original raw value.

To construct both the scores $\mathcal{S}_{m,k}$ and weights Ψ_m for each of the criteria, two standardization procedures are used. The first, re-scales the scores based on the weight of each score with respect to the total score over all topics. This is given in the form

$$\acute{\mathcal{C}1}_k^m = \mathcal{C}_k^m \times \frac{\sum_{j=1, j \neq k}^{K} \mathcal{C}_j^m}{\sum_{j=1}^{K} \mathcal{C}_j^m}. \tag{10}$$

The fraction in Equation (10) is a normalized weight of the topic distance.

The second standardization procedure is the score range procedure that uses the minimum and maximum values as scaling points for standardization. This is given by

$$\acute{\mathcal{C}2}_k^m = \frac{\mathcal{C}_k^m - \mathcal{C}_{\min}^m}{\mathcal{C}_{\max}^m - \mathcal{C}_{\min}^m} \tag{11}$$

where $\mathcal{C}_{\min}^m$ ($\mathcal{C}_{\max}^m$) is the minimum (maximum) distance value measured by the distance measure m under the criterion $\mathcal{C}$. While the first standardization rescales the raw measures to a relatively smaller range, the score range procedure bounds the resulted scores between zero and one. The former will be used as the topic score in the final TSR while the latter is used as the weight.

4.2 Intra-criterion Weighted Linear Combination

Before computing the rank of topic significance, it is required to combine the different distance measures within each J/I criterion into a single figure. Thus, the second phase of the topic ranking performs a Weighted Linear Combination (WLC) of the standardized scores of the distance measures as given in Equation (9). The weights Ψ_m in the equation determine the contributions of the distance measures in the total score. In this work, all the attributes under each criterion are assumed to weigh equally. Thus, the intra-criterion WLC is given by the mean score of the three distance measures.

So, given the standardized scores of the three distance measures under criterion $\mathcal{C}$ for topic k, i.e. $\acute{\mathcal{C}}_k^{KL}$, $\acute{\mathcal{C}}_k^{COR}$, and $\acute{\mathcal{C}}_k^{COS}$, then, the WLC score of the criterion $\mathcal{C}$ for topic k is given by

$$\mathcal{S}_k^c = \frac{\acute{\mathcal{C}}_k^{\mathrm{KL}} + \acute{\mathcal{C}}_k^{\mathrm{COR}} + \acute{\mathcal{C}}_k^{\mathrm{COS}}}{3}. \tag{12}$$

Substituting the two standardized scores $\acute{\mathcal{C}1}_k^m$ and $\acute{\mathcal{C}2}_k^m$ in Equation (12) results in two scores $\mathcal{S}1_k^c$ and $\mathcal{S}2_k^c$. Under each standardized procedure, a topic will have three intra-WLC scores $\mathcal{S}_k^u$, $\mathcal{S}_k^v$, and $\mathcal{S}_k^b$ based on the uniformity, vacuousness, and the background criteria, respectively.

4.3 Inter-criterion Weighted Combination

In this phase, a Weighted Combination[2] (WC) is performed over the scores computed in phase two. This involves assigning a weight, $\hat{\Psi}_c$, to each criterion $\mathcal{C}$ in order to adjust its contribution in the final ranking. However, two different WC techniques are used to combine the scores and the weights.

The first WC technique is based on Equation (10) and uses the standardized score of the background criterion as a weight for the uniformity and vacuousness scores as follows

$$\hat{S}_k = \hat{\mathcal{S}1}_k^b \left(\acute{\Psi}_u \mathcal{S}1_k^u + \acute{\Psi}_v \mathcal{S}1_k^v \right) \tag{13}$$

where $\acute{\Psi}_u$ ($\acute{\Psi}_v$) is the weight of the Uniformity (Vacuousness) criterion in the score and $\hat{\mathcal{S}1}_k^b$ is the rank (or weight) of the topic background. The rank is computed by substituting the intra-criterion score of topic background, $\acute{\mathcal{S}1}_k^b$, for $\acute{\mathcal{C}}_k^m$ in Equation (11). So, the background indicator is used to weigh the uniformity and vacuousness of the topic's word distribution by the uniformity of its distribution over the documents.

The second technique is performed over the intra-criteria scores that are based on the score range standardization procedure (Equation 11). This is done by a simple application of the WLC in Equation (9) as follows

$$\hat{\Psi}_k = \Psi_u \mathcal{S}2_k^u + \Psi_v \mathcal{S}2_k^v + \Psi_b \mathcal{S}2_k^b \tag{14}$$

where Ψ_c is the weight of the criterion $\mathcal{C}$ in the score $\hat{\Psi}_k$. These weights are assumed to sum to 1 so that the total score remains bounded between zero and one.

4.4 The Final Topic Significance Score

To compute the final rank, the score in Equation (13) is considered the total topic score while the normalized weight in Equation (14) is used as the weight of the topic score. Thus, the final rank of the topic significance is given by

$$TSR_k = \hat{\Psi}_k \times \hat{S}_k. \tag{15}$$

5 Experimental Design

The proposed post analysis to rank the significance of the topics in probabilistic models is evaluated on synthetic and real data. An LDA Gibbs sampler topic model is first used to learn the model parameters using the corpus of documents. The resulted topics are evaluated against the ground truth for the simulated data and the 20Newsgroups, and subjectively by investigating the topics and checking their significance for all the datasets. All experiments were implemented using

[2] Since the weights and scores are constructed from the computed distances, this phase (and the one that follows) are no longer linear.

Table 4. Topic distributions of the simulated data

TopicID	k1	k2	k3
TopicName	River	Bank	Factory
Topic %age	33%	34%	33%
↓Dictionary	$p(w_i\|k1)$	$p(w_i\|k2)$	$p(w_i\|k3)$
river	0.37	0	0
stream	0.41	0	0
bank	0.22	0.28	0
money	0	0.3	0.07
loan	0	0.2	0
debt	0	0.12	0
factory	0	0	0.33
product	0	0	0.25
labor	0	0	0.25
news	0.05	0.05	0.05
reporter	0.05	0.05	0.05

a modified version of the "Matlab Topic Modeling Toolbox", authored by Mark Steyvers and Tom Griffiths[3].

The weights of the different criteria in Equations (13) and (14) are first tuned using four different sets of documents that were generated from the synthetic data under different settings of K. The weights that resulted in the best ranking based on the ground truth are then fixed for the real datasets. Hence, the weights of the topic uniformity and vacuousness in Equation (13) ($\acute{\Psi}_u$ and $\acute{\Psi}_v$) are set to 0.6 and 0.4, respectively, while they are assigned to equal values, $\Psi_u, \Psi_v = 0.25$, in Equation (14) and the background weight Ψ_b is set to 0.5.

There are four datasets that are used in the experiments.

Simulated Data. The synthetic dataset consists of 6 samples of 16 documents that have been generated from three static equally weighted topic distributions. On average, the document size was 16 words. Table 4 shows the dictionary and topic distributions of the data. The dataset is configured such that shared words (background words) exists between subsets of topics, e.g. money and bank, and among all the topics, e.g. news and reporters. Given each sample of documents, LDA was run to estimate the topics.

In some experiments, fake junk topics were deliberately injected before computing the topic ranks. These topics were randomly sampled from the J/I topic distributions $\boldsymbol{\Omega}^{\mathcal{U}}$ and $\boldsymbol{\Omega}^{\mathcal{V}}$ and are denoted as Utopic and Vtopic, respectively.

20Newsgroups. The 20 Newsgroups data set is a collection of approximately 20,000 newsgroup documents, partitioned across 20 different newsgroups[4]. Some of the newsgroups are very closely related to each other (e.g. comp.sys.ibm. pc.hard ware / comp.sys.mac.hardware), while others are highly unrelated (e.g misc.forsale / soc.religion.christian). Data preprocessing included the removal of stop and rare words. The final dataset consisted of 1,359,612 word tokens and a dictionary size of 46191 terms.

[3] The Topic Modeling Toolbox is available at: psiexp.ss.uci.edu/research/programs_data/toolbox.htm

[4] The dataset is available at http://people.csail.mit.edu/jrennie/20Newsgroups/.

Table 5. TSR of simulated data without (left) and with (right) injected J/I topics

Without injected J/I topics				
ID	Topic Distribution	$\hat{S}$	$\hat{\Psi}$	TSR
3	river stream	3.263	0.766	2.491
4	factory labor production	3.602	0.653	2.377
1	money loan debt	2.260	0.486	1.09
5	reporter bank	0.502	0.3673	0.191
2	bank news	0.212	0.246	0.05

With injected J/I topics				
ID	Topic Distribution	$\hat{S}$	$\hat{\Psi}$	TSR
4	river stream	3.4	0.8	2.7
5	production factory labor	3.3	0.8	2.7
3	money loan debt	2.4	0.7	1.6
2	bank	1.5	0.5	0.8
1	reporter news	0.9	0.7	0.7
	Vtopic	1.7	0.2	0.3
	Utopic	0.14	0.03	0

NIPS Proceedings. The NIPS set consists of the full text of the 13 years of proceedings from 1988 to 2000 Neural Information Processing Systems (NIPS) Conferences[5]. The data was preprocessed for down-casing, removing stopwords and numbers, and removing the words appearing less than five times in the corpus. The data set contains 1,740 research papers, 13,649 unique words, and 2,301,375 word tokens in total.

5.1 Experimental Results

The Topic Significance Ranking algorithm was first evaluated on the simulated data. Table 5 lists the topics discovered by LDA with K set to 5 along with their total TSR rankings. The listed TSR is the average rank of the topic over six different samples. The topics are ordered by their significance index. It can be seen that the proposed ranking method is able to properly rank the topics based on their true significance. Both fake and true junk topics such as Topic 5 and 2 had the lowest ranks, while legitimate topics such as Topics 3 (k1), 4 (k3) and 1 (k2) have gained the highest ranks.

The rank of the topic, in general, depends on the amounts of background words that its distribution carries. For example, Topic k2 (money bank loan) is ranked lower than the other topics because both the "money" and "bank" terms are shared between more than one theme.

The proposed TSR is also tested using the 20Newsgroups dataset. Table 6 lists the distribution of topics that received the highest (lowest) TSR index. To determine the class of each topic and compare the results with the ground truth, an F1 measure (pF1) is computed based on a "probabilistic Contingency Table" (pCT) of size $C \times K$, where C is the number of classes. The table is constructed based on the document-topic distribution and the document labels. By considering the X topics with the highest probability under each document, an entry $pCT(i, j)$ in the table is the average probability that the topic j appeared in documents of class i. Then, the F1 measure is computed based on the contingency table pCT. A class is then assigned to the topic that has the highest pF1 measure . Table 6 lists the classes under which each topic had the highest pCT entry. The class that is assigned to the topic, i.e. the pF1 measure of the class

[5] The original dataset is available at the NIPS Online Repository. http://nips.djvuzone.org/txt.html.

document under that topic is the highest, are marked by *. Nearly half of the topics, e.g. 10, 32, and 34, did not get a high pF1 index for any class, while three topics (7, 12, and 23) had the highest pF1 measures for, and hence assigned to, more than one class.

It can be seen that the TSR rank matches the pF1 measure in 6 of the 10 highest ranked topics and 6 of the lowest ranked topics. For example, the distribution of topic 4 is focused on the "space" theme which matches the pF1 measure of the class "space" that has assigned the highest index for the class under that particular topic. Furthermore, the TSR is able to highly rank topics that better represent a class even though the corresponding pF1 index is not the highest. Topics 11 and 37 are examples of such topics. Although the classes "talk.politics.misc" and "comp.graphics" were assigned to topics 35 and 25, respectively, based on the pF1 measure, the word-distribution of topics 11 and 37 better represent the class semantics, see Table 6. Thus, the TSR does a better unsupervised judgment based on the topic distributions only.

When examining the lowest ranked topics, the most insignificant topics had a very low and approximately identical pF1 indexes for most, and sometimes all, of the classes. The word distribution of these topics clearly include a large set of background words, e.g. email header terminology (re, edu, ca), greeting words (topics 32 and 10), verbs (topic 16), and names of people and organizations (topic 34). The rest of the list included topics that had a high pF1 measure for one or more class, like topics 18, 19, and 21, or had been assigned a class, like topics 25 and 35. While an explanation regarding the latter topics was given earlier, the former topics illustrate another interesting observations. First, topic 18 contains the background words of the two religion related classes. As the TSR is high for topic 12 which better describes the underlying theme in more specific terms, it correctly identifies topic 18 as a background topic by assigning a low rank to it. The same explanation can be given for topic 25.

On the other hand, the class "misc.forsale" introduced a different behavior. First, the class is clearly heterogenous by its nature and involves a lot of shared words with other classes, particularly autos, electronics and computer-related classes. Thus, the topic is expected to have a large variance and heavy tailed distribution which makes it dominated for lower significance ranking. In addition, it can be seen that topic 19 provides a closer description for the class than topic 35. In fact, 294 documents (50.5%) of the class had topic 19 as the highest topic, compared to 11 documents for topic 35. However, the pF1 measure of topic 19 is less than the pF1 of topic 35 because the average relative importance of the latter topic in the class documents (0.5) is higher than the former topic (0.2). Although topic 35 is focused on political themes, the words "black" and "white" could be responsible of attracting the "forsale" documents into this topic. As a result, the topic's vacuous significance index is clearly affected.

The proposed TSR showed similar outcomes when tested on NIPS dataset. Table 7 lists the NIPS topics that gained the highest and lowest 10 indexes. The most significant topics clearly correspond to genuine themes of NIPS. Examples include reinforcement learning (30), speech recognition (41), image processing

Table 6. The 20Newsgroups: distribution and class of the 10 highest and lowest ranked topics

ID	Topic	class (pF1 measure)	TSR
	Highest Ranked Topics		
12	god jesus christ christian bible christians hell faith lord paul believe	soc.religion.christian*(0.121) talk.religion.misc*(0.112)	19.06
11	president money think going stephanopoulos tax don insurance pay care working clinton jobs bill	talk.politics.misc(0.1)	16.76
39	file output program entry section check line build ok read	comp.windows.x(0.073) comp.sys.ibm.pc.hw(0.071)	16.1
37	edu software image graphics ftp version pub data images package	comp.graphics(0.074) comp.windows.x(0.072)	14.17
20	turkish armenian armenians war turkey armenia soviet today greek genocide history	talk.politics.mideast*(0.198)	13.05
27	window server display widget mit application motif set manager sun	comp.windows.x*(0.176)	12.99
17	la period st play pts power pp chicago gm flyers buffalo van mon	rec.sport.hockey(0.156)	12.91
6	god believe say true truth question exist reason evidence religion existence argument atheism atheists	alt.atheism*(0.106) soc.religion.christian(0.111)	12.57
4	space nasa gov earth launch moon orbit satellite shuttle henry lunar	sci.space*(0.144)	12.40
23	drive scsi mb disk hard card system bit mac drives speed bus mhz apple	comp.sys.ibm.pc.hw*(0.134) comp.sys.mac.hw*(0.105)	12.146
	Lowest Ranked Topics		
32	thanks mail uk ac help advance fax university looking email hi appreciated		2.89
10	edu writes article ca apr news uiuc think don cso heard sorry		3.53
24	com writes article apr netcom hp opinions att ibm mark wrote		4.02
34	org edu chris david john scott mil navy ed jeff robert		4.06
16	don think re want going say things ll thing let ve doesnot maybe		4.14
35	black white edu virginia sex article sexual cover gay writes	misc.forsale*(0.128) talk.politics.misc*(0.118)	4.27
18	church think catholic thought true mean christian order group religion	talk.religion.misc(0.097) soc.religion.christian(0.086)	4.33
25	problem problems find line try ve help tried lines don	comp.graphics*(0.185) comp.sys.mac.hw(0.108)	4.407
19	price buy offer sale sell interested cd shipping printer asking sound condition apple cost computer	misc.forsale(0.084) comp.sys.mac.hardware(0.074)	4.47
21	writes science think system theory objective moral don morality article	sci.med(0.111) alt.atheism(0.106)	4.59

(43), and neuroscience (10). On the other hand, common terms across NIPS publications have been grouped by LDA in distinguished topics and have received the lowest significance rankings, see Table 7.

To verify the proposed approach to compute the ranks based on the judgments of a variety of distance measures, TSR rankings based on individual distance measures are constructed and compared to the proposed TSR. This is achieved by ignoring the intra-criterion WLC phase and directly combining the standardized distances for each of the individual measures separately. The resulted TRS ranks are called TSR-KL, TSR-COS, and TSR-COR for the KL-divergence, cosine dissimilarity, and coefficient correlation based ranks, respectively. Table 8 shows the TSR rankings for the topics of the simulated data with injected J/I topics. The topics are ordered by the TSR-KL rank. Given the true densities of the topics (Table 4), it can be seen that ranks from individual measures do

Table 7. The NIPS: distribution and ranks of the 10 highest and lowest ranked topics

ID	Topic	$\hat{S}$	$\hat{\Psi}$	TSR
	Highest Ranked Topics			
30	state action policy function reinforcement actions optimal time algorithm	23.4	0.7	16.8
10	firing spike cell cells neurons time potential membrane rate neuron	22.9	0.7	16.6
41	speech recognition word system training hmm words context speaker acoustic	23.0	0.7	16.2
17	cells cell visual orientation cortex receptive cortical spatial field fields	21.6	0.7	16.1
31	analog circuit chip figure current output vlsi voltage input circuits	22.9	0.7	15.8
29	control motor trajectory arm forward feedback movement inverse hand	20.7	0.7	15.1
43	image images visual pixel vision pixels figure edge features texture	19.6	0.7	14.3
44	motion direction velocity field moving flow directions eeg time optical	18.7	0.7	13.9
24	node nodes tree rules rule trees structure set representation connectionist	18.7	0.7	13.5
18	neurons synaptic input activity synapses connections inhibitory figure excitatory	18.9	0.6	12.7
	Lowest Ranked Topics			
27	case order general simple form work theory fact section terms	3.7	0.2	0.7
11	rate convergence results values number large random size constant fixed	9.5	0.33	3.2
34	method problem function optimal methods estimation solution parameter based	9.8	0.3	3.5
46	performance set training results test table number data method experiments	11.3	0.3	3.7
35	system time data systems real block large applications computer user	10.9	0.3	3.9
14	network neural net systems information architecture processing work	10.3	0.5	4.8
21	input output layer inputs training weights outputs network back hidden	11.9	0.5	6.2
16	noise information distribution correlation variance gaussian function density	12.5	0.5	6.3
12	local space figure points map point dimensional regions global region	12.1	0.5	6.3
32	learning algorithm weight gradient error weights descent time update	15.0	0.5	7.1

Table 8. Synthetic data: the TSR based on individual distance measures compared to the combined TSR

ID	Topic	**TSR-KL**	TSR- COS	TSR- COR	TSR
3	money loan debt	**2.05**	0.32	0.71	1.60
4	river stream	**2.01**	0.37	0.60	2.70
2	bank	**1.90**	0.36	0.58	0.80
1	reporter news	**1.83**	0.34	1.48	0.70
5	production factory labor	**1.82**	0.39	0.89	2.70
	Vtopic	**0.85**	0.45	0.46	0.30
	Utopic	**0.40**	0.42	0.89	0.00

not always provide the correct judgment regarding the semantic significance of the topics. In fact, the injected J/I topics "Vtopic" and "Utopic" have received the highest ranks under TSR-COS, while topic "Utopic" was the highest ranked topic under TSR-COR. In addition, based on the TSR-KL, topic 5, which corresponds to the genuine theme $k3$, was ranked lower than other insignificant topics, topic 1 (reporter news) and topic 2 (bank). Clearly, the intra-criterion WLC of the distance measures strengthens the judgments of the individual measures and provides a better representation of the topics' semantic significance.

Similarly, testing on the 20Newsgroups has revealed similar findings. Table 9 lists the 10 highest significant topics from the 20Newsgroups based on the TSR-KL rank. The table also shows the order of these topics under the cosine dissimilarity (TSR-COS) ranking and the proposed TSR ranking. It can be seen that the TSR-KL have introduced three topics to the list that are clearly not significant based on their distribution and the pF1 measure. The TSR-COS agrees with the TSR-KL in two of these J/I topics and introduces another insignificant topic (topic 33: de ma pa um em ei el rs sg mu di) to the list. In addition, topics

Table 9. The 20Newsgroups: the 10 highest ranked topics based on the TSR-KL

ID	Topic	class(pF1 measure)	TSR	TSR-COS
37	edu software image graphics ftp version pub data images package	comp.graphics(0.126) comp.windows.x(0.110)	4	8
39	file output program entry section check line build ok read	comp.sys.ibm.pc.hw(0.99) comp.windows.x(0.97)	3	1
23	drive scsi mb disk hard card system bit mac drives speed bus mhz memory controller pc board ram data apple	comp.sys.ibm.pc.hw*(0.208) comp.sys.mac.hw*(0.159)	10	9
31	list mail internet send information posting group email faq news address message usenet		12	3
20	turkish armenian armenians war turkey armenia soviet today greek genocide history	talk.politics.mideast*(0.360)	5	14
27	window server display widget mit application motif set manager sun	comp.windows.x*(0.252)	6	4
16	don think re want going say things ll thing let ve doesnot maybe		36	39
11	president money think going stephanopoulos tax don insurance pay care working clinton jobs bill	talk.politics(0.181)	2	11
12	god jesus christ christian bible christians hell faith lord paul believe	talk.religion.misc*(0.206) soc.religion.christian*(0.195)	1	18
8	db didn told home saw say don says going took re started happened building room wanted wife		15	6

such as 5, 12, and 16 illustrate how the combined rank provide a better judgment about the topic significance when compared to the individual measures.

6 Conclusion

In order to overcome the uncertainty that surrounds the outcome of a generative model, this paper presents a novel unsupervised analysis of Probabilistic Topic Models (PTM) for Topic Significance Ranking (TSR) to automatically distinguish genuine topics from Junk and Insignificant (J/I) topics. The proposed solution measures the distance of a topic from a set of J/I topic distributions using three different distance measures. A descriptive Topic Significance Ranking is constructed by applying 4 levels of Weighted Combination decision strategy. To the best of our knowledge, this work is the first attempt to automatically evaluate the inferred topics of PTM to judge their semantic significance.

The proposed ranking approach was evaluated on simulated and real datasets. The results are evaluated against the ground truth, when exists, and subjectively by examining the topic distribution. The outcomes confirm the potential of the proposed method as it is able to correctly highly rank the true topics while J/I topics received low figures. The approach was also verified against less complex ranking systems that depend on the judgment of a single distance measure.

To extend this work, we plan first to investigate the sensitivity of the approach to the applied combination techniques and to the weight settings. In addition, analyzing the effect of the number of components on the resulted rank is also considered. Consequently, the rank can be extended to be used as an indicator to adjust the setting of this critical parameter. The use of other J/I definition criteria and/or distance measures is also under consideration. In addition, further analysis of the use of the TSR in visualizing the evolution of topics in streaming text is planned.

References

1. AlSumait, L., Barbará, D., Domeniconi, C.: Online LDA: Adaptive Topic Model for Mining Text Streams with Application on Topic Detection and Tracking. In: Proceedings of IEEE International Conference on Data Mining, ICDM 2008 (2008)
2. Bishop, C.M.: Pattern Recognition and Machine Learning. Springer, Heidelberg (2006)
3. Bouyssou, D., Marchant, T., Pirlot, M., Tsoukias, A., Vincke, P.: Evaluation and Decision Models with Multiple Criteria. Springer, Heidelberg (2006)
4. Blei, D.M., Ng, A.Y., Jordan, M.I.: Latent Dirichlet Allocation. The Journal of Machine Learning Research 3, 993–1022 (2003)
5. Chemudugunta, C., Smyth, P., Steyvers, M.: Modeling General and Specific Aspects of Documents with a Probabilistic Topic Model. In: Proceedings of Neural Information Processing Systems (2007)
6. Griffiths, T.L., Steyvers, M.: Finding scientific topics. In: Proceeding of the National Academy of Sciences, pp. 5228–5235 (2004)
7. Joachims, T.: A Statistical Learning Model of Text Classification with Support Vector Machines. In: Proceedings of the Conference on Research and Development in Information Retrieval, SIGIR (2001)
8. Steyvers, M., Griffiths, T.L.: Probabilistic Topic Models. In: Landauer, T., McNamara, D., Dennis, S., Kintsch, W. (eds.) Latent Semantic Analysis: A Road to Meaning. Lawrence Erlbaum, Mahwah (2005)
9. Tan, P., Steinbach, M., Kumar, V.: Introduction to Data Mining. Pearson Addison Wesley, London (2006)
10. Wang, X., McCallum, A.: Topics over Time: A Non-Markov Continuous-Time Model of Topical Trends. In: ACM SIGKDD international conference on Knowledge discovery in data mining (2006)

Communication-Efficient Classification in P2P Networks

Hock Hee Ang, Vivekanand Gopalkrishnan, Wee Keong Ng, and Steven Hoi

Nanyang Technological University, 50 Nanyang Avenue, Singapore

Abstract. Distributed classification aims to learn with accuracy comparable to that of centralized approaches but at far lesser communication and computation costs. By nature, P2P networks provide an excellent environment for performing a distributed classification task due to the high availability of shared resources, such as bandwidth, storage space, and rich computational power. However, learning in P2P networks is faced with many challenging issues; viz., scalability, peer dynamism, asynchronism and fault-tolerance. In this paper, we address these challenges by presenting CEMPaR—a communication-efficient framework based on cascading SVMs that exploits the characteristics of DHT-based lookup protocols. CEMPaR is designed to be robust to parameters such as the number of peers in the network, imbalanced data sizes and class distribution while incurring extremely low communication cost yet maintaining accuracy comparable to the best-in-the-class approaches. Feasibility and effectiveness of our approach are demonstrated with extensive experimental studies on real and synthetic datasets.

1 Introduction

In recent years, peer-to-peer (P2P) networks have become increasingly popular on the Internet. Due to the greatly improved availability and accessibility, P2P networks are also emerging as excellent platforms for performing distributed data mining tasks such as P2P data classification [1,2,3,4,5]. Distributed data mining is important and useful to a broad range of real world applications. For example, in a P2P content sharing system, user preferences such as types of files shared can be mined to optimize delivery, and also to provide targeted advertising. In media annotation tasks [6], users typically only produce tag information for their own repositories. However, by employing P2P classification, peers are able to collaboratively auto-annotate their repositories (at least partially) by learning from the annotations of other peers.

While its potential is immense, mining in a P2P network is significantly more difficult than mining on a centralized dataset. P2P classification has a number of challenges [7] including *scalability* (Is the algorithm able to produce an acceptable solution within an acceptable time given the large number of peers and large amount of data?), *peer dynamism* (Is the algorithm robust enough to handle the availability and unavailability of data as peers connect and disconnect from the

W. Buntine et al. (Eds.): ECML PKDD 2009, Part I, LNAI 5781, pp. 83–98, 2009.

network?), *asynchronism* (Is the algorithm able produce an acceptable solution without performing global synchronization?).

Existing classification works in the P2P environment [1,2,3,4,5] have incurred high communication cost either during model construction or the prediction phase. This affects the scalability of the algorithms as the global data size and the number of peers increases. In addition, these algorithms may not be robust as their classification accuracy or computation and communication costs vary widely across different situations; e.g., in networks with unbalanced data class and size distribution.

In our previous work, we presented AllCascade [1], an approximate P2P classification algorithm based on cascading Reduced SVM (RSVM) [8] that greatly reduces the size of generated models. While it achieves high accuracy, AllCascade also incurs high communication and computation costs. In order to improve its efficiency, we presented RandBag [2], an approach based on bagging. However, RandBag is a non-deterministic approximation solution where accuracy and cost (computation and communication) fluctuate from peer to peer and under different situations.

In this paper, we observe that DHT-based P2P networks [9] have certain properties that may be exploited to address the above problems. Based on these properties, we design CEMPaR, a **C**ommunication **E**fficient **M**ultiple **Pa**rameter **R**obust framework. CEMPaR is a highly accurate P2P classification framework that (a) produces deterministic prediction, (b) reduces redundancy in classification model propagation, (c) achieves fault tolerance for a slight increase in communication cost, and (d) balances computation loads.

To the best of our knowledge, this is the first work in the area of P2P classification that takes advantage of DHT-based P2P network. Through theoretical and extensive empirical validation, we demonstrate that the proposed approach is scalable, tolerant of peer dynamism, invariant to imbalanced distribution of data size and class labels, and yields robust performance under varying conditions. We also show, over several real and synthetic datasets, that the proposed approach achieves comparable accuracy with the best-of-breed approaches for significantly lower computation and communication costs.

The rest of this paper is organized as follows. Background and related work are discussed in Section 2. The proposed approach is presented in Section 3, and experimentally validated in Section 4. Finally, conclusions and directions for future work are presented in Section 5.

2 Background and Related Work

A P2P network consists of N interconnected heterogeneous peers $\{p_1, p_2, \ldots, p_N\}$, where each peer p_i holds a set of training data instances $\ell_i(\mathbf{x}_i, y_i)$. Each instance is described by a d-dimensional data vector $\mathbf{x}_i \in \mathbb{R}^d$, and belongs to a specific class $y_i \in \mathcal{Y}$. The objective of P2P classification is to *efficiently* learn from the training data of all peers ($\ell = N\ell_i$) in order to accurately predict the class label of unlabeled data instances.

DHT-based P2P Networks. DHT-based P2P networks are popular as they provide efficient message routing for resource discovery. These approaches generally use consistent hashing—they assign each peer a unique identifier in the identifier ring space. Chord [10]—a DHT-based lookup protocol, assigns identifiers in the range from 0 to 2^b, where b is the number of bits for the identifier key. Using consistent hashing, identifier assignments remain unaffected by dynamic peers who join/leave arbitrarily. Moreover, there is a high probability that peers are well distributed in the identifier-ring space. As for data, they are hashed in a similar process and allocated to the node whose identifier is closest to (but not smaller than) the generated key. Each peer indexes a small number (b) of other peers' physical addresses. A resource can be found (or message routed) using its key by recursively looking up peers' indexes. This efficient divide-and-conquer approach of the identifier-ring space requires a number of hops at most logarithmic to the size of the network. The following is a key property of DHT protocols.

Property 1. On a given DHT with a circular identifier key space (e.g., Chord), whenever a message is sent to key i, if the peer with the key exists, the message will be delivered to the peer; otherwise, it will be routed to the peer with the next sequentially larger key (i.e., $peer(k + x)$ where $x > 0$ and x is minimum).

2.1 P2P Classification

With the number of peers in a P2P network exceeding the hundreds or thousands, P2P systems can be characterized as a massively distributed system requiring very high scalability. Moreover, the data of peers may change frequently and peers may join or leave the network anytime. Hence, P2P classification must be dynamic and fault tolerant. Global synchronization is also not possible due to the size of the network, latency and bandwidth cost [7].

Existing P2P classification approaches typically either perform local [4] or distributed [1,2,5] learning. Local learning performs training locally without incurring any communication during the training phase. Luo *et al.* [4] proposed building local classifiers using Ivotes [11] and performed prediction using a communication optimal distributed voting protocol. Unlike training, the prediction process requires the propagation of unseen data to most, if not all peers. This incurs huge communication cost if predictions are frequent.

Distributed learning approaches not only build models from the local training data, but also collaboratively learn from other peers. As a trade-off to the communication cost incurred during training, the cost of prediction can be significantly reduced. Siersdorfer and Sizov [5] classified Web documents by propagating SVM models built from local data among neighboring peers. Predictions are performed only on the collected models, which incur no communication cost.

To reduce communication cost and improve classification accuracy, we have proposed AllCascade [1] in an earlier work that performs a cascading of RSVM. RSVM is able to significantly reduce the size of the local model. However, AllCascade requires massive propagation of the local models and the cascading computation is repeated in all peers, wasting resources due to duplications.

To reduce duplication, an improvement to AllCascade with bagging (RandBag) [2] was proposed. By locally cascading random k peers' models and distributed voting with v random peers' cascaded model, we simulated the effects of bagging and reduce the duplication in training. The selection of k and v allows one to control the trade-off between training and testing communication cost while achieving satisfactory accuracy.

Due to its random components, RandBag may yield different prediction results for the same test data from different peers at the same point in time (non-deterministic). The sizes of collected data will also vary widely from peer to peer, resulting in an unbalanced load. Moreover, optimization of the selection process is impossible due to the large number of choices ($\binom{N}{k}$ and $\binom{N}{v}$) and the communication cost involved.

3 CEMPaR Framework

This section presents our proposed approach which exploits the advantages of DHT lookup protocols to perform efficient and robust learning in P2P networks.

3.1 Communication Structure Overlay

To reduce peer interactions considerably, we introduce the notion of a *super-peer* in the P2P network. The super-peers are dynamically selected from peers in the P2P network such that each super-peer is a representative of a subset of peers in the P2P network. Using super-peers, one is able to significantly reduce the huge amount of P2P communication among peers. However, the difficulty is that peers may not be able to locate their associated super-peers since peers in a P2P network usually know only a small number of their own neighbors. To address this challenge, we propose to apply DHT-based P2P network protocols [9] (e.g., Chord [10]) to facilitate the tasks of resource discovery and communication. Below, we present an efficient communication overlay scheme built upon DHT-based network protocols.

In our approach, the entire identifier ring space of a DHT-based network is equally split into g groups (with consideration to peer distribution, load balancing, and ease for super-peer assignment) where *group* is formally defined below:

Definition 1. *(Group) A group* $\boldsymbol{G}$ *is a contiguous subset of the identifier ring space. Given that the identifier ring space is evenly splitted, the number of identifiers contained in each group is* $|\boldsymbol{G}| = 2^b/g$, *i.e.,* $\boldsymbol{G} = [(2^b/g * i), (2^b/g * (i+1)))$ *for group* $i \in [0, g)$, *where* g *the number of groups and* b *the number of bits for an identifier key.*

For each group, we need to assign a super-peer amongst the peers to represent and manage the group. The super-peer assignment should be easily managed and efficient for discovery by peers. By exploiting the property of DHT-based network as shown in Property 1, we suggest a simple yet effective super-peer assignment approach that is formally defined below:

Definition 2. *(**Super-peer**) Given a set of peers whose identifiers $\boldsymbol{P}$ are a subset of an identifier group $\boldsymbol{G}$ in a DHT-based network, i.e., $\boldsymbol{P} \subset \boldsymbol{G}$, a super-peer is the peer with the smallest identifier in* $\mathbf{P}$: $s_p = \min_{i \in \boldsymbol{P}} i$.

The above approach enables CEMPaR to easily and deterministically locate super-peers using the DHT lookup service. In particular, by simply sending a message to the smallest identifier of a group, we guarantee that the message will be sent to the super-peer of the associated group, as shown in Theorem 1.

Theorem 1. *For a set of peers* $\mathbf{P} \subset \mathbf{G}$, *a message sent to the smallest identifier in group* $\mathbf{G}$, *denoted as* s_g, *will always be delivered to the super-peer of* $\mathbf{P}$ *with time complexity of* $O(\lg N)$.

Hence, in our approach, we allow only super-peers to receive models from other peers and to make predictions on unseen test data for a classification task.

Remark. We note that the assumption that each group has at least one peer whose identifier lies in the group's identifier range may not always be satisfied; e.g., when the number of peers in the P2P network is less than the number of groups. However, it only affects the condition that the key of the super-peer must lie in group $\mathbf{G}$, it does not affect the delivery of a message to the super-peer, as the identifier key overflows to the next group whose super-peer may be the super-peer of more than one group. This exceptional situation is rectified by the relocation process (discussed later) when a peer with an identifier $id \in \mathbf{G}$ joins.

By introducing the concepts of *group* and *super-peer*, we develop an efficient communication scheme in CEMPaR that resolves two critical tasks: (1) discovery of super-peers and (2) communication between peers. In particular, we offer two efficient solutions that are built upon the DHT-based protocols below.

DiscoverSP(*gid*, *irv*)—This function uses the underlying DHT lookup protocol to route the message containing information request vector *irv* and sender's physical address to the super-peer of group id *gid* using the first identifier of the group. *irv* encodes the sender's request for information such as physical address of receiver, mean vector, class counts, and etc. This function incurs $O(\lg N)$ messages which is optimal as opposed to a linear search of the identifier ring costing $O(N)$ (c.f. [10]). Hence, this function is best used when the physical address of the recipient is unknown. The size of the message to be sent is very small, which includes 1 byte for *irv*, and 5 bytes for the sender's physical address.

SendMsg(*rip*, *data*)—This aims to send the message containing sender's physical address, a set of content *data*; e.g., model, class count, mean vector, replica list, and etc, directly to the recipient's physical address *rip*. As the message is sent directly to the recipient, it is optimal ($O(1)$) and is best for sending large data. The size of the message to be sent includes 5 bytes for the sender's physical address and the size of the content.

3.2 Learning Modules

We now discuss core learning modules for performing training and prediction tasks in our framework.

3.2.1 Local Model Construction

We adopt RSVM [8] for training local models for peers in each group since RSVM produces the training model containing support vectors that are at most s percent of the total training data for a local dataset [1,2], which in turn caps the overall communication and learning cost.

3.2.2 Model Propagation and Cascading

Following the local model construction is model propagation and cascading in which peers send local models to super-peers and super-peers collect models from peers and update the cascaded classification models.

One key issue for model propagation is to determine which super-peer should a peer propagate its local model. A naïve way is to simply send the model to all super-peers. Apparently, this is inefficient due to intensive communication and computation cost. Ideally, we wish the local model be sent to the best super-peers that results in the best global classification performance. Unfortunately, in a P2P network, optimizing global classification performance is often intractable.

In practice, we want the cascaded models of the super-peers to be as diverse as possible as in ensemble classification, the best classification performance is often achieved when the models are diverse [12]. In addition, learning from previous experience [2], we want to ensure that every super-peer maintains an (approximate) equal class and data size distribution. This is achieved by balancing the load distribution and maintaining the natural class distribution of data on each super-peer. Although natural class distribution may not produce the best classification results [13], it provides an overview of the global class distribution to allow cost-sensitive learning. Finally, we also aim to reduce the overall redundancy in computation and communication cost.

To this end, we propose a greedy approach for model propagation. When a peer p is ready for propagation, it first collects information from all super-peers, including the number of collected instances (for each class) and the mean vector of the collected data (for each class) for each super-peer, via *DiscoverSP*. With the collected data, for each class type, the super-peer with the smallest instance count will be chosen, and in the mean time, the instance from the local model that is closest to the mean vector of the selected super-peer will be assigned. This process repeats until all instances in all classes have been assigned. As the instances are assigned in a disjoint manner, we avoid duplicate communication cost. Finally, the assigned data are sent to super-peers by using *SendMsg* since peer p has already obtained the physical address of all super-peers via *DiscoverSP*.

Note that to minimize the discrepancy of class count when multiple peers are performing model propagation, peers can first calculate the class count to be assigned and send the counts to the super-peer via *SendMsg* before the assignment of the support vectors. Once all support vectors are assigned, they are propagated to the respective super-peers. Finally, the model propagation algorithm is summarized in Algorithm 1.

Once the super-peers have received the models, in addition to merging the newly collected instances with the super-peer's cascaded model, each super-peer

Algorithm 1. Model Propagation for peer p_i.

input: number of groups g, local support vectors $\mathbf{SV}_i$
for $j \leftarrow 0$ **to** $g-1$ **do**
    $\mathbf{MV}$, $\mathbf{CC}$, $\mathbf{IP} \leftarrow$ DiscoverSP(j, *irv*:¡mean vector ($\mathbf{MV}$), class count ($\mathbf{CC}$)¿);
foreach *class label* y *in* $\mathcal{Y}$ **do**
    while SV_i^y *not* $\emptyset$ **do**
        $j \leftarrow$ group with the least count of class y in $\mathbf{CC}$;
        $sv \leftarrow$ closest support vector in $\mathbf{SV}_i^y$ to $\mathbf{MV}_j$;
        remove sv from $\mathbf{SV}_i$ and add it to $\mathbf{SV}_j$;
        update $\mathbf{CC}_j$;
for $j \leftarrow 0$ **to** $g-1$ **do**
    SendMsg($\mathbf{IP}_j$,$\mathbf{SV}_j$);

Algorithm 2. Model Cascading for super-peer s_i.

input : received model RM_j, collected data $\mathbf{CD}_i$, cascaded model CM_i, mean vector $\mathbf{MV}_i$, class count $\mathbf{CC}_i$
output: CM_i, $\mathbf{MV}_i$, $\mathbf{CC}_i$
$\mathbf{CD}_i \leftarrow$ combine received model RM_j with $\mathbf{CD}_i$;
$CM_i \leftarrow$ train SVM on local cascade model $CM_i \cup$ received model RM_j ;
foreach *class label* y *in* $\mathcal{Y}$ **do**
    update $\mathbf{CC}_i^y$ and $\mathbf{MV}_i^y$;

also updates the mean vector for the set of instances of each class and the instance count of each class. The model cascading algorithm is summarized in Algorithm 2.

Remark. In a stable network, the communication cost for the above model propagation process is only $O(m)$ where m is the total number of support vectors of all local models. In practice, as a P2P network is in nature highly dynamic, additional cost might be incurred to ensure correctness and robustness. We will discuss issues of relocation and replication in subsequent parts.

3.2.3 Prediction

During prediction, since only g super-peers are performing data collection and cascading the models, peers that need to predict unseen data simply sends the test instances to these super-peers and then aggregate the votes returned by the super-peers. However, it will incur heavy computational load on the super-peers.

We propose to replicate super-peers' cascaded models (c.f. Section 3.3). With the replicas, peers requesting prediction first request the replica list (containing physical addresses of replicas) of super-peers via *DiscoverSP*. Then, for every replica, by sending a *ping* message and with the reply from the replica (via *SendMsg*), a round trip time (RTT) is obtained. The RTT measures the network distance from the initiating peer to the replica. The initiating peer will then send the test instances to the nearest replica of each group. Once the replicas has finished predicting the test instances, they will send their predictions back to the initiating peer via *SendMsg*.

Algorithm 3. Prediction.

input : test instance $\mathbf{t}_i$, number of groups g
output: prediction y_i
for $j \leftarrow 0$ **to** $g-1$ **do**
    RL $\leftarrow$ DiscoverSP(j,*irv*:¡replica list (**RL**)¿);
    foreach *replica* $r \in$ ***RL*** **do**
        SendMsg(r,"ping");
PL $\leftarrow$ select nearest replica of each group;
for $j \leftarrow 0$ **to** $g-1$ **do**
    $\mathbf{V}_j \leftarrow$ SendMsg($\mathbf{PL}_j$,$\mathbf{t}_i$);
$y_i \leftarrow$ select class with most votes in $\mathbf{V}$;

The initiating peer then aggregates all votes to make the final prediction once all replies are received. Communication is efficient as all the messages sent are based on optimized communication functions. Finally, in order to reduce the communication of pinging replicas, caching of the RTT could be done for use in subsequent prediction. The prediction algorithm is summarized in Algorithm 3.

3.3 Maintenance Modules

3.3.1 Relocation

Since peers in P2P networks are dynamic, to ensure that the newly elected super-peer always hold the group's latest cascaded model, it is necessary to relocate the group's cascaded model to the newly elected super-peer. As peers' joining and leaving are tracked by Chord, no effort on our part is needed to track the peer changes.

The relocation process is as follows. When a new peer joins the group with the smallest identifier key within the group, it is elected as the new super-peer and the data from the old super-peer are relocated to the new super-peer. As the DHT network provides the physical address, in addition to the identifier key of a new peer, the cascaded model can be relocated in an efficient manner via the *SendMsg* function.

Although there are other options to ensure correctness of super-peers' models, relocation of models is preferred as it does not require any modification to the underlying P2P DHT protocol and it keeps the propose approach simple and efficient.

3.3.2 Replication

In order for the proposed approach to be fault tolerant while reducing the load of super-peers (during predictions), replication of the cascaded models is needed. To create the replicas, we adopt a concept similar to the proposed proximity routing of Chord, where each super-peer replicates the cascaded models to r successive peers. Since the identifier assignment of the peers is totally random, it is very likely that peers reside in different geographical locations; thus, speeding up the access of neighboring peers. In addition, with multiple peers having the cascaded models, the load of the super-peer for both communication and computation can be reduced substantially. Moreover, it is unlikely that all the replicas from the same group fail simultaneously.

On the failure of a predecessor, a peer sends an election message to its super-peer's key. Given that the failed predecessor is a super-peer, the successive replica will receive the election message and then assume the role of a super-peer until departure from the network or the entrance of a new super-peer.

To ensure that r successive peers hold the latest cascaded model, replication is performed by a super-peer when significant changes in the cascaded model is made or if any of the r successive peers changes (tracked by Chord). A check on the successive peers' data is made before replicating to reduce redundancy.

Since Chord already maintains a list of sibling nodes (successive in the key values) with their physical address, tracking the changes of these successive peers is trivial by simply retrieving the list from Chord (which does the actual job). Replication to the new successive peers (either newly joined or to replace a peer that has left) will be handled by the super-peer. In addition, we note that peers will always retain their collected data so as to reduce data propagation in the event of any change in the super-peer or replicas. All communications are done in an efficient manner using SendMsg as the physical address of all replicas are known.

3.4 Complexity Analysis

Here we analyze the complexity of computation and communication cost. To simplify the analysis, we assume the size of the local data ℓ_i of all peers $p_i, i \in 1, \ldots, N$ is equal. We also assume that every peer will be using the same percentage $s << 1$ ($s = 0.01$ for our case) of their local dataset for building the RSVM which will result in a model size of at most $m_i = s\ell_i$ for a peer p_i and the maximum size of the cascaded models of the entire P2P network will be $m = \sum_1^N m_i$.

3.4.1 Time Complexity Analysis

The time cost for model training is mainly composed of local model construction, model partitioning, model cascading, and computation of mean vectors. The time cost for the prediction phase mainly comprises the cost for super-peers to make their predictions on test instances. A summary of the time complexity of the training and prediction phases in CEMPaR follows.

With the following computational costs: (1) local model construction—$O(\ell_i m_i^2)$ for each peer, (2) model partitioning—$O(gm_i)$ for each peer, (3) computation of mean vector—$O(m/g)$ for each super-peer, and (4) model cascading—$O((m/g)^3)$ for each super-peer. The worst case time complexity of training is therefore $O((m/g)^3)$ given that all super-peers compute in parallel.

With SVM, the prediction cost incurred by each super-peer is $O(mt/g)$. Hence, this is the worst case time complexity when all super-peers predict in parallel.

3.4.2 Communication Cost Analysis

For communication cost, we note that compared to the transmission of data instances, the cost of sending non-instance message such as *irv* (1 byte) and physical address (5 bytes) is negligible. Therefore, we only examine the cost of sending data instances. A summary of the communication complexity of training and prediction phases in CEMPaR follows.

Given the following communication cost: (1) mean vector—$O(Ng)$, (2) local model with relocation—$O(m)$, and (3) replication—$O(rm)$. The communication cost for training is therefore $O((r+1)m)$.

As for each prediction task, the peer only needs to send the test instances to the g super-peers/replicas; hence, the communication cost is $O(gt)$

Given that the computation cost of training for RandBag is $O((km/N)^3)$ [2], it is comparable with CEMPaR, depending on the value of k (number of models to collect for RandBag) and g, and is lower than approaches like centralized SVM and RSVM and AllCascade whose computation cost of training is at least $O(m^3)$. In addition, the communication cost of CEMPaR and RandBag ($O(km)$) differs by k and $r+1$, and it is noted that k is required to be at least linear to the number of peers N to ensure satisfactory while it can be observed that when $r = \lg N$, the chance of any group failing for CEMPaR is less than 0.01 when $N = 500$ (good fault tolerance property). Hence, with consideration to both the computation and communication cost, CEMPaR will be the preferred approach.

Finally, in terms of the prediction cost, AllCascade incurs no communication cost although its training cost is very high. In addition, though the communication cost of prediction for RandBag ($O(vt)$ where v is the number of voting peers) and CEMPaR only differs by the factors g and v, from our empirical evaluation, we observed that the value of g can be significantly smaller than v to achieve satisfactory performance.

4 Experimental Results

We perform extensive experiments to show that our proposed approach: (a) incurs significantly lower communication cost compared with other P2P classification algorithms, (b) achieves accuracy comparable with other approaches, and (c) is robust with respect to imbalanced data and class distribution, number of peers, groups and failure of peers.

4.1 Experimental Setup

We simulate real world P2P problems with large sized datasets using the multi-class Covertype dataset [14] and the Synthetic Classification Data Set Generator (SCDS) [1]. We created a Binary Covertype dataset from Covertype, with only 2 classes (class 2 against the rest). Using SCDS, we generated two datasets with 1,000,000 instances each: Binary SCDS with 2 classes and 32 continuous attributes, of which 4 are relevant, and Multi-class SCDS with 6 classes and 32 continuous attributes, of which 10 are relevant. In addition, 20 percent of the attribute values and 20 percent of class labels are wrongly assigned to represent noise in data. We used 500 peers for both Covertype datasets, and 900 peers for both SCDS datasets so that each peer roughly has around 1000 data instances. The dataset column of Table 1 summarizes the datasets used.

[1] `http://www.datasetgenerator.com`

SVM takes too long to train on large datasets, so we used SVM RSVM [8] as the baseline centralized classification algorithm (implemented in C++). The same RSVM code is also used in our proposed approach for building the local model while the C-SVM [15,8] is used for model cascading. For approaches using RSVM, one percent of the local data was used by every peer so that it has sufficient data for building a representative model. All approaches also used the RBF kernel and the γ and C values were chosen using the model selection tool provided with LIBSVM based on one percent of stratified sampled data from each of the datasets. Unless otherwise stated, for RandBag, the number of cascading models k and number of voting peers v were set as 10% of the number of peers N, and for CEMPaR, the number of groups g was chosen as 10. For all experiments, 10-fold cross validation were performed and in addition, for each fold, 50 independent runs were executed for RandBag and CEMPaR.

In order to compute the communication cost, we used the OverSim P2P network simulator [16] with CEMPaR built on top of the Chord protocol [10] using default settings. The P2P network initializes without any peer, and peers were made to join one at a time until the maximum number of peers was reached and the network was allowed to stabilize. This scheme was chosen to capture all the intermediate overhead cost that might be incurred. Communication cost is typically measured in terms of total size of data transmitted in the network and not actual time taken [3]. Therefore, we report communication cost as total instances sent which is the dominating component of data transmitted. Executables for the experiments are available at `http://www.cais.ntu.edu.sg/~vivek/pubs/cempar09`.

Experiments were performed on a cluster of 16 machines, each with two Intel Dual Core Xeon 3.0GHz processors, 4-GB RAM, connected by gigabit ethernet.

4.2 Accuracy

In these experiments, every dataset was equally partitioned among the peers, and the class label was randomly distributed. The accuracy of the competing approaches is presented in Table 1 and tested for statistical difference using the Mann-Whitney-Wilcoxon (MWW) test with $P >= 0.05$. First, we observe that CEMPaR is comparable to the centralized RSVM. Although centralized RSVM is more accurate than CEMPaR on the Binary and Multiclass Covertype datasets, the difference is not significant according to MWW. On the other hand, CEMPaR is significantly superior on the Binary and Multiclass SCDS datasets by the same test. Second, we note that CEMPaR is generally comparable to the other state-of-art P2P approaches. We find that AllCascade performs slightly better than our method (significant only for Binary and Multiclass SCDS). This is reasonable because AllCascade takes all local models for training the final cascaded model while we only take a small portion of local models. Our approach is however significantly more efficient than AllCascade in terms of both time and communication efficiency. Compared with SVM ensemble, results of CEMPaR are always significantly different, and better in three out of four datasets. Compared with the RandBag approach, we can see that our approach is significantly better than RandBag ($k = v = g$) and slightly worse than RandBag ($k = v = 0.1N$). However, the latter setting incurs

Table 1. Classification accuracy (equally partitioned data, random class distribution)

Dataset (Instances, Attributes, Classes)	Centralized RSVM	SVM Ensemble	All-Cascade	RandBag $k = v = 0.1N$	RandBag $k = v = g$	CEMPaR
Binary Covertype (581K, 54, 2)	71.97%	52.35%	72.93%	69.29%	61.93%	68.99%
Multiclass Covertype (581K, 54, 7)	67.16%	46.41%	65.60%	67.27%	61.53%	64.27%
Binary SCDS (1M, 32, 2)	91.28%	92.01%	91.85%	91.82%	85.68%	91.62%
Multiclass SCDS (1M, 32, 6)	57.03%	58.99%	63.21%	60.92%	54.25%	60.60%

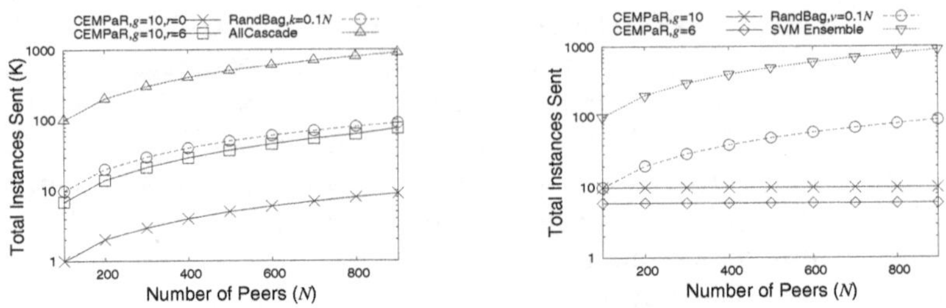

(a) Model Propagation (b) Test Propagation ($t = 1$)

Fig. 1. Communication cost vs number of peers

significantly higher communication cost for RandBag. This again validates the effectiveness of our method; i.e., excellent communication efficiency and competitive classification performance.

4.3 Communication Cost

We used the binary SCDS dataset, and varied the number of peers from 100 to 900 with each peer having roughly 1000 instances. The total communication cost incurred by various approaches on this problem is presented in Figure 1. We observe that, for both model propagation and prediction (test data propagation) tasks, our proposed approach incurs significantly lesser communication cost compared to other approaches that need to perform the same tasks. Since SVM ensemble does not perform model propagation and AllCascade does not perform test propagation, they are not depicted in the corresponding plots. However, it may be noted that their communication costs; viz., test propagation cost for SVM Ensemble and model propagation cost for AllCascade are far greater than the proposed approach (around two orders of magnitude).

4.4 Sensitivity to Parameters

We studied the robustness of the competing approaches by varying data sizes and class distributions on peers, number of peers, groups and failure of peers, and evaluating the effect on classification accuracy and communication costs.

4.4.1 Data Size Distribution

We assigned the multi-class Covertype data to the peers by sampling the data *size* from exponential, normal and uniform distributions. As the results show (in

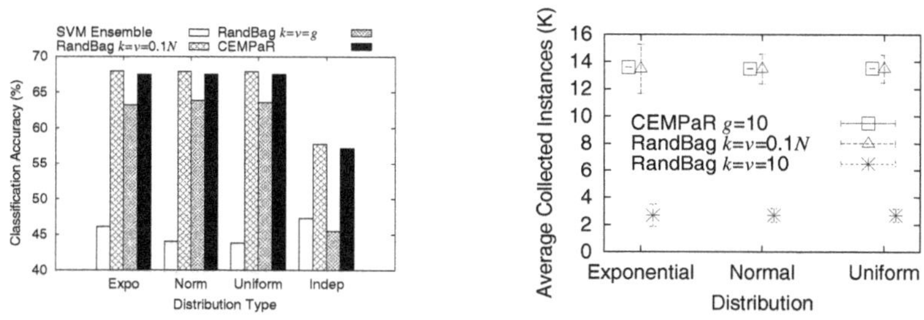

(a) Accuracy (b) Communication Cost

Fig. 2. Effect of size distribution on accuracy and communication. For Expo, Norm and Uniform, $N = 500$, Multi-class Covertype. For Indep, $N = 900$, Multi-class SCDS.

Figure 2(a) under Expo, Norm, Uniform), distribution of the local peer data does not have any significant effect on classification accuracy of any of the approaches.

However, the effect on communication cost (see Figure 2(b)) reveals an interesting point. The bar and line plot shows the average and standard deviation (s.d.) of the number of instances collected by each peer respectively, when peer data sizes are selected from non-equal distribution (exponential, normal and uniform). We notice that the s.d. of the instances collected for RandBag is significantly larger than that of our proposed approach. In other words, for RandBag, the number of instances collected by each peer varies widely, thus creating an uneven load distribution. Whereas in our proposed approach, the number of instances collected by each super-peer does not differ much (very small s.d.), thus evenly distributing the load of model cascading among all super-peers, demonstrating the effectiveness of the load distribution mechanism.

4.4.2 Class Distribution

The multi-class SCDS dataset was distributed equally among 900 peers with each peer's data belonging to 2 out of the 6 classes. This is a similar setting to that of the photo annotation problem in P2P environment [6]. Comparative accuracy results (see Figure 2(a) under Indep), show that contrary to the SVM ensemble approach, both our approach and RandBag achieve an accuracy comparable to centralized RSVM (57.06%). This demonstrates that cascade approaches are resilient to the class distribution of data, as representatives of many peers are merged together, thus providing a solution based on all classes' data. Class distribution has no impact on communication cost.

4.4.3 Number of Peers (N)

Studying the results of Section 4.3, we find that the communication cost of model propagation (Figure 1(a)) for all cascade approaches grows linearly to the number of peers N, or more specifically, the amount of training data in the P2P network. CEMPaR has the smallest factor, followed by RandBag and AllCascade. In addition, note that the model propagation cost of CEMPaR ($r =$ 0 and 6) includes the relocation cost, which contributes toward less than 12% of the total propagation cost (for 900 peers).

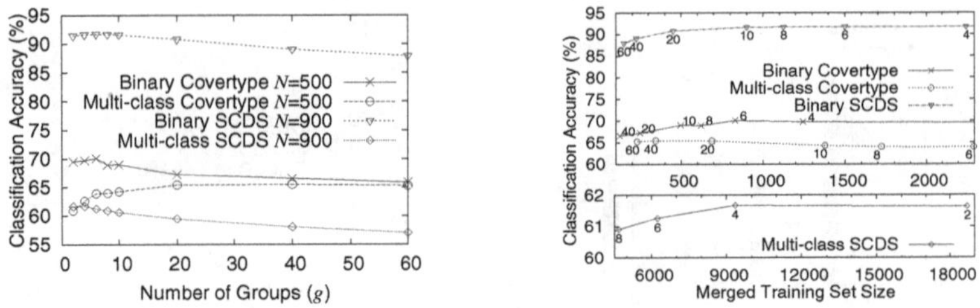

(a) Number of Groups (b) Merged Training Size

Fig. 3. Effect of number of groups and merged training size on classification accuracy

With respect to the communication cost for prediction (of a single test instance) which is shown in Figure 1(b), CEMPaR incurs a cost linear to the number of groups g (6 and 10) and constant to the number of peers, RandBag and SVM Ensemble incur linear communication cost with respect to the number of peers p where the factor of the cost for RandBag is v/N and for SVM Ensemble it is close to 1.

4.4.4 Number of Groups (g)

For this experiment, we varied the number of groups for all datasets. From the classification accuracy plots (see Figure 3(a)), we observe that for all datasets, as the number of groups increases, the classification accuracy first increases then starts to decrease steadily. For both the Binary datasets, the best accuracy is achieved for 6 groups, whereas for the multi-class SCDS dataset it is 4 and for the multi-class Covertype dataset, it is between 20 to 40 (with little differences).

To get a better understanding, we present in Figure 3(b) the merged training size of each super-peer for the different number of groups. We observed that in general, as the size of the merged training set increases, accuracy also increases, dropping slightly only when the number of groups approaches 2. However, the multi-class Covertype dataset behaves slightly different where the accuracy peaks at a smaller merged training size. This could be due to several reasons such as diversity of the cascade models, so further investigations are needed to draw a conclusion. For all datasets, we observe that the difference in accuracy between the optimal and worst number of groups is less than 5%. From these findings, we conclude that a sub-optimal choice on the number of groups does not have a significant impact on the accuracy.

Note that the number of groups affects the relocation cost and is observed that as the number of groups increases, the deviation of the relocation cost reduces and starts to stabilize. Due to space constraint results are not presented.

As expected and as shown earlier (Figure 1(b)), the prediction cost (test propagation) grows linearly with the number of groups. However, we observe from the results that choosing the number of groups as a log function of the number of peers greatly reduces the communication cost of prediction while achieving satisfactory classification accuracy.

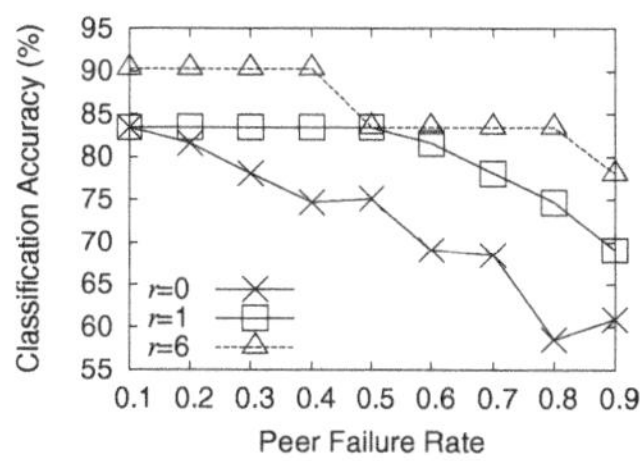

Fig. 4. Effect of peer failure rate on accuracy ($N = 500$, $g = 10$)

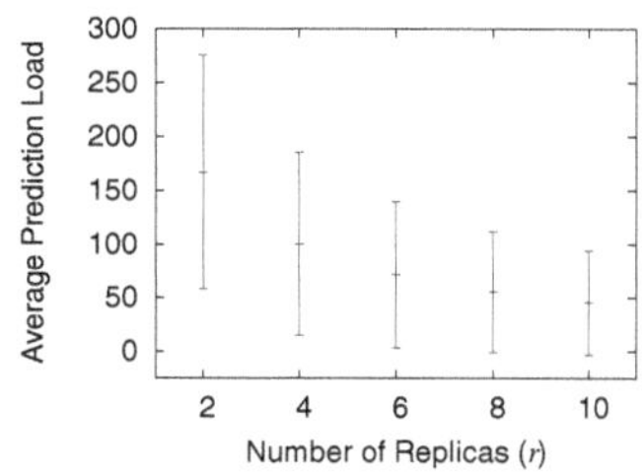

Fig. 5. Effect of number of replicas on prediction load ($g = 10$, $t = 500$)

4.4.5 Number of Replicas (r)

From Figure 1(a), we observe that the model propagation costs grows linearly with the number of replicas. Since the number of replicas does not affect test prediction cost (which only depends on g), here we study the effect of r on fault-tolerance of the group, and on the load balancing during prediction. We first simulated massive peer failure and varied the number of replicas to study the probability of group failure. We also report the number of prediction tasks performed by each super-peer/replica. Results of the fault tolerance and load balancing experiments are presented in Figure 4 and 5 respectively.

Figure 4 demonstrates that with the increase in peer failure rate, classification accuracy decreases. However, with the increase in the number of replicas, the effect of the accuracy reduction is reduced. Figure 5 shows a bar and line plot which demonstrate that as the number of replicas increases, the load handled by each replica decreases exponentially. Results show that with only 6 replicas, it is possible to maintain high accuracy with failure rate of up to 40%.

5 Conclusions

This paper proposes CEMPaR—a P2P classification framework that incurs low communication costs and is extremely robust to data and network parameters. CEMPaR utilizes the DHT networking protocol to efficiently and dynamically elect super-peers, which are then used to build the classification model. The resultant ensemble classifier which is based on cascading SVMs yields accuracy comparable to the centralized learning algorithms, while incurring significantly lesser communication cost than the existing P2P classification approaches. CEMPaR also manages to achieve good prediction load balancing and is fault-tolerant at the expense of very little model replication among other peers.

While this paper demonstrates the benefits of using an SVM-based cascaded learning approach, it must be noted that the CEMPaR framework can accommodate various learning strategies, including ensembles of different classifiers. It would be interesting to study whether other learning approaches could retain the benefits of the cascading SVM approach, especially robustness to data and class distribution. Since CEMPaR uses DHT lookup for efficiency, it is restricted to DHT-based P2P networks. However, given the popularity of this protocol, this is not much of a limitation. In future, we would like to explore the relationship between the diversity of data and the number of groups chosen for optimal accuracy and communication benefits. In addition, we would like to study and address the issues of concept drift and peer data privacy.

References

1. Ang, H.H., Gopalkrishnan, V., Hoi, S.C.H., Ng, W.-K.: Cascade RSVM in peer-to-peer networks. In: Daelemans, W., Goethals, B., Morik, K. (eds.) ECML PKDD 2008, Part I. LNCS (LNAI), vol. 5211, pp. 55–70. Springer, Heidelberg (2008)
2. Ang, H.H., Gopalkrishnan, V., Hoi, S.C.H., Ng, W.K., Datta, A.: Classification in P2P networks by bagging cascade RSVMs. In: VLDB Workshop on DBISP2P, pp. 13–25 (2008)
3. Gorodetskiy, V., Karsaev, O., Samoilov, V., Serebryakov, S.: Agent-based service-oriented intelligent P2P networks for distributed classification. In: Hybrid Information Technology, pp. 224–233 (2006)
4. Luo, P., Xiong, H., Lü, K., Shi, Z.: Distributed classification in peer-to-peer networks. In: ACM SIGKDD, pp. 968–976 (2007)
5. Siersdorfer, S., Sizov, S.: Automatic document organization in a P2P environment. In: ECIR, pp. 265–276 (2006)
6. Volkmer, T., Smith, J.R., Natsev, A.P.: A web-based system for collaborative annotation of large image and video collections: an evaluation and user study. In: ACM Multimedia, pp. 892–901 (2005)
7. Datta, S., Bhaduri, K., Giannella, C., Wolff, R., Kargupta, H.: Distributed data mining in peer-to-peer networks. IEEE Internet Computing, Special issue on Distributed Data Mining 10(4), 18–26 (2006)
8. Lin, K., Lin, C.: A study on reduced support vector machines. IEEE Transactions on Neural Networks 14(6), 1449–1459 (2003)
9. Balakrishnan, H., Kaashoek, M.F., Karger, D.R., Morris, R., Stoica, I.: Looking up data in P2P systems. Communications of the ACM 46(2), 43–48 (2003)
10. Stoica, I., Morris, R., Karger, D.R., Kaashoek, M.F., Balakrishnan, H.: Chord: A scalable peer-to-peer lookup service for internet applications. In: SIGCOMM, pp. 149–160 (2001)
11. Breiman, L.: Pasting small votes for classification in large databases and on-line. Machine Learning 36(1-2), 85–103 (1999)
12. Polikar, R.: Esemble based systems in decision making. IEEE Circuits and Systems Magazine 9(3), 21–45 (2006)
13. Weiss, G.M., Provost, F.: The effect of class distribution on classifier learning: An empirical study. Technical report, Department of Computer Science, Rutgers University (2001)
14. Asuncion, A., Newman, D.: UCI machine learning repository (2007)
15. Chang, C.C., Lin, C.J.: LIBSVM: a library for support vector machines (2001), `http://www.csie.ntu.edu.tw/~cjlin/libsvm`
16. Baumgart, I., Heep, B., Krause, S.: Oversim: A flexible overlay network simulation framework. In: IEEE Global Internet Symposium, pp. 79–84 (2007)

A Generalization of Forward-Backward Algorithm

Ai Azuma and Yuji Matsumoto

Nara Institute of Science and Technology
{ai-a,matsu}@is.naist.jp

Abstract. Structured prediction has become very important in recent years. A simple but notable class of structured prediction is one for sequences, so-called sequential labeling. For sequential labeling, it is often required to take a summation over all the possible output sequences, when estimating the parameters of a probabilistic model for instance. We cannot make the direct calculation of such a summation from its definition in practice. Although the ordinary forward-backward algorithm provides an efficient way to do it, it is applicable to limited types of summations. In this paper, we propose a generalization of the forward-backward algorithm, by which we can calculate much broader types of summations than the existing forward-backward algorithms. We show that this generalization subsumes some existing calculations required in past studies, and we also discuss further possibilities of this generalization.

1 Introduction

Many learning tasks in the real world involve complex structures, where there are multiple, interrelated labels to be assigned. A simple but notable class that involves structures is sequential labeling, where dependencies between labels constitute a linear chain. A lot of classification or pattern recognition tasks on natural language sentences, genes and the like can be formulated as sequential labeling.

For sequential labeling, it is often required to take a summation over all the possible output sequences. For example, the expectation-maximization (EM) algorithm for Hidden Markov Models (HMMs) [1] requires expected frequencies of hidden variables measured on the current model. Another example is the calculation of the normalization constant, or partition function, for Conditional Random Fields (CRFs) [2]. Gradient of the log-likelihood objective for CRFs also involves model expectations of features. These calculations are originally defined as summations over all the possible output sequences.

Such a naive definition of a summation is not suitable for practical computing purposes in its own, because the number of the possible output sequences is proportional to the exponential of the length of the input sequence in general. The forward-backward algorithm provides a solution to get around this difficulty when output sequences are well-structured. "Well-structuredness" means that

W. Buntine et al. (Eds.): ECML PKDD 2009, Part I, LNAI 5781, pp. 99–114, 2009.

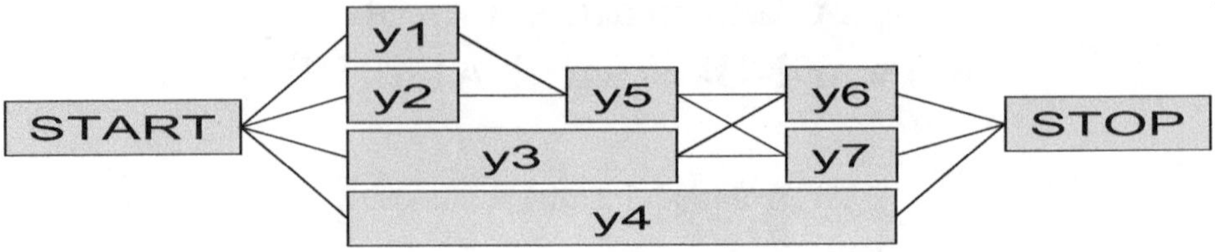

Fig. 1. An example trellis

output sequences can be encoded in a trellis, in which the number of nodes and arcs is proportional to a polynomial of the length of the input sequence. For example, the trellis shown in Fig.1 encodes seven sequences. This algorithm plays a crucial role in the aforementioned calculations. However, we would argue that the scope of applicability of the ordinary forward-backward algorithm is quite limited. The ordinary forward-backward algorithm is applicable to either one of the following forms,

$$\sum_{\boldsymbol{y}\in\mathcal{Y}(\boldsymbol{x})}\left(\prod_{c\in\mathcal{C}(\boldsymbol{y})}\phi(\boldsymbol{x},c)\right), \tag{1}$$

or

$$\sum_{\boldsymbol{y}\in\mathcal{Y}(\boldsymbol{x})}\left(\prod_{c\in\mathcal{C}(\boldsymbol{y})}\phi(\boldsymbol{x},c)\right)\left(\sum_{c\in\mathcal{C}(\boldsymbol{y})}f(\boldsymbol{x},c)\right), \tag{2}$$

where $\mathcal{Y}(\boldsymbol{x})$ denotes the set of all the possible sequences for a given input sequence $\boldsymbol{x}$, $\mathcal{C}(\boldsymbol{y})$ denotes the set of labels and adjacent pairs of labels appearing in $\boldsymbol{y}$, ϕ and f are complex-valued functions. Typically, ϕ represents a potential function of the distribution, and f represents an indicator function or a feature function.

We now want to derive an efficient algorithm applicable to much broader types of summations than (1) and (2). The term "efficient" here means that the computational cost is proportional not to the exponential of the length of the input sequence, but to a polynomial of it. As a result of the generalization proposed in this paper, we will derive an efficient algorithm applicable to the following form.

$$\sum_{\boldsymbol{y}\in\mathcal{Y}(\boldsymbol{x})}\left(\prod_{c\in\mathcal{C}(\boldsymbol{y})}\phi(\boldsymbol{x},c)\right)\left(\sum_{c\in\mathcal{C}(\boldsymbol{y})}f_1(\boldsymbol{x},c)\right)^{n_1}\cdots\left(\sum_{c\in\mathcal{C}(\boldsymbol{y})}f_K(\boldsymbol{x},c)\right)^{n_K}, \tag{3}$$

where $f_1,\ldots,f_K$ are K complex-valued functions, and $n_1,\ldots,n_K$ are non-negative integers.

In the next section, we will formalize our algorithm and show its validity. We will also give arguments in support of our estimation of its computational cost. In Sect.3, we will show that some specializations of the proposed algorithm lead

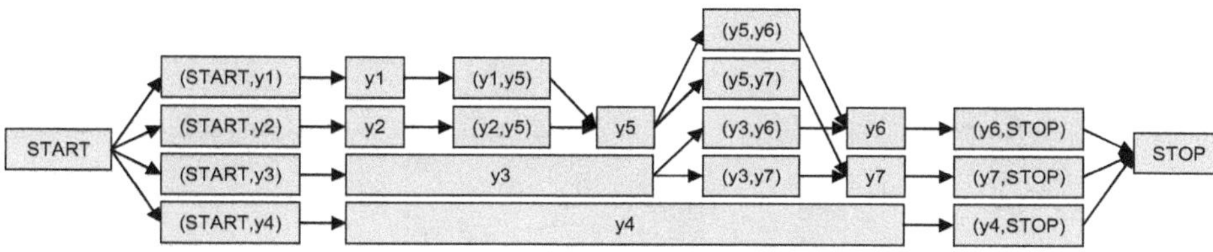

Fig. 2. The directed acyclic graph for the trellis shown in Fig.1

to some specific algorithms introduced in past studies. In Sect.4, we will show further possibilities of the algorithm. In Sect.5, we conclude this paper.

2 A Generalization of Forward-Backward Algorithm

In this section, we will formalize a generalization of the forward-backward algorithm. In order to write down definitions and proofs, we utilize directed acyclic graphs (DAG) instead of trellises. Consider a DAG $G = (V, E)$. A node in G represents either a label or an adjacent label pair in the trellis, i.e., $V := \bigcup_{\boldsymbol{y} \in \mathcal{Y}(\boldsymbol{x})} \mathcal{C}(\boldsymbol{y})$. An arc $(u, v) \in E$ exists if and only if, for an adjacent label pair $\langle y, y' \rangle$ in the trellis, $u = y \wedge v = \langle y, y' \rangle$ or $u = \langle y, y' \rangle \wedge v = y'$. For the trellis shown in Fig.1, the resulting DAG is shown in Fig.2. Hereafter, it is assumed that G has only one node whose in-degree is zero, denoted by src(G). It is also assumed that G has only one node whose out-degree is zero, denoted by snk(G). We abbreviate src(G) and snk(G) to src and snk respectively if G is obvious from the context. It is straightforward to extend the following discussion to DAGs with multiple source and sink nodes. Let $\Psi(G)$ be the set of directed paths whose starting node is src(G) and end node is snk(G). In this setting, summations over all the sequences in a trellis $\sum_{\boldsymbol{y} \in \mathcal{Y}(\boldsymbol{x})}$ is now rewritten as summations over the set of directed paths $\Psi(G)$, i.e., $\sum_{\pi \in \Psi(G)}$. We can also replace any function defined on $\mathcal{C}(\boldsymbol{y})$, i.e., $f(c)$ $(c \in \mathcal{C}(\boldsymbol{y}))$ with functions defined on V, i.e., $f(v)$ $(v \in V)$. For the sake of simplicity, we treat a directed path π in G as equivalent to the set of nodes appearing on π.

2.1 Standard Formulation

In this subsection, we will formalize an efficient algorithm to calculate summations of the form (3). Note that the summation $\sum_{\boldsymbol{y} \in \mathcal{Y}(G)}$ in the original formulation now becomes $\sum_{\pi \in \Psi(G)}$ in the new formulation. Readers are requested to read them as interchangeable.

Theorem 1 (Generalized Forward Algorithm). *Consider a DAG $G = (V, E)$. Let f_k $(k = 1, \ldots, K)$ be K complex-valued functions defined on V, and ϕ be a complex-valued function defined on V. For a directed path π in G, let $F_k(\pi) := \sum_{v \in \pi} f_k(v)$ $(k = 1, \ldots, K)$, and $\Phi(\pi) := \prod_{v \in \pi} \phi(v)$. Then, for non-negative integers $n_1, \ldots, n_K$,*[1]

[1] In this paper, we define $0^0 := 1$.

$$\sum_{\pi \in \Psi(G)} \Phi(\pi) {F_1}^{n_1}(\pi) \cdots {F_K}^{n_K}(\pi) = \alpha_{n_1,\ldots,n_K}(\mathrm{snk}) \ , \tag{4}$$

where α is defined recursively along a topological order of V, as follows.

$$\begin{aligned} \alpha_{n_1,\ldots,n_K}(\mathrm{src}) &:= \phi(\mathrm{src}) {f_1}^{n_1}(\mathrm{src}) \cdots {f_K}^{n_K}(\mathrm{src}) \ , \\ \alpha_{n_1,\ldots,n_K}(v) &:= \phi(v) \sum_{m_1=0}^{n_1} \left[\binom{n_1}{m_1} {f_1}^{m_1}(v) \cdots \sum_{m_K=0}^{n_K} \left[\binom{n_K}{m_K} {f_K}^{m_K}(v) \right.\right. \\ &\quad \left.\left. \times \sum_{v' \in \mathrm{prev}(v)} \alpha_{n_1-m_1,\ldots,n_K-m_K}(v') \right] \cdots \right] \qquad (v \neq \mathrm{src}) \ . \end{aligned} \tag{5}$$

Here, $\binom{n}{m}$ is the binomial coefficient, and prev(v) *denotes the set of preceding nodes of v, i.e.,* prev$(v) := \{x | (x, v) \in E\}$.

Proof. Hereafter, we will prove the following claim (6) from the recursive definition of α (5).

$$\alpha_{n_1,\ldots,n_K}(v) = \sum_{\pi \in \Psi_{\mathrm{src} \to v}(G)} \Phi(\pi) {F_1}^{n_1}(\pi) \cdots {F_K}^{n_K}(\pi) \qquad ({}^{\forall} v \in V) \ , \tag{6}$$

where $\Psi_{u \to v}(G)$ $(u, v \in V)$ is the set of directed paths whose starting node is u and end node is v. We utilize mathematical induction along a topological order of V.

For $v = \mathrm{src}$, (6) is obviously true from the definition of α (5).

Hereafter, for the sake of simplicity, we just refer to the case where there is only one function for F_k, i.e., $K = 1$. We also restate F_1 and n_1 as F and n respectively.

For a given $v \in V$, if we assume that (6) is true for ${}^{\forall} v' \in \mathrm{prev}(v)$, then

$$\begin{aligned} \alpha_n(v) &= \phi(v) \sum_{m=0}^{n} \left[\binom{n}{m} f^m(v) \sum_{v' \in \mathrm{prev}(v)} \alpha_{n-m}(v') \right] \\ &\qquad \text{(from the definition of } \alpha \text{ in (5))} \\ &= \phi(v) \sum_{m=0}^{n} \left[\binom{n}{m} f^m(v) \sum_{v' \in \mathrm{prev}(v)} \sum_{\pi \in \Psi_{\mathrm{src} \to v'}(G)} \Phi(\pi) F^{n-m}(\pi) \right] \\ &\qquad \text{(from the inductive assumption (6) for } {}^{\forall} v' \in \mathrm{prev}(v)) \\ &= \sum_{v' \in \mathrm{prev}(v)} \sum_{\pi \in \Psi_{\mathrm{src} \to v'}(G)} \left[\phi(v) \Phi(\pi) \sum_{m=0}^{n} \binom{n}{m} f^m(v) F^{n-m}(\pi) \right] \\ &= \sum_{v' \in \mathrm{prev}(v)} \sum_{\pi \in \Psi_{\mathrm{src} \to v'}(G)} \phi(v) \Phi(\pi) \left(f(v) + F(\pi) \right)^n \\ &\qquad (\because \text{ binomial theorem}) \\ &= \sum_{\pi \in \Psi_{\mathrm{src} \to v}(G)} \Phi(\pi) F^n(\pi) \ . \end{aligned} \tag{7}$$

So (6) appears to be true for v. Thus, (6) is true for $^\forall v \in V$, including the case $v = \text{snk}$, which is equivalent to (4) because $\Psi_{\text{src}\to\text{snk}}(G) = \Psi(G)$. We can also provide the proof of (6) for the case where there are multiple indices for α in a similar manner. □

This theorem shows that recursive calculation of α leads to the calculation of the left-hand side of (4). We call this recursive calculation $(n_1, \ldots, n_K)$*-th order forward algorithm*. It is worth noting that the specialization for the case $K = 0$, i.e., the (0)-th order forward algorithm is equivalent to the forward procedure of the ordinary forward-backward algorithm. Of course α_0 is equivalent to the ordinary forward variable (often denoted by α), too.

From the definition of α in (5), it is easy to show that the time complexity is proportional to $(|V| + |E|)\,(n_1+1)^2 \cdots (n_K+1)^2$, assuming that $\binom{n}{m}$ and $f_k{}^{n_k}(v)$ can be calculated in constant time.

In the ordinary forward-backward algorithm, we also calculate backward variables, that are often denoted by β. $\alpha(v)\beta(v)/\phi(v)$ is the summation of $\Phi(\pi)$ over the sequences appearing in the sub-trellis constrained on a node v. In other words, we can regard $\alpha(v)\beta(v)/\phi(v)$ as the marginal of $\Phi(\pi)$ on v. So, for a function $\acute{f}(v)$, summing up $\acute{f}(v)\alpha(v)\beta(v)/\phi(v)$ on all nodes leads to the summation of $\Phi(\pi)\sum_{v\in\pi}\acute{f}(v)$. This formulation is useful in the case where $\acute{f}$ is sparse, i.e., $\acute{f}(v) = 0$ for most of $v \in V$. Likewise, we can define generalized backward variables $\beta_{n_1,\ldots,n_K}$ in a similar manner to the definition of α in Th.1. A generalized forward variable $\alpha_{n_1,\ldots,n_K}(v)$ is equal to the summation of $\Phi(\pi)F_1{}^{n_1}(\pi)\cdots F_K{}^{n_K}(\pi)$ in the sub-trellis prior to v (see (6)), and a generalized β is equal to the summation of the same form in the sub-trellis posterior to v. Thus, a proper combination of generalized α and β yields the summation of $\Phi(\pi)F_1{}^{n_1}(\pi)\cdots F_K{}^{n_K}(\pi)$ over the sub-trellis constrained on v. In other words, such a combination can be regarded as the marginal of $\Phi(\pi)F_1{}^{n_1}(\pi)\cdots F_K{}^{n_K}(\pi)$ on v. So, we can make a similar discussion on a sparse function $\acute{f}(v)$ in our generalization. This argument is formally summarized in the following theorem.

Theorem 2 (Generalized Forward-backward Algorithm). *In addition to the definitions of* $G, f_k, F_k, \phi, \Phi, \alpha$ *in Th.1, let* $\acute{f}$ *be a complex-valued function defined on* V*, and, for a directed path* π *in* G*, let* $\acute{F}(\pi) := \sum_{v\in\pi}\acute{f}(v)$*. Then,*

$$\sum_{\pi\in\Psi(G)} \Phi(\pi)F_1{}^{n_1}(\pi)\cdots F_K{}^{n_K}(\pi)\acute{F}(\pi)$$
$$= \sum_{v\in V}\left[\acute{f}(v)\sum_{m_1=0}^{n_1}\left[\binom{n_1}{m_1}\cdots\sum_{m_K=0}^{n_K}\left[\binom{n_K}{m_K} \times \alpha_{m_1,\ldots,m_K}(v)\beta_{n_1-m_1,\ldots,n_K-m_K}(v)\right]\cdots\right]\right], \quad (8)$$

where β *is recursively defined along a reversed topological order of* V*, as follows.*

$$\beta_{n_1,\ldots,n_K}(\text{snk}) := \begin{cases} 1 & (n_1, \cdots, n_K = 0)\ , \\ 0 & \text{otherwise}\ , \end{cases}$$

$$\beta_{n_1,\ldots,n_K}(v) := \sum_{v' \in \mathrm{next}(v)} \left[\phi(v') \sum_{m_1=0}^{n_1} \left[\binom{n_1}{m_1} f_1{}^{m_1}(v') \cdots \times \sum_{m_K=0}^{n_K} \left[\binom{n_K}{m_K} f_K{}^{m_K}(v') \beta_{n_1-m_1,\cdots,n_K-m_K}(v') \right] \cdots \right] \right] \quad (v \neq \mathrm{snk}) \ , \tag{9}$$

where $\mathrm{next}(v)$ *denotes the set of succeeding nodes of* v*, i.e.*, $\mathrm{next}(v) := \{x | (v, x) \in E\}$.

Some readers may be confused by the definition of β in (9) because the definitions of α (in (5)) and β are asymmetric by design. In the ordinary forward-backward algorithm, α and β are usually defined in a symmetric manner, and combined in the product $\alpha(v)\beta(v)/\phi(v)$ because both $\alpha(v)$ and $\beta(v)$ have the same factor $\phi(v)$ redundantly. However, such a symmetric definition makes the combination of α and β in (8) quite complex. So we define β as in (9).

With the following relation

$$\sum_{\pi \in \Psi(G)} \Gamma(\pi) \acute{F}(\pi) = \sum_{v \in V} \left[\acute{f}(v) \sum_{\pi \in \Psi_v(G)} \Gamma(\pi) \right] \ , \tag{10}$$

where Γ is a complex-valued function defined on $\Psi(G)$ and $\Psi_v(G) := \{\pi \mid \pi \in \Psi(G), v \in \pi\}$, we give proof of Th.2.

Proof (of Th.2). Just like the proof of Th.1, we can prove the following claim.

$$\beta_{n_1,\ldots,n_K}(v) = \sum_{\pi \in \Psi_{v \leftarrow \mathrm{snk}}(G)} \Phi(G) F_1{}^{n_1}(\pi) \cdots F_K{}^{n_K}(\pi) \quad ({}^\forall v \in V, v \neq \mathrm{snk}) \ , \tag{11}$$

where $\Psi_{v \leftarrow \mathrm{snk}}(G) := \bigcup_{v' \in \mathrm{next}(v)} \Psi_{v' \rightarrow \mathrm{snk}}(G)$. Hereafter, for the sake of simplicity, we just refer to the case where $K = 1$. Obviously, we can split a directed path containing v into two parts. One is prior to v, and the other is posterior to v. In other words, for ${}^\forall v \in V$, $\Psi_v(G)$ is the direct product of $\Psi_{\mathrm{src} \rightarrow v}(G)$ and $\Psi_{v \leftarrow \mathrm{snk}}(G)$. So,

$$\begin{aligned}
&\sum_{\pi \in \Psi(G)} \Phi(\pi) F^n(\pi) \acute{F}(\pi) \\
&\quad = \sum_{v \in V} \left[\acute{f}(v) \sum_{\pi \in \Psi_v(G)} \Phi(\pi) F^n(\pi) \right] \\
&\qquad (\because (10) \text{ with } \Gamma(\pi) = \Phi(\pi) F^n(\pi)) \\
&\quad = \sum_{v \in V} \left[\acute{f}(v) \sum_{\pi \in \Psi_{\mathrm{src} \rightarrow v}(G)} \sum_{\pi' \in \Psi_{v \leftarrow \mathrm{snk}}(G)} \Phi(\pi \cup \pi') F^n(\pi \cup \pi') \right] \\
&\quad = \sum_{v \in V} \left[\acute{f}(v) \sum_{\pi \in \Psi_{\mathrm{src} \rightarrow v}(G)} \sum_{\pi' \in \Psi_{v \leftarrow \mathrm{snk}}(G)} \Phi(\pi) \Phi(\pi') \left(F(\pi) + F(\pi') \right)^n \right]
\end{aligned}$$

$$= \sum_{v \in V} \left[\acute{f}(v) \sum_{\pi \in \Psi_{\mathrm{src} \to v}(G)} \sum_{\pi' \in \Psi_{v \leftarrow \mathrm{snk}}(G)} \left[\Phi(\pi)\Phi(\pi') \times \sum_{m=0}^{n} \binom{n}{m} F^{m}(\pi) F^{n-m}(\pi') \right] \right]$$

$$(\because \text{ binomial theorem})$$

$$= \sum_{v \in V} \left[\acute{f}(v) \sum_{m=0}^{n} \binom{n}{m} \left(\sum_{\pi \in \Psi_{\mathrm{src} \to v}(G)} \Phi(\pi) F^{m}(\pi) \right) \times \left(\sum_{\pi' \in \Psi_{v \leftarrow \mathrm{snk}}(G)} \Phi(\pi') F^{n-m}(\pi') \right) \right] \quad (12)$$

$$= \sum_{v \in V} \left[\acute{f}(v) \sum_{m=0}^{n} \binom{n}{m} \alpha_m(v) \beta_{n-m}(v) \right]$$

$$(\because (6), (11)) .$$

□

This theorem shows that, by recursive calculation of α and β, and storing them for $^\forall v \in V$, we can calculate the summation appearing in the left-hand side of (8). We call this calculation $(n_1, \ldots, n_K)$*-th order forward-backward algorithm.* It is worth noting that the (0)-th order forward-backward algorithm is equivalent to the ordinary forward-backward algorithm for the form of (2).

In the same way as Th.1, the time complexity is proportional to $(|V| + |E|)(n_1 + 1)^2 \cdots (n_K + 1)^2$. The space complexity is proportional to $|V|(n_1 + 1) \cdots (n_K + 1)$ because we need to store α and β for $^\forall v \in V$.

2.2 Fourier-Transformed Formulation

In both Th.1 and Th.2, we can find that the definitions of α and β involve a sort of convolution. It is well known that a convolution is transformed into element-wise product under the Fourier transform. So, we can derive formulation with element-wise products instead of convolutions under the Fourier transform.

First, let us give a brief description of convolution and its relationship with the Fourier transform. Consider two complex-valued sequences $\{a_n\}, \{b_n\}$ $(n = 0, \ldots, 2N-1)$. Both have N trailing zeros in them, i.e., $a_n = 0, b_n = 0$ $(n = N, \ldots, 2N-1)$. Let a sequence $\{c_n\}$ be as follows.

$$c_n := \sum_{m=0}^{n} a_m b_{n-m} \qquad (n = 0, \ldots, 2N-1) . \quad (13)$$

Then,

$$\mathcal{F}(\{c_n\}) = \mathcal{F}(\{a_n\}) \circ \mathcal{F}(\{b_n\}) , \quad (14)$$

where $\mathcal{F} : \mathbb{C}^{2N} \to \mathbb{C}^{2N}$ is the discrete Fourier transform [2], and $\circ$ denotes an element-wise product. The relation (14) is often referred to as the convolution theorem.

Convolutions in the definitions of α and β are of the form

$$c_n := \sum_{m=0}^{n} \binom{n}{m} a_m b_{n-m} \ , \tag{15}$$

which is slightly different from the ordinary convolution in (13). However, with the definition of the binomial coefficient, we can rephrase (15) as

$$\frac{c_n}{n!} = \sum_{m=0}^{n} \frac{a_m}{m!} \cdot \frac{b_{n-m}}{(n-m)!} \ , \tag{16}$$

in which the sequence $\left\{\frac{c_n}{n!}\right\}$ is calculated by an ordinary convolution of $\left\{\frac{a_n}{n!}\right\}$ and $\left\{\frac{b_n}{n!}\right\}$.

Based on the convolution theorem mentioned above, we will derive the Fourier-transformed versions of Th.1 and Th.2. For the sake of simplicity, we just refer to the case where $K = 1$.

Theorem 3 (Fourier-transformed Generalized Forward Algorithm). *In addition to the definitions of $\phi, f := f_1, \alpha$ in Th.1, let*

$$\left\{\tilde{f}_n(v)\right\} := \mathcal{F}\left(\left\{\frac{f^0(v)}{0!}, \ldots, \frac{f^{N-1}(v)}{(N-1)!}, 0, \ldots, 0\right\}\right) \qquad ({}^\forall v \in V) \ , \tag{17}$$

and

$$\begin{aligned} \tilde{\alpha}_n(\mathrm{src}) &:= \phi(\mathrm{src})\tilde{f}_n(\mathrm{src}) \ , \\ \tilde{\alpha}_n(v) &:= \phi(v)\tilde{f}_n(v) \sum_{v' \in \mathrm{prev}(v)} \tilde{\alpha}_n(v') \qquad (v \neq \mathrm{src}) \ . \end{aligned} \tag{18}$$

Then, for $0 \leqq n \leqq N-1$ and ${}^\forall v \in V$,

$$\frac{\alpha_n(v)}{n!} = \left\{\mathcal{F}^{-1}\left(\{\tilde{\alpha}_n(v)\}\right)\right\}_n \ . \tag{19}$$

Proof. (Sketch) From the Fourier transform for the definition of α in (5), linearity of the Fourier transform and the convolution theorem (14), we can easily show the proof. □

Theorem 4 (Fourier-transformed Generalized Forward-backward Algorithm). *In addition to the definitions of $\phi, f, \tilde{f}, \tilde{\alpha}$ in Th.3, let*

$$\Phi(\pi) := \prod_{v \in \pi} \phi(v), \ \ F(\pi) := \sum_{v \in \pi} f(v), \ \ \acute{F}(\pi) := \sum_{v \in \pi} \acute{f}(v) \tag{20}$$

for a directed path π in G, and

[2] We use normalization constant 1 for $\mathcal{F}$, and $\frac{1}{2N}$ for the inverse discrete Fourier transform $\mathcal{F}^{-1}$.

$$\begin{aligned}
\tilde{\beta}_n(\text{snk}) &:= 1 \ , \\
\tilde{\beta}_n(v) &:= \sum_{v' \in \text{next}(v)} \phi(v') \tilde{f}_n(v') \tilde{\beta}_n(v') \ , \\
\tilde{S}_n &:= \sum_{v \in V} \acute{f}(v) \tilde{\alpha}_n(v) \tilde{\beta}_n(v) \ .
\end{aligned} \tag{21}$$

Then, for $0 \leqq n \leqq N-1$,

$$\frac{1}{n!} \sum_{\pi \in \Psi(G)} \Phi(\pi) F^n(\pi) \acute{F}(\pi) = \left\{ \mathcal{F}^{-1} \left(\left\{ \tilde{S}_n \right\} \right) \right\}_n \ . \tag{22}$$

Proof. (Sketch) The same as the proof of Th.3, with the Fourier transform of (8). □

The argument above can be extended to the case where there are multiple indices. We can apply the Fourier transform only to an arbitrary subset of indices, and we may leave the remaining indices as they are.

Using the fast Fourier transform (FFT), the proportionality factor of an index n_k in the time complexity is reduced from $(n_k+1)^2$ to $(n_k+1)\log(n_k+1)$.

3 Application of the Proposed Algorithm

In this section, we will show that some specializations of the generalization derived above correspond to some calculations studied in past research. Our intuition behind the following derivations is simple. Problems all boil down to how we can transform the summation at hand into an expression of the form (3). Once we obtain a formula of the form (3), our theorems immediately offer an efficient algorithm for calculating it.

In the following subsections, we assume that the distribution over output sequences is modeled by CRFs, which is defined as follows.

$$P(\pi | \boldsymbol{x}; \boldsymbol{\lambda}) = \frac{1}{Z(\boldsymbol{\lambda}; \boldsymbol{x})} \sum_{\pi \in \Psi(G)} \left(\prod_{v \in \pi} \phi(\boldsymbol{x}, v) \right) \tag{23}$$

with

$$Z(\boldsymbol{\lambda}; \boldsymbol{x}) := \sum_{\pi \in \Psi(G)} \left(\prod_{v \in \pi} \phi(\boldsymbol{x}, v) \right) \ , \tag{24}$$

$$\phi(\boldsymbol{x}, v) := \exp\left(\sum_k \lambda_k f_k(\boldsymbol{x}, v) \right) \ . \tag{25}$$

We also assume that all the expectations and covariances are taken under (23).

3.1 Covariance for Conditional Random Fields

For CRFs, it is sometimes necessary to calculate the covariance of feature functions. For feature functions $F_1(\pi), F_2(\pi)$ $(\pi \in \Psi(G))$, the covariance is

$$\mathrm{Cov}\left[F_1(\pi), F_2(\pi)\right] = E\left[F_1(\pi)F_2(\pi)\right] - E\left[F_1(\pi)\right] E\left[F_2(\pi)\right] \ . \tag{26}$$

If both F_1 and F_2 are of the form $\sum_{v \in \pi} f(v)$, the second term of (26) can be calculated either by the (1)-st order forward algorithm or the (0)-th order forward-backward algorithm, and the first term of (26) can be calculated either by the $(1,1)$-th order forward algorithm or the (1)-st order forward-backward algorithm.

Covariance of feature functions is useful for optimization of CRFs. Because it has rich information of curvature of the log-likelihood objective function, i.e.,

$$\frac{\partial^2}{\partial \lambda_i \partial \lambda_j} \mathcal{L}(\boldsymbol{\lambda}) = -\mathrm{Cov}[F_i(\boldsymbol{x}, \pi), F_j(\boldsymbol{x}, \pi)] \ , \tag{27}$$

where $\mathcal{L}$ is the log-likelihood objective function and $F_k(\boldsymbol{x}, \pi) := \sum_{v \in \pi} f_k(\boldsymbol{x}, v)$. For example, S. V. N. Vishwanathan et al. [3] used stochastic gradient methods to accelerate the optimization of the log-likelihood objective for CRFs. In stochastic gradient methods, it is required to calculate the Hessian-vector product. Although they use automatic differentiation to calculate the Hessian-vector product, we can derive an alternative algorithm for it as a specialization of the generalized forward-backward algorithm. For a vector $\boldsymbol{v}$, Hessian-vector product is

$$\begin{aligned} \left\{\boldsymbol{H}(\mathcal{L}) \cdot \boldsymbol{v}\right\}_i &= -\sum_j \mathrm{Cov}[F_i(\boldsymbol{x}, \pi), F_j(\boldsymbol{x}, \pi)] v_j \\ &= E[F_i(\boldsymbol{x}, \pi)] E[\boldsymbol{F}(\boldsymbol{x}, \pi) \cdot \boldsymbol{v}] - E[F_i(\boldsymbol{x}, \pi) \left(\boldsymbol{F}(\boldsymbol{x}, \pi) \cdot \boldsymbol{v}\right)] \ , \end{aligned} \tag{28}$$

where $\boldsymbol{H}$ is the Hessian matrix. The (1)-st order forward-backward algorithm provides an efficient algorithm to calculate the second term in (28). That algorithm is equivalent to the calculation by the automatic differentiation as far as the computational cost is concerned. [3]

As another practical application of the covariance, we can take differentiation of more complicated objective functions than the log-likelihood objective. For example, the objective function used in Jiao et al. [4] has the conditional entropy of CRFs.

$$\mathcal{H}(\boldsymbol{\lambda}) = E[\log P(\pi | \boldsymbol{x}; \boldsymbol{\lambda})] = E[\boldsymbol{F}(\boldsymbol{x}, \pi) \cdot \boldsymbol{\lambda}] - \log Z(\boldsymbol{\lambda}; \boldsymbol{x}) \ . \tag{29}$$

[3] Generally speaking, the algorithm implicitly derived by the automatic differentiation for the ordinary forward-backward algorithm is equivalent to the (1)-st order forward-backward algorithm. In addition, we conjecture that the higher-order automatic differentiation for the ordinary forward-backward algorithm is essentially equivalent to the generalized forward-backward algorithm proposed in this paper.

The ordinary ((0)-th order) forward-backward algorithm can calculate the gradient of the second term. The gradient of the first term is

$$\frac{\partial}{\partial \lambda_k} E[\boldsymbol{F}(\boldsymbol{x}, \pi) \cdot \boldsymbol{\lambda}] = \mathrm{Cov}\left[\boldsymbol{F}(\boldsymbol{x}, \pi) \cdot \boldsymbol{\lambda}, F_k(\boldsymbol{x}, \pi)\right] \ . \tag{30}$$

The discussion of the calculation of (30) is the same as the one for the second term of (28).

In more general formulation, for the expectation of a sufficient statistic $\bar{F}(\pi)$ ($\pi \in \Psi(G)$),

$$\frac{\partial}{\partial \lambda_k} E[\bar{F}(\pi)] = \mathrm{Cov}\left[\bar{F}(\pi), F_k(\boldsymbol{x}, \pi)\right] \ . \tag{31}$$

(31) falls within the applicable scope of the generalization proposed in this paper, assuming $\bar{F}$ is of the form $\bar{F}(\pi) = \sum_{v \in \pi} \bar{f}(v)$.

3.2 Hamming Loss for Conditional Random Fields

Kakade et al. [5] proposed to use Hamming loss objective function in parameter estimations involved in sequential labeling. For a correctly annotated input-output sequence pair $(\hat{\boldsymbol{x}}, \hat{\boldsymbol{y}})$, the objective to be optimized is

$$C_1(\boldsymbol{\lambda}; \hat{\boldsymbol{x}}, \hat{\boldsymbol{y}}) := \frac{1}{T} \sum_t \log P(\hat{y}_t | \hat{\boldsymbol{x}}; \boldsymbol{\lambda}) \ . \tag{32}$$

With notations used in this paper, we can rephrase this objective as follows.

$$C_1(\boldsymbol{\lambda}; \hat{\boldsymbol{x}}, \hat{\boldsymbol{y}}) = \frac{1}{T} \sum_t \log \sum_{\pi \in \Psi(G)} P(\pi | \hat{\boldsymbol{x}}; \boldsymbol{\lambda}) \delta_{\hat{\pi}_t}(\pi) \ , \tag{33}$$

where $\hat{\pi}_t$ is the node that corresponds to $\hat{y}_t$. T is the number of labels in $\hat{\boldsymbol{y}}$, and $\delta_{\hat{\pi}_t}(\pi) := \sum_{v \in \pi} \delta_{v, \hat{\pi}_t}$ is the indicator for the condition $\hat{\pi}_t \in \pi$. Its gradient is

$$\begin{aligned} \frac{\partial C_1}{\partial \lambda_k} = \frac{1}{T} \sum_t & \frac{\sum_{\pi \in \Psi(G)} P(\pi | \boldsymbol{x}; \boldsymbol{\lambda}) F_k(\boldsymbol{x}, \pi) \delta_{\hat{\pi}_t}(\pi)}{\sum_{\pi \in \Psi(G)} P(\pi | \boldsymbol{x}; \boldsymbol{\lambda}) \delta_{\hat{\pi}_t}(\pi)} \\ & - \sum_{\pi \in \Psi(G)} P(\pi | \boldsymbol{x}; \boldsymbol{\lambda}) F_k(\boldsymbol{x}, \pi) \ . \end{aligned} \tag{34}$$

The second term is of the form which can be calculated by the ordinary ((0)-th order) forward-backward algorithm.

To calculate the first term in (34), Kakade et al. [5] derived a sort of the forward-backward algorithm, which is as efficient as the ordinary forward-backward algorithm. We can show that a specialization of the generalization proposed in this paper also leads to an equivalent algorithm.

With following definitions

$$c(v) := \begin{cases} 1 & (v \text{ represents a correct label}) \ , \\ 0 & (\text{otherwise}) \ , \end{cases} \tag{35}$$

$$\alpha_0(v) := \begin{cases} \phi(v) & (v = \mathrm{src}) \ , \\ \phi(v) \sum_{v' \in \mathrm{prev}(v)} \alpha_0(v') & (\text{otherwise}) \ , \end{cases} \tag{36}$$

$$\alpha_1(v) := \begin{cases} \phi(v) \frac{c(v)}{\alpha_0(v)\beta_0(v)} & (v = \mathrm{src}) \ , \\ \frac{c(v)}{\alpha_0(v)\beta_0(v)} \alpha_0(v) + \phi(v) \sum_{v' \in \mathrm{prev}(v)} \alpha_1(v') & (\text{otherwise}) \ , \end{cases} \tag{37}$$

$$\beta_0(v) := \begin{cases} 1 & (v = \mathrm{snk}) \ , \\ \sum_{v' \in \mathrm{next}(v)} \phi(v')\beta_0(v') & (\text{otherwise}) \ , \end{cases} \tag{38}$$

$$\beta_1(v) := \begin{cases} 0 & (v = \mathrm{snk}) \ , \\ \frac{c(v)}{\alpha_0(v)\beta_0(v)} \beta_0(v) + \sum_{v' \in \mathrm{next}(v)} \phi(v')\beta_1(v') & (\text{otherwise}) \ , \end{cases} \tag{39}$$

and noting the following relation

$$\begin{aligned} \Delta(\pi) &:= \sum_t \frac{\delta_{\hat{\pi}_t}(\pi)}{\alpha_0(\hat{\pi}_t)\beta_0(\hat{\pi}_t)} = \sum_t \frac{\sum_{v \in \pi} \delta_{v,\hat{\pi}_t}}{\alpha_0(\hat{\pi}_t)\beta_0(\hat{\pi}_t)} \\ &= \sum_{v \in \pi} \left(\sum_t \frac{\delta_{v,\hat{\pi}_t}}{\alpha_0(\hat{\pi}_t)\beta_0(\hat{\pi}_t)} \right) = \sum_{v \in \pi} \frac{c(v)}{\alpha_0(v)\beta_0(v)} \ , \end{aligned} \tag{40}$$

we can derive an efficient algorithm to calculate the first term in (34) as follows.

$$\begin{aligned} &\frac{1}{T} \sum_t \frac{\sum_{\pi \in \Psi(G)} P(\pi|\boldsymbol{x};\boldsymbol{\lambda}) F_k(\boldsymbol{x},\pi) \delta_{\hat{\pi}_t}(\pi)}{\sum_{\pi \in \Psi(G)} P(\pi|\boldsymbol{x};\boldsymbol{\lambda}) \delta_{\hat{\pi}_t}(\pi)} \\ &\quad = \frac{1}{T} \sum_t \frac{\sum_{\pi \in \Psi(G)} P(\pi|\boldsymbol{x};\boldsymbol{\lambda}) F_k(\boldsymbol{x},\pi) \delta_{\hat{\pi}_t}(\pi)}{\alpha_0(\hat{\pi}_t)\beta_0(\hat{\pi}_t)} \\ &\quad (\because \text{(0)-th order forward-backward algorithm for the denominator}) \\ &\quad = \frac{1}{T} \sum_{\pi \in \Psi(G)} \left(P(\pi|\boldsymbol{x};\boldsymbol{\lambda}) F_k(\boldsymbol{x},\pi) \sum_t \frac{\delta_{\hat{\pi}_t}(\pi)}{\alpha_0(\hat{\pi}_t)\beta_0(\hat{\pi}_t)} \right) \\ &\quad = \frac{1}{T} \sum_{\pi \in \Psi(G)} P(\pi|\boldsymbol{x};\boldsymbol{\lambda}) F_k(\boldsymbol{x},\pi) \Delta(\pi) \\ &\quad = \frac{1}{T} \sum_{v \in V} f_k(\boldsymbol{x},v) \left(\alpha_1(v)\beta_0(v) + \alpha_0(v)\beta_1(v)\right) \\ &\quad (\because \text{(1)-st order forward-backward algorithm}) \ . \end{aligned} \tag{41}$$

Note that $\alpha_1(v)$ and $\beta_1(v)$ in this notation respectively correspond to $\omega^{\mathrm{f}}_{i+1}$ and ω^{b}_i used in [5]. Of course both sides are also equivalent concerning the computational cost.

3.3 Generalized Expectation Criteria for Conditional Random Fields

Mann et al. [6] proposed a semi-supervised training method called generalized expectation criteria to improve accuracy with unlabeled data. The objective used in [6] has an additional term

$$-\theta D(\hat{P}||\tilde{P}) \ , \tag{42}$$

where θ is a given hyper-parameter, D is the KL divergence, $\hat{P}$ is a given target distribution and $\tilde{P}$ is the conditional distribution of labels given a feature $f_m(\boldsymbol{x}, t)$ at time t, i.e., [4]

$$\tilde{P} := \tilde{P}(y_t | f_m(\boldsymbol{x}, t) = 1; \boldsymbol{\lambda}) \ . \tag{43}$$

We now focus on the parameter gradient. Instead of the derivation shown in [6], we first rephrase $\tilde{P}(y|f_m(x,t) = 1; \boldsymbol{\lambda})$ as

$$\tilde{P}(\ell(v)|f_m(v) = 1; \boldsymbol{\lambda}) = \frac{1}{U_m} \sum_{\boldsymbol{x} \in \boldsymbol{U}_m} \sum_{\pi \in \Psi(G)} P(\pi|\boldsymbol{x}; \boldsymbol{\lambda}) \Delta_{f_m, y}(\pi) \ , \tag{44}$$

where $\ell(v)$ is the label assigned to v, $\boldsymbol{U}_m$ is the set of sequences where f_m is present for some nodes, and

$$\Delta_{f_m, y}(\pi) = \sum_{v \in \pi} f_m(v) \delta_{\ell(v), y} \ . \tag{45}$$

Noting the following relation

$$\Delta_{f_m}(\pi) := \sum_l \frac{\hat{P}}{\tilde{P}} \Delta_{f_m, l}(\pi) = \sum_{v \in \pi} \frac{\hat{P}(\ell(v)|f_m(v) = 1)}{\tilde{P}(\ell(v)|f_m(v) = 1; \boldsymbol{\lambda})} f_m(v) \tag{46}$$

we get

$$\frac{\partial}{\partial \lambda_k} D(\hat{P}||\tilde{P}) = -\frac{1}{U_m} \sum_{\boldsymbol{x} \in \boldsymbol{U}_m} \sum_l \frac{\hat{P}}{\tilde{P}} \frac{\partial}{\partial \lambda_k} E\left[\Delta_{f_m, l}(\pi)\right] \ , \tag{47}$$

and

$$\sum_l \frac{\hat{P}}{\tilde{P}} \frac{\partial}{\partial \lambda_k} E\left[\Delta_{f_m, l}(\pi)\right] = -E[\Delta_{f_m}(\pi) F_k(\boldsymbol{x}, \pi)] + E[\Delta_{f_m}(\pi)] E[F_k(\boldsymbol{x}, \pi)] \ . \tag{48}$$

Here, the first term can be calculated by the (1)-st order forward-backward algorithm, the second term by the ordinary ((0)-th order) forward-backward algorithm.

Note that this calculation can drastically reduce the computational cost compared with the algorithm proposed in [6]. Computational cost needed by our formulation is scaled by up to a constant factor compared with the ordinary forward-backward algorithm, whereas [6] takes the forward-backward algorithm for each label so computational cost is proportional to the number of labels.

[4] In the same way as their paper, we assume that all output sequences are same in length and f_m is binary.

4 Further Possibilities

In the previous section, we show some specializations of our generalized forward-backward algorithm. However, we have not fully utilized the potential capacity of the higher order generalization of the forward-backward algorithm we proposed in this paper. To show a sample application of the higher order generalization, we will describe the expected F-measure optimization.

4.1 Expected F-Measure Optimization

Jansche [7] utilized the expected f-measure as an objective function in optimization of logistic regression models. He has just described on single label, non-structured case. Here, let this criteria be sequentially-extended. We define

$$\mathcal{F}_\gamma(\boldsymbol{\lambda}; \hat{\boldsymbol{x}}) := E[\mathcal{F}_\gamma(\pi)] \ , \tag{49}$$

where the expectation is taken under (23), and $\mathcal{F}_\gamma(\pi)$ is the label-wise f-measure for a given output sequence π, i.e.,

$$\mathcal{F}_\gamma(\pi) := \frac{(1+\gamma^2)l(\pi)}{|\pi| + \gamma^2 L} \ . \tag{50}$$

L is the number of labels in the correctly annotated answer sequence, and $l(\pi)$ is the number of labels in the output sequence π that coincide with the correct labels. $|\pi|$ is the total number of labels appearing in π. Hereafter, we refer only to the case $\gamma = 1$. Let $\mathcal{F}(\pi) := \mathcal{F}_1(\pi)$. For the case where all the possible output sequences have the same number of labels (length), optimization concerning (49) – calculations of value and gradient – requires up to the $(1,1)$-th order forward-backward algorithm. On the other hand, for the case where the lengths of the output sequences are different as in [8], because the denominator in (49) is variable in the summation, calculations of the value and the gradient of (49) are not trivial at all.

However, with the Taylor expansion of the reciprocal function in (49), we get

$$\begin{aligned} E[\mathcal{F}(\pi)] &= \frac{1}{Z(\boldsymbol{\lambda}; \boldsymbol{x})} \sum_{\pi \in \Psi(G)} \Phi(\pi) \frac{2l(\pi)}{|\pi| + L} \\ &= \frac{2}{Z(\boldsymbol{\lambda}; \boldsymbol{x})(L + x_0)} \sum_{\pi \in \Psi(G)} \left[\Phi(\pi) l(\pi) \sum_{m=0}^{\infty} \left(\frac{|\pi| - x_0}{-L - x_0} \right)^m \right] \\ &= \frac{2}{Z(\boldsymbol{\lambda}; \boldsymbol{x})(L + x_0)} \sum_{m=0}^{\infty} \sum_{\pi \in \Psi(G)} \Phi(\pi) l(\pi) \left(\frac{|\pi| - x_0}{-L - x_0} \right)^m , \end{aligned} \tag{51}$$

where x_0 is a complex value that satisfies the condition $\left| \frac{|\pi| - x_0}{-L - x_0} \right| < 1 \ (^\forall \pi \in \Psi(G))$ (such a value really exists at any time). By truncating the higher terms in the Taylor series, we can approximate the value.

$$E[\mathcal{F}(\pi)] \approx \frac{2}{Z(\boldsymbol{\lambda}; \boldsymbol{x})(L + x_0)} \sum_{m=0}^{M} \sum_{\pi \in \Psi(G)} \Phi(\pi) l(\pi) \left(\frac{|\pi| - x_0}{-L - x_0} \right)^m \ . \tag{52}$$

The summation $\sum_{\pi \in \Psi(G)}$ in (52) is of the form to which the $(m, 1)$-th order forward algorithm is applicable. We can also derive an approximation of the gradient of (52).

$$\frac{\partial}{\partial \lambda_k} E[\mathcal{F}(\pi)] = E[\mathcal{F}(\pi) F_k(\boldsymbol{x}, \pi)] - E[\mathcal{F}(\pi)] E[F_k(\boldsymbol{x}, \pi)] \tag{53}$$

The first term in (53) can be approximated by the $(m, 1)$-th forward-backward algorithm in a similar manner to (52). Therefore, we can optimize the objective with numerical optimization routines.

For a given constant M, these calculations are as efficient as the ordinary forward-backward algorithm, scaled by up to a constant factor. For trellises in which the length of a node is up to 5, we have experimentally confirmed that $M \sim 15$ proves to be sufficient to approximate the objective and the gradient with enough precision under the double-precision floating-point arithmetic.

5 Conclusion

In this paper, we proposed a generalization of the forward-backward algorithm, which is applicable to much broader types of summations over all possible sequences than the ordinary forward-backward algorithm. We also show that our theorems in this paper can offer efficient algorithms for some calculations required in past studies, even for the summation of a non-linear function using a Taylor expansion.

We are now investigating other possibilities of applications of our generalization, including optimization of much more complex probabilistic models than log-linear models on sequential labeling tasks (we assume CRFs with polynomial kernels [9] [10], for example).

While we proposed the generalization only for linear chains, it is straightforward to extend this generalization to trees. However, it will be much more challenging to find a counterpart in the case of loopy structures. We are going to study such a possibility.

References

1. Dempster, A.P., Laird, N.M., Rubin, D.B.: Maximum likelihood from incomplete data via the em algorithm. J. Roy. Stat. Soc. Ser. B 39(1), 1–38 (1977)
2. Lafferty, J., McCallum, A., Pereira, F.: Conditional random fields: Probabilistic models for segmenting and labeling sequence data. In: Proc. of ICML 2001, pp. 282–289 (2001)
3. Vishwanathan, S.V.N., Schraudolph, N.N., Schmidt, M.W., Murphy, K.P.: Accelerated training of conditional random fields with stochastic gradient methods. In: Proc. of ICML 2006, pp. 969–976 (2006)
4. Jiao, F., Wang, S., Lee, C.-H., Greiner, R., Schuurmans, D.: Semi-supervised conditional random fields for improved sequence segmentation and labeling. In: Proc. of COLING-ACL 2006, July 2006, pp. 209–216 (2006)

5. Kakade, S., Teh, Y.W., Roweis, S.: An alternate objective function for Markovian fields. In: Proc. of the ICML 2002, vol. 19 (2002)
6. Mann, G.S., McCallum, A.: Generalized expectation criteria for semi-supervised learning of conditional random fields. In: Proc. of ACL 2008: HLT, June 2008, pp. 870–878 (2008)
7. Jansche, M.: Maximum expected f-measure training of logistic regression models. In: Proc. of HLT/EMNLP 2005, October 2005, pp. 692–699 (2005)
8. Sarawagi, S., Cohen, W.W.: Semi-markov conditional random fields for information extraction. In: NIPS, vol. 17, pp. 1185–1192 (2004)
9. Altun, Y., Smola, A.J., Hofmann, T.: Exponential families for conditional random fields. In: Proc. of UAI 2004, pp. 2–9 (2004)
10. Lafferty, J., Zhu, X., Liu, Y.: Kernel conditional random fields: Representation and clique selection. In: Proc. of ICML 2004 (2004)

Mining Graph Evolution Rules

Michele Berlingerio[1], Francesco Bonchi[2], Björn Bringmann[3], and Aristides Gionis[2]

[1] ISTI - CNR, Pisa, Italy
[2] Yahoo! Research, Barcelona, Spain
[3] Katholieke Universiteit Leuven, Belgium

Abstract. In this paper we introduce *graph-evolution rules*, a novel type of frequency-based pattern that describe the evolution of large networks over time, at a local level. Given a sequence of snapshots of an evolving graph, we aim at discovering rules describing the local changes occurring in it. Adopting a definition of support based on *minimum image* we study the problem of extracting patterns whose frequency is larger than a minimum support threshold. Then, similar to the classical association rules framework, we derive graph-evolution rules from frequent patterns that satisfy a given minimum confidence constraint. We discuss merits and limits of alternative definitions of support and confidence, justifying the chosen framework. To evaluate our approach we devise *GERM* (Graph Evolution Rule Miner), an algorithm to mine all graph-evolution rules whose support and confidence are greater than given thresholds. The algorithm is applied to analyze four large real-world networks (i.e., two social networks, and two co-authorship networks from bibliographic data), using different time granularities. Our extensive experimentation confirms the feasibility and utility of the presented approach. It further shows that different kinds of networks exhibit different evolution rules, suggesting the usage of these local patterns to globally discriminate different kind of networks.

1 Introduction

With the increasing availability of large social-network data, the study of the temporal evolution of graphs is receiving a growing attention. While most research so far has been devoted to analyze the change of global properties of evolving networks, such as the diameter or the clustering coefficient, not much work has been done to study graph evolution at a microscopic level. In this paper, we consider the problem of searching for patterns that indicate local, structural changes in dynamic graphs. Mining for such local patterns is a computationally challenging task that can provide further insight into the increasing amount of evolving-network data.

Following a frequent pattern-mining approach, we introduce the problem of extracting *Graph Evolution Rules* (*GER*), which are rules that satisfy given constraints of minimum support and confidence in evolving graphs. An example

W. Buntine et al. (Eds.): ECML PKDD 2009, Part I, LNAI 5781, pp. 115–130, 2009.

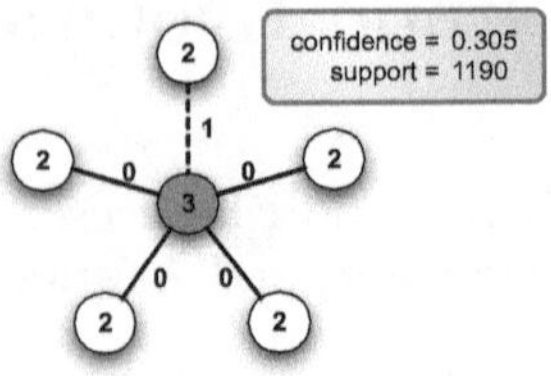

Fig. 1. A Graph Evolution Rule extracted from the DBLP co-authorship network

of a real *GER* extracted form the DBLP co-authorship network is given in Fig. 1: nodes are authors, with an edge between two nodes if they co-authored a paper.

In this specific example, the node labels represent a class of degree of the node: the higher the label the higher the degree of the node. It is important to note that the label refers to the degree of the node in the input graph, not in the rule. In particular the label 3 indicates a node with degree > 50. In general, node labels may represent any property of a node. The labels on the edges instead are more important as they represent the (relative) year in which the first collaboration between two authors was established. Intuitively (later we provide all the needed definitions) the rule might be read as a sort of local evidence of *preferential attachment*, as it shows a researcher with a large degree (label 3) that at time t is connected to four medium degree researchers (labels 2), and that at time $t+1$ will be connected to another medium degree researcher. The definition, extraction and subsequent empirical analysis of such *Graph Evolution Rules* (*GER*) constitute the main body of our work.

The remainder of the paper is organized as follows: Section 2 describes the problem under investigation and defines the novel kind of pattern we are interested in. In Section 3 we describe the details of our algorithm. We report on our experimental results in Section 4 and present related work in Section 5. Finally, in Section 6 we discuss possible future research directions and in Section 7 we provide our conclusions.

2 Patterns of Graph Evolution

2.1 Time-Evolving Graphs

We start by describing how we *conceptually* represent an evolving graph, and subsequently discuss how to *actually* represent the graph in a more compact format. As usual the terminology $G = (V, E, \lambda)$ is used to denote a graph G over a set of nodes V and edges $E \subseteq V \times V$, with a labeling function $\lambda : V \cup E \mapsto \Sigma$, assigning to nodes and edges labels from an alphabet Σ. These labels represent properties, and for simplicity we assume that they do not change with time. As an example, in a social network where nodes model its members, node properties may be *gender*, *country*, *college*, etc., while an edge property can be the kind of connection between two users. The evolution of the graph over time is conceptually represented by a series of undirected graphs $G_1, \ldots, G_T$, so that $G_t = (V_t, E_t)$ represents the graph at time t. Since $G_1, \ldots, G_T$ represent different

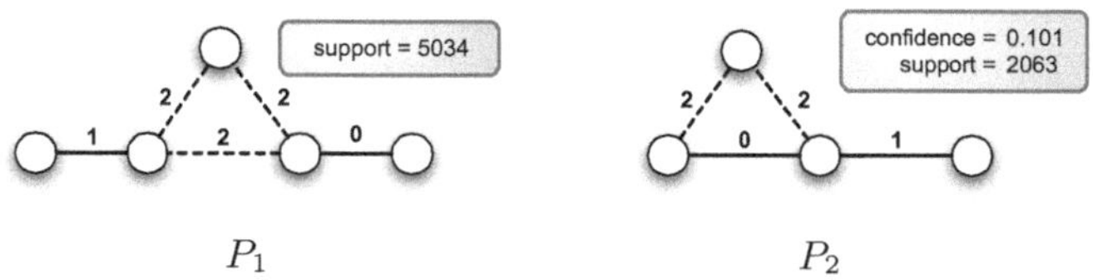

Fig. 2. Relative time patterns extracted from two different samples of the DBLP co-authorship network: respectively 1992-2002 for (P_1), and 2005-2007 (P_2). Dataset details are given in Sec. 4.1.

snapshots of the same graph, we have $V_t \subseteq V$ and $E_t \subseteq E$. For simplicity of presentation, we assume that as the graph evolves, nodes and edges are only added and never deleted: i.e., $V_1 \subseteq V_2 \subseteq \ldots V_T$ and $E_1 \subseteq E_2 \subseteq \ldots E_T$.

It is worth noting that the number of edge deletions in social networks is so small to be negligible when analyzing the temporal evolution of networks. However, in our framework we can handle also deletions by slightly changing the matching operator as described in Section 6.

Our mining algorithm represents the dataset by simply collapsing all the snapshots $G_1, \ldots, G_T$ in one undirected graph G, in which edges are *time-stamped* with their first appearance. Thus, we have $G = (V, E)$ with $V = \bigcup_{t=1}^{T} V_t = V_T$ and $E = \bigcup_{t=1}^{T} E_t = E_T$. To each edge $e = (u, v)$ a time-stamp $t(e) = \arg\min_j\{E_j \mid e \in E_j\}$ is assigned. Note that time-stamps on the nodes may be ignored as a node always comes with its first edge and hence this information is implicitly kept in edge time-stamps. Overall, a time-evolving graph is described as $G = (V, E, t, \lambda)$, with t assigning time-stamps to the set of edges E.

2.2 Patterns

Consider a time-evolving graph G, as defined above. Intuitively a pattern P of G is a subgraph of G that in addition to matching edges of G also matches their time-stamps, and if present, the properties on the nodes and edges of G.

Definition 1 (Absolute-time pattern)
Let $G = (V, E, t, \lambda)$ and $P = (V_P, E_P, t_P, \lambda_P)$ be graphs, where G is the time-evolving dataset and P a pattern. We assume that P is connected. An occurrence *of P in G is a function $\varphi : V_P \mapsto V$ mapping the nodes of P to the nodes of G such that for all $u, v \in V_P$:*

$$\begin{aligned} &i)\ (u,v) \in E_P \Rightarrow (\varphi(u), \varphi(v)) \in E, \\ &ii)\ (u,v) \in E_P \Rightarrow t(\varphi(u), \varphi(v)) = t(u,v),\ \text{and} \\ &iii)\ \lambda_P(v) = \lambda(\varphi(v)) \wedge \lambda_P((u,v)) = \lambda((\varphi(u), \varphi(v))) \end{aligned}$$

In case no labels are present for edges or nodes, the last condition (iii) is ignored. Two examples of patterns from the DBLP co-authorship network are shown in Fig. 2. Those examples motivate us to make two important decisions. First, since our goal in this paper is to study patterns of evolution we naturally focus on patterns that refer to more than one snapshots such as the examples in Fig. 2.

In other terms we are not interested in patterns where all edges have the same time-stamp. The second decision is based on the following observation. Consider pattern P_1: arguably, the essence of the pattern is the fact that two distinct pairs of connected authors, one collaboration created at time 0, and one at time 1, are later (at time 2) connected by a collaboration involving one author from each pair, plus a third author. We would like to account for an occurrence of that event even if it was taking place at times, say, 16, 17 and 18. To capture this intuition we define *relative-time patterns.*

Definition 2 (Relative-time Pattern). *Let G and P be a graph and pattern as in Definition 1. We say that P* occurs in G at relative time *if there exists a $\Delta \in \mathbb{R}$ and a function $\varphi : V_P \mapsto V$ mapping the nodes of P to the nodes in G such that $\forall u, v \in V_P$*

$$
\begin{aligned}
&i)\ (u,v) \in E_P \Rightarrow (\varphi(u), \varphi(v)) \in E, \\
&ii)\ (u,v) \in E_P \Rightarrow t(\varphi(u), \varphi(v)) = t(u,v) + \Delta, \text{ and} \\
&iii)\ \lambda_P(v) = \lambda(\varphi(v)) \wedge \lambda_P((u,v)) = \lambda((\varphi(u), \varphi(v)))
\end{aligned}
$$

The difference between Definitions 1 and 2 is only in the second condition. As a result of Definition 2, we obtain naturally forming equivalence classes of structurally isomorphic relative time patterns that differ only by a constant on their edge time-stamps. To avoid the resulting redundancies in the search space of all relative time patterns we only pick one representative pattern for each equivalence class, namely *the one where the lowest time-stamp is zero.*

In the remainder of this paper we focus on relative time patterns, as they subsume the absolute time case: they are both more interesting and more challenging to mine.

2.3 Support

Next we discuss the support measure we use. Let $\mathcal{G}_\Sigma$ be the set of all graphs over an alphabet Σ. We define support as a function $\sigma : \mathcal{G}_\Sigma \times \mathcal{G}_\Sigma \mapsto \mathbb{N}$. Given a host-graph G and a pattern P, the value of $\sigma(P, G)$ reflects the support of the pattern in the host-graph.

Defining a concept of support for the single graph setting is a non-trivial task, which has received attention recently [14,8,4,5]. The most important property that a definition of support must satisfy is anti-monotonicity, that is, for all graphs G, P and P', where P is a subgraph of P', it must hold that $\sigma(P, G) \geq \sigma(P', G)$. This property is exploited by pattern miners to prune the search space. Anti-monotonicity holds trivially in the transactional setting, but is more tricky for the single-graph setting. For instance, while the total number of occurrences of a pattern is intuitively a meaningful measure, it is not anti-monotonic. As an example consider Fig. 4(b): the number of occurrences in the host graph Y of the pattern indicated as "body" is 1, while the number of occurrences of its supergraph indicated as "head" is 2, thus violating anti-monotonicity.

A first feasible support measure was proposed in [14] followed by a refinement published in [8]. Both measures rely on solving a maximum independent

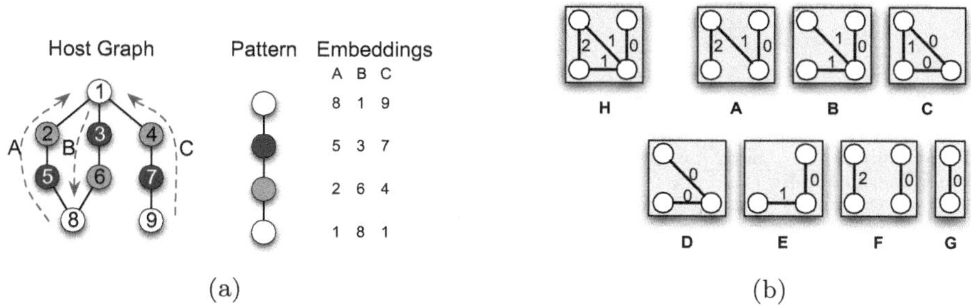

Fig. 3. (a): a graph with three different occurrences of a pattern evaluates to $\sigma = 2$. (b): a graph H with relative edge labels and all possible relative subgraphs A, B, C, D, E, F, G

set problem *MIS* which is NP-complete. We employ the *minimum image based support* measure recently introduced in [4] which does not require solving a *MIS*. This measure is based on the number of unique nodes in the graph $G = (V_G, E_G)$ that a node of the pattern $P = (V_P, E_P)$ is mapped to, and defined as follows:

Definition 3 (Support)

$$\sigma(P, G) = \min_{v \in V_P} \mid \{\varphi_i(v) : \varphi_i \textit{ is an occurrence of } P \textit{ in } G\} \mid$$

By taking the node in P as reference which is mapped to the least number of unique nodes in G, the anti-monotonicity of the measure is ensured. An example of minimum image based support is given in Fig. 3(a). Even if the pattern has 3 occurrences in the host graph, it has support $\sigma = 2$. In fact the lower white node of the pattern can only be mapped to nodes 1 and 8 in the host graph.

The advantage of this definition over other definitions introduced [14,8] is twofold. From a practical point of view it is computationally easier to calculate since it does not require the computation of all possible occurrences of a pattern-graph in the host-graph. Additionally it does not require to solve a maximal independent set problem for each candidate pattern. From a theoretical perspective we know that this definition is an upper bound for the according overlap based definitions [4,8]. Hence the support according to this definition is closer to the real number of occurrences in the graph.

2.4 Rules and Confidence Measure

The support of a pattern can provide insight into how often such an event may happen compared to other specific changes, but not how likely is a certain sequence of steps. To acquire this information we need to decompose a pattern into the particular steps and subsequently determine the *confidence* for each transition. Each step can be considered as a rule *body*→*head* with both *body* and *head* being patterns as defined in the previous section. Unfortunately, this does not yet solve our problem, but rather introduces two important questions:

measure	head	body	confidence
minimum image support	1	1	1.00
number of occurrences	2	6	0.33

(a)

measure	head	body	confidence
minimum image support	1	1	1.00
number of occurrences	2	1	2.00

(b)

Fig. 4. Two example host-graphs X and Y illustrating different problems with support and confidence notions

1. How to decompose a pattern into *head* and *body*?
2. What are reasonable definitions of confidence?

Regarding the decomposition consider pattern H in Fig. 3(b). An occurrence of H implies an occurrence of all its sub-patterns $A - G$. Similarly to the definition of association rules all $A - G$ can be considered candidate-*body* in order to form a graph evolution rule with pattern H as *head.* Fortunately, most of those possibilities can be discarded immediately. First, we are interested in evolution and hence only care about rules describing edges emerging in the future. This allows us to discard bodies A, C, D, E, and F thus only leaving B and G. Furthermore, the step should be as small as possible to allow for a high granularity wherefore we would drop candidate-body G in the example, leaving B as body for the head H. Following the same reasoning, G would be the only choice as body for B as head. Similar the other rules in the example are $E \rightarrow A, D \rightarrow C, G \rightarrow E$. The natural body thus would be the head discarding all the edges from the last time-step of the target-pattern. More formally:

Definition 4 (Graph Evolution Rule). *Given a pattern head P_H the body P_B is defined as: $E_B = \{e \in E_H \mid t(e) < \max_{e^* \in E_H}(t(e^*))\}$ and $V_B = \{v \in V_H \mid deg(v, E_B) > 0\}$, where $deg(v, E_B)$ denotes the degree of v with respect to the edges in E_B. Moreover we constrain P_B to be* connected. *Finally, the support of a graph evolution rule is the support of its head.*

This definition yields a unique body for each head and therefore a unique confidence value for each head. This allows us to represent the rules by the head only. Note that the definition disallows disconnected graphs as body due to the lack of a support-definition for disconnected graphs. As a consequence *not all frequent patterns can be decomposed into graph evolution rules.*

Consider for instance pattern P_1 in Fig. 2: after removing all edges with the highest time-stamp, and discarding disconnected nodes, the graph that remains still contains two disconnected components (the one-edge component with label 1, and the one with label 0). Since the support is not defined for such disconnected pattern, P_1 can not be decomposed to be a *GER*. On the other hand, P_2 can be decomposed: in fact after removing all edges with the maximum time-stamp, and subsequently the disconnected node, we obtain a connected graph that will become the body of the rule for which P_2 is the head. Note that a

GER can be represented in two different ways: either explicitly as two patterns (*body*→*head*), or implicitly by representing only the head as P_2 in Fig. 2. This is possible since there is a unique body for each head.

Finally, we have to choose a reasonable definition of *confidence* of a rule. Following the classic association rules framework, a first choice is to adopt the ratio of head and body supports as confidence. With the support being anti-monotonic this yields a confidence value which is guaranteed to be between zero and one. However, Fig. 4(a) shows that this definition may in some cases lack a reasonable semantic interpretation. In the upper host-graph X we find three possible ways to close a triangle given the edges from time-stamp 7. The confidence of 1 suggests that all of these will close to form triangles, while the graph shows that only one actually does. To overcome this counterintuitive result, we investigated if the ratio of number of occurrences of head and body can be employed to solve this issue. While this definition of confidence allows for more reasonable semantics for the case in Fig. 4(a), it has the clear disadvantage that, due to the lack of anti-monotonicity, it may yield confidence values larger than 1, as in Fig. 4(b). In our experiments we compare the two alternative definitions showing that the minimum-image-based support is an effective and useful concept, while the occurrence-based definition has unpredictable behavior. Moreover, while the support is already available as it is computed for extracting the frequent patterns, the occurrence based confidence needs a separate and costly computation.

3 Mining Graph Evolution Rules

GERM is an adaptation of the algorithm in [4], which was devised to prove the feasibility of the minimum image based support measure, and which in turn, was an adaptation of gSpan [22]. Thus, *GERM* inherits the main characteristics from those algorithms. In particular, *GERM* is based on a DFS traversal of the search space, which leads to very low memory requirements. Indeed, in all the experiments that we performed the memory consumption was negligible.

We next describe in detail how to adapt gSpan to *GERM* whereas the main changes are in the *SubgraphMining* method shown as Algorithm 1. The first key point is that we mine patterns in large single graphs, while gSpan was developed to extract patterns from sets of graphs. The part most involved in adapting gSpan is the *support computation* in line 7. Thus we start from the implementation of [4], where gSpan support calculation is replaced by the minimum image based support computation, without the need for changing the core of the algorithm.

One of the key elements in gSpan is the use of the *minimum DFS code*, which is a canonical form introduced to avoid multiple generations of the same pattern.

We need to change this canonical form in order to enable *GERM* to mine patterns with relative time-stamps (cf. line 1). As explained after Definition 2, we only want one representative pattern per equivalence class, namely the

Algorithm 1. *SubgraphMining*($\mathbb{GS}, \mathbb{S}, s$)

```
if s ≠ min(s) then return // using our canonical form
S ← S ∪ s
generate all s' potential children with one edge growth
Enumerate(s)
for all c, c is s' child do
    // using definition 3 based on definition 1 or definition 2
    if support(c) ≥ minSupp then
        s ← c
        SubgraphMining(GS, S, s)
```

one with the lowest time-stamp being zero. This is achieved by modifying the canonical form such that the first edge in the canonical form is always the one with the lowest time-stamp, as compared to gSpan where the highest label is used as a starting node for the canonical form. Any pattern grown from such a pattern by extending the canonical form will have the same lowest time-stamp, which we set to zero by a simple constraint on the first edge. Hence we guarantee to extract only one pattern per equivalence class which dramatically increases performance and eliminates redundancy in the output.

Note that when matching a pattern to the host-graph we implicitly fix a value of Δ, representing the time gap between the pattern and the host graph. In order to complete the match all remaining edges must adhere to this value of Δ. If all the edges match with the Δ set when matching the first edge, the pattern is discovered to match the host-graph with that value of Δ.

Another important issue is to be able to deal with large real-world graphs, in which several nodes have high degree (the degree distribution in our datasets follows a power law). In typical applications of frequent-subgraph mining in the transactional setting, such as biology and chemistry, the graphs are typically of small size and they are not high-degree nodes. Dealing with large graphs and high degrees give rise to increased computational complexity of the search. In particular, having nodes with large degree increases the possible combinations that have to be evaluated for each subgraph-isomorphism test. We thus equip *GERM* with a user-defined constraint specifying the maximum number of edges in a pattern. This constraint allows to deal more efficiently with the DFS strategy by reducing the search space. Our experiments confirm that the total running time is much more influenced by the maximum-edge constraint than by the minimum support threshold.

4 Experimental Results

In this section, we report our experimental analysis. The *GERM* algorithm was implemented in C++. All the experiments were conducted on a Linux cluster equipped with 8 Intel Xeon processors at 1.8Ghz and 16Gb of RAM.

Table 1. Dataset statistics: Number of nodes and edges and resulting average degree for the total graph as well as for the largest connected component (LCC) out of all connected components (CC). Further the growth rate in terms of edges: total growth as ratio between the graph size at the final and the initial time-stamps, and average growth rate per time-stamp.

							LCC			Growth Rates	
Dataset	**Date**	**|V|**	**|E|**	**avg deg**	**T**	**#CC**	**|V|**	**|E|**	**avg deg**	**total**	**avg**
flickr-month	03-05	147	241	1.64	24	16	74	182	2.43	60347	2.832
flickr-week	02-05	149	246	1.64	76	16	76	186	2.45	246331	0.241
y!360-month	04-05	177	205	1.16	10	17	110	155	1.40	68470	5.150
y!360-week	04-05	177	205	1.16	41	17	110	155	1.40	68470	0.834
arxiv92-01	92-01	709	289	4.08	10	6	49	260	5.32	803	1.691
dblp92-02	92-02	129	277	2.15	11	13	83	220	2.63	25	0.408
dblp03-05	03-05	109	233	2.15	3	14	53	153	2.88	3	0.871
dblp05-07	05-07	135	290	2.15	3	16	72	201	2.76	3	0.749
		×1000	×1000			×1000	×1000	×1000			

4.1 Datasets

We conducted experiments on four real-world datasets: two social networks (Flickr and Y!360) and two bibliographic networks (DBLP and arXiv). Table 1 reports statistics on the resulting graphs.

Flickr (`http://www.flickr.com/`)**:** Flickr is a popular photo-sharing portal. We sampled a set of Flickr users with edges representing mutual friendship and edge time-stamp the moment when the bidirectional contact was established. We generated one graph with monthly and one with weekly granularity.

Y!360 (`http://360.yahoo.com/`)**:** Yahoo! 360° is an online service for blogging. Again we sampled a set of users and proceed as in the Fickr dataset. In this case the monthly and weekly datasets contain exactly the same time period.

DBLP (`http://www.informatik.uni-trier.de/~ley/db/`)**:** This dataset is based on a recent snapshot of DBLP, which has yearly time granularity. Vertices represent authors and edges represent the co-authorship relation. The edge time stamps represent the year of the first co-authorship. Three different graph snapshots were extracted, for three different years.

arXiv (`http://arxiv.org/`)**:** Another co-authorship graph dataset, extracted from the arXiv repository. The resulting graphs *arxiv92-01* contain co-authorship relations that emerged in the years 1992 to 2001 with a yearly time granularity.

As discussed in Section 2, our framework allows to have vertex and edge labels that represent additional information. We experiment with node labels that are based on two graph-theoretic measures: the *degree* and the *closeness centrality*. These measures change as the graph evolves. To obtain static labels the measures are computed on the whole graph, corresponding to the last time stamp and then they are discretized in 5 bins.

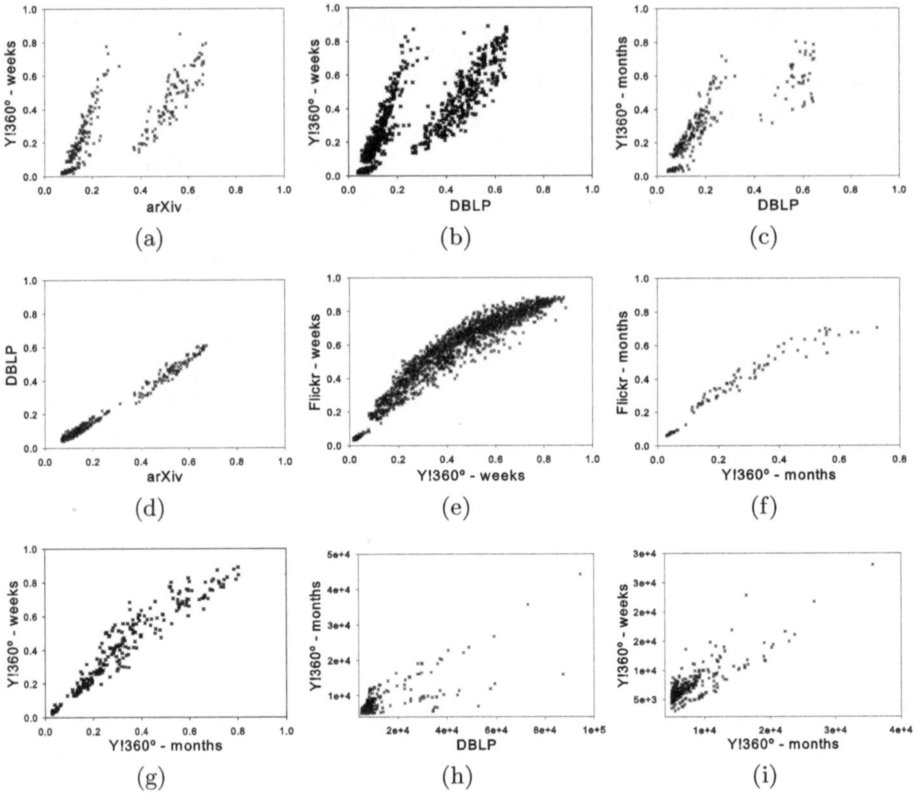

Fig. 5. (a)–(g): comparison of confidence of graph evolution rules in different networks. (h) and (i): comparison of support of patterns in different networks.

4.2 Results

We analyze the experimental results with regard to the following questions:

Q1 Do the extracted patterns and rules characterize the studied network?
Q2 Do different time granularities influence the confidence of the rules?
Q3 How do the different confidence definitions compare?
Q4 How do the parameters and the type of dataset influence the number of derivable rules, the number of patterns obtained, and the running time?

Q1: Discriminative analysis. Fig. 5 compares different pairs of datasets in terms of the confidence of extracted rules. For each pair-wise comparison we show the scatter plot of the confidence of the rules that are (*i*) most frequent in each dataset, and (*ii*) common in both datasets. The plots allow for several interesting observations. First, Fig. 5(a), (b) and (c) show that the confidence of the extracted rules are different between a co-authorship network (arXiv and DBLP) and a social network (Y!360).[1] On the other hand, the confidence of

[1] Using Flickr instead of Y!360 gives similar results.

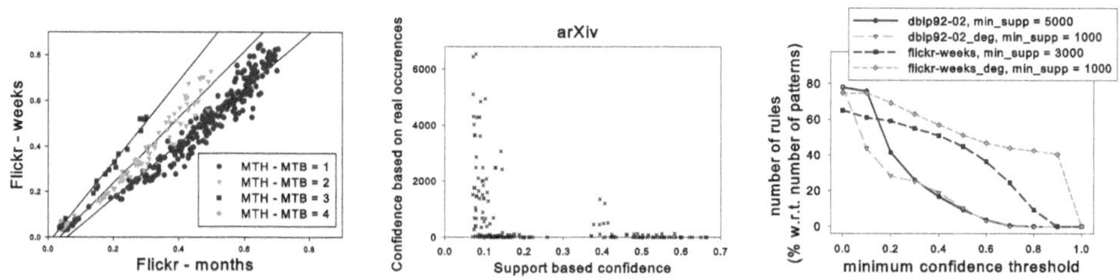

(a) (b) (c)

Fig. 6. (a): confidence comparison between monthly and weekly granularity. (b): scatter plot comparing the two different definitions of confidence discussed in Section 2.4. (c) number of valid rules as percentage of the number of frequent patterns, for varying confidence.

rules are similar between the two co-authorship networks (figure (d)) and the two social networks (figures (e)-(g)). Thus, the plots confirm our claim that graph evolution rules characterize the different types of networks.

Fig. 5 (h) and (i) compare the same two datasets as in figures (c) and (g), but using the measure of support instead of confidence. One notices that the measure of support cannot be used to characterize different types of networks.

Q2: Time-granularity. Fig. 6(a) is the scatter plot of confidence of rules extracted from the same network but with different time granularity. The colors/shapes in the plot correspond to the difference between maximum time stamp on an edge in the head (MTH) and maximum time stamp on an edge in the body (MTB) of the rule. We notice that the rules form three clear clusters (shown with different colors/shapes and the corresponding regression lines) that correspond to the different between the maximum time stamp in the the head and the body of each rule. In particular, we see that rules that have higher weekly confidence than monthly correspond to a larger difference MTH$-$MTB. This observation can be explained if we think about confidence as prediction: the difference MTH$-$MTB can be thought as the temporal gap that must be bridged by a prediction task, and clearly predicting further in the future is more difficult (i.e., lower confidence).

Q3: Confidence measures. Fig. 6(b) shows that the two confidence measures disagree. A more thorough investigation shows that all the rules with an occurrence-based confidence exceeding 200 have the most simple body: one single edge. Furthermore, all those rules span 3 or 4 time-steps from body to head. Given that all those rules share the same simplistic body, which can be matched anywhere, a prediction task, especially at 3 or 4 steps into the future, is doomed to fail. On the other hand, the support-based confidence, assigns score below 0.2 to all rules with the simplistic body – declaring them almost meaningless – thus indicating that support-based confidence is a more appropriate measure to use.

Q4: Influence of parameters. We performed further experimentation on the number of extracted rules and the running time of the algorithm. Fig. 6(c) shows the number of valid graph evolution rules as percentage of the number of frequent

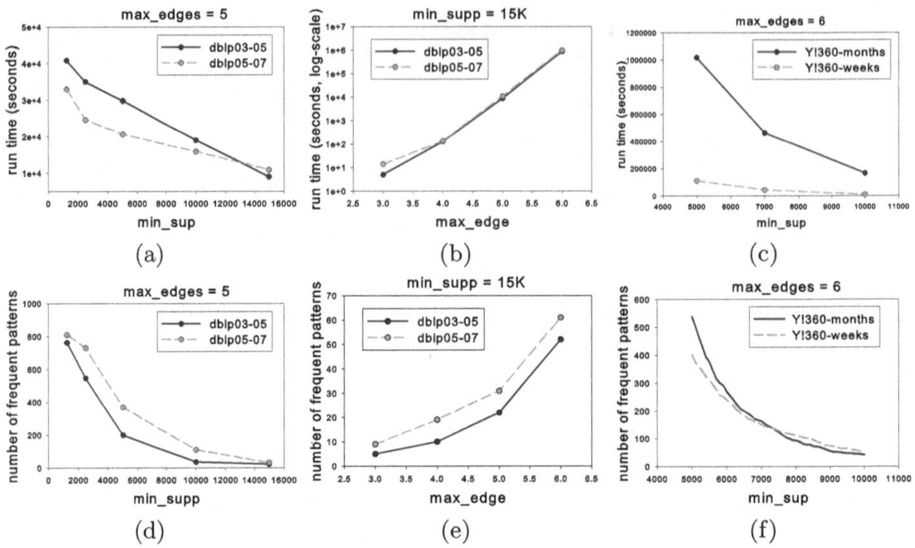

Fig. 7. Running time and number of patterns found with varying *min. support* and *max. edge* thresholds

patterns found for various thresholds of minimum confidence. This is done on one bibliographic and one social network, with and without node labels. In all cases, the number of rules is close to 80% of the number of frequent patterns. The results reaffirm the observation from Fig. 6(a), namely, that rules extracted from a dataset with weekly granularity have higher confidence than rules extracted from a dataset with yearly granularity.

Fig. 7(e) shows that the number of extracted patterns grows as a function of the number of edges allowed in the pattern. This behavior is expected as the number of possible graphs increases exponentially with the number of edges. Fig. 7(d) shows another typical result: lowering the support threshold allows for more complex patterns that contain more edges, and thus a similar increase in the number of extracted patterns.

Fig 7(a)-(b) show that the running time is affected more by the maximum-edge than the minimum-support constraint. While the increase is almost linear with decreasing minimum support, the running time grows exponentially with an increasing maximum edge size.

A more interesting observation can be made from Fig. 7(c) and (f), in which we compare the Y!360 graph for the two different time granularities. The weekly graph has 41 edge labels and is more *diverse* than the monthly graph, which has only 10. Comparing the two datasets, we see that while the running times are very different, the number of extracted patterns is almost the same. The explanation of this apparent discrepancy is the following: With respect to the number of patterns, notice that, on one hand, more edge labels allow for more patterns to be found, on the other hand, less edge labels allow patterns to be found more repeatedly. Thus for a fixed number of edges, there are more high-support patterns in the graph with

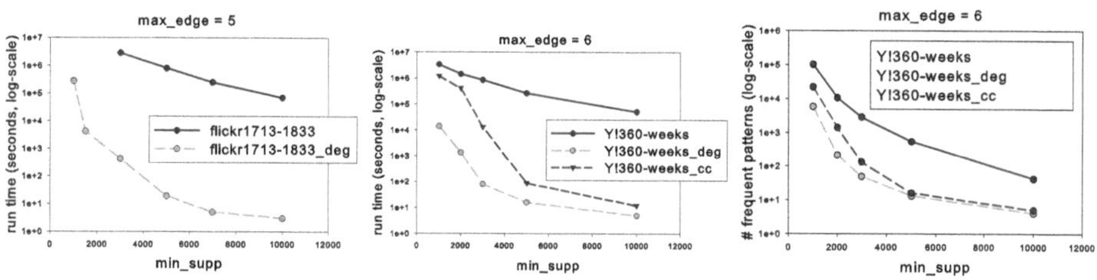

(a) (b) (c)

Fig. 8. Running time and number of patterns found on networks with labelled nodes with varying level of minimum support

the few edge labels, and more low-support patterns in the graph with the many edge labels. With respect to the running times, more patterns of smaller size can be found in the graph with the many edge labels, and for those small patterns the subgraph-isomorphism problem is easier to solve. Furthermore, it is easier to encounter a non-matching edge earlier, thus being able to terminate earlier the search-branch for the subgraph-isomorphism.

Similar results hold for graphs with labelled nodes, as shown in Fig. 8. However, in the case of node labels, the introduced diversity has the effect of reducing the number of patterns found, and the running time.

5 Related Work

Several papers have focused on the global evolution of networks. For instance, Backstrom et al. [2] studied the evolution of communities in social networks, and Leskovec et al. [16] discovered the shrinking diameter phenomena on time-evolving networks. On the other hand, studying network evolution at a more local level, Leskovec et al. [15] used a methodology based on the maximum-likelihood principle and they showed that edge locality plays a critical role in the evolution of networks.

Other recent papers present algorithmic tools for the analysis of evolving networks. Tantipathananandh et al. [20] focus on assessing the community affiliation of users and how it changes over time. Sun et al. [18], apply the MDL principle to the discovery of communities in dynamic networks. The main difference of the work of Sun et al. [18] from previous work such as [1,19] is that they develop a parameter-free framework. However, as in [20], the focus lies on identifying approximate clusters of users and their temporal changes. Ferlez et al. [7] use the MDL principle for monitoring the evolution of a network.

Desikan and Srivastava [6] study the problem of mining temporally evolving web graphs. Three levels of interest are defined: single node, subgraphs and whole graph analysis, each of them requiring different techniques. Inokuchi and Washio [11] propose a fast method to mine frequent subsequences from graph-sequence data defining a formalism to represent changes of subgraphs over time. However the time in which the changes take place is not specified in the patterns. Liu et al. [17] identify subgraphs changing over time by means of vertex-importance scores and vertex-closeness changes in subsequent snapshots of the graphs. The most relevant subgraphs are hence not the most frequent, but the most significant based on the two defined measures. The paper that is most related to our work is the one by Borgwardt et al. [3] who represent the history of an edge as a sequence of 0's and 1's representing the absence and presence of the edge respectively. Then conventional graph-mining techniques are applied to mine frequent patterns. However, there are several differences to our approach. First, the employed mining algorithm GREW is not complete, but heuristic. Further, the overlap-based support measure used requires solving an maximal independent set problem for which a greedy algorithm is used. Another computational issue with their approach is the extension of an edge in the so-called inter-asynchronous FDS case. Accordingly the size of the networks analyzed in the paper is rather small.

Various proposals for mining frequent patterns in the single graph context [14,21,8,4] were discussed in Section 2. A recent paper by Calders et al. [5] introduces a new measure named *minimum clique partition*, which analogous to the maximal independent set is based on the notion of an overlap graph and thus requires solving an NP-complete problem. They prove that support measures based maximal independent set and minimum clique partition are the minimal and the maximal possible meaningful overlap measures, and show that [12] introduced a function which is sandwiched between these two measures; computable in polynomial time. However, any of those measures requires computing an overlap graph for each candidate pattern, which is a costly operation in itself due to requiring enumerating all occurrences of a pattern.

6 Extensions and Future Work

In this section we discuss briefly how to relax some of the restrictions of our problem definition.

Consider first the pattern H in Fig. 3(b). Imagine that, for a particular dataset, it is the case that when there is a star of size 3 an edge between two peripheral nodes appear. Pattern H captures partly this phenomenon, but is too "specific" as it emphasizes that the star was formed in particular time instances before the appearance of the last edge. A more general pattern would be to replace the time-stamp of the last edge with T, and the time-stamp of all the edges in the star with the constraint "$< T$", which will have to be satisfied when tested with the time-stamps of the host graph. We plan to extend our algorithm to experiment with this idea as a continuation of our work.

For sake of presentation, in Section 2 we assumed that graphs can only grow in time. However, our approach can be easily extended to handle edge-deletions if an edge can appear and disappear at most once. The extension would consider two time-stamps t_I (time of insertion) and t_D (time of deletion) on each edge instead of the single time t. By modifying definitions 1 and 2 condition (ii) to $\forall (u,v) \in E_P$ it is $t_I(\varphi(u),\varphi(v)) = t_I(u,v) + \Delta$ and $t_D(\varphi(u),\varphi(v)) = t_D(u,v) + \Delta$.

We did not implement the above matching since two out of four datasets (arXiv and DBLP) are by definition only growing (no deletions), and deletions are rare in the other two. As future work we plan to incorporate deletions and study networks with a higher likelihood of such events.

In our approach, we have not considered node or edge relabelling so far. Considering node and edge relabeling is very interesting, as in many graphs, such as social networks, the properties of nodes and edges change over time. For example, in social-network analysis it would be interesting to study the change of leadership in communities and its effects.

Besides all the above, which are possible extensions to the type of patterns we are able to mine, we would like to go further by leveraging the concept of rule confidence, and designing a paradigm that will allow us to *predict* graph evolution, and that, together with *GERM*, will provide helpful tools analyzing datasets of dynamic graphs.

7 Conclusions

Following a frequent pattern mining approach, we defined relative time patterns and introduced introduced the problem of extracting *Graph Evolution Rules*, satisfying given constraints of minimum support and confidence, from an evolving input graph. While providing the problem definition we discussed alternative definitions of support and confidence, their merits and limits. We implemented *GERM* an effective solution to mine Graph Evolution Rules, and extensively test it on four large real-world networks (two social networks, and two co-authorship networks), using different time granularities. Our experiments confirmed the feasibility and the utility of our framework and allowed for interesting insights. In particular we showed that Graph Evolution Rules with their associated concept of confidence, indeed characterize the different types of networks.

Availability. The executable code of the GERM software is freely available at: `http://www-kdd.isti.cnr.it/~berlingerio/so/gm/`.

References

1. Aggarwal, C.C., Yu, P.S.: Online analysis of community evolution in data streams. In: SDM (2005)
2. Backstrom, L., Huttenlocher, D., Kleinberg, J., Lan, X.: Group formation in large social networks: membership, growth, and evolution. In: KDD (2006)
3. Borgwardt, K.M., Kriegel, H.-P., Wackersreuther, P.: Pattern mining in frequent dynamic subgraphs. In: ICDM (2006)
4. Bringmann, B., Nijssen, S.: What is frequent in a single graph? In: Washio, T., Suzuki, E., Ting, K.M., Inokuchi, A. (eds.) PAKDD 2008. LNCS (LNAI), vol. 5012, pp. 858–863. Springer, Heidelberg (2008)
5. Calders, T., Ramon, J., Van Dyck, D.: Anti-monotonic overlap-graph support measures. In: ICDM (2008)

6. Desikan, P., Srivastava, J.: Mining temporally changing web usage graphs. In: Mobasher, B., Nasraoui, O., Liu, B., Masand, B. (eds.) WebKDD 2004. LNCS (LNAI), vol. 3932, pp. 1–17. Springer, Heidelberg (2006)
7. Ferlez, J., Faloutsos, C., Leskovec, J., Mladenic, D., Grobelnik, M.: Monitoring network evolution using MDL. In: ICDE (2008)
8. Fiedler, M., Borgelt, C.: Subgraph support in a single graph. In: Workshop on Mining Graphs and Complex Data, MGCS (2007)
9. Hoeffding, W.: Probability inequalities for sums of bounded random variables. Journal of the American Statistical Association 58(301), 13–30 (1963)
10. Holder, L.B., Cook, D.J., Djoko, S.: Substucture discovery in the SUBDUE system. In: AAAI KDD Workshop (1994)
11. Inokuchi, A., Washio, T.: A fast method to mine frequent subsequences from graph sequence data. In: ICDM (2008)
12. Knuth, D.E.: The sandwich theorem. Electronic Journal of Combinatorics 1, 1 (1994)
13. Kuramochi, M., Karypis, G.: Frequent subgraph discovery. In: ICDM (2001)
14. Kuramochi, M., Karypis, G.: Finding frequent patterns in a large sparse graph. Data Mining and Knowledge Discovery 11(3), 243–271 (2005)
15. Leskovec, J., Backstrom, L., Kumar, R., Tomkins, A.: Microscopic evolution of social networks. In: KDD (2008)
16. Leskovec, J., Kleinberg, J.M., Faloutsos, C.: Graphs over time: densification laws, shrinking diameters and possible explanations. In: KDD (2005)
17. Liu, Z., Yu, J.X., Ke, Y., Lin, X., Chen, L.: Spotting significant changing subgraphs in evolving graphs. In: ICDM (2008)
18. Sun, J., Faloutsos, C., Papadimitriou, S., Yu, P.S.: Graphscope: parameter-free mining of large time-evolving graphs. In: KDD (2007)
19. Sun, J., Tao, D., Faloutsos, C.: Beyond streams and graphs: dynamic tensor analysis. In: KDD (2006)
20. Tantipathananandh, C., Berger-Wolf, T., Kempe, D.: A framework for community identification in dynamic social networks. In: KDD (2007)
21. Vanetik, N., Shimony, S.E., Gudes, E.: Support measures for graph data. Data Mining Knowledge Discovery 13(2), 243–260 (2006)
22. Yan, X., Han, J.: gSpan: Graph-based substructure pattern mining. In: ICDM (2002)
23. Zhu, F., Yan, X., Han, J., Yu, P.S.: gPrune: A constraint pushing framework for graph pattern mining. In: Zhou, Z.-H., Li, H., Yang, Q. (eds.) PAKDD 2007. LNCS (LNAI), vol. 4426, pp. 388–400. Springer, Heidelberg (2007)

Parallel Subspace Sampling for Particle Filtering in Dynamic Bayesian Networks

Eva Besada-Portas[1], Sergey M. Plis[1], Jesus M. de la Cruz[2], and Terran Lane[1]

[1] University of New Mexico, Albuquerque NM 87131, USA
[2] Universidad Complutense de Madrid, 28040 Madrid, Spain

Abstract. Monitoring the variables of real world dynamic systems is a difficult task due to their inherent complexity and uncertainty. Particle Filters (PF) perform that task, yielding probability distribution over the unobserved variables. However, they suffer from the curse of dimensionality problem: the number of particles grows exponentially with the dimensionality of the hidden state space. The problem is aggravated when the initial distribution of the variables is not well known, as happens in global localization problems. We present a new parallel PF for systems whose variable dependencies can be factored into a Dynamic Bayesian Network. The new algorithms significantly reduce the number of particles, while independently exploring different subspaces of hidden variables to build particles consistent with past history and measurements. We demonstrate this new PF approach on some complex dynamical system estimation problems, showing that our method successfully localizes and tracks hidden states in cases where traditional PFs fail.

1 Introduction

Estimating the hidden state of a multivariate dynamical system remains a large challenge. While the Dynamic Bayesian Network (DBN) formalism provides us an excellent way to *represent* such systems, performing *inference* in these models remains difficult, and approximations are usually necessary. Among the most popular approximate inference methods for DBNs are *particle filters* (PFs): point-mass approximations to the hidden state distribution, which are updated via Monte Carlo sampling steps [1].

In spite of their popularity, PFs can be sensitive and tricky to apply to new problems. One of the core challenges arises from the familiar curse of dimensionality: in even modest dimensionality spaces, the chances are high that many or most particles will fall into near-zero probability regions of the state space, leading to serious particle depletion and quickly driving the PF off track. This problem is evident, for example, in multi-object tracking tasks [2]. The difficulty is exacerbated by poor initial particle distributions that are unlikely to choose any high-probability particles. The dimensionality/depletion difficulty can be reduced by careful choice of initial particle distribution and sampling and reweighting distributions, but doing so requires extensive domain knowledge engineering.

In this paper, we propose a principled sampling framework to overcome some of the dimensionality drawbacks for PFs applied to DBNs with factored state spaces and multiple conditionally independent observations (e.g. multi-object tracking). Our approach

W. Buntine et al. (Eds.): ECML PKDD 2009, Part I, LNAI 5781, pp. 131–146, 2009.

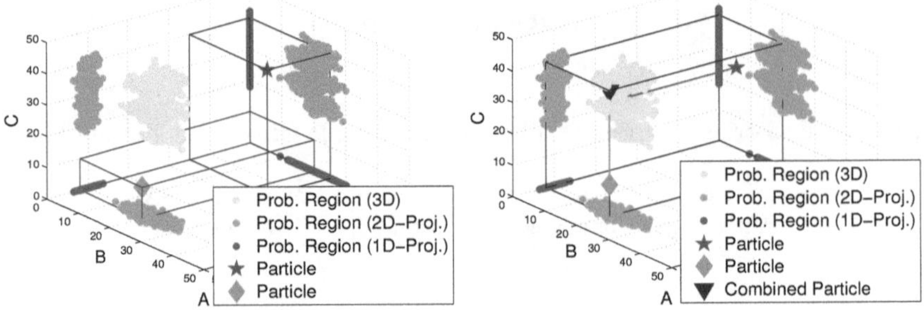

(a) Low probability (b) High probability

Fig. 1. Combining the information of two low probability particles (Figure 1a) to create a higher probability one (Figure 1b)

requires less prior knowledge about domain dynamics and fewer ad hoc corrections than other methods [3,4].

The key insight is displayed in Figure 1a. Here we see a state space of three variables, $\{A,B,C\}$, with the support of the target distribution displayed as a bright yellow cloud centered at approximately $[A=40,B=20,C=30]$. Two particles (red star and green diamond) fall far from the target distribution and are assigned zero probability. However, the red star particle has significant probability in the projection into the $\{C\}$ subspace (dark olive cloud), while the green diamond particle has non-negligible probability in the A,B subspace (dark olive cloud). Our method amounts to carefully combining the useful dimensions of these particles (A and B from the green diamond particle and C from the red star) into a hybrid particle (Figure 1b, blue triangle) that falls within our target distribution. The trick is doing so in a way that correctly accounts for both the observation likelihoods associated with different subspaces and the particles' previous state histories.

Of course, the basic insight is not, itself, new [5,6,7,8]. What we add to previous work is a principled design of the PF importance sampling (IS) and weighted resampling (WR) steps that provides a firm probabilistic foundation for combining such particles. Doing so has three benefits. First, it allows a form of "parallel" filtering that works with separate, lower-dimensional subspaces of variables simultaneously (Section 3). Second, it helps overcome poor particle initialization, improving state localization and convergence when the initial particle sample is far from the target distribution (Section 5). And finally, it allows us to get away with simultaneously fewer particles and less prior knowledge/less sophisticated sampling distributions than existing methods (Section 4).

2 Particle Filter Fundamentals

Our core contribution is to modify the standard PF for DBNs [9] by substituting its proposal distribution for another that lets it sample the subspaces defined for different subsets of hidden variables and introducing the weight update equation related with the new proposal. In this section we introduce an uniform notation and formalism in which to state the standard PF and our modifications to it. In the process, we also introduce a sub-contribution: the "instantaneous" PF, a variation of the standard PF that emerges to estimate only the current state of the hidden variables instead of the state of all the hidden variables at any time.

2.1 Definitions and Notation

In this paper, a capital letter U represents a random variable, a boldface capital letter $\boldsymbol{U}$ – a set of random variables, a lowercase letter u – the specific value of the corresponding random variable U, and a lowercase bold letter $\boldsymbol{u}$ – an assignment of the values to the variables of its set $\boldsymbol{U}$. We also define $\mathcal{P}(\boldsymbol{U})$ as a partition of $\boldsymbol{U}$.

A BN is an annotated directed graph that encodes the probability distribution of a set of random variables $\boldsymbol{V}$. DBNs model the distribution of a sequence of sets of variables. A set of variables belonging to k^{th} time slice is represented as $\boldsymbol{V}_k$ and the total set of variables up to the current time slice t is $\boldsymbol{V}_{0:t}$. To distinguish hidden and observed variables, we use X, $\boldsymbol{X}_t$ and $\boldsymbol{X}_{0:t}$ to denote the former, and Y, $\boldsymbol{Y}_t$ and $\boldsymbol{Y}_{0:t}$ for the latter. The graph is completely defined by the sets of parents of all its variables: $\mathrm{Pa}_k(V)$ represents the subset of the parents of variable V that belong to time slice k and $\mathrm{pa}_k(V)$ their assignment to particular values. Similarly, we also define the set of children of a variable and their assignments by $\mathrm{Ch}_k(V)$ and $\mathrm{ch}_k(V)$.

Probability distributions will be represented as $\mathrm{p}(\cdot)$, $\mathrm{q}(\cdot)$ and $\mathrm{r}(\cdot)$: the first related with the probabilities of the variables of the system and the others with the proposal distributions used to sample the particles from. The operation $a \sim \mathrm{q}(\cdot)$ represents sampling a according to $\mathrm{q}(\cdot)$.

A PF approximates the probability $\mathrm{p}(\boldsymbol{X}|\boldsymbol{y})$ of a set of hidden variables $\boldsymbol{X}$ given the values of the measurements $\boldsymbol{Y}$ by the point mass distribution $\sum_{i=1}^{N} w^{(i)} \delta(\boldsymbol{X} - \boldsymbol{x}^{(i)})$, where $\delta(\cdot)$ is Dirac delta function, $\boldsymbol{x}^{(i)}$ are the values of the variables in $\boldsymbol{X}$ in the i-th particle, $w^{(i)}$ their weights, and N the number of particles. Additionally, $x^{(i)}$, $\boldsymbol{x}_{0:t}^{(i)}$ and $\mathrm{pa}_t^{(i)}(V)$ represent the assignments of variable X, the variables in $\boldsymbol{X}_{0:t}$ and the variables in $\mathrm{Pa}_t(V)$ to the values they have in the i-th particle, and $w_t^{(i)}$ stands for the weight of the particle when we have unrolled the DBN up to the time stamp t.

2.2 Importance Sampling (IS) and Weighted Resampling (WR)

To develop a new PF we can modify the two basic operations that are normally combined to obtain the values of the particles $\boldsymbol{x}^{(i)}$ and the weights $w^{(i)}$ [10]:

- Importance sampling is used to (1) create new particles by means of a proposal distribution $\mathrm{q}(\boldsymbol{X}|\boldsymbol{y})$ that generates their values ($\boldsymbol{x}^{(i)} \sim \mathrm{q}(\boldsymbol{X}|\boldsymbol{y})$) and (2) calculate their weights as $w^{(i)} = \mathrm{p}(\boldsymbol{x}^{(i)}|\boldsymbol{y})/\mathrm{q}(\boldsymbol{x}^{(i)}|\boldsymbol{y})$.
- Weighted resampling is used to redistribute an existing set of weighted particles $(\boldsymbol{x}^{(i)}, w^{(i)})$ according to the resampling function $\mathrm{r}(\boldsymbol{X})$, without changing the approximated distribution. The new set of weighted particles $(\boldsymbol{x}'^{(i)}, w'^{(i)})$ is obtained as $\boldsymbol{x}'^{(i)} = \boldsymbol{x}^{(j)}$ and $w'^{(i)} = w^{(j)}/\mathrm{r}(\boldsymbol{x}^{(j)})$ with $j \sim r(\boldsymbol{x}^{(j)})$.

In short, IS is used to create the particles while WR is carried out to increment the number of the particles in the regions of high interest and reduce them in the others.

2.3 PF for DBN

Note that we have not specified which random variables are included in the sets $\boldsymbol{X}$ and $\boldsymbol{Y}$. Although it is often non stated explicitly in the literature, different PFs emerge from different choices of hidden variables $\boldsymbol{X}$ and observed values $\boldsymbol{y}$. When $\boldsymbol{X}=\boldsymbol{X}_{0:t}$ the PF is responsible of estimating the probability of the trajectory $\boldsymbol{X}_{0:t}$ while when $\boldsymbol{X}=\boldsymbol{X}_t$ is in charge of estimating the probability of the state $\boldsymbol{X}_t$ at the current time slice t. To distinguish the two problems, we name the PF filters that solve the first problem "trajectory" PF and the second "instantaneous". The set of observed variables $\boldsymbol{Y}$ is in both cases $\boldsymbol{Y}=\boldsymbol{Y}_{0:t}$.

The IS operation of both filters can be carried out sequentially exploiting the independence assumption imposed by the structure of the DBN. In the paper, we only consider DBNs whose variables have parents only belonging to the current or previous time slice ($\forall V\in \boldsymbol{V}_t \wedge \forall k \notin \{t,t-1\}\ \mathrm{Pa}_k(V)=\emptyset$). With this restriction, the structure of the DBN factors the joint probability of the set of hidden and observation variables up to time slice t as in Eq. (1). The dependence of each variable on its parents imposes an ancestral ordering in the evaluation of the probabilities that needs to be considered by the PFs.

$$\begin{aligned} \mathrm{p}(\boldsymbol{X}_{0:t},\boldsymbol{Y}_{0:t}) &= \mathrm{p}(\boldsymbol{X}_t,\boldsymbol{Y}_t|\boldsymbol{X}_{0:t-1},\boldsymbol{Y}_{0:t-1})\,\mathrm{p}(\boldsymbol{X}_{0:t-1},\boldsymbol{Y}_{0:t-1}) \\ \mathrm{p}(\boldsymbol{X}_t,\boldsymbol{Y}_t|\boldsymbol{X}_{0:t-1},\boldsymbol{Y}_{0:t-1}) &= \prod_{X\in\boldsymbol{X}_t}\mathrm{p}(X|\mathrm{Pa}_{t-1}(X),\mathrm{Pa}_t(X))\prod_{Y\in\boldsymbol{Y}_t}\mathrm{p}(Y|\mathrm{Pa}_{t-1}(Y),\mathrm{Pa}_t(Y)) \end{aligned} \tag{1}$$

Although the two types of PFs work with different $\boldsymbol{X}$, both are used to obtain the values of the current state variables $\boldsymbol{X}_t$, because the trajectory $\boldsymbol{X}_{0:t}$ includes them. Additionally, for some types of proposals $\mathrm{q}(\boldsymbol{X}|\boldsymbol{y})$, they are equivalent. However, the search space of the first is significantly bigger, and it usually needs more particles, as [11] shows for the case of HMMs.

"Trajectory" PF. Its weights $w_t^{(i)}$ can be calculated with expression (2), which exploits the factorization of the DBN, assumes that $\mathrm{q}(\boldsymbol{x}_{0:t-1}^{(i)}|\boldsymbol{y}_{0:t})=\mathrm{q}(\boldsymbol{x}_{0:t-1}^{(i)}|\boldsymbol{y}_{0:t-1})$ and considers that instead of sampling from $\mathrm{q}(\boldsymbol{x}_t^{(i)}|\boldsymbol{x}_{0:t-1}^{(i)},\boldsymbol{y}_{1:t})$ we can do it from $\mathrm{q}(\boldsymbol{x}_t^{(i)}|\boldsymbol{x}_{0:t-1},\boldsymbol{y}_{1:t})$.

$$\begin{aligned} w_t^{(i)} &= \frac{\mathrm{p}\left(\boldsymbol{x}_{0:t}^{(i)}|\boldsymbol{y}_{0:t}\right)}{\mathrm{q}\left(\boldsymbol{x}_{0:t}^{(i)}|\boldsymbol{y}_{0:t}\right)} \propto \frac{\mathrm{p}\left(\boldsymbol{x}_{0:t}^{(i)},\boldsymbol{y}_{0:t}\right)}{\mathrm{q}\left(\boldsymbol{x}_{0:t}^{(i)}|\boldsymbol{y}_{0:t}\right)} = \frac{\mathrm{p}\left(\boldsymbol{x}_t^{(i)},\boldsymbol{y}_t|\boldsymbol{x}_{0:t-1}^{(i)},\boldsymbol{y}_{0:t-1}\right)}{\mathrm{q}\left(\boldsymbol{x}_t^{(i)}|\boldsymbol{x}_{0:t-1}^{(i)},\boldsymbol{y}_{0:t}\right)}\frac{\mathrm{p}\left(\boldsymbol{x}_{0:t-1}^{(i)},\boldsymbol{y}_{0:t-1}\right)}{\mathrm{q}\left(\boldsymbol{x}_{0:t-1}^{(i)}|\boldsymbol{y}_{0:t-1}\right)} \\ &\propto \frac{\prod_{Y\in\boldsymbol{Y}_t}\mathrm{p}\left(y|\mathrm{pa}_{t-1}^{(i)}(Y),\mathrm{pa}_t^{(i)}(Y)\right)\prod_{X\in\boldsymbol{X}_t}\mathrm{p}\left(x^{(i)}|\mathrm{pa}_{t-1}^{(i)}(X),\mathrm{pa}_t^{(i)}(X)\right)}{\mathrm{q}\left(\boldsymbol{x}_t^{(i)}|\boldsymbol{x}_{0:t-1},\boldsymbol{y}_{1:t}\right)} w_{t-1}^{(i)} \end{aligned} \tag{2}$$

Different proposals and combinations of IS and WR create different "trajectory" PF. The IS step of KLPF [9] uses $\mathrm{q}(\boldsymbol{x}_t^{(i)}|\boldsymbol{x}_{0:t-1},\boldsymbol{y}_{1:t})=\prod_{X\in\boldsymbol{X}_t}\mathrm{p}(x^{(i)}|\mathrm{pa}_{t-1}^{(i)}(X),\mathrm{pa}_t^{(i)}(X))$, and so $w_t^{(i)}=\prod_{Y\in\boldsymbol{Y}_t}\mathrm{p}(y|\mathrm{pa}_{t-1}^{(i)}(Y),\mathrm{pa}_t^{(i)}(Y))w_{t-1}^{(i)}$. According to the ancestral ordering of the variables, for each $X\in\boldsymbol{X}_t$ its value $x^{(i)}\sim\mathrm{p}(X|\mathrm{pa}_{t-1}^{(i)}(X),\mathrm{pa}_t^{(i)}(X))$ and for each $Y\in\boldsymbol{Y}_t$ multiplies $w_t^{(i)}$ by its corresponding likelihood $\mathrm{p}(y|\mathrm{pa}_{t-1}^{(i)}(Y),\mathrm{pa}_t^{(i)}(Y))$. The WR step of KLPF is carried out once all $V \in \boldsymbol{V}_t$ are processed, using $\mathrm{r}(\boldsymbol{x}^{(i)}) = w_t^{(i)}$, what makes the resampled particles have all the same weight. The PF in [12] uses the same factorization and sequential weight updates as KLPF, but it alternates, according to the ancestral ordering, IS for each $X \in \boldsymbol{X}_t$ and WR with $\mathrm{r}(\boldsymbol{x}^{(i)}) = \mathrm{p}(y|\mathrm{pa}_{t-1}^{(i)}(Y), \mathrm{pa}_t^{(i)}(Y))$ after the likelihood of each measurement $Y \in \boldsymbol{Y}_t$ is used to multiply the $w_t^{(i)}$. That is, KLPF updates completely $w_t^{(i)}$ taking into account all the variables of time slice t and does WR after it, while the PF in [12] updates the $w_t^{(i)}$ for each variable and performs WR after each measurement appears in the ancestral ordering. The other "trajectory" PFs listed in Section 1 combine KLPF with other techniques.

"Instantaneous" PF. To the best of authors' knowledge, there are no "instantaneous" PFs designed for general DBNs. However, its $w_t^{(i)}$ can also be calculated by extending the ideas that [11] utilizes for HMMs to the general DBN case. Equation (3), our first contribution in this paper, calculates the weights by exploiting the factorization of the DBN and marginalizing the hidden variables of the previous time slice. To simplify the expression, it is possible to take out of the summation those variables that don't have hidden parents belonging to the previous time slice ($V \in \boldsymbol{V}_t$ s.t. $\mathrm{Pa}_{t-1}(V) \cap \boldsymbol{X}_{t-1} = \emptyset$).

$$
\begin{aligned}
w_t^{(i)} &= \frac{\mathrm{p}\left(\boldsymbol{x}_t^{(i)}|\boldsymbol{y}_{0:t}\right)}{\mathrm{q}\left(\boldsymbol{x}_t^{(i)}|\boldsymbol{y}_{0:t}\right)} \propto \frac{\mathrm{p}\left(\boldsymbol{x}_t^{(i)}, \boldsymbol{y}_t|\boldsymbol{y}_{0:t-1}\right)}{\mathrm{q}\left(\boldsymbol{x}_t^{(i)}|\boldsymbol{y}_{0:t}\right)} \propto \frac{\int \mathrm{p}\left(\boldsymbol{x}_t^{(i)}, \boldsymbol{y}_t, \boldsymbol{x}_{t-1}|\boldsymbol{y}_{0:t-1}\right) d\boldsymbol{x}_{t-1}}{\mathrm{q}\left(\boldsymbol{x}_t^{(i)}|\boldsymbol{y}_{0:t}\right)} \\
&\propto \frac{\int \mathrm{p}\left(\boldsymbol{x}_t^{(i)}, \boldsymbol{y}_t|\boldsymbol{x}_{t-1}, \boldsymbol{y}_{0:t}\right) \mathrm{p}(\boldsymbol{x}_{t_1}|\boldsymbol{y}_{0:t-1})\, d\boldsymbol{x}_{t-1}}{\mathrm{q}\left(\boldsymbol{x}_t^{(i)}|\boldsymbol{y}_{0:t}\right)} \\
&\propto \frac{\sum_{j=1}^{N}\left(w_{t-1}^{(j)} \prod_{Y \in \boldsymbol{Y}_t} p\left(y|\mathrm{pa}_{t-1}^{(j)}(Y), \mathrm{pa}_t^{(i)}(Y)\right) \prod_{X \in \boldsymbol{X}_t} p\left(x^{(i)}|\mathrm{pa}_{t-1}^{(j)}(X), \mathrm{pa}_t^{(i)}(X)\right)\right)}{q\left(\boldsymbol{x}_t^{(i)}|\boldsymbol{y}_{0:t}\right)}
\end{aligned}
\tag{3}
$$

Although a factorization of the proposal distribution is also possible, the sequential way proposed for updating $w_t^{(i)}$ in the "trajectory" PF proposed [12] is no longer valid for the hidden variables that have to stay inside the summation.

3 A Parallel Sampling Framework for DBN

In a case with N hidden variables where there is an observation for each of these variables, the likelihood of a state is a product of the likelihoods of each hidden variable in the state. A likely particle should be likely according to each and all of its variables, otherwise its weight is going to be negligible. An easy situation arises when all hidden variables are independent and their joint probability can be factored into a product of its marginals. In this case, we can simply solve N PF in parallel and obtain a solution. The problem arises when there are inter-dependencies in the hidden state, i.e. we have a DBN of N hidden interacting random variables and their N conditionally independent observations. If our prior is slightly wrong in even one of the hidden variables it is difficult to obtain a particle with a non-negligible weight. Thus a wrong start usually leads to the hidden state representation completely going off track.

Existing approaches to addressing the problem can be roughly split into three categories: 1) tracking only a subset of hidden variables while handling its complement through additional probabilistic techniques [3,4], 2) sequentially alternating IS and WR sub-steps for the variables belonging to each time slice of the DBN [10,12], and (3) parallel creation of particles in subsets of hidden variables with their subsequent recombination by varying means [5,6,7,8].

The PFs presented in Section 2.3, sample the values of the current time hidden variables $\boldsymbol{X}_t$ from the values that their parents have inside the same particle. So the probability of each particle is highly dependent on the values that all the variables have inside that particle and therefore on the initial sampling distribution. In problems with many hidden variables, if no particle is "good" initially or at some point, there is little chance for the presented PFs to recover.

However, the values of some variables of some particles that have zero probability can be inside of the non-zero probability subspace associated with those variables. Additionally, different particles can have different subsets of variables inside different non-zero probability subspaces. This scenario is illustrated by Fig. 1a, where the light yellow region represents the non-zero probability zone of a set of three variables $\{A,B,C\}$, the darker olive surfaces and lines its projections into the different possible subspaces, and the red star and green diamond the position of two particles with zero probability but whose $\{C\}$ and $\{A,B\}$ variables are respectively inside probable subspaces. The core idea of our PF framework consists on using the information of probable areas of different particles to build non-zero probable particles by parallel sampling different subspaces of hidden variables. Figure 1b shows with a blue triangle a probable particle obtained combining the probable regions of the red and green particles.

However, to create probabilistic consistent PFs we need to combine IS and WR steps, and select the $\mathrm{q}(\cdot)$ and $\mathrm{r}(\cdot)$ functions. As the main distinction between the weight update operation of the "trajectory" and "instantaneous" PFs appears in the numerator, we can also use the same proposal for both types of filters. So, the parallel PF framework presented in this paper is based on 1) the proposal presented in the following section, that independently samples in parallel different subsets of hidden variables, and 2) in the calculus of the $w_t^{(i)}$ of the particles considering this proposal and Eq. (2) or (3).

3.1 A Parallel Sampling Proposal

Independently sampling different subspaces of hidden variables belonging to each time slice $\boldsymbol{X}_t$ can be achieved factoring the proposal $\mathrm{q}(\boldsymbol{x}_t^{(i)}|\cdot,\boldsymbol{y}_{1:t})$ that appears in the denominator of Eq. (2) or (3) in as many proposals as subsets $\boldsymbol{X}$ exist in a given disjoint partition $\mathcal{P}(\boldsymbol{X}_t)$ of $\boldsymbol{X}_t$:

$$\mathrm{q}\left(\boldsymbol{x}_t^{(i)}|\cdot,\boldsymbol{y}_{1:t}\right) = \prod_{\boldsymbol{X}\in\mathcal{P}(\boldsymbol{X}_t)} \mathrm{q}\left(\boldsymbol{x}^{(i)}|\cdot,\boldsymbol{y}_{1:t}\right) \tag{4}$$

The sampling proposal of each subset has to select highly probable values of their corresponding variables given all the past information. This behavior can be approximated with a mixture model of the product of the transition priors of all the variables belonging to each subset. However, to increment the flexibility of our PF, we substitute the transition prior by a proposal of each variable given its parents. This lets the user decide whether to use the transition prior ($\mathrm{q}(x|\mathrm{pa}_{t-1}^{(j)}(X),\mathrm{pa}_t^{(i)}(X)) = \mathrm{p}(x|\mathrm{pa}_{t-1}^{(j)}(X),\mathrm{pa}_t^{(i)}(X))$) or another distribution that will let the PF explore other regions of the space. The complete proposal is presented in Eq. (5), where $\alpha_j^{\boldsymbol{X}}$ stands for the weight of each component of the mixture related with the subspace defined by the variables in the subset $\boldsymbol{X}$.

$$\mathrm{q}\left(\boldsymbol{x}_t^{(i)}|\cdot,\boldsymbol{y}_{1:t}\right) = \prod_{\boldsymbol{X}\in\mathcal{P}(\boldsymbol{X}_t)} \left(\sum_{j=1}^{N}\left(\alpha_j^{\boldsymbol{X}}\prod_{X\in\boldsymbol{X}}\mathrm{q}\left(x|\mathrm{pa}_{t-1}^{(j)}(X),\mathrm{pa}_t^{(i)}(X)\right)\right)\right) \tag{5}$$

The selection of the $\alpha_j^{\boldsymbol{X}}$ is fundamental because it let us change the importance of each component of the mixture. To make it consistent with the past measurements we can make it proportional to the product of the likelihoods related with the hidden variables of the subspace defined by the subset $\boldsymbol{X}$. However, to increment the flexibility of the system, we substitute the likelihoods for other distributions that relate the observed variables with their parents as it is presented in equation (6).

$$\alpha_j^{\boldsymbol{X}} = \prod_{Y\in\mathrm{Ch}_t(X)\wedge X\in\boldsymbol{X}} \mathrm{q}\left(y|\mathrm{pa}_{t-1}^{(j)}(Y),\mathrm{pa}_t^{(i)}(Y)\right) \tag{6}$$

However, we can't sample from the mixture model if the weights depend on values of the different components of the mixture. To overcome this difficulty, we can follow the approach of Auxiliary PFs [13,11]: we will assign to each hidden variables $X\in\mathrm{Pa}_t(Y)$ in Eq. (6) the mean value that will be obtained from sampling from $\mathrm{q}(x|\mathrm{pa}_{t-1}^{(j)}(X),\mathrm{pa}_t^{(i)}(X))$ instead of a sampled one.

Equation (5) and (6) define completely our parallel proposal. However, it is important to highlight that the parallel sampling idea proposed by (4) can be extended to other types of factorization. For instance, we could extend the idea of substituting the transition priors and likelihoods by adding some extra components to the mixture that will let us sample from regions of space no related with any particle at the previous time slice or modifying $\alpha_j^{\boldsymbol{X}}$ to create more particles in certain regions of the space, according with some prior kwnowlegde of the problem.

3.2 Defining the Partition of the Hidden and Observation Variables

The structure of the DBN imposes some restriction on the disjoint partition of hidden variables. Additionally, Eq. (6) imposes some constraints too in the partition of the observed variables. Both are closely related.

For the partition of the hidden variables, the isochronal connected hidden variables and the isochronal hidden variables whose child is the same observed variable need to be in the same partition. That is, the basic hidden partition fulfills that $\forall\boldsymbol{X}\in\mathcal{P}(\boldsymbol{X}_t)B\in\boldsymbol{X}\,\mathrm{if}\,X\in\boldsymbol{X}\wedge(X\in\mathrm{Pa}_t(B)\vee(Y\in\boldsymbol{Y}_t\wedge B\in\mathrm{Pa}_t(Y)\wedge X\in\mathrm{Pa}_t(Y)))$. Any other partition is formed by the union of subsets of the basic one.

Regarding the partition of the observed variables, we need to divide them according to the selected partition of the hidden variables, grouping the measurements with its hidden parents. So, the basic observed partition fulfills $\forall\boldsymbol{Y}\in\mathcal{P}(\boldsymbol{Y}_t)\exists\boldsymbol{X}\in\mathcal{P}(\boldsymbol{X}_t)\mathrm{s.t.}\forall Y\in\boldsymbol{Y}\exists X\in\boldsymbol{X}\mathrm{s.t.}Y\in\mathrm{Ch}_t(X)$.

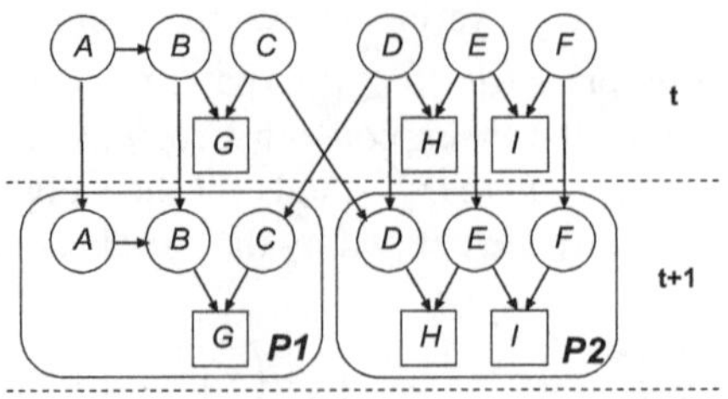

Fig. 2. Example of partition for a DBN

An example is presented in Fig. 2, where there is a DBN with 6 hidden variables $\{A,B,C,D,E,F\}$ and 3 measurements $\{G,H,I\}$ per time slice. This network has two basic hidden partitions (P1 and P2), represented with the big rounded squares. Variables A and B belong to P1 because $A \in \mathrm{Pa}_t(B)$. C also belongs to P1, because G is a common child of B and C. D, E and F belong to P2 because they are related through their common children H and I. The measurements are divided according to the hidden partitions: G belongs to the observed partition associated with P1 because it is a child of B and C, and H and I to the other because they are children of D, E and F.

3.3 The Complete Parallel Framework

The first step, before carrying out the filtering steps of our framework consists on defining the disjoint partitions, taking into account the restrictions presented in Sec. 3.2.

Once the subsets are defined, and the original particles created with an initialization proposal, we can sample the subspaces in parallel with the proposal defined by Eq. (5) and (6). In short, we get the mean values of the hidden variables and calculate the $\alpha_j^{\boldsymbol{X}}$ of each subset taking into account the ancestral ordering. We use $\alpha_j^{\boldsymbol{X}}$ to select a component of the mixture and sample the hidden values from that component.

To calculate the $w_t^{(i)}$ of the particles created with the parallel proposal, we use either (2) or (3). The numerator can be zero when the values of the hidden variables in the current time slice are non probabilistically "consistent" with the ones of the previous time slice and/or the measurements. The denominator can't, because we have generated it with the proposal. So, the final $w_t^{(i)} \geq 0$. When $w_t^{(i)} = 0$ for all the particles, as all of them are equally non probable, we accept them with the same probability. This automatically resets our PF to the values proposed by $\mathrm{q}(\cdot)$ when the PF is lost (all $w_t^{(i)} = 0$).

Additionally, we can also include a WR step after calculating $w_t^{(i)}$, or and create a Serial and Parallel sampling PF that alternate substeps of our Importance Parallel Sampling step with WR steps for the groups we sample in serial. However, the "instantaneous" PF can't include a WR step before we have sampled all the variables that can't be extracted from the numerator's summation in (3).

The computational cost of our PF framework depends on the expressions used for calculating the numerator ("trajectory" or "instantaneous" case) and denominator (the parallel sampling proposal) of (2) or (3). How to minimize the cost is out of the scope of this article, although we believe that the use of N-body learning can reduce it significantly as implied by [11]. Additionally, we can use parallel computing architectures for calculating the probabilities values and sample groups of hidden variables.

4 Related Work

This section compares the behavior of our PFs with some of the existing ones, starting with the PFs that combine the values of different subsets of hidden variables to create new particles and following with other two techniques closely related with our PF. Table 1 highlights the most relevant parts of our comparison.

Among the first group of PFs, the Factored PF in [5] is developed for systems with discrete hidden variables. It approximates the distribution as a product of distributions of subsets of hidden variables. The probability of each subset is represented as a weighted set of particles, and the inference is carried out building particles of the whole set of hidden variables, performing a PF step with the whole particles, and marginalizing the whole PF distribution to obtain the point-mass distribution of each subset. Our PF, valid for DBNs with continuous and discrete variables, follows a different path: it keeps the weighted particle distribution of all hidden variables, samples from different subsets according to $\alpha_j^{\boldsymbol{X}}$, and builds the whole particle with the weights calculated by (2) or (3).

The Hybrid-PF for the specific problem of [6] combines the Factored PF for discrete variables with the Rao-Blackwellised technique for continuous ones, and includes a look-ahead step to sample the most probable particles according to the current measurements before building the whole particle. Our proposal is more general, doesn't distinguish discrete and continuous variables, and the look-ahead step is implicitly implemented with our selection of $\alpha_j^{\boldsymbol{X}}$.

The Genetic PF for HMM in [7] includes a Genetic Algorithm (GA) crossover and mutation step, placed after having sampled the values of the hidden variable with the transition prior and before calculating the weights with the version of (2) for HMM. This lets the PF explore the whole space, although the calculated $w_t^{(i)}$ don't consider the two new steps. So, its $w_t^{(i)}$ represents only the probability of a hidden variable at the current time given the current time measurement, although the probability of sampling areas of the space consistent with previous measurements is higher.

The Hierarchical PF in [8] consists in multiple PF steps connected in serial and in parallel. Although structurally the closest to our parallel PFs, it doesn't show how to update $w_t^{(i)}$ after the parallel connection, while our PFs provides the sampling distribution as well as the updating weights equations.

It is important to highlight the connection of our PFs with the "instantaneous" PF for HMM [11]. Its proposal, extended to DBNs, will be equal to ours when all the hidden and observed variables belong to the same partition $(\mathcal{P}(\boldsymbol{Y}_t) = \boldsymbol{Y}_t) \wedge (\mathcal{P}(\boldsymbol{X}_t) = \boldsymbol{X}_t)$.

Finally, the serial PF presented in [10,12], that sequentially alternates IS and WR sub-steps for the variables belonging to each time slice of the DBN, can be used in combination with our PFs to build general Serial/Parallel PFs because the two approaches complement each other: the serial approach is beneficial for DBNs which have many isochronal links of hidden variables and can only be partitioned into a few sets, while the parallel approach is beneficial for DBNs which are naturally partitioned in many

Table 1. Comparison of our PFs with others (D. Discrete, C. Continuous)

PF	Foundation	Variables	DBN type	Probabilistic Consistent weight update
Ours	Parallel sampling proposal	D. + C.	Highly parallel	Yes
[5]	Factored posterior + PF for particles created from the posterior	D.	Any	No
[6]	[5] + Rao-Blackwellised	D. + C.	Special	No
[7]	Normal PF + genetic mutation and crossover	D. + C.	HMM	No
[8]	Serial and Parallel sampling	D. + C.	Any	No
[11]	Instantaneous PF	D. + C.	HMM	Yes
[12]	Alternating IS and WR inside time slice	D. + C.	Highly serial	Yes

subsets. Due to this distinction, in the experimental section, to show the performance of our framework, we select DBNs that show the advantages of our approach. As the experiments show there is an abundance of problems of this kind.

5 Experimental Results

The results presented in this paper compare our PFs with the KLPF because it is the generic algorithm for DBNs[1].

The first set of experiments, performed with simulated data compare the performance of the PFs using the Root Square Mean Error (RMSE) between the mean value of the particles estimates for each PF and the true value of the hidden variables at the last step of the simulation. In the second set, carried out with real world data, we use the different PFs to score the structure of a DBN given the measurements. In both cases, our PFs perform better than KLPF.

5.1 Simulated Experiments

This section compares our PFs, working with a limited number of particles, with KLPF in two complex real problems modeled by different DBNs. We have selected them because the structures of their DBNs, with multiple hidden variables that can be sampled in parallel, let them benefit significantly from our PFs. In fact, the second one is a simplification of our original problem: sea rescue where an Unmanned Air Vehicle (UAV) has to track several objects that are in the water and whose initial positions is not completely known. The first problem is a modification demonstrating our PFs on a more complex DBN. In both cases, we want to localize a set of O mobile objects given the information provided by the existing sensors. To be able to distinguish the variables related with the l-th object, in this section some variables names have a subindex (V_l).

In the first problem, each mobile object (M_l) is repelled by another (M_k) with a common unknown force (F), and the position of each mobile is observed by a different

[1] A preliminar analysis, excluded for space limitation, showed that the Serial PF by [10,12] worked even worst than KLPF due to the highly parallel structure of our DBNs.

sensor (S_l). The PFs have to estimate the probability of the hidden variables (M_l, F) given the measurements (S_l) for the DBN whose structure is defined by $\mathrm{Pa}_{t-1}(M_l) = \{M_l, M_{rem(l+1,O+1)}, F\}$, $\mathrm{Pa}_{t-1}(F) = \{F\}$ and $\mathrm{Pa}_t(S_l) = \{M_l\}$. The structure of this DBN let us divide the hidden variables in $O+1$ groups, each with a hidden variable and assign to the weights of each mixture the likelihoods of the measurement associated with each hidden variable ($\alpha_j^{\{F\}} = 1$ and $\alpha_j^{\{M_l\}} = \mathrm{p}(s_l|\mathrm{pa}_t^{(i)}(S_l))$). We also make the proposal of each hidden variable equal to its prior transition model. Although each hidden variable belongs to a different group, the difficulty of this problem comes from the fact that they are highly coupled by the $\mathrm{Pa}_{t-1}(M_l)$ relation.

The second problem is an abstraction of a sea rescue problem with UAVs. The objects (M_l) move with the direction of the sea current and wind (E) and the sensors, which are onboard the UAV (U), are only able to detect (D_l) the mobiles that are within a circular area of the UAV and provide their position (S_l) when detected.The PFs have to estimate the probability of the hidden variables (M_l, E) given the observed ones (U, D_l, S_l) for the DBN whose structure is defined by $\mathrm{Pa}_{t-1}(M_l) = \{M_l, E\}$, $\mathrm{Pa}_{t-1}(E) = \{E\}$, $\mathrm{Pa}_t(D_l) = \{M_l, U\}$ and $\mathrm{Pa}_t(S_l) = \{M_l, D_l\}$. Again, the structure of this DBN let us divide the hidden variables in $O+1$ groups, each with a hidden variable and assign to the weights of each mixture the product of the likelihoods of the measurements associated with each hidden variable ($\alpha_j^{\{E\}} = 1$ and $\alpha_j^{\{M_l\}} = \mathrm{p}(d_l|\mathrm{pa}_t^{(i)}(D_l))\mathrm{p}(s_l|\mathrm{pa}_t^{(i)}(S_l))$). We also make the proposal of each M_l equal to its prior transition model. However, to let our PFs explore a bigger region of E, its proposal distribution is the prior transition model with extra noise. Although this problem is less coupled than the first one, it is also difficult due to the use of multiple measurements per hidden variable and the fact that when the object is not inside the circular area its position is not observed.

For each problem, we run to types of experiments. In both types, we measure at the last step of the simulation the Root Mean Square Error (RMSE) between the mean value of the particles estimates for each PF and the true value of the hidden variables.

In the first type, we calculate the RMSE over the same experiment 50 times for the KLPF, and our "trajectory" and "instantaneous" PFs, initializing the particles in each experiment and PF randomly using the same initial proposal. We then use the RMSE to clasify the experiments in "convergent" and "divergent" according with a selected threshold. Table 2a and 2b show the number of convergent runs and the mean over convergent runs of RMSE of the three PFs for each problem. Our PFs, which obtain similar solutions, hasn't been tested with more than 100 particles and the RMSE is not defined (n/d) when no run converges. For the first problem, Table 2a shows that while our PF convergence increases with N, no runs of KLPF converge. The bad results of KLPF are due to the high dependence of the hidden variables on each other, which is especially problematic for a global localization problem where the particles are created randomly in all the regions of the space. However, as our PFs sample independently from the more likely regions of each group of variables, the estimated positions of the mobiles can converge more easily to their real values. For the second problem, Table 2b shows that the convergence of all PFs increases with N, although our PFs have a higher velocity of growth. Hidden variables are not strongly interdependent and particles are initialized randomly in a smaller region of the space increasing chances of starting with

Table 2. First experiment: Number of convergent runs (conv.) and RMSE (mean over convergent runs) of the two PFs for experiments of problems 1 and 2 with L=10

(a) problem 1

# particles		20	50	100	500	1000
KLPF	conv.	0	0	0	0	0
	RMSE	n/d	n/d	n/d	n/d	n/d
"Trajectory" PF	conv.	7	30	42		
	RMSE	810	662	414		
"Instant." PF	conv.	12	27	36		
	RMSE	830	620	410		

(b) problem 2

# particles		20	50	100	500	1000
KLPF	conv.	4	7	8	22	23
	RMSE	832	1650	457	362	392
"Trajectory" PF	conv.	24	37	46		
	RMSE	892	483	382		
"Instant." PF	conv.	24	34	46		
	RMSE	956	429	382		

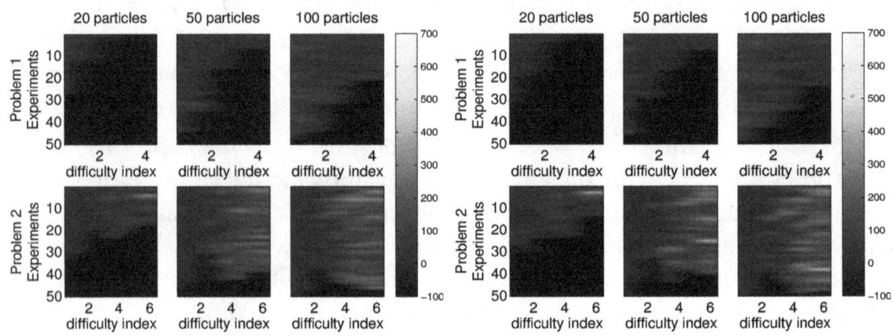

(a) "Trajectory" PF (b) "Instantaneous" PF

Fig. 3. Second experiment: Comparing KLPF with the "trajectory" and "instantaneous" PFs with the same initial particles

a good particle. This leads to improved performance of KLPF and superior performance of our PFs, both respect to the results observed in problem 1.

In the second type of experiments, we calculate the RMSE for the KLPF and our two PFs, with the same initial particles for all of them with the purpose of comparing how the filters work under the same initial conditions. The complete experiment consists on running the three filters for different number of particles (20, 50 and 100), with different initialization region proposals (4 for problem 1 and 6 for problem 2), and the same initial particles. For the convergent runs[2] of each of our PFs, we represent in Figure 3 the quotient of the RMSE of KLPF by the RMSE of one of our PFs ("Trajectory" in 3a and "Instantaneous" in 3b) for each problem (figure row), number of particles (figure column), initialization regions (x-axis, incrementing the size of the region, and so the difficulty of the problem with the number) and experiment number (y-axis). Quotients bigger than 1 show that our corresponding PF was better while smaller worst. The divergent runs are represented in black (set to -100). For problem 1, we can observe how incrementing the number of particles helps each of our PFs and how incrementing

[2] We use the same convergence criterium as in experiment 1.

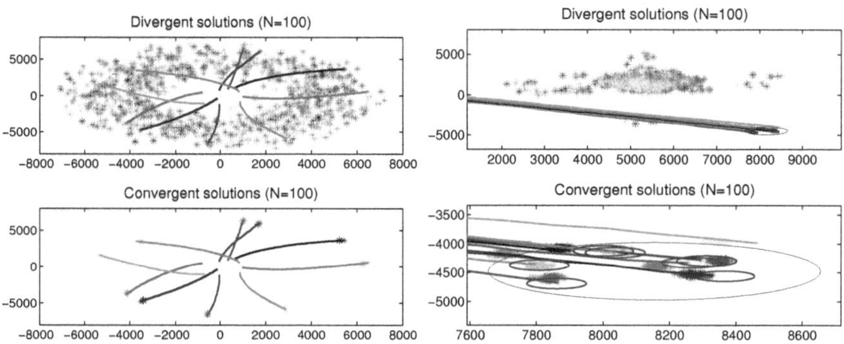

(a) Problem 1 (b) Problem 2

Fig. 4. Particle positions at the final step (pluses) and real trajectory (lines) of the 10 mobile objects (each with a different grey level) of a convergent and divergent run of the PFs. The big circle in Problem 2 represents the area around the UAV while the little ones are centered around M_l for the detected mobiles.

the size of the region makes it worst. As none runs of the KLPF converge, each of our PFs outperforms the KLPF for all its convergent runs (quotient is always bigger than 1). For the second problem, when the difficulty index is small, KLFP and each of our PFS obtain similar RMSE, but as the difficulty index grows, the convergent runs of our corresponding PF get better than KLPF. Incrementing the number of particles also increments the number of convergent runs of our PFs. So, in short, our PFs deals better than KLPF with poorly initialized particles and complex problems.

Finally, Fig. 4 shows examples of convergent and divergent solutions for each of the problems. Fig. 4a, for problem 1, shows the position of the mobiles in the particles at the final iteration (pluses) and the real trajectory (lines) for a typical run of KLPF (that always diverges) and of our PFs. Fig. 4b, for problem 2, represents the position of the mobiles in the particles at the final iteration (pluses) and the real trajectory (lines); the circular area around the UAV (big circle) and an area around the measured position $s_{l,t}$ of the detected mobiles (small circles). Note the different axes used in each case, due to the wider distribution of the particles for the divergent example.

5.2 Real Data

For the real data experiment we have used a functional Magnetic Resonance Imaging (fMRI) dataset [14]. FMRI signal represents the Blood Oxygenation Level Dependent (BOLD) response measured in a small cubic region of the brain (voxel). BOLD response is itself governed by the underlying hidden neural activity. As the model of the BOLD signal generation from neural activity we have used the hemodynamic forward model based on a coupled system of ordinary differential equations [15].

The number of voxel measurements collected per time point in a typical fMRI experiment is too large to model directly. Thus voxel time courses were averaged on a per Region of Interest (ROI) basis. ROIs were selected according to the widely used Talairach database [16].

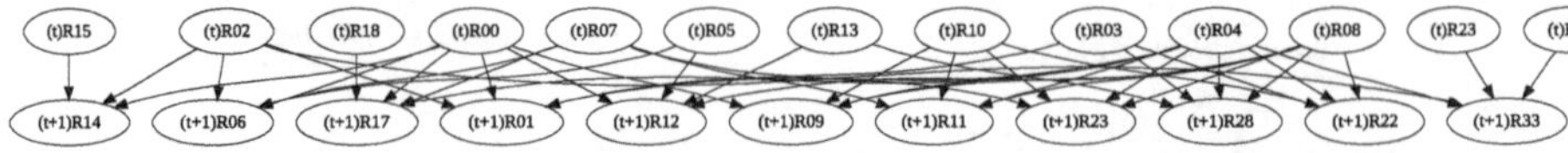

Fig. 5. The network for a subset of regions of interest (ROIs) from the Talairach anatomical atlas database. ROI names are replaced by short labels since they are not significant fr our work. Note that each ROI is observed through its indirect measurement by fMRI.

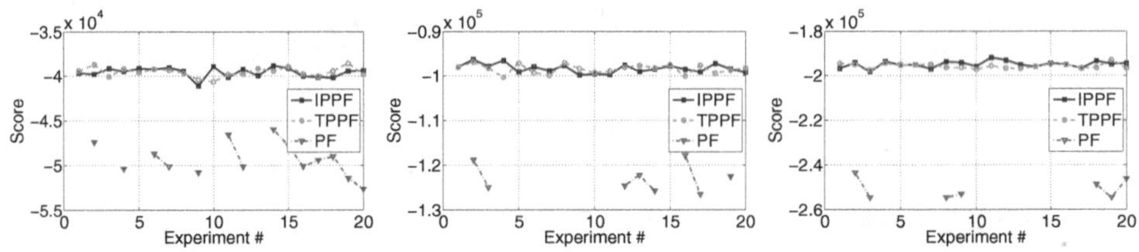

(a) 20 particles (b) 50 particles (c) 100 particles

Fig. 6. Cross validation log likelihood score plots for the instantaneous (IPPF) and trajectory (TPPF) parallel PFs, as well as regular particle filter (PF). Not displayed points mean that kernel density estimator did not produce a valid result.

In order to obtain the underlying structure of the DBN, we have used the approach that treats fMRI data as fully observed and quantizes it into categorical representation [14]. Among the discovered DBN families we have used several most significant ones according to the cross validation procedure (t-test with p-value of 0.05). This resulted in a DBN of 42 hidden variables per slice. Figure 5 shows a small portion of the hidden structure of the DBN we use. The observation nodes that are present for each ROI in a time slice are not shown.

The two novel parallel PF algorithms as well as the KLPF were run on this dataset for 50 time points (100 seconds with 2 seconds fMRI sampling rate). Since the ground truth for neural activity of ROIs is unknown in this dataset, for evaluation we have used cross validation log likelihood based on kernel density estimators with Gaussian kernels and automatically chosen variance value using a cross validation procedure [17].

The results for 20 runs using 20, 50 and 100 particles are shown in Figure 6. Parallel PFs show higher score (better) although are not much different in their performance. KLPF has considerably lower mean score as well as a higher variance. Note how the score decreases as the number of particles grows. This is an expected behavior with Parzen window estimators, that is due to the smoother and generally wider estimates in the cases of more data that lead to lower probability values. Thus in general this result supports the one obtain on simulated data.

6 Conclusions

This paper presents two PFs to estimate the trajectory or current state of the hidden variables of a DBN. It consists of a highly configurable proposal that samples in parallel the values of different subsets of hidden variables to build a whole particle whose weight is updated according to the sampling proposal and the estimation problem.

Our PFs explore different subspaces of the state space in parallel while performing the sampling step, and the whole space as a block while updating the weights. Inside each subspace the exploration is carried out by means of a mixture distribution, whose weights and components are selected by the user to control the regions of the space it wants to explore. For the whole space step, either the "trajectory" update weight function by Koller and Lenser [9] or the "instantaneous" update proposed in this paper can be used. In our experiments, our two filters show similar performances.

Our tests show that our PFs usually better sample the state space when the particles are initialized in spread regions, hence it is superior to KLPF in keeping track of the hidden state while needing smaller number of particles.

Finally, in combination with the Serial PF method proposed in [12], we can build a Parallel & Serial PF that updates the weights according to the sampling functions.

Acknowledgments

This work was supported by the Spanish Grants DPI2006-15661-C02-01 and CAM S-0505/DPI 0391. Further, Dr. Besada-Portas was supported by the Spanish post-doctoral Grant EX-2007-0915, Dr. Plis by NIMH Grant 1 R01 MH076282-01, and Dr. Lane by NSF Grant IIS-0705681. The authors also thank the Aula Sun-UCM for providing access to their computational resources for doing parts of the experiments.

References

1. Doucet, A., Freitas, N., Gordon, N. (eds.): Sequential Monte Carlo methods in practice. Springer, Heidelberg (2001)
2. MacCormick, J., Isard, M.: Partitioned sampling, articulated objects, and interface-quality hand tracking. In: Vernon, D. (ed.) ECCV 2000. LNCS, vol. 1843, pp. 3–19. Springer, Heidelberg (2000)
3. Doucet, A., de Freitas, N., Murphy, K., Russell, S.: Rao-blackwellised particle filtering for dynamic bayesian networks. In: 16th Conference on Uncertainty in Artificial Intelligence, pp. 176–183 (2000)
4. Vaswani, N.: Particle filters for infinite (or large) dimensional state spaces-part 2. In: IEEE ICASSP (2006)
5. Ng, B., Peshkin, L.: Factored particles for scalable monitoring. In: Proceedings of the 18th Conference on Uncertainty in Artificial Intelligence, pp. 370–377. Morgan Kaufmann, San Francisco (2002)
6. Das, S., Lawless, D., Ng, B., Pfeffer, A.: Factored particle filtering for data fusion and situation assessments in urban environments. In: 8th International Conference on Information Fusion (July 2005)
7. Park, S., Hwang, J., Rou, K., Kim, E.: A new particle filter inspired by biological evolution: Genetic Filter. Proceeding of World Academy of Science, Engineering and Technology 21, 57–71 (2007)
8. Brandao, B.C., Wainer, J., Goldenstein, S.K.: Subspace hierarchical particle filter. In: XIX Brazilian Symposium on Computer Graphics and Image Processing (2006)
9. Koller, D., Lerner, U.: Sampling in factored dynamic systems. In: Sequential Monte Carlo in Practice, pp. 445–464 (2001)

10. Maccormick, J., Blake, A.: A probabilistic exclusion principle for tracking multiple objects. Intenational Journal of Computer Vision 39(1), 57–71 (2000)
11. Klaas, M., de Freitas, N., Doucet, A.: Toward practical n^2 Monte Carlo: the Marginal Particle Filter. In: Proceedings of UAI 2005, Arlington, Virginia, pp. 308–331. AUAI Press (2005)
12. Rose, C., Saboune, J., Charpillet, F.: Reducing particle filtering complexity for 3D motion capture using dynamic bayesian networks. In: 23th AAAI Conference on Artificial Intelligence (2008)
13. Pitt, M.K., Shephard, N.: Filtering via simulation: Auxiliary Particle Filters. Journal of the American Statistical Association 94(446), 590–599 (1999)
14. Burge, J., Lane, T., Link, H., Qiu, S., Clark, V.P.P.: Discrete dynamic bayesian network analysis of fMRI data. Human Brain Mapping (November 2007)
15. Friston, K.J., Harrison, L., Penny, W.: Dynamic Causal Modelling. NeuroImage 19(4), 1273–1302 (2003)
16. Lancaster, J.L., Woldorff, M.G., Parsons, L.M., Liotti, M., Freitas, C.S., Rainey, L., Kochunov, P.V., Nickerson, D., Mikiten, S.A., Fox, P.T.: Automated Talairach atlas labels for functional brain mapping. Human Brain Mapping 10(3), 120–131 (2000)
17. Hofmann, R., Tresp, V.: Discovering structure in continuous variables using bayesian networks. In: Touretzky, D.S., Mozer, M.C., Hasselmo, M.E. (eds.) Advances in Neural Information Processing Systems, vol. 8, pp. 500–506. MIT Press, Cambridge (1996)

Adaptive XML Tree Classification on Evolving Data Streams

Albert Bifet and Ricard Gavaldà

Universitat Politècnica de Catalunya, Barcelona, Spain
{abifet,gavalda}@lsi.upc.edu

Abstract. We propose a new method to classify patterns, using closed and maximal frequent patterns as features. Generally, classification requires a previous mapping from the patterns to classify to vectors of features, and frequent patterns have been used as features in the past. Closed patterns maintain the same information as frequent patterns using less space and maximal patterns maintain approximate information. We use them to reduce the number of classification features. We present a new framework for XML tree stream classification. For the first component of our classification framework, we use closed tree mining algorithms for evolving data streams. For the second component, we use state of the art classification methods for data streams. To the best of our knowledge this is the first work on tree classification in streaming data varying with time. We give a first experimental evaluation of the proposed classification method.

1 Introduction

Pattern classification and frequent pattern discovery have been important tasks over the last decade. Nowadays, they are becoming harder, as the size of the pattern datasets is increasing, data often comes from sequential, streaming sources, and we cannot assume that data has been generated from a static distribution. If we want accuracy in the results of our algorithms, we have to consider that the distribution that generates data may vary over time, often in an unpredictable and drastic way.

Tree Mining is becoming an important field of research due to the fact that XML patterns are tree patterns and that XML is becoming a standard for information representation and exchange over the Internet. XML data is growing and it will soon constitute one of the largest collection of human knowledge. Other applications of tree mining appear in chemical informatics, computer vision, text retrieval, bioinformatics, and Web analysis. XML tree classification has been done traditionally using information retrieval techniques considering the labels of nodes as bags of words. With the development of frequent tree miners, classification methods using frequent trees appeared [21, 14, 8, 12]. Recently, closed frequent miners were proposed [7, 18, 1], and using them for classification tasks is the next natural step.

W. Buntine et al. (Eds.): ECML PKDD 2009, Part I, LNAI 5781, pp. 147–162, 2009.

Closure-based mining on relational data has recently provided some interesting algorithmic developments. In this paper we use closure-based mining to reduce drastically the number of attributes in tree classification tasks. Moreover, we show how to use maximal frequent trees to reduce even more the number of attributes needed in tree classification, in many cases without loosing accuracy.

We propose a general framework to classify XML trees based on subtree occurrence. It is composed of a Tree XML Closed Frequent Miner, and a classifier algorithm. We propose specific methods for dealing with adaptive data streams. In [5] a new approach was proposed for mining closed patterns adaptively from data streams that change over time, and it was used for closed *unlabeled* rooted trees. In this work, we use the extension to the more challenging case of *labeled* rooted trees.

The rest of the paper is organized as follows. We discuss related work in Section 2. Section 3 gives background and Section 4 introduces our closure operator and its properties needed for our classification algorithm. Section 5 shows the tree classification framework and it introduces the adaptive closed frequent mining method. Experimental results are given in Section 6, and some conclusions in Section 7.

2 Related Work

Zaki and Aggarwal presented XRules in [21]. Their classification method mines frequent trees in order to create classification rules. They do not use closed frequent trees, only frequent trees. XRules is cost-sensitive and uses Bayesian rule based class decision making. They also proposed methods for effective rule prioritization and testing.

Kudo and Matsumoto presented a boosting method for tree classification in [14]. Their method consists of decision stumps that uses significant frequent subtrees as features and a Boosting algorithm which employs the subtree-based decision stumps as weak learners. With Maeda they extended this classification method to graphs in [13].

Other works use SVMs defining tree Kernels [8, 12]. Tree kernel is one of the convolutions kernels, and maps the example represented in a labeled ordered tree into all subtree spaces. The feature space uses frequent trees and not closed trees.

Garriga et al. [10] showed that when considering labeled itemsets, closed sets can be adapted for classification and discrimination purposes by conveniently contrasting covering properties on positive and negative examples. They formally proved that these sets characterize the space of relevant combinations of features for discriminating the target class.

Chi et al. proposed CMTreeMiner [7], the first algorithm to discover all closed and maximal frequent labeled induced subtrees without first discovering all frequent subtrees. CMTreeMiner shares many features with CloseGraph [19]. Termier et al. proposed DryadeParent [18], based on the hooking principle first introduced in Dryade. They claim that the branching factor and depth of the frequent patterns to find are key factors in the complexity of tree mining algorithm and that DryadeParent outperforms CMTreeMiner, on datasets where the frequent patterns have a high branching factor.

In [5], we proposed a new approach for mining closed frequent patterns adaptively from data streams that change over time, and we applied it to unlabelled frequent tree mining.

To the best of our knowledge this is the first work defined for classifying trees and mining labeled closed frequent trees in streaming data that evolve with time, and the first one in using closed and maximal frequent trees for feature reduction.

3 Preliminaries

Following Formal Concept Analysis usage, we are interested in (possibly infinite) sets endowed with a partial order relation. Elements of these sets are generically called *patterns*.

The set of all patterns will be denoted with $\mathcal{T}$, but actually all our developments will proceed in some finite subset of $\mathcal{T}$ which will act as our universe of discourse.

Given two patterns t and t', we say that t is a *subpattern* of t', or t' is a *super-pattern* of t, if $t \preceq t'$. Two patterns t, t' are said to be *comparable* if $t \preceq t'$ or $t' \preceq t$. Otherwise, they are incomparable. Also we write $t \prec t'$ if t is a proper subpattern of t' (that is, $t \preceq t'$ and $t \neq t'$).

The input to our data mining process is a dataset $\mathcal{D}$ of transactions, where each transaction $s \in \mathcal{D}$ consists of a transaction identifier, *tid*, and a transaction pattern. The dataset is a finite set in the standard setting, and a potentially infinite sequence in the data stream setting. Tids are supposed to run sequentially from 1 to the size of $\mathcal{D}$. From that dataset, our universe of discourse $\mathcal{U}$ is the set of all patterns that appear as subpattern of some pattern in $\mathcal{D}$.

Figure 1 shows a finite dataset example of trees.

As is standard, we say that a transaction s *supports* a pattern t if t is a subpattern of the pattern in transaction s. The number of transactions in the dataset $\mathcal{D}$ that support t is called the *support* of the pattern t. A subpattern t is called *frequent* if its support is greater than or equal to a given threshold *min_sup*. The frequent subpattern mining problem is to find all frequent subpatterns in a given dataset. Any subpattern of a frequent pattern is also frequent and, therefore, any superpattern of a nonfrequent pattern is also nonfrequent (the *antimonotonicity* property).

3.1 Frequent Pattern Compression

We define a frequent pattern t to be *closed* if none of its proper superpatterns has the same support as it has. Generally, there are much fewer closed patterns than frequent ones. In fact, we can obtain all frequent subpatterns with their support from the set of frequent closed subpatterns with their supports. So, the

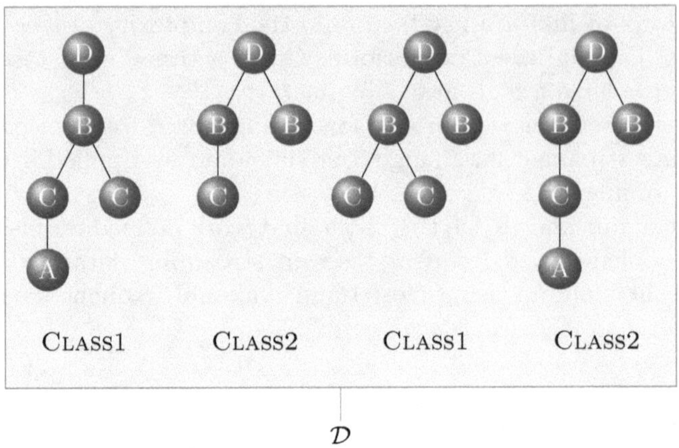

Class	Trees
CLASS1	$((0, D), (1, B), (2, C), (3, A), (2, C))$
CLASS2	$((0, D), (1, B), (2, C), (1, B))$
CLASS1	$((0, D), (1, B), (2, C), (2, C), (1, B))$
CLASS2	$((0, D), (1, B), (2, C), (3, A), (1, B))$

Fig. 1. A dataset example of 4 tree transactions

set of frequent closed subpatterns maintains the same information as the set of all frequent subpatterns.

The closed trees for the dataset of Figure 1 are shown in the Galois lattice of Figure 2.

We define a frequent pattern t to be *maximal* if none of t's proper superpatterns is frequent. All maximal patterns are closed, but not all closed patterns are maximal, so there are more closed patterns than maximal. Note that we can obtain all frequent subpatterns from the set of maximal frequent subpatterns, but not their support. So, the set of maximal frequent subpatterns maintains approximately the same information as the set of all frequent subpatterns.

4 Classification Using Compressed Frequent Patterns

The pattern classification problem is defined as follows. A set of examples of the form (t, y) is given, where y is a discrete class label and t is a pattern. The goal is to produce from these examples a model $\hat{y} = f(t)$ that will predict the classes y of future pattern examples with high accuracy.

We use the following approach: we convert the pattern classification problem into a vector classification learning task, transforming patterns into vectors of attributes. Attributes will be frequent subpatterns, or a subset of these frequent subpatterns.

Suppose $\mathcal{D}$ has d frequent subpatterns denoted by $t_1, t_2, \ldots, t_d$. We map any pattern t to a vector x of d attributes: $x = (x_1, ..., x_d)$ such that for each attribute i, $x_i = 1$ if $t_i \preceq t$ or $x_i = 0$ otherwise.

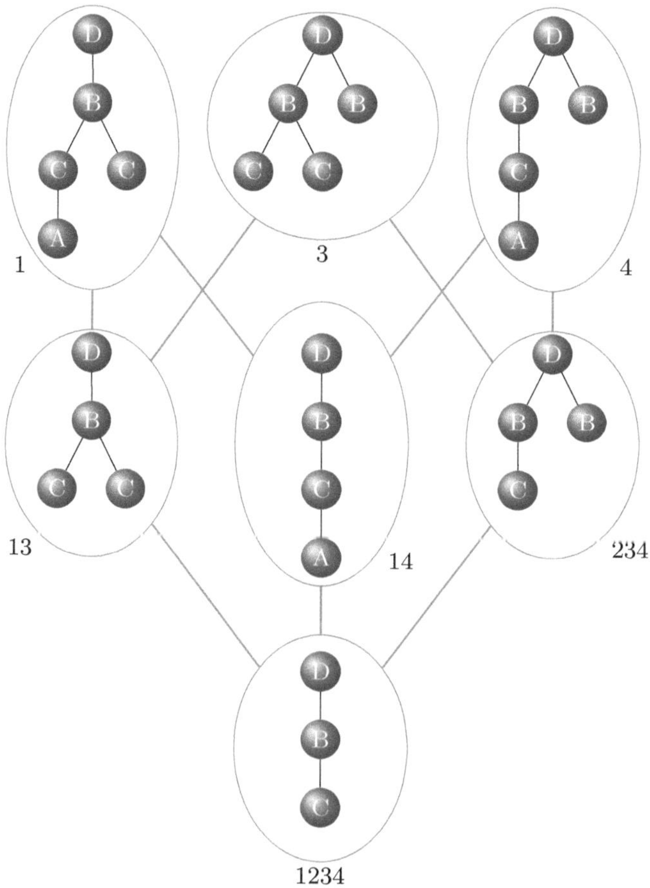

Fig. 2. Example of Galois Lattice of Closed trees

As the number of frequent subpatterns is huge, we perform a feature selection process, selecting a subset of these frequent subpatterns, maintaining the same information, or approximate. Figures 3 and 4 show frequent trees and its conversion to vectors of attributes. Note that closed trees have the same information as frequent trees, but maximal trees loose some support information, as mentioned in Section 3.1.

4.1 Closed Frequent Patterns

Recall that if X is a set with a partial order $\leq$, a *closure operator* on X is a function $C : X \to X$ that satisfies the following for all x in X: $x \leq C(X), C(x) = C(C(x))$, for all $y \in X$, $x \leq y$ implies $C(x) \leq C(y)$. A Galois connection is provided by two functions, relating two lattices in a certain way. Here our lattices are not only plane power sets of the transactions but also plain power

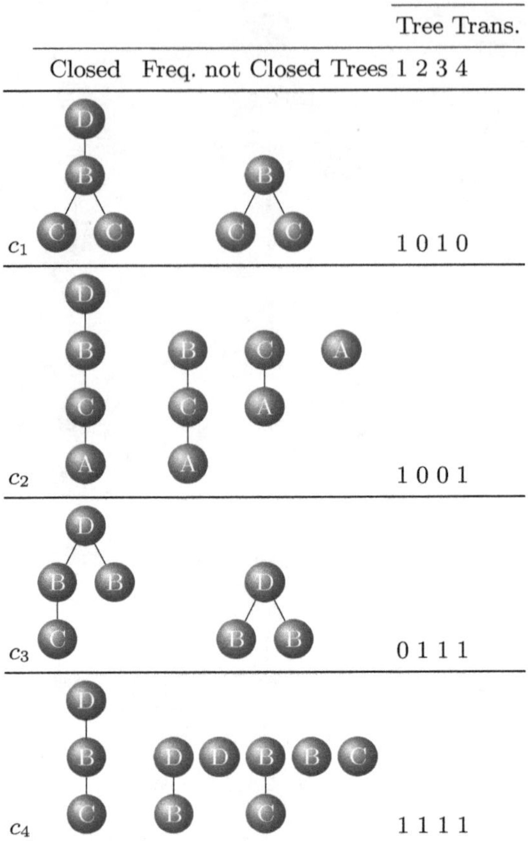

Fig. 3. Frequent trees from dataset example ($min\_sup = 30\%$), and their corresponding attribute vectors

sets of the corresponding solutions. On the basis of the binary relation $t \preceq t'$, the following definition and proposition are rather standard.

Definition 1. *The Galois connection pair is defined by:*

- *For finite* $A \subseteq \mathcal{D}$, $\sigma(A) = \{t \in \mathcal{T} \mid \forall t' \in A\,(t \preceq t')\}$
- *For finite* $B \subset \mathcal{T}$, *not necessarily in* $\mathcal{D}$, $\tau_{\mathcal{D}}(B) = \{t' \in \mathcal{D} \mid \forall t \in B\,(t \preceq t')\}$

Proposition 1. *The composition* $\Gamma_{\mathcal{D}} = \sigma \circ \tau_{\mathcal{D}}$ *is a closure operator on the subsets of* $\mathcal{D}$.

Theorem 1. *A pattern t is closed for* $\mathcal{D}$ *if and only if it is maximal in* $\Gamma_{\mathcal{D}}(\{t\})$.

In other words, Theorem 1 states each closed set is uniquely defined through its maximal elements. On this basis, our algorithms can avoid duplicating calculations and redundant information by just storing the maximal patterns of each closed set. We denote $\Delta(t)$ as the set of maximal patterns of each closed set of $\Gamma_{\mathcal{D}}(\{t\})$.

	Frequent Trees													
	c_1		c_2				c_3		c_4					
Id	c_1	f_1^1	c_2	f_2^1	f_2^2	f_2^3	c_3	f_3^1	c_4	f_4^1	f_4^2	f_4^3	f_4^4	f_4^5
1	1	1	1	1	1	1	0	0	1	1	1	1	1	1
2	0	0	0	0	0	0	1	1	1	1	1	1	1	1
3	1	1	0	0	0	0	1	1	1	1	1	1	1	1
4	0	0	1	1	1	1	1	1	1	1	1	1	1	1

	Closed Trees				Maximal Trees			
Id Tree	c_1	c_2	c_3	c_4	c_1	c_2	c_3	**Class**
1	1	1	0	1	1	1	0	CLASS1
2	0	0	1	1	0	0	1	CLASS2
3	1	0	1	1	1	0	1	CLASS1
4	0	1	1	1	0	1	1	CLASS2

Fig. 4. Closed and maximal frequent trees from dataset example ($min\_sup = 30\%$), and their corresponding attribute vectors

We can relate the closure operator to the notion of closure based on support, as previously defined, as follows: t is closed for $\mathcal{D}$ if and only if: $\Delta_{\mathcal{D}}(\{t\}) = \{t\}$.

Following the standard usage on Galois lattices, we consider now implications of the form $A \to B$ for sets of patterns A and B from $\mathcal{U}$. Specifically, we consider the following set of rules: $A \to \Gamma_{\mathcal{D}}(A)$. Alternatively, we can split the consequents into $\{A \to t \mid t \in \Gamma_{\mathcal{D}}(A)\}$.

It is easy to see that $\mathcal{D}$ obeys all these rules: for each A, any pattern of $\mathcal{D}$ that has as subpatterns all the patterns of A has also as subpatterns all the patterns of $\Gamma_{\mathcal{D}}(A)$.

Proposition 2. *Let t_i be a frequent pattern for $\mathcal{D}$. A transaction pattern t satisfies $t_i \preceq t$, if and only if it satisfies $\Delta_{\mathcal{D}}(t_i) \preceq t$.*

We use Proposition 2 to reduce the number of attributes on our classification task, using only closed frequent patterns, as they keep the same information. The attribute vector of a frequent pattern will be the same as its closed pattern attribute vector. Figure 4 shows the attribute vectors for the dataset of Figure 1.

4.2 Maximal Frequent Patterns

Maximal patterns are patterns that do not have any frequent superpattern. All maximal patterns are closed patterns. If min_sup is zero, then maximal patterns are transaction patterns. We denote by $M_1(t), M_2(t), \ldots, M_m(t)$ the maximal superpatterns of a pattern t. We are interested in the implications of the form $t_c \to (M_1(t) \vee M_2(t) \vee \ldots \vee M_m(t))$ where t_c is a closed pattern.

Proposition 3. *Let t_c be a closed non-maximal frequent pattern for $\mathcal{D}$. Let $M_1(t_c), M_2(t_c), \ldots, M_m(t_c)$ be the maximal superpatterns of pattern t_c. A transaction pattern t satisfies $t_c \prec t$, if and only if at least one of the maximals superpattern $M_i(t_c)$ of pattern t_c satisfies $M_i(t_c) \preceq t$.*

Proof. Suppose that pattern t_c satisfies $t_c \prec t$ but no maximal superpattern $M_i(t_c)$ satisfies $M_i(t_c) \preceq t$. Then, pattern t_c has no frequent superpattern. Therefore, it is maximal, contradicting the assumption.

Suppose, for the other direction, that a maximal superpattern $M_i(t_c)$ of t_c satisfies $M_i(t_c) \preceq t$. Then, as t_c is a $M_i(t_c)$ subpattern, $t_c \preceq M_i(t_c)$, and it holds that $t_c \preceq M_i(t_c) \preceq t$.

For non-maximal closed patterns, the following set of rules holds if t_c is not a transaction pattern:

$$t_c \rightarrow \bigvee M_i(t_c)$$

Note that for a transaction pattern t_c that it is closed and non-maximal, there is no maximal superpattern $M_i(t_c)$ of pattern t_c that satisfies $M_i(t_c) \preceq t_c$. If there are no closed non-maximal transaction patterns, we do not need to use all closed patterns as attributes, since non-maximal closed patterns may be derived from maximal patterns.

Using Proposition 3, we may reduce the number of attributes on our classification task, using only maximal frequent patterns, as they keep much of the information as closed frequent patterns.

5 XML Tree Classification Framework on Data Streams

In this section we specialize the previous approach to the case of labelled trees, such as XML trees.

Trees are connected acyclic graphs, *rooted trees* are trees with a vertex singled out as the root, and *unranked* trees are trees with unbounded arity. We say that $t_1, \ldots, t_k$ are the *components* of tree t if t is made of a node (the root) joined to the roots of all the t_i's. We can distinguish between the cases where the components at each node form a sequence (ordered trees) or just a set (*unordered trees*). We will deal with rooted, unranked trees. We assume the presence of labels on the nodes.

An *induced subtree* of a tree t is any connected subgraph rooted at some node v of t that its vertices and edges are subsets of those of t. An *embedded subtree* of a tree t is any connected subgraph rooted at some node v of t that does not break the ancestor-descendant relationship among the vertices of t. We are interested in induced subtrees. Formally, let s be a rooted tree with vertex set V' and edge set E', and t a rooted tree t with vertex set V and edge set E. Tree s is an *induced subtree* (or simply a *subtree*) of t (written $t' \preceq t$) if and only if 1) $V' \subseteq V$, 2) $E' \subseteq E$, and 3) the labeling of V' is preserved in t. This notation can be extended to sets of trees $A \preceq B$: for all $t \in A$, there is some $t' \in B$ for which $t \preceq t'$.

Our XML Tree Classification Framework has two components:

- An XML closed frequent tree miner, for which we could use any incremental algorithm that maintains a set of closed frequent trees.
- A Data stream classifier algorithm, which we will feed with tuples to be classified online. Attributes in these tuples represent the occurrence of the current closed trees in the originating tree, although the classifier algorithm need not be aware of this.

In this section, we describe the two components of the framework, the XML closed frequent tree miner, and the data stream classifier.

5.1 Adaptive Tree Mining on Evolving Data Streams

Using a methodology based on Galois Lattice Theory, we use three closed tree mining algorithms: an increment IncTreeMiner, a sliding-window based one, WinTreeMiner, and an algorithm that mines closed trees adaptively from data streams. It is basically an adaptation of the theoretical framework developed in [5], which deals with quite general notion of pattern and subpattern, to the case of labeled rooted trees.

For maximal frequent trees, the following properties hold:

- adding a tree transaction to a dataset of trees $\mathcal{D}$, may increase or decrease the number of maximal trees for $\mathcal{D}$.
- adding a transaction with a closed tree to a dataset of trees $\mathcal{D}$, may modify the number of maximal trees for $\mathcal{D}$.
- deleting a tree transaction from a dataset of trees $\mathcal{D}$, may increase or decrease the number of maximal trees for $\mathcal{D}$.
- deleting a tree transaction that is repeated in a dataset of trees $\mathcal{D}$ from it, may modify the number of maximal trees for $\mathcal{D}$.
- a non maximal closed tree may become maximal if
    - it was not frequent and now its support increases to a value higher or equal to min_sup
    - all of its maximal supertrees become non-frequent
- a maximal tree may become a non maximal tree if
    - its support decreases below min_sup
    - a non-frequent closed supertree becomes frequent

We could check if a closed tree becomes maximal when

- removing closed trees because they do not have enough support
- adding a new closed tree to the dataset
- deleting a closed tree from the dataset

We use three tree mining algorithms adapting the general framework for patterns presented in [5]:

- IncTreeMiner, an incremental closed tree mining algorithm,
- WinTreeMiner, a sliding window closed tree mining algorithm
- AdaTreeMiner, an adaptive closed tree mining algorithm

The batches are processed using the non-incremental algorithm explained in [3]. AdaTreeMiner is a new tree mining method for dealing with concept drift, using `ADWIN` [4], an algorithm for detecting change and dynamically adjusting the length of a data window. `ADWIN` is parameter- and assumption-free in the sense that it automatically detects and adapts to the current rate of change. Its only parameter is a confidence bound δ, indicating how confident we want to be in the algorithm's output, inherent to all algorithms dealing with random processes.

Also important for our purposes, `ADWIN` does not maintain the window explicitly, but compresses it using a variant of the exponential histogram technique in [9]. This means that it keeps a window of length W using only $O(\log W)$ memory and $O(\log W)$ processing time per item, rather than the $O(W)$ one expects from a naïve implementation. When windows tend to be large, this usually results in substantial memory savings.

We propose two strategies to deal with concept drift:

- AdaTreeMiner1: Using a sliding window, with an `ADWIN` estimator deciding the size of the window
- AdaTreeMiner2: Maintaining an `ADWIN` estimator for each closed set in the lattice structure.

In the second strategy, we do not delete transactions. Instead, each `ADWIN` monitors the support of a closed pattern. When it detects a change, we can conclude reliably that the support of this pattern seems to be changing in the data stream in recent times.

5.2 Data Stream Classifier

The second component of the framework is based on MOA. **M**assive **O**nline **A**nalysis (MOA) [11, 6] is a framework for online learning from continuous supplies of examples, such as data streams. It is closely related to the well-known WEKA project, and it includes a collection of offline and online as well as tools for evaluation. In particular, it implements boosting, bagging, and Hoeffding Trees, both with and without Naïve Bayes classifiers at the leaves.

We use bagging and boosting of decision trees, because these ensemble methods are considered the state-of-the-art classification methods.

6 Experimental Evaluation

We tested our algorithms on synthetic and real data. All experiments were performed on a 2.0 GHz Intel Core Duo PC machine with 2 Gigabyte main memory, running Ubuntu 8.10.

6.1 Tree Classification

We evaluate our approach to tree classification on both real and synthetic classification data sets.

Table 1. Comparison of classification algorithms. Memory is measured in MB. The best individual accuracy is indicated in boldface.

Bagging	Time	Acc.	Mem.
AdaTreeMiner1	161.61	**80.06**	4.93
AdaTreeMiner2	212.57	65.78	4.42
WinTreeMiner W=100,000	192.01	72.61	6.53
WinTreeMiner W= 50,000	212.09	66.23	11.68
IncTreeMiner	212.75	65.73	4.4

Boosting	Time	Acc.	Mem.
AdaTreeMiner1	236.31	**79.83**	4.8
AdaTreeMiner2	326.8	65.43	4.25
WinTreeMiner W=100,000	286.02	70.15	5.8
WinTreeMiner W= 50,000	318.19	63.94	9.87
IncTreeMiner	317.95	65.55	4.25

For synthetic classification, we use the tree generation program of Zaki [20], available from his web page. We generate two mother trees, one for each class. The first mother tree is generated with the following parameters: the number of distinct node labels $N = 200$, the total number of nodes in the tree $M = 1,000$, the maximal depth of the tree $D = 10$ and the maximum fanout $F = 10$. The second one has the following parameters: the number of distinct node labels $N = 5$, the total number of nodes in the tree $M = 100$, the maximal depth of the tree $D = 10$ and the maximum fanout $F = 10$.

A stream is generated by mixing the subtrees created from these mother trees. In our experiments, we set the total number of trees in the dataset to be $T = 1,000,000$. We added artificial drift changing labels of the trees every $250,000$ samples, so closed and maximal frequent trees evolve over time. We use bagging of 10 Hoeffding Trees enhanced with adaptive Naïve Bayes leaf predictions, as classification method. This adaptive Naïve Bayes prediction method monitors the error rate of majority class and Naïve Bayes decisions in every leaf, and chooses to employ Naïve Bayes decisions only where they have been more accurate in past cases.

Table 1 shows classification results. We observe that AdaTreeMiner1 is the most accurate method, and that the accuracy of WinTreeMiner depends on the size of the window.

For real datasets, we use the Log Markup Language (LOGML) dataset from Zaki et al. [16, 21], that describes log reports at their CS department website. LOGML provides a XML vocabulary to structurally express the contents of the log file information in a compact manner. Each user session is expressed in LOGML as a graph, and includes both structure and content.

The real CSLOG data set spans 3 weeks worth of such XML user-sessions. To convert this into a classification data set they chose to categorize each user-session into one of two class labels: edu corresponds to users from an "edu"

Table 2. Comparison of tree classification algorithms. Memory is measured in MB. The best individual accuracies are indicated in boldface (one per row).

BAGGING		Maximal						Closed					
		Unordered			Ordered			Unordered			Ordered		
	# Trees	Att.	Acc.	Mem.	Att.	Acc.	Mem.	Att.	Acc.	Mem.	Att.	Acc.	Mem.
CSLOG12	15483	84	**79.64**	1.2	77	79.63	1.1	228	78.12	2.54	183	78.12	2.03
CSLOG23	15037	88	**79.81**	1.21	80	79.8	1.09	243	78.77	2.75	196	78.89	2.21
CSLOG31	15702	86	**79.94**	1.25	80	79.87	1.17	243	77.6	2.73	196	77.59	2.19
CSLOG123	23111	84	**80.02**	1.7	78	79.97	1.58	228	78.91	4.18	181	78.91	3.31

BOOSTING		Maximal						Closed					
		Unordered			Ordered			Unordered			Ordered		
	# Trees	Att.	Acc.	Mem.	Att.	Acc.	Mem.	Att.	Acc.	Mem.	Att.	Acc.	Mem.
CSLOG12	15483	84	**79.46**	1.21	77	78.83	1.11	228	75.84	2.97	183	77.28	2.37
CSLOG23	15037	88	79.91	1.23	80	**80.24**	1.14	243	77.24	2.96	196	78.99	2.38
CSLOG31	15702	86	**79.77**	1.25	80	79.69	1.17	243	76.25	3.29	196	77.63	2.62
CSLOG123	23111	84	79.73	1.69	78	**80.03**	1.56	228	76.92	4.25	181	76.43	3.45

domain, while other class corresponds to all users visiting the CS department from any other domain. They separate each week's logs into a different data set (CSLOGx, where x stands for the week; CSLOG12 is the combined data for weeks 1 and 2). Notice that the edu class has much lower frequency rate than other.

Table 2 shows the results on bagging and boosting using 10 Hoeffding Trees enhanced with adaptive Naïve Bayes leaf predictions. The results are very similar for the two ensemble learning methods. Using maximals and closed frequent trees, we obtain results similar to [20]. Comparing maximal trees with closed trees, we see that maximal trees use 1/4 to 1/3rd of attributes, 1/3 of memory, and they perform better.

6.2 Closed Frequent Tree Labeled Mining

As far as we know, CMTreeMiner is the state-of-art algorithm for mining induced closed frequent trees in databases of rooted trees. The main difference with our approach is that CMTreeMiner is not incremental and only works with bottom-up subtrees, and our method works with both bottom-up and top-down subtrees.

For synthetic data, we use the same dataset as in [7] and [20] for rooted ordered trees. The synthetic dataset T8M is generated by the tree generation program of Zaki [20], used for the evaluation on tree classification. In brief, a mother tree is generated first with the following parameters: the number of distinct node labels $N = 100$, the total number of nodes in the tree $M = 10,000$, the maximal

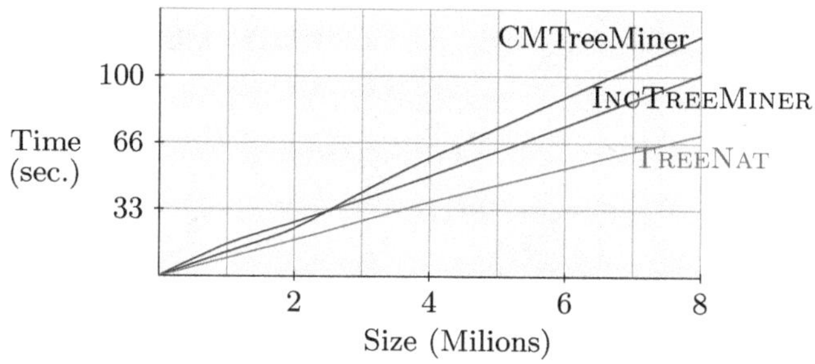

Fig. 5. Time used on ordered trees, T8M dataset

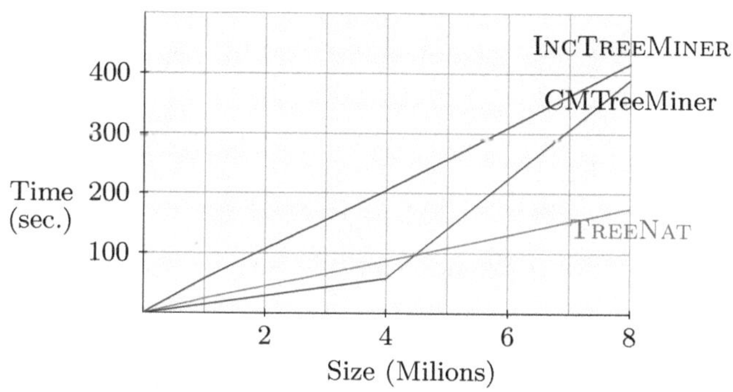

Fig. 6. Time used on unordered trees, TN1 dataset

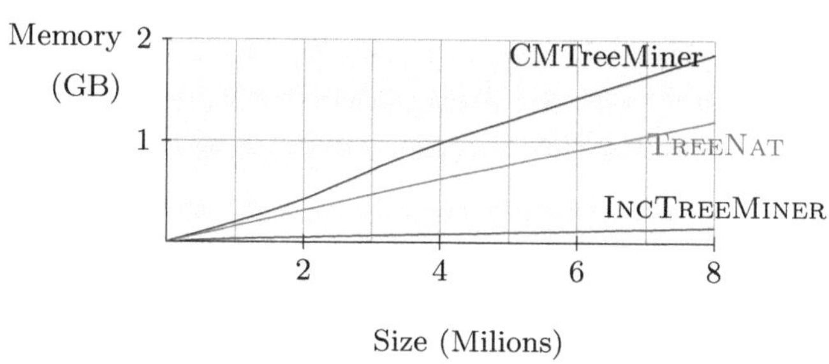

Fig. 7. Memory used on ordered trees, T8M dataset

depth of the tree $D = 10$ and the maximum fanout $F = 10$. The dataset is then generated by creating subtrees of the mother tree. In our experiments, we set the total number of trees in the dataset to be from $T = 0$ to $T = 8,000,000$.

The results of our experiments on synthetic data are shown in Figures 5,6,7, and 8. We observe that as the data size increases, the running times of IncTreeMiner and CMTreeMiner become closer, and that IncTreeMiner uses much less memory than CMTreeMiner. CMTreeMiner failed in our experiments

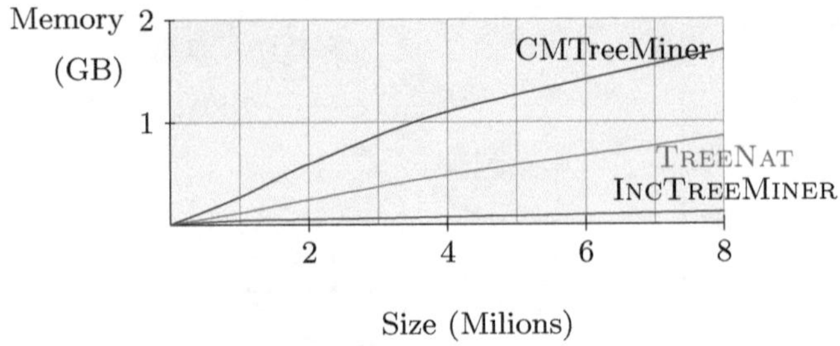

Fig. 8. Memory used on unordered trees on T8M dataset

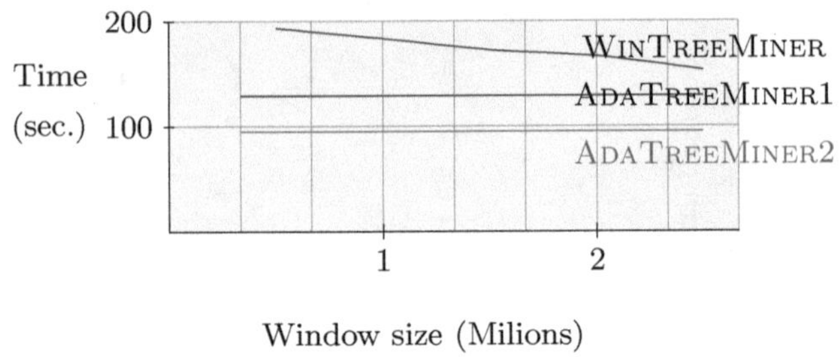

Fig. 9. Time used on ordered trees on T8M dataset varying window size

when dataset size reached 8 million trees: not being an incremental method, it must store the whole dataset in memory all the time *in addition* to the lattice structure, in contrast with our algorithms.

In Figure 9 we compare WinTreeMiner with different window sizes to AdaTreeMiner on T8M dataset. We observe that the two versions of AdaTreeMiner outperform WinTreeMiner for all window sizes.

7 Conclusions

The scheme for classification based on our methods, efficiently selects a reduced number of attributes, and achieves higher accuracy (even more in the more selective case in which we keep only attributes corresponding to maximal trees). Our approach to tree mining outperforms CMTreeMiner in time and memory consumption when the number of trees is huge, because CMTreeMiner is not an incremental method and it must store the whole dataset in memory all the time.

Song et al.[17] introduced the concept of relaxed frequent itemset using the notion of relaxed support. We can see relaxed support as a mapping from all possible dataset supports to a set of relaxed intervals. Relaxed closed mining is a powerful notion that reduces the number of closed subpatterns. We introduced the concept of logarithmic relaxed frequent pattern in [5]. Future work will be to apply this notion to our classification method by introducing an attribute for each relaxed frequent closed pattern, instead of one for each closed frequent pattern. Also, we would like to apply these classification methods to other kinds of patterns.

And finally, we would like to extend our work to generators [15]. In [2] the authors were interested in implications of trees of the form $G \rightarrow Z$, where G is a generator of Z. When $\Gamma(G) = Z$ for a set of trees $G \neq Z$ and G is minimal among all the candidates with closure equal to Z, we say that G is a generator of Z. Generator based representations contain the same information as the frequent closed ones. In the literature, there is no method for finding generators of trees in evolving data streams. As future work, we would like to compare classification using tree generators, with the classification methods presented in this paper.

Acknowledgments

Partially supported by the EU PASCAL2 Network of Excellence (FP7-ICT-216886), and by projects SESAAME-BAR (TIN2008-06582-C03-01), MOISES-BAR (TIN2005-08832-C03-03). Albert Bifet has been supported by a FI grant through the Grups de Recerca Consolidats (SGR) program of Generalitat de Catalunya.

References

[1] Arimura, H., Uno, T.: An output-polynomial time algorithm for mining frequent closed attribute trees. In: ILP, pp. 1–19 (2005)

[2] Balcázar, J.L., Bifet, A., Lozano, A.: Mining implications from lattices of closed trees. In: Extraction et gestion des connaissances (EGC 2008), pp. 373–384 (2008)

[3] Balcázar, J.L., Bifet, A., Lozano, A.: Mining frequent closed rooted trees. Accepted for publication in Machine Learning Journal (2009)

[4] Bifet, A., Gavaldà, R.: Learning from time-changing data with adaptive windowing. In: SIAM International Conference on Data Mining (2007)

[5] Bifet, A., Gavaldà, R.: Mining adaptively frequent closed unlabeled rooted trees in data streams. In: 14th ACM SIGKDD International Conference on Knowledge Discovery and Data Mining (2008)

[6] Bifet, A., Holmes, G., Pfahringer, B., Kirkby, R., Gavaldà, R.: New ensemble methods for evolving data streams. In: 15th ACM SIGKDD International Conference on Knowledge Discovery and Data Mining. ACM, New York (2009)

[7] Chi, Y., Xia, Y., Yang, Y., Muntz, R.: Mining closed and maximal frequent subtrees from databases of labeled rooted trees. Fundamenta Informaticae XXI, 1001–1038 (2001)

[8] Collins, M., Duffy, N.: New ranking algorithms for parsing and tagging: kernels over discrete structures, and the voted perceptron. In: ACL 2001, pp. 263–270 (2001)

[9] Datar, M., Gionis, A., Indyk, P., Motwani, R.: Maintaining stream statistics over sliding windows. SIAM Journal on Computing 14(1), 27–45 (2002)

[10] Garriga, G.C., Kralj, P., Lavrač, N.: Closed sets for labeled data. J. Mach. Learn. Res. 9, 559–580 (2008)

[11] Holmes, G., Kirkby, R., Pfahringer, B.: MOA: Massive Online Analysis (2007), http://sourceforge.net/projects/moa-datastream
[12] Kashima, H., Koyanagi, T.: Kernels for semi-structured data. In: ICML, pp. 291–298 (2002)
[13] Kudo, T., Maeda, E., Matsumoto, Y.: An application of boosting to graph classification. In: NIPS (2004)
[14] Kudo, T., Matsumoto, Y.: A boosting algorithm for classification of semi-structured text. In: EMNLP, pp. 301–308 (2004)
[15] Li, J., Li, H., Wong, L., Pei, J., Dong, G.: Minimum description length principle: Generators are preferable to closed patterns. In: AAAI (2006)
[16] Punin, J., Krishnamoorthy, M., Zaki, M.: LOGML: Log markup language for web usage mining. In: WEBKDD Workshop, with SIGKDD (2001)
[17] Song, G.-j., Yang, D.-q., Cui, B., Zheng, B., Liu, Y., Xie, K.-Q.: CLAIM: An efficient method for relaxed frequent closed itemsets mining over stream data. In: Kotagiri, R., Radha Krishna, P., Mohania, M., Nantajeewarawat, E. (eds.) DASFAA 2007. LNCS, vol. 4443, pp. 664–675. Springer, Heidelberg (2007)
[18] Termier, A., Rousset, M.-C., Sebag, M., Ohara, K., Washio, T., Motoda, H.: DryadeParent, an efficient and robust closed attribute tree mining algorithm. IEEE Trans. Knowl. Data Eng. 20(3), 300–320 (2008)
[19] Yan, X., Han, J.: CloseGraph: mining closed frequent graph patterns. In: KDD 2003, pp. 286–295. ACM Press, New York (2003)
[20] Zaki, M.J.: Efficiently mining frequent trees in a forest. In: KDD 2002 (2002)
[21] Zaki, M.J., Aggarwal, C.C.: XRules: an effective structural classifier for xml data. In: KDD 2003, pp. 316–325. ACM Press, New York (2003)

A Condensed Representation of Itemsets for Analyzing Their Evolution over Time

Mirko Boettcher[1], Martin Spott[2], and Rudolf Kruse[1]

[1] University of Magdeburg, Faculty of Computer Science, 39106 Magdeburg, Germany
{miboettc,kruse}@iws.cs.uni-magdeburg.de
[2] Intelligent Systems Research Centre, BT Group plc, Adastral Park, Ipswich IP5 3RE, United Kingdom
martin.spott@bt.com

Abstract. Driven by the need to understand change within domains there is emerging research on methods which aim at analyzing how patterns and in particular itemsets evolve over time. In practice, however, these methods suffer from the problem that many of the observed changes in itemsets are temporally redundant in the sense that they are the side-effect of changes in other itemsets, hence making the identification of the fundamental changes difficult. As a solution we propose temporally closed itemsets, a novel approach for a condensed representation of itemsets which is based on removing temporal redundancies. We investigate how our approach relates to the well-known concept of closed itemsets if the latter would be directly generalized to account for the temporal dimension. Our experiments support the theoretical results by showing that the set of temporally closed itemsets is significantly smaller than the set of closed itemsets.

1 Introduction

In many application areas data is being collected over a long time. Due to its temporal nature such data not only captures influences, like market forces or the launch of a new product, but also reflects the changes of the underlying domain. Often, change can mean a risk (like a shrinking subgroup of target customers) or an opportunity (like an evolving market niche). In either case, it is in many domains not only imperative to detect change in order to survive or to win but it is also essential for successful decision making to analyze and act upon it.

As a response to this need there is increasing research interest in methods which aim at analyzing the changes within a domain by describing and modeling how the results of data mining—models and patterns—evolve over time. The term change mining has been coined as an umbrella term for such methods [1]. Change mining approaches have been proposed for a variety of patterns and models. Nonetheless, many studies focus on analyzing change in the context of itemsets, not only because itemsets are rather comprehensible itself but also because their evolution can be represented in a convenient and interpretable way.

W. Buntine et al. (Eds.): ECML PKDD 2009, Part I, LNAI 5781, pp. 163–178, 2009.

It is generally assumed that itemsets cannot change in their symblic representation but only in quantitative measures which describe them, the most common of which is support, i.e. the frequency of occurrence within a data set. The evolution of itemsets is therefore captured in time series (also called histories), whereby most approaches only utilize support histories [2,3,4,5,6]. Nevertheless, they can often be adapted to other measures.

Consider, as an example, survey data which contains information about used telecommunication services, like broadband or phone, and the social background of customers, like their gender. Itemset change mining is applied to this dataset to discover evolving customer segments in a sociographical context. Assume that the following itemset which specifies one customer segment has been discovered: XY : BROADBAND=YES, GENDER=MALE. Figure 1 shows its support history describing the size of this segment relative to all customers over 20 periods. Change mining approaches would, for example, employ statistical tests [4,6], pattern matching in time series [2] or heuristics [5] to detect that this history shows many characteristic features which might be of interest to the expert: a trend turning point and a declining and inclining trend to the left, respectively to the right of it.

Generally, changing itemsets hint at changes in the underlying domain. Such changes may indicate that an intervening action is required, for instance, to rectify a problem [3]. On the other hand, an itemset which always remains stable in its support can be expected to describe an *invariant* of the domain. Invariants, however, are of less interest to experts because they are almost always known and usually do not indicate a serious problem [7]. Still, they may be of interest if the domain under consideration is in its basic underlying principles has not been fully understood, yet.

Besides being only of secondary interest, invariants also lead to a drastically larger number of itemsets with changes in their histories. To continue our example, assume that the fraction of males among all broadband users is an invariant of the domain. In this case, the change of X : BROADBAND=YES will be qualitatively the same as the one of XY : BROADBAND=YES, GENDER=MALE. Both histories will only differ by a scaling factor which, in turn, is the invariant. This is also shown in Figure 1. In the context of change mining it can be said that both itemsets are *temporally redundant* with respect to each other.

Temporal redundancy is very common in practice. Many of the changes observed in itemset histories are simply the effect of a combination of changes in other itemset histories with domain invariants. For an expert it will be a very tedious task to identify the fundamental changes in which he is primarily interested. This problem is even worsened by the vast number of itemsets which are discovered. For this reason it is desirable to obtain a *condensed representation* which captures the fundamental set of changing itemset histories and allows to reconstruct the shape of all other itemsets histories that are necessary for change mining. To our knowledge this problem has up to now neither been addressed in the field of change mining nor in the area of condensed representations of itemsets.

Our goal in this paper is twofold: first of all we want to contribute to the field of change mining by providing a condensed representation of itemsets which incorporates the temporal dimension. More precisely, we introduce *temporally closed itemsets* as an approach to reduce the number of itemsets by accounting for temporal redundancies. The set of temporally closed itemsets is minimal in the sense that the shape of every other itemset's history can be reconstructed from it. Here we follow the argument of Agrawal and Psaila [2] and Chakrabarti et al [3] that an itemset's relevance is primarily dictated by its qualitative change over time and not by its actual support value. Secondly, we want to tighten the link between itemset mining and itemset change mining by investigating how temporally closed itemsets relate to the well-known concept of closed itemsets. In particular, we show that the set of temporally closed itemsets is a subset of the set of closed itemsets if the notion of the latter is directly generalized to the temporal dimension. The subset property is the result of removing those redundancies from the set of closed itemsets which are only visible when itemsets are analyzed over time.

The remainder of this paper is organized as follows. In Section 2 we discuss related work. Section 3 and Section 4.1 introduce the necessary background on frequent itemset mining and closed itemsets. In Section 4.2 we discuss how closed itemsets can be straightforwardly generalized to be applicable to sequences of time periods in order to have a benchmark for our approach. In Section 5 we define temporal redundancy by introducing the concept of *temporally derivable itemsets*, which we will subsequently use in Section 6 as basis for the definition of the set of *temporally closed itemsets*. Section 7 discusses how the set of temporally closed itemsets can be discovered. Section 8 shows the experimental results we obtained.

2 Related Work

The approach described in this paper is related to two so far rather distinct fields of association mining: change mining and condensed representations. For this reason we will first provide an overview over existing change mining methods for associations, followed by some background on condensed representations.

Several methods have been proposed in the area of association mining which aim to discover interesting changes in histories of itemsets and association rules, respectively. Agrawal et al [2] proposed a query language for shapes of histories. Liu et al [4] showed how trend, semi-stable and stable rules can be distinguished using a statistical approach. In [3] the temporal description length of an itemset is introduced which rates support changes by using methods from information theory. Frameworks to monitor and analyse changes in support and confidence are described in [5,1]. None of these publications discusses how the set of discovered itemsets can be effectively reduced such that the shape of all other itemsets can still be derived, nor do they discuss how existing reduction techniques for itemsets can be extended towards the temporal dimension. Liu et al proposed a method to detect so-called *fundamental rule changes* that aims to identify changes in support and confidence of association rules which cannot be explained by other changes [8]. The authors provide heuristic criteria for solving this task. However, their approach differs to our approach of temporally closed itemsets because it can only be applied to histories of two periods length, whereas much longer histories are the norm when analyzing change. An extension to many periods is not straightforward due to the form of the underlying statistical test.

Several approaches have been proposed which lead to a condensed representation of the set of discovered itemsets such that all other itemsets can be derived from the representation. Three such techniques are: closed itemsets [9,10], counting inference [11] and deduction rules [12]. From the perspective of analyzing the change of itemsets over time these methods treat each element of a sequence of temporally ordered data sets independently from each other. For this reason, they do not have the capability to detect redundancies which are only visible if itemsets are analysed over time. Of these condensed representation approaches, closed itemsets are related to our approach. We will discuss them in more detail in Section 4.1.

3 Itemsets and Support Histories

Formally, itemset discovery is applied to a data set of *transactions*. Every transaction T is a subset of a set of items L. A subset $X \subseteq L$ is called *itemset*. It is said that a transaction T *supports* an itemset X if $X \subseteq T$. If $X \subset Y$ holds for two itemsets X and Y we will say that X is *more general* than Y because X puts less restrictions on the underlying transaction set. Likewise, we say that Y is *more specific* than X. Furthermore, we define $XY := X \cup Y$ for simplicity.

The statistical significance of an itemset X is measured by its *support* $\text{supp}(X)$ which estimates $P(X \subseteq T)$, or short $P(X)$. It is said that an itemset is *frequent* if its support is greater than or equal to a user-defined minimum support value supp_{min}. The *downward closure property* of itemsets states that for two itemsets $Y \supset X$ the support of X is greater or equal to the one of Y.

As other authors we define the change of an itemset by the change of its support over time. The time series of support values is called *support history*. Formally, let D be a time-stamped data set and $[t_0,\ t_n]$ the minimum time span that covers all its tuples. The interval $[t_0,\ t_n]$ is divided into $n > 1$ non-overlapping periods $T_i := [t_{i-1}, t_i]$, such that the corresponding subsets $D_i \subset D$ each have a size $|D_i| \gg 1$. After carrying out frequent itemset discovery for each D_i, $i = 1, \ldots, n$ the support of each itemset X is now related to a specific time period T_i. We will indicate this by using the notation $\text{supp}_i(X)$. An itemset X which has been discovered in all periods is therefore described by n support values. Imposed by the order of time the values form sequences $(\text{supp}_1(X), \ldots \text{supp}_n(X))$ which are also called *support histories*.

4 Closed Itemsets

4.1 Definition

Closed itemsets are a subset of itemsets from which all other itemsets can be derived without further mining. The formal underpinnings of closed itemset algorithms can be found in the theory of lattices and Galois connection closures [9]. Still, their meaning is rather intuitive: a closed itemset is the largest itemset common to a set of transactions. All non-closed itemsets have the same support as their closure, which is the smallest closed itemset containing them. This means that non-closed sets can be regarded redundant.

Formally, a closed itemset is defined as follows (cf. [9]):

Definition 1 (Closed Itemset). *An itemset X is a closed itemset iff there exists no proper superset $Y \supset X$ such that* $\text{supp}(X) = \text{supp}(Y)$.

Equivalently, an itemset X is called *non-closed* iff there exists a proper superset Y such that $\text{supp}(X) = \text{supp}(Y)$. The largest of such supersets is also called the *closure* of X.

As already mentioned in Section 2 closed itemsets can only be applied to a single data set. As a result, they do not account for redundancies imposed by the temporal dimension, where we deal with a sequence of data sets.

4.2 Generalization to a Sequence of Time Periods

To analyze the approach proposed in this paper it is desirable to have an established condensed representation with which it can be compared, both theoretically and experimentally. Although closed itemsets are probably the most widely used condensed representation they cannot directly be used for this purpose because they were not developed with the temporal dimension in mind. For this reason the definition of closed itemsets given in the previous section has to be generalized from a single data set corresponding to one time period to a sequence of time periods.

On the one hand, from such a generalization one would expect that it is straightforward in the sense that it resembles the original definition of closed itemsets as good as possible. On the other hand, the generalization should be sufficient for the purpose of change analysis: an itemset should only be regarded as *non-closed over a sequence of time periods* if no change information is lost.

As an example how change information can be lost consider a non-closed itemset whose closure varies across periods. Should this non-closed itemset be regarded as closed or as non-closed over a sequence of time periods? As laid out in the Introduction a notion of redundancy in the context of change analysis should be based on invariants of the domain. If an itemset with varying closure would be regarded as non-closed over a sequence of time periods and thus not presented to a user this change information is lost. Consequently, it is reasonable to regard itemsets whose closure varies across periods as closed in the context of a generalization of closed itemsets towards sequences of time periods.

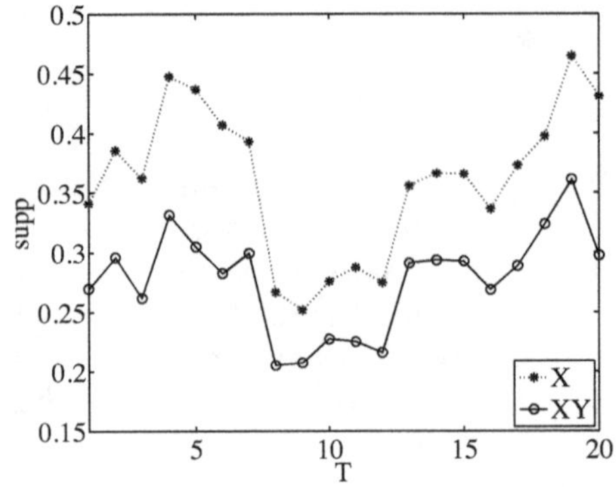

Fig. 1. Histories of the itemsets XY and X showing that $X \hookrightarrow XY$

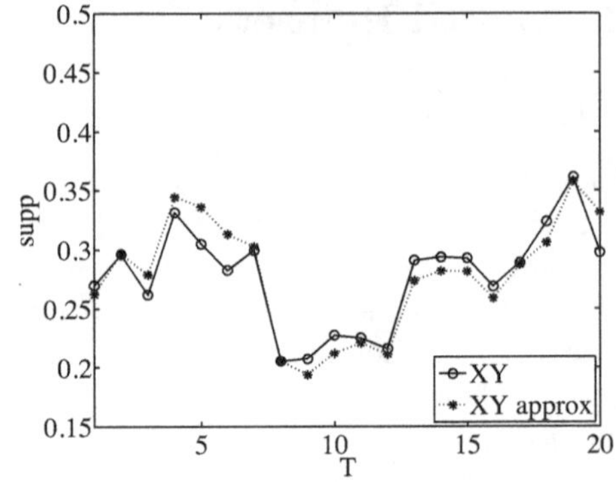

Fig. 2. Approximated history of XY using the history of X

Generally, two requirements have to be met by a temporal generalization of closed itemsets. In the first place, an itemset X should be non-closed over a sequence of time periods only if it is non-closed in all periods. Secondly, following the discussion in Section 1 the underlying reason for non-closedness should be an invariant, i.e. the superset Y with equal support (the closure Y of X) should be the same in every time period.

A direct temporal generalization of non-closed itemsets which meets the requirements above is the following: An itemset is *non-closed over a sequence of time periods* $T_i, i = 1, \ldots, n$ iff there exists an itemset $Y \supset X$ such that for all periods $\text{supp}_i(X) = \text{supp}_i(Y)$ $i = 1, \ldots, n$. Equivalently, itemsets which are *closed over a sequence of time periods* are defined as follows:

Definition 2 (Closed over a Sequence). *An itemset X is closed over the sequence of time periods $\{T_1, \ldots, T_n\}$ iff there exists no itemset $Y \supset X$ such that* $\text{supp}_i(X) = \text{supp}_i(Y), i = 1, \ldots, n$.

The above definition means that the set of itemsets which are closed over a sequence of time periods contains all itemsets which are closed in at least one period. Additionally, it also contains itemsets X which are non-closed in each individual time period but for which their closure differs across periods. It should be noted, however, that the latter is an extremely rare constellation as it has already been reported by authors in the field of incremental itemset mining (cf. [13]). Indeed, we did not observe non-closed itemsets with temporally varying closure in the data sets we used for our experiments.

If the context is clear we will for brevity refer to the *direct temporal generalization of closed itemsets* introduced in this section simply as closed itemsets.

5 Temporally Derivable Itemsets

As laid out in the Introduction, the aim is to find a set of itemsets which is non-redundant in the sense that it is the minimal set necessary to derive the shape of the history of all remaining itemsets. We therefore first have to define what makes a history of an itemset XY derivable from the history of X and thus the itemset XY *temporally derivable*:

Definition 3 (Temporally Derivable Itemset). *Let* $XY, X \neq \emptyset$ *be an itemset and* $(\text{supp}_1(XY), \ldots, \text{supp}_n(XY))$ *its support history. The itemset* XY *is temporally derivable with regard to an itemset* X*, denoted* $X \hookrightarrow XY$*, iff for each* $XZ, Z \subseteq Y$ *with support history* $(\text{supp}_1(XZ), \ldots, \text{supp}_n(XZ))$ *there exists a constant* $\epsilon, 0 < \epsilon \leq 1$ *such that* $\text{supp}_i(XY) = \epsilon\, \text{supp}_i(XZ), i = 1, \ldots, n$.

The main idea behind the definition is that the history of an itemset and hence the itemset itself is temporally derivable if it has the same shape as the history of a more general itemset apart from a scaling factor ϵ. To emphasize the scaling factor ϵ we will sometimes use the notation $X \overset{\epsilon}{\hookrightarrow} Y$. From the criterion $\text{supp}_i(XY) = \epsilon\, \text{supp}_i(X), i = 1, \ldots, n$ used within the definition it directly follows that ϵ decreases with increasing size of Y, i.e. for $X \overset{\epsilon_1}{\hookrightarrow} Y$ and $X \overset{\epsilon_2}{\hookrightarrow} Z$ with $Y \subset Z$ it is $\epsilon_2 \leq \epsilon_1$. The criterion $\text{supp}_i(XY) = \epsilon\, \text{supp}_i(X), i = 1, \ldots, n$ can also be rewritten as $\epsilon = \text{supp}_i(XY)/\text{supp}_i(X) = P(XY \mid T_i)/P(X \mid T_i) = P(Y \mid XT_i)$. This means, the probability of Y is required to be constant over time given X, so the fraction of transactions containing Y additionally to X constantly grows in the same proportion as X. In other words, the confidence (represented by the scaling factor ϵ) of the rule $X \rightarrow Y$ does not change over time. Such time-invariant properties, however, often represent domain knowledge known to a user. Thus, a user would be able to infer the history of XY if he knows the one of X. In the opposite direction, he could also derive the history of X from the one of XY.

Figures 1 and 2 show an example of a temporally derivable itemset taken from the customer survey data used for our experiments, cf. Section 8. For reasons of data protection, the underlying itemset cannot be revealed. For illustration, the reader is referred to the example given in the Introduction, instead. Figure 1 shows the support histories of the less specific itemset at the top and the more specific itemset below, both over 20 time periods. The shape of the two histories is obviously very similar and it turns out that the history of the more specific itemset XY can approximately be determined using the more general one X by applying a scaling factor. As shown in Figure 2, the reconstruction is not exact. The reason for this is noise. As a result, a statistical test is employed in Section 7.2 to test for temporal derivability. Obviously, the history of the less specific itemset could be determined from the more specific in the same way. In the following we will show several properties of temporally derivable itemsets which we will use later on in this paper:

Lemma 1. *All itemsets are temporally derivable with regard to themselves, i.e.* $X \hookrightarrow X$.

Proof. Lemma 1 follows directly from Definition 3.

Lemma 2. *If* $X \overset{\epsilon_1}{\hookrightarrow} Y$ *and* $Y \overset{\epsilon_2}{\hookrightarrow} Z$ *then* $X \overset{\epsilon_1 \epsilon_2}{\hookrightarrow} Z$*, i.e. derivability is transitive.*

Proof. By Definition 3 it is $\epsilon_1 \sup_i(X) = \sup_i(Y)$ and $\epsilon_2 \sup_i(Y) = \sup_i(Z)$. Substitution yields $\epsilon_1\epsilon_2 \sup_i(X) = \sup_i(Z)$ and thus $X \overset{\epsilon_1\epsilon_2}{\hookrightarrow} Z$.

6 Temporally Closed Itemsets

If XY is temporally derivable from X then XY and Y have histories with qualitatively the same shape. Further, if we assume that the relevance of an itemset is primarily determined by the qualitative changes represented in its history both itemsets, X and XY, would have the same interestingness. For example, in Figure 1 both histories show all characteristic features that would make them interesting for a user: a trend turning point and a declining, respectively inclining, trend left and right from it. Hence, if one is known the other can be regarded as temporally redundant.

Commonly, sequences of itemsets $X_1 \hookrightarrow X_2 \ldots \hookrightarrow X_n$ temporally derivable from each other are discovered. Thereby, we assume that this sequence is maximal in the sense that there exists no $Y \subset X_1$ or $Z \supset X_n$ such that $Y \hookrightarrow X_1$ or $X_n \hookrightarrow Z$, respectively. From such a sequence we will define the maximum element X_n as being non-redundant and treat the others as redundant. We will call such non-redundant itemsets *temporally closed itemsets* because they are related to closed itemsets as we prove later in this section.

Definition 4 (Temporally Closed Itemset). *An itemset X is temporally closed iff there exists no itemset $Y \supset X$ such that $X \hookrightarrow Y$.*

Apparently, from the above sequence $X_1 \hookrightarrow X_2 \ldots \hookrightarrow X_n$ the minimum element X_1 could also have been chosen as the non-redundant element. Nevertheless, the choice of the maximum X_n as the basis for the definition of temporally closed itemsets provides the advantage that in this way they can be related to closed itemsets and thus extending this established notion by temporal considerations.

To analyze how temporally closed itemsets relate to the temporal generalization of closed itemsets discussed in Section 4.2 the definition of an itemset which is closed over a sequence of time periods needs to be linked to the notion of temporal derivability. By comparing Definition 2 with Definition 3 it can be seen that the link between the two concepts can be expressed as follows:

Lemma 3. *An itemset X is closed over the sequence of time periods $\{T_1, \ldots, T_n\}$ iff there exists no itemset $Y \supset X$ such that $X \overset{1}{\hookrightarrow} Y$.*

Proof. Follows directly from the definition of a temporally derivable itemset (cf. Definition 3).

We now have the necessary tools to prove the central theorem of this paper which shows that temporally closed itemsets are a subset of the direct temporal generalization of closed itemsets introduced in Section 4.2, i.e. a subset of the itemsets which are closed over a sequence of time periods.

Theorem 1. *Let C be the set of all closed itemsets over the sequence of time periods $\{T_1, \ldots, T_n\}$ and TC be the set of temporally closed itemsets. Then, it is $TC \subseteq C$.*

Proof.

$$\begin{aligned} X \in TC &\overset{Def. 4}{\iff} \nexists Y \supset X : \exists \epsilon \in (0,1] : X \overset{\epsilon}{\hookrightarrow} Y \\ &\implies \nexists Y \supset X : X \overset{1}{\hookrightarrow} Y \\ &\overset{Lemma\ 3}{\iff} X \in C \end{aligned}$$

From $X \in TC \Rightarrow X \in C$ it follows that $TC \subseteq C$.

The following counterexample shows that TC can indeed be a proper subset of C. Consider the itemsets $X_1 \subset X_2 \subset X_3 \subset X_4$ with $X_1, X_3 \in C$. Further, assume that $X_1 \overset{0.5}{\hookrightarrow} X_2 \overset{1}{\hookrightarrow} X_3 \overset{0.5}{\hookrightarrow} X_4$. Using Lemma 2 it is $X_1 \hookrightarrow X_4$ and $X_3 \hookrightarrow X_4$. Using Definition 4 it follows that $X_1 \notin TC$ and $X_3 \notin TC$.

This means, every temporally closed itemset is also closed over a sequence of time periods but not every itemset which is closed over a sequence is also a temporally closed one. The counterexample shows that a temporally closed itemset can be temporally derivable from multiple closed itemsets. As we will see in our experiment results in Section 8 temporally closed itemsets form a (almost always proper) subset of directly generalized closed itemsets in which temporal redundancies have been removed. The set of temporally closed itemsets can in fact be significantly smaller than the set of closed ones as we will demonstrate in our experimental evaluation in Section 8. At the same time, temporally closed itemsets are lossless in the sense that they can be used to uniquely determine the shape of the histories of all remaining itemsets.

7 Discovery Procedure

To obtain the set of temporally closed itemsets we use a two step approach in which we first generate a set of candidate itemsets and then test every candidate whether it is temporally closed, or not. This two step procedure will be detailed in the following.

7.1 Candidate Generation

A naive approach to obtain a candidate set would be to consider the set of itemsets which are frequent in every time period. Because this set is usually vast the subsequent testing step would be very time consuming. A more efficient approach is to restrict the candidates to the set C of frequent itemsets which are closed over a sequence of time periods. According to Theorem 1 this set is a superset of the set of temporally closed itemsets TC. It is, however, by several factors smaller than the set of all frequent itemsets.

Given a time-stamped data set D and a segmentation $T_i := [t_{i-1}, t_i]$ of the covered time span $[t_0,\ t_n]$, D is divided into the corresponding subsets $D_i \subset D$. From this sequence of temporally ordered, disjunct data sets $D_1, \ldots, D_n$ the candidate set C can be obtained in three steps.

Mining Closed Itemsets in each Period. An algorithm for closed itemset mining is applied to each data set which yields a sequence of closed itemset sets $C_1, \ldots, C_n$.

Merging the Sets of Closed Itemsets. The closed itemset sets $C_1, \ldots, C_n$ have to be inserted into the candidate set C. We start by setting $C := C_1$ and converting C into a data structure in which each node represents an itemset and maintains a list of parents (largest subsets) and children (smallest supersets). We then subsequently insert the remaining C_i in C whereby the list of parents and childrens in each node is being maintained. Assume that $C_1, \ldots, C_{j-1}$ have already been inserted in C and that $X \in C_j$ is the next itemset to be inserted. The following scenarios are possible:

- *X already exists in C:* In this case the support of X in the data set D_j is inserted at position j in the support history of X in C.
- *X does not exist in C:* In this case it is checked whether there exists a $Y \in C$ such that $X \subset Y$. If no such proper superset exists, X was infrequent in earlier periods and is not inserted because a complete support history of it cannot be obtained anymore. If a proper superset exists X was frequent but non-closed in earlier periods and thus is inserted into C. The support of X in the data set D_j is inserted at position j in the support history of X in C.

Complexity. Assume that m is the length of the longest itemset, that the number of different items is p, and that all C_i have equal size. The initial computational effort to create the data structure for C is $O(|C|^2)$. The effort to search an itemset X in C is bound by the length m of the longest possible path in C. To insert an itemset X into C after the smallest proper superset has been found all largest proper subsets and all smallest proper supersets of X in C need to be accessed and their parent list, respectively children list, updated. This can be accomplished in $O(m+p)$ because every itemset cannot have more than m parent nodes and more than p child nodes. Under the assumption that always a fraction α of itemsets in C_i needs to be inserted into C the overall complexity of inserting the sets $C_2, \ldots, C_n$ into C therefore is $O(|C_i|^2 + mn|C_i| + \alpha n(m+p)|C_i|)$. A bottleneck is the quadratic effort for creating the initial data structure. This effort, however, can be avoided by using in the previous step a closed itemset miner like CHARM-L [10] which outputs the data structure directly.

Handling Incomplete Histories. It needs to be checked whether C contains itemsets with incomplete histories. An incomplete history can occur either because an itemset was non-closed or infrequent in some but not all periods, respectively. We iterate over each element $X \in C$ starting with the largest itemsets. We visit itemsets with no children always first and visit all others in order of size. The following two scenarios for itemsets X with incomplete histories are possible:

- *The itemset X has no children:* This implies that in at least one period either X or its closure were infrequent. In this case it is not possible to obtain a

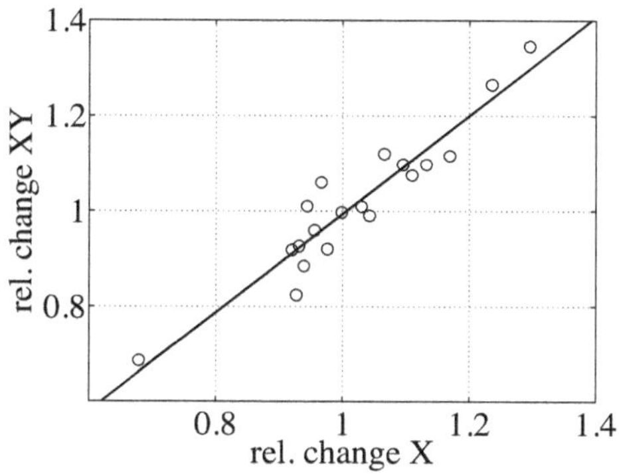

Fig. 3. Scatter plot of the relative changes of the support histories shown in Figure 1. The fitted regression line is $\Delta\,\mathrm{supp}(XY) = 1.0332 \cdot \Delta\,\mathrm{supp}(X) - 0.0396$ and the correlation coefficient $r \approx 0.9545$.

complete support history of X, thus the itemset is removed from C and the children list of its parents updated accordingly.

– *The itemset X has children:* This implies that the itemset X was non-closed in at least one period. Assuming that its support is missing in period j then one of the children of X must have a support value for this period, either because it is the closure of X in j or because the support value was completed in an earlier step. Due to X being non-closed in at least one period, there may exist subsets $Z \subset X$ which are non-closed in every period but have a different closure across periods (one of which is X). Following the discussion in Section 4.2 these itemsets are closed over a sequence of time periods and thus also part of the candidate set C. For this reason, each itemset $Z \subset X$ for which no itemset $X' \in C$ exists such that $X' \subset Z \subset X$ is also inserted into C.

Complexity. We use the same notation as in the complexity analysis before. Additionally we assume that a fraction β of itemsets in C is infrequent and that a fraction γ is non-closed in some but not all periods, respectively. In the first scenario for each deleted itemset the children list of each parent needs to be updated. Because each itemset can have at most m parents the computational effort thus is $O(m\beta|C|)$. In the second scenario, for each of the $\gamma|C|$ itemsets each direct child needs to be accessed. Since the number of children is bound by p the computational effort therefore is $O(p\gamma|C|)$. Further the set NC of itemsets which were non-closed in each period but had a varying closure need to be inserted. As in the previous step the complexity of inserting an itemset is $O(m+p)$. Taking into account that every itemset with complete history needs to be accessed once the overall complexity therefore is $O((1+(m-1)\beta+(p-1)\gamma)|C|+(2p+m)|NC|)$. We do not have an estimation for β, γ and $|NC|$ but in our experiments we neither encountered itemsets with varying closedness nor with varying closure. This gives rise to the assumption that γ and $|NC|$ will be very small in practice. Concerning β we observed that typically $10-20\%$ of the itemsets in C had an incomplete history due to infrequency in some periods.

7.2 Testing for Temporal Closedness

To check whether an itemset $X \in C$ is temporally non-closed we need to test whether an itemset XY exists which can be temporally derived from X. This, in turn, means we have to test whether ϵ in $\text{supp}_i(XY) = \epsilon\,\text{supp}_i(X), i = 1, \ldots, n$ is constant over time. Due to data usually being noisy as we showed in Figure 2, we will not check this criterion directly, but instead statistically test its validity. Also, we rewrite the criterion in an equivalent form to account for the order of values over time in the histories. Our experiments have shown that direct use of the criterion counterintuitively marked some histories as temporally derivable when they were noisy.

Let $\Delta_i\,\text{supp}(X) := \frac{\text{supp}_i(X)}{\text{supp}_{i-1}(X)}$ be the relative change in support for itemset X between two periods T_{i-1} and $T_i, i = 2, \ldots, n$. Then, the above criterion holds, iff $\Delta_i\,\text{supp}(XY) = \Delta_i\,\text{supp}(X)$ for any $i = 2, \ldots, n$. This means, if the itemset XY is temporally derivable from X then the relative changes in the history of XY are equal to the temporally related relative changes in the history of the itemset X.

Imagine $\Delta_i\,\text{supp}(X)$ and $\Delta_i\,\text{supp}(XY)$ in a plotted graph, whereby – as implied by Definition 3 – $\Delta_i\,\text{supp}(XY)$ is the dependent quantity. If $\Delta_i\,\text{supp}(XY) = \Delta_i\,\text{supp}(X)$ holds, then all points in the plot should be on a straight line with slope 1 and intercept 0. In practice, however, this equality will rarely hold due to noise. As a solution, we model the underlying relationship as $\Delta_i\,\text{supp}(XY) = \Delta_i\,\text{supp}(X) + \gamma$ where γ is a random error with zero mean and unknown, but low variance.

Under the assumption that the dependency of $\Delta_i\,\text{supp}(XY)$ from $\Delta_i\,\text{supp}(X)$ can be generally described by $\Delta_i\,\text{supp}(XY) = a \cdot \Delta_i\,\text{supp}(X) + b + \gamma$, we fit a regression line $\Delta\,\text{supp}(XY) = \hat{a} \cdot \Delta\,\text{supp}(X) + \hat{b}$. The parameters $\hat{a}$ and $\hat{b}$ are estimates for a and b and obtained by minimizing the regression error. We then test if $\Delta_i\,\text{supp}(X)$ is statistically equal to $\Delta_i\,\text{supp}(XY)$ by carrying out the following two steps:

1. Based on the estimates $\hat{a}$ and $\hat{b}$ we test the hypothesis that the true parameters of the model are $a = 1$ and $b = 0$ using a standard t-test.
2. Additionally, we test whether the variance of γ is small, i.e. whether the $(\Delta_i\,\text{supp}(X), \Delta_i\,\text{supp}(XY))$ are sufficiently close to the regression line, by setting a threshold $\tilde{r}$ for Pearson's correlation coefficient r.

Figure 3 illustrates the testing procedure. It shows the scatter plot of the relative changes of the support histories from Figure 1. The fitted regression line is $\Delta\,\text{supp}(XY) = 1.0332 \cdot \Delta\,\text{supp}(X) - 0.0396$ and the correlation coefficient $r \approx 0.9545$. The above test procedure using a significance level of 0.05 and $\tilde{r} = 0.95$ shows that XY is indeed temporally derivable from the history of X.

Complexity. The complexity of this step is apparently $O(|C|)$ because each itemset's parent can be directly accessed through the parent list.

8 Experimental Results

As Theorem 1 as the central result of this publication states temporally closed itemsets form a subset of those itemsets which are closed over a sequence of time periods. The set of itemsets which are closed over a sequence of time periods also forms the candidate set to be tested for temporal closedness. For this reason, the question to be answered experimentally is how much the set of temporally closed itemsets is smaller than the set of itemsets which are closed over a sequence.

For our experiments we chose two data sets. One data set, here called CRS, is extracted from the data-warehouse of a telecommunication company. The other data set we extracted from the IPUMS project[1] [14] which is dedicated to collecting, harmonizing and freely distributing census data.

The CRS data set contains answers of customers to a survey collected over a period of 20 weeks. Each record is described by 19 nominal attributes with a domain size between 2 and 9. We transformed the data set into a transaction set by recoding every (attribute, attribute value) combination as an item. Then we split the transaction set into 20 subsets, each corresponding to a period of one week. The subsets contain between 385 and 547 transactions.

The data set we extracted from IPUMS contains census data of the USA collected during the years 2001–2006. Due to the data set being vast we restricted the data to the states New Jersey, New York, and Pennsylvania. From the available attributes we selected 15 concerning the person himself (e.g. age, race, gender), the house they are living in (e.g. number of bedrooms, year of built), and their profession (e.g. travel time, avg. hours worked per week, net income). Numeric attributes were converted into nominal ones using uniform binning. The domain size of the attributes varies between 2 and 9. We split the data set year-wise resulting in six data sets each containing between 130364 (for 2002) and 397788 (for 2006) records. We applied the same preprocessing steps as for the CRS data.

For each data set we then obtained the candidate set to be tested for temporal closedness. To each subset of each data set we applied a closed itemset miner[2] using 11 different minimum support thresholds in steps of 0.01 in the range from $\text{supp}_{\text{min}} = 0.05$ to $\text{supp}_{\text{min}} = 0.15$. For each value of supp_{min} we then generated from the obtained 20 sets of closed itemsets for CRS, respectively 6 sets for IPUMS, the candidate sets as described in Section 7.

For both data sets we observed that the closed itemsets were the same in every period, i.e. closed itemsets did not turn into non-closed ones and vice versa. This indicates that such events are very rare. It also implies that for both data sets the candidate set is equal to the set of closed itemsets in each period.

We tested the elements of the candidate sets for temporally closed itemsets by applying Definition 4 in combination with the test procedure in Section 7.2. We also investigated the number of itemsets which are closed over the sequence of

---

[1] http://usa.ipums.org/usa/

[2] We used the frequent closed itemset miner contained within the *apriori* software package by Ch. Borgelt. It can be obtained from http://borgelt.net/fpm.html

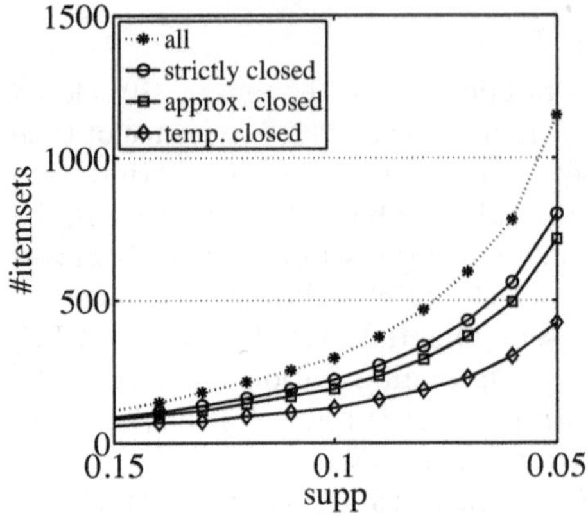

Fig. 4. Results for the CRS data set

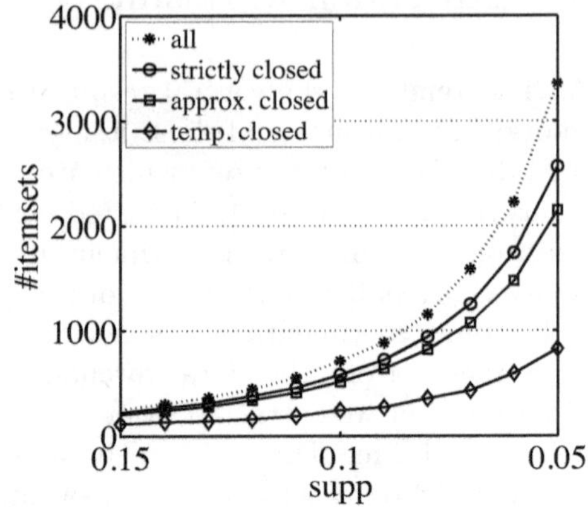

Fig. 5. Results for the IPUMS data set

time periods. Here, we employed two approaches. The first one uses the original definition which requires strict equality of support values (cf. Definition 2). To rule out the effects of low quality data we also tested for *approximate closedness*, i.e. we regarded an itemset as non-closed if its support value is approximately the one of a more general itemset. Here, we applied the test from Section 7.2 to the candidate set extended by an additional test for $\epsilon > 0.98$ because for *strict closedness* it must be $\epsilon = 1$ (cf. Lemma 3).

To compare the number of itemsets returned by our approach with the overall number of frequent itemsets (i.e. closed and non-closed frequent itemsets) we also applied a frequent itemset miner to both data sets using the same support threshold as for the other experiments. We only kept those itemsets which were frequent in every period.

The experimental results for the CRS data set are shown in Figure 4 and for the IPUMS data set in Figure 5 . Both figures show the the number of all frequent itemsets discovered and the number of itemsets which are temporally closed, approximately closed, and strictly closed. As can be seen, the approach of temporally closed itemsets leads to a significant reduction in the number of itemsets compared to both closed itemset approaches. For example, for $\text{supp}_{\text{min}} = 0.05$ mining only for closed itemsets reduces the CRS result set to roughly 69% and the IPUMS result set to roughly 76% of its initial size while the temporally closed itemset approach leads to a reduction of 36% and 24%, respectively. This means, for the CRS data the set of temporally closed itemsets is by a factor of 1.7 smaller than the set of strictly closed itemsets. For the IPUMS data this factor is with 3.1 even better.

Figure 6 and Figure 7 show for $\text{supp}_{\text{min}} = 0.05$ how the factor ϵ is distributed which maps the history of a non-temporally closed itemset to the smallest temporally closed itemset derivable from it. As we may expect from the results in Figure 4 and Figure 5 the range of ϵ is spread over a large range approximately $[0.75, 1]$ for the CRS data and $[0.4, 1]$ for the IPUMS data. The bar on the very right side in

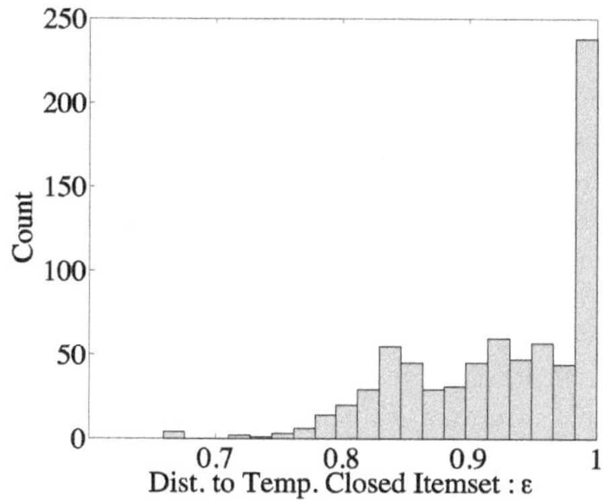

Fig. 6. Histogram of the distance ε of non-temporally closed itemsets to the corresponding temporally closed one for the CRS data.

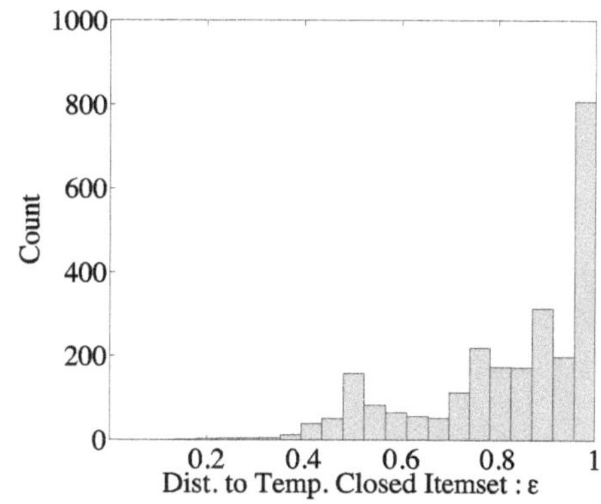

Fig. 7. Histogram of the distance ε of non-temporally closed itemsets to the corresponding temporally closed one for the IPUMS data.

each histogram ($\epsilon \approx 1$) roughly indicates the number of itemsets that would have been discarded by the generalized closed itemset approach described in Section 4.2. Our approach, in contrast, would discard every itemset shown in the histogram and thus reduce the set of closed itemsets by a very large extent.

9 Conclusion

Many businesses collect huge volumes of time-stamped data about all kinds of processes. This data reflects changes in the underlying domain. It is crucial for the success of most businesses to detect these changes, and finally to adapt or react to them. As a response to this need there is emerging research on data mining methods which aim at understanding change within a domain by analyzing how patterns evolve over time. Several studies have been conducted on analyzing how itemsets change over time. However, these approaches do not account for temporal redundancies, i.e. the problem that many of the observed changes are simply the side-effect of other changes.

In this paper we solved this problem by introducing temporally closed itemsets as a condensed representation of itemsets which is, on the one hand, free of temporal redundancies but which, on the other hand, still contains all the information needed for change analysis. Based on temporally closed itemsets it is possible to derive the shape of the history of all other itemsets. In particular, we showed that temporally closed itemsets are a subset of the set of closed itemsets if the definition of the latter would be directly generalized to be applicable to sequences of time periods. Our experiments not only demonstrated that temporally closed itemsets do exist in real-world data. We also showed that the set of temporally closed itemsets can be smaller than the set of closed itemsets by a factor of two to three and by orders of magnitude smaller than the set of initially discovered itemsets.

References

1. Böttcher, M., Spiliopoulou, M., Höppner, F.: On exploiting the power of time in data mining. SIGKDD Explorations Newsletter 10(2), 3–11 (2008)
2. Agrawal, R., Psaila, G.: Active data mining. In: Fayyad, U.M., Uthurusamy, R. (eds.) Proceedings of the 1st ACM SIGKDD International Conference on Knowledge Discovery and Data Mining, Montreal, Quebec, Canada, pp. 3–8. AAAI Press, Menlo Park (1995)
3. Chakrabarti, S., Sarawagi, S., Dom, B.: Mining surprising patterns using temporal description length. In: Proceedings of the 24th International Conference on Very Large Databases, pp. 606–617. Morgan Kaufmann Publishers Inc., San Francisco (1998)
4. Liu, B., Ma, Y., Lee, R.: Analyzing the interestingness of association rules from the temporal dimension. In: Proceedings of the IEEE International Conference on Data Mining, pp. 377–384. IEEE Computer Society Press, Los Alamitos (2001)
5. Spiliopoulou, M., Baron, S., Günther, O.: Efficient monitoring of patterns in data mining environments. In: Kalinichenko, L.A., Manthey, R., Thalheim, B., Wloka, U. (eds.) ADBIS 2003. LNCS, vol. 2798, pp. 253–265. Springer, Heidelberg (2003)
6. Böttcher, M., Spott, M., Nauck, D., Kruse, R.: Mining changing customer segments in dynamic markets. Expert Systems with Applications 36(1), 155–164 (2009)
7. Berger, C.R.: Slippery slopes to apprehension: Rationality and graphical depictions of increasingly threatening trends. Communication Research 32(1), 3–28 (2005)
8. Liu, B., Hsu, W., Ma, Y.: Discovering the set of fundamental rule changes. In: Proceedings of the 7th ACM SIGKDD International Conference on Knowledge Discovery and Data Mining, pp. 335–340 (2001)
9. Pasquier, N., Bastide, Y., Taouil, R., Lakhal, L.: Efficient mining of association rules using closed itemset lattices. Information Systems 24(1), 25–46 (1999)
10. Zaki, M.J., Hsiao, C.J.: Efficient algorithms for mining closed itemsets and their lattice structure. IEEE Transactions on Knowledge and Data Engineering 17(4), 462–478 (2005)
11. Bastide, Y., Taouil, R., Pasquier, N., Stumme, G., Lakhai, L.: Mining frequent patterns with counting inference. SIGKDD Explorations Newsletter 2(2), 66–75 (2000)
12. Calders, T., Goethals, B.: Mining all non-derivable frequent itemsets. In: Elomaa, T., Mannila, H., Toivonen, H. (eds.) PKDD 2002. LNCS (LNAI), vol. 2431, pp. 74–85. Springer, Heidelberg (2002)
13. Chi, Y., Wang, H., Yu, P.S., Muntz, R.R.: Catch the moment: maintaining closed frequent itemsets over a data stream sliding window. Knowledge and Information Systems 10(3), 265–294 (2006)
14. Ruggles, S., Sobek, M., Alexander, T., Fitch, C.A., Goeken, R., Hall, P.K., King, M., Ronnander, C.: Integrated public use microdata series: Version 4.0, machine-readable database. Minnesota population center, Minneapolis (producer and distributor) (2008)

Non-redundant Subgroup Discovery Using a Closure System

Mario Boley and Henrik Grosskreutz

Fraunhofer IAIS, Schloss Birlinghoven, Sankt Augustin, Germany
{mario.boley,henrik.grosskreutz}@iais.fraunhofer.de

Abstract. Subgroup discovery is a local pattern discovery task, in which descriptions of subpopulations of a database are evaluated against some quality function. As standard quality functions are functions of the described subpopulation, we propose to search for equivalence classes of descriptions with respect to their extension in the database rather than individual descriptions. These equivalence classes have unique maximal representatives forming a closure system. We show that minimum cardinality representatives of each equivalence class can be found during the enumeration process of that closure system without additional cost, while finding a minimum representative of a single equivalence class is **NP**-hard. With several real-world datasets we demonstrate that search space and output are significantly reduced by considering equivalence classes instead of individual descriptions and that the minimum representatives constitute a family of subgroup descriptions that is of same or better expressive power than those generated by traditional methods.

1 Introduction

Subgroup discovery [2,12,17] is a local pattern discovery task: descriptions of subpopulations of a database are evaluated against some real-valued quality function, and those descriptions exceeding some given minimum quality are returned to the user. The quality functions commonly used in this course like Piatetsky-Shapiro, binomial test, or Gini-index (see [12] for a list) are *functions of the extension* of a subgroup description. Traditional subgroup discovery algorithms, however, search in the space of subgroup descriptions, usually conjunctions of attribute/value equality constraints, rather than in the space of extensions. Since many descriptions can have an identical extension on the given data, this may lead to (i) many redundant evaluations of the quality function and (ii) to a result set that contains multiple descriptions of the same subpopulation.

In contrast we propose to consider extension based *equivalence classes* of subgroup descriptions rather than individual descriptions. Thereby our algorithms implicitly search in the space of subgroup extensions, and, consequently, they have a potentially reduced search space and return at most one description of each extension, *a representative*, to the user. For this purpose we propose to use descriptions with a *minimum number of constraints*, i.e., a minimum representative. This choice is motivated by the common hypothesis that (i) short

W. Buntine et al. (Eds.): ECML PKDD 2009, Part I, LNAI 5781, pp. 179–194, 2009.

descriptions are easier to interpret than long descriptions and (ii) in scenarios in which subgroups are used as building blocks for global models short descriptions lead to better generalization. In summary we consider the following computational problem:

*Problem 1 (*NON-REDUNDANT-SUBGROUP-DISCOVERY*). Given* a dataset, a quality function q, and a minimum quality threshold q^*, *list* a result set $\mathcal{R}$ of subgroup descriptions that satisfies

completeness, i.e., for all subgroup descriptions H with $q(H) \geq q^*$ there is an $H' \in \mathcal{R}$ that has the same extension as H,
non-redundancy, i.e., for all pairs of distinct subgroup descriptions $H, H' \in \mathcal{R}$ the extensions of H and H' are distinct, and
representative minimality, i.e., for all listed descriptions $H \in \mathcal{R}$ there is no shorter description H' having the same extension.

In addition we also discuss the standard problem variations of mining only representatives of the *top-k* quality equivalence classes as well as mining only classes having a representative not exceeding a given *length-limit.*

Results and Contribution. We formalize extension based equivalence classes and show that they theoretically can subsume an exponential number of individual subgroup descriptions. Thus, searching equivalence classes rather than individual descriptions has the potential to reduce search space and output tremendously. Indeed, as we show in an empirical study, a significant reduction can also be observed on ten well-known real-world datasets.

While each equivalence class has a unique maximal representative, we show that the number of minimal and minimum representatives can grow exponentially in the length of that maximal description, and it is **NP**-hard, given some description, to compute an equivalent description of minimum length. This result is complemented by our observation that a simple greedy strategy approximates a minimum representative within a logarithmic factor.

We use this approximation technique together with the fact that the unique maximal representatives are forming a closure system to develop a first algorithmic solution to Problem 1 that builds on any of the known algorithms that can enumerate closure systems combined with an anti-monotone pruning condition. As an alternative we present an algorithm that directly traverses the equivalence classes via the exact minimum representatives. This approach bypasses the hardness of computing minimum representatives by building them inductively from one another. It comes, however, at the cost of additional memory requirements.

In a concluding empirical evaluation we compare both approaches to each other and to traditional exhaustive subgroup discovery. We assess good performance, as well as a surprisingly good predictive power given that the method has not been optimized towards this goal.

Prior Work. Addressing output redundancy is a concern of subgroup discovery research right from the start (see, e.g., [8]). More recent approaches include successive weighted covering [13] and the removal of irrelevant descriptions [7]. For the first method it is important to note that it can be combined with our's rather than being an excluding alternative. The second approach is closely related to this work in that it also uses the closure system of maximal representatives. While we are interested in listing minimum representatives of each equivalence class, their focus is on discarding irrelevant classes.

Traditional subgroup miners usually exploit an optimistic estimator of their quality function in order to make the approach of searching in the space of descriptions feasible. Hence they create an anti-monotone search space that contains the family of all interesting descriptions. While our algorithms use equivalence classes instead of individual descriptions, all the optimistic estimator techniques including recent findings [10] can still be applied. Consequently, our algorithms always use a condensed version of traditional method's search spaces.

Methodologically our work is directly related to formal concept analysis (e.g., [6]) and closed set mining [3,14,16] because the maximal representatives together with their equivalence classes form a concept lattice. While our algorithms build on closure system enumeration techniques, we are only interested in minimum representatives. This is in contrast to closed set mining where algorithms seek to list all maximal (closed) members or minimal members (generators or free sets).

2 Basic Definitions

Throughout this work we denote "elementary objects" by non-capital letters, e.g., e, sets of elementary objects by capital letters, e.g, E and families, i.e., sets, of sets by calligraphic letters, e.g., $\mathcal{E}$. In particular the power set of some set E is denoted by $\mathcal{P}(E)$. The symbol "$\subset$" denotes the strict subset relation between sets. For a family $\mathcal{S}$ the terms **minimal** and **maximal** refer to the subset relation, i.e., a set $S \in \mathcal{S}$ is a minimal element of $\mathcal{S}$ if there no strict subset $S' \subset S$ that is also an element of $\mathcal{S}$. In contrast, the term **minimum** is used with respect to the cardinality of a set, i.e., $S \in \mathcal{S}$ is called minimum element of $\mathcal{S}$ if there is no $S' \in \mathcal{S}$ with $|S'| < |S|$.

Subgroup Descriptions. Let $\mathcal{A} = A_1, \ldots, A_n$ be a sequence of n sets we refer to as **attributes**. A **data record** over $\mathcal{A}$ is an n-tuple $D = (a_1, \ldots, a_n) \in A_1 \times \cdots \times A_n$, for which we denote its i-th component by $D(i) = a_i$. A **dataset** $\mathcal{D}$ over $\mathcal{A}$ is a multiset of data records over $\mathcal{A}$. Note that we do not consider labeled data respectively target attributes at this points. Labels are introduced in the paragraph about subgroup quality below.

The subgroup description language considered in this work is the language of conjunctions of attribute/value equality constraints. We formalize this as follows: a **constraint** over $\mathcal{A}$ is an expression $(A_i = v)$ with $i \in \{1, \ldots, n\}$ and $v \in A_i$. The set of all such constraints is denoted $\boldsymbol{C}_{\mathcal{A}}$. The family of **subgroup descriptions** over $\mathcal{A}$, i.e., the language of conjunctions of such constraints, is then $\mathcal{L}_{\mathcal{A}} = \mathcal{P}(\boldsymbol{C}_{\mathcal{A}})$. In the following we drop the index $\mathcal{A}$ whenever it is clear from the context.

The semantic of conjunctions arises through the following interpretation of subgroup descriptions: let $\mathcal{D}$ be a dataset over $\mathcal{A}$. A datarecord $D \in \mathcal{D}$ is said to

support a subgroup description $H \in \mathcal{L}$, denoted $D \models H$, if for all $(A_i = v) \in H$ it holds that $D(i) = v$. Then the **extension** of H in $\mathcal{D}$, denoted by $\mathcal{D}[H]$, is the submultiset of $\mathcal{D}$ containing the data records that support H. Extensions are anti-monotone with respect to the subset relation, i.e., it holds that $H \subseteq H' \Rightarrow \mathcal{D}[H] \supseteq \mathcal{D}[H']$.

Subgroup Quality. For the purpose of this work we simply regard a **quality function** as a map $q: \mathcal{L} \to \mathbb{R}$ and an **optimistic estimator** for q as a map $\hat{q}: \mathcal{L} \to \mathbb{R}$ satisfying for all $H \subseteq H' \in \mathcal{L}$ that $\hat{q}(H) \geq q(H')$. Given a **quality threshold** q^* the family of **interesting subgroup descriptions** is $\mathcal{R} = \{H \in \mathcal{L} : q(H) \geq q^*\}$. Usually, of course, a quality function depends on the given data. For instance for a **binary labeled dataset** $\mathcal{D}$, i.e., a dataset with associated labels $l(D) \in \{+, -\}$ for all $D \in \mathcal{D}$ a commonly used quality function (and the one used in our experiments) is the **binomial test quality function**

$$q(H) = \sqrt{\frac{|\mathcal{D}[H]|}{|\mathcal{D}|}} \left(\frac{|\mathcal{D}^+[H]|}{|\mathcal{D}[H]|} - \frac{|\mathcal{D}^+|}{|\mathcal{D}|} \right) ,$$

where $\mathcal{D}^+ = \{D \in \mathcal{D} : l(D) = +\}$ denotes the dataset of all +-labeled data records. As optimistic estimator for the binomial test quality function we used $\hat{q}(H) = \sqrt{|\mathcal{D}[H]| / |\mathcal{D}|}\,(1 - |\mathcal{D}^+| / |\mathcal{D}|)$. In the following, however, the concrete form of q and $\hat{q}$ is not considered. We regard them as given blackboxes that "encapsulate" the data and rely only on the following two requirements:

1. For a given **minimum quality threshold** $q^* \in \mathbb{R}$ the **search space** $\mathcal{S} = \{H \in \mathcal{L} : \hat{q}(H) \geq q^*\}$ defined by $\hat{q}$ is **anti-monotone**, i.e., for all $H \subseteq H' \subseteq C$ it holds that $H' \in \mathcal{S}$ implies $H \in \mathcal{S}$.
2. The maps $q(H)$ and $\hat{q}(H)$ both are *functions of the extension* of H in the dataset, i.e., $\mathcal{D}[H] = \mathcal{D}[H']$ implies $q(H) = q(H')$.

The first requirement follows from the definition of optimistic estimators, and the second is true for the usually employed quality functions and their estimators.

3 Extension Equivalence and Compression

In this section we formally introduce equivalence classes of subgroup descriptions and investigate their potential in reducing search space and output of subgroup discovery. Unless explicitly mentioned otherwise, for the remainder of this article we assume that $\mathcal{D}$ is a dataset of size m over n attributes $\mathcal{A}$. The central notion of equivalence is:

Definition 1 (Description Equivalence). *Two subgroup descriptions $H, H' \in \mathcal{L}$ are **equivalent** (with respect to the dataset $\mathcal{D}$), denoted by $H \equiv H'$, if they have an identical extension on $\mathcal{D}$, i.e., $\mathcal{D}[H] = \mathcal{D}[H']$.*

Clearly, $\equiv$ is an equivalence relation on $\mathcal{L}$. For a $H \in \mathcal{L}$ we denote its **equivalence class** with respect to $\equiv$, i.e., $\{H' \in \mathcal{L} : H \equiv H'\}$, by $[H]$. For a family

Table 1. Construction 1 with $n = 6$

	A_1	A_2	A_3	A_4	A_5	A_6	1
D_1	0	0	1	1	1	1	−
D_2	1	1	0	0	1	1	−
D_3	1	1	1	1	0	0	−
D_4	1	1	1	1	1	1	+

of subgroup descriptions $\mathcal{H} \subseteq \mathcal{L}$ we denote by $\mathcal{H}^{\equiv}$ the equivalence classes it contains, i.e., $\mathcal{H}^{\equiv} = \{[H] \colon H \in \mathcal{H}\}$.

In order to investigate the potential reduction of search space and output, we now give a general dataset construction that intuitively reflects "worst-case siutations" for traditional subgroup discovery. It leads to several theoretical observations.

Construction 1. *For even positive integers $n \in \mathbb{N}$ we define the dataset $\mathcal{D}_n$ over n binary attributes $\mathcal{A} = (A_1, \ldots, A_n)$ with $A_i = \{0, 1\}$ for $i \in \{1, \ldots, n\}$ by $\mathcal{D}_n = (D_1, \ldots, D_{n/2+1})$ with $D_{n/2+1} = (1, \ldots, 1)$ and*

$$D_i(j) = \begin{cases} 0, & \text{if } j \in \{2i-1, 2i\} \\ 1, & \text{otherwise} \end{cases} ,$$

for $i = 1, \ldots, n/2$.

Table 1 illustrates this construction for $n = 6$ annotated with binary labels. For $q^* = 3/8$ only one equivalence class is interesting with respect to the binomial test quality function, i.e, the one containing the descriptions H with $\mathcal{D}[H] = \{D_4\}$. Thus, one solution to Problem 1 for this data is $\{(A_1 = 1), (A_3 = 1), (A_5 = 1)\}$. In contrast there are $3^3 = 27$ alternative descriptions of this extension: for each of the pairs $\{A_1, A_2\}$, $\{A_3, A_4\}$, and $\{A_5, A_6\}$ choose one or both attributes to be constraint to 1. Generally the datasets $\mathcal{D}_n$ witness that the compression rate achieved by considering equivalence classes instead of individual description can grow exponentially in the number of attributes (and data records).

Theorem 1. *For all positive integers $n \in \mathbb{N}$ there is a dataset $\mathcal{D}$ of size $n/2+1$ over n attributes $\mathcal{A}$ and a quality threshold q^* such that the compression rates $|\mathcal{R}| \,/\, |\mathcal{R}^{\equiv}|$ and $|\mathcal{S}| \,/\, |\mathcal{S}^{\equiv}|$ are in $O(\exp(n/2))$.*

In order to investigate the extend of compression that can be achieved in practice, we conducted experiments on ten real-world datasets, which are introduced in more detail in Section 7. Table 2 shows sizes of result families $\mathcal{R}$, compressed results families $\mathcal{R}^{\equiv}$, search spaces $\mathcal{S}$, and compressed search spaces $\mathcal{S}^{\equiv}$ for different quality thresholds q^*. The threshold $t100$ varies among the datasets: it is equal to the quality of the 100th highest quality subgroup description (note that because of ties in the quality $|\mathcal{R}_{t100}|$ can still be greater than 100). The threshold ϵ is equal to the smallest positive number distinguishable from zero in double precision. The results for $q^* = t100$ give a differentiated impression: ranging from tremendious compression rates of 749.000 (soybean) to no compression

Table 2. Uncompressed and compressed result families and search spaces

	credi.	lung-.	lymph	mush.	nurse.	sick	soybe.	splice	tic-t.	vote
$\lvert\mathcal{R}_{t100}\rvert$	100	38K	124	168	100	128	749K	100	113	101
$\lvert\mathcal{R}^{\equiv}_{t100}\rvert$	83	1	17	12	100	1	1	99	113	101
$\lvert\mathcal{S}_{t100}\rvert$	148K	456M	12K	3458	103K	>100M	>100M	398K	6067	3505
$\lvert\mathcal{S}^{\equiv}_{t100}\rvert$	87K	159K	4176	890	69K	8M	1M	395K	5824	3465
$\lvert\mathcal{R}_{\epsilon}\rvert$	6119K	>100M	1078K	>100M	11K	>100M	>100M	>100M	65K	3610K
$\lvert\mathcal{R}^{\equiv}_{\epsilon}\rvert$	175K	103K	19K	105K	11K	2M	2M	>100M	23K	82K
$\lvert\mathcal{S}_{\epsilon}\rvert$	17M	>100M	26M	>100M	192K	>100M	>100M	>100M	129K	11M
$\lvert\mathcal{S}^{\equiv}_{\epsilon}\rvert$	385K	183K	45K	228K	115K	9M	3M	>100M	43K	227K

(vote). For decreasing thresholds, however, a significant compression arises for all datasets: while (with one exception) it is tractable to search through all equivalence classes with a potentially positive quality ($q^* = \epsilon$), this is infeasible on most datasets for exhaustive enumeration due to the large number of equivalent descriptions.

4 Border Elements

Border elements, i.e., maximal and minimal members of an equivalence class $[H]$, play a special role. They contain all information necessary to check whether some given description is a member of $[H]$. Among the minimal members one can find minimum representatives, one of which we desire as representative for its class. In this section we state some basic but important mathematical and computational properties of the border elements.

The first observation is that every equivalence class has a unique maximal (most specific) element. It is given by the map σ introduced in the lemma below.

Lemma 2 (Pasquier et al. [14]). *For all subgroup descriptions $H \in \mathcal{L}$ it holds that $\sigma(H)$ given by*

$$\sigma(H) = \{(A_i{=}v) \colon 1 \leq i \leq n, \forall D \in \mathcal{D}[H]\,,\, D(i) = v\}\ .$$

is the unique maximal element of $[H]$, i.e., (i) $H \equiv \sigma(H)$, (ii) for all $H' \in \mathcal{L}$ with $H' \supset \sigma(H)$ it holds that $H \not\equiv H'$, and (iii) $\sigma(H)$ is unique with (i) and (ii).

While each equivalence class $[H]$ has a unique maximal element $\sigma(H)$, there can be more than one minimal (most general) element of $[H]$. In fact the number of minimal and even that of the minimum representatives can be exponential in the cardinality of $\sigma(H)$. Again, the datasets $\mathcal{D}_n$ from Construction 1 witness this statement: for all $I \subseteq \{1, \ldots, n/2\}$ the description H_I defined by

$$H_I = \{(A_{2i}{=}1) \colon i \in I\} \cup \{(A_{2i-1}{=}1)) \colon \{1, \ldots, n\} \setminus I\}$$

is a minimum description of the extension $\{D_{n+1}\}$, and there are $2^{n/2}$ such descriptions. We can conclude:

Theorem 3. *For all positive integers $n \in \mathbb{N}$ there is a dataset $\mathcal{D}$ of size $n/2+1$ over n attributes $\mathcal{A}$ such that there is an equivalence class with $O(\exp(n/2))$ minimum representatives.*

For Problem 1 we are only interested in constructing one minimum representative per interesting equivalence class. As an isolated task, however, this is intractable. This again contrasts the maximal representatives, which can be computed in time $O(nm)$. In particular the **NP**-hard MIN-SET-COVER problem—*given* a family of subsets $\mathcal{F} \subseteq \mathcal{P}(E)$ with $\bigcup \mathcal{F} = E$, *compute* a minimum subfamily $\mathcal{F}' \subseteq \mathcal{F}$ with $\bigcup \mathcal{F}' = E$—polynomially reduces to finding a minimum description. In addition the reduction preserves solution sizes. Thus, even the inapproximability result from [5] carries over to our problem.

Theorem 4. Given *a subgroup description $H \in \mathcal{L}$, it is*

(a) **NP**-*hard to* compute *an equivalent subgroup description $G \in [H]$ of minimum length, i.e., $|G| = \min\{|H'|\colon H' \in [H]\}$ and*
(b) hard[1] *to compute an approximation $G' \in [H]$ in polynomial time that satisfies $|G'| \leq |G|(1-\epsilon)\ln m$ for all $\epsilon > 0$ where $m = |\mathcal{D} \setminus \mathcal{D}[H]|$ is the number of data records not supporting H.*

Proof. We prove both statements by giving a polynomial time transformation of MIN-SET-COVER instances $\mathcal{F} \subseteq \mathcal{P}(E)$ to a dataset $\mathcal{D}$ over attributes $\mathcal{A}$ and a subgroup description $H \in \mathcal{L}_{\mathcal{A}}$ such that (i) extension equivalent descriptions $H' \in [H]$ correspond to set covers $\mathcal{F}_{H'} \subseteq \mathcal{F}$ of E of same size, i.e., $|H'| = |\mathcal{F}_{H'}|$ and (ii) $|\mathcal{D} \setminus \mathcal{D}[H]|$ is equal to the size of the set cover ground set $|E|$. Part (a) then follows from the **NP**-hardness of MIN-SET-COVER and (b) from the result of [5]. Let $E = \{1, \ldots, m\}$ and $\mathcal{F} = \{S_1, \ldots, S_n\}$ be a set cover instance. Set $\mathcal{A} = \{A_1, \ldots, A_n\}$ with $A_i = \{0, 1\}$ and $\mathcal{D} = D_1, \ldots, D_{m+1}$ with

$$D_i(j) = \begin{cases} 1, & \text{if } i \notin S_j \\ 0, & \text{otherwise} \end{cases}$$

for $i = \{1, \ldots, m\}$ and $D_{m+1} = (1, \ldots, 1)$. Furthermore, choose $H \in \mathcal{L}$ as $H = \{(A_i = 1)\colon 1 \leq i \leq n\}$. Let $H' \in [H]$ be a subgroup description equivalent to H. Then it follows from the definitions that $H' \subseteq \sigma(H)$ and $\mathcal{D}'[H'] = \{\}$ where $\mathcal{D}' = \mathcal{D} \setminus \mathcal{D}[H]$ denote the datarecords that are not supporting H. It follows for $\mathcal{F}_{H'} \subseteq \mathcal{F}$ defined by $\mathcal{F}_{H'} = \{S_i\colon (A_i = 1) \in H'\}$ that

$$\begin{aligned} &\forall i \in E,\, \exists j,\, i \in S_j \wedge S_j \in \mathcal{F}_{H'} \\ \Leftrightarrow &\forall i \in E,\, \exists (A_j = 1) \in H',\, D_i(j) = 0 \\ \Leftrightarrow &\forall i \in E,\, D_i \notin \mathcal{D}'[H'] \ . \end{aligned}$$

That is $H' \equiv H$ if and only if $\mathcal{F}_{H'}$ is a cover of E. Moreover, $|H'| = |\mathcal{F}_{H'}|$ as required. □

[1] Here, hardness means: "as hard as computing a solution to an **NP**-hard problem in time $n^{O(\log \log n)}$ for instances of size n."

5 Closure System Traversal and Greedy Approximation

Our first algorithmic approach towards solving Problem 1 is motivated by the observation that the transformation used within the proof of Theorem 4 can be reversed. Hence, finding a minimum equivalent representative of a given description H is in fact equivalent to MIN-SET-COVER. Incoporating that inverse transformation into the well-known greedy algorithm for MIN-SET-COVER yields the procedure:

1. **set** $\bar{\mathcal{D}} \leftarrow \mathcal{D} \setminus \mathcal{D}[H]$, $G \leftarrow \emptyset$
2. **while** $\bar{\mathcal{D}}[G] \neq \emptyset$ **set** $G \leftarrow G \cup \{\arg\min_{c \in \sigma(H)} |\bar{\mathcal{D}}[G \cup \{c\}]|\}$
3. **return** G

Slavík [15] found that the **greedy approximation factor** for MIN-SET-COVER is $g(m) = \ln m - \ln \ln m + 0.78$ for instances with a ground set E of size m. Taking into account our transformation we have the following result.

Lemma 5. *Given a subgroup description H, a minimum representative of $[H]$ can be approximated in time $O(|\sigma(H)|\, m)$ within $g(m)$ where $m = |\mathcal{D} \setminus \mathcal{D}[H]|$ is the number of data records not supporting H.*

Thus, any algorithm that traverses the search space of equivalence classes can be combined with the greedy algorithm to approximately solve Problem 1. Several known algorithms are identified as applicable for that task by another observation: the maximal representatives form a closure system.

Lemma 6 (Pasquier et al. [14]). *The map σ is a closure operator, i.e., it satisfies for all $H, H' \in \mathcal{L}$ that $H \subseteq \sigma(H)$ (extensivity), $H \subseteq H' \Rightarrow \sigma(H) \subseteq \sigma(H')$ (monotonicity), and $\sigma(H) = \sigma(\sigma(H))$ (idempotence).*

There are several efficient algorithms listing all closed sets of a closure operator like the divide and conquer algorithm from formal concept analysis [9]. Adapting a closed frequent itemset miner like LCM [16] to our task is even more natural: we plug in optimistic estimate pruning instead of frequency pruning, and instead of single items we have single constraints. Since there are at most nm valid constraints for a dataset of size m over n attributes, together with the performance of LCM we get the result:

Theorem 7. *Problem 1 can be solved in time $O(|\mathcal{S}^{\equiv}|\, n^2 m^2)$ and space $O(nm)$ if the representative minimality condition is relaxed to: for all listed descriptions $H \in \mathcal{R}$ there is no description H' having the same extension with $|H| > g(|\mathcal{D} \setminus \mathcal{D}[H]|)\, |H'|$.*

Note that in case of a constant number of attribute values the bound on the number of constraints boils down to $O(n)$, and consequently the time complexity in the theorem is improved by a factor m. The theoretical approximation guarantee of the greedy algorithm, although optimal with regard to Theorem 4, may appear somewhat weak. In practice, however, the worst-case bound is virtually never attained, and the greedy result is usually close to optimum (see

Section 7). Moreover, note that the potentially expensive greedy algorithm has to be called only for the returned result equivalence classes and not during the actual traversal of the closure system.

In order to adress the *top-k* problem variant, i.e., to list only representatives of k highest quality classes, only minor changes are necessary: instead of directly printing the interesting subgroup descriptions, collect them in a priority queue with capacity k. In this scenario the search space can be reduced significantly by adjusting the q^* threshold whenever a new subgroup description is added to the result queue (and the queue is full).

On the other hand, there is no easy way to include a *length-limit* for additional pruning, i.e., when we are only interested in reprentatives containing no more than l literals, this cannot be exploited for reducing the search space. The reason is that, even if exact minimum reprentatives would be on hand, in general LCM (or any other common closed set miner) does not list the closed sets in ascending order with respect to their minimum equivalent descriptions.

6 Inductive Minimum Representative Construction

In this section we present an alternative algorithmic approach for non-redundant subgroup discovery that, intuitively, is based on a breadth-first traversal of the directed graph containing as vertices all equivalence classes that lie in the search space and edges between any two classes $[H] \neq [H']$ such that there is a constraint $c \in \boldsymbol{C}$ with $(H \cup \{c\}) \in [H']$ (note that the existence of such a constraint is independent of the chosen representatives H, H'). It turns out that minimum representatives of an equivalence class $[H]$ correspond to shortest paths from $[\emptyset]$ to $[H]$ in that graph. Thus, beside having the weakness of a significant memory overhead because all visited classes have to be kept in memory in order to guarantee a non-redundant traversal, this strategy has two major advantages:

1. it generates minimum representatives of each equivalence class without additional cost and,
2. as it visits equivalence classes in ascending order with respect to the minimum cardinality of their members, it allows for pruning based on a length-limit.

The straightforward implementation of that graph traversal has the serious drawback that it reaches vertices via many redundant ways: a minimum representative G induces $|G|!$ different paths to $[G]$—one for each of its orderings. This effect can be significantly reduced by chosing some arbitrary but fixed order $\{c_1, \ldots, c_N\}$ of the constraint set $\boldsymbol{C}$. Thereby the expressions "$\max H$" and "$\min H$" are defined for non-empty descriptions $H \in \mathcal{L}$ by referring to the constraint in H with the maximum or minimum index, respectively. Algorithm 1 below uses this order to reduce its traversal paths to those that are in descending order. To describe its behavior in more detail and prove its correctness we define a modified **lexicographical order** on the descriptions $\mathcal{L}$, denoted by "$\prec_L$", given by:

$$H \prec_L H' \Leftrightarrow |H| < |H'| \vee (|H| = |H'| \wedge \max(H \Delta H') \in H') \ .$$

Using this strict linear order we can specify the elements that Algorithm 1 enumerates among the potentially many minimum representatives.

Definition 2 (Canonical Minimum Representative). *The* ***canonical minimum representative*** *of an equivalence class* $[H]$*, denoted by* $\mu(H)$*, is the unique minimum representative of* $[H]$ *that is minimal with respect to* $\prec_L$*.*

These canonical minimum representatives have the important property that they can be built from one another inductively via their suffixes.

Lemma 8. *Let* $G \neq \emptyset$ *be a non-empty canonical minimum representative of its equivalence class* $[G]$*, i.e.,* $G = \mu(G)$*. Then* $G' = G \setminus \{\min G\}$ *is the canonical minimum representative of* $[G']$*.*

Proof. Assume there is a $G'' \in [G']$ with $G'' \prec_L G'$. Then $G'' \cup \{\min G\} \prec_L G' \cup \{\min G\} = G$. But as

$$\begin{aligned} \mathcal{D}[G'' \cup \{\min G\}] &= \mathcal{D}[G''] \cap \mathcal{D}[\{\min G\}] \\ &= \mathcal{D}[G'] \cap \mathcal{D}[\{\min G\}] \\ &= \mathcal{D}[G' \cup \{\min G\}] = \mathcal{D}[G] \ , \end{aligned}$$

i.e., $(G'' \cup \{\min G\}) \in [G]$, this contradicts $G = \mu(G)$. □

Now we can prove the correctness and time complexity of Algorithm 1.

Algorithm 1. Inductive Minimum Representative Construction

Require: ordered ground set of constraints $\boldsymbol{C} = \{c_1, \ldots, c_N\}$,
extension closure operator σ,
quality function q with optimistic estimator $\hat{q}$, and quality threshold q^*
Output: family $\{\mu(H) \colon q(H) \geq q^*\}$ in lexicographical order

init $\mathcal{Q}$ as empty queue and $\mathcal{V}$ as empty prefix tree
enqueue $(\emptyset, \sigma(\emptyset), \boldsymbol{C})$ on $\mathcal{Q}$
while $\mathcal{Q} \neq \emptyset$ **do**
  dequeue front element $(G, S, \boldsymbol{A})$ of $\mathcal{Q}$
  if $q(S) \geq q^*$ **then print** G
  $\boldsymbol{A}' \leftarrow \{c \in \boldsymbol{A} \setminus S \colon c < \min G,\ \hat{q}(G \cup \{c\}) \geq q^*\}$
  for all $c_i \in \boldsymbol{A}'$ in ascending order of their index **do**
    $G' \leftarrow G \cup \{c_i\}$
    $S' \leftarrow \sigma(G')$
    if $S' \notin \mathcal{V}$ **then**
      add S' to $\mathcal{V}$
      enqueue $(G', S', \boldsymbol{A}')$ on $\mathcal{Q}$

Theorem 9. *Algorithm 1 exactly and non-redundantly lists* $\mu(H)$ *for all* $[H] \in \mathcal{R}^{\equiv} = \{[H] : q(H) \geq q^*\}$ *in lexicographical order in time* $O(|\mathcal{S}^{\equiv}|\, n^2 m^2)$ *and space* $O(|\mathcal{S}^{\equiv}|\, nm)$ *where* $\mathcal{S}^{\equiv} = \{[H] : \hat{q}(H) \geq q^*\}$ *is the search space of all potentially interesting equivalence classes.*

Proof. For each dequeued tuple at most $|\boldsymbol{C}|$ augmentations are evaluated involving a computation of σ and a visited check. The prefix-tree lookup is performed in time $|S| \leq |\boldsymbol{C}|$ and σ is computed in time nm. Also the space dominant data structure $\mathcal{V}$ contains at most one element of size at most $|\boldsymbol{C}|$ for each dequeued tuple. As $|\boldsymbol{C}|$ is bounded by nm the claim follows if we can show that in lexicographical order for each $[H] \in \mathcal{S}^{\equiv}$ a tuple $(\mu(H), \sigma(H), \boldsymbol{A})$ is enqueued and only tuples of this form are enqueued, i.e., if the following three properties hold:

(i) If a tuple $(G, S, \boldsymbol{A})$ is enqueued before $(G', S', \boldsymbol{A'})$ then $G \prec_L G'$.
(ii) For all $[H] \in \mathcal{S}^{\equiv}$ a tuple $(G, S, \boldsymbol{A})$ with $G = \mu(H)$ and $S = \sigma(H)$ is enqueued, and $\boldsymbol{A} \supseteq \{c \in \boldsymbol{C} \colon (G \cup \{c\}) \in \mathcal{S}^{\equiv},\, c < \min G\}$.
(iii) all enqueued tuples $(G, S, \boldsymbol{A})$ are of the form $G = \mu(H)$ and $S = \sigma(H)$ for some $[H] \in \mathcal{S}^{\equiv}$.

Property (i) is implied by the breadth-first strategy and the following observation: if $G_1 \prec_L G_2$ then all descriptions G'_1 generated from G_1 are lexicographically smaller than all descriptions G'_2 generated from G_2.

Assume that (ii) is violated for some $[H]$. Then choose a class $[H]$ that violates (ii) with a minimal $G' = \mu(H)$. As $(\emptyset, \sigma(\emptyset), \boldsymbol{C})$ is enqueued in line 2, it holds that $G' \neq \emptyset$. By Lemma 8, $G = G' \setminus \{\min G'\}$ is lexicographically minimal in $[G]$. The anti-monotinicity of the search space and $G \subset G'$ imply that (ii) holds for $[G]$. In particular a tuple $(G, S, \boldsymbol{A})$ is enqueued with $(\min G') \in \boldsymbol{A}$ because $\min G' < \min G$ (for the same reason and because of the anti-monotinicity of $\mathcal{S}^{\equiv}$, the augmentation set $\boldsymbol{A'}$ satisfies $\boldsymbol{A'} \supseteq \{c \in \boldsymbol{C} \colon (G' \cup \{c\}) \in \mathcal{S}^{\equiv},\, c < \min G'\}$). Thus, G' is generated subsequently in line 8. Then $\sigma(G')$ does not pass the visited check in line 10. This implies that $[G']$ has already been visited, say via $G'' \in [G']$. It follows from (*i*) that $G'' \prec_L G'$ contradicting $G' = \mu(G')$.

For (iii) observe that $S = \sigma(G)$ for all enqueued tuples by the generation of S in line 9. Now assume that $G \neq \mu(S)$ for an enqueued tuple $(G, S, \boldsymbol{A})$. Then there is an $G' \in [S]$ with $G' \prec_L G$. By the anti-monotinicity of the search space and (ii) a tuple $(G', S', \boldsymbol{A'})$ is enqueued, and by (i) it is enqueued before $(G, S, \boldsymbol{A})$. In the same iteration $S' = S$ is added $\mathcal{V}$. Consequently, $(G, S, \boldsymbol{A})$ can not be enqueued as it does not pass the visited check in line 10—a contradiction. □

There are some additional speedups that do not affect the worst-case time complexity. For the top-k scenario the same changes as for the closure/greedy approach can be applied. Furthermore, if at a node $(G, S, \boldsymbol{A})$ with $c_i, c_j \notin S$ it holds that $c_j \in \sigma(G \cup \{c_i\})$ then it follows $\sigma(G \cup \{c_i, c_j\}) = \sigma(G \cup \{c_i\})$ by monotonicity of σ. In this case the augmentation element c_i can be removed from $\boldsymbol{A'}$ of the child $(G \cup \{c_j\}, \sigma(G \cup \{c_j\}), \boldsymbol{A'})$ in case $c_i \prec_L c_j$ as it would redundantly generate the same equivalence class again. Furthermore, the sorting of the constraint set can have a substantial impact on the computation time. It is, however, a non-trivial problem to find an optimal sorting (see [9] for a comparison of different sorting strategies for divide and conquer closed set listing).

7 Empirical Evaluation

In this section we empirically compare non-redundant subgroup discovery with minimum representatives to traditional subgroup discovery. This includes an evaluation of both proposed algorithmic solutions for Problem 1. We considered ten datasets from the UCI Machine Learning Repository [1], which are presented along with their most important properties in Table 3. All numerical attributes where discretized using minimal entropy discretization. As representative traditional subgroup miner we used the state-of-the-art algorithm Dpsubgroup [10]. All involved algorithms were implemented in Java and will be published on the author's webpage. For the sake of a better comparision we used a simplified reimplementation of the LCM algorithm and not the implementation published by its author. The quality function was the binomial test quality function combined with the optimistic estimator introduced in Section 2. All experiments were performed on a Core 2 Duo E8400 @ 3Ghz running a Sun SE 6u10 Java Virtual Machine with 1.5 GB of Java heap space under Windows XP.

Computation Time. Table 4 contains the computation times that correspond to the compression experiments presented in Section 3 for Dpsubgroup (dpsg), LCM/greedy (lcm/gr), and Algorithm 1 (imr). The threshold $t100$, i.e., the quality of the 100th best subgroup description, is explicitly stated. The results essentially reflect the already observed search space reduction. Although, even for

Table 3. Datasets

dataset	credi.	lung-.	lymph	mush.	nurse.	sick	soybe.	splice	tic-t.	vote
label	bad	1	maln.	pois.	recm.	sick	EI	bspot.	pos.	repub.
n	15	56	18	22	8	39	60	35	9	16
m	1000	32	148	8124	12960	3772	3190	638	958	435
$\|C\|$	58	159	50	117	27	66	133	287	27	48

Table 4. Computation time (in seconds unless stated differently); "oom" and ">12h" for computations that ran out of memory or out of time, respectively

dataset	credi.	lung-.	lymph	mush.	nurse.	sick	soybe.	splice	tic-t.	vote
q^*	0.094	0.336	0.244	0.267	0.029	0.177	0.223	0.190	0.061	0.306
dpsg	**2**	84.4m	0.5	**0.6**	**1.2**	4.3h	10.6h	23	0.3	0.4
lcm/gr	3.2	**23**	0.3	1.0	2.3	**18.3m**	123	38	**0.2**	**0.2**
imr	3.6	**23**	**0.2**	0.9	2.4	oom	**115**	**20**	**0.2**	0.3
q^*					ϵ					
dpsg	242	>12h	457	>12h	**2**	>12h	>12h	>12h	**1**	127
lcm/gr	184	95	6.5	53m	85.5	**7h**	**41m**	>12h	15	59
imr	**26**	**60**	**4**	**39**	5	oom	oom	oom	2	**19**

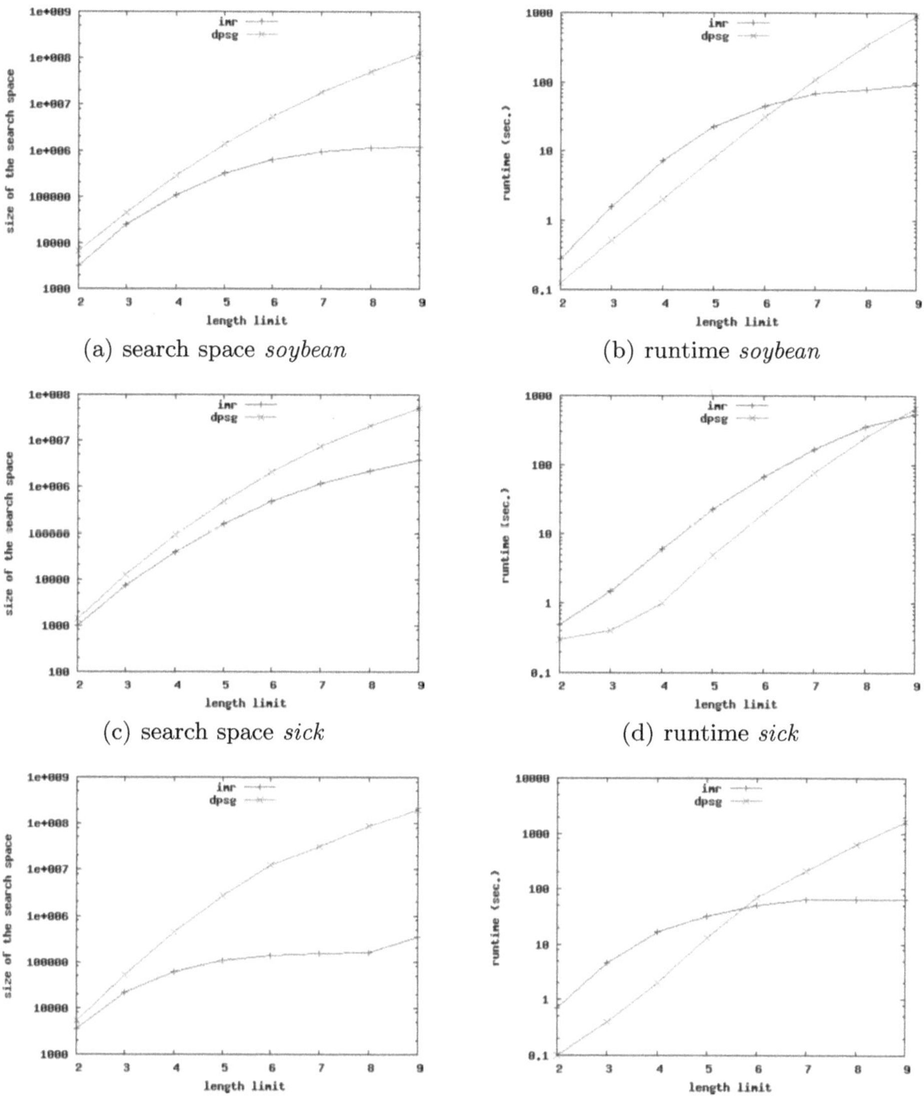

(a) search space *soybean* (b) runtime *soybean*

(c) search space *sick* (d) runtime *sick*

(e) search space *lung-.* (f) runtime *lung-.*

Fig. 1. Logscale search space and runtime comparison of Algorithm 1 (imr) vs. Dpsubgroup (dpsg) for increasing length limits

datasets on which a compression is achieved traditional subgroup discovery is not directly outperformed. This is illustrated in Figure 1 in which the development of search spaces and computation time for increasing length-limits are shown. The Dpsubgroup algorithm is only beaten by Algorithm 1 for sufficiently large differences in the search space. This behavior is due to the sophisticated data structures (fptrees [2,11]) Dpsubgroup uses in contrast to our algorithm.

Table 5. Greedy performance for $q^* = \epsilon$

	credi.	lung-.	lymph	mush.	nurse.	sick	soybe.	splice	tic-t.	vote
avg. min.	5.588	4.914	4.532	5.776	5.328	?	?	?	5.005	5.991
avg. apx.	5.603	5.032	4.562	5.862	5.328	9.062	6.591	?	5.041	6.039
max. dif.	5 vs. 3	7 vs. 4	5 vs. 3	6 vs. 4	no diff	?	?	?	6 vs. 4	8 vs. 5
time frac.	0.9	0.07	0.54	0.99	0.97	0.96	0.78	?	0.93	0.59

A further noteworthy fact is that unless Algorithm 1 ran out of memory (the oom entries) it always outperforms LCM/greedy. This motivates a more detailed investigation of the latter approach.

Greedy Performance. We analyze the perfomance of the greedy algorithm within the lcm/greedy approach in two respects: (a) the length of the produced representatives and (b) the greedy algorithm's fraction of the computation time for the experiments with $q^* = \epsilon$. Note that this is an extremely unfavorable case for the lcm/greedy approach because of the large number of interesting equivalence classes, each of which requires one greedy call. The results are listed in Table 5. We note that (a) for all datasets the length of the subgroup descriptions obtained using the greedy algorithm is only marginally greater than the minimum length and (b) the computation time of LCM/greedy was dominated by the greedy algorithm. Without the greedy approximations LCM even slightly outperformed Algorithm 1 on most datasets.

Predictive Power. While it is out of scope of this article to evaluate the claim that selecting minimum representatives improves understandability for users, their power as building blocks of global prediction models can be supported with standard accuracy experiments. Although not their primary intention, subgroup description families $\mathcal{R}$ are sometimes evaluated (see [13]) by investigating the area under the ROC curve (AUC) of the set of global models $\{h_i\colon 1 \leq i \leq |\mathcal{R}|\}$, where h_i classifies a given data point as positive if it supports any of the i highest quality subgroup descriptions from $\mathcal{R}$. For each dataset we compared the predictive quality of the

- top-20 subgroup description (sgd),
- minimum representatives of the top-20 equivalence classes (min-repr),
- and additionally the hypothesis of the rule learner J-RIP, a Java implementation of RIPPER [4].

A rule learner was chosen as additional benchmark because, among supervised learning methods, the nature of the hypotheses it produces is most similar to the subgroup based models. Table 6 shows the average AUC of ten cross validations using five folds. For performance reasons, we executed all subgroup discoveries with a depth limit of 5. Using equivalence classes instead of subgroup descriptions always resulted in a higher or equal AUC. Perhaps surprisingly, in addition, the hypotheses build from the minimum representatives also outperformed the rule learner hypotheses on the majority of datasets.

Table 6. Average AUC over ten 5-fold cross-validations

dataset	credi.	lung-.	lymph	mush.	nurse.	sick	splice	soybe.	tic-t.	vote
RIPPER	0.619	0.729	0.774	**1.0**	**0.815**	0.916	0.969	**0.917**	0.975	**0.959**
sgd	0.628	0.708	0.757	0.890	0.813	0.908	**0.987**	0.792	**0.999**	0.972
min-repr	**0.633**	**0.731**	**0.831**	0.946	0.813	**0.928**	**0.987**	0.856	**0.999**	0.972

8 Conclusion

Discussion. Beside the results stated in the introduction our experiments primarily revealed the following trade-off in using non-redundant subgroup discovery based on equivalence classes: while one gains a significant compression of the search space and output, this is sometimes outweighted by the fact that one loses the sophisticated data structures of traditional methods. There are, however, many datasets/quality thresholds managable by our algorithms, that were completely intractable before due to an exponential explosion of the search space. This opens up new opportunitics for some datasets like an exhaustive enumeration of all positive quality equivalence classes with a subsequent global optimization step.

Future Work. It is important to note that the results of this article easily generalize to *more expressive constraint languages* as long as they guarantee unique maximal representatives and extension anti-monotonicity with respect to some specialization relation. This is for instance the case for interval constraints, which are appealing in the presence of ordinal attributes. In constrast to the standard attribute/value equality constraints of the form "$A_i = v$" an intervall constraint for a real-valued attribute is of the form "$A_i \in [l, u]$". In the presence of more expressive constraint languages the equivalence class search space becomes even more important as there is an increasing number of ways to (redundantly) describe one and the same extension. This motivates a future study investigating the combination of non-redundant subgroup discovery and other constraint languages.

Acknowledgment

We thank the anonymous reviewers for their helpful comments. This research was partially supported by the European Commission under the project *RACWeB* (No. 045101).

References

1. Asuncion, A., Newman, D.J.: UCI machine learning repository (2007)
2. Atzmüller, M., Puppe, F.: SD-map – A fast algorithm for exhaustive subgroup discovery. In: Fürnkranz, J., Scheffer, T., Spiliopoulou, M. (eds.) PKDD 2006. LNCS (LNAI), vol. 4213, pp. 6–17. Springer, Heidelberg (2006)

3. Calders, T., Rigotti, C., Boulicaut, J.f.: A survey on condensed representations for frequent sets. In: Constraint Based Mining and Inductive Databases, pp. 64–80. Springer, Heidelberg (2005)
4. Cohen, W.W.: Fast effective rule induction. In: ICML, pp. 115–123 (1995)
5. Feige, U.: A threshold of ln n for approximating set cover. J. ACM 45(4), 634–652 (1998)
6. Ganter, B., Wille, R.: Formal Concept Analysis: Mathematical Foundations. Springer, Heidelberg (1999)
7. Garriga, G.C., Kralj, P., Lavrač, N.: Closed sets for labeled data. J. Mach. Learn. Res. 9, 559–580 (2008)
8. Gebhardt, F.: Choosing among competing generalizations. Knowledge Acquisition 3(4), 361–380 (1991)
9. Gély, A.: A generic algorithm for generating closed sets of a binary relation. In: Ganter, B., Godin, R. (eds.) ICFCA 2005. LNCS (LNAI), vol. 3403, pp. 223–234. Springer, Heidelberg (2005)
10. Grosskreutz, H., Rüping, S., Wrobel, S.: Tight optimistic estimates for fast subgroup discovery. In: Daelemans, W., Goethals, B., Morik, K. (eds.) ECML PKDD 2008, Part I. LNCS (LNAI), vol. 5211, pp. 440–456. Springer, Heidelberg (2008)
11. Han, J., Pei, J., Yin, Y.: Mining frequent patterns without candidate generation. In: SIGMOD Conference, pp. 1–12 (2000)
12. Klösgen, W.: Explora: A multipattern and multistrategy discovery assistant. In: Advances in Knowledge Discovery and Data Mining, pp. 249–271. AAAI Press, Menlo Park (1996)
13. Lavrac, N., Kavsek, B., Flach, P., Todorovski, L.: Subgroup discovery with CN2-SD. Journal of Machine Learning Research 5, 153–188 (2004)
14. Pasquier, N., Bastide, Y., Taouil, R., Lakhal, L.: Efficient mining of association rules using closed itemset lattices. Inf. Syst. 24(1), 25–46 (1999)
15. Slavík, P.: A tight analysis of the greedy algorithm for set cover. Journal of Algorithms 25(2), 237–254 (1997)
16. Uno, T., Asai, T., Uchida, Y., Arimura, H.: An efficient algorithm for enumerating closed patterns in transaction databases. In: Discovery Science, pp. 16–31 (2004)
17. Wrobel, S.: An algorithm for multi-relational discovery of subgroups. In: Komorowski, J., Żytkow, J.M. (eds.) PKDD 1997. LNCS, vol. 1263, pp. 78–87. Springer, Heidelberg (1997)

PLSI: The True Fisher Kernel and beyond

IID Processes, Information Matrix and Model Identification in PLSI*

Jean-Cédric Chappelier and Emmanuel Eckard

School of Computer and Communication Sciences
École Polytechnique Fédérale de Lausanne, Switzerland

Abstract. The Probabilistic Latent Semantic Indexing model, introduced by T. Hofmann (1999), has engendered applications in numerous fields, notably document classification and information retrieval. In this context, the Fisher kernel was found to be an appropriate document similarity measure. However, the kernels published so far contain unjustified features, some of which hinder their performances. Furthermore, PLSI is not generative for unknown documents, a shortcoming usually remedied by "folding them in" the PLSI parameter space.

This paper contributes on both points by (1) introducing a new, rigorous development of the Fisher kernel for PLSI, addressing the role of the Fisher Information Matrix, and uncovering its relation to the kernels proposed so far; and (2) proposing a novel and theoretically sound document similarity, which avoids the problem of "folding in" unknown documents. For both aspects, experimental results are provided on several information retrieval evaluation sets.

1 Introduction

Ten years ago, the "Probabilistic Latent Semantic Indexing" (PLSI) model [10,11,12] opened the road to the representation of documents as mixture proportions of so-called "latent topics". This model proved useful and led to several applications on textual data [6,14,18,26,27], audio data [1] and images [5,16,19,20,24].

In this context, the cosine similarity, originally used without much theoretical justification to evaluate the semantic similarities between documents, was replaced by the Fisher kernel similarity [11], which has a better theoretical basis. However, the kernels published so far contain unjustified features, some of which hinder their performances. The first contribution of this paper, detailed in Sect. 3.1 consists in the introduction of a new, rigorous development of this Fisher kernel, that uncovers its canonical form, and shows how it relates to the kernel originally proposed by Hofmann [11].

* Work supported by projects 200021-111817 and 200020-119745 of the Swiss National Science Foundation.

W. Buntine et al. (Eds.): ECML PKDD 2009, Part I, LNAI 5781, pp. 195–210, 2009.

Moreover, one major shortcoming of PLSI comes from its tendency to overfit [2,4,23] and its non-generative nature for new document models.[1] In the context of information retrieval, this is usually remedied by "folding in" the queries into the PLSI parameter space [9,10]. A number of extensions and alternatives have been proposed to address these issues: latent Dirichlet allocation [4], undirected PLSI [28], correlated topic models [3], rate adapting Poisson models [7]; but they come at the price of an increased complexity, especially regarding the runtime cost for the learning algorithms. The second contribution of this paper, detailed in Sect. 4, targets the "folding-in" phase for queries by introducing a new, theoretically grounded, document–query similarity that entirely avoids it. This novel approach is compared to Fisher kernel similarities for PLSI.

Finally, the third contribution of this paper, detailed in Sect. 5, lies in new experimental results on a large collection coming from the TREC–AP evaluation corpus. Up to authors' knowledge, it is the first time that PLSI is evaluated on an IR corpus of over 7000 documents and one million word occurrences.

2 PLSI Document Model and Similarity

PLSI is a latent topic-based model for textual document classification and Information Retrieval [10,12]. Documents are modeled as occurrences of successive random choices of document–term couples (d, w), knowing some topic $z \in Z$: iteratively, a topic z is chosen with probability $P(z)$; a term w and a document model d are then chosen, with probabilities $P(w|z)$ and $P(d|z)$ respectively. In PLSI, w and d are assumed to be independent knowing z; the probability of occurrence of a pair (d, w) thus being

$$P(d, w) = \sum_{z \in Z} P(z)\, P(w|z)\, P(d|z) \ . \quad (1)$$

A document realization $\widehat{d_0}$ from a collection C is modeled as the set of all (d, w) pairs sharing the same document model d_0: $\widehat{d_0} = \{(d_0, w) \in C\}$. As done in the PLSI literature to simplify expressions, we will henceforth not distinguish between document model d and its concrete realization $\widehat{d}$ in the collection.

The parameters of PLSI are $\theta = \{P(z), P(w|z), P(d|z)\}$, for all possible z, w and d in the model.[2] These parameters can be estimated over a document collection through (tempered) Expectation-Maximization (EM) [10,12].

Regarding document similarity measures for PLSI, Fisher kernels yielded significant improvements in performance over the original formulations that used the usual cosine measure [11].

---

[1] Notice, however, that PLSI is indeed a generative model for new occurrences of *already known* document models.

[2] Other equivalent parameterizations are also possible; for instance using $P(d)$, $P(w|z)$ and $P(z|d)$.

However, the original Fisher kernel for PLSI derived by Hofmann [11] was later found to neglect the contribution of the Fisher information matrix $G(\theta)$,[3] and to contain a normalization by document length $|d|$. Variants of the Fisher kernel were thus introduced [21]:

- the observation about renormalization by document length yielded the development of a Fisher kernel not normalized by $|d|$;
- from the observation about the information matrix stemmed the development of a "*DFIM*" kernel (for "Diagonal Fisher Information Matrix"), which takes the diagonal components of $G(\theta)$ into account. By contrast, whenever required, we shall name "*IFIM*" (Identity Fisher Information Matrix) the kernels that do not.

3 IID Processes Perspective on PLSI Fisher Kernels

3.1 IID Derivation of the Fisher Kernel

The Fisher kernel provides a similarity measure among instances of probabilistic models [13]: for two instances X and Y of a given family of stochastic models $P(X|\theta)$ parameterized with θ, it is defined as

$$K(X,Y) = U_X(\theta)^T G(\theta)^{-1} U_Y(\theta) \ ,$$

where $U_X(\theta)$ is the gradient of the log-likelihood: $U_X(\theta) = \nabla_\theta \log P(X|\theta)$, and the Fisher information matrix $G(\theta)$ is the covariance of $U_X(\theta)$:

$$G(\theta) = \mathbf{E}_X[U_X(\theta)\, {U_X}^T(\theta)] \ .$$

Lemma 1. *The Fisher kernel between two instances X_1^n and Y_1^m of an independent and identically-distributed (i.i.d.) stochastic process is the sum of the Fisher kernels between the individual random variables X_i and Y_j, divided by the number of variables in the processes:*

$$K(X_1^n, Y_1^m) = \frac{1}{n}\frac{1}{m}\sum_{i=1}^{n}\sum_{j=1}^{m} K(X_i, Y_j) \ .$$

We refer to Appendix A for the proof.

The Fisher kernel for PLSI can be derived in a manner which stems directly from the definition of the PLSI model as an i.i.d. process of (d, w) pairs probabilized by (1).

In this context, $X_i = (d, w)$ and $Y_i = (q, w')$ (for two document models d and q, and terms w and w' from vocabulary V). Through Lemma 1, the Fisher kernel for PLSI as an i.i.d. process is:

$$K^{\mathrm{IID}}(d,q) = \frac{1}{|d|}\frac{1}{|q|}\sum_{w\in V}\sum_{w'\in V} n(d,w)\, n(q,w')\, K\Big((d,w),(q,w')\Big) \ , \quad (2)$$

[3] Hofmann identified $G(\theta)$ with the identity matrix through a reparametrization suited for multinomial models; however, PLSI is neither a multinomial, nor in an exponential family, and $G(\theta)$ may significantly differ from identity in such cases.

where $n(d,w)$ is the number of occurrences of word w in query d, $|d| = \sum_w n(d,w)$ is the length of document d, and $K\left((d,w),(q,w')\right)$ is the "atomic kernel" between a pair of terms w and w' belonging to documents d and q respectively. This "atomic kernel" can be written as (see Appendix B for details): $K\left((d,w),(q,w')\right) =$

$$\frac{P(d|z)}{P(d,w)}\frac{P(q|z)}{P(q,w')}\cdot\sum_z\left[P(w|z)P(w'|z)\alpha(z)+\delta_{w,w'}\gamma(w,z)\right] , \tag{3}$$

with $\delta_{w,w'} = 1$ if $w = w'$ and 0 otherwise, and

$$\alpha(z) = \begin{cases} P(z) & \text{in IFIM,} \\ \left(\sum_{d\in C}\sum_{w\in V} n(d,w)\left(\frac{P(w|z)P(d|z)}{P(d,w)}\right)^2\right)^{-1} & \text{in DFIM,} \end{cases}$$

$$\text{and } \gamma(w,z) = \begin{cases} P(w|z)P^2(z) & \text{in IFIM,} \\ \left(\sum_{d\in C} n(d,w)\left(\frac{P(d|z)}{P(d,w)}\right)^2\right)^{-1} & \text{in DFIM.} \end{cases}$$

Injecting (3) into (2) leads to

$$K^{\mathrm{IID}}(d,q) = \sum_{w\in V}\sum_{w'\in V}\widehat{P}(w|d)\cdot\widehat{P}(w'|q)\cdot\frac{P(d|z)}{P(d,w)}\frac{P(q|z)}{P(q,w)}\cdot$$
$$\cdot\sum_z\left[P(w|z)P(w'|z)\cdot\alpha(z)+\delta_{w,w'}\gamma(w,z)\right] , \tag{4}$$

where $\widehat{P}(w|d) = \frac{n(d,w)}{|d|}$.

3.2 Relation between $K^{\mathrm{IID}}(d,q)$ and $K^{\mathrm{H}}(d,q)$

Hofmann's original development of the Fisher kernel [11], $K^H(d,q) =$

$$\sum_{z\in Z}\frac{P(z|d)P(z|q)}{P(z)} + \sum_{w\in V}\widehat{P}(w|d)\widehat{P}(w|q)\sum_{z\in Z}\frac{P(z|d,w)P(z|q,w)}{P(w|z)} ,$$

uses the assumption that

$$\sum_{w\in V}\frac{\widehat{P}(w|d)}{P(w|d)}P(w|z) \simeq 1 . \tag{5}$$

Introducing

$$\zeta(d,z) = \sum_{w\in V} n(d,w)\frac{P(w|z)}{P(d,w)} , \tag{6}$$

Equation (4) becomes:

$$K^{\mathrm{IID}}(d,q) = \frac{1}{|d|}\frac{1}{|q|}\sum_{z\in Z}P(d|z)P(q|z)\cdot$$
$$\cdot\left[\alpha(z)\zeta(d,z)\zeta(q,z)+\sum_{w\in V}\frac{n(d,w)}{P(d,w)}\frac{n(q,w)}{P(q,w)}\gamma(w,z)\right] . \tag{7}$$

Injecting (5) into (6) entails that $\zeta(d,z) \simeq \frac{|d|}{P(d)}$. Furthermore, noticing that

$$P(d) \simeq \frac{|d|}{|C|}, \qquad \text{where } |C| = \sum_{w \in V} \sum_{d \in C} n(d,w) \ , \tag{8}$$

is experimentally verified to a precision inferior to 1%, we can state that $\zeta(d,z) \simeq |C|$. In the IFIM case, this leads to

$$K^{\mathrm{IID}}(d,q) \simeq K^{\mathrm{H}}(d,q) \ .$$

Thus, at the price of assumptions (5) and (8), the Hofmann IFIM kernel can be seen as an approximation of the IID one.

In the DFIM case, the equivalence is as not exact, as the DFIM α and γ are not the same in the IID kernel and Hofmann's:

	Hofmann	IID
$\alpha^{-1}(z)$	$\sum_d \frac{P(z\|d)^2}{P(z)}$	$\sum_d \sum_w n(d,w) \left(\frac{P(w\|z)P(d\|z)}{P(d,w)}\right)^2$
$\gamma^{-1}(w,z)$	$\sum_d \widehat{P}^2(w\|d) \left(\frac{P(d\|z)}{P(d,w)}\right)^2$	$\sum_d n(d,w) \left(\frac{P(d\|z)}{P(d,w)}\right)^2$

3.3 Implementation of the IID Kernel

In practice, (7) is more efficiently computed by taking into account that $\alpha(z)$ depends only on z (and neither on d nor on w): $\alpha(z)$ and $\zeta(d,z)$ can be precomputed once for all for the entire corpus. $\gamma(w,z)$ and $\zeta(q,z)$ are best computed for each query, as to take advantage of the limited number of different terms present in a query: $\gamma(w,z)$ is computed only for $w \in q$ (i.e. $w \in V$ such as $n(w,q) > 0$). Such processing is an order of magnitude quicker using these precomputations.

Furthermore, the computation of all Fisher kernels can be decomposed into two independent parts K_z and K_w which stem from the contributions of the latent categories and of the terms, respectively; for instance using (7): $K^{\mathrm{IID}}(d,q) = K_z^{\mathrm{IID}}(d,q) + K_w^{\mathrm{IID}}(d,q)$, where

$$K_z^{\mathrm{IID}}(d,q) = \frac{1}{|d|}\frac{1}{|q|} \sum_{z \in Z} P(d|z)P(q|z)\,\alpha(z)\,\zeta(d,z)\,\zeta(q,z) \ ,$$

and

$$K_w^{\mathrm{IID}}(d,q) = \frac{1}{|d|}\frac{1}{|q|} \sum_{z \in Z} \left[P(d|z)P(q|z) \cdot \sum_{w \in V} \frac{n(d,w)}{P(d,w)} \frac{n(q,w)}{P(q,w)} \gamma(w,z) \right] \ .$$

4 Avoiding the Folding-In of Queries

The second contribution of this paper consists in the development of a document similarity measure for PLSI that entirely removes the so-called "folding-in" phase for unknown document models q (typically the queries in Information Retrieval), and all the problems related to the learning of new parameters $P(q|z)$, as well as their adequacy with already existing ones: changing from $P(q|z) = 0$ to $P(q|z) > 0$ should imply the rescaling of all the $P(d|z)$ according to $\sum_\delta P(\delta|z) = 1$, i.e. by a factor $1/(1 + P(q|z)) \simeq 1 - P(q|z)$.

To remove folding-in entirely, we follow the Language-Model approach [22,29] and consider queries, not as new document models for which new parameters $P(q|z)$ must be learned, but rather as new occurrences of already learned document models.

The retrieval problem thus simply turns into model identification (rather than learning of new model): for a given query q, which are the (already learned) models d best representative of q?

One usual way to address such a question is to minimize the Kullback-Leibler divergence between the empirical distribution (q) and the model distribution (d) [15]:

$$\mathcal{S}_{\mathrm{KL}}(d,q) = -\mathrm{KL}\left(\widehat{P}(w|q), P(w|d)\right) = \sum_{w \in q \cap d} \widehat{P}(w|q) \log \frac{P(w|d)}{\widehat{P}(w|q)} , \tag{9}$$

where $w \in q \cap d$ denotes all the words appearing in q (i.e. $n(q,w) > 0$) such that $P(d,w) > 0$.

Notice that this formulation uses $\widehat{P}(w|q) = n(q,w)/|q|$, for which no learning is required, as opposed to $P(w|q)$, for which unknown parameters $P(q|z)$ have to be estimated (usually done through "folding in").

5 Experiments

We are thus faced with 13 different document similarity measures: $\mathcal{S}_{\mathrm{KL}}(d,q)$, which does not require query folding-in (Sect. 4), and 12 Fisher kernel variants: the two IFIM models K^{H} and K^{IID}, and the corresponding DFIM kernels $K^{\mathrm{DFIM\text{-}H}}$ and $K^{\mathrm{DFIM\text{-}IID}}$; as well as, separately, the K_z and K_w components of all these kernels (as defined in Sect. 3.3). Across $K^{\mathrm{(DFIM\text{-})H}}$ and $K^{\mathrm{(DFIM\text{-})IID}}$, the latter provide 8 different kernels.

The following questions arise regarding these 13 kernels:

1. Are there significant differences between all the possible variants of the Fisher kernel and which one is the best?
2. Can the new approaches here proposed compete with the Hofmann kernel variant?
3. How do these compare to the IR state-of the art model, BM25 [25], especially on a large document collection?

To address these questions in line with previously published work on PLSI, we experimented on the standard Information Retrieval benchmarks from the SMART collection[4]: CACM, CISI, MED, CRAN and TIME. We furthermore

[4] `ftp://ftp.cs.cornell.edu/pub/smart/`

Table 1. Characteristics of the document collections used for evaluation

	CACM	CRAN	TIME	CISI	MED	AP89_01XX
# Terms (stems)	4 911	4 063	13 367	5 545	7 688	13 379
# occurrences ($\|C\|$)	90 927	120 973	114 850	87 067	76 571	1 321 482
Documents						
#	1 587	1 398	425	1 460	1 033	7 466
avg. $\|d\|$	56.8	85.1	268.6	56.7	73.8	177.2
Queries						
#	64	225	83	112	30	50
avg. $\|q\|$	12.7	8.9	8.2	37.7	11.4	79.3

explored the limits of PLSI learning tractability, and experimented on a significantly bigger corpus[5] consisting of a subpart of the TREC–AP 89 corpus [8]. For tractability reasons, we kept only the 7466 first documents of this collection,[6] and queries 1 to 50. The main characteristics of the evaluation corpora are given in Table 1.

For experiments on the SMART collection, 6 runs with different learning initial conditions were performed for all the models, and for different numbers of topics: $|Z| \in \{1, 2, 8, 16, 32, 64, 128\}$, totalling 2730 experiments. For the TREC-AP part, due to its size, a single run was performed for each $|Z| \in \{1, 32, 48, 64, 80, 128\}$, totaling 78 experiments.

For all the experiments, stemming was performed using the Porter algorithm of Xapian.[7] Evaluation results were obtained using the standard `trec_eval` tool.[8] We here use the standard Mean Average Precision (MAP) to display the results: we plot the MAP against the number of latent topics $|Z|$, averaged over all experiments and with error bars corresponding to 1-standard deviation. The conclusions are exactly the same using either 5-point precision or R-precision, except on MED; we come back to this latter point at the end of this section.

The main results out of these experiments, summarized in Table 2, are:

1. The kernels K^{IID} and K^{H} have very similar performances, which experimentally validates the theoretical result that K^{H} approximates K^{IID} (see Fig. 1). $K^{\mathrm{DFIM\text{-}IID}}$ and $K^{\mathrm{DFIM\text{-}H}}$ are a bit less similar, due to the slightly different form that $G(\theta)$ takes. However, K^{IID} is more computationally demanding than K^{H}, at least one order of magnitude slower.

[5] Over 5 times as many documents and 10 times as many word occurrences as in the SMART collection.

[6] Documents AP890101-0001 to AP890131-0311. The EM learning for e.g. $|Z| = 128$ took 45 hours of CPU time and used 6.7 Gb of RAM on a dedicated computer server with one octo core 2-GHz Intel Xenon processor and 32 Gb of memory.

[7] `http://xapian.org/`

[8] `http://trec.nist.gov/trec_eval/`

Table 2. Main results and conclusions out of 2808 experiments over 13 models on 6 corpora

		CACM	CRAN	TIME	CISI	MED	AP89_01XX
Results	BM25 MAP	**31.4**	**42.4**	**69.2**	12.3	52.3	19.7
	Best PLSI model MAP	30.0	39.6	60.8	**20.2**	**53.8**	**21.6**
	Best PLSI model is:	$K_w^{\text{DFIM-H}}$	$\mathcal{S}_{\text{KL}}$	$K_w^{\text{DFIM-H}}$	K_w^{H}	K^{H}	$K_w^{\text{DFIM-H}}$
	for $\|Z\| =$	32	128	8	8	32	48
	K_w^{H} MAP	**30.0**	33.6	55.6	**20.2**	49.8	16.5
	$K_w^{\text{DFIM-H}}$ MAP	23.2	37.0	**60.8**	15.6	45.5	**21.6**
	$\mathcal{S}_{\text{KL}-128}$ MAP	22.9	**39.6**	49.1	19.5	**52.8**	11.4
Concl.	PLSI $>$ BM25?	No	No	No	**YES**	**yes**	**yes**
	$\mathcal{S}_{\text{KL}-128}$ w.r.t. Fisher kernels	$<$	$\boxed{>}$	$<$	$\simeq$	$\simeq$	$<$
	DFIM $G(\theta)$ helps? (on K_w)	Yes	**Yes**	**Yes**	No	No	**Yes**

2. We can confirm that the DFIM Fisher kernels for PLSI outperform by their original IFIM versions (Fig. 1). They are furthermore dominated by their K_w component.
3. K_z deteriorates performances in general: used alone, it performs poorly; furthermore the performances of K^{H} decrease as the role of K_z becomes more important for growing $|Z|$: starting from K_w at low $|Z|$, the performances of K^{H} reach down K_z at higher $|Z|$(Fig. 3).
   On the other hand, K_w alone is always good, if not the best (Figs. 2 and 3).
4. $K_w^{\text{DFIM-H}}$ and $K^{\text{DFIM-H}}$ have similar behaviors (Fig. 3). The reason is that the normalizing role of $G(\theta)$ makes $K_z \ll K_w$ for DFIM kernels.
5. $\mathcal{S}_{\text{KL}}$ has a growing performance with $|Z|$, the number of latent-topics (Fig. 2). We had to stop at $|Z| = 128$ for tractability reasons (for both learning and evaluation running times).
6. $\mathcal{S}_{\text{KL}}$ can outperform the best Fisher kernel on CRAN and reaches similar performances on MED and CISI (Fig. 2). It should however be emphasized that the former does not require any folding-in phase for queries as the latter does.
7. The best PLSI-based kernels can perform better than the state-of-the-art BM25 model, especially on corpora which could be considered semantically more difficult: CISI, where few words are shared between queries and documents (thus particularly suited to test the extend to which a retrieval models is robust to synonymy, or "topics".[9]), MED (specialized vocabulary), and TREC–AP.

The only collection where conclusions should be more nuanced is MED: the different models do not behave in the same way at different recall values (Fig. 4); some are better at low recall and others are better at higher recall. Global measures as MAP or R-Prec cannot represent such nuances.

[9] CISI is remarkable in the sense that some query-document matches are expected between queries and documents that do not share any significant term.

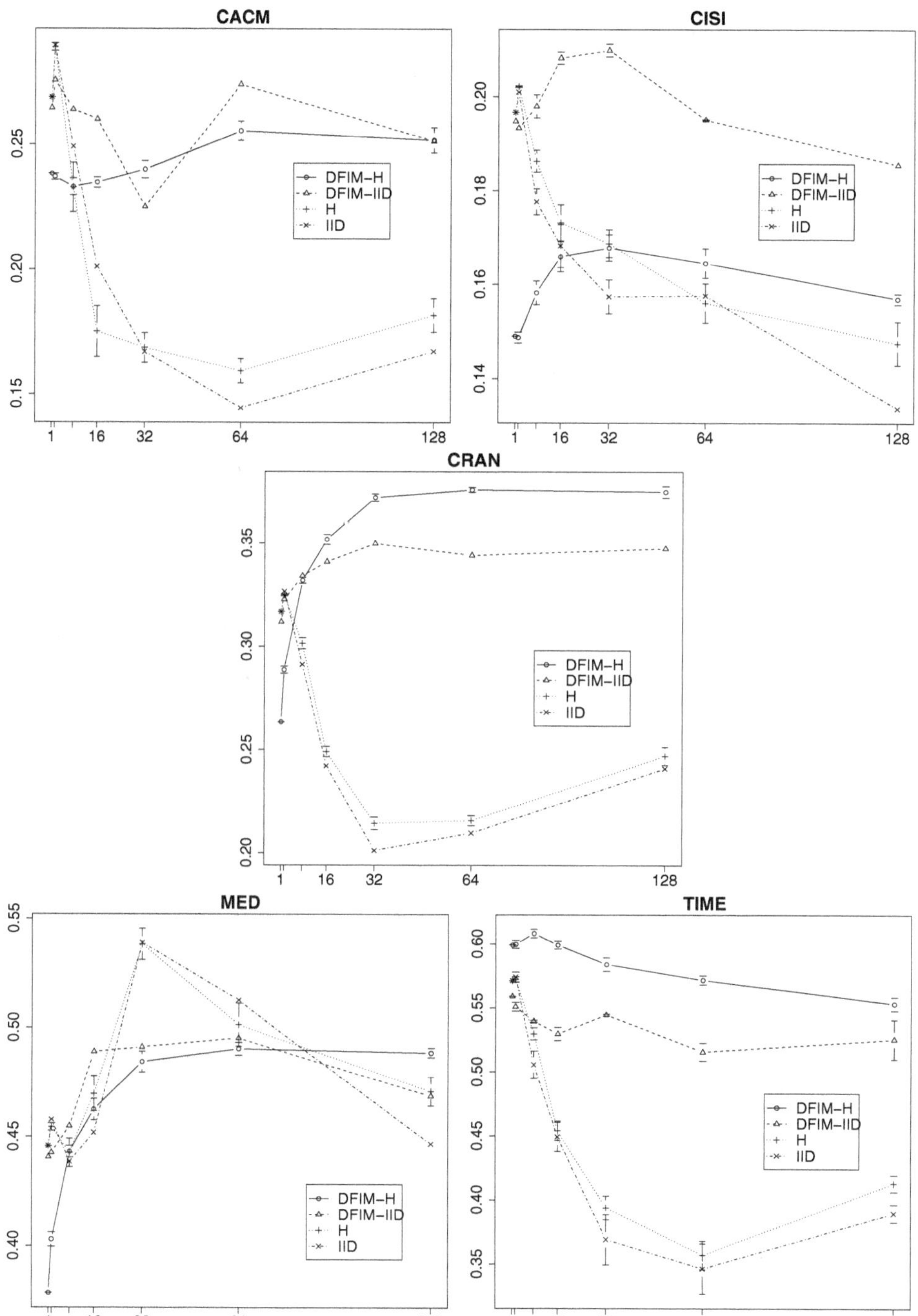

Fig. 1. Behaviors (MAP vs $|Z|$) of the K^{IID} and K^{H} kernels, both for IFIM and DFIM variants

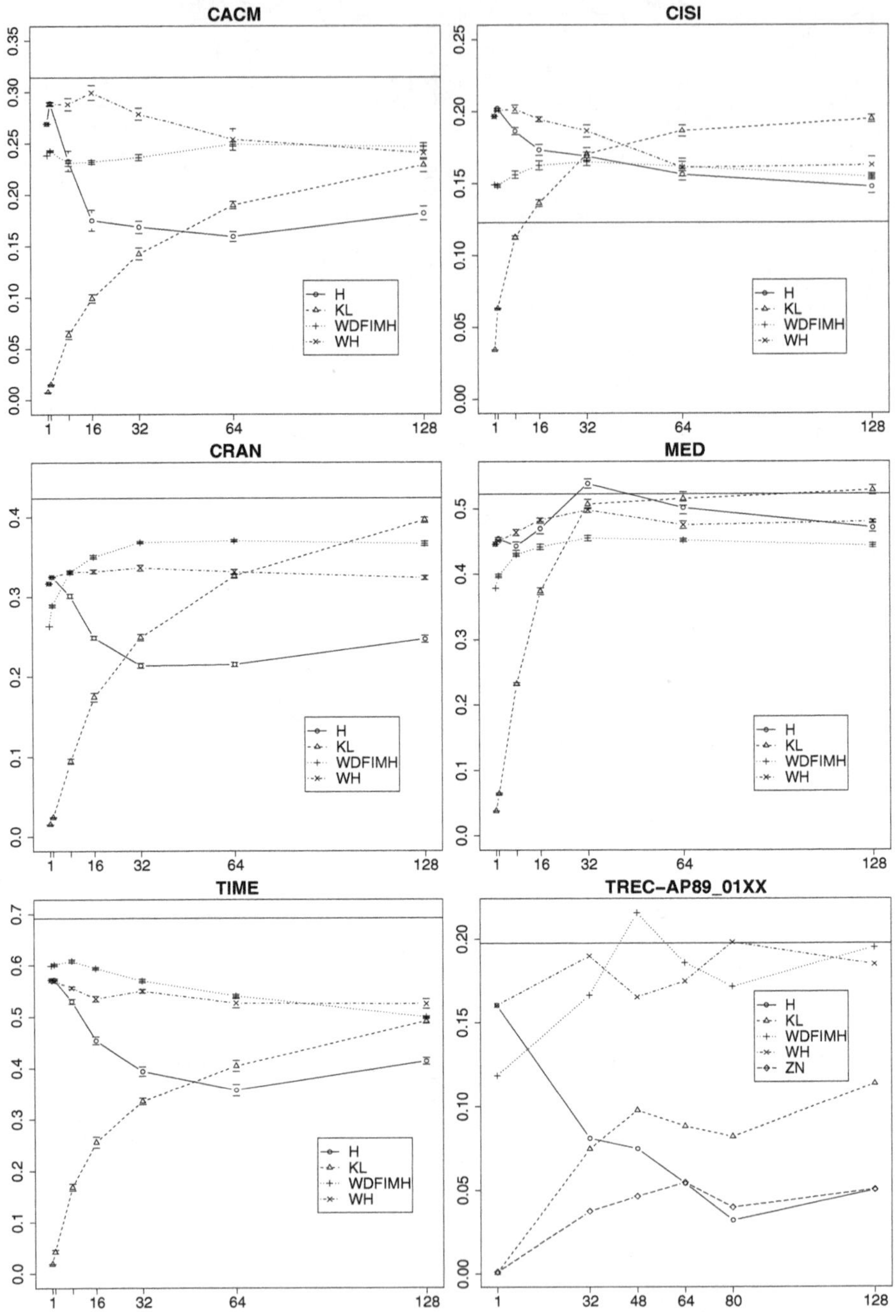

Fig. 2. Results (MAP vs $|Z|$) obtained on the six corpora considered for different models: K^{H} (H), $\mathcal{S}_{\mathrm{KL}}$ (KL), $K_w^{\mathrm{DFIM\text{-}H}}$ (WDFIMH), and K_w^{H} (WH). The horizontal bar represents the MAP of state-of-the-art BM25 model (which does not depend on $|Z|$).

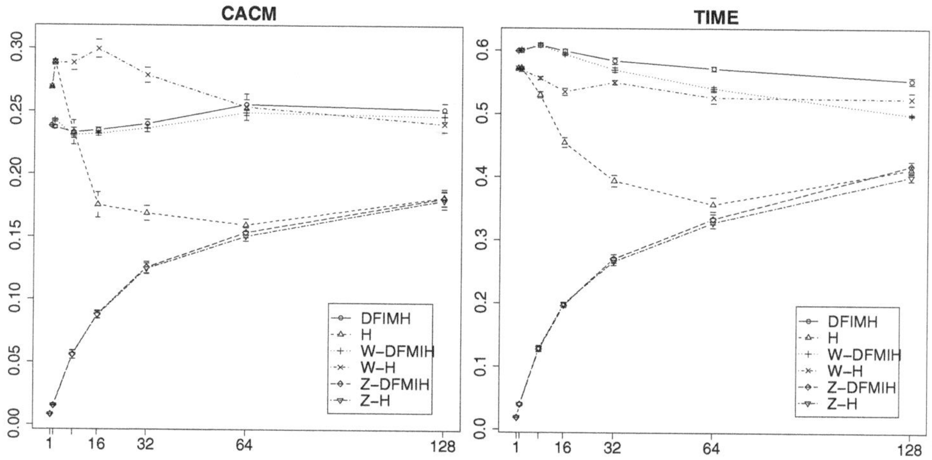

Fig. 3. Two typical examples comparing different variants of the Fisher kernel for PLSI (MAP vs $|Z|$): $K^{\text{DFIM-H}}$ (DFIMH), K^{H} (H), $K_w^{\text{DFIM-H}}$ (W-DFIMH), K_w^{H} (W-H), $K_z^{\text{DFIM-H}}$ (Z-DFIMH), and K_z^{H} (Z-H). This illustrates that the latent-topic part K_z performs poorly in comparison to the word part K_w, and impairs the combined kernels: notice how K^{H} starts from K_w^{H} at low $|Z|$, and degrades to K_z^{H} as $|Z|$ grows.

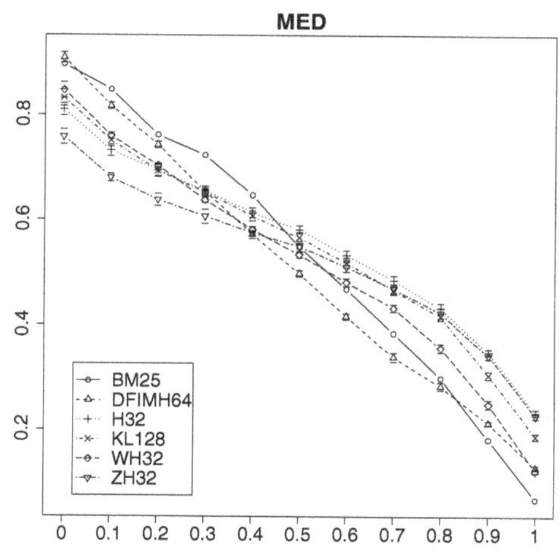

Fig. 4. Precision vs Recall curves on MED for BM25, $K^{\text{DFIM-H}}$ for $|Z|$=64 (DFIMH64), K^{H} for $|Z|$=32 (H32), $\mathcal{S}_{\text{KL}}$ for $|Z|$=128 (KL128), K_w^{H} for $|Z|$=32 (WH32), and K_z^{H} for $|Z|$=32 (YH32)

6 Conclusion

This paper offers two contributions: the first one is a rigorous development of the Fisher kernel for PLSI models, by which we uncover the kernel that properly takes into account the i.i.d. nature of PLSI, thereby explaining the underlying reasons for normalization by document and query length in Hofmann's kernel; and restore the contribution of the Fisher Information Matrix into these kernels. The second contribution is a theoretically grounded similarity for PLSI which entirely avoids query folding-in. Furthermore, we were able to perform an Information Retrieval evaluation of PLSI on a collection much larger than the SMART collections on which it is usually evaluated.

Regarding the questions we wanted to address experimentally, we can conclude:

1. There are significant differences between all the possible variants of Fisher kernel for PLSI and the best variant is globally $K_w^{\text{DFIM-H}}$.
   Regarding the role of the Fisher information matrix, its normalizing impact improves the results on bigger collections (TIME, CRAN, TREC–AP89).
   Regarding the role of topics and terms components, K_z should clearly be neglected in favor of K_w; at least for tractable numbers of topics (small $|Z|$).
   The rigorous IID kernels $K_w^{\text{(DFIM-)IID}}$ offer performances similar to those of $K_w^{\text{(DFIM-)H}}$; this is to be expected, as the Hofmann's kernel family turns out to be an approximation of the IID one. Since the kernels $K_w^{\text{(DFIM-)IID}}$ are computationally more expensive, $K_w^{\text{(DFIM-)H}}$ should be preferred in practice.
2. The new approach which avoids query folding-in can compete with the best Fisher kernel variants, especially for high number of topics.
3. These models (either KL divergence or Fisher kernels) can compete with BM25, especially on corpora which could be considered semantically more difficult as CISI, MED and TREC–AP.

We thus experimentally confirm that topic-based models as PLSI could be interesting for information retrieval in not too large but semantically difficult document collections, where documents and queries do not necessarily share at lot of terms. In such cases, we would recommend $K_w^{\text{DFIM-H}}$ as similarity measure, or $\mathcal{S}_{\text{KL}}$, (9), when sufficiently large numbers of latent topics are tractable. The advantage of the latter is the lack of folding-in phase for queries.

The overall conclusion, however, is that PLSI is not well-suited for large scale ad hoc IR, mainly because it is not a fully generative model. This model simply does not scale with the number of documents. Furthermore, as sophisticated as it can be, PLSI hardly outperforms the state-of-the-art BM25 model, at a complexity price that is not worth paying. PLSI might be interesting and better than state-of-the-art models for large numbers of latent-topics, but the former limitation (huge number of parameters) makes such levels intractable in practice. It may indeed be the case, especially for bigger document collections, that a *much* larger $|Z|$ improves the performances of K_z or $\mathcal{S}_{\text{KL}}$, but, due to the non-scalability of PLSI, such levels won't finish training in practice, especially for the bigger document collections where they would be interesting.

References

1. Ahrendt, P., Goutte, C., Larsen, J.: Co-occurrence models in music genre classification. In: IEEE Int. Workshop on Machine Learning for Signal Processing (2005)

2. Bast, H., Weber, I.: Insights from viewing ranked retrieval as rank aggregation. In: Proc. of Int. Workshop on Challenges in Web Information Retrieval and Integration (WIRI 2005), pp. 232–239 (2005)
3. Blei, D., Lafferty, J.: A correlated topic model of *Science*. Annals of Applied Statistics 1(1), 17–35 (2007)
4. Blei, D.M., Ng, A.Y., Jordan, M.I.: Latent Dirichlet allocation. Journal of Machine Learning Research 3, 993–1022 (2003)
5. Bosch, A., Zisserman, A., Munoz, X.: Scene classification via plsa. In: Proc. of the European Conf. on Computer Vision (2006)
6. Gaussier, E., Goutte, C., Popat, K., Chen, F.: A hierarchical model for clustering and categorising documents. In: Proc. of 24th BCS-IRSG Europ. Coll. on IR Research, pp. 229–247 (2002)
7. Gehler, P.V., Holub, A.D., Welling, M.: The rate adapting Poisson model for information retrieval and object recognition. In: Proc. 23rd Int. Conf. on Machine Learning, pp. 337–344 (2006)
8. Harman, D.: Overview of the fourth Text REtrieval Conference (TREC–4). In: Proc. of the 4th Text REtrieval Conf., pp. 1–23 (1995)
9. Hinneburg, A., Gabriel, H.-H., Gohr, A.: Bayesian folding-in with Dirichlet kernels for PLSI. In: Proc. of the 7th IEEE Int. Conf. on Data Mining, pp. 499–504 (2007)
10. Hofmann, T.: Probabilistic latent semantic indexing. In: Proc. of 22nd Annual Int. ACM SIGIR Conf. on Research and Development in Information Retrieval, pp. 50–57 (1999)
11. Hofmann, T.: Learning the similarity of documents: An information-geometric approach to document retrieval and categorization. In: Advances in Neural Information Processing Systems, vol. 12, pp. 914–920 (2000)
12. Hofmann, T.: Unsupervised learning by probabilistic latent semantic analysis. Machine Learning 42(1), 177–196 (2001)
13. Jaakkola, T., Haussler, D.: Exploiting generative models in discriminative classifiers. In: Advances in Neural Information Processing Systems, vol. 11, pp. 487–493. MIT Press, Cambridge (1999)
14. Jin, X., Zhou, Y., Mobasher, B.: Web usage mining based on probabilistic latent semantic analysis. In: Proc. of 10th Int. Conf. on Knowledge Discovery and Data Mining, pp. 197–205 (2004)
15. Lafferty, J., Zhai, C.: Document language models, query models, and risk minimization for information retrieval. In: Proc. ACM SIGIR Conf. on Research and Development in Information Retrieval (2001)
16. Lienhart, R., Slaney, M.: Plsa on large-scale image databases. In: Proc. of the 2007 Int. Conf. on Acoustics, Speech and Signal Processing, IEEE (ICASSP 2007), vol. 4, pp. 1217–1220 (2007)
17. McLachlan, G., Peel, D.: Finite Mixture Models. Wiley, Chichester (2000)
18. Mei, Q., Zhai, C.: A mixture model for contextual text mining. In: Proc. of 12th Int. Conf. on Knowledge Discovery and Data Mining, pp. 649–655 (2006)
19. Monay, F., Gatica-Perez, D.: Plsa-based image auto-annotation: Constraining the latent space. In: Proc. ACM Int. Conf. on Multimedia, ACM MM (2004)
20. Monay, F., Gatica-Perez, D.: Modeling semantic aspects for cross-media image indexing. IEEE Transactions on Pattern Analysis and Machine Intelligence 29 (2007)
21. Nyffenegger, M., Chappelier, J.-C., Gaussier, E.: Revisiting Fisher kernels for document similarities. In: Proc. of 17th European Conf. on Machine Learning, pp. 727–734 (2006)

22. Ponte, J.M., Croft, W.B.: A language modeling approach to information retrieval. In: 21st SIGIR Conf. on Research and Development in Information Retrieval, pp. 275–281 (1998)
23. Popescul, A., Ungar, L.H., Pennock, D.M., Lawrence, S.: Probabilistic models for unified collaborative and content-based recommendation in sparse-data environments. In: Proc. of the 17th Conf. in Uncertainty in Artificial Intelligence, pp. 437–444 (2001)
24. Quelhas, P., Monay, F., Odobez, J.-M., Gatica-Perez, D., Tuytelaars, T., Gool, L.V.: Modeling scenes with local descriptors and latent aspects. In: Proc. of ICCV 2005, vol. 1, pp. 883–890 (2005)
25. Robertson, S.E., Walker, S., Jones, S., Hancock-Beaulieu, M., Gatford, M.: Okapi at TREC–3. In: Proc. of the 3rd Text REtrieval Conf. (1994)
26. Steyvers, M., Smyth, P., Rosen-Zvi, M., Griffiths, T.: Probabilistic author-topic models for information discovery. In: Proc. 10th Int. Conf. on Knowl. Discovery and Data Mining, pp. 306–315 (2004)
27. Vinokourov, A., Girolami, M.: A probabilistic framework for the hierarchic organisation and classification of document collections. Journal of Intelligent Information Systems 18(2/3), 153–172 (2002)
28. Welling, M., Rosen-Zvi, M., Hinton, G.: Exponential family harmoniums with an application to information retrieval. In: Advances in Neural Information Processing Systems, vol. 17, pp. 1481–1488 (2005)
29. Zhai, C.: Statistical language models for information retrieval a critical review. Found. Trends Inf. Retr. 2(3), 137–213 (2008)

A Proof of Lemma 1

Let us consider X_1^n a n-element long instance of an i.i.d. stochastic process. To simplify notations, we shall henceforth write $X = X_1^n$. Its log-likelihood is expressed by $\log P(X) = \sum_{i=1}^n \log P(X_i)$. Linearity of derivation operators entails that the corresponding Fisher score is written $U_\theta(X) = \sum_{i=1}^n U_\theta(X_i)$.

The Fisher information matrix for n-event instances of this i.i.d. stochastic process is defined as $G_X(\theta) = E_X\left[U_\theta(X)\, U_\theta(X)^T\right]$; and for a single instance: $G_1(\theta) = E_{X_i}\left[U_\theta(X_i)\, U_\theta(X_i)^T\right]$ (which is independent of X_i since the process is i.i.d.). It can be written as:

$$
\begin{aligned}
G_X(\theta) &= E_X\left[\left(\sum_{i=1}^n U_\theta(X_i)\right)\left(\sum_{i=j}^m U_\theta(X_j)\right)^T\right] \\
&= \sum_{i=1}^n \left(E_X\left[U_\theta(X_i) U_\theta(X_i)^T\right] + \sum_{j\neq i} E_X\left[U_\theta(X_i) U_\theta(X_j)^T\right] \right) \\
&= \sum_{i=1}^n \left(G_{X_i}(\theta) + E_{X_i}\left[U_\theta(X_i)\right] \sum_{j\neq i} \underbrace{E_{X_j}\left[U_\theta(X_j)\right]^T}_{=0} \right) = \sum_{i=1}^n G_1(\theta) \\
&= n \cdot G_1(\theta) \ .
\end{aligned}
$$

The "natural gradient" ϕ_X is obtained from the ordinary gradient U_X via $\phi_X = {G_X}^{-1} U_X$ [13]. In the case of i.i.d. stochastic processes, the previous results lead to

$$\phi_X = {G_X}^{-1} U_X = \frac{1}{n} \cdot \sum_{i=1}^{n} {G_1}^{-1} U_{X_i} = \frac{1}{n} \cdot \sum_{i=1}^{n} \phi_{X_i} \ .$$

Eventually, the Fisher kernel between two instances X_1^n and Y_1^m of an i.i.d. process is given by

$$K(X_1^n, Y_1^m) = {\phi_X}^T G_1 \phi_Y = \left(\frac{1}{n} \sum_{i=1}^{n} {G_1}^{-1} U_{X_i} \right)^T G_1 \left(\frac{1}{m} \sum_{i=j}^{m} {G_1}^{-1} U_{Y_j} \right)$$
$$= \frac{1}{n \cdot m} \sum_{i=1}^{n} \sum_{i=j}^{m} U_{X_i}^T {G_1}^{-1} U_{Y_j} = \frac{1}{n \cdot m} \sum_{i=1}^{n} \sum_{i=j}^{m} K(X_i, Y_j) \ .$$

B Development of the PLSI Atomic Fisher Kernel

B.1 Fisher Score $U_{(d,w)}$

The Fisher score for PLSI is written $U_{(d,w)}(\theta) = \nabla_\theta \log P(d,w)$. Derivations are performed with respect to $P(z)$, $P(d|z)$ and $P(w|z)$ respectively for all terms w, documents d and categories z. Let us write $\tilde{d}$ and $\tilde{w}$ the indices of the document and term with respect to which derivations of $P(d,w)$ are performed. Then,

$$U_{(d,w)}(\theta) = \nabla_\theta \log P(d,w) = \begin{pmatrix} \frac{\partial \log P(w,d)}{\partial P(z)} \\ \frac{\partial \log P(w,d)}{\partial P(\tilde{w}|z)} \\ \frac{\partial \log P(w,d)}{\partial P(\tilde{d}|z)} \end{pmatrix} = \begin{pmatrix} \frac{P(w|z)P(d|z)}{P(w,d)} \\ \delta_{\tilde{w}w} \frac{P(d|z)P(z)}{P(w,d)} \\ \delta_{\tilde{d}d} \frac{P(w|z)P(z)}{P(w,d)} \end{pmatrix} ,$$

where $\delta_{\tilde{w}w} = 1$ if $\tilde{w} = w$ and 0 else (and similarly for d).

Note that two terms $\frac{\partial \log P(w,d_i)}{\partial P(\tilde{d}|z)}$ and $\frac{\partial \log P(w,d_j)}{\partial P(\tilde{d}|z)}$ are both non-zero if and only if $i = j$, that is if a document is compared to itself. Since this case is trivial, the terms of $U_{(d,w)}(\theta)$ that stem from the derivation w.r.t. $P(\tilde{d}|z)$ can be ignored in the kernel.

At this stage, for the sake of consistency with Hofmann's derivation, a square root reparametrization can be introduced:

$$\theta(z) \to \rho(z) = 2\sqrt{P(z)} \quad \text{and} \quad \theta(w|z) \to \rho(w|z) = 2\sqrt{P(w|z)} \ .$$

In the DFIM case, this reparametrization eventually cancels out; however, in the IFIM case, its contribution remains, making it a necessary step in the derivation. With this reparametrization, $\frac{\partial P(z)}{\partial \rho(z)} = \frac{1}{2}\rho(z) = \sqrt{P(z)}$ and $\frac{\partial P(w|z)}{\partial \rho(\tilde{w}|z)} = \delta_{\tilde{w}w} \frac{1}{2} \rho(w|z) = \delta_{\tilde{w}w} \sqrt{P(w|z)}$. Hence,

$$U_{(d,w)}(\rho) = \begin{pmatrix} \sqrt{P(z)} \frac{P(w|z)P(d|z)}{P(d,w)} \\ \delta_{\tilde{w}w} \sqrt{P(w|z)} \frac{P(d|z)P(z)}{P(w,d)} \end{pmatrix} .$$

B.2 Fisher Information Matrix $G(\theta)$

The Fisher information matrix is written

$$G(\theta) = E_{(d,w)}\left[U_{(d,w)}(\theta)\, U_{(d,w)}{}^T(\theta)\right] \; .$$

Let us consider the parts of the matrix G which stem from $U_{(d,w)}(z) = \frac{\partial \log P(d,w)}{\partial P(z)}$ (noted G_z) and from $U_{(d,w)}(\tilde{w}|z) = \frac{\partial \log P(d,w)}{\partial P(\tilde{w}|z)}$ (noted G_w). The diagonal of $G(\theta)$ can be approximated by (e.g. [17]):

$$G_z(z) = \sum_{(d,w)\in C} U_{(d,w)}(z)^2, \qquad G_w(\tilde{w}, z) = \sum_{(d,w)\in C} U_{(d,w)}(\tilde{w}|z)^2 \; ,$$

which leads to

$$G_z(z) = \sum_{d\in C}\sum_{w\in d} n(d,w)\left(\frac{P(w|z)P(d|z)}{P(d,w)}\right)^2 \qquad \text{and}$$

$$G_w(w,z) = \sum_{d\in C} n(d,w)\left(\frac{P(w|z)P(z)}{P(d,w)}\right)^2 \; .$$

B.3 Fisher Kernel

The expressions for $U_{(d,w)}$ and G, can be assembled into the Fisher kernel $K\big((d,w_d),(q,w_q)\big) = U_{(d,w_d)}\, G^{-1}\, U_{(q,w_q)}$, written as

$$\begin{aligned}
K\big((d,w_d),(q,w_q)\big) &= K_z\big((d,w_d),(q,w_q)\big) + K_w\big((d,w_d),(q,w_q)\big), \text{ where} \\
K_z\big((d,w_d),(q,w_q)\big) &= \sum_z U_{(d,w_d)}\, G_z(z)^{-1}\, U_{(q,w_q)} \; , \text{ and} \\
K_w\big((d,w_d),(q,w_q)\big) &= \sum_z\sum_{\tilde{w}} U_{(d,w_d)}\, G_w(\tilde{w},z)^{-1}\, U_{(q,w_q)} \\
&= \delta_{w_d w_q} \sum_z U_{(d,w_d)}\, G_w(w_d,z)^{-1}\, U_{(q,w_q)} \; .
\end{aligned}$$

Eventually, $K\Big((d,w_d),(q,w_q)\Big) =$

$$\sum_z \Bigg[\frac{P(w_d|z)P(d|z)}{P(d,w_d)}\frac{P(w_q|z)P(q|z)}{P(q,w_q)} P(z)G_z(z)^{-1} + \delta_{w_d,w_q}\frac{P(d|z)P(q|z)P^2(z)}{P(d,w_d)P(q,w_d)} P(w_d|z)G_w(w_d,z)^{-1}\Bigg] \; .$$

Note the normalizing role of $G(\theta)$: all the terms not depending on d in U_d will cancel out, as they can be factorized in $G_z(z)$ and $G_w(w,z)$, respectively: $P(z)$ cancels out in K_z and $P(w|z)P^2(z)$ cancels out in K_w.

Semi-supervised Document Clustering with Simultaneous Text Representation and Categorization

Yanhua Chen, Lijun Wang, and Ming Dong

Machine Vision and Pattern Recognition Lab
Department of Computer Science, Wayne State University
Detroit, MI 48202, USA
{chenyanh,ljwang,mdong}@wayne.edu

Abstract. In order to derive high quality information from text, the field of text mining has advanced swiftly from simple document clustering to co-clustering with words and categories. However, document co-clustering without any prior knowledge or background information is a challenging problem. In this paper, we propose a Semi-Supervised Non-negative Matrix Factorization (SS-NMF) framework for document co-clustering. Our method computes new *word-document* and *document-category* matrices by incorporating user provided constraints through simultaneous distance metric learning and modality selection. Using an iterative algorithm, we perform tri-factorization of the new matrices to infer the document, category and word clusters. Theoretically, we show the convergence and correctness of SS-NMF co-clustering and the advantages of SS-NMF co-clustering over existing approaches. Through extensive experiments conducted on publicly available data sets, we demonstrate the superior performance of SS-NMF for document co-clustering.

Keywords: Semi-supervised co-clustering, Non-negative matrix factorization.

1 Introduction

Document clustering is the task of automatically organizing text documents into meaningful clusters (groups) such that documents in the same cluster are similar, and are dissimilar from the ones in other clusters. It is one of the most important tasks in text mining and has received extensive attention in the data mining community. A number of different techniques [1,2,3,4] were proposed in the literature for clustering documents.

With the rapid development of the Internet and computational technologies in the past decade, the field of text mining has advanced swiftly from simple document clustering to more demanding tasks such as the production of granular taxonomies, sentiment analysis, and document summarization, in the hope of deriving higher quality information from text. These new applications in text mining typically involve multiple interrelated types of objects (e.g., categories, documents and words). Consequently, co-clustering was proposed in the literature [5,6]. In the heart of word-document co-clustering, the similarity between documents is defined by their word representations while the similarity between words is defined by their appearances in documents. In other words, document similarity and word similarity are defined in a reinforcing manner. In such a way, document and word can be grouped at the same time, leading to

W. Buntine et al. (Eds.): ECML PKDD 2009, Part I, LNAI 5781, pp. 211–226, 2009.

simultaneous document clustering and text representation. Similarly, high-order co-clustering uses the information contained in categories, documents, and words together, and is able to discover a hidden global structure in the heterogeneous text data [7,8]. This global structure, integrating document clustering with simultaneous text representation and categorization, provides us a better understanding of the roles and interactions of words, documents and categories in text analysis, which is highly valuable in many applications, and not achievable when clustering each data type independently.

However, current co-clustering methods are mostly developed based on the spectral graph model, and thus inapplicable to large text data sets. Moreover, they are completely unsupervised. Accurately co-clustering documents without domain dependent background information is still a challenging task. In this paper, we propose a Semi-Supervised NMF (SS-NMF) based framework to incorporate prior knowledge into document co-clustering. Under the proposed SS-NMF co-clustering methodology, a user is able to provide constraints on a few documents specifying whether they "must" (*must-link*) or "cannot" (*cannot-link*) be clustered together. Our goal is to improve the quality of document co-clustering by learning a distance metric based on these constraints. Using an iterative algorithm, we perform tri-factorizations of the new *word-document* and *document-category* matrices, obtained with the learnt distance metric, to infer the document clusters while simultaneously deriving the text representation (word clusters) and categorization (category clusters). The major contribution of this work is summarized as follows,

1. We propose a novel algorithm for document co-clustering based on NMF. Computationally, NMF co-clustering is more efficient and flexible than spectral methods, and can provide more meaningful clustering results.

2. To the best of our knowledge, this is the first work on semi-supervised data co-clustering providing significance of each modality. Through distance metric learning and modality selection, prior knowledge is integrated into document co-clustering, making *must-link* documents as tight as possible and *cannot-link* documents as loose as possible.

3. From a theoretical perspective, our approach is mathematically rigorous. The convergence and correctness are proved. In addition, we show that our work provides a general framework for data co-clustering. Existing approaches such as the well-established spectral co-clustering algorithms can be considered as special cases of our method.

The rest of the paper is organized as follows. We review related work in Section 2. The proposed SS-NMF co-clustering algorithm is derived in Section 3. Our theoretical analysis on the correctness and convergence of the algorithm and on the advantages over spectral co-clustering approaches are presented in Section 4. Experimental results appear in Section 5. Finally, we conclude in Section 6.

2 Related Work

In this section, we briefly review related work in co-clustering (documents, words, and categories) and semi-supervised clustering.

In general, co-clustering approaches can be divided into two representative categories: information theory-based models and graph theoretic methods. In the former category, Dhillon et al. [9] presented a pairwise co-clustering algorithm to maximize the mutual information between the clustered random variables subject to the constraints on the number of row and column clusters. Later, Gao et al.[10] extended this method for high-order co-clustering. However, there is no sound objective function and theoretical proof on the effectiveness and correctness of these algorithms. On the other hand, graph theoretic approaches have a well-defined objective function. Spectral learning, such as Bipartite Spectral Graph Partitioning (BSGP) [5], was proposed and applied to co-cluster documents and words. With the similar philosophy, Gao et al. proposed Consistent Bipartite Graph Co-partitioning for high-order co-clustering to do hierarchical taxonomy preparation [7]. Recently, Rege et al. proposed to directly minimize the isoperimetric ratio of the weighted bipartite or high-order graph[11,12]. Experimental results on word-document and word-document-category co-clustering show that their approaches outperform the spectral methods in terms of the quality, speed, and stability. More recently, Long et al. [8] proposed Spectral Relational Clustering (SRC), in which they formulated heterogenous co-clustering as collective factorization on related matrices and derived a spectral algorithm to cluster multi-type interrelated data objects simultaneously. SRC provides more flexibility by lifting the requirement of one-to-one association in graph-based co-clustering. However, as a spectral method, it requires solving an eigen-problem, which computationally is not efficient to deal with large text data sets.

Semi-supervised clustering uses class labels or pairwise constraints on examples to aid unsupervised clustering. Two sources of supervised information are usually available to a semi-supervised clustering method: class labels or some pairwise constraints (*must-link* or *cannot-link*) as *a prior*. Existing methods for semi-supervised clustering based on source information generally fall into two categories: *semi-supervised clustering with labels* and *semi-supervised clustering with constraints* methods. In constraint-based approaches, the clustering algorithm itself is modified so that the available labels or constraints are used to bias the search for an appropriate clustering of the data [13]. In distance-based approaches, an existing clustering algorithm that uses a distance measure is employed; however, the distance measure is first trained to satisfy the labels or constraints in the supervised data [14]. Recent research in semi-supervised clustering tends to combine the constraint-based with distance-based approaches. Noticeable efforts on semi-supervised clustering algorithm include: Semi-Supervised Kernel K-means [15], Semi-Supervised Spectral Normalize Cuts [16] and SS-NMF [17,18]. In [19], it is shown that SS-NMF provides a unified framework for semi-supervised clustering. Many existing algorithms can be considered as special cases of SS-NMF. However, until now all semi-supervised methods are only applicable to homogeneous data clustering.

Even though the research on document co-clustering and semi-supervised clustering has attracted substantial attention in the past years, there has been no mathematically rigorous approach for semi-supervised data co-clustering. In the following, we will derive a theoretically sound algorithm based on SS-NMF, and apply it for document co-clustering.

3 SS-NMF Co-clustering

In this section, we first propose a SS-NMF model for general data co-clustering. Then, we narrow down to document clustering with simultaneous text representation and categorization, and discuss 1) how to incorporate prior knowledge through distance metric learning and modality selection, and 2) how to efficiently infer document, word and category clusters simultaneously using matrix factorization.

3.1 Model Formulation

Nonnegative Matrix Factorization (NMF) is a group of algorithms in multivariate analysis and and linear algebra where a matrix $\mathbf{X}$ is factorized into two nonnegative matrices, $\mathbf{F}$ and $\mathbf{G}$. It is initially proposed for "parts-of-whole" decomposition [20], and later extended to a general framework for data clustering [21]. It can model widely varying data distributions and do both hard and soft clustering. Let $\mathbf{X} = (\mathbf{x}_1, ..., \mathbf{x}_n) \in \mathbb{R}^{d \times n}$ be the data matrix of nonnegative elements. NMF factorizes $\mathbf{X}$ into two non-negative matrices,

$$\mathbf{X} \approx \mathbf{F}\mathbf{G}^T, \tag{1}$$

where $\mathbf{F} \in \mathbb{R}^{d\times k}$ is cluster centroid, $\mathbf{G} \in \mathbb{R}^{n \times k}$ is cluster indicator, and k is the number of clusters. The factorizations are typically obtained by the least square minimization.

Given a Heterogenous Relational Data (HRD) set, $\mathcal{X}_1 = \{x_{11}, ..., x_{1n_1}\}$,..., $\mathcal{X}_c = \{x_{c1}, ..., x_{cn_c}\}$,..., $\mathcal{X}_l = \{x_{l1}, ..., x_{ln_l}\}$, each representing one data type, our goal is to simultaneously cluster $\mathcal{X}_1$ into k_1 disjoint clusters, ..., and $\mathcal{X}_l$ into k_l disjoint clusters. To derive a solution to the co-clustering problem under matrix factorization framework, we first model HRD as a set of related matrices, i.e., a relation matrix $\mathbf{R}^{(pq)} \in \mathbb{R}^{n_p \times n_q}$ is used to represent the relations between $\mathcal{X}_p$ and $\mathcal{X}_q$ $(1 \leq p, q \leq l)$. Then, we can formulate the task of co-clustering as a optimization problem with nonnegative tri-factorization of $\mathbf{R}^{(pq)}$,

$$J = \min_{\mathbf{G}^{(p)} \geq 0, \mathbf{G}^{(q)} \geq 0, \mathbf{S}^{(pq)} \geq 0} \sum_{1 \leq p,q \leq l} \|\mathbf{R}^{(pq)} - \mathbf{G}^{(p)}\mathbf{S}^{(pq)}\mathbf{G}^{(q)}\|^2 \tag{2}$$

where $\mathbf{G}^{(p)} \in \mathbb{R}^{n_p \times k_p}$ and $\mathbf{G}^{(q)} \in \mathbb{R}^{k_q \times n_q}$ are the cluster indicator matrices, and $\mathbf{S} \in \mathbb{R}^{k_p \times k_q}$ is the cluster association matrix which gives the relation among the clusters of different data types.

In semi-supervised document co-clustering, supervision is typically provided as two sets of pairwise constraints derived from given labels on the documents: *must-link* constraints $M = \{(\mathbf{x}_i, \mathbf{x}_j)\}$ and *cannot-link* constraints $C = \{(\mathbf{x}_i, \mathbf{x}_j)\}$, where $(\mathbf{x}_i, \mathbf{x}_j) \in M$ implies that $\mathbf{x}_i$ and $\mathbf{x}_j$ are labeled as belonging to the same cluster, while $(\mathbf{x}_i, \mathbf{x}_j) \in C$ implies that $\mathbf{x}_i$ and $\mathbf{x}_j$ are labeled as belonging to different clusters. Figure 1 shows the triplet data (e.g., categories, documents and words), which is a basic element of general HRD. If we can successfully co-cluster such triplet data, the corresponding technique can be easily extended to structures involving more data types. In Figure 1, the relations between words and documents, and documents and categories are denoted by a *word-document* matrix $\mathbf{R}^{(12)}$ and a *document-category* matrix $\mathbf{R}^{(23)}$, respectively. The edges marked with M indicate the *must-link* constraints M, while the edges marked with C denote *cannot-link* constraints C. The dotted line shows the optimal clustering result. Note that in the following discussions, we will focus on the triplet co-clustering, or more specifically, word-document-category co-clustering. However, the derived algorithm is in general applicable to structures with more than three data types.

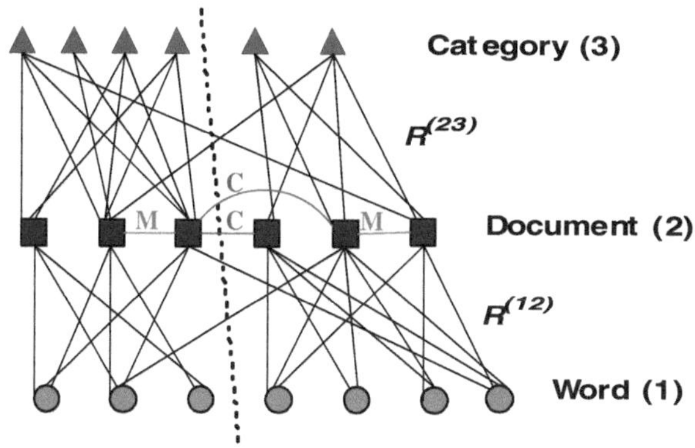

Fig. 1. Word-Document-Category co-clustering with must-link and cannot-link constraints

3.2 SS-NMF for Triplet Data

We now present the SS-NMF based triplet co-clustering algorithm. For simultaneous text representation and categorization, documents have to be clustered together with both words and categories. Let $\mathbf{R}^{(12)}$ and $\mathbf{R}^{(23)}$ denote the *word-document* and *document-category* matrix, respectively. The goal of SS-NMF triplet co-clustering is to iteratively cluster rows and columns of $\mathbf{R}^{(12)}$, and rows and columns of $\mathbf{R}^{(23)}$, subject to the *M* and *C* constraints on the documents. The first step in triplet co-clustering is to obtain the new matrix $\widetilde{\mathbf{R}}$ between different data types. In other words, we need to learn a distance metric $\mathbf{L}$ for each relation based on *must-link* and *cannot-link* constraints such that the clustering result on the central type (e.g., documents) is globally optimized. Specifically, a distance metric $\mathbf{L}^{(pq)}$ (where $(pq) \in \{(12), (23)\}$) over each relation of the form $d(\mathbf{x}_i^{(pq)}, \mathbf{x}_j^{(pq)}) = \sqrt{(\mathbf{x}_i^{(pq)} - \mathbf{x}_j^{(pq)})^T \mathbf{L}^{(pq)} (\mathbf{x}_i^{(pq)} - \mathbf{x}_j^{(pq)})}$ will be learnt, such that $(\mathbf{x}_i^{(pq)}, \mathbf{x}_j^{(pq)}) \in M$ are moved closer to each other while $(\mathbf{x}_i^{(pq)}, \mathbf{x}_j^{(pq)}) \in C$ are moved further away. That is, we solve the following optimization problem,

$$\max \quad g(\mathbf{L}^{(pq)}) = \frac{\sum_{(\mathbf{x}_i^{(pq)}, \mathbf{x}_j^{(pq)}) \in C} ||\mathbf{x}_i^{(pq)}, \mathbf{x}_j^{(pq)}||_{\mathbf{L}^{(pq)}}}{\sum_{(\mathbf{x}_i^{(pq)}, \mathbf{x}_j^{(pq)}) \in M} ||\mathbf{x}_i^{(pq)}, \mathbf{x}_j^{(pq)}||_{\mathbf{L}^{(pq)}}} \tag{3}$$

where $||.||$ is the Frobenius matrix norm. This maximization problem is equivalent to the generalized Semi-Supervised Linear Discriminate Analysis (SS-LDA) problem as follows,

$$J = \min \frac{trace(\mathbf{L}^{(pq)} \mathbf{W}_M{}^{(pq)})}{trace(\mathbf{L}^{(pq)} \mathbf{B}_C^{(pq)})} \tag{4}$$

where $\mathbf{W}_M$ is within-distance matrix from must-link constraints, $\mathbf{B}_C$ is between-distance matrix from cannot-link constraints, and can be solved accordingly [14]. Moreover, triplet co-clustering has an additional layer of complexity. Because categories and words can play a different role in the grouping of documents, we have to consider the issue of modality selection. To this end, we introduce a factor, $\mathbf{a} = [\alpha^{(12)}, \alpha^{(23)}]$, to denote the relative importance of "word" and "category". Note that the modality selection and distance metric learning are strongly dependent. This suggests that these two objectives must be achieved simultaneously. In Algorithm 1, we propose an iterative algorithm to learn the optimal distance metrics $\mathbf{L}^{(12)}, \mathbf{L}^{(23)}$ and modality importance factor **a** for the given constraints. Based on the learnt distance metrics $\mathbf{L}^{(12)}$ and $\mathbf{L}^{(23)}$, we compute two new relational data matrices, $\widetilde{\mathbf{R}}^{(12)}$ and $\widetilde{\mathbf{R}}^{(23)}$. To achieve triplet co-clustering, we need to perform non-negative tri-factorization of new relational matrix as follows,

$$J = \min_{\substack{\mathbf{G}^{(1)}\geq 0, \mathbf{G}^{(2)}\geq 0, \mathbf{G}^{(3)}\geq 0\\ \mathbf{S}^{(12)}\geq 0, \mathbf{S}^{(23)}\geq 0}} (\|\widetilde{\mathbf{R}}^{(12)} - \mathbf{G}^{(1)}\mathbf{S}^{(12)}(\mathbf{G}^{(2)})\|^2 + \|\widetilde{\mathbf{R}}^{(23)} - \mathbf{G}^{(2)^T}\mathbf{S}^{(23)}(\mathbf{G}^{(3)})\|^2) \tag{5}$$

Our main idea is to iteratively update the cluster structures for each data type in Equation (5). The details are given in Algorithm 2.

Algorithm 1. Simultaneous Distance Metric Learning and Modality Selection

INPUT: Original relational matrices $\mathbf{R}^{(12)}, \mathbf{R}^{(23)}$, central type $\mathcal{X}_2$ with must-link constraint M, and cannot-link constraint C
OUTPUT: Optimal distance metric $\mathbf{L}^{(12)}$, $\mathbf{L}^{(23)}$ and modality importance factor **a**
METHOD:

1. Construct target relation $\widetilde{\mathbf{M}}$ based on constraints M and C, where each element $\widetilde{m}_{ij}$ is 1 if $(\mathbf{x}_i, \mathbf{x}_j) \in M$, and 0 if $(\mathbf{x}_i, \mathbf{x}_j) \in C$
2. Obtain the initial distance metrics $\mathbf{L}^{(12)}$ and $\mathbf{L}^{(23)}$ by SS-LDA with constraints M and C
3. Set the number of iterations t=0
   (a) Learn new relational matrices $\widetilde{\mathbf{R}}^{(12)}$ and $\widetilde{\mathbf{R}}^{(23)}$
   (b) Formulate matrices $\mathbf{M}^{(12)} = (\widetilde{\mathbf{R}}'^{(12)})^T\widetilde{\mathbf{R}}'^{(12)}$ and $\mathbf{M}^{(23)} = \widetilde{\mathbf{R}}'^{(23)}(\widetilde{\mathbf{R}}'^{(23)})^T$, where $\widetilde{\mathbf{R}}'^{(12)}$ and $\widetilde{\mathbf{R}}'^{(23)}$ contain only samples of $\mathcal{X}_2$ with constraints
   (c) Optimize the following function to obtain modality importance factor **a**
   $$\mathbf{a}^{opt} = arg\min_{\alpha} \|\widetilde{\mathbf{M}} - \alpha^{(12)}\mathbf{M}^{(12)} + \alpha^{(23)}\mathbf{M}^{(23)}\|^2$$
   (d) Learn the new distance metrics $\mathbf{L}^{(12)}$ and $\mathbf{L}^{(23)}$ for $\alpha^{(12)}\widetilde{\mathbf{R}}^{(12)}$ and $\alpha^{(23)}\widetilde{\mathbf{R}}^{(23)}$ by SS-LDA
4. If $\mathbf{a}_{t+1} - \mathbf{a}_t > \varepsilon$, set $t = t + 1$ and go to steps (a)-(d); otherwise, stop and output the optimal distance metrics $\mathbf{L}^{(12)}, \mathbf{L}^{(23)}$ and modality importance factor **a**

4 Theoretical Analysis

4.1 Algorithm Convergence and Correctness

We now prove the theoretical convergence and correctness of SS-NMF co-clustering algorithm. Motivated by [6], we render the proof based on optimization theory, auxiliary function and several matrix inequalities.

Correctness. First, we prove the correctness of the algorithm, which can be stated as,

Proposition 1. *If the solution converges based on the updating rules in Equations (6)-(10), the solution satisfies the KKT optimality condition.*

Algorithm 2. SS-NMF for Triplet Co-Clustering

INPUT: Original relational matrices $\mathbf{R}^{(12)}$ and $\mathbf{R}^{(23)}$, new distance metrics $\mathbf{L}^{(12)}$ and $\mathbf{L}^{(23)}$
OUTPUT: Cluster indicator matrices $\mathbf{G}^{(1)}$, $\mathbf{G}^{(2)}$, and $\mathbf{G}^{(3)}$, cluster association matrices $\mathbf{S}^{(12)}$ and $\mathbf{S}^{(23)}$
METHOD:

1. Obtain new relational matrices through projection: $\widetilde{\mathbf{R}}^{(12)} = \sqrt{\mathbf{L}^{(12)}}\mathbf{R}^{(12)}$ and $\widetilde{\mathbf{R}}^{(23)} = \sqrt{\mathbf{L}^{(23)}}\mathbf{R}^{(23)}$
2. Initialize $\mathbf{G}^{(1)}, \mathbf{G}^{(2)}, \mathbf{G}^{(3)}, \mathbf{S}^{(12)}, \mathbf{S}^{(23)}$ with non-negative values.
3. Iterate for each i and h until *convergence*

   (a) Cluster indicator matrices:

$$\mathbf{G}_{ih}^{(1)} \leftarrow \mathbf{G}_{ih}^{(1)} \frac{(\widetilde{\mathbf{R}}^{(12)}\mathbf{G}^{(2)^T}\mathbf{S}^{(12)^T})_{ih}}{(\mathbf{G}^{(1)}\mathbf{S}^{(12)}\mathbf{G}^{(2)}\mathbf{G}^{(2)^T}\mathbf{S}^{(12)^T})_{ih}} \tag{6}$$

$$\mathbf{G}_{ih}^{(2)} \leftarrow \mathbf{G}_{ih}^{(2)} \frac{(\mathbf{S}^{(12)^T}\mathbf{G}^{(1)^T}\widetilde{\mathbf{R}}^{(12)}) + (\widetilde{\mathbf{R}}^{(23)}\mathbf{G}^{(3)^T}\mathbf{S}^{(23)^T})^T}{(\mathbf{S}^{(12)^T}\mathbf{G}^{(1)^T}\mathbf{G}^{(1)}\mathbf{S}^{(12)}\mathbf{G}^{(2)}) + (\mathbf{G}^{(2)^T}\mathbf{S}^{(23)}\mathbf{G}^{(3)}\mathbf{G}^{(3)^T}\mathbf{S}^{(23)^T})^T} \tag{7}$$

$$\mathbf{G}_{ih}^{(3)} \leftarrow \mathbf{G}_{ih}^{(3)} \frac{(\mathbf{S}^{(23)^T}\mathbf{G}^{(2)}\widetilde{\mathbf{R}}^{(23)})_{ih}}{(\mathbf{S}^{(23)^T}\mathbf{G}^{(2)}\mathbf{G}^{(2)^T}\mathbf{S}^{(23)}\mathbf{G}^{(3)})_{ih}} \tag{8}$$

   (b) Cluster association matrices:

$$\mathbf{S}_{ih}^{(12)} \leftarrow \mathbf{S}_{ih}^{(12)} \frac{(\mathbf{G}^{(1)^T}\widetilde{\mathbf{R}}^{(12)}\mathbf{G}^{(2)^T})_{ih}}{(\mathbf{G}^{(1)^T}\mathbf{G}^{(1)}\mathbf{S}^{(12)}\mathbf{G}^{(2)}\mathbf{G}^{(2)^T})_{ih}} \tag{9}$$

$$\mathbf{S}_{ih}^{(23)} \leftarrow \mathbf{S}_{ih}^{(23)} \frac{(\mathbf{G}^{(2)}\widetilde{\mathbf{R}}^{(23)}\mathbf{G}^{(3)^T})_{ih}}{(\mathbf{G}^{(2)}\mathbf{G}^{(2)^T}\mathbf{S}^{(23)^T}\mathbf{G}^{(3)}\mathbf{G}^{(3)^T})_{ih}} \tag{10}$$

Proof. Following the standard theory of constrained optimization, we introduce the Lagrangian multipliers λ_1, λ_2, λ_3 , λ_4, λ_5 and λ_6 to minimize the Lagrangian function,

$$\begin{aligned} &L(\mathbf{G}^{(1)}, \mathbf{G}^{(2)}, \mathbf{G}^{(3)}, \mathbf{S}^{(12)}, \mathbf{S}^{(23)}, \lambda_1, \lambda_2, \lambda_3, \lambda_4, \lambda_5, \lambda_6) \\ &= \|\widetilde{\mathbf{R}}^{(12)} - \mathbf{G}^{(1)}\mathbf{S}^{(12)}\mathbf{G}^{(2)}\|^2 + \|\widetilde{\mathbf{R}}^{(23)} - \mathbf{G}^{(2)^T}\mathbf{S}^{(23)}\mathbf{G}^{(3)}\|^2 \\ &\quad - \mathrm{Tr}(\lambda_1\mathbf{C}^{(1)^T}) - \mathrm{Tr}(\lambda_2\mathbf{S}^{(12)^T}) - \mathrm{Tr}(\lambda_3\mathbf{C}^{(2)^T}) \\ &\quad - \mathrm{Tr}(\lambda_4\mathbf{C}^{(2)}) - \mathrm{Tr}(\lambda_5\mathbf{S}^{(23)^T}) - \mathrm{Tr}(\lambda_6\mathbf{C}^{(3)^T}) \end{aligned} \tag{11}$$

Based on the KKT complementarity conditions $\frac{\partial L}{\partial \mathbf{G}^{(1)}} = 0$, $\frac{\partial L}{\partial \mathbf{S}^{(12)}} = 0$, $\frac{\partial L}{\partial \mathbf{G}^{(2)}} = 0$, $\frac{\partial L}{\partial \mathbf{S}^{(23)}} = 0$ and $\frac{\partial L}{\partial \mathbf{G}^{(3)}} = 0$, we obtain the following five equations,

$$\begin{aligned} &2\widetilde{\mathbf{R}}^{(12)}\mathbf{G}^{(2)^T}\mathbf{S}^{(12)^T} - 2\mathbf{G}^{(1)}\mathbf{S}^{(12)}\mathbf{G}^{(2)}\mathbf{G}^{(2)^T}\mathbf{S}^{(12)^T} + \lambda_1 = 0 \\ &2\mathbf{G}^{(1)^T}\widetilde{\mathbf{R}}^{(12)}\mathbf{G}^{(2)^T} - 2\mathbf{G}^{(1)^T}\mathbf{G}^{(1)}\mathbf{S}^{(12)}\mathbf{G}^{(2)}\mathbf{G}^{(2)^T} + \lambda_2 = 0 \\ &2\mathbf{S}^{(12)^T}\mathbf{G}^{(1)^T}\widetilde{\mathbf{R}}^{(12)} - 2\mathbf{S}^{(12)^T}\mathbf{G}^{(1)^T}\mathbf{G}^{(1)}\mathbf{S}^{(12)}\mathbf{G}^{(2)} + \lambda_3 + \\ &(2\widetilde{\mathbf{R}}^{(23)}\mathbf{G}^{(3)^T}\mathbf{S}^{(23)^T} - 2\mathbf{G}^{(2)^T}\mathbf{S}^{(23)}\mathbf{G}^{(3)}\mathbf{G}^{(3)^T}\mathbf{S}^{(23)^T})^T + \lambda_4 = 0 \\ &2\mathbf{G}^{(2)}\widetilde{\mathbf{R}}^{(23)}\mathbf{G}^{(3)^T} - 2\mathbf{G}^{(2)}\mathbf{G}^{(2)^T}\mathbf{S}^{(23)}\mathbf{G}^{(3)}\mathbf{G}^{(3)^T} + \lambda_5 = 0 \\ &2\mathbf{S}^{(23)^T}\mathbf{G}^{(2)}\widetilde{\mathbf{R}}^{(23)} - 2\mathbf{S}^{(23)^T}\mathbf{G}^{(2)}\mathbf{G}^{(2)^T}\mathbf{S}^{(23)}\mathbf{G}^{(3)} + \lambda_6 = 0 \end{aligned}$$

We apply the Hadamard multiplication on both sides of above five equations by $\mathbf{G}^{(1)}$, $\mathbf{S}^{(12)}$, $\mathbf{G}^{(2)}$, $\mathbf{S}^{(23)}$, and $\mathbf{G}^{(3)}$, respectively. Using KKT conditions of

$$\lambda_1 \odot \mathbf{G}^{(1)} = 0 \qquad \lambda_2 \odot \mathbf{S}^{(12)} = 0 \qquad \lambda_3 \odot \mathbf{G}^{(2)} = 0$$
$$\lambda_4 \odot \mathbf{G}^{(2)} = 0 \qquad \lambda_5 \odot \mathbf{S}^{(23)} = 0 \qquad \lambda_6 \odot \mathbf{G}^{(3)} = 0$$

where $\odot$ denotes the Hadamard product of two matrices and letting $\lambda_3 = \lambda_4$, we can prove that if $\mathbf{G}^{(1)}$, $\mathbf{S}^{(12)}$, $\mathbf{G}^{(2)}$, $\mathbf{S}^{(23)}$, and $\mathbf{G}^{(3)}$ are a local minimizer of the objective function in Equation (11), the following five equations are satisfied,

$$((\widetilde{\mathbf{R}}^{(12)}{\mathbf{G}^{(2)}}^T{\mathbf{S}^{(12)}}^T) - (\mathbf{G}^{(1)}\mathbf{S}^{(12)}\mathbf{G}^{(2)}{\mathbf{G}^{(2)}}^T{\mathbf{S}^{(12)}}^T)) \odot \mathbf{G}^{(1)} = 0$$
$$(({\mathbf{G}^{(1)}}^T\widetilde{\mathbf{R}}^{(12)}{\mathbf{G}^{(2)}}^T) - ({\mathbf{G}^{(1)}}^T\mathbf{G}^{(1)}\mathbf{S}^{(12)}\mathbf{G}^{(2)}{\mathbf{G}^{(2)}}^T)) \odot \mathbf{S}^{(12)} = 0$$
$$(({\mathbf{S}^{(12)}}^T{\mathbf{G}^{(1)}}^T\widetilde{\mathbf{R}}^{(12)} + (\widetilde{\mathbf{R}}^{(23)}{\mathbf{G}^{(3)}}^T{\mathbf{S}^{(23)}}^T)^T) - ({\mathbf{S}^{(12)}}^T{\mathbf{G}^{(1)}}^T\mathbf{G}^{(1)}$$
$$\mathbf{S}^{(12)}\mathbf{G}^{(2)} + ({\mathbf{G}^{(2)}}^T\mathbf{S}^{(23)}\mathbf{G}^{(3)}{\mathbf{G}^{(3)}}^T{\mathbf{S}^{(23)}}^T)^T)) \odot \mathbf{G}^{(2)} = 0$$
$$((\mathbf{G}^{(2)}\widetilde{\mathbf{R}}^{(23)}{\mathbf{G}^{(3)}}^T) - (\mathbf{G}^{(2)}{\mathbf{G}^{(2)}}^T\mathbf{S}^{(23)}\mathbf{G}^{(3)}{\mathbf{G}^{(3)}}^T)) \odot \mathbf{S}^{(23)} = 0$$
$$(({\mathbf{S}^{(23)}}^T\mathbf{G}^{(2)}\widetilde{\mathbf{R}}^{(23)}) - ({\mathbf{S}^{(23)}}^T\mathbf{G}^{(2)}{\mathbf{G}^{(2)}}^T\mathbf{S}^{(23)}\mathbf{G}^{(3)})) \odot \mathbf{G}^{(3)} = 0$$

Based on the above five equations, we derive the proposed updating rules of Equations (6)-(10). If the updating rules converge, the solution satisfies the KKT optimality condition. Proof is completed.

Convergence. Next, we prove the convergence of the algorithm. In Proposition 2, we show that the objective function decreases monotonically under the five updating rules of Equations (6)-(10) . This can be done by making use of an auxiliary function similar to that used in [21,22].

Proposition 2. *If any four of five matrices $\mathbf{G}^{(1)}$, $\mathbf{S}^{(12)}$, $\mathbf{G}^{(2)}$, $\mathbf{S}^{(23)}$, and $\mathbf{G}^{(3)}$ are fixed, $J = \|\widetilde{\mathbf{R}}^{(12)} - \mathbf{G}^{(1)}\mathbf{S}^{(12)}\mathbf{G}^{(2)}\|^2 + \|\widetilde{\mathbf{R}}^{(23)} - {\mathbf{G}^{(2)}}^T\mathbf{S}^{(23)}\mathbf{G}^{(3)}\|^2$ decreases monotonically under the updating rules of Equations (6)-(10).*

Proof. Due to the space constraints, we give the proof of convergence for one updating rule (e.g., the rule in Equation (6)) and skip the others. However, the proof of all five updating rules is similar to each other. The mathematical derivation below can be applied to other rules as well.

So, we need to show: If $\mathbf{S}^{(12)}$, $\mathbf{G}^{(2)}$, $\mathbf{S}^{(23)}$, and $\mathbf{G}^{(3)}$ are fixed matrices, then $J(\mathbf{G}^{(1)}) = \|\widetilde{\mathbf{R}}^{(12)} - \mathbf{G}^{(1)}\mathbf{S}^{(12)}\mathbf{G}^{(2)}\|^2 + \|\widetilde{\mathbf{R}}^{(23)} - {\mathbf{G}^{(2)}}^T\mathbf{S}^{(23)}\mathbf{G}^{(3)}\|^2$ decreases monotonically under the updating rule of Equation (6).

First, a function $F({\mathbf{G}^{(1)}}^{(t+1)}, {\mathbf{G}^{(1)}}^{(t)})$ is called an auxiliary function of $J({\mathbf{G}^{(1)}}^{(t+1)})$ if it satisfies the following two conditions: $F({\mathbf{G}^{(1)}}^{(t+1)}, {\mathbf{G}^{(1)}}^{(t)}) \geq J({\mathbf{G}^{(1)}}^{(t+1)})$ and $F({\mathbf{G}^{(1)}}^{(t+1)}, {\mathbf{G}^{(1)}}^{(t)}) = J({\mathbf{G}^{(1)}}^{(t+1)})$ for any ${\mathbf{G}^{(1)}}^{(t+1)}, {\mathbf{G}^{(1)}}^{(t)}$.

We define ${\mathbf{G}^{(1)}}^{(t+1)} = arg\min F({\mathbf{G}^{(1)}}^{(t+1)}, {\mathbf{G}^{(1)}}^{(t)})$. By constructing an appropriate auxiliary function, we can prove the following equation,

$$J({\mathbf{G}^{(1)}}^{(t)}) = F({\mathbf{G}^{(1)}}^{(t)}, {\mathbf{G}^{(1)}}^{(t)}) \geq F({\mathbf{G}^{(1)}}^{(t+1)}, {\mathbf{G}^{(1)}}^{(t)}) \geq J({\mathbf{G}^{(1)}}^{(t+1)})$$

Thus, $J({\mathbf{G}^{(1)}}^{(t)})$ is monotonic decreasing (non-increasing). The proof is completed.

4.2 Advantages of SS-NMF

We now show that NMF provides a general framework for data co-clustering by establishing the relationship between NMF and other well-known spectral high-order co-clustering algorithms, i.e., Spectral Relational Clustering (SRC) [8]. In fact, this algorithm can be considered as a special case of NMF co-clustering.

SRC is proposed in [8] for high-order document co-clustering. It iteratively embeds each type of data into low dimensional spaces and benefits from the interactions in the hidden structure of different data types. The underlying objective function is,

$$\min_{\mathbf{G}^{(1)T}\mathbf{G}^{(1)}=\mathbf{I},\mathbf{G}^{(2)T}\mathbf{G}^{(2)}=\mathbf{I},\mathbf{G}^{(3)T}\mathbf{G}^{(3)}=\mathbf{I}} (\|\mathbf{R}^{(12)} - \mathbf{G}^{(1)}\mathbf{S}^{(12)}\mathbf{G}^{(2)}\|^2 + \|\mathbf{R}^{(23)} - \mathbf{G}^{(2)^T}\mathbf{S}^{(23)}\mathbf{G}^{(3)}\|^2)$$

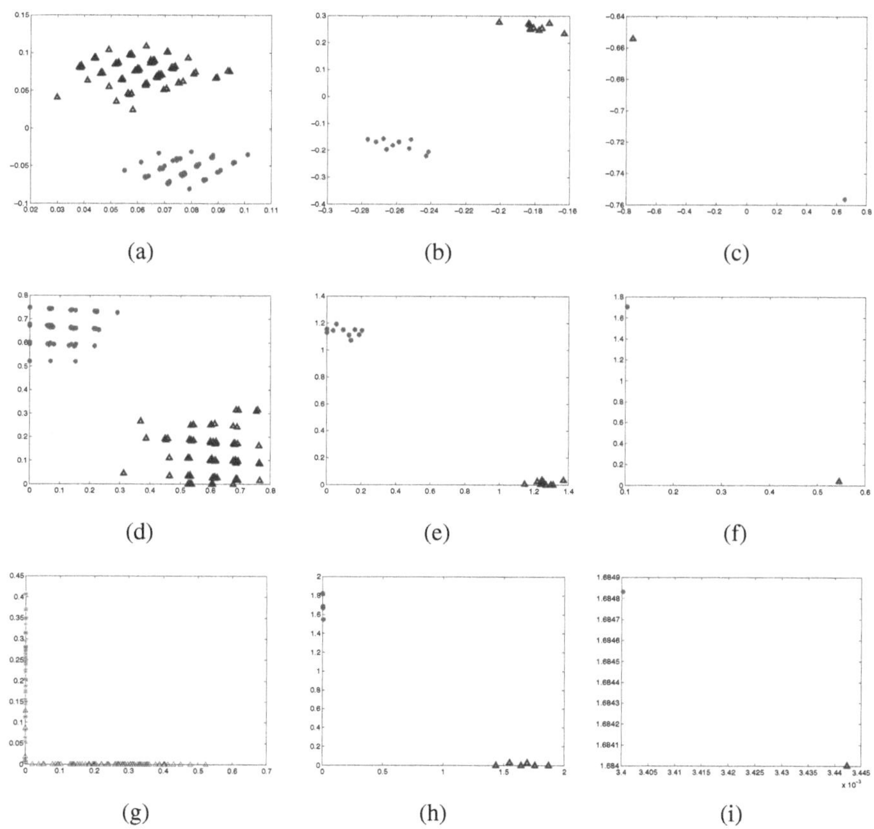

Fig. 2. (a)-(c): Clustering results by SRC in the subspace of the first two singular vectors of $\mathbf{G}^{(1)}$,$\mathbf{G}^{(2)}$ and $\mathbf{G}^{(3)}$. There is no direct relationship between the axes and the clusters. (d)-(f): Clustering results by NMF in the subspace of the two column vectors of $\mathbf{G}^{(1)}$,$\mathbf{G}^{(2)}$ and $\mathbf{G}^{(3)}$. The data points from the two clusters are distributed closely to the two axes. (g)-(i): Clustering results by SS-NMF (with 5% constraints) in the subspace of the two column vectors of $\mathbf{G}^{(1)}$,$\mathbf{G}^{(2)}$ and $\mathbf{G}^{(3)}$. The data points from the two clusters are distributed exactly along the two axes.

On the other hand, NMF-based high-order co-clustering is to minimize the following function,

$$\min_{\mathbf{G}^{(1)}\geq 0,\mathbf{G}^{(2)}\geq 0,\mathbf{G}^{(3)}\geq 0,\mathbf{S}^{(12)}\geq 0,\mathbf{S}^{(23)}\geq 0} (\|\mathbf{R}^{(12)} - \mathbf{G}^{(1)}\mathbf{S}^{(12)}\mathbf{G}^{(2)}\|^2 + \|\mathbf{R}^{(23)} - \mathbf{G}^{(2)^T}\mathbf{S}^{(23)}\mathbf{G}^{(3)}\|^2)$$

It is clear that the major difference between NMF co-clustering and SRC lies in the fact that SRC requires the cluster indictor matrices be orthogonal, while NMF co-clustering relaxes this requirements to be near-orthogonal. This relaxation can provide us more meaning clustering results.

The advantage of NMF or SS-NMF over SRC can best be illustrated using an example. In the following example, the synthetic data set has 200 words, 20 documents, and 2 categories, each having two clusters of equal size. More specifically, we have two relational matrices: $\mathbf{R}^{(12)}$ of size 200×20 and $\mathbf{R}^{(23)}$ of size 20×2, both binary matrices with 2-by-2 block structures generated by the Bernoulli distribution. $\mathbf{R}^{(12)}$ is generated based on the block structure $\begin{bmatrix} 0.9 & 0.1 \\ 0.2 & 0.8 \end{bmatrix}$ and $\mathbf{R}^{(23)}$ is based on the block structure $\begin{bmatrix} 0.8 & 0.2 \\ 0.1 & 0.9 \end{bmatrix}$.

Unlike SRC, NMF or SS-NMF maps the data into a non-negative latent semantic space which is not required to be orthogonal. Panels (a)-(c), (d)-(f) and (g)-(i) in Figure 2 show the clustering results by SRC, NMF and SS-NMF, in which two clusters are denoted by stars and triangles, respectively. For NMF or SS-NMF, we plot the data points in the subspace of two column vectors of $\mathbf{G}^{(1)}$, $\mathbf{G}^{(2)}$ and $\mathbf{G}^{(3)}$, while for SRC the subspace of the first two singular vectors is used. Note that for either NMF or SS-NMF, each data point takes a non-negative value on both axes. In the NMF subspace, each axis corresponds to a cluster, and all the data points belonging to the same cluster are nicely located closely to the axis. In the SS-NMF subspace, the data points belonging to the same cluster are almost spread along the axis. This indicates that SS-NMF can provide better clustering accuracy than unsupervised NMF because the cluster label for a data point is determined by finding the axis with which the data point has the largest projection value. On the other hand, in the SRC subspace, we observe no direct relationship between the axes (singular vectors) and the clusters.

5 Experiments and Results

In this section, we empirically demonstrated the performance of SS-NMF in co-clustering documents, words and categories by comparing it with well-established co-clustering algorithms. Through these comparisons, we showed the relative position of SS-NMF with respect to existing approaches to document co-clustering[1]. All algorithms were implemented using MATLAB 7.0.

[1] At present, there is no other existing work on semi-supervised co-clustering with constraints, so a comparison is not feasible.

5.1 Data Description and Preprocessing

We have primarily utilized the data set used in [23] [2]. Data sets $oh5$ and $oh15$ are from OHSUMED collection, a subset of MEDLINE database, which contains $233,445$ documents indexed using $14,321$ unique categories. Data set WAP is from the WebACE Project, and each document corresponds to a web page listed in the subject hierarchy of Yahoo!. Data set $re0$ is from $Reuters - 21578$ text categorization collection (distribution 1.0). We also used $Newsgroup$ data which contains about 2000 articles from 20 newsgroups [24] [3]. In our experiments, we mixed up some of the data sets mentioned above. Table 1 gives the details of the data sets for word-document-category co-clustering.

Table 1. Data sets for text (word-document-category) co-clustering

Name	*Data sets*	*Data structure*	*No. of categories*	*No. of clusters*	*No. of documents*
HT1	*oh15,re0*	{*Adenosine-Diphosphate,Aluminum,Cell-Movement*}, {*cpi,money*}	*2*	*5*	*899*
HT2	*oh15,re0*	{*Blood-Coagulation-Factors,Enzyme-Activation,Staphylococcal-Infections*}, {*Jobs,reserves*}	*2*	*5*	*461*
HT3	*oh15,re0*	{*Aluminum,Blood-Coagulation-Factors,Blood-Vessels*} {*housing,retail*}	*2*	*5*	*256*
HT4	*oh5,re0*	{*Aluminum,Cell-Movement,Staphylococcal-Infections*}, {*cpi,wpi*}	*2*	*5*	*391*
HT5	*WAP,re0*	{*media,film,music*}, {*cpi,jobs*}	*2*	*5*	*404*
HT6	*Newsgroup*	{*rec.sport.baseball,rec.sport.hockey*}, {*talk.politics.guns,talk.politics.mideast,talk.politics.misc*}	*2*	*5*	*500*
HT7	*Newsgroup*	{*comp.graphics,comp.os.ms-windows.misc*}, {*rec.autos,rec.motorcycles*}, {*sci.crypt,sci.electronics*}	*3*	*6*	*300*

We used term frequency to build *word-document* matrix and carry out feature selection to choose the top 1000 words by the mutual information. The *Document-category* matrix is constructed by computing the probability of each document belonging to each category. The following technique is used: (1) For each class of documents, select the top 1000 words based on mutual information. (2) For each document, if any of the top 1000 word occurs, the amount of occurrence is 1, otherwise 0. (3) The probability of one document belonging to a category is the ratio of the sum of occurrence of the top 1000 words in this document to 1000. Thus, every element of *document-category* matrix is in the range $[0, 1]$. In addition, for semi-supervised clustering, we defined the percentage (%) of pairwise constraints with respect to all the possible document pairs, which is $\binom{\text{total docs}}{2}$. The document constrains are generated by randomly selecting documents from each class of the data set.

5.2 Evaluation Method

We evaluated the clustering results using the accuracy rate AC. The AC metric measures how accurately a learning method assigns labels $\hat{y}_i$ to the ground truth y_i, and is defined as,

[2] http://www.cs.umn.edu/ ~han/data/tmdata.tar.gz

[3] http://www.cs.uiuc.edu/homes/dengcai2/Data/TextData.html

$$AC = \frac{\sum_{i=1}^{n} \delta(y_i, \hat{y_i})}{n}. \quad (12)$$

where n denotes the total number of documents/categories in the experiment, and δ is the delta function that equals one if $\hat{y_i} = y_i$, otherwise zero. Since an iterative algorithm is not guaranteed to find the global minimum, it is beneficial to run the algorithm several times with different initial values and choose the average of all the test runs as the final accuracy value. In our experiments, for each given cluster number k, we conducted 10 test runs and final AC value is the average of all the 10 test runs.

5.3 Word-Document-Category Co-clustering

We conducted experiments to co-cluster words, documents and categories, and compared the performance of SS-NMF with SRC [8] and unsupervised NMF.

Co-clustering Accuracy. Table 2 shows document clustering accuracy obtained by SRC, unsupervised NMF, and SS-NMF with 15% constraints, respectively. It is obvious that SS-NMF outperforms SRC or unsupervised NMF in all the data sets. In general, SRC performs the worst amongst the three. Its accuracy on data set HT7 with 3 categories and 6 document clusters is only 19%, while SS-NMF provides an accuracy over 63%. Also from Table 2, we observed that SS-NMF can achieve high clustering accuracy, over 80% in 5 out of 7 data sets. Note that we have at least five document clusters in each of these data sets, so comparatively the baseline accuracy is only 20%. In Figure 3, we plotted the *AC* value against increasing percentage of pairwise constraints for SS-NMF. It is obvious to see that SRC and unsupervised NMF are consistently outperformed by SS-NMF with varying amounts of constrains across all the data sets. In addition, when more prior knowledge is available, the performance of SS-NMF clearly gets better.

In the left panel of Table 3, we reported the accuracy of text categorization by SRC, unsupervised NMF, and SS-NMF. For all the data sets, the *AC* value of SS-NMF is either the best or the closely-followed second best amongst the three methods. This result shows that even though the original document-category matrix is biased in the distance metric learning towards the constraints on the documents, SS-NMF still can provide a highly competitive results on category clustering.

For co-clustering, we obtained the clusters of words simultaneously with the clusters of documents and categories. However, for text representation, there is no ground truth available to compute an *AC* value. Here, we selected the "top" 10 words based on mutual information for each word cluster associated with a category cluster, and listed them in the right panel of Table 3. These words can be used to represent the underlying "concept" of the corresponding category cluster.

Modality Selection. As described in Section 3.2, distance metric and modality importance are learnt iteratively in Algorithm 1. First, modality selection can provide additional information on the relative importance of various relations (e.g., "word" and "category") for the grouping the central data type (e.g., "document"). Moreover, from a technical point of view, it also acts like feature selection when computing the new relational data matrix. Table 4 lists the modality importance for the two relations: *word-document* and *document-category* in SS-NMF with 1% constraints. A higher value in

Table 2. Comparison of document clustering accuracy between SRC, unsupervised NMF and SS-NMF with 15% constraints on word-document-category co-clustering

Name	*SRC*	*NMF*	*SS-NMF*
HT1	*0.4772*	*0.5250*	*0.8509*
HT2	*0.4989*	*0.6529*	*0.8243*
HT3	*0.3359*	*0.5391*	*0.6875*
HT4	*0.4450*	*0.5601*	*0.8261*
HT5	*0.6411*	*0.6386*	*0.8267*
HT6	*0.4989*	*0.5780*	*0.8620*
HT7	*0.1900*	*0.4333*	*0.6467*

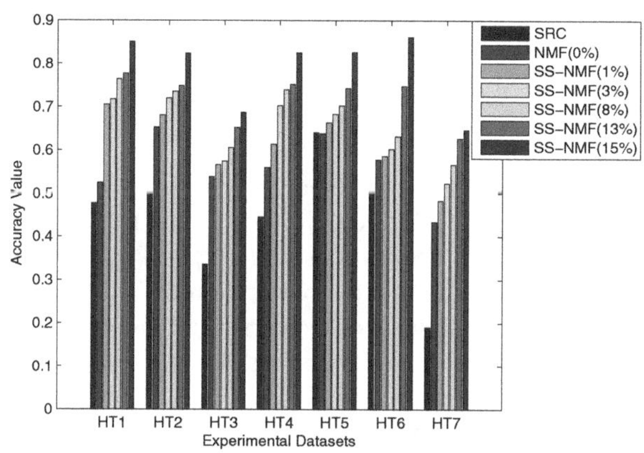

Fig. 3. Comparison of document clustering accuracy between SRC, unsupervised NMF, and SS-NMF with different amounts of constraints for word-document-category co-clustering

Table 3. Text categorization: clustering accuracy of categories and Text representation: top ten words for each category

Name	*SRC*	*NMF*	*SS-NMF*	*Representative words for each category*
HT1	*0.8*	*0.8*	*0.8*	{*via,coverag,calcium,purif,modifi,increm,identif,receiv,explant,delta*} {*market,pct,bank,rate,monei,billion,dollar,mln,dlr,currenc*}
HT2	*0.8*	*0.6*	*0.8*	{*studi,activ,patient,suggest,protein,increas,result,effect,treat,infect*} {*januari,pct,februari,reserv,unemploy,billion,bank,fell,mln,rose*}
HT3	*0.4*	*0.8*	*0.6*	{*increas,patient,activ,perform,suggest,studi,effect,examin,result,factor*} {*februari,adjust,fall,sale,depart,retail,fell,season,level,month*}
HT4	*0.4*	*0.8*	*0.8*	{*cell,treatment,determin,site,bone,neutrophil,single,anim,change,differ*} {*consum,statist,index,inflat,rise,compar,base,month,increas,rose*}
HT5	*0.8*	*0.4*	*0.8*	{*pm,star,film,hollywood,set,releas,octob,director,time,million*} {*rise,price,rose,statist,unemploy,inflat,compar,consum,januari,increas*}
HT6	*0.8*	*0.6*	*0.8*	{*disregard,jai,pyramid,winner,aaron,baltimor,dean,leaf,ban,stanlei*} {*sahak,ohanus,melkonian,appressian,serazuma,armenian,serdar,escap,turkish,sdpa*}
HT7	*0.8*	*0.5*	*0.7*	{*mac,color,al,push,bit,sse,lower,size,traffic,screen*} {*licenc,egreeneast,clipper,drink,claim,biker,safeti,clean,dod,motorcycl*} {*vga,univ,pub,servic,educ,bill,robert,school,technic,game*}

Table 4. Modality importance: words v.s. categories

Name	*word-document*	*document-category*
HT1	*0.9996*	*0.3884*
HT2	*0.9999*	*0.4331*
HT3	*0.6837*	*0.9949*
HT4	*0.7607*	*0.7233*
HT5	*0.2479*	*0.9998*
HT6	*0.9999*	*0.1751*
HT7	*0.2390*	*0.9990*

the table indicates more importance. It is clear that the significance of words and categories are quite distinct in different data sets. Specifically, *word-document* relation seems to play an more important role for document clustering in data sets HT1, HT2, HT4 and HT6, while *document-category* relation is more important in the rest. This information provides us a better understanding of the underlying process that generates the document clusters.

Time Complexity. Finally, we compared the computational speed of SRC, unsupervised NMF and SS-NMF for document co-clustering. The time complexity of SRC is $\mathcal{O}(t(l \max(n_d, n_w)^3 + kn_d n_f)$, SS-NMF is $\mathcal{O}(tl(n_d^3 + kn_d n_f))$, and unsupervised NMF is $\mathcal{O}(tlkn_d n_f)$, where t is the number of iterations, l is the number of data types, $k = \max(k_d, k_f)$ is maximum number of clusters in all data types (e.g., word, document or category), n_d is the number of documents, and n_f is the maximum number of words or categories for all modalities. Normally, $n_f = n_w$ since n_w(number of words)¿ n_c (number of categories). So, given t, l and k , the actual computational speed is usually determined by n_d or n_w. Figure 4(a) illustrates the computational speed of SRC, SS-NMF and unsupervised NMF, with increasing number of documents for a fixed n_w, while Figure 4(b) shows the computational speed with increasing number of words for a fixed n_d. The experiments were performed on a machine with Dual 3GHz Intel Xeon processors and 2GB RAM.

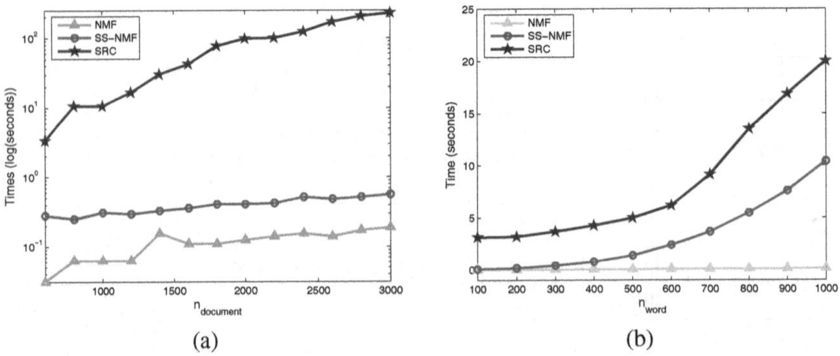

Fig. 4. Computational speed comparison for SRC , Unsupervised NMF, and SS-NMF. The time required by each of the algorithms are displayed in log(seconds) for increasing $n_{document}$ (a), and in seconds for increasing n_{word} (b).

Amongst the three, unsupervised NMF is the quickest as it uses an efficient iterative algorithm to compute the cluster indicator and cluster association matrices. SS-NMF ranks the second and its time gradually increases as the number of samples or features increases. The difference between SS-NMF and unsupervised NMF is mainly due to the additional computation required to learn the new distance metric through SS-LDA, in which we need to solve a generalized eigen-problem. We observed that in Figure 4(a), the computing time for SS-NMF is close to unsupervised NMF since both have linear complexity with n_d when n_w is fixed. On the other hand as shown in Figure 4(b), time for SS-NMF increases more quickly ($\mathcal{O}(tln_w^3)$) when n_c is fixed. In both cases, SRC is the slowest comparatively. Even though SRC is completely unsupervised, it needs to solve a computationally more expensive constrained eigen-decomposition problem and require additional k-means post-processing to infer the clusters. In short, SS-NMF approach provides the efficient way for document co-clustering.

6 Conclusions

In this paper, we presented SS-NMF co-clustering: a novel semi-supervised approach for document clustering with simultaneous text representation and categorization. In SS-NMF co-clustering model, users are able to provide supervision in terms of *must-link* and *cannot-link* pairwise constraints on the documents, which are used to derive new *word-document* and *document-category* matrices through distance metric learning and modality selection. Tri-factorization of the new matrices is then performed to obtain the grouping of documents, words and categories. We demonstrated that SS-NMF outperforms existing methods in document co-clustering on publicly available text data sets.

Acknowledgment

This research was partially funded by U. S. National Science Foundation under grants IIS-0713315 and CNS-0751045, and by the 21st Century Jobs Fund Award, State of Michigan, under grant 06-1-P1-0193.

References

1. Liu, X., Gong, Y., Xu, W., Zhu, S.: Document clustering with Cluster Refinement and Model Selection Capabilites. In: Proc. of ACM SIGIR, pp. 191–198 (2002)
2. Willett, P.: Recent Trends in Hierarchic Document Clustering: a Critical Review. Information Process Management 24(5), 577–597 (1988)
3. Ding, C., He, X., Zha, H., Simon, H.: A Min-max Cut Algorithm for Graph Partitioning and Data Clustering. In: Proc. of IEEE ICDM, pp. 107–114 (2001)
4. Xu, W., Liu, X., Gong, Y.: Document Clustering based on Non-negative Matrix Factorization. In: Proc. of ACM SIGIR, pp. 267–273 (2003)
5. Dhillon, I.S.: Co-Clustering Documents and Words Using Bipartite Spectral Graph Partitioning. In: Proc. of ACM SIGKDD, pp. 269–274 (2001)

6. Long, B., Zhang, Z., Yu, P.S.: Co-clustering by Block Value Decomposition. In: Proc. of ACM SIGKDD, pp. 635–640 (2005)
7. Gao, B., Liu, T.-Y., Cheng, Q., Feng, G., Qin, T., Ma, W.-Y.: Hierarchical Taxonomy Preparation for Text Categorization using Consistent Bipartite Spectral Graph Copartitioning. IEEE Transactions on Knowledge and Data Engineering 17(9), 1263–1273 (2005)
8. Long, B., Zhang, Z., Wu, X., Yu, P.S.: Spectral Clustering for Multi-type Relational Data. In: Proc. of ICML, pp. 585–592 (2006)
9. Dhillon, I.S., Mallela, S., Modha, D.S.: Information-theoretic Co-clustering. In: Proc. of ACM SIGKDD, pp. 89–98 (2003)
10. Gao, B., Liu, T.-Y., Mao, W.-Y.: Star-structured High-order Heterogeneous Data Co-clustering based on Consistent Information Theory. In: Proc. of IEEE ICDM, pp. 880–884 (2006)
11. Rege, M., Dong, M., Fotouhoi, F.: Co-clustering Documents and Words using Bipartite Isoperimetric Graph Partitioning. In: Proc. of IEEE ICDM, pp. 532–541 (2006)
12. Rege, M., Dong, M., Hua, J.: Graph Theoretical Framework for Simultaneously Integrating Visual and Textual Features for Efficient Web Image Clustering. In: Proc. of WWW, pp. 317–326 (2008)
13. Hiu, M., Law, C., Topchy, A., Jain, A.K.: Model-based Clustering with Probabilistic Constraints. In: Proc. of SIAM ICDM, pp. 641–645 (2005)
14. Xing, E.P., Ng, A.Y., Jordan, M.I., Russell, S.: Distance Metirc Learning with Application to Clustering with Side-information. In: Proc. of NIPS, pp. 362–371 (2001)
15. Kulis, B., Basu, S., Dhillon, I.S., Mooney, R.: Semi-supervised Graph Clustering:a Kernel Approach. In: Proc. of ICML, pp. 457–464 (2005)
16. Ji, X., Xu, W.: Document Clustering with Prior Knowledge. In: Proc. of ACM SIGIR, pp. 405–412 (2006)
17. Chen, Y., Rege, M., Dong, M., Hua, J.: Incorporating User Provided Constraints into Document Clusteirng. In: Proc. of IEEE ICDM, pp. 577–582 (2007)
18. Chen, Y., Rege, M., Dong, M., Fotouhi, F.: Deriving Semantics for Image Clustering from Accumulated User Feedbacks. In: Proc. of ACM MM, pp. 313–316 (2007)
19. Chen, Y., Rege, M., Dong, M., Hua, J.: Non-negative Matrix Factorization for Semi-supervised Data Clustering. Journal of Knowledge and Information Systems 17(3), 355–379 (2008)
20. Lee, D.D., Seung, H.S.: Learning the Parts of Objects by Non-negative Matrix Factorization. Nature 401, 788–791 (1999)
21. Ding, C., Li, T., Peng, W., Park, H.: Orthogonal Nonnegative Matrix Tri-factorizations for Clustering. In: Proc. of ACM SIGKDD, pp. 126–135 (2006)
22. Lee, D.D., Seung, H.S.: Algorithms for Non-negative Matirx Factroization. In: Proc. of NIPS, pp. 362–371 (2001)
23. Han, E.-H., Karypis, G.: Centroid-based document classification: Analysis and experimental results. In: Zighed, D.A., Komorowski, J., Żytkow, J.M. (eds.) PKDD 2000. LNCS (LNAI), vol. 1910, pp. 424–431. Springer, Heidelberg (2000)
24. Lang, K.: News weeder: Learning to Filter Networks. In: Proc. of ICML, pp. 331–339 (1995)

One Graph Is Worth a Thousand Logs: Uncovering Hidden Structures in Massive System Event Logs

Michal Aharon, Gilad Barash, Ira Cohen, and Eli Mordechai

HP-Labs Israel, Technion City, Haifa, Israel
firstname.lastname@hp.com

Abstract. In this paper we describe our work on pattern discovery in system event logs. For discovering the patterns we developed two novel algorithms. The first is a sequential and efficient text clustering algorithm which automatically discovers the templates generating the messages. The second, the PARIS algorithm (Principle Atom Recognition In Sets), is a novel algorithm which discovers patterns of messages that represent processes occurring in the system. We demonstrate the usefulness of our analysis, on real world logs from various systems, for debugging of complex systems, efficient search and visualization of logs and characterization of system behavior.

1 Introduction

Almost every piece of software writes messages into event logs. Systems composed of many components, such as web services, complex enterprise applications and even complex printing presses collect such events from their many components into log files. These logs are meant to be utilized both in the development stages and in normal operations for debugging and understanding of the behavior of the complex system. In fact, studies have shown these reasons to be the main impetus of companies in collecting logs in the first place [1].

While these logs hold vast amount of information describing the behavior of applications and systems, finding relevant information within the logs can be a very challenging task; even modest systems can log thousands of messages per second. Most users still use the good old unix "grep" for finding potentially relevant messages in event logs. Existing commercial tools, such as those from Splunk, LogLogic and Xpolog, collect and join logs from different sources and provide a more convenient search through the logs. However, the indexing provided by these tools still does not lead to automation in leveraging the logs for tasks such as automated problem debugging, process identification or visualization of the information in the logs.

There are two main fundamental challenges in transforming the logs into machine readable form, enabling automated analysis.

First, a "translation" of the text based events (messages) in the logs into a dictionary of event types: The number of templates creating events is limited, but the actual number of distinct events observed is very large, and grows quickly as a function of time. This is because the actual messages in the logs include various parameters that change from instance to instance of the same type of message (e.g., "login user

W. Buntine et al. (Eds.): ECML PKDD 2009, Part I, LNAI 5781, pp. 227–243, 2009.

$name"). Unfortunately, in many systems, the event templates generating the messages are not available, making it challenging to compute representative statistics enabling automated analysis over the events.

Second, a pattern finding mechanism to provide a compressed/concise representation of processes represented in the logs is a major challenge. Many processes spawn multiple messages into logs, e.g., a failure of a process can cause multiple messages in different logs representing the output of different software components creating interleaved sequences of events. Automated systems benefit greatly from identification and representation of such groups, as opposed to individual messages, as it reduces noise, compresses the data and provides more accurate representation of processes in the system.

Solving both of these problems requires the design of efficient machine learning algorithms. Existing research in the area focused mainly on the second problem – analysis of log patterns from event log streams [10][11][12][13][14] . These works focus on discovery of temporal patterns or correlation of event statistics to problems, but either assume a known dictionary [10][11][12][13] or assume access to the source code for its discovery [14]. Most of these works generally ignore the complexities introduced when collecting logs from various components in a complex system, namely, interleaving of sequences of events, asynchronous events and high dimensionality.

In this paper we describe our solution to the above challenges and demonstrate the solution on a number of common use cases. Our solution includes two novel algorithms. The first is an efficient sequential text clustering algorithm for creating the dictionary representing the event types. The second, the PARIS algorithm (Principle Atom Recognition In Sets), automatically discovers principle patterns (atoms) of log event types, representing processes in the system. We use the output of these algorithms to visualize logs, enable fast debugging of system problems, and search in the logs via a highly compressed lossless representation of the information in the logs.

The paper is organized as follows. We start with a problem description in the next section. We then describe the two main contributions of this paper; an online dictionary creation algorithm for semi-structured machine generated events, followed by the PARIS algorithm. We discuss the use cases for our log analysis algorithms and demonstrate the results of these algorithms on several large scale datasets collected from enterprise applications. Next, we describe the related work followed by a discussion and conclusions.

2 Problem Description

Logs are semi-structured events generated automatically when software components output messages describing actions, warnings or errors during their operation. A log event typically has a timestamp, representing the time at which the software wrote the event, and at least a text message describing the event. In some log events additional fields appear, often describing severity level, source method/function, etc.

Table 1. Log entry examples from an enterprise application. Marked words represent variables in the templates. Sequences of a similar process are marked 1,2,3 in the first column.

1	2008-02-06 14:35:16	unexpected failure while trying to ping user session #55555 the session authentication has failed
1	2008-02-06 14:35:17	failed to retrieve the meta data of project 'null0' the session authentication has failed.
1	2008-02-06 14:35:19	failed to get licenses for project session the session authentication has failed.
1	2008-02-06 14:35:19	error processing request from 192.111.22.33 data starts with 0 \00000023\0 conststr download
2	2008-02-06 14:35:39	unexpected failure while trying to ping user session #44444 the session authentication has failed
2	2008-02-06 14:35:40	failed to retrieve the meta data of project 'null1' the session authentication has failed.
3	2008-02-06 14:35:41	unexpected failure while trying to ping user session #33333 the session authentication has failed
3	2008-02-06 14:35:41	failed to get licenses for project session the session authentication has failed.
2	2008-02-06 14:50:08	failed to get licenses for project session the session authentication has failed.
2	2008-02-06 14:50:09	error processing request from 192.111.22.33 data starts with 0 \00000014\0 conststr download
3	2008-02-06 14:50:11	Failed to retrieve the meta data of project 'null3' the session authentication has failed.
3	2008-02-06 14:50:14	error processing request from 192.111.22.33 data starts with 0 \00000512\0 conststr download

Consider the examples of log entries from a running application shown in Table 1.

Even though the events in this log contain 10 distinct messages, it is easy to see that there are only four message templates that generated the messages in this log. Variables words (IP address, user session #, project name, conststr) generated the 10 distinct messages shown. If the templates were known, it would be easy to map each message to its generating templates; however, such templates are rarely known in practice. In addition, the number of log events with distinct messages in the log files we collected represented between 10-70% of the total number of log events. With millions of log events this number becomes too high for any type of automated analysis on the event log time sequence. The problem then becomes to "compress" the log events so that events with similar messages are grouped together. In this paper we pose this problem as a clustering one, with the goal of recovering the message templates generating the log events.

A second type of behavior is observed in logs when a system reaches a certain state; causing different software components to output log entries, sometimes in an ordered sequence, and sometimes unordered. The log entries in Table 1 show three sequences of log entries (marked as 1,2,3 in the first column), each caused by the same application state – a failure to authenticate a user session. While some of the event types always occur when an authentication failure occurs (the first three), the fourth message, 'error processing request from …' occurs in other states as well. It is desirable to capture such processes and represent them as one for better characterization of the system behavior. The second problem is thus to automatically discover such event sequences from the massive logs. Clearly, this second problem requires a solution to the first one, so that sequences can be compared and matched. A further challenge in solving this problem stems from the fact that the logs entries may represent multiple system states at the same time, leading to interleaved sequences (e.g., sequence 2 and 3 above), each representing the same or a different state; e.g., a user session authentication failure may occur at the same time as a failure to update a database – the logs would represent interleaved events from the two failures, without clear indication of the source of the failure. Any solution must be capable of discovering the different sequences despite the interleaving.

Formally, we represent each log entry, e, by the tuple (t,msg), where t is the time-stamp of the message and msg is the message text, represented by the word vector: $msg = w_1, w_2, \ldots, w_n$, where w_i is the word in the i'th position, while n is the number of words in a message. Each w_i represents a word from all the words present in the logs, and the set of log entries is E. The first problem is to discover a set of message clusters $C = c_1, c_2, c_3, \ldots, c_k$, where $k << |E|$, and map each event e(t,msg) to one of the clusters, leading to the new representation of each event as (t, c_i). The second problem is to automatically discover repeating event sequences, even when interleaving between sequences occurs, and when event clusters appear in multiple different types of sequences.

3 Online Dictionary Creation Algorithm

To create the dictionary, mapping the text messages to a typically much smaller set of message clusters, we leverage the fact that messages produced by the same template are usually identical in many of the words, with differences only at various parameters. Additionally, word ordering is important, therefore any similarity function needs to take word ordering into account. In our implementation we used the order sensitive cosine similarity between the messages:

$$\textbf{(Eq. 1)}\ < msg_1, msg_2 > = \frac{n_{12}}{\sqrt{n_1 \cdot n_2}},$$

where n_{12} is the number of identical words comparing *each word position* of msg_1 and msg_2, and n_1, n_2 are the number of words in each message.

An alternative to the cosine similarity defined above is an edit distance, or variations of it, allowing for insertions and deletions. However, we found that in practice there is little need for it, and using it adds a significant computational overhead.

Additionally, any algorithm is required to meet the following:

1. Efficient: Given the massive amount of log events, any algorithm must process the logs very quickly, keeping up at least with the rate of incoming messages. Our algorithm is linear time in terms of number of messages. Our implementation of the algorithm is able to process 25,000 messages per second, keeping up with all systems we have encountered so far.
2. Produce value immediately: Our algorithm is an online algorithm, producing new clusters as needed as more messages are observed.
3. Consistency of clusters: Two messages that belonged to the same cluster at time t, cannot belong to conflicting clusters at time t+1, otherwise it can result in conflicting conclusions depending on t. Our algorithm builds a forest of cluster trees, and ensures that messages are always a part of the same tree in the forest, thus maintaining global consistency.
4. Similar messages with different semantics or frequent parameter values should be in separate clusters: It occurs that two messages are almost identical, but the semantics can be very different – e.g., “network is up” vs. “network is down” (often the message can be much longer, with the only difference being

the words "up" and "down"). Additionally, a message could be "login user $name" could have many instances where $name = "root", and the rest with varying names, implying that user "root" should probably be considered separately from the others. To produce the desired result in such cases, our algorithm splits clusters based on the entropy of word positions in the messages and the words within each position to maintain **high** entropy among the members of a cluster in the word positions considered parameters.

Our online dictionary creation algorithm is shown in Table 2.

The algorithm begins with an empty set of clusters. Each new event is compared to a representative message of the existing clusters in the order in which the clusters were created, and is assigned to the first cluster to which the similarity threshold is exceeded; this ensures the satisfaction of the consistency requirement. If the similarity threshold is not surpassed for any of the existing clusters, a new cluster is created and the event message is used as the representative message of the new cluster.

Table 2. Dictionary creation algorithm

```
Algorithm Dictionary Creation
Parameters: MinimumSimilarityThreshold, MinimumSamplesForSplit,
MinimumWordPositionEntropy, MinimumWordPercentForSplit
Initialize Clusters = {ø};
CREATING ROOT CLUSTERS AND ASSIGNING EVENTS TO CLUSTERS
for every log event e(t,msg) in E Do{
    for every Cluster{i} in Clusters {
        if cosine(msg,Clusters{i}) > MinimumSimilarityThreshold {
            Add e(t,msg) as member of Clusters{i};
            Add e(t,msg) to relevant leaf in Clusters{i} tree;
            break;
        };
    };
    if no cluster matched e(t,msg)
        Add msg as new cluster in Clusters;
}
SPLITTING CLUSTERS (check occurs every X events)
for every leaf cluster in Clusters{
    if number of events assigned to the cluster > MinimumSamplesForSplit{
        Compute the entropy of every word position in cluster h(j), j=1,..., max msg length
        Find i=argmin_j h(j) such that h(j)>MinimumWordPositionEntropy;
        Find all words for which p_ki •100 > MinimumWordPercentForSplit
        Split to leaf clusters according to the words above and another for all other words
        Update membership of all events e(t,msg) in the split cluster
    }
}
```

The second step of the algorithm considers splitting a cluster if the following conditions are met:

1. There is a minimum number of events that belonged to the cluster
2. A word position has an entropy smaller than a splitting threshold (but not zero) and at least one word in that word position appeared in x% of the messages.

The entropy of a word position is computed as: $h(j) = -\sum_{k=1}^{n} p_{kj} \cdot \log(p_{kj})$, where n is the number of words in the dictionary, p_{kj} is the probability that word k appears in position j, computed as $p_{kj} = {n_{kj}}/{n_c}$, where n_{kj} is the number of times word k appeared in position j, and n_c is the number of messages belonging to the cluster.

When the two conditions are met, a cluster is split into at least two clusters, and possibly more if more words that pass the x% threshold. We use x = 10% in our experiment, and the minimum number of messages in a cluster as 1000.

The output of the algorithm is a forest of cluster trees, in which the branches of the tree represent splits based on the entropy criterion, and the tree roots are based on the cosine similarity criterion. In terms of efficiency, the algorithm performs a single pass over the data, preserving word counts for splits as it reads the messages. Creating the root of the forest is purely online, while the splitting phase is performed periodically on select clusters such that new messages are not held up for long. We show in our experiments that despite the heuristic nature of the algorithm, it achieves clustering results that are both accurate in terms of recovering message templates, and also very similar to batch clustering on the same data.

4 PARIS Algorithm

The PARIS (for Principal Atoms Recognition In Sets) algorithm is designed for identification of sets of events that tend to occur together. We assume that in each point in time several processes occur in the system, each generates its own set of log messages and the full log is a union of the individual log sets. We also do not assume that the messages generated by a process are ordered since many systems are a-synchronic making the ordering meaningless. PARIS gets as input the full log, and identifies the individual sets of messages that belong to one process or failure. Therefore, PARIS actually provides an alternative representation of the full log as a collection of atoms, where each atom is a known set of messages produced by one process or failure. Such an alternative representation provides several advantages,

1. Event suppression and filtering – one atom is used instead a set of log messages, which allows efficient representation of the data.
2. Analysis – each atom stands for one process or failure, and therefore representation by atoms is more meaningful than the full representation. Note that unlike atoms, a single log message can appear in several different processes or failures, and therefore its occurrence does not necessarily describe any specific system state.
3. Reduction of noise – lets assume a process occurs every once in a while in the system, and causes a set of log messages. PARIS will identify this set as an atom that describes the process. Let's also assume in a new time window we witness most of these log messages but not all of them. This inconsistency can be either noise (other messages were deleted, delayed or were wrongly assigned to a different time window), or as an intrinsic problem in the system

(e.g. the process was stopped in the middle). In both cases, PARIS representation can easily detect this inconsistency – fix it, or alert.

PARIS – Problem Description

Let $D_1, D_2, \ldots D_N$ be N different sets of elements, each consists of a finite set of values taken from a finite alphabet $v_1, v_2, \ldots, v_T$ (in our case the alphabet is the set of log messages ids, given as a result of the dictionary creation algorithm). We assume the content of each set D_i consists of at most L smaller sets, denoted also as principal atoms. The set of all principal atoms is denoted as A, and consists of K elements $A = \{A_1, A_2, \ldots, A_K\}$. Each principal atom A_j holds a set of values $A_j = \{v_{j1}, v_{j2}, \ldots, v_{jw}\}$. The atoms are not necessarily distinct, nor do they consist of all available values. A representation of the set D_i using A is denoted as $F(A,R_i)$, where R_i is a set of indices, and $F(A,R_i) = \bigcup_{j \in R_i} A_j$. We denote by F(A,R) the set of all representations, $F(A,R) = \left\{ F(A,R_i) \mid 1 \le i \le |R| \right\}$. The PARIS algorithm aims to find a set of atoms A and a set of representations R that minimize the sum of distances between $F(A,R_i)$ and D_i for $1 \le i \le N$. Therefor, the cost function can be defined as

$$\text{(Eq 2)} \quad \{A,R\} = \arg\min_{A,R} \sum_{i=1}^{N} d(D_i, F(A,R_i)) \ s.t.\ \forall_i |R_i| \le L,\ |A| \le k,$$

where d is a distance metric between sets. PARIS's execution and results heavily depend on the definition of the distance function d. The simplest distance function we defined counts the number of elements that are not common in the two sets, and normalizes it by the size of D_i,

$$d(D_i, F(A,R_i)) = \frac{|D_i \otimes F(A,R_i)|}{|D_i|},$$

where $\otimes$ is the XOR operator that returns the set of elements that appear in one set and not in the other. However, in many cases we found this distance function to be too simplified, as it does not consider cases in which only part of the atom appears in D_i due to, for example, missing log entries. We defined a slightly modified distance function that uses a slack parameter r,

$$d^r(D_i, F(A,R_i)) = d(D_i, B(F(A,R_i), D_i, r)),$$

Where

$$B(F(A,R_i), D_i, r) = F(\tilde{A}, R_i),$$

and $\tilde{A}_i = \arg\min_{\tilde{A}} d(D_i, F(\tilde{A}, R_i))$ for $\tilde{A}_i \subseteq A_i$, $|\tilde{A}_i| \ge r \cdot |A_i|$. That is, we allow the distance metric to consider only a portion r of the elements of each atom in the representation reducing the penalty for mismatches. When $r=1$, the function d^r is identical to d.

Implementing PARIS

As the solution of (2) is combinatorial in its nature, we designed an iterative scheme for minimizing it. In each iteration there are two stages: representation stage and atom optimization stage.

- Representation stage: we fix the set **A**, and solve for all i $R_i = \arg\min_{R_i} d^r(D_i, F(A, R_i))$. This is done greedily by adding one atom after the other to the representation. In each stage the added atom is the one that best minimizes the distance. The process stops when the distance is no longer minimized, or when $|R_i| = L$.
- Atom optimization stage: we change A in order to minimize the distance of the representation to D. We do so one atom after the other. When optimizing A_i, we fix all other atoms, and consider only the data sets that include A_i in their representation (other data sets will not effect A_i). For each such data set we define **representation error set** E_i,

  $$E_i = \{D_i / F(A, \{R_i / i\})\},$$

  where the operator '/' stands for set subtraction. E_i actually holds all the elements in D_i that are not yet represented by all other atoms except A_i. We then re-define A_i in order to better approximate all these representation error sets.

In each iteration the cost function (2) is reduced, and therefore the algorithm is promised to converge to a minimum solution. As the solution is local, several heuristic operations are done in order to force a more global solution (such as replacing identical atoms or atoms that contain other atoms). Other versions or PARIS are designed mainly to omit the need for the initial setting of K and L, and to better avoid local minima solutions.

Validating PARIS Using Synthetic Data

A necessary and important stage in validating the algorithm is to execute it on synthetic data that was generated directly according to the model assumptions. For this, we randomly selected a set of K 'true' atoms from an alphabet of size 200 (each atom included 8 elements). Then we assumed a fixed r and generated the input sets D. Each set D_i was generated by a union of random L atoms. From each of the atoms only r of the elements were taken. In addition, some noise was added to each set by switching on/off n elements. Finally, we executed the algorithm after supplying it with the parameters (k, L, r), and with the input sets D. The success rate of the algorithm was measured by the number of successful atom restorations out of k.

In the first experiment we set, K=50, L=3, N=3000, and varying r (between 0.4 and 1). The rates of successful restorations with noise levels of 0, 2, 4, 10 are described in the blue, red and green graphs in Figure 1.

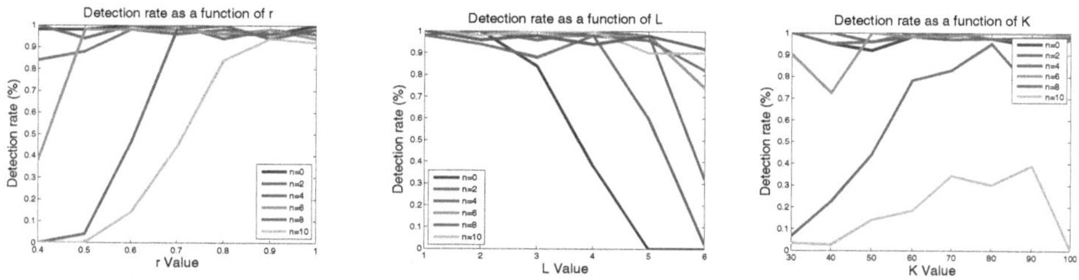

(a) (b) (c)

Fig. 1. (a-c) PARIS synthetic experiments

Another experiment was done with a fixed r=0.6, and L varies between 1 and 7, with noise levels of 0,2,4,6,8,10. The results are presented in Figure 1(b). A third experiment was done with fixed r=0.6, and L=3, while K varies between 30 and 100 in jumps of 10 and with noise levels of 0,2,4,6,8,10. The results are presented in Figure 1(c). These results demonstrate the ability of PARIS to successfully extract the original atoms, even in the presence of noise.

However, when applying PARIS on real data, the setting of K, L and r is done heuristically, using reasonable assumptions and knowledge of the system and the origin of the data. In the log experiments we used the following: k=80, L=3, r=0.51.

5 Use Cases for the Log Analysis Algorithms

The two algorithms described in the previous section transform the system event logs from semi-structured text to machine readable forms. The obvious question is what are the use cases that justify running these algorithms on system logs? In this section we describe three such use cases which we have seen in our experience to be useful for operations of large Enterprise IT systems. There could be additional use cases which we have not encountered yet.

The first use case, and also the most straightforward one, is to use the transformed event logs to aid in diagnosis of system problems. In most IT applications, indications of problems stem from abnormal measurement values in monitors that are critical to the business, such as transaction response time or throughput. When those indicate a problem, the operators needs to discover the root cause through mounds of data; monitors of system behavior, such as CPU utilization, memory, network, etc, and system event logs. In recent years, work in this area showed how to aid in the diagnosis using machine learning methods such as Bayesian network classifiers, clustering [7][8][9] using real-valued monitors. Adapting these methods to use the output of our algorithm is straightforward; the rate or binary representation of each event type from the dictionary, and/or atoms from the output of PARIS, is computed over time windows of 1-5 minutes, matching the frequency of the monitors used to detect problems. This turns the log events to temporal measurements enabling the application of existing learning technologies to help classify and describe problem periods.

A second use case is visualization of the system event logs over time for gaining better understanding of the system operation. While this use case is similar to the first in many respects, visualization of the log events over time produces views that enable quick understanding of system operation, such as reboots, normal periodic processes (e.g., database partition), and processes that are running amok, not causing any detectable problem at the application level yet. Whereas in the first use case the diagnosis of a specific problem that occurred is a supervised learning problem, this use case is unsupervised, leveraging visualization and additional unsupervised techniques to detect anomalies or behavioral patterns from the logs.

A third use case is the use of the output of the algorithm for efficient indexing of the logs, reducing both space requirements and speeding up search through the logs significantly over standard indexing. The clusters serve as the index to each message, coupled with the varying words, to produce a very fast and small index representing exactly all event logs.

We show examples of each use case in the experimental section.

6 Experimental Results

For our experiments we collected five log message datasets taken from systems in different fields of the IT world, in accordance with the use cases mentioned above. In order to demonstrate the generality of the algorithm, we chose one hardware log (that of a printing press), one Windows Server event log (which represents an infrastructure environment) and two enterprise business application logs. We were careful to select logs in which system problems were discovered, as well as logs from normal processing time of the systems. We ran each of the logs through the dictionary creation algorithm, transforming them into sets of clusters, unique messages and word dictionaries.

In addition to clustering the messages into distinct clusters, the dictionary creation algorithm also kept track of the distinct message strings, number of distinct non-numeric words in the logs of each system, as well as statistical information about the messages, such as the average number of words and the median number of words in each message.

Table 3. Result Summary for Log Datasets

Source	Time Frame	Number of messages	Number of unique messages	Number of clusters	Number of distinct words	Median message length	Index size reduction
Business App 1	Start: 2008-01-10 06:00:01 End: 2008-09-03 11:14:16	4,210,513	153,619	4,193	112,112	25	90%
Printer Press	Start: 2008-04-14 22:09:10 End: 2008-04-14 22:12:51	11,204	5,631	204	1,796	10	50%
Window s Events	Start: 2006-09-05 01:55:19 End: 2009-06-19 02:38:45	66,102	25,340	476	14,550	26	40%
Business App 2	Start: 2009-01-25 13:19:31 End: 2009-03-23 09:06:33	483,768	70,102	1,115	42,057	10	90%

Table 3 summarizes the results of running each of the datasets through the algorithm. For every data set it shows the timeframe of the messages in the log, how many messages were processed (number of messages), how many distinct messages were in the logs (number of unique messages), the number of clusters they were grouped into (number of clusters), the number of distinct words that were discovered in the log (not including numeric strings), and the median number of words in a message.

Index Size Reduction and Space Compression

Special attention must be given to the last column of Table 3 Result Summary for Log Datasets. As mentioned previously in the third use case, the algorithm efficiently indexes the logs, both reducing space requirements and speeding up search through the logs significantly over standard indexing. The last column of Table 3 demonstrates this. It shows the percentage by which the index size was reduced in the representation of the logs. In the cases of the business applications, the reduction in size was up to 90%! This compression in representation of the logs is done on two levels. The first consists of keeping track of unique messages and the number of times they occur rather than keeping track of every instance of every message. The second level of reduction consists of using the cluster as an index to the unique messages, coupled with keeping only the subsequent varying words of every unique pattern message, to produce a very fast and small index representing exactly all event logs. The 90% reduction is the saving of the clustering step compared to the first step of reduction – an index based on the unique messages. A 99% reduction is achieved over the naïve index which doesn't keep track of the unique messages.

Figure 2 further demonstrates the reduction in size that the algorithm provides for the Business Application 1 data, showing that over a 6 month period the number of unique messages (labeled as "Distinct Messages") grows at a steep rate while the number of clusters (labeled as "Template Messages") stays relatively very small.

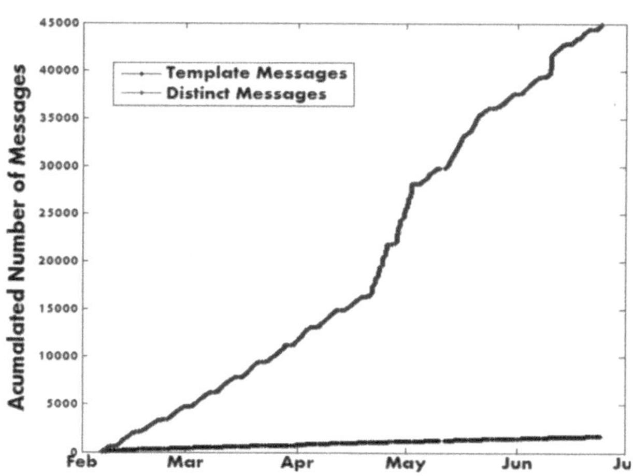

Fig. 2. Growth of distinct (unique) messages vs. cluster (Template) messages created by the algorithm

Dictionary Accuracy

In order to gauge the precision of our clustering methodology, we utilized the event logs from Windows to perform an accuracy test of our algorithm. Windows event messages include an event id for every message, a form of ground truth clusters preordained by the Windows server. We analyzed our algorithm's clusters to see what types of Windows event id's are included in each of them, and to see how well our cluster ids correspond to the Windows event id designated by the Windows server. We measure cluster accuracy using clustering purity and normalized mutual information (NMI) [15] compared to the ground truth given by the windows event ids.

For the 25,340 unique messages in the Windows logs, there were a total of 104 distinct event ids and a total of 467 clusters created by the algorithm[1]. The algorithm clustered the events with 96.7% purity in comparison to the Windows event ids, and NMI of 0.41. These numbers show that there was a great match between the clusters and the windows event ids, although the algorithm produced four times as many event types as existed in the data. Compared to the number of unique messages though (over 25,000), this number is quite small. It is also noteworthy that some Windows events had different event ids despite being comprised of the exact same text. This ambiguity in the source data can also contribute to the lack of purity in the clusters.

We also compared our clustering algorithm to the result of a batch clustering algorithm on the windows event data. As a batch clustering algorithm we used the hierarchical cluster tree algorithm with single linkage clustering [15], adjusting it to produce the same number of clusters as our online algorithm. The NMI between the results of our clustering ands the batch algorithm was 0.88, while the purity was at 98%, indicating the result of our online heuristic algorithm is nearly the same as the batch algorithm.

Example 1: Root cause analysis for application performance debugging

The following example demonstrates the use case of using the log analysis algorithm for diagnosis of performance problems in a transactional business application using the system logs. During routine monitoring of performance, a spike in transaction response time was noticed (see figure 3).

In addition, the performance monitors showed that the database CPU usage and the application server CPU usage were increasing, despite the fact that the number of connections to the application server was decreasing.

The monitor indicated multiple symptoms of a performance problem occurring in the system, but no indication of what might be causing it. The different components that compose the environment all have corresponding logs with error messages that occurred at the time of the spike, but no clear way to disseminate them. It is only when processing the system logs through the algorithms that a clear picture of the environment is composed using that information and error messages from multiple sources are put into context of timeline and order (see figure 4).

Once processed through the algorithms, error messages from multiple sources were visualized together, indicating order of appearance and cluster classification.

[1] The algorithm ran with similarity threshold of 0.6.

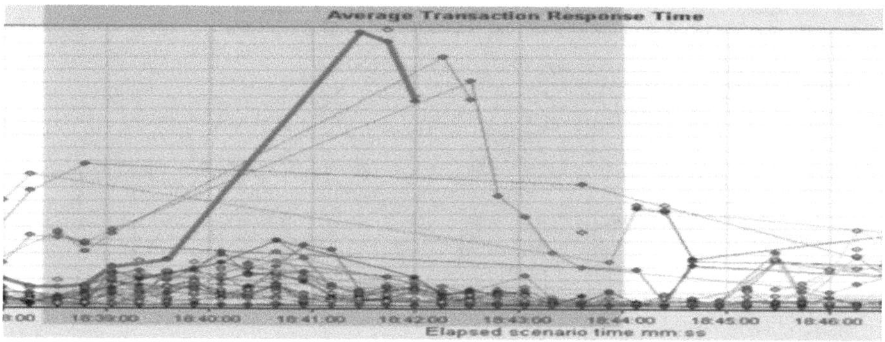

Fig. 3. Spike in transaction response time. The spike represents a performance problem for the application.

Analyzing the messages in this fashion assisted in isolating the root problem that caused the system errors – a java IllegalStateException message: java.lang.IllegalStateException: getAttribute: Session already invalidated. Once identified, explanations and solutions were easily found and applied (e.g., http://forums.sun.com/thread.jspa?threadID=5129793).

Example 2: Error/Behavior Detection in IT Management Software

The second example we provide is that of an IT Management system whose dashboard did not report any data about system health, despite the data being collected. As opposed to the previous case, in this situation there were no accompanying symptomatic behaviors by components in the system.

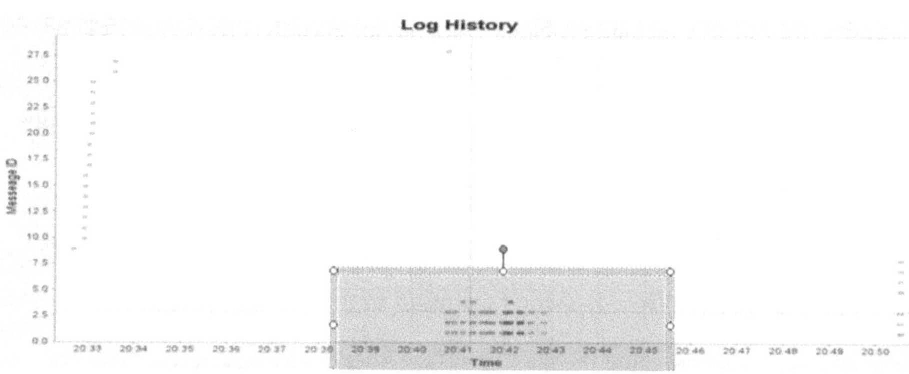

Fig. 4. Error messages at time of system errors, visualized after dictionary creation processing. The y-axis refers to the IDs generated by the dictionary creation algorithm (clusters). The x-axis is the time. Every dot in the graph represents an actual message from the log, showing its mapping to one of the discovered clusters. The period shaded period corresponds to the period with the spike in response time. The corresponding log error messages relevant to the problem are clearly visible in the graph.

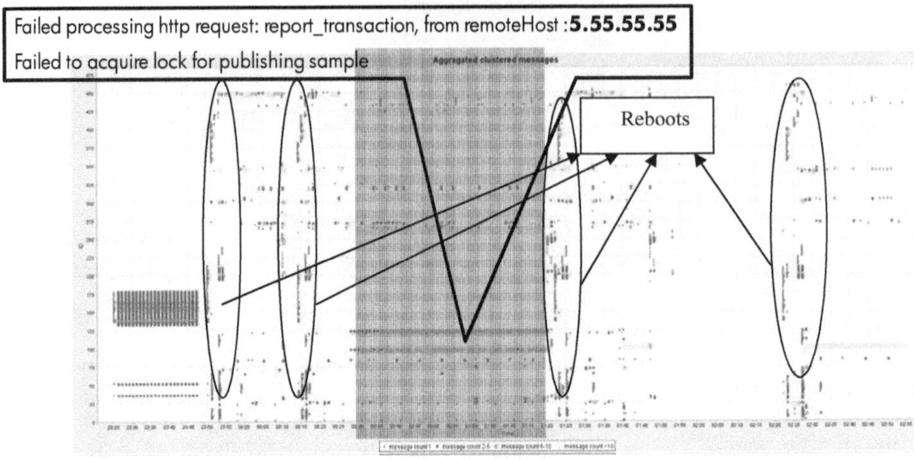

Fig. 5. Visualization of the messages from IT Management System logs following the dictionary creation step. The Region marked in red is the time when a problem occurred in the system; the marked message clusters that appear in this period were deemed to describe the root cause of the problem. Regions marked with ovals demonstrate the pattern of events when the system was performing a restart.

Once again a multitude of logs were processed by the algorithms and, when visualized, provided some insight into the problem in the system. The time-series graph in Figure 5 displays the results of the log analysis process. The area shaded in red is the timeframe in which the problem with application occurred, and upon examination of the graph, the problem is isolated given the two error messages, which appear only during this problem, stating that the report transaction failed to publish its results to the dashboard due to a locking problem. Once again, once the problem is isolated, the corresponding resources can get to work on fixing it.

From the visualization, further information about the system's behavior can be obtained, both normal and abnormal. For example, the points surrounded by ellipses represent groups of messages that recur regularly, which upon investigation are identified as belonging to the normal shutdown and startup processes of the server.

PARIS Results: Identifying Processes

We have shown in the previous section the result of the PARIS algorithm on artificial data. We applied the PARIS algorithm on the Business App 1 data set, using 15 minute windows and representing the data in the window as a binary vector; a vector representing at time window t is a vector of length N (the dictionary size), with "1"s in the indices of event clusters that appeared at least once during the time window. While no complete ground truth exists regarding the different processes in the system, we resort to experts to determine the validity of our output.

Table 4 below shows examples of three atoms, showing the corresponding event clusters in each. These atoms represent three failure types in the system creating the logs.

We see that there are shared events in Atom 78 and Atom 14 (highlighted), although the failure modes in the system are different in these two cases. The messages in Atom 79 represent a third failure mode, representing the messages occurring after a session authentication failure. When we looked at the timeline of when this atom appeared, we saw it repeating. Some investigation showed that it is due to an old script that was being executed at a regular interval that went undetected for months until our analysis

Viewing the timeline of when the atoms appeared, we saw that some failure types sometimes occur concurrently, due to the concurrent nature of the system, which serves multiple users performing various different tasks at the same time.

Unfortunately, we do not have ground truth data similar to the windows event logs, as such ground truth requires deep knowledge of the applications. However, the anecdotal evidence presented above do validate both the need for an algorithm such as PARIS and that PARIS can produce meaningful results in complex system event logs.

Table 4. Example Atoms corresponding to three failure types in the system

Failure type 1: Atom 78	Failure Type 2: Atom 14	Failure Type 3: Atom 79
failed to get coverage, failed to get list of entities keys	failed to cover req invalid values in removecoverage input frec parameter	failed to post common settings the project session authentication has failed.
failed to complete the action cannot create new test duplicate test name 'xxx'	failed to complete the action cannot create new test duplicate test name 'xxx'	failed to retrieve the meta data of project 'xxx' the project session authentication has failed
failed to get test value unexpected failure in getvaluepostprocess	failed to get test value unexpected failure in getvaluepostprocess	failed to get licenses for xx project session the project session authentication has failed.
failed to post design steps values cannot add new design step to test #xxx test #xxx not found	failed to post design steps values cannot add new design step to test #xxx test #xxx not found	failed to get the specified common settings the project session authentication has failed.
	failed to post req, failed to build hierarchical item descriptor	error creating request from x \xxx\xx conststr gettestsetvalue,
		failed to lock object the project session authentication has failed.
		exception in getting sequence value login has timed out.

7 Related Work

Methods for analyzing system log events have gained the attention of machine learning researchers in recent years. Most works assume a known message dictionary and apply time series analysis methods to the logs. In this category, among others, are [11][12][10], who developed methods to mine temporal event patterns, using HMMs, temporal graphs and other temporal models to extract important patterns. These works assume a known dictionary mapping the messages to a finite set of types, and also do not account for interleaved log sequences, which appear in large scale transactional systems.

[13] introduce a method for reducing the dimensionality of large scale logs by using dimensionality reduction and clustering of groups of log events. However, their method is designed to work in support center settings, where a support engineer receives chunks of logs only when the system is known to have failures, and limits the log types collected. Xu et al [14] also analyze logs based on log message groups and

use PCA to detect anomalous messages in the logs, analyzing logs continuously during operations. Their solution to the dictionary creation problem is to analyze the source code of the monitored applications, identifying possible message templates from the code and validating them in the logs. In contrast, our method assumes no access to the source code, using only the logs themselves to discover the possible templates, making our method applicable to a wider range of systems. Our PARIS algorithm combines the advantages of both the temporal pattern mining approaches and the batch approaches (such as PCA). It does not ignore time in the analysis, but tries to represent efficiently each time window with a set of atoms, representing processes, thus accounting for both groups of messages and interleaving of sequences.

The PARIS algorithm can also be compared to other existing methods in signal processing and machine learning. The simple version of PARIS is the set variation of the overcomplete dictionary learning algorithm K-SVD [2] in signal processing with real-valued data. K-SVD discovers a dictionary matrix so that each signal in the input data can be well represented by a **sparse linear combination of its atoms**. Its efficiency in image and video processing was proved in applications such as denoising of images and videos [3][4], image completion [2] and compression [5]. Similar to the latent variables in Latent Dirichlet Allocation [6], PARIS's atoms can be viewed as representing concept clusters, system processes in our case, allowing multiple concept clusters to represent each time period; this property overcomes the interleaving problem present in the logs.

8 Discussion and Summary

In this paper we introduced two novel methods for extracting patterns and summarizing complex system event logs. While our work shows promise on real logs, there are various open issues and future work opportunities.

While our dictionary creation algorithm is online and efficient, in some systems it may become necessary to analyze logs locally on the various distributed system machines as it may be too expensive to transfer logs over the network to a single central node. The challenge becomes to generate one consistent dictionary for all logs being analyzed on various machines, so that logs of the same type (e.g., Apache logs from various web servers) produce a single dictionary.

Second, our use of the PARIS algorithm on the logs currently uses a fixed size time window. However, for some processes, shorter or longer windows may be required, and the boundaries of the current windows are arbitrary and may split processes. An adaptive windowing approach may be required for more accurate process identification. Additionally, further validation is required for the PARIS algorithm on logs with more ground truth data.

Our heuristic dictionary creation algorithm performed well on system event logs and would most likely fair well on texts generated automatically from a finite set of templates. While we used the word order sensitive cosine similarity, it can be easily modified to ignore word order and use other similarity measures. We have also been experimenting with the PARIS algorithm to extracts concept classes from general text corpus, with initial results showing promise.

To conclude, the title of this paper, "one graph is worth a thousand logs", was coined by the quality engineer who used our system to quickly diagnose a performance problem with his application. This paper provides the necessary machine learning methods that transform massive amounts of system event logs into a form that enables meaningful visualizations and automated analysis that indeed help operators understand the complex behavior of the systems they manage.

References

[1] Shenk, J.: Demanding More from Log Management Systems, A SANS Whitepaper (June 2008), `http://www.sans.org/reading_room/analysts_program/LogMgt_June08.pdf`
[2] Aharon, M., Elad, M., Bruckstein, A.M.: The K-SVD: An Algorithm for Designing of Overcomplete Dictionaries for Sparse Representation. The IEEE Trans. On Signal Processing 54(11), 4311–4322 (2006)
[3] Elad, M., Aharon, M.: Image Denoising Via Sparse and Redundant representations over Learned Dictionaries. The IEEE Trans. on Image Processing 15(12), 3736–3745 (2006)
[4] Protter, M., Elad, M.: Image Sequence Denoising Via Sparse and Redundant Representations. IEEE Trans. on Image Processing 18(1), 27–36 (2009)
[5] Bryt, O., Elad, M.: Compression of Facial Images Using the K-SVD Algorithm. Journal of Visual Communication and Image Representation 19(4), 270–283 (2008)
[6] Blei, D.M., Ng, A.Y., Jordan, M.I.: Latent dirichlet allocation. Journal of machine Learning Research 3 (2003)
[7] Cohen, I., Goldszmidt, M., Kelly, T., Symons, J., Chase, J.S.: Correlating instrumentation data to system states: A building block for automated diagnosis and control. In: Proc. 6th USENIX OSDI, San Francisco, CA (December 2004)
[8] Cohen, I., Zhang, S., Goldszmidt, M., Symons, J., Kelly, T., Fox, A.: Capturing, indexing, clustering, and retrieving system history. In: Proc. 20th ACM SOSP (2005)
[9] Powers, R., Cohen, I., Goldszmidt, M.: Short term performance forecasting in enterprise systems. In: SIGKDD 2005 (2005)
[10] Peng, W., Perng, C., Li, T., Wang, H.: Event Summarization for System Management. In: SIGKDD 2007 (2007)
[11] Hellerstein, J.L., Ma, S., Perng, C.-S.: Discovering actionable patterns in event data. IBM System Journal 41(3), 475 (2002)
[12] Li, T., Liang, F., Ma, S., Peng, W.: An integrated framework on mining logs Files for computing system management. In: SIGKDD 2005 (2005)
[13] Sabato, S., Yom-Tov, E., Tsherniak, A., Rossetm, S.: Analyzing System Logs: A New View of What's Important. In: Second Workshop on Tackling Computer Systems Problems with Machine Learning Techniques, SysML 2007 (2007)
[14] Xu, W., Huang, L., Fox, A., Patterson, D., Jordan, M.: Mining Console Logs for Large-Scale System Problem Detection. In: SysML 2008 (2008)
[15] Manning, C.D., Raghavan, P., Schütze, H.: Introduction to Information Retrieval. Cambridge University Press, Cambridge (2008)

Conference Mining via Generalized Topic Modeling

Ali Daud[1], Juanzi Li[1], Lizhu Zhou[1], and Faqir Muhammad[2]

[1] Department of Computer Science & Technology, 1-308, FIT Building, Tsinghua University, Beijing, China, 100084
[2] Department of Mathematics & Statistics, Allama Iqbal Open University, Sector H-8, Islamabad, Pakistan, 44000
ali_msdb@hotmail.com, ljz@keg.cs.tsinghua.edu.cn, dcszlz@tsinghua.edu.cn, aioufsd@yahoo.com

Abstract. Conference Mining has been an important problem discussed these days for the purpose of academic recommendation. Previous approaches mined conferences by using network connectivity or by using semantics-based intrinsic structure of the words present between documents (modeling from document level (DL)), while ignored semantics-based intrinsic structure of the words present between conferences. In this paper, we address this problem by considering semantics-based intrinsic structure of the words present in conferences (*richer semantics*) by modeling from conference level (CL). We propose a generalized topic modeling approach based on Latent Dirichlet Allocation (LDA) named as Conference Mining (ConMin). By using it we can discover topically related conferences, conferences correlations and conferences temporal topic trends. Experimental results show that proposed approach significantly outperformed baseline approach in discovering topically related conferences and finding conferences correlations because of its ability to produce less sparse topics.

Keywords: Richer Semantics, Conference Mining, Generalized Topic Modeling, Unsupervised Learning.

1 Introduction

With the emergence of the Web, automatic acquirement of useful information from the text has been a challenging problem, when most of the information is implicit within the entities (e.g. documents, researchers, conferences, journals) and their relationships. For example, various conferences are held every year about different topics and huge volume of scientific literature is collected about conferences in digital libraries. It provides us with many challenging discovery tasks useful from researchers' point of view. For example, a new researcher can be interested in obtaining authoritative conferences of specific research area to do literature review or a group of researchers would like to know about conferences related to their research area for submitting papers.

Previous approaches used for conference mining problem can be categorized into two major frameworks 1) graph connectivity based approaches as a basis for representation and analysis of relationships between conferences [24,25] on the basis of co-authorship and publishing in the same venue and 2) topic modeling based approaches

W. Buntine et al. (Eds.): ECML PKDD 2009, Part I, LNAI 5781, pp. 244–259, 2009.

which make use of latent topic layer between words and documents to capture the semantic correlations between them. Recently one of the topic modeling approaches argued that conferences and authors are interdependent and should be modeled together [20]. Consequently, a unified topic modeling approach Author-Conference-Topic1 (ACT1) was proposed, which can discover topically related authors and conferences on the basis of semantics-based structure of the words by considering conferences information. Above mentioned frameworks based on graph connectivity ignored the semantics-based information. While, recent topic modeling approach viewed conferences information just as a stamp (token), which became the reason of ignoring implicit semantics-based text structure present between the conferences. We think this information is very useful and important for mining conferences.

In this paper, we will consider semantics-based text structure present between the conferences explicitly. We generalized previous topic modeling approach [20] idea of mining conferences from a single document "Constituent-Document" (*poorer semantics* because of only some semantically related words are present in one document) to all publications of conference "Super-Document" (*richer semantics* because of many semantically related words are present in all documents of one conference). It can provide grouping of conferences in different groups on the basis of latent topics (semantically related probabilistic cluster of words) present between the conferences. We propose a Latent Dirichlet Allocation (LDA) [4] based ConMin approach which can discover topically related conferences. We used discovered topics to find associations between conferences by using sKL divergence and shown temporal topic trends of conferences. We empirically showed that ConMin approach clearly achieve better results than ACT1 approach for conference mining and solution provided by us produced quite intuitive and functional results.

The novelty of work described in this paper lies in the; formalization of the key conference mining issues, proposal of generalized topic modeling (ConMin) approach to deal with the issues by capturing *richer semantics*, and experimental verification of the effectiveness of our approach on real-world dataset. To the best of our knowledge, we are the first to deal with the aforementioned conference related discovery issues directly (not through authors generated topics like ACT1) by proposing a generalized topic modeling approach from DL to CL.

The rest of the paper is organized as follows. In Section 2, we formalize the key conference related mining issues. Section 3 illustrates our proposed approach for modeling conferences with its parameter estimation details. In Section 4, dataset, parameters settings, performance measures, baseline approach with empirical studies and discussions about the results are given; applications of proposed approach are provided at the end of this section. Section 5 provides related work and section 6 brings this paper to the conclusions and future work.

Note that in the rest of the paper, we use the term constituent-document, accepted paper, and document interchangeably. Additionally "super-document" means all the documents of one conference.

2 Problem Setting

Our work is focused on mining conferences through their accepted papers. Each conference accepts many papers every year. To our interest, each publication contains

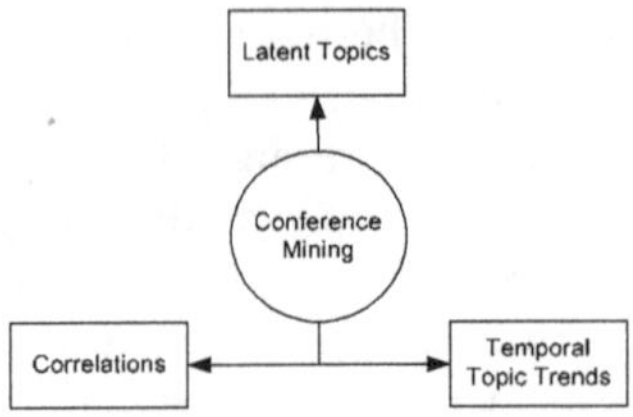

Fig. 1. Conferences related discovery issues

title which covers most of the highly related sub research areas. Conferences with their accepted papers on the basis of latent topics can be mined. Figure 1 provides a pictorial look of conference related mining issues discussed here.

We denote a conference (Super-Document) c as a vector of N_c words based on all accepted papers (Constituent-Documents) by the conference and formalize conference mining problem as three subtasks. Intuition behind considering conference as super-document is based on thinking that semantics at super-document level are richer as compared to semantics at a single document (Constituent-Document).

1) Discovery and Ranking of Conferences related to Topics: Given a conference c with N_c words, find the latent topics Z of conference. Formally for a conference, we need to calculate the probability $p(z|c)$, where z is a latent topic and c is a conference.
   Predict Z topics for a conference: Given a new conference c (not contained previously in the corpus) with W_c words, predict the topics contained in the conference.
2) Discovery of Conferences Correlations: Given two conferences c_1 and c_2 with N_{c1} and N_{c2} words respectively, find the correlations between conferences.
3) Discovery of Conferences Temporal Topic Trends: Given a conference c with N_c words for every year, access the temporal topic likeliness of a conference.

3 Conference Modeling

In this section, before describing our ConMin approach, we will first describe how documents are modeled with topics using topic model LDA, followed by modeling of conferences with authors' topics (ACT1 approach).

3.1 Modeling Documents with Topics (LDA)

Fundamental topic modeling assumes that there is a hidden topic layer $Z = \{z_1, z_2, z_3, \ldots, z_i\}$ between the word tokens and the documents, where z_i denotes a latent topic and each document d is a vector of N_d words $\mathbf{w}_d$. A collection of D documents is defined by $D = \{\mathbf{w}_1, \mathbf{w}_2, \mathbf{w}_3, \ldots, \mathbf{w}_d\}$ and each word w_{id} is chosen from a vocabulary of size V. LDA [4] is a state-of-the-art topic modeling approach which makes use of latent topic layer to capture semantic dependencies between the words. First, for each document d, a multinomial distribution θ_d over topics is randomly sampled from a Dirichlet distribution with parameter α. Second, for each word w, a topic z is chosen from this topic distribution. Finally, the word w is generated by randomly sampling from a topic-specific multinomial distribution Φ_z. The generating probability of word w from document D for LDA is given as:

$$P(w|d,\theta,\phi) = \sum_{z=1}^{T} P(w|z,\phi_z)P(z|d,\theta_d) \quad (1)$$

3.2 Modeling Conferences with Authors Topics (ACT1 (DL) Approach)

Recently, LDA is extended to discover topically related conferences indirectly by using topics of documents generated by authors [20]. In ACT1 model, each author is represented by the probability distribution θ_d over topics and each topic is represented as a probability distribution Φ_z over words and Ψ_z over conferences for each word of a document for that topic. The generative probability of the word w with conference c for author r of a document d is given as:

$$P(w,c|r,d,\phi,\Psi,\theta) = \sum_{z=1}^{T} P(w|z,\phi_z)P(c|z,\Psi_z)P(z|r,\theta_r) \quad (2)$$

3.3 Modeling Conferences with Topics (ConMin (CL) Approach)

The basic idea of topic modeling that words and documents can be modeled by considering latent topics became the intuition of modeling the words and conferences directly through latent topics. We generalize this idea from DL [4] to CL by considering documents as sub-entities of a conference. In our approach a conference is viewed as a composition of the words of its all accepted publications. Symbolically, for a conference c we can write it as: $C = \{\mathbf{d_1} + \mathbf{d_2} + \mathbf{d_3} + \ldots + \mathbf{d_i}\}$, where d_i is one document in a conference.

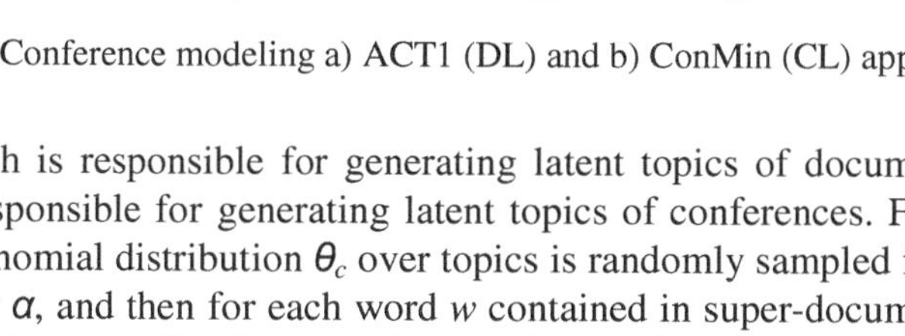

Fig. 2. Conference modeling a) ACT1 (DL) and b) ConMin (CL) approaches

DL approach is responsible for generating latent topics of documents, while CL approach is responsible for generating latent topics of conferences. For each conference c, a multinomial distribution θ_c over topics is randomly sampled from a Dirichlet with parameter α, and then for each word w contained in super-document, a topic z is chosen from this topic distribution. Finally, the word w is generated by randomly sampling from a topic-specific multinomial distribution Φ_z with parameter β.

The generative process is as follows:

1. For each conference $c = 1,\ldots, C$
   Choose θ_c from Dirichlet (α)

2. For each topic $z = 1,\ldots, T$
   Choose Φ_z from Dirichlet (β)
3. For each word $w = 1,\ldots, N_c$ of conference c
   Choose a topic z from multinomial (θ_c)
   Choose a word w from multinomial (Φ_z)

Figure 3 shows the generating probability of the word w from the conference c is given as:

$$P(w|c,\theta,\phi) = \sum_{z=1}^{T} P(w|z,\phi_z)P(z|c,\theta_c) \quad (3)$$

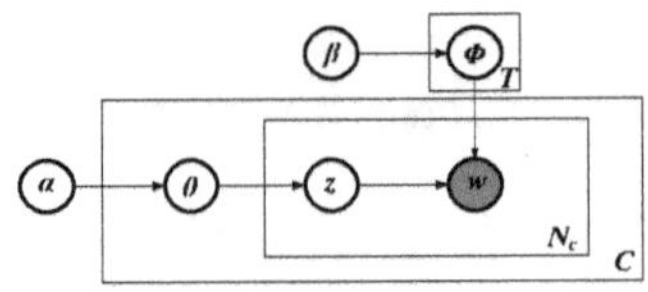

Fig. 3. ConMin approach (generalized smoothed LDA)

We utilize Gibbs sampling [1] for parameter estimation in our approach which has one latent variable z and the conditional posterior distribution for z is given by:

$$P(z_i = j|\mathbf{z}_{-i},\mathbf{w}) \propto \frac{n_{-i,j}^{(wi)}+\beta}{n_{-i,j}^{(\cdot)}+W\beta} \frac{n_{-i,j}^{(ci)}+\alpha}{n_{-i,\cdot}^{(ci)}+Z\alpha} \quad (4)$$

where $z_i = j$ represents the assignments of the ith word in a conference to a topic j. $\mathbf{z}_{-i}$ represents all topic assignments excluding the ith word, and $\mathbf{w}$ represents all words in the dataset. Furthermore, $n_{-i,j}^{(wi)}$ is the total number of words associated with topic j, excluding the current instance, and $n_{-i,j}^{(ci)}$ is the total number of words from conference c assigned to topic j, excluding the current instance. "." Indicates summing over the column where it occurs and $n_{-i,j}^{(\cdot)}$ stands for number of all words that are assigned to topic z excluding the current instance.

During parameter estimation, the algorithm only needs to keep track of W x Z (words by topic) and Z x C (topic by conference) count matrices. From these count matrices, topic-word distribution $\boldsymbol{\Phi}$ and conference-topic distribution $\boldsymbol{\theta}$ can be calculated as:

$$\phi_{zw} = \frac{n_{-i,j}^{(wi)}+\beta}{n_{-i,j}^{(\cdot)}+W\beta} \quad (5)$$

$$\theta_{cz} = \frac{n_{-i,j}^{(ci)}+\alpha}{n_{-i,\cdot}^{(ci)}+Z\alpha} \quad (6)$$

where, ϕ_{zw} is the probability of word w in topic z and θ_{cz} is the probability of topic z for conference c. These values correspond to the predictive distributions over new words w and new topics z conditioned on w and z.

4 Experiments

4.1 Dataset

We downloaded five years publication dataset of conferences from DBLP [8,14] by only considering conferences for which data was available for years 2003-2007. In total, we extracted 90,124 publications for 261 conferences and combined them into a super-document separately for each conference. We then preprocessed corpus by a) removing stop-words, punctuations and numbers b) down-casing the obtained words, and c) removing words that appear less than three times in the corpus. This led to a vocabulary size of V=10,902 and a total of 571,439 words in the corpus. Figure 4 shows quite smooth yearly data distribution for number of publications in the conferences.

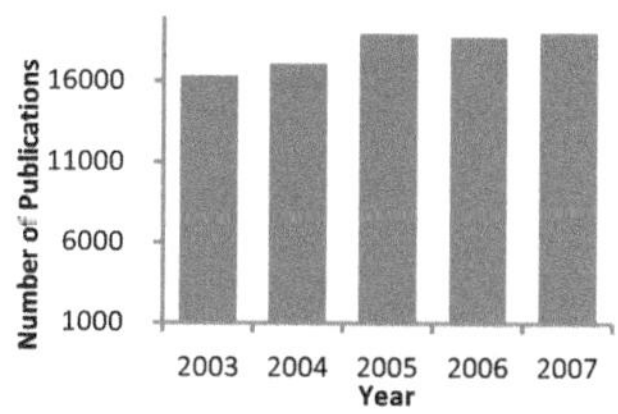

Fig. 4. Histogram illustrating data distribution

4.2 Parameter Settings

One can estimate the optimal values of hyper-parameters α and β (figure 3) by using Expectation Maximization (EM) method [11] or Gibbs sampling algorithm [10]. EM algorithm is susceptible to local maxima and computationally inefficient [4], consequently Gibbs sampling algorithm is used. For some applications topic models are sensitive to the hyper parameters and need to be optimized. For application in this paper, we found that our topic model based approach is not sensitive to the hyper parameters. In our experiments, for 200 topics Z the hyper-parameters α and β were set at 50/Z and .01 respectively. The numbers of topics Z were fixed at 200 on the basis of human judgment of meaningful topics plus measured perplexity [2] on 20% held out test dataset for different number of topics Z from 2 to 300. We ran five independent Gibbs sampling chains for 1000 iterations each. All experiments were carried out on a machine running Windows XP 2006 with AMD Athlon I Dual Core Processor (1.90 GHz) and 1 GB memory. The run time per each chain was 1.26 hours.

4.3 Performance Measures

Perplexity is usually used to measure the performance of latent-topic based approaches; however it cannot be a statistically significant measure when they are used for information retrieval [Please see [2] for details]. In our experiments, at first we used average entropy to measure the quality of discovered topics, which reveals the purity of topics. Entropy is a measure of the disorder of system, less intra-topic entropy is usually better. Secondly, we used average Symmetric KL (sKL) divergence [19] to measure the quality of topics, in terms of inter-topic distance. sKL divergence is used here to measure the relationship between two topics, more inter-topic sKL divergence (distance) is usually better.

To measure the performance in terms of precision and recall [2] is out of question due to unavailability of standard dataset and use of human judgments cannot provide appropriate (unbiased) answers for performance evaluation. Consequently, we used a simple error rate method to evaluate the performance in terms of conferences ranking. We discovered top 9 conferences related to top most conference (e.g. for ConMin "XML Databases" topic it is XSym) in each topic by using sKL divergence [please see table 1]. We compared these top 9 conferences with topically discovered top 10 conferences and calculated error rate with respect to their absence or presence in the topically ranked conferences list.

$$Entropy\ of\ (Topic) = -\sum_z P(z) log_2[P(z)] \tag{7}$$

$$sKL(i,j) = \sum_{z=1}^{T}\left[\theta_{iz} log\frac{\theta_{iz}}{\theta_{jz}} + \theta_{jz} log\frac{\theta_{jz}}{\theta_{iz}}\right] \tag{8}$$

4.4 Baseline Approach

We compared proposed ConMin with ACT1 and used same number of topics for comparability. The numbers of Gibbs sampler iterations used for ACT1 are 1000 and parameter values same as the values used in [20]. We used the same machine which was used for proposed approach; run time per each chain for ACT1 was 3.00 hours almost double than proposed approach. It shows that ConMin approach is also better in terms of time complexity.

4.5 Results and Discussions

The effect of topic sparseness on the model performance is studied both qualitatively and quantitatively. Firstly, we provide qualitative comparison between ConMin and ACT1 approaches. We discovered and probabilistically ranked conferences related to specific area of research on the basis of latent topics. Table 1 illustrates 7 different topics out of 200, discovered from the 1000th iteration of a particular Gibbs sampler run. The words associated with each topic for ConMin approach are strongly semantically related (less sparse) than that of ACT1, as they are assigned higher probabilities (please see prob. column in table 1). So, they make compact topics in the sense of conveying a semantic summary of a specific area of research [Please see figure 5 to see quantitative comparison of topic compactness]. Additionally it is observed that because of topic sparseness topically related conferences are also sparse (not from the specific area of research).

Consequently the conferences associated with each topic for ConMin are also more precise than ACT1, as they are assigned high probabilities (please see prob. Column in table 1). Only higher probabilities assigned to topic words and conferences is not extremely convincing, so we also investigated the bad impact of topic sparseness due to lower probabilities on the performance of baseline approach. For example, from top ten conferences six conferences related to "XML Databases" topic discovered by ACT1 are VLDB, SIGMOD, ICDE, Xsym, ADBIS, WIDM which are related to databases research area and other four ECOOP, SEKE, CAISE and KI are more related to software engineering and artificial intelligence research areas. While for ConMin topic “XML Databases” all the conferences are related to only databases research area. Similarly for “Data Mining” topic top ten conferences discovered by ConMin are more precise then ACT1 as for ACT1 SAC (Cryptography), CCGRID (Cluster Computing and Grid), ACM SenSys (Embedded Networked and Sensor Systems), ICDCS (Distributed Computing Systems) and ISISC (Information Security and Cryptology) are not actually related to data mining research area, additionally ACT1 is unable to find PAKDD, PKDD, DAWAK and DS for “Data Mining” topic among top ten conferences but they are well-known conferences in this field. One can see that PKDD and PAKDD are discovered by ACT1 for “Web Search” topic, which mismatches with the real world data. Similar kind of problem is encountered by ACT1 for other topically related conferences. It concludes that sparser the topics the discovered conferences will also be sparse which will result in poor performance of the approach.

Here it is obligatory to mention that top 10 conferences associated with a topic are not necessarily most well-known conferences in that area, but rather are the conferences that tend to produce most words for that topic in the corpus. However, we see that top ranked conferences for different topics are in fact top class conferences of that area of research for proposed approach. For example for topic 28 “Bayesian Networks” and topic 117 “XML Databases” top ranked conferences are more or less the

Table 1. An illustration of 7 discovered topics (top ConMin approach, bottom ACT1 approach). Each topic is shown with the top 10 words and conferences. The titles are our interpretation of the topics.

Topic 117 (ConMin)		Topic 164 (ConMin)		Topic 63 (ConMin)		Topic 138 (ConMin)		Topic 190 (ConMin)		Topic 28 (ConMin)		Topic 0 (ConMin)	
“XML Databases”		“Semantic Web”		“Information Retrieval”		“Digital Libraries”		“Data Mining”		"Bayesian Networks"		"Web Search"	
Word	*Prob.*	*Word*	*Prob.*	*Word*	*Probability*	*Word*	*Prob.*	*Word*	*Prob.*	*Word*	*Prob.*	*Word*	*Prob.*
xml	0.121514	semantic	0.125522	retrieval	0.157699	digital	0.234255	mining	0.147924	Bayesian	0.083057	web	0.328419
query	0.059027	web	0.12249	information	0.112182	libraries	0.099236	data	0.107059	networks	0.057923	search	0.02874
databases	0.0547	owl	0.03093	query	0.05448	library	0.09544	clustering	0.056024	inference	0.042624	content	0.024066
database	0.052969	rdf	0.029718	relevance	0.037277	metadata	0.031998	frequent	0.044513	time	0.028964	semantic	0.024066
processing	0.050199	ontologies	0.023048	feedback	0.029392	access	0.020611	patterns	0.036455	belief	0.028418	xml	0.019565
queries	0.045179	annotation	0.01941	search	0.022583	collections	0.01573	time	0.027054	causal	0.024593	language	0.018007
relational	0.03237	end	0.016378	user	0.020074	collection	0.013019	streams	0.02667	continuous	0.0235	pages	0.017314
efficient	0.025447	data	0.01274	language	0.017924	image	0.012477	pattern	0.022066	graphical	0.022954	information	0.015929
management	0.020773	large	0.010921	xml	0.017565	educational	0.012477	high	0.021298	structured	0.021315	user	0.014717
schema	0.020081	networks	0.010921	term	0.017207	oai	0.011935	privacy	0.017077	graphs	0.019676	collaborative	0.014544
Conference	*Prob.*	*Conference*	*Prob.*	*Conference*	*Prob.*	*Conference*	*Prob.*	*Conference*	*Prob.*	*Conference*	*Prob.*	*Conference*	*Prob.*
Xsym	0.413636	ISWC	0.330486	SIGIR	0.242417	JCDL	0.293113	SDM	0.251071	UAI	0.227882	WWW	0.234292
VLDB	0.199081	ASWC	0.326289	ECIR	0.194643	ECDL	0.27024	KDD	0.213337	AAAI	0.049531	LA-WEB	0.214421
SIGMOD	0.197517	WWW	0.040461	CIKM	0.086882	ELPBU	0.086239	ICDM	0.198849	NIPS	0.048314	WISE	0.213057
ICDE	0.192734	WIDM	0.014888	SPIRE	0.053974	MKM	0.04002	PKDD	0.196895	ICML	0.046224	WIDM	0.192592
IDEAS	0.1875	PODS	0.01374	SEBD	0.037998	DOCENG	0.025634	PAKDD	0.187208	ECML	0.044391	ICWS	0.159733
ADBIS	0.179348	ICCS	0.010382	ECDL	0.036844	Hypertext	0.017996	DAWAK	0.15004	Cana. AI	0.030308	WI	0.157155
SEBD	0.17217	ACSAC	0.009259	MMM	0.032828	SBBD	0.012186	DS	0.072158	ICTAI	0.016417	Hypertext	0.114341
BNCOD	0.165171	CAISE	0.008955	ICWS	0.029954	ECOOP	0.010417	IDEAS	0.066027	SDM	0.016065	ICWL	0.09839
ADC	0.164414	PSB	0.00837	WAIM	0.027234	SIGCSE	0.008574	ICDE	0.0647	EC	0.014017	ICWE	0.073778
PODS	0.162534	CADE	0.008267	ELPBU	0.022441	ECIR	0.008135	SSDBM	0.061772	AUSAI	0.012357	ASWC	0.0631
Topic 117 (ACT1)		**Topic 164 (ACT1)**		**Topic 63 (ACT1)**		**Topic 138 (ACT1)**		**Topic 190 (ACT1)**		**Topic 28 (ACT1)**		**Topic 0 (ACT1)**	
“XML Databases”		**“Semantic Web”**		**“Information Retrieval”**		**“Digital Libraries”**		**“Data Mining”**		**"Bayesian Networks"**		**"Web Search"**	
Word	*Prob.*	*Word*	*Prob.*	*Word*	*Probability*	*Word*	*Prob.*	*Word*	*Prob.*	*Word*	*Prob.*	*Word*	*Prob.*
data	0.03135	semantic	0.056959	retrieval	0.035258	digital	0.056555	data	0.029013	Bayesian	0.017148	web	0.065414
xml	0.031176	web	0.05335	information	0.020689	libraries	0.026451	mining	0.021635	learning	0.01287	search	0.017745
query	0.023387	ontology	0.025683	search	0.018469	library	0.021862	clustering	0.020054	networks	0.011704	based	0.016747
database	0.01802	based	0.016861	based	0.016387	based	0.012868	patterns	0.008459	models	0.010926	semantic	0.015748
web	0.013	ontologies	0.012851	web	0.015277	information	0.00938	learning	0.007668	inference	0.006649	services	0.007512
system	0.012135	owl	0.011247	text	0.01167	metadata	0.008279	based	0.007668	probabilistic	0.00626	data	0.006514
processing	0.011789	services	0.010846	document	0.011392	evaluation	0.006994	classification	0.007141	based	0.005871	information	0.006514
based	0.011096	rdf	0.010045	query	0.010976	web	0.00681	preserving	0.006351	markov	0.004705	approach	0.005765
relational	0.010231	approach	0.008842	relevance	0.009588	collections	0.00681	streams	0.006087	graphical	0.004705	queries	0.005765
management	0.010231	service	0.008441	evaluation	0.007646	search	0.006627	privacy	0.005824	information	0.004705	query	0.005516
Conference	*Prob.*	*Conference*	*Prob.*	*Conference*	*Prob.*	*Conference*	*Prob.*	*Conference*	*Prob.*	*Conference*	*Prob.*	*Conference*	*Prob.*
VLDB	0.450054	ASWC	0.496074	SIGIR	0.651289	JCDL	0.609793	SDM	0.695489	UAI	0.978935	WWW	0.986798
SIGMOD	0.378506	ISWC	0.49582	ECIR	0.249118	ECDL	0.379536	ICDM	0.185296	NIPS	0.001382	CIKM	0.001388
ICDE	0.150949	ICWS	0.000534	CIKM	0.080613	WISE	0.00207	KDD	0.102877	ISAAC	0.000724	ECIR	0.000711
Xsym	0.014415	KI	0.00028	SPIRE	0.014316	SBBD	0.00116	VLDB	0.002225	AUSAI	0.000724	PKDD	0.000711
ECOOP	0.000233	IEAAIE	0.00028	DAWAK	0.000179	SODA	0.000705	ICDE	0.00186	PODS	0.000724	SPIRE	0.000711
SEKE	0.000233	INFOCOM	0.00028	PKDD	0.000179	DOCENG	0.00025	SAC	0.000766	SIGIR	0.000724	TCVG	0.000372
WIDM	0.000233	LA-WEB	0.00028	WISE	0.000179	CASES	0.00025	CCGRID	0.000766	AINA	0.000066	TAB. AUX	0.000372
CAISE	0.000021	ADBIS	0.000025	KI	0.000016	ACL	0.00025	SenSys	0.000401	CAISE	0.000066	PAKDD	0.000372
KI	0.000021	AGILE	0.000025	ADBIS	0.000016	ECOOP	0.00025	ICDCS	0.000401	KI	0.000066	KI	0.000034
ADBIS	0.000021	XP	0.000025	AGILE	0.000016	CAISE	0.000023	ISISC	0.000401	ADBIS	0.000066	ADBIS	0.000034

best conferences of artificial intelligence and databases fields, respectively. Both topics also show deep influence of Bayesian networks on artificial intelligence and move from simple databases to XML database, respectively. We think, characteristically in top class conferences submitted papers are very carefully judged for the relevance to the conference research areas which results in producing more semantically related words; this is why top class conferences are ranked higher.

Proposed approach discovers several other topics related to data mining such as neural networks, multi-agent systems and pattern matching, also other topics that span the full range of areas encompassed in the dataset. A fraction of non-research topics, perhaps 10-15%, are also discovered that are not directly related to a specific area of research, as the words present in those topics were actually used as a glue between scientific terms. In addition to qualitative comparison between ConMin and ACT1, we also provide quantitative comparison to explain the effect of topics sparseness on the performance of approach. Figure 5 (a) shows the average entropy of topic-word distribution for all topics measured by using equation 7. Lower entropy curve of proposed approach for different number of topics Z = 50, 100, 150, 200, 250, 300 shows its effectiveness for obtaining less sparse topics which resulted in its better ranking performance shown in table 1. Figure 5 (b) shows the average distance of topic-word distribution between all pairs of the topics measured by using equation 8. Higher sKL divergence curve for different number of topics Z = 50, 100, 150, 200, 250, 300 confirms the effectiveness of the proposed approach for obtaining compact topics as compared to baseline approach.

From the curves in figure 5 (a) and figure 5 (b) it is clear that ConMin approach outperformed ACT1 approach for different number of topics. The performance difference for different number of topics is pretty much even, which corroborate that proposed approach dominance is not sensitive to the number of topics.

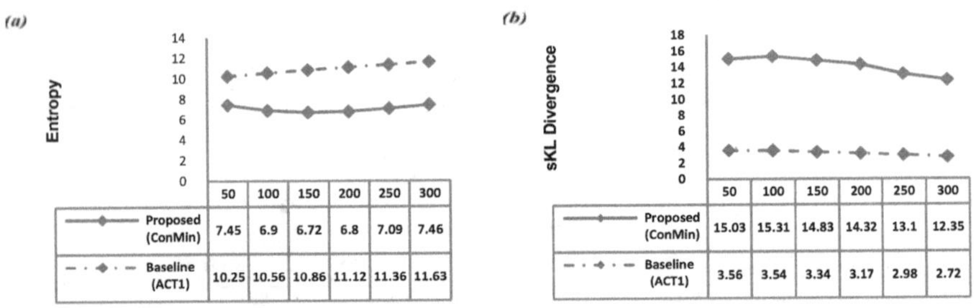

Fig. 5. a) Average Entropy curve as a function of different number of topics, lower is better and b) Average sKL divergence curve as a function of different number of topics, higher is better

Now we provide comparison in terms of error rate. Table 2 shows top 9 conferences discovered related to the first conference of each topic for ConMin and ACT1 approaches by using sKL divergence. For example, in case of "XML Databases" topic ADC, ADBIS, IDEAS, BNCOD, VLDB, SIGMOD, PODS, DASFAA and DEXA are top 9 conferences correlated with "Xsym" for ConMin approach.

The highlighted blocks in table 2 shows that similar results are found for discovered topics in table 1 and sKL divergence calculated for top most conference. For example, in case of ConMin approach top 10 conferences shown in table 1 for "XML Databases" topic has 7 conferences in common, which are ADC, ADBIS, IDEAS, BNCOD, VLDB, SIGMOD and PODS. From top 9 related conferences for seven selected topics (same is the case with non selected topics) shown in the table 2 the error rate (ER) for ConMin is less than ACT1, except digital libraries topic and ConMin approach has 30.16 % less average error rate than ACT1. It shows the bad effect of topics sparseness on conferences ranking performance of ACT1, and its inability to discover better results in comparison with proposed approach.

Table 2. An illustration of 7 topics sparseness effect on ranking in terms of error rate (ER). Here acronyms are XML Databases (XMLDB), Semantic Web (SeW), Information Retrieval (IR), Digital Libraries (DiL), Data Mining (DM), Bayesian Networks (BN) and Web Search (WS).

ConMin Approach							ACT1 Approach						
XMLDB	**SeW**	**IR**	**DiL**	**DM**	**BN**	**WS**	**XMLDB**	**SeW**	**IR**	**DiL**	**DM**	**BN**	**WS**
ADC	ASWC	ECIR	ECDL	ICDM	ICML	WI	SIGMOD	ISWC	ECIR	ECDL	KDD	EC	Hypertext
ADBIS	ER	CIKM	ELPBU	PAKDD	ECML	LA-WEB	ICDE	LA-WEB	CIKM	WISE	ICDM	ICML	SPIRE
IDEAS	LA-WEB	NLDB	Hypertext	KDD	NIPS	WISE	Xsym	KI	SPIRE	SBBD	SEDB	ALT	LISA
BNCOD	ISTA	ACL	WWW	PAKDD	AAAI	ICWS	ADA	ADA	WISE	ISI	ICDE	PODS	MATES
VLDB	WI	ICWS	ICWL	DS	ALT	CIKM	Ada-Eu	Xsym	MKM	ECOOP	VLDB	ADA	SGP
SIGMOD	SEBD	WWW	SIGIR	ECML	COLT	WAIM	ISTA	PPDP	DOCENG	DOCENG	ISISC	COLT	ICSOC
PODS	WWW	WISE	DOCENG	DAWAK	Cana. AI	WIDM	SDM	FSTCS	TableAUX	SODA	ADA	ISAAC	SIGIR
DASFAA	CAISE	KDD	ECIR	IDEAL	SDM	Hypertext	ICFP	ECOOP	ISSAC	CASES	SAC	Xsym	ICWS
DEXA	WIDM	MMM	LA-WEB	ICML	ICTAI	JCDL	APLAS	ICWS	RCLP/LPAR	ADA	SAM	PPDP	FC
ER=22.22	ER=55.55	ER=55.55	ER=44.44	ER=33.33	ER=22.22	ER=33.33	ER=66.66	ER=66.66	ER=55.55	ER=33.33	ER=33.33	ER=77.77	ER=88.88
Average Error Rate = 30.15							**Average Error Rate = 60.31**						

4.6 Applications of Proposed Approach

4.6.1 Topics for New Conferences

One would like to quickly access the topics for new conferences which are not contained in the training dataset by offline trained model. Provided parameter estimation Gibbs sampling algorithm requires significant processing time for large number of conferences. It is computationally inefficient to rerun the Gibbs sampling algorithm for every new conference added to the dataset. For this purpose we apply equation 4 only on the word tokens in the new conference each time temporarily updating the count matrices of (word by topic) and (topic by conference). The resulting assignments of words to topics can be saved after a few iterations (20 in our simulations which took only 2 seconds for one new conference). Table 3 shows this type of inference. To show predictive power of our approach we treated two conferences as test conferences one at a time, by training model on remaining 260 conferences to discover latent topics. Discovered topics are then used to predict the topics for words of the test conference.

Predicted words associated with each topic are quite intuitive, as they provide a summary of a specific area of research and are true representatives of conferences. For example, KDD conference is one of the best conferences in the area of Data Mining. Top five predicted topics for this conference are very intuitive, as "Data Mining", "Classification and Clustering", "Adaptive Event Detection", "Data Streams" and "Time Series Analysis" all are prominent sub-research areas in the field of data mining and knowledge discovery. Topics predicted for SIGIR conference are also intuitive and precise, as they match well with conference sub-research areas. Comparatively ACT1 (DL) approach is unable to directly predict topics for new conferences.

Table 3. An illustration of top five predicted topics for SIGIR and KDD conferences; each topic is shown with its probability, title (our interpretation of the topics) and top 10 words

SIGIR		
Topic Words	**Title**	**Probability**
retrieval, search, similarity, query, based, clustering, classification, relevance, document, evaluation	Information Retrieval	.2001
information, based, text, document, approach, documents, web, user, content, structured	Web based Information	.1340
language, text, extraction, semantic, disambiguation, question, word, answering, relations, natural	Intelligent Question Answering	.0671
web, search, collaborative, xml, user, pages, information, mining, content, sites	Web Search	.0415
models, probabilistic, random, structure, graph, exploiting, conditional, hidden, probability, markov	Probabilistic Models	.0361
KDD		
Topic Words	**Title**	**Probability**
mining, clustering, data, patterns, discovery, frequent, association, rules, algorithm, rule	Data Mining	.1819
classification, data, feature, selection, clustering, support, vector, machine, machines, Bayesian	Classification and Clustering	.0809
based, approach, model, multi, algorithm, method, efficient, analysis, detection, adaptive	Adaptive Event Detection	.0652
data, streams, stream, similarity, semantic, queries, incremental, adaptive, distributed, trees	Data Streams	.0618
time, high, large, efficient, dimensional, series, method, scalable, correlation, clusters	Time Series Analysis	.0584

In addition to the quantitative and qualitative evaluation of topically related conferences, we also quantitatively illustrate the predictive power of proposed approach in predicting words for the new conferences. For this purpose, perplexity is derived for conferences by averaging results for each conference over five Gibbs samplers. The perplexity for a test set of words W_c, for conference c of test data C_{test} is defined as:

$$perplexity\ (C_{test}) = exp\left[-\frac{logp(W_c)}{N_c}\right] \tag{9}$$

Figure 6 shows the average perplexity for different number of topics for AAAI, SIGIR, KDD and VLDB conferences, which fairly indicate the stable predictive power of proposed approach after 50 topics for all conferences.

4.6.2 Conference Correlations

ConMin and ACT1 both approaches can be used for automatic correlation discovery [19] between conferences, which can be utilized to conduct joint conferences in the future. To illustrate how it can be used in this respect, distance between conferences i and j is calculated by using equation 8 for topics distribution conditioned on each of the conferences distribution.

We calculated the dissimilarity between the conferences by using equation 8, smaller dissimilarity values means higher correlation between the conferences. For similar pairs less dissimilarity value and for dissimilar pairs higher dissimilarity value indicate better performance of our approach.

Table 4 shows correlation between 8 pairs of conferences, with every two pairs in order from top to down have at least one conference in common making four (A, B, C, D) common pairs. Common conference pairs show the effectiveness of our approach in discovering more precise conferences correlations. For example, common pair A has ASWC (Asian Semantic Web Conference) conference common in pairs (1, 2). Dissimilarity value between pair 1 (pretty much related conferences Asian Semantic Web Conference and International Semantic Web Conference) is smaller for ConMin .176 than that of ACT1 2.75, and dissimilarity value between pair 2 (related conferences to normal extent) is smaller for ConMin 3.16 than that of ACT1 3.61, which shows that ConMin can find correlations better. Common pair B has ECIR (European Conference on Information Retrieval) common in pairs (3, 4). Dissimilarity value between pair 3 is smaller for ConMin 1.13 than that of ACT1 1.89 because both are IR related conferences, while dissimilarity value between pair 4 is greater for ConMin 4.03 than that of ACT1 1.58 because ECIR is top ranked conference for IR topic in table 1 and JCDL (Joint Conference on Digital Libraries) is top ranked conference for topic Digital Libraries in table 1 for both approaches, which shows that ConMin can better disambiguate which conference is related to which conference and to which extent. On the other hand according to ACT1 approach ECIR is more related to JCDL 1.58 than SIGIR (Special Interest Group Conference on Information Retrieval) 1.89 which is against the real world situation. The results for pairs C and D represent same situation as pair B, which proves overall authority of ConMin approach on ACT1 in capturing semantics-based correlations between conferences.

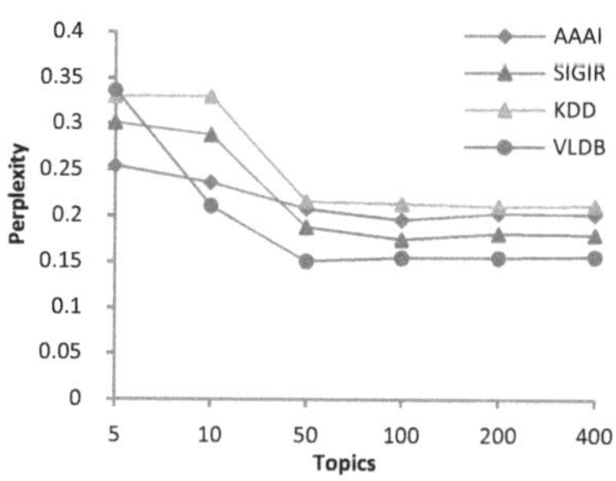

Fig. 6. Measured perplexity for new conferences

Table 4. *sKL* divergence for pairs of Conferences of ConMin and ACT1

Common Pairs	Pairs	Conferences	T=200 ConMin	T=200 ACT1
A	1	ASWC	.176	2.75
		ISWC		
	2	ASWC	3.16	3.61
		WWW		
B	3	ECIR	1.13	1.89
		SIGIR		
	4	ECIR	4.03	1.58
		JCDL		
C	5	SDM	1.49	2.31
		KDD		
	6	SDM	3.91	1.25
		UAI		
D	7	PODs	2.28	3.33
		VLDB		
	8	PODs	7.68	3.16
		ISWC		

4.6.3 Conferences Temporal Topic Trends

In most of the cases, conferences can be dominated by different topics in different years, which can provide us with topic drift for different research areas in different conferences. We used yearly data from (2003-2007) to analyze these temporal topic trends. Using 200 topics *Z*; for each conference corpus was partitioned by year, and for each year all of the words were assigned to their most likely topic using ConMin approach. It provided us the probability of topics assigned to each conference for a given year. The results provide interesting and useful indicators of temporal topic status of conferences. Figure 7 shows the results of plotting topics for SIGIR and KDD, where each topic is indicated in the legend with the five most probable words. Temporal conference trends can be captured by Topics over Time [22] and Dynamic Topic Models [5], but we are not focusing on that here.

The left plot shows the super dominant continuing topic "Information Retrieval" and other four topics having very low and steady likeliness trend for SIGIR conference. The right plot shows the ongoing dominancy of "Data Mining" topic and steady increase in the popularity of topics "Information Retrieval" and "Vector based Learning" for KDD (Knowledge Discovery in Databases) conference. As a whole, both conferences are dominated by one topic over the years, which is also one of the

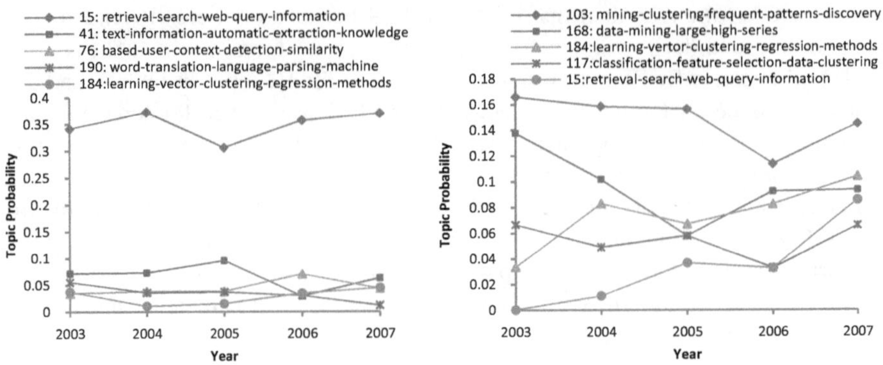

Fig. 7. Temporal topic trends of conferences

judgment criteria of the excellence of the conference and ongoing popularity of that topic. Here, it is necessary to mention that the probability for each topic per year of a conference only indicates probabilities assigned to topics by our approach, and makes no direct assessment of the quality or importance of the particular sub-area of a conference. Nonetheless, despite these caveats, obtained results are quite informative and indicate understandable temporal status of research topics in the conferences. Comparatively, ACT1 (DL) approach is unable to directly discover temporal topic trends.

5 Related Work

Automatic extraction of topics from text is performed by [15,16] to cluster documents into groups based on similar semantic contents. Clustering provides a good way to group similar documents, but clustering is inherently limited by the fact that each document is only associated with one cluster. For this reason soft clustering representation techniques are mandatory, which can allow documents composed of multiple topics to relate to more than one cluster on the basis of latent topics.

Probabilistic Latent Semantic Indexing (PLSI) [11] was proposed as a probabilistic alternative to projection and clustering methods. While PLSI produced impressive results on a number of document modeling problems, the number of parameters in the model grows linearly with the size of the corpus, which leads to serious problems of model over fitting and it was not clear how to assign a probability to a document outside the corpus.

Consequently, a more general probabilistic topic model LDA was proposed [4]. LDA assumes that each word in the document is generated by a latent topic and explicitly models the words distribution of each topic as well as the prior distribution over topics in the document. We generalized LDA to model conferences directly instead of indirectly modeling conferences like ACT1 [20] which modeled conferences through topics generated by the authors, and obtained more precise results.

Entities are modeled as graphs and related groups of entities were discovered either by network linkage information [17] or by iterative removal of edges between graphs [9,18,21]. Collaborative filtering [6,7] is employed to discover related groups of entities. They recommended items to the users on the basis of similarity between users and items. Content-based filtering [3] can also be used to recommend items on the basis of correlations between the content of the items and the users' preferences. This method creates a profile for each item or user to characterize their nature.

Previously, topics of conferences are extracted on the basis of keyword frequency from paper titles for related conferences finding [24] and specific area conferences are suggested by using pair-wise random walk algorithm [25], without considering semantic information present in the text. Differently, a topic modeling approach is used to discover topically related conferences [20]. Aforementioned approaches were incapable of considering implicit semantic information based text structure present between conferences. While, in real world co-occurrence of words and conferences; instead of co-occurrence of words and documents, can provide more appropriate semantics-based conferences correlations.

Traditionally, Kernighan-Lin algorithm and spectral bisection method [12,17] used the network linkage information between the entities to find the relationships between them. Both approaches are useless if there is no network connectivity information. Differently, correlations between authors and topics are discovered by using semantic information presented in the text [19]. Recently, Eclipse Developers correlations are discovered by using KL Divergence [13]. Here, we used sKL Divergence to discover semantics-based correlations between conferences.

Temporal topic trends of computer science were discovered in Citeseer documents [16,19] by utilizing clustering and semantics-based text information. Recently, Dynamic Topic model and Topics over Time [5,22] are used to find the general topic trends in the field of computer science. A Bayesian Network was proposed on the basis of authors to understand the research field evolution and trends [23]. Here, we used ConMin to discover topic trends specific to conferences without using authors' information, these topics are also representative of general topic trends in computer science field.

6 Conclusions and Future Work

This study deals with the problem of conference mining through capturing rich semantics-based structure of words present between conferences. We conclude that our generalization from DL to CL is significant; as proposed generalized approach's discovered and probabilistically ranked conferences (can also be applied to journals datasets such as HEP or OHSUMED) related to specific knowledge domains are better than baseline approach. While, predicted topics for new conferences are practical and meaningful. Proposed approach was also proved effective in finding conferences correlations when compared with the baseline approach. We demonstrated the effectiveness of proposed approach by applying it for analyzing temporal topic trends, which provide useful information. CL (capturing conference-level semantic structure) approach can handle the problem of DL (not capturing conference-level semantic structure) approach and provides us with dense topics. We studied the effect of generalization on topics denseness and concluded that sparser topics will results in poor performance of the approach. Empirical results show better performance of proposed approach on the basis of *richer semantics* as compared to baseline approach. Even though our approach is quite simple, nonetheless it reveals interesting information about different conference mining tasks.

Possible future direction of this work is use of authors' information in addition to already used information for discovering research community. As we think, the research community discovered from CL will be more precise than that of DL due to topics denseness.

Acknowledgements. The work is supported by the National Natural Science Foundation of China under Grant (90604025, 60703059), Chinese National Key Foundation Research and Development Plan under Grant (2007CB310803) and Higher Education Commission (HEC), Pakistan. We are thankful to Jie Tang and Jing Zhang for sharing their codes, valuable discussions and suggestions.

References

1. Andrieu, C., Freitas, N.D., Doucet, A., Jordan, M.: An Introduction to MCMC for Machine Learning. Journal of Machine Learning 50, 5–43 (2003)
2. Azzopardi, L., Girolami, M., van Risjbergen, K.: Investigating the Relationship between Language Model Perplexity and IR Precision-Recall Measures. In: Proc. of the 26th ACM SIGIR Conference on Research and Development in Information Retrieval, Toronto, Canada, July 28-August 1 (2003)
3. Balabanovic, M., Shoham, Y.: Content-Based Collaborative Recommendation. Communications of the ACM, CACM (1997)
4. Blei, D.M., Ng, A.Y., Jordan, M.I.: Latent Dirichlet Allocation. Journal of Machine Learning Research 3, 993–1022 (2003)
5. Blei, D.M., Lafferty, J.: Dynamic Topic Models. In: Proc. of 23rd International Conference on Machine Learning (ICML), Pittsburgh, Pennsylvania, USA, June 25-29 (2006)
6. Breese, J., Heckerman, D., Kadie, C.: Empirical Analysis of Predictive Algorithms for Collaborative Filtering. In: Proc. of the International Conference on Uncertainty in Intelligence (UAI), pp. 43–52 (1998)
7. Deshpande, M., Karypis, G.: Item-based Top-n Recommendation Algorithms. ACM Transactions on Information Systems 22(1), 143–177 (2004)
8. DBLP Bibliography database, `http://www.informatik.uni-trier.de/~ley/db/`
9. Girvan, M., Newman, M.E.J.: Community Structure in Social and Biological Networks. In: Proc. of the National Academy of Sciences, USA, vol. 99, pp. 8271–8276 (2002)
10. Griffiths, T.L., Steyvers, M.: Finding scientific topics. In: Proc. of the National Academy of Sciences, pp. 5228–5235 (2004)
11. Hofmann, T.: Probabilistic Latent Semantic Analysis. In: Proc. of the 15th Annual Conference on Uncertainty in Artificial Intelligence (UAI), Stockholm, Sweden, July 30-August 1 (1999)
12. Kernighan, B.W., Lin, S.: An Efficient Heuristic Procedure for Partitioning Graphs. Bell System Technical Journal 49, 291–307 (1970)
13. Linstead, E., Rigor, P., Bajracharya, S., Lopes, C., Baldi, P.: Mining Eclipse Developer Contributions via Author-Topic Models. In: 29th International Conference on Software Engineering Workshops, ICSEW (2007)

14. Ley, M.: The DBLP Computer Science Bibliography: Evolution, Research Issues, Perspectives. In: Proc. of the International Symposium on String Processing and Information Retrieval (SPIRE), Lisbon, Portugal, September 11-13, 2002, pp. 1–10 (2002)
15. McCallum, A., Nigam, K., Ungar, L.H.: Efficient Clustering of High-dimensional Data Sets with Application to Reference Matching. In: Proc. of the 6th ACM SIGKDD Conference on Knowledge Discovery and Data Mining, Boston, MA, USA, August 20-23, 2000, pp. 169–178 (2000)
16. Popescul, A., Flake, G.W., Lawrence, S., et al.: Clustering and Identifying Temporal Trends in Document Databases. In: IEEE Advances in Digital Libraries (ADL), pp. 173–182 (2000)
17. Pothen, A., Simon, H., Liou, K.P.: Partitioning Sparse Matrices with Eigenvectors of Graphs. SIAM Journal on Matrix Analysis and Applications 11, 430–452 (1990)
18. Radicchi, F., Castellano, C., Cecconi, F., et al.: Dening and Identifying Communities in Networks. In: Proc. of the National Academy of Sciences, USA (2004)
19. Rosen-Zvi, M., Griffiths, T., Steyvers, M.: Smyth. P.: The Author-Topic Model for Authors and Documents. In: Proc. of the 20th International Conference on Uncertainty in Artificial Intelligence (UAI), Banff, Canada, July 7-11 (2004)
20. Tang, J., Zhang, J., Yao, L., Li, J., Zhang, L., Su, Z.: ArnetMiner: Extraction and Mining of Academic Social Networks. In: Proc. of the 14th ACM SIGKDD International Conference on Knowledge Discovery and Data Mining (SIGKDD), Las Vegas, USA, August 24-27 (2008)
21. Tyler, J.R., Wilkinson, D.M., Huberman, B.A.: Email as Spectroscopy: Automated Discovery of Community Structure within Organizations. In: Proc. of the International Conference on Communities and Technologies, pp. 81–96 (2003)
22. Wang, X., McCallum, A.: Topics over time: A non-markov continuous-time model of topical trends. In: Proc. of the 12th ACM SIGKDD International Conference on Knowledge Discovery and Data Mining, Philadelphia, USA, August 20-23 (2006)
23. Wang, J.-L., Xu, C., Li, G., Dai, Z., Luo, G.: Understanding Research Field Evolving and Trend with Dynamic Bayesian Networks. In: Zhou, Z.-H., Li, H., Yang, Q. (eds.) PAKDD 2007. LNCS (LNAI), vol. 4426, pp. 320–331. Springer, Heidelberg (2007)
24. Zaiane, O.R., Chen, J., Goebel, R.: DBconnect: Mining Research Community on DBLP Data. In: Joint 9th WEBKDD and 1st SNA-KDD Workshop, San Jose, California, USA, August 12 (2007)
25. Zhang, J., Tang, J., Liang, B., et al.: Recommendation over a Heterogeneous Social Network. In: Proc. of the 9th International Conference on Web-Age Information Management (WAIM), ZhangJiaJie, China, July 20-22 (2008)

Within-Network Classification Using Local Structure Similarity

Christian Desrosiers and George Karypis

Department of Computer Science & Engineering,
University of Minnesota, Twin Cities
{desros,karypis}@cs.umn.edu

Abstract. Within-network classification, where the goal is to classify the nodes of a partly labeled network, is a semi-supervised learning problem that has applications in several important domains like image processing, the classification of documents, and the detection of malicious activities. While most methods for this problem infer the missing labels collectively based on the hypothesis that linked or nearby nodes are likely to have the same labels, there are many types of networks for which this assumption fails, e.g., molecular graphs, trading networks, etc. In this paper, we present a collective classification method, based on relaxation labeling, that classifies entities of a network using their local structure. This method uses a marginalized similarity kernel that compares the local structure of two nodes with random walks in the network. Through experimentation on different datasets, we show our method to be more accurate than several state-of-the-art approaches for this problem.

Keywords: Network, semi-supervised learning, random walk.

1 Introduction

Networked data is commonly used to model the relations between the entities of a system, such as hyperlinks connecting web pages, citations relating research papers, and calls between telephone accounts. In such models, entities are represented by nodes whose label gives their type, and edges are relations between these entities. As is it often the case, important information on the nature of certain entities and links may be missing from the network. The task of recovering the missing types of entities and links (i.e. node and edge labels) based on the available information, known as within-network classification, is a semi-supervised learning problem key to several applications like image processing, classifying document and web pages, classifying protein interaction and gene expression data, part-of-speech tagging, detecting malicious or fraudulent activities, and recommending items to consumers.

Classification methods for this problem suffer from two important limitations. First, many of these methods are based on the principle that nodes that are close to each other in the network are likely to have the same type, a principle known as *homophily*. While evidence suggests that homophily applies to certain

W. Buntine et al. (Eds.): ECML PKDD 2009, Part I, LNAI 5781, pp. 260–275, 2009.

networks, such as social networks [1], there are many types of networks for which this principle fails. For instance, in molecules, nearby atoms are no more likely to have the same type than distant ones. In such networks, the type of a node is instead dictated by underlying rules which may be learned by considering the relations of this node with other ones. Also, while other methods classify nodes based on their neighborhood these methods consider the distribution of labels in this neighborhood but not its structure. As we will see, the local structure of a node in the network contains important information that can improve its classification.

1.1 Contributions

This paper makes two contributions to the problem of within-network classification. First, it introduces a novel collective classification framework that combines the iterative approach of relaxation labeling with the power of similarity kernels. Unlike existing relational classifiers, which only consider the distribution of labels in the neighborhood of a node, this framework allows to use complex similarity measures between nodes. Secondly, while methods based on random walks have recently been proposed for the within-network classification problem [4,8,20], such methods evaluate the similarity between nodes using their proximity in the network. Following the success of structural kernels on the problem of graph classification [3,7,12], we present a new similarity measure inspired by marginalized graph kernels [10], that evaluates the local structure similarity between two nodes with random walks. This similarity measure upgrades marginalized kernels by considering the uncertainty of labels and the degree of nodes in the network.

The rest of this paper is organized as follows. In Section 2, we present an overview of existing methods for within-network classification. We then describe our method in Section 3, and evaluate it experimentally on several datasets in Section 4. Finally, we conclude with a short summary of our approach and contributions.

2 Related Work

Unlike traditional machine learning approaches, methods for the classification of networked data must deal with additional challenges that result from the interdependence of node classes. To overcome this problem, it has been recognized that the type of the nodes should be inferred simultaneously instead of individually, a technique known as collective classification [15]. Collective inference methods for the within-network classification problem can generally be divided in two groups: exact and approximate inference methods. Exact inference methods for this problem focus on learning the joint probability distribution of node labels. Among the best known methods of this group are those using Markov Random Fields (MRF) [11], where the joint distribution is defined as the product of potential functions that operate on the cliques of the network. Various extensions to MRFs, that also take into account observed attribute data, have also been developed. Among these are Conditional Random Fields [11], Relational Markov Networks [18] and Markov Logic Networks [6].

Due to the computational complexity of exact inference, approximation methods, such as the one presented in this paper, are normally used for the within-network classification problem. Among these is Gibbs sampling [9], which approximates the joint distribution by generating samples for the unknown labels. The main drawback of this method is its slow convergence, especially for large networks [2]. Related approaches are Relaxation Labeling (RL) [5] and Loopy Belief Propagation [19], where a vector containing the label probabilities of each node of unknown label is maintained. In RL, these vectors are initialized with apriori probabilities, either given or obtained from the data, and, at each subsequent iteration, are recomputed using a given relational classifier, until convergence or a maximum number of iterations is reached. Nodes of unknown label are then given the label of greatest probability. Unlike RL methods, Iterative Classification (IC) approaches [13,16] assign, at every iteration, a label to each node of unknown label, using a given relational classifier. To facilitate convergence, the amount of classified nodes at each iteration can be gradually increased during the process. Although our classification approach could also be used within an IC framework, we have found the updated label probabilities of RL to have better convergence properties.

2.1 Relational Classifiers

As pointed out in [15], the performance of iterative classification methods, such as RL and IC, greatly depends on the relational classifier used. A classifier strongly based on homophily, the Weighted-Vote Relational Neighbor (WVRN) [14], computes the label probability of a node as a weighted sum of the probabilities of neighbor nodes of having the same label. This simple classifier was found to work well with a RL method in the classification of documents and web pages.

Another classifier is the Class-Distribution Relational Neighbor (CDRN) [15]. This classifier assigns to each node v of known label a vector whose k-th element contain the sum, over each neighbor u of v, of the probability of u to have label k. A reference vector is then obtained for each label k as the average of the vectors belonging to nodes with known label k, and the probability of a node to have label k is defined as the similarity (L_1, L_2, cosine, etc.) between its vector and the reference vector of label k. While our classification approach is also based on node similarity, it is more general than CDRN which only considers the distribution of labels in the neighborhood of a node.

Two other relational classifiers are the Network-Only Bayes (NOB) classifier [5] and the Network-Only Link-Based (NOLB) [13] classifier. The former, which was originally used with an RL method to classify documents employs a naive Bayes approach to compute the label probability of a node, assumed to be independent given the labels of its neighbors. Finally, the NOLB classifier learns a multiclass logistic regression model using the label distribution (raw or normalized counts, or aggregation of these values) in the neighborhood of nodes with known labels. In [13], this classifier was used within an IC method to classify documents. As with CDRN, these methods do not use the local structure of a node in its classification.

3 A Novel Classification Approach

Our method is composed of two parts: a classification approach based on relaxation labeling and a node structure similarity inspired by marginalized kernels.

3.1 Relaxation Labeling Framework

Although the methods presented in this paper could be extended to the multivariate case of the within-network classification problem, we will focus on the univariate case.

We model relational data as a partially labeled graph $G = (V, E, W, L_V, L_E, l)$ where V is a set of nodes, E a set of edges between the nodes of V, $W \subset V$ is the set of nodes for which the true labels are known, L_V and L_E are respectively the sets of node and edge labels, and l is a function that maps each node and edge to a label of the corresponding set. We denote by l_u the label of a node u and $l_{u,v}$ the label of an edge (u, v). Denoting U the set of unlabeled nodes of G, i.e. $U = V \setminus W$, the problem consists in assigning to each $u \in U$ a label in L_V based on the labels of nodes in W.

As with other RL methods, our approach works by iteratively updating the label probabilities of each unlabeled node using a relational classifier, until convergence. Let K_u be a random variable modeling the label of a node u. Our relational classifier is based on the assumption that the probability $\mathrm{P}(K_u = k)$ of a node u to have label k is determined by the membership of u to a certain "node class", modeled by random variable C_u whose possible values are the nodes of V. Thus, $\mathrm{P}(C_u = v)$ represents the probability of node u to "behave" in the same way as node v, or more simply, the similarity between u and v. Following this model, $\mathrm{P}(K_u = k)$ can be obtained by marginalizing C_u:

$$\begin{aligned}
\mathrm{P}(K_u = k) &= \sum_{v \in V} \mathrm{P}(K_u = k, C_u = v) \\
&= \sum_{v \in V} \mathrm{P}(K_u = k | C_u = v)\, \mathrm{P}(C_u = v) \\
&\propto \sum_{v \in V} \mathrm{P}(K_v = k)\, \mathrm{P}(C_u = v).
\end{aligned}$$

Using the shorthand notation $\pi_{v,k} = \mathrm{P}(K_v = k)$ and letting $\sigma_{u,v} \propto \mathrm{P}(C_u = v)$ denote the similarity between u and v, this expression becomes

$$\pi_{u,k} = \frac{1}{Z} \sum_{v \in V} \pi_{v,k}\, \sigma_{u,v},$$

where Z is a normalization constant. Assuming the label probabilities are all binary, i.e. $\pi_{v,k} = \delta\,(l_v = k)$, and letting $V_k \subseteq V$ be the nodes that have label k, the model simplifies to

$$\pi_{u,k} = \frac{1}{Z} \sum_{v \in V_k} \sigma_{u,v}.$$

This expression underlines an important drawback, where the probability of a label k is proportional the number of nodes that have k as label. In cases where there is an important bias in the distribution of labels, such as those reported in the experimental section, this causes all the unlabeled nodes to get the most frequent label. To alleviate this problem, we use a different formulation where

$$\pi_{u,k} = \frac{1}{Z} \cdot \frac{\sum_{v \in V} \pi_{v,k}\, \sigma_{u,v}}{\sum_{v \in V} \pi_{v,k}}.$$

If we consider, once again, the label probabilities as binary, this expression becomes

$$\pi_{u,k} = \frac{1}{Z} \cdot \frac{1}{|V_k|} \sum_{v \in V_k} \sigma_{u,v},$$

i.e., $\pi_{u,k}$ is proportional to the average similarity with nodes of label k.

Finally, this model is augmented with two parameters $\alpha \geq 0$ and $\beta \geq 0$ which respectively encode the importance given to label uncertainty and node similarity:

$$\pi_{u,k} = \frac{1}{Z} \cdot \frac{\sum_{v \in V} \pi_{v,k}^{\alpha}\, \sigma_{u,v}^{\beta}}{\sum_{v \in V} \pi_{v,k}^{\alpha}}. \tag{1}$$

Our classification approach can be summarized with the following iterative process. First, we initialize the label probability of unlabeled nodes using apriori probabilities, either known or approximated from the labeled nodes. Then, at each iteration, we use (1) to update the label probabilities $\pi_{u,k}$ of each unlabeled node $u \in U$ and label $k \in L_V$, using the values of the previous iteration. These values are then normalized to make sure that they constitute a probability distribution. This process is repeated until the label probabilities converge, i.e. the average change is inferior to a given threshold $\epsilon > 0$, or we reach a given number of iteration $T_{\max}$. Finally, we assign to each unlabeled node $u \in U$ the label k of highest probability $\pi_{u,k}$.

Note that this approach can also be used to classify the unlabeled edges of a graph G. The idea is to transform G by replacing each edge $(u, v) \in E$ by a new node uv and two new edges, (u, uv) and (uv, v). The graph obtained in this way has $|V| + |E|$ nodes with labels from a set of $|L_V| + |L_E|$ nodes labels, and $2|E|$ edges with the same label.

3.2 Random Walk Structure Similarity

Our approach to evaluate the local structure similarity of two nodes is based on marginalized graph kernels [10], which compute similarities as the probability of generating the same sequence of labels in two parallel random walks.

While a more general approach using product graphs has been proposed to compute the structural similarity between graphs [7], the probabilistic framework of marginalized kernels is better suited to cope with the label uncertainties of our RL method. We should also mention that other types of kernels have been proposed to measure the similarity between *nodes*, such as exponential, diffusion and regularization kernels [17], and kernels based on random walks [4,8,20]. However, these kernels are mostly based on the proximity of the nodes in the graph, not their structural similarity.

Our similarity measure differs from marginalized kernels in two respect. First, it evaluates the similarity between two nodes of a same graph, instead of between two different graphs. Accordingly, the similarity between two nodes u and u' is defined as the probability of generating the same sequence with random walks starting at u and u'. Secondly, the labels of some nodes are only known as a probability. To cope with this problem, we make the label generation stochastic such that label k is generated at node v with probability $\pi_{v,k}$.

Since we do not demonstrate the semi-definite positiveness of our proposed kernels and because our classification framework is meant to be very flexible, the term *kernel* should be considered as a *general similarity measure* throughout the rest of the paper.

3.3 Derivation of the Similarity Kernel

Denote by $\mathrm{P}_t(v|u)$ the probability that the walk jumps from a node u to an adjacent node v and $\mathrm{P}_e(v)$ the probability that the walk stops at node v, satisfying the constraint that

$$\mathrm{P}_e(u) + \sum_{v \in V} \mathrm{P}_t(v|u) \; = \; 1. \tag{2}$$

Following these definitions, the probability of visiting a sequence of nodes $\mathbf{v} = (\mathbf{v}_0, \ldots, \mathbf{v}_n)$ in a random walk starting at node $\mathbf{v}_0$ is

$$\mathrm{P}(\mathbf{v}) = \left(\prod_{i=1}^{n} \mathrm{P}_t(\mathbf{v}_i|\mathbf{v}_{i-1}) \right) \mathrm{P}_e(\mathbf{v}_i).$$

Let $\mathrm{P}_l(k|v)$ and $\mathrm{P}_l(k|u,v)$ denote, respectively, the probability of generating label $k \in L_V$ at node v and the probability of generating label $k' \in L_E$ while traversing edge (u,v). Given node sequence $\mathbf{v}$, the conditional probabilities of generating the sequences of node labels $\mathbf{s}$ and edge labels $\mathbf{q}$ are

$$\mathrm{P}(\mathbf{s}|\mathbf{v}) \; = \; \prod_{i=1}^{n} \mathrm{P}_l(\mathbf{s}_i|\mathbf{v}_i)$$

$$\mathrm{P}(\mathbf{q}|\mathbf{v}) \; = \; \prod_{i=1}^{n} \mathrm{P}_l(\mathbf{q}_i|\mathbf{v}_{i-1}, \mathbf{v}_i)$$

Let $\mathcal{W}_u^{(n)}$ be the set of possible sequences of $n+1$ nodes visited in a random walk starting at node u. The marginalized probability of a sequence $\mathbf{s}$, given a start node $u = \mathbf{v}_0$, is obtained by summing over all sequences of $\mathcal{W}_u^{(n)}$:

$$\begin{aligned}\mathrm{P}(\mathbf{s},\mathbf{q}|u) &= \sum_{n=1}^{\infty}\sum_{\mathbf{v}\in\mathcal{W}_u^{(n)}} \mathrm{P}(\mathbf{s}|\mathbf{v})\mathrm{P}(\mathbf{q}|\mathbf{v})\mathrm{P}(\mathbf{v})\\ &= \sum_{n=1}^{\infty}\sum_{\mathbf{v}}\left(\prod_{i=1}^{n}\mathrm{P}_t(\mathbf{v}_i|\mathbf{v}_{i-1})\mathrm{P}_l(\mathbf{s}_i|\mathbf{v}_i)\mathrm{P}_l(\mathbf{q}_i|\mathbf{v}_{i-1},\mathbf{v}_i)\right)\mathrm{P}_e(\mathbf{v}_i).\end{aligned}$$

Denote by $\mathcal{S}^{(n)}$ and $\mathcal{Q}^{(n)}$ the set containing, respectively, all sequences of n node labels and edge labels, the probability of generating the same sequence in two parallel random walks starting at nodes u and u' is given by

$$\begin{aligned}\sigma_{u,u'} &= \sum_{n=1}^{\infty}\sum_{\mathbf{s}\in\mathcal{S}^{(n)}}\sum_{\mathbf{q}\in\mathcal{Q}^{(n)}} \mathrm{P}(\mathbf{s},\mathbf{q}|u)\mathrm{P}(\mathbf{s},\mathbf{q}|u')\\ &= \sum_{n=1}^{\infty}\sum_{\mathbf{s},\mathbf{q}}\sum_{\mathbf{v}\in\mathcal{W}_u^{(n)}}\sum_{\mathbf{v}'\in\mathcal{W}_{u'}^{(n)}}\left(\prod_{i=1}^{n}\mathrm{P}_t(\mathbf{v}_i|\mathbf{v}_{i-1})\mathrm{P}_t(\mathbf{v}'_i|\mathbf{v}'_{i-1})\mathrm{P}_l(\mathbf{s}_i|\mathbf{v}_i)\mathrm{P}_l(\mathbf{s}_i|\mathbf{v}'_i)\right.\\ &\qquad\left.\mathrm{P}_l(\mathbf{q}_i|\mathbf{v}_{i-1},\mathbf{v}_i)\mathrm{P}_l(\mathbf{q}_i|\mathbf{v}'_{i-1},\mathbf{v}'_i)\right)\mathrm{P}_e(\mathbf{v}_n)\mathrm{P}_e(\mathbf{v}'_n)\\ &= \sum_{n=1}^{\infty}\sum_{\mathbf{s},\mathbf{q}}\sum_{\mathbf{v},\mathbf{v}'}\left(\prod_{i=1}^{n} a(\mathbf{v}_{i-1},\mathbf{v}'_{i-1},\mathbf{v}_i,\mathbf{v}'_i,\mathbf{s}_i,\mathbf{q}_i)\right)\mathrm{P}_e(\mathbf{v}_n)\mathrm{P}_e(\mathbf{v}'_n),\end{aligned}$$

where

$$a(\mathbf{v}_{i-1},\mathbf{v}'_{i-1},\mathbf{v}_i,\mathbf{v}'_i,\mathbf{s}_i,\mathbf{q}_i)=\mathrm{P}_t(\mathbf{v}_i|\mathbf{v}_{i-1})\mathrm{P}_t(\mathbf{v}'_i|\mathbf{v}'_{i-1})\mathrm{P}_l(\mathbf{s}_i|\mathbf{v}_i)\mathrm{P}_l(\mathbf{s}_i|\mathbf{v}'_i)\mathrm{P}_l(\mathbf{q}_i|\mathbf{v}_{i-1},\mathbf{v}_i)\mathrm{P}_l(\mathbf{q}_i|\mathbf{v}'_{i-1},\mathbf{v}'_i).$$

The computation of $\sigma_{u,u'}$ can be greatly simplified using the following recurrence: the probability of generating the same sequence of n labels two parallel random walks starting at nodes u and u', written $r_{u,u'}^{(n)}$, can be obtained from the probability of visiting nodes v and v', respectively from u and u', and the probability of generating the same sequences of $n-1$ node and edge labels, starting at nodes v and v'. This recurrence can be written as

$$r_{u,u'}^{(n)} = \begin{cases}\sum_{v,v'\in V}\sum_{k\in L_V}\sum_{k'\in L_E} a(u,u',v,v',k,k')\; r_{v,v'}^{(n-1)} & ,n\geq 1\\ \mathrm{P}_e(u)\;\mathrm{P}_e(u') & ,n=0\end{cases}$$

The probability of generating the same sequences of at most N labels starting from nodes u and u', written $R_{u,u'}^{(N)}$, is then

$$\begin{aligned}R_{u,u'}^{(N)} &= \sum_{n=1}^{N} r_{u,u'}^{(n)}\\ &= \sum_{n=1}^{N}\sum_{v,v'\in V}\sum_{k\in L_V}\sum_{k'\in L_E} a(u,u',v,v',k,k')\; r_{v,v'}^{(n-1)}\\ &= \sum_{v,v'}\sum_{k,k'} a(u,u',v,v',k,k')\sum_{n=1}^{N} r_{v,v'}^{(n-1)}\\ &= \sum_{v,v'}\sum_{k,k'} a(u,u',v,v',k,k')\left(\mathrm{P}_e(v)\mathrm{P}_e(v') + R_{v,v'}^{(N-1)}\right),\end{aligned}$$

where $R_{u,u'}^{(0)} = 0$ for all u, u'. We then have $\sigma_{u,u'} = \lim_{N\to\infty} R_{u,u'}^{(N)}$.

Denote by N_u the neighbors of node u and let $d_u = |N_u|$ be the degree of u. Setting the termination probabilities of u to a constant $\mathrm{P}_e(u) = \gamma$, and letting the transition probabilities be uniform over the neighbors of u, following the constraint of (2), we have $\mathrm{P}_t(v|u) = (1-\gamma)/d_u$ if $v \in N_u$ and 0 otherwise. Furthermore, using $\mathrm{P}_l(k|v) = \pi_{v,k}$ and $\mathrm{P}_l(k'|u,v) = \delta(l_{u,v} = k')$ as node and edge label probabilities, the formulation of the kernel becomes

$$R^{(N)}_{u,u'} = \frac{(1-\gamma)^2}{d_u d_{u'}} \sum_{v \in N_u} \sum_{v' \in N_{u'}} \sum_{k \in L_V} \delta\left(l_{u,v} = l_{u',v'}\right) \pi_{v,k} \pi_{v',k} \left(\gamma^2 + R^{(N-1)}_{v,v'}\right). \quad (3)$$

Other than being computed between pairs of nodes instead of graphs, this expression differs from the one of [10] by the fact that the label probabilities are also marginalized. To compute the kernel, we use a bottom-up iterative approach, where we use (3) to compute the probabilities $R^{(N)}$ based on $R^{(N-1)}$. We repeat this process for increasing values of N, until the similarity values converge, i.e. the average change is smaller than a given ϵ, or N reaches a given limit $N_{\max}$.

3.4 Exploiting Node Degrees

A problem with the kernel definition of (3) is that it does not consider the difference between the degrees of two nodes u and v, while evaluating their similarity. To illustrate this, suppose we limit the walk length in (3) to $N_{\max} = 1$, i.e. we consider only the direct neighbors of u and v. Moreover, suppose that the label of every node is known, i.e. $\pi_{u,k} = \delta(l_u = k)$. Under these constraints, the similarity kernel becomes

$$\sigma_{u,v} = \frac{(1-\gamma)^2\gamma^2}{d_u d_v} \sum_{k \in L_V} n_{u,k}\, n_{v,k},$$

where $n_{u,k} \leq d_u$ denotes the number of neighbors of u that have label k. Thus, this simplified kernel simply compares ratios of neighbors having each label k, similar to what is done in the CDRN classifier. Using this formulation, the similarity between the nodes u and v of Figure 1 (a)-(b) is equal to the self-similarity of these nodes: $\sigma_{u,u} = \sigma_{v,v} = \sigma_{u,v} = \frac{1}{2}(1-\gamma)^2\gamma^2$.

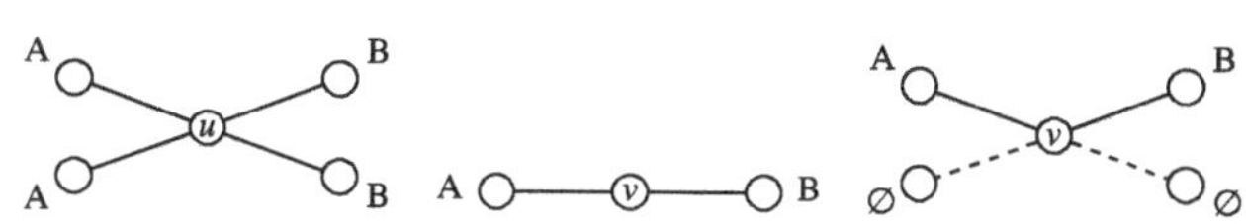

Fig. 1. (a)-(b) The neighborhood of two nodes u, v and (c) the transformed neighborhood of v

In order to consider the difference in the degrees, we modify the kernel formulation as

$$R_{u,u'}^{(N)} = \frac{(1-\gamma)^2}{\max\{d_u, d_{u'}\}^2} \sum_{v \in N_u} \sum_{v' \in N'_u} \sum_{k \in L_V} \delta\left(l_{u,v} = l_{u',v'}\right) \pi_{v,k} \pi_{v',k} \left(\gamma^2 + R_{v,v'}^{(N-1)}\right) \quad (4)$$

This modification to the kernel can be interpreted in the parallel random walks framework as follows. If the degree of the node visited by a walk is less than the degree of the node visited by the other walk, temporary edges are added from this node to a dummy node of label $\varnothing \notin L_V$, such that both nodes have the same degree. With the same probability as the true neighbors, the random walk can jump to this dummy node, after which the probability of generating the same sequence becomes null. Figure 1(c) illustrates this idea for nodes u, v of (a) and (b). Using this new formulation, the similarity values for nodes u and v, again limiting the walk length to $N_{\max} = 1$, are $\sigma_{u,u} = \sigma_{v,v} = \frac{1}{2}(1-\gamma)^2\gamma^2 \geq \frac{1}{4}(1-\gamma)^2\gamma^2 = \sigma_{u,v}$.

3.5 Convergence and Complexity

While the convergence of the similarity kernels defined above is shown in Appendix A, the collective classification method presented in this paper, as most RL methods, is not guaranteed to converge since the node structure similarities $\sigma_{u,v}$ vary from one iteration to the next. However, by limiting the number of allowed iterations to $T_{\max}$, we can still obtain a solution in the non-converging case. Furthermore, while the classification process can be expensive in the worst-case, i.e. $O\left(T_{\max} N_{\max} d_{\max}^2 |L_V||V|^2\right)$, its complexity is closer to $O(|V|^2)$ in practice due to four reasons: 1) there are much less node labels than nodes, 2) the nodes of many real-life graphs have a low bounded degree (e.g., molecular graphs), 3) the relevant structural information of a node is contained within a short distance, and 4) the RL algorithm normally converges in a few iterations, regardless of $|V|$.

4 Experimental Evaluation

In this section, we test our framework on the problem of classifying the unlabeled nodes of a partly labeled graph.

4.1 Experimental Setting

We tested our classification approach on five datasets. The first three datasets, which are available online at the IAM Graph Database Repository[1], were originally used for the prediction of mutagenicity, AIDS antiviral activity, and protein function. The first two model chemical compounds as undirected graphs where the nodes represent atoms, node labels are the chemical symbols of these

[1] `http://www.iam.unibe.ch/fki/databases/iam-graph-database`

Table 1. Properties of the datasets

Property	Mutagen.	AIDS	Protein	Cornell	Texas
Nb. graphs	4,337	2,000	600	1	1
Avg. nodes	30.3	15.7	32.6	351	338
Avg. edges	30.8	16.2	62.1	1392	986
Node labels	14	38	3	6	6
Edge labels	3	3	5	1	1
Freq. class	44.3%	59.3%	49.4%	41.5%	48.1%

atoms, and edges are covalent bonds between atoms. Edge labels give the valency of these bonds. The third dataset models proteins into undirected graphs using their secondary structure, such that nodes are secondary structure elements (SSE) labeled as helix, sheet, or turn. Every node is connected with an edge to its three nearest neighbors[2] in space, and edges are labeled with their structural type.

Finally, the last two datasets, which were created for the WebKB project, contain graphs modeling the links between Web pages collected from computer science departments of the Cornell and Texas Universities. These two datasets, available online[3], have often been used to benchmark within-network classification methods, as in [15]. While the link information is sometimes converted into a co-citation graph, we evaluate our approach directly on the original Web page link graph. Furthermore, we consider the multiclass classification problem where pages can have one of six types: *student, faculty, staff, department, course* and *project*. Finally, while they are used in the evaluation of other methods, the edges weights representing the number of links between two Web pages, are ignored by our methods.

Table 1 gives some properties of these datasets: the number of graphs, the average number of nodes and edges of these graphs, their number of node and edge labels, and the percentage of nodes having the most frequent class label.

As suggested in [15], we compare our approach with the classification methods implemented in NetKit-SRL[4]. This toolkit provides a general framework for within-network classification that allows the user to choose any combination of collective inference approach, i.e. RL, IC or Gibbs sampling, and relational classifier, i.e. WVRN, CDRN, NOB or NOLB (using either raw or normalized counts of neighbors with a given label). For additional information, the reader may refer to Section 2 or to [15]. Although we have tested every possible combination of collective classification approach and relational classifier, we have kept, for each classifier, the approach which worked best. Including the two methods proposed in this paper, i.e. our RL framework with the similarity kernels of (3) and (4), a total of 7 methods, described in Table 2, are tested.

[2] Note that a node can have more than three neighbors since the relation "nearest-neighbor" is not symmetric.

[3] `http://netkit-srl.sourceforge.net/data.html`

[4] `http://netkit-srl.sourceforge.net/`.

Table 2. Tested classification methods

Method	Description
RL-WVRN:	RL with WVRN
RL-CDRN:	RL with CDRN (cosine similarity on normalized counts)
IC-NOB:	IC with NOB
IC-NOLB-count:	IC with NOLB (raw counts)
IC-NOLB-norm:	IC with NOLB (normalized counts)
RL-RW:	Our RL with the kernel of (3).
RL-RW-deg:	Our RL with the kernel of (4).

The five datasets were used differently in our experiments. For the first three ones, which contain many small graphs, we randomly sampled six sets of 100 graphs and then merged the graphs of each of these sets into larger test graphs, considering the small graphs as individual components of the larger ones (1500 to 3500 nodes depending on the dataset). We then randomly selected one of these test graphs to tune the parameters of the tested methods and used the five others to evaluate their performance. For each of these five test graphs, 10 runs were performed, where we randomly selected a subset of nodes from which we removed the labels. We then computed the F1-score using the precision and recall obtained for each class, weighted by the number of nodes in these classes, and averaged this value over the 5×10 classification runs. For the graphs of the last two datasets, parameters were tuned using another WebKB dataset modeling the links between Web pages of the University of Washington. As with the other datasets, 10 runs were performed on each of these two graphs and the F1-scores were averaged over these runs.

4.2 Results

Figures 2 gives the F1-scores obtained by the seven tested methods on the five datasets, for decreasing percentages of labeled nodes. Note that we have used a different range of labeled nodes for the two WebKB graphs (10% to 80% instead of 2.5% to 50%), since these graphs have much less nodes than the other ones.

We can see that our structure similarity kernel approach that considers node degrees, i.e. RL-RW-deg, outperforms the other classification methods for datasets where the type of a node is well correlated with its local structure (i.e., Mutagenicity and AIDS datasets), especially when a small portion of nodes are labeled. Moreover, within the classification methods of Netkit-SRL, the IC method based on the multiclass regression using the raw counts, i.e. IC-NOLB-count, provides results comparable with RL-RW-deg when the labels of a sufficient number of nodes are known. However, as the number of labeled nodes reduces, this method fails to learn a proper regression model and its performance drops. Finally, the methods based on homophily, such as RL-WDRN, perform poorly on this type of data.

Our classification approach considering node degrees also works well on other types of data, such as the Web page link graphs, where it is comparable to

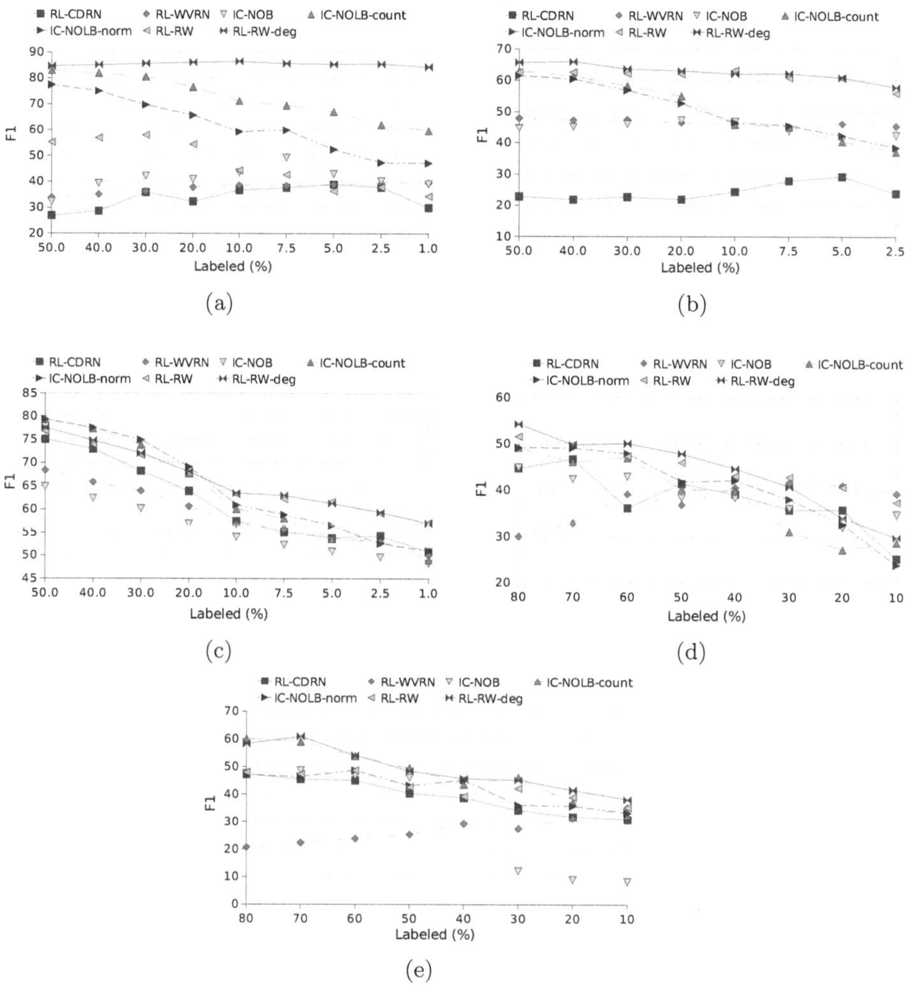

Fig. 2. F1-scores obtained for the tested datasets: (a) Mutagenicity, (b) AIDS, (c) Protein, (d) Cornell and (e) Texas

the best NetKit-SRL method. For the Cornell dataset, however, it appears that WDRN works the best when a few labels are known. In this case, this method simply assigns the most frequent label to unlabeled nodes, which gives decent results due to the biased distribution of labels (see Table 1). As more node labels are known, the classification relies increasingly on proximity which actually degrades the performance of this method.

Comparing our two similarity kernels, we observe a variation in the results obtained for the different types of data. Thus, while RL-RW-deg is significantly better than RL-RW on the Mutagenicity data, the performance of these two methods is comparable on the AIDS data. This is due to the fact valency of an atom, i.e. degree of a node, is a good indicator of the type of this atom, but this information is noisy in the AIDS data since bonds to hydrogen atoms have been omitted. For the Protein and WebKD datasets, however, the degree of a node provides a weaker signal for classification, and both approaches give comparable results.

4.3 Influence of Parameters and Runtimes

Although the results presented in this section were obtained on the same datasets as for the validation, these results were not used to tune our methods.

Figure 3(a) gives the average accuracy of RL-RW-deg on the Mutagenicity data (using a percentage of labeled nodes of 50%) for different values of parameters α and β, which control the impact of label uncertainty and similarity in our relational classifier. We notice that the accuracy can sometimes be improved by increasing the importance of nodes with uncertain labels w.r.t. nodes of known label, i.e. using $\alpha < 1$. This could be explained by the fact that using such values provides a smoother convergence of the method. This could also explain the poor results of the RL-CDRN method, which corresponds to using $\alpha \to \infty$ in our framework (assuming the random walk length is limited to 1).

The impact of the random walk termination probability γ on the classification of the AIDS data (using a percentage of labeled nodes of 50%) is shown in Figure 3(b). To illustrate how this parameter influences the length of the random walks, we varied the maximum walk length $N_{\max}$ of the kernel. When the walk lengths are the least limited, i.e. $N_{\max} = 6$, we notice that the accuracy is reduced when the termination probability γ increases. We also see that the greatest gain in accuracy occurs for $N_{\max} = 2$, suggesting that most of the structural information of a node, for this data, is contained within a short distance of this node.

The last analysis focuses on the times required to run our methods on a machine equipped with two 2.60GHz i686 processors and 1Gb of RAM. Figure 4 gives the mean runtimes of RL-RW-deg on the Mutagenicity data (using a percentage of labeled nodes of 50%), for different values of kernel parameter γ. As a reference, we also give the runtime of RL-CDRN, the slowest Netkit-SRL classification method

	RL β						
RL α	0.5	1.0	1.5	2.0	2.5	3.0	3.5
0.25	87.42	87.48	87.75	88.22	88.42	88.68	89.28
0.50	87.42	87.55	88.02	88.42	88.42	88.82	89.41
0.75	86.95	87.95	88.15	87.88	88.02	88.68	88.88
1.00	86.42	87.82	87.75	87.08	85.82	83.69	81.89
1.25	86.22	86.88	84.82	82.36	77.03	72.44	67.44
1.50	84.75	83.02	76.23	69.04	59.19	44.67	43.81

(a)

Kernel	Kernel γ				
$N_{\max}$	0.1	0.3	0.5	0.7	0.9
1	62.54	62.54	62.54	62.54	62.54
2	66.92	66.92	67.49	65.80	62.76
3	65.69	67.15	66.25	64.45	62.76
4	65.46	67.82	66.02	64.00	62.76
5	67.71	67.82	66.02	64.00	62.76
6	66.92	67.60	66.02	64.00	62.76

(b)

Fig. 3. Impact of the parameters on the classification accuracy: (a) RL parameters α and β on the Mutagenicity data, and (b) kernel parameters $N_{\max}$ and γ on the AIDS data

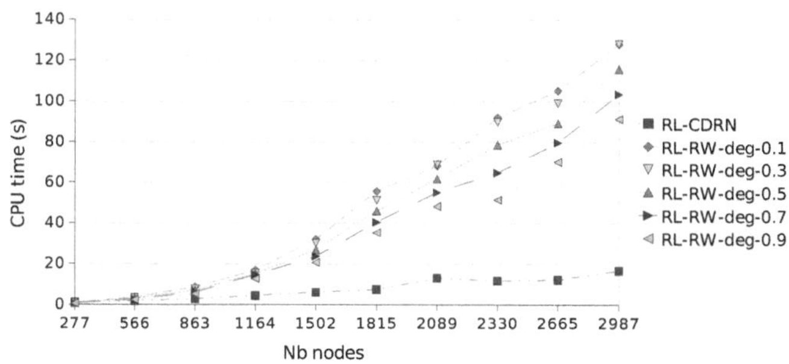

Fig. 4. Runtime (in seconds) of our approach on the Mutagenicity data

for this data. While our method is noticeably slower than methods based only on direct neighbors, such as RL-CDRN, it performed the classification within only 2 minutes for networks with 3,000 nodes, suggesting it could be used for even larger networks.

5 Conclusion

This paper presented a novel approach for the problem of within-network classification. Unlike other methods for this problem, which are based on the principle that nearby nodes in the network are likely to have the same type, or which use only the distribution of labels in the neighborhood of a node, this approach classifies a node based on its local structure similarity with other nodes in the network. Furthermore, a new method was proposed to evaluate the structural similarity between nodes in a partly labeled graph. This method, which uses random walks, extends marginalized graph kernels by considering the label uncertainty and the degree of nodes in the network. Our classification approach was tested on real-life data from the several fields, and the experimental results have shown our method to outperform several state-of-the-art methods when the type of a node is correlated to its structure, and perform as well as these methods with other types of data.

Acknowledgements. This work was supported by NSF ACI-0133464, IIS-0431135, NIH RLM008713A, and by the Digital Technology Center at the University of Minnesota.

References

1. Barabasi, A., Jeong, H., Neda, Z., Ravasz, E., Schubert, A., Vicsek, T.: Evolution of the social network of scientific collaborations. Physica A 311(3-4), 590–614 (2002)
2. Besag, J.: On the statistical analysis of dirty pictures. Journal of the Royal Statistical Society 48(3), 259–302 (1986)
3. Borgwardt, K., Ong, C., Schönauer, S., Vishwanathan, S., Smola, A., Kriegel, H.-P.: Protein function prediction via graph kernels. Bioinformatics 21(1), 47–56 (2005)

4. Callut, J., Francoisse, K., Saerens, M., Dupont, P.: Semi-supervised classification from discriminative random walks. In: Daelemans, W., Goethals, B., Morik, K. (eds.) ECML PKDD 2008, Part I. LNCS (LNAI), vol. 5211, pp. 162–177. Springer, Heidelberg (2008)
5. Chakrabarti, S., Dom, B., Indyk, P.: Enhanced hypertext categorization using hyperlinks. In: SIGMOD 1998: Proc. of the 1998 ACM SIGMOD Int. Conf. on Management of data, pp. 307–318. ACM Press, New York (1998)
6. Domingos, P., Richardson, M.: Markov logic: A unifying framework for statistical relational learning. In: Proc. of the ICML 2004 Workshop on Statistical Relational Learning and its Connections to Other Fields, pp. 49–54 (2004)
7. Gaertner, T., Flach, P., Wrobel, S.: On graph kernels: Hardness results and efficient alternatives. In: Proc. of the 16th Annual Conf. on Computational Learning Theory, August 2003, pp. 129–143. Springer, Heidelberg (2003)
8. Gallagher, B., Tong, H., Eliassi-Rad, T., Faloutsos, C.: Using ghost edges for classification in sparsely labeled networks. In: KDD 2008: Proc. of the 14th ACM SIGKDD Int. Conf. on Knowledge discovery and data mining, pp. 256–264. ACM Press, New York (2008)
9. Geman, S., Geman, D.: Stochastic relaxation, gibbs distributions, and the bayesian restoration of images. In: Neurocomputing: foundations of research, pp. 611–634 (1988)
10. Kashima, H., Tsuda, K., Inokuchi, A.: Marginalized kernels between labeled graphs. In: Proc. of the 12th In. Conf. on Machine Learning, pp. 321–328. AAAI Press, Menlo Park (2003)
11. Lafferty, J.D., McCallum, A., Pereira, F.C.N.: Conditional random fields: Probabilistic models for segmenting and labeling sequence data. In: ICML 2001: Proc. of the 18th Int. Conf. on Machine Learning, pp. 282–289. Morgan Kaufmann Publishers Inc., San Francisco (2001)
12. Li, X., Zhang, Z., Chen, H., Li, J.: Graph kernel-based learning for gene function prediction from gene interaction network. In: BIBM 2007: Proc. of the 2007 IEEE Int. Conf. on Bioinformatics and Biomedicine, Washington, DC, USA, pp. 368–373. IEEE Computer Society Press, Los Alamitos (2007)
13. Lu, Q., Getoor, L.: Link-based classification. In: Fawcett, T., Mishra, N., Fawcett, T., Mishra, N. (eds.) Proc. 12th Int'l Conf. Machine Learning (ICML), pp. 496–503. AAAI Press, Menlo Park (2003)
14. Macskassy, S.A., Provost, F.: A simple relational classifier. In: Proc. of the 2nd Workshop on Multi-Relational Data Mining (MRDM 2003), pp. 64–76 (2003)
15. Macskassy, S.A., Provost, F.: Classification in networked data: A toolkit and a univariate case study. Journal of Machine Learning Research 8, 935–983 (2007)
16. Neville, J., Jensen, D.: Iterative classification in relational data. In: Proc. Workshop on Statistical Relational Learning, AAAI, pp. 13–20. AAAI Press, Menlo Park (2000)
17. Smola, A., Kondor, R.: Kernels and regularization on graphs. In: Warmuth, M., Schölkopf, B. (eds.) Proc. of the 2003 Conf. on Computational Learning Theory (COLT) and Kernels Workshop, pp. 144–158 (2003)
18. Taskar, B., Abbeel, P., Koller, D.: Discriminative probabilistic models for relational data. In: UAI 2002, Proc. of the 18th Conf. in Uncertainty in Artificial Intelligence, pp. 485–492. Morgan Kaufmann, San Francisco (2002)
19. Yedidia, J.S., Freeman, W.T., Weiss, Y.: Constructing free-energy approximations and generalized belief propagation algorithms. IEEE Transactions on Information Theory 51(7), 2282–2312 (2005)

20. Zhu, X., Ghahramani, Z., Lafferty, J.: Semi-supervised learning using gaussian fields and harmonic functions. In: Proc. of the 12th Int. Conf. on Machine Learning (ICML), pp. 912–919 (2003)

A Proof of Convergence

Proposition 1. *The kernel defined by (3) converges for any* $0 < \gamma \leq 1$.

Proof. Consider the probability of generating the same sequence of n labels in two random walks starting at u and u', as defined by the following equation:

$$r^{(n)}_{u,u'} = \frac{(1-\gamma)^2}{d_u d_{u'}} \sum_{v \in N_u} \sum_{v' \in N_{u'}} \sum_k \delta\left(l_{u,v} = l_{u',v'}\right) \pi_{v,k}\, \pi_{v',k}\, r^{(n-1)}_{v,v'}.$$

Since $r^{(n)}_{u,u'}$ is computed by summing and multiplying non-negative terms, by induction, it is also non-negative. Furthermore, this value can be bounded from above as

$$\begin{aligned} r^{(n)}_{u,u'} &\leq \frac{(1-\gamma)^2}{d_u d_{u'}} \sum_{v,v'} r^{(n-1)}_{v,v'} \sum_{k \in L_V} \sum_{k' \in L_V} \pi_{v,k}\pi_{v',k'} \\ &= \frac{(1-\gamma)^2}{d_u d_{u'}} \sum_{v,v'} r^{(n-1)}_{v,v'} \left(\sum_k \pi_{v,k}\right)\left(\sum_{k'} \pi_{v',k'}\right) \\ &= \frac{(1-\gamma)^2}{d_u d_{u'}} \sum_{v,v'} r^{(n-1)}_{v,v'}. \end{aligned}$$

Let $r^{(n-1)}_{\max} = \max_{v,v' \in V} r^{(n-1)}_{v,v'}$, we have

$$\begin{aligned} r^{(n)}_{u,u'} &= (1-\gamma)^2\, r^{(n-1)}_{\max} \\ &\leq (1-\gamma)^{2n}\, r^{(0)}_{\max} \qquad \text{(by induction)} \\ &= (1-\gamma)^{2n}\, \gamma^2. \end{aligned}$$

Finally, the probability of generating the same sequence of any length between u and u' is bounded by

$$\begin{aligned} \sigma_{u,u'} &= \sum_{n=1}^{\infty} r^{(n)}_{u,u'} \\ &\leq \gamma^2 \sum_{n=1}^{\infty} (1-\gamma)^{2n} \\ &= \frac{\gamma^2(1-\gamma)^2}{1-(1-\gamma)^2}, \end{aligned}$$

where we have used the fact that the series is geometric and $(1-\gamma)^2 < 1$.

Proposition 2. *The kernel defined by (4) converges for any* $0 < \gamma \leq 1$.

Proof. This can be shown using the same approach as with Proposition 1.

Multi-task Feature Selection Using the Multiple Inclusion Criterion (MIC)

Paramveer S. Dhillon[1], Brian Tomasik[2], Dean Foster[3], and Lyle Ungar[1]

[1] CIS Department, University of Pennsylvania, Philadelphia, PA 19104, U.S.A.
[2] Computer Science Department, Swarthmore College, PA 19081, U.S.A.
[3] Statistics Department, University of Pennsylvania, Philadelphia, PA 19104, U.S.A.

Abstract. We address the problem of joint feature selection in multiple related classification or regression tasks. When doing feature selection with multiple tasks, usually one can "borrow strength" across these tasks to get a more sensitive criterion for deciding which features to select. We propose a novel method, the Multiple Inclusion Criterion (MIC), which modifies stepwise feature selection to more easily select features that are helpful across multiple tasks. Our approach allows each feature to be added to none, some, or all of the tasks. MIC is most beneficial for selecting a small set of predictive features from a large pool of potential features, as is common in genomic and biological datasets. Experimental results on such datasets show that MIC usually outperforms other competing multi-task learning methods not only in terms of accuracy but also by building simpler and more interpretable models.

1 Introduction

We consider the problem of feature selection for a set of related tasks which are expected to partially share common sets of predictive features. When one is trying to predict a set of related responses ("tasks"), be they multiple clinical outcomes for patients or growth rates under different conditions for yeast strains, it may be possible to "borrow strength" by sharing information between the models for the different responses.

The problem of building shared models for multiple related tasks is popularly known as "Multi-Task Learning" or "Transfer Learning" and has been studied extensively [1,2,3,4,5,6,7]. To give a couple examples: [2] do joint empirical risk minimization and treat the multi-response problem by introducing a low-dimensional subspace which is common to all the response variables. [7] construct a multivariate Gaussian prior with a full covariance matrix for a set of "similar" supervised learning tasks and then use semidefinite programming (SDP) to combine these estimates and learn a good prior for the current learning task. Some traditional methods such as neural networks also share parameters between the different tasks [1]. However, none of the above methods does feature selection. This limits their applicability in domains such as genomics, where often only a handful of the thousands of potential features are predictive so that feature selection is very important.

W. Buntine et al. (Eds.): ECML PKDD 2009, Part I, LNAI 5781, pp. 276–289, 2009.

There has been some work on feature selection for multi-task learning. [8] uses maximum-entropy discrimination to select a single subset of features across multiple SVM regression or classification problems that share a common set of potential features. Several other papers work within the framework of regularized regression, taking the penalty term to be an ℓ_1 norm over features of an ℓ_p norm over the coefficients for each feature (an "$\ell_1 - \ell_p$" penalty). [9] consider the case $p = \infty$, while [4,10] use $p = 2$. [4] show that the general subspace selection problem can be formulated as an optimization problem involving the trace norm. [10] focus on the case where the trace norm is not required and instead use a homotopy-based approach to evaluate the entire regularization path efficiently [11]. The idea behind $\ell_1 - \ell_p$ penalties is that when $p > 1$, the cost of making a coefficient nonzero is smaller for features that are shared across more tasks. Indeed, for either $p = 2$ or $p = \infty$, these algorithms tend in practice to yield nonzero coefficients for all of the tasks associated with features that get selected.

Our approach is different. Unlike the above algorithms, we not only select features but select tasks for each feature. This amounts to an $\ell_0 - \ell_0$ penalty ("two-level" sparsity), with one ℓ_0 penalty on features and the other on the associated tasks.[1] Optimization with an exact ℓ_0 penalty requires subset selection, known to be NP-hard [12], but a close solution can be found by stepwise search. Though approximate, stepwise ℓ_0 methods generally yield sparser models than exact ℓ_1 methods [13].

In this paper, we use the information-theoretic Minimum Description Length (MDL) principle [14] to derive an efficient coding scheme for multitask stepwise regression that we call the Multiple Inclusion Criterion (MIC). MIC gives improved performance in prediction for multiple related tasks when there are many candidate features of which only a few are predictive and the predictive features are shared by different tasks. Because it allows each feature to be considered relevant for only a subset of the tasks, MIC achieves sparser models than methods which always share features across all tasks.

The rest of the paper is organized as follows. After outlining the notation used, we describe the MIC coding scheme and its properties in detail. We then present experimental results on synthetic and real datasets, and conclude.

2 Multiple Inclusion Criterion (MIC) for Feature Selection

2.1 Notation Used

The symbols used throughout this section are defined in the Table 1. All the values in the table are given by data except m^*, which is unknown. In particular, we have an $n \times h$ response matrix $\mathbf{Y}$, with a shared $n \times m$ feature matrix $\mathbf{X}$.

[1] Another case of two-level sparsity could be an $\ell_1 - \ell_1$ penalty, but as [10] note, this is equivalent to a single ℓ_1 penalty over the entire set of features and fails to share information across tasks.

Table 1. Symbols used and their definitions

Symbol	Meaning
n	Number of observations
m	Number of candidate features
m^*	Number of beneficial features
h	Total number of tasks
k	Number of tasks into which a feature has been added
j	Index of feature
ν	Index of observation

2.2 A Brief Overview of MIC

In general, penalized likelihood methods like MIC aim to minimize an objective function of the form

$$\text{score} = -\log(\text{likelihood of } \mathbf{Y} \text{ given } \mathbf{X}) + F \times q, \tag{1}$$

where q is the current number of features in the model. As [15] notes, various penalties F have been proposed, including $F = 1$, corresponding to AIC (Akaike Information Criterion), $F = \frac{\ln n}{2}$, corresponding to BIC (Bayesian Information Criterion), and $F = \ln m$, corresponding to RIC (Risk Inflation Criterion—similar to a "Bonferroni correction") [16].

Each of these penalties can be interpreted within the framework of the Minimum Description Length (MDL) principle [14]. MDL envisions a "sender," who knows $\mathbf{X}$ and $\mathbf{Y}$, and a "receiver," who knows only $\mathbf{X}$. In order to transmit $\mathbf{Y}$ using as few bits as possible, the sender encodes not the raw $\mathbf{Y}$ matrix but instead a model for $\mathbf{Y}$ given $\mathbf{X}$, followed by the residuals of $\mathbf{Y}$ about that model. The length S of this message, in bits, is called the *description length* and is the sum of two components. The first is S_E, the number of bits for encoding the residual errors, which according to standard MDL is given by the negative log-likelihood of the data given the model; note that this is the first term of (1). The second component, S_M, is the number of bits used to describe the model itself and can be seen as corresponding to the second term of (1). We use the phrase *total description length* (TDL) to denote the combined length of the message for all h tasks.

In the setting of multiple responses,[2] we select features for the h tasks simultaneously to minimize S. Thus, when we evaluate a feature for addition into the model, we want to maximize the reduction of TDL ΔS^k incurred by adding that feature to a subset k of the h tasks ($1 \le k \le h$):

$$\Delta S^k = \Delta S_E^k - \Delta S_M^k$$

[2] We interchangeably use the terms "multiple responses" and "multiple tasks," though our setting is more specific than that of standard multitask problems in which each task may have a different feature matrix; we assume all tasks share the same feature matrix.

where $\Delta S_E^k > 0$ is the reduction in residual-error coding cost due to the data likelihood increase given the new feature, and $\Delta S_M^k > 0$ is the increase in model cost to encode the new feature.[3]

As will be seen in Section 2.3, MIC's model cost includes a component for coding feature coefficients that resembles the AIC or BIC penalty, plus a component for specifying which features are present in the model that resembles the RIC penalty.

In Section 2.4 we compare three approaches to stepwise regression with multiple tasks: (1) a feature is added to all tasks or none; (2) features are added independently to each task (i.e., no transfer); or (3) features can be added to a subset of tasks but the tasks share strength. We first describe a code for approach (3) below.

2.3 Coding Schemes for MIC

Code ΔS_{jE}^k. Let **E** be the residual error matrix:

$$\mathbf{E} = \mathbf{Y} - \hat{\mathbf{Y}},$$

where $\mathbf{Y}$ and $\hat{\mathbf{Y}}$ are the $n \times h$ response and prediction matrices, respectively.

ΔS_{jE}^k is the decrease in negative log-likelihood that results from adding feature j to some subset k of the h tasks. If all the tasks were independent, then ΔS_{jE}^k would simply be the sum of the changes in negative log-likelihood for each of the h models separately. However, we may want our model to allow for nonzero covariance among the tasks. This is particularly true for stepwise regression, because in the first iterations of a stepwise algorithm, the effects of features not present in the model show up as part of the "noise" error term, and if two tasks share a feature not yet in the model, the portion of the error term due to that feature will be the same.

Thus, letting ϵ_ν, $\nu = 1, 2, \ldots, n$, denote the error for the νth row of **E**, we assume $\epsilon_\nu \overset{\text{i.i.d.}}{\sim} \mathcal{N}(0, \Sigma)$, with Σ an $h \times h$ covariance matrix. In other words,

$$P(\epsilon_\nu) = \frac{1}{\sqrt{(2\pi)^h |\Sigma|}} \exp\left(-\frac{1}{2}\epsilon_\nu^T \Sigma^{-1} \epsilon_\nu\right)$$

in which $(\cdot)^T$, $(\cdot)^{-1}$, and $|\cdot|$ are the matrix transpose, inverse, and determinant, respectively. Therefore,

$$\begin{aligned} S_E &= -\log \prod_{\nu=1}^{n} P(\epsilon_\nu) \\ &= \frac{n}{2}\log\left((2\pi)^h |\Sigma|\right) + \frac{1}{2\ln 2}\sum_{\nu=1}^{n} \epsilon_\nu^T \Sigma^{-1} \epsilon_\nu, \end{aligned} \tag{2}$$

[3] ΔS_E^k is always greater than zero, because even a spurious feature will slightly increase the data likelihood.

where the $\frac{1}{\ln 2}$ factor appears because we use logarithm base 2 (here and throughout the remainder of the paper). Note that the superscript k in ΔS^k_{jE} indicates that the reduction is incurred by adding a new feature to k tasks, but the calculation ΔS^k_{jE} is over all h tasks; i.e., the whole residual error $\mathbf{E}$ is taken into account.

Code $\boldsymbol{\Delta S^k_{jM}}$. To describe ΔS^k_{jM} when a feature is added, MIC uses a three-part coding scheme:

$$\Delta S^k_{jM} = \ell_I + \ell_H + \ell_{\hat{\theta}},$$

where ℓ_I is the number of bits needed to describe which feature is being added, ℓ_H is the cost of specifying the subset (k) of the h task models in which to include the feature, and $\ell_{\hat{\theta}}$ is the description length of the k nonzero feature coefficients. We now consider different coding schemes for ℓ_I, ℓ_H, and $\ell_{\hat{\theta}}$.

Code ℓ_I. For most data and feature sets, little is known *a priori* about which features will be beneficial.[4] We therefore assume that if a feature x_j is beneficial, its index j is uniformly distributed over $\{1, 2, \ldots, m\}$. This implies $\ell_I = \log m$ bits to encode the index, reminiscent of the RIC penalty for equation (1).

RIC often uses no bits to code the coefficients of the features that are added, based on the assumption that m is so large that the $\log m$ term dominates. This assumption is not valid in the multiple response setting, where the number of models h could be large. If a feature is added to k of the h tasks, the cost of encoding the k coefficients may be a major part of the cost. We describe the cost to code a coefficient below.

Code $\ell_{\hat{\theta}}$. This term corresponds to the number of bits required to code the value of the coefficient of each feature. We could use either AIC or the more conservative BIC to code the coefficients. As explained below, we use 2 bits for each coefficient, similar to AIC.

Given a model, MDL chooses the values of the coefficients that maximize the likelihood of the data. [18] proposes approximating $\hat{\theta}$, the Maximum Likelihood Estimate (MLE), using a grid resolved to the nearest standard error. That is, instead of specifying $\hat{\theta}$, we imagine encoding a rounded-integer value of $\hat{\theta}$'s z-score $\hat{z}$, where $\hat{\theta} = \hat{\theta}_0 + \hat{z}\ \mathrm{SE}(\hat{\theta})$, with $\hat{\theta}_0$ being the default, null-hypothesis value (here, 0) and $\mathrm{SE}(\hat{\theta})$ being the standard error of $\hat{\theta}$.

We assume a "universal prior" distribution for $\hat{z}$, in which half of the probability is devoted to the null value $\hat{\theta}_0$ and the other half is concentrated near $\hat{\theta}_0$ and decays slowly. In particular, for $\hat{\theta} \neq \hat{\theta}_0$, the coding cost is $2 + \log^+ |\hat{z}|$ $+ 2\log^+ \log^+ |\hat{z}|$ bits, where $\log^+$ denotes logarithm base 2 if the argument is positive, else 0. This prior distribution makes sense in hard problems of feature selection where beneficial features are just marginally significant. Since $\hat{z}$ is quite small in such hard problems, the 2 bits will dominate the other two terms. In fact, we simply assume $\ell_{\hat{\theta}} = 2$.

[4] Following [17], we define a "beneficial" feature as one which, if added to the model, would reduce error on a hypothetical infinite test set.

Code ℓ_H. In order to specify the subset of task models that include a given feature, we encode two pieces of information: First, how many of the tasks (k) have the feature? Second, which subset of k tasks are those?

One way to encode k is to use $\log h$ bits to specify an integer in $\{1, 2, \ldots, h\}$; this implicitly corresponds to a uniform prior distribution on k. However, since we generally expect that smaller values of k are more likely, we instead use coding lengths inspired by the "idealized universal code for the integers" of [19] and [18]: The cost to code k is $\log^* k + c_h$, where $\log^* k = \log^+ k + \log^+ \log^+ k + \log^+ \log^+ \log^+ k + \ldots$, and c_h is the constant required to normalize the implied probability distribution over $\{1, 2, \ldots, h\}$. $c_\infty \approx \log 2.865 \approx 1.516$ [18], but for $h \in \{5, \ldots, 1000\}$, $c_h \approx 1$.

Given k, there are $\binom{h}{k}$ possible subsets of tasks. Taking the subsets to have some pre-specified ordering, we can list the index of our desired subset within that ordering using $\log \binom{h}{k}$ bits.

Thus, in total, we have

$$\ell_H = \log^* k + c_h + \log \binom{h}{k}.$$

2.4 Comparison of the Coding Schemes

The preceding discussion outlined a coding scheme for what we might call "Partially Dependent MIC," or "Partial MIC," in which models for different tasks can share some or all features.

As suggested at the end of Section 2.2, we can also consider a "Fully Dependent MIC," or "Full MIC," scheme in which each feature is shared across all or none of the task models. This amounts to a restricted Partial MIC in which $k = 0$ or $k = h$ for each feature. The advantage comes in not needing to specify the subset of tasks used, saving ℓ_H bits for each feature in the model; however, Full MIC may need to code more coefficient values than Partial MIC.

A third coding scheme is simply to specify each task model in isolation from the others. We call this the "RIC" approach, because each model pays $\log m$ bits for each feature to code its index; this is equivalent, up to the base of the logarithm, to the $F = \ln m$ penalty in equation 1. (However, we include an additional cost of $\ell_{\hat{\theta}}$ bits to code a coefficient.) If the sum of the two costs is sufficiently less than the bits saved by the increase of the data likelihood from adding the feature to the model, the feature will be added to the model. RIC assumes that the beneficial features are not significantly shared across tasks.

We compare the relative coding costs under these three coding schemes for the case where we evaluate a hypothetical feature, x_j, that is beneficial for k tasks and spurious for the remaining $h - k$ tasks. Suppose that Partial MIC and RIC both add the feature to only the k beneficial tasks, while Full MIC adds it to all h tasks. We assume that if the feature is added, the three methods save approximately the same number of bits in encoding residual errors, ΔS_E^k. This would happen if, say, the additional $h - k$ coefficients that Full MIC adds to its models save a negligible number of residual-coding bits (because those features

Table 2. Costs in bits for each of the three schemes to code a model with $k = 1$, $k = \frac{h}{4}$, and $k = h$ nonzero coefficients. $m \gg h \gg 1$, $\ell_I = \log m$, $\ell_{\hat{\theta}} = 2$, and for $h \in \{5, \ldots, 1000\}$, $c_h \approx 1$. Examples of these values for $m = 2{,}000$ and $h = 20$ appear in brackets; the smallest of the costs appears in bold.

k	Partial MIC		Full MIC		RIC	
1	$\log m + c_h + \log h + 2$	[18.4]	$\log m + 2h$	[51.0]	$\log m + 2$	**[13.0]**
$\frac{h}{4}$	$\log m + \log^* \left(\frac{h}{4}\right) + c_h + \log \binom{h}{h/4} + \frac{h}{2}$	**[39.8]**	$\log m + 2h$	[51.0]	$\frac{h}{4} \log m + \frac{h}{2}$	[64.8]
h	$\log m + \log^* h + c_h + 2h$	[59.7]	$\log m + 2h$	**[51.0]**	$h \log m + 2h$	[259.3]

are spurious) and if the estimate for Σ is sufficiently diagonal that the negative log-likelihood calculated using (2) for Partial MIC approximately equals the sum of the negative log-likelihoods that RIC calculates for each response separately.

Table 2 shows that RIC and Partial MIC are the best and the second best coding schemes when $k = 1$, and that their difference is on the order of $\log h$. Full MIC and Partial MIC are the best and the second best coding schemes when $k = h$, and their difference is on the order of $\log^* h$. Partial MIC is best for $k = \frac{h}{4}$.

2.5 Stepwise Search Method

To find a model that minimizes TDL, we use a modified greedy stepwise search as shown in Algorithm 1. For each feature, we evaluate the change in TDL that would result from adding that feature to the model with the optimal number of associated tasks. We add the best feature and then recompute the changes in TDL for the remaining features. This continues until there are no more features that would reduce TDL if added. The number of evaluations of features for possible addition is thus $\mathcal{O}(mm_s)$, where m_s is the number of features eventually added.

To select the optimal number of task models (k) in which to include a given feature, we again use a stepwise-style search. In this case, we evaluate the reduction in TDL that would result from adding the feature to each task, add the feature to the best task, recompute the reduction in TDL for the remaining tasks, and continue.[5] However, unlike a normal stepwise search, we continue this process until we have added the feature to all h task models. The reason for this is two-fold. First, because we want to borrow strength across tasks, we need to avoid overlooking cases where the correlation of a feature with any single task is insufficiently strong to warrant addition, yet the correlations with all of the tasks are. Second, the $\log \binom{h}{k}$ term in Partial MIC's coding cost does not increase monotonically with k, so even if adding the feature to an intermediate number of

[5] A stepwise search that re-evaluates the quality of each task at each iteration is necessary because, if we take the covariance matrix Σ to be nondiagonal, the values of the residuals for one task may affect the likelihood of residuals for other tasks. If we take Σ to be diagonal, as we do in Section 3, then an $\mathcal{O}(h)$ search through the tasks without re-evaluation suffices.

Algorithm 1. Partial MIC

Include the intercept (feature number 1) in all h response models.
$remaining\_features = \{2, \ldots, m\}$.
$keep\_adding\_features =$ true.
while $keep\_adding\_features$ **do**
  for j in $remaining\_features$ **do**
    // Find the best subset of response models to which to add feature j.
    for $k = 1$ to h **do**
      Try including feature j in the best k response models. (We greedily assume that the best k responses are the union of the best $k-1$ responses with the remaining response that, if included, would most increase likelihood.)
      Compute ΔS_{jE}^{k}, the decrease in data residual cost, and ΔS_{jM}^{k}, the resulting increase in model-coding cost, relative to not including feature j in any response models.
    end for
    Let k_j be the value of k that maximizes $\Delta S_{jE}^{k} - \Delta S_{jM}^{k}$.
    $\Delta S_j := \Delta S_{jE}^{k_j} - \Delta S_{jM}^{k_j}$.
  end for
  Let j^* be the feature j that maximizes ΔS_j, the reduction in TDL for adding feature j.
  if $\Delta S_{j^*} > 0$ **then**
    Add feature j^* to the appropriate k_{j^*} response models.
    $remaining\_features = remaining\_features - \{j^*\}$.
  else
    $keep\_adding\_features =$ false.
  end if
end while

tasks does not look promising, adding it to all of them might still be worthwhile. Thus, when evaluating a given feature, we compute the description length of the model $\mathcal{O}(h^2)$ times. Since we need to identify the optimal k for each feature evaluation, the entire algorithm requires $\mathcal{O}(h^2 m m_s)$ evaluations of TDL.

While not shown explicitly in Algorithm 1, we use two branch-and-bound-style optimizations to cut this cost significantly in practice:

1. Before searching through subsets of responses to find the optimal subset for each feature, we make an $\mathcal{O}(m)$ sweep through the features to compute an upper bound on the decrease in TDL that could result from adding that feature as

$$\text{(decrease in TDL if the feature is added to all } h \text{ response models)} - \log m. \tag{3}$$

   Here, the first term is an upper bound on the benefit of adding the feature to the optimal number of response models (since adding a feature can only make a model fit better), and the second term underestimates the model cost of adding the feature, regardless of how many response models would actually be used. We sort the features in decreasing order by this upper

bound, and when we reach features whose upper bounds are less than the best actual decrease in TDL observed so far, we terminate the search early.
2. For the stepwise search over responses, we can bound from above the potential benefit of adding the feature to k response models as

$$\begin{aligned}&\text{(decrease in TDL if the feature is added to all } h \text{ response models)}\\&-\left(\log^* k + c_k + \log\binom{h}{k} + 2k\right),\end{aligned} \tag{4}$$

where the subtracted term represents the coding cost of including the feature in k response models. We can stop the search early when no higher value of k has an upper bound that exceeds the best reduction in TDL seen so far for any feature's response subset.[6]

Although we did not attempt to do so, it may be possible to formulate MIC using a *regularization path*, or *homotopy*, algorithm of the sort that have become popular for performing ℓ_1 regularization without the need for cross-validation (e.g., [20]). If possible, this would be significantly faster than stepwise search.

3 Experimental Results

This section evaluates the MIC approach on three synthetic datasets, each of which is designed to match the assumptions of, respectively, the Partial MIC, Full MIC, and RIC coding schemes described in Section 2.4. We also test on two biological data sets, a Yeast Growth dataset [21], which consists of real-valued growth measurements of multiple strains of yeast under different drug conditions, and a Breast Cancer dataset [22], which involves predicting prognosis, ER status, and three other descriptive variables from gene-expression values for different cell lines.

We compare the three coding schemes of Section 2.4 against two other multitask algorithms: "AndoZhang" [2] and "BBLasso" [10], as implemented in the Berkeley Transfer Learning Toolkit [23]. We did not compare MIC with other methods from the toolkit as they all require the data to have additional structure, such as *meta-features* [6,7], or expect the features to be frequency counts, such as for the Hierarchical Dirichlet Processes algorithm. Also, none of the neglected methods does feature selection.

For AndoZhang we use 5-fold CV to find the best value of the parameter that [2] call h (the dimension of the subspace Θ, not to be confused with h as we use it in this paper). We tried values in the range $[1, 100]$ as is done in [2].

MIC as presented in Section 2.3 is a regression algorithm, but AndoZhang and BBLasso are both designed for classification. Therefore, we made each of our responses binary 0/1 values before applying MIC with a regular regression

[6] We say "no higher value of k" rather than "the next higher value of k" because (4) does not decrease monotonically with k, due to the $\log\binom{h}{k}$ quantity.

likelihood term. Once the features were selected, however, we used logistic regression applied to just those features to obtain MIC's actual model coefficients.[7]

As noted in Section 2.3, MIC's negative log-likelihood term can be computed with an arbitrary $h \times h$ covariance matrix Σ among the h tasks. On the data sets in this paper, we found that estimating all h^2 entries of Σ could lead to overfitting, so we instead took Σ to be diagonal. Informal experiments showed that estimating Σ as a convex combination of the full and diagonal estimates could also work well.

3.1 Evaluation on Synthetic Datasets

We created synthetic data according to three separate scenarios—called Partial, Full, and Independent. For each scenario, we generated a matrix of continuous responses as

$$\mathbf{Y}_{n\times h} = \mathbf{X}_{n\times m}\mathbf{w}_{m\times h} + \epsilon_{n\times h}$$

where m = 2,000 features, h = 20 responses, and n = 100 observations. Then, to produce binary responses, we set to 1 those response values that were greater than or equal to the average value for their column and set to 0 the rest; this produced a roughly 50-50 split between 1's and 0's because of the normality of the data. Each nonzero entry of w was i.i.d. $\mathcal{N}(0, 1)$, and entry of ϵ was i.i.d. $\mathcal{N}(0, 0.1)$, with no covariance among the ϵ entries for different tasks. Each task had $m^* = 4$ beneficial features, i.e., each column of w had 4 nonzero entries.

The scenarios differed according to the distribution of the beneficial features in w.

- In the Partial scenario, the first feature was shared across all 20 responses, the second was shared across the first 15 responses, the third across the first 10 responses, and the fourth across the first 5 responses. Because each response had four features, those responses $(6 - 20)$ that did not have all of the first four features had other features randomly distributed among the remaining features (5, 6, ..., 2000).
- In the Full scenario, each response shared exactly features $1 - 4$, with none of features $5 - 2000$ being part of the model.
- In the Independent scenario, each response had four random features among $1, \ldots, 2000$.

For the synthetic data, we report precision and recall to measure the quality of feature selection. This can be done both at a coefficient level (Was each nonzero coefficient in w correctly identified as nonzero, and vice versa?) and at an overall feature level (For features with *any* nonzero coefficients, did we correctly identify them as having nonzero coefficients for any of the tasks, and vice versa?). Note that Full MIC and BBLasso always make entire rows of their estimated w matrices nonzero and so tend to have larger numbers of nonzero coefficients. Table 3 shows the performance of each of the methods on five instances of the

[7] This contrasts with BBLasso, which in the default Transfer Learning Toolkit implementation uses the original regression coefficients (mostly zeros).

Table 3. Test-set accuracy, precision, and recall of MIC and other methods on 5 instances of various synthetic data sets generated as described in Section 3.1. Standard errors are reported over each task; that is, with 5 data sets and 20 tasks per data set, the standard errors represent the sample standard deviation of 100 values divided by $\sqrt{100}$ (except for feature-level results, which apply only to entire data sets and so are divided by $\sqrt{5}$). Baseline accuracy, corresponding to guessing the majority category, is roughly 0.5. *Note:* AndoZhang's NA values are due to the fact that it does not explicitly select features.

Method	True Model	Partial MIC	Full MIC	RIC	BBLasso	AndoZhang
			Partial Synthetic Dataset			
Test error	0.07 ± 0.00	$\mathbf{0.10 \pm 0.00}$	0.17 ± 0.01	0.12 ± 0.01	0.19 ± 0.01	0.50 ± 0.02
Coeff. precision	1.00 ± 0.00	0.84 ± 0.02	0.26 ± 0.01	0.84 ± 0.02	0.04 ± 0.00	NA
Coeff. recall	1.00 ± 0.00	0.77 ± 0.02	0.71 ± 0.03	0.56 ± 0.02	0.81 ± 0.02	NA
Feature precision	1.00 ± 0.00	0.99 ± 0.01	0.97 ± 0.02	0.72 ± 0.05	0.20 ± 0.03	NA
Feature recall	1.00 ± 0.00	0.54 ± 0.05	0.32 ± 0.03	0.62 ± 0.04	0.54 ± 0.01	NA
			Full Synthetic Dataset			
Test error	0.07 ± 0.00	$\mathbf{0.08 \pm 0.00}$	$\mathbf{0.08 \pm 0.00}$	0.11 ± 0.01	0.09 ± 0.00	0.45 ± 0.02
Coeff. precision	1.00 ± 0.00	0.98 ± 0.01	0.80 ± 0.00	0.86 ± 0.02	0.33 ± 0.03	NA
Coeff. recall	1.00 ± 0.00	1.00 ± 0.00	1.00 ± 0.00	0.63 ± 0.02	1.00 ± 0.00	NA
Feature precision	1.00 ± 0.00	0.80 ± 0.00	0.80 ± 0.00	0.36 ± 0.06	0.33 ± 0.17	NA
Feature recall	1.00 ± 0.00	1.00 ± 0.00	1.00 ± 0.00	1.00 ± 0.00	1.00 ± 0.00	NA
			Independent Synthetic Dataset			
Test error	0.07 ± 0.00	0.17 ± 0.01	0.36 ± 0.01	$\mathbf{0.13 \pm 0.01}$	0.35 ± 0.01	0.49 ± 0.00
Coeff. precision	1.00 ± 0.00	0.95 ± 0.01	0.06 ± 0.01	0.84 ± 0.02	0.02 ± 0.00	NA
Coeff. recall	1.00 ± 0.00	0.44 ± 0.02	0.15 ± 0.02	0.58 ± 0.02	0.43 ± 0.02	NA
Feature precision	1.00 ± 0.00	1.00 ± 0.00	1.00 ± 0.00	0.83 ± 0.02	0.30 ± 0.05	NA
Feature recall	1.00 ± 0.00	0.44 ± 0.02	0.14 ± 0.02	0.58 ± 0.03	0.42 ± 0.06	NA

Partial, Full, and Independent synthetic data sets. On the Partial data set, *Partial MIC* performed the best, closely followed by *RIC*; on the Full synthetic data, *Full MIC* and *Partial MIC* performed equally well; and on the Independent synthetic data, the *RIC* algorithm performed the best closely followed by *Partial MIC*. It is also worth noting that the best-performing methods tended to have the best precision and recall on coefficient selection. The performance trends of the three methods are in consonance with the theory of Section 2.4.

The table shows that only in one of the three cases does one of these methods compete with MIC methods. BBLasso on the Full synthetic data shows comparable performance to the MIC methods, but even in that case it has a very low feature precision, since it added many more spurious features than the MIC methods.

3.2 Evaluation on Real Datasets

This section compares the performance of MIC methods with AndoZhang and BBLasso on a Yeast dataset and Breast Cancer dataset. These are typical of biological datasets in that only a handful of features are predictive from thousands

Table 4. Accuracy and number of coefficients and features selected on five folds of CV for the Yeast and Breast Cancer data sets. For the Yeast data, $h = 20$, $m = 6{,}715$, $n = 104$. For the Breast Cancer data, $h = 5$, $m = 5{,}000$, $n = 100$. Standard errors are over the five CV folds; i.e., they represent (sample standard deviation) / $\sqrt{5}$. *Note:* AndoZhang's NA values are due to the fact that it does not explicitly select features.

Method	Partial MIC	Full MIC	RIC	BBLasso	AndoZhang
			Yeast Dataset		
Test error	$\mathbf{0.38 \pm 0.04}$	0.39 ± 0.04	0.41 ± 0.05	0.43 ± 0.03	0.39 ± 0.03
Num. coeff. sel.	22 ± 4	64 ± 4	9 ± 1	1268 ± 279	NA
Num. feat. sel.	4 ± 0	3 ± 0	9 ± 1	63 ± 14	NA
			Breast Cancer Dataset		
Test error	$\mathbf{0.33 \pm 0.08}$	0.37 ± 0.08	0.36 ± 0.08	$\mathbf{0.33 \pm 0.08}$	0.44 ± 0.03
Num. coeff. sel.	3 ± 0	11 ± 1	2 ± 0	61 ± 19	NA
Num. feat. sel.	2 ± 0	2 ± 0	2 ± 0	12 ± 4	NA

of potential features. This is precisely the case in which MIC outperforms other methods. MIC not only gives better accuracy but does so by choosing fewer features than BBLasso's $\ell_1 - \ell_2$-based approach.

Yeast Dataset. Our Yeast dataset comes from [21]. It consists of real-valued growth measurements of 104 strains of yeast ($n = 104$ observations) under 313 drug conditions. In order to make computations faster, we hierarchically clustered these 313 conditions into 20 groups using correlation as the similarity measure. Taking the average of the values in each cluster produced $h = 20$ real-valued responses (tasks), which we then binarized into two categories: values at least as big as the average for that response (set to 1) and values below the average (set to 0).[8] The features consisted of 526 markers (binary values indicating major or minor allele) and 6,189 transcript levels in rich media for a total of $m = 6{,}715$ features.

Table 4 shows test errors from 5-fold CV on this data set. As can be seen from the table, Partial MIC performs better than BBLasso and AndoZhang. RIC and Full MIC perform slightly worse than Partial MIC, underscoring the point that it is preferable to use a more general MIC coding scheme compared to Full MIC or RIC. The latter methods have strong underlying assumptions, which cannot always correctly capture sharing across tasks. Like Partial MIC, AndoZhang did well on this data set; however, because the algorithm scales poorly with large numbers of tasks, the computation took 39 days (at least using a naïve extension of the default Transfer Learning Toolkit implementation).

Breast Cancer Dataset. Our second data set pertains to Breast Cancer, containing data from five of the seven data sets used in [22]. It contains 1,171

[8] The split was not exactly 50-50, as the data were sometimes skewed, with the mean falling above or below the median.

observations for 22,268 RMA-normalized gene-expression values. We considered five associated responses (tasks); two were binary—prognosis ("good" or "poor") and ER status ("positive" or "negative")—and three were not—age (in years), tumor size (in mm), and grade (1, 2, or 3). We binarized the three non-binary responses into two categories: Response values at least as high as the average, and values below the average. Finally we scaled the dataset down to $n = 100$ and $m = 5{,}000$ (the 5,000 features with the highest variance), to save computational resources. Table 4 shows test errors from 5-fold CV on this data set. As is clear from the table, Partial MIC and BBLasso are the best methods here. But as was the case with other datasets, BBLasso puts in more features, which is undesirable in domains (like biology and medicine) where simpler and hence more interpretable model are sought.

4 Conclusion

We proposed a novel coding scheme, MIC, for feature selection in the presence of multiple related tasks. MIC is a penalized-likelihood method based on the principle of Minimum Description Length (MDL). Sharing across tasks happens at two levels in MIC. Firstly, we (optionally) build a shared covariance matrix for the tasks (important when the various tasks are nearly identical), and secondly, we use a shared coding scheme to specify which of the tasks (models) each feature is added to. Among many attractive properties of the method is its immunity to overfitting and the absence of any free parameters that might require cross validation, unlike ℓ_1 regularization methods. Results on synthetic and real data demonstrate that MIC performs better than state-of-the-art Multi-Task Learning Methods, not only in terms of accuracy but also in terms of simplicity of the resulting models. MIC works best when there are many features of which only a few are relevant, a situation that occurs commonly in computational biology and bioinformatics applications.

References

1. Caruana, R.: Multitask learning. In: Machine Learning, pp. 41–75 (1997)
2. Ando, R., Zhang, T.: A Framework for Learning Predictive Structures from Multiple Tasks and Unlabeled Data. The Journal of Machine Learning Research 6, 1817–1853 (2005)
3. Jacob, L., Bach, F., Vert, J.P.: Clustered multi-task learning: A convex formulation. In: Koller, D., Schuurmans, D., Bengio, Y., Bottou, L. (eds.) Advances in Neural Information Processing Systems, vol. 21 (2009)
4. Argyriou, A., Evgeniou, T., Pontil, M.: Convex multi-task feature learning. Mach. Learn. 73(3), 243–272 (2008)
5. Ben-David, S., Borbely, R.S.: A notion of task relatedness yielding provable multiple-task learning guarantees. Mach. Learn. 73(3), 273–287 (2008)

6. Lee, S.I., Chatalbashev, V., Vickrey, D., Koller, D.: Learning a meta-level prior for feature relevance from multiple related tasks. In: ICML 2007: Proceedings of the 24th international conference on Machine learning, pp. 489–496. ACM, New York (2007)
7. Raina, R., Ng, A.Y., Koller, D.: Constructing informative priors using transfer learning. In: ICML 2006, pp. 713–720. ACM, New York (2006)
8. Jebara, T.: Multi-task feature and kernel selection for SVMs. In: ICML 2004, ACM Press, New York (2004)
9. Turlach, B., Venables, W., Wright, S.: Simultaneous variable selection. Technometrics 47(3), 349–363 (2005)
10. Obozinski, G., Taskar, B., Jordan, M.I.: Joint covariate selection and joint subspace selection for multiple classification problems. Statistics and Computing (2009)
11. Efron, B., Hastie, T., Johnstone, L., Tibshirani, R.: Least angle regression. Annals of Statistics 32, 407–499 (2004)
12. Natarajan, B.: Sparse approximate solutions to linear systems. SIAM journal on computing 24, 227 (1995)
13. Lin, D., Pitler, E., Foster, D.P., Ungar, L.H.: In defense of ℓ_0. In: Workhsop on Feature Selection at International Conference on Machine Learning, ICML 2008 (2008)
14. Rissanen, J.: Hypothesis selection and testing by the mdl principle. The Computer Journal 42, 260–269 (1999)
15. George, E., Foster, D.: Calibration and empirical Bayes variable selection. Biometrika 87(4), 731–747 (2000)
16. Foster, D.P., George, E.I.: The risk inflation criterion for multiple regression. The Annals of Statistics 22(4), 1947–1975 (1994)
17. Zhou, J., Foster, D., Stine, R., Ungar, L.: Streamwise feature selection. The Journal of Machine Learning Research 7, 1861–1885 (2006)
18. Rissanen, J.: A universal prior for integers and estimation by minimum description length. Annals of Statistics 11(2), 416–431 (1983)
19. Elias, P.: Universal codeword sets and representations of the integers. IEEE Transactions on Information Theory 21(2), 194–203 (1975)
20. Friedman, J.: Fast Sparse Regression and Classification (2008)
21. Litvin, O., Causton, H.C., Chen, B., Pe'er, D.: Special feature: Modularity and interactions in the genetics of gene expression. Proceedings of the National Academy of Sciences of the United States of America (February 2009) PMID: 19223586
22. van't Veer, L.J., Dai, H., van de Vijver, M.J., He, Y.D., Hart, A.A., Mao, M., Peterse, H.L., van der Kooy, K., Marton, M.J., Witteveen, A.T., Schreiber, G.J., Kerkhoven, R.M., Roberts, C., Linsley, P.S., Bernards, R., Friend, S.H.: Gene expression profiling predicts clinical outcome of breast cancer. Nature 415(6871), 530–536 (2002)
23. Kao, W.C., Rakhlin, A.: Transfer learning toolkit (2007), `http://multitask.cs.berkeley.edu`

Kernel Polytope Faces Pursuit

Tom Diethe and Zakria Hussain

Department of Computer Science, University College London
{t.diethe,z.hussain}@cs.ucl.ac.uk

Abstract. Polytope Faces Pursuit (PFP) is a greedy algorithm that approximates the sparse solutions recovered by ℓ_1 regularised least-squares (Lasso) [4,10] in a similar vein to (Orthogonal) Matching Pursuit (OMP) [16]. The algorithm is based on the geometry of the polar polytope where at each step a basis function is chosen by finding the maximal vertex using a path-following method. The algorithmic complexity is of a similar order to OMP whilst being able to solve problems known to be hard for (O)MP. Matching Pursuit was extended to build kernel-based solutions to machine learning problems, resulting in the sparse regression algorithm, Kernel Matching Pursuit (KMP) [17]. We develop a new algorithm to build sparse kernel-based solutions using PFP, which we call Kernel Polytope Faces Pursuit (KPFP). We show the usefulness of this algorithm by providing a generalisation error bound [7] that takes into account a natural regression loss and experimental results on several benchmark datasets.

Keywords: Polytope Faces Pursuit, Orthogonal Matching Pursuit, Pseudo-dimension, Sample Compression Bounds, Regression, Kernel methods.

1 Introduction

Estimating a sparsely represented function from a set of training examples is a classical problem in regression. Sparsity of representation is an important issue, for reasons of computational efficiency and for its influence on generalisation performance [5,6]. Suppose we are given a sequence of observations $\mathbf{X} = [\mathbf{x}_1, \ldots, \mathbf{x}_m], \mathbf{x}_i \in \mathbb{R}^n$ with corresponding outputs $\mathbf{y} \in \mathbb{R}$. We would like to find a sparse set of weights $\mathbf{w} \in \mathbb{R}^n$ such that the regression loss (e.g. $\|\mathbf{y} - \mathbf{X}'\mathbf{w}\|_2^2$) is minimised. It is thereotically possible to directly enforce sparsity through the ℓ_0 optimisation problem

$$\min_{\mathbf{w}} \|\mathbf{w}\|_0 \quad \text{s.t.} \quad \mathbf{y} = \mathbf{X}'\mathbf{w} \tag{1}$$

Finding this ℓ_0 solution is known to be $NP-$hard. However the equivalent ℓ_1 optimisation problem

$$\min_{\mathbf{w}} \|\mathbf{w}\|_1 \quad \text{s.t.} \quad \mathbf{y} = \mathbf{X}'\mathbf{w} \tag{2}$$

W. Buntine et al. (Eds.): ECML PKDD 2009, Part I, LNAI 5781, pp. 290–301, 2009.

is a convex optimisation problem and can be solved using general purpose solvers. A reformulation of this that directly minimises the regression loss is the Least Absolute Shrinkage and Selection Operator (LASSO) [15], which is given by

$$\min_{\mathbf{w}} \|\mathbf{y} - \mathbf{X}'\mathbf{w}\|_2^2 + \lambda \|\mathbf{w}\|_1 , \tag{3}$$

i.e. a form of ℓ_1-penalised least squares. Basis pursuit [3] uses the Least Angle Regression Solver (LARS) to solve the LASSO problem. The algorithm requires the computation of the full regularisation path through the LARS. However, for large scale problems, this optimisation becomes inefficient and in some cases intractible.

Matching Pursuit was proposed in the signal-processing community as an algorithm that decomposes any signal into a linear expansion of waveforms that are selected from a redundant dictionary of functions [8]. It is a general, greedy, sparse function approximation scheme with the squared error loss, which iteratively adds new functions (i.e. basis functions) to the linear expansion.

If we take as dictionary of functions of the form $K(\cdot, x_i)$ where x_i is the input part of a training example, then the linear expansion has essentially the same form as a Support Vector Machine. Matching Pursuit and its variants were developed primarily in the signal-processing and wavelets community, but there are many interesting links with the research on kernel-based learning algorithms developed in the machine learning community.

Kernel Matching Pursuit (KMP) [17] was developed as a method to estimate a function from training examples in the presence of noise in the context of a Reproducing Kernel Hilbert Space (RKHS). The general idea is to decompose the function to learn on a sparse-optimal set of spanning functions. Unlike Basis Pursuit (BP), which finds the exact ℓ_1 solution, the implementation does not rely on the LASSO formulation or the LARS. We are then effectively finding an approximation to the regression problem in the RKHS defined by the kernel function,

$$\begin{aligned} &\min_{\boldsymbol{\alpha}} \|\boldsymbol{\alpha}\|_1 \\ \text{s.t.} \quad &\mathbf{y} = \mathbf{K}\boldsymbol{\alpha} \end{aligned} \tag{4}$$

where $\boldsymbol{\alpha}$ resembles the dual weight vector found by other kernel methods such as Kernel Ridge Regression [12].

More recently, connections have been made between Matching Pursuit, Kernel-PCA, Sparse Kernel Feature analysis, and how greedy algorithms of this kind can be used to compress the design matrix in SVMs to allow handling of very large data sets [14,13].

Further investigation of the criteria under which ℓ_0/ℓ_1 equivalence holds led to consideration of the $d-$dimensional *polytope* (the $d-$dimensional generalisation of a polygon) [4]. Using this geometric interpretation, a greedy algorithm called Polytope Faces Pursuit (PFP) has been proposed [9] which adopts a path-following approach through the relative interior faces of the polar polytope. In order to generalise this to its kernelised form, we begin by converting (4) into its standard form

$$\min_{\boldsymbol{\alpha}} \|\tilde{\boldsymbol{\alpha}}\|_1 \quad (5)$$
$$\text{s.t.} \quad \mathbf{y} = \tilde{\mathbf{K}}\tilde{\boldsymbol{\alpha}}, \quad \tilde{\boldsymbol{\alpha}} \geq 0$$

where $\tilde{\mathbf{K}} = [\mathbf{K}, -\mathbf{K}]$ and $\tilde{\boldsymbol{\alpha}}$ has $2m$ nonnegative components, with the standard weight vector recoverable by $\alpha_i = \tilde{\boldsymbol{\alpha}}_i - \tilde{\boldsymbol{\alpha}}_{i+m}$ [2]. The corresponding *dual* of this linear program is

$$\max_{\mathbf{c}} \mathbf{y}'\mathbf{c} \quad (6)$$
$$\text{s.t.} \quad \tilde{\mathbf{K}}'\mathbf{c} \leq 1$$

which has an optimal dual weight vector $\mathbf{c}$ which coincides with the optimum $\boldsymbol{\alpha}$ of the primal formulation. Note that in the sense of kernel methods, the formulation (5) is already in the dual space, so the weight vector $\mathbf{c}$ is in fact the dual of the dual. The greedy approach to the solution of (6) then follows the same approach as taken in standard PFP [9]. This will be described further in section 2.2

We present experimental results on real world datasets, which show that KPFP is competitive with the KMP, Kernel Ridge Regression (KRR) and the LARS solver for LASSO on datasets derived from the UCI and STATLOG repositories.

2 Kernel Polytope Faces Pursuit

2.1 Preliminaries

Assume we have a sample S containing examples as paired inputs $\mathbf{x} \in \mathbb{R}^n$ and outputs $y \in \mathbb{R}$. Let $\mathbf{X} = (\mathbf{x}_1, \ldots, \mathbf{x}_m)'$ be the input vectors stored in matrix $\mathbf{X}$ as row vectors, where $'$ denotes the transpose of vectors or matrices. For simplicity we always assume that the examples are already projected into the kernel defined feature space, so that the kernel matrix $\mathbf{K}$ has entries $\mathbf{K}[i,j] = \langle \mathbf{x}_i, \mathbf{x}_j \rangle$. In the analysis section we will explicitly denote the feature map $\phi(\mathbf{x})$ for some vector $\mathbf{x}$. The notation $\mathbf{K}[:,i]$ will denote the ith column of the matrix $\mathbf{K}$. When given a set of indices $\mathbf{i} = \{i_1, \ldots, i_k\}$ (say) then $\mathbf{K}[\mathbf{i},\mathbf{i}]$ denotes the square matrix defined solely by the index set $\mathbf{i}$.

For analysis purposes we assume that the training examples are generated i.i.d. according to an unknown but fixed probability distribution that also governs the generation of the test data. Expectation over the training examples (empirical average) is denoted by $\hat{\mathbb{E}}[\cdot]$, while expectation with respect to the underlying distribution is denoted $\mathbb{E}[\cdot]$.

2.2 Algorithm

We now derive the Kernel Polytope Faces Pursuit (KPFP) algorithm, which is a generalisation of Polytope Faces Pursuit (PFP) to Reproducing Kernel Hilbert Spaces (RKHS).

Table 1. Notation

$\mathbf{X}$	Input matrix
$\mathbf{y}$	Output vector
$\hat{\mathbf{y}}$	Estimates of outputs
$\mathbf{K}$	Kernel matrix with entries $\mathbf{K}_{i,j} =< \phi(x_i), \phi(x_j) >$
$\tilde{\mathbf{K}}$	$[\mathbf{K}, -\mathbf{K}]$
$\boldsymbol{\alpha}$	Dual (sparse) weight vector
$\mathbf{r}$	Vector of residuals
$\mathbf{c}$	Double-dual weight vector
k_{max}	Sparsity parameter
$\mathbf{A}^{\dagger}$	pseudo-inverse of matrix $\mathbf{A}$
$\mathbf{0}, \mathbf{1}$	Vector of zeroes and ones respectively
$\mathbf{I}$	Identity matrix

At each step the approach to the solution of this problem is to identify the optimal vertex which is the maximiser of $\mathbf{y}'\mathbf{c}$, which is similar to the way in which KMP builds up its solution. However the difference is that at each step, the path is constrained on the polytope face F given by the vertex of the previous step. This is achieved by projecting $\mathbf{y}$ into a subspace parallel to F to give $\mathbf{r} = (\mathbf{I} - \mathbf{Q})\mathbf{y}$ where $\mathbf{Q} = \frac{\mathbf{K}[:,i]\mathbf{K}[:,i]'}{|\mathbf{K}[:,i]|^2}$. Since $\boldsymbol{\alpha} = \mathbf{K}[:,i]'^{\dagger}\mathbf{y}$ and $\hat{y} = \mathbf{K}[:,i]\boldsymbol{\alpha}$ we have $\mathbf{r} = \mathbf{y} - \mathbf{K}[:,i]\boldsymbol{\alpha} = \mathbf{y} - \hat{\mathbf{y}}$ meaning that $\mathbf{r}$ is the residual from the approximation at step i.

The second step, which is where the main difference between (O)MP and PFP arises, involves projecting within the face F that has just been found, rather than from the origin. This is done by projecting along the residual $\mathbf{r}$. Therefore to find the next face at each step we find the maximum *scaled* correlation

$$\mathbf{i}_i = \arg\max_{i \notin \mathbf{i}} \tilde{\mathbf{K}}[:,i]'\mathbf{r}/(1 - \tilde{\mathbf{K}}[:,i]'\mathbf{c}) \tag{7}$$

where we only consider bases such that $\tilde{\mathbf{K}}[:,i]'\mathbf{r} > 0$.

We then proceed by removing any constraints that violate the condition that $\tilde{\boldsymbol{\alpha}}$ contains any negative entries. This is achieved by finding $j \in \mathbf{i}$ such that $\tilde{\boldsymbol{\alpha}}_j < 0$, removing j from $\mathbf{i}$ and removing the face from the current solution. We then recaluclate $\tilde{\boldsymbol{\alpha}}$ and continue until $\boldsymbol{\alpha}_j \geq 0, \forall\, j$. Although this step is necessary to provide exact solutions to (6), it may be desirable in some circumstances to remove this step due to the fact that our primal space is in fact the dual space of an RKHS. This would result in faster iterations but less sparse solutions. In section 3 we compare the performance of the algorithm with and without this step. The full algorithm is given in Algorithm 1.

2.3 Generalisation Error Bound

For the generalisation error bound we assume that the data are generated iid from a fixed but unknown probability distribution P over the joint space $\mathcal{X} \times \mathcal{Y}$. Given the *true error* of a function f:

$$\mathrm{err}(f) = \mathbb{E}_{(x,y)\sim P} L(f(x), y),$$

Require: kernel $\mathbf{K}$, sparsity parameter $k > 0$, training outputs $\mathbf{y}$
Initialise $\tilde{\mathbf{K}} = [\mathbf{K}, -\mathbf{K}]$, $\tilde{\alpha} = [\,]$, $\boldsymbol{\alpha} = [\,]$, $\hat{\mathbf{y}} = \mathbf{0}$, $\tilde{A} = [\,]$, $\mathbf{r} = \mathbf{y}$, $\mathbf{c} = \mathbf{0}$
for $i = 1$ to k **do**
  Find face $\mathbf{i}_i = \arg\max_{i \notin \mathbf{i}} \tilde{\mathbf{K}}[:, i]'\mathbf{r}/(1 - \tilde{\mathbf{K}}[:, i]'\mathbf{c})$ where $\tilde{\mathbf{K}}[:, i]'\mathbf{r} > 0$
  Add constraint: $\tilde{\mathbf{A}} = [\tilde{\mathbf{A}}, \tilde{\mathbf{K}}[:, \mathbf{i}_i]]$
  Update $\mathbf{B} = (\tilde{\mathbf{A}})^{\dagger}$, $\tilde{\alpha} = \mathbf{B}\mathbf{y}$
  (Optional) Release violating constraints:
  while $\exists\ \tilde{\alpha}_j < 0, \forall\ j$ **do**
    Remove face j: $\tilde{\mathbf{A}} = \tilde{\mathbf{A}} \setminus \tilde{\mathbf{K}}[:, j]$, $\mathbf{i} = \mathbf{i} \setminus \{j\}$
    Update $\mathbf{B} = \tilde{\mathbf{K}}[:, \mathbf{i}_i]^{\dagger}$, $\boldsymbol{\alpha} = \mathbf{B}\mathbf{y}$
  end while
  Set $\mathbf{c} = \mathbf{B}'\mathbf{1}$, $\hat{\mathbf{y}} = \tilde{\mathbf{A}}\tilde{\alpha}$, $\mathbf{r} = \mathbf{y} - \hat{\mathbf{y}}$
end for
Calculate $\boldsymbol{\alpha}_i = \tilde{\alpha}_i + \tilde{\boldsymbol{\alpha}}_{i+m}$
Ensure: final set $\mathbf{i}$, (sparse) dual weight vector $\boldsymbol{\alpha}$, predicted outputs $\hat{y}$

Algorithm 1. Kernel Polytope Faces Pursuit

where $L(\hat{y}, y)$ is the loss between the predicted $\hat{y}$ and true y, the *empirical error* of f given S:

$$\mathrm{err}_S(f) = \frac{1}{m} \sum_{i=1}^{m} L_S(f(x_i), y_i)$$

and the estimation error $\mathrm{est}(f)$

$$\mathrm{est}(f) = |\mathrm{err}(f) - \mathrm{err}_S(f)|,$$

we would like to find an upper bound for $\mathrm{est}(f)$.

We use Theorem 17.1 from [1] (using the ℓ_∞ covering number as opposed to the ℓ_1 covering number) which states that:

$$P^m\{\exists f \in F : |\mathrm{err}(f) - \mathrm{err}_S(f)| \geq \epsilon\} \leq 4\mathcal{N}_\infty(\epsilon/16, F, 2m)\exp(-\epsilon^2 m/32),$$

where $\mathcal{N}_\infty(\epsilon, F, m)$ is the ℓ_∞ *covering number.* [1] This covering number can be upper bounded using Theorem 12.2 from [1]:

$$\mathcal{N}_\infty(\epsilon, F, m) \leq \left(\frac{emR}{\epsilon d}\right)^d,$$

where R is the support of the distribution and d denotes the *pseudo-dimension.* As with KMP [7] the KPFP also has VC-dimension (pseudo-dimension) k, when k is the number of basis vectors chosen. However, in contrast to the KMP bound of [7] we use the pseudo-dimension to apply a natural regression loss function, the so-called squared error:

$$L(f(x), y) = (f(x) - y)^2.$$

[1] Note that the $\ell_\infty \geq \ell_1$ covering number is always an upper bound on the ℓ_1 covering number.

Therefore there is no need to fix a bandwidth parameter as was the case with the bound of [7] *i.e.*, no need to map the regression loss into a classification one. We follow the proof technique of [7] but instead apply the sample compression technique over pseudo-dimension bounds, resulting in a slightly more involved proof.

Theorem 1. *Let* $f \in F : \mathbf{X} \mapsto [0,1]$ *be the function output by any* sparse (dual) kernel regression *algorithm which builds regressors using basis vectors,* m *the size of the training set* S *and* k *the size of the chosen basis vectors* $\mathbf{i}$. *Let* $\bar{S} = S \setminus S_{\mathbf{i}}$ *denote the examples outside of the set* $S_{\mathbf{i}}$. *Assume without loss of generality that the last* k *examples in* S *form the set* $S_{\mathbf{i}}$. *Let* R *be the radius of the ball containing the support of* S, *then with* $1-\delta$ *confidence we can upper bound the true error* $\mathrm{err}(f)$ *of function* f *given any training set* S *by,*

$$\mathrm{err}(f) \leq \mathrm{err}_{\bar{S}}(f) + \frac{\sqrt{32^2 + 128(m-k)\left(k \ln \frac{em}{k} + k \ln 32e(m-k)R + 1 + \ln \frac{4km}{\delta}\right)} - 32}{2(m-k)}.$$

Proof. First consider a fixed k for the indices $\mathbf{i}$. Assume that the first $m-k$ points from S are drawn independently and apply Theorem 17.1 (and Theorem 12.2) from [1] to obtain the bound

$$P^m\{\bar{S} : |\mathrm{err}(f) - \mathrm{err}_{\bar{S}}(f)| \geq \epsilon\} \leq 4\left(\frac{32e(m-k)R}{\epsilon k}\right)^k \exp\left(\frac{-\epsilon^2(m-k)}{32}\right). \quad (8)$$

Given that we would like to choose k basis vectors from m choices we have $\binom{m}{k}$ different ways of selecting them. Multiplying the rhs of Equation 8 by $\binom{m}{k}$ and setting it equal to δ we get:

$$\begin{aligned} P^m\{S : \exists f \in \mathrm{span}\{S_{\mathbf{i}}\} \text{ s.t. } |\mathrm{err}(f) - \mathrm{err}_{\bar{S}}(f)| \geq \epsilon\} \\ \leq 4\binom{m}{k}\left(\frac{32e(m-k)R}{\epsilon k}\right)^k \exp\left(\frac{-\epsilon^2(m-k)}{32}\right). \quad (9) \end{aligned}$$

Next by setting the rhs of Equation (9) to δ, taking logarithms and rearranging we get

$$\frac{\epsilon^2(m-k)}{32} = k \ln \frac{em}{k} + k \ln 32e(m-k)R - \ln \epsilon + \ln k + \ln \frac{4}{\delta}.$$

We would like to write this bound in terms of ϵ and use the following result [11] which states that for any $\alpha > 0$, $\ln \epsilon \leq \ln \frac{1}{\alpha} - 1 + \alpha\epsilon$. Substituting this result with $\alpha = 1$ (a smaller α can be used but would make the bound less neat) we get

$$\epsilon^2(m-k) = 32\left(k \ln \frac{em}{k} + k \ln 32e(m-k)R - \ln 1 + 1 - \epsilon + \ln k + \ln \frac{4}{\delta}\right),$$

which yields the following quadratic equation:

$$(m-k)\epsilon^2 + 32\epsilon - 32\left(k \ln \frac{em}{k} + k \ln 32e(m-k)R + 1 + \ln \frac{4k}{\delta}\right) = 0.$$

Therefore, solving for ϵ gives the result[2] when we further apply the bound m times. □

This bound can be specialised to a Gaussian kernel[3] that uses the mean squared error loss.

Corollary 1. *For a Gaussian kernel and using all the definitions from Theorem 1 we can upper bound the loss of KPFP by:*

$$\mathrm{err}(f) \leq \frac{1}{m-k} \sum_{i=1}^{m-k} L_{\bar{S}}(f(x_i), y_i) + \frac{\sqrt{32^2 + 128(m-k)\left(k \ln \frac{em}{k} + k \ln 32e(m-k) + 1 + \ln \frac{4km}{\delta}\right)} - 32}{2(m-k)}.$$

Remark 1. The consequences of Theorem 1 (and Corollary 1) is that although the pseudo-dimension can be infinite even in cases where learning is successful[4], we will generate a bound that is *always* finite. Also, this is the first bound for KMP and KPFP (proposed in this paper) to use *the natural regression loss* in order to upper bound generalisation error. The bound is naturally trading off empirical error with complexity – as the training error decreases the bound gets smaller, and as the number of basis vectors (complexity) increase the bound gets larger. A good trade-off is to find small training error whilst using a small number of basis vectors. Clearly, the KMP and KPFP algorithms try and optimise this trade-off, and the bound suggests that this will result in good generalisation.

It is quite obvious that the output of the function class $F : \mathbf{X} \mapsto [0,1]$ is not bounded between 0 and 1 in most 'real world' regression scenarios. Therefore, we can give a more practically useful bound for a function class $F : \mathbf{X} \mapsto [-B, B]$ where the outputs are bounded in the range of $[-B, B] \in \mathbb{R}$.

Corollary 2. *Let $\|\mathbf{w}\|_2 \leq B \in \mathbb{R}$ and $\|\mathbf{x}_i\|_2 \leq 1, i = 1, \ldots, m$. Let $f \in F : \mathbf{X} \mapsto [-B, B]$ be the function output by any* sparse (dual) kernel regression *algorithm which builds regressors using basis vectors, m the size of the training set S and k the size of the chosen basis vectors* $\mathbf{i}$. *Let $\bar{S} = S \setminus S_{\mathbf{i}}$ denote the examples outside of the set $S_{\mathbf{i}}$. Assume without loss of generality that the last k examples in S form the set $S_{\mathbf{i}}$. Let R be the radius of the ball containing the support of S,*

---

[2] Only solve the quadratic equation for the positive quadrant.

[3] We use the Gaussian kernel in the experiments.

[4] Note that the pseudo-dimension is a generalisation of the VC-dimension and hence the same problems of infinite VC-dimension also apply to the pseudo-dimension.

then with $1 - \delta$ *confidence we can upper bound the true error* $\mathrm{err}(f)$ *of function* f *given any training set* S *by,*

$$\mathrm{err}(f) \leq \mathrm{err}_{\bar{S}}(f) + 2B\frac{\sqrt{32^2 + 128(m-k)\left(k \ln \frac{em}{k} + k \ln 32e(m-k)R + 1 + \ln \frac{4km}{\delta}\right)} - 32}{2(m-k)}.$$

Proof. Denote the function class $\tilde{F} = \left\{\frac{f \pm B}{2B} : f \in F\right\} : \mathbf{X} \mapsto [0,1]$. Therefore, given any function $\tilde{f} \in \tilde{F}$ Theorem 1 holds. Furthermore, for any function class $F : \mathbf{X} \mapsto [-B, B]$ we have:

$$2B\mathrm{err}(\tilde{f}) \leq 2B\mathrm{err}_{\bar{S}}(\tilde{f}) + 2B\frac{\sqrt{32^2 + 128(m-k)\left(k \ln \frac{em}{k} + k \ln 32e(m-k)R + 1 + \ln \frac{4km}{\delta}\right)} - 32}{2(m-k)},$$

which completes the proof when we make the substitutions $\mathrm{err}(f) = 2B\mathrm{err}(\tilde{f})$ and $\mathrm{err}_{\bar{S}}(f) = 2B\mathrm{err}_{\bar{S}}(\tilde{f})$. □

3 Experiments

We present a comparison on 9 benchmark datasets derived from the UCI, StatLib, and Delve benchmark repositories. Details of the datasets are given in Table 2. We analyse the performance of KPFP, KMP, Kernel Ridge Regression (KRR) and LASSO using the Least Angle Regression Solver (LARS) using Radial Basis Function (RBF) kernels. We used 10 randomised splits into training and test sets. For each of the datasets we used cross-validation (c.v.) to select the optimal RBF kernel width parameter for KRR. We then used this kernel as input to the KMP, LARS and KPFP algorithms. For both KMP and KPFP the initial sparsity level k was set in training by a heuristic method to the lesser of 100 or the number of training examples. Since the train and test error curves for both KMP and KPFP tend to follow each other well, we used the index of minimum training error as the final sparsity value. The means and standard deviations of the generalisation error for each method and dataset are given in Table 3.

The results show that overall the sparse methods (KMP,KPFP,LARS) all perform better than KRR. It is interesting to compare the performance of KPFP with and without the release of violating constraints (KPFPv and KPFP respectively). KPFPv performs nearly as well as KMP on all datasets except for `cpusmall`, whilst requiring fewer bases in the final solutions. On the other hand, KPFPv results in solutions that are the least sparse of the three methods, but results in the lowest generalisation error. LARS which gives an exact solution to the LASSO problem performs the worst here, showing that the exact solution is not necessarily the optimal one for generelisation. The key to the performance of all of these methods is in selecting the appropriate stopping point k. This is

Table 2. Number of examples and dimensions of each of the 9 benchmark datasets

Dataset	# examples	# dimensions
abalone	4177	8
bodyfat	252	14
cpusmall	8192	12
housing	506	13
mpg	392	7
mg	1385	6
pyrim	74	27
space_ga	3107	6
triazines	186	60

Table 3. Mean MSE (μ) and standard deviations (σ) for 9 benchmark datasets for Kernel Ridge Regression (KRR), Kernel Matching Pursuit (KMP), LASSO using the Least Angle Regression Solver (LARS) and Kernel Polytope Faces Pursuit with and without violation release (KPFPv,KPFP). The total number of wins over all splits of the data for each algorithm is given in the last row.

Dataset	**KRR**		**KMP**			**LARS**			**KPFPv**			**KPFP**		
	μ	σ	μ	σ	k	μ	σ	k	μ	σ	k	μ	σ	k
abalone	8.70	1.79	5.70	2.56	49.2	21.64	28.80	5.4	6.07	1.16	7.3	**4.82**	0.24	37.7
bodyfat	0.00	0.00	0.00	0.00	49.1	0.01	0.02	5.7	0.00	0.00	30.1	**0.00**	0.00	129.7
cpusmall	216.35	64.04	15.66	2.51	24.0	519.06	95.45	10.3	69.97	2.51	13.4	**12.50**	1.51	54.2
housing	72.19	19.59	**21.93**	7.17	50.3	56.84	19.35	8.9	34.16	8.19	21.9	23.22	6.67	150.8
mpg	39.47	24.57	20.70	14.37	50.6	42.05	48.27	7.7	13.11	3.35	11.5	**10.98**	1.97	161.1
mg	0.04	0.01	**0.02**	0.00	49.0	0.11	0.19	4.4	0.02	0.00	7.6	0.02	0.00	48.7
pyrim	0.02	0.01	0.02	0.02	24.3	0.02	0.01	11.6	0.02	0.01	17.8	**0.01**	0.01	39.0
space_ga	0.03	0.01	0.02	0.00	49.9	0.05	0.05	4.8	0.02	0.00	6.0	**0.02**	0.00	38.2
triazines	0.02	0.01	0.03	0.02	50.9	0.02	0.00	11.3	**0.02**	0.00	34.4	0.02	0.00	109.7
wins	3		34			6			9			39		

quite difficult to achieve in KMP, as the algorithm tends to overfit quite quickly, and there is no obvious criteria for stopping. For example, if cross-validation were used to select k, the resulting value would be too low, as the number of bases would selected from a smaller validation set. In our experiments we found that by selecting an intial k through a heuristic method and then choosing the minimiser of the training error resulted in the best compromise. In KPFP and KPFPv the optimal value for k is more easily achieved, as the training and test error curves tend to follow each other quite well, and we also have an (optional) stopping parameter θ_{max}. In fact, the value of θ to which θ_{max} is compared also follows the error curves. We found that by taking the minimiser of θ as the number of bases was a reliable way of estimating k.

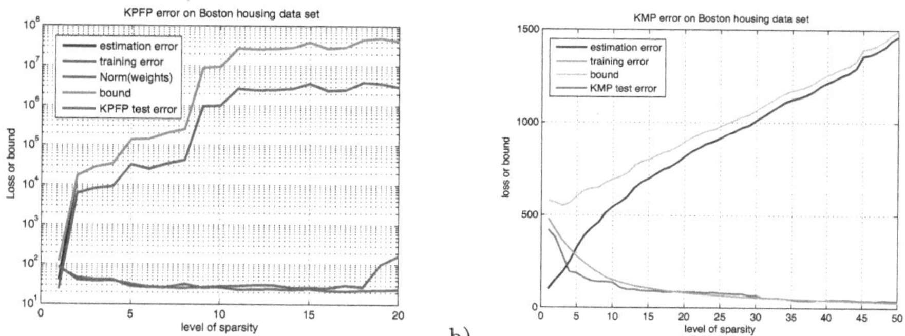

a) b)

Fig. 1. a) Plot of generalisation error bound for different values of k using RBF kernels for the 'Boston housing' data set. The log of the generalisation error is shown on the y axis. The plot shows the empirical error of the set $\bar{S}$ (denoted training error, in green), the estimation error (in blue), the norm of the weight vector (in red), the bound value which is calculated from these three values (in cyan), and the generalisation (true) error (in magenta). Note that the empirical error follows the true error very well, which justifies its' use in the setting of the sparsity parameter. However the bound value is swamped by the norm of the weight vector (needed according to Corrollory 2), and as such is not useful. b) The bound values for the KMP algorithm. Note that in this case the bound (which is valid for this algorithm too) is more useful, simply because the norm of the weight vector does not blow up as quickly.

3.1 Bound Experiments

Finally we present results of the performance of the bound. Figure 3.1 shows typical plots of the bound. For Figure 3.1 (b) the number of training examples chosen were 450 and the number of test examples were 56, with the Gaussian width parameter set to $\sigma = 0.035$. The bound values tend to fall as basis vectors are added, before rising again as the complexity of the solution rises. Hence the first minimum of the bound value could serve as an appropriate point to stop the algorithm. This is clearly much more efficient than using cross-validation to select the value of k, the number of basis vectors to use. However in our experiments this resulted in stopping too early, resulting in underfitting. Further refinement of the bound may improve its performance in this respect.

4 Conclusions

Polytope Faces Pursuit (PFP) is a greedy algorithm that approximates the sparse solutions recovered by ℓ_1 regularised least-squares (LASSO) [4,10] in a similar vein to (Orthogonal) Matching Pursuit (OMP) [16]. The algorithm is based on the geometry of the polar polytope where at each step a basis function is chosen by finding the maximal vertex using a path-following method. The algorithmic complexity is of a similar order to OMP whilst being able to solve problems known to be hard for (O)MP.

We extended the PFP algorithm to a kernel verison, which we called Kernel Polytope Faces Pursuit (KPFP). We showed the utility of this algorithm by providing a novel generalisation error bound which used the natural regression loss and pseduo-dimension in order to upper bound its loss. The experimental results were also encouraging and showed that KPFP was competitive against the KMP and Kernel Ridge Regression.

A future research direction is to tighten the bound and use it to find the number of basis vectors during training.

Acknowledgements

We would like to thank John Shawe-Taylor for his helpful comments regarding the generalisation analysis. The research leading to these results has received funding from the European Community's Seventh Framework Programme (FP7/2007–2013) under *grant agreement* no. 216529, Personal Information Navigator Adapting Through Viewing, PinView.

References

1. Anthony, M., Bartlett, P.: Neural Network Learning: Theoretical Foundations. Cambridge University Press, Cambridge (1999)
2. Chen, S.: Basis Pursuit. PhD thesis, Department of Statistics, Stanford University (November 1995)
3. Chen, S.S., Donoho, D.L., Saunders, M.A.: Atomic decomposition by basis pursuit. SIAM Journal on Scientific Computing 20(1), 33–61 (1998)
4. Donoho, D.: Neighborly polytopes and sparse solution of underdetermined linear equations. Technical report, Department of Statistics, Stanford Univ., Stanford, CA (2005)
5. Floyd, S., Warmuth, M.: Sample compression, learnability, and the Vapnik-Chervonenkis dimension. Machine Learning 21(3), 269–304 (1995)
6. Graepel, T., Herbrich, R., Shawe-Taylor, J.: Generalisation error bounds for sparse linear classifiers. In: Proceedings of the Thirteenth Annual Conference on Computational Learning Theory, pp. 298–303 (2000)
7. Hussain, Z., Shawe-Taylor, J.: Theory of matching pursuit. In: Koller, D., Schuurmans, D., Bengio, Y., Bottou, L. (eds.) Advances in Neural Information Processing Systems, vol. 21, pp. 721–728 (2009)
8. Mallat, S., Zhang, Z.: Matching pursuit with time-frequency dictionaries. IEEE Transactions on Signal Processing 41(12), 3397–3415 (1993)
9. Plumbley, M.D.: Polar polytopes and recovery of sparse representations (2005)
10. Plumbley, M.D.: Recovery of sparse representations by polytope faces pursuit. In: Rosca, J.P., Erdogmus, D., Príncipe, J.C., Haykin, S. (eds.) ICA 2006. LNCS, vol. 3889, pp. 206–213. Springer, Heidelberg (2006)
11. Shawe-Taylor, J., Anthony, M., Biggs, N.L.: Bounding sample size with the Vapnik-Chervonenkis dimension. Discrete Applied Mathematics 42(1), 65–73 (1993)
12. Shawe-Taylor, J., Cristianini, N.: Kernel Methods for Pattern Analysis. Cambridge University Press, Cambridge (2004)

13. Smola, A.J., Bartlett, P.: Sparse greedy gaussian process regression. Advances in neural information processing systems 13 (2001)
14. Smola, A.J., Schölkopf, B.: Sparse greedy matrix approximation for machine learning. In: Proceedings of 17th International Conference on Machine Learning, pp. 911–918. Morgan Kaufmann, San Francisco (2000)
15. Tibshirani, R.: Regression selection and shrinkage via the lasso. Technical report, Department of Statistics, University of Toronto (June 1994), `ftp://utstat.toronto.edu/pub/tibs/lasso.ps`
16. Tropp, J.A., Gilbert, A.C., Strauss, M.J.: Algorithms for simultaneous sparse approximation. part i: Greedy pursuit. Signal Processing, special issue Sparse approximations in signal and image processing 86, 572–588 (2006)
17. Vincent, P., Bengio, Y.: Kernel matching pursuit. Machine Learning 48(1-3), 165–187 (2002)

Soft Margin Trees

Jorge Díez, Juan José del Coz, Antonio Bahamonde, and Oscar Luaces

Artificial Intelligence Center, University of Oviedo at Gijón, Asturias, Spain
www.aic.uniovi.es

Abstract. From a multi-class learning task, in addition to a classifier, it is possible to infer some useful knowledge about the relationship between the classes involved. In this paper we propose a method to learn a hierarchical clustering of the set of classes. The usefulness of such clusterings has been exploited in bio-medical applications to find out relations between diseases or populations of animals. The method proposed here defines a distance between classes based on the margin maximization principle, and then builds the hierarchy using a linkage procedure. Moreover, to quantify the goodness of the hierarchies we define a measure. Finally, we present a set of experiments comparing the scores achieved by our approach with other methods.

1 Introduction

In many Machine Learning and Data Mining applications, users are not only interested in learning good classifiers but also in gaining some insight into the application domain. This is the case, for instance, of learning techniques for association rule, clustering or feature selection.

Given a dataset of labeled examples, in this paper we present a method to cluster the set of classes. This learning task has received little attention in the literature; typically, clustering deals with examples instead of classes. However, once we have classes attached to examples, their clusters draw valuable information about the similarities and differences between classes.

In *Medicine*, [1,2] report methods for clustering SAGE (Serial Analysis of Gene Expression) data to detect similarities and dissimilarities between different types of cancer on the subcellular level. There are also many interesting applications in *Genetics of Populations*. The aim is to discover relationships between species or breeds according to their genetic descriptions. Thus, [3] studied 50 indigenous cattle breeds from Africa; [4] clustered a total of 1272 termites representing 56 genetically distinct colonies in central North Carolina; [5] investigated the genetic structure and variation of 21 populations of cattle in northern Eurasia and the neighbouring Near Eastern regions; [6], in order to facilitate the assessments of epidemiological risks, showed the genetic structure of human populations using genotypes at 377 autosomal microsatellite loci in 1056 individuals from 52 populations.

Roughly speaking the clustering of classes has been faced in two ways. The straightforward approach represents each class using a single feature vector,

W. Buntine et al. (Eds.): ECML PKDD 2009, Part I, LNAI 5781, pp. 302–314, 2009.

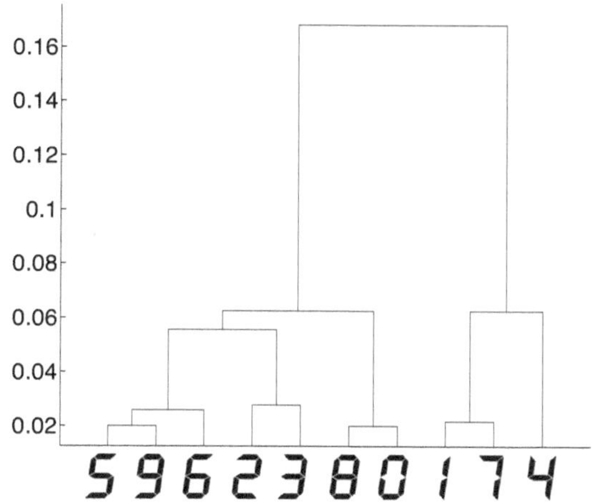

Fig. 1. Hierarchical clustering of classes obtained on led dataset. This domain contains 7 boolean attributes, representing the 7 light-emitting diodes, and 10 classes, the set of decimal digits. To make the task more difficult, the problem has some noise. Each attribute value has the 10% probability of having its value inverted.

usually the centroid of the examples of the class in the input space [7]. Then, the collection of these vectors is clustered using algorithms as k-means, Self-Organized Maps [8] or a hierarchical clustering. The main drawback of this approach is that representing a class by a single vector may imply an important loss of information.

The second approach proceeds in two stages. First, a clustering is obtained using the complete dataset of individual examples. Then, in an *ad hoc* way, the users analyze the groups so obtained. Typically, the individual examples are represented in a 2-D map using visualizations tools like Self-Organized Maps [9,10,1]; from this map an application-specific discussion *infers* a graph that represents the relationship between classes [11]. Using these types of methods, knowledge discovery depends heavily on the capability of users to interpret the clustering of individual examples. Using this approach, it is a difficult task to find the true relationship between all classes of a classification task.

The method proposed in this paper builds a hierarchical clustering [12] of the set of classes. The core idea is to define a metric between classes. Given that the starting data is a classification learning task, the metric of classes is defined from a bundle of binary classifiers. Then an agglomerative hierarchical clustering method will provide a binary tree or *dendrogram* with classes placed at leaves. The dendrogram can be broken at different levels to yield different clusterings of the set of classes.

The organization of the classes so learned is a meaningful tree that will be easily interpreted by an expert of the domain. Figure 1 depicts the relationships of classes of the well-know *led* dataset. Notice that the topmost split of classes separates those that share the bottom led (left hand side) from those in which that led is off, classes $1, 7$, and 4. As one could hope, the closest classes are the groups $\{8, 0\}$, $\{5, 9, 6\}$, $\{1, 7\}$, and $\{2, 3\}$.

The organization of the paper is the following. In the next section we present an overview of related work. Then, we introduce the central idea of the paper, the definition of the distance of classes that uses an algorithm based on the margin-maximization principle. Additionally, we present a measure to compare different hierarchical clusterings of classes. In Section 5, we report some experimental results supporting our approach. We shall show that our method is able to build meaningful hierarchical clusters of classes, significantly better than those produced by other methods. We close the paper drawing some conclusions.

2 Related Work

There are two kinds of related work. The papers that present biomedical applications, quoted in the Introduction, describe mainly *ad hoc* methods to cluster the set of classes. On the other hand, there are a number of approaches that build clusters trying to assemble multi-class classifiers based on Support Vector Machines (SVM) [13]. This section is devoted to explain the relation of these papers with the method presented here.

First of all, let us recall that there are two types of approaches for multi-class classification using SVM. One is considering all data in a single optimization problem [14,15]. The other is decomposing the multi-class task into a series of binary SVMs, such as One-vs-All (OVA) [13], One-vs-One (OVO) [16], and using a directed acyclic graph (DAG) [17]. As none of these methods significantly outperforms the others, this is an active research subject.

Lately, some authors [7,18,19,20] have proposed decomposition algorithms which follow a similar approach: learn a decision tree (a special case of DAG) based on a hierarchy of classes. First, these methods build a dendrogram of classes; and then, a binary SVM is learned for each internal node of that hierarchy in order to separate the examples of each subset of classes (see Figures 1 and 2). The classification procedure goes from the root to leaves guided by the predictions of SVMs classifiers at internal nodes. All these methods basically differ in the way they build the hierarchy of classes.

In [7] the authors propose a method to construct a binary tree. It proceeds top-down, and at each node it uses a *k-means* clustering of the centroids of classes to divide the set of classes into two groups.

In [19] the authors present the so called Dendrogram-based Support Vector Machines (DSVM). In order to build the dendrogram, DSVM computes the centroid of each class, and then uses an agglomerative hierarchical algorithm. We shall include this method in the Experimental Results section.

In [18], the algorithm Half-Against-Half (HAH) is presented. After some discussion about different methods to build the hierarchy structure (generating it at random or using prior knowledge), the authors propose to use a hierarchical clustering algorithm based on the mean distance between classes. Since the authors do not state clearly the concrete hierarchical clustering algorithm used, we shall not include this method in Section 5.

However, the work that inspired this paper is [20]. In fact, our approach can be seen as a generalization of it. The authors present a multi-class classifier for high-dimensional input spaces, called *Margin Trees Classifier*, that achieves an accuracy comparable to that of OVO SVM on 7 cancer microarray data sets. The algorithm is somehow similar to those cited before since it uses a hierarchy of classes to build a decision tree classifier. The authors report a study comparing different procedures to build the hierarchy: complete linkage, single linkage and a greedy algorithm. They conclude proposing the complete linkage.

However, the main idea introduced by Tibshirani and Hastie is that they use the margin to compute the distance between two classes. In their approach, the dimension of the input space is always greater than the number of examples. Therefore, all classes are separable. In the next section we present a generalization to the non-separable case of this method, applying the basic principles of margin maximization.

In our opinion, another important contribution of [20] is that the authors are the first to remark the additional interpretability of the model obtained in biological tasks. Despite they describe their method only as a multi-class classifier, they also point out the utility of the cluster of classes in real applications. In this paper we want to explore this idea and we shall focus in the method as a clustering of classes.

3 Soft Margin Trees

Let $\mathcal{X}$ be an input space, and $\mathcal{Y} = \{C_1, ..., C_k\}$ a finite set of classes. We consider a multi-classification task given by a training set $S = \{(\boldsymbol{x}_1, y_1), \ldots, (\boldsymbol{x}_n, y_n)\}$ drawn from an unknown distribution $Pr(X, Y)$ from the product $\mathcal{X} \times \mathcal{Y}$. Within this context, our learning task is to build a dendrogram T in which each class labels exactly one leave, T has k leaves, and $k-1$ internal nodes.

In Figure 1 you can see the dendrogram obtained applying the method described here to the led dataset [21]. Here, y-axis represents the distance between clusters grouped together.

In order to define an algorithm for agglomerative hierarchical clustering we require two elements:

1. A linkage scheme to recalculate inter-cluster distances when the two most similar or near clusters are merged.
2. A method to calculate a symmetric dissimilarity matrix D based on pairwise dissimilarities or distances. The value in the l-th row, m-th column is the distance between classes C_l and C_m. Notice that it will be necessary to apply this method $\binom{k}{2}$ times to calculate D.

In the following subsections we describe our proposals in these elements of the method for clustering classes.

3.1 Linkage Method

An agglomerative hierarchical clustering algorithm starts defining one cluster for each class. Then, the algorithm proceeds iteratively joining together (*linking*) the two closest clusters. Differences between linkage methods arise from different ways to define distances between clusters. Thus, using different linkage methods can produce different dendrograms. Here, due to the lack of space, we can not study the effect of using different linkage methods.

Following the conclusions drawn by [20], we shall use the *complete linkage* method. In their experimental results all methods produce the same accuracy, but complete linkage gives rise to more balanced trees. This kind of trees are more interpretable than those obtained by other methods. In the experiments reported in Section 5 all hierarchical clustering algorithms compared use the complete linkage method. They only differ in the way they compute the dissimilarity matrix.

3.2 A Margin-Based Metric for Classes

In [20], the distance between two separable classes is defined as the margin between them. To compute the distance between classes C_l and C_m, the class labels (y_i) of the examples of those classes are relabeled (y'_i) as $+1$ and -1, respectively. The hyperplane of maximum margin that separates the nearest examples of those classes is defined by a weight vector $\boldsymbol{w}$ and a bias b that can be obtained solving the following optimization problem:

$$\begin{aligned}(\boldsymbol{w}, b) &= \operatorname*{argmax}_{\|\boldsymbol{w}\|=1} (M) \\ &\text{s.t. } y'_i(\langle \boldsymbol{w}, \boldsymbol{x}_i \rangle + b) \geq M, \quad \forall y_i \in C_l \cup C_m.\end{aligned} \tag{1}$$

Finally, in Margin Trees, the distance between classes C_l and C_m is the margin:

$$D(l, m) = 2 \cdot M. \tag{2}$$

Actually, the optimization problem in Equation 1 is equivalent, see for instance [22,23], to a typical hard-margin SVM formulation:

$$\begin{aligned}&\min \ \frac{1}{2}||\boldsymbol{w}||^2 \\ &\text{s.t. } \ y'_i(\langle \boldsymbol{w}, \boldsymbol{x}_i \rangle + b) \geq 1, \quad \forall y_i \in C_l \cup C_m.\end{aligned} \tag{3}$$

In this case, the margin between both classes is equal to:

$$D(l, m) = \frac{2}{||\boldsymbol{w}||}. \tag{4}$$

Our proposal is to use a *soft*-margin SVM formulation instead of a hard-margin one:

$$\begin{aligned} \min \ & \frac{1}{2}||\boldsymbol{w}||^2 + C \sum_{\forall y_i \in C_l \cup C_m} \xi_i, \\ \text{s.t. } & y_i'(\langle \boldsymbol{w}, \boldsymbol{x}_i \rangle + b) \geq 1 - \xi_i, \\ & \xi_i \geq 0, \qquad \forall y_i \in C_l \cup C_m, \end{aligned} \tag{5}$$

where C is a regularization parameter. The distance between classes C_l and C_m is defined by the following expression:

$$D(l,m) = \frac{1}{\frac{1}{2}||\boldsymbol{w}||^2 + C \sum_{\forall y_i \in C_l \cup C_m} \xi_i}. \tag{6}$$

The idea is that when classes are very different, the classifier will be given by a simple model (i.e., $||\boldsymbol{w}||$ will be low) and/or the number of misclassified examples or points inside the margin will be small ($\sum \xi_i \rightarrow 0$), so the distance (Equation 6) will be high. When classes are similar, the model will be complex (the norm of weight vector will be high) and/or there will be a lot of misclassified examples or points inside the margin ($\sum \xi_i \gg 0$); in this case the distance will be small. Thus, the optimization problem in Equation 5 minimizes an expression that captures faithfully the differences between two classes.

It must be noted that, in order to ensure that all distances of matrix D share an identical scale, the regularization parameter C must be the same in the $\binom{k}{2}$ SVM binary classifiers needed to calculate all pairwise distances.

The differences between distances of Margin Trees and those proposed in this paper may be quite subtle in linearly separable cases. Then, both depend only on the norm of weight vectors, but these vectors may differ. However, in general, our metric has a couple of advantages over that of Margin Trees: 1) it can be applied to non-separable problems, and 2) even in a separable case, the soft-margin approach can find a different solution because it takes into account, at the same time, the complexity of the model and the expected loss. In fact, these are the same benefits that can be obtained using the soft margin approach instead of the hard margin in traditional SVM.

Figure 2 shows these ideas. There are some examples in a 2-dimensional space labeled in four classes linearly separable. At a first glance, it seems that there are two groups of classes: {1,3} and {2,4}. Applying the Margin Trees algorithm (left side panel), Equations 1 or 3, the two most similar classes are 1 and 2; a couple of outlayers make the hard margin quite small. The separating hyperplane and margin frontiers are displayed in Figure 2. On the other hand, our method (right side panel), Equation 5, takes advantage of the possibility to misclassify some examples or to place them inside the margin. Therefore, the method proposed in this paper captures the relationship between those classes better than Margin Trees.

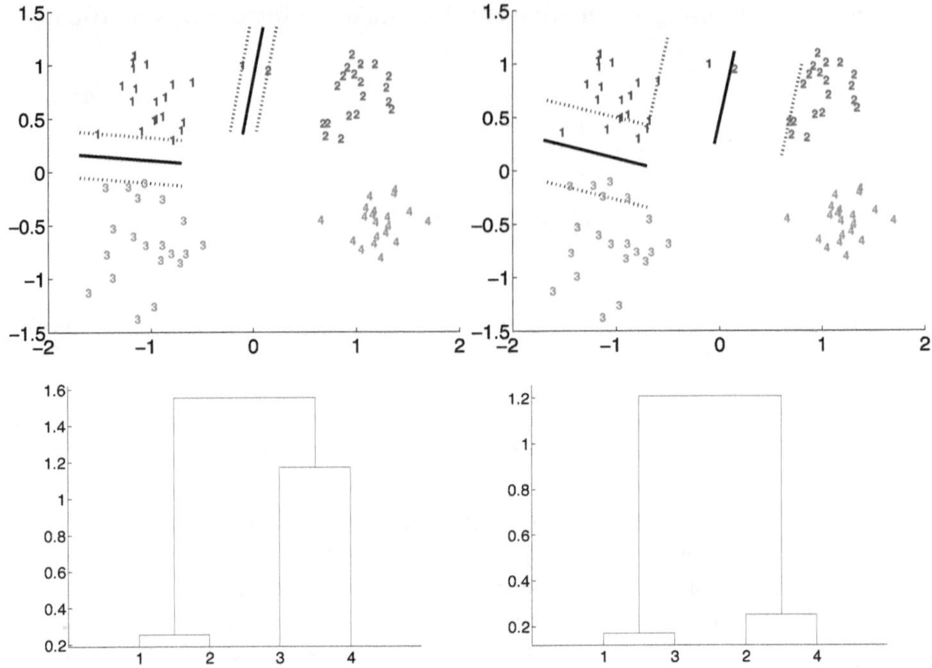

Fig. 2. Example of a Margin Tree (left panel) and a Soft Margin Tree (right panel). Separating hyperplanes and hierarchical clustering of classes are displayed. The method proposed in this paper captures the relationship between those classes better than Margin Trees.

4 A Method and a Metric to Compare Hierarchical Clustering of Classes

A very important aspect of learning methods is to measure the quality of the obtained knowledge. In supervised learning this task is accomplished by a set of well established metrics. They allow users to compare different techniques and to estimate the future performance of predictive models. But unsupervised learning algorithms, like clustering, are difficult to compare.

The learning task of clustering a set of classes is an unsupervised task, we do not know the *correct* organization of classes. However, we can take advantage of having a dataset with examples labeled by those classes.

Given a hierarchical clustering of k classes, as was mentioned in Section 2, it is possible to build a classifier learning $k-1$ binary SVMs. Let $h : \mathcal{X} \to \mathcal{Y}$ be such a classifier. A first approach to define a metric for the quality of clusterings of classes could be to measure the accuracy of h. In fact, if the hierarchy is *good*, the accuracy of h must be high.

Table 1. Description of the datasets used in the experiments. They are divided in two parts. The first is a collection of classification tasks non-linearly separable drawn from the UCI repository [21]. The second part are cancer datasets that were previously used in [20].

Dataset	#classes	#examples	#features
zoo	7	101	16
glass	6	214	9
ecoli	8	336	7
dermatology	6	366	33
vehicle	4	846	18
vowel	11	990	11
led	10	1000	7
yeast	10	1484	8
car	4	1728	6
image	7	2310	19
landsat	6	6435	36
brain	5	42	5597
lymphoma	3	62	4026
srbct	4	63	2308
stanford	14	261	7452
9 tumors	9	60	7131
11 tumors	11	174	12533
14 tumors	14	190	16063

But this is an incomplete view of the problem. It is more important to study what happens when the classifier fails, than to calculate only the accuracy. A hierarchical clustering of classes represents the similarity between those classes. When h missclassifies an example $\boldsymbol{x}_i$, if the hierarchy is good, the predicted class $h(\boldsymbol{x}_i)$, must be near (in the hierarchy) to the true class y_i .

Therefore, we propose to measure the distance in the hierarchy between predictions and true classes; that is, the number of arcs in the dendrogram T placed between the leaves labeled by those classes. We called this measure *Prediction Distance (PD)*:

$$PD(h(\boldsymbol{x}_i), y_i, T) = \#\texttt{arcs}(h(\boldsymbol{x}_i), y_i, T). \tag{7}$$

For instance, the Prediction Distance of an example of class 5 (see Figure 1) classified as class 9 is 2, but if prediction were class 2, PD would rise to 5. The farthest classes from class 5 are 1 and 7, both at distance 8. Notice that the maximum distance between two classes in a dendrogram of k classes is k.

Averaging this metric over a set of test examples we shall measure the goodness of a hierarchy T.

5 Experimental Results

In order to evaluate the benefits of our approach we conducted a battery of experiments. The aim is to show that our approach is able to build meaningful hierarchical clusters of classes, significantly better than those produced by other methods.

As it was indicated in Section 3.1, each algorithm for hierarchical clustering employed in the experiments used the complete linkage method to build the tree. Therefore, the differences between the algorithms arise from the way they compute the matrixes of distances between each pair of classes. We considered four approaches:

- *Random.* A random symmetrical distance matrix is computed. Since *PD* is an unbounded measure, we used *Random* method to obtain an upper bound on all datasets. It is expected that the other approaches perform significantly better.
- *DSVM* [19]. Dendrogram-based Support Vector Machines computes the distance matrix by means of the Euclidean distance between the centroids of each pair of classes.
- *Margin Trees (MT).* The distance matrix is computed using the Margin Trees idea from [20], showed in Equations 3 and 4.
- *Soft Margin Trees (SMT).* The distance matrix is computed applying Equations 5 and 6.

Once we have built a hierarchical clustering of classes, a multi-class classifier may be learned with a binary SVM attached to each internal node. We shall report the accuracy of these classifiers ("0/1" errors) in addition to the Prediction Distance (*PD*) (Equation 7). Although the quality of the trees can be measured only by means of the prediction distance, we include the accuracy to compare their predictive power with the *SVM* multiclass.

The datasets used in the experiments are described in Table 1. There are datasets with more number of examples than features and vice versa. The first group is formed by datasets obtained from the UCI repository [21]. We selected those datasets that fulfill the following rules: continuous or ordinal attribute values, no more than 40 attributes, no more than 10000 examples, and excluding datasets with missing values and with less than 4 classes. The second group is composed by the datasets used in [20]. They are 7 datasets which target is to classify cancer patients from gene expressions captured by microarrays.

All the scores reported were estimated by means of a 5-fold cross validation repeated 2 times. We did not use the 10-fold procedure, since in certain datasets there are too few examples in some classes. The *SVM* implementation used was *libsvm* [24] with the linear kernel in all cases. Cancer datasets can be separated by a linear-SVM, so the utility of using different kernels it is not clear.

For each hierarchical system considered (*Random, DSVM, MT and SMT*), we used the same regularization parameter C in all models learned at each node in the tree. To select this parameter, we utilized a 2-fold cross validation repeated 5 times on training data searching within $C \in [10^{-2}, \ldots, 10^{2}]$. The aim in this

Table 2. Cross validation results both in "0/1" errors and *PD*

	SVM	*Random*		*DSVM*		*MT*		*SMT*	
Dataset	0/1	0/1	*PD*	0/1	*PD*	0/1	*PD*	0/1	*PD*
zoo	4.98	5.48	0.2336	4.50	0.2750	*MT* classifier only works	with separable datasets	4.98	0.1788
glass	33.20	40.44	1.7156	35.76	1.1292			37.40	1.1502
ecoli	11.75	16.58	0.6888	14.01	0.3873			13.71	0.3782
dermatology	3.55	3.14	0.1312	4.36	0.1036			3.82	0.0928
vehicle	20.33	22.99	0.7400	20.63	0.4817			20.63	0.4805
vowel	19.19	54.60	3.2303	26.77	1.1354			29.70	1.3449
led	27.70	36.95	2.0790	27.80	1.1950			27.95	1.1860
yeast	41.61	46.36	2.2685	42.79	1.4340			43.43	1.3672
car	14.35	19.50	0.6754	15.11	0.4751			17.30	0.4161
image	4.07	12.53	0.5890	10.13	0.3812			4.29	0.1126
landsat	13.05	18.91	0.7876	14.96	0.5829			13.18	0.4051
brain	13.19	14.44	0.5278	14.58	0.5611	13.33	0.5125	13.33	0.5125
lymphoma	0.00	0.00	0.0000	0.00	0.0000	0.00	0.0000	0.00	0.0000
srbct	1.60	1.60	0.0564	0.83	0.0167	0.83	0.0250	0.83	0.0250
stanford	5.36	5.35	0.3759	6.12	0.2700	5.93	0.3162	5.93	0.3162
9 tumors	48.33	51.67	2.4667	45.83	2.3083	44.17	2.3083	44.17	2.3083
11 tumors	10.08	9.81	0.5603	9.81	0.5492	10.09	0.5206	10.09	0.5206
14 tumors	30.53	32.11	2.0368	31.58	1.9026	28.68	1.5289	28.68	1.5289
Average	16.83	21.80	1.0646	18.09	0.7327			17.75	0.6847

grid search was to optimize the *PD*. In the case of the *SMT*, the C parameter found by the search was used both for computing the distance matrix and to build the classifiers of each internal node.

Following [25], we used the Wilcoxon signed ranks test to compare the performance of the classifiers by pairs when the measurements are "0/1" errors or *PD*.

In Table 2, we show the results obtained in the experiments. Considering the quality of the hierarchies measured by *PD*, we can see that *SMT* achieved the best results. The differences are significant with $p < 0.01$ for *Random*, and with $p < 0.03$ for *DSVM*. The *MT* algorithm was not applied on UCI datasets since they are not separable. To summarize the results, in Table 3 we show the number of victories, defeats and ties between each pair of algorithms. We can see that *SMT* is the unique algorithm that never loses *versus Random* in all datasets; and it only ties ones, in *lymphoma* dataset where all algorithms obtain the best performance.

Additionally, it is important to remark that *SMT* and *MT* achieve the same results in those datasets in which the number of features is higher than the number of examples.

With respect to "0/1" errors, Table 2 shows that the results attained by *SVM* multiclass are better in general (as it happens in [20]). Actually, *SVM* is significantly better than *Random* and *DSVM* with $p < 0.01$, and better than *SMT* with $p < 0.04$. Comparing multi-class classifiers attached to hierarchies of

Table 3. Summary of *PD* scores. Number of wins (w), losses (l), and ties (t) between pairs of algorithms considered. In the case of *DSVM versus SMT* (4/12/2), it must be read as follows: *DSVM* is better than our approach *SMT* 4 times, worse 12 times, and in 2 datasets they obtain equal results.

w/l/t	*MT*	*SMT*	*Random*
DSVM	2/3/2	4/12/2	15/2/1
MT		0/0/7	6/0/1
SMT			17/0/1

classes, *SMT* is significantly better than *Random* with $p < 0.01$. These results show that these methods have less predictive power than *SVM*, although they produce valuable descriptive models represented in the hierarchies of classes.

Notice that sometimes an algorithm A obtains better "0/1" errors, but a worse *PD* than other algorithm B. Even in that case, these scores mean that the hierarchy learned by A is worse than that produced by B. The reason is that although the hypothesis h_A learned by A missclassifies less examples than the corresponding hypothesis h_B of B, the errors of h_A are far in the corresponding hierarchy from true classes. On the other hand, the more frequent misclassifications of h_B are close to the true classes.

6 Conclusions

In this paper we presented a new method for clustering a set of classes of a multi-class learning task. These clusterings have shown their usefulness in fields such as *medicine* or *genetics of populations*. The aim in all these cases is not only to learn an accurate classifier but also to gain some insight into the application domain.

Our approach computes a distance matrix between classes using a method based on the soft margin of each pair of classes. Then, using a complete linkage method, it is possible to build a dendrogram where the leaves are labeled by classes. We tested *Soft Margin Trees* with other algorithms for clustering classes and it was shown that *SMT* produces significantly better hierarchical clusters of classes than those produced by other methods.

Acknowledgments

The research reported here is supported in part under grant TIN2008-06247 from the MICINN (Ministerio de Ciencia e Innovación of Spain).

References

1. Patra, J., Ang, E.L., Meher, P., Zhen, Q.: A new som-based visualization technique for dna microarray data. In: IJCNN 2006. International Joint Conference on Neural Networks, pp. 4429–4434 (2006)

2. Ng, R., Sander, J., Sleumer, M., et al.: Hierarchical cluster analysis of SAGE data for cancer profiling. In: Proceedings of BIOKDD 2001 Workshop on Data Mining in Bioinformatics, pp. 65–72 (2001)
3. Hanotte, O., Bradley, D.G., Ochieng, J.W., Verjee, Y., Hill, E.W., Rege, J.E.O.: African Pastoralism: Genetic Imprints of Origins and Migrations. Science 296(5566), 336–339 (2002)
4. Vargo, E.: Hierarchical analysis of colony and population genetic structure of the eastern subterranean termite, reticulitermes flavipes, using two classes of molecular markers. Evolution 57(12), 2805–2818 (2003)
5. Li, M., Tapio, I., Vilkki, J., Ivanova, Z., Kiselyova, T., Marzanov, N., Cinkulov, M., Stojanovic, S., Ammosov, I., Popov, R., Kantanen, J.: The genetic structure of cattle populations (Bos taurus) in northern Eurasia and the neighbouring Near Eastern regions: implications for breeding strategies and conservation. Molecular Ecology 16(18), 3839–3853 (2007)
6. Rosenberg, N.A., Pritchard, J.K., Weber, J.L., Cann, H.M., Kidd, K.K., Zhivotovsky, L.A., Feldman, M.W.: Genetic Structure of Human Populations. Science 298(5602), 2381–2385 (2002)
7. Vural, V., Dy, J.G.: A hierarchical method for multi-class support vector machines. In: ICML 2004: Proceedings of the twenty-first international conference on Machine learning, pp. 105–112. ACM Press, New York (2004)
8. Kohonen, T.: Self-organizing maps. Springer-Verlag New York, Inc., Secaucus (1997)
9. Flexer, A.: On the use of self-organizing maps for clustering and visualization. In: Principles of Data Mining and Knowledge Discovery, pp. 80–88. Springer, Heidelberg (1999)
10. Vesanto, J.: Som-based data visualization methods. Intelligent Data Analysis 3, 111–126 (1999)
11. Nikkilä, J., Törönen, P., Kaski, S., Venna, J., Castrén, E., Wong, G.: Analysis and visualization of gene expression data using Self-Organizing Maps. Neural Networks 15(8-9), 953–966 (2002)
12. Johnson, S.: Hierarchical clustering schemes. Psychometrika 32(3), 241–254 (1967)
13. Vapnik, V.: Statistical Learning Theory. John Wiley, New York (1998)
14. Weston, J., Watkins, C.: Multi-class support vector machines. In: Verleysen, M. (ed.) Proceedings of the 6th European Symposium on Artificial Neural Networks (ESANN), D. Facto Press, Brussels (1999)
15. Crammer, K., Singer, Y.: On the algorithmic implementation of multiclass kernel-based vector machines. Journal of Machine Learning Research 2, 265–292 (2001)
16. Kreßel, U.: Pairwise classification and support vector machines. In: Schölkopf, B., Burges, C.J.C., Smola, A.J. (eds.) Advances in Kernel Methods – Support Vector Learning, pp. 255–268. MIT Press, Cambridge (1999)
17. Platt, J.C., Cristianini, N., Shawe-taylor, J.: Large margin dags for multiclass classification. In: Advances in Neural Information Processing Systems, pp. 547–553. MIT Press, Cambridge (2000)
18. Lei, H., Govindaraju, V.: Half-against-half multi-class support vector machines. In: Oza, N.C., Polikar, R., Kittler, J., Roli, F. (eds.) MCS 2005. LNCS, vol. 3541, pp. 156–164. Springer, Heidelberg (2005)
19. Benabdeslem, K., Bennani, Y.: Dendogram based svm for multi-class classification. In: 28th International Conference on Information Technology Interfaces, pp. 173–178 (2006)
20. Tibshirani, R., Hastie, T.: Margin Trees for High-dimensional Classification. Journal of Machine Learning Research 8, 637–652 (2007)

21. Asuncion, A., Newman, D.J.: UCI machine learning repository. School of Information and Computer Sciences. University of California, Irvine, California, USA (2007)
22. Boser, B.E., Guyon, I., Vapnik, V.: A training algorithm for optimal margin classifiers. Computational Learing Theory, 144–152 (1992)
23. Cristianini, N., Shawe-Taylor, J.: An Introduction to Support Vector Machines and other kernel-based learning methods. Cambridge University Press, Cambridge (2000)
24. Wu, T., Lin, C., Weng, R.: Probability Estimates for Multi-class Classification by Pairwise Coupling. The Journal of Machine Learning Research 5, 975–1005 (2004)
25. Demšar, J.: Statistical Comparisons of Classifiers over Multiple Data Sets. Journal of Machine Learning Research 7, 1–30 (2006)

Feature Weighting Using Margin and Radius Based Error Bound Optimization in SVMs

Huyen Do, Alexandros Kalousis, and Melanie Hilario

University of Geneva,
Computer Science Department,
7 Route De Drize, 1227, Carouge, Switzerland
{Huyen.Do,Alexandros.Kalousis,Melanie.Hilario}@unige.ch

Abstract. The Support Vector Machine error bound is a function of the margin and radius. Standard SVM algorithms maximize the margin within a given feature space, therefore the radius is fixed and thus ignored in the optimization.

We propose an extension of the standard SVM optimization in which we also account for the radius in order to produce an even tighter error bound than what we get by controlling only for the margin.

We use a second set of parameters, $\boldsymbol{\mu}$, that control the radius introducing like that an explicit feature weighting mechanism in the SVM algorithm. We impose an l_1 constraint on $\boldsymbol{\mu}$ which results in a sparse vector, thus performing feature selection. Our original formulation is not convex, we give a convex approximation and show how to solve it. We experiment with real world datasets and report very good predictive performance compared to standard SVM.

Keywords: Feature Weighting, Support Vector Machine, convex optimization.

1 Introduction

Support Vector Machines have been proven one of the most successful machine learning tools over the last decade. Their wide acceptance from the machine learning community has to do by the excellent performance that they achieve over different learning problems—excellent performance that is founded on sound theoretical concepts and namely the fact that they control directly their error bound. In the classification setting they work by establishing a class separating hyperplane in some feature space, typically given by a kernel function. The theory shows that their error bound is a function of both the margin, γ, informally the distance of the nearest data points from the hyperplane, and the radius, R, of the smallest sphere enclosing the data; more precisely, the function depends on the ratio R^2/γ^2.

However, even though this double dependency is well known, all SVM algorithms optimize the error bound by focusing only on the margin and ignoring the radius. One could argue that for a given feature space the radius of the smallest

W. Buntine et al. (Eds.): ECML PKDD 2009, Part I, LNAI 5781, pp. 315–329, 2009.

sphere enclosing the data remains fixed and thus can be ignored. However it is quite easy to show that under a very simple scenario, for example by weighting or selecting features, the radius changes. The question that then arises is what is the best way to control that change, i.e. best in the sense of the error bound optimization. This question has been partially explored in the context of feature selection; for example [1] and [2] directly try to exploit this dual dependency of the error bound in order to determine which features to remove and which to retain. SVMRFE [3] also implicitly changes the radius since it removes features; however since the algorithm is based on the recursive application of a standard SVM, its cost function obviously disregards the radius.

Nevertheless it still remains a challenge to exploit all the facets of the generalization bound in the learning process. In this paper we move one step ahead in this direction and propose an SVM algorithm that optimizes the error bound by controlling both for the margin and the radius. To do so we extend the standard margin-based cost function of SVM (which controls the margin by controlling the size of the norm of the normal vector, $\mathbf{w}$, of the maximum margin hyperplane), to include also the radius. We control the latter by introducing a second vector of parameters, $\boldsymbol{\mu}$, which in fact performs feature weighting; we call our algorithm MR-SVM (Margin-Radius SVM). The proposed algorithm includes a sparsity constraint on $\boldsymbol{\mu}$ which is based on the l_1 norm; this forces many features to have a zero weight, thus performing also feature selection.

The paper is organized as follows. In section 2 we briefly review previous work on SVM that deal with both the margin and the radius, mainly in the context of feature selection. In section 3 we discuss the various error bounds that motivate the use of the radius and present the standard SVM framework. We give our main contribution in section 4 and present some experiments on several benchmark datasets in section 5. Finally, we conclude in section 6 where we also present some ideas for future work.

2 Related Work

The idea of optimizing the error bound of SVM by controlling both the margin and the radius has received relatively limited attention. This despite the fact that one would expect, at least in principle, to be able to produce tighter error bounds when we control for both, thus achieving better generalization performance. Somehow naturally the only two works of which we are aware that move in that direction fall within the feature selection research domain.

Weston et al. in [1] start from exactly the same idea, i.e. that the generalization performance of SVM depends on the ratio R^2/γ^2 and not only on the margin γ and propose a feature selection algorithm which tries to optimize that ratio. To do so they introduce a binary valued vector $\boldsymbol{\sigma} \in \{0,1\}^d$, where each $\sigma_i, i := 1,\ldots,d$ corresponds to one of the dimensions of the input space. The $\boldsymbol{\sigma}$ controls which features are included or removed through a componentwise multiplication with the learning instances. They then express the ratio as a function of $\boldsymbol{\sigma}$, i.e. $f(\boldsymbol{\sigma}) = \frac{R^2}{\gamma^2}(\boldsymbol{\sigma})$. Their goal is to find that vector $\boldsymbol{\sigma}^*$

which minimizes $f(\boldsymbol{\sigma})$. However the original formulation of the problem requires searching over all possible feature subsets which is a combinatorial problem that is only tractable by greedy search methods. The authors propose an alternative approach in which they relax $\boldsymbol{\sigma}$ to be a real valued vector and formulate the problem as an approximation of integer programming which they solve by using gradient descent by computing the gradient $\partial f(\boldsymbol{\sigma})/\partial \boldsymbol{\sigma}$. There is no guarantee that the algorithm will reach a global minimum. Moreover since they view the problem as a feature selection problem they parametrize the ratio in a manner that includes or removes features on the basis of the original input space, i.e. the parametrization performed by $\boldsymbol{\sigma}$ is not done in the feature space. However the radius is computed in the feature space and obviously tighter error bounds can be achieved if the parametrization is done in the feature space, in the same way that the parametrization of the margin is done in the feature space and not in the original input space.

Rakotomamonjy in [2] examines a number of different SVM-based functions and feature ranking approaches to perform feature selection. Among the different functions he examines we find the radius-margin ratio as it was given in the previous paragraph, i.e. $f(\boldsymbol{\sigma})$. He explores two approaches to establish a ranking of the features from $f(\boldsymbol{\sigma})$, which he calls zero-order and first-order approaches. In the zero-order approach the importance of a feature is given by the value of $f(\boldsymbol{\sigma})$ when that feature has been removed. In other words the importance of feature i is given by $f(\boldsymbol{\sigma}^{(-i)})$, where $\sigma_{j\neq i}^{(-i)} = 1$ and $\sigma_i^{(-i)} = 0$. In the first-order approach the importance of a feature i is given by the value of the partial derivative $\partial f(\boldsymbol{\sigma})/\partial \sigma_i$ calculated at $\sigma_i = 1$. The two approaches are incorporated within a stepwise selection method to determine which features to retain. We note here that this work does not try to optimize $f(\boldsymbol{\sigma})$ but simply uses it as a way to determine the importance of the different features in the input space and decided which ones to retain and which to eliminate. Moreover the initial value of $f(\mathbf{1})$, where $\mathbf{1}$ is the vector of ones, may not be the optimal value and the method removes those features that have the smallest effect on the value of $f(\mathbf{1})$.

In both approaches the parametrization of the radius-margin ratio is achieved via the parametrization of the kernel function which is written in the form $K(\boldsymbol{\sigma}.\mathbf{x}_i, \boldsymbol{\sigma}.\mathbf{x}_j)$, where the operator . denotes the componentwise multiplication. [4] examined the more general problem of tuning any number of parameters of a kernel and proposed a framework that relies on the use of theoretical error bounds to perform that tuning. One of the error bounds they examined was the radius-margin ratio. Obviously the $K(\boldsymbol{\sigma}.\mathbf{x}_i, \boldsymbol{\sigma}.\mathbf{x}_j)$ form can be addressed within their framework, and in fact they did use their method and this kernel parametrization to perform feature selection and scaling. As in the two previous works, feature scaling and weighting is performed in the input space. The framework they proposed is based on a two step iterative optimization procedure. The first step solves a standard SVM problem in which the set of parameters $\boldsymbol{\sigma}$ is kept fixed. The second step uses gradient descent to update the parameters $\boldsymbol{\sigma}$ in the direction that minimizes the theoretical error bound, i.e. the radius-margin ratio. However the radius-margin ratio can have many local minima, see for example [5], and a simple gradient method may easily get trapped in one of them.

As it is apparent from the above discussion, one of the learning tasks in which SVMs have been used often is that of feature selection. Probably the most popular algorithm of this family is SVM-RFE, introduced by [3], which is a backwards feature elimination procedure that removes features based on the values of their weights as these are determined by a linear kernel SVM. In fact the work of [2] extends this work. The linear kernel has been proven one of the most popular kernels for SVM not only because of its very good predictive performance, but also because of its understandability; its weights directly reflect the importance of the corresponding features, provided that features have been normalized. These explain why it is the kernel of choice not only for the feature selection task but also more generally for the task of classification in problems with high dimensionality.

The same trend also appears in regression problems where some of the most successful algorithms for high dimensional problems are algorithms that produce linear models by imposing sparsity constraints on the vector of weights of the coefficients, e.g. Lasso, [6], LARS, [7], and more recently Adaptive Lasso, [8]. All of them use the l_1 norm to control the coefficients of the linear regression models, which is known to produce sparse weight vectors. In an interesting twist Adaptive Lasso uses two sets of weights: a first set of weights controls the importance of the different features and the second is the set of lasso coefficients learned on the weighted feature set. The author proposes to derive the first set of weights from the solution of the Ordinary Least Squares regression [9]. Note here that unlike the different regression algorithms mentioned the standard SVM does not impose strong sparsity constraints such as the one imposed by the l_1 norm.

3 SVM, Margin and Radius Based Error Bounds

There are a number of theorems in statistical learning that bound the expected classification error of the thresholded linear classifier given by the maximum margin hyperplane by quantities that are related to its margin and the radius of the smallest sphere that encloses the data. Below we give two of them that correspond to the cases of linearly separable and non-separable training sets.

Consider a mapping $\mathbf{x} \mapsto \mathbf{\Phi}(\mathbf{x})$ in which a training instance $\mathbf{x} \in \mathcal{X}$ is mapped to an inner product feature space $\mathcal{H}$. Moreover the inner product of $\mathcal{H}$ is given by $K(.,.)$, a kernel function defined in the original input space $\mathcal{X}$, i.e. $K(\mathbf{x}_i, \mathbf{x}_j) = \langle \mathbf{\Phi}(\mathbf{x}_i), \mathbf{\Phi}(\mathbf{x}_j) \rangle$.

Theorem 1. *[4], Given a training set $S = \{(\mathbf{x}_1, y_1), ..., (\mathbf{x}_l, y_l)\}$ of size l, a feature space $\mathcal{H}$ and a hyperplane $(\mathbf{w}, b)$, the margin $\gamma(\mathbf{w}, b, S)$ and the radius $R(S)$ are defined by*

$$\gamma(\mathbf{w}, b, S) = \min_{(\mathbf{x}_i, y_i) \in S} \frac{y_i(\langle \mathbf{w}, \mathbf{\Phi}(\mathbf{x}_i) \rangle + b)}{\|\mathbf{w}\|}$$

$$R(S) = \min_{\mathbf{a}} \max_{i} \|\mathbf{\Phi}(\mathbf{x}_i) - \mathbf{a}\|$$

The maximum margin algorithm $L_l : (\mathcal{X} \times \mathcal{Y})^l \rightarrow \mathcal{H} \times \mathbb{R}$ *takes as input a training set of size* l *and returns a hyperplane in feature space such that the margin* $\gamma(\mathbf{w}, b, S)$ *is maximized. Note that assuming the training set is separable means that* $\gamma > 0$. *Under this assumption, for all probability measures* P *underlying the data* S, *the expectation of the misclassification probability*

$$p_{err}(\mathbf{w}, b) = P(sign(\langle \mathbf{w}, \mathbf{\Phi}(\mathbf{X}) \rangle + b) \neq Y)$$

has the bound

$$E\{p_{err}(L_{l-1}(Z))\} \leq \frac{1}{l} E \left\{ \frac{R^2(Z)}{\gamma^2(L_l(Z), Z)} \right\}$$

The expectation is taken over the random draw of a training set Z *of size* $l-1$ *for the left hand side and* l *for the right hand side.*

The following theorem gives a similar result for the error bound of the linearly non-separable case.

Theorem 2. *[10], Consider thresholding real-valued linear functions* $\mathcal{L}$ *with unit weight vectors on an inner product space* $\mathcal{H}$ *and fix* $\gamma \in \mathcal{R}^+$. *There is a constant* c, *such that for any probability distribution* $\mathcal{D}$ *on* $\mathcal{H} \times \{-\infty, \infty\}$ *with support in a ball of radius* R *around the origin, with probability* $1 - \delta$ *over* l *random examples* S, *any hypothesis* $f \in \mathcal{L}$ *has error no more than:*

$$err(f)_{\mathcal{D}} \leq \frac{c}{l}(\frac{R^2 + \|\boldsymbol{\xi}\|_2^2}{\gamma^2} log^2 l + log\frac{1}{\delta}), \tag{1}$$

where $\boldsymbol{\xi} = \boldsymbol{\xi}(f, S, \gamma)$ *is the margin slack vector with respect to* f *and* γ.

It is clear from both theorems that the bound on the expected error depends not only on the margin but also on the radius of the data. The expected error is bounded in the linearly separable and non-separable cases by functions of the ratios R^2/γ^2 and $(R^2 + \|\boldsymbol{\xi}\|_2^2)/\gamma^2$ respectively.

The standard soft margin SVM builds exactly on these results, namely theorem 2, by learning maximal margin hyperplanes while controlling for the l_2 norm of the slack vector in an effort to optimize the error bound given in equation 1. Maximizing for the margin is equivalent to minimizing the norm of the normal vector, $\mathbf{w}$, of the separating hyperplane, thus the optimization problem that is solved by the standard soft margin SVM is given by:

$$\begin{aligned} &\min_{\boldsymbol{\xi}, \mathbf{w}, b} \ \frac{1}{2}\langle \mathbf{w}, \mathbf{w} \rangle + \frac{C}{2}\sum_{i=1}^{l} \xi_i^2 \\ &s.t \ \ y_i(\langle \mathbf{w}, \mathbf{\Phi}(\mathbf{x}_i) \rangle + b) \geq 1 - \xi_i, i = 1, \ldots, l \end{aligned} \tag{2}$$

C is a regularization parameter that controls the trade off between the size of the margin and the norm of the slack variables vector; the latter is directly related to the total number of training set missclassifications. Obviously this approach to error bound optimization focuses exclusively on the margin and ignores the radius; under this problem formulation the latter is fixed and optimizing the cost function of equation 2 is equivalent to optimizing the error bound given in equation 1. Usually we solve the dual optimization problem of equation 2; this is given by:

$$\max_{\boldsymbol{\alpha}} W(\boldsymbol{\alpha}) = \sum_{i}^{l} \alpha_i - \frac{1}{2}\sum_{ij}^{l} \alpha_i \alpha_j y_i y_j (K(\mathbf{x}_i, \mathbf{x}_j) + \frac{1}{C}\delta_{ij})$$

$$s.t \quad \sum_{i=1}^{l} y_i \alpha_i = 0, \alpha_i \geq 0, i = 1, ..., l.$$

where $\boldsymbol{a}$ is the vector of Langrange multipliers, and δ_{ij} is the Kronecker δ, defined to be 1 if $i = j$ and 0 otherwise. The decision function of SVM is $f(\mathbf{x}) = sign(\langle \mathbf{w}^*, \boldsymbol{\Phi}(\mathbf{x})\rangle + b^*) = sign(\sum_i^l y_i \alpha_i^* K(\mathbf{x}_i, \mathbf{x}) + b^*)$ where $\mathbf{w}^*$, b^* and α_i^* are the solutions of 2 and its dual form.

The radius of the smallest sphere that contains all instances $\mathbf{x}_i$ in the $\mathcal{H}$ feature space defined by the $\boldsymbol{\Phi}(\mathbf{x})$ mapping is computed by the following formula [11]:

$$\min_{R, \boldsymbol{\Phi}(\mathbf{x}_0)} \quad R^2 \tag{3}$$
$$s.t. \quad \|\boldsymbol{\Phi}(\mathbf{x}_i) - \boldsymbol{\Phi}(\mathbf{x}_0)\|^2 \leq R^2, \forall i$$

where $\boldsymbol{\Phi}(\mathbf{x}_0)$ is the center of the sphere.

4 Margin and Radius Based SVM

In this section we will show how it is possible to control not only for the margin but also for the radius in an effort to achieve better error bounds. Consider the feature space $\mathcal{H}$ given by the mapping function $\boldsymbol{\Phi}(\mathbf{x}) = (\Phi_1(\mathbf{x}), \Phi_2(\mathbf{x}), ..., \Phi_d(\mathbf{x})) \in \mathcal{R}^d$, where $\mathbf{x} \in \mathcal{X}$; let $\boldsymbol{\mu} \in R^d$ be a vector of parameters whose role will be to perform feature weighting in the feature space. Then the feature space $\mathcal{H}_{\boldsymbol{\mu}}$ given by the mapping $\boldsymbol{\Phi}_{\boldsymbol{\mu}}(\mathbf{x}) = \boldsymbol{\Phi}(\mathbf{x}).\sqrt{\boldsymbol{\mu}}$ is a feature space the radius of which is no longer constant but can be controlled by the weighting vector $\boldsymbol{\mu}$ ($\sqrt{\boldsymbol{\mu}}$ denotes the componentwise square root of $\boldsymbol{\mu}$). Therefore, if we use the standard SVM, which maximizes only the margin, we will not be fully optimizing the generalization bound given in equation 1. To exploit the full potential of the radius-margin error bound we should also learn the vector $\boldsymbol{\mu}$. Note that in the standard SVM formulation $\boldsymbol{\mu} = \mathbf{1}$, i.e. all features of the feature space have equal importance or weight. For reasons that will become apparent below, and without loss of generality, we will work in the normalized feature space $\mathcal{H}'$, i.e. $\boldsymbol{\Phi}'(\mathbf{x}) = \frac{\boldsymbol{\Phi}(\mathbf{x})}{\|\boldsymbol{\Phi}(\mathbf{x})\|}$; this normalization is also achieved by the kernel $K'(\mathbf{x}, \mathbf{y}) = \frac{K(\mathbf{x},\mathbf{y})}{\sqrt{K(\mathbf{x},\mathbf{x})K(\mathbf{y},\mathbf{y})}}$, where $K(\mathbf{x}, \mathbf{y})$ is the kernel associated with the $\boldsymbol{\Phi}(\mathbf{x})$ mapping. To simplify notation in the following we will be using $\mathcal{H}$, $\boldsymbol{\Phi}(\mathbf{x})$, and $K(\mathbf{x}, \mathbf{y})$, instead of $\mathcal{H}'$, $\boldsymbol{\Phi}'(\mathbf{x})$, and $K'(\mathbf{x}, \mathbf{y})$.

We will now discuss a number of inequalities that relate the radius of the $\mathcal{H}_{\boldsymbol{\mu}}$ space to the radii of the spaces defined by each one of its features. Consider the one-dimensional space given by $\mathcal{H}_k$, $\Phi_k(\mathbf{x})$, which we get by projecting $\mathcal{H}$ on its kth dimension. Its radius is given by $R_k = (\max_j(\Phi_k(\mathbf{x}_j)) - \min_j \Phi_k(\mathbf{x}_j))/2$. Let $R^2_{\boldsymbol{\mu}}$ be the radius of the feature space $\mathcal{H}_{\boldsymbol{\mu}}$. The following inequalities hold:

$$\max_k(\mu_k R_k^2) \leq R^2_{\boldsymbol{\mu}} \leq \sum_{k=1}^d \mu_k R_k^2 \leq \max_k(R_k^2) \leq 1 \qquad (4)$$
$$s.t. \ \sum_k^d \mu_k = 1, \mu_k \geq 0, \ \forall k$$

The inequality, $\max_k(R_k^2) \leq 1$, holds because the feature space $\mathcal{H}$ is normalized. The rest of the proof is given in the appendix.

We will now describe how we can optimize the radius-margin error bound given in equation 1 by learning also the $\boldsymbol{\mu}$ vector. A straightforward way to do so is to modify the cost function of the standard SVM so that it also includes the radius. We define the the following optimization problem in the $\mathcal{H}_{\boldsymbol{\mu}}$ feature space:

$$\min_{\mathbf{w},b,\boldsymbol{\xi},\boldsymbol{\mu}} \ \frac{1}{2}\langle \mathbf{w}, \mathbf{w}\rangle R^2_{\boldsymbol{\mu}} + \frac{C}{2}\sum_{i=1}^{l} \xi_i^2 \qquad (5)$$
$$s.t. \ y_i(\langle \mathbf{w}, \sqrt{\boldsymbol{\mu}}.\boldsymbol{\Phi}(\mathbf{x}_i)\rangle + b) \geq 1 - \xi_i, \forall i$$

By rewriting $\mathbf{w} := \sqrt{\boldsymbol{\mu}}.\mathbf{w}$ we also have $w_k := \sqrt{\mu_k} w_k$ and then we can rewrite equation 5 as:

$$\min_{\mathbf{w},b,\boldsymbol{\xi},\boldsymbol{\mu}} \ \frac{1}{2}\sum_{k}^{d} \frac{\langle w_k, w_k\rangle}{\mu_k} R^2_{\boldsymbol{\mu}} + \frac{C}{2}\sum_{i}^{l} \xi_i^2 \qquad (6)$$
$$s.t. \ y_i(\sum_k^d \langle w_k, \Phi_k(\mathbf{x}_i)\rangle + b) \geq 1 - \xi_i,$$
$$\sum_{k=1}^d \mu_k = 1, \mu_k \geq 0, \forall k$$

The constraint $\sum_{k=1}^d \mu_k = 1$ is added so that we get a unique solution. If $\mu_k = 0$ we will see that from the dual form given later we have $w_k = 0$; in this case, we use the convention that $\frac{0}{0} = 0$. Note the similarity of the two sets of parameters, $\mathbf{w}$ and $\boldsymbol{\mu}$, to the two sets of weights used by the Adaptive Lasso algorithm of [8], however here we simultaneously optimize over both sets of parameters within the same optimization problem and not independently as it is done in [8].

We will denote the cost function of equation 6 by $F(\mathbf{w}, b, \boldsymbol{\xi}, \boldsymbol{\mu})$. This optimization problem is not a convex optimization problem because the cost function F is not convex. From the set of inequalities given in equation 4 we have $R^2_{\boldsymbol{\mu}} \leq \sum_k^d \mu_k R_k^2 \leq 1$ and from this we can get:

$$F(\mathbf{w}, b, \boldsymbol{\xi}, \boldsymbol{\mu}) = \frac{1}{2}\sum_k^d \frac{\langle w_k, w_k\rangle}{\mu_k} R_{\boldsymbol{\mu}}^2 + \frac{C}{2}\sum_i^l \xi_i^2 \leq \tag{7}$$
$$\frac{1}{2}\sum_k^d \frac{\langle w_k, w_k\rangle}{\mu_k} \sum_k^d \mu_k R_k^2 + \frac{C}{2}\sum_i^l \xi_i^2 =$$
$$(\frac{1}{2}\sum_k^d \frac{\langle w_k, w_k\rangle}{\mu_k} + \frac{C}{2\sum_k^d \mu_k R_k^2}\sum_i^l \xi_i^2)\sum_k^d \mu_k R_k^2 \leq$$
$$(\frac{1}{2}\sum_k^d \frac{\langle w_k, w_k\rangle}{\mu_k} + \frac{C}{2\sum_k^d \mu_k R_k^2}\sum_i^l \xi_i^2) = \widetilde{F}(\mathbf{w}, b, \boldsymbol{\xi}, \boldsymbol{\mu})$$

The tighter the bound $R_{\boldsymbol{\mu}}^2 \leq \sum_k^d \mu_k R_k^2$ is, the closer $\widetilde{F}$ will be to the original error bound given by F. $\widetilde{F}(\mathbf{w}, b, \boldsymbol{\xi}, \boldsymbol{\mu})$ is a convex function so instead of the original soft margin optimization problem given in formula 6 we propose to solve the following upper bounding convex optimization problem:

$$\min_{\mathbf{w}, b, \boldsymbol{\xi}, \boldsymbol{\mu}} \frac{1}{2}\sum_k^d \frac{\langle w_k, w_k\rangle}{\mu_k} + \frac{C}{2\sum_k^d \mu_k R_k^2}\sum_i^l \xi_i^2 \tag{8}$$
$$s.t. \;\; y_i(\sum_k^d \langle w_k, \Phi_k(x_i)\rangle + b) \geq 1 - \xi_i,$$
$$\sum_{k=1}^d \mu_k = 1, \mu_k \geq 0, \forall k$$

The l_1 norm constraint on $\boldsymbol{\mu}$ has the potential to result in a sparser solution, i.e. many of the μ_i could be zero, which can not be obtained using standard SVM. The dual function of the optimization problem of equation 8 is:

$$W_s(\boldsymbol{\alpha}, \boldsymbol{\mu}) = -\frac{1}{2}\sum_{ij}^l \alpha_i\alpha_j y_i y_j \sum_k^d \mu_k \langle \Phi_k(\mathbf{x}_i), \Phi_k(\mathbf{x}_j)\rangle \tag{9}$$
$$+\sum_i^l \alpha_i - \frac{\sum_k^d \mu_k R_k^2}{2C}\langle \boldsymbol{\alpha}, \boldsymbol{\alpha}\rangle$$
$$= -\frac{1}{2}\sum_{ij}^l \alpha_i\alpha_j y_i y_j (\sum_k^d \mu_k \langle \Phi_k(\mathbf{x}_i), \Phi_k(\mathbf{x}_j)\rangle$$
$$+\frac{\sum_k^d \mu_k R_k^2}{C}\delta_{ij}) + \sum_i^l \alpha_i$$

The dual optimization problem is:

$$\max_{\boldsymbol{\alpha},\boldsymbol{\mu}} \; W_s(\boldsymbol{\alpha},\boldsymbol{\mu}) \tag{10}$$
$$s.t. \; \sum_i^l \alpha_i y_i = 0,$$
$$\alpha_i \geq 0, \forall i$$
$$\sum_{ij}^l \alpha_i \alpha_j y_i y_j \langle \Phi_k(\mathbf{x}_i), \Phi_k(\mathbf{x}_j)\rangle = \frac{R_k^2 C \sum_i^l \xi_i^2}{(\sum_k \mu_k R_k^2)^2}, \forall k,$$

As it is obvious the cost function and the constraints are not expressed in the form of a kernel function on the feature space $\mathcal{H}$ but instead require access to its explicit representation. This limits for the moment the application of the method that we propose here only to features spaces for which we have access to their explicit form, e.g. linear or polynomial feature spaces. In the next section we will show how we can solve this optimization problem.

4.1 Algorithm

The dual function 9 is quadratic with respect to $\boldsymbol{\alpha}$ and linear with respect to $\boldsymbol{\mu}$. One way to solve the optimization problem 8 is by using a two step iterative algorithm such as the ones described in [4,12]. Following such a two step approach, in the first step we will solve a quadratic problem that optimizes over $(\mathbf{w}, b)$, while keeping $\boldsymbol{\mu}$ fixed; as a consequence the resulting dual function is a simple quadratic function of $\boldsymbol{\alpha}$ which can be optimized easily. In the second step we will solve a linear problem that optimizes over $\boldsymbol{\mu}$. More precisely the formulation of the optimization problem with the two-step approach takes the following form:

$$\min_{\boldsymbol{\mu}} \; J(\boldsymbol{\mu}) \tag{11}$$
$$s.t. \; \sum_{k=1}^d \mu_k = 1, \mu_k \geq 0, \forall k$$

where

$$J(\boldsymbol{\mu}) = \begin{cases} \min_{\mathbf{w},b} \; \frac{1}{2}\sum_k^d \frac{\langle w_k, w_k\rangle}{\mu_k} + \frac{C}{\sum_k^d \mu_k R_k^2} \sum_i^l \xi_i^2 \\ s.t. \quad y_i(\sum_k^d \langle w_k, \Phi_k(\mathbf{x}_i)\rangle + b) \geq 1 - \xi_i \end{cases} \tag{12}$$

To solve the outer optimization problem, i.e. $\min_{\boldsymbol{\mu}} J(\boldsymbol{\mu})$, we use gradient descent.

At each iteration, we fix $\boldsymbol{\mu}$, compute the value of $J(\boldsymbol{\mu})$ and then compute the gradient of $J(\boldsymbol{\mu})$ with respect to $\boldsymbol{\mu}$. The dual function of equation 12 is the $W_s(\boldsymbol{\alpha}, \boldsymbol{\mu})$ function already given in equation 9. Since $\boldsymbol{\mu}$ is fixed we now optimize only over $\boldsymbol{\alpha}$ (the resulting dual optimization problem is much simpler compared to the original soft margin dual optimization problem given in formula 10):

$$\max_{\boldsymbol{\alpha}} W_s(\boldsymbol{\alpha}, \boldsymbol{\mu})$$
$$s.t. \sum_i^l \alpha_i y_i = 0, \alpha_i \geq 0, \forall i$$

which has the same form as the SVM quadratic optimization problem, the only difference is that the C parameter here is equal to $\frac{C}{\sum_k^d \mu_k R_k^2}$.

For the strong duality, at the optimal solution $\boldsymbol{\alpha}^*$, the value of $W_s(\boldsymbol{\alpha}, \boldsymbol{\mu})$, and thus the $J(\boldsymbol{\mu})$ value, is given by:

$$W_s(\boldsymbol{\alpha}^*, \boldsymbol{\mu}) = -\frac{1}{2}\sum_{ij}^l \alpha_i^* \alpha_j^* y_i y_j (\sum_k^d \mu_k \langle \Phi_k(\mathbf{x}_i), \Phi_k(\mathbf{x}_j)\rangle$$
$$+ \frac{\sum_k^d \mu_k R_k^2}{C}\delta_{ij}) + \sum_i^l \alpha_i^*$$

The last step of the algorithm is to compute the gradient of the $J(\boldsymbol{\mu})$ function, formula 12, with respect to $\boldsymbol{\mu}$. As [4] and [12] have pointed out, we can use the theorem of Bonnans and Shapiro, [13], to compute gradients of such functions. Hence:

$$\frac{\partial J(\boldsymbol{\mu})}{\partial \mu_k} = -\frac{1}{2}\sum_{ij}^l \alpha_i^* \alpha_j^* y_i y_j (\langle \Phi_k(\mathbf{x}_i), \Phi_k(\mathbf{x}_j)\rangle + \frac{R_k^2}{C}\delta_{ij})$$

To compute the optimal step in the gradient descent we used line search. The complete two-step procedure is given in algorithm 1.

Algorithm 1. MR-SVM

Initialize $\mu_k^1 = \frac{1}{d}$ for $k = 1, ..., d$
repeat
  Set $R_{\boldsymbol{\mu}}^2 = \sum_k^d \mu_k^t R_k^2$
  compute $J(\boldsymbol{\mu}^t)$ as the solution of a quadratic optimization problem with given μ_k^t

  compute $\frac{\partial J}{\partial \mu_k}$ for $k = 1, ..., d$
  compute optimal step γ_t
  $\boldsymbol{\mu}^{t+1} \leftarrow \boldsymbol{\mu}^t + \gamma_t \frac{\partial J(\boldsymbol{\mu})}{\partial \boldsymbol{\mu}}$
until *stopCriterion* is *true*

4.2 Computational Complexity

At each step of the iteration we have to compute the solution of a standard SVM, with a fixed $\boldsymbol{\mu}$ and C equal to $\frac{C}{\sum_k^d \mu_k R_k^2}$, which has a complexity $O(n^3)$ where n

is number of instances. Moreover when $\boldsymbol{\mu}$ is updated we have to recompute the approximation of $R^2_{\boldsymbol{\mu}}$, a computation that is linear in the number of features, $O(d)$, where d is number of features.

5 Experiments

We experimented with 12 different datasets. Six of them, *Ionosphere, Liver, Sonar, Wdbc, Wpbc, Musk1*, were taken from the UCI repository, and six, *Colon-Cancer, CentralNervousSystem, FemaleVsMale, Leukemia, stroke, ovarian*, [14], from the domain of genomics and proteomics. A short description of the datasets is given in Table 1. In the experiments we limit ourselves to the linear feature space, although as we mentioned previously any feature space for which we have access to its explicit form can be used. We compare the performance of the standard SVM with the linear kernel, STD-SVM, to that of MR-SVM. We tuned the hyperparameter C by inner cross-validation choosing from the set of values $\{0.1, 1, 10, 100, 500, 1000\}$. We terminate MR-SVM when the duality gap is smaller than 0.01. We estimated the classification error using 10-fold cross validation. In table 2 we give the results including: the classification errors of both algorithms, the number of non-zero weight features selected by MR-SVM, the percentage that these non-zero weight features represent with respect to the total number of features, and the result of McNemar's test of significance with a significance level of 0.05 (if MR-SVM is significantly better than standard SVM we denote that by +, if it is significantly worse by -, and if they are equivalent by =).

As it is obvious from the results the performance of MR-SVM is much better than that of standard SVM. Its classification performance is significantly better in seven out of the 12 datasets and significantly worse only in two of them. Remember here that the two algorithms operate in exactly the same space, i.e.

Table 1. Datasets Description

Dataset	#Inst.	#Features	#Class1	#Class2
Ionosphere	351	34	126	225
Liver	345	6	145	200
Sonar	208	60	97	111
Wdbc	569	32	357	212
Wpbc	198	34	151	47
Musk1	476	166	269	207
ColonCancer	62	2000	40	22
CentralNervous	60	7129	21	39
FemaleVSMale	134	1524	67	67
Leukemia	72	7128	25	47
Stroke	208	171	101	107
Ovarian	253	385	62	91
Prostate	322	390	253	69

Table 2. Results of the experiments. The average errors of standard SVM and MR-SVM are reported, together with the average number of non-zero weight features established by MR-SVM and the respective number of total features. The last column gives the results of the McNemar's significance test, + means that MR-SVM is significantly better than the standard SVM.

DATASET	STD-SVM	MR-SVM	$\#\mu_i \neq 0$	$\frac{\#\mu_i \neq 0}{\#Features}$	SIG.
IONOSPHERE	11.71	11.14	28.3	83.23	+
LIVER	32.35	32.35	5.9	98.33	=
SONAR	24.5	23.00	44.3	73.88	+
WDBC	1.96	2.50	16.7	52.18	-
WPBC	18.95	17.37	24.8	72.94	+
MUSK1	13.62	13.83	68.1	41.02	=
COLONCANCER	15.00	16.67	2000	100	=
CENTRALNERVOUS	38.33	31.67	6442	90.36	+
FEMALEVSMALE	13.08	11.54	1524	100	+
LEUKEMIA	1.43	1.43	6922	97.10	=
STROKE	28.00	26.00	76.8	44.91	+
OVARIAN	04.80	3.60	53.2	13.81	+
PROSTATE	18.44	20.00	93.07	23.86	-

the one that corresponds to the linear kernel. Their only difference is that MR-SVM directly optimizes for both the radius and the margin and not just the margin as the standard SVM does. This has the potential, at least in theory, to produce lower error bounds than those we would get by optimizing only for the margin—a fact that seems to be confirmed by the performance results we get.

Apart from the very good classification performance that MR-SVM achieves we also get for many of the datasets, though not for all of them, a significant reduction of the number of features actually used in the learned model, many of them are assigned a μ_i that has a value of zero due to the use of the l_1 constraint. The mean value of features retained is around 68.6% of the total number of features over the different datasets.

6 Discussion and Future Work

In this paper we present an extension of the standard SVM that incorporates in its cost function not only the margin but also the radius of the smallest sphere that encloses the data, thus implementing directly well known error bounds from statistical learning theory. Our experimental results show that indeed optimizing the error bound accounting for both the radius and the margin leads to much better classification performance than when we optimize only for the margin as it is typically done in the standard SVM algorithms. We want to further verify the preliminary results reported here by experimenting with more datasets; if these are verified, and provided that we are able to provide a kernelized version, the proposed algorithm has a potential of very wide application.

A possible way of producing a kernelized version of our algorithm is by using kernel PCA [15] and expressing the original feature space by projecting it on its PCA components. In this case we do not need the explicit representation of the feature space, and we will work on the representation given by the projections on the PCA dimensions of the feature space. Since these can be computed explicitly we will be able to directly apply our algorithm on the transformed space.

Apart from the remarkable predictive performance, a further advantage of the algorithm comes as a result of the way we control the radius through the introduction of the $\boldsymbol{\mu}$ vector of parameters. This vector weights the different features in the feature space; moreover due to the incorporated sparsity constraint on $\boldsymbol{\mu}$ the algorithm has the potential to produce sparser linear models than those of the standard SVM. The result is that we do not only get a feature weighting mechanism but we also perform direct feature selection. However we would like to explore further the l_1 constraint defined on the $\boldsymbol{\mu}$ vector in order to see whether it is possible to control the desired level of sparsity in the final models. Remember that we set that constraint to one so that we get a unique solution. For some of the datasets this resulted in quite sparse solutions retaining as few as 14% or 24% of the original features, yet in others it retained all of them. We want to understand better the conditions under which this constraint becomes active and removes a significant number of features. Moreover we would like to explore the option of regularizing this norm as it is done in other methods, such as Lasso, in order to get even sparser solutions.

Finally we should mention that it is straightforward to use our algorithm, in its present form, within the SVM-RFE algorithm in order to replace the standard linear kernel SVM used there. Its advantage over the standard SVM will be the fact that at each iteration we can potentially remove a larger number of features the weight of which will be zero due the sparsity constraint and with a better predictive performance as our results indicate.

Acknowledgments. The work reported in this paper was partially funded by the European Commission through EU projects DropTop (FP6-037739), DebugIT (FP7-217139) and e-LICO (FP7-231519).

References

1. Weston, J., Mukherjee, S., Chapelle, O., Pontil, M., Poggio, J., Vapnik, V.: Feature selection for svms. Advances in Neural Information Processing Systems 13, 668–674 (2000)
2. Rakotomamonjy, A.: Variable selection using svm-based criteria. Journal of Machine Learning Research 3, 1357–1370 (2003)
3. Guyon, I., Weston, J., Barnhill, S., Vapnik, V.: Gene selection for cancer classification using suppor vector machine. Machine Learning 46, 389–422 (2002)
4. Chapelle, O., Vapnik, V., Bousquet, O., Mukherjee, S.: Choosing multiple parameters for support vector machines. Machine Learning 46(1-3), 131–159 (2002)
5. Duan, K., Keerthi, S.S., Poo, A.N.: Evaluation of simple performance measures for tuning svm hyperparameters. Neurocomputing 51, 41–59 (2002)

6. Tibshirani, R.: Regression shrinkage and selection via the lasso. Roal statistics 58, 276–288 (1996)
7. Efron, B., Hastie, T., Tibshirani, R.: Least angle regression. Annals of statistics (2003)
8. Zou, H.: The adaptive lasso and its oracle properties. Journal of the American statistical association 101, 1418–1429 (2006)
9. Hastie, T., Tibshirani, R., Friedman, J.: The elements of statistical learning theory. Springer, Heidelberg (2001)
10. Cristianini, N., Shawe-Taylor, J.: An introduction to Support Vector Machines. Cambridge University Press, Cambridge (2000)
11. Vapnik, V.: Statistical learning theory. Wiley Interscience, Hoboken (1998)
12. Bach, F., Rakotomamonjy, A., Canu, S., Grandvalet, Y.: SimpleMKL. Journal of Machine Learning Research (2008)
13. Bonnans, J., Shapiro, A.: Optimization problems with perturbation: A guided tour. SIAM Review 40(2), 202–227 (1998)
14. Kalousis, A., Prados, J., Hilario, M.: Stability of feature selection algorithms: a study on high dimensional spaces. Knowledge and Information Systems 12(1), 95–116 (2007)
15. Shawe-Taylor, J., Cristianini, N.: Kernel methods for pattern analysis. Cambridge University Press, Cambridge (2004)
16. Leo Liberti, N.M.: Global OPtimization - From Theory to Implementation. Springer, Heidelberg (2006)
17. Collobert, R., Weston, J., Bottou, L.: Trading convexity for scalability. In: Proceedings of the 23th Conference on Machine Learning (2006)
18. Stephen Boyd, L.V. (ed.): Convex optimization. Cambridge University Press, Cambridge (2004)

Appendix

Proof of inequality (4). If $K_{\boldsymbol{\mu}}(\mathbf{x}, \mathbf{x}')$ is the kernel function associated with the $\mathbf{\Phi}_{\boldsymbol{\mu}}(\mathbf{x})$ mapping then the computation of the radius in the dual form is given by [15]:

$$\max_{\beta_i \beta_j} R^2 = \sum_{i}^{l} \beta_i K(\mathbf{x}_i, \mathbf{x}_i) - \sum_{ij}^{l} \beta_i \beta_j K(\mathbf{x}_i, \mathbf{x}_j) \tag{13}$$

$$s.t. \sum_{i}^{l} \beta_i = 1, \beta_i \geq 0$$

We also introduce the kernels $K_{\boldsymbol{\mu}}$ and K_k as follows: $K_{\boldsymbol{\mu}}(\mathbf{x}_i, \mathbf{x}_j) = \langle \boldsymbol{\Phi}_{\boldsymbol{\mu}}(\mathbf{x}_i), \boldsymbol{\Phi}_{\boldsymbol{\mu}}(\mathbf{x}_j) \rangle = \langle \sqrt{\boldsymbol{\mu}}.\boldsymbol{\Phi}(\mathbf{x}_i), \sqrt{\boldsymbol{\mu}}\boldsymbol{\Phi}(\mathbf{x}_j) \rangle = \sum_{k=1}^{d} \mu_k \langle \Phi_k(\mathbf{x}_i), \Phi_k(\mathbf{x}_j) \rangle = \sum_{k=1}^{d} \mu_k K_k(\mathbf{x}_i, \mathbf{x}_j)$, i.e., $K_{\boldsymbol{\mu}}$ is the kernel, inner product, in the weighted feature space, $\mathcal{H}_{\boldsymbol{\mu}}$, and $K_k(\mathbf{x}_i, \mathbf{x}_j) = \langle \Phi_k(\mathbf{x}_i), \Phi_k(\mathbf{x}_j) \rangle$, is the trivial kernel, inner product, on the kth dimension, $\mathcal{H}_k$, of the non-weighted feature space $\mathcal{H}$. Remember that the solution $\boldsymbol{\mu}$ satisfies: $\sum_{k=1}^{d} \mu_k = 1$.

If β^* is the optimal solution of (13) when $K = K_{\boldsymbol{\mu}}$, and $\hat{\beta}^k$ is the optimal solution of (13) when $K = K_k$, i.e. :

$$R_{\boldsymbol{\mu}}^2 = \sum_{k=1}^{d} \mu_k (\sum_{i=1}^{l} \beta_i^* K_k(\mathbf{x}_i, \mathbf{x}_i) - \sum_{i,j=1}^{l} \beta_i^* \beta_j^* K_k(\mathbf{x}_i, \mathbf{x}_j))$$

$$R_k^2 = \sum_{i=1}^{l} \hat{\beta^k}_i K_k(\mathbf{x}_i, \mathbf{x}_i) - \sum_{i,j=1}^{l} \hat{\beta^k}_i \hat{\beta^k}_j K_k(\mathbf{x}_i, \mathbf{x}_j)$$

then

$$\sum_{i=1}^{l} \beta_i^* K_k(\mathbf{x}_i, \mathbf{x}_i) - \sum_{i,j=1}^{l} \beta_i^* \beta_j^* K_k(\mathbf{x}_i, \mathbf{x}_j) \leq$$

$$\sum_{i=1}^{l} \hat{\beta^k}_i K_k(\mathbf{x}_i, \mathbf{x}_i) - \sum_{i,j=1}^{l} \hat{\beta^k}_i \hat{\beta^k}_j K_k(\mathbf{x}_i, \mathbf{x}_j)$$

Therefore: $R_{\boldsymbol{\mu}}^2 \leq \sum_{k=1}^{d} \mu_k R_k^2$. This still obviously holds true when K is linear kernel.

Proof of convexity of R-MKL (Eq.8). To prove that 8 is convex, it is enough to show that functions $\frac{x^2}{\mu}$, where $x \in \mathbb{R}$, $\mu \in \mathbb{R}^+$; and $\frac{\xi^2}{\sum_k^M \alpha_k \mu_k}$, where $\xi \in \mathbb{R}, \mu_k, \alpha_k \in \mathbb{R}^+$, are convex. The former is quadratic-over-linear function which is convex. The later is convex because its epigraph is a convex set [18].

Margin and Radius Based Multiple Kernel Learning

Huyen Do, Alexandros Kalousis, Adam Woznica, and Melanie Hilario

University of Geneva, Computer Science Department
7, route de Drize, Battelle batiment A, 1227 Carouge, Switzerland
{Huyen.Do,Alexandros.Kalousis,Adam.Woznica,Melanie.Hilario}@unige.ch

Abstract. A serious drawback of kernel methods, and Support Vector Machines (SVM) in particular, is the difficulty in choosing a suitable kernel function for a given dataset. One of the approaches proposed to address this problem is Multiple Kernel Learning (MKL) in which several kernels are combined adaptively for a given dataset. Many of the existing MKL methods use the SVM objective function and try to find a linear combination of basic kernels such that the separating margin between the classes is maximized. However, these methods ignore the fact that the theoretical error bound depends not only on the margin, but also on the radius of the smallest sphere that contains all the training instances. We present a novel MKL algorithm that optimizes the error bound taking account of both the margin and the radius. The empirical results show that the proposed method compares favorably with other state-of-the-art MKL methods.

Keywords: Learning Kernel Combination, Support Vector Machines, convex optimization.

1 Introduction

Over the last few years kernel methods [1,2], such as Support Vector Machines (SVM), have proved to be efficient machine learning tools. They work in a feature space implicitly defined by a positive semi-definite kernel function, which allows the computation of inner products in feature spaces using only the objects in the input space.

The main limitation of kernel methods stems from the fact that in general it is difficult to select a kernel function, and hence a feature mapping, that is suitable for a given problem. To address this problem several several attempts have been recently made to learn kernel operators directly from the data [3,4,5,6,7,8,9,10,11,12]. The proposed methods differ in the objective functions (e.g. CV risk, margin based, alignment, etc.) as well as in the classes of kernels that they consider (e.g. combination of finite or infinite set of basic kernels).

The most popular approach in the context of kernel learning considers a finite set of predefined *basic kernels* which are combined so that the margin-based

W. Buntine et al. (Eds.): ECML PKDD 2009, Part I, LNAI 5781, pp. 330–343, 2009.

objective function of SVM is optimized. The learned kernel K is a linear combination of basic kernels K_i, i.e. $K(\boldsymbol{x},\boldsymbol{x}') = \sum_{i=1}^{M} \mu_i K_i(\boldsymbol{x},\boldsymbol{x}')$, $\mu_i \geq 0$, where M is the number of basic kernels, and $\boldsymbol{x}$ and $\boldsymbol{x}'$ are input objects. The weights μ_i of the kernels are included in the margin-based objective function. This setting is commonly referred to as the Multiple Kernel Learning (MKL).

The MKL formulation has been introduced in [3] as a semi-definite programming problem, which scaled well only for small problems. [7] extended that work and proposed a faster method based on the conic duality of MKL and solved the problem using Sequential Minimal Optimization (SMO). [5] reformulated the MKL problem as semi-infinite linear problem. In [6] the authors proposed an adjustment in the cost function of [5] to improve predictive performance. Although the MKL approach to kernel learning has some limitations (e.g. one has to choose the basic kernels), it is widely used because of its simplicity, interpretability and good performance.

The MKL methods that use the SVM objective function do not exploit the fact that the error bound of SVM depends not only on the separating margin, but also on the radius of the smallest sphere that encloses the data. In fact even the standard SVM algorithms do not exploit the latter, because for a given feature space the radius is fixed. However in the context of MKL the radius is not fixed but is a function of the weights of the basic kernels.

In this paper we propose a novel MKL method that takes account of both radius and margin to optimize the error bound. Following a number of transformations, these problems are cast in a form that can be solved by the two step optimization algorithm given in [6].

The paper is organized as follows. In Section 2 we introduce the general MKL framework. Next, in Section 3 we discuss the various error bounds that motivate the use of the radius. The main contribution of the work is presented in Section 4 where we propose a new method for multiple kernel learning that aims to optimize the margin- and radius-dependent error bound. In Section 5 we present the empirical results on several benchmark datasets. Finally, we conclude with Section 6 where we also present pointers to future work.

2 Multiple Kernel Learning Problem

Consider a mapping of instances $\mathbf{x} \in \mathcal{X}_i$, to a new feature space $\mathcal{H}_i$

$$\mathbf{x} \rightarrow \mathbf{\Phi}_i(\mathbf{x}) \in \mathcal{H}_i \tag{1}$$

This mapping can be performed by a kernel function $K_i(\mathbf{x},\mathbf{x}')$ which is defined as the inner product of the images of two instances $\mathbf{x}$ and $\mathbf{x}'$ in $\mathcal{H}_i$, i.e. $K_i(\mathbf{x},\mathbf{x}') = \langle \mathbf{\Phi}_i(\mathbf{x}), \mathbf{\Phi}_i(\mathbf{x}')\rangle$; $\mathcal{H}_i$ may have even infinite dimensionality. Typically, the computation of the inner product in $\mathcal{H}_i$ is done implicitly, i.e. without having to compute explicitly the images $\mathbf{\Phi}_i(\mathbf{x})$ and $\mathbf{\Phi}_i(\mathbf{x}')$.

2.1 Original Problem Formulation of MKL

Given a set of training examples $S = \{(\mathbf{x}_1, y_1), ..., (\mathbf{x}_l, y_l)\}$ and a set of basic kernel functions, $\boldsymbol{Z} = \{K_i(\mathbf{x},\mathbf{x}')|i := 1, \ldots M\}$, the goal of MKL is to optimize a cost function $Q(f(\boldsymbol{Z},\boldsymbol{\mu})(\mathbf{x},\mathbf{x}'),S)$ where $f(\boldsymbol{Z},\boldsymbol{\mu})(\mathbf{x},\mathbf{x}')$ is some positive semi-definite function of the set of the basis kernels, parametrized by $\boldsymbol{\mu}$; most often a linear combination of the form:

$$f(\boldsymbol{Z},\boldsymbol{\mu})(\mathbf{x},\mathbf{x}') = \sum_{i=1}^{M} \mu_i K_i(\mathbf{x},\mathbf{x}'), \mu_i \geq 0, \sum_{i}^{M} \mu_i = 1 \quad (2)$$

To simplify notation we will denote $f(\boldsymbol{Z},\boldsymbol{\mu})$ by $K_{\boldsymbol{\mu}}$. In the remaining part of this work we will only focus on the normalized versions of K_i, defined as:

$$K_i(\mathbf{x},\mathbf{x}') := \frac{K_i(\mathbf{x},\mathbf{x}')}{\sqrt{K_i(\mathbf{x},\mathbf{x}) \cdot K_i(\mathbf{x}',\mathbf{x}')}}. \quad (3)$$

If $K_{\boldsymbol{\mu}}$ is a linear combination of kernels then its feature space $\mathcal{H}_{\boldsymbol{\mu}}$ is given by the mapping:

$$\mathbf{x} \rightarrow \boldsymbol{\Phi}_{\boldsymbol{\mu}}(\mathbf{x}) = (\sqrt{\mu_1}\boldsymbol{\Phi}_1(\mathbf{x}), ..., \sqrt{\mu_M}\boldsymbol{\Phi}_M(\mathbf{x}))^T \in \mathcal{H}_{\boldsymbol{\mu}} \quad (4)$$

where $\boldsymbol{\Phi}_i(\mathbf{x})$ is the mapping to the $\mathcal{H}_i$ feature space associated with the K_i kernel, as this was given in Formula 1.

In previous work within the MKL context the cost function, Q, has taken different forms such as the Kernel Target Alignment, which measures the "goodness" of a kernel for a given learning task [9], or the typical SVM cost function combining classification error and the margin [3,5,6], or as in [4] any of the above with an added regularization term for the complexity of the combined kernel.

3 Margin and Radius Based Error Bounds

There are a number of theorems in statistical learning that bound the expected classification error of the thresholded linear classifier, that corresponds to the maximum margin hyperplane, by quantities that are related to the margin and the radius of the smallest sphere that encloses the data. Below we give two of them that are applicable on linearly separable and non-separable training sets, respectively.

Theorem 1. *[10], Given a training set* $S = \{(\mathbf{x}_1, y_1), ..., (\mathbf{x}_l, y_l)\}$ *of size* l*, a feature space* $\mathcal{H}$ *and a hyperplane* $(\mathbf{w}, b)$*, the margin* $\gamma(\mathbf{w}, b, S)$ *and the radius* $R(S)$ *are defined by*

$$\gamma(\mathbf{w}, b, S) = \min_{(\mathbf{x}_i, y_i) \in S} \frac{y_i(\langle \mathbf{w}, \boldsymbol{\Phi}(\mathbf{x}_i)\rangle + b)}{\|\mathbf{w}\|}$$
$$R(S) = \min_{\mathbf{a}} \max_{i} \|\boldsymbol{\Phi}(\mathbf{x}_i) - \mathbf{a}\|$$

The maximum margin algorithm $L_l : (\mathcal{X} \times \mathcal{Y})^l \rightarrow \mathcal{H} \times \mathbb{R}$ *takes as input a training set of size* l *and returns a hyperplane in feature space such that the margin* $\gamma(\mathbf{w}, b, S)$ *is maximized. Note that assuming the training set is separable means* *that $\gamma > 0$. Under this assumption, for all probability measures P underlying the data S, the expectation of the misclassification probability*

$$p_{err}(\mathbf{w}, b) = P(sign(\langle \mathbf{w}, \mathbf{\Phi}(\mathbf{X})\rangle + b) \neq Y)$$

has the bound

$$E\{p_{err}(L_{l-1}(Z))\} \leq \frac{1}{l} E\left\{\frac{R^2(Z)}{\gamma^2(L_l(Z), Z)}\right\}$$

The expectation is taken over the random draw of a training set Z of size $l-1$ for the left hand side and l for the right hand side.

The following theorem gives a similar result for the error bound of the linearly non-separable case.

Theorem 2. *[13], Consider thresholding real-valued linear functions $\mathcal{L}$ with unit weight vectors on an inner product space $\mathcal{H}$ and fix $\gamma \in \mathcal{R}^+$. There is a constant c, such that for any probability distribution $\mathcal{D}$ on $\mathcal{H} \times \{-\infty, \infty\}$ with support in a ball of radius R around the origin, with probability $1-\delta$ over l random examples S, any hypothesis $f \in \mathcal{L}$ has error no more than:*

$$err(f)_{\mathcal{D}} \leq \frac{c}{l}\left(\frac{R^2 + \|\boldsymbol{\xi}\|_2^2}{\gamma^2} log^2 l + log\frac{1}{\delta}\right), \tag{5}$$

where $\boldsymbol{\xi} = \boldsymbol{\xi}(f, S, \gamma)$ is the margin slack vector with respect to f and γ.

It is clear from both theorems that the bound on the expected error depends not only on the margin but also on the radius of the data, being a function of the R^2/γ^2 ratio. Nevertheless standard SVM algorithms can ignore the dependency of the error bound on the radius because for a fixed feature space the radius is constant and can be simply ignored in the optimization procedure. However in the MKL scenario where the $\mathcal{H}_{\boldsymbol{\mu}}$ feature space is not fixed but depends on the parameter vector $\boldsymbol{\mu}$ the radius is no longer fixed but it is a function of $\boldsymbol{\mu}$ and thus should not be ignored in the optimization procedure. The radius of the smallest sphere that contains all instances in the $\mathcal{H}$ feature space defined by the $\mathbf{\Phi}(\mathbf{x})$ mapping is computed by the following formula [14]:

$$\min_{R, \Phi(\mathbf{x}_0)} \quad R^2 \tag{6}$$
$$s.t. \quad \|\mathbf{\Phi}(\mathbf{x}_i) - \mathbf{\Phi}(\mathbf{x}_0)\|^2 \leq R^2, \forall i$$

It can be shown that if $K_{\boldsymbol{\mu}}$ is a linear combination of kernels, of the form given in Formula 2, then for the $R_{\boldsymbol{\mu}}$ radius of its $\mathcal{H}_{\boldsymbol{\mu}}$ feature space the following inequalities hold:

$$\max_i(\mu_i R_i^2) \leq R_{\boldsymbol{\mu}}^2 \leq \sum_{i=1}^{M} \mu_i R_i^2 \leq \max_i(R_i^2), \tag{7}$$
$$s.t. \sum_i^M \mu_i = 1$$

where R_i is the radius of the component feature space $\mathcal{H}_i$ associated with the K_i kernel. The proof of the above statement is given in the appendix.

4 MKL with Margin and Radius Optimization

In the next sections we will show how we can make direct use of the dependency of the error bound both on the margin and the radius in the context of the MKL problem in an effort to decrease even more the error bound than what is possible by optimizing only over the margin.

4.1 Soft Margin MKL

The standard l_2-soft margin SVM is based on theorem 2 and learns maximal margin hyperplanes while controlling for the l_2 norm of the slack vector in an effort to optimize the error bound given in equation 5; as already mentioned previously the radius although it appears in the error bound is not considered in the optimization problem due to the fact that for a given feature space it is fixed. The exact optimization problem solved by the l_2-soft margin SVM is [13]:

$$\min_{\mathbf{w},b,\boldsymbol{\xi}} \frac{1}{2}\langle \mathbf{w}, \mathbf{w}\rangle + \frac{C}{2}\sum_{i=1}^{l} \xi_i^2 \tag{8}$$
$$s.t. \ y_i(\langle \mathbf{w}, \boldsymbol{\Phi}(\mathbf{x}_i)\rangle + b) \geq 1 - \xi_i, \forall i$$

The solution hyperplane $(\mathbf{w}^*, b^*)$ of this problem realizes the maximum margin classifier with geometric margin $\gamma = \frac{1}{\|\mathbf{w}^*\|}$.

When instead of a single kernel we learn with a combination of kernels $K_{\boldsymbol{\mu}}$ then the radius of the resulting feature space $\mathcal{H}_{\boldsymbol{\mu}}$ depends on the parameters $\boldsymbol{\mu}$ which are also learned. We can profit from this additional dependency and optimize not only for the margin but also for the radius, as Theorems 1 and 2 suggest, in the hope of reducing even more the error bounds than what would be possible by just focusing on the margin.

A straightforward way to do so is to alter the cost function of the above optimization problem so that it also includes the radius. Thus we define the primal form of soft margin MKL optimization problem as follows:

$$\min_{\mathbf{w},b,\boldsymbol{\xi},\boldsymbol{\mu}} \frac{1}{2}\langle \mathbf{w}, \mathbf{w}\rangle R_{\boldsymbol{\mu}}^2 + \frac{C}{2}\sum_{i=1}^{l} \xi_i^2 \tag{9}$$
$$s.t. \ y_i(\langle \mathbf{w}, \boldsymbol{\Phi}(\mathbf{x}_i)\rangle + b) \geq 1 - \xi_i, \forall i$$

Accounting for the form $\boldsymbol{\Phi}_{\boldsymbol{\mu}}$ of the feature space $\mathcal{H}_{\boldsymbol{\mu}}$, as it is given in equation 4, this optimization problem can be rewritten as:

$$\min_{\mathbf{w},b,\boldsymbol{\xi},\boldsymbol{\mu}} \frac{1}{2}\sum_{k}^{M}\langle \mathbf{w}_k, \mathbf{w}_k\rangle R_{\boldsymbol{\mu}}^2 + \frac{C}{2}\sum_{i=1}^{l} \xi_i^2 \tag{10}$$
$$s.t. \ y_i(\sum_{k}^{M}\langle \mathbf{w}_k, \sqrt{\mu_k}\boldsymbol{\Phi}_k(\mathbf{x}_i)\rangle + b) \geq 1 - \xi_i, \forall i$$

where the $\mathbf{w}$ is the same as that of Formula 9 and equal to $(\mathbf{w}_1, \ldots, \mathbf{w}_M)$, $R^2_{\boldsymbol{\mu}}$ can be computed by equation 6. By letting $\mathbf{w} := \sqrt{\boldsymbol{\mu}}.\mathbf{w}$ then $\mathbf{w}_k := \sqrt{\mu_k}\mathbf{w}_k$ we can rewrite equation 10 as[1]:

$$\min_{\mathbf{w},b,\boldsymbol{\xi},\boldsymbol{\mu}} \frac{1}{2}\sum_k^M \frac{\langle \mathbf{w}_k, \mathbf{w}_k \rangle}{\mu_k} R^2_{\boldsymbol{\mu}} + \frac{C}{2}\sum_i^l \xi_i^2 \tag{11}$$
$$s.t.\ y_i(\sum_k^M \langle \mathbf{w}_k, \boldsymbol{\Phi}_k(\mathbf{x}_i)\rangle + b) \geq 1 - \xi_i,$$
$$\sum_{k=1}^M \mu_k = 1, \mu_k \geq 0, \forall k$$

The non-negativity of $\boldsymbol{\mu}$ is required to guarantee that the kernel combination is a valid kernel function; the constraint $\sum_{k=1}^M \mu_k = 1$ is added to make the solution interpretable (kernel with bigger weight can be interpreted as more important one) and to get a specific solution (note that if $\boldsymbol{\mu}$ is solution of 11 (without the constraint $\sum_{k=1}^M \mu_k = 1$), then $\lambda\boldsymbol{\mu}, \lambda \in \mathbb{R}^+$ is also its solution).

We will denote the cost function of equation 11 by $F(\mathbf{w}, b, \boldsymbol{\xi}, \boldsymbol{\mu})$. F is not a convex function, this is probably the main reason why in current MKL algorithms the radius is simply removed from the original cost function, therefore they do not really optimize the generalization error bound.

From the set of inequalities given in equation 7 we have $R^2_{\boldsymbol{\mu}} \leq \sum_k^M \mu_k R_k^2$ and from this we can get:

$$\begin{aligned} F(\mathbf{w}, b, \boldsymbol{\xi}, \boldsymbol{\mu}) = & \quad \tfrac{1}{2}\sum_k^M \tfrac{\langle \mathbf{w}_k, \mathbf{w}_k\rangle}{\mu_k} R^2_{\boldsymbol{\mu}} + \tfrac{C}{2}\sum_i^l \xi_i^2 & \leq \\ & \tfrac{1}{2}\sum_k^M \tfrac{\langle \mathbf{w}_k, \mathbf{w}_k\rangle}{\mu_k} \sum_k^M \mu_k R_k^2 + \tfrac{C}{2}\sum_i^l \xi_i^2 & = \\ & (\tfrac{1}{2}\sum_k^M \tfrac{\langle \mathbf{w}_k, \mathbf{w}_k\rangle}{\mu_k} + \tfrac{C}{2\sum_k^M \mu_k R_k^2}\sum_i^l \xi_i^2)\sum_k^M \mu_k R_k^2 & \leq \\ & (\tfrac{1}{2}\sum_k^M \tfrac{\langle \mathbf{w}_k, \mathbf{w}_k\rangle}{\mu_k} + \tfrac{C}{2\sum_k^M \mu_k R_k^2}\sum_i^l \xi_i^2) & = \widetilde{F}(\mathbf{w}, b, \boldsymbol{\xi}, \boldsymbol{\mu}) \end{aligned} \tag{12}$$

The last inequality holds because in the context we examine we have $\sum_k^M \mu_k R_k^2 \leq 1$. This is a result of the fact that we work with the normalized feature spaces, using the normalized kernels as these were defined in equation 3, thus we have $R_k^2 \leq 1$ and since $\sum_k^M \mu_k = 1$ it holds that: $\sum_k^M \mu_k R_k^2 \leq 1$. Since $\widetilde{F}$ is an upper bound of F and moreover it is convex[2] we are going to use it as our objective function. As a result, we propose to solve, instead of the original soft margin optimization problem given in equation 11, the following upper bounding convex optimization problem:

---

[1] Note that if $\mu_k = 0$ then from the dual form we have $\mathbf{w}_k = 0$. In this case, we use the convention that $\frac{0}{0} = 0$.

[2] The convexity of this new function can be easily proved by showing that the Hessian matrix is positive semi-definite.

$$\min_{\mathbf{w},b,\boldsymbol{\xi},\boldsymbol{\mu}} \frac{1}{2}\sum_{k}^{M}\frac{\langle \mathbf{w}_k, \mathbf{w}_k\rangle}{\mu_k} + \frac{C}{2\sum_k^M \mu_k R_k^2}\sum_i^l \xi_i^2 \tag{13}$$
$$s.t. \;\; y_i(\sum_k^M \langle \mathbf{w}_k, \boldsymbol{\Phi}_k(\mathbf{x}_i)\rangle + b) \geq 1 - \xi_i,$$
$$\sum_{k=1}^{M}\mu_k = 1, \mu_k \geq 0, \forall k$$

The dual function of this optimization problem is:

$$\begin{aligned} W_s(\boldsymbol{\alpha},\boldsymbol{\mu}) &= -\frac{1}{2}\sum_{ij}^{l}\alpha_i\alpha_j y_i y_j \sum_k^M \mu_k K_k(\mathbf{x}_i,\mathbf{x}_j) \\ &+ \sum_i^l \alpha_i - \frac{\sum_k^M \mu_k R_k^2}{2C}\langle \boldsymbol{\alpha},\boldsymbol{\alpha}\rangle \\ &= -\frac{1}{2}\sum_{ij}^{l}\alpha_i\alpha_j y_i y_j(\sum_k^M \mu_k K_k(\mathbf{x}_i,\mathbf{x}_j) + \frac{\sum_k^M \mu_k R_k^2}{C}\delta_{ij}) + \sum_i^l \alpha_i \end{aligned} \tag{14}$$

where δ_{ij} is the Kronecker δ defined to be 1 if $i = j$ and 0 otherwise. The dual optimization problem is given as:

$$\max_{\boldsymbol{\alpha},\boldsymbol{\mu}} W_s(\boldsymbol{\alpha},\boldsymbol{\mu}) \tag{15}$$
$$s.t. \;\; \sum_i \alpha_i y_i = 0,$$
$$\alpha_i \geq 0, \forall i$$
$$\sum_{ij}^{l}\alpha_i\alpha_j y_i y_j K_k(\mathbf{x}_i,\mathbf{x}_j) = \frac{R_k^2 C \sum_i \xi_i^2}{(\sum_k \mu_k R_k^2)^2}, \forall k$$

In the next section we will show how we can solve this new optimization problem.

4.2 Algorithm

The dual function 14 is quadratic with respect to $\boldsymbol{\alpha}$ and linear with respect to $\boldsymbol{\mu}$. One way to solve the optimization problem 13 is to use a two step iterative algorithm such as the ones described in [6], [10]. Following such a two step approach, in the first step we will solve a quadratic problem that optimizes over $(\mathbf{w}, b)$, while keeping $\boldsymbol{\mu}$ fixed; as a consequence the resulting dual function is a simple quadratic function of $\boldsymbol{\alpha}$ which can be optimized easily. In the second step we will solve a linear problem that optimizes over $\boldsymbol{\mu}$.

More precisely, the formulation of the optimization problem with the two-step approach takes the following form:

$$\min_{\boldsymbol{\mu}} \; J(\boldsymbol{\mu}) \qquad (16)$$
$$s.t. \; \sum_{k=1}^{M} \mu_k = 1, \mu_k \geq 0, \forall k$$

where

$$J(\boldsymbol{\mu}) = \begin{cases} \min_{\mathbf{w},b} \frac{1}{2} \sum_k^M \frac{\langle \mathbf{w}_k, \mathbf{w}_k \rangle}{\mu_k} + \frac{C}{2\sum_k^M \mu_k R_k^2} \sum_i^l \xi_i^2 \\ s.t. \quad y_i(\sum_k^M \langle \mathbf{w}_k, \boldsymbol{\Phi}_k(\mathbf{x}_i) \rangle + b) \geq 1 - \xi_i \end{cases} \qquad (17)$$

To solve the outer optimization problem, i.e. $\min_{\boldsymbol{\mu}} J(\boldsymbol{\mu})$, we use gradient descent method. At each iteration, we fix $\boldsymbol{\mu}$, compute the value of $J(\boldsymbol{\mu})$ and then compute the gradient of $J(\boldsymbol{\mu})$ with respect to $\boldsymbol{\mu}$. The dual function of Formula 17 is the $W_s(\boldsymbol{\alpha}, \boldsymbol{\mu})$ function already given in Formula 14. Since $\boldsymbol{\mu}$ is fixed we now optimize only over $\boldsymbol{\alpha}$ (the resulting dual optimization problem is much simpler compared to the original soft margin dual optimization problem given in Formula 15):

$$\max_{\boldsymbol{\alpha}} \; W_s(\boldsymbol{\alpha}, \boldsymbol{\mu})$$
$$s.t. \; \sum_i \alpha_i y_i = 0, \alpha_i \geq 0, \forall i$$

which has the same form as the SVM quadratic optimization problem, the only difference is that the C parameter here is equal to $\frac{C}{\sum_k^M \mu_k R_k^2}$.

For the strong duality, at the optimal solution $\boldsymbol{\alpha}^*$, the values of dual cost function and primal cost function are equal. Thus the value of $W_s(\boldsymbol{\alpha}, \boldsymbol{\mu})$, and the $J(\boldsymbol{\mu})$ value, is given by:

$$W_s(\boldsymbol{\alpha}^*, \boldsymbol{\mu}) = -\frac{1}{2} \sum_{ij}^{l} \alpha_i^* \alpha_j^* y_i y_j (\sum_k^M \mu_k K_k(\mathbf{x}_i, \mathbf{x}_j) + \frac{\sum_k^M \mu_k R_k^2}{C} \delta_{ij}) + \sum_i^l \alpha_i^*$$

The last step of the algorithm is to compute the gradient of the $J(\boldsymbol{\mu})$ function, Formula 17, with respect to $\boldsymbol{\mu}$. As [6] have pointed out, we can use the theorem of Bonnans and Shapiro [15] to compute gradients of such functions. Hence, the gradient is in the following form:

$$\frac{\partial J(\boldsymbol{\mu})}{\partial \mu_k} = -\frac{1}{2} \sum_{ij}^{l} \alpha_i^* \alpha_j^* y_i y_j (K_k(\mathbf{x}_i, \mathbf{x}_j) + \frac{R_k^2}{C} \delta_{ij})$$

To compute the optimal step in the gradient descent we used line search. The complete two-step procedure is given in Algorithm 1.

Algorithm 1. *R*-MKL

Initialize $\mu_k^1 = \frac{1}{M}$ for $k = 1, ..., M$
repeat
  Set $R_\mu^2 = \sum_k^M \mu_k^t R_k^2$
  compute $J(\boldsymbol{\mu}^t)$ as the solution of a quadratic optimization problem with $K := \sum_k^M \mu_k^t K_k$
  compute $\frac{\partial J}{\partial \mu_k}$ for $k = 1, ..., M$
  compute optimal step γ_t
  $\boldsymbol{\mu}^{t+1} \leftarrow \boldsymbol{\mu}^t + \gamma_t \frac{\partial J(\mu)}{\partial \mu}$
until *stopCriteria* is *true*

4.3 Computational Complexity

At each step of the iteration we have to compute the solution of a standard SVM, with kernel $K = \sum_{k=1}^M \mu_k K_k$, and C equal to $\frac{C}{\sum_k^M \mu_k R_k^2}$, which is a quadratic programming problem with a complexity of $O(n^3)$ where n is number of instances. Moreover, when $\boldsymbol{\mu}$ is updated we have to recompute the approximation of $R_{\boldsymbol{\mu}}^2$; the complexity of this procedure is linear in the number of kernels, $O(M)$.

5 Experiments

We experimented with ten different datasets. Six of them were taken from the UCI repository (*Ionosphere, Liver, Sonar, Wdbc, Wpbc, Musk1*), while four come from the domain of genomics and proteomics (*ColonCancer, CentralNervousSystem, FemaleVsMale, Leukemia*) [16]; these four are characterized by small sample and high dimensionality morphology. A short description of the datasets is given in Table 1. We experimented with two different types of basic kernels, i.e. polynomials and Gaussians, and performed two sets of experiments. In the first set of experiments we used both types of kernels and in the second one we focused only on Gaussians kernels. For each set of experiments the total number of basic kernels was 20; for the first set we used polynomial kernels of degree one, two, and three and 17 Gaussians with bandwidth δ that ranged from 1 to 17 with a step of one; for the second set of experiments we only used Gaussian kernels with bandwidth δ that ranged from 1 to 20 with a step of one.

We compared our MKL algorithm (denoted as *R*-MKL) with two state-of-the-art MKL algorithms: Support Kernel Machine (SKM) [7], and SimpleMKL [6]. We estimate the classification error using 10-fold cross validation. For comparison purposes we also provide the performances of the best single kernel (BK) and the majority classifier (MC); the latter always predicts the majority class. The performance of the BK is that of the best single kernel estimated also by 10-fold cross validation, since it is the best result after seeing the performance of all individual kernels on the available data it is optimistically biased. We tuned the parameter C in an inner-loop 10-fold cross-validation choosing the values from the set $\{0.1, 1, 10, 100\}$. All algorithms terminate when the duality gap is smaller than 0.01. All input kernel matrices are normalized by equation 3.

Table 1. Short description of the classification datasets used

DATASET	#INST	#ATTR	#CLASS1	#CLASS2
IONOSPHERE	351	34	126	225
LIVER	345	6	145	200
SONAR	208	60	97	111
WDBC	569	32	357	212
WPBC	198	34	151	47
MUSK1	476	166	269	207
COLONCANCER	62	2000	40	22
CENTRALNERVOUS	60	7129	21	39
FEMALEVSMALE	134	1524	67	67
LEUKEMIA	72	7128	25	47

We compared the significance level of the performance differences of the algorithms with McNemar's test [17], where the level of significance is set to 0.05. We also established a ranking schema of the examined MKL algorithms based on the results of the pairwise comparisons [18]. More precisely, if an algorithm is significantly better than another it is credited with one point; if there is no significant difference between two algorithms then they are credited with 0.5 points; finally, if an algorithm is significantly worse than another it is credited with zero points. Thus, if an algorithm is significantly better than all the others for a given dataset it has a score of two. We give the full results in Tables 2, 3. For each algorithm we report a triplet in which the first element is the estimated classification error, the second is the number of selected kernels, and the last is the above described rank.

Our kernel combination algorithm does remarkably well in the first set of experiments, Table 2, in which it is significantly better than both other algorithms in four datasets and significantly worse in two; for the four remaining datasets there are no significant differences. Note that in the cases of *Wpbc* and

Table 2. Results for the first experiments, where both polynomial and Gaussian kernels are used. Each triplet x,y,z gives respectively the classification error, the number of selected kernels, and the number of significance point that the algorithm scores for the given experiment set and dataset. Columns BK and MC give the errors of the best single kernel and the majority classifier, respectively.

D. SET	SKM	SIMPLE	R-MKL	BK	MC
IONOS.	04.00,02,1.5	03.71,02,1.5	04.86,02,0	05.71	36.00
LIVER	33.82,05,1.5	33.53,13,1.5	36.18,03,0	30.29	42.06
SONAR	15.50,01,1.0	15.50,01,1.0	15.50,01,1	17.50	46.00
WDBC	11.25,03,1.0	13.04,18,0.0	03.75,18,2	08.57	37.32
WPBC	23.68,17,1.0	23.68,01,1.0	23.68,01,1	23.68	23.68
MUSK1	11.70,07,1.0	13.40,18,0.0	06.60,01,2	04.47	43.83
COLON.	18.33,18,1.0	18.33,18,1.0	16.67,18,1	11.67	35.00
CENTNE.	35.00,17,1.0	35.00,17,1.0	35.00,17,1	31.67	35.00
FEMALE.	33.85,20,1.0	38.92,18,0.0	20.00,18,2	22.31	60.00
LEUKE.	07.14,18,0.5	07.14,18,0.5	02.86,18,2	02.86	34.29

Table 3. Results for the second set of experiments, where only Gaussian kernels are used. The table contains the same information as the previous one.

D. Set	SKM	Simple	R-MKL	BK	MC
Ionos.	04.86,02,1.0	05.43,04,1.0	05.14,03,1.0	05.14	36.00
Liver	33.53,03,1.0	33.53,20,1.0	33.53,16,1.0	34.71	42.06
Sonar	15.50,01,1.0	15.50,01,1.0	15.50,01,1.0	17.00	46.00
Wdbc	37.32,03,1.0	37.32,19,1.0	37.32,20,1.0	37.32	37.32
Wpbc	23.68,20,1.0	23.68,19,1.0	23.68,20,1.0	23.68	23.68
Musk1	43.83,14,1.0	43.83,19,1.0	43.83,20,1.0	43.83	43.83
Colon.	NA	35.00,20,1.5	35.00,20,1.5	35.00	35.00
CentNe.	35.00,20,1.0	35.00,20,1.0	35.00,20,1.0	35.00	35.00
Female.	60.00,20,1.0	60.00,20,1.0	60.00,20,1.0	60.00	60.00
Leuke.	34.29,20,1.0	34.29,20,1.0	34.29,20,1.0	34.29	34.29

CentralNervousSystem all algorithms have a performance that is similar to that of the majority classifier, i.e. the learned models do not have any discriminatory power. By examining the classification performances of the individual kernels on these datasets we see that none of them had a performance that was better than that of the majority classifier; this could explain the bad behavior of the different kernel combination schemata. Overall, for this set of experiments R-MKL gets 12 significance points over the different datasets, SKM 10.5, and SimpleMKL 7.5. The performance improvements of R-MKL over the two other methods are are quite impressive on those datasets on which R-MKL performs well; more precisely its classification error is around 30%, 50%, and 40%, of that of the other algorithms for *Wdbc*, *Musk1*, and *Leukemia* datasets respectively.

In the second set of experiments,Table 3, all methods perform very poorly in seven out of ten datasets; their classification performance is similar to that of the majority classifier. In the remaining datasets, with the exception of *ColonCancer* for which SKM failed (we used the implementation provided by the authors of the algorithm and it returns with some errors), there is no significant difference between the three algorithms. The collectively bad performance in the last seven databases is explained by the fact that none of the basic kernels had a classification error that was better than that of the majority classifier. Overall, for this set of experiments SKM scores 9 points, and Simple MKL and R-MKL score 10.5 points each.

Comparing the number of selected kernels by the different kernel combination methods using a paired t-test (significance level of 0.05) revealed no statistically significant differences between the three algorithms on both sets of experiments.

In an effort to get an empirical estimation of the quality of the approximation of the radius that we used to make the optimization problems convex, we computed the approximation error defined as $\frac{\sum_k^M \mu_k R_k^2 - R_\mu^2}{R_\mu^2}$. We computed this error over the different folds of the ten-fold cross-validation for each dataset. The average approximation error over the different datasets was 0.0056. We also computed this error over 1000 random values of $\boldsymbol{\mu}$ for each dataset and the average error was 0.0104. Thus, the empirical evidence seems to indicate that the $R_{\mu}^2 \leq \sum_k^M \mu_k R_k^2$ bound is relatively tight, at least for the datasets we examined.

6 Conclusion and Future Work

In this paper we presented a new kernel combination method that incorporates in its cost function not only the margin but also the radius of the smallest sphere that encloses the data. This idea is a direct implementation of well known error bounds from statistical learning theory. To the best of our knowledge this is the first work in which the radius is used together with the margin in an effort to minimize the generalization error. Even though the resulting optimization problems were non-convex and we had to use an upper bound on the radius to get convex forms, the empirical results were quite encouraging. In particular, our method competed with other state-of-the-art methods for kernel combination, thus demonstrating the benefit and the potential of the proposed technique. Finally, we mention that it is still a challenging research direction to fully exploit the examined generalization bound.

In future work we would like to examine optimization techniques for directly solving the non-convex optimization problem presented in Formula 11. In particular, we will examine whether it is possible to decompose the cost function as a sum convex and concave functions, or to represent it as *d.m functions* (difference of two monotonic functions) [19,20]. Additionally, we plan to analyze the bound $R_{\mu}^2 \leq \sum_k^M \mu_k R_k^2$ and see how it relates with the real optimal value.

Acknowledgments. The work reported in this paper was partially funded by the European Commission through EU projects DropTop (FP6-037739), DebugIT (FP7-217139) and e-LICO (FP7-231519). The support of the Swiss NSF (Grant 200021-122283/1) is also gratefully acknowledged.

References

1. Shawe-Taylor, J., Cristianini, N.: Kernel methods for pattern analysis. Cambridge University Press, Cambridge (2004)
2. Schölkopf, B., Smola, A.J.: Learning with Kernels: Support Vector Machines, Regularization, Optimization, and Beyond. MIT Press, Cambridge (2001)
3. Lanckriet, G., Cristianini, N., Bartlett, P., Ghaoui, L.E.: Learning the kernel matrix with semidefinite programming. Journal of Machine Learning Research 5, 27–72 (2004)
4. Ong, C.S., Smola, A.J., Williamson, R.C.: Learning the kernel with hyperkernels. Journal of Machine Learning Research 6, 1043–1071 (2005)
5. Sonnenburg, S., Ratsch, G., Schafer, C.: A general and efficient multiple kernel learning algorithm. Journal of Machine Learning Research 7, 1531–1565 (2006)
6. Bach, F., Rakotomamonjy, A., Canu, S., Grandvalet, Y.: SimpleMKL. Journal of Machine Learning Research (2008)

7. Bach, F.R., Lanckriet, G.R.G., Jordan, M.I.: Multiple kernel learning, conic duality, and the smo algorithm. In: ICML 2004: Proceedings of the twenty-first international conference on Machine learning, p. 6. ACM, New York (2004)
8. Lanckriet, G., Bie, T.D., Cristianini, N.: A statistical framework for genomic data fusion. Bioinformatics 20 (2004)
9. Cristianini, N., Shawe-Taylor, J., Elisseeff, A.: On kernel-target alignment. Journal of Machine Learning Research (2002)
10. Chapelle, O., Vapnik, V., Bousquet, O., Mukherjee, S.: Choosing multiple parameters for support vector machines. Machine Learning 46(1-3), 131–159 (2002)
11. Crammer, K., Keshet, J., Singer, Y.: Kernel design using boosting. In: Advances in Neural Information Processing Systems, vol. 14. MIT Press, Cambridge (2002)
12. Bousquet, O., Herrmann, D.: On the complexity of learning the kernel matrix. In: Advances in Neural Information Processing Systems, vol. 14. MIT Press, Cambridge (2003)
13. Cristianini, N., Shawe-Taylor, J.: An introduction to Support Vector Machines. Cambridge University Press, Cambridge (2000)
14. Vapnik, V.: Statistical learning theory. Wiley Interscience, Hoboken (1998)
15. Bonnans, J., Shapiro, A.: Optimization problems with perturbation: A guided tour. SIAM Review 40(2), 202–227 (1998)
16. Kalousis, A., Prados, J., Hilario, M.: Stability of feature selection algorithms: a study on high dimensional spaces. Knowledge and Information Systems 12(1), 95–116 (2007)
17. McNemar, Q.: Note on the sampling error of the difference between correlated proportions or percentages. Psychometrika 12, 153–157 (1947)
18. Kalousis, A., Theoharis, T.: Noemon: Design, implementation and performance results for an intelligent assistant for classifier selection. Intelligent Data Analysis Journal 3, 319–337 (1999)
19. Leo Liberti, N.M. (ed.): Global Optimization - From Theory to Implementation. Springer, Heidelberg (2006)
20. Collobert, R., Weston, J., Bottou, L.: Trading convexity for scalability. In: Proceedings of the 23th Conference on Machine Learning (2006)
21. Stephen Boyd, L.V. (ed.): Convex optimization. Cambridge University Press, Cambridge (2004)

Appendix

Proof of Inequality 7. If $K(\mathbf{x}, \mathbf{x}')$ is the kernel function associated with the $\mathbf{\Phi}(\mathbf{x})$ mapping then the computation of the radius in the dual form is given in [1]:

$$\max_{\beta_i \beta_j} R^2 = \sum_i^l \beta_i K(\mathbf{x}_i, \mathbf{x}_i) - \sum_{ij}^l \beta_i \beta_j K(\mathbf{x}_i, \mathbf{x}_j) \tag{18}$$

$$s.t. \sum_i^l \beta_i = 1, \beta_i \geq 0$$

If $\boldsymbol{\beta^*}$ is the optimal solution of (18) when $K = K_{\boldsymbol{\mu}} = \sum_1^M \mu_k K_k$, and $\hat{\boldsymbol{\beta}}^{\boldsymbol{k}}$ is the optimal solution of (18) when $K = K_k$, i.e. :

$$R^2_{\boldsymbol{\mu}} = \sum_{k=1}^{M} \mu_k (\sum_{i=1}^{l} \beta_i^* K_k(\mathbf{x}_i, \mathbf{x}_i) - \sum_{i,j=1}^{l} \beta_i^* \beta_j^* K_k(\mathbf{x}_i, \mathbf{x}_j))$$

$$R^2_k = \sum_{i=1}^{l} \hat{\beta^k}_i K_k(\mathbf{x}_i, \mathbf{x}_i) - \sum_{i,j=1}^{l} \hat{\beta^k}_i \hat{\beta^k}_j K_k(\mathbf{x}_i, \mathbf{x}_j)$$

then

$$\sum_{i=1}^{l} \beta_i^* K_k(\mathbf{x}_i, \mathbf{x}_i) - \sum_{i,j=1}^{l} \beta_i^* \beta_j^* K_k(\mathbf{x}_i, \mathbf{x}_j) \leq$$
$$\sum_{i=1}^{l} \hat{\beta^k}_i K_k(\mathbf{x}_i, \mathbf{x}_i) - \sum_{i,j=1}^{l} \hat{\beta^k}_i \hat{\beta^k}_j K_k(\mathbf{x}_i, \mathbf{x}_j)$$

Therefore: $R^2_{\boldsymbol{\mu}} \leq \sum_{k=1}^{M} \mu_k R^2_k$

Proof of convexity of R-MKL (Eq.13). To prove that 13 is convex, it is enough to show that functions $\frac{x^2}{\mu}$, where $x \in \mathbb{R}$, $\mu \in \mathbb{R}^+$, and $\frac{\xi^2}{\sum_k^M \alpha_k \mu_k}$, where $\xi \in \mathbb{R}, \mu_k, \alpha_k \in \mathbb{R}^+$ are convex. The first is quadratic-over-linear function which is convex. The second is convex because its epigraph is a convex set [21].

Inference and Validation of Networks

Ilias N. Flaounas[1], Marco Turchi[2], Tijl De Bie[2], and Nello Cristianini[1,2]

[1] Department of Computer Science, Bristol University
Merchant Venturers Building, Woodland Road, Bristol, BS8-1UB, United Kingdom
[2] Department of Engineering Mathematics, Bristol University
Queen's Building, University Walk, Bristol, BS8-1TR, United Kingdom
`http://patterns.enm.bris.ac.uk`

Abstract. We develop a statistical methodology to validate the result of network inference algorithms, based on principles of statistical testing and machine learning. The comparison of results with reference networks, by means of similarity measures and null models, allows us to measure the significance of results, as well as their predictive power. The use of Generalised Linear Models allows us to explain the results in terms of available ground truth which we expect to be partially relevant. We present these methods for the case of inferring a network of News Outlets based on their preference of stories to cover. We compare three simple network inference methods and show how our technique can be used to choose between them. All the methods presented here can be directly applied to other domains where network inference is used.

Keywords: Network inference, Network validation, News Outlets network.

1 Introduction

Network Inference is a ubiquitous problem, found in fields as diverse as genomics, epidemiology or social sciences. Elements of a set (e.g., genes, people or news outlets) are connected by links that represent relations between them (e.g., co-expression, social contact, similar reporting bias, etc). While we can often observe the state of the network nodes, the underlying topology of the network is hidden, and must be inferred based on a finite set of observations of node-states.

Several methods have been proposed to infer this underlying network structure, in different communities and under different conditions. Examples include gene regulatory networks [1], biochemical regulatory networks[2] and protein interaction networks[3]. We focus on the general problem of testing, or validating, the result of this inference. Since often ground truth is missing, validation against related but different networks, or against networks inferred in different ways, is the only option or the only viable alternative to costly experiments.

In this paper, we present and study general methods to assess the results of Network Inference algorithms, from a statistical and machine learning point of view, and we demonstrate them on a challenging test case: the inference and

W. Buntine et al. (Eds.): ECML PKDD 2009, Part I, LNAI 5781, pp. 344–358, 2009.

validation of a network of News Outlets, based on content similarity information. All the principles and methods are however general, and can be applied to different domains.

We argue that network inference algorithms need to satisfy two key properties. First, the inferred network needs to be stable, meaning that networks inferred on independent data must be similar to each other. Second, it must be related to any available independent ground truth known or assumed to affect the network topology. Both these properties can be verified by testing if the inferred network is similar to a reference network.

For the first property, the reference network would be a network inferred based on independent data. For example for the network of News Outlets, we show that our network inference algorithm produces networks that are significantly similar to each other, when operating on independent datasets. This stability strongly indicates that the algorithm is capturing a signal, not noise.

For the second property, the reference network would be a network constructed based on independent ground truth data. For example for the network of News Outlets we show how the inferred network is significantly related to other—directly observable—networks of news outlets, such as those based on geographic, linguistic and media-type similarity.

Hence, both properties are verified by assessing if the inferred network is related to a reference network. More specifically, we want to verify if the inferred network is related significantly stronger to the reference network than a random network would be related to it. This is formalized in statistics by means of the key notion of statistical significance of a pattern, as expressed by a p-value. In order for this to be defined, we need to make two choices: a test statistic that quantifies how related the inferred network is to the reference network, and a null model for the inferred network.

The test statistic can be defined by quantifying the similarity of the inferred network to the reference network considered, and we will discuss various options for this similarity measure. To define the null model we will make use of two established approaches for random network generation. The choices we explore in this paper are exemplary, and other choices may be more appropriate in other applications. We will discuss the implications of these design choices, by comparing three different network inference algorithms for the News Outlet network inference application.

Finally, as a separate validation from a machine learning perspective, we investigate if we can predict the inferred network topology based on independent information. In particular, we show how Generalised Linear Models can be used to 'explain' the inferred network in terms of any known ground truth networks as discussed above.

Our approaches can be readily transferred to social sciences and genomics, where the availability of ground truth is the key problem when validating network inference.

2 Network Validation

The analysis of patterns found in data can generally be validated in two different ways: either by assessing their significance, or by measuring their predictive power. In the first case, we are interested in measuring the probability that a similar pattern could be found in randomly generated data. In the second case, we are interested in measuring the extent to which patterns found in a subset of the data, can be found in an independent subset of the data. Of course the two approaches have many relations, but in this study we will simply address them separately. We will call them respectively 'Hypothesis Testing' and 'Predictive Power' approach.

Notations. A network $\mathcal{G} = (\mathcal{N}, \mathcal{E})$ is comprised of the set of nodes $\mathcal{N} = \{N_1, N_2, ...N_n\}$ and the set of edges $\mathcal{E} \subseteq \mathcal{N} \times \mathcal{N}$, $n = |\mathcal{N}|$ is the total number of nodes of the network and $e = |\mathcal{E}|$ is the total number of edges.

2.1 Hypothesis Testing

The key idea of hypothesis testing is to quantify the probability that a the value of a test statistic evaluated on observed data could have been found also in random data. In our strategy to evaluate network inference algorithms, the test statistic is the similarity to a chosen reference network, and we denote it as $t_{\mathcal{G}_R}$ where $\mathcal{G}_R$ stands for the reference network. We denote as $\mathcal{G}_I$ the network inferred by the inference algorithm. The null hypothesis H_0 is that the inferred network is sampled from some underlying distribution. Hence, the hypothesis test boils down to quantifying the probability that a random network is at least as similar to a chosen reference network as the inferred network:

$$p = P_{\mathcal{G} \sim H_0}(t_{\mathcal{G}_R}(\mathcal{G}) \geq t_{\mathcal{G}_R}(\mathcal{G}_I)) \tag{1}$$

For most null hypotheses, it would be impractical to compute the p-value exactly. However, it can be reliably estimated by sampling a large number K of networks from the null hypothesis. Then the p-value is measured as the fraction of those for which the test statistic defined as the similarity to a reference network is smaller than that for the inferred network, or more precisely:

$$p \approx \frac{\#\{\mathcal{G} : t_{\mathcal{G}_R}(\mathcal{G}) \geq t_{\mathcal{G}_R}(\mathcal{G}_I)\} + 1}{K + 1} \tag{2}$$

If the p-value is small, this means that the inferred network is more similar to the reference network than expected by chance. Then the null hypothesis is rejected, as it is unlikely given the evidence. When the null hypothesis is rejected in this way, the alternative must hold. This means that the inferred network is significantly different from random networks sampled from the null hypothesis in its similarity to the reference network, supporting the inference method used to infer the network. In case that distances are used as test statistics instead of similarities, the inequality signs in p-value equations should be inverted. Often a significance threshold α is chosen, and the null hypothesis is rejected if $p < \alpha$, where α is selected depending on the application.

In the following two subsections we will address the issue of selection of a test statistic and null models.

Test Statistics. As a test statistic we will use the similarity of the inferred network to a reference network, which we expect to be in some sense related to it. In this section we will discuss both the similarity measure and the choice of reference networks.

In literature several approaches have been proposed for the measurement of similarity between networks [4,5,6]. In the case of comparing two networks $\mathcal{G}_A$ and $\mathcal{G}_B$ that have the same set of nodes $\mathcal{N}$, one can compare the topological properties or their link structure. Perhaps the simplest comparison involves counting how many pairs of nodes have the same linkage status (connected or disconnected). The edges of each network can be considered as independent sets of elements and the comparison of networks is reduced to a comparison of sets of edges. Jaccard distance can be used to compare two sets as a measure of dissimilarity between them[7]. It is obtained by dividing the difference of the sizes of the union and the intersection of two sets by the size of the union:

$$JD(\mathcal{E}_A, \mathcal{E}_B) = \frac{|\mathcal{E}_A \cup \mathcal{E}_B| - |\mathcal{E}_A \cap \mathcal{E}_B|}{|\mathcal{E}_A \cup \mathcal{E}_B|} \tag{3}$$

This quantity ranges between zero and one, with a value of zero indicating identical networks, and a value of one indicating no shared edges.

Of course one could define other measures, that consider less local properties, for example one could count how many triplets of nodes have the same connectivity status (in this way counting common network motifs [8]), or one could ignore the specifics of network topology, and focus on the distances between nodes represented by it. So a comparison of the all-pairs distance-matrix for each network could lead to a useful similarity measure.

In this study, as test statistic we will use the Jaccard distance from a reference network.

The choice of reference network is a very important one, as often in network inference applications we only have access to indirect evidence of the network topology (this being one of the key motivations for modern network inference). We have however often access to other networks (or sub-networks) for which the ground truth can be assumed to be known, and which we expect to be somewhat related to the network we are investigating.

If the data can be divided into independent sets, for example, we should assume that networks inferred on different parts of the data should be significantly similar. This can lead both to a bootstrap process, but also to the analysis of temporal data, as we will discuss in Sect. 3.3. Observing significant similarity to independently generated networks can provide strong support for a hypothesis, also in the case of networks generated with completely different types of data, as we will demonstrate using geographic, linguistic and media-type similarity, to test the significance of a network of news-media outlets.

Null Models. Every statistical test aims at answering the following question: what is the probability that a pattern like the one currently analysed is the result of chance? Of course this quantity (p-value) can be computed only after a random process has been specified, to formalise the notion of 'chance'. The Null Model has the crucial role of providing a baseline comparison to assess the significance of inferred patterns.

In the case of validating network patterns, we will need to specify a model of random network generation. If the observed similarity to a reference network is found also in randomly generated graphs, then we cannot conclude that we have found a significant pattern in the given data. In this study we will present two methods of random network generation, although many others are possible.

Erdös-Rényi Model
The first model is the celebrated $G(n,p)$ Erdös - Rényi model[9]. A graph of n nodes is generated by connecting nodes randomly. Two nodes have an independent probability p to be connected. This probability defines the density of the graph. Indeed the expected number of edges e is:

$$e = \tbinom{n}{2}p \tag{4}$$

and the distribution of the degree of any particular node N is binomial:

$$P(deg(N) = l) = \binom{n-1}{l} p^k (1-p)^{n-1-l} \tag{5}$$

Switching Randomisation
Although the Erdös-Rényi model is very natural and simple to analyse, it leads to topologies that are often very different from topologies observed in real world situations. For example, it does not exhibit the power-law in degree distributions that is often found in social networks. To remedy this, one can define a random-network generation model that—by construction—has the same degree distribution as the inferred network, and yet is randomly sampled from the space of possible networks. Such models can be created by a switching approach[10]. This method starts from a given graph and randomises it by switching edges between nodes. If the pairs of nodes A-B and C-D are connected, the model will switch the connections to create the edges A-D and B-C. The number of iterations is arbitrary but an adequate number is considered 100 times the number of edges [11].

2.2 Predictive Power

We are interested in the possibility of predicting the network topology based on other observable properties of the network. If some ground truth information about the inferred network is known, it is expected to be able to 'explain' the existence of some edges of the network. If more than one ground truth components are available this knowledge can be combined in order to improve the understanding of the inferred network. The combination of ground truth elements can be made using Generalized Linear Models (GLMs).

Generalized Linear Models. J. Nelder and R. Wedderburn introduced GLMs as a way to provide a unified framework for various non-linear or non-normal linear variations of regression[12]. GLM splits the model for the observed data Y_i into a random and a systematic component through a function called the link function. Under GLM Y_i is assumed to be generated from a distribution function of the exponential family [13]. The mean μ of the distribution depends on the independent variables, $\mathbf{X}$, through:

$$E(\mathbf{Y}) = \mu = g^{-1}(\eta) = g^{-1}(\mathbf{X}\beta) \quad (6)$$

where $E(\mathbf{Y})$ is the expected value of $\mathbf{Y}$; η is the linear predictor which is a linear combination of unknown parameters β; g is the link function; and the elements of $\mathbf{X}$ are typically measured by experimenters. The variance of the distribution is a function of the mean that can also follow the same exponential family distribution. The unknown parameters are easily estimated by maximum likelihood or other techniques.

Network Topology Prediction. The quality of the GLM models and the accepted ground truth components can be measured based on their power to predict the topology of the inferred network. Our aim is to measure the ability of the GLM model to predict the existence of an edge of the network. Using a methodology similar to this found in supervised classification we separate the network into a training and test sub-network. The training network is used to calculate the GLMs parameters. These parameters are combined with the accepted ground truth and are used to predict the structure of the test network. A generally accepted accuracy measurement is the Area Under Curve (AUC) based on the ROC analysis of the predictions on the test set. The separation into train and test sub-networks is performed multiple times under a cross-validation scheme in order to reduce bias.

3 Experimental Study

We will illustrate the validation methodology on the specific task of inferring the network of news outlets that are connected by the same bias in choosing stories to cover. This case study has many points in common with standard network inference tasks, for example gene regulation networks (while being easier to interpret): ground truth is not directly observed, side information is available, data is noisy, and so on. In this section we also introduce three increasingly complex network inference algorithms and use our methodology to compare their output.

3.1 Content-Based Inference of Media Outlets Network

In this application, we are interested in linking news outlets that have similar interests in choosing stories to cover. In this research a news story is defined as a set of news articles that cover the same event, practically found as a cluster.

We analyse a set of 1,017,348 articles gathered over a period of 12 consecutive weeks starting from October 1st, 2008, from 543 online news outlets, distributed over 32 different countries, in 22 different languages, including 7 different media types (e.g., newspapers, blogs, etc). This dataset was created as part of a separate project, which will not be discussed here[14]. While many of the outlets of interest offer their content in English language, we machine-translated the content of the others into English, by using Moses software [15,16].

Articles are preprocessed using stop-word removal, stemming and are vectorised using the TF-IDF representation[17]. They are then clustered in order to form the stories that will be used as the base for the network inference. The distance of two articles is measured using the cosine similarity[17]. The clustering algorithm identified on average 974 stories per day.

We used the Best Reciprocal Hit (BRH) clustering method, borrowed from the field of bioinformatics [18]. The choice of clustering algorithm is not central to the discussion of this study.

3.2 Three Network Inference Algorithms

We compared three network inference algorithms, all connecting pairs of nodes that have a sufficiently high level of similarity. While other inference methods are possible, we focused on this approach here for simplicity. Since we will use real valued similarity measures, we will also to choose a threshold in order to derive the linkage structure, and this threshold will control the density of the resulting graph.

We will assume we have an Outlet-by-Story matrix, indicating which outlets carried each given news story. There are 543 outlets, and 81,816 stories in total. Every outlet is hence described by an indicator vector in 'story-space'.

Method A. The simplest approach is to connect two outlets if they share some minimum number of stories. If the threshold is set to one, every pair of outlets that share at least one story are connected. In other words, the similarity measure between outlets is the scalar product between their indicator-vectors in story-space. This approach can easily lead to very dense networks since many stories are shared by the majority of outlets.

Method B. A more sophisticated approach would apply weights to the candidate edges of the network. A popular weighing scheme is based on the TF-IDF. Under this scheme each outlet correspond to a document and each story to a term. The frequency of story j that belong to outlet k is

$$f_j^k = \frac{s_j^k}{s^k} \tag{7}$$

where the nominator is the number of times the story appears to the outlet k and s^k is the total number of stories of outlet k. The corresponding inverse outlet frequency i_j^k is defined as

$$i_j^k = \log \frac{n}{n_j} \tag{8}$$

where n is the total number of outlets and n_j is the number of outlets that have story j. Thus, a vector of size J, that is the total number of different stories, is assigned to each outlet, one weight for each story:

$$w_j^k = f_j^k \cdot i_j^k, j = 1, 2..J \tag{9}$$

The similarity of two outlets N_a and N_b can now be measured as their cosine similarity:

$$sim(N_a, N_b) = \sum_{t=1}^{J} w_t^a w_t^b \tag{10}$$

Method C. Another method which is similar to the previous one is weighting each story with a weight f_j based on the frequency of the story, independently of the outlet that publish it:

$$f_j = \frac{1}{n_j} \tag{11}$$

where n_j is the number of outlets that have story j. Stories that are found in the majority of media receive a small weight and stories found in few media receive higher weight. The maximum weight is $1/2$ since we consider as stories clusters that have articles of at least two different outlets, and the minimum weight is $1/n$. If we normalise the above measure to the range of zero to one we get

$$f_j' = \frac{2(n - n_j)}{(n - 2) \cdot n_j} \tag{12}$$

where n is the total number of outlets and we consider two as the minimum number of outlets that can belong to a cluster. This way measure of similarity between two outlets N_a and N_b is defined as:

$$sim'(N_a, N_b) = \frac{\sum_{t=1}^{J} f_t' y_a(j) y_b(j)}{\sum_{t=1}^{J} y_a(t) y_b(t)} \tag{13}$$

where $y_k(j)$ is one if outlet k has story j and zero otherwise.

3.3 Results

In this section we present the application of our methodology for inferring and validating the News Outlets network. We show that the network presents stability in time using independent datasets, that using some ground truth knowledge we can select the appropriate inference algorithm, and that finally we can predict the network structure.

Stability in Time. Figure 1 presents the stability of the network for the 12 consecutive weeks. The dataset of each week is independent of the data of the other weeks and the first week is used only as reference network. The threshold of the three network inference methods was set to produce networks of the same density of $\sim$ 5000 edges per week.

To determine the stability of the network inference algorithm, we test whether an inferred network's similarity to the inferred network from the previous week is significant. In order to do this, we carry out two hypothesis tests: one for each of the possible null models discussed in 2.1. As test statistic we used the Jaccard Distance. We sampled $K = 1000$ networks and estimating the p-value as the fraction of those for which the Jaccard distance with the reference network from the previous week is smaller than for the inferred network.

The networks inferred by each method are significantly similar ($p < 0.001$) to those inferred the previous week with the same method, under both null models.

Comparison to Related Networks. In this test, we compare the network inferred by each of the three Methods, with three other networks, obtained using independent (but possibly related) information. The first one (Location Network) linking outlets with the same geographic location; the second one (Language Network) linking outlets written in the same language; the third one (Media Type) liking outlets of the same type (e.g., newspaper, magazine, broadcast, blog, etc). These networks of outlets are formed of several disjoint cliques, and we expect some of them to relate to the news-choice preference of an outlet. Clearly, a story that is important and publishable for UK media may be uninteresting to the French media. Language is also an important factor, independently of the location of the outlet. For example, we measured that the number of stories that mention the word 'Pope' in Spanish-language media in the USA is three times larger than in English-language media in the same country. About the influence of media type, it is worth mentioning that certain stories may reported in blogs before they appear in mainstream traditional media.

Figure 1 presents the comparison of the content-based network inferred by Methods A, B, C to the 'Location' network. In this case only the Methods B and C are significant with $p < 0.001$. Figure 2 presents the case of the 'Language' where only Method C yields significant patterns ($p < 0.001$) over all weeks' datasets. Finally Fig. 2 illustrates the 'Media-Type' case where Method A and Method C yield significant results ($p < 0.001$). Only method C present significant results for all reference networks over all independent datasets and the two null models that were used to make comparisons. Note that although the distances between networks seem relative small, they are highly significant.

Selecting Inference Method. The selection of the inference method will be made based on their ability to create significant results. We selected a significance level of 0.001 and based our decision on this. Table 1 compares the three methods for the 11 weeks' independent datasets (the first week was used only as reference network). The numbers represent the number of weeks that the methods presented significant results with $p < 0.001$ for the different reference networks and the two null models. Only Method C presents significant results for all the performed tests and datasets, performing at least as good as the two

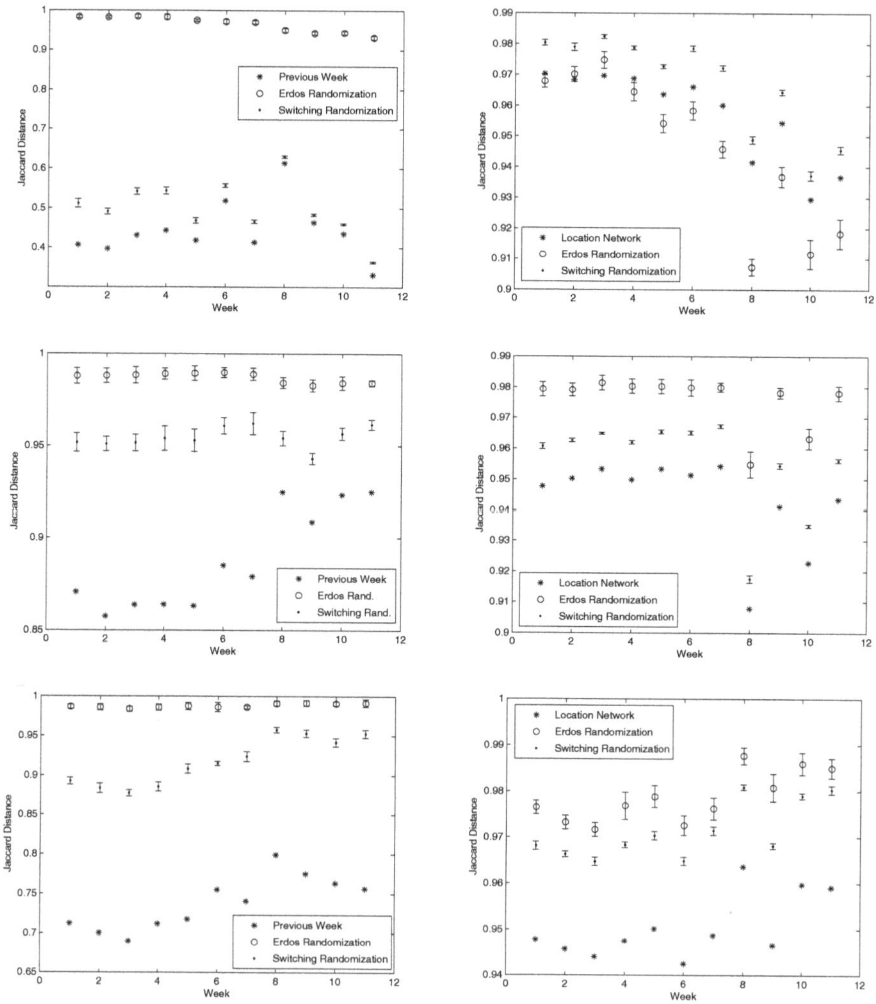

Fig. 1. Network stability on sequential weeks (on the left) and 'Location' reference network (on the right) for the three network reconstruction methods. Errorbars are +/− 3 standard deviations over the mean value.

other methods investigated on all tests carried out. We can therefore conclude that Method C is better than both Methods A and B.

In Figure 3 we report the network of the media outlets obtained with Method C. To the best of our knowledge, this is the first map of this kind to be published. The visualisation of the network was made by using the Cytoscape software [19].

Prediction of Edges. We investigated the ability of prediction of an edge of media outlet network based on the GLM analysis and the three available ground truth reference networks. For the GLM analysis we adopted the normal distribution for Y_i and the identity link function where $\mu = X\beta$. The accuracy

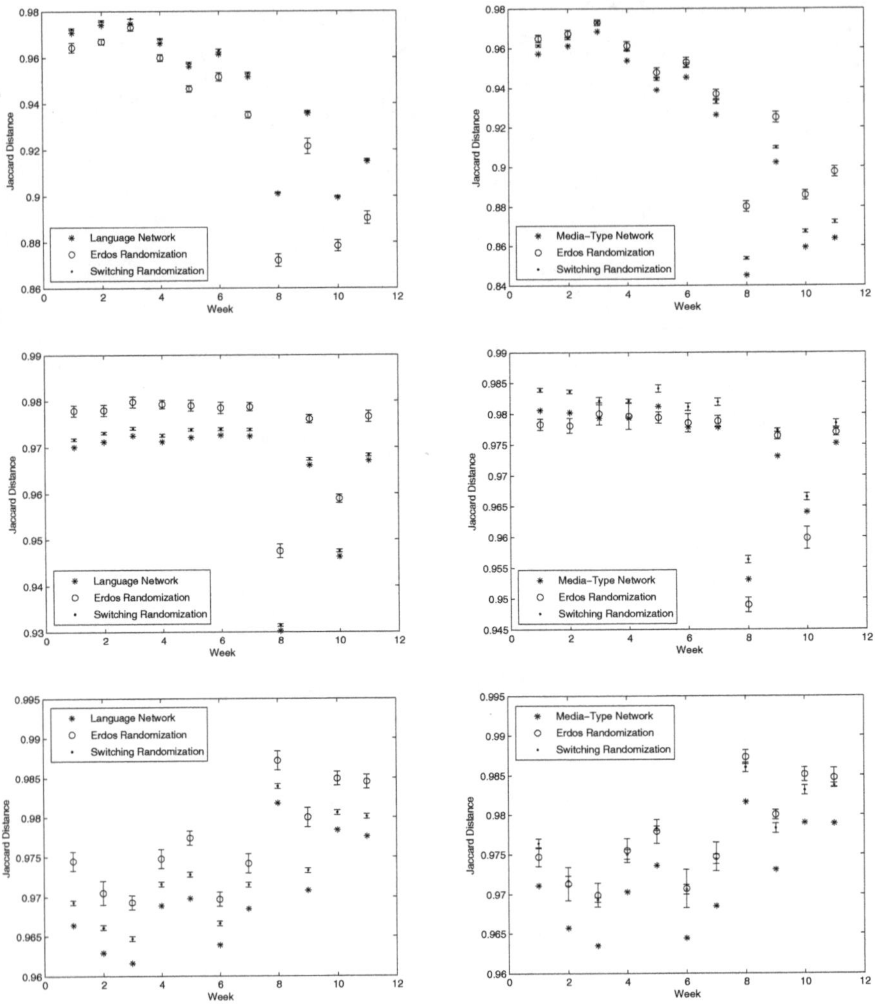

Fig. 2. Comparison to the 'Language' (on the left) and 'Media-Type' (on the right) reference networks. Errorbars are +/− 3 standard deviations over the mean value.

Table 1. The number of weeks that each Method presented significant results with $p < 0.001$. ER stands for the Erdös and SR for the Switching Random Graphs.

	Previous week		Location		Language		Media-Type	
	ER	SR	ER	SR	ER	SR	ER	SR
Method A	11	11	1	11	0	9	11	11
Method B	11	11	11	11	11	11	2	11
Method C	11	11	11	11	11	11	11	11

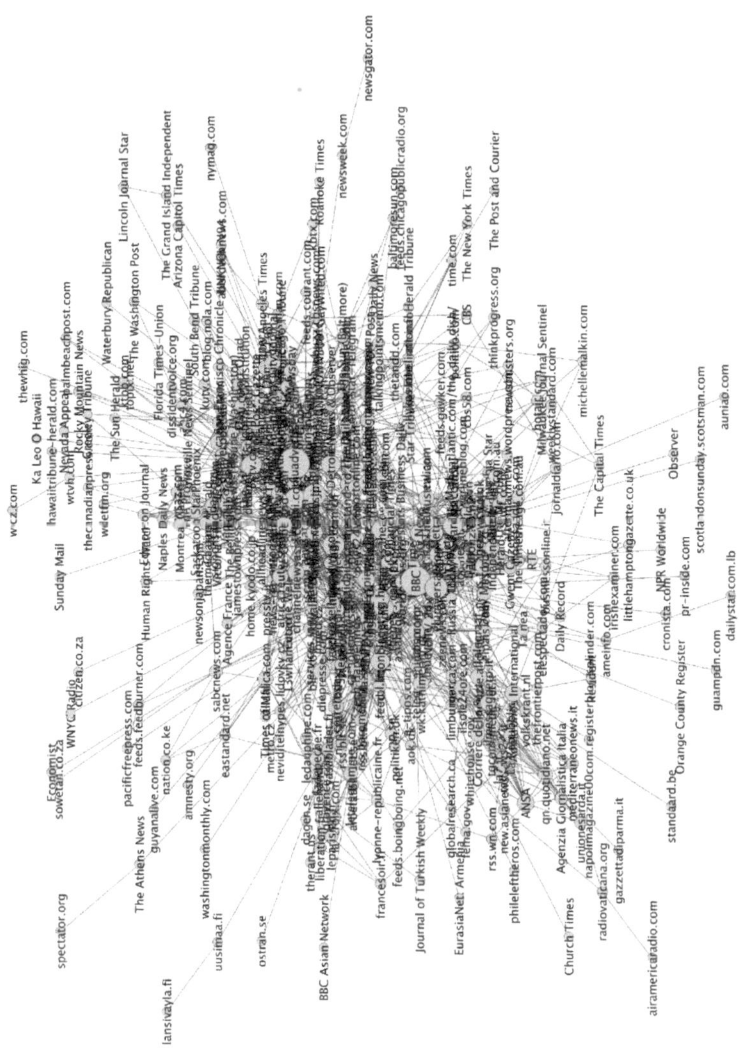

Fig. 3. A snapshot of the News Media Network. A high threshold is set to produce more sparse graph for visualisation reasons. This graph contains 351 nodes and 1612 edges. Singletons are omitted. The node sizes are proportional to their degrees.

was measured as the AUC for a 100-fold cross-validation scheme for different densities of the inferred network. For each inference method two figures are presented: The first illustrates the accuracy of each ground component if it was used by itself for predictions compared to using all three, and the second the accuracy of using pairs of components compared to using all three. Figure 4 present the edge prediction results for the three network inference methods and under different scenarios: Using all three ground truth networks combined, using each one of them separately and using all pairwise combinations of them. The

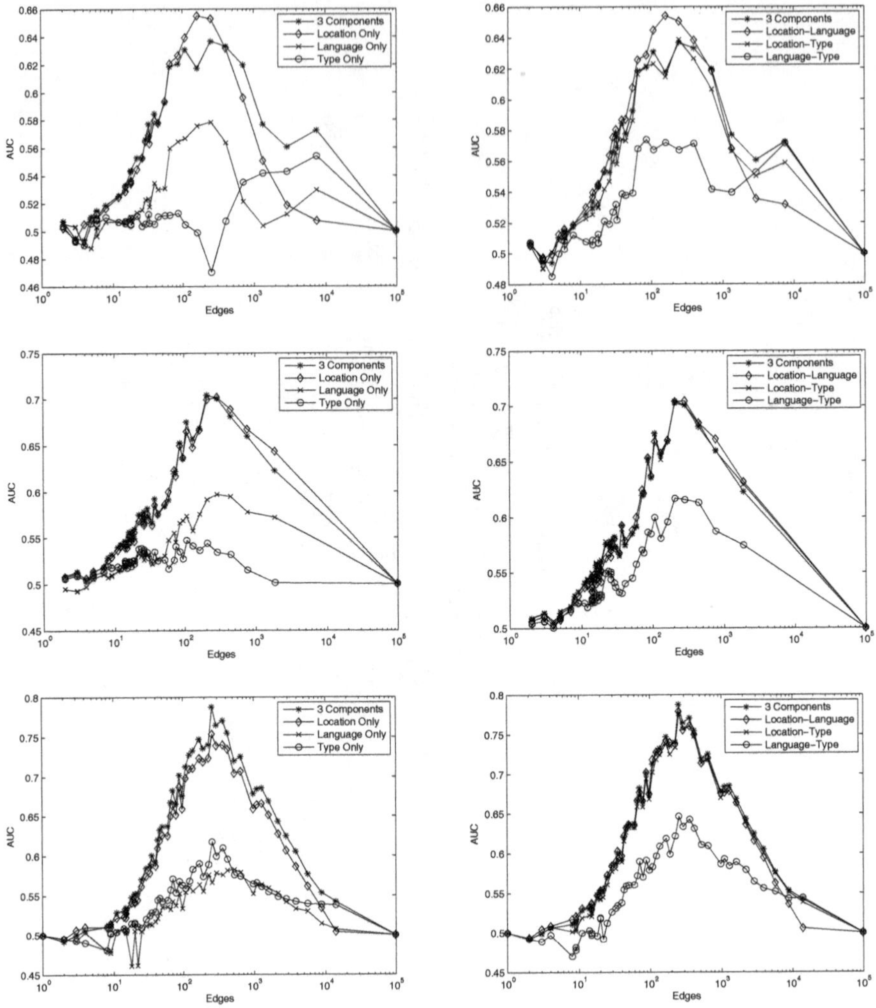

Fig. 4. AUC accuracy for edge prediction using Method A on the top row, Method B in the middle row and Method C in the bottom row, based on GLM analysis over different network densities

best prediction accuracy, 77.11%, over all different network densities, is reached using all three ground truth networks and the Method C.

4 Conclusions

The validation of network inference results in terms of statistical assumptions or in terms of related networks, indicates how to handle the common case where ground truth is difficult to obtain. Concepts from statistical testing can directly provide a framework for assessing and comparing results, algorithms and also datasets.

Importantly, when one can distinguish in a principled way between two algorithms, then one can also search the hypothesis space for the best possible network. Future work in this direction will include the design of network inference algorithms that directly optimise the stability and significance of the output, instead of just choosing between existing heuristic algorithms.

Acknowledgements. The authors want to thank Omar Ali and Phil Naylor, respectively for their contribution in data gathering and computer infrastructure support and the entire 'Pattern Analysis and Intelligent Systems' group at the University of Bristol for discussions. This work is partially supported by the European Commission through the PASCAL2 Network of Excellence (FP7-216866), and the IST project SMART (FP6-033917). Ilias Flaounas is supported by Alexander S. Onassis Public Benefit Foundation and Nello Cristianini is supported by a Royal Society Wolfson Merit Award.

References

1. D'haeseleer, P., Liang, S., Somogyi, R.: Genetic network inference: from co-expression clustering to reverse engineering. Bioinformatics 16, 707–726 (2000)
2. Ma'ayan, A.: Insights into the Organization of Biochemical Regulatory Networks Using Graph Theory Analyses. J. Biol. Chem. 284, 5451–5455 (2009)
3. Paris, L., Bazzoni, G.: The Protein Interaction Network of the Epithelial Junctional Complex: A System-Level Analysis. Mol. Biol. Cell 19, 5409–5421 (2008)
4. Pelillo, M.: Replicator Equations, Maximal Cliques, and Graph Isomorphism. Neural Computation 11(8), 1933–1955 (1999)
5. Bunke, H.: Error Correcting Graph Matching: On the Influence of the Underlying Cost Function. IEEE Transactions on Pattern Analysis and Machine Intelligence 21, 917–922 (1999)
6. Fernández, M.-L., Valiente, G.: A graph distance metric combining maximum common subgraph and minimum common supergraph. Pattern Recognition Letters 22, 753–758 (2001)
7. Jaccard, P.: Ètude comparative de la distribution florale dans une portion des Alpes et des Jura. Bulletin del la Societ. Vaudoise des Sciences Naturelles 37, 547–579 (1901)
8. Milo, R., Shen-Orr, S., Itzkovitz, S., Kashtan, N., Chklovskii, D., Alon1, U.: Network Motifs: Simple Building Blocks of Complex Networks. Science 298, 824–827 (2002)
9. Erdös, P., Rényi, A.: On Random Graphs. Publicationes Mathematicae 6, 290–297 (1959)
10. Rao, A.R., Jana, R., Bandyopadhya, S.: A Markov chain Monte Carlo method for generating random (0, 1)-matrices with given marginals. Indian J. of Statistics 58, 225–242 (1996)
11. Milo, R., Kashtan, N., Itzkovitz, S., Newman, M.E.J., Alon, U.: On the uniform generation of random graphs with prescribed degree sequences (2003) Arxiv cond-mat/0312028

12. Nelder, J., Wedderburn, R.: Generalized Linear Models. Journal of the Royal Statistical Society 135 Series A (General), 370–384 (1972)
13. McCullagh, P., Nelder, J.: Generalized Linear Models. Chapman and Hall, London (1989)
14. Turchi, M., Flaounas, I., Ali, O., De Bie, T., Snowsill, T., Cristianini, N.: Found In Translation. In: Buntine, W., et al. (eds.) ECML/PKDD 2009. LNCS, vol. 5781. Springer, Heidelberg (2009), `http://patterns.enm.bris.ac.uk/publications/found-in-translation` (accepted for publication)
15. Koehn, P., Hoang, H., et al.: Moses: Open Source Toolkit for Statistical Machine Translation. Annual Meeting-Association for Computational Linguistics 45 (2007)
16. Koehn, P., Och, F.J., Marcu, D.: Statistical phrase-based translation. In: 2003 Conference of the North American Chapter of the Association for Computational Linguistics on Human Language Technology, pp. 48–54. Association for Computational Linguistics, Morristown (2003)
17. Liu, B.: Web Data Mining, Exploring Hyperlinks, Contents, and Usage Data. Springer, Heidelberg (2007)
18. Hirsh, A., Fraser, H.: Protein dispensability and rate of evolution. Nature 411, 1046–1049 (2001)
19. Shannon, P., Markiel, A., Ozier, O., Baliga, N.S., Wang, J.T., Ramage, D., Amin, N., Schwikowski, I.T.: Cytoscape: A Software Environment for Integrated Models of Biomolecular Interaction Networks. Genome Res. 13, 2498–2504 (2003)

Binary Decomposition Methods for Multipartite Ranking

Johannes Fürnkranz[1], Eyke Hüllermeier[2], and Stijn Vanderlooy[3]

[1] Department of Computer Science, TU Darmstadt
juffi@ke.tu-darmstadt.de
[2] Department of Mathematics and Computer Science, Marburg University
eyke@mathematik.uni-marburg.de
[3] Department of Knowledge Engineering, Maastricht University
s.vanderlooy@maastrichtuniversity.nl

Abstract. Bipartite ranking refers to the problem of learning a ranking function from a training set of positively and negatively labeled examples. Applied to a set of unlabeled instances, a ranking function is expected to establish a total order in which positive instances precede negative ones. The performance of a ranking function is typically measured in terms of the AUC. In this paper, we study the problem of multipartite ranking, an extension of bipartite ranking to the multi-class case. In this regard, we discuss extensions of the AUC metric which are suitable as evaluation criteria for multipartite rankings. Moreover, to learn multipartite ranking functions, we propose methods on the basis of binary decomposition techniques that have previously been used for multi-class and ordinal classification. We compare these methods both analytically and experimentally, not only against each other but also to existing methods applicable to the same problem.

1 Introduction

There are several connections between "learning to rank", a topic of increasing interest in machine learning research, and conventional classifier learning. First, some ranking problems such as *label ranking* [16] can be seen as direct extensions of (multi-class) classification. Second, many learning methods for ranking essentially reduce the original problem to a standard classification problem. Third, aspects of ranking and sorting are also of interest for classification itself. A notable case is ROC analysis, which evaluates the ability of classifiers to sort positive and negative instances in terms of the area under the ROC curve, abbreviated as AUC [7]. The problem to learn classifiers with high AUC (instead of low error rate) is called *bipartite ranking*.

In this paper, we are interested in extending bipartite ranking from the binary to the multi-class case. This problem, that we shall refer to as *multipartite ranking*, is closely related to *ordinal classification*, that is, classification with a totally ordered set of classes. Yet, just like in bipartite ranking, the goal is not to learn a good classifier, but a good ranker, that is, a function that systematically

W. Buntine et al. (Eds.): ECML PKDD 2009, Part I, LNAI 5781, pp. 359–374, 2009.

ranks "high" classes ahead of "low" classes. Note that a ranking is in a sense a refinement of the order information provided by an ordinal classifier, as the latter does not distinguish between objects within the same category. As an example, compare the task of two reviewers: one has to make an ordinal prediction for each paper (reject, weak reject, weak accept, accept), and the other has to rank the papers according to quality. Obviously, the latter approach provides more detailed information that can be used not only for partitioning papers into the above four categories.

To solve the multipartite ranking problem, we explore the use of binary decomposition techniques. Such techniques have already been applied quite successfully in conventional classification, ordinal classification, and label ranking [1,8,16], but have not yet been explored in prior work on multipartite ranking [20,22]. Our main result is that, compared to existing methods applicable for the multipartite ranking problem, binary decomposition techniques are at least competitive in terms of predictive accuracy, and presumably even superior, while being computationally much more efficient.

In the next section, we introduce the multipartite ranking problem, discuss its relation to bipartite ranking and ordinal classification, and also address the question of how to evaluate multipartite ranking functions. In Section 3, we propose two methods for multipartite ranking which are based on binary decomposition techniques. Section 4 is devoted to an experimental analysis of these methods. The paper ends with some concluding remarks in Section 5.

2 Ordinal Classification and Multipartite Ranking

In this section, we introduce the problem of multipartite ranking and discuss corresponding performance metrics. Beforehand, we recall the related problems of ordinal classification and bipartite ranking.

2.1 Ordinal Classification

In ordinal classification, also called ordinal regression in statistics, the set of class labels $\mathcal{L} = \{\lambda_1, \lambda_2 \dots \lambda_m\}$ is endowed with a natural (total) order relation $\lambda_1 \prec \lambda_2 \prec \cdots \prec \lambda_m$. This distinguishes ordinal from conventional classification, where $\mathcal{L}$ is an unordered set.

From a learning point of view, the ordinal structure of $\mathcal{L}$ is *additional* information that a learner should try to exploit, and this is what existing methods for ordinal classification essentially seek to do [8,4]. In fact, the problem of ordinal classification is in a sense in-between classification and regression, two problems that have been extensively studied. Like in classification, the output space is finite, and like in regression, the elements of this space are ordered.

Thus, it is hardly surprising that both classification and regression algorithms have been used to tackle ordinal classification problems, even though both approaches are obviously problematic: A simple classification method will neglect information about the class order, whereas a regression method will make too strong assumptions, because it does not only exploit the order relation but assumes meaningful distances between output values [19]. To avoid these problems, new algorithms have been developed in the machine learning field in recent years, which are able to exploit class order information in a meaningful way [13,8,5,4].

2.2 Bipartite Ranking

In the problem of *bipartite ranking*, training data consists of a set of positively and negatively labeled instances, just like in conventional binary classification. However, instead of learning a classifier that can be used for assigning instances to one of the two classes, the goal is to learn a ranking function $f(\cdot)$ that can be used for ordering a set of instances from most likely positive to most likely negative. Thus, given a set X of instances with unknown class labels, the learned ranking function outputs a linear order of these instances. Typically, this is accomplished by *scoring* the instances, i.e., $f(\cdot)$ is implemented as an $\mathbb{X} \to \mathbb{R}$ mapping that assigns a real-valued score $f(\boldsymbol{x})$ to each instance $\boldsymbol{x}$ from an instance space $\mathbb{X}$. The instances are then ranked according to their respective scores.

The most commonly used metric to evaluate a predicted ranking is the area under the ROC curve (AUC) [7]:

$$\mathrm{AUC}(f, X) = \frac{1}{|P||N|} \sum_{\boldsymbol{x} \in P} \sum_{\boldsymbol{x}' \in N} S(f(\boldsymbol{x}'), f(\boldsymbol{x})) \ , \tag{1}$$

where $P \subset X$ and $N \subset X$ are the positive and negative instances in X (hence $X = P \cup N$, $P \cap N = \emptyset$). The mapping $S(\cdot, \cdot)$ outputs 1 when the positive instance is ranked before the negative one, and 0 in the reverse case. An output of 1/2 is given when the instances are assigned the same score.

2.3 Multipartite Ranking

Given that class labels are ordered, as in ordinal classification, the idea of bipartite ranking can obviously be generalized from the binary to the multi-class case. Given a set of instances X with class labels in $\mathcal{L} = \{\lambda_1, \lambda_2 \dots \lambda_m\}$, the goal is to order them in such a way that, ideally, the instances from λ_m precede those from λ_{m-1}, which in turn precede those from class λ_{m-2}, etc. Subsequently, we shall refer to the problem of learning a corresponding ranking function from a training set $T = \{(\boldsymbol{x}_i, \ell_{\boldsymbol{x}_i})\}_{i=1}^n \subset \mathbb{X} \times \mathcal{L}$ of labeled instances as *multipartite ranking.*

A common approach to learning the scoring function f consists of turning the original training data into a set of *order constraints* on $f(\cdot)$, and then finding a function that is as much as possible in agreement with these constraints. More specifically, each pair of observed examples $(\boldsymbol{x}_i, \ell_{\boldsymbol{x}_i})$ and $(\boldsymbol{x}_j, \ell_{\boldsymbol{x}_j})$ with $\ell_{\boldsymbol{x}_i} \succ \ell_{\boldsymbol{x}_j}$ gives rise to a constraint $f(\boldsymbol{x}_i) > f(\boldsymbol{x}_j)$. To make the problem amenable to existing learning algorithms, the idea is to express such constraints as classification examples. Suppose, for example, that $f(\cdot)$ is a linear function $\boldsymbol{x} \mapsto \langle \alpha, \boldsymbol{x} \rangle$. Then,

$$f(\boldsymbol{x}) > f(\boldsymbol{x}') \quad \Leftrightarrow \quad \langle \alpha, \boldsymbol{x} \rangle - \langle \alpha, \boldsymbol{x}' \rangle > 0 \quad \Leftrightarrow \quad \langle \alpha, \boldsymbol{x} - \boldsymbol{x}' \rangle > 0 \ ,$$

which is equivalent to saying that $\boldsymbol{z} = \boldsymbol{x} - \boldsymbol{x}'$ should be classified as positive (or $-\boldsymbol{z}$ as negative) by a standard binary classifier [13]. The number of pairwise constraints, and hence the size of the training data for the binary classifier, will typically be much larger than the original training data (cf. Section 3.5).

2.4 Evaluation Metrics for Multipartite Ranking

To evaluate the performance of a predicted multipartite ranking of a set X of instances, different metrics have been proposed in the literature. An obvious generalization of the AUC, which estimates the probability that a randomly chosen positive instance is ranked higher than a randomly chosen negative instance, is to consider the probability that a randomly chosen pair of instances from different classes is ranked correctly by $f(\cdot)$, i.e.

$$\mathbf{P}\left(f(\boldsymbol{x}) < f(\boldsymbol{x}') \,|\, \lambda_{\boldsymbol{x}} \prec \lambda_{\boldsymbol{x}'}\right) .$$

Assuming that the training examples are drawn independently from an underlying probability distribution on $\mathbb{X} \times \mathcal{L}$, an unbiased estimate of this probability is obtained by the C-index

$$C(f, X) = \frac{1}{\sum_{i<j} n_i n_j} \sum_{1 \leq i < j \leq m} \sum_{(\boldsymbol{x}, \boldsymbol{x}') \in X_i \times X_j} S(f(\boldsymbol{x}'), f(\boldsymbol{x})) , \tag{2}$$

where X_i is the subset of instances $\boldsymbol{x} \in X$ whose true class is λ_i, and $n_i = |X_i|$. The C-index is commonly used as a metric of concordance in statistics [11]. It is essentially equivalent to the pairwise ranking error introduced in [13]. Obviously, the AUC defined in (1) is a special case of (2) with $X_1 = P$ and $X_2 = N$.

A related metric is the Jonckheere-Terpstra statistic [14], which is closely related to a multi-class extension of the AUC that has been proposed in [12]:

$$U(f, X) = \frac{2}{m(m-1)} \sum_{1 \leq i < j \leq m} \mathrm{AUC}(f, X_i \cup X_j) . \tag{3}$$

The key difference between (2) and (3) is that in (2), the contribution of a class is proportional to the size of the class, while each class has the same weight in (3). In fact, (2) can be written as a weighted sum of pairwise AUCs:

$$C(f, X) = \frac{1}{\sum_{i<j} n_i n_j} \sum_{1 \leq i < j \leq m} n_i n_j \, \mathrm{AUC}(f, X_i \cup X_j) \tag{4}$$

An alternative proposal for extending the AUC has recently been made in [22]. This alternative is motivated by the observation that decomposing the evaluation of a multipartite ranking into several pairwise evaluations may arguably come along with a certain loss of information and, moreover, can violate desirable transitivity properties. For example, pairwise AUCs can violate stochastic transitivity: $\mathrm{AUC}(f, X_1 \cup X_2) \geq 1/2$, $\mathrm{AUC}(f, X_2 \cup X_3) \geq 1/2$ but $\mathrm{AUC}(f, X_1 \cup X_3) < 1/2$.

Therefore, to evaluate rankings in a more global way, the idea is to consider the probability

$$\mathbf{P}\left(f(\boldsymbol{x}_1) < f(\boldsymbol{x}_2) < \ldots < f(\boldsymbol{x}_m) \,|\, \ell_{\boldsymbol{x}_1} = \lambda_1, \ldots, \ell_{\boldsymbol{x}_m} = \lambda_m\right) ,$$

where the $(\boldsymbol{x}_i, \ell_{\boldsymbol{x}_i})$, $i = 1 \ldots m$, are again drawn independently from an underlying probability distribution on $\mathbb{X} \times \mathcal{L}$. This gives rise to the following evaluation metric for the ranking of a finite set X of instances:

$$W(f, X) = \frac{1}{\prod_{i=1}^{m} n_i} \sum_{\boldsymbol{x}_1 \in X_1, \ldots, \boldsymbol{x}_m \in X_m} \mathbb{I}(f(\boldsymbol{x}_1) < f(\boldsymbol{x}_2) < \ldots < f(\boldsymbol{x}_m)) , \quad (5)$$

where the indicator function $\mathbb{I}(\cdot)$ maps truth values of predicates to $\{0, 1\}$. Thus, the basic idea is to select m instances, one from each class, and to check whether they are correctly ordered or not. The metric (5) is simply the average over all possible m-tuples that can be verified in this way.

Despite having some potential advantages, we believe that (5) does also exhibit some questionable properties. In particular, evaluating a ranking of m elements in terms of a simple 0/1 loss, as done by the indicator function $\mathbb{I}(\cdot)$, does not distinguish between very poor rankings (e.g., a reversal of the correct ranking) and rankings that are "almost correct" (in which, for example, only two adjacent classes are swapped). To take an extreme example, suppose that $X_1 = \{\boldsymbol{x}_1\}$ and $X_2 = \{\boldsymbol{x}_2\}$ each only contain a single instance, and that the predicted ranking swaps these two instances ($f(\boldsymbol{x}_1) > f(\boldsymbol{x}_2)$), while the instances from all other classes are always ordered correctly by $f(\cdot)$. Still, (5) will yield the worst evaluation $W(f, X) = 0$.

This strong sensitivity toward small mistakes could in principle be avoided by replacing the indicator function $\mathbb{I}(\cdot)$ in (5) with a more tolerant metric, for example the (normalized) sum of concordant pairs

$$\frac{2}{m(m-1)} \sum_{1 \le i < j \le m} \mathbb{I}(f(\boldsymbol{x}_i) < f(\boldsymbol{x}_j)) .$$

This would obviously yield a metric closely related to the pairwise variants (2) and (3). Still, when expressing the metric thus obtained in terms of pairwise AUCs, another questionable property of (5) becomes obvious. In fact, one will again obtain a weighted combination of pairwise AUCs, just like (4), but now the AUC of classes λ_i and λ_j is weighted by $\prod_{i=1}^{m} n_i/(n_i n_j)$. In other words, the weighing is now *inversely* related to the size of the classes. This is because, to produce all m-tuples in (5), a pair of instances $(\boldsymbol{x}_i, \boldsymbol{x}_j)$ is combined with all instances from all other classes. In our above example, where $|X_1| = |X_2| = 1$, the instance pair $(\boldsymbol{x}_1, \boldsymbol{x}_2)$ has an extreme influence, since it will necessarily appear in *all* m-tuples.

Subsequently, we shall focus on the pairwise evaluation metrics (2) and (3), which are intuitively appealing and widely adopted in the literature.

2.5 From Multipartite Ranking to Ordinal Classification

We end this section with a few comments on the relationship between multipartite ranking and ordinal classification, the problem that we also started with at the beginning of the section. Obviously, an ordinal classification function can be used as a ranking function. In fact, note that an ordinal classifier becomes a scoring function by interpreting its predictions, namely class labels, as scores (i.e., predicting class λ_i for instance $\boldsymbol{x}$ is considered as scoring $\boldsymbol{x}$ by i). Needless to say, however, this type of scoring will produce a large number of ties, which is why one cannot expect ordinal classifiers to be good rankers.

Conversely, a ranking function $f(\cdot)$ can be turned into an ordinal classifier by *thresholding*: Given $m+1$ threshold values t_i, $i = 1 \dots m+1$, class λ_i is predicted for an instance $\boldsymbol{x}$ if the score $f(\boldsymbol{x})$ is between t_i and t_{i+1}. The main problem here is to find optimal thresholds, i.e., thresholds that lead to an optimal classification performance [21]. However, this problem is beyond the scope of this paper.

3 Binary Decomposition for Multipartite Ranking

As mentioned in Section 2.3, existing ranking methods are mostly based on the idea of transforming the original ranking problem into a binary classification problem. Roughly speaking, each pair of instances (from different classes) gives rise to a training example. It is worth mentioning that these methods are indeed applicable though not specifically designed for multipartite ranking: Their input is a set of order constraints on instances ($f(\boldsymbol{x}_i) > f(\boldsymbol{x}_j)$), which *can* originate from class information ($\ell_{\boldsymbol{x}_i} \succ \ell_{\boldsymbol{x}_j}$), but may also come from other sources; in fact, the existence of classes is not even assumed. In this section, we propose an alternative strategy which is specialized to the multipartite ranking problem and exploits class information in a more explicit way.

3.1 Exploiting Class Information by Binary Decomposition

Instead of transforming the original problem to a *single* binary classification problem, we decompose it into several such problems, resorting to binary decomposition techniques that have already been used successfully in multi-class classification. A decomposition technique specifically designed for ordinal classification has been proposed in [8]. Since ordinal classification and multipartite ranking are closely related (cf. Section 2.5), the idea to adapt this method to the latter problem suggests itself, and we will do so in Section 3.2. Another promising decomposition technique is the all-pairs learning scheme that has already been used in conventional and ordinal classification, as well as in other types of ranking problems such as label ranking [9,10,16,15]. In fact, this technique is especially motivated by (3) and (4), which show that the quality of a multipartite ranking is in direct correspondence with the quality of the associated bipartite rankings. Thus, it should, in principle, be possible to optimize the former by optimizing the latter, and this is what the all-pairs approach is seeking to do. We shall elaborate on this approach in Section 3.3.

One key advantage of binary decomposition is that the related problems are simpler and usually much smaller than a single binary problem, as they avoid the combinatorial explosion caused by considering all pairs of the original training examples (cf. Section 3.5). Roughly speaking, by exploiting class information, a large number of order constraints can be satisfied in an implicit way: A classifier that separates n_0 negative from n_1 positive instances, and which is hence trained on $n_0 + n_1$ examples, automatically satisfies $n_0 n_1$ order constraints.

On the other hand, binary decomposition also produces an additional problem, namely the need to *aggregate* the predictions of the different models into a single ranking: Binary decomposition techniques learn a *set* of models $\mathcal{M}_i$ on different subproblems. Given a query instance $\boldsymbol{x}$, this instance is submitted to all models, and the predictions $\mathcal{M}_i(\boldsymbol{x})$ have to be combined into an overall prediction. In the context of multipartite ranking, aggregation can essentially be realized at two different levels: (1) *Aggregation of rankings:* In this case, each model $\mathcal{M}_i$ produces a ranking, and these rankings are combined into a *consensus ranking.* (2) *Aggregation of scoring functions:* If all models $\mathcal{M}_i$ are based on scoring functions $f_i(\cdot)$, a second option is to combine these functions into a single function $f(\cdot)$, which means combining the scores $f_i(\boldsymbol{x})$ into a single score $f(\boldsymbol{x})$. Computationally, the first alternative is more complex, since, depending on the concrete criterion used, optimal rank aggregation may become very expensive. We shall therefore adopt the second approach.

3.2 The Approach of Frank and Hall

A simple and intuitively appealing approach to ordinal classification has been proposed by Frank and Hall [8]. The idea of this method, subsequently referred to as F&H, is to decompose the original problem involving m classes $\mathcal{L} = \{\lambda_1, \lambda_2 \ldots \lambda_m\}$ into $m-1$ binary problems. The i-th problem is defined by the "meta-classes" $C_- = \{\lambda_1, \lambda_2 \ldots \lambda_i\}$ and $C_+ = \{\lambda_{i+1}, \lambda_{i+2} \ldots \lambda_m\}$ playing the role, respectively, of the negative and positive class.

Let $\mathcal{M}_i$, $i = 1 \ldots m-1$, denote the model learned on this problem. Given a query instance $\boldsymbol{x}$, a prediction $\mathcal{M}_i(\boldsymbol{x})$ is interpreted as an estimation of the probability $\mathbf{P}(\ell_{\boldsymbol{x}} \succ \lambda_i)$ that the class of $\boldsymbol{x}$, denoted $\ell_{\boldsymbol{x}}$, is in C_+. Consequently, the models must guarantee outputs in the unit interval. From these probabilities, a probability distribution on $\mathcal{L}$ is derived as follows:

$$\begin{aligned}
\mathbf{P}(\ell_{\boldsymbol{x}} = \lambda_1) &= 1 - \mathbf{P}(\ell_{\boldsymbol{x}} \succ \lambda_1) \\
\mathbf{P}(\ell_{\boldsymbol{x}} = \lambda_i) &= \max\{\,\mathbf{P}(\ell_{\boldsymbol{x}} \succ \lambda_{i-1}) - \mathbf{P}(\ell_{\boldsymbol{x}} \succ \lambda_i), 0\,\}, \quad i = 2, \ldots, m-1 \qquad (6) \\
\mathbf{P}(\ell_{\boldsymbol{x}} = \lambda_m) &= \mathbf{P}(\ell_{\boldsymbol{x}} \succ \lambda_{m-1})
\end{aligned}$$

Eventually, the class with the highest probability is predicted.

Coming back to the problem of multipartite ranking, recall that we seek to aggregate the scoring functions associated with individual models into an overall scoring function that defines the ranking. Here, these functions are given by the predictions of the models $\mathcal{M}_i$, that is, $f_i(\boldsymbol{x}) = \mathcal{M}_i(\boldsymbol{x})$, and each of them induces an individual ranking. An obvious aggregation function is given by

$$f(\boldsymbol{x}) = \sum_{i=1}^{m-1} f_i(\boldsymbol{x}) = \sum_{i=1}^{m-1} \mathcal{M}_i(\boldsymbol{x}) \ . \tag{7}$$

This aggregation is meaningful as it systematically assigns higher scores to instances from higher classes. For example, an instance $\boldsymbol{x}$ of class λ_1 will be in the negative meta-class of all $m-1$ binary classifiers, i.e., all $\mathcal{M}_i$ should return a low score $f_i(\boldsymbol{x})$, and hence the cumulative score $f(\boldsymbol{x})$ will be low.

Formally, it is worth mentioning that (7) is in direct correspondence with the *expected class index* of an instance, given that the models $\mathcal{M}_i$ yield reasonable probability estimations and are consistent in the sense that $\mathcal{M}_i(\boldsymbol{x}) \geq \mathcal{M}_{i+1}(\boldsymbol{x})$. In fact, the mapping $i \mapsto \mathcal{M}_i(\boldsymbol{x})$ is nothing else than a decumulative distribution function, and hence, with $\mathbb{E}(i)$ the expected class index of $\boldsymbol{x}$:

$$f(\boldsymbol{x}) = \sum_{i=1}^{m-1} \mathbf{P}(\ell_{\boldsymbol{x}} \succ \lambda_i) = \sum_{i=1}^{m-1} \sum_{j=i+1}^{m} \mathbf{P}(\ell_{\boldsymbol{x}} = \lambda_j) = \sum_{i=1}^{m-1} (i-1)\mathbf{P}(\ell_{\boldsymbol{x}} = \lambda_i) = \mathbb{E}(i) - 1$$

3.3 Learning by Pairwise Comparison

Learning by pairwise comparison (LPC), also known as all-pairs or round robin learning [9], is a popular binarization technique for multi-class classification. LPC trains a separate model $\mathcal{M}_{i,j}$ for each pair of classes $(\lambda_i, \lambda_j) \in \mathcal{L} \times \mathcal{L}$, $1 \leq i < j \leq m$; thus, a total number of $m(m-1)/2$ models is needed. At classification time, a query $\boldsymbol{x}$ is submitted to all models, and each prediction $\mathcal{M}_{i,j}(\boldsymbol{x})$ is interpreted as a vote for a label. More specifically, assuming scoring classifiers that produce normalized scores $f_{i,j} = \mathcal{M}_{i,j}(\boldsymbol{x}) \in [0,1]$, the *weighted voting* technique interprets $f_{i,j}$ and $f_{j,i} = 1 - f_{i,j}$ as weighted votes for classes λ_i and λ_j, respectively, and predicts the class λ^* with the highest sum of weighted votes, i.e., $\lambda^* = \arg\max_i \sum_{j \neq i} f_{i,j}$.

LPC has been used successfully for conventional multi-class classification, but has also been shown to produce strong results for ordinal classification [10,15]. To derive a ranking function from the ensemble of models $\mathcal{M}_{i,j}$, we again need a proper scoring function. Imitating the derivation of (7), a reasonable candidate is the sum of the predictions "in favor of a higher class", that is

$$f(\boldsymbol{x}) = \sum_{1 \leq i < j \leq m} f_{j,i}(\boldsymbol{x}) = \sum_{1 \leq i < j \leq m} \mathcal{M}_{j,i}(\boldsymbol{x}) \ . \tag{8}$$

A variant of this aggregation, motivated by the fact that the C-index (2) is a *weighted* combination of the pairwise AUCs, is

$$f(\boldsymbol{x}) = \sum_{1 \leq i < j \leq m} p_i p_j \, f_{j,i}(\boldsymbol{x}) = \sum_{1 \leq i < j \leq m} p_i p_j \, \mathcal{M}_{j,i}(\boldsymbol{x}) \ , \tag{9}$$

where p_i is the probability of class λ_i estimated by the relative frequency in the training data. In fact, one may expect that (9) is more suitable to optimize the C-index (2), while (8) might be preferable when using (3) as evaluation metric.

3.4 Comparing LPC and F and H

A critical issue of the LPC approach is the so-called "non-competence" problem. Even though the pairwise models $\mathcal{M}_{i,j}$ are trained only on examples from classes λ_i and λ_j, they have to be queried by all instances at prediction time. Thus, if $\boldsymbol{x}$ neither belongs to λ_i nor to λ_j, then $\mathcal{M}_{i,j}$ is actually not "competent" and the prediction $\mathcal{M}_{i,j}(\boldsymbol{x})$ becomes arguably questionable.

This problem is well-known from standard classification, and is also relevant for other binary decomposition techniques which are based on a generalized version of error correcting output codes that allows classifiers to be trained on proper subsets of the label set [1]. Despite this problem, LPC is known to perform extremely well for classification and often outperforms other decomposition techniques that do not suffer from the non-competence problem, including the one-vs-rest decomposition and the F&H method. A key advantage of LPC is a *simplification effect* produced by the all-pairs decomposition: Two-class problems are maximally simple from a learning point of view, and the predictions of models trained on these problems are hence more accurate. Since methods like one-vs-rest and F&H train each model on *all* classes, it is true that they only produce competent models. On the other hand, however, the problems are more difficult (typically requiring more complex decision boundaries), and hence the model predictions presumably less accurate. Besides, there is another potential advantage of LPC, namely a kind of *redundancy* due to the training of a larger (namely quadratic) number of models. Thanks to this redundancy, is is easier to compensate for prediction errors of individual models.

For the following reason, however, one may expect the non-competence problem to be more severe for multipartite ranking than for classification: In classification, one score is derived for each class label λ_i by combining the predictions $\mathcal{M}_{i,j}(\boldsymbol{x})$, $1 \leq i \neq j \leq m$. Thus, at least the computation of the score for the true label $\ell_{\boldsymbol{x}}$ does not involve any incompetent prediction, and as long as these predictions are correct, the true class will be the winner of the voting scheme, regardless of all other (non-competent) predictions. In multipartite ranking, on the other hand, the score of an instance $\boldsymbol{x}$ depends on *all* models $\mathcal{M}_{i,j}$, and most of them are not competent for $\boldsymbol{x}$.

Fortunately, due to the ordinal structure, there is still reason to hope that even the non-competent models will make reasonable predictions: Given that the ordinal structure of the set of class labels $\mathcal{L}$ is also reflected in the topology of the instance space $\mathbb{X}$, an assumption which is implicitly made by all ordinal classification methods [15], a model $\mathcal{M}_{i,j}$ will *indirectly* also be trained for classes λ_k, $i \neq k \neq j$. For example, when the model $M_{2,3}$ is queried with an instance $\boldsymbol{x}$ whose true class is λ_4, a vote for the higher class λ_3 (which is supposed to be "closer" to λ_4) should be more likely than a vote for the lower class λ_2.

In summary, LPC has the disadvantage of partially non-competent models, but the advantage of simplicity and redundancy. Which of these effects will dominate in practice is a question that cannot be answered in general, especially since the answer will strongly depend on the data set and learning algorithm used. First insights will be offered by our empirical study presented in Section 4.

3.5 Computational Complexity

The F&H method trains $m-1$ models on the complete set of training examples whose size is $|T| = n$. Thus, the total number of training examples for the transformed problem is $(m-1)n$, and when using a base learner with complexity $O(k^\alpha)$, the overall complexity for training an F&H ranking model is $O(mn^\alpha)$.

LPC requires a total of $(m-1)n$ training examples, too, because each original example $(\boldsymbol{x}_i, \ell_{\boldsymbol{x}_i})$ is used in exactly $m-1$ binary models $\mathcal{M}_{i,j}$. Correspondingly, the training complexity is $O(m^2n^\alpha)$ for a base learner with complexity $O(k^\alpha)$. It should be noted, however, that the pairwise problems, which involve only the instances from two classes, are typically much smaller than n. For example, for a uniform class distribution where each class has $\approx n/m$ examples, the complexity is $O(m^{2-\alpha}n^\alpha)$. Thus, for base learners with super-linear complexity ($\alpha > 1$), LPC is typically even more efficient than F&H.

For methods that construct a single binary classification problem, with one order constraint for each pair of instances with different class labels, the number of training examples can become as large as $O(n^2)$. With a classifier whose complexity is $O(k^\alpha)$, the overall complexity is hence $O(n^{2\alpha})$ and increases much faster with the size of the training set T than for the decomposition methods.

However, it is worth mentioning that, for support vector machines, a quadratic growth in the number of examples can be avoided under certain conditions. In [18], Joachims proposes a sophisticated cutting plane algorithm for SVM training that exploits the sparsity of a data set and scales with $O(s \cdot n \cdot \log(n))$, where s is the average number of non-zero features. Unfortunately, an implementation of this approach (`http://svmlight.joachims.org/`) is offered only for the case of two ranks ($m = 2$), which precludes its use in our experimental analysis.

4 Experimental Analysis

In this section, we provide an extensive empirical evaluation and comparison between methods for multipartite ranking. Our primary interest is to show that a reduction to multiple binary classifiers is at least competitive to state-of-the-art ranking methods in terms of predictive accuracy, while being much more efficient from a computational point of view. Besides, we are also interested in comparing the two decomposition methods, LPC and F&H, amongst each other.

4.1 Data Sets, Ranking Methods, and Experimental Setup

Due to a lack of ordinal benchmark data sets, several previous studies including [8,10,15] have resorted to discretized regression data sets for experimental purposes. This is reasonable and has the advantage that, by changing the discretization, ordinal data sets can be produced in a quite flexible manner. We have used twenty-one regression data sets from the UCI repository [2]. These data sets vary strongly in size and number and type of features, and hence, they are representative for a wide range of data that may occur in practice; see Table 1. To obtain an ordinal class attribute, the numerical output attributes were

Table 1. The twenty-five data sets used in the experiments (name, number of instances, number of numerical features, number of nominal features). The last four data sets are truly ordinal, and the number of classes is given in brackets after the name.

#	name	size	num	nom	#	name	size	num	nom
1	Abalone	4177	7	1	14	House 8L	22784	8	0
2	Ailerons	13750	40	0	15	House 16H	22754	16	0
3	Auto Mpg	398	40	0	16	Kinematics	8192	8	0
4	Auto Price	159	14	1	17	MV Artificial	40768	7	3
5	Bank 8FM	8192	8	0	18	Pumadyn 8NH	8192	8	0
6	Bank 32NH	8192	32	0	19	Pumdayn 32H	8192	32	0
7	Boston Housing	506	12	1	20	Servo	167	0	4
8	California Housing	20640	8	0	21	Stocks	950	9	0
9	CPU Act	8192	21	0					
10	CPU Small	8192	12	0	22	ERA (9)	1000	40	0
11	Delta Ailerons	7129	5	0	23	ESL (9)	488	40	0
12	Elevators	16599	18	0	24	Eucalyptus (5)	736	14	5
13	Friedman Artificial	40768	10	0	25	LEV (5)	1000	40	0

discretized into $m = 5$ classes using equal-frequency binning. It is worth noting that this did not always lead to classes of equal size, since in many data sets, one or more values of the numerical output attribute occurs many times. We only report results for $m = 5$, because other values produced quite similar results. We complemented the discretized regression data sets with four truly ordinal classification data sets taken from [3], namely those that have reasonable size and at least five classes.

As a baseline, we selected SVMRank [17], a state-of-the-art method for ranking problems which is based on support vector learning (`http://svmlight.joachims.org/`). We used this method in its default setting with a linear kernel. F&H and LPC were implemented with logistic regression as a base learner, which comes down to fitting a linear model and using the logistic link function ($\text{logit}(x) = \log(x/(1-x))$) to derive $[0,1]$-valued scores, the type of model output requested by both methods. Essentially, all three methods are therefore based on linear models so that the comparison is fair from this point of view. For LPC, we tried both aggregation strategies, the unweighted version (8) and the weighted variant (9); we shall refer to these methods as LPC-U and LPC-W, respectively. As a side remark, we mention that we also tried the classifier versions of F&H and LPC (ordering instances by the predicted class). As expected, however, these were consistently outperformed by their ranking variants, and therefore we excluded them from the analysis that we will present below.

We report averages of several test statistics over five repetitions of a stratified ten-fold cross validation, each time with a different random permutation of the data. The standard deviations were very small (often of the magnitude 10^{-4}) and, for the ease of exposition, are therefore omitted in the tables.

4.2 Experimental Results

The ranking performance of the methods in terms of the C-index (2) and the Jonckheere-Terpstra statistic (3) are shown in Table 3. To analyze the results, we followed the two-step statistical test procedure recommended in [6], consisting of a Friedman test of the null hypothesis that all learners have equal performance and, in case this hypothesis is rejected, a Nemenyi test to compare learners in a pairwise way. Both tests are based on the average ranks over all data sets, which are shown at the bottom of the table. The ranks on the individual data sets are indicated by the numbers in brackets (the best method has rank 1, the worst rank 4; average ranks are assigned in case of ties).

The results show that F&H is significantly better than the rest, while no significant difference is detected among the others (the critical rank difference is 0.88 for a significance level of 10%). Note that, as expected, the performance of LPC-U increases in comparison to LPC-W when using (3) instead of (2) as a performance metric. Overall, however, these differences are small.

The main conclusion from this experiment is that the binary decomposition methods are quite competitive to SVMRank in terms of ranking performance, and in fact, the ranking variant of F&H seems to be even superior. (We are somewhat cautious here, since the computational complexity of SVMRank prevented from a truly thorough parameter optimization.) In terms of computational complexity, the decomposition methods are much more efficient, outperforming SVMRank by several orders of magnitude on some data sets. In fact, training of a single SVMRank model may easily take hours to days, and running the experiments for this method was only feasible on a parallel computer. Obviously, this precludes a direct comparison of run-times. We can mention, though, that the run-times of F&H and LPC are comparable to the run-times of their corresponding classification variants, and for both methods, 5 repetitions of a 10-fold cross validation never took more than one hour of CPU time.

As to the direct comparison of F&H and LPC, we complemented the ranking results by the performance of the respective classifier-versions in terms of classification accuracy; see Table 2. This study essentially confirms previous results [10,15], which have shown that LPC, even though it is not specifically designed

Table 2. Classification accuracy for the classifier-variants of LPC and F&H

#	LPC	F&H	#	LPC	F&H	#	LPC	F&H	#	LPC	F&H
1	.5248	.5157	8	.5490	.5463	14	.5040	.4814	20	.7425	.7199
2	.6084	.6118	9	.7195	.7183	15	.4967	.4603	21	.8307	.7840
3	.6836	.6846	10	.6686	.6676	16	.4044	.4161	22	.4036	.4062
4	.5987	.5902	11	.5716	.5587	17	.9149	.8837	23	.7729	.7741
5	.7897	.7885	12	.5727	.5707	18	.4860	.4575	24	.6378	.6619
6	.4851	.4855	13	.5547	.5547	19	.3863	.3933	25	.5972	.6022
7	.6608	.6561									

Table 3. Ranking performance of the different methods

#	C-index				Jonckheere-Terpstra statistic			
	LPC-U	LPC-W	F&H	SVMRank	LPC-U	LPC-W	F&H	SVMRank
1	.8398 (3.0)	.8401 (2.0)	.8417 (1.0)	.8361 (4.0)	.8383 (2.5)	.8383 (2.5)	.8401 (1.0)	.8352 (4.0)
2	.9171 (1.0)	.9170 (2.0)	.9167 (4.0)	.9168 (3.0)	.9121 (1.0)	.9120 (2.0)	.9118 (3.0)	.9117 (4.0)
3	.9468 (4.0)	.9470 (3.0)	.9495 (2.0)	.9583 (1.0)	.9471 (4.0)	.9473 (3.0)	.9498 (2.0)	.9590 (1.0)
4	.9142 (3.0)	.9137 (4.0)	.9217 (2.0)	.9342 (1.0)	.9130 (3.0)	.9125 (4.0)	.9210 (2.0)	.9367 (1.0)
5	.9764 (3.5)	.9764 (3.5)	.9774 (1.0)	.9766 (2.0)	.9764 (3.5)	.9764 (3.5)	.9774 (1.0)	.9766 (2.0)
6	.8422 (3.5)	.8422 (3.5)	.8431 (1.0)	.8427 (2.0)	.8418 (3.5)	.8418 (3.5)	.8427 (1.0)	.8423 (2.0)
7	.9129 (4.0)	.9130 (3.0)	.9242 (1.0)	.9241 (2.0)	.9132 (3.5)	.9132 (3.5)	.9245 (1.0)	.9240 (2.0)
8	.8782 (2.5)	.8782 (2.5)	.8784 (1.0)	.8734 (4.0)	.8782 (2.5)	.8782 (2.5)	.8784 (1.0)	.8734 (4.0)
9	.9249 (4.0)	.9250 (3.0)	.9528 (1.0)	.9518 (2.0)	.9238 (4.0)	.9239 (3.0)	.9519 (1.0)	.9509 (2.0)
10	.9105 (4.0)	.9108 (3.0)	.9349 (1.0)	.9324 (2.0)	.9090 (4.0)	.9093 (3.0)	.9334 (1.0)	.9310 (2.0)
11	.8722 (1.0)	.8721 (2.0)	.8715 (3.0)	.8704 (4.0)	.8594 (1.0)	.8592 (2.0)	.8585 (3.0)	.8575 (4.0)
12	.8696 (2.0)	.8697 (1.0)	.8686 (3.0)	.8640 (4.0)	.8695 (2.0)	.8696 (1.0)	.8686 (3.0)	.8639 (4.0)
13	.8839 (3.5)	.8839 (3.5)	.8853 (1.0)	.8849 (2.0)	.8839 (3.5)	.8839 (3.5)	.8853 (1.0)	.8849 (2.0)
14	.8317 (1.5)	.8317 (1.5)	.8243 (3.0)	.8169 (4.0)	.8317 (1.5)	.8317 (1.5)	.8244 (3.0)	.8170 (4.0)
15	.8239 (1.0)	.8238 (2.0)	.8164 (3.0)	.8123 (4.0)	.8239 (1.5)	.8239 (1.5)	.8164 (3.0)	.8123 (4.0)
16	.7741 (2.5)	.7741 (2.5)	.7753 (1.0)	.7734 (4.0)	.7741 (2.5)	.7741 (2.5)	.7753 (1.0)	.7734 (4.0)
17	.9780 (2.5)	.9780 (2.5)	.9797 (1.0)	.9505 (4.0)	.9780 (2.5)	.9780 (2.5)	.9797 (1.0)	.9505 (4.0)
18	.7892 (2.5)	.7892 (2.5)	.7895 (1.0)	.7806 (4.0)	.7892 (2.5)	.7892 (2.5)	.7895 (1.0)	.7806 (4.0)
19	.6721 (3.5)	.6721 (3.5)	.6724 (2.0)	.6732 (1.0)	.6721 (3.5)	.6721 (3.5)	.6724 (2.0)	.6732 (1.0)
20	.9031 (4.0)	.9052 (3.0)	.9272 (1.0)	.9219 (2.0)	.9058 (4.0)	.9068 (3.0)	.9271 (1.0)	.9231 (2.0)
21	.9325 (2.0)	.9324 (3.0)	.9636 (1.0)	.9320 (4.0)	.9324 (2.5)	.9324 (2.5)	.9635 (1.0)	.9320 (4.0)
22	.7415 (3.5)	.7415 (3.5)	.7418 (2.0)	.7434 (1.0)	.7263 (3.0)	.7262 (4.0)	.7265 (2.0)	.7277 (1.0)
23	.9645 (4.0)	.9648 (2.0)	.9654 (1.0)	.9647 (3.0)	.9662 (3.0)	.9664 (2.0)	.9672 (1.0)	.9655 (4.0)
24	.8908 (4.0)	.8954 (3.0)	.9346 (2.0)	.9395 (1.0)	.8895 (4.0)	.8936 (3.0)	.9341 (2.0)	.9384 (1.0)
25	.8635 (4.0)	.8647 (3.0)	.8660 (2.0)	.8679 (1.0)	.8775 (1.0)	.8748 (4.0)	.8757 (3.0)	.8764 (2.0)
avg. rank	2.96	2.72	1.68	2.64	2.78	2.78	1.68	2.76

Table 4. C-index statistics of the individual models M_{ij} and M_i for LPC and F&H, respectively. We report the average min, mean, median, and max. For comparison, we also add the performance of the complete (aggregated) model.

#	LPC					F&H				
	min	mean	median	max	overall	min	mean	median	max	overall
1	.6553	.8011	.8185	.8356	.8401	.8185	.8295	.8311	.8372	.8417
2	.8917	.9099	.9121	.9156	.9170	.9110	.9135	.9138	.9154	.9167
3	.7851	.8718	.8703	.9113	.9470	.8546	.9034	.9153	.9283	.9495
4	.7253	.8169	.8342	.8882	.9137	.8103	.8663	.8752	.9043	.9217
5	.9433	.9657	.9732	.9763	.9764	.9726	.9746	.9747	.9764	.9774
6	.7936	.8266	.8310	.8378	.8422	.8239	.8351	.8379	.8407	.8431
7	.7887	.8450	.8653	.8971	.9130	.8741	.8906	.8912	.9059	.9242
8	.8533	.8656	.8688	.8732	.8782	.8606	.8664	.8665	.8721	.8784
9	.8746	.9042	.8797	.9492	.9250	.9435	.9476	.9484	.9502	.9528
10	.8576	.8933	.8930	.9280	.9108	.9243	.9273	.9273	.9304	.9349
11	.8477	.8590	.8590	.8662	.8721	.8580	.8649	.8668	.8681	.8715
12	.7719	.8313	.8396	.8631	.8697	.7991	.8435	.8563	.8623	.8686
13	.8753	.8834	.8844	.8852	.8839	.8820	.8841	.8847	.8852	.8853
14	.7804	.8101	.8136	.8262	.8317	.7945	.8131	.8165	.8251	.8243
15	.7309	.7878	.7956	.8104	.8238	.7866	.8018	.8024	.8156	.8164
16	.7450	.7631	.7661	.7723	.7741	.7650	.7691	.7691	.7730	.7753
17	.7056	.8760	.8996	.9424	.9780	.8186	.8976	.9176	.9368	.9797
18	.5884	.7367	.7647	.7814	.7892	.7147	.7581	.7689	.7797	.7895
19	.5544	.6401	.6638	.6683	.6721	.6701	.6706	.6702	.6719	.6724
20	.6968	.7850	.7913	.8457	.9052	.7322	.8274	.8438	.8900	.9272
21	.5115	.7420	.7618	.8650	.9324	.7154	.8295	.8478	.9069	.9636
22	.7043	.7345	.7383	.7419	.7415	.7392	.7413	.7418	.7423	.7418
23	.8960	.9393	.9492	.9622	.9648	.9521	.9596	.9619	.9625	.9654
24	.7205	.8007	.7827	.8939	.8954	.8412	.8757	.8735	.9146	.9346
25	.7726	.8465	.8526	.8673	.8647	.8441	.8590	.8622	.8674	.8660

for ordinal problems, is at least competitive and often superior to F&H: The classification version of LPC wins 15 times against the classification variant of F&H and looses only 9 times; see Table 2. All the more interesting is the question why F&H performs better for the ranking problem.

In Section 3.3, some possible answers to this question have already been anticipated. To get more insight, we looked at the ranking performance of the *individual* models. More precisely, we let each of the models $\mathcal{M}_i$ in F&H and $\mathcal{M}_{i,j}$ in LPC rank all examples, and compared this ranking to the true ordinal classification of the examples. Table 4 shows the minimum, mean, median, and maximum C-index among these models for both LPC and F&H. It can be seen that F&H is consistently better in terms of the minimum, i.e., the worst model $\mathcal{M}_i$ is still better than the worst model $\mathcal{M}_{i,j}$. This may not be too surprising, because the pairwise approach learns a much higher number of models. However, the same can also be observed for the mean and the median (with a single exception), and even in terms of the maximum, F&H is better in 20 of the 25 data sets. Less surprisingly, the variability among the LPC models $\mathcal{M}_{i,j}$ is higher than among the F&H models $\mathcal{M}_i$, as can be seen from the difference between the maximum and the minimum. We take these results as strong evidence for the dominance of the non-competence problem of LPC models, as discussed in Section 3.3, and consider this problem as a reasonable explanation for the superiority of F&H.

Worth mentioning is finally a kind of ensemble effect revealed by the results in Table 4: For both methods, LPC and F&H, the overall performance of the aggregated model is consistently better than the best performance of an individual model.

5 Conclusions

We have elaborated on the use of binary decomposition methods in the context of multipartite ranking, a generalization of bipartite ranking to the multi-class case. The use of such methods is motivated by their successful application to related problems, including multi-class classification, ordinal classification, and label ranking. Our results have shown that decomposition methods are competitive, if not even superior, to state-of-the-art ranking methods in terms of predictive accuracy, while being much more efficient from a computational point of view. In any case, they offer a viable alternative to hitherto existing methods.

In future work, we plan to further improve the performance of the methods proposed in this paper, for example by means of alternative aggregation schemes. Moreover, only simple classification methods such as logistic regression have been used as base learners so far. Such methods seek to maximize classification accuracy in the first place, and hence optimize ranking performance (in terms of AUC) only indirectly. Therefore, the use of AUC-optimizing classifiers as base learners is likely to yield improved results.

Acknowledgements. This work was supported by the *German Science Foundation (DFG)* and the *Dutch Organisation for Scientific Research (NWO)*.

References

1. Allwein, E., Schapire, R., Singer, Y.: Reducing multiclass to binary: A unifying approach for margin classifiers. JMLR 1, 113–141 (2001)
2. Asuncion, A., Newman, D.: UCI machine learning repository (2007), `http://www.ics.uci.edu/~mlearn/MLRepository.html`
3. Ben David, A.: Ordinal real-world data sets repository (2008), `http://www.cs.waikato.ac.nz/ml/weka/index_datasets.html`
4. Cardoso, J., da Costa, J.P.: Learning to classify ordinal data: The data replication method. JMLR 8, 1393–1429 (2007)
5. Chu, W., Keerthi, S.: New approaches to support vector ordinal regression. In: Proceedings ICML, pp. 145–152. ACM, New York (2005)
6. Demsar, J.: Statistical comparisons of classifiers over multiple data sets. JMLR 7, 1–30 (2006)
7. Fawcett, T.: An introduction to ROC analysis. Pattern Recognition Letters 27(8), 861–874 (2006)
8. Frank, E., Hall, M.: A simple approach to ordinal classification. In: Flach, P.A., De Raedt, L. (eds.) ECML 2001. LNCS (LNAI), vol. 2167, pp. 145–156. Springer, Heidelberg (2001)
9. Fürnkranz, J.: Round robin classification. JMLR 2, 721–747 (2002)
10. Fürnkranz, J.: Round robin ensembles. Intelligent Data Analysis 7(5), 385–404 (2003)
11. Gnen, M., Heller, G.: Concordance probability and discriminatory power in proportional hazards regression. Biometrika 92(4), 965–970 (2005)
12. Hand, D., Till, R.: A simple generalization of the area under the ROC curve for multiple class problems. Machine Learning 45(2), 171–186 (2001)
13. Herbrich, R., Graepel, T., Obermayer, K.: Large margin rank boundaries for ordinal regression. In: Advances in Large Margin Classifiers, pp. 115–132. MIT Press, Cambridge (2000)
14. Higgins, J.: Introduction to Modern Nonparametric Statistics. Duxbury Press (2004)
15. Hühn, J., Hüllermeier, E.: Is an ordinal class structure useful in classifier learning? Int. J. Data Mining, Modelling and Management 1(1), 45–67 (2009)
16. Hüllermeier, E., Fürnkranz, J., Cheng, W., Brinker, K.: Label ranking by learning pairwise preferences. Artificial Intelligence 172, 1897–1917 (2008)
17. Joachims, T.: Optimizing Search Engines Using Clickthrough Data. In: Proceedings ACM SIGKDD, Edmonton, Alberta, Canad, pp. 133–142. ACM Press, New York (2002)
18. Joachims, T.: Training linear SVMs in linear time. In: Proceedings ACM SIGKDD, Philadelphia, PA, USA, pp. 217–226. ACM Press, New York (2006)
19. Kramer, S., Widmer, G., Pfahringer, B., De Groeve, M.: Prediction of ordinal classes using regression trees. Fundamenta Informaticae 34(1-2), 1–15 (2000)
20. Rajaram, S., Agarwal, S.: Generalization bounds for k-partite ranking. In: Proceedings of the NIPS 2005 Workshop on Learning to Rank, Whistler, BC, Canada, pp. 28–23 (2005)
21. Shashua, A., Levin, A.: Ranking with large margin principle: Two approaches. In: Advances in NIPS, vol. 15, pp. 937–944. MIT Press, Cambridge (2002)
22. Waegeman, W., Baets, B.D., Boullart, L.: ROC analysis in ordinal regression learning. Pattern Recognition Letters 29(1), 1–9 (2008)

Leveraging Higher Order Dependencies between Features for Text Classification

Murat C. Ganiz[1,3], Nikita I. Lytkin[2], and William M. Pottenger[2,3]

[1] Department of Computer Science
Lehigh University, USA
[2] Department of Computer Science
Rutgers, The State University of New Jersey, USA
[3] DIMACS
Rutgers, The State University of New Jersey, USA

Abstract. Traditional machine learning methods only consider relationships between feature values within individual data instances while disregarding the dependencies that link features across instances. In this work, we develop a general approach to supervised learning by leveraging higher-order dependencies between features. We introduce a novel Bayesian framework for classification named Higher Order Naive Bayes (HONB). Unlike approaches that assume data instances are independent, HONB leverages co-occurrence relations between feature values across different instances. Additionally, we generalize our framework by developing a novel data-driven space transformation that allows any classifier operating in vector spaces to take advantage of these higher-order co-occurrence relations. Results obtained on several benchmark text corpora demonstrate that higher-order approaches achieve significant improvements in classification accuracy over the baseline (first-order) methods.

Keywords: machine learning, text classification, higher order learning, statistical relational learning, higher order naive bayes, higher order support vector machine.

1 Introduction

A well known problem in real-world applications of machine learning methods is that expert labeling of large amounts of data for training a classifier is prohibitively expensive. Often in practice, only a small amount of labeled data is available for training. Traditional methods of classification treat individual data instances independently. In this case, however, a small training set makes an adequate estimation of the model parameters of a classifier very challenging. Prior work has demonstrated the value of leveraging explicit [1,2,3] as well as implicit [4,5] link information within data in order to provide a richer data representation for model estimation.

In Sect. 4.1, we build on a graph-based data representation from our prior work [4] and introduce a novel Bayesian framework for classification named Higher Order

W. Buntine et al. (Eds.): ECML PKDD 2009, Part I, LNAI 5781, pp. 375–390, 2009.

Naive Bayes (HONB). Unlike approaches that assume data instances are independent, HONB leverages co-occurrence relations between feature values across different instances. We term these implicit co-occurrence relations higher-order paths. Features (e.g., words in documents of a text collection) are richly connected by such higher-order paths, and a model built by HONB exploits this rich connectivity.

We further generalize our framework in Sect. 4.2 by developing a novel data-driven space transformation that allows any classifier operating in vector spaces to take advantage of relational dependencies captured by higher-order paths between features.

We evaluate the proposed methods[1] on several benchmark text corpora across a wide range of training set sizes in Sect. 5. In that section, we also draw comparisons with the results reported in [6], where a word clustering approach was proposed for dealing with small and sparse training data for text classification.

The paper is organized as follows. Related work is discussed in Sect. 2. In Sect. 3, we provide the necessary background for development of the proposed methods described in Sect. 4. Experimental results are presented in Sect. 5. A discussion of the effects of leveraging higher-order dependencies for text classification is provided in Sect. 6. Concluding remarks are made in Sect. 7.

2 Related Work

Our motivation for using higher-order dependencies for classification stems from advances in the areas of link mining [7] and information retrieval. In addition to (or sometimes instead of) using the more traditional data representation by feature vectors characterizing each data instance independently of the others, link-based approaches [3,8,9] to collective classification leverage explicit dependencies, or links, within networked data [9]. Several studies [1,2,3] have shown that collective classification can achieve significant reductions in classification errors by performing inferences about multiple data instances simultaneously. However, such methods are context-dependent and are therefore not designed to classify single data instances. This restriction limits the domain of applicability of link-based classifiers.

In this work, we propose classification methods that leverage higher-order dependencies in the form of implicit links between instances and features of the training data. Unlike collective classifiers, methods presented in this work maintain the ability to classify single data instances without requiring any additional context information.

In [5], we gave a mathematical proof supported by empirical results of the dependence of Latent Semantic Indexing (LSI) [10], a technique often used in text mining and information retrieval, on higher-order relations and in particular on higher-order term co-occurrences expressed via higher-order paths. A higher-order path is exemplified in Fig. 1 (reproduced from [5]) in the context of text data. Figure 1 depicts three documents, D1, D2 and D3, each containing two terms represented by

[1] "Systems and Methods for Data Transformation for Supervised Machine Learning," M.C. Ganiz, N.I. Lytkin and W.M. Pottenger, U.S. Patent Pending 61/185255.

the letters A, B, C and D. Shown below the three documents in Fig. 1 is a higher-order path that links term A with term D through B and C. This path contains three edges and is therefore referred to as a third-order path.

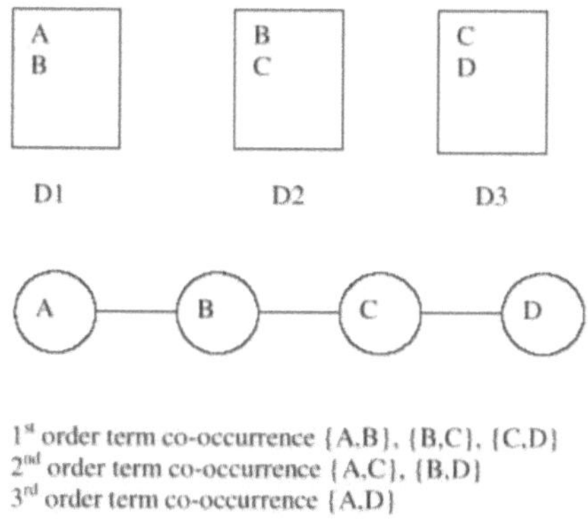

Fig. 1. Higher-order co-occurrences [5]

Higher-order relations play an important role in many other systems for text mining and information retrieval. In [11], higher-order associations were used for rule mining in textual data. Higher-order co-occurrences were used in [12] for solving a component of the problem of lexical choice, which identifies synonyms in a given context. In another effort, [13] used second-order co-occurrences for improving the run-time performance of LSI. Higher-order co-occurrences have also been used in other applications including word sense disambiguation [14] and stemming [15].

It should be noted that our framework extends beyond the textual domain. In other words, instances D1, D2 and D3 from Fig. 1 need not be text documents – they may be records in a database, or instances in a labeled training dataset. Likewise, features A, B, C, etc. need not be terms – they may be values in a database record, or feature-value pairs in a data instance.

In fact, in [4] we have successfully used higher-order paths for capturing dependencies amongst data instances when modeling time series data for anomaly detection in the Border Gateway Protocol, which constitutes the backbone of the Internet's routing infrastructure. The approach presented in this work builds on the data representation introduced in our prior work [4] and also described in Sect. 3.2.

3 Background

In this section we review parts of the Bayesian learning theory (Sect. 3.1) and the data representation (Sect. 3.2) underlying the development of approaches presented in Sect. 4. Although the methods discussed in this work are not limited to a particular application domain, here we restrict our attention to textual data. We assume the input space to be an n-dimensional binary vector space where each document is represented by a vector whose non-zero coordinates correspond to terms present in the document.

Formally, let $W = \{w_1, \ldots, w_n\}$ denote the set of binary features that correspond to terms in the vocabulary. Any document d can therefore be represented by an n-dimensional binary vector $w(d) = (w_1(d), \ldots, w_n(d))$, where

$$w_i(d) = \begin{cases} 1, & \text{if document } d \text{ contains term } w_i \\ 0, & \text{otherwise.} \end{cases}$$

For convenience of further exposition, we will simply write d to denote the corresponding vector $w(d)$.

3.1 Bayesian Learning Theory

Given a document d and two classes c_1 and c_2, the Bayes discriminant function can be written as

$$f(d) = \log \frac{P(c_1|d)}{P(c_2|d)} = \log \frac{P(d|c_1)P(c_1)}{P(d|c_2)P(c_2)}, \tag{1}$$

where

$$P(d|c_j) = P(w_1(d), \ldots, w_n(d)|c_j), \ \forall j \in \{1, 2\}, \tag{2}$$

is the conditional likelihood of document d belonging to class c_j, and $P(c_j)$ is the prior probability of class c_j.

By assumption of mutual independence of terms given a class, the conditional likelihood (2) becomes

$$P(w_1(d), \ldots, w_n(d)|c_j) = \prod_{i=1}^{n} P(w_i(d)|c_j), \tag{3}$$

where $P(w_i(\cdot)|c_j)$ denotes the conditional probability mass function of term w_i in class c_j. The Bayes discriminant function (1) therefore becomes

$$\begin{aligned} f(d) = & \sum_{i:w_i(d)=1} \log \tfrac{P(w_i|c_1)}{P(w_i|c_2)} + \\ & \sum_{i:w_i(d)=0} \log \tfrac{1-P(w_i|c_1)}{1-P(w_i|c_2)} + \log \tfrac{P(c_1)}{P(c_2)}, \end{aligned} \tag{4}$$

where $P(w_i|c_j)$ denotes the conditional probability of occurrence of term w_i in documents of class c_j, i.e., $P(w_i|c_j) = P(w_i(\cdot) = 1|c_j)$. The log likelihood ratios in (4) are, essentially, term weighting factors that attain large absolute values for terms that are strong discriminators between a pair of classes.

This framework is not limited to binary classification problems. The Bayes discriminant function for a general K-class classification task can be written as

$$g(d) = \arg \max_{j=1,\ldots,K} P(c_j|d). \tag{5}$$

Given a set of class labels $C = \{c_1, \ldots, c_K\}$ and the corresponding (training) set D_j of documents representing class c_j for each $j \in \{1, \ldots, K\}$, conditional probabilities $P(w_i|c_j)$ are estimated by

$$P(w_i|c_j) = \frac{1 + \sum_{d \in D_j} w_i(d)}{2 + |D_j|}, \tag{6}$$

which is the ratio of the number of documents that contain term w_i in class c_j to the total number of documents in class c_j. The constants in numerator and denominator in (6) are introduced according to Laplace's rule of succession in order to avoid zero-probability terms.

The prior class probabilities $P(c_j)$ are estimated by

$$P(c_j) = \frac{|D_j|}{\sum_{k=1}^{K} |D_k|}. \tag{7}$$

Together, equations (3) and (5–7) comprise the well-known Naive Bayes classifier.

3.2 Higher Order Data Representation

The data representation on which we build in the following section was initially used in our prior work [4] for anomaly detection in time series data. In the representation employed herein, a set D of documents is considered as a bipartite graph $G = (V_D \cup V_W, E)$. Vertices in V_D correspond to documents, while vertices in V_W correspond to terms. Two vertices $d \in V_D$ and $w \in V_W$ are connected by an edge $(d, w) \in E$ iff document d contains the term w.

A higher-order path in dataset D is a chain subgraph of G. For example, a chain w_i—d_l—w_k—d_r—w_j, which we will denote by $(w_i, d_l, w_k, d_r, w_j)$, encodes the fact that document $d_l \in D$ contains terms w_i and w_k, while document $d_r \in D$ contains terms w_k and w_j. The order of a path is determined by the number of document vertices the path spans. Thus, path $(w_i, d_l, w_k, d_r, w_j)$ captures a second-order co-occurrence between terms w_i and w_j, realized through term w_k shared by distinct documents d_l and d_r.

Higher-order paths simultaneously capture term co-occurrences within documents as well as term sharing patterns across documents, and in doing so provide a much richer data representation than the traditional feature vector form. As we will see in Sect. 5, the richness of representation becomes crucial when the small size of a training dataset prohibits traditional methods from accurately estimating model parameters.

4 Approach

In this section we present two novel methods of leveraging higher-order dependencies between features for supervised machine learning. In Sect. 4.1, we build on the Naive Bayes classifier by introducing a higher-order classifier termed Higher Order Naive Bayes. Our approach, however, can be generalized beyond the probabilistic machine learning methods. In Sect. 4.2, we introduce a data-driven space transformation that allows any learner that operates in vector spaces to leverage higher-order dependencies in data.

4.1 Higher Order Naive Bayes

Higher-order paths as defined in Sect. 3.2 allow us to extract rich relational information between features in a dataset. We incorporate this information into a Bayesian learning framework by modifying the parameter estimation equations (6) and (7) to depend on the counts of higher-order paths as follows.

Let $\varphi(w_i, D)$ denote the number of higher-order paths that contain term w_i given the dataset D, and let $\Phi(D)$ denote the total number of higher-order paths in D. The parameter estimation equations of the proposed Higher Order Naive Bayes classifier are:

$$\hat{P}(w_i|c_j) = \frac{1 + \varphi(w_i, D_j)}{2 + \Phi(D_j)}, \tag{8}$$

and

$$\hat{P}(c_j) = \frac{\Phi(D_j)}{\sum_{k=1}^{K} \Phi(D_k)}. \tag{9}$$

4.2 Leveraging Higher Order Dependencies by a Vector Space-Based Classifier

In this section, we present a novel data transformation that allows any classifier operating in vector spaces to take advantage of higher-order dependencies between features. We describe our approach for the case of binary classification. This, however, does not limit the applicability of the proposed approach, because numerous methods for multi-class classification based on binary classifiers have been proposed (see [16] for an overview). The proposed data transformation proceeds as follows.

Given two sets D_j and D_k of (training) documents from classes c_j and c_k, resp., the class conditional term probabilities (8) are computed. Let us denote the corresponding conditional log likelihood ratios as

$$\phi_i^{(1)} = \log \frac{\hat{P}(w_i|c_j)}{\hat{P}(w_i|c_k)}, \tag{10}$$

and

$$\phi_i^{(0)} = \log \frac{1 - \hat{P}(w_i|c_j)}{1 - \hat{P}(w_i|c_k)}. \tag{11}$$

Each binary document vector $d = (d_1, \ldots, d_n)$, $d \in D_j \cup D_k$, is then transformed into a real vector $\hat{d} = (\hat{d}_1, \hat{d}_2, \ldots, \hat{d}_n)$, where

$$\hat{d}_i = \begin{cases} \frac{\phi_i^{(1)}}{\sqrt{|\phi_i^{(1)}|}}, & \text{if } d_i = 1, \phi_i^{(1)} \neq 0 \\ \frac{\phi_i^{(0)}}{\sqrt{|\phi_i^{(0)}|}}, & \text{if } d_i = 0, \phi_i^{(0)} \neq 0 \\ 0, & \text{otherwise.} \end{cases} \tag{12}$$

Finally, the resulting dataset $\hat{D}_j \cup \hat{D}_k$ is used as input for training a binary classifier for classes c_j and c_k.

Data transformation (12) assigns weights that are high in absolute values for highly discriminative terms present in a document. The normalizing factors[2] in (12) moderate the spread of values of each feature in order to allow less discriminative terms to retain a certain level of influence over the classification. This level of influence depends on the discriminative power of a term as measured by (10) and (11). An illustration of the effect of the normalizing factors can be found in Sect. 6.

5 Experimental Results

Experimental evaluation was carried out on six widely-used text corpora. Four of these datasets were RELIGION, SCIENCE, POLITICS and COMP subsets of the 20 News Groups (20NG) [17] benchmark data. These particular subsets were selected to allow us to draw comparisons of our results with the related work [6], which will be discussed later in this section. Our preprocessing procedures closely followed those in [6]. First, all cross-postings in the 20NG data were removed. Then, for each dataset we performed stop word removal, stemming and removal of all terms that occurred in fewer than three documents in the dataset. The remaining terms were ranked by Information Gain. The top 2000 terms were selected. Finally, 500 documents were sampled at random from each class to comprise the 20NG datasets used in our experiments.

The other two datasets, Citeseer and Cora, are collections of scholarly research articles preprocessed by [18]. The Citeseer dataset contained 3312 documents with a vocabulary of 3703 terms. The Cora dataset comprised of 2708 documents with a vocabulary of 1433 terms. A summary description of all six datasets is provided in Table 1.

Naive Bayes (NB) was used as the baseline for evaluation of the proposed Higher Order Naive Bayes (HONB) classifier. Support Vector Machine (SVM) [19] was chosen as the base classifier for evaluation of the data transformation

Table 1. Six datasets used in the experiments

Dataset	Classes
RELIGION (3)	alt.atheism, soc.religion.christian, talk.religion.misc
SCIENCE (4)	sci.crypt, sci.electronics, sci.med, sci.space
POLITICS (3)	talk.politics.guns, talk.politics.mideast, talk.politics.misc
COMP (5)	comp.graphics, comp.os.ms-windows.misc, comp.sys.ibm.pc.hardware, comp.sys.mac-hardware, comp.windows.x
Citeseer (6)	AI, Agents, DB, HCI, IR, ML
Cora (6)	Case_Based, Genetic_Algorithms, Neural_Networks, Probabilistic_Methods, Reinforcement_Learning, Theory

[2] It is possible to omit the normalizing factors in (12). However, we have found experimentally that the normalized transformation, on average, yields slightly higher classification accuracies.

proposed in Sect. 4.2. SVM with the linear kernel has been shown [20] to perform well on text classification problems. The linear kernel allows us to observe the direct impact of leveraging higher-order dependencies, without any additional data transformations as performed implicitly by other kernel functions. Multi-class problems were addressed using the "one-against-one" classification scheme [21]. Under this scheme, a binary SVM classifier is constructed for every pair of classes. A data instance is then classified by each binary classifier and the final classification is determined by the majority vote over the assigned class labels. We refer to a SVM classifier constructed on the transformed data as Higher Order SVM (HOSVM).

Experiments described in this section were done using second-order paths for HONB and HOSVM. A $O\left(n^2(m+n)\right)$ algorithm for obtaining the counts of second-order paths in a dataset with m instances in n dimensions can be found in [22]. We have conducted additional experiments using third-order paths, but the results did not differ significantly.

Figure 2 shows mean classification accuracies obtained by varying training set size from 5% up to 60%. For each training set size, eight trials were performed. On each trial, a set of documents were randomly sampled from each class for training, while the rest were used for testing. On every trial, all terms that did not appear in any of the training documents were disregarded. The classifiers were then trained in the corresponding subspace of the original term space.

As can be seen from Figs. 2(a)–2(d), HONB and HOSVM[3] consistently outperformed first-order classifiers NB and SVM[3], resp., on the RELIGION, SCIENCE, POLITICS and COMP datasets. All of these accuracy improvements were statistically significant at the 5% level for (HONB, NB) pair of classifiers and in 83% of the cases for the (HOSVM, SVM) pair.

A slightly different pattern of performance can be observed on the Citeseer (Fig. 2(e)) and Cora (Fig. 2(f)) datasets. While classification accuracy of each method monotonically increased with increasing training set size, HONB drastically outperformed NB when training sets were small, reaching close to accuracy levels of SVM-based methods. HOSVM performed 1.5-2% better than SVM 66% of the time, but never worse overall. These results are especially encouraging in that they suggest that higher-order methods were able to construct robust models even when training data was particularly scarce. We speculate that changes in the distribution of highly-discriminative terms across classes as a result of increasing training set size allowed NB to outperform HONB on the Citeseer and Cora datasets once the amount of training data reached a certain point. We are currently investigating this hypothesis and intend to report our findings as part of our future work.

Consistent and statistically significant accuracy improvements attained by higher-order classifiers on small training sets led us to explore this aspect

[3] When selecting the value of the soft margin cost parameter C for SVM, we considered the set $\{10^{-4}, 10^{-3}, \ldots, 10^{4}\}$ of possible values. On every trial, we picked the smallest value of C which resulted in the highest accuracy obtained on the *training* set.

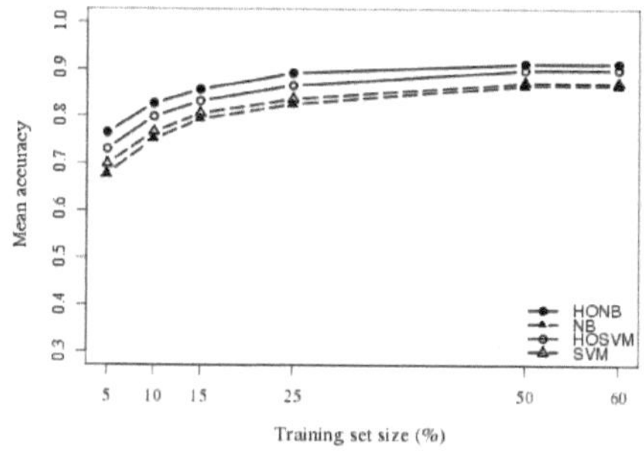

(a) RELIGION

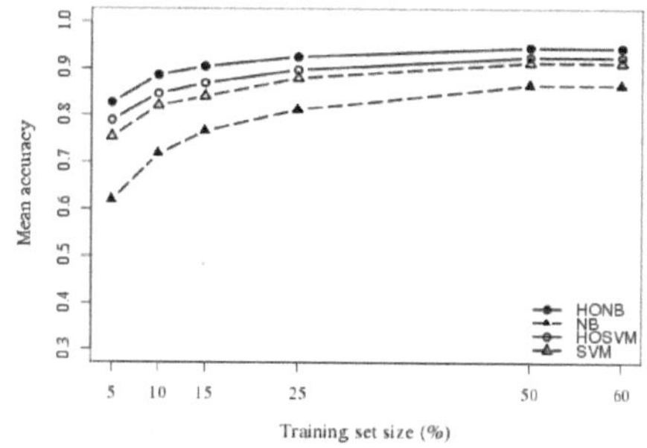

(b) SCIENCE

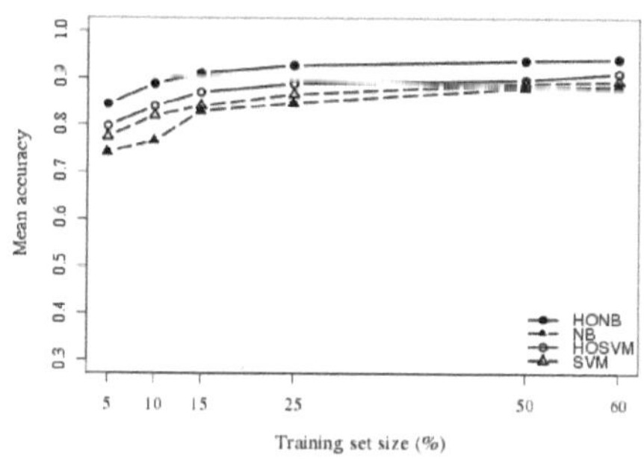

(c) POLITICS

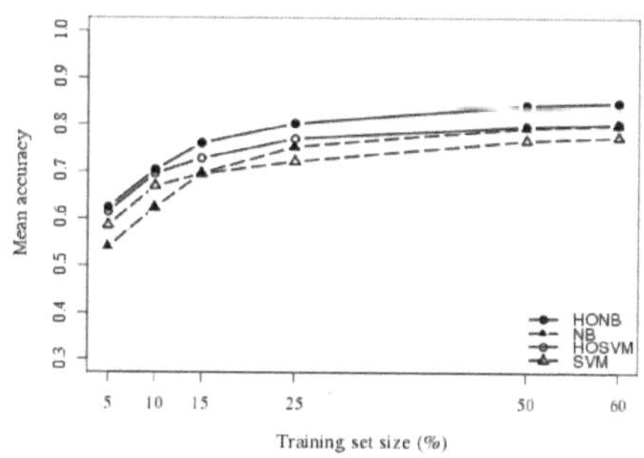

(d) COMP

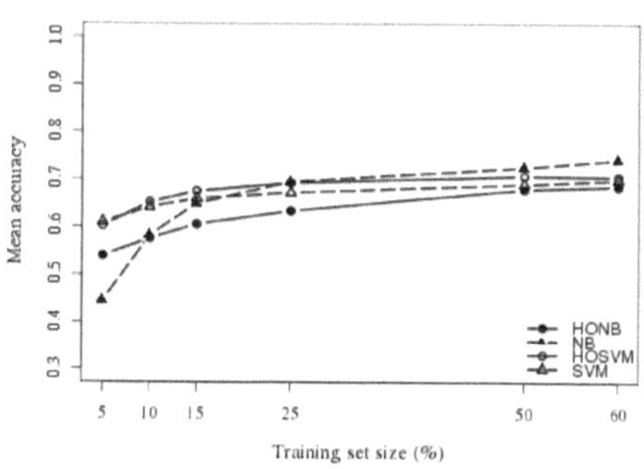

(e) Citeseer

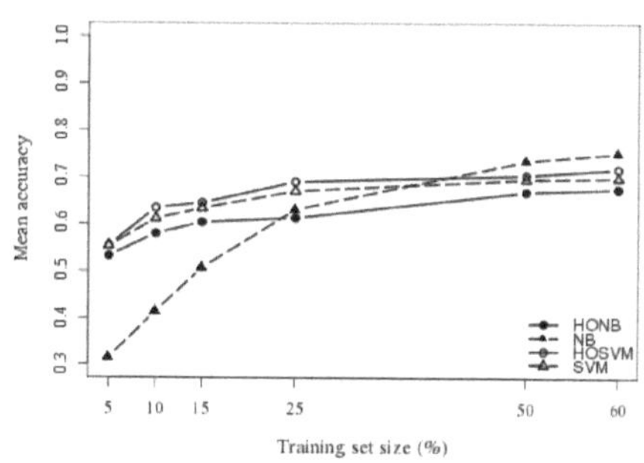

(f) Cora

Fig. 2. Scalability across training set size

further. In order to simulate a real-world scenario where only a few labeled data instances are available, and to illustrate accuracy improvements in a setting comparable with the related work [6], we focused our attention on 5% training samples. In case of the 20NG datasets, for instance, this corresponded to training on 25 documents per class and testing on the other 475 documents per

Table 2. Mean classification accuracies

	NB		HONB		SVM		HOSVM	
Dataset	Acc.	St. dev.	Acc.	St. dev.	Acc.	St. dev.	Acc.	St. dev.
RELIGION (3)	0.66	0.051	**0.741**	0.03	0.699	0.022	**0.723**	0.023
SCIENCE (4)	0.632	0.071	**0.833**	0.043	0.751	0.029	**0.792**	0.039
POLITICS (3)	0.73	0.042	**0.832**	0.01	0.763	0.03	**0.793**	0.047
COMP (5)	0.539	0.051	**0.623**	0.036	0.585	0.022	**0.614**	0.031
Citeseer (6)	0.444	0.025	**0.539**	0.015	**0.609**	0.01	0.602	0.01
Cora (6)	0.315	0.006	**0.532**	0.021	0.553	0.021	**0.554**	0.026

class. Classification accuracies averaged over eight trials are reported in Table 2. Highest accuracies for pairs (NB, HONB) and (SVM, HOSVM) of classifiers are highlighted in bold. The corresponding standard deviations are also reported in Table 2.

The obtained results indicate that leveraging higher-order dependencies in the RELIGION, SCIENCE, POLITICS and COMP datasets lead to significant improvements in classification accuracies of both Bayesian and SVM-based approaches. The improvements of HONB over NB and of HOSVM over SVM on those datasets are statistically significant at the 5% level. The only exception is the (HOSVM, SVM) pair on the POLITICS dataset. Although the difference in SVM and HOSVM accuracies on the POLITICS dataset was significant at level $\alpha = 0.158$, HOSVM outperformed SVM on seven out of eight trials on that data by an average of 3%.

HONB performed particularly well on the 20NG data, outperforming NB by 11.7% and SVM by 5.8% on average. HONB also outperformed HOSVM by an average of 2.7%, although the differences in accuracy between the two classifiers were not statistically significant at the 5% level. HOSVM consistently outperformed SVM by an average of 3.1%.

On the Citeseer dataset, HONB outperformed NB by about 9% (significant at the 5% level), while HOSVM and SVM performed at the same level. On the Cora dataset, NB's average classification accuracy was 31.5%, which is the lowest accuracy across all datasets and classifiers considered in this work. Such low accuracy and almost zero standard deviation (Table 2) resulted from NB assigning all test documents to the majority class Neural_Networks. By exploiting valuable higher-order dependencies, HONB attained approximately 53% average accuracy, which is not significantly different at the 5% level from the roughly 55% accuracy of SVM and HOSVM.

In another set of experiments, we expanded our reach to compare the performance of higher-order classifiers to other approaches that also attempt to achieve better generalization performance on sparse input data. These related research efforts use clustering as a dimensionality reduction technique. In general they cluster words into groups and these groups are used as features in text classification. The authors of [6] observed that one advantage of using word clusters can be seen when word statistics (i.e., NB parameter estimates) are relatively hard to estimate. In order to demonstrate this, they conducted experiments with

Table 3. Mean classification accuracies reported by [6]

Dataset	mNB	NBWC
RELIGION (3)	0.525	0.553
SCIENCE (4)	0.65	0.725
POLITICS (3)	0.62	0.67
COMP (5)	0.473	0.508

Table 4. Mean classification accuracies of a pure higher-order classifier

Dataset	HONB	pure HONB
RELIGION (3)	0.741	0.745
SCIENCE (4)	0.833	0.842
POLITICS (3)	0.832	0.836
COMP (5)	0.623	0.649

small numbers of labeled training documents from the subsets listed in Table 1 of the 20NG data. To provide a more objective comparison, we focused on the relative improvement over respective base models. In [6] the authors used multinomial NB (mNB) as their base model and improved on the performance of this model using word clusters as features instead of just words. The latter algorithm is indicated as NBWC in Table 3. Similarly, we used binomial NB as our base model and improved on the performance of this model using higher-order paths instead of individual documents to estimate parameters.

Tables 2 and 3 show that compared to NBWC, HONB achieved much better performance than the corresponding base model for all datasets (4.7% average difference between NBWC and mNB versus 11.7% average difference between HONB and NB). Additionally, our results are statistically significant.

In order to further verify the value of leveraging higher-order dependencies within the data, additional experiments were conducted. In the first experiment, when estimating the conditional probabilities (8) and (9), we used only the "pure" second-order paths, i.e., paths that involved terms which did not co-occur together in any single document of the training set. The results of these experiments are reported in Table 4 comparing performance of HONB with that of the pure HONB. The differences in performance of these classifiers were minimal and not statistically significant.

In the second experiment, prior to training an SVM classifier with the linear kernel, a data transformation analogous to (12) was performed using the first-order conditional term probabilities (6) instead of higher-order probabilities (8). The resulting approach is referred to as NBSVM. Mean classification accuracies attained by NBSVM[3] are shown in Table 5. Comparison of Tables 2 and 5 makes it clear that NBSVM performed worse than both SVM and HOSVM. These results indicate that taking advantage of higher-order dependencies was indeed crucial for achieving the performance improvements attained by HONB and HOSVM.

Table 5. Mean classification accuracies of NBSVM

Dataset	NBSVM	HOSVM
RELIGION (3)	0.678	0.723
SCIENCE (4)	0.745	0.792
POLITICS (3)	0.759	0.793
COMP (5)	0.576	0.614

We have also conducted experiments with the Radial Basis Function (RBF) kernel for the HOSVM and SVM classifiers. The results were consistent with the findings of [20]. Namely, there were no significant differences between classification accuracies attained with the linear kernel and those attained with the RBF kernel.

In summary, higher-order approaches HONB and HOSVM outperformed their first-order counterparts. Consistent performance improvements were observed on the 20NG data across a wide range of training set sizes. On the Citeseer and Cora datasets, HONB outperformed NB when training set sizes were below 10% and 25%, respectively. Under the conditions of small (5%) training samples from the Citeseer dataset, HOSVM performed the same as SVM, which outperformed HONB by 7%. HONB, SVM and HOSVM produced very similar results on the Cora dataset. It is interesting to note that while HOSVM scaled better across datasets, HONB did not require any parameter tuning and performed exceptionally well on the 20NG data.

6 Discussion

In order to gain a deeper understanding of the effect of using higher-order paths for estimation of conditional term probabilities (8), let us consider Fig. 3 generated based on one of the 5% (25 documents per class) training samples from the RELIGION dataset using two of its classes, "alt.atheism" and "soc.religion.christian". For every term w_i, Fig. 3(a) presents a plot of the conditional log probability ratio (10) obtained from higher-order probabilities (8) (horizontal axis) versus the log ratio obtained from first-order probabilities (6) (vertical axis). Notice the differences in scales of values on the axes of Fig. 3(a): $[-20, 20]$ for higher-order log ratios versus $[-4, 4]$ for first-order log ratios. Additionally, three distinct groups of terms appeared as a result of using higher-order paths for estimating the model parameters. We found that terms that fell into the right (left) most group are highly-discriminative terms that appeared in documents of only one of the classes in the *training* set. Figure 3(a) reveals that due to drastic difference in scales of values of higher- and first-order log ratios, highly-discriminative terms exert much stronger influence on classification in HONB than in NB. In other words, highly-discriminative terms contribute more heavily to the Bayesian discriminant function (4) under HONB than under NB.

Similarly, Fig. 3(b) shows a plot of the conditional log probability ratios (11) of non-occurrence of a term in a HONB model versus a NB model. An important feature in this figure is the one order of magnitude difference in values

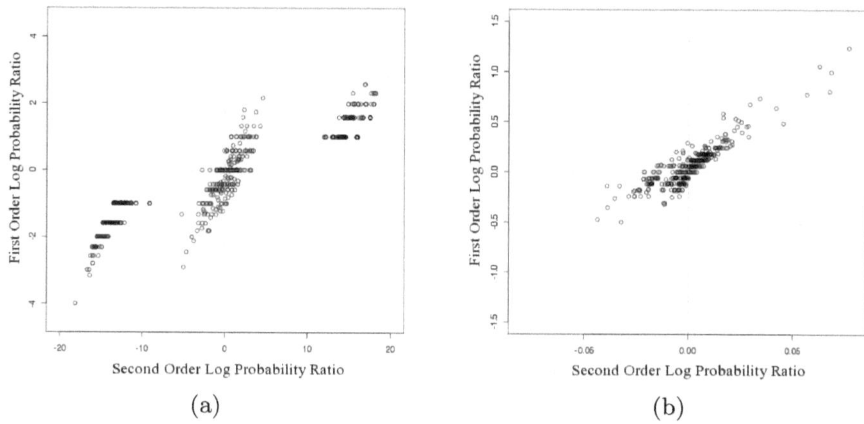

Fig. 3. Conditional log probability ratios obtained from higher-order probabilities (8) (horizontal axes) versus the log ratios obtained from first-order probabilities (6) (vertical axes) on "alt.atheism" and "soc.religion.christian" classes of one of the 5% (25 documents per class) training samples from the RELIGION dataset

on the axes: $[-0.1, 0.1]$ for higher-order log ratios on the horizontal axis versus $[-1.5, 1.5]$ for first-order log ratios on the vertical axis. Together, Figs. 3(a) and 3(b) indicate that while both NB and HONB take into account presence of terms as well as their absence, HONB tends to place more emphasis on the presence of terms in a document being classified.

As was noted in Sect. 4.2, the normalizing factors in (12) were introduced in order to allow terms that may appear in multiple classes, but are still good discriminators as measured by the log likelihood ratios (10) and (11), to have a non-negligible impact during document classification by HOSVM. The effect of these normalizing factors can be seen by comparing Figs. 3(a) and 3(b) with Figs. 4(a) and 4(b).

On the vertical axes in Figs. 4(a) and 4(b) are plotted the same first-order conditional log probability ratios as in Figs. 3(a) and 3(b), respectively. Plotted on the horizontal axes of Figs. 4(a) and 4(b) are the higher-order conditional log probability ratios shown on the horizontal axes of Figs. 3(a) and 3(b), respectively, and normalized as in (12). Note the change in scales of the horizontal axes once normalization has been applied: $[-20, 20]$ before normalization (Fig. 3(a)) versus $[-4, 4]$ after, and $[-0.1, 0.1]$ before normalization (Fig. 3(b)) versus $[-0.3, 0.3]$ after. It is this change in scales coupled with the increased spread of the middle group of terms along the horizontal axis that allowed good discriminator terms from the middle group in Fig. 3(a) to increase their relative influence during document classification by HOSVM.

Although it is trivial to identify strongly discriminative features in a given training set, the question remains of how to weight those features for pattern classification. Methods proposed in this work address this question by leveraging higher-order dependencies between features.

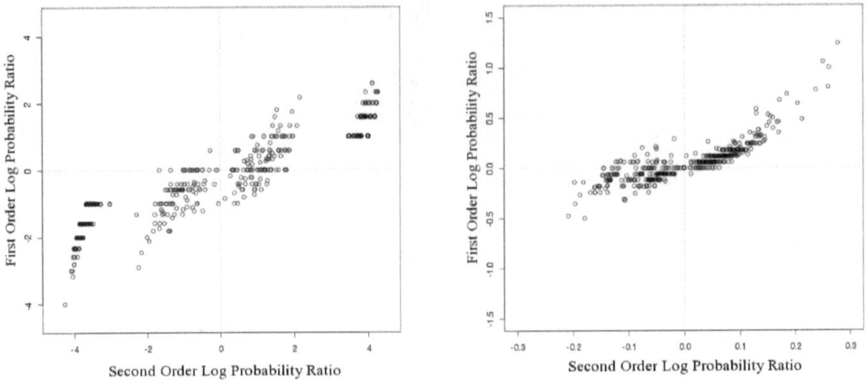

(a) (b)

Fig. 4. Conditional log probability ratios obtained from higher-order probabilities (8) and normalized as in (12) for use by HOSVM (horizontal axes) versus the log ratios obtained from first-order probabilities (6) (vertical axes) on the same dataset as in Fig. 3.

7 Conclusions and Future Work

In prior work [5], we gave a mathematical proof supported by empirical results of the dependence of LSI on higher-order paths. In this work, a general approach to leveraging higher-order dependencies for supervised learning was developed. We presented a novel classification method termed Higher Order Naive Bayes (HONB). We further generalized our framework by developing a new data transformation that allows any classifier operating in vector spaces to take advantage of the rich relational information captured by higher-order paths.

Higher-order paths allow a classifier to operate on a much richer data representation than the conventional feature vector form. This is especially important when working with small and sparse training sets where accurate parameter estimation of traditional (first-order) models becomes very challenging [6]. Experimental results affirmed the value of leveraging higher-order paths, resulting in significant improvements in classification accuracies on benchmark text corpora across a wide range of training set sizes. In addition, we compared our results with those reported in [6], where term clusters were used (instead of terms) as features for classification by Naive Bayes (NBWC). Our experiments demonstrated that HONB consistently achieved better performance over the baseline Naive Bayes (NB) classifier than did NBWC.

In ongoing efforts, we are developing methods for leveraging higher-order dependencies in various domains including nuclear detection, as well as in large datasets. We are also investigating extensions of the proposed framework to a number of other supervised learning approaches, and to unsupervised machine learning.

Acknowledgments

The authors wish to thank Rutgers University, the National Science Foundation and the Department of Homeland Security. This material is based upon work partially supported by the National Science Foundation under Grant Numbers 0703698 and 0712139 as well as by the Department of Homeland Security Science & Technology Directorate. Any opinions, findings, and conclusions or recommendations expressed in this material are those of the authors and do not necessarily reflect the views of the National Science Foundation, the Department of Homeland Security or Rutgers University.

The authors also wish to thank several colleagues at DIMACS, Rutgers University, including Dr. Ilya B. Muchnik for valuable discussions regarding the higher-order path counting algorithm, Dr. Alexey V. Nefyodov for helpful suggestions on the presentation of this work and Shenzhi Li for her many insightful comments during the development of the Higher Order Learning framework.

We are also grateful for the help of other co-workers, family members and friends. Co-author W. M. Pottenger also gratefully acknowledges the continuing help of his Lord and Savior, Yeshua the Messiah (Jesus the Christ) in his life and work.

References

1. Chakrabarti, S., Dom, B., Indyk, P.: Enhanced hypertext categorization using hyperlinks. SIGMOD Rec. 27(2), 307–318 (1998)
2. Neville, J., Jensen, D.: Iterative classification in relational data. In: Proc. AAAI, pp. 13–20. AAAI Press, Menlo Park (2000)
3. Taskar, B., Segal, E., Koller, D.: Probabilistic classification and clustering in relational data. In: Proceedings of the Seventeenth International Joint Conference on Artificial Intelligence, pp. 870–878 (2001)
4. Ganiz, M.C., Kanitkar, S., Chuah, M.C., Pottenger, W.M.: Detection of interdomain routing anomalies based on higher-order path analysis. In: ICDM 2006: Proceedings of the Sixth International Conference on Data Mining, pp. 874–879. IEEE Computer Society, Los Alamitos (2006)
5. Kontostathis, A., Pottenger, W.M.: A framework for understanding latent semantic indexing (LSI) performance. Inf. Process. Manage. 42(1), 56–73 (2006)
6. Slonim, N., Tishby, N.: The power of word clusters for text classification. In: 23rd European Colloquium on Information Retrieval Research (2001)
7. Getoor, L., Diehl, C.P.: Link mining: a survey. SIGKDD Explor. Newsl. 7(2), 3–12 (2005)
8. Lu, Q., Getoor, L.: Link-based classification. In: Fawcett, T., Mishra, N. (eds.) ICML, pp. 496–503. AAAI Press, Menlo Park (2003)
9. Neville, J., Jensen, D.: Dependency networks for relational data. In: Fourth IEEE International Conference on Data Mining, 2004. ICDM 2004, pp. 170–177 (2004)
10. Deerwester, S., Dumais, S.T., Furnas, G.W., Landauer, T.K., Harshman, R.: Indexing by latent semantic analysis. Journal of the American Society for Information Science 41, 391–407 (1990)

11. Li, S., Wu, T., Pottenger, W.M.: Distributed higher order association rule mining using information extracted from textual data. SIGKDD Explor. Newsl. 7(1), 26–35 (2005)
12. Edmonds, P.: Choosing the word most typical in context using a lexical co-occurrence network. In: Proceedings of the 35th Annual Meeting of the Association for Computational Linguistics, pp. 507–509 (1997)
13. Zhang, X., Berry, M.W., Raghavan, P.: Level search schemes for information filtering and retrieval. Inf. Process. Manage. 37(2), 313–334 (2001)
14. Schütze, H.: Automatic word sense discrimination. Comput. Linguist. 24(1), 97–123 (1998)
15. Xu, J., Croft, W.B.: Corpus-based stemming using cooccurrence of word variants. ACM Trans. Inf. Syst. 16(1), 61–81 (1998)
16. Scholkopf, B., Smola, A.J.: Learning with Kernels: Support Vector Machines, Regularization, Optimization, and Beyond. MIT Press, Cambridge (2001)
17. Lang, K.: Newsweeder: Learning to filter netnews. In: Proceedings of the Twelfth International Conference on Machine Learning, pp. 331–339 (1995)
18. Sen, P., Getoor, L.: Link-based classification. Technical Report CS-TR-4858, University of Maryland (February 2007)
19. Vapnik, V.: Statistical Learning Theory. John Wiley, Chichester (1998)
20. Joachims, T.: Learning to Classify Text Using Support Vector Machines: Methods, Theory and Algorithms. Kluwer Academic Publishers, Norwell (2002)
21. Kreßel, U.H.G.: Pairwise classification and support vector machines. In: Advances in kernel methods: support vector learning, pp. 255–268. MIT Press, Cambridge (1999)
22. Ganiz, M.C., Lytkin, N.I., Pottenger, W.M.: Leveraging higher order dependencies between features for text classification. Technical Report 2009-16, DIMACS, Rutgers University (June 2009)

Syntactic Structural Kernels for Natural Language Interfaces to Databases

Alessandra Giordani and Alessandro Moschitti

Department of Computer Science and Engineering
University of Trento
Via Sommarive 14, 38100 POVO (TN) - Italy
{agiordani,moschitti}@disi.unitn.it

Abstract. A core problem in data mining is to retrieve data in a easy and human friendly way. Automatically translating natural language questions into SQL queries would allow for the design of effective and useful database systems from a user viewpoint. Interesting previous work has been focused on the use of machine learning algorithms for automatically mapping natural language (NL) questions to SQL queries.

In this paper, we present many structural kernels and their combinations for inducing the relational semantics between pairs of NL questions and SQL queries. We measure the effectiveness of such kernels by using them in Support Vector Machines to select the queries that correctly answer to NL questions. Experimental results on two different datasets show that our approach is viable and that syntactic information under the form of pairs of syntactic tree fragments (from queries and questions) plays a major role in deriving the relational semantics between the two languages.

Keywords: Natural Language Processing; Kernel Methods; Support Vector Machines.

1 Introduction

In the last decade many natural language interfaces to database (NLIDBs) have been proposed to translate the human intent into machine-readable instructions [1,2,3,4,5,6,7,8]. Despite this, little progress has been made in developing an interface that can be used by any untrained user without manual annotation and intervention. The problem of automatically translating natural language (NL) questions into SQL queries is an interesting and appealing research in data mining. For example, solving this problem would suggest the role of syntax for mapping NLs to artificial languages; this would have a direct impact in the field of information systems. Unfortunately computational linguistics and artificial intelligence research [9] has shown that such mapping problem cannot be addressed with a deep semantic approach, thus a concrete solution should rely on shallow and statistical methods.

W. Buntine et al. (Eds.): ECML PKDD 2009, Part I, LNAI 5781, pp. 391–406, 2009.

In this paper, we propose a set of structural kernels, e.g. Sequence and Tree Kernels [10,11,12,13], and Support Vector Machines (SVMs) to map NL into SQL.

First, starting from a set of correct pairs of questions and the related SQL queries, available for our target DBs, we design an algorithm to produce incorrect pairs and additional correct pairs, i.e. negative and positive examples, respectively.

Second, we model a representation of the above question/query pairs in terms of syntactic structures, i.e. we build pairs of syntactic parse trees automatically derived by off-the-shelf natural language parsers and the straightforward application of SQL grammar.

Third, we train SVMs with the above data, where the structural representation of the pairs is encoded by means of different types of kernels, i.e. linear, polynomial, string and tree kernels and their combinations. This allows us to automatically exploit the associative patterns between NL and SQL syntax to detect correct and incorrect pairs from an operational semantics viewpoint.

Finally, given a new question and the set of available queries (i.e. the repository of queries asked to the target DB), we produce the set of pairs containing such question and then we use SVMs to rank pairs in terms of *correctness*. We select the top scored pair as the query that answers the given question.

The new contributions with respect to our previous research on Natural Language Interface to Databases [14] are the following:

- We propose and study sequence kernels to provide a pair representation that is shallower than the one based on deep syntactic parsing. Although, they prove to not be essential for the design of the most accurate model, their comparison with polynomial kernels gives some indications on the role of feature pair spaces.
- We experimented with a large number of kernels showing that, in contrast with our previous findings, complex kernels relevantly improve the simple space of term (word) pairs.
- We applied our semi-automatic algorithm to a second dataset of correct question and query pairs, namely RESTQUERIES [7], to design a new dataset for classification. We made it available[1] along with the one derived by GEOQUERIES.

The use of RESTQUERIES allowed us to (1) assess the high effectiveness of product kernels and the feature pair spaces, which, even in their simple form (e.g. word pairs), highly improve the traditional linear kernel; (2) show that syntactic information is very important since it improves the best model by about 10 absolute percent points; and (3) find out that complex kernels such as the polynomial expansion of pairs of tree fragments and bigrams can produce the highest results.

In the remainder, Section 2 shows our proposed algorithm to generate a training set of question and query pairs used by the kernel-based classifier as described

[1] `http://disi.unitn.it/~moschitt/corpora.htm`

in Section 3. Section 4 discusses the experimental setup and results, Section 5 reviews some state of the art NLIDBs, to which we compare, and finally, Section 6 draws conclusions.

2 Dataset Generation

The goal of this research is the development of a NLIDB that maps NL questions into SQL queries based on a machine learning approach; consequently, we need to have training data, i.e. a set of positive and negative examples. In practical cases, we can assume to have a set of positive examples consisting of correct question/query pairs[2], i.e. such that the execution of the query retrieves a correct answer for the question. Assuming the availability of negative examples is a more strong assumption since providing the correct query for an user information need is a more natural task than providing the incorrect solution.

Therefore, to create negative examples, we may use the initial set of questions and queries in the correct pairs and randomly pair them. Unfortunately, this may generate false negatives since different questions may have more than one answer (and vice-versa), thus a manual verification of such pairs is required. To reduce such costly manual intervention, we can exploit the semantic equivalency between the pairs' members and its transitivity closure to pair questions to their correct queries. This allow us to extend the set of positive examples.

The semantic equivalency can be calculated by means of a clustering algorithm, which groups questions that represent the same information need with queries that correctly retrieve it. An approach to effectively detect such equivalence is the generalization of questions and queries, e.g. *What are some good restaurants in Berkeley* becomes *What are some good restaurants in a city*.

In the next sections, we describe our approach to automatically generate the target dataset. The main steps are: (1) generalize question and query instances, (2) cluster the generalized pairs, (3) generate all the true positives by pairing questions and queries belonging to the same clusters, and (4) annotate as true negatives all remaining pairings between questions and queries of distinct clusters. This also requires a limited manual intervention.

2.1 Generalizing Pairs

The aim of pair generalization is to make the detection of semantically equivalent questions and queries easy. Our approach consists in considering questions or queries having similar structures instantiated by the same semantic concepts. The latter are generalizations of important domain terms occurring both in the question and in the related query. For example, terms like *Berkeley*, *San Francisco*, etc., are substituted with the concept *city*. Note that: (a) we can identify

[2] For example, correct pairs may be defined when databases are designed and validated. Also, we may ask the DB operator to collect the set of queries that she/he designed in response of typical specific (questions) asked by DB users.

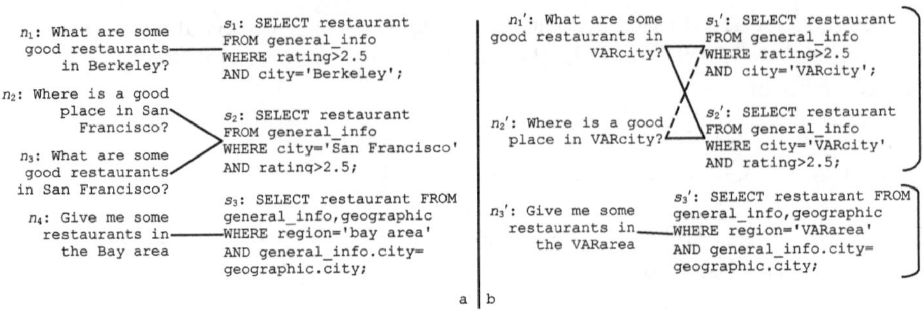

Fig. 1. Example of the initial corpus (*A, on the left*) and the generalized version (*B, on the right*). The latter is divided in two clusters (*identified by the two brackets*).

concepts by extracting the column names (in the database) that naturally store domain terms; and (b) concepts are expressed in the *WHERE* condition of the given SQL queries.

An example of the generalization phase is shown in Figure 1, which reports questions and queries of a restaurant domain. More in detail, on the left there is a set of four pairs containing four distinct questions and their three related queries (connected by lines) whereas on the right four generalized pairs are shown. In the question and query pair $\langle n_1, s_1 \rangle$, where n_1: "What are some good restaurants in Berkeley?" and s_1: `SELECT restaurant FROM general_info WHERE rating>2.5 AND city='Berkeley'`, since *Berkeley* is associated with the column *city*, its occurrences in n_1 and s_1 are substituted with the concept/variable *VARcity*.

We note that, after substituting instances with variables, both n_1 and n_3 are generalized into n'_1, which can then be paired with two distinct SQL queries, i.e. s'_1 and s'_2. This is correct since there can be more SQL queries that correctly retrieve an answer to an NL question. We can define them to be *semantically equivalent*, i.e. $s'_1 \equiv s'_2$. Conversely, there can be many NL questions that map to the same query, e.g. $n_2 \equiv n_3$[3].

It is worth noting that with the generalization process, we introduce redundancy that we eliminate by removing duplicated questions and queries. Thus, the output dataset is usually smaller than the initial one. However the number of training examples will be larger, not only because of the introduction of negatives but also due to the automatic discovering of new positives.

2.2 Pair Clustering and Final Dataset Annotation

Once the pairs have been generalized, we cluster them according to their semantic equivalence so that we can automatically derive new positive examples by swapping their members. We define semantically equivalent pairs those correct pairs with (a) equivalent NL questions, i.e. whose generalized version is the same or (b) equivalent SQL queries. Given that two equivalent queries must retrieve

[3] It is worth noting that in this equivalence is true in this domain but in other domains can be false, since *good places* not necessarily refers to restaurants.

the same result set, we can automatically test their equivalence by simply executing them. Unfortunately, this is just a necessary condition (e.g. two different queries can have the same answer) therefore we manually evaluate new pairings obtained applying this condition.

Note that automatically detecting semantic equivalence of natural language questions with perfect accuracy is a hard task, so we consider as semantically equivalent either identical questions (after generalization) or those associated with semantic equivalent queries. We also apply transitivity closure to both members of pairs to extend the set of equivalent pairs.

For example, in Figure 1.b s'_1 and s'_2 retrieve the same results so we verify that they are semantically equivalent queries and we assign them to the same cluster (CL1), i.e. information need about good restaurants in a city (with a rating larger than 2.5 stars). Alternatively, we can also consider that n'_1 and n'_2 are both paired with s'_2 to derive that they are equivalent, avoiding the human intervention. Concerning s'_3, it retrieves a result set different form the previous one so we can automatically assign it to a different cluster (CL2), i.e. involving questions about restaurants in a region. Note that, once n'_2 is shown to be semantically equivalent to n'_1, we can pair them with s'_1 to create the new pair (indicated by the dashed line) $\langle n'_2, s'_1 \rangle$. Indeed the negative example set is $\langle n'_3, s'_1 \rangle, \langle n'_3, s'_2 \rangle, \langle n'_1, s'_3 \rangle, \langle n'_2, s'_3 \rangle$.

3 Kernel Methods for Question/Query Representation

Kernel Methods refer to a large class of learning algorithms based on inner product vector spaces, among which Support Vector Machines (SVMs) are one of the most well-known algorithms. The main idea is that the parameter model vector $\boldsymbol{w}$ generated by SVMs (or by other kernel-based machines) can be rewritten as $\sum_{i=1..l} y_i \alpha_i \boldsymbol{x}_i$, where y_i is equal to 1 for positive and -1 for negative examples, $\alpha_i \in \Re$ with $\alpha_i \geq 0$, $\forall i \in \{1, .., l\}$ $\boldsymbol{x}_i$ are the training instances. Therefore we can express the classification function as $\text{Sgn}(\sum_{i=1..l} y_i \alpha_i \boldsymbol{x}_i \cdot \boldsymbol{x} + b) = \text{Sgn}(\sum_{i=1..l} y_i \alpha_i \phi(o_i) \cdot \phi(o) + b)$, where $\boldsymbol{x}$ is a classifying object, b is a threshold and the product $K(o_i, o) = \langle \phi(o_i) \cdot \phi(o) \rangle$ is the kernel function associated with the mapping ϕ.

Note that it is not necessary to apply the mapping ϕ, we can use $K(o_i, o)$ directly. This allows, under the Mercer's conditions [15] for defining abstract functions which generate implicit feature spaces. The latter allow for an easier feature extraction and the use of huge feature spaces (possibly infinite), where the scalar product (i.e. $K(\cdot, \cdot)$) is implicitly evaluated.

In the following section, we first propose a structural representation of the question and query pairs, then we report the two more adequate kernels for syntactic structure representation, i.e. the Syntactic Tree Kernel (STK) [11], which computes the number of syntactic tree fragments and the Extended Syntactic Tree Kernel (STK_e) [16], which includes leaves in STK. In the last subsection we show how to engineer new kernels from them.

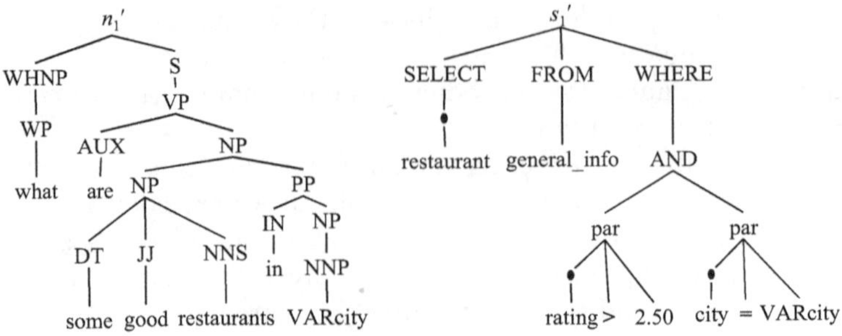

Fig. 2. Question/Query Syntactic trees

3.1 Representing Question and Queries Pairs

In Data Mining and Information Retrieval the so-called bag-of-words (BOW) has been shown to be effective to represent textual documents, e.g. [17,18]. However, in case of questions and queries we deal with small textual objects in which the semantic content is expressed by means of few words and poorly reliable probability distributions. In these conditions the use of syntactic representation improves BOW and should be always used, [13,19,20,21,22,23].

Therefore, in addition to BOW, we represent questions and queries using their syntactic trees[4]. As shown in Figure 2 for questions (a) we use the Charniak's syntactic parser [25] while for queries (b) we implemented an ad-hoc SQL parser. The latter builds a SQL parse tree for each query following its syntactic derivation according to MySQL grammar. The grammar has been slightly modified to accommodate the usage of the symbol • for the production of *items* in the SELECT clause and in WHERE conditions. In such an SQL tree, the internal nodes are only the SQL keywords of the query plus the special symbol • whereas the leaves are names of tables and columns of the database, category variables or operators. Note that, although we eliminated comma and dot from the grammar, it is still possible to obtain the original SQL query, by just performing a preorder traversal of the tree.

To represent the above structures in a learning algorithm we use tree kernels described in the following section.

3.2 Tree Kernels

The main underlying idea of tree kernels is to compute the number of common substructures between two trees T_1 and T_2 without explicitly considering the whole fragment space. Let $\mathcal{F} = \{f_1, f_2, \ldots, f_{|\mathcal{F}|}\}$ be the set of tree fragments and $\chi_i(n)$ an indicator function equal to 1 if the target f_i is rooted at node n and equal to 0 otherwise. A tree kernel function over T_1 and T_2 is defined as $TK(T_1, T_2) = \sum_{n_1 \in N_{T_1}} \sum_{n_2 \in N_{T_2}} \Delta(n_1, n_2)$, where N_{T_1} and N_{T_2} are the sets of

[4] Early work on the use of syntax for text categorization were based on part-of-speech tags, e.g. [24].

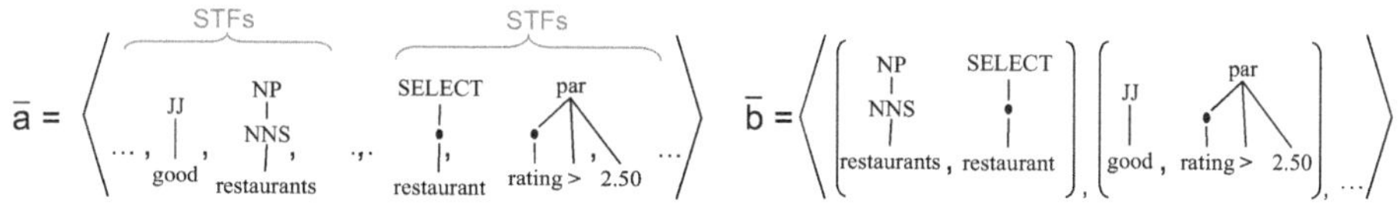

Fig. 3. Feature spaces for the tree pair in Figure 2 a) joint space STK+STK b) Cartesian product STK×STK

nodes in T_1 and T_2, respectively, and $\Delta(n_1, n_2) = \sum_{i=1}^{|\mathcal{F}|} \chi_i(n_1)\chi_i(n_2)$. The Δ function is equal to the number of common fragments rooted in nodes n_1 and n_2, and thus, depends on the fragment type. We report its algorithm for the evaluation of the number of syntactic tree fragments (STFs).

A syntactic tree fragment (STF) is a set of nodes and edges from the original tree which is still a tree and with the constraint that any node must have all or none of its children. This is equivalent to state that the production rules contained in the STF cannot be partial. To compute the number of common STFs rooted in n_1 and n_2, the STK uses the following Δ function [11]:

1. if the productions at n_1 and n_2 are different then $\Delta(n_1, n_2) = 0$;
2. if the productions at n_1 and n_2 are the same, and n_1 and n_2 have only leaf children (i.e. they are pre-terminal symbols) then $\Delta(n_1, n_2) = \lambda$;
3. if the productions at n_1 and n_2 are the same, and n_1 and n_2 are not pre-terminals then
$\Delta(n_1, n_2) = \lambda \prod_{j=1}^{l(n_1)} (1 + \Delta(c_{n_1}(j), c_{n_2}(j)))$,

where $l(n_1)$ is the number of children of n_1, $c_n(j)$ is the j-th child of the node n and λ is a decay factor penalizing larger structures.

Figure 3.a shows some STFs of the NL and SQL trees in Figure 2. STFs satisfy the constraint that grammatical rules cannot be broken. For example, *[VP [AUX NP]]* is a STF, which has two non-terminal symbols, *AUX* and *NP*, as leaves whereas *[VP [AUX]]* is not a STF.

STK does not include individual nodes as features. As shown in [16] using its extension (STK_e) we can include at least the leaves, (which in constituency trees correspond to words) by simply inserting the following step 0 in the algorithm above [16]:

0. if n_1 and n_2 are leaf nodes and their labels are identical then $\Delta(n_1, n_2) = \lambda$;

3.3 String Kernels

The String Kernels that we consider count the number of substrings containing gaps (i.e. some of the symbols of the original string are skipped) shared by two sequences. We adopted the efficient algorithm described in [15,10,26,27]. Characters in the sequences can be substituted with any set of symbols. In our study we preferred to use words obtaining word sequences. For example, given the query: `Select restaurant from general_info` sample substrings, extracted by the Sequence Kernel (SK), are: `Select restaurant`, `Select from general_info`, `Select general_info`, `Select from`, etc. It is worth noting that: (i) longer subsequences receive lower weights, (ii) some words can be omitted, i.e. gaps and (iii) gaps determine a weight since an exponential decay factor is applied, where the exponent is the number of words and gaps between the first and last words.

3.4 Kernel Engineering for Pair Representation

Kernel engineering [28,29,30] can be carried out by combining basic kernels with additive or multiplicative operators or by designing specific data objects, e.g. the tree representation for the SQL syntax, to which standard kernels are applied. Since our data is a set of pairs, we need to represent the members of a pair and their interdependencies. For this purpose, given two kernel functions, $k_1(.,.)$ and $k_2(.,.)$, and two pairs, $p_1 = \langle n_1, s_1 \rangle$ and $p_2 = \langle n_2, s_2 \rangle$, a first approximation is given by summing the kernels applied to the components: $K(p_1, p_2) = k_1(n_1, n_2) + k_2(s_1, s_2)$. This kernel will produce the union of the feature spaces of questions and queries. For example, the explicit vector representation of the space of STK of the pair in Figure 2 is shown in Figure 3.a. The Syntactic Tree Fragments of the question will be in the same space of the Syntactic Tree Fragments of the query.

In theory a more effective kernel is the product $k(n_1, n_2) \times k(s_1, s_2)$ since it generates pairs of fragments as features, where the overall space is the Cartesian product of the used kernel spaces. For example Figure 3.b shows pairs of STF fragments, which are essential to capture the relational semantics between the syntactic tree subparts of the two languages [31]. In particular, the second fragment pair of the figure may suggest that the adjective phrase *good* expresses similar semantics of the syntactic construct `WHERE rating>2.5`. In other words, the above pair feature suggests that the whole query may be a correct translation of the given question.

As additional feature and kernel engineering, we also exploit the ability of the polynomial kernel to add feature conjunctions. By simply applying the function $(1 + K(p_1, p_2))^d$, we can generate conjunction up to d features. Thus, we can obtain tree fragment conjunctions and conjunctions of pairs of tree fragments.

The next section will show the results using different kernel combination for pair representation.

4 The Experiments

We ran several experiments to evaluate the accuracy of our approach in automatically selecting correct SQL queries for NL questions, where the selection of the correct query is modeled as a ranking problem. The ranker is constituted by SVMs and by the kernels described in Section 3. To show the generality of our approach we created two different datasets by applying our algorithm described in Section 2 to two different corpora.

4.1 Setup

We address the problem of finding a query whose result answers to a question according to the following ranking problem. Given a question $n \in \mathcal{N}$ and the complete set of the available queries $\mathcal{S}$, we classify the set of all possible pairs $P(n) = \{\langle n, s \rangle : s \in \mathcal{S}\}$. Then we use the classification score to rank the element of $P(n)$ and select the pair with the highest score[5].

To learn the classifier we used SVM-Light-TK[6], which extends the SVM-Light optimizer [18] with tree kernels. i.e. Syntactic Tree Kernel (STK) and its extension (STK_e) as described in Section 3. We implemented the String Kernel [10] (word sequence kernel [26]) and modeled many different combinations described in the next section. We used the default parameters, i.e. the cost and trade-off parameters = 1 (for normalized kernels) and $\lambda = 0.4$ (see Sec. 3.2).

To generate our datasets we applied our algorithm described in Section 2 to GeoQueries250 and RestQueries corpora[7].

The first corpus is about geography questions. After the generalization process the initial 250 pairs of questions/queries were reduced to 155 pairs containing 154 NL questions and 79 SQL queries. We found 76 clusters, from which we generated 165 positive and 12,001 negative examples for a total of 154 × 79 pairs. Such dataset will be referred to as Geo.

The second dataset regards questions about restaurants. The initial 250 pairs were generalized by 197 pairs involving 126 NL questions and 77 SQL queries. We clustered these pairs in only 26 groups, which lead to 852 positive examples and 9,702 negatives. Such dataset will be referred to as Rest.

To evaluate the results of our mapping models, we applied standard 10-fold cross validation and measure the average accuracy and the Std Dev. of selecting the correct query for each question.

4.2 Results on Geo Dataset

We tested several models for ranking based on different kernel combinations whose results are reported on Table 1 and Table 2. The first column of Table 1 lists kernel combination by means of product and sum between pairs of basic kernels used for the question and the query, respectively. The latter column shows the average accuracy (over 10 folds) ± Std. Dev.

More in detail, our basic kernels are: (1) linear kernel (LIN) built on the bag-of-stems (BOS) of the questions or of the query; (2) a polynomial kernel of degree 3 on the above BOSs (POLY); (3) the Syntactic Tree Kernel (STK) on the parse tree of the question or the query and (4) STK extended with leaf features (STK_e). Note that we can also sum or multiply different kernels, e.g. POLY×STK.

---

[5] More effective approaches have been proposed [11,32].

[6] `http://disi.unitn.it/~moschitt/Tree-Kernel.htm`

[7] Questions in both corpora were originally collected from a web-based interface and manually translated into logical formulas in Prolog by Mooney's group [7]. Popescu et al. [2] manually converted them into SQL. Thanks to our clustering algorithm we discovered and fixed many errors and inconsistencies in SQL queries.

Table 1. Kernel combination accuracies (± Std. Dev) for GEO dataset

Combination	Accuracy
LIN + LIN	**57.3**±10.4
LIN × LIN	**70.7**±12.0
POLY × POLY	71.9±11.5
STK × STK	70.3±9.3
STK_e × STK_e	70.1±10.9
LIN × STK	74.6±9.6
LIN × STK_e	**75.6**±13.1
POLY × STK	73.8±9.5
POLY × STK_e	73.5±10.4
STK × LIN	64.7±11.5
STK_e × LIN	68.3±9.6
STK × POLY	65.4±10.9
STK_e × POLY	68.3±9.6

Table 2. Advanced kernel combination accuracies (± Std. Dev) for GEO dataset

Advanced Kernels	Accuracy
STK^2+$POLY^2$	72.7±9.7
STK_e^2+$POLY^2$	73.2±11.4
$(1+LIN^2)^2$	**73.6**±9.4
$(1+POLY^2)^2$	73.2±10.9
$(1+STK^2)^2$	69.4±10.0
$(1+STK_e^2)^2$	70.0±12.2
$(1+LIN^2)^2+STK^2$	75.6±8.3
$(1+POLY^2)^2+STK^2$	72.6±10.5
$(1+LIN^2)^2$+LIN×STK	**75.9**±9.6
$(1+POLY^2)^2$+POLY×STK	73.2±10.9
POLY×STK+STK^2+$POLY^2$	73.9±11.5
POLY×STK_e+STK_e^2+$POLY^2$	75.3±11.5
LIN×STK+STK^2+LIN^2	74.5±9.1
LIN×STK_e+STK_e^2+LIN^2	74.9±11.8

An examination of the reported figures suggests that: first, the basic traditional model based on linear kernel and BOS, i.e. LIN + LIN, provides an accuracy of only 57.3%, which is greatly improved by LIN×LIN=LIN^2, i.e. by 13.5 points [8]. The explanation is that the sum cannot express the relational feature pairs coming from questions and queries, thus LIN cannot capture the underlying shared semantics between them. It should be noted that only kernel methods allow for an efficient and easy design of LIN^2; the traditional approach would have required to build the Cartesian product of the question BOS by query BOS. This can be very large, e.g. 10K features for both spaces lead to a pair space of 100M features.

Second, the feature pair space is essential since the accuracy of all kernels implementing the union spaces of question and query representations[9] is much lower than the baseline model for feature pairs, i.e. LIN^2.

Third, if we include conjunctions in the BOS representation by using POLY, we improve the LIN model, i.e. 71.9% vs. 70.8%. $POLY^2$ is also better than STK^2 since it includes individual term/word bigrams that are not included by STK.

Next, the lower accuracy provided by STK^2 and STK_e^2 suggests that syntactic models can improve BOS although too many (possibly incorrect) syntactic features (generated by the syntactic parser) make the model unstable. This consideration leads us to experiment with the model LIN × STK and LIN × STK_e, which combine words of the questions with syntactic constructs of SQL queries.

[8] Although the Std. Dev. associated with the model accuracy is high, the one associated with the distribution of difference between the model accuracy is much lower, i.e. 5%.

[9] Given the limited space, we could not report the results of these poorly accurate kernels.

Table 3. Kernel combination accuracies (± Std. Dev) for REST dataset

Combination	Accuracy
LIN + LIN	**20.9**±11.9
LIN × LIN	**37.1**±16.2
POLY × POLY	74.5±14.0
STK × STK	71.8±10.8
STK_e × STK_e	62.5±11.6
LIN × STK	**79.1**±11.5
LIN × STK_e	77.2±12.8
POLY × STK	**82.3**±11.8
POLY × STK_e	78.0±12.2
STK × LIN	38.3±13.4
STK × POLY	45.5±11.8
SK_3 × SK_3	67.4±11.1
SK_3 × STK	**81.7**±13.3
SK_3 × STK_e	78.5±11.5

Table 4. Advanced kernel combination accuracies (± Std. Dev) for REST dataset

Advanced Kernels	Accuracy
STK^2+$POLY^2$	78.6±11.9
STK_e^2+$POLY^2$	73.3±10.2
$(1+LIN^2)^2$	52.5±10.2
$(1+POLY^2)^2$	74.5±14.0
$(1+STK^2)^2$	74.5±14.0
$(1+STK_e^2)^2$	62.5±11.6
$(1+SK_3^2)^2$	69.8±10.0
$(1+POLY\times STK)^2$	**84.7**±11.5
$(1+POLY^2)^2+STK^2$	78.7±12.1
$(1+POLY^2)^2+POLY\times STK$	78.1±13.8
$POLY\times STK+STK^2+POLY^2$	78.6±11.9

They produce statistically significant higher results (at 90% of confidence), i.e. 74.6% and 75.6%. This suggests that the syntactic parse tree of the SQL query is very reliable (indeed, it is obtained with 100% of accuracy) while the natural language parse tree, although accurate, introduces noise that degrades the overall feature representation. As a consequence it is more effective to use words only in the representation of the first member of the pairs.

This is also shown by the last four lines of Table 1, showing the low accuracies obtained when relying on NL syntactic parse trees and SQL BOSs. Thus the only viable possibility to improve LIN × STK_e was to use the polynomial kernel in the combination POLY × STK_e. The slightly lower outcome shows that POLY is equivalent to LIN.

Moreover, we experimented with very advanced kernels built on top of feature pair spaces as shown in Table2. For example, we sum different pair spaces, STK_e^2 and $POLY^2$, and we apply the polynomial kernel on top of pair spaces by creating conjunctions, over feature pairs. This operation tends to increase too much the cardinality of the space and makes it ineffective. However, using the simplest initial space, i.e. LIN, to build pair conjunctions, i.e. $(1+LIN^2)^2$, we obtain a very interesting and high result, i.e. 73.6%. Using the joint space of the kernel above and kernel products, we can still improve our models, e.g. $(1+LIN^2)^2+LIN\times STK$.

This suggests that kernel methods have the potentiality to describe relational problems using simple building blocks although new theory describing the degradation of kernels when the space is too complex is required.

Finally, to study the stability of our complex kernels, we compared the learning curve of the baseline model, i.e. LIN+LIN, with the those of the best models, i.e. LIN×STK_e and $STK^2+(1+LIN^2)^2$. Figure 4 shows that surprisingly, complex

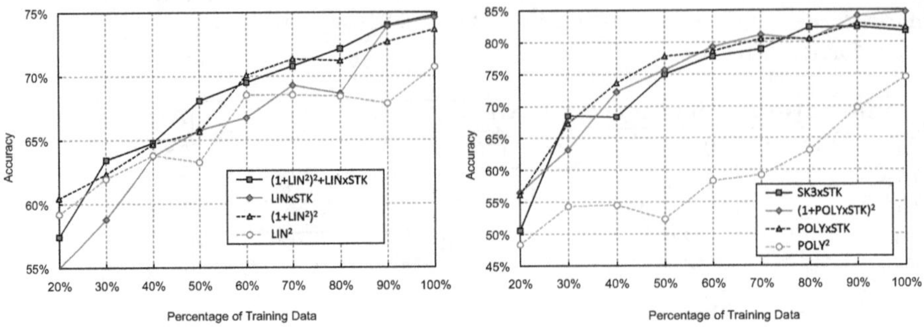

Fig. 4. Learning curves for GEOQUERIES and RESTQUERIES corpora

kernels are not only more accurate but also more stable, i.e. their accuracy increases smoothly according to the availability of training data.

4.3 Results on Rest Dataset

The previous results, although interesting, show that syntactic information plays a minor role and that complex kernels do not significantly improve our question translator (similar findings were derived in the preliminary experiments in [14]).

To verify such hypothesis, we experimented with the second dataset whose results, reported in tables Table 3 and Table 4, provide more interesting data.

First, the baseline model, i.e. LIN + LIN, produces very low accuracy, i.e. only 20.9%, which is highly improved by LIN^2, also showing a very low result, i.e. 37.1%, although.

Second, surprisingly, including conjunctions in the BOS representation, i.e. by using $POLY^2$, the accuracy of LIN^2 doubled (74.5%). Moreover, if we combine POLY with syntactic SQL subtrees, i.e. POLY×STK, we obtain another relevant improvement, i.e. 82.3%. This confirms that it is better to use stems in the representation of the first member of the pairs and syntactic parse trees in the second member.

Third, given that n-gram based text representation technique has shown to outperform bag-of-words approaches [10], we experimented with String Kernel (SK). The results confirm that sequences of three words (SK_3^2) better represent questions than BOS (see SK_3^2 vs. LIN^2). Nevertheless, the conjunctions of POLY are more effective.

Next, we experimented with advanced kernel combinations. The results, listed in table Table 4, show that the advanced polynomial kernel combination $(1+ POLY \times STK)^2$ outperforms[10] the best kernel combination, POLY × STK.

Finally, Figure 4 illustrates the learning curve of the best kernels, i.e. POLY × STK, the advanced polynomial kernel applied on top of it, i.e. $(1+POLY\times STK)^2$,

[10] Although the Std. Dev. associated with the model accuracy is high, the one associated with the distribution of difference between the model accuracy is much lower (about 2%). Considering also that we used 10 folds, we could verify that the first is better than the second at 90% of confidence limit.

$POLY^2$ and $SK_3 \times STK$. The plots show that the best kernels including the syntactic information are superior to the very accurate and rich kernels based on only BOS.

5 Related Work and Discussion

In this section we discuss some NLIDBs that have been tested on GeoQueries[11] or RestQueries datasets. NLIDBs can be classified according to the approach used to retrieve an answer to a given question from a database. For sake of space limit, we only discuss a system for each different approach. For a complete review of other state-of-the-art systems, please refer to Chandra and Mihalcea [33].

Authoring systems rely on semantic grammar specified by an expert user (i.e. the author) to interpret question over a database. The author has to name database elements, tailor entries and define additional concepts. CatchPhrase [3] is an authoring system that has been evaluated on GeoQueries250. In particular, two students were asked to author the system to cover 100 of the 250 questions each. Then the remaining questions were split in 2 test sets and translated by the system first into logical queries in tuple calculus representation and then into SQL queries. The average accuracy was 69%.

Many systems instead adopt a machine learning approach to induce semantic grammars from corpora of correct pairs of questions and queries, e.g. Krisp [1]. This takes pairs of sentences and their computer-executable meaning representations as training input to find a mapping between sentences and Prolog assertions using an SVM classifier. For each production in the meaning representation language the model is learned using string subsequence kernels. Then the classification is used to compositionally represent a natural language sentence in its meaning representation. The reported experiments using standard 10-fold cross validation show an accuracy of 70% and of 75% on GeoQueries880 and GeoQueries250, respectively.

In Precise [2], the derivation of semantic interpretation of ambiguous phrases is reduced to a graph matching problem. Precise finds valid mapping(s) from a complete tokenization of a given question to a set of database elements and then converts them into a SQL query (queries). The system achieves 100% Precision on a subset of questions while rejecting semantically intractable questions for a final Recall of 77.5% and 95% for GeoQueries880 and RestQueries respectively.

The performance of the above mentioned systems were originally measured according to Precision and Recall since they do not to generate a correct answer in particular output conditions. In contrast our approach allows for always having one answer, therefore it can be more appropriately measured with accuracy. We note that our approach is comparable to Krisp obtaining the similar outcome,

[11] GeoQueries250 is a subset of GeoQueries880 dataset whose questions are also available in other languages. Since our learning algorithm is language independent, we plan to experiment with other natural languages but also with GeoQueries880 so we report others' result also on this dataset.

i.e. 75.6% vs 75% (of Krisp), on GeoQueries250. Regarding the comparison with Precise, it should be noted that we corrected several errors in the SQL testset prepared in [2] (many of queries did not return the correct values and others were syntactically incorrect).

It is worth noting that our system, in contrast with previous generative approaches, retrieves the best matching query among the given set of all possible queries. One could argue that we cannot find a correct answer to a given unseen NL question if the SQL query is not present in the initial dataset. However, we can rely on query logs which reliably represent frequent and required queries asked to DBs.

There exist other systems [4,5,7,8] that were tested on GeoQueries880 with different experimental-setup, so results are not directly comparable.

With respect to other natural language tasks that employ tree kernels, several models have been proposed, e.g. [11,34,35,36,37,38,39,40,41,42].

6 Conclusions

In this paper, we approach the problem of deriving a shared semantic between natural language and programming language by automatically learning a model based the syntactical representation of the training examples. In our experiments we consider pairs of NL questions and SQL queries as training examples. These are annotated by means of our algorithm starting from a given initial annotation. In particular we experimented with the annotation available in GeoQueries250 and RestQueries corpora. We generated new datasets adding new positive pairs, creating negatives example set and also fixing some errors. Our datasets are publicly available so that other systems can be compared with our benchmark corpora.

To represent syntactic/semantic relationships expressed by training pairs, we encode such pairs in SVM by means of kernel functions. We designed innovative combinations between different kernels for structured data applied to pairs of objects, that, to the best of our knowledge, represent a novel approach to describe relational semantics between NL and SQL languages.

Experimental results show a promising accuracy, which can be largely improved, e.g. by model tuning. This suggests that our approach is viable to mine semantic relations between natural language and SQL.

In the future we would like to extend this research by focusing on advanced shallow semantic approaches such as predicate argument structures [43].

References

1. Kate, R.J., Mooney, R.J.: Using string-kernels for learning semantic parsers. In: Proceedings of the 21st ICCL and 44th Annual Meeting of the ACL, Sydney, Australia, July 2006, pp. 913–920. Association for Computational Linguistics (2006)
2. Popescu, A.M., Etzioni, A.O., Kautz, A.H.: Towards a theory of natural language interfaces to databases. In: Proceedings of the 2003 International Conference on Intelligent User Interfaces, Miami, pp. 149–157. Association for Computational Linguistics (2003)

3. Minock, M., Olofsson, P., Näslund, A.: Towards building robust natural language interfaces to databases. In: Kapetanios, E., Sugumaran, V., Spiliopoulou, M. (eds.) NLDB 2008. LNCS, vol. 5039, pp. 187–198. Springer, Heidelberg (2008)
4. Zettlemoyer, L.S., Collins, M.: Learning to map sentences to logical form: Structured classification with probabilistic categorial grammars. In: UAI, pp. 658–666 (2005)
5. Wong, Y.W., Mooney, R.: Learning for semantic parsing with statistical machine translation. In: Proceedings of the Human Language Technology Conference of the NAACL, Main Conference, New York City, USA, June 2006, pp. 439–446. Association for Computational Linguistics (2006)
6. Dale, R., Somers, H.L., Moisl, H. (eds.): 9. In: Database Interfaces, pp. 209–240. Marcel Dekker Inc., New York (2000)
7. Tang, L.R., Mooney, R.J.: Using multiple clause constructors in inductive logic programming for semantic parsing. In: Proceedings of the 12th European Conference on Machine Learning, Freiburg, Germany, pp. 466–477 (2001)
8. Ge, R., Mooney, R.: A statistical semantic parser that integrates syntax and semantics. In: Proceedings of the Ninth Conference on Computational Natural Language Learning (CoNLL-2005), Ann Arbor, Michigan, June 2005, pp. 9–16. Association for Computational Linguistics (2005)
9. Winograd, T.: Understanding Natural Language. Academic Press, New York (1972)
10. Lodhi, H., Taylor, J.S., Cristianini, N., Watkins, C.J.C.H.: Text classification using string kernels. In: NIPS, pp. 563–569 (2000)
11. Collins, M., Duffy, N.: New ranking algorithms for parsing and tagging: Kernels over discrete structures, and the voted perceptron. In: Proceedings of ACL 2002 (2002)
12. Vishwanathan, S.V.N., Smola, A.J.: Fast kernels for string and tree matching. In: Advances in Neural Information Processing Systems, vol. 15, pp. 569–576. MIT Press, Cambridge (2003)
13. Moschitti, A.: Efficient convolution kernels for dependency and constituent syntactic trees. In: Fürnkranz, J., Scheffer, T., Spiliopoulou, M. (eds.) ECML 2006. LNCS (LNAI), vol. 4212, pp. 318–329. Springer, Heidelberg (2006)
14. Giordani, A., Moschitti, A.: Semantic mapping between natural language questions and sql queries via syntactic pairing. In: NLDB 2009: Proceedings of the 13th international conference on Natural Language and Information Systems (2009)
15. Shawe-Taylor, J., Cristianini, N.: Kernel Methods for Pattern Analysis. Cambridge University Press, Cambridge (2004)
16. Zhang, D., Lee, W.S.: Question classification using support vector machines. In: Proceedings of the 26th annual international ACM SIGIR conference on Research and development in informaion retrieval, pp. 26–32. ACM Press, New York (2003)
17. Salton, G.: Recent trends in automatic information retrieval. In: SIGIR 1986, Proceedings of the 9th Annual International ACM SIGIR Conference on Research and Development in Information Retrieval, Pisa, Italy, September 8-10, 1986, pp. 1–10. ACM, New York (1986)
18. Joachims, T.: Making large-scale SVM learning practical. In: Schölkopf, B., Burges, C., Smola, A. (eds.) Advances in Kernel Methods (1999)
19. Moschitti, A., Quarteroni, S., Basili, R., Manandhar, S.: Exploiting syntactic and shallow semantic kernels for question/answer classification. In: Proceedings of ACL 2007, Prague, Czech Republic (2007)
20. Moschitti, A., Quarteroni, S.: Kernels on linguistic structures for answer extraction. In: Proceedings of ACL 2008: HLT, Short Papers, Columbus, Ohio (2008)
21. Chali, Y., Joty, S.: Improving the performance of the random walk model for answering complex questions. In: Proceedings of ACL 2008: HLT, Short Papers, Columbus, Ohio, pp. 9–12 (2008)

22. Shen, D., Lapata, M.: Using semantic roles to improve question answering. In: Proceedings of EMNLP-CoNLL (2007)
23. Surdeanu, M., Ciaramita, M., Zaragoza, H.: Learning to rank answers on large online QA collections. In: Proceedings of ACL 2008: HLT, Columbus, Ohio (2008)
24. Basili, R., Moschitti, A., Pazienza, M.: A text classifier based on linguistic processing. In: Proceedings of IJCAI 1999, Machine Learning for Information Filtering (1999)
25. Charniak, E.: A maximum-entropy-inspired parser. In: Proceedings of NAACL 2000 (2000)
26. Cancedda, N., Gaussier, E., Goutte, C., Renders, J.M.: Word sequence kernels. J. Mach. Learn. Res. 3, 1059–1082 (2003)
27. Moschitti, A.: Kernel methods, syntax and semantics for relational text categorization. In: Proceeding of CIKM 2008, NY, USA (2008)
28. Moschitti, A., Bejan, C.: A semantic kernel for predicate argument classification. In: Proceedings of CoNLL 2004, Boston, MA, USA (2004)
29. Moschitti, A., Coppola, B., Pighin, D., Basili, R.: Engineering of syntactic features for shallow semantic parsing. In: Proceedings of ACL 2005 Workshop on Feature Engineering for Machine Learning in NLP, USA (2005)
30. Moschitti, A., Pighin, D., Basili, R.: Tree kernels for semantic role labeling. Computational Linguistics 34(2), 193–224 (2008)
31. Moschitti, A., Zanzotto, F.: Fast and effective kernels for relational learning from texts. In: Ghahramani, Z. (ed.) Proceedings of the 24th Annual International Conference on Machine Learning, ICML 2007 (2007)
32. Moschitti, A., Pighin, D., Basili, R.: Semantic role labeling via tree kernel joint inference. In: Proceedings of CoNLL-X, New York City (2006)
33. Chandra, Y., Mihalcea, R.: Natural language interfaces to databases, University of North Texas, Thesis, M.S. (2006)
34. Kudo, T., Matsumoto, Y.: Fast Methods for Kernel-Based Text Analysis. In: Hinrichs, E., Roth, D. (eds.) Proceedings of ACL, pp. 24–31 (2003)
35. Cumby, C., Roth, D.: Kernel Methods for Relational Learning. In: Proceedings of ICML 2003, Washington, DC, USA, pp. 107–114 (2003)
36. Culotta, A., Sorensen, J.: Dependency Tree Kernels for Relation Extraction. In: ACL 2004, Barcelona, Spain, pp. 423–429 (2004)
37. Kudo, T., Suzuki, J., Isozaki, H.: Boosting-based parse reranking with subtree features. In: Proceedings of ACL 2005, US (2005)
38. Toutanova, K., Markova, P., Manning, C.: The Leaf Path Projection View of Parse Trees: Exploring String Kernels for HPSG Parse Selection. In: Proceedings of EMNLP 2004, Barcelona, Spain (2004)
39. Kazama, J., Torisawa, K.: Speeding up Training with Tree Kernels for Node Relation Labeling. In: Proceedings of EMNLP 2005, Toronto, Canada, pp. 137–144 (2005)
40. Shen, L., Sarkar, A., Joshi, A.k.: Using LTAG Based Features in Parse Reranking. In: EMNLP, Sapporo, Japan (2003)
41. Zhang, M., Zhang, J., Su, J.: Exploring Syntactic Features for Relation Extraction using a Convolution tree kernel. In: Proceedings of NAACL, New York City, USA, pp. 288–295 (2006)
42. Zhang, D., Lee, W.: Question classification using support vector machines. In: Proceedings of SIGIR 2003, Toronto, Canada. ACM Press, New York (2003)
43. Giuglea, A.M., Moschitti, A.: Semantic role labeling via framenet, verbnet and propbank. In: Proceedings of ACL 2006, Sydney, Australia (2006)

Active and Semi-supervised Data Domain Description

Nico Görnitz, Marius Kloft, and Ulf Brefeld

Machine Learning Group
Technische Universität Berlin
Franklinstr. 28/29, 10587 Berlin, Germany
{goernitz,mkloft,brefeld}@cs.tu-berlin.de

Abstract. Data domain description techniques aim at deriving concise descriptions of objects belonging to a category of interest. For instance, the support vector domain description (SVDD) learns a hypersphere enclosing the bulk of provided unlabeled data such that points lying outside of the ball are considered anomalous. However, relevant information such as expert and background knowledge remain unused in the unsupervised setting. In this paper, we rephrase data domain description as a semi-supervised learning task, that is, we propose a semi-supervised generalization of data domain description (SSSVDD) to process unlabeled *and* labeled examples. The corresponding optimization problem is non-convex. We translate it into an unconstraint, continuous problem that can be optimized accurately by gradient-based techniques. Furthermore, we devise an effective active learning strategy to query low-confidence observations. Our empirical evaluation on network intrusion detection and object recognition tasks shows that our SSSVDDs consistently outperform baseline methods in relevant learning settings.

1 Introduction

Data domain description techniques aim to devise concise descriptions of observed data. The task is to find minimal regions in feature space containing all data points that belong to the category of the observed data. Observations that do not fall into this region deviate from the normality and are rejected.

Data domain description techniques are therefore frequently being applied to outlier and anomaly detection problems where a model of normality is devised from available observations. Anomality of new objects is measured by their distance (in some metric space) from the learned model of normality, historically also known as "the sense of self" [7].

In network intrusion detection, the main merit of anomaly detection techniques is their ability to detect previously unknown attacks. One might think that the collective expertise amassed in the computer security community rules out major outbreaks of "genuinely novel" exploits. Unfortunately, a wide-scale deployment of efficient tools for obfuscation, polymorphic mutation and encryption results in an exploding variability of attacks. Although being only "marginally

W. Buntine et al. (Eds.): ECML PKDD 2009, Part I, LNAI 5781, pp. 407–422, 2009.

novel", such attacks quite successfully defeat signature-based detection tools. This reality brings one-class anomaly detection back into the research focus of the security community [14,15,11,28,27,19,20]. Until now, anomaly detection is usually being regarded as an unsupervised learning task for good reasons: Firstly, the rejection class cannot be sampled per definition as it comprises rare and unlikely events. Secondly, outliers are frequently too diverse to be modeled by only a single rejection class.

Nevertheless, data domain description techniques exhibit appealing properties for dealing with multiple, non-stationary class-distributions in settings where shifting distributions can be modeled by all means. For instance, domain descriptions have been successfully applied to multi-class classification problems with temporally varying numbers of categories such as event detection tasks and object recognition systems. Instead of maintaining expensive multi-class classifiers that have to be retrained using all available data once a new category is added, one simply learns a single domain description for every (new) category of interest.

We claim that an unsupervised learning setting for data domain descrption is often too restricted for practical applications. Firstly, these methods have to be trained solely on normal data which is hardly possible without already *knowing* the labelings. Although state-of-the-art techniques prove robust against injecting a few instances of the rejection class into the training data [2,24], knowing the class ratios is often crucial for accurate parameter adjustments. Secondly, one often knows the categories of certain training instances, be it manually labeled or recently seen instances. Such expert knowledge cannot be exploited in unsupervised settings and the learned models are sub-optimal in the sense that they leave out important information.

In this paper, we rephrase data domain description as a semi-supervised learning task, that is, we present semi-supervised data domain description (SSSVDD) that allows for processing unlabeled as well as labeled data to include expert and prior knowledge. Our model learns a minimal enclosing hypersphere in feature space that contains the normal data where point-wise errors are relaxed by slack variables. The inclusion of examples of the rejection class turns the optimization problem non-convex. As a remedy, we translate the optimization into an unconstraint, continuous problem with fewer parameters. It can therefore be optimized faster, and the retrieved local minima are substantially better on average [3]. The SSSVDD contains the unsupervised data domain description [24] as a special case that is obtained when no label information is used in the training process.

Furthermore, we devise an active learning strategy to query low-confidence decisions, hence guiding the user in the labeling process. Active learning selects an instance to be labeled by the user from the pool of unlabeled data. The selection process is designed to find unlabeled examples in the pool which – once labeled – lead to the maximal improvement of the hypothesis. Thus, the SSSVDD is initially trained solely on unlabeled examples and then subsequently refined by incorporating labeled examples that have been queried by the active learning rule. The training process can be terminated at any time, for instance when the desired predictive performance is obtained.

Empirical results on network intrusion detection and object recognition tasks show the benefit of casting data domain description into a semi-supervised learning framework: The SSSVDD significantly outperforms appropriate baseline methods for all learning settings. This effect is significantly enhanced by active learning. Our active learning strategy not only reduces the manual labeling effort for the practitioner, it also allows for automatically identifying novel network attacks for the intrusion detection tasks.

Our paper is structured as follows. Section 2 reviews related work and Section 3 introduces the classical data domain description. We extend the latter to a semi-supervised learning method in Section 4 where we also discuss optimization issues. Section 5 introduces our active learning strategy and Section 6 reports on empirical results. Section 7 concludes.

2 Related Work

Data domain description is usually regarded as an unsupervised or one-class classification task. Prominent approaches comprise k-nearest neighbors [2] or other distance based methods [9], quadratic programming [24], and statistical methods [30,25]. In this paper, we rephrase data domain description as a semi-supervised task (see [32] for an overview).

Active learning for anomaly detection has been studied by [22,17,1]. [1] take a max-margin approach and propose to query points that lie close to the decision hyperplane and violate the margin criterion in order to minimize the error rate. By contrast, the approach by [17] aims at detecting rejection categories in the data using as few queries as possible. Finally, the approach taken in [22] combines the former two active learning strategies to find interesting regions in feature space *and* to decrease the error-rate simultaneously.

Furthermore, there are several extensions of unsupervised data domain descriptors allowing for the inclusion of labeled examples. For instance, [8,12,26,31] present fully-supervised variants of the classical support vector data description (SVDD) [24]. However, the objective functions are no longer convex and the proposed optimizations in dual space may suffer from duality gaps. Another variant proposed in [23] is trained on unlabeled and instances belonging to the rejection class. Although this approach seems promising, it also suffers from non-convexity of the objective.

3 Support Vector Data Description

In this section, we briefly review the classical support vector domain description (SVDD) [24]. We are given a set of n *normal* inputs $\boldsymbol{x}_1, \ldots, \boldsymbol{x}_n \in \mathcal{X}$ and a function $\phi : \mathcal{X} \rightarrow \mathcal{F}$ extracting features out of the inputs. For instance, $\boldsymbol{x}_i$ may refer to the i-th recorded request and $\phi(\boldsymbol{x}_i)$ may encode the vector of bigrams occurring in $\boldsymbol{x}_i$.

The goal of the SVDD is to find a concise description of the normal data such that anomalous data can be easily identified as outliers. In the underlying one-class scenario, this translates to finding a minimal enclosing hypersphere (i.e., center $\mathbf{c}$ and radius R) that contains the normal input data [24], see Figure 1 (left). Given the function

$$f(\boldsymbol{x}) = \|\phi(\boldsymbol{x}) - \mathbf{c}\|^2 - R^2,$$

the boundary of the ball is described by the set $\{\boldsymbol{x} : f(\boldsymbol{x}) = 0 \wedge \boldsymbol{x} \in \mathcal{X}\}$. That is, the parameters of f are to be chosen such that $f(\boldsymbol{x}) < 0$ for normal data and $f(\boldsymbol{x}) > 0$ for anomalous points. The center $\mathbf{c}$ and the radius R can be computed accordingly by solving the following optimization problem [24]

$$\begin{aligned} \min_{R,\mathbf{c},\boldsymbol{\xi}} \quad & R^2 + \eta \sum_{i=1}^{n} \xi_i \\ \text{s.t.} \quad & \forall_{i=1}^{n} : \|\phi(\boldsymbol{x}_i) - \mathbf{c}\|^2 \le R^2 + \xi_i \\ & \forall_{i=1}^{n} : \xi_i \ge 0. \end{aligned} \tag{1}$$

The trade-off parameter η adjusts point-wise violations of the hypersphere. That is, a concise description of the data might benefit from omitting some data points in the computation of the solution. Discarded data points induce slack that is absorbed by variables ξ_i. Thus, in the limit $\eta \to \infty$, the hypersphere will contain all input examples irrespectively of their utility for the model and $\eta \to 0$ implies $R \to 0$ and the center $\mathbf{c}$ reduces to the centroid of the data.

The above optimization problem can be translated into an equivalent dual formulation by exploiting the identity $\mathbf{c} = \sum_{i=1}^{n} \alpha_i \phi(\boldsymbol{x}_i)$. We arrive at the dual SVDD optimization problem [24],

$$\begin{aligned} \max_{\boldsymbol{\alpha}} \quad & \sum_{i=1}^{n} \alpha_i k(\boldsymbol{x}_i, \boldsymbol{x}_i) - \sum_{i,j=1}^{n} \alpha_i \alpha_j k(\boldsymbol{x}_i, \boldsymbol{x}_j) \\ \text{s.t.} \quad & \sum_{i=1}^{n} \alpha_i = 1 \quad \text{and} \quad 0 \le \alpha_i \le \eta \quad \forall i = 1, \ldots, n. \end{aligned}$$

Once optimal parameters $\boldsymbol{\alpha}^*$ are found these are used as plug-in estimates to compute the anomaly score for new and unseen instances. A new observation $\bar{\boldsymbol{x}}$ is accepted if

$$k(\bar{\boldsymbol{x}}, \bar{\boldsymbol{x}}) - 2 \sum_{i=1}^{n} \alpha_i^* k(\boldsymbol{x}_i, \bar{\boldsymbol{x}}) + \sum_{i,j=1}^{n} \alpha_i^* \alpha_j^* k(\boldsymbol{x}_i, \boldsymbol{x}_j) \le R^2.$$

[8,12,26,23] propose extensions of the SVDD to incorporate labeled data into the learning process. The corresponding optimization problems are however not convex and the dual solution might suffer from a duality gap.

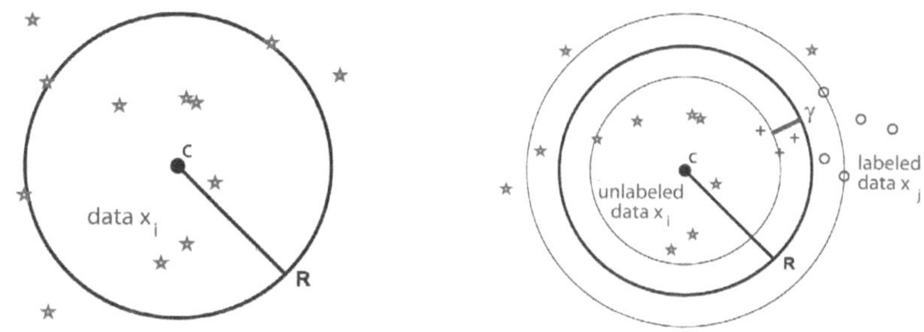

Fig. 1. Left: An exemplary solution of the SVDD. Right: Illustration of SSSVDD that incorporates unlabeled (green) as well as labeled data of the normal class (red) and the rejection category (blue).

4 Semi-supervised Data Domain Description

In this section, we derive our semi-supervised data domain description. In addition to n normal observations $\boldsymbol{x}_1, \ldots, \boldsymbol{x}_n \in \mathcal{X}$ we are also given m labeled pairs $(\boldsymbol{x}^*_{n+1}, y^*_{n+1}), \ldots, (\boldsymbol{x}^*_{n+m}, y^*_{n+m}) \subset \mathcal{X} \times \{+1, -1\}$, where we associate normal data with the positive class and outliers as the negative class. As in the previous section, we aim at finding a model $f(\boldsymbol{x}) = ||\phi(\boldsymbol{x}) - \mathbf{c}||^2 - R^2$ that generalizes well on unseen data, however, the model is now devised on the basis of labeled and unlabeled data. A straight-forward extension of the SVDD in Equation (1) using both, labeled and unlabeled examples, is given by

$$\begin{aligned}
\min_{R,\gamma,\mathbf{c},\boldsymbol{\xi}} \quad & R^2 - \kappa\gamma + \eta_u \sum_{i=1}^{n} \xi_i + \eta_l \sum_{j=n+1}^{n+m} \xi^*_j \\
\text{s.t.} \quad & \forall_{i=1}^{n} : \|\phi(\boldsymbol{x}_i) - \mathbf{c}\|^2 \leq R^2 + \xi_i \\
& \forall_{j=n+1}^{n+m} : y^*_j \left(\|\phi(\boldsymbol{x}^*_j) - \mathbf{c}\|^2 - R^2\right) \leq -\gamma + \xi^*_j \qquad (2) \\
& \forall_{i=1}^{n} : \xi_i \geq 0, \\
& \forall_{j=n+1}^{n+m} : \xi^*_j \geq 0.
\end{aligned}$$

The optimization problem has additional constraints for the labeled examples that have to fulfill the margin criterion with margin γ. Trade-off parameters κ, η_u, and η_l balance margin-maximization and the impact of unlabeled and labeled examples, respectively. To avoid cluttering the notation unnecessarily, we omit the obvious generalization of allowing different trade-offs η_l^+ and η_l^- for positively and negatively labeled instances, respectively. The additional slack variables ξ^*_j are bound to labeled examples and allow for point-wise relaxations of margin violations by labeled examples. The solution of the above optimization problem is illustrated in Figure 1 (right).

The inclusion of negatively labeled data turns the above optimization problem non-convex and optimization in the dual is prohibitive. As a remedy, we translate

Equation (2) into an unconstraint, continuous problem [3,33]. For the above problem, it is possible to resolve the slack terms:

$$\xi_i = \ell\left(R^2 - ||\phi(\boldsymbol{x}_i) - \mathbf{c}||^2\right)$$
$$\xi_j^* = \ell\left(y_j^*\left(R^2 - ||\phi(\boldsymbol{x}_j^*) - \mathbf{c}||^2\right) - \gamma\right)$$

where $\ell(t) = \max\{-t, 0\}$ is the common hinge loss where we explicitely deal with the margin γ in the argument t because γ is part of the optimization. We can now pose optimization problem (2) as a simple minimization problem *without* constraints as follows,

$$\min_{R,\gamma,\mathbf{c}} \quad R^2 - \kappa\gamma + \eta_u \sum_{i=1}^{n} \ell\left(R^2 - ||\phi(\boldsymbol{x}_i) - \mathbf{c}||^2\right) + \eta_l \sum_{j=n+1}^{n+m} \ell\left(y_j^*\left(R^2 - ||\phi(\boldsymbol{x}_j^*) - \mathbf{c}||^2\right) - \gamma\right). \tag{3}$$

Note that the optimization problems in Equations (2) and (3) are equivalent so far. We now substitute the Huber loss for the hinge loss to obtain a smooth and differentiable function that can be optimized with gradient-based techniques. The Huber loss $\ell_{\Delta,\epsilon}$ is displayed in Figure 2 and given by

$$\ell_{\Delta,\epsilon}(t) = \begin{cases} \Delta - t & : \ t \leq \Delta - \epsilon \\ \frac{(\Delta+\epsilon-t)^2}{4\epsilon} & : \ \Delta - \epsilon \leq t \leq \Delta + \epsilon \\ 0 & : \ \text{otherwise} \end{cases}$$
$$\ell'_{\Delta,\epsilon}(t) = \begin{cases} -1 & : \ t \leq \Delta - \epsilon \\ -\frac{1}{2}(\frac{\Delta-t}{\epsilon} + 1) & : \ \Delta - \epsilon \leq t \leq \Delta + \epsilon \\ 0 & : \ \text{otherwise}\ . \end{cases} \tag{4}$$

For notational convenience, we focus on the Huber loss for $\ell_{\Delta=0,\epsilon}(t)$ and move margin dependent terms into the argument t. Using the Huber loss $\ell_{0,\epsilon}$, computing the gradients of the slack variables ξ_i associated with unlabeled examples with respect to the primal variables R and $\mathbf{c}$ yields

$$\frac{\partial \xi_i}{\partial R} = 2R\ell'_\epsilon(R^2 - ||\phi(\boldsymbol{x}_i) - \mathbf{c}||^2)$$
$$\frac{\partial \xi_i}{\partial \mathbf{c}} = 2(\phi(\boldsymbol{x}_i) - \mathbf{c})\ell'_\epsilon(R^2 - ||\phi(\boldsymbol{x}_i) - \mathbf{c}||^2).$$

The derivatives of their counterparts ξ_j^* for the labeled examples with respect to R, γ, and $\mathbf{c}$ are given by

$$\frac{\partial \xi_j^*}{\partial R} = 2y_j^* R\ell'_\epsilon\left(y_j^*\left(R^2 - ||\phi(\boldsymbol{x}_j^*) - \mathbf{c}||^2\right) - \gamma\right)$$
$$\frac{\partial \xi_j^*}{\partial \gamma} = -\kappa\ell'_\epsilon\left(y_j^*\left(R^2 - ||\phi(\boldsymbol{x}_j^*) - \mathbf{c}||^2\right) - \gamma\right)$$
$$\frac{\partial \xi_j^*}{\partial \mathbf{c}} = 2y_j^*(\phi(\boldsymbol{x}_j^*) - \mathbf{c})\ell'_\epsilon\left(y_j^*\left(R^2 - ||\phi(\boldsymbol{x}_j^*) - \mathbf{c}||^2\right) - \gamma\right).$$

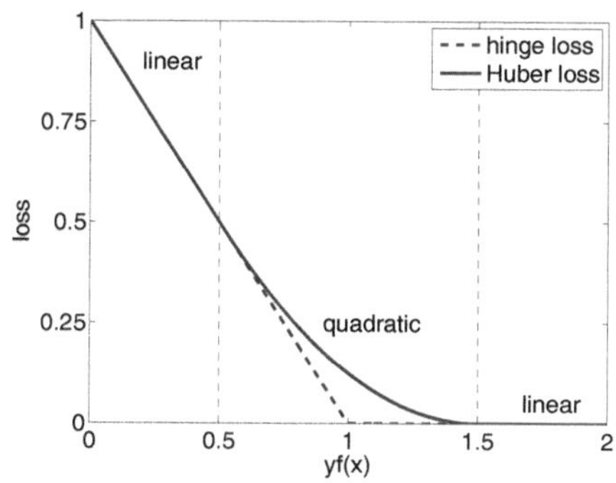

Fig. 2. The differentiable Huber loss $\ell_{\delta=1,\epsilon=0.5}$

Substituting the partial gradients, we resolve the gradient of Equation 3 with respect to the primal variables:

$$\frac{\partial EQ3}{\partial R} = 2R + \eta_u \sum_{i=1}^{n} \frac{\partial \xi_i}{\partial R} + \eta_l \sum_{j=n+1}^{n+m} \frac{\partial \xi_j^*}{\partial R}, \tag{5}$$

$$\frac{\partial EQ3}{\partial \gamma} = -\kappa + \eta_l \sum_{j=n+1}^{n+m} \frac{\partial \xi_j^*}{\partial \gamma}, \tag{6}$$

$$\frac{\partial EQ3}{\partial \boldsymbol{c}} = \eta_u \sum_{i=1}^{n} \frac{\partial \xi_i}{\partial \boldsymbol{c}} + \eta_l \sum_{j=n+1}^{n+m} \frac{\partial \xi_j^*}{\partial \boldsymbol{c}}. \tag{7}$$

The above equations can be plugged directly into off-the-shelf gradient-based optimization tools to optimize Equation (3) in the input space for the identity $\phi(\boldsymbol{x}) = \boldsymbol{x}$. However, predictive power is often related to (possibly) non-linear mappings ϕ of the input data into some high-dimensional feature space. An application of the representer theorem (see Appendix) shows that the center $\boldsymbol{c}$ can be expanded as

$$\boldsymbol{c} = \sum_i \alpha_i \phi(\boldsymbol{x}_i) + \sum_j \alpha_j y_j^* \phi(\boldsymbol{x}_j^*). \tag{8}$$

According to the chain rule, the gradient of Equation (3) with respect to the $\alpha_{i/j}$ is given by

$$\frac{\partial EQ3}{\partial \alpha_{i/j}} = \frac{\partial EQ3}{\partial \boldsymbol{c}} \frac{\partial \boldsymbol{c}}{\partial \alpha_{i/j}}.$$

Using Equation (8), the partial derivatives $\frac{\partial \boldsymbol{c}}{\partial \alpha_{i/j}}$ resolve to

$$\frac{\partial \boldsymbol{c}}{\partial \alpha_i} = \phi(\boldsymbol{x}_i) \quad \text{and} \quad \frac{\partial \boldsymbol{c}}{\partial \alpha_j} = y_j^* \phi(\boldsymbol{x}_j^*), \tag{9}$$

respectively. Applying the cain-rule to Equations (5),(6),(7) and (9) gives the gradients of Equation (3) with respect to the $\alpha_{i/j}$. The final objective function allowing for the use of kernel functions can be stated as

$$\min_{R,\gamma,\alpha} \quad R^2 - \kappa\gamma + \eta_u \sum_{i=1}^{n} \ell_\epsilon \left(R^2 - k(\boldsymbol{x}_i, \boldsymbol{x}_i) + (2\boldsymbol{e}_i - \alpha)' K \alpha\right)$$
$$+ \eta_l \sum_{j=n+1}^{n+m} \ell_\epsilon \left(y_j^* \left(R^2 - k(\boldsymbol{x}_j^*, \boldsymbol{x}_j^*) + (2\boldsymbol{e}_j^* - \alpha)' K \alpha\right) - \gamma\right), \quad (10)$$

where kernel K is given by $K(\boldsymbol{x}, \boldsymbol{x}') = \langle \phi(\boldsymbol{x}), \phi(\boldsymbol{x}') \rangle$ and $\boldsymbol{e}_1, \ldots, \boldsymbol{e}_{n+m}$ is the standard base of $\mathbb{R}^{n+m}$. By rephrasing the problem as an unconstrained optimization problem, its intrinsic complexity has not changed. However, the local minima of Optimization Problems (3) and (10) can now easily be found with gradient-based techniques such as conjugate gradient descent. In general, unconstrained optimization is also easier to implement than constrained optimization. We will observe the benefit of this approach in the following.

5 Active Learning

The SSSVDD is initially trained solely on unlabeled examples and then subsequently refined by incorporating labeled examples that have been queried by the active learning rule. We now devise an active learning strategy to query low-confidence decisions, hence guiding the user in the labeling process. Our active learning strategy selects an instance of the unlabeled data pool to be labeled by the user. The selection process is designed to find the unlabeled example in the pool which – once labeled – leads to the maximal improvement of the actual model.

We begin with a commonly used active learning strategy which simply queries borderline points. The strategy is sometimes called *margin strategy* and can be expressed by asking the user to label the point $\boldsymbol{x}'$ that is closest to the decision hypersphere [1,29]

$$\boldsymbol{x}' = \operatorname*{argmin}_{\boldsymbol{x} \in \{\boldsymbol{x}_1, \ldots, \boldsymbol{x}_n\}} \frac{|f(\boldsymbol{x})|}{\Omega} = \operatorname*{argmin}_{\boldsymbol{x} \in \{\boldsymbol{x}_1, \ldots, \boldsymbol{x}_n\}} \frac{|R^2 - \|\phi(\boldsymbol{x}) - \mathbf{c}\|^2|}{\Omega}, \quad (11)$$

where Ω is a normalization constant and given by $\Omega = \max_i |f(\boldsymbol{x}_i)|$.

However, when dealing with many non-stationary outlier and/or attack categories, it is beneficial to identify novel reject classes as soon as possible. We translate this into an active learning strategy as follows. Let $A = (a_{ij})_{i,j=1,\ldots,n+m}$ be an adjacency matrix, for instance obtained by a k-nearest-neighbor approach, where $a_{ij} = 1$ if $\boldsymbol{x}_i$ is among the k-nearest neighbors of $\boldsymbol{x}_j$ and 0 otherwise. We introduce an extended labeling $\bar{y}_1 \ldots, \bar{y}_{n+m}$ for all examples by defining $\bar{y}_i = 0$ for unlabeled instances and retaining the labels for labeled instances, i.e.,

$\bar{y}_j = y_j$. Using these pseudo labels, Equation (12) returns the unlabeled instance according to

$$x' = \underset{x_i \in \{x_1,\ldots,x_n\}}{\operatorname{argmin}} \frac{1}{2k} \sum_{j=1}^{n+m} (\bar{y}_j + 1)\, a_{ij}. \tag{12}$$

The above strategy explores unknown regions in feature space and subsequently deepens the learned knowledge by querying clusters of potentially similar objects to allow for good generalizations.

Nevertheless, using Equation (12) alone may result in querying points lying close to the center of the hypersphere or far from its boundary. These points will hardly contribute to an improvement of the hypersphere. In other words, only a combination of both strategies (11) and (12) guarantees the active learning to query points of interest. Our final active learning strategy is therefore given by

$$x' = \underset{x_i \in \{x_1,\ldots,x_n\}}{\operatorname{argmin}} = \delta \frac{|f(x)|}{\Omega} + \frac{1-\delta}{2k} \sum_{j=1}^{n+m} (\bar{y}_j + 1)\, a_{ij} \tag{13}$$

for $\delta \in [0,1]$. The combined strategy queries instances that are close to the boundary of the hypersphere *and* lie in potentially anomalous clusters with respect to the k-nearest neighbor graph. Depending on the actual value of δ, the strategy jumps from cluster to cluster and thus helps to identify interesting regions in feature space. For the special case of no labeled points our combined strategy reduces to the margin strategy.

Usually, an active learning step is followed by an optimization step of the SSSVDD taking into account the newly labeled data. This procedure is of course time-consuming and can be altered for practical settings, for instance by querying a couple of points before performing a model update. Irrespectively of the actual implementation, alternating between active learning and updating the model can be repeated until a desired predictive performance is obtained.

6 Empirical Results

In this section, we empirically evaluate the SSSVDD and the active learning strategies and compare their performances to appropriate strawmen. The baselines SVDD and SVDD^{neg} [23] are implemented in Matlab and optimized by SMO [18]. Additional baselines for the object recognition tasks are binary SVMs. SSSVDDs are optimized by conjugate gradient descent. Parameters of the active learning strategy are set to $k = 10$, $\alpha = 0.1$ for simplicity. We experiment on network intrusion and object recognition tasks.

6.1 Intrusion Detection

For the intrusion detection experiments we use HTTP traffic recorded within 10 days at Fraunhofer Institute FIRST. The data set comprises 145,069 unmodified

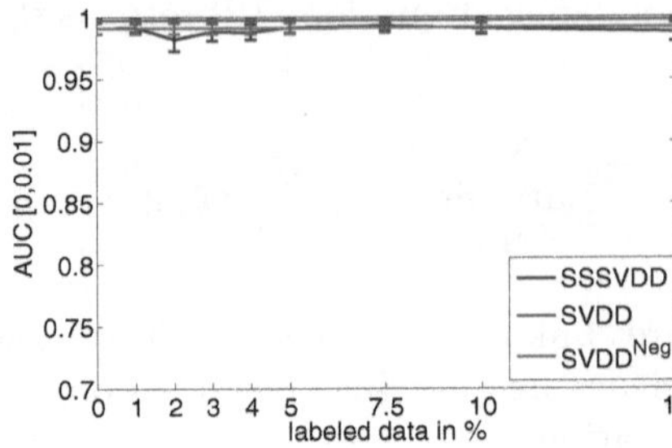

Fig. 3. Results for normal vs. malicious

connections of average length of 489 bytes. We refer to the FIRST data as the *normal pool*. The *malicious pool* contains 27 real attack classes generated using the Metasploit framework [16]. It covers 15 buffer overflows, 8 code injections and 4 other attacks including HTTP tunnels and cross-site scripting. Every attack is recorded in 2 – 6 different variants using virtual network environments and decoy HTTP servers.

To study the robustness of the different approaches in a more realistic scenario we also study techniques to obfuscate malicious content by adapting attack payloads to mimic benign traffic in feature space [6]. As a consequence, the extracted features do not deviate from a model of normality and the classifier is likely to be fooled by the attack. For our purposes it already suffices to study a simple cloaking technique by adding common HTTP headers to the payload while the malicious body of the attack remains unaltered. We apply this technique to the malicious pool and refer to the obfuscated set of attacks as *cloaked pool*.

We focus on two scenarios: normal vs. malicious and normal vs. cloaked data. For both settings, the respective byte streams are translated into a bag-of-3-grams representation. For each experiment, we randomly draw 966 training examples from the normal pool and 34 attacks either from the malicious or the cloaked pool, depending on the scenario. Holdout and test sets are also drawn at random and consist of 795 normal connections and 27 attacks, each. We make sure that attacks of the same attack class occur either in the training, or in the test set but not in both. We report on 10 repetitions with distinct training, holdout, and test sets and measure the performance by the area under the ROC curve in the false-positive interval $[0, 0.01]$ ($\text{AUC}_{0.01}$)

Figure 3 shows the results for normal vs. malicious data pools, where the x-axis depicts the percentage of randomly drawn labeled instances. Irrespectively of the amount of labeled data, the malicious traffic is detected by all methods equally well as the intrinsic nature of the attacks is well captured by the bag-of-3-grams representation. There is no significant difference between the classifiers. However, our next experiment shows the fragility of the of these results in the presence of simple cloaking techniques. Simply obfuscating the attacks by copying normal headers into the malicious payload leads to dramatically different results.

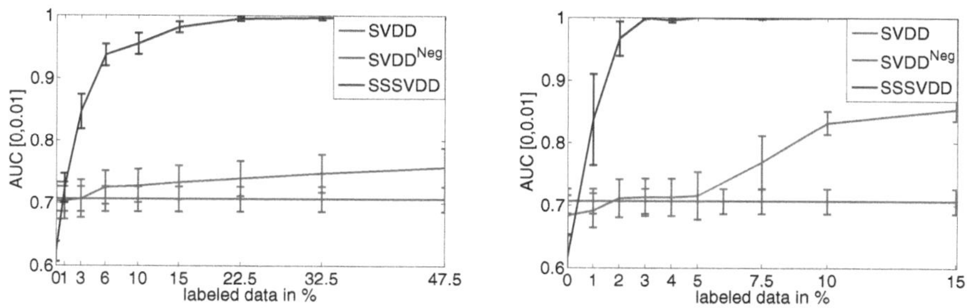

Fig. 4. Results for normal vs. cloaked. Left: Random samling. Right: Active learning.

Figure 4 (left) displays the results for normal vs. cloaked data. First of all, the performance of the unsupervised SVDD drops to only 70%. We obtain a similar result for the SVDD^{neg}; incorporating cloaked attack information into the training process of the SVDD leads to an increase of about 5% which is far from any practical value. Notice that the SVDD^{neg} cannot make use of labeled data of the normal class. Thus, its moderate ascent in terms of the number of labeled examples is credited to the class ratio of 966/34 for the random labeling strategy. The bulk of additional information cannot be exploited and has to be left out. By contrast, the semi-supervised SSSVDD includes all labeled data into the training process and clearly outperforms the two baselines. For only 5% labeled data, the SSSVDD easily beats the best baseline and for randomly labeling 30% of the available data it separates almost perfectly between normal and cloaked malicious traffic.

Nevertheless, labeling 30% of the data is not realistic for practical applications. We thus explore the benefit of active learning for inquiring label information of borderline and low-confidence points. Figure 4 shows the results for normal vs. cloaked data where the labeled data for SVDD^{neg} and SSSVDD is chosen according to the active learning strategy in Equation (13). The unsupervised SVDD that does not make use of labeled information remains at an $AUC_{0.01}$ of 70%. Compared to the results for a random labeling strategy (Figure 4, left), the performance of its counterpart SVDD^{neg} increases significantly. The ascent of the SVDD^{neg} is now steeper and yields 85% for 15% labeled data. However, the SSSVDD also improves for active learning and dominates the baselines. Using active learning, we need to label only 3% of the data for attaining an almost perfect separation, compared to 25% for a random labeling strategy. Our active learning strategy effectively boosts the performance and reduces the manual labeling effort significantly.

Figure 5 details the impact of our active learning strategy in Equation (13). We compare the number of outliers detected by the combined strategy with the margin-based strategy in Equation (11) (see also [1]) and by randomly drawing instances from the unlabeled pool. As a sanity check, we also included the theoretical outcome for random sampling. The results show that the combined strategy effectively detects malicious traffic much faster than the margin-based strategy.

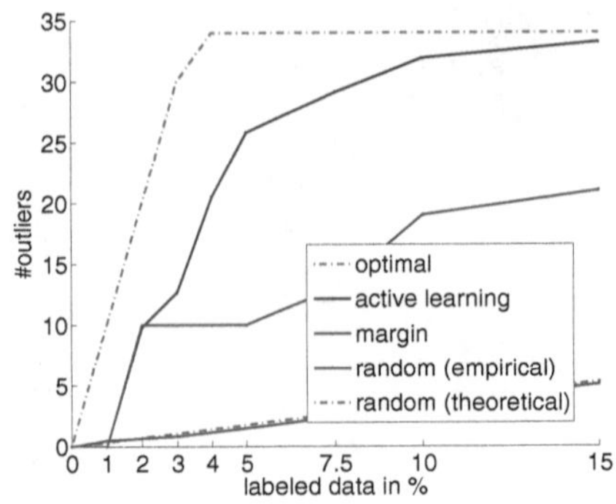

Fig. 5. Number of outliers found by active learning

6.2 Object Recognition

For our object recognition experiments we use the classification data of the VOC 2008 challenge [5]. The data set comprises 8780 images and 20 object classes. An image is annotated with a class label if at least one object from that class is detectable in the image. We use the training and holdout sets for our experiments which contain 4340 images.

For computational reasons, we focus on the three randomly drawn classes *aeroplane* (198 instances), *bird* (286 instances), and *dog* (266 instances), exemplary images are displayed in Figure 6. From the pool, we draw 375 instances randomly as independent test set while the remaining 375 examples are used for model selection over 10 repetitions. In each run, we randomly draw 10 labeled images of each class, 148 unlabeled instances, and 187 holdout examples.

We employ pyramid histograms [10] of visual words [4] (PHOW) for pyramid levels 0,1,2 over the grey channel. We obtain a feature vector for every image by concatenating histograms of all levels. For the grey channel, 1200 visual words are computed by k-means clustering on SIFT features [13] from randomly drawn images of each class. The underlying SIFT features are extracted from a dense grid of pitch ten.

Figure 7 compares regular support vector machines (SVMs) with SSSVDDs where both approaches apply margin-based active learning (Equation (11)) and the combined strategy in Equation (13) for detecting query points. For only a few labeled data points and many unlabeled examples (which cannot be utilized

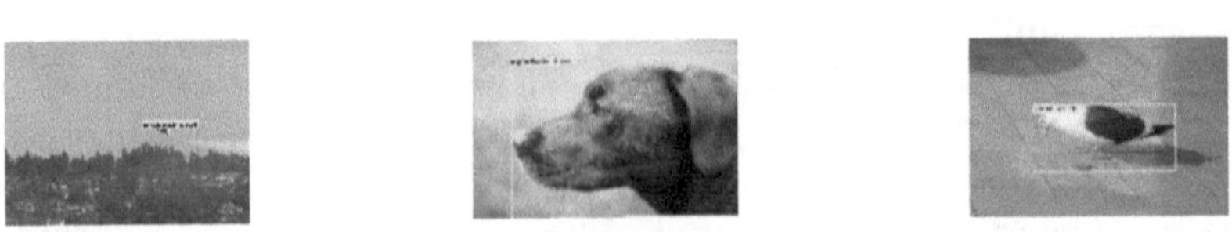

Fig. 6. Exemplary images from the VOC2008 object recognition data set. From left to right: aeroplane, dog, and bird.

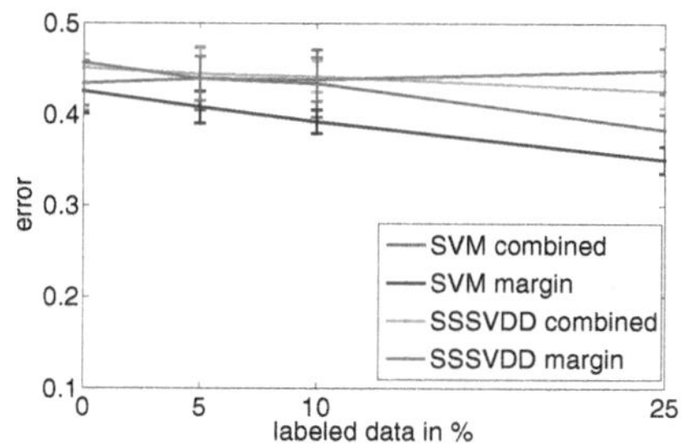

Fig. 7. Error rates for the VOC2008 data set

by SVMs), both approaches perform comparably. However, for increasing percentages of labeled data, the task becomes more and more a binary problem for which the SVM is well suited. For 25% labeled data, the SVM beats the SSSVDD significantly. Nevertheless, SSSVDD proves robust when labeled data is scarce and expensive to obtain; unlabeled examples are effectively exploited to augment sparse labelings.

7 Conclusion

In this paper, we proposed to view data domain description as a semi-supervised learning problem to allow for the inclusion of prior and expert knowledge. We generalized support vector data description to a semi-supervised learning algorithm (SSSVDD). Since the objective function of the SSSVDD is not convex, we translated the optimization problem into an unconstraint, continuous problem which can be optimized with efficient gradient-based techniques. Furthermore, we proposed a novel active learning strategy to guide the user in the labeling process of the unlabeled data by querying instances that are not only close to the boundary of the hypersphere but also likely members of novel rejection categories.

Empirically, we showed on network intrusion detection and object recognition tasks that rephrasing the unsupervised problem setting as a semi-supervised task is worth the effort. For instance in the network intrusion detection task, SSSVDDs prove robust in scenarios where the performance of baseline approaches deteriorate due to obfuscation techniques. Moreover, we observe the effectiveness of our active learning strategy which significantly improves the quality of the SSSVDD and spares practitioners from labeling unnecessarily many data points.

Acknowledgements. We thank Konrad Rieck for providing the kernels for the HTTP traffic and Christina Müller and Shinichi Nakajima for helping us with the object recognition task. This work was supported in part by the German Bundesministerium für Bildung und Forschung (BMBF) under the project ReMIND (FKZ 01-IS07007A) and by the FP7-ICT Programme of the European Community, under the PASCAL2 Network of Excellence, ICT-216886.

References

1. Almgren, M., Jonsson, E.: Using active learning in intrusion detection. In: Proc. IEEE Computer Security Foundation Workshop (2004)
2. Angiulli, F.: Condensed nearest neighbor data domain description. In: Advances in Intelligent Data Analysis VI (2005)
3. Chapelle, O., Zien, A.: Semi-supervised classification by low density separation. In: Proceedings of the International Workshop on AI and Statistics (2005)
4. Csurka, G., Bray, C., Dance, C., Fan, L.: Visual categorization with bags of keypoints. In: Workshop on Statistical Learning in Computer Vision, ECCV, Prague, Czech Republic, May 2004, pp. 1–22 (2004)
5. Everingham, M., Van Gool, L., Williams, C.K.I., Winn, J., Zisserman, A.: The PASCAL Visual Object Classes Challenge 2008 (VOC2008) Results (2008), `http://www.pascal-network.org/challenges/VOC/voc2008/`
6. Fogla, P., Sharif, M., Perdisci, R., Kolesnikov, O., Lee, W.: Polymorphic blending attacks. In: Proceedings of USENIX Security Symposium (2006)
7. Forrest, S., Hofmeyr, S.A., Somayaji, A., Longstaff, T.A.: A sense of self for unix processes. In: Proc. of IEEE Symposium on Security and Privacy, Oakland, CA, USA, pp. 120–128 (1996)
8. Hoi, C.-H., Chan, C.-H., Huang, K., Lyu, M., King, I.: Support vector machines for class representation and discrimination. In: Proceedings of the International Joint Conference on Neural Networks (2003)
9. Knorr, E.M., Ng, R.T.: Algorithms for mining distance-based outliers in large datasets. In: Proceedings of the 24th International Conference on Very Large Data Bases (1998)
10. Lazebnik, S., Schmid, C., Ponce, J.: Beyond bags of features: Spatial pyramid matching for recognizing natural scene categories. In: IEEE Computer Society Conference on Computer Vision and Pattern Recognition, New York, USA, vol. 2, pp. 2169–2178 (2006)
11. Lee, W., Stolfo, S.J.: A framework for constructing features and models for intrusion detection systems. ACM Transactions on Information Systems Security 3, 227–261 (2000)
12. Liu, Y., Zheng, Y.F.: Minimum enclosing and maximum excluding machine for pattern description and discrimination. In: ICPR 2006: Proceedings of the 18th International Conference on Pattern Recognition, Washington, DC, USA, 2006, pp. 129–132. IEEE Computer Society Press, Los Alamitos (2006)
13. Lowe, D.: Distinctive image features from scale invariant keypoints. International Journal of Computer Vision 60(2), 91–110 (2004)
14. Mahoney, M.V., Chan, P.K.: Learning nonstationary models of normal network traffic for detecting novel attacks. In: Proc. of ACM SIGKDD International Conference on Knowledge Discovery and Data Mining (KDD), pp. 376–385 (2002)
15. Mahoney, M.V., Chan, P.K.: Learning rules for anomaly detection of hostile network traffic. In: Proc. of International Conference on Data Mining (ICDM) (2003)
16. Maynor, K., Mookhey, K., Cervini, J.F.R., Beaver, K.: Metasploit toolkit. Syngress (2007)
17. Pelleg, D., Moore, A.: Active learning for anomaly and rare-category detection. In: Proc. Advances in Neural Information Processing Systems (2004)
18. Platt, J.C.: Fast training of support vector machines using sequential minimal optimization. In: Advances in kernel methods: support vector learning (1999)

19. Rieck, K., Laskov, P.: Detecting unknown network attacks using language models. In: Büschkes, R., Laskov, P. (eds.) DIMVA 2006. LNCS, vol. 4064, pp. 74–90. Springer, Heidelberg (2006)
20. Rieck, K., Laskov, P.: Language models for detection of unknown attacks in network traffic. Journal in Computer Virology 2(4), 243–256 (2007)
21. Schölkopf, B., Smola, A.J.: Learning with Kernels. MIT Press, Cambridge (2002)
22. Stokes, J.W., Platt, J.C.: Aladin: Active learning of anomalies to detect intrusion. Technical report, Microsoft Research (2008)
23. Tax, D.M.J.: One-class classification. PhD thesis, Technical University Delft (2001)
24. Tax, D.M.J., Duin, R.P.W.: Support vector data description. Machine Learning 54, 45–66 (2004)
25. Thottan, M., Ji, C.: Anomaly detection in ip networks. IEEE Transactions on Signal Processing 51(8), 2191–2204 (2003)
26. Wang, J., Neskovic, P., Cooper, L.N.: Pattern classification via single spheres. In: Computer Science: Discovery Science, DS (2005)
27. Wang, K., Parekh, J.J., Stolfo, S.J.: Anagram: A content anomaly detector resistant to mimicry attack. In: Zamboni, D., Krügel, C. (eds.) RAID 2006. LNCS, vol. 4219, pp. 226–248. Springer, Heidelberg (2006)
28. Wang, K., Stolfo, S.J.: Anomalous payload-based network intrusion detection. In: Jonsson, E., Valdes, A., Almgren, M. (eds.) RAID 2004. LNCS, vol. 3224, pp. 203–222. Springer, Heidelberg (2004)
29. Warmuth, M.K., Liao, J., Rätsch, G., Mathieson, M., Putta, S., Lemmen, C.: Active learning with support vector machines in the drug discovery process. Journal of Chemical Information and Computer Sciences 43(2), 667–673 (2003)
30. yan Yeung, D., Chow, C.: Parzen-window network intrusion detectors. In: Proceedings of the Sixteenth International Conference on Pattern Recognition, pp. 385–388 (2002)
31. Yuan, C., Casasent, D.: Pseudo relevance feedback with biased support vector machine. In: Proceedings of the International Joint Conference on Neural Networks (2004)
32. Zhu, X.: Semi–supervised learning in literature survey. Technical Report 1530, Computer Sciences, University of Wisconsin-Madison (2005)
33. Zien, A., Brefeld, U., Scheffer, T.: Transductive support vector machines for structured variables. In: Proceedings of the International Conference on Machine Learning (2007)

Appendix

In this section, we show the applicability of the representer theorem for semi-supervised support vector domain descriptions.

Theorem 1 (Representer Theorem [21]). *Let $\mathcal{H}$ be a reproducing kernel Hilbert space with a kernel $k : \mathcal{X} \times \mathcal{X} \to \mathbb{R}$, a symmetric positive semi-definite function on the compact domain. For any function $L : \mathcal{R}^n \to \mathbb{R}$, any nondecreasing function $\Omega : \mathbb{R} \to \mathbb{R}$. If*

$$J^* := \min J(f)_{f \in \mathcal{H}} := \min f \in \mathcal{H}\{\Omega\left(||f||^2_{\mathcal{H}}\right) + L\left(f(\boldsymbol{x}_1), \ldots, f(\boldsymbol{x}_n)\right)\}$$

is well-defined, then there exist $\alpha_1, \ldots, \alpha_n \in \mathbb{R}$, such that

$$f(\cdot) = \sum_{i=1}^{n} \alpha_i k(\boldsymbol{x}_i, \cdot) \tag{14}$$

achieves $J(f) = J^*$. Furthermore, if Ω is increasing, then each minimizer of $J(f)$ can be expressed in the form of Eq. (14).

Lemma 1. *The representer theorem can be applied to Equation* (3).

Proof. Recall the primal SSSVDD objective function which is given by

$$\begin{aligned} J(R, \gamma, \mathbf{c}) =& R^2 - \kappa\gamma + \eta_u \sum_{i=1}^{n} \ell\left(R^2 - ||\phi(\boldsymbol{x}_i) - \mathbf{c}||^2\right) \\ &+ \eta_l \sum_{j=n+1}^{n+m} \ell\left(y_j^*\left(R^2 - ||\phi(\boldsymbol{x}_j^*) - \mathbf{c}||^2\right) - \gamma\right). \end{aligned}$$

Substituting $T := R^2 - ||c||^2$ leads to the new objective function

$$\begin{aligned} J(T, \gamma, \mathbf{c}) =& ||c||^2 + T - \kappa\gamma + \eta_u \sum_{i=1}^{n} \ell\left(T - ||\phi(\boldsymbol{x}_i)||^2 + 2\phi(\boldsymbol{x}_i)'\mathbf{c}\right) \\ &+ \eta_l \sum_{j=n+1}^{n+m} \ell\left(y_j^*\left(T - ||\phi(\boldsymbol{x}_j^*)||^2 + 2\phi(\boldsymbol{x}_j^*)'\mathbf{c}\right) - \gamma\right). \end{aligned}$$

Expanding the center $\mathbf{c}$ in terms of labeled and unlabeled input examples is now covered by the representer theorem. After the optimization, T can be easily re-substituted to obtain the primal variables R, γ, and $\mathbf{c}$. □

A Matrix Factorization Approach for Integrating Multiple Data Views

Derek Greene and Pádraig Cunningham

School of Computer Science & Informatics, University College Dublin
{derek.greene,padraig.cunningham}@ucd.ie

Abstract. In many domains there will exist different representations or "views" describing the same set of objects. Taken alone, these views will often be deficient or incomplete. Therefore a key problem for exploratory data analysis is the integration of multiple views to discover the underlying structures in a domain. This problem is made more difficult when disagreement exists between views. We introduce a new unsupervised algorithm for combining information from related views, using a *late integration* strategy. Combination is performed by applying an approach based on matrix factorization to group related clusters produced on individual views. This yields a projection of the original clusters in the form of a new set of "meta-clusters" covering the entire domain. We also provide a novel model selection strategy for identifying the correct number of meta-clusters. Evaluations performed on a number of multi-view text clustering problems demonstrate the effectiveness of the algorithm.

1 Introduction

In many data analysis tasks there will naturally exist several different ways to describe the same set of data objects. This leads to the availability of multiple distinct representations or "views" that encode patterns relevant to the domain [1]. The question then arises, how can we integrate these representations in a way that allows us to effectively identify and explore these patterns? For some data exploration applications, we may have access to a set of views that are entirely *compatible* – the same patterns will occur across all views. The problem then becomes the identification of a single consensus model describing the patterns common to all views [2]. In other cases significant discord may exist between the data in different views [3]. An effective data integration procedure must then reconcile these disagreements, identifying common patterns, while also preserving those that are unique to each view.

In this paper we propose a simple but effective algorithm for combining data from multiple views, based on a *late integration* strategy [4]. The proposed approach, referred to as Integration by Matrix Factorization (IMF), takes representative clusterings generated independently on each available view, constructs an intermediate matrix representation of those clusterings, and applies a factorization procedure to this representation to reconcile the groups arising from the individual views. The factorization procedure preserves the contribution of the

W. Buntine et al. (Eds.): ECML PKDD 2009, Part I, LNAI 5781, pp. 423–438, 2009.

original clusters to the new groups, thereby highlighting the contribution made by each of the views. In addition we propose an entropy-based model selection procedure for automatically identifying the number of groups. To evaluate our approach we consider the problem of organizing topical news stories, represented by related text documents distributed across multiple views. These evaluations indicate that IMF can address common issues arising in real-world integration problems – such as disagreement between views, noisy views, and missing data.

This paper is organized as follows. Section 2 provides a brief overview of existing techniques for matrix factorization and fusing data from different sources. In Section 3 we discuss various issues that frequently arise when integrating multiple datasets in practice, and describe the proposed algorithm in detail. In Section 4 we present an empirical evaluation of the algorithm on synthetically-generated multi-view text datasets, followed by an evaluation on a real-world integration problem in Section 5. The paper finishes with some conclusions and suggestions for future work in Section 6.

2 Related Work

2.1 Matrix Factorization

Lee & Seung [5] proposed *Non-negative Matrix Factorization* (NMF), an unsupervised approach for dimensionality reduction, which approximates a data matrix as a product of factors that are constrained so that they will not contain negative values. By modeling each object as the additive combination of a set of non-negative basis vectors, a readily interpretable clustering of the data can be produced without further post-processing. These basis vectors are not required to be orthogonal, which facilitates the discovery of overlapping groups. The factorization process itself involves minimizing the difference between the original data and the approximation, most commonly by iteratively applying a pair of multiplicative update rules until the process converges to a local minimum [5].

2.2 Ensemble Clustering

In an unsupervised ensemble learning scenario we have access to a collection of "base clusterings", consisting of different clusterings generated on data originating from the same source. These clusterings represent the members of the ensemble. The primary aim of *ensemble clustering* [6] is to aggregate the information provided the ensemble members to produce a more accurate, stable clustering. A variety of strategies have been proposed to combine an ensemble to produce a single solution. For instance, the most widely-used strategy has been to consider information derived from the base clusterings to determine the level of *co-association* between each pair of objects in a dataset. Once a pairwise co-association matrix has been constructed, a standard algorithm such as single-linkage agglomerative clustering [7] or multi-level graph partitioning [6] is applied to produce a consensus clustering. The latter formulation was referred to by the authors as the Cluster-based Similarity Partitioning Algorithm (CSPA).

Rather than merely examining the pairwise relations between data objects, several authors have suggested examining the relations between the actual clusters contained in all base clusterings. Strehl & Ghosh [6] proposed the HyperGraph Partitioning Algorithm (HGPA), which involves transforming disjoint base clusterings to a hypergraph representation. Each node in the hypergraph represents a data object, and hyperedges are defined by the base cluster binary membership vectors. Subsequently a consensus clustering is produced by partitioning the hypergraph using the METIS algorithm [8].

The task of aggregating multiple clusterings can also be viewed as a *cluster correspondence* problem, where similar clusters from different base clusterings are matched together to produce a single "average clustering". Strehl & Ghosh [6] described a solution, referred to as the Meta-CLustering Algorithm (MCLA), which involves constructing a hypergraph where each hyperedge represents a cluster. The edges of the graph are then divided into a balanced k-way partition. Based on this edge partition, a majority voting scheme is used to assign data objects into the final clusters. The correspondence problem has been tackled by a number of other authors using *cumulative voting* ensemble clustering schemes, which are based on the assumption that there will be a direct relationship between individual clusters across all the base clusterings [9].

2.3 Data Integration

Blum & Mitchell [1] initially proposed the application of machine learning techniques in a *multi-view* setting, a problem which arises in domains where the data objects will naturally have several different representations. A useful broad distinction between techniques in this area was described by Pavlidis *et al.* [4], who identified three general data integration strategies: *early integration* involves the direct combination of data from several views into a single dataset before learning; *intermediate integration* involves computing separate similarity matrices on the views and producing a combined pairwise representation which is then passed to the learning algorithm; and *late integration* involves applying an algorithm to each individual view and subsequently combining the results.

While theoretical work in this area has largely focused on supervised learning problems, researchers have also considered the problem of producing clusterings from several different data sources. For instance, Bickel & Scheffer [2] proposed multi-view extensions of existing partitional and agglomerative clusterings algorithms. These algorithms were applied to the problem of clustering web pages, as represented by both textual information and hyperlinks. A general two-stage framework for reconciling discordant views in an unsupervised setting was described by Berthold & Patterson [3]. Other approaches have included minimizing the disagreement between views by casting the integration problem as an instance of bipartite spectral clustering [10], or as a semi-supervised clustering task where pairwise constraints generated from one view are used to influence the clustering process in another [11]. Both of these approaches are naturally limited to scenarios involving pairs of views.

3 Methods

3.1 Motivation

Given a set of views $\{V_1, \ldots, V_v\}$, let $\{x_1, \ldots, x_n\}$ denote the complete set of data objects present in the domain (*i.e.* $V_1 \cup V_2 \cdots \cup V_v$). The data integration task involves producing a complete clustering of the n objects to uncover all significant underlying "patterns" or groups present in the domain. In practice such integration tasks will often encounter one or more of the following issues:

Diversity of representation: In some views a feature-based representation will be available for data objects, while in other views only relation-based representations will be available, often in the form of graphs or networks.

Incomplete views: A representation for a data object in each view will not always be available. Rather, each view will often contain a subset of the total set of data objects in the domain.

Missing patterns: Patterns may be present in the data in one view, but largely or entirely absent from another view. As a consequence the number of patterns in each view will also vary.

Disagreement between views: The assignment of data objects to patterns may be inconsistent between views. Such disagreements can arise due to the unique characteristics of problem domain, or can simply be the result of noise within a view.

With the above requirements in mind, we propose a factorization-based formulation of the *late integration* [4] strategy for exploring domains were two or more related views exist. This approach, referred to as Integration by Matrix Factorization (IMF), takes representative clusterings generated independently on each individual view (using an algorithm appropriate for that view), constructs an intermediate representation of the clusterings, and decomposes this representation to reconcile the groups arising from the individual views. The fact that IMF operates on previously generated clusterings alone, rather than any specific representation of the original data, neatly avoids the diversity of representation issue. Late integration brings a number of additional benefits: the ability to harness parallel computing resources by processing large data views separately, the aggregation of information from views where privacy issues arise (*e.g.* financial, legal or commercially-sensitive data), and the facility to reuse knowledge available in existing legacy clusterings [6]. Later in Section 4 we demonstrate that the IMF algorithm can also address the other key integration issues of incomplete views, missing patterns, and disagreement between the set of available views.

3.2 Integration by Matrix Factorization

Intermediate representation. Formally we have access to a set of representative clusterings $\mathbb{C} = \{\mathcal{C}_1, \ldots, \mathcal{C}_v\}$, one per view, where $\mathcal{C}_h$ indicates the set of k_h clusters $\{c_h^1, \ldots, c_h^{k_h}\}$ generated on the view V_h. The sum $l = \sum_{i=1}^{v} k_i$ is the total number of clusters generated on all views. The clusterings may be generated by an algorithm that produces a disjoint partition (*e.g.* standard k-means or the kernelized equivalent), probabilistic clusters (*e.g.* EM clustering), or arbitrary non-negative membership weights (*e.g.* NMF). Hierarchical clusterings can be combined by applying a suitable cut-off strategy to produce a disjoint partition. However, for the remainder of this paper we focus on disjoint clusterings.

The constituent clusterings in $\mathbb{C}$ can be represented by a set of non-negative membership matrices $\mathbb{M} = \{\mathbf{M}_1, \ldots, \mathbf{M}_v\}$, where $\mathbf{M_h} \in \mathbb{R}^{n \times k_h}$ represents the cluster membership of objects in $\mathcal{C}_h$ generated on view V_h. For objects which are not present or clustered in a given view, the corresponding row in the membership matrix of the clustering for that view will contain zero values. By transposing the matrices in $\mathbb{M}$ and stacking them vertically, we can construct a matrix of clusters $\mathbf{X} \in \mathbb{R}^{l \times n}$. Each row in $\mathbf{X}$ now corresponds to an individual cluster from the clusterings in $\mathbb{C}$, while each column corresponds to a data object in the original domain. Following the discussion in [6], conceptually we can view this representation as the adjacency matrix of a hypergraph consisting of n vertices and l weighted hyperedges. Alternatively we can interpret the columns of $\mathbf{X}$ as an embedding of the original objects in a new l-dimensional space.

Factorization process. The goal of the integration process is to project the clusters in $\mathbb{C}$ to a set of $k' < l$ new basis vectors or "meta-clusters", where k' represents the number of underlying patterns present in the domain. These meta-clusters represent the additive combinations of clusters generated on one or more different views. Clusters generated on the same view can also be grouped together. This may be desirable in cases where a pattern has been incorrectly split in a view, or where the constituent clusterings are generated at a higher resolution than is required for the integrated solution.

Formally, the process involves producing an approximation of $\mathbf{X}$ in the form of the product of two non-negative factors:

$$\mathbf{X} \approx \mathbf{PH} \quad \text{such that} \quad \mathbf{P} \geq 0\,, \mathbf{H} \geq 0$$

where the rows of $\mathbf{P} \in \mathbb{R}^{l \times k'}$ represent the projection of the original clusters to a set of new basis vectors representing k' "meta-clusters". These meta-clusters can be additively combined using the coefficient values from the matrix $\mathbf{H} \in \mathbb{R}^{k' \times n}$ to reconstruct an approximation of the original set of clusters in $\mathbf{X}$. Furthermore, each column in $\mathbf{H}$ can be viewed as the membership of the original complete set of n data objects with respect to the k' meta-clusters.

To measure the reconstruction error between the original matrix $\mathbf{X}$ and the pair of factors $(\mathbf{P}, \mathbf{H})$ we can compute the Frobenius norm:

$$||\mathbf{X} - \mathbf{PH}||_{\mathsf{F}}^2 = \sum_{i=1}^{l} \sum_{j=1}^{n} [X_{ij} - (\mathbf{PH})_{ij}]^2 \qquad (1)$$

To minimize Eqn. 1 we iteratively apply the multiplicative update rules proposed by Lee & Seung [5]:

$$P_{ic} \leftarrow P_{ic} \frac{(\mathbf{X}\mathbf{H}^{\mathsf{T}})_{ic}}{(\mathbf{P}\mathbf{H}\mathbf{H}^{\mathsf{T}})_{ic}} \qquad H_{cj} \leftarrow H_{cj} \frac{(\mathbf{P}^{\mathsf{T}}\mathbf{X})_{cj}}{(\mathbf{P}^{\mathsf{T}}\mathbf{P}\mathbf{H})_{cj}}$$

The rules are applied until the change in the objective (Eqn. 1) between one iteration and the next is below an arbitrarily small value. The computational cost of each iteration is $O(lnk')$ when using dense matrix multiplication operations.

The additive nature of the factorization procedure can be useful in interpreting the results of the integrating process. Based on the values in the projection matrix $\mathbf{P}$, we can calculate a matrix $\mathbf{T} \in \mathbb{R}^{v \times k'}$ indicating the contribution of the view V_h to each meta-cluster:

$$T_{hf} = \frac{\sum_{c_f^j \in \mathcal{C}_f} P_{jf}}{\sum_{g=1}^{l} P_{gf}} \tag{2}$$

That is, the sum of the projection weights in $\mathbf{P}$ for the clusters generated on V_h, normalized with respect to the total projection weight for each meta-cluster (*i.e.* the column sums of $\mathbf{P}$). A value T_{hf} close to 0 indicates that the view V_h has made little contribution to the f-th meta-cluster, while a value close to 1 indicates that the view V_h has made the predominant contribution.

To illustrate the integration process, Figure 1 shows a simple problem involving objects $\{x_1, \ldots, x_7\}$ represented in two views. The corresponding clusterings $\mathbb{C} = \{\mathcal{C}_1, \mathcal{C}_2\}$ are transformed to the intermediate representation $\mathbf{X}$, and factorization is applied to yield the matrices $(\mathbf{P}, \mathbf{H})$. The entries in $\mathbf{P}$ illustrate how the clusters from these clusterings are combined to produce $k' = 3$ meta-clusters. The actual object membership weights for these meta-clusters are shown in $\mathbf{H}$. The contributions made by the two views to the meta-clusters are given by the entries on the rows of the matrix $\mathbf{T}$.

$$\begin{array}{l} \mathcal{C}_1 = \{\{x_1, x_2, x_3\}, \{x_4, x_5\}\} \\ \mathcal{C}_2 = \{\{x_6, x_7\}, \{x_1, x_2\}\} \end{array} \longrightarrow \mathbf{X} \quad \begin{array}{c|ccccccc} & x_1 & x_2 & x_3 & x_4 & x_5 & x_6 & x_7 \\ \hline c_1^1 & 1 & 1 & 1 & 0 & 0 & 0 & 0 \\ c_1^2 & 0 & 0 & 0 & 1 & 1 & 0 & 0 \\ \hdashline c_2^1 & 0 & 0 & 0 & 0 & 0 & 1 & 1 \\ c_2^2 & 1 & 1 & 0 & 0 & 0 & 0 & 0 \end{array}$$

$$\mathbf{X} \approx \mathbf{P}\mathbf{H} \longrightarrow \mathbf{P} \quad \begin{array}{c|ccc} c_1^1 & 1.2 & 0.0 & 0.0 \\ c_1^2 & 0.0 & 1.2 & 0.0 \\ \hdashline c_2^1 & 0.0 & 0.0 & 1.2 \\ c_2^2 & 0.9 & 0.0 & 0.0 \end{array} \qquad \mathbf{H}^{\mathsf{T}} \quad \begin{array}{c|ccc} x_1 & 1.0 & 0.0 & 0.0 \\ x_2 & 1.0 & 0.0 & 0.0 \\ x_3 & 0.5 & 0.0 & 0.0 \\ x_4 & 0.0 & 0.8 & 0.0 \\ x_5 & 0.0 & 0.8 & 0.0 \\ x_6 & 0.0 & 0.0 & 0.8 \\ x_7 & 0.0 & 0.0 & 0.8 \end{array}$$

$$\mathbf{T} \quad \begin{array}{c|ccc} V_1 & 0.6 & 1.0 & 0.0 \\ \hdashline V_2 & 0.4 & 0.0 & 1.0 \end{array}$$

Fig. 1. Example of IMF applied to clusterings from two views generated in a domain containing 7 data objects. A value of $k' = 3$ is used for the number of meta-clusters.

Factorization initialization. The sensitivity of NMF-like algorithms to the choice of initial factors has been noted by a number of authors [12,13]. While stochastic initialization is widely used in this context, ideally we would like to produce a single integrated clustering without requiring multiple runs of the integration process. Therefore to initialize the integration process, we populate the pair $(\mathbf{P}, \mathbf{H})$ by employing the deterministic NNDSVD strategy described by Boutsidis & Gallopoulos [13]. This strategy applies two sequential SVD processes to the matrix $\mathbf{X}$ to produce a pair of initial factors. In addition to being deterministic, NNDSVD is suitable in the context of integration as it has a tendency to produce comparatively sparse factors. As we shall see in the next section this is particularly desirable in the case of the projection matrix $\mathbf{P}$.

3.3 Model Selection

The selection of a suitable value for the number of meta-clusters k' is central to the data integration process. A value that is too low could force unrelated clusters to be grouped together, while a value that is too high could potentially cause the integration process to fail to merge related clusters from different views.

When selecting a model we consider the uncertainty of the mapping between clusters from different views, based on the uncertainty of the values in the projection matrix $\mathbf{P}$. Firstly we normalize the rows of $\mathbf{P}$ to unit length, yielding a normalized matrix $\hat{\mathbf{P}} \in [0, 1]$. In the ideal case each row in $\hat{\mathbf{P}}$ will contain a single value 1 and $(k' - 1)$ zeros, signifying that the corresponding base cluster has been perfectly matched to a single meta-cluster. This notion of uncertainty can be formally described in terms of the normalized entropy of the rows in $\hat{\mathbf{P}}$.

To illustrate this, we refer back to the previous example (Figure 1) of combining two clusterings. Figure 2 shows normalized project matrices corresponding to models for $k' = 3$ and $k' = 4$ respectively. In the latter case the integration procedure splits cluster c_1^1 between two meta-clusters (instead of matching it solely with c_2^2, which it subsumes as shown in Figure 1). Consequently the values in the first row of $\hat{\mathbf{P}}$ have a higher level of entropy. In contrast the matrix for $k' = 3$ shows a perfect match between each cluster from $\mathbb{C}$ and one of the three meta-clusters, suggesting that this model is more appropriate for the problem.

To identify an appropriate number of meta-clusters k', we can test values from a broad range $k \in [k_{min}, k_{max}]$ informed by the user's knowledge of the domain.

$$\underset{(k'=3)}{\hat{\mathbf{P}}} \quad \begin{matrix} c_1^1 \\ c_1^2 \\ c_2^1 \\ c_2^2 \end{matrix} \begin{bmatrix} 1.0 & 0.0 & 0.0 \\ 0.0 & 1.0 & 0.0 \\ 0.0 & 0.0 & 1.0 \\ 1.0 & 0.0 & 0.0 \end{bmatrix} \qquad \underset{(k'=4)}{\hat{\mathbf{P}}} \quad \begin{matrix} c_1^1 \\ c_1^2 \\ c_2^1 \\ c_2^2 \end{matrix} \begin{bmatrix} 0.4 & 0.0 & 0.0 & 0.6 \\ 0.0 & 1.0 & 0.0 & 0.0 \\ 0.0 & 0.0 & 1.0 & 0.0 \\ 1.0 & 0.0 & 0.0 & 0.0 \end{bmatrix}$$

Fig. 2. Example of model selection for data integration applied to clusterings from two views, using two candidate values for the number of meta-clusters k'

For each candidate k we construct $\hat{\mathbf{P}}$. For each row j in this matrix we calculate the normalized entropy of the projection values:

$$e(\hat{\mathbf{P}}_j) = -\frac{1}{\log k} \sum_{h=1}^{k} P_{jh} \log (P_{jh}) \tag{3}$$

An evaluation $s(k) \in [0, 1]$ for the suitability of the model with k meta-clusters is given by subtracting the mean row entropy from 1:

$$s(k) = 1 - \frac{1}{l} \sum_{j=1}^{l} e(\hat{\mathbf{P}}_j) \tag{4}$$

A value for the final number of meta-clusters k' can be chosen so as to maximize the value of Eqn. 4.

In practice we observe that Eqn. 4 will not have an expected value of zero when combining randomly generated clusterings, and will exhibit a bias toward higher values of k. We can readily address this by employing the widely-used adjustment technique described in [14] to correct for chance agreement:

$$\hat{s}(k) = \frac{s(k) - \bar{s}(k)}{1 - \bar{s}(k)} \tag{5}$$

The value $\bar{s}(k)$ is the expected evaluation score for a factorization containing k meta-clusters. In practice we can find an approximation for $\bar{s}(k)$ by applying the following for a sufficiently large number of runs: take a given intermediate matrix $\mathbf{X}$, randomly permute the values in the columns, apply factorization with parameter k, and recalculate Eqn. 4. The expected value is given by the mean of $s(k)$ across all permutations.

3.4 Ensemble Multi-view Integration

The "one-shot" integration scenario described in Section 3.2 assumes the availability of a definitive clustering for each view. However, in many cases a variety of different clusterings may be available for each view – either generated on different subsets of the data in a view, produced using different parameter values, or simply as a result of the use of a clustering algorithm that converges to different local minima (*e.g.* k-means with random initialization). In many cases ensemble clustering techniques can harness the diversity present in such collections of clusterings [6]. We can naturally apply the IMF approach in a multi-view ensemble clustering setting, where $\mathbb{C}$ contains multiple clusterings generated on each view. As we shall see in our evaluation in Section 5, IMF can often take advantage of the diversity in a multi-view ensemble to produce a superior clustering.

4 Evaluation on Synthetic Multi-view Data

To evaluate the ability of the IMF approach described in Section 3 to handle a number of key issues that arise in data integration problems, we applied IMF to cluster multiple different views artificially produced from single-view news text corpora. Since we can assume that a single news article consists of one or more segments of text (*e.g.* one or more consecutive paragraphs), we can construct views containing related segments of text. The overall task then becomes the identification of an accurate clustering of the entire collection based on the segment information provided in the different views. Using this synthetically generated data, we can examine the effectiveness of the proposed approach in the context of the requirements detailed in Section 3.1.

4.1 Dataset Construction

We make use of the *bbc* and *bbcsport* news corpora[1] which have been previously used in document clustering tasks [12], and produce multiple views for documents based on related text segments. The original *bbc* corpus contains a total of 2225 documents with 5 annotated topic labels, while the original *bbcsport* corpus contains a total of 737 documents also with 5 annotated labels. From each corpus we constructed new synthetic datasets with 2-4 views as follows:

1. We split each raw document into segments. This was done by separating the documents into paragraphs, and merging sequences of consecutive paragraphs until 1-4 segments of text remained, such that each segment was at least 200 characters long. Each segment is logically associated with the original document from which it was obtained.
2. The segments for each document were randomly assigned to views, with the restriction that at most one segment from each document was assigned to the same view.
3. Standard stemming, stop-word removal and TF-IDF normalization procedures were separately applied to the segments in the individual views.

Details of the six resulting multi-view datasets[2] are provided in Table 1. To quantify algorithm performance, we calculate the *normalized mutual information* (NMI) [6] between clusterings and the set of annotated label information provided for the original corpora. These annotations are derived from the categories assigned to the original online news articles. Since NMI evaluates disjoint clusterings, we convert weighted membership matrices to disjoint clusterings by assigning each document to the cluster for which it has the highest weight. Note that in all cases NMI scores are calculated relative to the entire corpus, rather than relative to the subset present in any individual view.

4.2 "One-Shot" Multi-view Integration

To examine the effectiveness of the IMF approach, we consider the scenario of combining a set of v clusterings, each coming from a different view. To provide a set of representative clusterings for our synthetic views, we apply spectral clustering followed by weighted kernel k-means as described in [15]. Since our focus

[1] Both available from `http://mlg.ucd.ie/datasets/bbc.html`
[2] Available from `http://mlg.ucd.ie/datasets/segment.html`

Table 1. Details of the synthetic multi-view text datasets

Datasets	View	Documents
bbc-seg2	1	2125
	2	2112
bbc-seg3	1	1828
	2	1832
	3	1845
bbc-seg4	1	1543
	2	1524
	3	1574
	4	1549

Collection	View	Documents
bbcsport-seg2	1	644
	2	637
bbcsport-seg3	1	519
	2	531
	3	513
bbcsport-seg4	1	400
	2	410
	3	437
	4	432

here is not on the generation of these constituent clusterings, for convenience we set the number of clusters to the correct number of labeled classes for the associated corpora. The representative clusterings also provide a reasonable baseline comparison. When applying IMF itself, we select a value for the parameter k' using the procedure proposed in Section 3.3, using a candidate range $k' \in [4, 12]$.

Incomplete views. As indicated by the figures in Table 1, the synthetic view construction methodology will naturally result in cases where documents will be represented by segments in some but not all of the views (*i.e.* the views are not complete). Therefore we can directly examine the ability of the proposed algorithm to deal with this scenario. A summary of the results of the "one-shot" experiments on the six synthetic datasets is given in Table 2. The mean and standard deviation of the NMI scores for the constituent clusterings are listed for comparison. In all cases the application of IMF produced integration clusterings that were significantly better than those generated on the individual views.

Table 2 also shows the number of meta-clusters k' automatically selected by the entropy-based criterion (Eqn. 4). While the model selection procedure did not always exactly attain the "correct" number of clusters (both the *bbc* and *bbcsport* corpora contain 5 annotated topics), this value did appear in the top three recommended choices for five of the six datasets.

Table 2. Accuracy (NMI) of the IMF approach on synthetic multi-view data, using one clustering per view

Dataset	k'	Base	IMF
bbc-seg2	4	0.77 ± 0.05	0.80
bbc-seg3	5	0.71 ± 0.01	0.83
bbc-seg4	5	0.60 ± 0.01	0.83
bbcsport-seg2	4	0.74 ± 0.05	0.80
bbcsport-seg3	10	0.54 ± 0.05	0.62
bbcsport-seg4	6	0.39 ± 0.04	0.56

Table 3. Mean accuracy (NMI) of the IMF approach on synthetic multi-view data *with missing patterns*, using one clustering per view

Dataset	Base	IMF
bbc-seg2	0.76 ± 0.02	0.79 ± 0.03
bbc-seg3	0.66 ± 0.01	0.85 ± 0.03
bbc-seg4	0.56 ± 0.02	0.82 ± 0.04
bbcsport-seg2	0.69 ± 0.05	0.78 ± 0.03
bbcsport-seg3	0.51 ± 0.06	0.63 ± 0.04
bbcsport-seg4	0.39 ± 0.04	0.52 ± 0.04

Missing patterns. To examine the behavior of IMF in scenarios were a pattern is entirely absent from a view, we took the synthetic datasets and removed all segments relating to a different randomly chosen label from each view (*i.e.* so that each view only contains segments pertaining to $k' - 1$ classes). We repeated this process for 20 runs, applying weighted kernel k-means on this data followed by IMF integration. For computational reasons, we use the same values of k' selected in the last set of experiments. Mean and standard deviation of NMI scores for these experiments are reported in Table 3. Again the IMF approach performs significantly better than the representative clusterings, and is successful in combining clusterings where an exact one-to-one mapping between the clusters in $\mathbb{C}$ does not necessarily exist. It is also worth noting that the NMI scores achieved are very close to those achieved when integrating clusterings generated on views with all patterns present (Table 2).

Disagreement between views. Next we examined the problem of discord between connected views. Specifically we considered the scenario where one view is considerably less informative than the others. In practice we selected one view at random and permuted 10% to 40% of the non-zero term values for each document, producing a noisy view on which a clustering was generated with spectral-initialized kernel k-means. IMF was then applied to integrate the noisy clustering together with the non-noisy clusterings from the $v - 1$ other views. In these experiments we set the value of k' to the "correct" number of class labels. We repeated the entire process for 30 runs and averaged the resulting NMI scores. Mean NMI scores for the base clusterings (both noisy and non-noisy) and the resulting integrated clusterings are given in Table 4.

A key test in this experiment is whether an integrated clustering can improve on its constituent clusterings in the presence of noisy views, rather than having performance equivalent to the "weakest link" among the views. As expected we observe that the meta-clusters produced by IMF remain significantly more accurate than the underlying constituent clusterings. Secondly, comparing the results to those in Table 2, we see that for 10-20% noise there is little decrease in clustering accuracy. For more extreme levels of noise, the IMF clustering on datasets derived from the *bbc* corpus remain reasonably accurate, while we see a greater effect on the datasets derived from the *bbcsport* corpus. In general these

Table 4. Mean accuracy (NMI) of the IMF approach on synthetic data using one clustering per view, where one of the views contains 10% to 40% noisy term values

Dataset	10% noise		20% noise		30% noise		40% noise	
	Base	IMF	Base	IMF	Base	IMF	Base	IMF
bbc-seg2	0.79	0.84	0.79	0.83	0.75	0.80	0.72	0.77
bbc-seg3	0.69	0.83	0.67	0.81	0.64	0.78	0.59	0.75
bbc-seg4	0.57	0.81	0.56	0.79	0.54	0.78	0.49	0.72
bbcsport-seg2	0.71	0.81	0.66	0.74	0.65	0.74	0.60	0.66
bbcsport-seg3	0.53	0.65	0.51	0.63	0.47	0.58	0.41	0.47
bbcsport-seg4	0.40	0.53	0.38	0.51	0.35	0.49	0.26	0.35

experiments suggest that the IMF approach is reasonably tolerant to disagreement between views, and cases where one view is weaker than the others.

5 Evaluation on Real-World Data

In this section we describe an evaluation of the proposed integration approach performed on a real-world multi-view document clustering task – namely that of clustering topical news stories where multiple reports of the same news story are available from different news sources. We constructed a new multi-view dataset[3], referred to as the *3sources* collection, from three well-known online news sources: BBC[4], Reuters[5], and The Guardian[6]. This dataset exhibits a number of common aspects of multi-view problems highlighted previously – notably that certain stories will not be reported by all three sources (*i.e.* incomplete views), and the related issue that sources vary in their coverage of certain topics (*i.e.* partially missing patterns).

In total we collected 948 news articles covering 416 distinct news stories from the period February–April 2009. Of these stories, 169 were reported in all three sources, 194 in two sources, and 53 appeared in a single news source. Each story was manually annotated with one or more of the six topical labels: *business*, *entertainment*, *health*, *politics*, *sport*, *technology*. These roughly correspond to the primary section headings used across the three news sources. To facilitate comparisons using the NMI measure, in our evaluation we consider only the dominant topic for each news story, yieldings a disjoint set of annotated classes as shown in Table 5.

5.1 "One-Shot" Multi-view Integration

For our first evaluation on the *3sources* data, we consider a "one-shot" integration process. Once again we generate a representative clustering on each view

[3] Available from `http://mlg.ucd.ie/datasets/3sources.html`
[4] `http://news.bbc.co.uk`
[5] `http://reuters.co.uk`
[6] `http://www.guardian.co.uk`

Table 5. The distribution of dominant topic labels for stories in the *3sources* collection. The overall total number of stories per label is given, as well as the number of articles present within each individual view.

Label	Overall	BBC	Guardian	Reuters
business	122	87	78	94
entertainment	70	53	41	43
health	57	45	24	27
politics	61	48	40	23
sport	90	81	76	71
technology	67	38	43	36

Table 6. Performance of IMF on the *3sources* collection, compared with clusterings generated on individual views

Algorithm/View	NMI	Assigned
Weighted kernel k-means (*BBC*)	0.65	85%
Weighted kernel k-means (*The Guardian*)	0.52	73%
Weighted kernel k-means (*Reuters*)	0.55	71%
Integration by Matrix Factorization	0.71	100%

using weighted kernel k-means [15], setting the value of k to the number of labels. A value $k' = 7$ for the number of meta-clusters was automatically selected using the entropy criterion described in Section 3.3. It is interesting to note that the additional cluster reflects the fact the integration procedure identifies two distinct clusters for "business and finance" – one cluster largely pertaining to reports on the global economic downturn, the other cluster containing stories directly related to business and finance, but also containing stories from *sport* and *entertainment* that have a business or financial dimension. The presence of this latter group reflects the actual overlapping nature of the topics in the collection. However, as noted earlier we focus on the disjoint labels during our evaluation to allow comparison with algorithms producing disjoint clusters.

Table 6 shows a comparison of the performance of the proposed approach to the clusterings produced on documents from the individual news sources. IMF out-performs the three weighted kernel clusterings, and the resulting integrated clustering is considerably more informative than those generated on the Guardian and Reuters views. We observed that including these "weaker" sources of information does not significantly impact upon the effectiveness of the data integration process.

5.2 Ensemble Multi-view Integration

In the second evaluation performed on this collection, we consider the multi-view ensemble clustering problem described in Section 3.4. To generate the individual ensemble members, we use standard k-means with random initialization and cosine similarity. The number of base clusters is set to the number of labels, and

Table 7. Accuracy (NMI) of IMF on the *3sources* collection, compared with four well-known ensemble clustering algorithms. The NMI scores for the constituent base clusterings are also listed.

Algorithm	Mean	Min	Max
Base	0.37	0.02	0.66
IMF	**0.78**	0.70	0.80
CSPA	0.62	0.61	0.64
HGPA	0.47	0.36	0.60
MCLA	0.72	0.65	0.79
Voting	0.74	0.67	0.79

the parameter value $k' = 7$ determined in Section 5.1 is used for the number of meta-clusters. As well as examining the performance of IMF, for comparison we also applied several well-known ensemble clustering algorithms from the literature: the CSPA co-association clustering approach [6], the HGPA and MCLA hypergraph-based methods [6], and the correspondence clustering cumulative voting method described in [9]. In total we conducted 30 runs, each involving the generation and integration of 100 different base clusterings per view.

Table 7 summarizes the experimental results for the four approaches under consideration, in terms of average, minimum and maximum NMI scores achieved over the 30 runs. The range of NMI scores for the complete set of 9000 base clusterings is also given. On average 76% of the total number of news stories was assigned in each base clustering, reflecting the incomplete nature of the views. Even though the information provided by the base clustering was often of very poorly quality, we see that most of the integration algorithms performed reasonably well, with the exception of the HGPA procedure. On average the IMF approach was the most successful of the techniques under consideration, suggesting that it frequently availed of the information provided by the "weak" but diverse clusterings generated on the three views.

Effect of ensemble size. An important issue in ensemble clustering that is often neglected is the effect of ensemble size (*i.e.* the number of base clusterings in the ensemble) on clustering performance. For larger datasets, even with the availability of parallel computing resources, the number of clusterings that can reasonably be generated can often be strictly limited. As the size of the ensemble decreases, the ensemble multi-view clustering task approaches the one-shot integration task. However we still may face the problem of having access to only a set of weak or unrepresentative clusterings. Therefore it is desirable to employ an integration approach that will be effective when given a relatively small set of potentially weak base clusterings.

To examine this issue, we compare the behavior of the IMF algorithm with that of the four alternative approaches used above, as the number of ensemble members increases. Specifically we consider ensemble sizes $\in [5, 100]$ of k-means clusterings generated on the *3sources* collection, with an approximately equal number of clusterings per view. To account for variability in the results we repeat

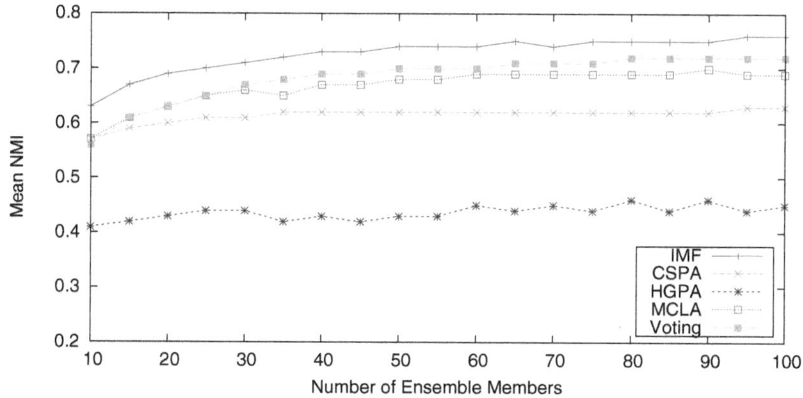

Fig. 3. Plot of mean clustering accuracy (NMI) scores for IMF compared to popular ensemble clustering algorithms, for ensembles of varying size generated on the *3sources* collection

the process over 30 trials with different sets of base clusterings. As before, a value of $k' = 7$ was used as the number of final clusters. The results of the comparison are shown in Figure 3. We observe that IMF shows superior clustering accuracy in comparison to the alternative integration algorithms, particularly for smaller ensemble sizes – an NMI score of at least 0.70 was generally obtained after 25 clusterings have been added.

6 Conclusion

In this paper we presented a simple but effective approach for performing unsupervised data integration on two or more connected views. Experiments on synthetic and real-world multi-view text datasets yielded encouraging results, both in tasks where a single representative clustering was available for each view, and in cases where a larger diverse collection of clusterings was available for integration. In the latter task the proposed IMF approach out-performed a number of popular ensemble clustering algorithms. An additional aspect of IMF is that the additive nature of the factorization process provides an aid in interpreting the output of the integration process.

Acknowledgments. This work is supported by Science Foundation Ireland Grant Nos. 05/IN.1/I24 and 08/SRC/I140 (Clique: Graph & Network Analysis Cluster).

References

1. Blum, A., Mitchell, T.: Combining labeled and unlabeled data with co-training. In: Proc. 11th Annual Conference on Computational learning theory, pp. 92–100 (1998)
2. Bickel, S., Scheffer, T.: Multi-view clustering. In: Proc. 4th IEEE International Conference on Data Mining, pp. 19–26 (2004)

3. Berthold, M., Patterson, D.: Towards learning in parallel universes. In: Proc. 2004 IEEE International Conference on Fuzzy Systems, vol. 1 (2004)
4. Pavlidis, P., Weston, J., Cai, J., Noble, W.: Learning Gene Functional Classifications from Multiple Data Types. Journal of Computational Biology 9(2), 401–411 (2002)
5. Lee, D.D., Seung, H.S.: Learning the parts of objects by non-negative matrix factorization. Nature 401, 788–791 (1999)
6. Strehl, A., Ghosh, J.: Cluster ensembles - a knowledge reuse framework for combining multiple partitions. JMLR 3, 583–617 (2002)
7. Jain, A.K., Fred, A.: Data clustering using evidence accumulation. In: Proc. 16th International Conference on Pattern Recognition., vol. 4, pp. 276–280 (2002)
8. Karypis, G., Kumar, V.: A fast and high quality multilevel scheme for partitioning irregular graphs. SIAM Journal on Scientific Computing 2096(1), 359–392 (1998)
9. Dimitriadou, E., Weingessel, A., Hornik, K.: A combination scheme for fuzzy clustering. International Journal of Pattern Recognition and Artificial Intelligence 16(7), 901–912 (2002)
10. de Sa, V.: Spectral clustering with two views. In: ICML Workshop on Learning With Multiple Views (2005)
11. Zeng, E., Yang, C., Li, T., Narasimhan, G.: On the Effectiveness of Constraints Sets in Clustering Genes. In: Proc. 7th IEEE International Conference on Bioinformatics and Bioengineering (BIBE 2007), pp. 79–86 (2007)
12. Greene, D., Cunningham, P.: Producing accurate interpretable clusters from high-dimensional data. In: Jorge, A.M., Torgo, L., Brazdil, P.B., Camacho, R., Gama, J. (eds.) PKDD 2005. LNCS (LNAI), vol. 3721, pp. 486–494. Springer, Heidelberg (2005)
13. Boutsidis, C., Gallopoulos, E.: SVD based initialization: A head start for non-negative matrix factorization. Pattern Recognition (2008)
14. Hubert, L., Arabie, P.: Comparing partitions. Journal of Classification, 193–218 (1985)
15. Dhillon, I.S., Guan, Y., Kulis, B.: Kernel k-means: spectral clustering and normalized cuts. In: Proc. 2004 ACM SIGKDD International Conference on Knowledge Discovery and Data Mining, pp. 551–556 (2004)

Transductive Classification via Dual Regularization

Quanquan Gu and Jie Zhou

State Key Laboratory on Intelligent Technology and Systems
Tsinghua National Laboratory for Information Science and Technology (TNList)
Department of Automation, Tsinghua University, Beijing 100084, China
gqq03@mails.tsinghua.edu.cn, jzhou@tsinghua.edu.cn

Abstract. Semi-supervised learning has witnessed increasing interest in the past decade. One common assumption behind semi-supervised learning is that the data labels should be sufficiently smooth with respect to the intrinsic data manifold. Recent research has shown that the features also lie on a manifold. Moreover, there is a duality between data points and features, that is, data points can be classified based on their distribution on features, while features can be classified based on their distribution on the data points. However, existing semi-supervised learning methods neglect these points. Based on the above observations, in this paper, we present a dual regularization, which consists of two graph regularizers and a co-clustering type regularizer. In detail, the two graph regularizers consider the geometric structure of the data points and the features respectively, while the co-clustering type regularizer takes into account the duality between data points and features. Furthermore, we propose a novel transductive classification framework based on dual regularization, which can be solved by alternating minimization algorithm and its convergence is theoretically guaranteed. Experiments on benchmark semi-supervised learning data sets demonstrate that the proposed methods outperform many state of the art transductive classification methods.

1 Introduction

In many practical machine learning problems, the acquisition of sufficient labeled data is often expensive and/or time consuming. On the contrary, in many cases, large number of unlabeled data are far easier to obtain. Consequently, semi-supervised learning [1] [2], which aims to learn from both labeled and unlabeled data points, has received significant attention in the past decade. In general, semi-supervised learning can be categorized into two classes: (1) Transductive learning [3] [4] [5] [6] [7]: to estimate the labels of the given unlabeled data; and (2) Inductive learning [8]: to induce a decision function which has a low error rate on the whole sample space. In our study, we focus on transductive classification.

Many transductive classification methods have been proposed up to now [1] [2], among which graph based method [3] [4] [5] [6] [7] is one of the most popular

W. Buntine et al. (Eds.): ECML PKDD 2009, Part I, LNAI 5781, pp. 439–454, 2009.

approaches. One common assumption behind graph based transductive classification is that the data labels should be sufficiently smooth with respect to the intrinsic data manifold, i.e. *Cluster Assumption* [1] [3] [4]. This assumption can be achieved by graph regularization [9]. In detail, it models the whole data set as an undirected weighted graph, whose vertices correspond to the data points, and edges reflect the affinity between pairwise data points. Some of the vertices on the graph are labeled, while the remainder are unlabeled, and the goal of graph based transductive classification is to predict the labels of those unlabeled data points such that the predicted labels are sufficiently smooth with respect to the data graph. Other assumptions include *Local Learning Assumption* [6] [10], which says the label of each point can be predicted by the points in its neighborhood, and *local linear embedding assumption* [11] [5], which says if a data point can be reconstructed from its neighbors, then its label can be reconstructed from the labels of its neighbors by the same reconstruction coefficients.

The motivation of our work is twofold. First, recent research has shown that not only the data points are sampled from some low dimensional manifold embedded in the high dimensional ambient space [11] [12], namely data manifold, but also the features lie on a manifold [13] [14], namely feature manifold. Second, there is a duality between data points and features, i.e. data points can be classified based on their distribution on features while features can be classified based on their distribution on the data points. This is originally proposed in co-clustering literature [15] [16] [17], which suggests that clustering of features of a data matrix can lead to the improvement of data clustering. To demonstrate the usefulness of the duality between data points and features, we give an illustrative example in Fig. 1. As far as we know, existing semi-supervised learning methods fail to consider these points mentioned above together, which may further improve the performance of semi-supervised learning.

In this paper, we present a dual regularization. It consists of two graph regularizers and a co-clustering type regularizer. The graph regularizers explore the geometric structure of the data points and features respectively, while the co-clustering type regularizer utilizes the duality between the data points and features. Furthermore, we propose a novel framework for transductive classification based on dual regularization, which can be solved by alternating minimization algorithm and its convergence is theoretically guaranteed. Encouraging experimental results on benchmark semi-supervised learning data sets illustrate that the proposed methods outperform many existing transductive classification algorithms.

It is worth noting that in [18] [19], the authors proposed a co-clustering type regularization for transductive learning. However, in their method, besides partial supervision in the data points, partial supervision in the form of feature labels is also assumed, which is hardly obtained in many applications. Similar idea has also been applied for clustering [20]. Our method is different from theirs, since we do not require the supervision on the feature side. Furthermore, graph regularizers are adopted in our dual regularization, which considers the geometric structure of data points and features.

D1: NMF for webpage clustering D2: LSI for text classification

D3: PCA for face classification D4: Kmeans for image clustering

(a) A synthetic data set

	clustering	classification	face	image	text	webpage
D1	1	0	0	0	0	1
D2	0	1	0	0	1	0
D3	0	1	1	0	0	0
D4	1	0	0	1	0	0

(b) The document word matrix

Fig. 1. An illustrative example. Suppose we have 4 documents, i.e. D1, D2, D3 and D4. The true label of D1 and D2 is "Data mining", while the true label of D3 and D4 is "Computer vision". And suppose there are 6 words in the vocabulary, i.e. "clustering, classification, face, image, text, webpage". We assume D1 and D3 are labeled, while D2 and D4 are unlabeled and need to be predicted. If we only rely on the knowledge of the document side, then D2 is closer to D3 than D1, and D4 is closer to D1 than D3. So we will wrongly classify D2 as "Computer vision" and D4 as "data mining". However, if we has the clusters information of the words, we may obtain 3 clusters, i.e. "clustering, classification", "face, image" and "text, webpage", based on which D2 is closer to D1 than D3 while D4 is closer to D3 than D1. As a result, we will correctly classify D2 as "Data mining". Similar analysis can be conducted on the word side.

The remainder of this paper is organized as follows. In Section 2, we will briefly review graph-based transductive learning. In Section 3, we first present dual regularization, followed which we propose a novel transductive classification method based on dual regularization. The experiments on benchmark semi-supervised learning data sets are demonstrated in Section 4. Finally, we draw conclusions and point out the future work in Section 5.

2 A Brief Review of Graph Based Transductive Classification

Before we go any further, let's first briefly review the general framework of graph based transductive classification [3] [4] [5] [6] [7], since it is the foundation of this paper.

In the setting of transductive classification, we are given a data set $\mathcal{X} = \{\mathbf{x}_1, \ldots, \mathbf{x}_l, \mathbf{x}_{l+1}, \ldots, \mathbf{x}_n\} \subset \mathbb{R}^d$, and a label set $\mathcal{L} = \{1, 2, \ldots, c\}$, the first l points $\mathbf{x}_i, 1 \leq i \leq l$ are labeled as $y_i \in \mathcal{L}$ and the remaining points $\mathbf{x}_u, l+1 \leq u \leq n$ are unlabeled. Each $\mathbf{x}_i$ is drawn from a fixed but usually unknown distribution $p(\mathbf{x})$. Typical graph based transductive classification seeks c optimal classification functions $f^j, 1 \leq j \leq c$, by minimizing the following criterion

$$J = \sum_{j=1}^{c} \sum_{i=1}^{l} L(y_i, f^j(\mathbf{x}_i)) + \lambda \sum_{j=1}^{c} ||f^j||_I^2, \tag{1}$$

where $L(,)$ is some loss function, e.g. hinge loss or square loss, $||f^j||_I$ measures the smoothness of f^j with respect to the intrinsic data manifold, $\lambda > 0$ is the regularization parameter, which controls the balance between the loss and label smoothness.

Specifically, if we choose $L(,)$ as square loss, and $||f^j||_I$ as graph regularization, then Eq.(1) can be formulated as follows

$$\min_{\mathbf{F}} \operatorname{tr}((\mathbf{F} - \mathbf{Y})^T \mathbf{C}(\mathbf{F} - \mathbf{Y})) + \lambda \operatorname{tr}(\mathbf{F}^T \mathbf{L} \mathbf{F}), \tag{2}$$

where $\mathbf{F} = (\mathbf{f}^1, \mathbf{f}^2, \ldots, \mathbf{f}^c) \in \mathbb{R}^{n \times c}$ with $\mathbf{f}^j = [f^j(\mathbf{x}_1), f^j(\mathbf{x}_2), \ldots, f^j(\mathbf{x}_n)]^T$ is the class assignment matrix, $\mathbf{Y} \in \mathbb{R}^{n \times c}$ is the label matrix with $Y_{ij} = 1$ if $\mathbf{x}_i$ is labeled as $y_i = j$ and $Y_{ij} = 0$ otherwise, $\mathbf{L} \in \mathbb{R}^{n \times n}$ is called graph Laplacian [21], and $\mathbf{C} \in \mathbb{R}^{n \times n}$ is a diagonal matrix, with its ith diagonal element $C_{ii} = C_l > 0$ for $1 \leq i \leq l$, and $C_{ii} = C_u \geq 0$ for $l + 1 \leq i \leq n$, where C_l and C_u are two parameters.

It is easy to show that the solution of Eq.(2) is

$$\mathbf{F} = (\mathbf{C} + \lambda \mathbf{L})^{-1} \mathbf{C} \mathbf{Y} \tag{3}$$

And the predicted label of $\mathbf{x}_i, l + 1 \leq i \leq n$ is determined by

$$y_i = \arg \max_{1 \leq j \leq c} \mathbf{F}_{ij}, l + 1 \leq i \leq n \tag{4}$$

Most of the existing graph based transductive classification methods can be unified in Eq(2), and only differ in the setting of graph Laplacian $\mathbf{L}$ and/or the diagonal matrix $\mathbf{C}$. For example, [3] chose $\mathbf{L}$ as *combinational graph Laplacian* and $C_l = \infty, C_u = 0$. [4] set $\mathbf{L}$ as *normalized graph Laplacian* and $C_l = C_u = 1$. Both of the graph Laplacians mentioned above correspond to *Cluster Assumption* [1], while [6] selected $\mathbf{L}$ as *local learning graph Laplacian* which is based on *Local Learning Assumption*, and $C_l = 1, C_u = 0$. And [5] selected $\mathbf{L}$ as *local linear embedding graph Laplacian* and $C_l = C_u = 1$.

For convenience, we present in Table 1 the notation used in the rest of this paper.

Table 1. Notation used in this paper

Notation	Description	Notation	Description
n	number of data points	$\mathbf{F}$	class assignment matrix of size $n \times c$
d	number of features	$\mathbf{f}_{i\cdot}$	ith row of $\mathbf{F}$
c	number of data classes	$\mathbf{f}_{\cdot i}$	ith column of $\mathbf{F}$
m	number offeature clusters	$\mathbf{G}$	feature partition matrix of size $d \times m$
$\mathcal{X}$	data set	$\mathbf{g}_{i\cdot}$	ith row of $\mathbf{G}$
$\mathbf{X}$	data matrix of size $d \times n$	$\mathbf{g}_{\cdot i}$	ith column of $\mathbf{G}$
$\mathbf{x}_{i\cdot}$	ith row of $\mathbf{X}$	$\mathbf{L}_F$	data graph Laplacian of size $n \times n$
$\mathbf{x}_{\cdot i}$	ith column of $\mathbf{X}$	$\mathbf{L}_G$	feature graph Laplacian of size $d \times d$

3 The Proposed Method

In this section, we will first present dual regularization. Then we will propose a transductive classification framework based on dual regularization, followed with the optimization algorithm as well as the proof of its convergence.

3.1 Dual Regularization

As we have mentioned above, we aim to explore the geometric structure on both the data point side and the feature side. To achieve this objective, we turn to graph regularization [9]. In detail, we construct two graphs i.e. data graph and feature graph, to explore the geometric structure of data manifold and feature manifold. In the following, we will introduce the construction of data graph and feature graph respectively. We will adopt *Cluster Assumption* [1] [3] [4] as a running example.

Data Graph. We construct a data graph $\mathcal{G}_F$ whose vertices correspond to $\{\mathbf{x}_{\cdot 1}, \ldots, \mathbf{x}_{\cdot n}\}$. According to *Cluster Assumption*, if data points $\mathbf{x}_{\cdot i}$ and $\mathbf{x}_{\cdot j}$ are close to each other, then their class labels $\mathbf{f}_{i\cdot}$ and $\mathbf{f}_{j\cdot}$ should be close as well. This is formulated as follows,

$$\frac{1}{2}\sum_{ij}||\frac{\mathbf{f}_{i\cdot}}{\sqrt{D^F_{ii}}} - \frac{\mathbf{f}_{j\cdot}}{\sqrt{D^F_{jj}}}||^2 W^F_{ij} \tag{5}$$

where W^F_{ij} is the affinity matrix of data graph measuring how close $\mathbf{f}_{i\cdot}$ and $\mathbf{f}_{j\cdot}$ will be, $D^F_{ii} = \sum_j W^F_{ij}$ is the diagonal degree matrix of data graph.

We define the data affinity matrix $\mathbf{W}^F$ as follows,

$$W^F_{ij} = \begin{cases} \exp\frac{-d(\mathbf{x}_{\cdot i},\mathbf{x}_{\cdot j})^2}{\sigma^2}, & \text{if } \mathbf{x}_{\cdot j} \in \mathcal{N}(\mathbf{x}_{\cdot i}) \text{ or } \mathbf{x}_{\cdot i} \in \mathcal{N}(\mathbf{x}_{\cdot j}) \\ 0, & \text{otherwise.} \end{cases} \tag{6}$$

where $\mathcal{N}(\mathbf{x}_{\cdot i})$ denotes the k-nearest neighbor of $\mathbf{x}_{\cdot i}$. $\exp\frac{-d(\mathbf{x}_{\cdot i},\mathbf{x}_{\cdot j})^2}{\sigma^2}$ is Gaussian similarity where $\sigma > 0$ is the width. $d(\mathbf{x}_{\cdot i}, \mathbf{x}_{\cdot j})$ denotes the distance between $\mathbf{x}_{\cdot i}$ and $\mathbf{x}_{\cdot j}$.

Eq.(5) can be further rewritten as

$$\begin{aligned} &\frac{1}{2}\sum_{i,j}||\frac{\mathbf{f}_{i\cdot}}{\sqrt{D^F_{ii}}} - \frac{\mathbf{f}_{j\cdot}}{\sqrt{D^F_{jj}}}||^2 W^F_{ij} \\ &= \mathrm{tr}(\mathbf{F}^T(\mathbf{I} - (\mathbf{D}^F)^{-\frac{1}{2}}\mathbf{W}^F(\mathbf{D}^F)^{-\frac{1}{2}})\mathbf{F}) \\ &= \mathrm{tr}(\mathbf{F}^T\mathbf{L}_F\mathbf{F}) \end{aligned} \tag{7}$$

where $\mathbf{F} \in \mathbb{R}^{n\times c}$ is the class assignment matrix of data points, and $\mathbf{L}_F = \mathbf{I} - (\mathbf{D}^F)^{-\frac{1}{2}}\mathbf{W}^F(\mathbf{D}^F)^{-\frac{1}{2}}$ is called *normalized graph Laplacian* (NLap) of the data graph $\mathcal{G}_F$. Eq.(7) reflects the label smoothness of the data points. The smoother the data labels are with respect to the underlying data manifold, the smaller the value of Eq.(7) will be.

Feature Graph. Similar with the construction of the data graph $\mathcal{G}_F$, we construct a feature graph $\mathcal{G}_G$ whose vertices correspond to $\{\mathbf{x}_{1\cdot}, \ldots, \mathbf{x}_{d\cdot}\}$. According to *Cluster Assumption* again, if features $\mathbf{x}_{i\cdot}$ and $\mathbf{x}_{j\cdot}$ are near, then their cluster labels $\mathbf{g}_{i\cdot}$ and $\mathbf{g}_{j\cdot}$ should be near as well. This is formulated as follows

$$\frac{1}{2}\sum_{ij}||\frac{\mathbf{g}_{i\cdot}}{\sqrt{D_{ii}^G}} - \frac{\mathbf{g}_{j\cdot}}{\sqrt{D_{jj}^G}}||^2 W_{ij}^G \tag{8}$$

where W_{ij}^G is the affinity matrix of feature graph measuring how close $\mathbf{g}_{i\cdot}$ and $\mathbf{g}_{j\cdot}$ will be, $D_{ii}^G = \sum_j W_{ij}^G$ is the diagonal degree matrix of feature graph.

Again we define the feature affinity matrix $\mathbf{W}^G$ as follows,

$$W_{ij}^G = \begin{cases} \exp \frac{-d(\mathbf{x}_{i\cdot},\mathbf{x}_{j\cdot})^2}{\sigma^2}, & \text{if } \mathbf{x}_{j\cdot} \in \mathcal{N}(\mathbf{x}_{i\cdot}) \text{ or } \mathbf{x}_{i\cdot} \in \mathcal{N}(\mathbf{x}_{j\cdot}) \\ 0, & \text{otherwise.} \end{cases} \tag{9}$$

where $\mathcal{N}(\mathbf{x}_{i\cdot})$ denotes the k-nearest neighbor of $\mathbf{x}_{i\cdot}$.

Eq.(8) can be further rewritten as

$$\begin{aligned} &\frac{1}{2}\sum_{i,j}||\frac{\mathbf{g}_{i\cdot}}{\sqrt{D_{ii}^G}} - \frac{\mathbf{g}_{j\cdot}}{\sqrt{D_{jj}^G}}||^2 W_{ij}^G \\ &= \mathrm{tr}(\mathbf{G}^T(\mathbf{I} - (\mathbf{D}^G)^{-\frac{1}{2}}\mathbf{W}^G(\mathbf{D}^G)^{-\frac{1}{2}})\mathbf{G}) \\ &= \mathrm{tr}(\mathbf{G}^T\mathbf{L}_G\mathbf{G}) \end{aligned} \tag{10}$$

where $\mathbf{G} \in \mathbb{R}^{d\times m}$ is the partition matrix of features, $\mathbf{L}_G = \mathbf{I} - (\mathbf{D}^G)^{-\frac{1}{2}}\mathbf{W}^G(\mathbf{D}^G)^{-\frac{1}{2}}$ is the *normalized graph Laplacian* (NLap) of the feature graph $\mathcal{G}_G$. Eq.(10) reflects the label smoothness of the features. The smoother the feature labels are with respect to the underlying feature manifold, the smaller the value of Eq.(10) will be.

Based on the two graph regularizers introduced above, we present a regularization as follows

$$\lambda\mathrm{tr}(\mathbf{F}^T\mathbf{L}_F\mathbf{F}) + \mu\mathrm{tr}(\mathbf{G}^T\mathbf{L}_G\mathbf{G}) + \eta||\mathbf{X} - \mathbf{G}\mathbf{S}\mathbf{F}^T||_F^2 \tag{11}$$

where $||\cdot||_F$ is the Frobenius norm, $\lambda, \mu, \eta \geq 0$ are regularization parameters. It consists of three terms. The first term is graph regularizer defined in Eq.(7), which is also the same as the second term in Eq.(2). The second term is graph regularizer defined in Eq.(10). The third term is the most important one. It is a co-clustering [17] type regularizer. This term reflects the approximation error of matrix tri-factorization for co-clustering. The smaller it is, the better the approximation will be. It establishes a bridge between the data points in the first term and the features in the second term, through which the label information of data points and features can be transferred from one to another, which may benefit the classification of data points. Eq.(11) is called *Dual Regularization.*

By now, we have presented dual regularization. It is worth noting that although in the derivation, we adopt *Cluter Assumption*, other kinds of assumptions, e.g. *Local Learning Assumption* can also be used. In other words, besides the *normalized graph Laplacian* (NLap), other kinds of graph Laplacians can also be utilized in Eq.(11), e.g. *local learning graph Laplacian* (LLL). For the detail of LLL, please refer to [6] [10].

3.2 Transductive Classification via Dual Regularization

Based on the dual regularization in Eq.(11) presented above, we propose a novel transductive classification framework as follows,

$$\begin{aligned} J_{TCDR} &= \mathrm{tr}((\mathbf{F}-\mathbf{Y})^T\mathbf{C}(\mathbf{F}-\mathbf{Y})) \\ &+ \lambda \mathrm{tr}(\mathbf{F}^T\mathbf{L}_F\mathbf{F}) + \mu \mathrm{tr}(\mathbf{G}^T\mathbf{L}_G\mathbf{G}) + \eta||\mathbf{X}-\mathbf{GSF}^T||_F^2, \end{aligned} \tag{12}$$

where $||\cdot||_F$ is the Frobenius norm, $\mathbf{F} \in \mathbb{R}^{n\times c}$ is the class assignment matrix of the data points, $\mathbf{G} \in \mathbb{R}^{d\times m}$ is the partition matrix of the features, $\mathbf{L}_F \in \mathbb{R}^{n\times n}$ is graph Laplacian for data points, and $\mathbf{L}_G \in \mathbb{R}^{d\times d}$ is graph Laplacian for features. $\lambda, \mu, \eta \geq 0$ are regularization parameters. We call Eq.(12) *Transductive Classification via Dual Regularization* (TCDR). TCDR provides a unified framework for transductive classification. Different settings of the graph Laplacians $\mathbf{L}_F$ and $\mathbf{L}_G$, along with the diagonal matrix $\mathbf{C}$ lead to various instantiations of TCDR. When letting $\mu = \eta = 0$ in Eq.(12), TCDR degenerates to traditional graph based transductive classification framework in Eq.(2). To this end, existing graph based transductive classification methods can be seen as the special case of TCDR.

By its definition, the elements in $\mathbf{F}$ and $\mathbf{G}$ can only take binary values, which makes the minimization in Eq.(12) very difficult, therefore we relax $\mathbf{F}$ and $\mathbf{G}$ into continuous nonnegative domain. Then the objective of TCDR in Eq.(12) turns out to be,

$$\begin{aligned} J_{TCDR} &= \mathrm{tr}((\mathbf{F}-\mathbf{Y})^T\mathbf{C}(\mathbf{F}-\mathbf{Y})) \\ &+ \lambda \mathrm{tr}(\mathbf{F}^T\mathbf{L}_F\mathbf{F}) + \mu \mathrm{tr}(\mathbf{G}^T\mathbf{L}_G\mathbf{G}) + \eta||\mathbf{X}-\mathbf{GSF}^T||_F^2, \\ &\text{s.t. } \mathbf{F} \geq 0, \mathbf{G} \geq 0, \end{aligned} \tag{13}$$

To make the objective in Eq.(13) lower bounded, we use L_2 normalization on columns of $\mathbf{F}$ and $\mathbf{G}$ in the optimization, and compensate the norms of $\mathbf{F}$ and $\mathbf{G}$ to $\mathbf{S}$.

3.3 Optimization

As we see, minimizing Eq.(13) is with respect to $\mathbf{F}, \mathbf{G}$ and $\mathbf{S}$. And we cannot give a closed-form solution. In the following, we will present an alternating scheme to optimize the objective. In other words, we will optimize the objective with respect to one variable when fixing the other variables. This procedure repeats until convergence.

Computation of S. Optimizing Eq.(13) with respect to $\mathbf{S}$ is equivalent to optimizing

$$J_1 = ||\mathbf{X} - \mathbf{GSF}^T||_F^2 \tag{14}$$

Setting $\frac{\partial J_1}{\partial \mathbf{S}} = 0$ leads to the following updating formula

$$\mathbf{S} = (\mathbf{G}^T\mathbf{G})^{-1}\mathbf{G}^T\mathbf{XF}(\mathbf{F}^T\mathbf{F})^{-1} \tag{15}$$

Computation of F. Optimizing Eq.(13) with respect to $\mathbf{F}$ is equivalent to optimizing

$$\begin{aligned} J_2 &= \mathrm{tr}((\mathbf{F}-\mathbf{Y})^T\mathbf{C}(\mathbf{F}-\mathbf{Y})) + \lambda \mathrm{tr}(\mathbf{F}^T\mathbf{L}_F\mathbf{F}) + \eta||\mathbf{X} - \mathbf{GSF}^T||_F^2, \\ &\text{s.t. } \mathbf{F} \geq 0, \end{aligned} \tag{16}$$

For the constraint $\mathbf{F} \geq 0$, we cannot get a closed-form solution of $\mathbf{F}$. In the following, we will present an iterative multiplicative updating solution. We introduce the Lagrangian multiplier $\boldsymbol{\alpha} \in \mathbb{R}^{n\times c}$, thus the Lagrangian function is

$$\begin{aligned} L(\mathbf{F}) = &\mathrm{tr}((\mathbf{F}-\mathbf{Y})^T\mathbf{C}(\mathbf{F}-\mathbf{Y})) \\ &+ \lambda \mathrm{tr}(\mathbf{F}^T\mathbf{L}_F\mathbf{F}) + \eta||\mathbf{X} - \mathbf{GSF}^T||_F^2 - \mathrm{tr}(\boldsymbol{\alpha}\mathbf{F}^T) \end{aligned} \tag{17}$$

Setting $\frac{\partial L(\mathbf{F})}{\partial \mathbf{F}} = 0$, we obtain

$$\alpha = 2\mathbf{CF} - 2\mathbf{CY} + 2\lambda\mathbf{R}_F\mathbf{F} - 2\eta\mathbf{A} + 2\eta\mathbf{FB} \tag{18}$$

where $\mathbf{A} = \mathbf{X}^T\mathbf{GS}$ and $\mathbf{B} = \mathbf{S}^T\mathbf{G}^T\mathbf{GS}$.

Using the Karush-Kuhn-Tucker condition [22] $\boldsymbol{\alpha}_{ij}\mathbf{F}_{ij} = 0$, we get

$$[\mathbf{CF} - \mathbf{CY} + \lambda\mathbf{L}_F\mathbf{F} - \eta\mathbf{A} + \eta\mathbf{FB}]_{ij}\mathbf{F}_{ij} = 0 \tag{19}$$

Introduce $\mathbf{L}_F = \mathbf{L}_F^+ - \mathbf{L}_F^-$, $\mathbf{A} = \mathbf{A}^+ - \mathbf{A}^-$ and $\mathbf{B} = \mathbf{B}^+ - \mathbf{B}^-$ where $\mathbf{A}_{ij}^+ = (|\mathbf{A}_{ij}| + \mathbf{A}_{ij})/2$ and $\mathbf{A}_{ij}^- = (|\mathbf{A}_{ij}| - \mathbf{A}_{ij})/2$ [23], we obtain

$$[\mathbf{CF} - \mathbf{CY} + \lambda\mathbf{L}_F^+\mathbf{F} - \lambda\mathbf{L}_F^-\mathbf{F} - \eta\mathbf{A}^+ + \eta\mathbf{A}^- + \eta\mathbf{FB}^+ - \eta\mathbf{FB}^-]_{ij}\mathbf{F}_{ij} = 0 \tag{20}$$

Eq.(20) leads to the following updating formula

$$\mathbf{F}_{ij} \leftarrow \mathbf{F}_{ij}\sqrt{\frac{[\mathbf{CY} + \lambda\mathbf{L}_F^-\mathbf{F} + \eta\mathbf{A}^+ + \eta\mathbf{FB}^-]_{ij}}{[\mathbf{CF} + \lambda\mathbf{L}_F^+\mathbf{F} + \eta\mathbf{A}^- + \eta\mathbf{FB}^+]_{ij}}} \tag{21}$$

Computation of G. Optimizing Eq.(13) with respect to $\mathbf{G}$ is equivalent to optimizing

$$\begin{aligned} J_3 &= \mu\mathrm{tr}(\mathbf{G}^T\mathbf{L}_G\mathbf{G}) + \eta||\mathbf{X} - \mathbf{GSF}^T||_F^2 \\ &\text{s.t. } \mathbf{G} \geq 0, \end{aligned} \tag{22}$$

Since $\mathbf{G} \geq 0$, we introduce the Lagrangian multiplier $\boldsymbol{\beta} \in \mathbb{R}^{d\times m}$, thus the Lagrangian function is

$$L(\mathbf{G}) = \mu \mathrm{tr}(\mathbf{G}^T \mathbf{L}_G \mathbf{G}) + \eta ||\mathbf{X} - \mathbf{G}\mathbf{S}\mathbf{F}^T||_F^2 - \mathrm{tr}(\boldsymbol{\beta}\mathbf{G}^T) \tag{23}$$

Setting $\frac{\partial L(\mathbf{G})}{\partial \mathbf{G}} = 0$, we obtain

$$\boldsymbol{\beta} = 2\mu \mathbf{L}_G \mathbf{G} - 2\eta \mathbf{P} + 2\eta \mathbf{G}\mathbf{Q} \tag{24}$$

where $\mathbf{P} = \mathbf{X}\mathbf{F}\mathbf{S}^T$ and $\mathbf{Q} = \mathbf{S}\mathbf{F}^T\mathbf{F}\mathbf{S}^T$.

Using the Karush-Kuhn-Tucker complementarity condition [22] $\boldsymbol{\beta}_{ij}\mathbf{G}_{ij} = 0$, we get

$$[\mu \mathbf{L}_G \mathbf{G} - \eta \mathbf{P} + \eta \mathbf{G}\mathbf{Q}]_{ij} \mathbf{G}_{ij} = 0. \tag{25}$$

Introduce $\mathbf{L}_G = \mathbf{L}_G^+ - \mathbf{L}_G^-$, $\mathbf{P} = \mathbf{P}^+ - \mathbf{P}^-$ and $\mathbf{Q} = \mathbf{Q}^+ - \mathbf{Q}^-$, we obtain

$$[\mu \mathbf{L}_G^+ \mathbf{G} - \mu \mathbf{L}_G^- \mathbf{G} - \eta \mathbf{P}^+ + \eta \mathbf{P}^- + \eta \mathbf{G}\mathbf{Q}^+ - \eta \mathbf{G}\mathbf{Q}^-]_{ij} \mathbf{G}_{ij} = 0. \tag{26}$$

Eq.(26) leads to the following updating formula

$$\mathbf{G}_{ij} \leftarrow \mathbf{G}_{ij} \sqrt{\frac{[\mu \mathbf{L}_G^- \mathbf{G} + \eta \mathbf{P}^+ + \eta \mathbf{G}\mathbf{Q}^-]_{ij}}{[\mu \mathbf{L}_G^+ \mathbf{G} + \eta \mathbf{P}^- + \eta \mathbf{G}\mathbf{Q}^+]_{ij}}} \tag{27}$$

3.4 Convergence Analysis

In this section, we will investigate the convergence of the updating formula in Eq.(21) and Eq.(27). We use the auxiliary function approach [24] to prove the convergence of the algorithm. Here we first introduce the definition of auxiliary function [24].

Definition 1. *[24] $Z(h, h')$ is an auxiliary function for $F(h)$ if the conditions*

$$Z(h, h') \geq F(h), Z(h, h) = F(h),$$

are satisfied.

Lemma 1. *[24] If Z is an auxiliary function for F, then F is non-increasing under the update*

$$h^{(t+1)} = \arg\min_h Z(h, h^{(t)})$$

Proof. $F(h^{(t+1)}) \leq Z(h^{(t+1)}, h^{(t)}) \leq Z(h^{(t)}, h^{(t)}) = F(h^{(t)})$

Lemma 2. *[23] For any nonnegative matrices $\mathbf{A} \in \mathbb{R}^{n\times n}$, $\mathbf{B} \in \mathbb{R}^{k\times k}$, $\mathbf{S} \in \mathbb{R}^{n\times k}$, $\mathbf{S}' \in \mathbb{R}^{n\times k}$, and $\mathbf{A}$, $\mathbf{B}$ are symmetric, then the following inequality holds*

$$\sum_{i=1}^{n}\sum_{p=1}^{k} \frac{(\mathbf{A}\mathbf{S}'\mathbf{B})_{ip}\mathbf{S}_{ip}^2}{\mathbf{S}'_{ip}} \geq tr(\mathbf{S}^T\mathbf{A}\mathbf{S}\mathbf{B})$$

Theorem 1. *Let*

$$J(\mathbf{F}) = tr(\mathbf{F}^T\mathbf{C}\mathbf{F} - 2\mathbf{F}^T\mathbf{C}\mathbf{Y} + \lambda\mathbf{F}^T\mathbf{L}_F\mathbf{F} - 2\eta\mathbf{A}\mathbf{F}^T + \mathbf{F}\mathbf{B}\mathbf{F}^T) \tag{28}$$

Then the following function

$$\begin{aligned} Z(\mathbf{F},\mathbf{F}') = & \sum_{ij}\frac{(\mathbf{C}\mathbf{F}')_{ij}\mathbf{F}_{ij}^2}{\mathbf{F}'_{ij}} - 2\sum_{ij}(\mathbf{C}\mathbf{Y})_{ij}\mathbf{F}'_{ij}(1+\log\frac{\mathbf{F}_{ij}}{\mathbf{F}'_{ij}}) \\ & + \lambda\sum_{ij}\frac{(\mathbf{L}_F^+\mathbf{F}')_{ij}\mathbf{F}_{ij}^2}{\mathbf{F}'_{ij}} - \lambda\sum_{ijk}(\mathbf{L}_F^-)_{jk}\mathbf{F}'_{ji}\mathbf{F}'_{ki}(1+\log\frac{\mathbf{F}_{ij}\mathbf{F}_{ik}}{\mathbf{F}'_{ij}\mathbf{F}'_{ik}}) \\ & - 2\eta\sum_{ij}\mathbf{A}_{ij}^+\mathbf{F}'_{ij}(1+\log\frac{\mathbf{F}_{ij}}{\mathbf{F}'_{ij}}) + 2\eta\sum_{ij}\mathbf{A}_{ij}^-\frac{\mathbf{F}_{ij}^2+\mathbf{F}'^2_{ij}}{2\mathbf{F}'_{ij}} \\ & + \sum_{ij}\frac{(\mathbf{F}'\mathbf{B}^+)_{ij}\mathbf{F}_{ij}^2}{\mathbf{F}'_{ij}} - \sum_{ijk}\mathbf{B}_{jk}^-\mathbf{F}'_{ij}\mathbf{F}'_{ik}(1+\log\frac{\mathbf{F}_{ij}\mathbf{F}_{ik}}{\mathbf{F}'_{ij}\mathbf{F}'_{ik}}) \end{aligned}$$

is an auxiliary function for $J(\mathbf{F})$*. Furthermore, it is a convex function in* $\mathbf{F}$ *and its global minimum is*

$$\mathbf{F}_{ij} = \mathbf{F}_{ij}\sqrt{\frac{[\mathbf{C}\mathbf{Y} + \lambda\mathbf{L}_F^-\mathbf{F} + \eta\mathbf{A}^+ + \eta\mathbf{F}\mathbf{B}^-]_{ij}}{[\mathbf{C}\mathbf{F} + \lambda\mathbf{L}_F^+\mathbf{F} + \eta\mathbf{A}^- + \eta\mathbf{F}\mathbf{B}^+]_{ij}}} \tag{29}$$

Proof. See Appendix.

Theorem 2. *Updating* $\mathbf{F}$ *using Eq.(21) will monotonically decrease the value of the objective in Eq.(13), hence it converges.*

Proof. By Lemma 1 and Theorem 1, we can get that $J(\mathbf{F}^0) = Z(\mathbf{F}^0,\mathbf{F}^0) \geq Z(\mathbf{F}^1,\mathbf{F}^0) \geq J(\mathbf{F}^1) \geq \ldots$ So $J(\mathbf{F})$ is monotonically decreasing. Since $J(\mathbf{F})$ is obviously bounded below, we prove this theorem.

Theorem 3. *Let*

$$J(\mathbf{G}) = tr(\mu\mathbf{G}^T\mathbf{L}_G\mathbf{G} - 2\eta\mathbf{G}^T\mathbf{P} + \mu\mathbf{G}\mathbf{Q}\mathbf{G}^T) \tag{30}$$

Then the following function

$$\begin{aligned} Z(\mathbf{G},\mathbf{G}') = & \sum_{ij}\frac{(\mathbf{L}_G^+\mathbf{G}')_{ij}\mathbf{G}_{ij}^2}{\mathbf{G}'_{ij}} - \sum_{ijk}(\mathbf{L}_G^-)_{jk}\mathbf{G}'_{ji}\mathbf{G}'_{ki}(1+\log\frac{\mathbf{G}_{ij}\mathbf{G}_{ik}}{\mathbf{G}'_{ij}\mathbf{G}'_{ik}}) \\ & - 2\eta\sum_{ij}\mathbf{P}_{ij}^+\mathbf{G}'_{ij}(1+\log\frac{\mathbf{G}_{ij}}{\mathbf{G}'_{ij}}) + 2\eta\sum_{ij}\mathbf{P}_{ij}^-\frac{\mathbf{G}_{ij}^2+\mathbf{G}'^2_{ij}}{2\mathbf{G}'_{ij}} \\ & + \mu\sum_{ij}\frac{(\mathbf{G}'\mathbf{Q}^+)_{ij}\mathbf{G}_{ij}^2}{\mathbf{G}'_{ij}} - \mu\sum_{ijk}\mathbf{Q}_{jk}^-\mathbf{G}'_{ij}\mathbf{G}'_{ik}(1+\log\frac{\mathbf{G}_{ij}\mathbf{G}_{ik}}{\mathbf{G}'_{ij}\mathbf{G}'_{ik}}) \end{aligned}$$

is an auxiliary function for $J(\mathbf{G})$. *Furthermore, it is a convex function in* $\mathbf{G}$ *and its global minimum is*

$$\mathbf{G}_{ij} = \mathbf{G}_{ij}\sqrt{\frac{[\mu\mathbf{L}_G^{-}\mathbf{G} + \eta\mathbf{P}^{+} + \eta\mathbf{G}\mathbf{Q}^{-}]_{ij}}{[\mu\mathbf{L}_G^{+}\mathbf{G} + \eta\mathbf{P}^{-} + \eta\mathbf{G}\mathbf{Q}^{+}]_{ij}}} \tag{31}$$

Proof. For the limit of space, we omit it here.

Theorem 4. *Updating* $\mathbf{G}$ *using Eq.(27) will monotonically decrease the value of the objective in Eq.(13), hence it converges.*

Proof. By Lemma 1 and Theorem 3, we can get that $J(\mathbf{G}^0) = Z(\mathbf{G}^0, \mathbf{G}^0) \geq Z(\mathbf{G}^1, \mathbf{G}^0) \geq J(\mathbf{G}^1) \geq \ldots$ So $J(\mathbf{G})$ is monotonically decreasing. Since $J(\mathbf{G})$ is obviously bounded below, we prove this theorem.

4 Experiments

In this section, we evaluate the proposed methods on many benchmark semi-supervised learning data sets. Two instantiations of TCDR are evaluated: (1)*normalized graph Laplacian* (NLap) + TCDR, which chooses $\mathbf{L}_F$ and $\mathbf{L}_G$ as NLap. Note that it is just the method derived as a running example in Section 3; and (2) *local learning graph Laplacian* (LLL) + TCDR, which chooses $\mathbf{L}_F$ and $\mathbf{L}_G$ as LLL [6].

4.1 Data Sets

In our experiments, we use 9 benchmark semi-supervised learning data sets, which can be found in [1] [7].

g241c & g241n[1]: Each data set contains two classes with 350 points in each class, and the data sets are generated in a way of violating the cluster assumptions or misleading class structures.

USPS, COIL & Digit1: The first two data sets are generated from the famous USPS and COIL databases, such that the resultant image data did not appear to be manifold explicitly. The digit1 data set is generated by transforming the image of digit 1, and the image data appears a manifold structure strongly.

Cornell, Texas, Wisconsin & Washington: All these four data sets are selected from the famous WebKB database[2], and the web pages are classified into 5 $\sim$ 6 categories.

Table 1 summarizes the characteristics of the data sets mentioned above. For more details about these data sets, please refer to [1].

[1] http://www.kyb.tuebingen.mpg.de/ssl-book/benchmarks.html
[2] http://www.cs.cmu.edu/ webkb/

Table 2. Description of a subset of UCI database

Datasets	#samples	#classes	#dimensions	Datasets	#samples	#classes	#dimensions
g241c	1500	2	241	cornell	826	6	4134
g241n	1500	2	241	texas	811	5	4029
USPS	1500	2	241	wisconsin	1210	6	4189
COIL	1500	6	241	washington	1165	6	4165
digit1	1500	2	241				

4.2 Methods and Parameter Settings

We compare our methods with some state of the art graph based transductive classification algorithms in the following.

Gaussian Field and Harmonic Function (GFHF) [3]: The width of the Gaussian similarity is set via the grid $\{2^{-3}\sigma_0^2, 2^{-2}\sigma_0^2, 2^{-1}\sigma_0^2, \sigma_0^2, 2\sigma_0^2, 2^2\sigma_0^2, 2^3\sigma_0^2\}$, where σ_0 is the mean distance between any two samples in the training set. And the size of $\mathcal{N}(\cdot)$ is searched by the grid $\{5, 10, 50, 80, n-1\}$

Learning with Local and Global Consistency (LLGC) [4]: The width of the Gaussian similarity and the size of $\mathcal{N}(\cdot)$ are also determined the same as that in GFHF, and the regularization parameter is set by searching the grid $\{0.1, 1, 10, 100\}$.

Transductive Classification with Local Learning Regularization (TCLLR) [6]: The neighborhood size for constructing local learning regularizer is searched by the grid $\{5, 10, 50, 80\}$, and the regularization parameter is set by searching the grid $\{0.1, 1, 10, 100\}$.

Transductive Classification via Dual Regularization (TCDR): In NLap+TCDR, we use *normalized graph Laplacian* on both the data point side and the feature side, and the width of the Gaussian similarity as well as the size of $\mathcal{N}(\cdot)$ are tuned the same as in GFHF. And we set $C_l = C_u = 1$. In LLL+TCDR, we use *local learning graph Laplacian* for both data points and features, and the neighborhood size is tuned the same as that in TLLR. And we set $C_l = 1, C_u = 0$. Besides, we set the number of feature clusters the same as the number of data classes for simplicity, i.e. $m = c$. And the regularization parameters, i.e. λ, μ, η are set by searching the grid $\{0.1, 1, 10, 100\}$.

For synthetic and image data sets, the distance between $\mathbf{x}_{\cdot i}$ and $\mathbf{x}_{\cdot j}$ (or $\mathbf{x}_{i\cdot}$ and $\mathbf{x}_{j\cdot}$) is computed as $d(\mathbf{x}_{\cdot i}, \mathbf{x}_{\cdot j}) = ||\mathbf{x}_{\cdot i} - \mathbf{x}_{\cdot j}||_2$ (or $d(\mathbf{x}_{i\cdot}, \mathbf{x}_{j\cdot}) = ||\mathbf{x}_{i\cdot} - \mathbf{x}_{j\cdot}||_2$).

For text data sets, we use TFIDF to weight the term-document matrix. The distance between two points $\mathbf{x}_{\cdot i}$ and $\mathbf{x}_{\cdot j}$ is defined as

$$d(\mathbf{x}_{\cdot i}, \mathbf{x}_{\cdot j}) = 1 - \frac{\langle \mathbf{x}_{\cdot i}, \mathbf{x}_{\cdot j} \rangle}{||\mathbf{x}_{\cdot i}|| \cdot ||\mathbf{x}_{\cdot j}||}. \tag{32}$$

And the distance between $\mathbf{x}_{i\cdot}$ and $\mathbf{x}_{j\cdot}$ can be computed analogously.

In order to compare these algorithms fairly, we randomly select $\{5\%, 10\%, \ldots, 45\%, 50\%\}$ data points as labeled samples, while the rest as unlabeled samples. Since the labeled set is randomly chosen, we repeat each experiment 20 times and calculate the average transductive classification accuracy. We run each

Table 3. Classification Accuracy with 10% labeled samples on the 9 data sets

Data Sets	g241c	g241n	USPS	COIL	digit1	cornell	texas	wisconsin	washington
GFHF	0.7998	0.7763	0.9377	0.9230	0.9813	0.7075	0.7151	0.7451	0.7860
LLGC	0.7980	0.8101	0.9633	0.9250	0.9781	0.7276	0.7442	0.7787	0.7906
TCLLR	0.8541	0.8551	0.9623	0.8730	0.9774	0.7439	0.7719	0.8308	0.8078
NLap+TCDR	0.8195	0.8207	**0.9742**	**0.9323**	**0.9822**	0.8064	0.7914	0.8376	0.8407
LLL+TCDR	**0.8770**	**0.8799**	0.9694	0.8855	0.9779	**0.8173**	**0.8209**	**0.8807**	**0.8936**

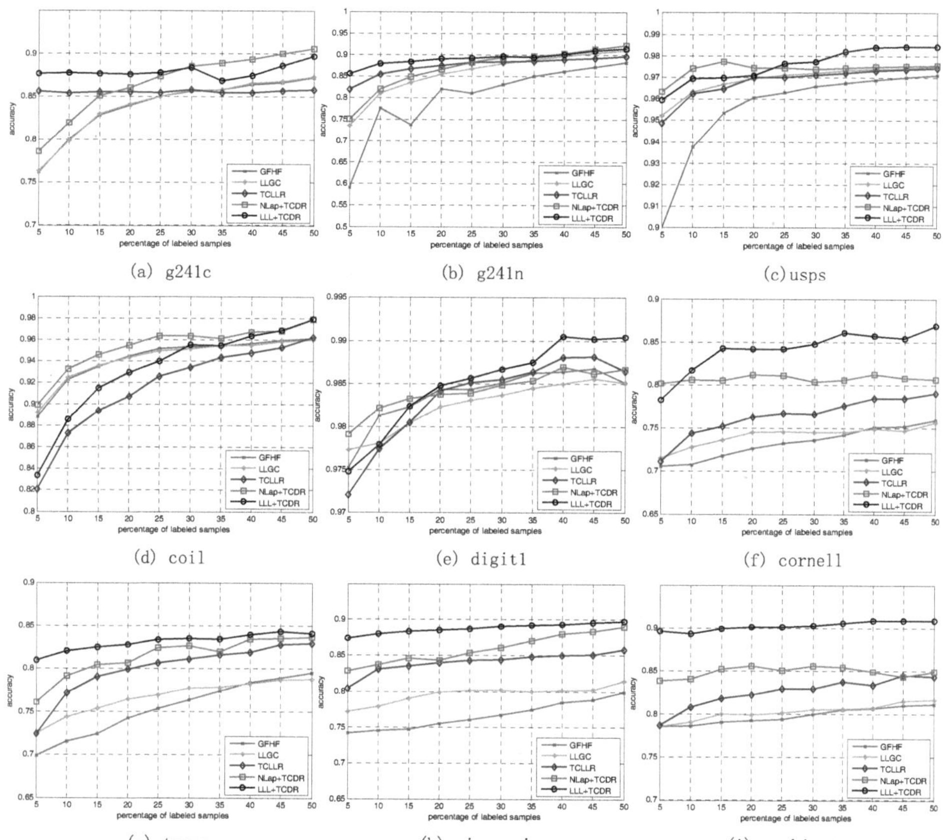

(a) g241c (b) g241n (c) usps (d) coil (e) digit1 (f) cornell (g) texas (h) wisconsin (i) washington

Fig. 2. Classification accuracy with respect to the proportion of labeled samples on the 9 data sets

algorithms under different parameter settings, and select the best average result to compare with each other.

4.3 Classification Results

The experimental results are shown in Fig. 2. In all figures, the x-axis represents the percentage of randomly labeled points, while the y-axis is the average transductive classification accuracy.

To illustrate the experimental results better, we also list the results of these algorithms on all the data sets with 10% labeled samples in Table 3.

It is obvious that TCDR outperforms other methods on all the data sets. In detail, we can see that NLap+TCDR outperforms LLGC consistently, while LLL+TCDR outperforms TCLLR consistently. This is due to that LLGC is the special case of NLap+TCDR and TCLLR is the special case of LLL+TCDR. And the consistent improvement indicates that clustering the features indeed benefits the classification of data points. In addition, it is worth noting that LLL+TCDR achieved higher classification accuracy than NLap+TCDR on text data sets. This indicates that *Local Learning Assumption* is more suitable for text classification than *Cluster Assumption.* The reason is probably that the data matrix of text data is very sparse, *normalized graph Laplacian* based on the distance defined in Eq.(32) may not be able to explore the geometric structure very well. In contrast, the *local learning graph Laplacian* can explore the geometric structure better in this case.

5 Conclusion and Future Work

In this paper, we present a dual regularization to explore the geometric structure in data manifold and feature manifold, along with the duality between data points and features. Furthermore, we propose a novel framework for transductive classification via dual regularization, which can be solved by alternating minimization algorithm and its convergence is theoretically guaranteed. Encouraging experimental results on benchmark semi-supervised learning data sets illustrate that the proposed methods outperform many existing approaches.

In the future work, we will devote to extending the dual regularization framework from transductive learning to inductive learning [8].

Acknowledgments

This work was supported by the National Natural Science Foundation of China (No.60721003, No.60673106 and No.60573062) and the Specialized Research Fund for the Doctoral Program of Higher Education. We thank the anonymous reviewers for their helpful comments.

References

1. Chapelle, O., Schölkopf, B., Zien, A. (eds.): Semi-Supervised Learning. MIT Press, Cambridge (2006)
2. Zhu, X.: Semi-supervised learning literature survey. Technical report, Computer Sciences, University of Wisconsin-Madison (2008)
3. Zhu, X., Ghahramani, Z., Lafferty, J.D.: Semi-supervised learning using gaussian fields and harmonic functions. In: ICML, pp. 912–919 (2003)
4. Zhou, D., Bousquet, O., Lal, T.N., Weston, J., Schölkopf, B.: Learning with local and global consistency. In: NIPS (2003)

5. Wang, F., Zhang, C.: Label propagation through linear neighborhoods. In: ICML, pp. 985–992 (2006)
6. Wu, M., Schölkopf, B.: Transductive classification via local learning regularization. In: AISTATS, pp. 628–635 (March 2007)
7. Wang, F., Li, T., Wang, G., Zhang, C.: Semi-supervised classification using local and global regularization. In: AAAI, pp. 726–731 (2008)
8. Belkin, M., Niyogi, P., Sindhwani, V.: Manifold regularization: A geometric framework for learning from labeled and unlabeled examples. Journal of Machine Learning Research 7, 2399–2434 (2006)
9. Smola, A.J., Kondor, R.: Kernels and regularization on graphs. In: Schölkopf, B., Warmuth, M.K. (eds.) COLT/Kernel 2003. LNCS (LNAI), vol. 2777, pp. 144–158. Springer, Heidelberg (2003)
10. Gu, Q., Zhou, J.: Local learning regularized nonnegative matrix factorization. In: IJCAI (2009)
11. Roweis, S.T., Saul, L.K.: Nonlinear dimensionality reduction by locally linear embedding. Science 290(5500), 2323–2326 (2000)
12. Niyogi, P.: Laplacian eigenmaps for dimensionality reduction and data representation. Neural Computation 15, 1373–1396 (2003)
13. Sandler, T., Blitzer, J., Talukdar, P., Pereira, F.: Regularized learning with networks of features. In: NIPS (2008)
14. Gu, Q., Zhou, J.: Co-clustering on manifolds. In: KDD (2009)
15. Dhillon, I.S.: Co-clustering documents and words using bipartite spectral graph partitioning. In: KDD, pp. 269–274 (2001)
16. Dhillon, I.S., Mallela, S., Modha, D.S.: Information-theoretic co-clustering. In: KDD, pp. 89–98 (2003)
17. Ding, C.H.Q., Li, T., Peng, W., Park, H.: Orthogonal nonnegative matrix t-factorizations for clustering. In: KDD, pp. 126–135 (2006)
18. Sindhwani, V., Hu, J., Mojsilovic, A.: Regularized co-clustering with dual supervision. In: NIPS, pp. 556–562 (2008)
19. Sindhwani, V., Melville, P.: Document-word co-regularization for semi-supervised sentiment analysis. In: ICDM, pp. 1025–1030 (2008)
20. Li, T., Ding, C.H.Q., Zhang, Y., Shao, B.: Knowledge transformation from word space to document space. In: SIGIR, pp. 187–194 (2008)
21. Chung, F.R.K.: Spectral Graph Theory. American Mathematical Society (February 1997)
22. Boyd, S., Vandenberghe, L.: Convex optimization. Cambridge University Press, Cambridge (2004)
23. Ding, C.H., Li, T., Jordan, M.I.: Convex and semi-nonnegative matrix factorizations. IEEE Transactions on Pattern Analysis and Machine Intelligence 99(1) (2008)
24. Lee, D.D., Seung, H.S.: Algorithms for non-negative matrix factorization. In: NIPS, pp. 556–562 (2000)

Appendix: Proof of Theorem 1

Proof. We rewrite Eq.(28) as

$$\begin{aligned} L(\mathbf{F}) = \operatorname{tr}(&\mathbf{F}^T\mathbf{C}\mathbf{F} - 2\mathbf{F}^T\mathbf{C}\mathbf{Y} + \lambda\mathbf{F}^T\mathbf{L}_F^+\mathbf{F} - \lambda\mathbf{F}^T\mathbf{L}_F^-\mathbf{F} \\ &- 2\mathbf{F}^T\mathbf{A}^+ + 2\mathbf{F}^T\mathbf{A}^- + \mathbf{F}\mathbf{B}^+\mathbf{F}^T - \mathbf{F}\mathbf{B}^-\mathbf{F}^T) \end{aligned} \tag{33}$$

By applying Lemma 2, we have

$$\mathbf{tr}(\mathbf{F}^T\mathbf{C}\mathbf{F}) \le \sum_{ij} \frac{(\mathbf{C}\mathbf{F}')_{ij}\mathbf{F}^2_{ij}}{\mathbf{F}'_{ij}}, \mathbf{tr}(\mathbf{F}^T\mathbf{L}^+_F\mathbf{F}) \le \sum_{ij} \frac{(\mathbf{L}^+_F\mathbf{F}')_{ij}\mathbf{F}^2_{ij}}{\mathbf{F}'_{ij}},$$
$$\mathbf{tr}(\mathbf{F}\mathbf{B}^+\mathbf{F}^T) \le \sum_{ij} \frac{(\mathbf{F}'\mathbf{B}^+)_{ij}\mathbf{F}^2_{ij}}{\mathbf{F}'_{ij}}$$

Moreover, by the inequality $a \le \frac{(a^2+b^2)}{2b}, \forall a,b > 0$, we have

$$\mathrm{tr}(\mathbf{F}^T\mathbf{A}^-) = \sum_{ij} \mathbf{A}^-_{ij}\mathbf{F}_{ij} \le \sum_{ij} \mathbf{A}^-_{ij}\frac{\mathbf{F}^2_{ij} + \mathbf{F}'^2_{ij}}{2\mathbf{F}'_{ij}}$$

To obtain the lower bound for the remaining terms, we use the inequality that $z \ge 1 + \log z, \forall z > 0$, then

$$\mathrm{tr}(\mathbf{F}^T\mathbf{A}^+) \ge \sum_{ij} \mathbf{A}^+_{ij}\mathbf{F}'_{ij}(1 + \log\frac{\mathbf{F}_{ij}}{\mathbf{F}'_{ij}}), \ \mathrm{tr}(\mathbf{F}^T\mathbf{L}^-_F\mathbf{F}) \ge \sum_{ijk}(\mathbf{L}^-_F)_{jk}\mathbf{F}'_{ji}\mathbf{F}'_{ki}(1 + \log\frac{\mathbf{F}_{ji}\mathbf{F}_{ki}}{\mathbf{F}'_{ji}\mathbf{F}'_{ki}})$$
$$\mathrm{tr}(\mathbf{F}^T\mathbf{C}\mathbf{Y}) \ge \sum_{ij}(\mathbf{C}\mathbf{Y})_{ij}\mathbf{F}'_{ij}(1 + \log\frac{\mathbf{F}_{ij}}{\mathbf{F}'_{ij}}), \ \mathrm{tr}(\mathbf{F}\mathbf{B}^-\mathbf{F}^T) \ge \sum_{ijk}\mathbf{B}^-_{jk}\mathbf{F}'_{ij}\mathbf{F}'_{ik}(1 + \log\frac{\mathbf{F}_{ij}\mathbf{F}_{ik}}{\mathbf{F}'_{ij}\mathbf{F}'_{ik}})$$

By summing over all the bounds, we can get $\mathbf{Z}(\mathbf{F}, \mathbf{F}')$, which obviously satisfies (1) $\mathbf{Z}(\mathbf{F}, \mathbf{F}') \ge J_{TCDR}(\mathbf{F})$; (2)$\mathbf{Z}(\mathbf{F}, \mathbf{F}) = J_{TCDR}(\mathbf{F})$

To find the minimum of $\mathbf{Z}(\mathbf{F}, \mathbf{F}')$, we take

$$\frac{\partial Z(\mathbf{F}, \mathbf{F}')}{\partial \mathbf{F}_{ij}} = 2\frac{(\mathbf{C}\mathbf{F}')_{ij}\mathbf{F}_{ij}}{\mathbf{F}'_{ij}} - 2(\mathbf{C}\mathbf{Y})_{ij}\frac{\mathbf{F}'_{ij}}{\mathbf{F}_{ij}} + 2\lambda\frac{(\mathbf{L}^+_F\mathbf{F}')_{ij}\mathbf{F}_{ij}}{\mathbf{F}'_{ij}} - 2\lambda(\mathbf{L}^-_F\mathbf{F}')_{ij}\frac{\mathbf{F}'_{ij}}{\mathbf{F}_{ij}}$$
$$- 2\mathbf{A}^+_{ij}\frac{\mathbf{F}'_{ij}}{\mathbf{F}_{ij}} + 2\mathbf{A}^-_{ij}\frac{\mathbf{F}_{ij}}{\mathbf{F}'_{ij}} + 2\frac{(\mathbf{F}'\mathbf{B}^+)_{ij}\mathbf{F}_{ij}}{\mathbf{F}'_{ij}} - 2(\mathbf{F}'\mathbf{B}^-)_{ij}\frac{\mathbf{F}'_{ij}}{\mathbf{F}_{ij}}$$

and the Hessian matrix of $Z(\mathbf{F}, \mathbf{F}')$

$$\frac{\partial^2 Z(\mathbf{F}, \mathbf{F}')}{\partial \mathbf{F}_{ij}\partial \mathbf{F}_{kl}} = \delta_{ik}\delta_{jl}(2\frac{(\mathbf{C}\mathbf{F}')_{ij}}{\mathbf{F}'_{ij}} + 2(\mathbf{C}\mathbf{Y})_{ij}\frac{\mathbf{F}'_{ij}}{\mathbf{F}^2_{ij}} + 2\lambda\frac{(\mathbf{L}^+_F\mathbf{F}')_{ij}}{\mathbf{F}'_{ij}} + 2\lambda(\mathbf{L}^-_F\mathbf{F}')_{ij}\frac{\mathbf{F}'_{ij}}{\mathbf{F}^2_{ij}}$$
$$+ 2\mathbf{A}^+_{ij}\frac{\mathbf{F}'_{ij}}{\mathbf{F}^2_{ij}} + 2\frac{\mathbf{A}^-_{ij}}{\mathbf{F}'_{ij}} + 2\frac{(\mathbf{F}'\mathbf{B}^+)_{ij}}{\mathbf{F}'_{ij}} + 2(\mathbf{F}'\mathbf{B}^-)_{ij}\frac{\mathbf{F}'_{ij}}{\mathbf{F}^2_{ij}})$$

which is a diagonal matrix with positive diagonal elements. Thus $Z(\mathbf{F}, \mathbf{F}')$ is a convex function of $\mathbf{F}$. Therefore, we can obtain the global minimum of $Z(\mathbf{F}, \mathbf{F}')$ by setting $\frac{\partial Z(\mathbf{F}, \mathbf{F}')}{\partial \mathbf{F}_{ij}} = 0$ and solving for $\mathbf{F}$, from which we can get Eq.(29).

Stable and Accurate Feature Selection

Gokhan Gulgezen*, Zehra Cataltepe**, and Lei Yu

Istanbul Technical University, Computer Engineering Department, Istanbul, Turkey
Binghamton University, Computer Science Department, Binghamton, NY, USA
gulgezen@itu.edu.tr, cataltepe@itu.edu.tr, lyu@cs.binghamton.edu

Abstract. In addition to accuracy, stability is also a measure of success for a feature selection algorithm. Stability could especially be a concern when the number of samples in a data set is small and the dimensionality is high. In this study, we introduce a stability measure, and perform both accuracy and stability measurements of MRMR (Minimum Redundancy Maximum Relevance) feature selection algorithm on different data sets. The two feature evaluation criteria used by MRMR, MID (Mutual Information Difference) and MIQ (Mutual Information Quotient), result in similar accuracies, but MID is more stable. We also introduce a new feature selection criterion, MID_α, where redundancy and relevance of selected features are controlled by parameter α.

Keywords: Feature Selection, Stable Feature Selection, Stability, MRMR (Minimum Redundancy Maximum Relevance).

1 Introduction and Previous Work

Many feature selection algorithms have been developed in the past with a focus on improving classification accuracy while reducing dimensionality. Traditionally, the relevance of a feature is the most important selection criterion because using highly relevant features improves classification accuracy [1]. A majority of feature selection algorithms concentrate on feature relevance [2]. In order to have a more compact feature subset with good generalization, the selected features need to be non-redundant. There are several studies on feature redundancy and how the trade-off between feature relevance and redundancy affects classification accuracy [3,4].

A relatively neglected issue is the *stability of feature selection* - the insensitivity of the result of a feature selection algorithm to variations in the training set. This issue is important in many applications with high-dimensional data, where feature selection is used as a knowledge discovery tool for identifying characteristic markers for the observed phenomena [5]. For example, in microarray data analysis, a feature selection algorithm may select largely different subsets of features (genes) under variations to the training data [6,7]. Such instability dampens the confidence of domain experts in investigating any of the various

* Supported by Tubitak master scholarship.
** Supported partially by Tubitak research project no 105E164.

W. Buntine et al. (Eds.): ECML PKDD 2009, Part I, LNAI 5781, pp. 455–468, 2009.

subsets of selected features for biomarker identification. It is worthy noting that stability of feature selection results should be investigated together with classification accuracy, because domain experts are not interested in a strategy that yields very stable feature sets, but leads to a bad predictive model (e.g., arbitrarily picking the same set of features under training data variation).

There exist very limited studies on the stability of feature selection algorithms. An early work in this direction was done by Kalousis et al. [6]. Their work compared the stability of a number of feature ranking and weighting algorithms under training data variation based on various stability measures on high-dimensional data, and demonstrated that different algorithms which performed similarly well for classification had a wide difference in terms of stability. More recently, two techniques were proposed to explicitly achieve stable feature selection without sacrificing classification accuracy: ensemble feature selection [8] and group-based feature selection [7].

The above studies have not addressed an important issue: how different tradeoff between relevance and redundancy affects the stability of feature selection algorithms. Our study attempts to address this issue by evaluating the stability of two different MRMR (Minimum Redundancy Maximum Relevance) feature evaluation criteria: MID (Mutual Information Difference) and MIQ (Mutual Information Quotient), which balance the two objectives, maximum relevance and minimum redundancy, in different ways. We theoretically and empirically show that MID produces more stable feature subsets. Furthermore, we introduce an extension of MID where relevance and redundancy of a feature may have different weights in feature evaluation. We show that for each data set, stability of MRMR can be controlled through the use of this weighting parameter.

The rest of the paper is organized as follows: In Section 2 we review the MRMR feature selection method, and the two criteria MID and MIQ which are used by MRMR for feature evaluation. Section 3 discusses the stability measure that we use. Section 4 discusses theoretically and practically why MID is more stable than MIQ and introduce the extension MID_α where the contribution of redundancy and relevance to feature score calculation is scaled by means of a parameter α. In Section 5 experimental results are given and Section 6 concludes the paper.

2 MRMR Feature Selection Algorithm

MRMR [9] is a filter based feature selection algorithm which tries to select the most relevant features with the target class labels and minimize the redundancy among those selected features simultaneously, the algorithm uses Mutual Information $I(X, Y)$ that measures the level of similarity between two discrete random variables X and Y:

$$I(X,Y) = \sum_{x \in X} \sum_{y \in Y} p(x,y) log(\frac{p(x,y)}{p_1(x)p_2(y)}) \quad (1)$$

where $p(x, y)$ is the joint probability distribution function of X and Y, and $p_1(x)$ and $p_2(y)$ are the marginal probability distribution functions of X and Y respectively.

For notational simplicity, we represent each feature f_i using the vector of N observations for that feature: $f_i = \left[f_i^1, f_i^2, f_i^3, ..., f_i^N\right]$. f_i is an instance of the discrete random variable F_i. $I(F_i, F_j)$ will be used to represent the mutual information between features i and j, where $i, j = 1, 2, ..., d$ and d is the input dimensionality which equals the number of features in the dataset. In order to measure relevance, MRMR algorithm again uses mutual information between target class label $h = \left[h^1, h^2, h^3, ..., h^N\right]$ and the feature i which will be denoted as $I(H, F_i)$.

Let S denote the feature set that we want to select and $|S|$ its cardinality. In order to make sure that the selected feature subset is the best subset, two conditions should be met. First one is the minimum redundancy condition:

$$W = \frac{1}{|S|^2} \sum_{F_i, F_j \in S} I(F_i, F_j) \tag{2}$$

and the other one is the maximum relevancy condition:

$$V = \frac{1}{|S|} \sum_{F_i \in S} I(F_i, H) \tag{3}$$

According to [9], the two simplest combinations of these two conditions are:

$$max(V - W) \tag{4}$$

$$max(V/W) \tag{5}$$

Because of the fact that obtaining the best subset that satisfies one of the above equations requires $O(N^{|S|})$ search, MRMR uses the following algorithm to solve this optimization problem. First feature is selected according to Eq. (3). After that the feature i that satisfies the conditions below in Eqs. (6) and (7) is selected at each step and the selected features remain in the feature set S. m is the number of features in feature set (number of selected features) and $\Omega_S = \Omega - S$ is the feature subset of all features except those already selected.

$$\min_{F_i \in \Omega_S} \frac{1}{|S|} \sum_{F_j \in S} I(F_i, F_j) \tag{6}$$

$$\max_{F_i \in \Omega_S} I(F_i, H) \tag{7}$$

The combination of Eqs. (6) and (7) according to Eqs. (4) and (5) result in two selection criteria in Table 1:

As it can be seen, Eq. (6) is equivalent to the condition in Eq. (2) and Eq. (7) is an approximation of Eq. (3). The complexity of the algorithm above is given to be $O(|S| \cdot N)$ in [9].

Table 1. Two different schemes to search for the next feature in MRMR optimization conditions

ACRONYM	FULL NAME	FORMULA
MID	Mutual information difference	$\max_{F_i \in \Omega_S} \left[I(F_i, H) - \frac{1}{\lvert S \rvert} \sum_{F_j \in S} I(F_i, F_j) \right]$
MIQ	Mutual information quotient	$\max_{F_i \in \Omega_S} \left\{ I(F_i, H) / \left[\frac{1}{\lvert S \rvert} \sum_{F_j \in S} I(F_i, F_j) \right] \right\}$

3 Stability Evaluation

In order to measure the stability of a feature selection algorithm, a measure of similarity between two sets of feature selection results is needed. We will use a method similar to the one proposed by [7]. Let $R_1 = \{F_i\}_{i=1}^{|R_1|}$ and $R_2 = \{F_j\}_{j=1}^{|R_2|}$ denote two sets of feature selection results and each F_i and F_j represent an individual feature. In order to evaluate stability between R_1 and R_2, [7] propose to model R_1 and R_2 together as a weighted complete bipartite graph $G = (V, E)$, with nodes $V = R_1 \cup R_2$, and edges $E = \{(F_i, F_j) \mid F_i \in R_1, F_j \in R_2\}$, and every edge (F_i, F_j) is associated with a weight $\omega(F_i, F_j)$. In our method, all weights are determined by calculating the symetrical uncertainty between pair of features F_i and F_j. This entropy based nonlinear correlation is called symmetrical uncertainty, SU, and is calculated in the following way:

$$SU_{i,j} = 2 \left[\frac{IG(F_i \mid F_j)}{H(F_i) + H(F_j)} \right] \tag{8}$$

As it is defined earlier, in this equation, F_i and F_j are random variables which refer to the i th and jth input features respectively and information gain, entropy and conditional entropy are defined as:

$$IG\,(X \mid Y) = H\,(X) - H\,(X \mid Y) \tag{9}$$

$$H\,(X) = \sum_{x \in X} p\,(x)\, log_2\,(p\,(x)) \tag{10}$$

$$H\,(X \mid Y) = \sum_{y \in Y} p\,(y) \sum_{x \in X} p\,(x \mid y)\, log_2\,(p\,(x \mid y)) \tag{11}$$

In our method $\omega\,(F_i, F_j)$ equals to $SU_{i,j}$ and the overall similarity between feature sets R_1 and R_2 is defined as:

$$Sim^M(R_1, R_2) = \frac{1}{|M|} \sum_{F_i, F_j \in M} \omega(F_i, F_j) \tag{12}$$

where M is a maximum matching in G. The problem of maximum weighted bipartite matching (also known as the assignment problem) is to find an optimal matching where the sum of the weights of all edges in the matching has a maximal value. There exist various efficient algorithms for finding an optimal solution. The purpose of using such a method is to assess the similarity between two sets of features by considering the similarity of feature values instead of features indices which makes sense when two feature subsets contain a large portion of different but highly correlated features. In order to find the optimal solution, Hungarian Algorithm is used. The algorithm is implemented by Alexander Melin from University of Tennessee and taken from the Matlab Central web site [10]. The algorithm is designed for finding minimum weight matching so in order to find maximum weight matching, the sign of the entries of the performance matrix is inversed.

Our stability measure differs from that of Yu and Ding in the following way. First of all, in their paper, they measure the similarity between two sets of feature groups not two sets of individual features. Second, each weight in a bipartite graph is decided by the correlation coefficient between the centers or the most representative features of the two feature groups. Our methodology is a special case of that of Yu and Ding where each F_i and F_j represent individual features and the similarity between them is decided by the symmetrical uncertainty.

4 Stability of MID versus MIQ

In this section, the stability of MID and MIQ techniques are compared both theoretically and experimentally.

4.1 Theoretical Analysis

As it is mentioned before, the MRMR algorithm uses two basic calculations, MID (Mutual information difference) and MIQ (Mutual information quotient), for selecting the next feature among Ω_S, the feature subset of all features except those already selected. MID is the difference of the mutual information between feature F_i and the class label h and the mean of the sum of mutual information values between feature F_i and F_j, which $F_j \in S$ and $j = 1, 2, \ldots, |S|$. MIQ is the ratio of the mutual information between feature F_i and the class label h to the mean of the sum of mutual information values between feature F_i and F_j, which $F_j \in S$ and $j = 1, 2, \ldots, |S|$. Since we have a limited number of samples, we can not obtain the real probability distribution of features or labels, therefore the mutual information values computed also contain a sampling error. Since both features and labels are discrete random variables, the feature-feature or feature-label mutual information computations contain similar types of error.

Let $V+\epsilon$ and $W+\delta$ be the random variables that correspond to the relevance and redundancy values computed over a sample of size N and let ϵ and δ be distributed as $N(0,\sigma_N^2)$.

The variance of MID and MIQ is a direct indication of the stability of these estimators, because as the variance increases different features could be selected at each step of feature selection.

The variance of MID is quite easy to compute:

$$var(MID) = var((V+\epsilon) - (W+\delta)) = 2\sigma_N^2 \tag{13}$$

The mean of MID equals the actual value of the difference between relevance and redundancy.

The mean and variance of the ratio MIQ is much harder to compute. First of all, if $W+\delta$ has a nonnegligible distribution around 0, then the ratio has a Cauchy component, which means the mean and variance are undefined and the second moment is infinite [11]. When both the numerator and the denominator are far from zero, then the ratio is normally distributed with mean V/W and unit variance. As seen in the next section in the experimental analysis, especially for small number of features, W is close to zero and the MIQ shows a large variance.

4.2 Experimental Analysis

Since we have only a finite number of samples, computation of the actual values of V and W are not possible. In order to estimate the mean and variances of V, W, MID and MIQ, we use bootstrapping. We apply MID and MIQ

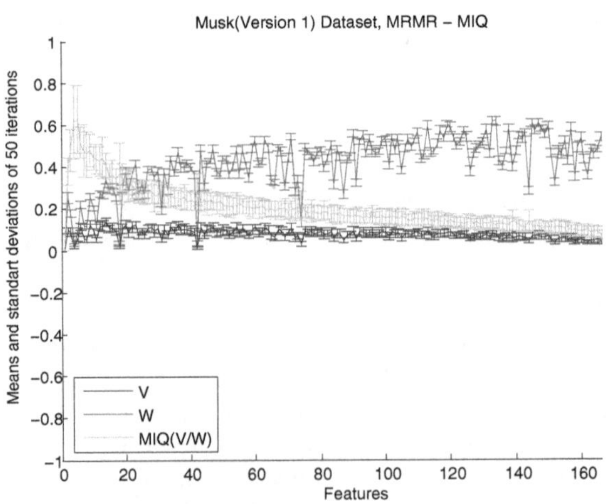

Fig. 1. Mean and standard deviation of the V, W and V/W (MIQ) calculations(y axis) while the number of selected features(x axis) varies for Musk (Version 1) dataset

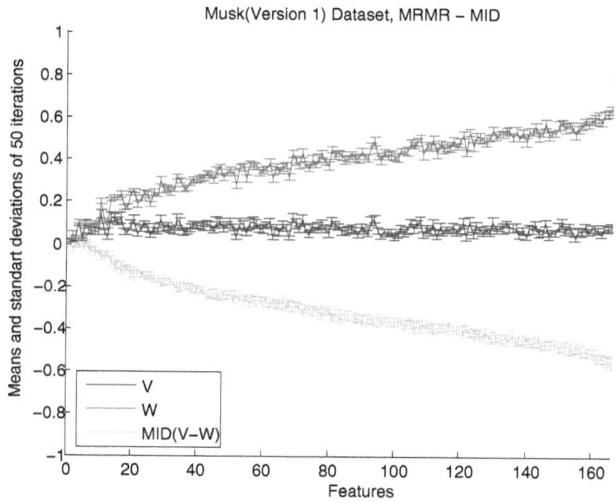

Fig. 2. Mean and standard deviation of the V, W and $V - W$ (MID) calculations(y axis) while the number of selected features(x axis) varies for Musk (Version 1) dataset

techniques respectively on the whole dataset and obtain feature sequences for each algorithm, then we bootstrap the data and do the mutual information calculations again on these new datasets according to feature sequences that are obtained before. We repeat the bootstrap evaluations 50 times and present mean and standart deviation of the values in each feature selection step for the Musk(Version 1) dataset from the UCI [12].

As seen in figures 1 and 2, the entropy values that are calculated by MID technique have smaller variance than the entropy values that are calculated by MIQ, which means difference of V and W gives more stable results.

4.3 MID_α Trading Off Stability and Accuracy

As seen at the previous section, MID gives more stable results than MIQ. We propose a modification to MID as follows: $MID_\alpha = \alpha V - (1 - \alpha) W$. We aim to control stability and accuracy by means of changing α. Various α values($\alpha = [0, 0.25, 0.5, 0.75, 1]$) are used in the Results section below.

5 Experiments

In this section we give details on the datasets used in the experiments, experimental setup and results.

5.1 Data Sets

Experiments were performed on 8 datasets. Seven of them were from the UCI [12]: Ionosphere, Sonar, Parkisons, Musk (Version 1), Multiple-features, Handwritten digits and Wine. The eighth dataset was the Audio Genre data set.

Audio genre dataset consists of the 5 least confused genres of the Tzanetakis data set [13]: Classical, Hiphop, Jazz, Pop and Reggae, each with 100 samples. Two different sets of audio features are computed. First Timbral, rhythmic content and pitch content features yielding 30 features are extracted using Marsyas toolbox [13]. Second, 20 features covering temporal and spectral properties are etracted using the databionic music miner framework given in [14].

All feature values in the datasets are discretized to 10 equal length bins between their maximum and minimum values. MRMR algorithm executions and stability calculations between feature sequences are performed on these discretized features.

Table 2 shows the number of features, instances and classes for the 8 datasets.

Table 2. Information about datasets

Dataset	Features	Instances	Classes
Ionosphere	34	351	2
Sonar	60	208	2
Parkinsons	23	195	2
Musk(Version 1)	166	476	2
Audio Genre	50	500	5
Multi-features Digits	649	2000	10
Handwritten Digits	64	3823	10
Wine	13	178	3

5.2 Experimental Setup

The stability of a feature selection algorithm is defined as the average similarity of various sets of results produced by the same feature selection algorithm under training data variations. Each subset of samples can be obtained by randomly sampling or bootstrapping the full set of samples. Lets say, we have a dataset D which contains N samples and we want to measure the stability of a specific algorithm. In order to do that, we bootstrap N samples from the dataset 10 times to obtain 10 training sets, $D_{train,i}$, $i = 1, 2, \ldots, q$ where $q = 10$. After each bootstrap process, samples that do not belong to $D_{train,i}$ are considered to be the test set $D_{test,i}$ for that iteration. In order to demonstrate the stability of an algorithm, first, a feature selection is performed using all training data and feature sequence R_i is obtained. Since we want to compare the change in the selected features when the sample size gets smaller, we draw samples of size $r * |D_{train,i}|$ from $D_{train,i}$ where we chose r from $r = \left[r^1, r^2, \ldots, r^j\right]$. In the experiments below we chose $j = 5$ and $r = [0.1, 0.25, 0.5, 0.75, 1]$. We obtain feature sequences R_i^j by implementing feature selection algorithm on smaller datasets that are obtained by different ratio values. As a result, for each ratio value, r^j, the stability of the algorithm over q subsets of samples is given by:

$$\frac{1}{q} \sum_{i=1}^{q} Sim^M \left(R_i, R_i^j\right) \tag{14}$$

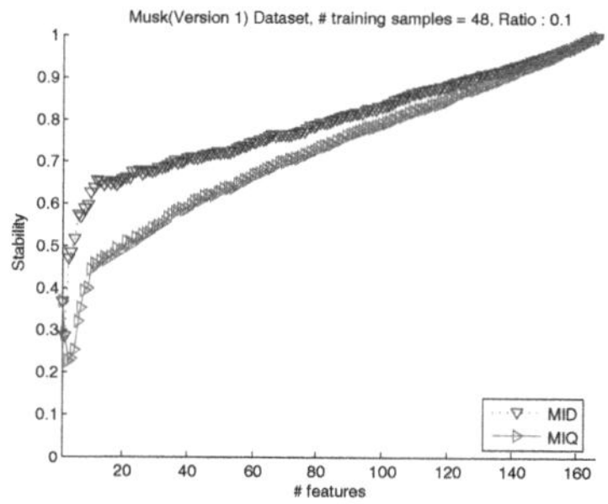

(a) Stability values of MID and MIQ while the number of selected features(x axis) varies when $r = 0.1$ for Musk (Version 1) dataset.

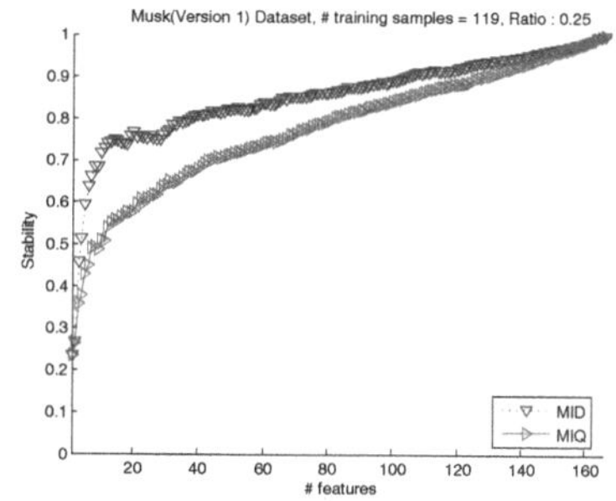

(b) Stability values of MID and MIQ while the number of selected features(x axis) varies when $r = 0.25$ for Musk (Version 1) dataset.

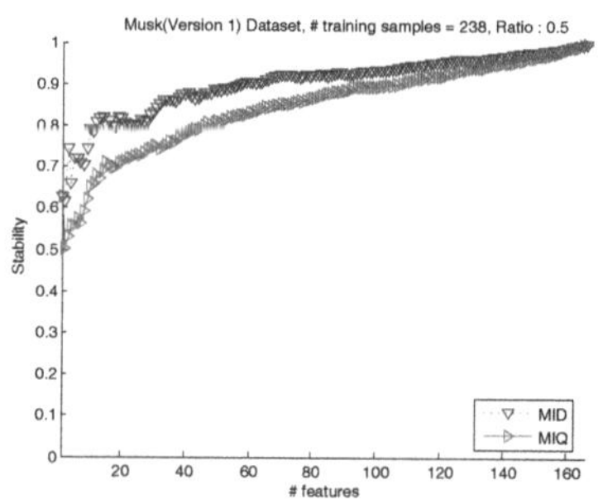

(c) Stability values of MID and MIQ while the number of selected features(x axis) varies when $r = 0.5$ for Musk (Version 1) dataset.

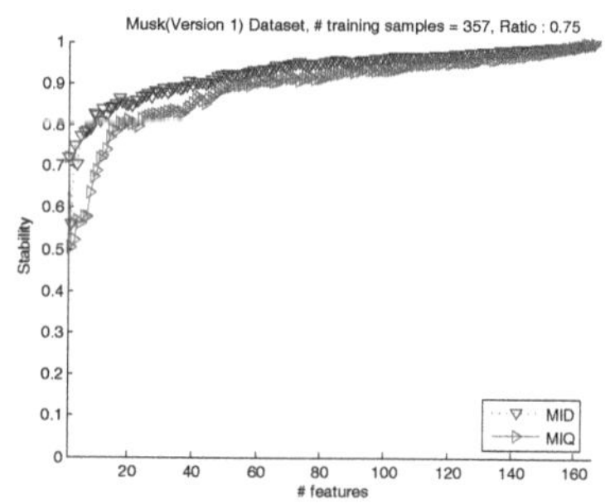

(d) Stability values of MID and MIQ while the number of selected features(x axis) varies when $r = 0.75$ for Musk (Version 1) dataset.

Fig. 3. Stability values of MID and MIQ for Musk (Version 1) dataset

In our experiments we do stability and accuracy computation of feature selection algorithms as the size of available training samples decrease. In [7] stability computation is performed only for a single bootstrap sample which corresponds to approximately $r = 0.632$ ([15] p. 333).

The classifiers that are used in experiments are k-nearest neighbors classifiers with the k value of 3. The support vector machines were also used, however their accuracies did not show a significant difference, hence they are not given here.

The stability comparison of all algorithms and the accuracy values of feature sequences computed on $D_{test,i}$ can be seen at the Results section.

5.3 Results

Stability of MID and MIQ : We first compared stability and accuracy of MID and MIQ on different datasets. In general MID is more stable than MIQ. The stability values of MID and MIQ for Musk (Version 1) dataset is shown

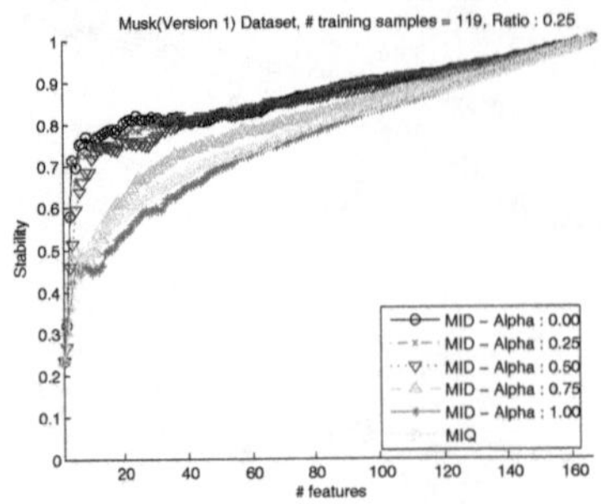

(a) Stability values while the number of selected features(x axis) varies when $r = 0.25$ for Musk (Version 1) dataset.

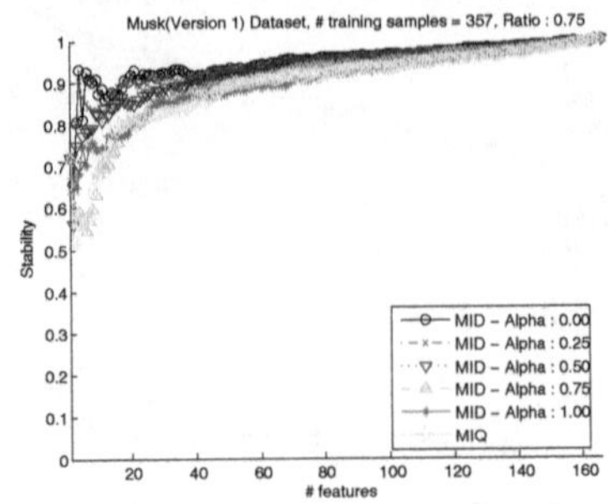

(b) Stability values while the number of selected features(x axis) varies when $r = 0.75$ for Musk (Version 1) dataset.

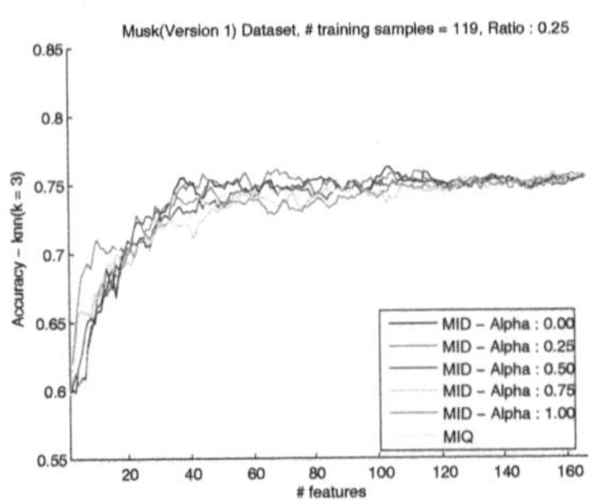

(c) Accuracy values computed with k-nn classifier($k = 3$), while the number of selected features(x axis) varies when $r = 0.25$ for Musk (Version 1) dataset.

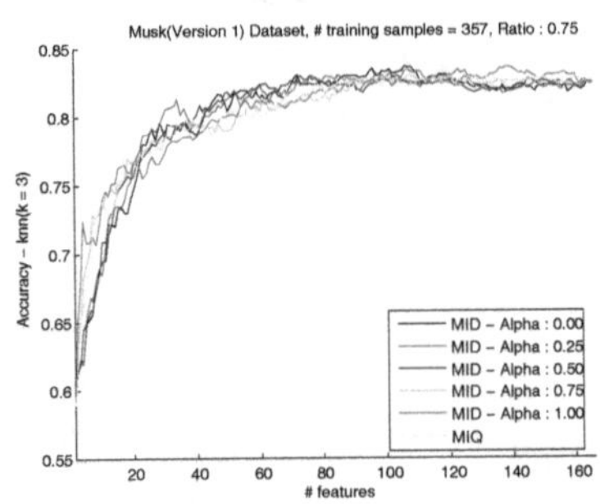

(d) Accuracy values computed with k-nn classifier($k = 3$), while the number of selected features(x axis) varies when $r = 0.75$ for Musk (Version 1) dataset.

Fig. 4. Stability and accuracy values for Musk (Version 1) dataset

in Figure 3. As seen in the figure, MID is always more stable than MIQ, both for different values of r and the number of features selected, for this dataset.

Stability of MID, MIQ and MID_α : In order to compare stability and accuracy of MID, MIQ and MID_α we performed experiments on all the datasets. We demonstrate the mean stability and accuracy values for the Musk (Version 1) and Multi-features datasets in Figures 4 and 5. The stability and accuracy plots for the other datasets are omitted due to space restrictions but are available at [16].

As seen in Figures 4 and 5, MID is again more stable than MIQ. However, it is not clear which value of α results in the most stable feature selection. In order to find the best value of parameter α in terms of stability and accuracy for each dataset, we propose comparing the mean ranks of each feature selection method MID_α for each number of selected features. For a certain number of selected features, we compute the rank of a method (smaller rank means better stability or accuracy),

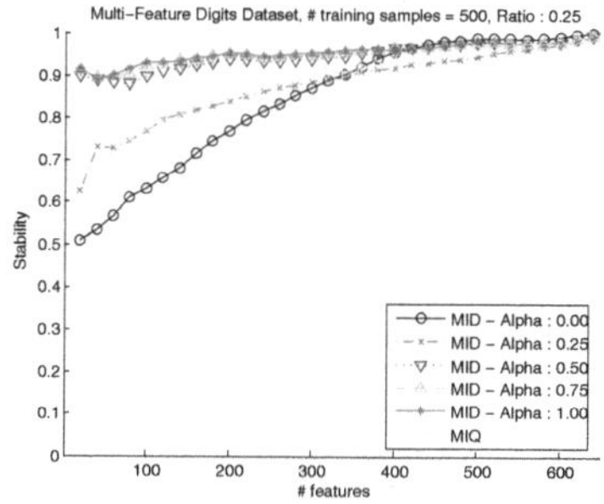

(a) Stability values while the number of selected features(x axis) varies when $r = 0.25$ for Multi-features dataset.

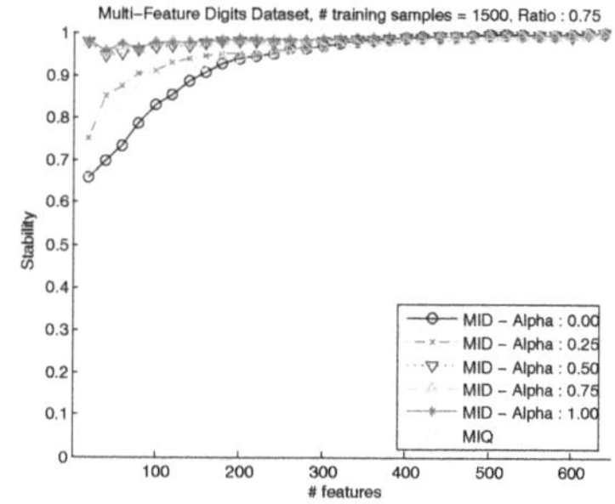

(b) Stability values while the number of selected features(x axis) varies when $r = 0.75$ for Multi-features dataset.

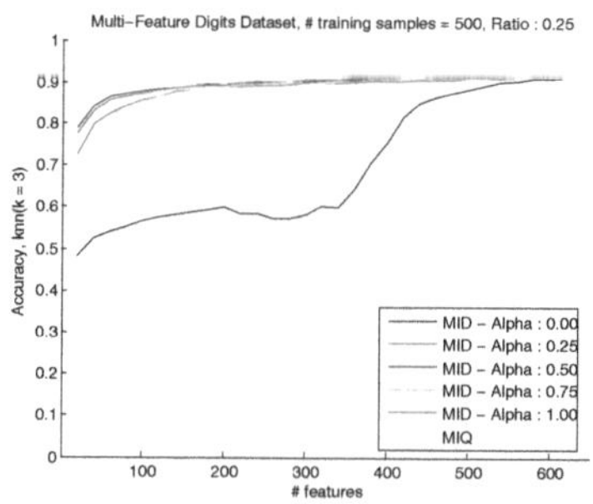

(c) Accuracy values computed with k-nn classifier($k = 3$), while the number of selected features(x axis) varies when $r = 0.25$ for Multi-features dataset.

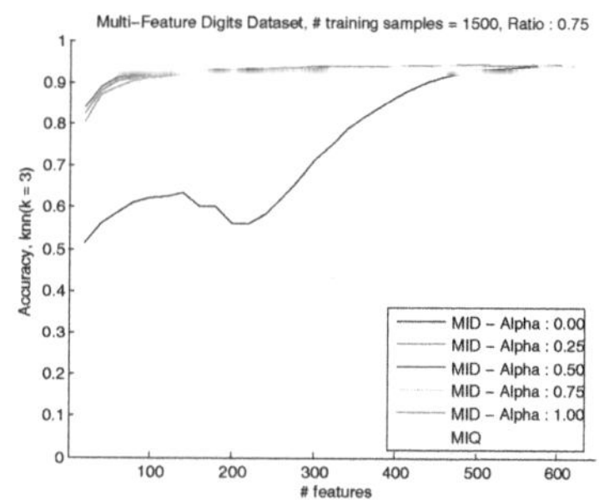

(d) Accuracy values computed with k-nn classifier($k = 3$), while the number of selected features(x axis) varies when $r = 0.75$ for Multi-features dataset.

Fig. 5. Stability and accuracy values for Multi-features dataset

Table 3. Rank of α values and their standard deviation according to stability for Musk (Version 1) dataset

Ratio r	$\alpha = 0$	$\alpha = 0.25$	$\alpha = 0.5$	$\alpha = 0.75$	$\alpha = 1$
0.1	1.7 ∓ 0.9	1.8 ∓ 0.8	2.7 ∓ 0.8	3.9 ∓ 0.7	4.7 ∓ 0.8
0.25	1.9 ∓ 0.8	1.4 ∓ 0.6	2.7 ∓ 0.6	4 ∓ 0.4	4.8 ∓ 0.8
0.5	1.7 ∓ 0.7	1.5 ∓ 0.6	2.8 ∓ 0.5	4.1 ∓ 0.5	4.8 ∓ 0.7
0.75	1.4 ∓ 0.7	1.8 ∓ 0.6	2.9 ∓ 0.6	4.1 ∓ 0.6	4.7 ∓ 0.7

Table 4. Rank of α values and their standard deviation according to accuracy(knn = 3) for Musk (Version 1) dataset

Ratio r	$\alpha = 0$	$\alpha = 0.25$	$\alpha = 0.5$	$\alpha = 0.75$	$\alpha = 1$
0.1	4.2 ∓ 1.1	1.3 ∓ 0.6	1.9 ∓ 0.6	3.3 ∓ 0.8	4.1 ∓ 0.7
0.25	4.3 ∓ 1	1.1 ∓ 0.3	2 ∓ 0.4	3.3 ∓ 0.5	4.2 ∓ 0.8
0.5	4.5 ∓ 0.8	1.1 ∓ 0.4	1.9 ∓ 0.3	3.3 ∓ 0.7	4.1 ∓ 0.7
0.75	4.9 ∓ 0.4	1.1 ∓ 0.4	2 ∓ 0.5	3.2 ∓ 0.7	3.7 ∓ 0.6

then we average them for all the number of features. If two methods result in the same stability or accuracy, then we give them the same rank.

Table 3 shows the averages and standard deviations of stability ranks of MID_α methods and Table 4 shows the same ranks but for accuracy for Musk (Version 1) dataset.

Table 5. Rank of α values and their standart deviation according to stability for Multi-features dataset

Ratio r	$\alpha = 0$	$\alpha = 0.25$	$\alpha = 0.5$	$\alpha = 0.75$	$\alpha = 1$
0.1	3.3 ∓ 1.8	4.4 ∓ 0.6	2.4 ∓ 0.6	2.2 ∓ 0.8	2.7 ∓ 1.5
0.25	3.4 ∓ 1.9	4.5 ∓ 0.5	2.8 ∓ 0.6	2.2 ∓ 0.7	2.1 ∓ 1.3
0.5	3.3 ∓ 1.8	4.5 ∓ 0.5	3.2 ∓ 0.6	2.1 ∓ 0.8	1.9 ∓ 0.9
0.75	3.2 ∓ 1.8	4.5 ∓ 0.5	3.3 ∓ 0.6	2.1 ∓ 0.8	1.9 ∓ 1

Table 6. Rank of α values and their standart deviation according to accuracy(knn = 3) for Multi-features dataset

Ratio r	$\alpha = 0$	$\alpha = 0.25$	$\alpha = 0.5$	$\alpha = 0.75$	$\alpha = 1$
0.1	4.8 ∓ 0.7	3.9 ∓ 0.4	2.4 ∓ 0.9	1.8 ∓ 0.6	1.7 ∓ 0.8
0.25	4.7 ∓ 0.7	3.8 ∓ 0.7	1.9 ∓ 0.9	2.1 ∓ 0.6	2.1 ∓ 0.9
0.5	4.6 ∓ 1	3.6 ∓ 0.7	1.9 ∓ 0.9	2.1 ∓ 0.8	2.1 ∓ 1
0.75	4.7 ∓ 0.7	3.4 ∓ 1	1.8 ∓ 0.9	2 ∓ 0.7	2.2 ∓ 1

Table 7. Rank of α values and their standart deviation according to stability for all datasets when $r = 0.25$

Dataset	$\alpha = 0$	$\alpha = 0.25$	$\alpha = 0.5$	$\alpha = 0.75$	$\alpha = 1$
Ionosphere	3.5 ∓ 1.2	3 ∓ 0.9	3.4 ∓ 1.1	3 ∓ 1.9	1.4 ∓ 0.6
Sonar	2.2 ∓ 0.8	1.7 ∓ 0.8	2.6 ∓ 1.2	4.2 ∓ 0.8	4 ∓ 1.4
Parkinsons	2.5 ∓ 1.1	2.2 ∓ 1.1	3 ∓ 1.4	3.6 ∓ 1.3	2.7 ∓ 1.9
Musk(Version 1)	1.9 ∓ 0.8	1.4 ∓ 0.6	2.7 ∓ 0.6	4 ∓ 0.4	4.8 ∓ 0.8
Audio Genre	3.7 ∓ 1.1	4.5 ∓ 0.9	3.2 ∓ 0.7	1.8 ∓ 0.8	1.4 ∓ 0.5
Multi-features Digits	3.4 ∓ 1.9	4.5 ∓ 0.5	2.8 ∓ 0.6	2.2 ∓ 0.7	2.1 ∓ 1.3
Handwritten Digits	2.9 ∓ 1.5	4.4 ∓ 1	2.4 ∓ 1	2.3 ∓ 1.1	2 ∓ 1
Wine	3.7 ∓ 1.3	4.1 ∓ 1.4	2.5 ∓ 1	1.4 ∓ 0.5	1.8 ∓ 0.9

Table 8. Rank of α values and their standart deviation according to accuracy(knn = 3) for all datasets when $r = 0.25$

Dataset	$\alpha = 0$	$\alpha = 0.25$	$\alpha = 0.5$	$\alpha = 0.75$	$\alpha = 1$
Ionosphere	2.7 ∓ 1.9	3.2 ∓ 1.2	3.5 ∓ 1.2	3.1 ∓ 1.1	1.9 ∓ 1
Sonar	3.1 ∓ 1.9	1.8 ∓ 0.7	2.5 ∓ 1.2	3.7 ∓ 1	3.7 ∓ 1
Parkinsons	3.6 ∓ 1.6	2.2 ∓ 1.2	2.9 ∓ 1.4	3.2 ∓ 1	2.4 ∓ 1.3
Musk(Version 1)	4.3 ∓ 1	1.1 ∓ 0.3	2 ∓ 0.4	3.3 ∓ 0.5	4.2 ∓ 0.8
Audio Genre	4.8 ∓ 0.6	3.9 ∓ 0.7	2.9 ∓ 0.4	1.7 ∓ 0.6	1.4 ∓ 0.5
Multi-features Digits	4.7 ∓ 0.7	3.8 ∓ 0.7	1.9 ∓ 0.9	2.1 ∓ 0.6	2.1 ∓ 0.9
Handwritten Digits	4.4 ∓ 0.9	3.8 ∓ 1	2 ∓ 0.9	2 ∓ 0.9	1.7 ∓ 0.8
Wine	2.5 ∓ 1.4	4.3 ∓ 1.3	3.1 ∓ 1.3	1.8 ∓ 0.7	2.2 ∓ 0.7

Table 5 shows the averages and standard deviations of stability ranks of MID_α methods and Table 6 shows the same ranks but for accuracy for Multi-features dataset.

We summarize the stability and accuracy ranks of MID_α for all datasets in tables 7 and 8 respectively, when $r = 0.25$. According to these tables, $\alpha = 0.5$, i.e. MID is not necessarily the best choice in terms of either stability or accuracy. Different datasets favor different values of α for best stability and accuracy. While Ionosphere, audio genre, multi features, handwritten digits and wine datasets prefers $\alpha > 0.5$ for best stability, others perform better when $alpha < 0.5$.

6 Conclusion

In this paper, we first devised a method to evaluate the stability of a feature selection method as the number of dataset used for feature selection gets smaller. Then, using our stability evaluation method, on 8 different datasets, we showed that among the two feature selection criteria, MID and MIQ of the MRMR feature selection method, MID gives more stable results, while the accuracy of both criteria are comparable. We also showed theoretically, why MID is more stable than MIQ. Finally, we suggested an improvement on the MID criterion, MID_α, where the contribution of relevance and redundancy on feature selection is controlled through a parameter α. For different data sets, we evaluated the stability and accuracy performance of MID_α and observed that for each dataset the α that results in the best stability and accuracy could be different. Understanding what value of α is the best based on the characteristics of a dataset is one of the future works we are planning on.

Acknowledgments

The authors would like to thank Istanbul Technical University, Computer Engineering Department, Data Mining and Pattern Recognition Laboratory members Baris Senliol and Yusuf Yaslan for useful discussions.

References

1. John, G.H., Kohavi, R., Pfleger, K.: Irrelevant Feature and The Subset Selection Problem. In: Proceedings of the Eleventh International Conference on Machine Learning, pp. 121–129 (1994)
2. Liu, H., Yu, L.: Toward Integrating Feature Selection Algorithms for Classification and Clustering. IEEE Transactions on Knowledge and Data Engineering 17(4), 491–502 (2005)
3. Ding, C., Peng, H.: Minimum Redundancy Feature Selection from Microarray Gene Expression Data. In: Proceedings of the Computational Systems Bioinformatics conference (CSB 2003), pp. 523–529 (2003)
4. Yu, L., Liu, H.: Efficient Feature Selection via Analysis of Relevance and Redundancy. Journal of Machine Learning Research 5, 1205–1224 (2004)
5. Pepe, M.S., Etzioni, R., Feng, Z., et al.: Phases of Biomarker Development for Early Detection of Cancer. J. Natl. Cancer Inst. 93, 1054–1060 (2001)
6. Kalousis, A., Prados, J., Hilario, M.: Stability of Feature Selection Algorithms: A Study on High-Dimensional Spaces. Knowledge and Information Systems 12, 95–116 (2007)
7. Yu, L., Ding, C., Loscalzo, S.: Stable Feature Selection via Dense Feature Groups. In: Proceedings of the 14th ACM International Conference on Knowledge Discovery and Data Mining (KDD 2008), pp. 803–811 (2008)
8. Saeys, Y., Abeel, T., Van de Peer, Y.: Robust feature selection using ensemble feature selection techniques. In: Daelemans, W., Goethals, B., Morik, K. (eds.) ECML 2008, Part II. LNCS (LNAI), vol. 5212, pp. 313–325. Springer, Heidelberg (2008)
9. Ding, I., Peng, H.C.: Minimum Redundancy Feature Selection from Microarray Gene Expression Data. In: Proc. Second IEEE Computational Systems Bioinformatics Conf., pp. 523–528 (2003)
10. Hungarian Algorithm by Alexander Melin, MATLAB CENTRAL Web Site, `http://www.mathworks.com/matlabcentral/fileexchange/11609`
11. Marsaglia, G.: Ratios of Normal Variables and Ratios of Sums of Uniform Variables. Journal of the American Statistical Association 60(309), 193–204 (1965)
12. UCI Machine Learning Repository, `http://www.ics.uci.edu/~mlearn/MLRepository.html`
13. Tzanetakis, G., Cook, P.: Musical Genre Classification of Audio Signals. IEEE Transactions on Speech and Audio Processing 10(5), 293–302 (2002)
14. Ding, I., Peng, H.C., Moerchen, F., Ultsch, A., Thies, M., Loehken, I.: Modelling Timbre Distance with Temporal Statistics From Polyphonic Music. IEEE Transactions on Speech and Audio Processing 14, 81–90 (2006)
15. Alpaydin, E.: Introduction to Machine Learning. The MIT Press, Cambridge (2004)
16. Gulgezen, G.: Stable and Accurate Feature Selection. M.Sc. Thesis, Istanbul Technical University, Computer Engineering Department (2009)

Efficient Sample Reuse in EM-Based Policy Search

Hirotaka Hachiya[1], Jan Peters[2], and Masashi Sugiyama[1]

[1] Tokyo Institute of Technology, Tokyo, 152-8552, Japan
{hachiya@sg,sugi@}cs.titech.ac.jp
[2] Max-Planck Institute for Biological Cybernetics, 72076 Tübingen, Germany
jan.peters@tuebingen.mpg.de

Abstract. Direct policy search is a promising reinforcement learning framework in particular for controlling in continuous, high-dimensional systems such as anthropomorphic robots. Policy search often requires a large number of samples for obtaining a stable policy update estimator due to its high flexibility. However, this is prohibitive when the sampling cost is expensive. In this paper, we extend an EM-based policy search method so that previously collected samples can be efficiently reused. The usefulness of the proposed method, called *Reward-weighted Regression with sample Reuse* (R^3), is demonstrated through a robot learning experiment.

1 Introduction

Model-free reinforcement learning is an important tool for solving real-world Markov decision problems online. However, due to its high flexibility, many data samples are usually required for obtaining good control policies. In practice, the cost of collecting rollout data is often prohibitively expensive and too time-consuming for real-world problems where thousands of trials would require weeks or months of experiments. For example, when a robot learns how to hit a ball in baseball or tennis, we need to let the robot hit a ball hundreds of times for obtaining a reliable policy-improvement direction; then this policy update steps need to be repeated many times for finally obtaining a good policy. As a result, robot engineers need to spend a lot of time and effort to "nurse" the vulnerable robot through frequent mechanical maintenance. As in many other real-world reinforcement learning problems, it is therefore highly important to reduce the number of training samples generated by the physical system and instead re-use them efficiently in future updates.

A lot of efforts have been made to *reuse* previously collected samples, in particular in the context of value function approximation. A basic technique for sample reuse is to use *importance sampling* [16] for which the bias is canceled out asymptotically. However, a naive use of importance sampling significantly increases the variance of the estimator and, therefore, it becomes highly unstable. To mitigate this problem, the *per-decision* importance-weighting technique has been introduced for variance reduction [10]. This technique cleverly makes

W. Buntine et al. (Eds.): ECML PKDD 2009, Part I, LNAI 5781, pp. 469–484, 2009.

use of a property of Markov decision processes and eliminates irrelevant terms in the importance sampling identity. However, the obtained estimator still tends to be unstable and, thus, the importance-sampling paradigm has not been in active use for real-world reinforcement learning tasks yet. For more stable and efficient variance reduction, an *adaptive* importance sampling scheme has recently been proposed [4]. This formulates the off-policy value function approximation problem as *covariate shift adaptation*, which is a statistical learning paradigm under non-stationarity: a 'flattening' parameter is introduced to the importance weight for trading the variance reduction with a slight bias increase—the bias-variance trade-off is optimally controlled based on *importance-weighted cross-validation* [15].

Due to the above efforts, reinforcement learning methods based on value function approximation can successfully reuse previously collected samples in a stable manner. However, it is not easy to deal with continuous actions in the value function based policy iteration framework; the *direct policy search* approach is more suitable for learning control policies with continuous actions, e.g., the *policy gradient* method [18,17], the *natural policy gradient* method [5,9], and the *policy search by expectation-maximization* [2,8]. Reusing data samples is even more urgent in policy search approaches as small policy updating steps can result into under-utilization of the data. While plain importance sampling techniques have also been applied in the direct policy search, they were shown to be unstable [13,7]. For stabilization purposes, heuristic techniques are often used in practice, e.g., samples with smaller importance weights are not used for learning [6]. However, to the best of our knowledge, systematic treatment of instability issues in the policy search with sample reuse is still an open research topic.

The purpose of this paper is to propose a new framework for systematically addressing this problem. In particular, we combine the policy search by expectation-maximization [2,8] with the covariate shift adaptation paradigm, and develop an efficient data-reuse algorithm for direct policy learning. The effectiveness of the proposed method, called *Reward-weighted Regression with sample Reuse* (R^3), is demonstrated by robot-control experiments.

2 Policy Search Framework

We consider the standard reinforcement learning framework in which an agent interacts with a Markov decision process. In this section, we review how Markov decision problems can be solved using the policy search by expectation-maximization [2]; for Gaussian models, this results in the reward-weighted regression (RWR) algorithm [8].

2.1 Markov Decision Problems

Let us consider a Markov decision problem specified by $(\mathcal{S}, \mathcal{A}, P_{\mathrm{T}}, P_{\mathrm{I}}, R, \gamma)$, where $\mathcal{S}$ is a set of (continuous) states, $\mathcal{A}$ is a set of (continuous) actions, $P_{\mathrm{T}}(\boldsymbol{s}'|\boldsymbol{s}, a)$ ($\in (0,1]$) is the transition probability-density from state $\boldsymbol{s}$ to next state $\boldsymbol{s}'$ when action a is taken, $P_{\mathrm{I}}(\boldsymbol{s})$ ($\in (0,1]$) is the probability of initial states, $R(\boldsymbol{s}, a, \boldsymbol{s}')$ (≥ 0) is an immediate reward for transition from $\boldsymbol{s}$ to $\boldsymbol{s}'$ by taking action a, and γ ($\in (0,1]$) is the discount factor for future rewards. Let $\pi(a|\boldsymbol{s};\boldsymbol{\theta})$ ($\in (0,1]$) be a stochastic policy with parameter $\boldsymbol{\theta}$, which represents the conditional probability density of taking action a given state $\boldsymbol{s}$. Let us denote a *trajectory* of the agent by $d \equiv (\boldsymbol{s}_1, a_1, \boldsymbol{s}_2, a_2, \ldots, \boldsymbol{s}_N, a_N, \boldsymbol{s}_{N+1})$, where N is the length of the trajectory. Let $\mathcal{R}(d)$ be the *return* (i.e., the sum of discounted rewards) along trajectory d:

$$\mathcal{R}(d) \equiv \sum_{n=1}^{N} \gamma^{n-1} R(\boldsymbol{s}_n, a_n, \boldsymbol{s}_{n+1}).$$

Let $P(d;\boldsymbol{\theta})$ be the probability of occurring trajectory d under $P_{\mathrm{I}}(\boldsymbol{s}_1)$, $P_{\mathrm{T}}(\boldsymbol{s}_{n+1}|\boldsymbol{s}_n, a_n)$, and $\pi(a_n|\boldsymbol{s}_n;\boldsymbol{\theta})$:

$$P(d;\boldsymbol{\theta}) \equiv P_{\mathrm{I}}(\boldsymbol{s}_1) \prod_{n=1}^{N} \pi(a_n|\boldsymbol{s}_n;\boldsymbol{\theta}) P_{\mathrm{T}}(\boldsymbol{s}_{n+1}|\boldsymbol{s}_n, a_n). \tag{1}$$

The *expected return* is denoted by $J(\boldsymbol{\theta})$:

$$J(\boldsymbol{\theta}) \equiv \int \mathcal{R}(d) P(d;\boldsymbol{\theta}) \mathrm{d}d.$$

The goal of the policy search is to learn the optimal policy parameter $\boldsymbol{\theta}^*$ that maximizes the expected return $J(\boldsymbol{\theta})$:

$$\boldsymbol{\theta}^* \equiv \arg\max_{\boldsymbol{\theta}} J(\boldsymbol{\theta}). \tag{2}$$

2.2 EM-Based Policy Search

Naively maximizing $J(\boldsymbol{\theta})$ is hard since $J(\boldsymbol{\theta})$ usually contains high non-linearity. The basic idea of the policy search by expectation-maximization is to iteratively update the policy parameter $\boldsymbol{\theta}$ by maximizing a lower bound of the target cost function [3]. More precisely, let $\boldsymbol{\theta}_L$ be the current policy parameter, where the subscript L indicates the iteration number. Then the use of *Jensen's inequality* allows us to obtain a lower bound of the *normalized expected return* $\log(J(\boldsymbol{\theta})/J(\boldsymbol{\theta}_L))$ as

$$\begin{aligned}\log \frac{J(\boldsymbol{\theta})}{J(\boldsymbol{\theta}_L)} &= \log \int \frac{\mathcal{R}(d)P(d;\boldsymbol{\theta})}{J(\boldsymbol{\theta}_L)} \mathrm{d}d = \log \int \frac{\mathcal{R}(d)P(d;\boldsymbol{\theta}_L)}{J(\boldsymbol{\theta}_L)} \frac{P(d;\boldsymbol{\theta})}{P(d;\boldsymbol{\theta}_L)} \mathrm{d}d \\ &\geq \int \frac{\mathcal{R}(d)P(d;\boldsymbol{\theta}_L)}{J(\boldsymbol{\theta}_L)} \log \frac{P(d;\boldsymbol{\theta})}{P(d;\boldsymbol{\theta}_L)} \mathrm{d}d \equiv J_L(\boldsymbol{\theta}).\end{aligned}$$

Note that $\mathcal{R}(d)P(d;\boldsymbol{\theta}_L)/J(\boldsymbol{\theta}_L)$ is treated as a probability density function in applying Jensen's inequality; this is why $\mathcal{R}(d)$ is assumed to be non-negative and the expected return is normalized.

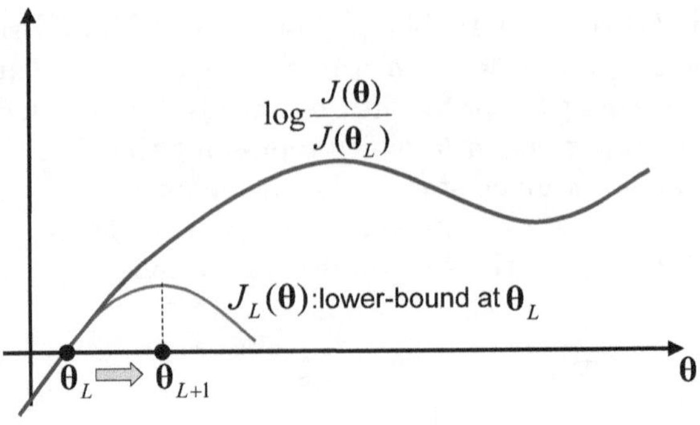

Fig. 1. Relation between the normalized expected return $\log \frac{J(\boldsymbol{\theta})}{J(\boldsymbol{\theta}_L)}$ and its lower bound $J_L(\boldsymbol{\theta})$. At $\boldsymbol{\theta}_L$, the lower bound touches $\log \frac{J(\boldsymbol{\theta})}{J(\boldsymbol{\theta}_L)}$ with zero value

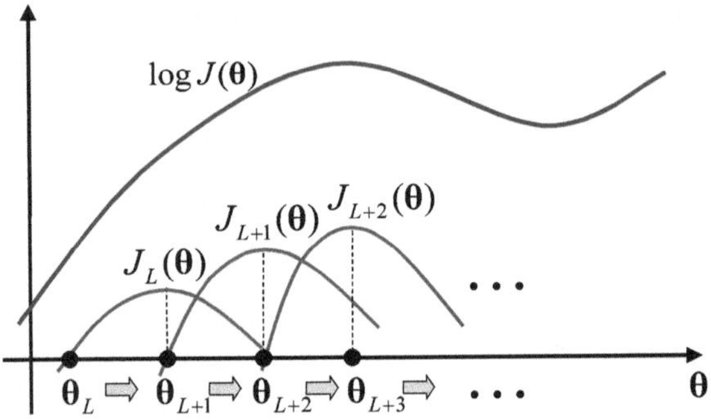

Fig. 2. The policy parameter $\boldsymbol{\theta}$ is updated iteratively by maximizing lower bounds

The expectation-maximization (EM) approach iteratively updates the parameter $\boldsymbol{\theta}$ by maximizing the lower bound $J_L(\boldsymbol{\theta})$:

$$\boldsymbol{\theta}_{L+1} \equiv \arg\max_{\boldsymbol{\theta}} J_L(\boldsymbol{\theta}). \tag{3}$$

Since $J_L(\boldsymbol{\theta}_L) = 0$, the lower bound $J_L(\boldsymbol{\theta})$ is *tight* (i.e., the lower bound touches the target function) at $\boldsymbol{\theta}_L$. Thus monotone non-decrease of the (un)normalized expected return is guaranteed (Fig.1):

$$J(\boldsymbol{\theta}_{L+1}) \geq J(\boldsymbol{\theta}_L).$$

This update is iterated until convergence (see Fig.2).

2.3 Reward-Weighted Regression

Let us employ the Gaussian policy model defined as

$$\pi(a|\boldsymbol{s};\boldsymbol{\theta}) = \frac{1}{\sigma\sqrt{2\pi}} \exp\Big(- \frac{(a - \boldsymbol{k}^\top \boldsymbol{\phi}(\boldsymbol{s}))^2}{2\sigma^2} \Big), \tag{4}$$

where $\boldsymbol{k} = (k_1, k_2, \ldots, k_b))^\top$ and σ are policy parameters, b is the number of basis functions, and $\boldsymbol{\phi}(\boldsymbol{s}) = (\phi_1(\boldsymbol{s}), \phi_2(\boldsymbol{s}), \ldots, \phi_b(\boldsymbol{s}))^\top$ are fixed basis functions. This model allows us to deal with one-dimensional action a and multi-dimensional state vector $\boldsymbol{s}$—multi-dimensional action vectors may be handled by concatenating one-dimensional models. The maximizer $\boldsymbol{\theta}_{L+1}$ of the lower bound $J_L(\boldsymbol{\theta})$ satisfies the following equation:

$$\begin{aligned} \frac{\partial}{\partial \boldsymbol{\theta}} J_L(\boldsymbol{\theta}) \Big|_{\boldsymbol{\theta}=\boldsymbol{\theta}_{L+1}} &= \int \frac{\mathcal{R}(d)P(d;\boldsymbol{\theta}_L)}{J(\boldsymbol{\theta}_L)} \frac{\partial}{\partial \boldsymbol{\theta}} \log P(d;\boldsymbol{\theta}) \Big|_{\boldsymbol{\theta}=\boldsymbol{\theta}_{L+1}} \mathrm{d}d \\ &= \int \frac{\mathcal{R}(d)P(d;\boldsymbol{\theta}_L)}{J(\boldsymbol{\theta}_L)} \sum_{n=1}^{N} \frac{\partial}{\partial \boldsymbol{\theta}} \log \pi(a_n|\boldsymbol{s}_n;\boldsymbol{\theta}) \Big|_{\boldsymbol{\theta}=\boldsymbol{\theta}_{L+1}} \mathrm{d}d = \boldsymbol{0}. \end{aligned} \tag{5}$$

A useful property of the Gaussian policy model is that the log-derivative of the policy model with respect to the parameters can be analytically computed as

$$\begin{aligned} \frac{\partial}{\partial \boldsymbol{k}} \log \pi(a|\boldsymbol{s};\boldsymbol{\theta}) &= \frac{a - \boldsymbol{k}^\top \boldsymbol{\phi}(\boldsymbol{s})}{\sigma^2} \boldsymbol{\phi}(\boldsymbol{s}), \\ \frac{\partial}{\partial \sigma} \log \pi(a|\boldsymbol{s};\boldsymbol{\theta}) &= \frac{(a - \boldsymbol{k}^\top \boldsymbol{\phi}(\boldsymbol{s}))^2 - \sigma^2}{\sigma^3}. \end{aligned}$$

Then the maximizer $\boldsymbol{\theta}_{L+1} = (\boldsymbol{k}_{L+1}^\top, \sigma_{L+1})^\top$ can be analytically obtained as

$$\begin{aligned} \boldsymbol{k}_{L+1} &= \Big(\int \mathcal{R}(d)P(d;\boldsymbol{\theta}_L) \sum_{n=1}^{N} \boldsymbol{\phi}(\boldsymbol{s}_n)\boldsymbol{\phi}(\boldsymbol{s}_n)^\top \Big)^{-1} \\ &\quad \times \Big(\int \mathcal{R}(d)P(d;\boldsymbol{\theta}_L) \sum_{n=1}^{N} a_n \boldsymbol{\phi}(\boldsymbol{s}_n) \Big), \\ \sigma_{L+1}^2 &= \Big(N \int \mathcal{R}(d)P(d;\boldsymbol{\theta}_L) \Big)^{-1} \\ &\quad \times \Big(\int \mathcal{R}(d)P(d;\boldsymbol{\theta}_L) \sum_{n=1}^{N} (a_n - {\boldsymbol{k}_{L+1}}^\top \boldsymbol{\phi}(\boldsymbol{s}_n))^2 \Big). \end{aligned}$$

The EM-based policy search for Gaussian models is called *reward-weighted regression* [8].

2.4 Learning from Episodic Data Samples

Suppose a dataset consisting of M episodes with N steps is available for each RWR iteration, where each episodic sample is generated as follows. Initially, the agent starts from a randomly selected state $\boldsymbol{s}_1$ following the initial-state probability density $P_{\mathrm{I}}(\boldsymbol{s}_1)$ and chooses an action based on the policy $\pi(a_n|\boldsymbol{s}_n;\boldsymbol{\theta}_L)$ at the L-th iteration. Then the agent makes a transition following $P_{\mathrm{T}}(\boldsymbol{s}_{n+1}|\boldsymbol{s}_n,a_n)$ and receives a reward r_n $(=R(\boldsymbol{s}_n,a_n,\boldsymbol{s}_{n+1}))$. This transition is repeated N times for M episodes—hence, the training data $\mathcal{D}^L$ gathered at the L-th iteration is expressed as $\mathcal{D}^L \equiv \{d_m^L\}_{m=1}^M$, where each episodic sample d_m^L is given by

$$d_m^L \equiv \{(\boldsymbol{s}_{m,n}^L, a_{m,n}^L, r_{m,n}^L, \boldsymbol{s}_{m,n+1}^L)\}_{n=1}^N.$$

Then the RWR solution $\boldsymbol{\theta}_{L+1} \equiv (\boldsymbol{k}_{L+1}^\top, \sigma_{L+1})^\top$ can be approximated using the L-th training data $\mathcal{D}^L$ as $\widehat{\boldsymbol{\theta}}_{L+1} \equiv (\widehat{\boldsymbol{k}}_{L+1}^\top, \widehat{\sigma}_{L+1})^\top$, where

$$\begin{cases} \widehat{\boldsymbol{k}}_{L+1} = \left(\sum_{m=1}^M \mathcal{R}(d_m^L) \sum_{n=1}^N \boldsymbol{\phi}(\boldsymbol{s}_{m,n}^L)\boldsymbol{\phi}(\boldsymbol{s}_{m,n}^L)^\top\right)^{-1} \\ \qquad \times \left(\sum_{m=1}^M \mathcal{R}(d_m^L) \sum_{n=1}^N a_{m,n}^L \boldsymbol{\phi}(\boldsymbol{s}_{m,n}^L)\right), \\ \widehat{\sigma}_{L+1}^2 = \left(N\sum_{m=1}^M \mathcal{R}(d_m^L)\right)^{-1} \left(\sum_{m=1}^M \mathcal{R}(d_m^L) \sum_{n=1}^N (a_{m,n}^L - \widehat{\boldsymbol{k}}_{L+1}^\top \boldsymbol{\phi}(\boldsymbol{s}_{m,n}^L))^2\right). \end{cases} \tag{6}$$

2.5 Importance Sampling

When the cost for gathering rollout samples is high, the number M of episodes should be kept small. As a result, the next policy parameter $\widehat{\boldsymbol{\theta}}_{L+1}$ suggested by RWR may not be sufficiently accurate. In order to improve the estimation accuracy, we could reuse the samples collected at the previous iterations $\{\mathcal{D}^l\}_{l=1}^L$.

If the policies remain unchanged by the RWR updates, just using Eq.(6) gives an *efficient* estimator[1] $\widehat{\boldsymbol{\theta}}_{L+1}^{\mathrm{NIW}} \equiv (\widehat{\boldsymbol{k}}_{L+1}^{\mathrm{NIW}\top}, \widehat{\sigma}_{L+1}^{\mathrm{NIW}})^\top$, where

$$\begin{aligned} \widehat{\boldsymbol{k}}_{L+1}^{\mathrm{NIW}} &= \left(\sum_{l=1}^L\sum_{m=1}^M \mathcal{R}(d_m^l) \sum_{n=1}^N \boldsymbol{\phi}(\boldsymbol{s}_{m,n}^l)\boldsymbol{\phi}(\boldsymbol{s}_{m,n}^l)^\top\right)^{-1} \\ &\quad \times \left(\sum_{l=1}^L\sum_{m=1}^M \mathcal{R}(d_m^l) \sum_{n=1}^N a_{m,n}^l \boldsymbol{\phi}(\boldsymbol{s}_{m,n}^l)\right), \\ (\widehat{\sigma}_{L+1}^{\mathrm{NIW}})^2 &= \left(N\sum_{l=1}^L\sum_{m=1}^M \mathcal{R}(d_m^l)\right)^{-1} \\ &\quad \times \left(\sum_{l=1}^L\sum_{m=1}^M \mathcal{R}(d_m^l) \sum_{n=1}^N \left(a_{m,n}^l - \widehat{\boldsymbol{k}}_{L+1}^{\mathrm{NIW}\top} \boldsymbol{\phi}(\boldsymbol{s}_{m,n}^l)\right)^2\right). \end{aligned}$$

[1] As the number of episodes goes to infinity, the estimated parameter converges to the optimal value and its variance achieves the Cramér-Rao lower bound [11].

The superscript 'NIW' stands for 'No Importance Weight'. However, since policies are updated in each RWR iteration, $\{\mathcal{D}^l\}_{l=1}^{L}$ generally follow different distributions for different policies and, therefore, the naive use of Eq.(6) will result in a biased estimator.

Importance sampling can be used for coping with this problem. The basic idea of importance sampling is to weight the samples drawn from a different distribution to match the target distribution. More specifically, from i.i.d. (*independent and identically distributed*) samples $\{d_m\}_{m=1}^{M}$ following $P(d;\boldsymbol{\theta}_l)$, the expectation of some function $g(d)$ over another probability density function $P(d;\boldsymbol{\theta}_L)$ can be consistently estimated by the *importance-weighted average*:

$$\frac{1}{M}\sum_{m=1}^{M} g(d_m)\frac{P(d_m;\boldsymbol{\theta}_L)}{P(d_m;\boldsymbol{\theta}_l)} \overset{M\rightarrow\infty}{\longrightarrow} \mathop{\mathbb{E}}_{P(d;\boldsymbol{\theta}_l)}\left[g(d)\frac{P(d;\boldsymbol{\theta}_L)}{P(d;\boldsymbol{\theta}_l)}\right]$$
$$= \int g(d)\frac{P(d;\boldsymbol{\theta}_L)}{P(d;\boldsymbol{\theta}_l)}P(d;\boldsymbol{\theta}_l)\mathrm{d}d = \mathop{\mathbb{E}}_{P(d;\boldsymbol{\theta}_L)}[g(d)]\,.$$

The ratio of two densities $P(d;\boldsymbol{\theta}_L)/P(d;\boldsymbol{\theta}_l)$ is called the *importance weight* for d.

This importance sampling technique can be employed in RWR for obtaining a consistent estimator $\widehat{\boldsymbol{\theta}}_{L+1}^{\mathrm{IW}} \equiv (\widehat{\boldsymbol{k}}_{L+1}^{\mathrm{IW}\top}, \widehat{\sigma}_{L+1}^{\mathrm{IW}})^\top$, where

$$\widehat{\boldsymbol{k}}_{L+1}^{\mathrm{IW}} = \Big(\sum_{l=1}^{L}\sum_{m=1}^{M}\mathcal{R}(d_m^l)w_{L,l}(d_m^l)\sum_{n=1}^{N}\boldsymbol{\phi}(\boldsymbol{s}_{m,n}^l)\boldsymbol{\phi}(\boldsymbol{s}_{m,n}^l)^\top\Big)^{-1}$$
$$\times\Big(\sum_{l=1}^{L}\sum_{m=1}^{M}\mathcal{R}(d_m^l)w_{L,l}(d_m^l)\sum_{n=1}^{N}a_{m,n}^l\boldsymbol{\phi}(\boldsymbol{s}_{m,n}^l)\Big),$$
$$(\widehat{\sigma}_{L+1}^{\mathrm{IW}})^2 = \Big(N\sum_{l=1}^{L}\sum_{m=1}^{M}\mathcal{R}(d_m^l)w_{L,l}(d_m^l)\Big)^{-1}$$
$$\times\Big(\sum_{l=1}^{L}\sum_{m=1}^{M}\mathcal{R}(d_m^l)w_{L,l}(d_m^l)\sum_{n=1}^{N}\Big(a_{m,n}^l-\widehat{\boldsymbol{k}}_{L+1}^{\mathrm{IW}\top}\boldsymbol{\phi}(\boldsymbol{s}_{m,n}^l)\Big)^2\Big).$$

Here, $w_{L,l}(d)$ denotes the importance weight defined by

$$w_{L,l}(d) \equiv \frac{P(d;\boldsymbol{\theta}_L)}{P(d;\boldsymbol{\theta}_l)}.$$

The superscript 'IW' abbreviates 'Importance Weight'. According to Eq.(1), the two probability densities $P(d;\boldsymbol{\theta}_L)$ and $P(d;\boldsymbol{\theta}_l)$ both contain $P_{\mathrm{I}}(\boldsymbol{s}_1)$ and $\{P_{\mathrm{T}}(\boldsymbol{s}_{n+1}|\boldsymbol{s}_n,a_n)\}_{n=1}^{N}$. Since they cancel out in the importance weight, we can compute the importance weight without the knowledge of $P_{\mathrm{I}}(\boldsymbol{s})$ and $P_{\mathrm{T}}(\boldsymbol{s}'|\boldsymbol{s},a)$ as

$$w_{L,l}(d) = \frac{\prod_{n=1}^{N}\pi(a_{,n}|\boldsymbol{s}_n;\boldsymbol{\theta}_L)}{\prod_{n=1}^{N}\pi(a_n|\boldsymbol{s}_n;\boldsymbol{\theta}_l)}.$$

Although the importance-weighted estimator is guaranteed to be consistent, it is not *efficient* even asymptotically. Therefore, the importance-weighted estimator tends to be unstable when the number of samples is rather small. The purpose of this paper is to propose a new framework for systematically addressing instability problems in the direct policy search.

3 Adaptive Importance Weight for Stable Policy Search

In this section, we propose a new policy search method called *Reward-weighted Regression with sample Reuse* (R^3) for efficient sample reuse.

3.1 Bias-Variance Trade-Off and Adaptive Importance Weight

When data samples $\{\mathcal{D}_l\}_{l=1}^{L}$ follow different distributions, the NIW estimator is biased even asymptotically. On the other hand, the IW estimator is asymptotically unbiased [14]. Thus, IW would generally have a smaller bias than NIW. However, IW has a larger variance than NIW, so there is the trade-off between the bias and variance.

In order to appropriately control the bias-variance trade-off, we introduce an *adaptive importance weighting* technique [14]. For a *flattening parameter* ν ($\in [0,1]$), the importance weight $w_{L,l}(d)$ is 'flattened' as $w_{L,l}^{\nu}(d)$ ($w_{L,l}(d)$ to the power of ν). Then we have $\widehat{\boldsymbol{\theta}}_{L+1}^{\mathrm{AIW}} \equiv (\widehat{\boldsymbol{k}}_{L+1}^{\mathrm{AIW}\top}, \widehat{\sigma}_{L+1}^{\mathrm{AIW}})^\top$, where

$$
\begin{cases}
\widehat{\boldsymbol{k}}_{L+1}^{\mathrm{AIW}} = \Big(\sum_{l=1}^{L}\sum_{m=1}^{M} \mathcal{R}(d_m^l) w_{L,l}^{\nu}(d_m^l) \sum_{n=1}^{N} \boldsymbol{\phi}(\boldsymbol{s}_{m,n}^l)\boldsymbol{\phi}(\boldsymbol{s}_{m,n}^l)^\top\Big)^{-1} \\
\qquad\times \Big(\sum_{l=1}^{L}\sum_{m=1}^{M} \mathcal{R}(d_m^l) w_{L,l}^{\nu}(d_m^l) \sum_{n=1}^{N} a_{m,n}^l \boldsymbol{\phi}(\boldsymbol{s}_{m,n}^l)\Big), \\
(\widehat{\sigma}_{L+1}^{\mathrm{AIW}})^2 = \Big(N \sum_{l=1}^{L}\sum_{m=1}^{M} \mathcal{R}(d_m^l) w_{L,l}^{\nu}(d_m^l)\Big)^{-1} \\
\qquad\times \Big(\sum_{l=1}^{L}\sum_{m=1}^{M} \mathcal{R}(d_m^l) w_{L,l}^{\nu}(d_m^l) \sum_{n=1}^{N} \Big(a_{m,n}^l - \widehat{\boldsymbol{k}}_{L+1}^{\mathrm{AIW}\top}\boldsymbol{\phi}(\boldsymbol{s}_{m,n}^l)\Big)^2\Big).
\end{cases}
\tag{7}
$$

'AIW' stands for 'Adaptive Importance Weight'. When $\nu = 0$, AIW is reduced to NIW. On the other hand, when $\nu = 1$, AIW is reduced to IW. In practice, an intermediate ν often produces a better estimator since the bias-variance trade-off can be better controlled.

In the above AIW method, the flattening parameter value ν can be different for each $\mathcal{D}^l$. However, for simplicity, we employ a single common value ν.

3.2 Flattening Parameter Selection

The performance of AIW depends on the choice of the flattening parameter ν, which allows us to control the bias-variance trade-off. Here, we show how the value of the flattening parameter can be optimally chosen using samples.

The goal of the policy search is to find the optimal policy that maximizes the expected return $J(\boldsymbol{\theta})$. Therefore, the optimal flattening parameter value ν_L^* at the L-th iteration is given by

$$\nu_L^* \equiv \arg\max_{\nu} J(\widehat{\boldsymbol{\theta}}_{L+1}^{\mathrm{AIW}}(\nu)). \tag{8}$$

Directly obtaining ν_L^* requires to compute the expected return $J(\widehat{\boldsymbol{\theta}}_{L+1}^{\mathrm{AIW}}(\nu))$ for each candidate of ν. To this end, we need to collect data samples following $\pi(a|\boldsymbol{s};\widehat{\boldsymbol{\theta}}_{L+1}^{\mathrm{AIW}}(\nu))$ for each ν, which is prohibitively expensive. In order to reuse samples generated by previous policies, we propose to use a variation of cross-validation called *importance-weighted cross-validation* (IWCV) [15].

The basic idea of IWCV is to split the training dataset $\mathcal{D}^{1:L} = \{\mathcal{D}^l\}_{l=1}^L$ into an 'estimation part' and a 'validation part'. Then the parameter $\widehat{\boldsymbol{\theta}}_{L+1}^{\mathrm{AIW}}(\nu)$ is learned from the estimation part and its expected return $J(\widehat{\boldsymbol{\theta}}^{\mathrm{AIW}}(\nu))$ is approximated using the validation part. Below, we explain in more detail how we apply IWCV to the selection of the flattening parameter ν in the current context. For simplicity, we assume that M is divisible by K, i.e., M/K is an integer.

Let us randomly divide the training dataset $\mathcal{D}^{1:L} = \{\mathcal{D}^l\}_{l=1}^L$ into K (we use $K = 5$ in the experiments) disjoint subsets $\{\mathcal{D}_k^{1:L}\}_{k=1}^K$ of the same size, where each $\mathcal{D}_k^{1:L}$ contains M/K episodic samples from every $\mathcal{D}^l$. Let $\widehat{\boldsymbol{\theta}}_{L+1,k}^{\mathrm{AIW}}(\nu)$ be the policy parameter learned from $\{\mathcal{D}_{k'}^{1:L}\}_{k' \neq k}$ (i.e., without $\mathcal{D}_k^{1:L}$) by AIW. We estimate the expected return of $\widehat{\boldsymbol{\theta}}_{L+1,k}^{\mathrm{AIW}}(\nu)$ using $\mathcal{D}_k^{1:L}$ as

$$\widehat{J}_{\mathrm{IWCV}}^k(\widehat{\boldsymbol{\theta}}_{L+1,k}^{\mathrm{AIW}}(\nu)) \equiv \frac{1}{\eta} \sum_{d^l \in \mathcal{D}_k^{1:L}} \mathcal{R}(d^l) w_{L,l}(d^l),$$

where d^l denotes an episodic sample from $\mathcal{D}^l$ and η is a normalization constant. An ordinary choice of η would be $\eta = LM/K$, but we use a 'stable' variant $\eta \equiv \sum_{d^l \in \mathcal{D}_k^{1:L}} w_{L,l}(d^l)$ [10].

The above procedure is repeated for all $k = 1, 2, \ldots, K$ and the average score is computed:

$$\widehat{J}_{\mathrm{IWCV}}(\widehat{\boldsymbol{\theta}}_{L+1}^{\mathrm{AIW}}(\nu)) \equiv \frac{1}{K} \sum_{k=1}^{K} \widehat{J}_{\mathrm{IWCV}}^k(\widehat{\boldsymbol{\theta}}_{L+1,k}^{\mathrm{AIW}}(\nu)).$$

This is the K-fold IWCV estimate of $J(\widehat{\boldsymbol{\theta}}_{L+1}^{\mathrm{AIW}}(\nu))$, which is shown to be unbiased [15].

We compute this K-fold IWCV score for each candidate value of the flattening parameter ν and choose the one that maximizes the IWCV score:

$$\widehat{\nu}_{\mathrm{IWCV}} \equiv \arg\max_{\nu} \widehat{J}_{\mathrm{IWCV}}(\widehat{\boldsymbol{\theta}}_{L+1}^{\mathrm{AIW}}(\nu)).$$

Note that the above IWCV scheme can be used also for choosing the type and number of basis functions $\{\phi_i(\boldsymbol{s})\}_{i=1}^b$ in the Gaussian policy model (4).

3.3 Numerical Examples

Here, we illustrate how the proposed method behaves using a one-dimensional ball-balancing simulation illustrated in Fig.3. The goal is to control the angle of the plate so that the ball is kept in the middle of the plate. The action space $\mathcal{A}$ consists of the angle α ($\in (-\pi/4, \pi/4)$) [rad] of the plate which is one-dimensional and continuous. The state space $\mathcal{S}$ is also continuous and a state vector $\boldsymbol{s} = (x, \dot{x})^\top$ consists of the distance x [m] between the centers of mass of the ball and the plate, and the velocity $\dot{x}$ [m/s] of the ball. The distance x and velocity $\dot{x}$ can be modeled using the following equations:

$$x_{t+1} = x_t + \dot{x}_{t+1}\Delta t,$$
$$\dot{x}_{t+1} = \dot{x}_t + \Delta t\Big(-\frac{f}{m}\dot{x}_t - 9.8\sin(a_t)\Big),$$

where $f = 0.5$ is the friction coefficient, $m = 3$ [kg] is the mass of the ball, a_t [rad] is the action chosen at time t, and $\Delta t = 0.05$ [s] is the duration of a time step. We assume that if an action a_t is chosen, the plate angle will be adjusted with a single time-step (this simulation is for illustration purposes; more realistic experiments will be shown in the next section). The reward function $R(s, a, s')$ is a quadratic function defined as

$$R'(\boldsymbol{s}, a, \boldsymbol{s}') = -\boldsymbol{s}'^\top \begin{pmatrix} 1 & 0 \\ 0 & 0.5 \end{pmatrix} \boldsymbol{s}' - 0.1a^2.$$

This reward function indicates that the agent will receive the maximum reward (i.e., zero) when the ball stops at the center of the plate ($\boldsymbol{s} = \mathbf{0}$). However, the above reward function takes negative values and therefore violates the non-negativity assumption imposed in the derivation of the EM algorithm (see Section 2). To cope with this problem, when the policy search is carried out, we convert the return $\mathcal{R}(d)$ as

$$\mathcal{R}'(d) = -\frac{1}{0.00001 + \mathcal{R}(d)}. \tag{9}$$

We use the Gaussian policy model with $\boldsymbol{\phi}(\boldsymbol{s}) = \boldsymbol{s}$. The discount factor was set to $\gamma = 0.95$.

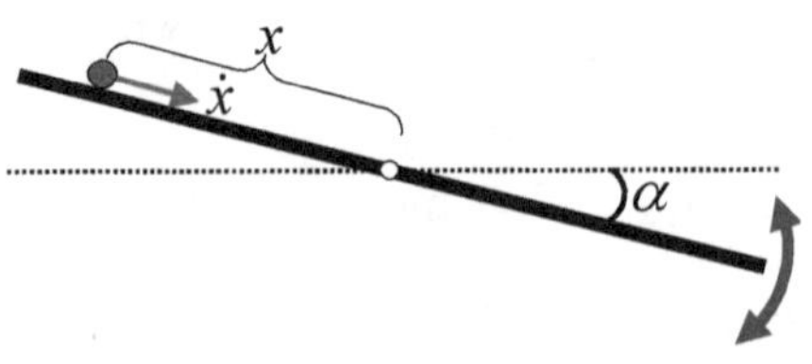

Fig. 3. Illustration of one-dimensional ball-balancing simulation. The goal is to control the angle α of the plate so that the ball is kept in the middle of plate.

First, let us illustrate how the flattening parameter ν influences the policy-parameter update. For this purpose, we compute $\widehat{\boldsymbol{\theta}}_{L+1}^{\mathrm{AIW}}(\nu)$ only from $\mathcal{D}^{L-1}$. The target policy parameter $\widehat{\boldsymbol{\theta}}_L = (\widehat{\boldsymbol{k}}_L^\top, \widehat{\sigma}_L)^\top$ is fixed to $\widehat{\boldsymbol{k}}_L = (0.5, 0.5)^\top$ and $\widehat{\sigma}_L = 0.2$. The sample-generation policy parameter $\widehat{\boldsymbol{\theta}}_{L-1}$ is chosen randomly as $\widehat{\boldsymbol{k}}_{L-1} = \widehat{\boldsymbol{k}}_L - (\beta_1, \beta_2)^\top$ and $\widehat{\sigma}_{L-1} = \widehat{\sigma}_L$, where β_1 and β_2 independently follow the uniform distribution on $[0, 0.2]$. Fig.4(a) depicts the true expected return $J(\widehat{\boldsymbol{\theta}}_{L+1}^{\mathrm{AIW}}(\nu))$ averaged over 20 trials as a function of the flattening parameter ν for $M = 5$ and 20.

The graph overall shows that when the number M of episodes is larger, the average expected return is also larger. However, the performance of NIW ($\nu = 0$) is not improved significantly when M is increased. The reason for this phenomenon is that increasing the number of samples does not contribute to reducing the estimation bias in NIW. The amount of variance reduction in NIW is limited as the variance is relatively small even when M is not large. On the other hand, the performance of IW ($\nu = 1$) is significantly improved as M is increased. This is because IW tends to have a large variance when M is small and increasing M highly contributes to reducing the variance. The graph also shows that neither NIW nor IW is the best in this simulation—intermediate values of ν (say, $0.2 \le \nu \le 0.4$) perform better than NIW and IW. Thus, given that ν is chosen optimally, AIW can outperform IW and NIW. Note that, although the amount of performance improvement by AIW over IW seems subtle in this one-step experiment, accumulation of this small gain through iterations actually causes big performance improvement (Fig.6).

Next, we illustrate how IWCV behaves. Fig.4(b) depicts the expected return $\widehat{J}_{\mathrm{IWCV}}(\widehat{\boldsymbol{\theta}}_{L+1}^{\mathrm{AIW}}(\nu))$ estimated by 5-fold IWCV, averaged over 20 trials as a function of the flattening parameter ν. The graphs show that IWCV roughly captures the trend of the true expected return for both cases ($M = 5$ and 20). Note that the orders of magnitude of the values in Fig.4(a) and in Fig.4(b) are different due to the importance weights. However, this does not cause a problem in model selection as long as relative profiles of the curves are similar.

Fig.5 depicts the expected return obtained by NIW ($\nu = 0$), IW ($\nu = 1$), and AIW+IWCV ($\nu \in \{0, 0.1, 0.2, \ldots, 1\}$ is selected in each trial using 5-fold IWCV) averaged over 20 trials as a function of the number M of episodes. This result indicates that the performance of NIW ($\nu = 0$) does not improve significantly when M is increased, implying that the estimation bias in NIW is crucial. On the other hand, the performance of IW ($\nu = 1$) is highly improved when M is increased. Thus, consistency of IW would be useful in this simulation. The proposed method, AIW+IWCV, tends to outperform both NIW and IW.

Finally, we illustrate how R^3 behaves in three different scenarios; through RWR iterations, ν is fixed to 0 (NIW), ν is fixed to 1 (IW), and ν is chosen by IWCV (R^3). Starting from a randomly-chosen initial policy-parameter $\widehat{\boldsymbol{\theta}}_1$, we let the agent collect samples $\mathcal{D}^L$ following the current policy $\pi(a|\boldsymbol{s}; \widehat{\boldsymbol{\theta}}_L)$, and then the policy parameter is updated using all samples $\{\mathcal{D}^l\}_{l=1}^L$. This is repeated over iterations. Fig.6 depicts the return averaged over 20 trials as a function of the

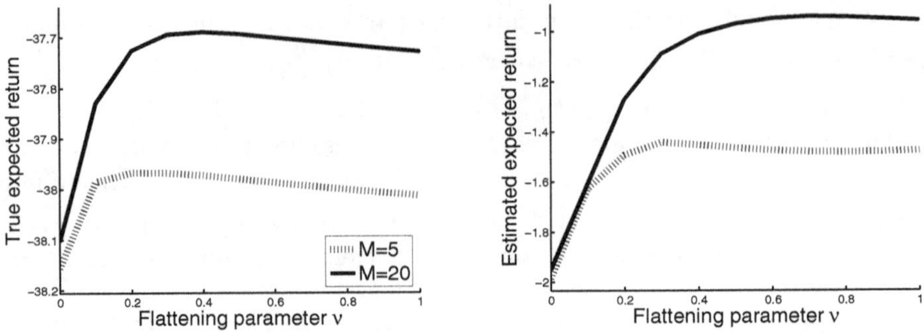

(a) True expected return (b) 5-fold IWCV estimate

Fig. 4. True expected return $J(\widehat{\boldsymbol{\theta}}^{\mathrm{AIW}}_{L+1}(\nu))$ and 5-fold IWCV estimate $\widehat{J}_{\mathrm{IWCV}}(\widehat{\boldsymbol{\theta}}^{\mathrm{AIW}}_{L+1}(\nu))$ averaged over 20 trials as a function of the flattening parameter ν in one-dimensional ball-balancing task

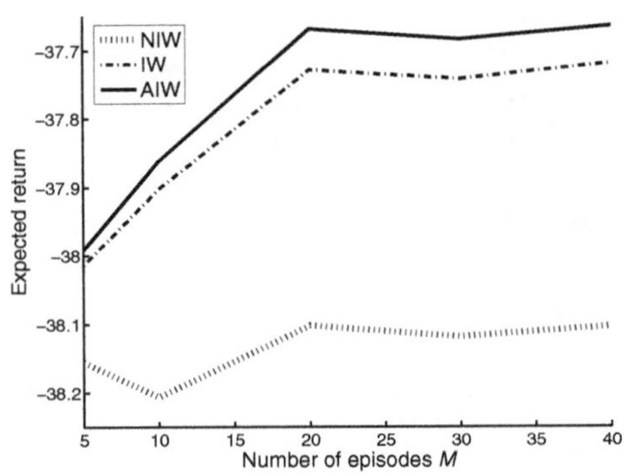

Fig. 5. Expected return obtained by NIW ($\nu = 0$), IW ($\nu = 1$), and AIW (ν is chosen by IWCV) averaged over 20 trials as a function of the number of episodes in one-dimensional ball-balancing task

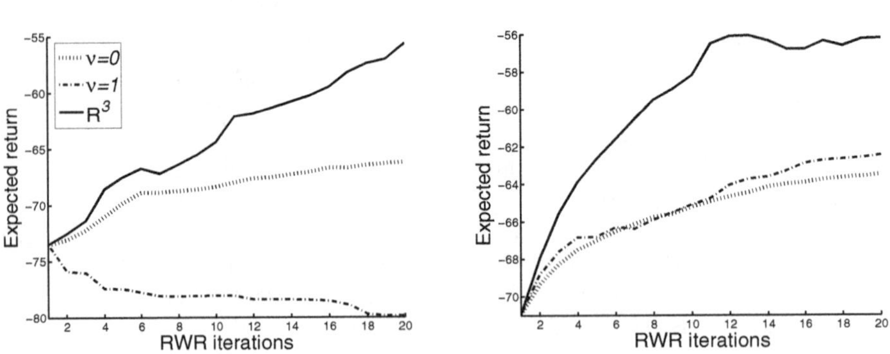

(a) $M = 5$ (b) $M = 20$

Fig. 6. The performance of policies learned in three scenarios: $\nu = 0$ (NIW), $\nu = 1$ (IW), and ν is chosen by IWCV (R^3) in one-dimensional ball-balancing task. The performance is measured by the expected return averaged over 20 trials.

number of RWR iterations; the number of episodes is $M = 5$ or 20, while the number of steps is fixed to $N = 20$.

The graphs show that R^3 tends to outperform the non-adaptive schemes. When ν is fixed to 0 (NIW), the performance of policy is stable for both cases. However, performance improvement is much slower than that for R^3. On the other hand, when ν is fixed to 1 (IW), the behavior depends on the number of samples; it works as well as NIW for $M = 20$, but the performance is poor and is not improved over iterations for $M = 5$ due to the high estimation variance.

Our original motivation to introduce R^3 was to reduce the number of samples for saving the sampling cost. When the sampling cost is high, it is preferable to keep M small, e.g., $M = 5$. In this case, the IW methods ($\nu = 1$) performs very poorly due to the huge estimation variance. On the other hand, the proposed R^3 method is stable even in the small sample case and its performance tends to be better than the naive RWR method ($\nu = 0$).

4 Applications

In this section, we evaluate the performance of our proposed method R^3 in a more complicated environment, i.e., a physically realistic ball-balancing task using a Barrett WAM^{TM} robot-arm simulator.

The 7 degree-of-freedom Barrett WAM^{TM} arm is mounted on the ceiling upside down and has a circular tray attached at the end-effector (see Fig.7). The goal is to control the robot so that the ball is always brought to the middle of the tray, similarly to the toy ball-balancing task in the previous section. However, unlike before, the angle of the tray cannot be directly controlled here as this is not feasible in a realistic scenario. Thus, achieving the goal is much harder.

The initial pose of the robot is set so that the tray is slightly tilted and the ball is rolling on the tray (see Fig.7). To simplify the problem, we control only two degrees of freedom: the wrist angle α_{roll} and the elbow angle α_{pitch}; all the remaining joints are fixed. Control of the wrist and elbow angles roughly corresponds to changing the *roll* and *pitch* of the tray, but not directly.

We design two separate control subsystems, each of which is in charge of roll and pitch control. Each subsystem has its own policy parameter $\boldsymbol{\theta}_*$, state space $\mathcal{S}_*$, and action space $\mathcal{A}_*$; '*' corresponds to 'roll' or 'pitch'. The state space $\mathcal{S}_*$ is continuous and consists of $(x_*, \dot{x}_*, \alpha_*)$, where x_* [m] is the distance between the centers of mass of the ball and the tray along each axis, $\dot{x}_*$ [m/s] is the velocity of the ball, and α_* [rad] is the joint angle. The action space $\mathcal{A}_*$ is continuous and corresponds to the target angle of the joint a_* [rad]. The reward function is defined as

$$R_*(\boldsymbol{s}, a, \boldsymbol{s}') = -(x'_*)^2 - (\dot{x}'_*)^2 - a_*^2.$$

Since this reward function is negative, the return is converted to a non-negative one by Eq.(9) when the policy search is carried out.

We explain how the control system is designed in more detail. As described in Fig.8, the control system has two feedback loops: joint-angle control using a *proportional and derivative (PD) controller* and motion control using an R^3

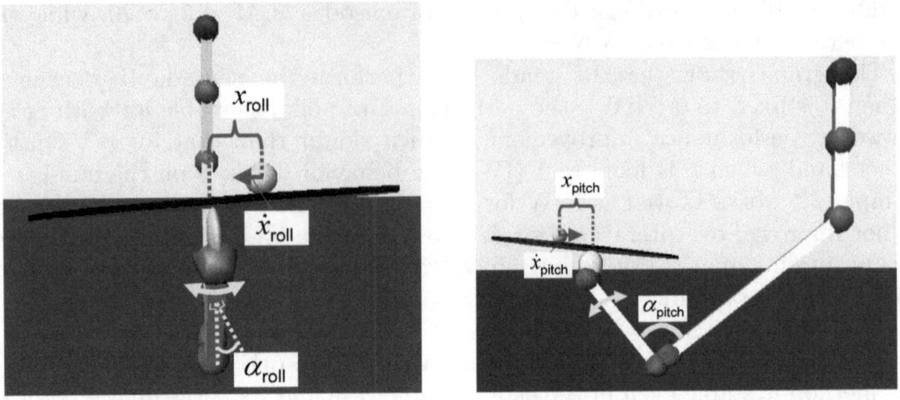

(a) The angle of wrist (b) The angle of elbow

Fig. 7. Ball-balancing task using a Barrett WAM^{TM} arm simulator. Two joints of the robots are controlled so as to keep the ball in the middle of the tray.

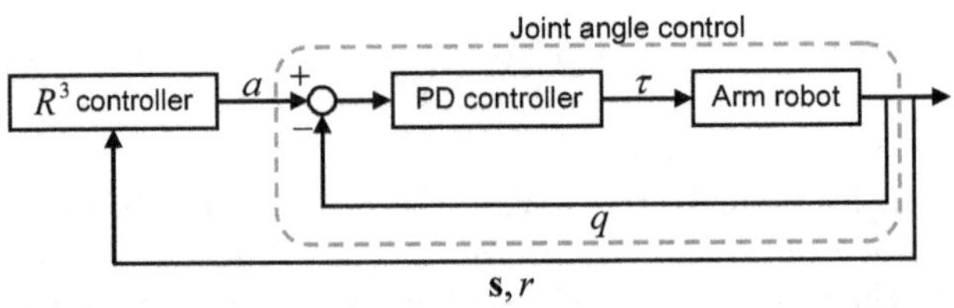

Fig. 8. The architecture of the Barrett WAM^{TM} robot-arm control system for ball balancing. The control system has two feedback loops, i.e., joint angle control by a PD controller and motion control by an R^3 controller.

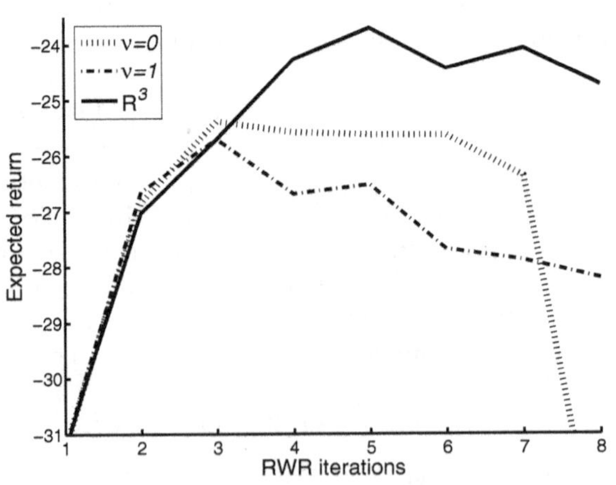

Fig. 9. The performance of learned policies when $\nu = 0$ (NIW), $\nu = 1$ (IW), and ν is chosen by IWCV (R^3) in ball balancing using a simulated Barrett WAM^{TM} arm system. The performance is measured by the expected return averaged over 10 trials.

controller. The PD controller outputs motor torque to achieve the desired joint angle a_* as

$$\tau_* = k_{\mathrm{p}}(\alpha_* - a_*) + k_{\mathrm{d}}\dot{\alpha}_*,$$

where $k_{\mathrm{p}} = 20$ is the proportional gain and $k_{\mathrm{d}} = 1$ is the derivative gain. On the other hand, the R^3 controller outputs the target joint angle obtained by current policy $\pi(a|\boldsymbol{s};\boldsymbol{\theta}_L)$; the policy parameter is learned by R^3 using the Gaussian policy model with linear basis function $\boldsymbol{\phi}(\boldsymbol{s}) = \boldsymbol{s}$.

We use the *Simulation Laboratory (SL) simulator* [12] for experiments, which is known to be highly realistic. The initial policy parameters and state[2] are randomly set as $\boldsymbol{k}_* = (\delta_*, \delta'_*, \delta''_*)^\top$, $\sigma_* = 0.2$, and $\boldsymbol{s}_* = (x_*, \dot{x}_*, \alpha_*)^\top = (\kappa_*, 0, 0.15)^\top$, where $\{\delta_*, \delta'_*, \delta''_*\}$ and κ_* are independently drawn the uniform distributions on $[0, 0.1]$ and $[0.045, 0.075]$, respectively. The dataset collected in each iteration consists of 5 episodes with 400 steps; the duration of an episode is 4 [s] in which[3] the sampling cycle and the R^3 control cycle are both 10 [ms]. We consider three scenarios here: sample reuse with $\nu = 0$ (fixed), $\nu = 1$ (fixed), and the proposed R^3 (ν is chosen by IWCV from $\{0.0, 0.2, 0.4, \ldots, 1.0\}$). The discount factor is set to $\gamma = 0.99$. Fig.9 depicts the averaged return over 10 trials as a function of the number of RWR iterations. The graph shows that R^3 nicely improves the performance. On the other hand, the performance using fixed $\nu = 0$ or $\nu = 1$ is saturated after the 2nd or 3rd iteration.

Overall, the proposed R^3 method is shown to be still promising in a complex robot-control task.

5 Conclusions

In real-world reinforcement learning problems, reducing the number of training samples is highly important as the sampling costs are often much higher than the computational cost. In this paper, we proposed a new framework of the direct policy search for efficient sample reuse. To overcome the instability problem caused by importance sampling, we proposed to combine reward-weighted regression with *adaptive* importance sampling techniques. The proposed method, called R^3, was shown to work well in experiments.

The proposed idea of using importance sampling techniques in the direct policy search is applicable to other policy search methods such as the *policy gradient* method [18,17], the *natural-policy gradient* method [5,9], and *policy search by dynamic programming* [1]. We will develop these variants and evaluate their performance in the future work.

Acknowledgements. HH acknowledges GCOE Computationism as a Foundation for the Sciences, and MS thanks MEXT Grant-in-Aid for Young Scientists (A), 20680007, SCAT, and AOARD.

---

[2] The angle of elbow α_{pitch} is offset by $\pi/2$.

[3] The time cycle of PD control is 2 [ms]. Thus, PD control is applied 5 times in each R^3 control.

References

1. Bagnell, J.A., Kakade, S., Ng, A.Y., Schneider, J.: Policy search by dynamic programming. In: Neural Information Processing Systems, vol. 16 (2003)
2. Dayan, P., Hinton, G.E.: Using expectation-maximization for reinforcement learning. Neural Computation 9(2), 271–278 (1997)
3. Dempster, A.P., Laird, N.M., Rubin, D.B.: Maximum likelihood from incomplete data via the em algorithm. Journal of the Royal Statistical Society B 39, 1–38 (1977)
4. Hachiya, H., Akiyama, T., Sugiyama, M., Peters, J.: Adaptive importance sampling with automatic model selection in value function approximation. In: Proceedings of the Twenty-Third National Conference on Artificial Intelligence (2008)
5. Kakade, S.: A natural policy gradient. In: Neural Information Processing Systems, vol. 14, pp. 1531–1538 (2002)
6. Kober, J., Peters, J.: Policy search for motor primitives in robotics. In: Neural Information Processing Systems, vol. 21 (2008)
7. Peshkin, C.R., Shelton, L.: Learning from scarce experience. In: Proceedings of International Conference on Machine Learning, pp. 498–505 (2002)
8. Peters, J., Schaal, S.: Reinforcement learning by reward-weighted regression for operational space control. In: Proceedings of the International Conference on Machine Learning (2007)
9. Peters, J., Vijayakumar, S., Shaal, S.: Natural actor-critic. In: Proceedings of the 16th European Conference on Machine Learning, pp. 280–291 (2005)
10. Precup, D., Sutton, R.S., Singh, S.: Eligibility traces for off-policy policy evaluation. In: Proceedings of International Conference on Machine Learning, pp. 759–766 (2000)
11. Rao, C.R.: Linear Statistical Inference and Its Applications. Wiley, Chichester (1973)
12. Schaal, S.: The SL Simulation and Real-Time Control Software Package. University of Southern California (2007)
13. Shelton, C.R.: Policy improvement for POMDPs using normalized importance sampling. In: Proceedings of Uncertainty in Artificial Intelligence, pp. 496–503 (2001)
14. Shimodaira, H.: Improving predictive inference under covariate shift by weighting the log-likelihood function. Journal of Statistical Planning and Inference 90(2), 227–244 (2000)
15. Sugiyama, M., Krauledat, M., Müller, K.-R.: Covariate shift adaptation by importance weighted cross validation. Journal of Machine Learning Research 8, 985–1005 (2007)
16. Sutton, R.S., Barto, A.G.: Reinforcement Learning: An Introduction. MIT Press, Cambridge (1998)
17. Sutton, R.S., Mcallester, M., Singh, S., Mansour, Y.: Policy gradient methods for reinforcement learning with function approximation. In: Advances in Neural Information Processing Systems, vol. 12, pp. 1057–1063. MIT Press, Cambridge (2000)
18. Williams, R.J.: Simple statistical gradient-following algorithms for connectionist reinforcement learning. Machine Learning 8, 229–256 (1992)

Applying Electromagnetic Field Theory Concepts to Clustering with Constraints

Huseyin Hakkoymaz[1], Georgios Chatzimilioudis[1], Dimitrios Gunopulos[1,2], and Heikki Mannila[3]

[1] Dept. of Computer Science, University of California, Riverside, CA, 92521, USA
{huseyin,gchatzim,dg}@cs.ucr.edu
[2] Dept. of Informatics and Telecommunications, Univ. of Athens, Greece
[3] HIIT, Helsinki University of Technology and University of Helsinki, Finland
heikki.mannila@cs.helsinki.fi

Abstract. This work shows how concepts from the electromagnetic field theory can be efficiently used in clustering with constraints. The proposed framework transforms vector data into a fully connected graph, or just works straight on the given graph data. User constraints are represented by electromagnetic fields that affect the weight of the graph's edges. A clustering algorithm is then applied on the adjusted graph, using k-distinct shortest paths as the distance measure. Our framework provides better accuracy compared to MPCK-Means, SS-Kernel-KMeans and Kmeans+Diagonal Metric even when very few constraints are used, significantly improves clustering performance on some datasets that other methods fail to partition successfully, and can cluster both vector and graph datasets. All these advantages are demonstrated through thorough experimental evaluation.

Keywords: Data Clustering, User Constraints, Electromagnetic Field Theory.

1 Introduction

The goal of clustering is to provide useful information by organizing data into groups (referred to as clusters). The use of labeled data is often important for the success of the clustering process and for the evaluation of the clustering accuracy. Consequently, learning approaches which use both labeled and unlabeled data have attracted the interest of the research community [3, 4, 6, 12, 20]. Such approaches incorporate user knowledge in the clustering technique, thus improving the clustering result. There are several ways to incorporate user knowledge in the clustering process. In this work we focus on the Clustering with User Constraints paradigm, where the user specifies constraints on groups of objects (typically pairs), and the goal is to produce a clustering that satisfies these constraints as much as possible.

In this paper, we present a novel way of applying user constraints for clustering; our approach is inspired by the Electromagnetic Field Theory in physics. We transform the dataset into a graph by linking k-nearest neighbors for each instance in the dataset, if the data are given as vectors. If the input is a distance matrix, no transformation is needed. Must-link and cannot-link constraints are then expressed naturally

W. Buntine et al. (Eds.): ECML PKDD 2009, Part I, LNAI 5781, pp. 485–500, 2009.

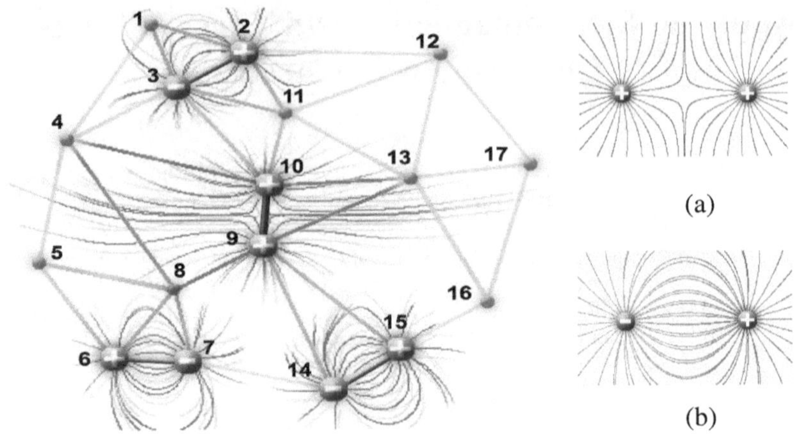

Fig. 1. The simulation of an EMF in a graph. (a)Like charges induces repulsive properties while (b) opposite charges induces attractive properties of other objects.

as magnetic fields between the nodes that are involved in the constraint. These fields impact edge weights based on the alignment of each edge compared to the magnetic field and its distance to the constraint axis. Using a graph representation yields the advantage that the edge weights can be adjusted without limitations, in contrast to Euclidean space where pairwise distances need to satisfy the triangle inequality. We exploit this liberty through a probabilistic model based on the nature of the constraint edges.

Contributions: We present a framework, called *EMC*, for clustering with constraints inspired by Electromagnetic Field Theory. The framework is general and assumes only that a distance function is available. The distance function does not need to satisfy the triangle inequality. If vector data is given, we transform it into a graph with minimal information loss; else if graph data is given we work directly on the given graph. We adjust the edge weights according to their proximity to the constraints in a probabilistic manner and run any clustering algorithm compatible with graphs. In our study, we use the K-Medoids algorithm [10], since it is less sensitive to outliers. For the distance metric to be used by the clustering algorithm, we propose *k simple-and-distinct shortest paths,* described in section 4.3. It allows for accurate clustering even with small number of constraints. Our method avoids the use of an objective function that strictly adheres to the given constraints, or a distance metric that is weighted according to the constraints and is applied globally.

Our experimental results confirm that our approach outperforms in accuracy current techniques including MPCK-Means [4], SS-Kernel-KMeans [12] and KMeans+ Diagonal Metric [20] even using very few constraints. Also, it can work on both vector and graph datasets.

2 Related Work

COP-KMeans [18] uses must-link and cannot-link constraints in an objective function to avoid incorrect assignment of data instances. It performs hard constrained clustering and computes the transitive closure of the constraints, thus suffering greatly from noisy constraints sensitivity. On the contrary, Xing et al. [20] proposed a distance metric learning algorithm which places a distance metric over the input space with the intention of assigning small distances between similar pairs. Distance metric learning aims to satisfy the maximum number of constraints by specifying different weights for different axes. The RCA algorithm [2] learns a Mahalanobis distance metric by using only must-link constraints. However, both approaches find one global metric which must be applied to all clusters.

MPCK-KMeans [4] integrates the strengths of both metric-based and constraint-based approaches in a principal manner. It learns individual distance-metrics for each cluster utilizing both unlabeled data and constraints. This allows different clusters to define their own space and have arbitrary boundaries. Its drawback: it suffers from noisy constraints.

HMRF-KMeans [3] is a probabilistic framework based on Hidden Markov Random Fields. Basu et al. have taken advantage of the available constraints in several ways. First, they estimate initial centroids using the constraints as the initialization step is crucial to accuracy of the KMeans algorithm. The model integrates constraint-based and distance-based approaches to maximize the joint likelihood of data and constraints while penalizing violated constraints. One weakness of the method is that it is applicable only to vector data, like all of the other mentioned so far.

Kulis et al. [12] extended HMRF-KMeans to a kernel-based framework which can handle both vector-based and graph-based data. They have established a connection between unweighted Kernel-KMeans, HMRF-KMeans and penalties for violated constraints. SS-Kernel-KMeans optimizes the kernel by preprocessing the similarity matrix with must-link and cannot-link constraints. Kernel methods are sensitive to the manual selection of the kernel's parameters. Yan and Domeniconi [22] proposed an adaptive method that estimates the optimal parameters. Our method adjusts edge weights more intuitively since it considers edges nearby the constraints as well, and this leads better performance. This model is based on the homophily concept [8], – nearby nodes tend to share same characteristics. Although the attraction/repulsion is practiced in unsupervised clustering [15], exploiting their use in semi-supervised clustering via the concept is completely new to our best knowledge.

Klein et al. [23] has based his algorithm on the same *homophily* intuition as our work, but can be applied only to vector data. Also, our method deals in a more comprehensive way with constraint satisfaction and triangular inequality restrictions.

3 Magnetically Affected Paths (MAP)

In this section, we describe the basic idea behind our approach in an intuitive way. We want to cluster a dataset of n points. We assume that a function defining the distance between any two points and a set of user defined constraints are given. Our goal is to cluster the points so that the constraints are satisfied as much as possible. We make no assumptions on the distance measure, but we assume that the constraints are on pairs of points, either Must-Link (the two points must be in the same cluster) or Cannot-Link (the two points cannot be in the same cluster). We use the user-defined constraints to "stretch" the space around the object that are accumulated around the constraints. Since we are dealing with graphs, our goal is to increase the weight of the paths connecting the objects that are in the neighborhood of a cannot-link constraint, and to decrease the weight of the paths connecting the objects that are in the neighborhood of a must-link constraint.

We realize the idea via the interesting analogy between electromagnetic field theory and graphs. We focus on the graph representation of a given dataset and assume that this graph has the characteristics of an electromagnetic field (EMF). In physics, an electric field is the property of the space in the vicinity of electric charges or in the presence of a time-varying magnetic field. The charges produce an electric field in space. This electric field exerts a force on other charged objects [14]. Opposite charged objects induce attractive properties, whereas like charged objects induce repulsive properties. To simulate these characteristics, we add charges to the nodes that take part in a pair-wise constraint. For must-link constraints, we add opposite charges to the node pair. The generated magnetic field decreases the weight of the affected edges. Cannot-link constraints (like charges) increase the weights of affected edge. The situation is illustrated in Figure 1, where must-link constraints are {*e(2,3), e(6,7), e(14,15)*} and the cannot-link constraints are {*e(9,10)*}.

We explore imaginary magnetic fields surrounding the pair of charged nodes. We check the nearby edges in the graph and identify the graph edges that are influenced by the magnetic field. Then, we reduce the weight of the affected edge or escalate it according to the constraint type. The magnitude of readjustment depends on the distance of the edge in regard to the magnetic field and its alignment in the field.

Before we proceed to the MAP, some definitions are necessary:

- *Constraint axis.* The straight line between the pair of nodes involved in a constraint
- *Reduction ratio rRatio(u,v)* is the decrement amount in edge weight w(u,v) due to a must-link constraint.
- *Escalation ratio eRatio(u,v)* is the increment amount in edge weight w(u,v) due to a cannot-link constraint.
- *Vertical distance vd(u,v).* The average distance of edge e(u,v) to the constraint axis.
- *Horizontal distance hd(u,v)* is defined as distance of edge e(u,v) to the mid-point of the *constraint axis.*

The effect of an EMF decreases as we get further away from the constraint axis (vertical distance). For must-link constraints, the horizontal distance has no effect on the reduction ratio. Cannot-link constraints utilize the horizontal distance of an edge to determine its probability of being in the separation region of two clusters. Intuitively, *the closer a regular edge e(u,v) is to the mid-point of a negative edge constraint, the higher the probability of it being an inter-cluster edge.* Based on this intuition, we apply the highest penalty to the edges overlapping with the mid-point of a constraint axis. The penalty is reduced as we go further away from the mid-point.

To summarize, MAP increases or decreases the weights of regular edges based on a probabilistic approach. Even though it classifies some of the edges incorrectly, overall re-adjustment of edge weights defines better distances in the graph domain.

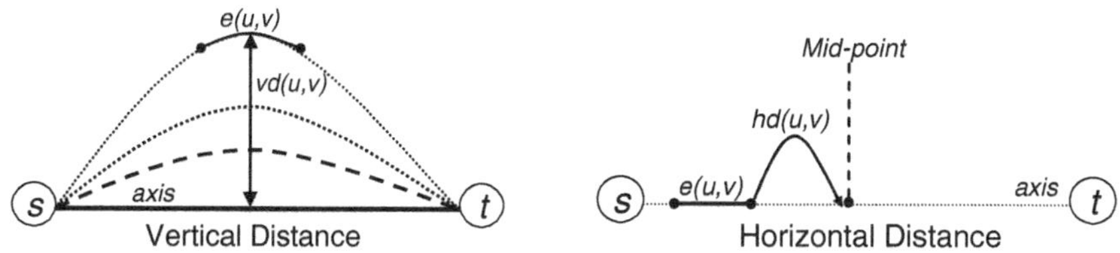

Fig. 2. Horizontal and vertical distances relative to the constraint axis (s and t are two nodes of a constraint)

4 EMC Framework

Here we describe our clustering framework that unifies MAP and clustering algorithms. The algorithm uses 3 steps to apply the MAP concept and cluster a given dataset: 1) *Graph Construction:* If given a vector-based dataset, it is converted into a graph by connecting k-nearest-neighbors. Must-link and cannot-link constraints are addressed as same or opposite charged nodes respectively. 2) *Weight Adjustment:* Identifies the edges that are affected by the given constraints and adjusts the edge weights accordingly. 3) *Clustering:* Runs an appropriate clustering algorithm to partition the adjusted graph.

4.1 Graph Construction

If the input is not a graph but a vector of data points Ds, our framework builds a graph reflecting the data with minimal loss of information. We list k nearest neighbors L_i for each object $x_i \in Ds$, according to their Euclidean distance, and add an edge between x_i and each $v \in L_i$. We assign the Euclidean distance $\|x_i - v\|_2$ as the edge weight of $e(x_i, v)$. k can be easily estimated from the graph size. Experiments show that k should be proportional to the dataset size $|Ds|$ for better accuracy.

We take all node pairs (s_i, t_i) involved in some constraint and charge s_i and t_i so that the force between them is equal to $\|s_i - t_i\|_2$. Remember, opposite charged pair of nodes create an attractive force (must-link constraint), whereas same charged pair of nodes create a repulsive force (cannot-link constraint).

Disconnected Components. Even if we set k to its optimal value, disconnected components might still exist. We identify all disconnected subgraphs, explore k nearest neighbors between subgraphs, and add an edge for each nearest neighbor connecting the disconnected components. This approach is similar to [12].

4.2 Weight Adjustment Algorithm

The weight adjustment phase applies the MAP concept to the graph in order to increase or decrease the edge weights. For each constraint $c(s,t)$, we extract a list L of edges $e(u,v) \in E$ which are affected by the constraint. Affected edges are labeled according to the following definition:

DEFINITION. *If an edge e(u,v) is in between nodes s and t of constraint c(s,t) and is not perpendicular to the constraint axis, then it is affected by constraint c(s,t).*

In our model, perpendicularity and betweenness is defined based on hop-count distance to the constraint pair nodes *s* and *t*. We run two breadth-first search algorithms starting at *s* and *t* separately and for each node we store entries *hc(u,s)* and *hc(u,t)*, the hop-count distance to *s* and to the *t* respectively. For a given edge *e(u,v)*, we check the hop-count entries of *u* and *v* to see whether the edge is affected by a constraint or not. Perpendicularity and betweenness is defined as the inverse behavior on *hc(u,s)*, *hc(u,t)* and *hc(v,s)*, *hc(v,t)* values for nodes *u* and *v*. In other words, if *hc(u,s)>hc(v,s)* and *hc(u,t)<hc(v,t)* or vise versa, then the edge is perpendicular to the constraint axis and between nodes *s* and *t*.

Once we identify the affected edges, we compute the escalation ratio for the cannot-link constraints or reduction ratio for the must-link constraints. In line with the main idea, we expect the effective escalation/reduction ratio on an affected edge to decrease as we get away from the constraint axis (vertical distance). For a cannot-link constraint, we expect an inversely proportional weight increase in regard to the distance of the edge to the mid-point of the constraint axis. To express this, we use a method similar to the validation process. Instead of hop counts, we find the shortest path distance to all edges starting at node *s* and *t*. Shortest path *dist(u,v)* is the sum of the weights of all edges that compose the shortest path. We compute the reduction ratio of *e(u,v)* for must-link constraints as follows:

$$rRatio(u,v) = norm(\frac{q_r}{r}) \tag{1}$$

Here, q_r is the weight for the must-link constraint and *r=(dist(u,s) + dist(u,t) + dist(v,s) + dist(v,t)) / w(s,t)* which is the dispersion ratio from the constraint *c(s,t)*. *norm()* is the normalizing function. To prevent extremely low values, normalization function maps the output to a higher interval.

Similarly, we can write the following equation to compute the escalation ratio due to a cannot-link constraint:

$$eRatio(u,v) = norm(\frac{q_e}{r \cdot \Delta}) \tag{2}$$

where q_e is the weight of the cannot-link constraint, Δ=(|*dist(u,s)-dist(u,t)*|+|*dist(v,s) - dist(v,t)*|)/*w(s,t)+c,* which is the average distance approximation function to the mid-point of the constraint axis to reflect the effect of horizontal distance as seen in Figure 2. The value of |*dist(u,s)-dist(u,t)*| becomes zero if node u has equal distances to the s and t. We add a constant value *c*=1 to the Δ so that in this case, no penalty is applied to *eRatio(u,v)*. If Δ increases, the effect of *eRatio(u,v)* reduces gradually.

After applying all constraints, we approximate the overall ratio of edge *e(u,v)* as:

$$tRatio(u,v) = \frac{\sum_{i=1}^{|C|} eRatio_i(u,v)}{|C|} - \frac{\sum_{j=1}^{|M|} rRatio_j(u,v)}{|M|} \tag{3}$$

In the last step, we adjust the edge weight as follows:

$$w_{new}(u,v) = w(u,v) \cdot \alpha^{tRatio(u,v)} \tag{4}$$

Algorithm. Weight_Adjustment_Algorithm

Input: G(V,E): graph with constraints

Output: G' (V,E'): graph with adjusted edge weights

P: proximity matrix

1. For each constraint *c(s,t)*
   a. Run breath-first search algorithm starting at node *s* and *t*; and record hop-counts for each node $v_i \in$V
   b. Run single shortest path algorithm starting at node *s* and *t*; and record shortest path distances for each node $v_i \in$V
   c. Identify affected edges using hop-counts and put them into list L
   d. Compute escalation/reduction ratio for each affected edge $e(u,v) \in$L
2. For each edge $e(u,v) \in$E
   a. Calculate overall ratio using following formula
   $$tRatio(u,v) = \frac{\sum_{i=1}^{|C|} eRatio_i(u,v)}{|C|} - \frac{\sum_{j=1}^{|M|} rRatio_j(u,v)}{|M|}$$
   b. Apply overall ratio to the edge weight $w_{new}(u,v) = w(u,v) \cdot \alpha^{tRatio(u,v)}$
3. For each node pair (v, u), calculate the distance $D(u,v) = \left(\sum_{q=1}^{k} \frac{1}{dist_q(u,v)} \right)^{-1}$.

Fig. 3. Pseudo-code of Weight Adjustment Algorithm

Empirically, we have observed that $1<\alpha<2$ is a good interval for the adjustment of the edge weights. It is obvious that if cannot-link constraints are dominant upon the must-link ones, then the edge weight increases. Note that vertical and horizontal distances are used just to compute the adjustment ratio and are not used in any way in the clustering process.

After adjusting all the edge weights, we have our final graph and the algorithm uses this graph to extracts distance matrix *D*. The distance between two node pairs is defined using *k*-shortest paths distance. Remember that *k* was the number of neighbors used in order to transform vector data into a graph representation. For vector data, we use the same *k* for the number of shortest paths since this is the maximum out-links number of an edge. Using multiple shortest paths as a distance between two nodes, works in practice better than a single shortest path. Involving more paths in the calculation of the distance involves also more constraints, which yields better accuracy even for a small number of constraints due to the homophily phenomenon. A naïve approach for k-shortest paths is using the Dijkstra's algorithm to discover *k* most significant paths one by one.

Each constraint affects the graph locally. Thus, we can focus on subgraphs that capture the relevant information we need rather than whole graph. We have adopted the idea of very small-connection subgraphs from [11]. The weight adjustment step asymptotically takes O((|*M*| + |*C*|) (|*E*| + |*V*| log|*V*|)) time, but in practice it is much faster. We explore k-shortest path for every node pairs. Extracting the proximity table requires O(k $|V|^2$ ($|E|$ + $|V|$ log$|V|$)) time. Thus, the weight adjustment phase takes overall O(k $|V|^2$ ($|E|$ + $|V|$ log$|V|$)).

4.3 Optimization for K-SD Shortest Path Algorithm

Finding the k-shortest simple paths for every pair of nodes is the bottleneck for the efficiency of our algorithm. Even though there is more than one k-shortest paths definition in the literature [5], we focus on k simple and distinct shortest paths in which no loop is allowed, i.e. all vertices on a path are distinct and no two paths share the same edge for a given source and destination pair.

We extend the single shortest path Dijkstra algorithm to k-shortest paths at a reasonable cost and refer to it as K-SD shortest path algorithm. The new algorithm is asymptotically only k times slower than the Dijkstra algorithm for one path. For each node, we define k entries to handle each lth path passing through, where $1 \leq l \leq k$. We initialize all entries to ∞ except the entries of source node s and nodes adjacent to the source. We set all s entries to 0. We assign monotonically increasing path labels i to each node u adjacent to s ensuring that each path is rooted at a different edge outgoing s. We update the lth entry of each node u to edge weight $w(s,u)$ and the parents of lth entry of the nodes to source node s where l is the path label pertaining to each adjacent node. Then, we initialize a minimum priority queue Q that contains all path

---

Algorithm. K-SD_Shortest_Path_Algorithm

Input: G'(V,E): graph with adjusted edge weights
s: source node
k: number of shortest path

Output: P: Distance matrix

1. Assign k path entries for each node such as $p_i(s,u)$
2. Initialize $p_i(s,u).length$ ← +∞ for each node $u \neq s$
3. For each node u adjacent to s
   a. Assign a monotonically increasing path id i to u
   b. Set $p_i(s,u).dist$ ← $w(s,u)$ and $parent_i(u)$ ← s
4. Let a min priority queue Q contain all path entries
5. while Q is not empty do
   a. Extract path entry pe ← Q.removeMin()
   b. Let i ← $pe.pathID$ and v ← $pe.node$
   c. Lock $parent_i(v)$ to prevent updates for v
   d. for each node u adjacent to v
      i. if v is locked for u, then continue
      ii. if $p_i(s,v).length+w(v,u)<p_i(s,u).length$, then
         $p_i(s,u).length$ ← $p_i(s,v).length+w(v,u)$
         $parent_i(u)$ ← v
         Update the value of $p_i(s,u)$ in queue Q

Fig. 4. Pseudo-code of K-SD Shortest Path Algorithm

entries. The rest of the algorithm works very similar to Dijkstra's shortest path algorithm except for a few additional restrictions in the relaxation routine.

When we remove the minimum entry from Q, we lock the parent of the entry in order to prevent further updates from this parent for other path entries, using bitmaps with each bit assigned to one neighbor node. In the relaxation procedure, first we check whether the current node is locked for destination node. If it is not, the algorithm allows it to relax the destination from current node. Otherwise, we simply proceed to the next available neighbor node. Another rule is that l^{th} entry of path entries at a node can be updated only by the lth path entry of the parent node. This limitation is necessary in order to force all paths to follow a different set of edges to the destination. We repeat the relaxation process until all path entries at every single node are updated and no more entries reside in the queue (Figure 4).

THEOREM 1. *K-SD Shortest Path algorithm identifies k simple and distinct shortest paths from a given source node to all other nodes in* $O(k^2\ |V|\ log|V|)$ *time.*

Proof: We assume the worst case scenario where we find all shortest path available, giving us $O(k\ |V|)$ entries in the queue. For each entry, we check all neighbor nodes and based on k-shortest path definition and our graph construction algorithm, we may have at most k neighbors. We may update the keys of all neighbors at each relaxation step, requiring $O(k\ log|V|)$ time. Therefore, our algorithm takes overall $O(k^2\ |V|\ log|V|)$ time to find all k simple and distinct shortest paths from a given source node s to all other nodes.

4.4 Multi-level Approach for Extracting Similarity Matrix

Even with the K-SD shortest path algorithm from the previous section, it takes $O(k^2 \cdot |V|^2 \cdot log|V|)$ to compute all pairs, which is inefficient for large datasets. Many state-of-the-art methods deal with this problem by using a multilevel approach. Our strategy is to partition the graph into smaller pieces, compute distance matrices for each partition, and correlate them via hubs to obtain the global solution [17].

We partition the graph into p equally sized subgraphs using METIS with the Kernighan-Lin objective and find local K-SD shortest path distances for each partition. Let D_i be the distance matrix for partition i. To establish a relationship between partitions and be able to estimate the global distances we use the hub concept discussed in [21]. The vertices that reside on the cut and bridge different clusters are considered hubs. We are interested only in high quality hubs, which have high degree and their edges are balanced amongst the different partitions it bridges. Unlike the hubs in SCAN, we assume hubs are special vertices belonging to all partitions that it bridges, as shown in Fig. 5. If needed, we add new edges to hubs to maintain the k-neighborhood.

Let H be the set of all hubs in the graph, H_i be the set of hubs in partition i, and m be the size of H. Notice that $m<<n$, which makes a fast correlation possible. We already have computed the distance between nodes in partition i and all hubs in H_i. Next, we find the pair-wise distances between hubs H so that we can compute the distance of two elements that belong to two different partitions. Since H_i hubs are also elements of another partition by definition, we get a fully connected graph of hubs, where edges represent the distance between two hubs. By running

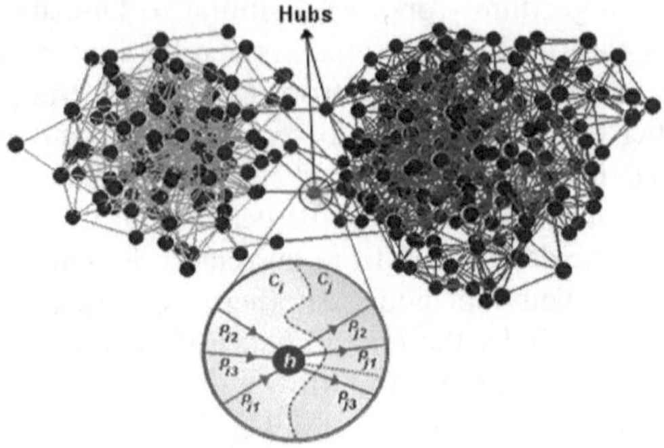

Fig. 5. The *Ionosphere* graph with two partitions and hubs connecting them, and the correlation of same paths for clusters C_i and C_j

Floyd-Warshall on this imaginary graph, we get a distance matrix *S* of all hubs in the graph. The matrix *S* will serve as an approximation of the KSD-shortest path distance between two nodes of different clusters.

For example, assume we have performed all the preparation for extracting the global distances in an efficient way. Let *s* be the source node, *t* be the destination node, and $S(h_i,h_j)$ be the distance between hubs h_i and h_j. Then, the KSD-shortest path problem can be expressed as the following optimization problem:

$$D(s,t) = min\{D_i(s,h_a) + S(h_a,h_b) + D_j(h_b,t)\} \quad \text{where } \forall h_a \in H_i \text{ and} \forall h_b \in H_j \} \tag{5}$$

We have the distances between *s* and all H_i elements in matrix D_i. We compute the distances from *s* to the other hubs in set $(H\text{-}H_i)$ through H_i using the *S* matrix. Each destination node checks only the hubs of its own partition in H_j to relax the shortest path distance to node *s*. It takes $O(|H_i|\ (|H| - |H_i|) + |V - V_i|\ |H_j|)$ to find *k*-shortest paths from *s* to all other nodes. Note that, $|V_i| \approx |V|\ /\ p$ and $|H_i| \approx |H|\ /\ p$. With this partitioning concept, our algorithm *(EMC)* has an overall time complexity of $O(\frac{|V|^2 k^2 \log(|V|)}{p} + |H|^3 + \frac{|V|\cdot|H|\cdot(|V|+|H|)}{p})$. If p=|V|, it finds only single shortest paths. In practice, this approach runs hundreds times faster compared to the naïve approach.

4.5 Clustering Algorithm

The framework allows us to use any graph-compatible clustering algorithm. The success of clustering essentially depends on the compatibility of the dataset and the clustering algorithm. Thus, the choice of the algorithm must be made cautiously.

We implement the K-Medoids algorithm, which utilizes the similarity matrix and can be applied on both vector and the graph data. The initial centroids play a significant role for the final clustering, so instead of random initialization, we take advantage of the given constraints. We take the transitive closure of the must-link constraints and define groups of nodes, which have to be clustered together. At this point, each group represents a set. We merge the closest two sets until we have *K* sets remaining. Eventually, we have *K* disconnected sets formed by constraints, which we use to initialize the medoids.

The algorithm starts by assigning every point $x_i \in Ds$ to the cluster that minimizes the distance between x_i and $\boldsymbol{\mu}_k$ where $\boldsymbol{\mu}_k$ is the cluster medoid of cluster k^*. Rather than Euclidean distance, we use the distance matrix extracted in the previous step for assignment. The algorithm re-estimates medoid $\boldsymbol{\mu}_k$ using the points assigned to cluster k^*. For each point x_i, we check the total distance to all other points and we assign the point with minimum distance as the cluster medoid. Then, we repeat the steps until algorithm converges or reaches to a pre-specified number of runs. The time complexity of the clustering process is $O(t\ K\ N^2)$ where t is the number of iterations, K is number of clusters and N is the size of the dataset.

5 Experiments

5.1 Experimental Setup

We experimented on two synthetic datasets and seven real datasets from UCI Machine Learning Repository [1]: *Soybean, Iris, Wine, Ionosphere, Balance, Breast Cancer and Satellite.* The properties of these dataset are summarized in Table 1. *N* is the number of instances, d is the number of dimensions, and K is the number of clusters in each dataset. We have measured the clustering accuracy as:

$$Accuracy = \sum_{i>j} \frac{1\{1\{c_i = c_j\} = 1\{\hat{c}_i = \hat{c}_j\}\}}{0.5N(N-1)} \tag{6}$$

where $1\{\cdot\}$ returns 1 if any pair of instances x_i and x_j are assigned correctly by the algorithm [20]. In each experimental setup, we have run the clustering algorithms for 50 times and reported the average accuracy ratios.

Must-link and cannot-link constraints are generated randomly at equal amounts and the total amount is varied proportional to the dataset size. For each run, we use the same constraint set for all algorithms. EMC parameters: we use α=1.6 (shown through preliminary experiments to give good results), and increase the nearest neighbors and number of shortest paths proportional to the dataset size. Weights of the positive/negative constraint edge, q_r/q_e, are set to 1.

Table 1. Datasets used in experiments and running time of the algorithms (*in seconds*)

		Soybean	*Iris*	*Wine*	*Ionosphere*	*Balance*	*Breast*	*Satellite*
Dataset details	*N*	47	150	178	351	625	683	4435
	d	35	4	13	34	4	9	36
	K	4	3	2	2	3	2	6
Running Times (secs)	*SS-Kernel*	~0.1	0.1	0.1	0.1	0.3	0.6	73.8
	MPCK	0.4	0.5	0.5	0.5	0.5	0.6	6.4
	Diagonal	0.2	0.3	0.6	1.3	3.6	5.1	304.3
	EMC	0.6	1.1	1.6	4.3	7.0	7.3	66.1

To visualize how our method works, we generated two synthetic datasets:

Gaussian: A set of 180 two-dimensional instances generated by Gaussian number generator, as shown in Figure 6. 120 instances in vertical and lower horizontal sets are labeled as class one. Upper horizontal set is label as class two.

ThreeCircles: Similar to *TwoCircles* data in [12], we have generated three layered circular data with 300 instances in 2 dimensions. Each circle represents one class with 100 data points in it.

We generated small amount of must-link and cannot link constraints (18 for *Gaussian* and 30 for *ThreeCircles*). Figure 6 shows final clustering of these datasets with high accuracy. For the circular data, Graph Construction process had a pre-clustering effect. As we have selected k-nearest neighbors for each point, there were not too many inter-cluster edges in the graph. In addition, the re-adjustment phase successfully wiped out the effects of these inter-cluster edges for the clustering phase. The rest of the experiments are run on the real datasets.

5.2 Effect of Parameters k and p

In this set of experiments we investigate the effect of the values of parameters k and p. k is the number of neighbors we pick, when converting vector data to graph data, and the number of shortest paths we use for our distance metric. For values below 5, we get heavily disconnected graphs, when converting vector data to graph data, thus experiments are not run for smaller values. We note that relatively small values of k (>4) are performing well enough and by using greater values we do not get significant improvements (Figure 9). On the other hand, greater values increase execution time. In summary, with small values of k we cannot fully take advantage of the k-shortest paths distance metric while for large values all edges are labeled as an affected edge, and consequently false escalation or reduction occurs for too many edges.

The number of partitions, p, has a significant effect on running time while preserving the accuracy (Figure 7). The algorithm runs up to 24x faster for the Breast dataset without significant loss of accuracy. Given that this is a small dataset, we have more gain in performance for larger datasets. After p=12, the accuracy starts to decline because the K-SD shortest path distance approximation does not keep up with very small-sized subgraphs. On the flip side, the running time starts increasing after p=16. The reason for this situation is the high number of hubs. The computation of matrix S starts to dominate the running time as it requires $O(|H|^3)$ time. One advantage of partitioning is that we no longer need to increase the value of parameter k and about five shortest paths are quite enough to compute distances accurately.

5.3 Effects of Only Must-Link or Cannot-Link Constraints

Next, we examine the effect of constraint types individually (Figure 8). When we use the must-link constraints alone, it reduces the weights of both inter-cluster and intra-cluster edges. When the reduction ratio on weights of intra-cluster edges is larger than inter-cluster edges, there is a small gain. Similarly, when cannot-link constraints are used alone, the weights of inter-cluster edges increase more than weights of intra-cluster edges. However, the gain in accuracy is greater than when using only positive edges. In that respect our algorithm's behavior differs from the one reported in [13],

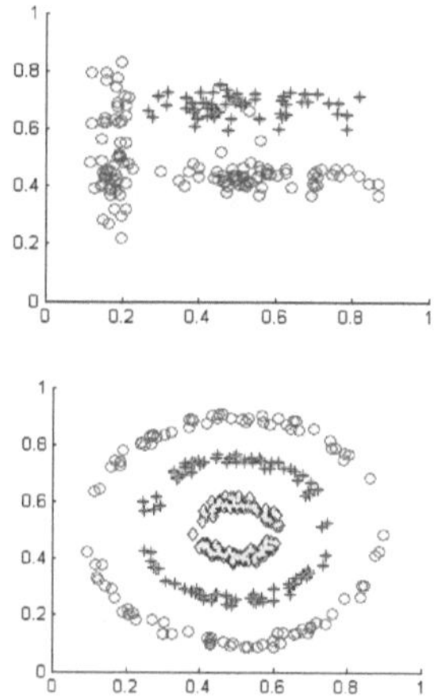

Fig. 6. EMC clusters (a) *Gaus-sian* and (b) *ThreeCircles* datasets with small set of constraints

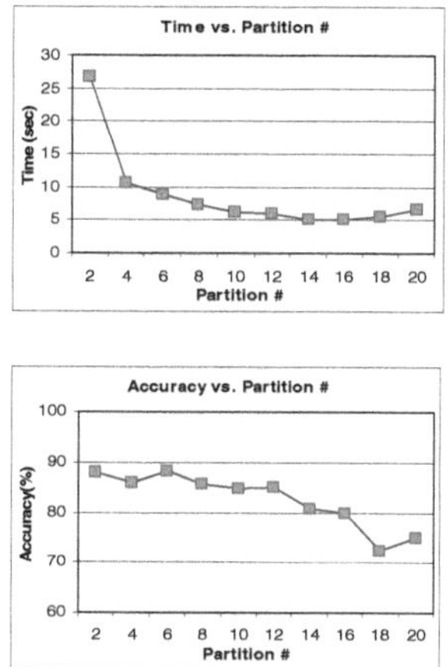

Fig. 7. Effect of partition # on clustering Breast dataset in terms of (a) running time and (b) accuracy

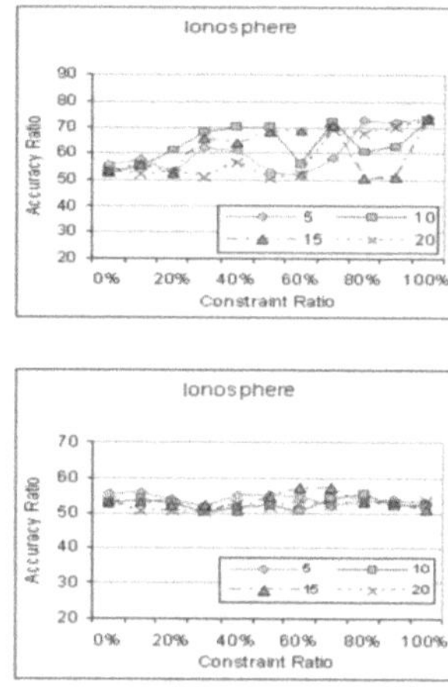

Fig. 8. The effect of only (a) negative edges and (b) only positive edges on Ionosphere dataset

and this shows that the informativeness of a constraint type depends on how it is applied. Furthermore, the accuracy ratio trend is not steady as the number of constraints is augmented. On the other hand, when used together, we get optimal results for the algorithm. On incorrectly validated edges, as seen in Figure 10, they cancel the effect of each other. We observed the same phenomena for other algorithms as well.

5.4 Comparison with Other Techniques

We used real datasets to compare our EMC algorithm with the MPCK-Means, SS-Kernel-KMeans and KMeans+Diagonal Metric algorithms, which are publicly available online. We used the same parameters for EMC: k=5 and p=|V|/80 (each subgraph has approximately 80 nodes. EMC outperforms MPCK-Means, SS-Kernel-KMeans and KMeans+Diagonal Metric algorithms on all datasets, except *Breast* and *Wine* *(Fig. 11)*. It runs better than SS-Kernel-KMeans and Kmeans+Diagonal Metric on *Breast* dataset and quite reasonable compared to the MPCK-Means. For the *Wine* dataset, graph-based methods such as SS-Kernel-KMeans and EMC do not improve the performance significantly and overall accuracy is very low compared to metric-based methods.

SS-Kernel-KMeans runs the min-cut objective while EMC tries to minimize the overall pairwise distance. The same way MPCK-Means and Kmeans+Diagonal Metric algorithms could not improve the clustering for *Balance* regardless of constraint amount, EMC fails to increase the accuracy for *Wine* dataset. In some experiments, we detect phenomena where the accuracy of the algorithm goes up and down slightly as we increase the number of constraints. As shown in [6], this is a general problem of randomly-chosen constraint sets, where some constraints reduce the clustering

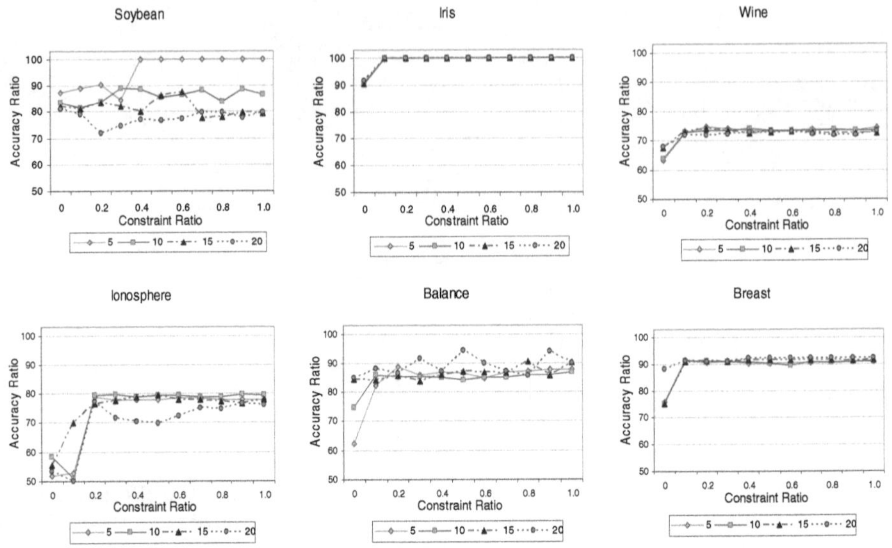

Fig. 9. Clustering results for EMC. The plots show the accuracy ratio achieved in respect to the constraints ratio used. The different lines stand for different values of k ranging from 5 to 20. Constraints amounts are $x{\cdot}N$ where x is the constraint ratio.

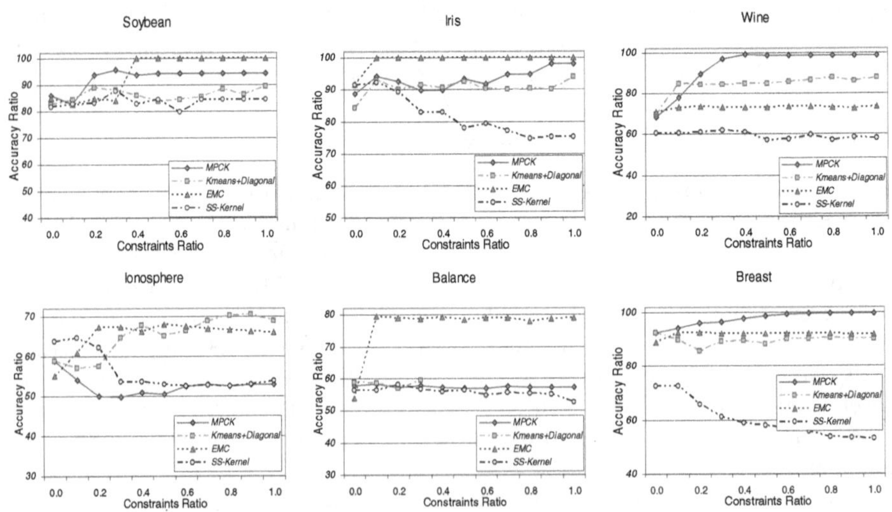

Fig. 10. EMC vs. MPCK-Means, KMeans+Diagonal and SS-Kernel-KMeans

performance. Thus, a learning metric or an edge weight re-adjustment method, is not always reliable for a small number of constraints. Compared to other methods, EMC is typically more trustworthy even when using few constraints.

5.5 Running Time Experiments

We have performed experiments on the running time of the algorithms. All experiments were carried out on 1.7 GHz Pentium IV machine with 512 MB memory. We have performed 10 experiments for each algorithm as we increase the constraint amount by $10\% \cdot N$, where N is the dataset size, for each experiment and reported the average running time of these experiments. The running time of EMC is comparable to the other algorithms tested. In addition, EMC scales almost linearly with the number of partitions in practice. It is almost linear because it processes equal-sized subgraphs and number of hubs affects the linearity in time complexity during merge process. Results are given in Table 1.

6 Conclusions

We have presented a framework that, when given a dataset of instances and user constraints, transforms vector data into a graph and improves the clustering algorithm distance metric by adjusting the edge weights based on user constraints. The most important contribution lies in the way the weights are adjusted, based on ideas from Electromagnetic Field Theory. Instead of modifying the distance metric, it alters the distances between objects in the graph domain. EMC algorithm allows us to cluster both vector-based and graph-based datasets and it works with distances only as well. In addition to K-Medoids, we can also integrate other clustering algorithms into the framework. We have shown that even when using a small amount of constraints, the algorithm improves the clustering accuracy significantly.

Acknowledgments. This work was partially supported by NSF IIS-0534781, NSF 0803410, ONR N00014-07-C-0311 AWARE, Health-e-Child and SemsorGrid4Env.

References

[1] Asuncion, A., Newman, D.J.: UCI Machine Learning Repository
[2] Bar-Hillel, A., Hertz, T., Shental, N., Weinshall, D.: Learning distance function using equivalence relations. In: ICML 2003, Washington DC (August 2003)
[3] Basu, S., Bilenko, M., Mooney, R.J.: A Probabilistic Framework for Semi-Supervised Clustering. In: KDD 2004, Seattle, WA (August 2004)
[4] Bilenko, M., Basu, S., Mooney, R.J.: Integrating Constraints and Metric Learning in Semi-Supervised Clustering. In: ICML 2004, Canada, July 2004, pp. 81–88 (2004)
[5] Brander, A., Sinclair, M.: A comparative study of k-shortest path algorithms. In: Proceedings of 11th UK Performance Engineering Workshop for Computer and Telecomm. Systems (1995)

[6] Davidson, I., Wagstaff, K., Basu, S.: Measuring Constraint-Set Utility for Partitional Clustering Algorithms. In: Proceedings of the 17th European Conference on Machine Learning, Berlin, Germany, September 18-22 (2006)
[7] Dhillon, I., Guan, Y., Kulis, B.: A fast kernel-based multilevel algorithm for graph clustering. In: Proceedings of ACM SIGKDD 2005, Chicago, Illinois, USA, August 21-24 (2005)
[8] Gallagher, B., Tong, H., Eliassi-Rad, T., Faloutsos, C.: Using ghost edges for classification in sparsely labeled networks. In: KDD 2008, Las Vegas, NV, USA, August 24-27 (2008)
[9] Karypis, G., Kumar, V.: A fast and high quality multilevel scheme for partitioning irregular graphs. SIAM Journal on Scientific Computing 20(1), 359–392 (1999)
[10] Kaufman, L., Rousseeuw, P.J.: Finding Groups in Data: an Introduction to Cluster Analysis. John Wiley & Sons, New York (1990)
[11] Koren, Y., North, S.C., Volinsky, C.: Measuring and Extracting Proximity in Networks. In: KDD 2006, Philadelphia, Pennsylvania, USA, August 20-23 (2006)
[12] Kulis, B., Basu, S., Dhillon, I., Mooney, R.: Semi-supervised graph clustering: a kernel approach. In: Proceedings of the 22nd international conference on Machine learning, Bonn, Germany, August 07-11, 2005, pp. 457–464 (2005)
[13] Law, M.H.C., Topchy, A.P., Jain, A.K.: Model-based clustering with probabilistic constraints. In: SDM 2005 (2005)
[14] Lutz, H., Stocker, H., Harris, J.W.: Handbook of Physics, 1st edn., pp. 439–444 (2002)
[15] Raytchev, B., Murase, H.: Unsupervised Face Recognition from Image Sequences Based on clustering with Attraction and Repulsion. In: CVPR 2001, vol. 2, p. 25 (2001)
[16] Suhir, E.: Applied Probability for Engineers and Scientists. McGraw-Hill, New York (1997)
[17] Tong, H., Faloutsos, C., Pan, J.: Fast Random Walk with Restart and Its Applications. In: ICDM 2006, Hong Kong (2006)
[18] Wagstaff, K., Cardie, C., Rogers, S., Schroedl, S.: Constrained K-Means clustering with background knowledge. In: ICML 2001, pp. 577–584 (2001)
[19] Weber, R., Schek, H., Blott, S.: A Quantitative Analysis and Performance Study for Similarity-Search Methods in High-Dimensional Spaces. In: VLDB 1998, New York, USA, pp. 194–205 (1998)
[20] Xing, E., Ng, A.Y., Jordan, M., Russell, S.: Distance metric learning, with app- lication to clustering with side-information. In: Advances in NIPS, vol. 15. MIT Press, Cambridge (2002)
[21] Xu, X., Yuruk, N., Feng, Z., Schweiger, T.A.J.: SCAN: a structural clustering algorithm for networks. In: KDD 2007, San Jose, California, USA, August 12-15 (2007)
[22] Yan, B., Domeniconi, C.: An Adaptive Kernel Method for Semi-supervised Clustering. In: Proc. of the 17th European Conference on Machine Learning, Berlin, Germany (September 2006)
[23] Klein, D., Kamvar, S.D., Manning, C.D.: From Instance-level Constraints to Space-Level Constraints: Making the Most of Prior Knowledge in Data Clustering. In: Proc. of 19th Int. Conf. on Machine Learning 2002, San Francisco, CA, USA (2002)

An ℓ_1 Regularization Framework for Optimal Rule Combination

Yanjun Han and Jue Wang

Laboratory of Complex Systems and Intelligence Science,
Institute of Automation, Chinese Academy of Sciences
Beijing, 100190, P.R. China
{yanjun.han,jue.wang}@ia.ac.cn

Abstract. In this paper ℓ_1 regularization is introduced into relational learning to produce sparse rule combination. In other words, as few as possible rules are contained in the final rule set. Furthermore, we design a rule complexity penalty to encourage rules with fewer literals. The resulted optimization problem has to be formulated in an infinite dimensional space of horn clauses R_m associated with their corresponding complexity $\mathcal{C}_m$. It is proved that if a locally optimal rule is generated at each iteration, the final obtained rule set will be globally optimal. The proposed meta-algorithm is applicable to any single rule generator. We bring forward two algorithms, namely, ℓ_1FOIL and ℓ_1Progol. Empirical analysis is carried on ten real world tasks from bioinformatics and cheminformatics. The results demonstrate that our approach offers competitive prediction accuracy while the interpretability is straightforward.

Keywords: Rule Learning, ℓ_1 regularization, Path Following Algorithm.

1 Introduction

Relational Learning [3] deals with tasks in which observations are given in the form of fragments connected by relations. In general, these observations cannot be represented by vectors in the Euclidean Space $\mathbb{R}^n$. For instance, in chemistry, an observation (i.e., a molecule) is described by atoms connected by bonds. However, substructures extracted from atoms and bonds are responsible features for this domain rather than atoms and bonds themselves. Therefore, fragments are often integrated into rules (features) R_m on which the learner is constructed. The learner, which combines different rules together, is denoted as $y = f(R_1, R_2, ...R_M)$.

Due to its powerful expressive ability, Inductive Logic Programming (ILP) became the earliest approach for relational learning [15]. Most algorithms in ILP attempt to find a set of rules covering the given data set. These rules are iteratively generated and then combined by logical disjunction. Formally, $y = R_1 \vee R_2 \vee ... \vee R_M$.

However, there is no statistical guarantee for the generalization performance of the algorithms purely based on ILP. This motivated the research of integrating ILP with kernel methods. Usually these algorithms [7] consist of two steps: an appropriate kernel is designed with the assistance of ILP beforehand, and then a support vector machine (SVM) is trained based on the predefined kernel. In [11], another dynamic approach

W. Buntine et al. (Eds.): ECML PKDD 2009, Part I, LNAI 5781, pp. 501–516, 2009.

kFOIL was brought forward. In kFOIL, the rules are dynamically generated and combined in an SVM, i.e., $y = SVM(R_1, R_2, ...R_M)$. Its interpretability is embodied in the learned rule set. In addition, approaches have been proposed to embed ILP into probabilistic structures, such as the Bayesian network [12].

In essence, these approaches except kernel predefining possess the same mechanism: the rules are iteratively generated by ILP algorithms and then combined in a specific learner. The performance of an algorithm depends on whether it can choose the relevant rules. In current research, various theory revision techniques, heuristic tricks, and experts' experience [3] are utilized to seek the important rules. However, there is no theoretical and thus objective results to guarantee the global optimality of the obtained rule set. That's just the motivation of our work.

In this paper, we bring out an ℓ_1 regularization framework to produce sparse rule combination, which is applicable to any horn clauses generator. The clauses are dynamically generated in the training process, and we establish a one-to-one correspondence between clauses and features. Therefore, the observations are dynamically mapped into the feature space induced by horn clauses. Due to the ℓ_1 regularization term, the most relevant features can be automatically picked out [22]. We adopt the hinge loss as the loss function which has a sound statistical foundation [20]. In addition, for the hinge loss with ℓ_1 regularization, it is proved that there exists an efficient path following algorithm through which the entire solution path can be obtained [16]. Each section on the path corresponds to a rule set as an approximation to the true model. Then we can apply cross validation to choose the model with the highest accuracy. This greatly facilitates model selection in relational learning. To promote rules with lower complexity, we modify ℓ_1 penalty with the rule complexity weight. The resulted optimization problem is an ℓ_1 SVM in an infinite dimensional space, and we design a path following meta-algorithm to obtain its entire solution path. If at each iteration the generated clause is locally optimal, then the obtained rule set is proved to be globally optimal. Moreover, we propose a simple calibration procedure for the case that the generated clause is only suboptimal. There is an interesting interpretation from the viewpoint of kernel learning. The interpretability of our approach is satisfactory in that the obtained clauses are linearly combined, i.e., $y = f_1(R_1)+f_2(R_2)+...+f_M(R_M)$. We propose two algorithms, i.e. ℓ_1FOIL and ℓ_1Progol, based on the path following meta-algorithm. The results on real world tasks demonstrate that our approach has competitive prediction accuracy.

Throughout the paper we use the following notations: $\mathcal{M}$ is the index set of the dimensions in the feature space. $\phi(x) \in \mathbb{R}^{\mathcal{M}}$ is the mapping of x into the feature space ($\phi_m(x) \in \mathbb{R}$ is the "*m-th*"coordinate of $\phi(x)$ for $m \in \mathcal{M}$). For a certain relational learning problem, $\mathcal{H}$ is the corresponding space of horn clauses. It is not hard to show that $|\mathcal{H}| \leq \aleph_0$ (finite or countably infinite). In our setting, there is a one-to-one correspondence between $\mathcal{H}$ and $\mathcal{M}$. Therefore, the *j-th* clause $h_j \in \mathcal{H}$ can be denoted by a $|\mathcal{M}|-$ length vector $\boldsymbol{e}_j$ whose *j-th* element is 1 and all other elements are 0. However, $\mathcal{H}$ is indexed dynamically in the rule constructing process rather than indexed beforehand.

The remainder of the paper is organized as follows: ℓ_1 regularization is introduced into relational learning in Section 2. In Section 3 a path following algorithm which generates the entire solution path is presented and the resulted rule set is proved to be optimal in the infinite space of horn clauses under certain condition. We interpret our approach from kernel learning viewpoint in section 4. In Section 5, the algorithm is evaluated on ten real world data sets. Finally, conclusions are drawn in Section 6.

2 Rule Complexity Penalty and Infinite Dimensional Rule Space

An ideal relational learning algorithm should possess the following two features: first, the relevant rules can be chosen from the vast ocean of candidates; second, the obtained rules are combined in an appropriate manner. The space of horn clauses for a certain relational learning task with n observations can be divided into 2^n equivalence classes, each of which corresponds to a set of rules whose output on the given observations are exactly the same. It seems that each equivalence class could be represented by any individual rule in it. However, this is far from true in that two rules with the same output on the training set may differ greatly on generalization performance. As will be analyzed below, it is more appropriate to differentiate the rules from the same equivalence class and formulate the optimal rule combining problem in infinite dimensional spaces. First of all, the relationship between features and rules has to be established.

2.1 The Correspondence between Features and Rules

Given: the background theory B, in the form of a set of horn clauses, i.e., clauses of the form $h \leftarrow b_1, ..., b_k$ where h and b_j are logical atoms. Background Knowledge depicts the basic facts and constraints in the given domain; a set of observations $D = \{x_i\}_{i=1}^n$, each of which is in the form of the ground facts and the corresponding label set is $\{y_i\}_{i=1}^n$; a set of rules $\{R_m\}_{m=1}^M \subset \mathcal{H}$, R_m are horn clauses.

The feature space is constructed as follows: If an observation x_i satisfies the rule R_m, its corresponding feature value $\phi_m(x_i)$ is set to 1, else to 0. Formally:

$$\phi_m(x_i) = \begin{cases} 1 & B \cup \{R_m\} \models x_i \\ 0 & B \cup \{R_m\} \not\models x_i \end{cases} \tag{1}$$

Therefore, there is a one-to-one correspondence between features and rules. Therefore, we will refer to them without differentiation. Nevertheless, the feature space is constructed dynamically in the learning process rather than determined beforehand.

2.2 ℓ_1 Penalty and Rule Complexity Penalty

For interpretability, the features are often linearly combined

$$f = \sum_{m \in \mathcal{M}} \beta_m \phi_m(x) + b, \quad |\mathcal{M}| \leq \aleph_0. \tag{2}$$

On the other hand, the ℓ_1 regularization is very popular in statistics and machine learning [17,22] because it encourages sparsity. This statistical terminology means that in the obtained linear model only the coefficients of a small number of features will be nonzero while the coefficients of other features will be exactly zero. Moreover, some theoretical research [9,22] indicate that under certain conditions the true model can be obtained by algorithms with ℓ_1 regularization. For linear rule combination, this implies that it is possible to pick out the rules involved in the underlying model through ℓ_1 regularization. Consequently, the problem is formulated as follows:

$$\min_{\boldsymbol{\beta},b} \sum_{i=1}^{n} L(y_i, \boldsymbol{\beta}^T \boldsymbol{\phi}(x_i) + b) \quad s.t. \ \|\boldsymbol{\beta}\|_1 \le t \tag{3}$$

where the loss function L is convex, $\|\boldsymbol{\beta}\|_1 = \sum_{m \in \mathcal{M}} |\beta_m|$ and $|\mathcal{M}| \le \aleph_0$. The above constraint form is equivalent to the following regularization form:

$$\min_{\boldsymbol{\beta},b} \sum_{i=1}^{n} L(y_i, \boldsymbol{\beta}^T \boldsymbol{\phi}(x_i) + b) + \lambda \|\boldsymbol{\beta}\|_1 \tag{4}$$

There is a one-to-one correspondence between λ and t [16]. In this paper, we will use the constraint form (3) which is more convenient for algorithm designing.

Features ϕ_m are generated dynamically in the training process. There are 2^n choices for ϕ_m, and thus it seems that finite space is enough to analyze the optimal rule combining problem (3). However, due to the characteristic of relational learning, an infinite dimensional space is more appropriate for this problem. There are three reasons: firstly, the rules in the same equivalence class may have different complexity (e.g,: an equivalence class may contains the rules of the form $\{w_1 x_1 + w_2 x_2 > 0\}, R_1 \rightarrow pos$, and an appropriate complexity measure for this kind of rules is $\|w\|_1 + 2$ in which 2 stands for the number of literals. Therefore, the complexity for these rules are real numbers and thus have potentially infinite values and rules with different complexity should be distinguished from each other; secondly, the rules in the same equivalence class may have different numbers of literals and shorter rules with less redundancy are preferred (e.g.: rule generators may yield rules of infinite length due to their rigidity, such as $x > 0, x > -1, x > -2, \cdots \rightarrow Positive(x)$); finally, the rules are dynamically generated and thus it is more convenient to assume they are from an infinite repository.

To encourage simpler rules, we substitute the ℓ_1 penalty $\|\boldsymbol{\beta}\|_1$ in (3) with rule complexity penalty $\|\boldsymbol{C} \odot \boldsymbol{\beta}\|_1$, where $\odot$ is the element-wise product of two vectors and C_m is the complexity of rule R_m. The resulted optimization problem is as follow:

$$\min_{\boldsymbol{\beta},b} \sum_{i=1}^{n} L(y_i, \boldsymbol{\beta}^T \boldsymbol{\phi}(x_i) + b) \quad s.t. \ \|\boldsymbol{C} \odot \boldsymbol{\beta}\|_1 = \sum_{m \in \mathcal{M}} |C_m \beta_m| \le t \tag{5}$$

Furthmore, (5) can be rewritten as follow:

$$\min_{\boldsymbol{\beta},b} \sum_{i=1}^{n} L(y_i, \boldsymbol{\beta}^T \tilde{\boldsymbol{\phi}}(x_i) + b) \quad s.t. \ \|\boldsymbol{\beta}\|_1 \le t \tag{6}$$

where $\tilde{\phi}_m(x_i) = \phi_m(x_i)/C_m$.

Therefore, optimization problems (3) and (5) are of the identical type. In this work, we measure the complexity of a clause by the number of literals, i.e., $C_m = \mathcal{N}_m$.

The effect of the rule complexity penalty consists of two parts: the output of rules are weighted according to its complexity and ℓ_1 regularization is performed on the weighted feature space. Moreover, for rules in each equivalence class their complexity may have infinite different values, and thus (6) is an infinite dimensional optimization problem.

2.3 The Optimality in Infinite Dimensional Rule Space and ℓ_1SVM

In [17] it is proved that the infinite dimensional problem (6) has a finite dimensional optimal solution. We restate and explain relevant theorems in [17] here.

Theorem 1. (Caratheodory's Convex Hull Theorem) *Let A be a finite set of points in $\mathbb{R}^n$. Then every $x \in co(A)$ can be expressed as a convex combination of at most $n+1$ points of A, where $co(A)$ is the set of all convex combinations of points in A.*

Theorem 1 can be readily extended to the case in which A is infinite. This theorem implies that for a given relational learning problem with n observations, it is possible to represent these observations with at most $n+1$ rules, as follows:

Theorem 2. (A sufficient condition for the existence of a sparse solution) *If the set $D = \{\phi_m(x) : m \in \mathcal{M}\} \subset \mathbb{R}^n$ is compact, then problem (6) has an optimal solution. Moreover, there exists an optimal solution supported on at most $n+1$ features in $\mathcal{M}$.*

Furthermore, a criterion is given to judge whether an $n+1$ dimensional solution is an optimal solution to (6).

Theorem 3. (The criterion for judging the optimality of a finite-supported solution) *If an optimal solution to (6) exists, and $\hat{\boldsymbol{\beta}}$ is a finite-supported candidate solution such that $\exists \mathcal{A} \subset \mathcal{M}, |\mathcal{A}| < \infty$ and $supp(\hat{\boldsymbol{\beta}}) = \mathcal{A}$. We can test its optimality using the following criterion:*
$\hat{\boldsymbol{\beta}}$ is optimal solution to (6) $\Leftrightarrow \forall \mathcal{B}$ s.t. $\mathcal{A} \subseteq \mathcal{B}, |\mathcal{B}| < \infty$, $\hat{\boldsymbol{\beta}}$ is optimal solution to:

$$\min_{\boldsymbol{\beta}} \sum_{i=1}^{n} L(y_i, \sum_{m \in \mathcal{B}} \beta_m \phi_m(x_i)) \quad s.t. \|\boldsymbol{\beta}\|_1 \leq \sum_{m \in \mathcal{A}} \hat{\beta}_m$$

This theorem implies that it suffices to show that a finite solution is optimal for any finite sub-problems containing it in order to prove its optimality for the original infinite problem. In the next section, we will utilize these theorems to prove that our algorithm indeed generate the optimal rule set.

In our work, the hinge loss is employed as the loss function L in (6), because there is sound theoretical foundation for SVM to guarantee its generalization performance [20]. Therefore, the obtained rules are linearly combined in the following ℓ_1SVM [24]:

$$\min_{\boldsymbol{\beta}, b} \sum_{i=1}^{n} \max(0, 1 - y_i(\boldsymbol{\beta}^T \tilde{\boldsymbol{\phi}}(x_i) + b)) \quad s.t. \ \|\boldsymbol{\beta}\|_1 \leq t. \tag{7}$$

where $\tilde{\boldsymbol{\phi}}(x_i) = \frac{1}{\mathcal{N}} \odot \boldsymbol{\phi}(\boldsymbol{x_i})$, $\mathcal{N}_m$ is the number of literals in R_m.

It is proved in [23] that in finite case the solution to (7) has an amazing piecewise linearity. Namely, $\boldsymbol{\beta}$ is a piecewise linear function of t. When t varies from zero to infinity, the entire path of β is generated. Each section of the path corresponds to a certain model $\hat{\beta}^*$, and this property greatly facilitates model selection. In this paper, we generalize ℓ_1SVM from finite feature spaces to the infinite space of horn clauses and design a similar path following algorithm to obtain the approximately optimal rule set.

3 The Algorithm

First of all, a nested algorithm integrating any rule generator and ℓ_1SVM is presented, and then we prove that an optimal rule set in the space of horn clauses can be generated by our algorithm if the optimal rule for each step is obtained.

Table 1. Details for optimization in step 2.5 and 2.6

2.5	$P(R): \min_g -\sum_{i\in\mathcal{L}} y_i(\sum_{j\in\mathcal{A}} g_j \frac{\phi_j(x_i)}{N_j} + g_R\phi_R(x_i)/N_R)$ s.t. $\begin{cases} \sum_{j\in\mathcal{A}} g_j \frac{\phi_j(x_i)}{N_j} + g_R \frac{\phi_R(x_i)}{N_R} = 0, \\ \text{for } i \in \mathcal{E} \\ \sum_{j\in\mathcal{A}} sign(\beta_j) g_j + \lvert g_R \rvert = 1 \\ g_j = 0, for j \notin \mathcal{A} \cup \{\mathcal{I}_R\} \end{cases}$
2.6	$Q(e): \min_g -\sum_{\mathcal{L}} y_i g^T \phi(x_i)/N$ s.t. $\begin{cases} \sum_{j\in\mathcal{A}} g_j \frac{\phi_j(x_i)}{N_j} = 0 \text{ ,for } i \in \mathcal{E} \setminus \{e\} \\ \sum_{j\in\mathcal{A}} sign(\beta_j) g_j = 1 \\ g_j = 0, \text{for } j \notin \mathcal{A} \end{cases}$

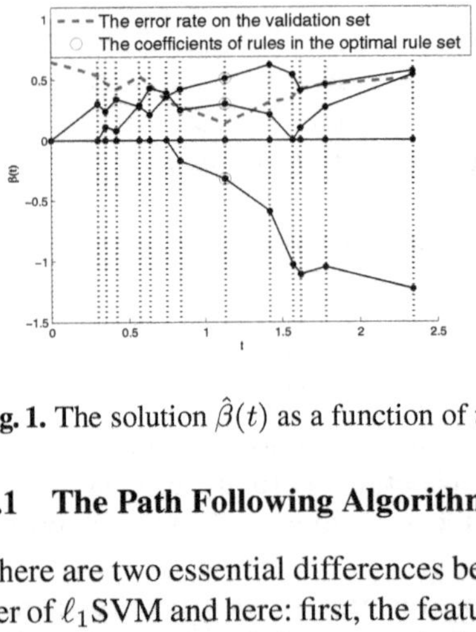

Fig. 1. The solution $\hat{\beta}(t)$ as a function of t

3.1 The Path Following Algorithm

There are two essential differences between the scenarios confronted in the seminal paper of ℓ_1SVM and here: first, the feature space induced by the horn clauses is potentially infinite rather than finite; second, the feature space here have to be dynamically constructed during the learning process rather than fixed beforehand. Therefore, the path following algorithm presented in [24] should be modified accordingly.

To describe the algorithm for finding the solution path, some concepts and definitions should be given first of all. A feature ϕ_m is called "active" when its corresponding coefficient β_m is nonzero. Features enter the active feature set $\mathcal{A} = \{j : \beta_j \neq 0\}$ successively when the value of t increases. When t is small, only the most relevant features can enter the active set $\mathcal{A}$; when t is large, $\mathcal{A}$ contains most of the features. The solution has a favorable piecewise linear property (Figure 1), i.e., $\forall m \in \mathcal{A} = \{j \mid \hat{\beta}_j \neq 0\}$,

Table 2. The Main Algorithm

1. **Initialize:**
   $\mathcal{HG}$: a Horn clause Generator. $\mathcal{N}_R$: the number of literals in R.
   The rule set $\mathcal{R} = \emptyset$,
   $R^* = \mathrm{argmin}_R\, \Delta L(\phi_R) = \mathrm{argmin}_{\mathcal{HG}\to R} - \sum_{\mathcal{L}} y_i\phi_R(x_i)/\mathcal{N}_R$, $\mathcal{R} = \{R^*\}$
   $\mathcal{A} = \{\mathcal{I}_{R^*}\}$, the index of R^* in the whole set of horn clauses.
   $g_j = 1, j \in \mathcal{A}$; $g_j = 0, j \notin \mathcal{A}$
   $\Delta L^* = \Delta L(R^*)$. $m = 0$, $\mathcal{A}^m = \mathcal{A}$, $P^m(R) = -\sum_{\mathcal{L}} y_i\phi_R(x_i)/\mathcal{N}_R$, $\beta^m = \mathbf{0}$, $m = m+1$
2. **While** $\Delta L^* \neq 0$
   2.1 $d_1 = \min\{d > 0 : (\beta + dg)_j = 0, j \in \mathcal{A}\}$
   2.2 $d_2 = \min\{d > 0 : 1 - y_i(\beta + dg)^T\phi(x_i)/\mathcal{N} = 0, i = 1, ..., n\}$
   2.3 set $d = min(d_1, d_2)$. $\beta = \beta + dg$.
   2.4 **If** $d = d_1$, **then** remove the variable attaining 0 from $\mathcal{A}$,
   If $d = d_2$, **then** add the observation entering the elbow to $\mathcal{E}$.
   2.5 $\mathcal{HG}$ is guided by : $\min_{\mathcal{HG}\to R} P(R)$, $P(R)$ is detailed in Table 1.
   The solution is (R^*, g^*), set $\phi(\cdot) = \{\phi_j(\cdot)\}_{j\in\mathcal{A}} \cup \phi_{R^*}(\cdot)$,
   $\Delta L_1 = -\sum_{\mathcal{L}} y_i {g^*}^T\phi(x_i)/\mathcal{N}$. ($\mathcal{N}_m$: the number of literals in R_m).
   2.6 $\min_{e\in\mathcal{E}} Q(e)$, $Q(e)$ is detailed in Table 1.
   The solution is (e^*, g^*), $\Delta L_2 = -\sum_{\mathcal{L}} y_i {g^*}^T\phi(x_i)/\mathcal{N}$.
   2.7 Let $\Delta L^* = min\{\Delta L_1, \Delta L_2\}$,
   If $\Delta L^* = \Delta L_1$, **then**
   invoke the **Calibration Procedure** (Section 3.3):
   $\mathcal{A}^m = \mathcal{A}$, $P^m(\cdot) = P(\cdot)$ (step 2.5), $\beta^m = \beta$, $R^m = R^*$, $m = m+1$.
   Update g and $\mathcal{A} \leftarrow \mathcal{A} \cup \{\mathcal{I}_{R^*}\}$. $\mathcal{R}_\mathcal{A} = \mathcal{R}_\mathcal{A} \cup \{R^*\}$
   If $\Delta L^* = \Delta L_2$, **then** update g and $\mathcal{E} \leftarrow \mathcal{E} \setminus \{e^*\}$, $\mathcal{L} \leftarrow \mathcal{L} \cup \{e^*\}$.
   If $\Delta L^* \leq 0$, **then** set $\Delta L^* = 0$.
   2.8 **Return** to 2.

$\hat{\beta}_m(t)$ is a piecewise linear function of t. Piecewise linear property only depends on the functional form of the loss function and the regularization term [16], and thus it still holds in the infinite dimensional space. Since the hinge loss $L(v) = max(0, 1-v)$ is non-smooth, the dataset has to be divided into three non-intersecting sets in order to carry out analysis, as follows:

$\mathcal{E} = \{i : 1 - y_i\beta^T\phi(x_i) = 0\}$, elbow set of observations;
$\mathcal{L} = \{i : 1 - y_i\beta^T\phi(x_i) > 0\}$, left of the elbow set;
$\mathcal{R} = \{i : 1 - y_i\beta^T\phi(x_i) < 0\}$, right of the elbow set (has no use in analysis);

Due to the nonsmoothness of the loss function and the ℓ_1 constraint, it is impossible to guide the optimization problem directly via gradient information. Intead, we have to solve the following optimization to get the steepest descent direction for L:

$$g_\beta = \underset{d_\beta}{\mathrm{argmin}} \quad \Delta L(\epsilon \boldsymbol{d}_\beta) = L(\boldsymbol{\beta} + \epsilon \boldsymbol{d}_\beta) - L(\boldsymbol{\beta}) = -\sum_{\mathcal{L}} y_i \epsilon \boldsymbol{d}_\beta^T \boldsymbol{\phi}(x_i)$$
$$s.t. \qquad \epsilon \boldsymbol{d}_\beta^T \boldsymbol{\phi}(x_i) = 0 \text{ for } i \in \mathcal{E}, \|\boldsymbol{d}_\beta\| = 1 \tag{8}$$

"$\boldsymbol{g}$" and "$\boldsymbol{d}$" stand for gradient and descent direction, respectively. $\boldsymbol{g}_\beta$ is the direction in which the loss function decreases fastest per unit increase of the ℓ_1 norm of the coefficient vector. If ϵ is a sufficiently small positive number, the observations in $\mathcal{R}$ and $\mathcal{L}$ still remains in their respective set when the coefficient vector grows from β to $\beta + \epsilon \boldsymbol{d}_{\beta}$. However, an observation may leave $\mathcal{E}$ and enter $\mathcal{L}$, and thus it is impossible to determine the incremental of the loss function correpsonding to the incremental of the coefficient vector $\epsilon \boldsymbol{d}_{\beta}$. Therefore, we add a contraint (the first constraint in (8)) to keep $\mathcal{E}$ unchanged. Furthermore, (8) can be simplified to the following form:

$$g_{\beta} = \underset{\boldsymbol{d}_{\beta}}{\operatorname{argmin}} - \sum_{\mathcal{L}} y_i \boldsymbol{d}_{\beta}^T \phi(x_i) \quad s.t. \quad \boldsymbol{d}_{\beta}^T \phi(x_i) = 0 \text{ for } i \in \mathcal{E}, \|\boldsymbol{d}_{\beta}\| = 1 \tag{9}$$

(9) is the basis for the optimization problem of step 2.5 and 2.6 (Table 2).

First of all, the rule generator algorithm is invoked to generate a feature (rule) that can decrease the loss the fastest per unit increase of ℓ_1 norm of the coefficient vector. The solution moves along that direction until one of the following two events happens: 1. A coefficient hits the non-differentiable points of the penalty, i.e., β_j becomes zero from nonzero for some j. Hence the corresponding rule is removed from $\mathcal{A}$. (step 2.1) 2. An observation hits the non-differentiable point of the loss, i.e., $1 - y_i \beta^T \phi(x_i)$ becomes zero from nonzero for some i. Hence this observation is added to $\mathcal{E}$ (step 2.2).

Then there are three types of actions one can take: 1. invoke the rule generator to yield a new rule and add it to $\mathcal{A}$. (step 2.5) 2. remove an observation from $\mathcal{E}$. (step 2.6) 3. do nothing (can be merged with the first action). The action that can cause the fastest decrease in the loss per unit increase of the ℓ_1 norm of the coefficient will be chosen.

3.2 The Optimality of the Algorithm

Now we will prove that if at each iteration an optimal rule for current step is generated, then the obtained rule set is an optimal solution to infinite dimensional optimization problem (6). This is an amazing property in that generally the global optimality of the rule set cannot be guaranteed just by the local optimality of single rules due to the greedy nature of rule generating process. This characteristic distinguishes our approach from existing rule combining methods.

Theorem 4. *Assume at any iteration of the algorithm, an optimal rule for step 1 and step 2.5 (**Table 2**) is obtained. For a certain iteration, assume we are after step 2.2 and the current rule set is $\mathcal{R}_{\mathcal{A}}$. Then the corresponding finitely-supported solution $\beta_{\mathcal{A}}$ is an optimal solution to (7) in infinite dimensional rule space with $t = \|\beta_{\mathcal{A}}\|_1$. Namely, $\mathcal{R}_{\mathcal{A}}$ is an optimal rule set with the constraint $t = \|\beta_{\mathcal{A}}\|_1$.*

Proof: Given a relational learning task and n observations $X = \{x_i\}_{i=1}^n$, the set of horn clauses can be divided into 2^n equivalence classes $H_p, p = 1, \cdots, 2^n$ according to the output on X. The corresponding feature value vector $\phi_p(X)$ is constructed as in (1) and the index set is $\mathcal{I}_p$. In this work, we measure the complexity of a rule R by its number of literals, i.e., $\mathcal{C} = \mathcal{N}_R$. $\forall h \in H_p$, if $\mathcal{N}_h$ is finite, then the feature vector set $D_p = \{\tilde{\phi}_i = \phi_p(X)/\mathcal{N}_{h_i}, i \in \mathcal{I}_p\}$ is finite. If $\exists h$, $\mathcal{N}_h$ is infinite, then D_p is a countable infinite set with a unique limit point $\mathbf{0}$ (a $n \times 1$ vector). Therefore, the overall feature vector set $D = \bigcup_{p=1}^{2^n} D_p$ is a bounded set with unique limit point $\mathbf{0}$. We construct a new feature vector set $D' = D \cup \{\mathbf{0}\}$. It is easy to show that the solution set for (7) with D'

equals that with D. Moreover, $D^{'}$ is a bounded closed set, and thus compact. According to Theorem 2, there is an optimal solution supported on at most $n+1$ features in $D^{'}$. Due to the equivalence between optimizing with D and $D^{'}$, this also holds for D. Next we will prove the optimality of the finitely-supported solution at any iteration. Assume at any iteration , an optimal rule for step 1 and step 2.5 (**Table 2**) is obtained. Therefore, there is no need to perform calibration at step 2.7. For finite D, the optimality of the algorithm has been proved in [23]. For infinite D, since the finite feature set result implies that for any finite $\mathcal{B}$ such that $\mathcal{A} \subseteq \mathcal{B} \subseteq \mathcal{M}$, the finitely-supported solution $\hat{\beta}_{\mathcal{A}}$ is also an optimal solution in feature set $\mathcal{B}$ given the restriction $\|\beta_{\mathcal{B}}\|_1 \leq \|\hat{\beta}_{\mathcal{A}}\|_1$. Theorem 3 and finite feature set result combined complete the proof. □

3.3 Suboptimality and Calibration Procedure

In practice the exact optimization in step 2.5 is prohibitive because the space of horn clauses is exponential to the number of observations or infinite. Therefore, at each step 2.5 a suboptimal rule is chosen instead of the theoretically optimal one. It is still not clear that to which extent the performance of the algorithm is influenced by this gap between theory and reality. We conjecture that this gap does not significantly impair the performance of our algorithm, since the difference between the optimal rule and the suboptimal rule lies in that the former corresponds to the steepest descent direction while the latter only corresponds to ordinary descent direction, and for convex objective function, an algorithm proceeds in an ordinary descent direction at each iteration can still converge at a slower rate. In experimental evaluation, we explored the impact of this gap on our algorithms' performance, and the results support our conjecture. We will carry out detailed theoretical analysis in the future work.

Nevertheless, there is a situation which has to be avoided: the "contradiction" on the solution path. Formally, for two iteration steps $m_2 > m_1$, rules selected at step 2.5 are R_{m_2} and R_{m_1} respectively, if R_{m_2} is better for the optimization problem P^{m_1} than R^{m_1}, then we say that an contradiction happens. A simple remedy is that once a new rule is generated, it is substituted into previous P_m to examine whether an contradiction exists, and algorithm is restarted at the earliest step when the contradiction happens.

3.4 Analysis of Computational Complexity

The computational cost of the algorithm consists of three parts: the first part is brought by the rule generator when searching for the optimal rule at step 2.5, the second part

Table 3. The Calibration Procedure

```
for n = 0 : (m − 1)
  if P^n(R*) < P^n(R^n)
    if n = 0, goto step 1
    else set m = n − 1, A = A^{n−1}, β = β^{n−1}, goto step 2.1
    end if
  end if
end for
```

is resulted in by the linear programming at step 2.5 and 2.6, and the third part is due to the computation for the entire solution path. The following notations are used in analysis: for *m-th* iteration, $|\mathcal{E}| = q$ and the size of current $\mathcal{A}$ is s_a. The number of the rules that have ever entered the active rule set $\mathcal{A}$ during the overall iterations is p, the complexity of the rule generator is $\mathcal{O}(C_f)$, and the number of invoking the rule generator to generate an approximate rule at step 2.5 is C_s.

The complexity for solving the inner linear programming at step 2.5 and 2.6 is $\mathcal{O}(s_a^3)$ [13]. If the training data is separable, then the number of joints on the path can be assumed to be $\mathcal{O}(n)$. Therefore, the overall complexity is $\mathcal{O}(((C_f + s_a^3)C_s + q \cdot s_a^3) \cdot n)$. In experiment, we find $C_f \propto n^2$, $C_s < 40$ and $p < n$. Since $q \leq min(n, p)$ and $s_a \leq p$, then the worst case complexity is $\mathcal{O}(40n^3 + np^4)$. Calibration is not considered in the above analysis. If calibration is performed for n_c times, then the worst case complexity is $\mathcal{O}(n_c(40n^3 + np^4))$. In experimental evaluation, it is observed n_c is about 2 on average when the beam search width is set to 5. Therefore, the worst case complexity with calibration is $\mathcal{O}(80n^3 + 2np^4)$.

4 The Kernel Learning Interpretation of the Algorithm

From the viewpoint of kernel learning, it can be shown that the ℓ_1 regularized rule learning algorithm is equivalent to multiple kernel learning [10].

Multiple kernel learning (MKL) attempts to learn the optimal kernel matrix from data. Different kernels reflect different facets of the data set. MKL aims to combine these kernels in an optimal manner such that the obtained kernel matrix can best represent the original data set, as follows:

$$\min_{\boldsymbol{d}} J(\boldsymbol{d}) \quad s.t.: \sum_{m=1}^{M} d_m = 1, d_m \geq 0 \tag{10}$$

where $J(\boldsymbol{d})$ is the optimal value of the following problem:

$$\begin{aligned} &\max_{\boldsymbol{\alpha}} -\frac{1}{2}\sum_{i,j} \alpha_i \alpha_j y_i y_j \sum_{m=1}^{M} d_m K_m(x_i, x_j) + \sum_i \alpha_i \\ &s.t. \sum_{i=1}^{n} \alpha_i y_i = 0, 0 \leq \alpha_i \leq t' \end{aligned} \tag{11}$$

Our ℓ_1 regularized algorithm can be proved to be equivalent to performing MKL with a linear kernel set induced by rules. First of all, the definition of linear kernel induced by rules is given as follows:

Definition 1. *Given a rule R_m, the linear kernel K_m induced by it is as follows:*

$$K_m(x_i, x_j) = < \tilde{\phi}_m(x_i), \tilde{\phi}_m(x_j) > \begin{cases} 1/N_m^2, B \cup \{R_m\} \models x_i \cap B \cup \{R_m\} \models x_j \\ 0, \quad otherwise \end{cases} \tag{12}$$

Theorem 5. *ℓ_1 SVM (7) is equivalent to the multiple kernel learning problem (11) with the linear kernel set $\{K_m\}_{m=1}^M$ constructed as in definition 1.*

Proof: Due to the block-sparsity as shown in [1], MKL (11) is equivalent to the following problem:

$$\min_{\beta} \sum_{i=1}^{n} L(y_i, \mathbb{B}^T \Phi(x_i)) \quad s.t. \|\mathbb{B}\|_{block} \leq t' \tag{13}$$

where L is the hinge loss, $\Phi = (\phi_1, ..., \phi_M)$, ϕ_m is the feature space induced by K_m, $\mathbb{B} = (\beta_1, ...\beta_M)$, β_m is the corresponding coefficient vector, $\|\mathbb{B}\|_{block} = \sum_{m=1}^{M} \|\beta_m\|_2$. According to definition 1, $\tilde{\phi}_m$ are single dimensional features, and thus (13) is simplified to (7). There is a one-to-one correspondence between t and t'. Consequently, (11) is equivalent to (7). □

The resulted kernel matrix is the optimal linear combination of the linear kernels induced by R_m, i.e., $K = \sum_{i=1}^{M} d_m^* K_m$. Our approach is actually a kernel matrix learning algorithm based on the obtained rules. Each rule provides a specific view, and these facets are finally integrated in the overall kernel matrix K. According to the kernel construction, the rules with lower complexity may have greater contribution to the final kernel matrix K. Therefore, as a kernel learning method, our approach is more flexible than the recent algorithms [7] which attempts to integrate ILP and kernel methods.

5 Experiments

As a general-purpose algorithm, our sparse rule combination approach is applicable to any rule generator such as FOIL, Progol, etc. In experiment, we realized two sparse rule combination algorithms respectively based on FOIL and Progol, denoted as ℓ_1FOIL and ℓ_1Progol. Both FOIL [15] and Progol [14] process data in a divide-and-conquer way, which consists of two iterative steps: the inner loop generates a single rule at each iteration, while the outer loop seeks for a set of rules to cover all positive observations. The difference lies in the way how a single rule is generated: FOIL can be considered as a first-order decision tree algorithm guided by FOIL-gain while Progol conducts a general-to-specific search in the theta-subsumption lattice of a single clause hypothesis through inverse entailment. The algorithms for comparison include **kFOIL** [11], **FOIL** [15], Rooted Kernel(**RKernel**,[19]), and Boosting FOIL (**BFOIL**). For **FOIL**, maximum number of clauses in a hypothesis was set to 25, maximum number of literals in a clause was set to 10, and a beam search with beam width 5 was performed instead of simple greedy search. For **BFOIL**, AdaBoost.M1 [5] was adopted for combining rules and the stepsize was fixed to 0.01 since some theoretical and experiment research [21,6] demonstrate this modification leads to better prediction accuracy, and the same parameters for FOIL were used. **kFOIL** is an algorithm that uses SVM as the score function for FOIL instead of the FOIL-gain. We used the same parameters as in **FOIL**. For the SVM part, a polynomial kernel of degree 2 was used. **RKernel** belongs to the family of graph kernel [7] algorithms in which structural information of data based on graph theory is utilized to construct predefined kernels for SVM. The walk length and the discount factor for **RKernel** were selected from $\{1,2,4,5,10\}$ and $\{0.1,0.5,0.8,1,2,10\}$ through cross validation. For ℓ_1FOIL and ℓ_1Progol, we also substituted simple greedy search with beam search with width 5, and there was no limitation on the number of clauses in a hypothesis and the number of literals in a clause because the complexity of the hypotheses would be controlled via ℓ_1 regularization. The LIBSVM implementation [2] was employed for the SVM part and the regularization constant C was selected from $\{0.1,1,10,100,1000,10000\}$ through cross validation. For all of the algorithms, the validation technique was utilized to determine when to stop training. In each fold of cross validation (10-fold cross validation was performed for all data sets), the data set was split to 3 parts, 70% for training, 20% for validating and 10% for testing.

5.1 Datasets

For comparison, we evaluated the above algorithms on ten real world classification datasets, including Mutagenesis [18], Alzheimer [8], NCTRER [4] and PTC[1]. On Mutagenesis (230 observations, MUT) the problem is to predict the mutagenicity of a set of compounds. Alzheimer consists of four independent datasets, each of which corresponds to a desirable property of drugs against Alzheimer's disease: inhibit amine reuptake (686 observations, ALR), low toxicity (886 observations, ALT), high acetyl cholinesterase inhibition (1326 observations, ALC) and good reversal of memory deficiency (642 observations, ALM). NCTRER (232 observations,NCT) is extracted from the EPA's DSSTox NCTRER database, the problem is to predict estrogens' binding activity for the estrogen receptor. On all of the above datasets, we used the atom and bond information only. PTC contains 417 compounds, each of which is labeled based on whether it is carcinogenic to female rats (349 observations, PFR), female mice (351 observations, PFM), male rats (336 observations, PMR), and male mice (344 observations, PMM). Following [19] and others, we treat any molecule labeled "CE", "SE" and "P" as positive, "NE" and "N" as negative, ignore other unsure classifications.

5.2 Results

The 10-fold cross-validation result of the prediction accuracy is shown in Table 4. As demonstrated therein, ℓ_1FOIL significantly outperforms the other algorithms on six out of ten tasks. In addition, on none of the tasks ℓ_1FOIL and ℓ_1Progol performed significantly worse than the other algorithms. It is worth noting that ℓ_1FOIL and ℓ_1Progol have similar performance on nearly all of the tasks although their rule generation mechanisms are totally different. This indicates that as a meta-algorithm the performance of our approach is insensitive to the choice of different rule generators. This phenomenon may be due to the following three reasons: firstly, in relational learning the true model for a certain task often consists of several rules, each of which accounts for part of observations; secondly, FOIL and Progol are highly expressive in that they both have very large hypothesis spaces, and thus the rules for the true model is probably contained

[1] http://www.predictive-toxicology.org/ptc/

Table 4. 10-fold cross validation predictive accuracy results. •/▲ indicate that the result for ℓ_1FOIL/ℓ_1Progol is significantly better than that of the corresponding method. On none of datasets, performance of ℓ_1FOIL or ℓ_1Progol is significantly worse (paired samples t-test, $\alpha = 0.05$). The second row for each dataset is the number of obtained rules on average.

Dataset	ℓ_1FOIL	ℓ_1Progol	kFOIL	FOIL	Rkernel	BFOIL
MUT	85.3 ± 2.8	86.9 ± 1.5	82.7 ± 10.3	75.3 ± 11.7 • ▲	85.4 ± 6.1	81.2 ± 7.4
	7.2 ± 2.1	8.4 ± 1.3	13.4 ± 5.2	7.5 ± 1.4	− − −	21.2 ± 2.2
ALR	92.6 ± 3.1	89.2 ± 4.5	87.9 ± 4.9•	76.5 ± 12.1 • ▲	83.1 ± 7.8 • ▲	87.3 ± 3.2•
	13.4 ± 4.8	15.7 ± 2.6	13.4 ± 1.9	12.7 ± 3.1	− − −	20.3 ± 3.7
ALT	91.6 ± 1.1	92.3 ± 2.4	89.5 ± 4.0	78.2 ± 5.2 • ▲	88.3 ± 8.1	86.3 ± 5.2
	12.1 ± 2.3	16.7 ± 3.6	22.3 ± 7.5	13.4 ± 4.1	− − −	16.9 ± 3.5
ALC	91.3 ± 2.4	88.2 ± 3.9	87.4 ± 3.8•	67.4 ± 5.5 • ▲	84.6 ± 6.8 • ▲	81.2 ± 7.8 • ▲
	8.4 ± 2.7	9.5 ± 3.6	15.6 ± 6.2	13.2 ± 3.4	− − −	17.1 ± 9.1
ALM	85.7 ± 4.2	89.3 ± 4.6	81.3 ± 3.5 • ▲	61.3 ± 12.8 • ▲	81.9 ± 3.6 • ▲	75.3 ± 2.1 • ▲
	11.3 ± 2.7	12.8 ± 5.4	7.9 ± 2.1	12.4 ± 5.3	− − −	21.4 ± 5.9
NTC	86.3 ± 2.3	85.2 ± 3.8	78.2 ± 8.7 • ▲	57.8 ± 6.3 • ▲	85.2 ± 7.2	79.2 ± 2.5 • ▲
	12.2 ± 2.2	13.1 ± 2.1	16.7 ± 7.5	14.3 ± 2.2	− − −	18.9 ± 4.1
PFR	75.2 ± 5.1	78.2 ± 3.4	82.2 ± 12.1•	57.9 ± 9.2 • ▲	67.3 ± 11.8•	73.3 ± 4.6▲
	17.3 ± 6.0	19.2 ± 2.6	23.5 ± 8.2	12.3 ± 4.9	− − −	13.9 ± 5.2
PFM	77.4 ± 7.1	78.6 ± 2.3	62.4 ± 7.2 • ▲	52.3 ± 10.1 • ▲	63.8 ± 6.1 • ▲	72.3 ± 6.4 • ▲
	18.7 ± 8.1	16.5 ± 3.2	21.3 ± 6.4	12.1 ± 3.1	− − −	19.4 ± 8.9
PMR	68.5 ± 3.4	66.5 ± 2.7	57.5 ± 5.2 • ▲	63.5 ± 6.4•	61.5 ± 7.3 • ▲	68.2 ± 8.4
	13.2 ± 5.5	17.2 ± 3.1	24.1 ± 2.1	13.6 ± 5.3	− − −	18.2 ± 7.2
PMM	70.9 ± 3.7	71.2 ± 5.2	72.1 ± 2.1	67.1 ± 9.2•	65.2 ± 8.9▲	63.8 ± 4.1•
	8.3 ± 2.1	11.6 ± 3.9	12.3 ± 2.4	7.9 ± 5.8	− − −	25.1 ± 3.2

in their hypothesis spaces; finally, the most relevant rules can be automatically picked out from the numerous candidates via ℓ_1 regularization. Therefore, the true model can be well approximated by the combination of expressive rule generators and ℓ_1 regularization. Besides the outstanding approximation ability, there are two more reasons for the success of our approach: firstly, due to ℓ_1 regularization the variance of the generalization performance is much less than the other algorithms; secondly, the entire solution path is generated and thus the risk of missing the right model is greatly reduced. Although boosting also performs rule combination, its performance is worse than our algorithms. The reason may be that due to the hinge loss our algorithm can pick out the support vectors and thus is less influenced by the noise than boosting. The comparison between **RKernel** and other algorithms is somewhat unfair in that there are two more tunable parameters for **RKernel**, i.e., the walk length and the discount factor. This partly explains its competitive performance in the evaluation. However, its interpretability is not as straightforward as that of our algorithms.

The 10-fold cross-validation result of the number of rules is also shown in Table 4. The number of rules generated by our algorithms is comparable to that by other algorithms. Although we claimed that sparse rule combination is achieved through our framework, the obtained rules is not necessarily fewer than that of other algorithms. The sparsity only implies that the rules involved in the true model may be chosen automatically. If for a certain task, the corresponding true model consists of many rules,

(a) Heteroatom: The atom at X is O, S or N. (b) Aromatic Ring (c) Phenol3nPhenyl: X can be any kind of atoms.

Fig. 2. Three Obtained Rules for NCTRER Dataset by ℓ_1FOIL

Table 5.

(a) The Influence of Beam Width on the Performance of ℓ_1 FOIL. A stands for prediction accuracy, R stands for the number of obtained rules and C stands for calibration times. The results are mean values of 10-fold cross validation.

Dataset	$Width = 1$			$Width = 3$			$Width = 5$		
	A	C	R	A	C	R	A	C	R
MUT	80.9	28.2	8.1	86.7	7.3	8.8	85.3	1.2	7.2
ALR	84.2	7.4	15.9	88.9	4.6	11.1	92.6	1.3	13.4
ALT	87.9	31.2	7.6	89.1	15.8	12.6	91.6	0	12.1
ALC	88.1	14.7	12.4	87.5	2.3	10.5	91.3	1.7	8.4
ALM	82.4	12.3	12.8	86.4	3.5	15.7	85.7	2.1	11.3
NCT	83.6	18.6	9.1	87.8	5.4	16.3	86.3	2.6	12.2
PFR	70.1	15.4	12.4	72.1	6.7	12.4	75.2	2.3	17.3
PFM	73.3	12.6	22.1	75.6	8.9	16.9	77.4	1.3	18.8
PMR	63.4	8.7	9.1	65.9	7.3	17.2	68.5	4.5	13.2
PMM	62.8	9.3	9.5	71.2	6.9	9.1	70.9	2.5	8.3

(b) The Influence of Rule Complexity Penalty on the Performance of ℓ_1 FOIL and ℓ_1 Progol. A stands for prediction accuracy and L stands for the number of literals in each rule. The results are mean values of 10-fold cross validation.

Dataset	$\mathcal{P}_a$FOIL		$\mathcal{P}_b$FOIL		$\mathcal{P}_a$Progol		$\mathcal{P}_b$Progol	
	A	L	A	L	A	L	A	L
MUT	85.3	2.4	83.1	3.3	86.9	2.6	85.4	5.1
ALR	92.6	4.7	93.2	6.4	89.2	3.2	91.3	4.7
ALT	91.6	3.6	92.7	7.5	92.3	4.1	89.7	4.9
ALC	91.3	3.2	88.9	4.8	88.2	3.2	91.2	5.6
ALM	85.7	5.4	80.4	3.5	89.3	3.7	90.9	4.8
NCT	86.3	4.2	82.1	4.6	85.2	2.3	86.6	3.2
PFR	75.2	2.6	73.5	4.5	78.9	3.2	82.3	5.8
PFM	77.4	3.8	76.9	5.4	78.6	5.7	77.1.9	4.3
PMR	68.3	3.4	72.2	3.6	66.5	4.8	70.2	5.3
PMM	70.3	2.3	68.1	4.3	71.2	3.1	68.3	4.1

then the performance of an algorithm apt to choose fewer rules cannot be expected to be satisfactory. For **RKernel**, the corresponding place is filled with "- - -" instead of a real number because it is based on kernel design rather than rule combination. For NCTRER dataset, three rules among the top ten rules (ordered by coefficients) are shown is Figure 2. This result is consistent with the study on NTCRER [4].

In Theorem 4, we proved that if at each iteration the optimal rule can be generated, then the obtained rule set is globally optimal. However, in practice only a suboptimal rule can be generated for each iteration. For beam search based rule generators, the approximity between the optimal rule and the suboptimal rule depends on the width of beam search. Therefore, the prediction accuracy, the calibration times (refer to section 3.3) and the number of rules can all be influenced by the beam width. We tried different search width (1, 3, 5) and the result is shown in Table 5(a). It can be observed that the calibration times decreases as the search width increases and this indicates that the chance of missing the optimal rule for a certain iteration decreases as the search width increases. The generalization performance improves as the search width increases as we expected. The number of obtained rules decreases as the search width increases. This implies that when the suboptimal rules chosen at each step are far from the optimal rules, more rules are needed to approximate the true model.

Another interesting question is how the size of rule set and prediction accuracy will be influenced by different designs of rule complexity penalties. For comparison , we evaluated ℓ_1FOIL and ℓ_1Progol based on two different rule complexity penalties: ($\mathcal{P}_a$) C_m is defined as the number of the literals in R_m ; ($\mathcal{P}_b$) all C_m are just set to 1, i.e., no additional rule complexity penalty. The result is shown in Table 5(b). The prediction accuracy under the two kinds of penalties are close to each other while the literals in each rule are fewer for the case of literal number penalty.

6 Conclusion and Future Work

In this paper we bring out an ℓ_1 regularized rule learning and combining approach for relational learning, which formally introduces ℓ_1 regularization into the infinite space of horn clauses. Our approach is applicable to any horn clause generators such as FOIL, Progol, etc. Moreover, due to the piecewise linearity of hinge loss the entire solution path can be generated for a given task. It is proved that if a locally optimal rule is generated at each iteration, then each cross-section of the path is an optimal rule set given the ℓ_1 constraint on coefficients. From the kernel learning viewpoint, our approach is equivalent to multiple kernel learning and the information carried in the obtained rules is optimally fused into the overall kernel matrix. The proposed algorithm is competitive with the other algorithms for comparison. In future, the approach presented here will be extended to tackle the regression tasks.

Acknowledgments. We gratefully acknowledge Kuijun Ma for her insightful remarks. This work was partially supported by National Basic Research Program of China under Grant No.2004CB318103 and National Natural Science Foundation of China under award No.60835002.

References

1. Bach, F.R., Lanckriet, G., Jordan, M.I.: Multiple Kernel Learning,Conic Duality, and the SMO Algorithm. In: 21st international conference on Machine learning. ACM Press, New York (2004)
2. Chang, C.C., Lin, C.J.: LIBSVM: a library for support vector machines (2001), `http://www.csie.ntu.edu.tw/~cjlin/libsvm`
3. Dzeroski, S., Lavrac, N.: An Introduction to Inductive Logic Programming, Relational Data Mining. Springer-Verlag New York, Inc., USA (2001)
4. Fang, H., Tong, W., Shi, L.M., Blair, R., Perkins, R., Branham, W., Hass, B.S., Xie, Q., Dial, S.L., Moland, C.L., Sheehan, D.M.: Structure-activity relationships for a large diverse set of natural, synthetic, and environmental estrogens. Chem. Res. Tox. 14, 280–294 (2001)
5. Freund, Y., Schapire, R.E.: A decision-theoretic generalization of on-line learning and an application to boosting. Journal of Computer and System Sciences 55(1), 119–139 (1997)
6. Friedman, J.H.: Greedy function approximation: A gradient boosting machine. Annals of Statistics 29, 1189–1232 (2001)
7. Gaertner, T., Lloyd, J.W., Flach, P.A.: Kernels and Distances for Structured Data. Machine Learning 57, 205–232 (2004)
8. King, R.S., Sternberg, M.J.E., Srinivasan, A.: Relating Chemical Activity to Structure: An Examination of "ILP" Successes. New Generation Computing 13(3-4), 411–433 (1995)
9. Knight, K., Fu, W.J.: Asymptotics for lass-type estimators. The Annals of Statistics 28(5), 1356–1378 (2000)

10. Lanckriet., G., Cristianini, N., Bartlett, P., El Ghaoui, L., Jordan., M.I.: Learning the Kernel Matrix with Semi-horn Programming. EECS Department, University of California, Berkeley (2002), http://www.eecs.berkeley.edu/Pubs/TechRpts/2002/5765.html,UCB/CSD-02-1206
11. Landwehr, N., Passerni, A., De Raedt, L., Frasconi, P.: kFOIL: Learning Simple Relational Kernels. In: Proc. of the 21st Natl. Conf. on Artificial Intelligence, AAAI Press, Boston (2006)
12. Landwehr, N., Kersting, K., De Raedt, L.: nFOIL:Integrating Naive Bayes and FOIL. Journal of Machine Learning Research (JMLR) 8, 481–507 (2007)
13. Megiddo, N.: Progress in Mathematical Programming: Interior-Point and Related Methods. Springer, New York (1988)
14. Muggleton, S.: Inverse Entailment and Progol. New Generation Computing 13(3 and 4), 245–286 (1995)
15. Quinlan, J.R.: Learning Logical Definitions from Relations. Machine Learning 5(3), 239–266 (1990)
16. Rosset, S., Zhu, J.: Piecewise linear regularized solution paths. Annals of Statistics 35, 1012–1030 (2007)
17. Rosset, S., Swirszcz, G., Srebro, N., Zhu, J.: ℓ_1 Regularization in Infinite Dimensional Feature Spaces. In: Bshouty, N.H., Gentile, C. (eds.) COLT. LNCS (LNAI), vol. 4539, pp. 544–558. Springer, Heidelberg (2007)
18. Srinivasan, A., Muggleton, S., Sternberg, M.J.E., King, R.D.: Theories for Mutagenicity: A Study in First-Order and Feature-Based Induction. Artificial Intelligence 85(1-2), 277–299 (1996)
19. Wachman, G., Khardon, R.: Learning from Interpretations: A Rooted Kernel for Ordered Hyper- graphs. In: 24th international conference on Machine learning. ACM Press, New York (2007)
20. Zhang, T.: Statistical behavior and consistency of classification methods based on convex risk minimization. The Annals of Statistics 32, 56–85 (2004)
21. Zhang, T., Yu, B.: Boosting with early stopping: convergence and consistency. Annals of Statistics 33, 2005 (2003)
22. Zhao, P., Yu, B.: On Model Selection Consistency of Lasso. J. Machine Learning Research 7, 2541–2563 (2006)
23. Zhu, J.: Flexible statistical modeling. Ph.D. Thesis. Stanford University (2003)
24. Zhu, J., Rosset, S., Hastie, T., Tibshirani, R.: 1-norm support vector machine. In: NIPS, vol. 16, MIT Press, Cambridge (2003)

A Generic Approach to Topic Models

Gregor Heinrich

Fraunhofer IGD + University of Leipzig
Darmstadt, Germany
heinrich@igd.fraunhofer.de

Abstract. This article contributes a generic model of topic models. To define the problem space, general characteristics for this class of models are derived, which give rise to a representation of topic models as "mixture networks", a domain-specific compact alternative to Bayesian networks. Besides illustrating the interconnection of mixtures in topic models, the benefit of this representation is its straight-forward mapping to inference equations and algorithms, which is shown with the derivation and implementation of a generic Gibbs sampling algorithm.

1 Introduction

Mixture models [1] are a powerful tool to model complex probabilistic distributions by convex sums of component densities, $p(x) = \sum_k p(z{=}k)p(x|\vartheta_k)$, where z is an index variable that indicates which component k the observation x originates from. Among the large class of such models, mixture models with discrete component densities $p(x|\vartheta_k)$ are of particular interest because in this case the component densities can serve as weighting functions for other mixtures, which themselves can have again discrete or non-discrete component densities. This fact makes it possible to construct models that consist of cascades or even networks of coupled discrete mixtures as generative structure underlying one or more observable mixtures with arbitrary (e.g., discrete or Gaussian) component densities.

Such a coupling of mixtures can be considered a defining characteristic of topic models, a class of probabilistic models that has become a central subject of research in text mining, computer vision, bioinformatics and other fields. Following the idea proposed by the seminal work on latent Dirichlet allocation (LDA [2]), topic models exploit the conjugacy of Dirichlet and multinomial/discrete distributions to learn discrete latent variables from discrete co-occurrence data (e.g., [3,4,5]) or from the co-occurrence of discrete and continuous features (e.g., [6]). Via the interrelation of the latent variables across different mixture levels, structures assumed in the data can be accounted for in specialised topic models, which renders the topic model approach a powerful and flexible framework.

However, the published work on topic models only defines this framework implicitly; authors tend to analyse and derive probabilistic properties and inference algorithms on a model-specific basis, typically using results from particular prior work. Although on the other hand frameworks for automatic inference in (more general) Bayesian networks exist that are in principle capable of handling topic models as special cases (e.g., WinBUGS [7], HBC [8], AutoBayes [9], or VIBES [10]), the generality of this software

W. Buntine et al. (Eds.): ECML PKDD 2009, Part I, LNAI 5781, pp. 517–532, 2009.

makes it difficult (1) to gain insights from the result of the automatic inference derivation process, and (2) to make performance improvements that may be possible for more restricted model structures, which is desirable especially because topic models have serious scalability issues. Such improvements may be based on the recent advances in massively parallel hardware, along with general-purpose programming platforms like OpenCL [11], or heterogeneous computing architectures including specialised FPGA processor designs, along with programming interfaces like the hArtes toolchain [12]. For such high-performance computing architectures, a generic approach to topic model inference may permit to reuse highly optimised kernels across models and therefore allow to focus optimisation effort.

Apart from theoretical interest, these practical considerations motivate a closer look on topic models with the intent to characterise their properties in a generic manner. Specifically, we generalise the probabilistic properties of topic models in Sec. 2. Motivated by this general characterisation, we propose a specialised representation of topic models in Sec. 3: mixture networks. Subsequently, as a basis for actual implementations we present a generic approach to inference in mixture networks in Sec. 4 for the case of Gibbs sampling, which has been implemented as a generic Gibbs sampling tool described in Sec. 5. We finish with conclusions and future work directions in Sec. 6.

2 Generalising Topic Models

In this section, we present a generic characterisation to topic models. As a basis for the following derivations, consider an arbitrary Bayesian network (BN [13]) with variables $U_n \in U$. Its likelihood can be generally formulated as:

$$p(U) = \prod_n p(U_n \,|\, \mathrm{pa}(U_n)) \tag{1}$$

where the operator $\mathrm{pa}(U_n)$ refers to the set of parents of some BN node that belongs to variable U_n.

Characteristics. As has been outlined in the Introduction, the first notable characteristic of topic models is their use of the conjugate Dirichlet and multinomial/discrete distributions. Focussing on discrete observations, such models can be structured entirely into "mixture levels", each of which consists of a set of multinomial components $\vec{\vartheta}_k \in \Theta \triangleq \{\vec{\vartheta}_k\}_{k=1}^K$ that are themselves drawn from Dirichlet priors with some set of hyperparameters $\vec{\alpha}_j \in A \triangleq \{\vec{\alpha}_j\}_{j=1}^J$. Based on one or more discrete values from parent nodes in the BN, a component k among the multinomial mixture is chosen and a discrete value x_i sampled from it, which is part of the observation sequence $X \triangleq \{x_i\}_{i \in I}$ with index sequence I. The corresponding generative process for one mixture level can be summarised as:

$$\begin{aligned} x_i \,|\, \vec{\vartheta}_k, k{=}g(\uparrow x_i, i) &\sim \mathrm{Mult}(x_i|\Theta, \uparrow x_i) \\ \vec{\vartheta}_k \,|\, \vec{\alpha}_j, j{=}f(\uparrow X) &\sim \mathrm{Dir}(\vec{\vartheta}_k|A, \uparrow X) \end{aligned} \tag{2}$$

where the component index k is some function of the incoming discrete values or their indices that maps to components of the local mixture level. For this, the parent variable operator $\uparrow x_i$ is introduced that collects all parent variables of x_i (excluding parameters: $\uparrow X = \mathrm{pa}(X)\backslash\Theta$), and the component selection function can consequently be expressed as $k = g(\uparrow x_i, i)$. Hyperparameter indices j can be chosen either to be global for all k, i.e., $j \equiv 1$, or similarly to the component indices, assigned to a group of components with some grouping function $j = f(\uparrow X)$. This grouping can be used to model clustering among components (see, e.g., [14,4]).

The generative process in Eq. 2 reveals the second characteristic of topic models: Mixture levels are solely connected via discrete parent variables ($\uparrow X$), which ensures a simple form of the joint likelihood of the model.[1] Based on Eq. 1, the complete topic model can be constructed from the mixture levels $\ell \in L$, yielding the likelihood:

$$p(X, \Theta|A) = \prod_{\ell \in L} p(X^\ell, \Theta^\ell \,|\, A^\ell; \uparrow X^\ell) \tag{3}$$

$$= \prod_{\ell \in L} \left[\prod_{i \in I} \mathrm{Mult}(x_i \,|\, \Theta, \uparrow x_i) \prod_{k=1}^{K} \mathrm{Dir}(p(\vec{\vartheta}_k \,|\, A, \uparrow X) \right]^{[\ell]} \tag{4}$$

where for simplicity we mark up variables specific to a level with a superscript ℓ, in brackets $[\cdot]^{[\ell]}$ for all their contents or for entire equations in the text. Without this markup, symbols are assumed model-wide sets of variables X, parameters Θ, hyperparameters A, etc. Eq. 4 shows the structure of the joint likelihood common for topic models: Multinomial observations factorise over the sequence of tokens generated by the model, and the Dirichlet priors factorise between the components.

Mixture level likelihood. Due to the conjugacy of the Dirichlet and multinomial/discrete distributions, the inner terms of Eq. 4 can be simplified further after a transformation from tokens with index $i \in I$ (part of a sequence) to counts over component dimensions with index over $t \in [1, T]$ (part of a "vocabulary"), each specific to a mixture level ℓ. For every ℓ, the following holds for the total count of co-occurrences between outcomes $x_i{=}t$ and mixture components $k{=}g(\uparrow x_i, i)$ responsible for them:

$$n_{k,t} = \sum_{i \in I} \delta(k - g(\uparrow x_i, i))\, \delta(t - x_i) \tag{5}$$

where $\delta(x)$ is the delta function, $\delta(x) = \{1 \text{ if } x{=}0, 0 \text{ otherwise}\}$. Using these counts, the likelihood of one mixture level becomes:

$$p(X^\ell, \Theta^\ell \,|\, A^\ell; \uparrow X^\ell) = \left[\prod_{k=1}^{K} \frac{1}{\Delta(\vec{\alpha}_j)} \prod_{t=1}^{T} \vartheta_{k,t}^{n_{k,t}+\alpha_{j,t}-1} \right]^{[\ell]} \tag{6}$$

where the product over t in Eq. 6 is the integrand of a Dirichlet integral and $\Delta(\vec{\alpha})$ is the partition function of the Dirichlet distribution, a T-dimensional generalisation of the beta function:

$$\Delta(\vec{a}) \triangleq \frac{\prod_{t=1}^{T} \Gamma(a_t)}{\Gamma(\sum_{t=1}^{T} a_t)} . \tag{7}$$

[1] With the hyperparameters dependent on $j = f(\cdot)$, formally there is an additional dependency between mixture levels, but this is dropped by assuming the set A known; common EM-type inference methods estimate hyperparameters independently inside their M-step; see Sec. 4.

Mixture level variants. The topic model framework is not restricted to the Dirichlet–multinomial type of mixture level that forms its core. Several possibilities exist to extend the framework and plug as levels into Eq. 3:

- *Symmetric hyperparameters* are a common variant to the standard vectors, see, e.g., the original LDA model [2]. The Dirichlet partition function simplifies to $\Delta_T(a) \triangleq \Gamma(a)^T/\Gamma(Ta)$.
- *Observed parameters* introduce known mixture proportions or fixed observations like labels (see, e.g., the author–topic model [3]), which leads to $p(X|\Theta;\uparrow X) = \prod_i \text{Mult}(x_i|\Theta,\uparrow x_i) = \prod_{k,t} \vartheta_{k,t}^{n_{k,t}}$ for a level, i.e., the Dirichlet vanishes in Eq. 3.
- *Infinite mixtures* allow model adaptation to data dimensionalities and typically use Dirichlet process (DP) mixtures or generalisations [15]. Mixture interrelations in topic models are handled typically using hierarichcal DPs [16]. Especially the stick-breaking representation of the DP and its finite approximations [17,18] promise to preserve a high similarity to finite-dimensional models.
- *Non-Dirichlet priors* like logistic–normal allow a more flexible combination of topics in particular mixture levels, as in the correlated topic model [19].
- *Non-discrete observation components* use, e.g., Gaussian distributions in the final mixture level (e.g., in Corr-LDA [6]). Parameters are preferrably drawn from conjugate priors to simplify inference. The likelihood of a non-discrete mixture level is $p(\tilde{X}|\tilde{\Theta},\tilde{A};\uparrow\tilde{X}) = \prod_i h(\tilde{x}_i|\tilde{\Theta},\uparrow\tilde{x}_i)\ \prod_k g(\tilde{\vartheta}_k|\tilde{\alpha}_j)$ with component distribution h and prior g.

To keep this paper focussed, we restrict ourselves to the first two variants mentioned, which already cover a vast body of models in the literature.

3 Mixture Networks

The dependency structure characteristic for topic models shown in Eq. 3 gives rise to the idea of a specialised graphical representation. In addition to the fact that in BN diagrams of more complex topic models, interrelations between mixtures are easily hidden in complex network structures, the introduction of a domain-specific graphical representation of topic models may help simplify derivation of their likelihood structure and inference equations.

To obtain such a representation, we use the two characteristics discussed in Sec. 2: Dirichlet–multinomial mixture levels and the connections between levels via discrete variables, which can be seen as nodes and edges in a new network structure. This leads to a representation of topic models as "mixture networks".

3.1 Definition

A mixture network (MN) is defined as a digraph $G(\mathcal{N},\mathcal{E})$ that consists of (1) a set of nodes, $\mathcal{N}$, where (a) an inner node represents a mixture level as described in Sec. 2, i.e., a sampling operation from a mixture component and (b) a terminal node represents an observable (discrete) value, as well as (2) a set of directed edges, $\mathcal{E} : \mathcal{N} \times \mathcal{N}$, where an

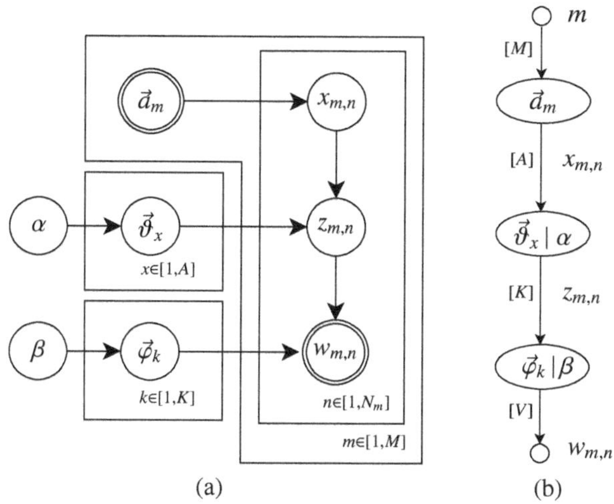

Fig. 1. The author–topic model, (a) Bayesian network and (b) mixture network

edge propagates a discrete value from its parent node to its child node. The child node then uses the value to choose one of its components.

Graphical notation. A graphical notation for mixture networks is proposed in Fig. 1(b) via the example of the author–topic model (ATM [3]), which models the topic association with authors with three mixture levels and whose BN is shown in Fig. 1(a). Opposed to BNs that visualise dependencies between random variables and express the repetitions of data points (plate notation), MNs focus on the interrelations between discrete mixtures (in the example: document–author, $\vec{a}_m$, author–topic, $\vec{\vartheta}_x|\alpha$, and topic–word, $\vec{\varphi}_k|\beta$, distributions), along with component numbers and dimensionalities ($[M]$, $[K]$ and $[V]$). The frequency of sampling a particular variable is encoded in subscripts (in the example: $\vec{\vartheta}_x$, $\vec{\varphi}_k$ and $w_{m,n}$ referring to author-, topic- and word-wise sampling schedules, respectively). Note that the top (inner) node does not indicate a hyperparameter because of its observed parameter (see mixture level variants in Sec. 2).

3.2 Example Models

In illustrate the applicability of the MN representation, the mixture network diagrams of some topic models from the literature are drawn in Fig. 2. Fig. 2(a) shows the MN of the model of latent Dirichlet allocation (LDA [2]), which served as a design paragon to all other models. It has two mixture levels: a document–topic mixture $\vec{\vartheta}_m|\alpha$ and a topic–term mixture $\vec{\varphi}_k|\beta$. The extension of this model gives illustrative insight into the organisation of mixtures to account for logical and semantic structure assumed in the data. A simple extension to the plain LDA model is the author–topic model shown in Fig. 2(b) as explained above.

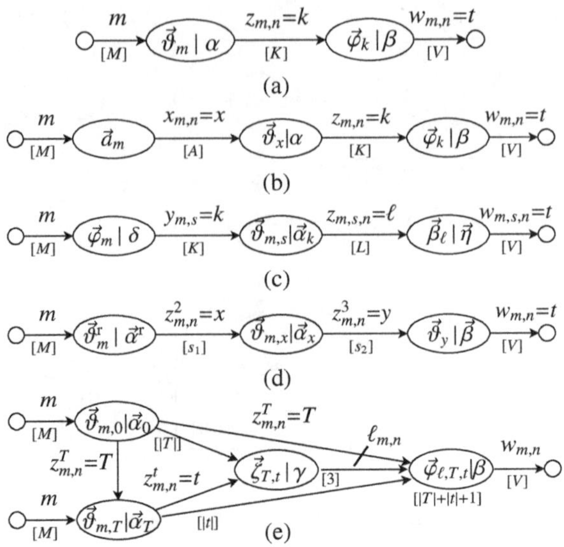

Fig. 2. Mixture networks of example models from the literature: (a) latent Dirichlet allocation, (b) author–topic model, (c) latent Dirichlet co-clustering model, (d) 4-level pachinko allocation, (e) hierarchical pachinko allocation (hPAM1)

Co-clustering. The model of latent Dirichlet co-clustering (LDCC [14]) in Fig. 2(c) uses aggregation to infer an additional logical layer of topics from the data: For each section s, a topic distribution $\vec{\vartheta}_{m,s}$ can be inferred, and document topic distributions $\vec{\varphi}_m$ index word-topics $z_{m,s,n}$ indirectly via section topics $y_{m,s}$, allowing a finer-grained handling of topic structure across documents with component selection function $g(\uparrow z_{m,s,n}, (m, s, n)) = (m, s)$. Further, segment topics $\vec{\vartheta}_{m,s}$ are coupled across documents via the topic-hyperparameters $\vec{\alpha}_y$ with component group function $f(\uparrow z) = k$.

Pachinko allocation. Another multi-level MN is the class of pachinko allocation models (PAM), of which the four-level variant as described in [4] is depicted in Fig. 2(d). For each word, a path through a topic hierarchy is sampled consisting of the indicators (z^1, z^2, z^3) where $z^1 = 1$ provides the root of the tree, associated with LDA-type document topics $\vec{\vartheta}^r_m$, and based on its sample, a document- and topic-dependent level $\vec{\vartheta}_{m,x}$ is sampled ($g(\uparrow z^3_{m,n}, (m, n)) = (m, x)$), finally indexing word-topics $\vec{\vartheta}_y$. Similar to the LDCC model, component grouping is used: $f(\uparrow z^3) = x$.

Hierarchical pachinko allocation [5] (hPAM) as shown in Fig. 2(e) is an example for a more complex model that allows a hierarchy of topic–term distributions: As in PAM, the topic hierarchy consists of document-specific root- and super- as well as global subtopics, but each node k in the hierarchy is associated with a topic–term distribution $\vec{\varphi}_k$, and for each word $w_{m,n}$, a complete topic path (root–super–sub) is sampled along with a level $\ell_{m,n}$ from $\vec{\zeta}_{T,t}$ specific to super- and sub-topics (hPAM1 in [5]). The topic sample on level $\ell_{m,n}$ selects from the set $k = \{1, 1 + T, 1 + |T| + t\}$ the component $\vec{\varphi}_k$ that finally generates the word.

Although with a different goal in mind, the concept of pachinko allocation models is closely related to the approach pursued with mixture networks because it allows to connect different levels of mixtures with great flexibility. In fact, MNs can be considered a generalisation of PAMs that allows free interconnection of nodes in general DAG structures with different types of mixture levels (observed, unobserved parameters) and with observable variables (edges or parameters) at arbitrary points in the network. By appropriate choice of index transformations $g^\ell(\uparrow x_i^\ell, i^\ell)$, even the component-dependent subtrees mentioned as the most flexible version of the PAM concept [4] may be realised with mixture networks.

4 Inference in Mixture Networks

Inference in the context of mixture networks refers to finding the parameters Θ and hyperparameters A given the observations. With the model variables X divided into sets of visible (observed) and latent (hidden) variables, $X = \{V, H\}$, this is typically a two part process of (1) Bayesian inference for the posterior distribution,

$$p(H, \Theta | V, A) = \frac{p(V, H, \Theta | A)}{p(V | A)}, \tag{8}$$

and (2) estimation of the hyperparameters, for which ML or MAP estimators are commonly sufficient because of the simpler search space.

As in many latent-variable models, determining the posterior Eq. 8 is generally intractable in mixture networks because of excessive dependencies between the latent variables H and parameters Θ in the marginal likelihood for the observations V in the denominator, $p(V|A) = \sum_H \int p(V, H, \Theta|A)\, \mathrm{d}\Theta$. To circumvent this intractability, approximate inference methods have been proposed, for topic models including mean-field variational Bayes [2], collapsed variational Bayes [20], expectation propagation [21] and collapsed Gibbs sampling [22].

For our purposes, a method is needed that has feasible complexity with reasonable accuracy even when it comes to modelling dependencies between variables. The full factorisation of variational mean-field distributions may be adverse for model fitting [23], and structured approaches become complicated quickly [10]. Expectation propagation on the other side has not been commonly used with more complex topic models. Thus, Gibbs sampling appears to be the most straight-forward method for a formulation of approximate inference for mixture networks.

4.1 Gibbs Sampling

Gibbs sampling [24] is an approximative inference method particularly suited for models where the marginals of the posterior can be expressed in closed form, in particular for high-dimensional discrete models. As a Markov-chain Monte Carlo (MCMC) method, Gibbs sampling uses a Markov chain that upon convergence approximately generates samples according to the posterior distribution. By sampling one dimension of the posterior at a time, Gibbs sampling avoids computationally complex Metropolis-Hastings acceptance calculations. One step in the Gibbs sampling inference approach thus corresponds to sampling dependent hidden variables h_i for each data token v_i from the full conditional distribution, $h_i \sim p(h_i|H_{\neg i}, V, \Theta, A)$, where $\cdot_{\neg i}$ refers to the complete set of tokens except i. Analogously, $\vec{\vartheta}_k$ must be sampled in such an approach [25].

With topic models, it has been shown, however, that collapsed approaches to Gibbs sampling, i.e., those that integrate out parameters Θ [26], lead to particularly good convergence behaviour [22] (which is attributed to the high independence of the remaining hidden variables). Therefore, the posterior considered for Gibbs sampling is $p(H|V, A) = \int p(H, \Theta|V, A)\, d\Theta$. The Markov state of the Gibbs sampler then reduces to H, and the resulting mixture network inference approach can be considered a form of stochastic EM algorithm [27] that trains the latent variables H in its E-step and hyperparameters A in its M-step.

To sample from posteriors of collapsed MNs, for each independent latent variable H^ℓ (generic variables, complete sequence: upper case) with tokens $h_i^\ell \in H^\ell \triangleq \{h_{i'}^\ell\}_{i' \in I^\ell}$ (tokens: lower case; convention: $h_i^\ell \equiv h_{i^\ell}^\ell$ unless otherwise noted), a separate full conditional distribution $p(h_i^\ell|H_{\neg i}^\ell, H^{\neg \ell}, V, A)$ must be formulated for each token $h_i^\ell \in H^\ell$ with $\cdot^{\neg \ell}$ used analogous to $\cdot_{\neg i}$. Typically, however, several hidden variables are dependent and need to be drawn as a block. Therefore, with dependency groups denoted by H^d with $H^\ell \subseteq H^d \subseteq H$ as sequences of groups of dependent tokens h_i^d, the full conditionals sought are: $p(h_i^d|H_{\neg i}^d, H^{\neg d}, V, A)$ for each group d and each token $i = i^d$. Remember that subscripts refer to sequence indices and superscripts to levels. Further note that h_i^d is a configuration of hidden variables that corresponds to a unique combination of components k^ℓ and outputs t^ℓ of the mixture levels involved.

Derivation. To find the full conditional distributions, we start from the joint likelihood, Eq. 3, and for a collapsed approach integrate out its parameters via Dirichlet integrals:

$$
\begin{aligned}
p(V, H|A) &= \prod_{\ell \in L} \left[\int \prod_{k=1}^{K} \frac{1}{\Delta(\vec{\alpha}_j)} \prod_{t=1}^{T} \vartheta_{k,t}^{n_{k,t}+\alpha_{j,t}-1}\, d\Theta \right]^{[\ell]} \\
&= \prod_{\ell \in L} \left[\prod_{k=1}^{K} \frac{\Delta(\vec{n}_k + \vec{\alpha}_j)}{\Delta(\vec{\alpha}_j)} \right]^{[\ell]}
\end{aligned} \tag{9}
$$

where the level-specific $\vec{n}_k$ are vectors of co-occurrence counts $n_{k,t}$.

This equation shows that the joint likelihood of the model variables is a product of ratios of Dirichlet partition functions for each component on each individual mixture level in the model. Interestingly, using the identity $\Gamma(a+n) = \Gamma(a) \prod_{c=0}^{n-1}(a+c)$ with real $a > 0$ and integer $n \geq 0$, we obtain a ratio of finite product sequences:

$$
\frac{\Delta(\vec{a} + \vec{n})}{\Delta(\vec{a})} = \frac{\prod_{t=1}^{T} \prod_{c=0}^{n_t-1} (a_t + c)}{\prod_{c=0}^{[\sum_t n_t]-1} ([\sum_{t=1}^{T} a_t] + c)}\,, \tag{10}
$$

which for a unit difference in a single element u, $\Delta(\vec{a}+\delta(t-u))/\Delta(\vec{a})$, reduces to $a_u / \sum_t a_t$. Note that with Eq. 10, we can alternatively expand Eq. 9 into products without any special functions, which comes at the cost of obtaining denominator terms in Eq. 9 specific to components k.

The next step to obtain full conditionals is to determine dependent edges $H^d \subseteq H$: Analogous to the "Bayes ball" algorithm in Bayesian networks, in MNs we can identify dependent edges by finding subgraphs that extend through nodes (1) whose component selection function $g(\uparrow x_i, i)$ contain hidden edge values or (2) have hidden outputs. In the examples given in Sec. 2, ATM, PAM and hPAM models have dependent edges; LDCC does not because the hidden variables are connected via hyperparameters (assumed given in the full conditional). Further, edges of nodes adjacent to subgraph H^d but independent of H^d are collected in a set $S^d \subset \{V, H\}$ with token sets s_i^d. We use the notation $\cdot^{\neg s}$ to denote the exclusion of S^d. With these definitions, full conditional distributions can be derived generically by applying the chain rule:

$$\begin{aligned} p(h_i^d | H_{\neg i}^d, H^{\neg d}, V, A) &= \frac{p(h_i^d, s_i^d | H_{\neg i}^d, S_{\neg i}^d, H^{\neg d, \neg s}, V^{\neg s}, A)}{p(s_i^d | H_{\neg i}^d, S_{\neg i}^d, H^{\neg d, \neg s}, V^{\neg s}, A)} \\ &\propto p(h_i^d, s_i^d | H_{\neg i}^d, S_{\neg i}^d, H^{\neg d, \neg s}, V^{\neg s}, A) \\ &= \frac{p(H, V | A)}{p(H_{\neg i}^d, S_{\neg i}^d, H^{\neg d, \neg s}, V^{\neg s} | A)} \\ &= \prod_{\ell \in \{H^d, S^d\}} \left[\prod_{k=1}^{K} \frac{\Delta(\vec{n}_k + \vec{\alpha}_j)}{\Delta(\vec{n}_{k, \neg i^d} + \vec{\alpha}_j)} \right]^{[\ell]} . \end{aligned} \tag{11}$$

Generic full conditionals. In Eq. 11, all terms except those with a count difference between numerator and denominator cancel out. The remainder of terms can be simplified by applying Eq. 10 with $\vec{a} = \vec{n}_{k^\ell, \neg i^d}^\ell + \vec{\alpha}_j^\ell$, and the resulting full conditional becomes a product of the following form if all mixture levels $\in \{H^d, S^d\}$ exclude only a single token with $\neg i^d$:

$$p(h_i^d | H_{\neg i}^d, H^{\neg d}, V, A) \propto \prod_{\ell \in \{H^d, S^d\}} \left[\frac{n_{k,t,\neg i^d} + \alpha_{j,t}}{\sum_{t=1}^{T} n_{k,t,\neg i^d} + \alpha_{j,t}} \right]^{[\ell]} . \tag{12}$$

The factors in Eq. 12 can be interpreted as posterior means of Dirichlet distributions with hyperparameters $\vec{\alpha}_j$ and observation counts $\vec{n}_{k,\neg i^d}$, $\langle \mathrm{Dir}(\cdot | \vec{n}_{k,\neg i^d} + \vec{\alpha}_j) \rangle$ on level ℓ. Although this form of full conditional factors is prevalent in a majority of topic models, with the scope of models considered in this paper alternative forms are possible:

- If $[g(\uparrow x_i, i)]^{[\ell]}$ contains no hidden edges, the denominator can be omitted (e.g., nodes with m as only component index).
- If one index i^d at a mixture level input aggregates a whole sequence of i^ℓ at its output, $\neg i^d$ corresponds to more than one token in the factor denominator in Eq. 11 (e.g., in LDCC, section topics $y_{m,s}$ aggregate word topic sequences $\{z_{m,s,n}\}_n$), which yields a factor analogous to Eq. 10:

$$\left[\frac{\prod_{t=1}^{T} \prod_{c=0}^{n_{k,t}-1} (c + \alpha_{j,t})}{\prod_{c=0}^{[\sum_{t=1}^{T} n_{k,t}]-1} (c + \sum_{t=1}^{T} \alpha_{j,t})} \right]^{[\ell]} . \tag{13}$$

- Finally, mixture levels with observed parameters have components $\vec{\vartheta}_k$ as factors. In this case, few non-zero elements in $\vec{\vartheta}_k$ support sparse representations, while symmetric non-zero values cancel out.

4.2 Parameter Estimation

Generally, estimation of parameters and hyperparameters is part an M-step dual to the Gibbs E-step in a stochastic EM procedure. It can be performed on a per-node basis in mixture networks.

Hyperparameters. In many topic models, hyperparameters are of decisive importance, e.g., to couple component groups or to model data dispersion. As there is no closed-form solution for estimation of Dirichlet parameters from count data, iterative or sampling-based approaches are commonly employed. Extending results from [28] yields the following fixed-point iterations for node-specific standard and symmetric Dirichlet distributions that result in maximum likelihood estimates:

$$\alpha_{j,t} \leftarrow \alpha_{j,t} \frac{\left(\sum_{\{k:f(k)=j\}} \Psi(n_{k,t} + \alpha_{j,t})\right) - K_j \Psi(\alpha_{j,t})}{\left[\sum_{\{k:f(k)=j\}} \Psi(\sum_{t=1}^{T} n_{k,t} + \alpha_{j,t})\right] - K_j \Psi(\sum_{t=1}^{T} \alpha_{j,t})}, \tag{14}$$

$$\alpha \leftarrow \alpha \frac{\left(\sum_{k=1}^{K} \sum_{t=1}^{T} \Psi(n_{k,t} + \alpha)\right) - KT\Psi(\alpha)}{T\left[\left(\sum_{k=1}^{K} \Psi([\sum_{t=1}^{T} n_{k,t}] + T\alpha)\right) - K\Psi(T\alpha)\right]}. \tag{15}$$

where $\Psi(x) = \mathrm{d}/\mathrm{d}x \log \Gamma(x)$ is the digamma function and level indicators ℓ are omitted. For the case $j \not\equiv 1$ we use $f(\uparrow X) = f(k)$ for notational simplicity. Each $\alpha_{j,t}$ then is estimated from K_j components for each of the J component groups. Estimators are initialised with a coarse-grained heuristic or a previous estimate and converge within few iterations.

Component parameters. Estimation of component parameters Θ is possible directly from the statistics of the collapsed state H and estimated hyperparameters A. Using the posterior mean of Dirichlet distributions given obervation counts $\vec{n}_k$ for each level ℓ, $\mathrm{Dir}(\vec{\vartheta}_k|\vec{\alpha}_j + \vec{n}_k) = \prod_i \mathrm{Mult}(x_i|\vec{\vartheta}_k) \cdot \mathrm{Dir}(\vec{\vartheta}_k|\vec{\alpha}_j)$, leads to the point estimate:

$$\vartheta_{k,t} = \frac{n_{k,t} + \alpha_{j,t}}{\sum_{t=1}^{T} n_{k,t} + \alpha_{j,t}} \tag{16}$$

where $\alpha_{j,t} \equiv \alpha$ for the symmetric case. Usually several samples $H^{(r)}, r \in [1, R]$ are taken from the stationary Markov chain with a sampling lag in between to ensure decorrelation. Finally parameters are averaged: $\vec{\vartheta}_k \approx R^{-1} \sum_r \vec{\vartheta}_k^{(r)}$.

4.3 Predictive Inference

In many applications, it is necessary to predict the topics of some query data set V' given the model $\mathcal{M}$ trained on the observations V. Regarding the information required to represent the model $\mathcal{M}$, two different types of node can be distinguished:

- *Topic nodes*, $\ell \in L^*$, represent mixtures whose components are not specific to documents, i.e., $g(\uparrow x_i, i) \equiv g(\uparrow x_i)$, and $\mathcal{M}$ contains their parameters $\Theta^* = \{\Theta^\ell\}_{\ell \in L^*}$,
- *Sequence nodes*, $\ell \in L'$, represent mixtures specific to documents, and $\mathcal{M}$ contains their hyperparameters $A' = \{A^\ell\}_{\ell \in L'}$ that allow to find parameters $\Theta' = \{\Theta^\ell\}_{\ell \in L'}$.

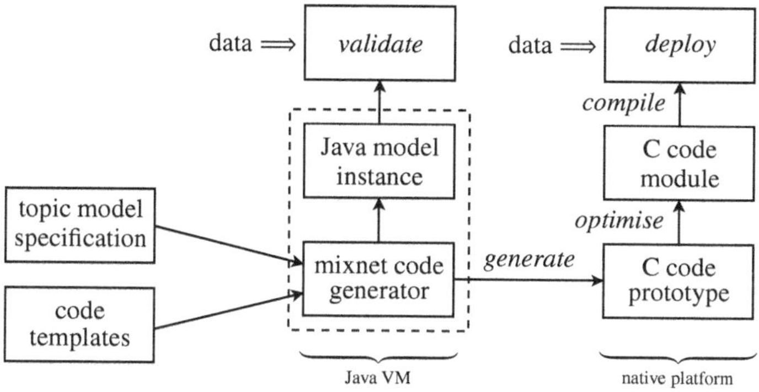

Fig. 3. Mixture network Gibbs sampler development workflow

Thus we can define $\mathcal{M} \triangleq \{\Theta^*, A'\}$, and finding the association of unseen data V' with a state H' can be achieved using Gibbs sampling with a predictive full conditional analogous to Eq. 11, only that now it is possible (1) to treat parameters of topic nodes Θ^* as observed and (2) to restrict sampling to the query state H' without M-step updates, which both accelerates convergence of H' compared to H:

$$p(h_i'^d | H_{\neg i}'^d, H'^{\neg d}, V', \mathcal{M}) \propto \prod_{\ell \in \{S^{*d}, H^{*d}\}} [\vartheta_{k,t}]^{[\ell]} \cdot \prod_{\ell \in \{S'^d, H'^d\}} \left[\prod_{k=1}^{K} \frac{\Delta(\vec{n}_k + \vec{\alpha}_j)}{\Delta(\vec{n}_{k,\neg i^d} + \vec{\alpha}_j)} \right]^{[\ell]} . \quad (17)$$

With this equation, all findings on generic full conditionals that were derived from the analogous Eq. 11 can be reused, including Eqs. 12 and 13. Parameters can be estimated again using Eq. 16.

5 Implementation

The coherence of Eqs. 12–17 across models leads to the conclusion that Gibbs sampler implementations can be achieved based on a small number of computation kernels. Few reusable kernels are desirable when targeting architectures that require high optimisation effort. In this section, a proof-of-concept implementation of the MN approach is outlined that, although it targets a CPU-based architecture, may be a basis for topic model implementation on massively parallel and FPGA-based architectures.

5.1 Generic Gibbs Samplers

The implementation of MN Gibbs samplers is based on a multi-stage workflow that allows to construct software modules with increasing levels of optimisation. This is intended to keep the interface for the researcher simple while retaining flexibility with respect to target architectures. An overview of the workflow is given in Fig. 3.

The central block in this process is the Java-based mixture network code generator, which is fed with a simple text script of a given MN (for example that shown in Fig. 4)

```
data:                          # input data
    w[m,n] : M * N[m] -> V     # word tokens (vocabulary size V)
state:                         # latent variables (E-step)
    x[m,n] : M * N[m] -> X     # supertopics (defines dimension X)
    y[m,n] : M * N[m] -> Y     # subtopic (defines dimension Y)
est:                           # estimated parameters (M-step)
    thetar : M * X             # document-supertopic level 1
    theta  : M * X * Y         # supertopic-subtopic level 2
    phi    : Y * V             # subtopic-term level 3
    alphar : 1                 # level 1 hparam (scalar)
    alpha  : X * Y             # level 2 hparam (with grouping)
    beta   : 1                 # level 3 hparam (scalar)
network:                       # format: parent_values >>
                               #     param[g(pav,i)] | hparam[f(pav)]
                               #     >> child_edge[sequence] = value
    m   >>   thetar[m]   | alphar    >>   x[m,n] = x
    x   >>   theta[m,x]  | alpha[x]  >>   y[m,n] = k
    k   >>   phi[k]      | beta      >>   w[m,n]
```

Fig. 4. Commented mixture network script for 4-level PAM

and allows two modes of operation: (a) The generator can create an instance of a Java-based Gibbs sampling class directly from the model script, e.g., for model validation purposes, and: (b) Based on a set of code templates, it generates C source code of the Gibbs sampler kernels that can then be further optimised and integrated with other code before it is compiled for the native computing platform.

In both cases, the generator applies the results of Sec. 4 to the information parsed from the script, creating Gibbs sampling algorithms as outlined in Fig. 5. Across different MN models, the design follows a stochastic EM approach that after initialisation loops over alternating sampling (E) and hyperparameter estimation (M) steps until convergence, after which samples can be drawn from the posterior. Important data structures in the generated code include the Markov state H, its count statistics as well as the arrays for multinomial sampling from the full conditional. The main computation kernels are those for full conditionals, Eq. 12 (including filling of the multinomial masses of $p(h_i^d|\cdot)$), for parameter estimation, Eqs. 14–16, as well as for convergence monitoring, which is described below. Currently, in addition to standard Dirichlet–multinomial nodes models can include observed nodes and parameters but are restricted to a single token sequence, which excludes aggregation as in the LDCC example.

Convergence monitoring and model quality. Gibbs sampling and other MCMC methods pose the general problem to determine when their Markov chain reaches a stationary state that allows to sample from the posterior distribution. With standard convergence diagnostics [29] difficult to apply to the high-dimensional discrete problem at hand, an alternative approach is to use some measure of model quality that reaches an optimum at convergence. Because of its generalisability and frequent use of similar approaches in topic model evaluation, the likelihood of test data given the trained model $\mathcal{M}$ (as defined in Sec. 4.3) has been chosen as quality measure, whose generalisation can be outlined as follows:

Algorithm `mixnetGibbs`(V, V')
Input: training and test observations V, V'
Global data: level-specific dimensions $K^H = \{K^\ell\}_{\ell \in H}$, $T^H = \{T^\ell\}_{\ell \in H}$, selection functions f and g, count statistics $N^\ell = [\{\vec{n}_k\}_{k=1}^K]^\ell$, $N^\ell \in N$ and their sums $\Sigma^\ell = [\{\sum_x n_{k,t}\}_{k=1}^K]^\ell$, $\Sigma^\ell \in \Sigma$ for each node with hidden parameters, memory for full conditional array $p(h_i^d|\cdot)$, likelihood $\mathcal{L}$
Output: topic associations H, parameters Θ and hyperparameters A
`// initialise`
for all nodes ℓ in topological order **do**
  random initialise hidden sequences $h_i^\ell \sim \text{Mult}(1/T^\ell)$, update counts N^ℓ and Σ^ℓ
`// Gibbs EM over burn-in period and sampling period`
while not (converged and R samples taken) **do**
  `// stochastic E step to sample collapsed state`
  for all dependency groups $H^d \subseteq H$ **do**
    for all joint tokens $h_i^d \in H^d$ **do**
      decrement counts N^d and sums Σ^d according to current state h_i^d
      assemble array for $p(h_i^d|H_{\neg i}^d, H^{\neg d}, V)$ acc. to Eq. 12
      sample new state $h_i^d \sim p(h_i^d|H_{\neg i}^d, H^{\neg d}, V)$
      increment counts N^d and sums Σ^d according to changed state h_i^d
  `// M step to estimate parameters`
  for all nodes ℓ **do**
    update hyperparameters A^ℓ acc. to Eqs. 14 and 15
  for all nodes ℓ **do**
    find parameters Θ^ℓ according to Eq. 16
  `// monitor convergence using test data likelihood`
  $\mathcal{L} \leftarrow$ **call** `testLik`(Θ, A, V') using Eqs. 16–18
  if $\mathcal{L}$ converged and L sampling iterations since last read out **then**
    `// different parameter read outs are averaged`
    $\bar{\Theta} \leftarrow \bar{\Theta} + \Theta$
`// Complete parameter average`
$\Theta = \bar{\Theta}/R$

Fig. 5. Generic Gibbs sampling algorithm

- For each sequence node of the network, the hidden state H' is trained on test data $V' = \{v_i'\}_i$ according to Eq. 17, resulting in predictive parameters $\Theta' = \{\Theta'^\ell\}_\ell$.
- For each test-data token v_i', the likelihood given parameters $\{\Theta', \Theta^*\}$ is calculated:

$$p(v_i' \mid \Theta', \Theta^*) = \sum_{h_i'} \prod_{\ell \in L} [\vartheta_{k,t}]^{[\ell]} \tag{18}$$

  where the sum over h_i' refers to marginalisation of all hidden variables. To calculate Eq. 18 efficiently, the mixture network is traversed level by level according to its generative process, multiplying the respective level parameters (elements of Θ'^ℓ or $\Theta^{*\ell}$) and summing over values of latent variables $h_i'^\ell$ not indexing components k^ℓ of child levels. Further, duplicate v_i' have identical likelihood.
- The log likelihood of held-out test documents is accumulated from the token likelihoods: $\mathcal{L}(V') = \sum_i \log p(v_i' \mid \Theta', \Theta^*)$.

As a variant, the test-set likelihood $\mathcal{L}(V')$ can be exponentiated and normalised with the number of tokens in the test data W' to obtain the perplexity: $\mathcal{P}(V') = \exp(-\mathcal{L}(V')/W')$, i.e., the inverse geometric mean of the token likelihoods. Both $\mathcal{L}(V')$ and $\mathcal{P}(V')$ are measures of how well a model is able to explain unseen data. Specifically, perplexity can be intuitively interpreted as the expected size of a vocabulary with uniform word distribution that the model would need to generate a token of the test data. A model that better captures co-occurrences in the data requires fewer possibilities to choose tokens given their context (document etc.). Due to the stochastic nature of the states H and H', values of $\mathcal{L}(V')$ and $\mathcal{P}(V')$ are not strictly monotonic over iterations. Thus, convergence of their moving-average process is used as indicator of Markov chain stationarity.

5.2 Validation

At this point, the focus of validation was on algorithms generated for a single-processor PC architecture, providing a basis for future investigation of specific high-performance architectures. In order to validate the implementation taken, generated and manually developed mixture network Gibbs samplers have been compared, including the examples LDA, ATM and PAM from Fig. 2 as well as several other models with two and three dependent hidden variables that handle labelled texts. Beside verification of the generated kernels, the code has been tested on the NIPS1-12[2] and Reuters-21578[3] data sets that in addition to text contain label information (authors, categories). Temporal performance achieved with the generated C-based algorithms came close to the respective manual implementations ($\Delta t < 2.5\%$). With equal seeds for random number generators, numerical behaviour turned out to be identical, considering Θ, A and $\mathcal{P}(V')$. Validation results are presented in further detail in a technical report [30].

6 Conclusions

We have presented a generic approach to topic models that covers a broad range of models in the literature. From their general characteristics, we have developed a representation of topic models as "mixture networks" along with a domain-specific graphical representation that complements Bayesian networks. Based on the mixture network representation, Gibbs sampling full conditionals were derived, which resulted in a generic Gibbs sampling algorithm and a "meta-Gibbs sampler" implementation based on code generation for specific models.

Future work can depart from these results in various directions. Extensions like the ones listed in Sec. 2 are desirable to widen the scope of the the mixture network approach, e.g., towards non-discrete observations as in the Corr-LDA model [6] and infinite mixtures with Dirichlet process priors [16]. Furthermore, the generic approach for Gibbs sampling may be applied analogously to collapsed variational Bayes [20].

The foremost research direction is, however, related to the actual motivation of this article discussed in the Introduction: to extend the code generation to high-performance computing architectures to help tackle the scalability issues common with topic models. The vision of this is a high-level language as a user front-end for implementations with optimised computing kernels. In addition to targeting computing platforms, improvements may be gained from heuristics like the statistically motivated acceleration of multinomial samplers proposed in [31], especially for the large sampling spaces of dependent latent variables.

[2] `http://www.cs.toronto.edu/~roweis/data.html`.

[3] `http://www.daviddlewis.com/resources/testcollections/reuters21578/`.

References

1. McLachlan, G., Peel, D.: Finite Mixture Models. Wiley, Chichester (2000)
2. Blei, D., Ng, A., Jordan, M.: Latent Dirichlet allocation. Journal of Machine Learning Research 3, 993–1022 (2003)
3. Steyvers, M., Smyth, P., Rosen-Zvi, M., Griffiths, T.: Probabilistic author-topic models for information discovery. In: The Tenth ACM SIGKDD International Conference on Knowledge Discovery and Data Mining (2004)
4. Li, W., McCallum, A.: Pachinko allocation: DAG-structured mixture models of topic correlations. In: ICML 2006: Proceedings of the 23rd international conference on Machine learning, pp. 577–584. ACM, New York (2006)
5. Mimno, D., Li, W., McCallum, A.: Mixtures of hierarchical topics with pachinko allocation. In: ICML 2007: Proceedings of the 24th International Conference on Machine Learning, pp. 633–640. ACM Press, New York (2007)
6. Barnard, K., Duygulu, P., Forsyth, D., de Freitas, N., Blei, D., Jordan, M.: Matching words and pictures. JMLR – Special Issue on Machine Learning Methods for Text and Images 3, 1107–1136 (2003)
7. Lunn, D., Thomas, A., Best, N., Spiegelhalter, D.: WinBUGS – a Bayesian modelling framework: concepts, structure, and extensibility. Statistics and Computing 10, 325–337 (2000)
8. Daumé, H.I.: HBC: Hierarchical Bayes Compiler (2007)
9. Gray, A.G., Fischer, B., Schumann, J., Buntine, W.L.: Automatic derivation of statistical algorithms: The EM family and beyond. In: NIPS, pp. 673–680 (2002)
10. Winn, J.M.: Variational Message Passing and its Applications. PhD thesis, University of Cambridge (2004)
11. Khronos OpenCL Working Group: The OpenCL Specification, version 1.0.29 (2008)
12. Rashid, M., Ferrandi, F., Bertels, K.: hArtes design flow for heterogeneous platforms. In: Proc. 10th International Symposium on Quality of Electronic Design (ISQED), pp. 330–338 (2009)
13. Wainwright, M.J., Jordan, M.I.: Graphical models, exponential families, and variational inference. Technical report, EECS Dept., University of California, Berkeley (2003)
14. Shafiei, M.M., Milios, E.E.: Latent Dirichlet co-clustering. In: ICDM 2006: Proceedings of the Sixth International Conference on Data Mining, pp. 542–551. IEEE Computer Society, Los Alamitos (2006)
15. Teh, Y.W., Jordan, M.I.: Hierarchical Bayesian nonparametric models with applications. In: Hjort, N., Holmes, C., Müller, P., Walker, S. (eds.) To appear in Bayesian Nonparametrics: Principles and Practice. Cambridge University Press, Cambridge (2009)
16. Teh, Y.W., Jordan, M.I., Beal, M.J., Blei, D.M.: Hierarchical Dirichlet processes. Journal of the American Statistical Association 101, 1566–1581 (2006)
17. Blei, D.M., Jordan, M.I.: Variational methods for the Dirichlet process. In: Proc. ICML (2006)
18. Ishwaran, H., James, L.F.: Gibbs sampling methods for stick breaking priors. Journal of the American Statistical Association 96, 161 (2001)
19. Blei, D., Lafferty, J.: A correlated topic model of science. Annals of Applied Statistics 1, 17–35 (2007)
20. Teh, Y.W., Newman, D., Welling, M.: A collapsed variational Bayesian inference algorithm for latent Dirichlet allocation. In: Advances in Neural Information Processing Systems, vol. 19 (2007)
21. Minka, T., Lafferty, J.: Expectation-propagation for the generative aspect model. In: Proc. UAI (2002)

22. Griffiths, T.L., Steyvers, M.: Finding scientific topics. Proceedings of the National Academy of Sciences 101, 5228–5235 (2004)
23. Dueck, D., Frey, B.J.: Probabilistic sparse matrix factorization. Technical report PSI TR 2004-023., U. Toronto, September 28 (2004)
24. Geman, S., Geman, D.: Stochastic relaxation, Gibbs distributions, and the Bayesian restoration of images. IEEE 6(6), 721–741 (1984)
25. Pritchard, J.K., Stephens, M., Donnelly, P.: Inference of population structure using multilocus genotype data. Genetics 155, 945–959 (2000)
26. Liu, J.S.: The collapsed Gibbs sampler in Bayesian computations with applications to a gene regulation problems. Journal of the American Statistical Association 89(427), 958–966 (1994)
27. Jank, W.: Stochastic variants of EM: Monte Carlo, quasi-Monte Carlo and more. In: Proc. American Statistical Association, Minneapolis, Minnesota (2005)
28. Minka, T.: Estimating a Dirichlet distribution. Web (2000)
29. Robert, C.P., Casella, G.: Monte Carlo Statistical Methods, 2nd edn. Springer, New York (2004)
30. Heinrich, G.: Generic topic models. Technical report 09RP008-FIGD, Fraunhofer IGD, Darmstadt (2009)
31. Porteous, I., Newman, D., Ihler, A., Asuncion, A., Smyth, P., Welling, M.: Fast collapsed Gibbs sampling for latent Dirichlet allocation. In: KDD 2008: Proceeding of the 14th ACM SIGKDD international conference on Knowledge discovery and data mining, pp. 569–577. ACM Press, New York (2008)

Feature Selection by Transfer Learning with Linear Regularized Models

Thibault Helleputte[1,2] and Pierre Dupont[1,2]

[1] University of Louvain, Computing Science and Engineering Dept., Reaumur Building, Place Sainte Barbe 2, B-1348 Louvain-la-Neuve, Belgium
[2] University of Louvain, Machine Learning Group
{Thibault.Helleputte,Pierre.Dupont}@uclouvain.be

Abstract. This paper presents a novel feature selection method for classification of high dimensional data, such as those produced by microarrays. It includes a partial supervision to smoothly favor the selection of some dimensions (genes) on a new dataset to be classified. The dimensions to be favored are previously selected from similar datasets in large microarray databases, hence performing inductive transfer learning at the feature level. This technique relies on a feature selection method embedded within a regularized linear model estimation. A practical approximation of this technique reduces to linear SVM learning with iterative input rescaling. The scaling factors depend on the selected dimensions from the related datasets. The final selection may depart from those whenever necessary to optimize the classification objective. Experiments on several microarray datasets show that the proposed method both improves the selected gene lists stability, with respect to sampling variation, as well as the classification performances.

1 Introduction

Classification of microarray data is a challenging problem as it typically relies on a few tens of samples but several thousand dimensions (genes). The number of microarray experiments needed to obtain robust models is generally orders of magnitude higher than the actual size of most datasets [1]. The number of available datasets is however continuously rising. Large databases like the NCBI's Gene Expression Omnibus (GEO) [2] or the EBI's ArrayExpress [3] offer tens of thousand microarray samples which are well formatted and documented. The construction of a large microarray dataset consisting of the simple juxtaposition of independent smaller datasets would be difficult or irrelevant due to differences either in terms of biological topics, technical constraints or experimental protocols.

Transfer learning techniques have been designed to overcome situations where too few samples for the task at hand are available, but where experience from slightly different tasks is available [4,5]. Knowledge is extracted from previous experience (source domains) to help solving the new problem (target domain).

W. Buntine et al. (Eds.): ECML PKDD 2009, Part I, LNAI 5781, pp. 533–547, 2009.

Samples of source and target domains are supposed to be drawn from similar but different distributions. This setting contrasts with semi-supervised learning where samples are supposed to be drawn from the same distribution but where some unlabeled samples are used to build the model [6].

Inductive transfer approaches require labeled data both for the source and target domains [7]. The target domain examples are unlabeled for transductive transfer methods [8] while fully unsupervised transfer techniques have been designed as well [9]. The transfer of knowledge between source and target domains may concern the examples [10,11,12], some model parameters [13,14] or, as in the present work, the feature space [7,15,16].

In multi-task learning [7] a common feature representation is learned at the same time for several tasks. Structural correspondence learning [15] uses features supposed to be relevant for several domains to generate new features for the target domain. An alternative method compares learning tasks by measuring a distance between relevance weights on a set of common features [16]. In contrast to those approaches for transferring feature representation, the transferred knowledge in our method can be automatically partly or fully dropped whenever it does not help to optimize the classification objective on the target domain. As a result, the specific choice of source datasets is not too critical. Benefits of transfer learning have been reported mainly as a gain in classification performances but, as detailed below, the proposed approach also improves the stability of the selected features.

In the particular context of microarray data, feature selection is commonly performed, both to increase the interpretability of the predictive model and possibly to reduce its cost [17,18]. In some cases feature selection has also been shown to improve classification accuracy [19]. *Biomarker selection* specifically refers to the identification of a small set of genes, also called a *signature*, related to a pathology or to an observed clinical outcome after a treatment. The lack of robustness of biomarker selection has been outlined [20]. A good signature is ideally highly stable with respect to sampling variation. In the context of biomarker selection from microarray data, high stability means that different sub-samples of patients lead to very similar sets of biomarkers. This is motivated by the assumption that the biological process explaining the outcome is mostly common among different patients.

Support Vector Machines (SVMs) are particularly convenient to classify high dimensional data with only a few samples. In their simplest form, SVMs simply reduce to maximal margin hyperplanes in the input space. Such models were shown to successfully classify microarray data either on the full input space [21] or combined with feature selection [22,23,24]. The latter approaches are *embedded* as the selection directly uses the classifier structure.

In the present work we rely on another embedded selection method with linear models, called $l1$-AROM [25]. This specific choice is motivated by the possibility to extend this approach in a simple yet efficient way to perform transfer learning by biasing the optimization procedure towards certain dimensions. We proposed recently such a *partially supervised* (PS) extension [26] but the favored dimensions were then defined from prior knowledge. In the context of microarray data, molecular biologists may indeed sometimes guess that a few genes should be considered *a priori* more relevant. In the present work, we do not use such prior knowledge but rather related datasets, hence performing inductive transfer learning at the feature level. The additional benefits are a fully automated feature selection procedure and the possibility to choose the number of features to be transferred independently of some expert knowledge. A practical approximation of this technique reduces to learn linear SVMs with iterative rescaling of the inputs. The rescaling factors depend here on previously selected features from existing datasets.

This initial feature selection on source domains is performed using a simple *univariate* t-test ranking while the final iterative selection is intrinsically *multivariate*. Using an initial univariate selection on the source domains is both computationally efficient and arguably a relevant starting point before transferring to a distinct target domain. As shown in our experiments this choice results in significant stability and classification performance improvements.

The rest of the paper is organized as follows. Section 2 briefly reviews the $l1$-AROM and $l2$-AROM feature selection techniques. Section 3 describes our partially supervised feature selection technique extending the AROM methods. Section 4 details how to use this technique to perform transfer learning. Experiments on microarray datasets are reported in section 5. Conclusions and future perspectives are discussed in section 6.

2 The AROM Methods

Given m examples $\mathbf{x}_i \in \mathbb{R}^n$ and the corresponding class labels $y_i \in \{\pm 1\}$ with $i = 1, ..., m$, a linear model $g(\mathbf{x})$ predicts the class of any point $\mathbf{x} \in \mathbb{R}^n$ as follows.

$$g(\mathbf{x}) = \text{sign}(\mathbf{w} \cdot \mathbf{x} + b) \tag{1}$$

Feature selection is closely related to a specific form of regularization of this decision function to enforce sparsity of the weight vector $\mathbf{w}$. Weston et al. [25] study in particular the zero-norm minimization subject to linear margin constraints:

$$\min_{\mathbf{w}} ||\mathbf{w}||_0 \text{ subject to } y_i(\mathbf{w} \cdot \mathbf{x}_i + b) \geq 1 \tag{2}$$

where $||\mathbf{w}||_0 = card\{w_j | w_j \neq 0\}$ and $card$ is the set cardinality. Since problem (2) is NP-Hard, a log $l1$-norm minimization is proposed instead.

$$\min_{\mathbf{w}} \sum_{j=1}^{n} \ln(|w_j| + \epsilon) \text{ subject to } y_i(\mathbf{w} \cdot \mathbf{x}_i + b) \geq 1 \tag{3}$$

where $0 < \epsilon \ll 1$ is added to smooth the objective when some $|w_j|$ vanishes. The natural logarithm in the objective facilitates parameter estimation with a simple gradient descent procedure. The resulting algorithm $l1$-AROM[1] iteratively optimizes the $l1$-norm of $\mathbf{w}$ with rescaled inputs.

[1] AROM stands for **A**pproximation of the ze**ro**-norm **m**inimization.

The $l2$-AROM method further approximates this optimization by replacing the $l1$-norm by the $l2$-norm. Even though such an approximation may result in a less sparse solution, it is very efficient in practice when $m \ll n$. Indeed, a dual formulation may be used and the final algorithm boils down to a linear SVM estimation with iterative rescaling of the inputs. A standard SVM solver can be iteratively called on properly rescaled inputs. A smooth feature selection occurs during this iterative process since the weight coefficients along some dimensions progressively drop below the machine precision while other dimensions become more significant. A final ranking on the absolute values of each dimension can be used to obtain a fixed number of features.

3 The Partially Supervised AROM Methods

Whenever some knowledge on the relative importance of each feature is available (either from actual prior knowledge or from a related dataset), the $l1$-AROM objective can be modified by adding a prior relevance vector $\boldsymbol{\beta} = [\beta_1, ..., \beta_n]^t$ defined over the input dimensions. Let $\beta_j > 0$ denote the relative prior relevance of the jth feature, the higher its value the more relevant the corresponding feature is *a priori* assumed. In practice, only a few dimensions can be assumed more relevant (e.g. $\beta_j > 1$) while the vast majority of remaining dimensions are not favored (e.g. $\beta_j = 1$). Section 5 further discusses the practical definition of $\boldsymbol{\beta}$. In contrast with semi-supervised learning, this is a form of partial supervision (PS) on the relevant *dimensions* rather than the labels.

The optimization problem of PS-$l1$-AROM is defined to penalize less the dimensions which are assumed *a priori* more relevant:

$$\min_{\mathbf{w}} \sum_{j=1}^{n} \frac{1}{\beta_j} \ln(|w_j| + \epsilon) \text{ subject to } y_i(\mathbf{w} \cdot \mathbf{x}_i + b) \geq 1 \tag{4}$$

It was recently shown how problem (4) can be reformulated as an *iterated* $l1$-norm optimization with margin constraints on rescaled inputs [26]:

$$\min_{\mathbf{w}'} \sum_{j=1}^{n} |w'_j| \text{ subject to } y_i(\mathbf{w}' \cdot (\mathbf{x}_i * \mathbf{w}_k * \boldsymbol{\beta}) + b) \geq 1 \tag{5}$$

where $*$ denotes the component-wise product and the initial weight vector is defined as $\mathbf{w}_0 = [1, \ldots, 1]^t$. At iteration $k+1$, problem (5) is solved given the previous weight vector $\mathbf{w}_k$ and the fixed relevance vector $\boldsymbol{\beta}$, and the process is iterated till convergence.

Similarly to the $l2$-AROM method presented in section 2, problem (5) can be approximated by replacing the $l1$-norm by the $l2$-norm. This modification results in PS-$l2$-AROM, a practical approach which is both easy to implement and computationally more efficient. The $l2$-norm formulation indeed reduces to estimate linear SVMs with iteratively rescaled margin constraints. The original $l2$-AROM method is obtained when $\beta_j = 1$ $(\forall j)$, in other words, without prior

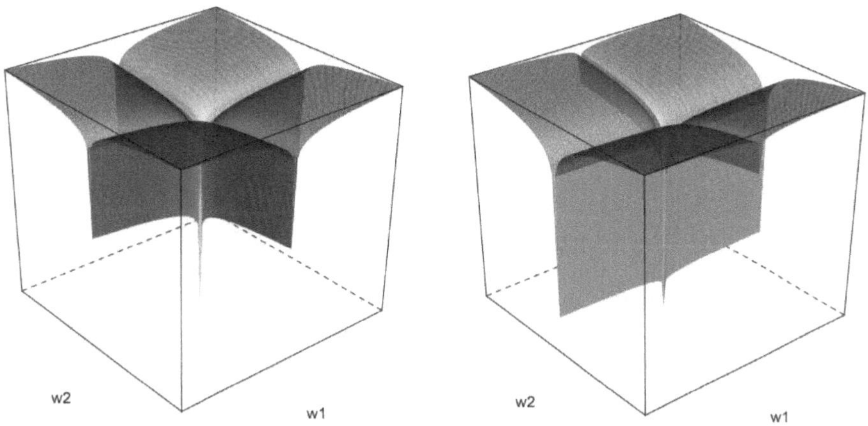

Fig. 1. 2D-representation of the Zero-Norm Approximation by $\sum_j \frac{1}{\beta_j} \ln |w_j|$. Left: without prior relevance ($\boldsymbol{\beta} = [1, 1]^t$). Right: with prior relevance ($\boldsymbol{\beta} = [5, 1]^t$).

preference between the input features. PS-$l2$-AROM further uses the relevance vector $\boldsymbol{\beta}$ to smoothly favor certain dimensions within the selection process.

Figure 1 illustrates why problem (3) is a good approximation to the zero-norm minimization. This objective is nearly flat on the whole space of parameters $\boldsymbol{w}$ except when a specific w_j tends towards zero. The objective is there strongly minimized. It also illustrates what happens if this objective is modified by introducing prior relevance on dimensions, as in problem (4). The objective function is again nearly flat everywhere but the gradient is now even smaller along a dimension corresponding to a larger β_j.

4 Transfer Learning with PS-$l2$-AROM

We discuss here how to use the PS-$l2$-AROM method for transfer learning. Let the *Target Domain* D_T be a set of samples $\mathbf{x}_i \in \mathbb{R}^n$, generated according to a distribution $P_T(\mathbf{x})$, and associated class labels $y_i \in \{\pm 1\}$ following $P_T(y|\mathbf{x})$, $i = 1, ..., m$. The task is to build a robust classification model $g_T(\mathbf{x})$ and to identify a discriminative signature $\mathbf{S}_T$ for D_T. It is generally possible to find related datasets, called *Source Domains* D_S, for which $P_S(y|\mathbf{x}) \approx P_T(y|\mathbf{x})$ but $P_S(\mathbf{x}) \neq P_T(\mathbf{x})$. For example, several microarray datasets are available on GEO [2] for which the class labels correspond to the same concepts, *cancer tissue* or *normal tissue*, but for which the gene expression distributions differ. There are many sources of divergence such as the type of biological samples, the RNA extraction protocol, the normalization steps applied to raw data, *etc.*

It is possible to use a partially supervised feature selection method as a transfer learning technique. The proposed approach is an inductive transfer learning technique since the class labels are known both for D_S and D_T. We assume that the data from all domains share a sufficiently large set of n features and, without loss of generality, that feature index j in D_S maps to the same index in D_T. The proposed approach simply uses any convenient feature selection on D_S to build an initial signature $\mathbf{S}_S$. A prior relevance vector $\boldsymbol{\beta} \in \mathbb{R}^n$ is defined from $\mathbf{S}_S$ to favor the actual selection of some dimensions on D_T:

$$\beta_j = \begin{cases} B \;\forall j \in \mathbf{S}_S \\ 1 \;\; \forall j \notin \mathbf{S}_S \end{cases} \tag{6}$$

where $B > 1$ corresponds to the weight to favor features belonging to $\mathbf{S}_S$. The choice of B is arbitrary but experiments reported in section 5.6 illustrate that the proposed method is not sensitive to a specific choice for a large range of possible values. The vector $\boldsymbol{\beta}$ is used to bias the selection of features on D_T via PS-l2-AROM to obtain a final signature $\mathbf{S}_T$. The selection on D_T is influenced by the knowledge extracted from D_S, i.e. a set of indexes of relevant features. Those transferred features are assumed relevant for D_S but not necessarily highly discriminative on D_T, since $P_S(\mathbf{x}) \neq P_T(\mathbf{x})$. Our modeling assumption is however that the features extracted from similar tasks provide useful information as compared to selecting features only from a single domain D_T. This assumption is confirmed by our practical experiments reported in section 5.

Any standard feature selection method can be used for the initial selection on D_S. We recommend in particular the use of a simple univariate technique such as a t-test ranking. It is computationally efficient and this initial selection is not meant to be highly accurate on D_S but to guide the detailed selection to be performed eventually on D_T.

5 Data and Experiments

This section describes and evaluates several practical ways of transfer learning based on the partially supervised feature selection implemented in PS-l2-AROM. Three prostate cancer microarray datasets are presented in section 5.1. We detail in section 5.2 the metrics used to assess the stability of selected features and classification performances. Baseline results are obtained with no transfer, that is by applying the l2-AROM feature selection technique on a given dataset. Improved stability and classification performances are obtained with a *single transfer*. This first protocol uses one related dataset as source domain to guide the selection on the target domain (section 5.3). Further improvements can be obtained with *multiple transfer* which combines several source domains (section 5.4). Experimental results are presented in section 5.5. Finally, the sensitivity to a specific choice of the prior weight value (the B parameter) is analyzed in section 5.6.

In a nutshell those experimental results show that transfer learning based on partially supervised feature selection always leads to a gain in stability as well as a systematic gain in classification performance for signature sizes of interest.

5.1 Microarray Data

Table 1 presents the main characteristics of the three prostate cancer microarray datasets used in this experimental section. For convenience, datasets will

Table 1. Microarray prostate datasets. Columns respectively show the dataset name, the number of normal samples, the number of tumor samples, the original number of features and the type of Affymetrix chips used.

dataset	Normal	Tumor	Features	Chip
SINGH	50	52	12,625	HGU95Av2
CHANDRAN	18	86	12,625	HGU95Av2
WELSH	9	25	12,626	HGU95A

be named after the first author of the publication along which they were made available (SINGH [27], CHANDRAN [28] and WELSH [29]). In the original publications, the task is almost the same for the three datasets: binary classification between tumor and normal tissues. In CHANDRAN, tumor samples are of two types: primary tumor and metastatic tumor. Tumor samples in WELSH correspond to 24 primary tumors and 1 lymph node metastasis. No precision is made about the type of tumor tissue in SINGH. The microarray technology used to produce those datasets is the same for SINGH and CHANDRAN, but is a bit older for WELSH (see table 1). Consequently, features (genes) present on each type of chip differ very slightly. Samples and RNA extraction were also performed according to different protocols. The internal normalization to produce one value for each feature also differ from set to set. All these differences make their simple combination in a larger dataset irrelevant. The three datasets are here reduced to the set of 12,600 features they share in common.

5.2 Evaluation Metrics

Stability measures to which extent k sets $\mathbf{S}$ of s selected features (gene signatures) share common features. Those sets can typically be produced by selecting features from different samplings of the data. Kuncheva [30] proposed such a stability index:

$$K(\{\mathbf{S}_1, \ldots, \mathbf{S}_k\}) = \frac{2}{k(k-1)} \sum_{i=1}^{k-1} \sum_{j=i+1}^{k} \frac{|\mathbf{S}_i \cap \mathbf{S}_j| - \frac{s^2}{n}}{s - \frac{s^2}{n}} \tag{7}$$

where n is the total number of features, and $\mathbf{S}_i$, $\mathbf{S}_j$ are two signatures built from different subsets of the training samples. The $\frac{s^2}{n}$ ratio in this formula corrects a bias due to the chance of selecting common features among two sets chosen at random. This correction motivates our use of this particular stability index. This index satisfies $-1 < K \leq 1$ and the greater its value the largest the number of commonly selected features in the various sets. A negative index for a set of signatures means that feature sharing is mostly due to chance.

Stability alone cannot characterize the quality of a subset of features. Indeed, if a large randomly chosen set of features were purely forced in every signature, the stability would be very high, but the model built on those features would likely have a poor classification performance. This performance is assessed here with the *Balanced Classification Rate*:

$$BCR = \frac{1}{2}\left(\frac{TP}{P} + \frac{TN}{N}\right) \quad (8)$$

where TP (resp. TN) is the number of positive (resp. negative) test samples correctly predicted as positive (resp. negative) among the P positive (resp. N negative) test samples. BCR is preferred to accuracy because microarray datasets often have unequal class priors. BCR is the average between *specificity* and *sensitivity*, two very common measures in the medical domain. BCR can also be generalized to multi-class problems more easily than ROC analysis.

5.3 Single Transfer

Our first experimental protocol uses ($k = 200$) random 90%-10% samplings from the target domain D_T. Each 90% fraction forms a training set. These samples are first normalized to zero median and unit standard deviation. Features are then selected via PS-l2-AROM and a linear soft-margin SVM is built on the selected dimensions. Each 10% fraction forms the associated test samples that are preprocessed according to the training normalization parameters. The vector β used for PS-l2-AROM is set by selecting a signature $\mathbf{S}_S$ on D_S with a t-test ranking[2]. The 50 top ranked features define $\mathbf{S}_S$, which is a common default signature size for biomarker selection. The feature selection is performed on D_T while favoring the genes from $\mathbf{S}_S$ according to:

$$\beta_j = \begin{cases} 10 \; \forall j \in \mathbf{S}_S \\ 1 \quad \forall j \notin \mathbf{S}_S \end{cases} \quad (9)$$

Here D_S is a single dataset different from D_T. Given the three datasets available, six combinations of D_S and D_T are tested. Stability over the 200 samplings and averaged BCR performances are reported. For comparison purposes, the same protocol is performed with no transferred knowledge, i.e. with $\beta_j = 1, \forall j \in 1, ..., n$.

5.4 Multiple Transfer

When several datasets are available as source domains it may be useful to combine the knowledge extracted from each of them to guide the feature selection on the target domain D_T. Such a multiple transfer protocol is described below with two source domains. The extension to more than two source domains is straightforward.

[2] This univariate filtering method ranks genes according to $\frac{\mu_{j+} - \mu_{j-}}{\sqrt{\sigma_{j+}^2/m_+ + \sigma_{j-}^2/m_-}}$, where μ_{j+} (resp. μ_{j-}) is the mean expression value of the gene j for the m_+ positively (resp. m_+ negatively) labeled samples, and σ_{j+}, σ_{j-} are the associated standard deviations.

A signature $\mathbf{S}_S$ of 50 features is extracted from the source domains D_{S_1,S_2} and applied to D_T via PS-$l2$-AROM. Features are ranked according to a t-test on each source dataset. The p best ranked features are selected from each source dataset and the intersection of those two signatures is computed. The parameter p is chosen such that the size of the intersection is 50. The rationale behind this choice is to transfer the same amount of knowledge in D_T as compared to the single transfer protocol. The values of p used for the three possible combinations of the available source datasets are 557 for (SINGH $\bigcap$ CHANDRAN), 275 for (WELSH $\bigcap$ SINGH) and 385 for (WELSH $\bigcap$ CHANDRAN). Those differences result from the fact that some combinations have more top ranked features in common than others. For example, the number of features needed to build an intersected signature of 50 genes is about twice as much for SINGH and CHANDRAN as compared to SINGH and WELSH. This could explain why using SINGH rather than CHANDRAN as D_S for WELSH as D_T gives better results in a single transfer protocol (see section 5.5).

5.5 Results

Figures 2 and 3 respectively show the BCR and the stability results for various signature sizes $|\mathbf{S}_T|$. For signatures significantly larger than the transferred knowledge ($|\mathbf{S}_S| = 50$) results are equivalent to baseline results (no transfer). In contrast, for signature sizes of practical interest (a few tens of genes), there is a large increase both in classification performance and stability.

For example, transferring knowledge from SINGH to WELSH (top of fig. 2) improves the average BCR from 79.6% (resp. 65.7%) for a signature size of 16 (resp. 10) genes up to more than 99.2% (resp. 98.7%). Those differences are statistically significant according to the corrected resampled t-test[3] proposed in [31]. BCR results on the CHANDRAN and SINGH target datasets follow the same trends.

Multiple transfer may further improve the BCR performances. This is in particular the case on the CHANDRAN dataset (center of fig. 2) with transferred knowledge from WELSH $\bigcap$ SINGH. For example, the BCR differences are statistically significant (p-value ≤ 0.025) between a single transfer and a multiple transfer with a final signature of 10 genes. In general, multiple transfer BCR results are always equivalent or better than single transfer results. Multiple transfer thus offers a more robust approach not requiring to carefully select which source domain need to be considered for a given target domain.

Transfer learning always improves the stability of the selected gene lists as illustrated in Fig. 3. The maximal stability is often reached around 50 features, which comes with no surprise since precisely 50 genes are favored during the selection on the target domain. However this maximal stability does not reach 100% which illustrates that the selected genes are not just those belonging to $\mathbf{S}_S$.

---

[3] Such a test corrects for the fact that the various test sets are not independent since they may overlap. The BCR differences are significant with a (likely conservative) p-value $= 1.9 \times 10^{-2}$ for 16 genes and a p-value $= 2.6 \times 10^{-4}$ for 10 genes.

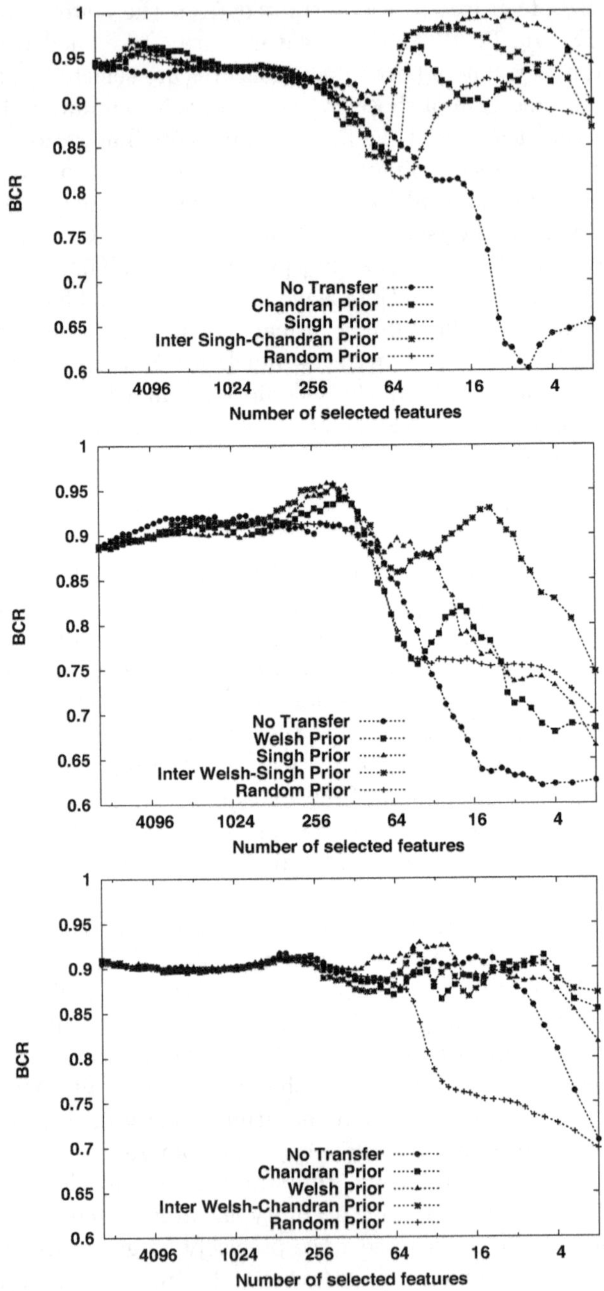

Fig. 2. Classification performances (Balanced Classification Rate) obtained on WELSH (top), CHANDRAN (center) and SINGH (bottom). `No Transfer` is the baseline for which features are selected on the target dataset without prior preference. The next two curves specify which dataset was used in a single transfer setting. The fourth curve refers to the multiple transfer setting.

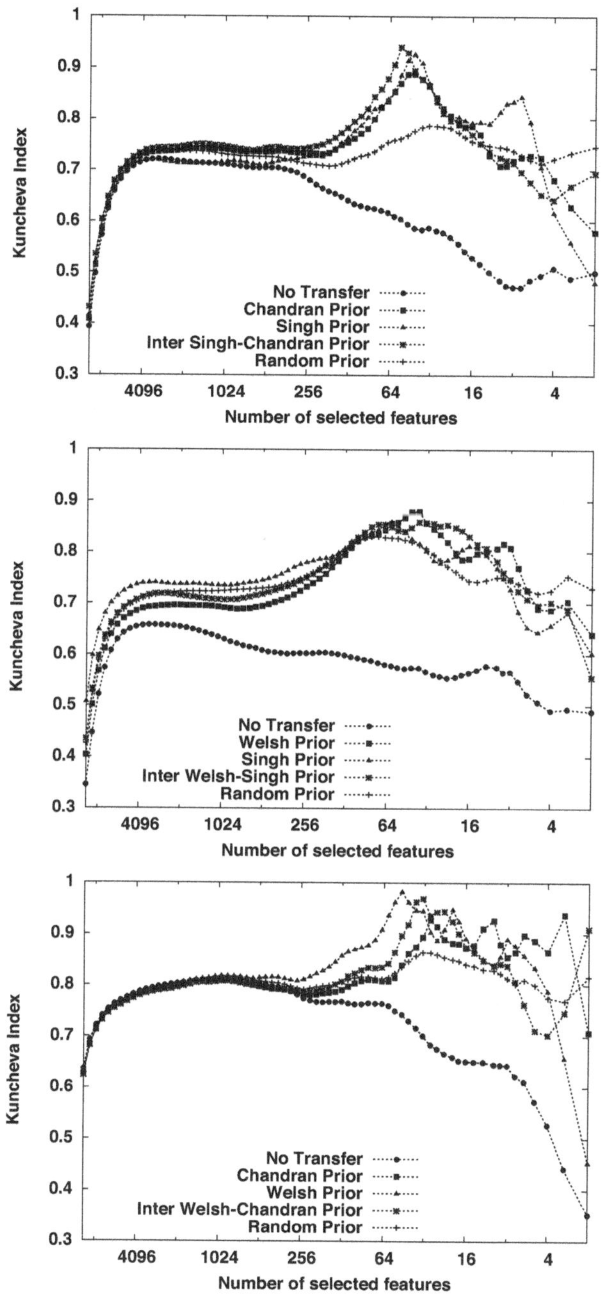

Fig. 3. Signature stability (Kuncheva index) obtained on WELSH (top), CHANDRAN (center) and SINGH (bottom). `No Transfer` is the baseline for which features are selected on the target dataset without prior preference. The next two curves specify which dataset was used in a single transfer setting. The fourth curve refers to the multiple transfer setting.

5.6 Impact of Prior Relevance Weight

In the experiments described in section 5.5, the value $B = 10$ was chosen to favor some dimensions via PS-$l2$-AROM. The influence of a specific choice of the B value on stability and classification performances is analyzed in this section. We detail experiments with multiple transfer since this approach offers the best results so far. Figure 4 displays stability and BCR results for $B = \{1, 2, 5, 10, 100, 1000\}$ on WELSH. The curves for $B = 1$ and $B = 10$ correspond to the previous settings respectively with no transfer and multiple transfer. Equivalent trends are observed on the other datasets (results not shown). The influence of the B value can be summarized as follows.

Results show that the higher the B value the stronger the stability peak around 50 features. This is a logical consequence of the design of PS-$l2$-AROM. The stability is not influenced for signature sizes $|\mathbf{S}_T|$ significantly larger than 50 features except for a very large $B = 1000$. For signature sizes smaller than 50 a better stability is obtained with B in the range $[10, 100]$. BCR results show a positive effect of the partial supervision as soon as B is greater than 5. The proposed approach is not highly sensitive to a specific choice of B in the range $[5, 100]$. The default value of $B = 10$ offers a reasonable choice overall. Hence the proposed approach does not require to carefully optimize the meta-parameter B in a nested validation loop.

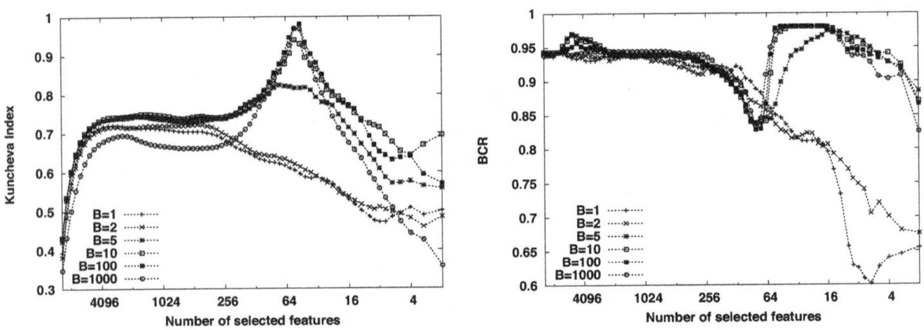

Fig. 4. Impact of the prior relevance weight on signature stability (Kuncheva index) and classification performances (Balanced Classification Rate) on WELSH

6 Conclusions and Perspectives

We address in this paper the problem of transfer learning for feature selection and classification of high dimensional data, such as those produced by microarray experiments. We propose a feature selection method on a *target domain* that can be partially supervised (PS) from features previously extracted from related *source domains*. Such knowledge can be acquired from public databases like GEO [2] or ArrayExpress [3]. The initial feature selection on the source domains is typically performed with a fast univariate technique. The purpose of this initial selection is to guide the selection process on the target domain.

We rely here on our recently proposed PS-$l2$-AROM method, a feature selection approach embedded within the estimation of a regularized linear model [26]. This algorithm reduces to linear SVM learning with iterative rescaling of the input features. The scaling factors depend here on the selected dimensions on the source domains. The proposed optimization procedure smoothly favors the preselected features but the finally selected dimensions may depart from those to optimize the classification objective under rescaled margin constraints.

Practical experiments on several microarray datasets illustrate that the proposed approach not only increases classification performances, a usual benefit of a sound transfer learning scheme, but also the stability of the selected dimensions with respect to sampling variation. We also show how a multiple transfer from various source domains can bring further improvements.

The proposed approach relies on a meta-parameter defining the prior weight of the favored dimensions during the partially supervised feature selection. We show experimentally that this method is not sensitive to a specific choice of this parameter for a large range of possible values. Distinct weight values for different features could also be considered in the future. One could for instance define those weights as a function of the p-values of the initial t-test. Here the t-test was applied on the source domain(s) but it could also be interesting to compute the t-test on the target domain itself, hence without transfer. The combination of a simple feature ranking method and the partially supervised feature selection could already improve stability and/or classification performances on a given dataset.

We rely on a simple univariate selection on the source domain(s). The purpose of this initial selection is indeed not to be highly accurate since the final and more refined selection is performed on a distinct target domain. Our current choice is a simple t-test ranking for this initial selection. Several alternatives could however be considered, including a multivariate embedded selection method such as L2-AROM. More importantly, the size of the initial signature extracted from the source domain(s) is currently fixed to 50 genes. This is a common default value for biomarker selection and it offers very good performances. It would however be interesting to investigate further the influence of this size on the quality of the final selection. A related issue would be the automatic selection of the best source domain(s) for a given target domain and the number of features to be extracted from each of them. Simple similarity measures between various datasets can probably help in this regard.

Our partially supervised feature selection is a general approach which does not depend, at least in principle, on how the favored dimensions are initially selected. The present work relies on other related datasets while our previous work used real *prior knowledge* from field experts [26]. A further and natural extension would combine the transferred knowledge which such prior knowledge whenever available.

Acknowledgment. Thibault Helleputte is funded by a FRIA grant (1.E091.07). All computations have been run on the FYNU's INGRID cluster thanks to the Center for Intensive Computation and Mass Storage (CISM) of the University of Louvain.

References

1. Ein-Dor, L., Zuk, O., Domany, E.: Thousands of samples are needed to generate a robust gene list for predicting outcome in cancer. PNAS 103(15), 5923–5928 (2006)
2. Edgar, R., Barrett, T.: Ncbi geo standards and services for microarray data. Nature Biotechnology 24, 1471–1472 (2006)
3. Parkinson, H., Kapushesky, M., Kolesnikov, N., Rustici, G., Shojatalab, M., Abeygunawardena, N., Berube, H., Dylag, M., Emam, I., Farne, A., Holloway, E., Lukk, M., Malone, J., Mani, R., Pilicheva, E., Rayner, T.F., Rezwan, F., Sharma, A., Williams, E., Bradley, X.Z., Adamusiak, T., Brandizi, M., Burdett, T., Coulson, R., Krestyaninova, M., Kurnosov, P., Maguire, E., Neogi, S.G., Rocca-Serra, P., Sansone, S.-A., Sklyar, N., Zhao, M., Sarkans, U., Brazma, A.: ArrayExpress update–from an archive of functional genomics experiments to the atlas of gene expression. Nucl. Acids Res. 37(suppl-1), D868–D872 (2009)
4. Silver, D.L., Bennett, K.P.: Guest editor's introduction: special issue on inductive transfer learning. Machine Learning 73, 215–220 (2008)
5. Pan, S.J., Yang, Q.: A survey on transfer learning. Technical Report HKUST-CS08-08, Department of Computer Science and Engineering, Hong Kong University of Science and Technology, Hong Kong, China (November 2008)
6. Chapelle, O., Schölkopf, B., Zien, A. (eds.): Semi-Supervised Learning. MIT Press, Cambridge (2006)
7. Argyriou, A., Evgeniou, T., Pontil, M.: Multi-task feature learning. In: NIPS, pp. 41–48 (2006)
8. Daumé III, H., Marcu, D.: Domain adaptation for statistical classifiers. Journal of Artificial Intelligence Research 26, 101–126 (2007)
9. Wang, Z., Song, Y., Zhang, C.: Transferred dimensionality reduction. In: Daelemans, W., Goethals, B., Morik, K. (eds.) ECML PKDD 2008, Part II. LNCS (LNAI), vol. 5212, pp. 550–565. Springer, Heidelberg (2008)
10. Liao, X., Xue, Y., Carin, L.: Logistic regression with an auxiliray data source. In: Proceedings of the 22nd International Conference on Machine Learning, Bonn, Germany, pp. 505–512 (2005)
11. Huang, J., Smola, A., Gretton, A., Borgwardt, K., Schölkopf, B.: Correcting sample selection bias by unlabeled data. In: Proceedings of the 19th Annual Conference on Neural Information Processing Systems, pp. 601–608. MIT Press, Cambridge (2007)
12. Dai, W., Yang, Q., Xue, G., Yu, Y.: Selft-thaught clustering. In: Proceedings of the 25th International Conference of Machine Learning, pp. 200–207 (2008)
13. Evgeniou, T., Pontil, M.: Regularized multi-task learning. In: Proceedings of the 10th ACM SIGKDD International Conference on Knowledge Discovery and Data Mining, pp. 109–117 (2004)
14. Lawrence, N., Platt, C.: Learning to learn with the informative vector machine. In: Proceedings of the 21st International Conference on Machine Learning, p. 65. ACM, New York (2004)
15. Blitzer, J., McDonald, R., Pereira, F.: Domain adaptation with structural correspondence learning. In: Proceedings of the 2006 Conference on Empirical Methods in Natural Language Processing, Sydney, Australia, July 2006, pp. 120–128. Association for Computational Linguistics (2006)
16. Mierswa, I., Wurst, M.: Efficient case based feature construction. In: Gama, J., Camacho, R., Brazdil, P.B., Jorge, A.M., Torgo, L. (eds.) ECML 2005. LNCS (LNAI), vol. 3720, pp. 641–648. Springer, Heidelberg (2005)

17. Guyon, I., Elisseef, A.: An introduction to variable and feature selection. Journal of Machine Learning Research 3, 1157–1182 (2003)
18. Saeys, Y., Inza, I., Larranaga, P.: A review of feature selection techniques in bioinformatics. bioinformatics 23(19), 2507–2517 (2007)
19. Krishnapuram, B., Carin, L., Hartemink, A.: 14: Gene Expression Analysis: Joint Feature Selection and Classifier Design. In: Kernel Methods in Computational Biology, pp. 299–317. MIT Press, Cambridge (2004)
20. Ein-Dor, L., Kela, I., Getz, G., Givol, D., Domany, E.: Outcome signature genes in breast cancer: is there a unique set? Bioinformatics 21 (2005)
21. Mukherjee, S.: 9: Classifying Microarray Data Using Support Vector Machines. In: A Practical Approach to Microarray Data Analysis, pp. 166–185. Springer, Heidelberg (2003)
22. Weston, J., Mukherjee, S., Chapelle, O., Pontil, M., Poggio, T., Vapnik, V.: Feature selection for SVMs. In: Advances in Neural Information Processing Systems, pp. 668–674 (2000)
23. Chapelle, O., Vapnik, V., Bousquet, O., Mukherjee, S.: Choosing multiple parameters for support vector machines. Machine Learning 46, 131–159 (2002)
24. Guyon, I., Weston, J., Barnhill, S., Vapnik, V.: Gene selection for cancer classification using support vector machines. Machine Learning 46(1-3), 389–422 (2002)
25. Weston, J., Elisseef, A., Schölkopf, B., Tipping, M.: Use of the zero-norm with linear models and kernel methods. Journal of Machine Learning Research 3, 1439–1461 (2003)
26. Helleputte, T., Dupont, P.: Partially supervised feature selection with regularized linear models. In: Proceedings of the 26th International Conference on Machine Learning (2009)
27. Singh, D., Febbo, P.G., Ross, K., Jackson, D.G., Manola, J., Ladd, C., Tamayo, P., Renshaw, A.A., D'Amico, A.V., Richie, J.P., Lander, E.S., Loda, M., Kantoff, P.W., Golub, T.R., Sellers, W.R.: Gene expression correlates of clinical prostate cancer behavior. Cancer Cell 1(2), 203–209 (2002)
28. Chandran, U., Ma, C., Dhir, R., Bisceglia, M., Lyons-Weiler, M., Liang, W., Michalopoulos, G., Becich, M., Monzon, F.: Gene expression profiles of prostate cancer reveal involvement of multiple molecular pathways in the metastatic process. BMC Cancer 7(1), 64 (2007)
29. Welsh, J.B., Sapinoso, L.M., Su, A.I., Kern, S.G., Wang-Rodriguez, J., Moskaluk, C.A., Frierson Jr., F.H., Hampton, G.M.: Analysis of Gene Expression Identifies Candidate Markers and Pharmacological Targets in Prostate Cancer. Cancer Res 61(16), 5974–5978 (2001)
30. Kuncheva, L.I.: A stability index for feature selection. In: Proceedings of the 25th International Multi-Conference: Artificial Intelligence and Applications, Anaheim, CA, USA, pp. 390–395. ACTA Press (2007)
31. Nadeau, C., Bengio, Y.: Inference for the generalization error. Machine Learning 52, 239–281 (2003)

Integrating Logical Reasoning and Probabilistic Chain Graphs

Arjen Hommersom, Nivea Ferreira, and Peter J.F. Lucas

Institute for Computing and Information Sciences,
Radboud University Nijmegen, Nijmegen, The Netherlands
{arjenh,nivea,peterl}@cs.ru.nl

Abstract. Probabilistic logics have attracted a great deal of attention during the past few years. While logical languages have taken a central position in research on knowledge representation and automated reasoning, probabilistic graphical models with their probabilistic basis have taken up a similar position when it comes to reasoning with uncertainty. The formalism of chain graphs is increasingly seen as a natural probabilistic graphical formalism as it generalises both Bayesian networks and Markov networks, and has a semantics which allows any Bayesian network to have a unique graphical representation. At the same time, chain graphs do not support modelling and learning of relational aspects of a domain. In this paper, a new probabilistic logic, chain logic, is developed along the lines of probabilistic Horn logic. The chain logic leads to relational models of domains in which associational and causal knowledge are relevant and where probabilistic parameters can be learned from data.

1 Introduction

There has been a considerable amount of work in the field of artificial intelligence during the past two decades on integrating logic and probability theory. This research was motivated by perceived limitations of both formalisms. Logic has for long acted as the common ground for almost all research on knowledge representation, reasoning and learning in artificial intelligence; yet, uncertainty cannot be handled easily in logic. Probabilistic graphical models take probability theory as their foundation; they have been proposed as formalisms for statistical learning and for reasoning with uncertainty. Although their associated graphical representation allows specifying relationship among objects in the domain of discourse such that it is possible to reason about their statistical dependences and independences, probabilistic graphical models are essentially propositional in nature, and they lack the representational richness of logics.

Several researchers have proposed probabilistic logics that merge those two types of languages in an attempt to redress their individual shortcomings. A variety of such languages is now available, each of them adopting a different view on the integration. Unfortunately, it appears that all of the available frameworks are still restricted in one way or the other. In particular, the available

W. Buntine et al. (Eds.): ECML PKDD 2009, Part I, LNAI 5781, pp. 548–563, 2009.

languages either support representing Bayesian-network-like independence information or Markov-network-like independence information. In this paper, we describe a probabilistic first-order language that is more expressive than similar languages developed earlier, in the sense that the probabilistic models that can be specified and reasoned about have Bayesian and Markov networks as special cases. This new probabilistic logic is called *chain logic*. This paper addresses the representation and reasoning aspects of chain logic as well as parameter learning of chain logic theories.

The organisation of this paper is as follows. In Section 2 we provide an overview of the basic notions of Horn clauses and chain graphs. Section 3 contains an introduction to the chain logic language, with details on its syntax and semantics. In Section 4, we focus on learning the parameters of chain logic theories. In Section 5 the most important related work is introduced and a detailed comparison to this is provided. Finally, Section 6 presents our conclusions.

2 Preliminaries

The work discussed in this paper builds upon two separate branches of research: (*i*) probabilistic graphical models, and (*ii*) abductive logic. We start by summarising the basic facts about probabilistic graphical models, in particular chain graph models. This is followed by a review of central notions from abductive logic. Both frameworks act as the foundation for chain logic as developed in the remainder of the paper.

2.1 Chain Graphs

A chain graph (CG) is a probabilistic graphical model that consists of labelled vertices, that stand for random variables, connected by directed and undirected edges. This representation allows chain graphs to be considered as a framework that generalises both directed acyclic graph probabilistic models, i.e., Bayesian networks, and undirected graph probabilistic models, i.e., Markov networks [4]. The definitions with respect to chain graphs given in this paper are in accordance with [5].

Let $G = (V, E)$ be a *hybrid graph*, where V denotes the set of *vertices* and E the set of *edges*, where an edge is either an *arc* (directed edge), or a *line* (undirected edge). Let indexed lower case letters, e.g., v_1 and v_2, indicate vertices of a chain graph. We denote an arc connecting two vertices by '$\rightarrow$' and a line by '$-$'. Consider two vertices v_1 and v_2. If $v_1 \rightarrow v_2$ then v_1 is a *parent* of v_2. If $v_1 - v_2$ then v_1 (v_2) is a *neighbour* of v_2 (v_1). The set of parents and neighbours of a vertex v are denoted by $\mathrm{pa}(v)$ and $\mathrm{ne}(v)$, respectively.

A *path* of length n in a hybrid graph $G = (V, E)$ is a sequence of distinct vertices $v_1, \ldots, v_{n+1}$, such that either $v_i - v_{i+1} \in E$, $v_i \rightarrow v_{i+1} \in E$, or $v_i \leftarrow v_{i+1} \in E$. A *directed path* is a path which includes at least one arc, and where all arcs have the same direction. A *cycle* is a path where the first and last vertex are the same. A *chain graph* is a hybrid graph with the restriction that no directed cycles exist.

If there is a line between every pair of vertices in a set of vertices, then this set is named *complete*. A *clique* is a maximally complete subset. Now, consider the graph obtained from a chain graph by removing all its arcs. What are left are vertices connected by lines, called *chain components*; the set of all chain components is denoted here by $\mathcal{C}$.

Associated to a chain graph $G = (V, E)$ is a joint probability distribution $P(X_V)$ that is faithful to the chain graph G, i.e., it includes all the independence information represented in the graph. This is formally expressed by the following *chain graph Markov property*:

$$P(X_V) = \prod_{C \in \mathcal{C}} P(X_C \mid X_{\mathrm{pa}(C)}) \tag{1}$$

with $V = \bigcup_{C \in \mathcal{C}} C$, and where each $P(X_C \mid X_{\mathrm{pa}(C)})$ factorises according to

$$P(X_C \mid X_{\mathrm{pa}(C)}) = Z^{-1}(X_{\mathrm{pa}(C)}) \prod_{M \in M(C)} \varphi_M(X_M) \tag{2}$$

given that $M(C)$ is the complete set in the moral graph[1] obtained from the subgraph $G_{C \cup \mathrm{pa}(C)}$ of G. The functions φ are non-negative real functions, called *potentials*; they generalise joint probability distributions in the sense that they do not need to be normalised. Finally, the normalising factor Z is defined as

$$Z(X_{\mathrm{pa}(C)}) = \sum_{X_C} \prod_{M \in M(C)} \varphi_M(X_M) \tag{3}$$

As a Bayesian network is a special case of a chain graph model, Equation (1) simplifies in that case to:

$$P(X_V) = \prod_{v \in V} P(X_v \mid X_{\mathrm{pa}(v)}) \tag{4}$$

which is the well-known factorisation theorem of Bayesian networks [5]. In this case, the chain components are formed by a family of random variables. Therefore, for each of those random variables the distribution is defined as the conditional probability function of this variable, given the value of its parents. Note that according to Equation (1), chain graphs can also be interpreted as a directed acyclic graph of chain components.

2.2 Abduction Logic

Abduction logic is defined as a special variant of function-free Horn logic, where the syntax of Horn clauses is slightly modified, and logical implication, '$\leftarrow$', is given a causal interpretation. *Abduction clauses* have the following form:

$$D \leftarrow B_1, \ldots, B_n : R_1, \ldots, R_m$$

[1] Moralisation encompasses: (1) adding lines between unconnected parents of a chain component, and (2) conversion of arcs into lines by ignoring their directions.

where the predicates of the atoms D and B_i are at least unary and the atoms R_j, called *templates*, express relationships among variables, where at least one variable appearing in the atoms D and B_i occurs in at least one template R_j. An example illustrating this representation is shown below (Example 1). Atoms that do not occur as head of a clause are called *assumables*. From a logical point of view, the ':' operator has the meaning of a conjunction; it is only included in the syntax to allow separating atoms that are templates from non-template atoms. The basic idea is to use atoms D and B_i to introduce specific variables, later interpreted as *random* variables, and the templates R_j to represent relations among those variables. Other variables can be introduced to define additional, logical relationships among objects, or to define generic properties.

Let T be a set of abduction clauses, called an *abductive theory* in this paper. Then, concluding a formula ψ from the theory is denoted by $T \vDash \psi$ (when using model theory) and $T \vdash \psi$ (when using deduction or proof theory).

Throughout this paper, we will write Ψ' as the set of ground instances of Ψ, where Ψ is a set of formulae. For example, $\mathcal{A}$ is the set of all assumables and we use $\mathcal{A}'$ to denote the set of ground instances of $\mathcal{A}$.

For abduction logic a special type of logical reasoning has been proposed, called *abduction*, which is defined in terms of model theory or deduction using so-called *explanations*: "a entails b" allows inferring a as an explanation of b. Given a set of atoms O, interpreted as *observations*, then these observations are explained in terms of the abductive theory and a set of assumables.

Definition 1. *An* explanation *of a set of atoms O based on the pair $\langle T, \mathcal{A}\rangle$ is defined as a set of ground assumables $E \subseteq \mathcal{A}'$ satisfying the following conditions:*

- *$T \cup E \vDash O$, and*
- *$T \cup E$ is consistent, i.e., $T \cup E \nvDash \bot$.*

A minimal explanation *E of O is an explanation whose proper subsets are not explanations of O. The set of all minimal explanations is denoted by $\mathcal{E}_T(O)$.*

Example 1. Suppose that we have the following piece of medical knowledge. Influenza (I) causes coughing (C), where coughing is known as a possible cause for hoarseness (H). In addition, coughing is known to be associated with dyspnoea (shortness of breath) (D), although a clear cause-effect relationship is missing. Dyspnoea restricts the oxygen supply to the blood circulation; the resulting low oxygen saturation of the blood will turn the skin to colour blue (B), which is a condition called cyanosis. This qualitative knowledge is represented by the causal network shown in Fig. 1. The associated abductive theory T is the following:

$$
\begin{aligned}
&I(x) \leftarrow: r_I(x)\\
&C(x) \leftarrow I(y) : r_{C,I}(x,y), r_{C,D}(x,z)\\
&D(x) \leftarrow I(y) : r_{C,I}(z,y), r_{C,D}(z,x)\\
&H(x) \leftarrow C(y) : r_{H,C}(x,y)\\
&B(x) \leftarrow D(y) : r_{B,D}(x,y)
\end{aligned}
$$

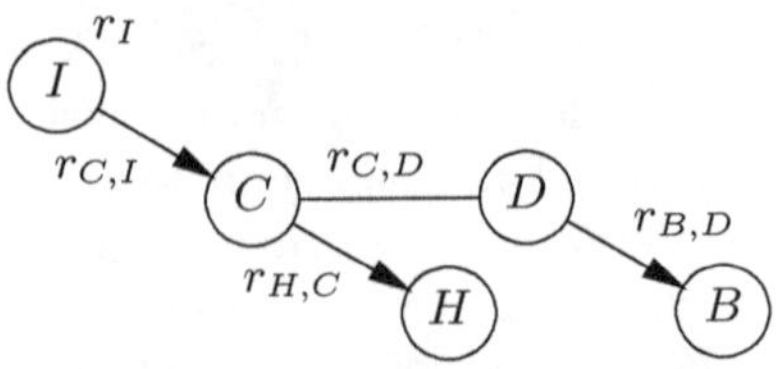

Fig. 1. Causal network model of causal and associational knowledge about influenza

where each of the variables has $\{f, t\}$ as domain. It now holds that:

$$T \cup \{r_I(t), r_{H,C}(t,t), r_{C,I}(t,t), r_{C,D}(t,t)\} \models H(t)$$

and $T \cup \{r_I(t), r_{H,C}(t,t), r_{C,I}(t,t), r_{C,D}(t,t)\} \nvDash \bot$.

The intuition behind the syntax of abduction clauses, such as $C(x) \leftarrow I(y) : r_{C,I}(x,y), r_{C,D}(x,z)$, is that $C(x) \leftarrow I(y)$ expresses the *potential* existence of a causal relation between the referred atoms, here $I(y)$ and $C(x)$. Note that $I(y)$ also appear in the clause $D(x) \leftarrow I(y) : r_{C,I}(z,y), r_{C,D}(z,x)$, following the fact that I is the parent of the chain component CD. Templates R_j, e.g. $r_{C,I}(x,y)$, expresses whether the relationship actually does or does not hold. When there are no atoms to the left of the ':' operator, such as in the clause $I(x) \leftarrow: r_I(x)$, the template represents a root node or an association with no parents.

3 Chain Logic

In this section, the chain logic language is formally defined. This paves the way for the next section where we will focus on learning.

3.1 Language Syntax

The formalism presented in this section is inspired by probabilistic Horn logic as introduced by Poole in [1]. For the sake of simplicity, we assume here finite domain specifications (infinite domains are briefly mentioned in Section 6). Furthermore, the unique names assumption holds for the different constants of the domain.

Chain logic (CL) extends abduction logic as described in Section 2.2 by interpreting templates as representing uncertain events. The actual definition of the uncertainty is done by means of a *weight* declaration. This is of the form

$$weight(a_1 : w_1, \ldots, a_n : w_n) \tag{5}$$

where a_i represents an atom and $w_i \in \mathbb{R}_0^+$. The set of atoms appearing in such declarations are the assumables $\mathcal{A}$. Here we require that the atoms in a weight declaration share the same variables. Furthermore, we require that a ground atom a – which is an instance of one of the assumables – does not appear as an instance of another assumable in another weight declaration. The weight declaration defines conjunctions of atoms that are mutually exclusive and exhaustive.

Therefore, together with the above elements, a CL specification also includes integrity constraint statements, i.e., clauses of the form

$$\bot \leftarrow a_i, a_j \tag{6}$$

for any pair a_i and a_j appearing in the same weight declaration where $i \neq j$. Such clauses are implicit in all of our given examples. We also allow the addition of another set of constraints referring to a pair of assumables appearing in different weight declarations, as seen in the example below.

Example 2. Consider the description given in Example 1. Uncertainty is defined by replacing the templates by potential functions. For the abductive theory in this example:

φ_{CI}	$\imath$	$\bar{\imath}$
c	8	2
$\bar{c}$	1	10

φ_{CD}	d	$\bar{d}$
c	18	2
$\bar{c}$	5	2

φ_{HC}	c	$\bar{c}$
h	0.6	0.1
$\bar{h}$	0.4	0.9

φ_{BD}	d	$\bar{d}$
b	0.3	0.001
$\bar{b}$	0.7	0.999

φ_{I}	
$\imath$	0.1
$\bar{\imath}$	0.9

This example can be represented in chain logic using the following abduction clauses:

$$\begin{array}{l} I(x) \leftarrow: \varphi_I(x) \\ C(x) \leftarrow I(y) : \varphi_{CI}(x,y), \varphi_{CD}(x,z) \\ D(x) \leftarrow I(y) : \varphi_{CI}(z,y), \varphi_{CD}(z,x) \\ H(x) \leftarrow C(y) : \varphi_{HC}(x,y) \\ B(x) \leftarrow D(y) : \varphi_{BD}(x,y) \\ \bot \leftarrow \varphi_{CI}(x,y), \varphi_{CD}(\bar{x},z) \end{array}$$

Furthermore, we can associate weights to the assumables according to the potential functions. For instance,

$$weight(\varphi_{CD}(t,t) : 18, \varphi_{CD}(t,f) : 2, \varphi_{CD}(f,t) : 5, \varphi_{CD}(f,f) : 2)$$

In order to be able to probabilistically interpret a CL theory T, a number of assumptions are added to those of abduction logic: (i) the theory is acyclic; (ii) the rules for every ground non-assumable represented in T' are covering, i.e., there is always a rule whose assumable holds; (iii) the bodies of the rules in T' for an atom are mutually exclusive; (iv) there is a set of ground assumables, one from every grounded weight declaration, consistent with T. As in Poole's probabilistic Horn logic, these assumptions are not intended to be enforced by the system: it is up to the modeller to comply to these requisites. Under this condition, we can then guarantee the probabilistic properties of the theory.

3.2 Semantics and Reasoning

The interpretation of chain logic theories T is done in terms of possible world semantics for the ground case.

Definition 2. *Let $\mathcal{P}$ be a set of predicates of the language. Then a possible world is a tuple $w = \langle D, \omega, \hat{p} \rangle$ where*

- *D is a set of ground terms of the language*
- *$\omega : \mathcal{A}' \to R_0^+$ is a function which assigns a weight to ground assumables $\mathcal{A}'$*
- *$\hat{p} : D^n \to \{true, false\}$ is a valuation function, for each $p \in \mathcal{P}$.*

Truth of formulae is then inductively defined as usual except that an atom a is false if there are clauses $a \leftarrow b_1$ and $a \leftarrow b_2$ in T and both b_1 and b_2 are true. Furthermore, we have:

$$\begin{aligned} w \models \quad & weight(a_1 : w_1, \ldots, a_n : w_n) \\ \textit{iff} \quad & \exists i\, w \models a_i \text{ and } \forall j \neq i\, w \not\models a_j \text{ and } \forall i\, \omega(a_i) = w_i \end{aligned}$$

which expresses that exactly one assumable is true in a weight declaration.

As a convenience, for arbitrary theories, we write $w \models T$, whenever for all groundings of T, denoted by T', we have $w \models T'$. The set of all possible worlds denoted W, for which we define a joint probability distribution.

Definition 3. *Let P_T be a non-negative real function of W that is defined as follows:*

$$P_T(w) = \begin{cases} \frac{1}{Z} \prod_{a \in \mathcal{A}'} \omega(a) & \text{if } w \models T \\ 0 & \text{otherwise} \end{cases}$$

where $Z = \sum_{w \in \{w \mid w \models T\}} \prod_{a \in \mathcal{A}'} \omega(a)$.

Clearly, the function P_T obeys the axioms of probability theory, as each weight is larger than or equal to 0 and, given that there is a set of consistent assumables consistent with T, there is at least one possible world for T, thus, it follows that $\sum_{w \in W} P_T(w) = 1$. Therefore, it is a joint probability distribution; P_T is sometimes abbreviated to P in the following. A probability for a formula conjunction φ can be derived by marginalising out the other atoms, i.e., $P(\varphi) = \sum_{w \models \varphi} P(w)$.

These definitions provide the means to reason logically, at the same time assigning probabilities to conjunctive formulae in the language. An alternative way to look at the reasoning process, however, is in terms of explanations of observations, as defined above, which will be considered next.

We define a *hypothesis* as a conjunction of ground instances, one for each assumable of a grounded weight declaration. The set of all such hypotheses is denoted by $\mathcal{H}$. The set of consistent hypotheses, with respect to T will be denoted by CH, i.e., $\text{CH} = \{H \in \mathcal{H} \mid T \cup H \not\models \bot\}$.

Proposition 1. *The joint probability distribution over the set of hypotheses is as follows:*

$$P_T(H) = \begin{cases} \frac{1}{Z} \prod_{a \in H} \omega(a) & \text{if } H \in \text{CH} \\ 0 & \text{otherwise} \end{cases}$$

where $Z = \sum_{H \in \text{CH}} \prod_{a \in H} \omega(a)$.

Given T, a minimal explanation E of some formula ψ is equivalent to a disjunction of hypotheses, i.e., $E \equiv \bigvee_i H_i$ with $H_i \in \mathcal{H}$. As all H_i are mutually exclusive, it follows that:

$$P_T(E) = P_T(\bigvee_i H_i) = \sum_i P_T(H_i)$$

which assigns a probability to minimal explanations. In order to assign a probability to a formula using explanations, we have the following result.

Theorem 1. *Under the assumptions mentioned in Section 3.1, if $\mathcal{E}_T(\psi)$ is the set of minimal explanations of the conjunction of atoms ψ from the chain logic theory T, then:*

$$P_T(\psi) = \sum_{E \in \mathcal{E}_T(\psi)} P(E)$$

Proof. This follows exactly the same line of reasoning of [1, page 53, proof of Theorem A.13]. □

This result shows that P is indeed a probability distribution over conjunctions of formulae if we use the definition of P_T above. Other probabilities can be calculated on the basis of these types of formulae, such as conditional probabilities. Below, we will sometimes refer to the resulting probability distribution by P_T in order to stress that we mean the probability calculated using Definition 3.

Example 3. Reconsider the uncertainty specification concerning influenza as described in Example 2. Consider here that we are interested in calculating the $P(B(t))$ (i.e., the probability of B being true). Recalling the definitions provided in Section 2.2, we obtain the minimal explanations for $B(t)$, i.e., $\mathcal{E}_T(B(t))$ as the set with the (8) members such as:

$$\{\varphi_{BD}(t,t), \varphi_{CD}(t,t), \varphi_{CI}(t,t), \varphi_I(t)\}$$
$$\{\varphi_{BD}(t,t), \varphi_{CD}(t,t), \varphi_{CI}(t,f), \varphi_I(f)\}$$
$$\vdots$$

We can then sum over the hypotheses that are consistent with these explanations.

$$\begin{aligned} P(B(t)) &= \textstyle\sum_{e \in \mathcal{E}_T(B(t))} P(e) \\ &= Z^{-1}(0.3 \cdot 18 \cdot 8 \cdot 0.1 + \ldots) = 27.7/Z \end{aligned}$$

Similarly, we can find that $P(B(f)) = 88.0/Z$, so $Z = 115.7$ and thus $P(B(t)) \approx 0.24$.

Abductive reasoning establishes the relevant variables for the computation of a marginal probability, i.e., it selects the portion of the chain graph that is relevant. Consider once again our running example. By asking if B is true, we obtain through the rule $B(x) \leftarrow D(y) : \varphi_{BD}(x,y)$ that D influences B. In terms of the graph, this means that we walk the arc in reverse direction, i.e., from effect to explaining cause, from B to D. We now look at the rules which have D as head, selecting $D(x) \leftarrow I(y) : \varphi_{CI}(z,y) \wedge \varphi_{CD}(z,x)$. From the potential φ_{CD} in this clause, the existence of the association between C and D is established. From the presence of potential φ_{CI} we – indirectly – recover and include also

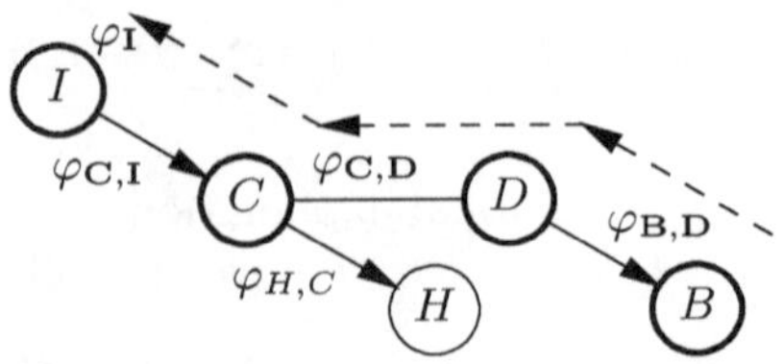

Fig. 2. The direction of reasoning about B is denoted with a dashed line. The nodes that represent variables that are abduced over and the variables that are relevant in the explanation are highlighted. Parts of the graph that are not relevant for the computation of $P(B)$ are not considered.

the influence of I on C. From the presence of predicate $I(y)$ we can proceed to including also the potential φ_I, which (as seen previously) is also important for the correct probabilistic computation. This process is graphically depicted in Fig. 2.

Example 4. Consider that we are interested in the probability of $P(I(t) \mid B(t))$. This probability can be obtained by the having $P(I(t) \wedge B(t))$ divided by $P(B(t))$. The calculation of $P(B(t))$ was shown in Example 3. By calculation the minimal explanations for $I(t) \wedge B(t)$, we obtain that $P(I(t) \wedge B(t)) = 4.5/Z \approx 0.04$, so it follows that $P(I(t) \mid B(t)) \approx \frac{0.04}{0.24} \approx 0.16$. Note that the prior probability for $I(t)$ is 0.1, so the evidence $B(t)$ has increased the probability for influenza.

3.3 Specification of Chain Graphs

In this section, we present the formal relation between chain graphs with discrete random variables and chain logic. For the sake of simplicity, we focus on chain graphs with binary variables, i.e., the set of constants is $\{t, f\}$, although the theory generalises to arbitrary arities. Complementary constants are denoted with a bar, i.e., $\bar{t} = f$ and $\bar{f} = t$.

The translation from a chain graph G to a chain logic theory T is as follows. First, introduce for each potential function φ_M a corresponding predicate φ_M and define a weight declaration containing $\varphi_M(c_0, \ldots, c_n) : w$ if $\varphi_M(X_M = (c_0, \ldots, c_n)) = w$, for all possible instantiations $(c_0, \ldots, c_n)$ of X_M. Second, we model the structure of the chain graph in chain logic. Consider a vertex v in G. For each component $C \in \mathcal{C}$ of G, there is a set of potential functions defined on the moral graph of the sub-graph $G_{C \cup \text{pa}(C)}$ which contains v or one of the parents of C. This set of potential functions is denoted by $\Phi_G(C, v)$. For every vertex v, we have the following formula in T:

$$V(x) \leftarrow \bigwedge\{V'(x_{v'}) \mid v' \in \text{pa}(C)\} : \\ \bigwedge\{\varphi_M(x_1, \ldots, x, \ldots, x_n) \mid \varphi_M \in \Phi_G(C, v)\}$$

and we ensure that each of the predicates defined for the same random variable shares that variable in the formula. However, this is not strictly necessary as different values for the same random variable in a component is also disallowed by the integrity constraints.

The integrity constraints are defined as follows. If we have two potential functions, namely an n-ary $\varphi_M(\ldots, v, \ldots)$ and an m-ary $\varphi'_M(\ldots, v, \ldots)$, i.e., which share a variable v *in the same chain component* (i.e., not between chain components), then we add the following formula to T:

$$\bot \leftarrow \varphi_M(x_0, \ldots, x, \ldots, x_n),\ \varphi'_M(x'_0 \ldots, \bar{x}, \ldots, x'_m)$$

for each variable that they share. As mentioned earlier, this ensures we do not generate explanations which have inconsistent assignments to the random variables within the same chain component.

In the following theorem, we establish that probabilities calculated from the chain logic theory correspond to the chain graph semantics.

Theorem 2. *Suppose* $v_1, \ldots, v_n$ *are vertices in a chain graph, with* T *as the corresponding chain logic theory by the translation described above, then:*

$$P(X_{v_1} = c_1, \ldots, X_{v_n} = c_n) = P_T(V_1(c_1), \ldots, V_n(c_n))$$

Proof. There is only one minimal explanation of $V_1(c_1) \wedge \cdots \wedge V_n(c_n)$, namely $\varphi_M(c_0^M, \ldots, c_m^M)$ for all potential functions in cliques in the moral graphs of chain components with their parents, such that the constants filled into the potential functions correspond to the values for each of the random variables.

The explanation describes exactly one hypothesis. Denote this as h. As the potential functions are related to exactly one component, we have the following equation:

$$\begin{aligned}
&\prod_{a \in h} \omega(a) \\
&= \prod_{C \in \mathcal{C}} \prod_{\varphi_j^C(c_0^j, \ldots, c_n^j) \in h} \varphi_j^C(X_{v_0^j} = c_0^j, \ldots, X_{v_n^j} = c_n^j) \\
&= \prod_{C \in \mathcal{C}} \prod_{M \in M(C)} \varphi_M(X_M)
\end{aligned} \tag{7}$$

where φ^C are potential functions defined for component C and $M(C)$ are the complete sets in the moral graph from the sub-graph $G_{C \cup \mathrm{pa}(C)}$.

Let $Z = \sum_{H \in \mathrm{CH}} \prod_{a \in H} \omega(a)$. Since there are no integrity constraints between variables in chain components (i.e., combinations of consistent potential functions which are in different chain components are consistent), we have that:

$$\begin{aligned}
Z &= \sum_{H \in \mathrm{CH}} \prod_{a \in H} \omega(a) \\
&= \prod_{C \in \mathcal{C}} \sum_{h \in \mathrm{CH}(C)} \prod_{\varphi_j^C} \varphi_j^C(X_{v_0^j} = c_0^j, \ldots, X_{v_n^j} = c_n^j) \\
&= \prod_{C \in \mathcal{C}} Z(X_{\mathrm{pa}(C)})
\end{aligned} \tag{8}$$

where CH(C) is the set of consistent hypotheses (w.r.t. T) restricted to the potential functions in that chain component. Then, the equivalence follows in the following way:

$$\begin{aligned}
&P(X_{v_1} = c_1, \ldots, X_{v_n} = c_n)\\
&=_{\text{(factorisation)}} \prod_{C \in \mathcal{C}} P(X_C \mid X_{\text{pa}(C)})\\
&=_{\text{(factorisation)}} \prod_{C \in \mathcal{C}} Z^{-1}(X_{\text{pa}(C)}) \prod_{M \in M(C)} \varphi_M(X_M)\\
&\quad \times \left(\prod_{C \in \mathcal{C}} Z^{-1}(X_{\text{pa}(C)})\right) \prod_{C \in \mathcal{C}} \prod_{M \in M(C)} \varphi_M(X_M)\\
&=_{\text{(Eq. 8)}} Z^{-1} \prod_{C \in \mathcal{C}} \prod_{M \in M(C)} \varphi_M(X_M))\\
&=_{\text{(Eq. 7)}} Z^{-1} \prod_{a \in w} \omega(a)\\
&=_{\text{(def. } P_T)} P_T(V_1(c_1), \ldots, V_n(c_n))
\end{aligned}$$

□

As we have shown in Section 3.2 that P_T adheres to the axioms of probability theory, chain graphs and the translated chain logic theory agree on all probabilities. This result shows that chain graphs can be translated to chain logic specifications. The converse is also true: all chain logic theories, which adhere to the assumptions of Section 3.1, correspond to a chain graph as a fully connected Markov network models and the associated probability distributions. This is not a minimal independence map of the underlying probability distribution in general, although conditional independence statements can be obtained by comparing explanations between formulae. As we are only able to represent direct causal links and indirect association, we conjecture that we have the same expressiveness in this logic as in chain graphs.

4 Learning Chain Graph Parameters

While observables and assumables make up the core of chain logic, determining the probabilistic parameters of assumables using observations stored in a database D is one of the essential tasks of learning in chain logic. Basically, the goal is to estimate weights of the assumables in a CL theory related to CGs. In general, it is not possible to easily estimate the potentials from data as they might have a complex dependency to the rest of the graph. However, if the individual components are triangulated, the factorisation can be stated in terms of marginal probabilities over the variables in a clique.

The proposed algorithm for determining the parameters is inspired by the use of a junction tree for probabilistic inference. The reason for this is that a junction tree provides sufficient information about the interactions between assumables, i.e., when they influence the same observables. Junction trees, with required properties such as the running intersection property, are only guaranteed to exist when the graph is triangulated, so we restrict ourselves to this case. Let $\mathcal{O}$ be the set of grounded non-assumables, which are the *observations*. As a convenience we write N_O with $O \subseteq \mathcal{O}$ for the number of tuples of D that contain O. In the following, we will assume that all $o \in \mathcal{O}$ are present in database D.

Algorithm 1. Learn CL parameters

Require: chain logic theory T, assumables $\mathcal{A}'$, observables $\mathcal{O}$, database D
for $a \in \mathcal{A}'$ **do**
  $\texttt{Effect}(a) \leftarrow \{o \in \mathcal{O} \mid \exists H \in \text{CH} : T \cup H \models o \text{ and } T \cup (H \setminus \{a\}) \not\models o\}$
  $\texttt{Rel}(a) \leftarrow \{o \in \mathcal{O} \mid \forall H \in \text{CH}\colon a \in H \wedge T \cup H \models \texttt{Effect}(a) \text{ implies } T \cup H \models o\}$
end for
$V \leftarrow \mathcal{A}'$
$E \leftarrow \{(a, a') \in \mathcal{A}' \times \mathcal{A}' \mid \texttt{Rel}(a) \cap \texttt{Rel}(a') \neq \varnothing\}$
$JG \leftarrow (V, E)$ with separator set $\texttt{Rel}(a) \cap \texttt{Rel}(a')$ associated to every edge (a, a')
let the weight of an edge in JG be the cardinality of its separator set
$J \leftarrow$ spanning tree of maximal weight of JG
DJ $\leftarrow$ any directed tree of J
for $a \in \mathcal{A}'$ **do**
  let S be the union of separators of a with its parents in DJ
  $\hat{\omega}(a) \leftarrow N_{\texttt{Effect}(a) \cup S}/N_S$
end for

The learning procedure is described in Algorithm 1. It will first identify effects of assumables (`Effect`), which are those observables that are implied by the assumables, possibly together with other assumables. Then, indirect relation (`Rel`) are identified, which are those observables that are always included in the derivation to the effects from an assumable. The latter can be used to build up a junction tree, where variables are instantiated for a particular value. From this structure, weights of the assumables can be learned. The properties of junction trees ensure that joint distribution corresponds to the relative frequency of observables, i.e., we have the following, general, result.

Theorem 3. *Given a chain logic theory T with associated (moralised, triangulated) chain graph G and database D, then after running Algorithm 1, the resulting weight declaration described by $\hat{\omega}$ and T will be such that:*

$$P_T(O) = \frac{N_O}{N_\varnothing} \tag{9}$$

for all possible observations $O \subseteq \mathcal{O}$.

Proof (sketch). If the assumable a models the potential function of some clique, then the set $\texttt{Rel}(a)$ contains the nodes of that clique. Furthermore, JG is isomorph to a junction graph of the underlying chain graph G. According to [6, Theorem 1], then J will be a junction tree of G. Because of the equivalence between reasoning in chain logic and chain graphs, we have:

$$P(x_V) = \prod_C \prod_{M \in M(C)} \hat{\omega}(\varphi_M)$$

(see Eq. 7). Since $\hat{\omega}(\varphi_M) = N_M/N_S$ where S is some separator, it follows that

$$\prod_C \prod_{M \in M(C)} \hat{\omega}(\varphi_M) = \frac{\prod_C \prod_{M \in M(C)} N_M}{R}$$

where R amounts to the product of the frequency of separators on the edges. Then, observe that all separators of the graph are there iff they are in R as each separator appears exactly once in an edge of a junction tree.

Finally, by application of [7, Lemma 1] and [8, proposition 12.3.2], this frequency coincides exactly with the frequency interpretation of $P(X_V)$, i.e., coincides with the relative frequency of the data. □

Example 5. Reconsider the chain graph of Fig. 2. By logical reasoning, it can be shown that: $\texttt{Rel}(\varphi_I(t)) = \{I(t)\}, \texttt{Rel}(\varphi_{CI}(t,t)) = \{C(t), I(t)\}$, etc, which gives the following graph (here shown with quantified variables to visualise the similarity to regular junction graphs):

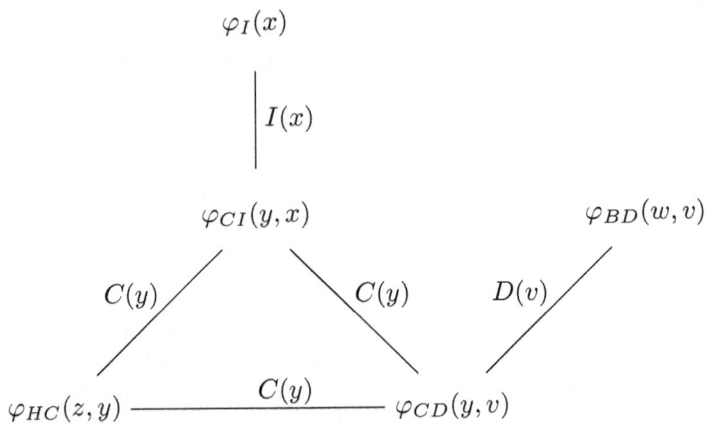

A tree can be obtained from this graph by removing any of the edges from the loop in the undirected graph. In fact, it does not matter which one to choose as, if there is a loop, then the separators are all the same set of variables. Equation 9 implies that for chain logic it is irrelevant which one is chosen, but some are more closely related to the original graph as others. For example, if we take $\varphi_I(x)$ as a root, then $\varphi_{CI}(y,x)$ is a conditional probability, as in the original graph. If, on the other hand we take $\varphi_I(x)$ as a leaf, then its weight will be 1 and thus $\varphi_{CI}(y,x)$ will be the *joint* probability of C and I, given its parent.

In order to obtain the interpretation of the original graph, we adapt Algorithm 1 by choosing DJ as the maximum weight spanning tree such that there is an arc from a to a' iff $s(a,a') \in \texttt{Effect}(a)$ and $s(a,a') \notin \texttt{Effect}(a')$, where $s(a,a')$ denotes $\texttt{Rel}(a) \cap \texttt{Rel}(a')$, i.e., whenever an assumable a' does not explain an observable it is related to, which means it must be conditioned on this observation. For triangulated chain graphs, it can be proven that such a tree exists.

Example 6. In the example above, we thus take the tree with arrows between φ_I and φ_{CI}, and between φ_{CI} and φ_{CD}, giving, e.g., the following tree:

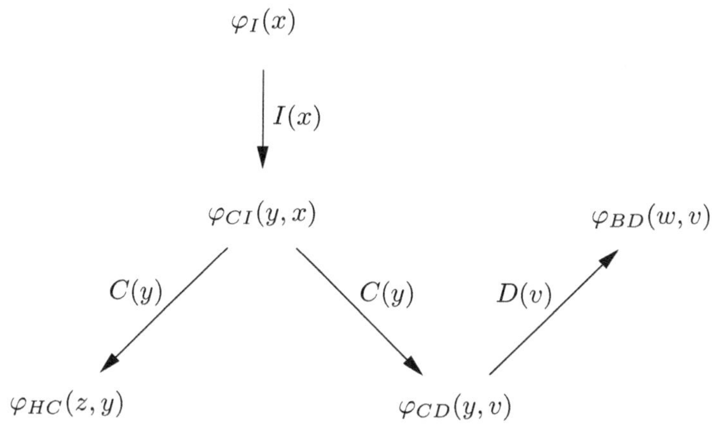

The learning algorithm will then, e.g., learn that:

$$\varphi_{CD}(x,y) = N_{\{C(x),D(y)\}}/N_{C(y)}$$

which corresponds to exactly the relative frequencies associated to variables in the original chain graph showing the relation between the two formalisms for learning.

Relational domains can be represented, and thus learned about using the same machinery, as long as the modeller ensures properties characterised by the class of triangulated chain graphs.

5 Comparison

Probabilistic Horn logic was originally proposed by Poole in [1]. It offers a framework that was shown to be as powerful as Bayesian networks, yet it has the advantage that it is a first-order language that integrates probabilistic and logical reasoning in a seamless fashion. Besides some changes in the terminology (such as using *weight* declarations in place of *disjoint* ones), the main differences in terms of syntax is the set of integrity constraints allowed and the probabilistic information captured in each formalism. Weights can sum up to any value, enabling the formalisation of potential functions instead of a (normalised) probability distribution. Furthermore, in our case, by allowing the use of extra integrity constraints, we are able to establish dependences among instantiations of hypotheses.

Those differences extend Poole's approach and allow us to obtain a more generic probabilistic model, being crucial for the representation of chain graph models. The graphical representation associated with a Bayesian network does not offer a unique way to represent the independence information, which makes the interpretation of Bayesian networks cumbersome. In contrast, an advantage of using chain graphs as underlying model is representing associations (e.g., *coughing* and *dyspnoea* in Example 1), which cannot be defined in Bayesian networks. In fact, chain graphs can capture the class of equivalent Bayesian networks. By using potential functions we can represent the quantitative influence between variables in a clique. The additional integrity constraints guarantee that instantiations of those potentials functions appear consistently in each explanation. Despite such differences, we still share with Poole's approaches some assumptions and similar results, for instance, with respect to the probability densities defined over the theory.

Bayesian logic programs [2] have similar limitations as probabilistic Horn logic; in addition, they are only proposed as formalisms to specify Bayesian networks in a logical way and reasoning is done in the generated Bayesian networks. Furthermore, the framework of Markov logic networks [3] has been proposed as a powerful language based on first-order logic to specify Markov networks. Yet, Markov networks are seen by researchers in probabilistic graphical models as the weakest type of such models, as much of the subtleties of representing conditional independence cannot be handled by Markov networks. In fact, formulae in Markov logic can only model associations between literals, whereas causal knowledge cannot be represented, for instance, between *coughing* and *hoarseness*. Furthermore, despite its expressive power, Markov logic is a generative language, i.e., specifications are translated into the corresponding graphical model on which reasoning is then performed in a standard fashion. The aim of the presented research was to design an expressive probabilistic logic that supports probabilistic reasoning and learning guided by the structure of the logic.

6 Final Considerations

In this paper we presented a simple, yet powerful, language for representing, reasoning with and learning generic chain graph models. Besides being able to incorporate both Bayesian and Markov network models as special cases, we maintain a strong relation between logical and probabilistic reasoning.

Our language still presents some restrictions. First, we use finite set of constants, which prohibits the use of continuous variables. For Markov logic networks, it has been shown that special cases of such networks can be extended to infinite domains by defining a Gibbs measure over sets of interpretations for logical formulae [9]. A similar approach could be taken here by defining a measure over the set of consistent states. Another limitation is the acyclicity assumption, which restricts the explicit representation of undirected graphs components. Even though we require certain assumptions for a sound probabilistic interpretation, weakening acyclicity seems feasible [10].

While we have shown in this paper that chain logic is powerful enough to define, reason, and learn about chain graphs, we have no strong reason to suspect that chain logic is restricted to this class of probabilistic graphical models. Although chain graphs form a fairly general class of probabilistic graphs, it might be the case that the language is applicable to a broader set of graphs. Also, modelling the independence implied in chain logic theories into a graphical model is an open question that will be investigated further.

With respect to learning, we have presented in this paper parameter learning of chain graph theories. Learning the structure of such graphs will be a subject of further research, but techniques from the inductive logic programming have been successful for learning Bayesian logic programs [11]. We believe similar ideas can be applied for learning chain logic theories.

Acknowledgements

The first and last author have been supported by the OCTOPUS project under the responsibility of the Embedded Systems Institute. This project is partially supported by the Netherlands Ministry of Economic Affairs under the Embedded Systems Institute program. The second author acknowledges the support through the B-Screen project funded by the Netherlands Organisation for Scientific Research (NWO) under BRICKS/FOCUS grant number 642.066.605.

References

1. Poole, D.: Probabilistic Horn abduction and Bayesian networks. AI Journal 64(1), 81–129 (1993)
2. Kersting, K., de Raedt, L.: Bayesian logic programs. Technical Report 151, Institute for Computer Science - University of Freiburg, CoRR cs.AI/0111058 (2001)
3. Richardson, M., Domingos, P.: Markov logic networks. Machine Learning 62, 107–136 (2006)
4. Pearl, J.: Probabilistic Reasoning in Inteligent Systems: Networks of Plausible Inference. Morgan Kaufmann, San Francisco (1988)
5. Lauritzen, S.L.: Graphical Models. Clarendon, Oxford (1996)
6. Jensen, F., Jensen, F.: Optimal junction trees. In: Uncertainty in Artificial Intelligence (1994)
7. Bouckaert, R.R., Studený, M.: Chain graphs: Semantics and expressiveness. In: Froidevaux, C., Kohlas, J. (eds.) ECSQARU 1995. LNCS, vol. 946, pp. 69–76. Springer, Heidelberg (1995)
8. Whittaker, J.: Graphical Models in Applied Multivariate Statistics. Wiley, Chichester (1990)
9. Singla, P., Domingos, P.: Markov logic in infinite domains. In: Proc. of UAI 2007, pp. 368–375. AUAU Press (2007)
10. Poole, D.: The independent choice logic for modelling multiple agents under uncertainty. AI Journal 94(1–2), 7–56 (1997)
11. Kersting, K., de Raedt, L.: Towards combining inductive logic programming with bayesian networks. In: Rouveirol, C., Sebag, M. (eds.) ILP 2001. LNCS (LNAI), vol. 2157, pp. 118–131. Springer, Heidelberg (2001)

Max-Margin Weight Learning for Markov Logic Networks

Tuyen N. Huynh and Raymond J. Mooney

The University of Texas at Austin, Austin TX 78712, USA
{hntuyen,mooney}@cs.utexas.edu

Abstract. Markov logic networks (MLNs) are an expressive representation for statistical relational learning that generalizes both first-order logic and graphical models. Existing discriminative weight learning methods for MLNs all try to learn weights that optimize the Conditional Log Likelihood (CLL) of the training examples. In this work, we present a new discriminative weight learning method for MLNs based on a max-margin framework. This results in a new model, Max-Margin Markov Logic Networks (M3LNs), that combines the expressiveness of MLNs with the predictive accuracy of structural Support Vector Machines (SVMs). To train the proposed model, we design a new approximation algorithm for loss-augmented inference in MLNs based on Linear Programming (LP). The experimental result shows that the proposed approach generally achieves higher F_1 scores than the current best discriminative weight learner for MLNs.

1 Introduction

Statistical relational learning (SRL) concerns the induction of probabilistic knowledge that supports accurate prediction for multi-relational structured data [1]. Markov Logic Networks (MLNs) are a recently developed SRL model that generalizes both full first-order logic and Markov networks [2]. An MLN consists of a set of weighted clauses in first-order logic. Rather than completely ruling out situations that violate these logical constraints, possible worlds simply become exponentially less likely as the total weight of violated clauses increases. MLNs have been successfully applied to a variety of real-world problems ranging from collective classification of web pages [3] to extraction of bibliographic information from scientific papers [4].

Existing discriminative training algorithms for learning MLN weights attempt to maximize the conditional log likelihood (CLL) of a set of *target predicates* given evidence provided by a set of *background predicates* [5,3,6]. If the goal is to predict accurate target-predicate probabilities, this approach is well motivated. However, in many applications, the actual goal is to maximize an alternative performance metric such as classification accuracy or F-measure. Max-margin methods are a competing approach to discriminative training that are well-founded in computational learning theory and have demonstrated empirical success in

W. Buntine et al. (Eds.): ECML PKDD 2009, Part I, LNAI 5781, pp. 564–579, 2009.

many applications [7]. They also have the advantage that they can be adapted to maximize a variety of performance metrics in addition to classification accuracy [8]. Max-margin methods have been successfully applied to structured prediction problems, such as in Max-Margin Markov Networks (M3Ns) [9] and structural SVMs [10]; however, until now, they have not been applied to an SRL model that generalizes first-order logic such as MLNs.

In this paper, we develop Max-Margin MLNs (M3LNs) by instantiating an existing general framework for max-margin training of structured models [11]. This requires developing a new algorithm for approximating the "loss-augmented" inference in MLNs. Extensive experiments in the two real-world MLN applications referenced above demonstrate that M3LNs generally produce improved results when the goal involves maximizing predictive accuracy metrics other than CLL.

The remainder of the paper is organized as follows. Section 2 provides some background on MLNs and structural SVMs. Section 3 presents the max-margin approach for weight learning in MLNs. Section 4 shows the experimental evaluation of the proposed approach. Section 5 and 6 discuss the related work and future work. Section 7 concludes the paper.

2 Background

2.1 MLNs and Alchemy

An MLN consists of a set of weighted first-order clauses. It provides a way of softening first-order logic by making situations in which not all clauses are satisfied less likely but not impossible [2]. More formally, let X be the set of all propositions describing a world (i.e. the set of all ground atoms), $\mathcal{F}$ be the set of all clauses in the MLN, w_i be the weight associated with clause $f_i \in \mathcal{F}$, $\mathcal{G}_{f_i}$ be the set of all possible groundings of clause f_i, and $\mathcal{Z}$ be the normalization constant. Then the probability of a particular truth assignment $\mathbf{x}$ to the variables in X is defined as [2]:

$$\begin{aligned} P(X=\mathbf{x}) &= \frac{1}{\mathcal{Z}} \exp\left(\sum_{f_i \in \mathcal{F}} w_i \sum_{g \in \mathcal{G}_{f_i}} g(\mathbf{x})\right) \\ &= \frac{1}{\mathcal{Z}} \exp\left(\sum_{f_i \in \mathcal{F}} w_i n_i(\mathbf{x})\right) \end{aligned} \quad (1)$$

where $g(\mathbf{x})$ is 1 if g is satisfied and 0 otherwise, and $n_i(\mathbf{x}) = \sum_{g \in \mathcal{G}_{f_i}} g(\mathbf{x})$ is the number of groundings of f_i that are satisfied given the current truth assignment to the variables in X.

There are two inference tasks in MLNs. The first one is to infer the Most Probable Explanation (MPE) or the most probable truth values for a set of unknown literals $\mathbf{y}$ given a set of known literals $\mathbf{x}$, provided as evidence (also called MAP inference). This task is formally defined as follows:

$$\arg\max_{\mathbf{y}} P(\mathbf{y}|\mathbf{x}) = \arg\max_{\mathbf{y}} \frac{1}{Z_x} \exp\left(\sum_i w_i n_i(\mathbf{x},\mathbf{y})\right)$$
$$= \arg\max_{\mathbf{y}} \sum_i w_i n_i(\mathbf{x},\mathbf{y}) \quad (2)$$

where Z_x is the normalization constant over all possible worlds consistent with $\mathbf{x}$, and $n_i(\mathbf{x},\mathbf{y})$ is the number of true groundings of clause f_i given the truth assignment $(\mathbf{x},\mathbf{y})$. MPE inference in MLNs is therefore equivalent to finding the truth assignment that maximizes the sum of the weights of satisfied clauses, a Weighted MAX-SAT problem. This is an NP-hard problem for which a number of approximate solvers exist, of which the most commonly used is MaxWalkSAT [12]. Recently, Riedel [13] proposed a more efficient and accurate MPE inference algorithm for MLNs called Cutting Plane Inference (CPI), which does not require grounding the whole MLN. However, the CPI method only works well for some classes of MLNs where the separation step of the CPI method returns a small set of constraints. In the worst case, it also constructs the whole ground MLN.

The second inference task in MLNs is computing the conditional probabilities of some unknown literals, $\mathbf{y}$, given some evidence $\mathbf{x}$. Computing these probabilities is also intractable, but there are good approximation algorithms such as MC-SAT [14] and lifted belief propagation [15].

Learning an MLN consists of two tasks: structure learning and weight learning. The weight learner can learn weights for clauses written by a human expert or automatically induced by a structure learner. There are two approaches to weight learning in MLNs: generative and discriminative. In discriminative learning, we know a priori which predicates will be used to supply evidence and which ones will be queried, and the goal is to correctly predict the latter given the former. Several discriminative weight learning methods have been proposed, all of which try to find weights that maximize the CLL (equivalently, minimize the negative CLL). In MLNs, the derivative of the negative CLL with respect to a weight w_i is the difference of the expected number of true groundings $E_w[n_i]$ of the corresponding clause f_i and the actual number according to the data n_i. However, computing the expected count $E_w[n_i]$ is intractable. The first discriminative weight learner [5] uses the voted perceptron algorithm [16] where it approximates the intractable expected counts by the counts in the MPE state computed by the MaxWalkSAT. Later, Lowd and Domingos [3] presented a number of first-order and second-order methods for optimizing the CLL. These methods use samples from MC-SAT to approximate the expected counts used to compute the gradient and Hessian of the CLL. Among them, the best performing is *preconditioner scaled conjugate gradient* (PSCG) [3]. This method uses the inverse diagonal Hessian as the preconditioner. Recently, Huynh and Mooney [6] proposed an efficient and accurate discriminative weight learner for MLNs when all clauses are non-recursive (A non-recursive clause is a clause that contains only one non-evidence literal). For information about previous work on structure learning see [17].

ALCHEMY [18] is an open source software package for MLNs. It includes implementations for all of the major existing algorithms for structure learning, generative weight learning, discriminative weight learning, and inference. Our proposed algorithm is implemented using ALCHEMY.

2.2 Structural Support Vector Machines

In this section, we briefly review the structural SVM problem and an algorithmic schema for solving it efficiently. For more detail, see [11]. In structured output prediction, we want to learn a function $h : \mathcal{X} \rightarrow \mathcal{Y}$, where $\mathcal{X}$ is the space of inputs and $\mathcal{Y}$ is the space of multivariate and structured outputs $\mathcal{Y}$, from a set of training examples S:

$$S = ((x_1, y_1), ..., (x_n, y_n)) \in (\mathcal{X} \times \mathcal{Y})^n$$

The goal is to find a function h that has low prediction error. This can be accomplished by learning a discriminant function $f : \mathcal{X} \times \mathcal{Y} \rightarrow \mathcal{R}$, then maximizing f over all $y \in \mathcal{Y}$ for a given input x to get the prediction.

$$h_w(x) = \arg\max_{y \in \mathcal{Y}} f_w(x, y)$$

The discriminant function $f_w(x, y)$ takes the form of a linear function:

$$f_w(x, y) = \mathbf{w}^T \Psi(x, y)$$

where $\mathbf{w} \in \mathcal{R}^N$ is a parameter vector and $\Psi(x, y)$ is a feature vector relating an input x and output y. The features need to be designed for a given problem so that they capture the dependency structure of y and x and the relations among the outputs y . Then, the goal is to find a weight vector $\mathbf{w}$ that maximizes the margin:

$$\gamma(x_i, y_i; w) = \mathbf{w}^T \Psi(x_i, y_i) - \max_{y_i' \in \mathcal{Y} \setminus y_i} \mathbf{w}^T \Psi(x_i, y_i')$$

To find such a weight vector, Joachims *et al.* [11] proposed to solve the following optimization problem, called the "1-slack" structural SVM problem:

Optimization Problem 1 (OP1): 1-Slack Structural SVMs [11]

$$\min_{\mathbf{w}, \xi \geq 0} \frac{1}{2} \mathbf{w}^T \mathbf{w} + C\xi$$

$$s.t.\ \forall (\bar{y_1}, ..., \bar{y_n}) \in \mathcal{Y}^n : \frac{1}{n} \mathbf{w}^T \sum_{i=1}^{n} \delta\Psi(\bar{y_i}) \geq \frac{1}{n} \sum_{i=1}^{n} \Delta(y_i, \bar{y_i}) - \xi$$

where $\delta\Psi(\bar{y_i}) = \Psi(x_i, y_i) - \Psi(x_i, \bar{y_i})$, and $\Delta(y_i, \bar{y_i})$ is the loss function. This optimization problem has an exponential number of constraints $|\mathcal{Y}|^n$, one for each possible combination of labels $(\bar{y_1}, ..., \bar{y_n}) \in \mathcal{Y}^n$, but it can be solved efficiently (provably linear runtime in the number of training examples) by Algorithm 1.

In each iteration, this algorithm solves a Quadratic Programming (QP) problem (line 4) to find the optimal weights corresponding to the current set of

Algorithm 1. Cutting-plane method for solving the "1-slack structural SVMs" [11]

1: **Input:** $S = ((x_1, y_1), ..., (x_n, y_n)), C, \varepsilon$

2: $\mathcal{W} \leftarrow \emptyset$

3: **repeat**

4:

$$(\mathbf{w}, \xi) \leftarrow \min_{\mathbf{w},\xi\geq 0} \frac{1}{2}\mathbf{w}^T\mathbf{w} + C\xi$$

$$s.t.\ \forall(\bar{y_1}, ..., \bar{y_n}) \in \mathcal{W} : \frac{1}{n}\mathbf{w}^T \sum_{i=1}^{n}[\Psi(x_i, y_i) - \Psi(x_i, \bar{y_i})] \geq \frac{1}{n}\sum_{i=1}^{n} \Delta(y_i, \bar{y_i}) - \xi$$

5: **for** $i = 1$ **to** n **do**

6: $\hat{y_i} \leftarrow \arg\max_{\hat{y}\in\mathcal{Y}}\{\Delta(y_i, \hat{y}) + \mathbf{w}^T\Psi(x_i, \hat{y})\}$

7: **end for**

8: $\mathcal{W} \leftarrow \mathcal{W} \cup \{(\hat{y_1}, ..., \hat{y_n})\}$

9: **until** $\frac{1}{n}\sum_{i=1}^{n} \Delta(y_i, \hat{y_i}) - \frac{1}{n}\mathbf{w}^T \sum_{i=1}^{n}[\Psi(x_i, y_i) - \Psi(x_i, \hat{y_i})] \leq \xi + \varepsilon$

10: return $(\mathbf{w}, \xi)$

constraints $\mathcal{W}$ and a loss-augmented inference problem (line 6), also called a separation oracle, to find the most violated constraint to add to $\mathcal{W}$. Since the QP problem does not depend on the structure of a particular problem (the $\Psi(x, y)$ and $\Delta(y, \bar{y})$), it can be solved by any QP solver. In contrast, for each specific problem, one needs to come up with an efficient way to solve the loss-augmented inference problem.

In summary, to apply structural SVMs to a new problem, one needs to design a new feature vector function $\Psi(x, y)$, choose a loss function $\Delta(y, \bar{y})$, and solve two argmax problems:

Prediction: $\arg\max_{y\in\mathcal{Y}} \mathbf{w}^T\Psi(x, y)$

Separation Oracle: $\arg\max_{\bar{y}\in\mathcal{Y}}\{\Delta(y, \bar{y}) + \mathbf{w}^T\Psi(x, \bar{y})\}$

3 Max-Margin Weight Learning for MLNs

3.1 Max-Margin Formulation

All of the current discriminative weight learners for MLNs try to find a weight vector $\mathbf{w}$ that optimizes the conditional log-likelihood $P(\mathbf{y}|\mathbf{x})$ of the query atoms $\mathbf{y}$ given the evidence $\mathbf{x}$. However, an alternative approach is to learn a weight vector $\mathbf{w}$ that maximizes the ratio:

$$\frac{P(\mathbf{y}|\mathbf{x}, \mathbf{w})}{P(\hat{\mathbf{y}}|\mathbf{x}, \mathbf{w})}$$

between the probability of the correct truth assignment $\mathbf{y}$ and the closest competing incorrect truth assignment $\hat{\mathbf{y}} = \arg\max_{\bar{\mathbf{y}}\in\mathbf{Y}\setminus\mathbf{y}} P(\bar{\mathbf{y}}|\mathbf{x})$. Applying equation 1 and taking the log, this problem translates to maximizing the margin:

$$\gamma(\mathbf{x}, \mathbf{y}; \mathbf{w}) = \mathbf{w}^T \mathbf{n}(\mathbf{x}, \mathbf{y}) - \mathbf{w}^T \mathbf{n}(\mathbf{x}, \hat{\mathbf{y}})$$
$$= \mathbf{w}^T \mathbf{n}(\mathbf{x}, \mathbf{y}) - \max_{\bar{\mathbf{y}} \in \mathbf{Y} \setminus \mathbf{y}} \mathbf{w}^T \mathbf{n}(\mathbf{x}, \bar{\mathbf{y}})$$

Note that, this translation holds for all log-linear models. For example, if we apply it to a CRF [19] then the result model is an M3N [9]. In fact, this translation is the connection between log-linear models and linear classifiers [20].

In turn, the max-margin problem above can be formulated as a "1-slack" structural SVM as follows:

Optimization Problem 2 (OP2): Max-Margin Markov Logic Networks

$$\min_{\mathbf{w}, \xi \geq 0} \frac{1}{2} \mathbf{w}^T \mathbf{w} + C\xi$$
$$s.t. \; \forall \bar{\mathbf{y}} \in Y : \mathbf{w}^T [\mathbf{n}(\mathbf{x}, \mathbf{y}) - \mathbf{n}(\mathbf{x}, \bar{\mathbf{y}})] \geq \Delta(\mathbf{y}, \bar{\mathbf{y}}) - \xi$$

So for MLNs, the number of true groundings of the clauses $\mathbf{n}(\mathbf{x}, \mathbf{y})$ plays the role of the feature vector function $\Psi(x, y)$ in the general structural SVM problem. In other words, each clause in an MLN can be viewed as a feature representing a dependency between a subset of inputs and outputs or a relation among several outputs.

As mentioned, in order to apply Algorithm 1 to MLNs, we need algorithms for solving the following two problems:

Prediction: $\arg\max_{\mathbf{y} \in Y} \mathbf{w}^T \mathbf{n}(\mathbf{x}, \mathbf{y})$

Separation Oracle: $\arg\max_{\bar{\mathbf{y}} \in Y} \{\Delta(\mathbf{y}, \bar{\mathbf{y}}) + \mathbf{w}^T \mathbf{n}(\mathbf{x}, \bar{\mathbf{y}})\}$

The prediction problem is just the (intractable) MPE inference problem discussed in section 2.1. We can use MaxWalkSAT to get an approximate solution, but we have found that models trained with MaxWalkSAT have very low predictive accuracy. On the other hand, recent work [21] has found that fully-connected pairwise Markov random fields, a special class of structural SVMs, trained with overgenerating approximate inference methods (such as relaxation) preserves the theoretical guarantees of structural SVMs trained with exact inference, and exhibits good empirical performance. Based on this result, we sought a relaxation-based approximation for MPE inference. We first present an LP-relaxation algorithm for MPE inference, then show how to modify it to solve the separation oracle problem for some specific loss functions.

3.2 Approximate MPE Inference for MLNs

MPE inference in MLNs is equivalent to the Weighted MAX-SAT problem, and there has been significant work on approximating this NP-hard problem using LP-relaxation [22,23]. The existing algorithms first relax and convert the Weighted MAX-SAT problem into a linear or semidefinite programming problem, then solve it and apply a randomized rounding method to obtain an approximate integral solution. These methods cannot be directly applied to MLNs, since they require the weights to be positive while MLN weights can be negative or infinite. So we modified the conversion used in these approaches to handle the case of negative and infinite weights.

Based on the evidence and the closed world assumption, a ground MLN contains only ground clauses (in clausal form) of the unknown ground atoms after removing all trivially satisfied and unsatisfied clauses. The following procedure translates the MPE inference in a ground MLN into an Integer Linear Programming (ILP) problem.

1. Assign a binary variable y_i to each unknown ground atom. y_i is 1 if the corresponding ground atom is $TRUE$ and 0 if the ground atom is $FALSE$.
2. For each ground clause C_j with infinite weight, add the following linear constraint to the ILP problem:

$$\sum_{i \in I_j^+} y_i + \sum_{i \in I_j^-} (1 - y_i) \geq 1$$

   where I_j^+, I_j^- are the sets of positive and negative ground literals in clause C_j respectively.
3. For each ground clause C_j with positive weight w_j, introduce a new auxiliary binary variable z_j, add the term $w_j z_j$ to the objective function, and add the following linear constraint to the ILP problem:

$$\sum_{i \in I_j^+} y_i + \sum_{i \in I_j^-} (1 - y_i) \geq z_j$$

   z_j is 1 if the corresponding ground clause is satisfied.
4. For each ground clause C_j with k ground literals and negative weight w_j, introduce a new auxiliary boolean variable z_j, add the term $-w_j z_j$ to the objective function and add the following k linear constrains to the ILP problem:

$$\begin{aligned} 1 - y_i \geq z_j, &\quad i \in I_j^+ \\ y_i \geq z_j, &\quad i \in I_j^- \end{aligned}$$

The final ILP has the following form:

Optimization Problem 3:

$$\begin{aligned}
\max_{y_i, z_i} & \sum_{C_j \in C^+} w_j z_j + \sum_{C_j \in C^-} -w_j z_j \\
s.t. & \sum_{i \in I_j^+} y_i + \sum_{i \in I_j^-} (1 - y_i) \geq 1 \quad \forall\, C_j \; where \; w_j = \infty \\
& \sum_{i \in I_j^+} y_i + \sum_{i \in I_j^-} (1 - y_i) \geq z_j \quad \forall C_j \in C^+ \\
& 1 - y_i \geq z_j \quad \forall\, i \in I_j^+ \; and \; C_j \in C^- \\
& y_i \geq z_j \quad \forall\, i \in I_j^- \; and \; C_j \in C^- \\
& y_i, z_j \in \{0, 1\}
\end{aligned}$$

Algorithm 2. The modified ROUNDUP procedure

```
Input: The LP solution y = {y_1, ..., y_n}
F ← ∅
for i = 1 to n do
    if y_i is integral then
        Remove all the ground clauses satisfied by assigning the value of y_i to the
        corresponding ground atom
    else
        add y_i to F
    end if
end for
repeat
    Remove the last item y_i in F
    Compute the sum w+ of the unsatisfied clauses where y_i appears as a positive
    literal
    Compute the sum w- of the unsatisfied clauses where y_i appears as a negative
    literal
    if w+ > w- then
        y_i ← 1
    else
        y_i ← 0
    end if
    Remove all the ground clauses satisfied by assigning the value of y_i to the cor-
    responding ground atom
until F is empty
return y
```

where C^+ and C^- are the set of clauses with positive and negative weights respectively. This ILP problem can be simplified by not introducing an auxiliary variable z_j for unit clauses, where we can use the variable y_i directly. This reduces the problem considerably, since ground MLNs typically contain many unit clauses (ALCHEMY combines all the non-recursive clauses containing the query atom into a unit clause whose weight is the sum of all the clauses' weights). Note that our mapping from a ground MLN to an ILP problem is a bit different from the one presented in [13] which generates two sets of constraints for every ground clause: one when the clause is satisfied and one when it is not. For a clause with positive weight, our mapping only generates a constraint when the clause is satisfied; and for a clause with negative weight, the mapping only imposes constraints when the clause is unsatisfied. The final ILP problem has the same solution with the one in [13], but it has fewer constraints since our mapping does not generate unnecessary constraints. We then relax the integer constraints $y_i, z_j \in \{0, 1\}$ to linear constraints $y_i, z_j \in [0, 1]$ to obtain an LP-relaxation of the MPE problem.

This LP problem can be solved by any general LP solver. If the LP solver returns an integral solution, then it is also the optimal solution to the original ILP problem. In our case, the original ILP problem is an NP-hard problem, so the LP solver usually returns non-integral solutions. Therefore, the LP solution needs to be rounded to give an approximate ILP solution. We first tried some of the randomized rounding methods in [23] but they gave poor results since the LP solution has a lot of fractional components with value 0.5. We then adapted a rounding procedure called ROUNDUP [24], a procedure for producing an upper-bound binary solution for a pseudo-Boolean function, to the case of pseudo-Boolean functions with linear constraints, which we found to work well. In each step, this procedure picks one fractional component and rounds it to 1 or 0. Hence, this process terminates in at most n steps, where n is the number of query atoms. Note that due to the dependencies between y_i's and z_j's (the linear constraints of the LP problem), this modified ROUNDUP procedure does not guarantee an improvement in the value of the objective function in each step like the original ROUNDUP procedure where all the variables are independent.

3.3 Approximation Algorithm for the Separation Oracle

The separation oracle adds an additional term, the loss term, to the objective function. So, if we can represent the loss as a linear function of the y_i variables of the LP-relaxation, then we can use the above approximation algorithm to also approximate the separation oracle. In this work, we consider two loss functions. The first one is the 0/1 loss function, $\Delta_{0/1}(\mathbf{y^T}, \mathbf{y})$ where $\mathbf{y^T}$ is the true assignment and $\mathbf{y}$ is some predicted assignment. For this loss function, the separation oracle is the same as the MPE inference problem since the loss function only adds a constant 1 to the objective function. Hence, in this case, to find the most violated constraint, we can use the LP-relaxation algorithm above or any other MPE inference algorithm. This 0/1 loss makes the separation oracle problem easier but it does not scale the margin by how different $\mathbf{y^T}$ and $\mathbf{y}$ are. It only requires a unit margin for all assignments $\mathbf{y}$ different from the true assignment $\mathbf{y^T}$. To take into account this problem, we consider the second loss function that is the number of misclassified atoms or the Hamming loss:

$$\begin{aligned}\Delta_{Hamming}(\mathbf{y^T}, \mathbf{y}) &= \sum_i^n [y_i^T \neq y_i] \\ &= \sum_i^n [(y_i^T = 0 \land y_i = 1) \lor (y_i^T = 1 \land y_i = 0)]\end{aligned}$$

From the definition, this loss can be represented as a function of the y_i's:

$$\Delta_{Hamming}(\mathbf{y^T}, \mathbf{y}) = \sum_{i:y_i^T=0} y_i + \sum_{i:y_i^T=1} (1 - y_i)$$

which is equivalent to adding 1 to the coefficient of y_i if the true value of y_i is 0 and subtracting 1 from the coefficient of y_i if the true value of y_i is 1. So we can use the LP-relaxation algorithm above to approximate the separation oracle with this Hamming loss function. Another possible loss function is $(1 - F_1)$ loss.

Unfortunately, this loss is a non-linear function, so we cannot use the above approach to optimize it. Developing algorithms for optimizing or approximating this loss function is an area for future work.

4 Experimental Evaluation

This section presents experiments comparing M3LNs to the current best discriminative weight learner for MLNs with recursive clauses, PSCG.

4.1 Datasets

We ran experiments on two large, real-world MLN datasets: WebKB for collective web-page classification, and CiteSeer for bibliographic citation segmentation. All the datasets and MLNs can be found at the ALCHEMY website[1].

The WebKB dataset consists of labeled web pages from the computer science departments of four universities. Different versions of this data have been used in previous work. To make a fair comparison, we used the version from [3], which contains 4,165 web pages and 10,935 web links. Each page is labeled with a subset of the categories: *course*, *department*, *faculty*, *person*, *professor*, *research project*, and *student*. The goal is to predict these categories from the words and links on the web pages. We used the same simple MLN from [3], which only has clauses relating words to page classes, and page classes to the classes of linked pages.

$Has(+word, page) \rightarrow PageClass(+class, page)$
$\neg Has(+word, page) \rightarrow PageClass(+class, page)$
$PageClass(+c1, p1) \wedge Linked(p1, p2) \rightarrow PageClass(+c2, p2)$

The plus notation creates a separate clause for each pair of word and page class, and for each pair of classes. The final MLN consists of 10,891 clauses, and a weight must be learned for each one. After grounding, each department results in an MLN with more than 100,000 ground clauses and 5,000 query atoms in a complex network. This also results in a large LP-relaxation problem for MPE inference.

For CiteSeer, we used the dataset and MLN used in [4]. The dataset has 1,563 citations and each of them is segmented into three fields: *Author*, *Title* and *Venue*. The dataset has four disconnected segments corresponding to four different research topics. We used the simplest MLN in [4], which is the isolated segmentation model. Despite its simplicity, after grounding, this model results in a large network with more than 30,000 query atoms and 110,000 ground clauses.

4.2 Metrics

We used F_1, the harmonic mean of recall and precision, to measure the performance of each algorithm. This is the standard evaluation metric in multiclass text categorization and information extraction. For systems that compute marginal probabilities rather than MPEs, we predict that an atom is true iff its probability is at least 0.5.

[1] http://alchemy.cs.washington.edu

Table 1. F_1 scores on WebKB

	Cornell	Texas	Washington	Wisconsin	Average
PSCG-MCSAT	0.418	0.298	0.577	0.568	0.465
PSCG-LPRelax	0.420	0.310	0.588	0.575	0.474
MM-$\Delta_{0/1}$-MaxWalkSAT	0.150	0.162	0.122	0.122	0.139
MM-$\Delta_{0/1}$-LPRelax	0.282	0.372	0.675	0.521	0.462
MM-$\Delta_{Hamming}$-LPRelax	**0.580**	**0.451**	**0.715**	**0.659**	**0.601**

Table 2. F_1 scores of different inference algorithms on WebKB

	Cornell	Texas	Washington	Wisconsin	Average
PSCG-MCSAT	0.418	0.298	0.577	0.568	0.465
PSCG-MaxWalkSAT	0.161	0.140	0.119	0.129	0.137
PSCG-LPRelax	0.420	0.310	0.588	0.575	0.474
MM-$\Delta_{Hamming}$-MCSAT	0.470	0.370	0.573	0.481	0.473
MM-$\Delta_{Hamming}$-MaxWalkSAT	0.185	0.184	0.150	0.154	0.168
MM-$\Delta_{Hamming}$-LPRelax	0.580	0.451	0.715	0.659	0.601

4.3 Methodology

We ran four-fold cross-validation (i.e. leave one university/topic out) on both datasets. For the max-margin weight learner, we used a simple process for selecting the value of the C parameter. For each train/test split, we trained the algorithm with five different values of C: 1, 10, 100, 1000, and 10000, then selected the one which gave the highest average F_1 score on training. The ϵ parameter was set to 0.001 as suggested in [11]. To solve the QP problems in Algorithm 1 and LP problems in the LP-relaxation MPE inference, we used the Mosek[2] solver. The PSCG algorithm was carefully tuned by its author. For MC-SAT and MaxWalkSAT, we used the default setting in Alchemy.

4.4 Results and Discussion

Table 1 and 3 present the performance of different systems on the WebKB and Citeseer datasets. Each system is named by the weight learner used, the loss function used in training, and the inference algorithm used in testing. For max-margin (MM) learner with margin rescaling, the inference used in training is the loss-augmented version of the one used in testing. For example, MM-$\Delta_{Hamming}$-LPRelax is the max-margin weight learner using the loss-augmented (Hamming loss) LP-relaxation MPE inference algorithm in training and the LP-relaxation MPE inference algorithm in testing.

Table 1 shows that the model trained by MaxWalkSAT has very low predictive accuracy. This result is consistent with the result presented in [13] which also found that the MPE solution found by MaxWalkSAT is not very accurate. Using the proposed LP-relaxation MPE inference improves the F_1 score from 0.139

[2] http://www.mosek.com/

Table 3. F_1 scores on CiteSeer

	Constraint	Face	Reasoning	Reinforcement	Average
PSCG-MCSAT	0.937	0.914	0.931	0.975	0.939
MM-$\Delta_{Hamming}$-LPRelax	0.933	0.922	0.924	0.958	0.934

Table 4. F_1 scores on CiteSeer with different parameter values

	Constraint	Face	Reasoning	Reinforcement	Average
PSCG-MCSAT-5	0.852	0.844	0.836	0.923	0.864
PSCG-MCSAT-10	0.937	0.914	0.931	0.973	0.939
PSCG-MCSAT-15	0.878	0.896	0.780	0.891	0.861
PSCG-MCSAT-20	0.850	0.859	0.710	0.784	0.801
PSCG-MCSAT-100	0.658	0.697	0.600	0.668	0.656
MM-$\Delta_{Hamming}$-LPRelax-1	0.933	0.922	0.924	0.955	0.934
MM-$\Delta_{Hamming}$-LPRelax-10	0.926	0.922	0.925	0.955	0.932
MM-$\Delta_{Hamming}$-LPRelax-100	0.926	0.922	0.925	0.954	0.932
MM-$\Delta_{Hamming}$-LPRelax-1000	0.931	0.918	0.925	0.958	0.933
MM-$\Delta_{Hamming}$-LPRelax-10000	0.932	0.922	0.919	0.968	0.935

to 0.462, the MM-$\Delta_{0/1}$-LPRelax system. Then the best system is obtained by rescaling the margin and training with our loss-augmented LP-relaxation MPE inference, which is the only difference between MM-$\Delta_{Hamming}$-LPRelax and MM-$\Delta_{0/1}$-LPRelax. The MM-$\Delta_{Hamming}$-LPRelax achieves the best F_1 score (0.601), which is much higher than the 0.465 F_1 score obtained by the current best discriminative weight learner for MLNs, PSCG-MCSAT.

Table 2 compares the performance of the proposed LP-relaxation MPE inference algorithm against MCSAT and MaxWalkSAT on the best trained models by PSCG and MM on the WebKB dataset. In both cases, the LP-relaxation MPE inference achieves much better F_1 scores than those of MCSAT and MaxWalkSAT. This demonstrates that the approximate MPE solution found by the LP-relaxation algorithm is much more accurate than the one found by the MaxWalkSAT algorithm. The fact that the performance of the LP-relaxation is higher than that of MCSAT shows that in collective classification it is better to use the MPE solution as the prediction than the marginal prediction.

For the WebKB dataset, there are other results reported in previous work, such as those in [9], but those results cannot be directly compared to our results since we use a different version of the dataset and test on a more complicated task (a page can have multiple labels not just one).

On the Citeseer results presented in Table 3, the performance of max-margin methods are very close to those of PSCG. However, its performance is much more stable. Table 4 shows the performance of MM weight learners and PSCG with different parameter values by varying the C value for MM and the number of iterations for PSCG. The best number of iterations for PSCG is 9 or 10. In principle, we should run PSCG until it converges to get the optimal weight vector. However, in this case, the performance of PSCG drops drastically on both training and testing after a certain number of iterations. For example, from Table 4 we can see that at 10 iterations PSCG achieves the best F_1 score of 0.939, but after 15 iterations, its F_1 score drops to 0.861 which is much worse than those of the max-margin weight learners. Moreover, if we let it run until 100 iterations, then its F_1 score is only 0.656. On the other hand, the performance of MM only varies a little bit with different values of C and we don't need to tune the number of iterations of MM. On this dataset, [4] achieved a F_1 score of 0.944 with the same MLN by using a version of the voted perceptron algorithm called Contrastive Divergence (CD) [25] to learn the weights. However, the performance of the CD algorithm is very sensitive to the learning rate [3], which requires a very careful tuning process to learn a good model.

Regarding training time, on both of the datasets, the max-margin weight learner is comparable to PSCG. It usually took about 200 iterations on the WebKB and less than 50 iterations on Citeseer to find the optimal weights, which resulted in a few hours of training for WebKB and less than an hour for Citeseer.

5 Related Work

Our work is related to various previous projects. Among them, M3N [9] is probably the most related. It is a special case of structural SVMs where the feature function $\Psi(x, y)$ is represented by a Markov network. When the Markov network can be triangulated and the loss function can be linearly decomposed, the original exponentially-sized QP can be reformulated as a polynomially-sized QP [9]. Then, the polynomially-sized QP can be solved by general QP solvers [26], decomposition methods [9], extragradient methods [27], or exponentiated gradient methods [28]. As mentioned in [9], these methods can also be used when the graph cannot be triangulated, but the algorithms only yield approximate solutions like our approach. However, these algorithms are restricted to the cases where a polynomially-sized reformulation exists [11]. That's why in this work we used the general cutting plane algorithm which imposes no restrictions on the representation. The ground MLN can be any kind of graph. On the other hand, since an MLN is a template for constructing Markov networks [2], the proposed model, M3LN, can also be seen as a template for constructing M3Ns. Hence, when the ground MLN can be triangulated and the loss is a linearly decomposable function, the algorithms developed for M3Ns can be applied. Our work is also closely related to the Relational Markov Networks (RMNs) [29]. However, by using MLNs, M3LNs are more powerful than RMNs in term of representation [2]. Besides, the objectives of M3LNs and RMNs are different. One tries to maximize the margin between the true assignment and other competing assignments, and one tries to maximize the conditional likelihood of the true assignment. Another related system is RUMBLE [30], a margin-based approach to first-order rule learning. In that work, the goal is to find a set of weighted rules that maximizes a quantity called margin minus variance. However, unlike M3LNs, RUMBLE only applies to independent binary classification problems and is unable to perform structured prediction or collective classification. In terms of applying the general structural SVM framework to a specific representation, our work is related to the work in [31] which used CRFs as the representation and graph cuts as the inference algorithm. In the context of discriminative learning, our work is related to previous work on discriminative training for MLNs [5,3,6,32]. We have mentioned some of them [5,3,6] in previous sections. The main difference between the work in [32] and ours is that we assume the structure is given and apply max-margin framework to learn the weights while [32] tries to learn a structure that maximizes the conditional likelihood of the data. Extending the max-margin framework to structure learning is an area for future work.

6 Future Work

Currently, our loss-augmented LP-relaxation MPE inference algorithm requires grounding the entire network first. It would be useful to develop new loss-augmented MPE inference algorithms which do not require grounding the full network. On the other hand, the max-margin framework allows optimizing different kinds of loss functions. Therefore, it would also be interesting to extend the current algorithm to optimize other loss functions. Besides, we also want to apply M3LNs to other structured prediction tasks and compare to other structured prediction models.

7 Conclusions

We have presented a max-margin weight learning method for MLNs based on the framework of structural SVMs. It resulted in a new model, M3LN, that has the representational expressiveness of MLNs and the predictive performance of SVMs. M3LNs can be trained to optimize different performance measures depending on the needs of the application. To train the proposed model, we developed a new approximation algorithm for loss-augmented MPE inference in MLNs based on LP-relaxation. The experimental results showed that the new max-margin learner generally has better or equally good but more stable predictive accuracy (as measured by F_1) than the current best discriminative MLN weight learner.

Acknowledgments

We thank the anonymous reviewers for their helpful comments. We also thank Daniel Lowd and Hoifung Poon for useful discussions and helping with the experiments. This research is sponsored by the DARPA and managed by the AFRL under contract FA8750-05-2-0283. The project is also partly support by ARO grant W911NF-08-1-0242. Most of the experiments were run on the Mastodon Cluster, provided by NSF Grant EIA-0303609. The first author also thanks the Vietnam Education Foundation (VEF) for its sponsorship.

References

1. Getoor, L., Taskar, B. (eds.): Statistical Relational Learning. MIT Press, Cambridge (2007)
2. Richardson, M., Domingos, P.: Markov logic networks. MLJ 62, 107–136 (2006)
3. Lowd, D., Domingos, P.: Efficient weight learning for Markov logic networks. In: Kok, J.N., Koronacki, J., Lopez de Mantaras, R., Matwin, S., Mladenič, D., Skowron, A. (eds.) PKDD 2007. LNCS (LNAI), vol. 4702, pp. 200–211. Springer, Heidelberg (2007)
4. Poon, H., Domingos, P.: Joint inference in information extraction. In: AAAI 2007, pp. 913–918 (2007)
5. Singla, P., Domingos, P.: Discriminative training of Markov logic networks. In: AAAI 2005, pp. 868–873 (2005)
6. Huynh, T.N., Mooney, R.J.: Discriminative structure and parameter learning for Markov logic networks. In: ICML 2008, pp. 416–423 (2008)
7. Cristianini, N., Shawe-Taylor, J.: An Introduction to Support Vector Machines and Other Kernel-based Learning Methods. Cambridge University Press, Cambridge (2000)
8. Joachims, T.: A support vector method for multivariate performance measures. In: ICML 2005, pp. 377–384 (2005)
9. Taskar, B., Guestrin, C., Koller, D.: Max-margin Markov networks. In: NIPS 2003 (2003)
10. Tsochantaridis, I., Joachims, T., Hofmann, T., Altun, Y.: Large margin methods for structured and interdependent output variables. JMLR 6, 1453–1484 (2005)
11. Joachims, T., Finley, T., Yu, C.N.: Cutting-plane training of structural SVMs. MLJ (2009), `http://www.springerlink.com/content/h557723w88185170`
12. Kautz, H., Selman, B., Jiang, Y.: A general stochastic approach to solving problems with hard and soft constraints. In: Dingzhu Gu, J.D., Pardalos, P. (eds.) The Satisfiability Problem: Theory and Applications, AMS, pp. 573–586 (1997)
13. Riedel, S.: Improving the accuracy and efficiency of MAP inference for Markov logic, UAI, 468–475 (2008)
14. Poon, H., Domingos, P.: Sound and efficient inference with probabilistic and deterministic dependencies. In: AAAI 2006, Boston, MA (July 2006)
15. Singla, P., Domingos, P.: Lifted first-order belief propagation. In: AAAI 2008, 1094–1099 (2008)
16. Collins, M.: Discriminative training methods for hidden Markov models: Theory and experiments with perceptron algorithms. In: CONLL 2002, pp. 1–8 (2002)
17. Domingos, P., Kok, S., Lowd, D., Poon, H., Richardson, M., Singla, P.: Markov logic. In: De Raedt, L., Frasconi, P., Kersting, K., Muggleton, S.H. (eds.) Probabilistic Inductive Logic Programming. LNCS (LNAI), vol. 4911, pp. 92–117. Springer, Heidelberg (2008)
18. Kok, S., Singla, P., Richardson, M., Domingos, P.: The Alchemy system for statistical relational AI. Technical report, Department of Computer Science and Engineering, University of Washington (2005), `http://www.cs.washington.edu/ai/alchemy`
19. Lafferty, J., McCallum, A., Pereira, F.: Conditional random fields: Probabilistic models for segmenting and labeling sequence data. In: ICML 2001, Williamstown, MA, pp. 282–289 (2001)
20. Collins, M.: Parameter estimation for statistical parsing models: Theory and practice of distribution-free methods. In: Harry Bunt, J.C., Satta, G. (eds.) New Developments in Parsing Technology. Kluwer, Dordrecht (2004)

21. Finley, T., Joachims, T.: Training structural SVMs when exact inference is intractable. In: ICML 2008, pp. 304–311 (2008)
22. Asano, T., Williamson, D.P.: Improved approximation algorithms for MAX SAT. J. of Algorithms 42(1), 173–202 (2002)
23. Asano, T.: An improved analysis of Goemans and Williamson's LP-relaxation for MAX SAT. Theoretical Computer Science 354(3), 339–353 (2006)
24. Boros, E., Hammer, P.L.: Pseudo-Boolean optimization. Discrete Applied Mathematics 123(1-3), 155–225 (2002)
25. Hinton, G.E.: Training products of experts by minimizing contrastive divergence. Neural Computation 14(8), 1771–1800 (2002)
26. Anguelov, D., Taskar, B., Chatalbashev, V., Koller, D., Gupta, D., Heitz, G., Ng, A.: Discriminative learning of Markov random fields for segmentation of 3D scan data. In: CVPR 2005, pp. 169–176 (2005)
27. Taskar, B., Lacoste-Julien, S., Jordan, M.I.: Structured prediction, dual extragradient and Bregman projections. JMLR 7, 1627–1653 (2006)
28. Collins, M., Globerson, A., Koo, T., Carreras, X., Bartlett, P.L.: Exponentiated gradient algorithms for conditional random fields and max-margin Markov networks. JMLR 9, 1775–1822 (2008)
29. Taskar, B., Abbeel, P., Koller, D.: Discriminative probabilistic models for relational data. In: UAI 2002, Edmonton, Canada, pp. 485–492 (2002)
30. Rückert, U., Kramer, S.: Margin-based first-order rule learning. MLJ 70(2-3), 189–206 (2008)
31. Szummer, M., Kohli, P., Hoiem, D.: Learning CRFs using graph cuts. In: Forsyth, D., Torr, P., Zisserman, A. (eds.) ECCV 2008, Part II. LNCS, vol. 5303, pp. 582–595. Springer, Heidelberg (2008)
32. Biba, M., Ferilli, S., Esposito, F.: Discriminative structure learning of Markov logic networks. In: Železný, F., Lavrač, N. (eds.) ILP 2008. LNCS (LNAI), vol. 5194, pp. 59–76. Springer, Heidelberg (2008)

Parameter-Free Hierarchical Co-clustering by n-Ary Splits

Dino Ienco, Ruggero G. Pensa, and Rosa Meo

University of Torino,
Department of Computer Science,
I-10149, Turin, Italy
{ienco,pensa,meo}@di.unito.it

Abstract. Clustering high-dimensional data is challenging. Classic metrics fail in identifying real similarities between objects. Moreover, the huge number of features makes the cluster interpretation hard. To tackle these problems, several co-clustering approaches have been proposed which try to compute a partition of objects and a partition of features simultaneously. Unfortunately, these approaches identify only a predefined number of flat co-clusters. Instead, it is useful if the clusters are arranged in a hierarchical fashion because the hierarchy provides insides on the clusters. In this paper we propose a novel hierarchical co-clustering, which builds two coupled hierarchies, one on the objects and one on features thus providing insights on both them. Our approach does not require a pre-specified number of clusters, and produces compact hierarchies because it makes $n-$ary splits, where n is automatically determined. We validate our approach on several high-dimensional datasets with state of the art competitors.

1 Introduction

Clustering is a popular data mining technique that enables to partition data into groups (clusters) in such a way that objects inside a group are similar to each other, and objects belonging to different groups are dissimilar [1]. When data are represented in a high-dimensional space, traditional clustering algorithms fail in finding an optimal partitioning because of the problem known as curse of dimensionality. Even though some distance metrics have been proposed to deal with high-dimensional data (e.g., cosine similarity) and feature selection tries to solve the problem by a reduction in the number of features, novel approaches emerged in the last years. One of the most appealing approach is co-clustering [2,3,4] whose solution provides contemporaneously a clustering of the objects and a clustering of the features. Furthermore, co-clustering algorithms are powerful because they exploit similarity measures on the clusters in one dimension of the problem in order to cluster the other dimension: that is, clusters of objects are evaluated by means of the clusters on the features and vice versa.

One of the classical aims of clustering is to provide a description of the data by means of an abstraction process. In many applications, the end-user is used to

W. Buntine et al. (Eds.): ECML PKDD 2009, Part I, LNAI 5781, pp. 580–595, 2009.

study natural phenomena by the relative proximity relationships existing among the analyzed objects. For instance, he/she compares animals by means of the relative similarity in terms of the common features w.r.t. a same referential example. Many hierarchical algorithms have the advantage that are able to produce a dendrogram which stores the history of the merge operations (or split) between clusters. As a result they produce a hierarchy of clusters and the relative position of clusters in this hierarchy is meaningful because it implicitly tells the user about the relative similarity between the cluster elements. This hierarchy is often immediately understandable: it constitutes a helpful conceptual tool to understand the inner, existing relationships among objects in the domain; it provides a visual representation of the clustering result and explains it. Furthermore, it provides a ready to use tool to organize the conceptual domain, to browse and search objects, discover their common features or differences, etc. It is a conceptual tool especially advisable if one cluster hierarchy - built on one dimension of the problem, the objects - gives insights to study the other dimension of the problem - the features - and gives information to produce the feature hierarchy. In this paper we propose a co-clustering algorithm that simultaneously produces a hierarchical organization in both the problem dimensions: the objects and the features. In many applications both hierarchies are extremely useful and are searched for: in text mining, for instance, documents are organized in categories grouping related documents. The resulting object hierarchy is useful because it gives a meaningful structure to the collection of documents. On the other side, keywords are organized in groups of synonyms or words with related meaning and this hierarchy provides a semantic network with meaningful insights on the relationships between keywords. In bioinformatics and in other applications, a similar discussion applies: genes or proteins are grouped in groups sharing a similar behavior while biological experiments by their related, involved functionalities. The key contributions of this paper are the following. We present a novel co-clustering algorithm, *HiCC*, that simultaneously produces two hierarchies of clusters: one on the objects and the other one on the features. As we will see with the aid of the experimental section, these hierarchies are meaningful to the end-user and are valid also under the viewpoint of several objective measures. Our algorithm is able to produce compact hierarchies because it produces $n-$ary splits in the hierarchy instead of the usual binary splits. This compactness improves the readability of the hierarchy. The third contribution is the adoption of an unusual cluster association measure in co-clustering: Goodman-Kruskal τ. It has been considered as a good choice in a comparative study on several evaluation function for co-clustering [5,6]. The last novelty is that our algorithm is parameter-less and it does not require the user to set a number of clusters for the two dimensions, which usually require a pre-processing stage and is not easy to set.

2 An Introductory Example to Co-clustering

Goodman-Kruskal τ [7] has been originally proposed as a measure of association between two categorical variables: it is a measure of proportional reduction in

	F_1	F_2	F_3
O_1	d_{11}	d_{12}	d_{13}
O_2	d_{21}	d_{22}	d_{23}
O_3	d_{31}	d_{32}	d_{33}
O_4	d_{41}	d_{42}	d_{43}

(a)

	CF_1	CF_2	
CO_1	t_{11}	t_{12}	T_{O_1}
CO_2	t_{21}	t_{22}	T_{O_2}
	T_{F_1}	T_{F_2}	T

(b)

Fig. 1. An example dataset (a) and a related co-clustering (b)

the prediction error of a dependent variable given information on an independent variable. We intend to use τ as measure of validation of a co-clustering solution that produces an association between two clusterings: the former on the values of the independent variable, the latter on the values of the dependent one.

We start our discussion by presenting a dataset example and one co-clustering on it. In Figure 1(a) we show a matrix whose rows represent the objects and the columns represent the features describing the objects themselves. d_{xy} is the frequency of the y-th feature in the x-th object.

Now consider table in Figure 1(b): it is the contingency table representing a co-clustering of the dataset of Figure 1(a). Symbol CO_i represents i-th cluster on objects while CF_j represents j-th cluster on features. T_{Oi} is the total counting for cluster CO_i, T_{Fj} is the total counting for cluster CF_j, and T is the global total. In this example we suppose CO_1 has aggregated objects O_1 and O_2 while CO_2 has aggregated objects O_3 and O_4. Similarly, CF_1 has aggregated features F_1 and F_2 while CF_2 contains only feature F_3. Co-cluster (CO_i, CF_j) is represented by the value t_{ij} stored in the cell at the intersection of the i-th object cluster and the j-th feature cluster. It has been computed by application of an aggregating function on features and on objects. The aggregation that produces co-cluster (CO_i, CF_j) is the following:

$$t_{ij} = \sum_{O_x \in CO_k} \sum_{F_y \in CF_l} d_{xy}$$

In order to review the meaning of τ for the evaluation of a co-clustering consider the table in Figure 1(b) whose cells at the intersection of the row CO_i with the column CF_j contain the frequencies of objects in cluster CO_i having the features in cluster CF_j.

$\tau_{CO|CF}$ determines the predictive power of the partitioning on features, *CF*, considered as an independent variable, for the prediction of the partitioning on objects, *CO*, considered as a dependent variable. The prediction power of *CF* is computed as a function of the error in the classification of *CO*.

The prediction error is first computed when we do not have any knowledge on the partitioning *CF*. E_{CO} denotes this error here. The reduction of this error allowed by *CF* is obtained by subtraction from E_{CO} of the error in the prediction of cluster in *CO* that we make when we have the knowledge of the cluster in *CF* (the independent variable) in any database example. $E_{CO}|CF$ denotes this latter error. The proportional reduction in prediction error of *CO* given *CF*, here called $\tau_{CO|CF}$, is computed by:

$$\tau_{CO|CF} = \frac{E_{CO} - E_{CO}|CF}{E_{CO}}$$

E_{CO} and $E_{CO}|CF$ are computed by a predictor which uses information from the cross-classification frequencies and tries to reduce as much as possible the prediction error. In the prediction, it also preserves the dependent variable distribution (relative frequencies of the predicted categories CO) in the following way: when no knowledge is given on CF, CO_i is predicted by the relative frequency T_{Oi}/T; otherwise, when CF_j is known for an example, CO_i is predicted with the relative frequency t_{ij}/T_{Fj}. Therefore, E_{CO} and $E_{CO}|CF$ are determined by:

$$E_{CO} = \sum_i \left(\frac{T - T_{Oi}}{T} \cdot T_{Oi}\right) \; ; E_{CO}|CF = \sum_i \sum_j \left(\frac{T_{Fj} - t_{ij}}{T_{Fj}} \cdot t_{ij}\right)$$

Analyzing the properties of τ, we can see that it satisfies many desirable properties for a measure of association between clusters. For instance, it is invariant by rows and columns permutation. Secondarily, it takes values between (0,1) (it is 0 iff there is independence between the object and feature clusters). Finally, it has an operational meaning: given an example, it is the relative reduction in the prediction error of the example's cluster in one of the problem dimensions, given the knowledge on the example's cluster in the other dimension, in a way that preserves the class distribution.

3 Original Algorithm

Before introducing our approach, we specify the notation used in this paper. Let $X = \{x_1, \ldots, x_m\}$ denote a set of m objects (rows) and $Y = \{y_1, \ldots, y_n\}$ denote a set of n features (columns). Let D denote a $m \times n$ data matrix built over X and Y, each element of D representing a frequency or a count. Let $R = \{r_1, \ldots, r_I\}$ denote a partition of rows of D and $C = \{c_1, \ldots, c_J\}$ denote a partition of columns of D. Given R and C, a contingency table is denoted by T, i.e., a $I \times J$ matrix such that each element $t_{ij} \in T$ ($i \in 1 \ldots I$ and $j \in 1 \ldots J$) is computed as specified in Section 2.

We build our algorithm on the basis of *τCoClust* [6], whose goal is to find a partition of rows R and a partition of columns C such that Goodman-Kruskal's $\tau_{C|R}$ and $\tau_{R|C}$ are optimized [5,8]. *τCoClust* is shown as algorithm 1.

In [8], the authors propose a heuristic to locally optimize the two coefficients. Given two co-clusterings matrices, T and T', the differential τ between T and T' is given by:

$$\Delta\tau_{C|R} = \tau_{C|R}(T) - \tau_{C|R}(T')$$

The local optimization strategy is sketched in Algorithm 2. For sake of brevity, we only present the optimization of the row partition. The optimization strategy of the column partition works in a similar way.

This algorithm is a stochastic optimization technique and it allows to obtain, starting from a given set of row clusters, another set of row clusters that optimizes the objective function. As first step, the algorithm chooses a cluster r_b randomly from the set of the initial row clusters R. Then it randomly picks an object x belonging to r_b. The original features of x are projected onto the current column clusters assignment. Then, the function tries to move x from the original cluster r_b to another cluster r_e s.t. $r_e \neq r_b$ and $r_e \in \{R \cup \emptyset\}$. Using function $\Delta_{\tau_{R|C}}(x, r_b, r_e, C)$, it verifies if there is an increment in the objective function as a result of this change. Then, it chooses cluster $r_{e_{min}}$ s.t. $r_{e_{min}} = \arg\min_{r_e} \Delta_{\tau_{R|C}}(x, r_b, r_e, C)$. Finally, the function updates the contingency table T and the row cluster partition (line 11). To perform this step, it first moves the object x from the cluster r_b to cluster $r_{e_{min}}$. Then it performs one or more of the following three actions:

1. if cluster r_b is empty after removing x, it deletes cluster r_b and updates the contingency table T consequently;
2. if cluster $r_{e_{min}}$ is the empty cluster, it adds a new cluster to the row partition and updates the contingency table T consequently;
3. if the two above mentioned cases do not apply, it simply updates the content of T in a consequent way.

Thanks to this strategy, the number of clusters may grow or decrease at each step, and their number only depends on the effective optimization of $\tau_{R|C}$. This makes this approach substantially different from the one described in [4] and in other state-of-the-art co-clustering approaches, where the number of co-clusters is fixed as a user-defined parameter. For a deep view on the efficient version of this algorithm see [8].

Algorithm 1. $\tau CoClust(D, N_{iter})$

```
Initialize R, C
T ← ContingencyTable(R, C, D)
while (t ≤ N_iter) do
    optimizeRowCluster(R, C, T)
    optimizeColumnCluster(R, C, T)
    t ← t + 1
end while
return (R, C)
```

4 Hierarchical Co-clustering

In this section we present our method, named HiCC (Hierarchical Co-Clustering by n-ary split). Let us introduce the notation.

4.1 Notations

Given the above described matrix D defined on the set of objects $X = \{x_1, \ldots, x_m\}$ and on the set of features $Y = \{y_1, \ldots, y_n\}$, the goal of our hierarchical

Algorithm 2. *optimizeRowCluster*(R, C, T)

$min_{\Delta\tau_{R|C}} = 0$
Randomly choose a cluster $r_b \in R$
Randomly choose an object $x \in r_b$
for all $r_e \in \{R \cup \emptyset\} s.t. r_b \neq r_e$ **do**
  if $(\Delta_{\tau_{R|C}}(x, r_b, r_e, C) < min_{\Delta\tau_{R|C}})$ **then**
    $e_{min} = e$
    $min_{\Delta\tau_{R|C}} = \Delta_{\tau_{R|C}}(x, r_b, r_e, C)$
  end if
end for
if $min_{\Delta\tau_{R|C}} \neq 0$ **then**
  Update R using $(x, r_b, r_{e_{min}})$ and modify T consequently
end if

Algorithm 3. *HiCC*(D, N_{iter})

$k \leftarrow 0$, $l \leftarrow 0$, $\mathcal{R} \leftarrow \emptyset$, $\mathcal{C} \leftarrow \emptyset$
$(R_k, C_l) \leftarrow \tau CoClust(D, N_{iter})$
$\mathcal{R} \leftarrow \mathcal{R} \cup R_k$, $\mathcal{C} \leftarrow \mathcal{C} \cup C_l$
while $(TERMINATION)$ **do**
  $R_{k+1} \leftarrow \emptyset$
  for all $r_{ki} \in R_k$ **do**
    $R'_i \leftarrow RandomSplit(r_{ki})$
    $t \leftarrow 0$
    $T_{ki} \leftarrow ContingencyTable(R'_i, C_l, D)$
    while $(t \leq N_{iter})$ **do**
      $optimizeRowCluster(R'_i, C_l, T_{ki})$
      $t \leftarrow t + 1$
    end while
    $R_{k+1} \leftarrow R_{k+1} \cup R'_i$
  end for
  $\mathcal{R} \leftarrow \mathcal{R} \cup R_{k+1}$
  $k \leftarrow k + 1$
  $C_{l+1} \leftarrow \emptyset$
  for all $c_{lj} \in C_l$ **do**
    $C'_j \leftarrow RandomSplit(c_{lj})$
    $t \leftarrow 0$
    $T_{lj} \leftarrow ContingencyTable(R_k, C'_j, D)$
    while $(t \leq N_{iter})$ **do**
      $optimizeColumnCluster(R_k, C'_j, T_{lj})$
      $t \leftarrow t + 1$
    end while
    $C_{l+1} \leftarrow C_{l+1} \cup C'_j$
  end for
  $\mathcal{C} \leftarrow \mathcal{C} \cup C_{l+1}$
  $l \leftarrow l + 1$
end while
return $(\mathcal{R}, \mathcal{C})$

co-clustering algorithm is to find a hierarchy of row clusters $\mathcal{R}$ over X, and a hierarchy of column clusters $\mathcal{C}$ over Y. Supposing that $\mathcal{R}$ has K levels, and that $\mathcal{C}$ has L levels, $\mathcal{R}$ and $\mathcal{C}$ are defined as $\mathcal{R} = \{R_1, \dots, R_K\}$ and $\mathcal{C} = \{C_1, \dots, C_L\}$ (where R_1 and C_1 are the clustering at the roots of the respective hierarchy). Each $R_k \in \mathcal{R}$ is a set of clusters denoted by $R_k = \{r_{k1}, \dots, r_{k|R_k|}\}$ where $|R_k|$ is the total number of clusters in R_k, $r_{ki} \subseteq X$, $\bigcup_i r_{ki} = X$ and $\forall i, j$ s.t. $i \neq j$ $r_{ki} \cap r_{kj} = \emptyset$. Similarly, $C_l = \{c_{l1}, \dots, c_{l|C_l|}\}$, and the other conditions hold for C_l too. Since $\mathcal{R}$ defines a hierarchy, each R_k must also satisfy the following conditions:

1. $\forall R_{k_1}, R_{k_2}$ s.t. $k_1 < k_2$, $|R_{k_1}| \leq |R_{k_2}|$;
2. $\forall R_k$ $(k > 1)$, $\forall r_{ki} \in R_k$, $\forall R_{k_0}$ $(k_0 < k)$, $\exists! r_{k_0 j \in R_{k_0}}$ s.t. $r_{ki} \subseteq r_{k_0 j}$;

Two similar conditions must hold for $\mathcal{C}$ too.

Our approach first computes the first level of the hierarchy (R_1 and C_1) using Algorithm 1. Then, it builds R_2 by fixing C_1 by optimization of $\tau_{R_2|C_1}$. In general, given a generic hierarchy level h, Algorithm 3 alternates the optimization of $\tau_{R_h|C_{h-1}}$ and $\tau_{C_h|R_h}$, finding the appropriate row cluster R_h and column cluster C_h, constrained by the two above mentioned conditions. We present now the details of our algorithm.

4.2 Algorithm Description

The whole procedure is presented in Algorithm 3. HiCC adopts a divisive strategy. At line 1 it initializes all our indices and structures. Using function $\tau CoClust$ (see Algorithm 1) it builds the first level of both hierarchies.

From line 4 to line 31 it builds the levels of the two hierarchies. From line 6 to line 13, for each row cluster $r_{ki} \in R_k$, the algorithm splits it into a new set of row clusters R'_i using *RandomSplit* function. This function first sets the cardinality R'_i randomly; then it randomly assigns each row in r_{ki} to a cluster of R'_i. Subsequently, it initializes a new contingency table T_{ki} related to the set R'_i of row clusters and to the set C_l of column clusters (we consider all the column clusters found at previous level). Without changing columns partition, the algorithm tries to optimize $\tau_{R'_i|C_l}$ using the *optimizeRowCluster* function (see Algorithm 2). It returns a new and optimized R'_i. After all r_{ki} have been processed, the algorithm adds the new level of the row hierarchy to $\mathcal{R}$ (line 16). Column clusters are then processed in the same way, using the row cluster assignment returned at the end of previous steps. In this way the two hierarchies grow until a TERMINATION condition is reached. The TERMINATION condition is satisfied when all leaves of the two hierarchies contain only one element. Obviously, some cluster may be split into singletons at higher levels than others. At the end, our algorithm returns both the hierarchies over rows ($\mathcal{R}$) and columns ($\mathcal{C}$).

As shown in [8,6], the local search strategy employed to update partitions, sometimes leads to some degradation of $\tau_{R|C}$ or $\tau_{C|R}$. This is due to the fact that an improvement on one partition may decrease the quality of the other one. The authors, however, showed that there is always a compensation effect such that, after an adequate number of iterations, the two coefficients become stable.

In our approach, a single cluster is partitioned while keeping the other partition fixed. This ensure that the corresponding coefficient increases at each iteration.

4.3 Complexity Discussion

HiCC complexity is influenced by three factors. The first one is the number of iterations. It influences the convergence of the algorithm. A deep study of the number of iterations is presented in [6]. The second factor is the number of row/column clusters in *optimizeRowCluster* and in *optimizeColCluster* functions. This number of clusters influences the swap step which tries to optimize the internal objective function moving one object from a cluster b to another cluster e. The number of possible different values that e can assume influences the speed of the algorithm, but this number varies during the optimization. The third factor is the depth of the two hierarchies, in particular the deeper one. This value influences the number of times the main loop from lines 4-31 of algorithm 3 is repeated. If we denote by N the number of iterations, by $\widetilde{c}$ the mean number of row/column clusters inside the optimization procedure, and by v the mean branching factor, we observe that the complexity of a split of r_{ki} and c_{lj} is equal as average to $O(N \times \widetilde{c})$. Each split is performed for each node of the two hierarchies except for the bottom level. The number of nodes in a tree with branching factor v is $\sum_{i=0}^{levels} v^i$ where $levels$ is the number of levels of the tree. We can expand this summation and we obtain $\frac{1-v^{1+levels}}{1-v}$. From the previous consideration we estimate the complexity of our approach $O\left(N \times \widetilde{c} \times \frac{1-v^{levels}}{1-v}\right)$. The worst case is verified when any split is binary and at each level $\widetilde{c} = n$, where n is the number of rows/columns. In this case, the complexity is $O(N \times (n-1) \times n)$, $(n-1)$ being the number of internal nodes in the hierarchy. In conclusion, in the worst case, the overall complexity is $O(Nn^2)$. In the experimental section, we show that our algorithm often splits into more than two clusters, thus reducing the number of internal nodes. Moreover, assumption $\widetilde{c} = n$ is very pessimistic. In general our algorithm runs in linear time with the number of iterations, and in subquadratic time with the number of objects/features. Notice also that in this work, the number of iterations is constant, but it could be adapted to the size of the set of objects, and then reduced at each split.

5 Experimental Validation

In this section we report on several experiments performed on real, high-dimensional, multi-class datasets. We compare our approach with Information-Theoretic Co-Clustering (ITCC) [4], a well-known co-clustering algorithm which minimizes the loss in mutual information. To evaluate our results, we use several objective performance parameters that measure the quality of clustering. Besides the precision of the hierarchical co-clustering we analyze also the hierarchies returned by our approach for both rows and columns. This analysis allows to emphasize the utility of the hierarchical structure w.r.t. standard flat

Table 1. Datasets characteristics

Dataset	n. instances	n. attributes	n. of classes
oh0	1003	3182	10
oh15	913	3100	10
tr11	414	6429	9
tr21	336	7902	6
re0	1504	2886	13
re1	1657	3758	25

co-clustering approaches. All experiments are performed on PC with a 2.6GHz Opteron processor, 4GB RAM, running Linux.

5.1 Datasets for Evaluation

To evaluate our results we use some of the datasets described in [9]. In particular we use:

- **oh0, oh15**: two samples from OHSUMED dataset. OHSUMED is a clinically-oriented MEDLINE subset of abstracts or titles from 270 medical journals over five-year period (1987-1991).
- **tr11, tr21**: two samples from TREC dataset. These data come from the Text REtrieval Conference archive.
- **re0, re1**: two samples from Reuters-21578 dataset. This dataset is widely used as test collection for text categorization research.

All datasets have more than five classes, which usually is a hard context for text categorization. The characteristics of datasets are shown in Table 1.

5.2 External Evaluation Measures

We evaluate the algorithm performance using three external validation indices. We denote by $\mathbf{C} = \{C_1 \dots C_J\}$ the partition built by the clustering algorithm on objects at a particular level, and by $\mathbf{P} = \{P_1 \dots P_I\}$ the partition inferred by the original classification. J and I are respectively the number of clusters $|\mathbf{C}|$ and the number of classes $|\mathbf{P}|$. We denote by n the total number of objects.

The first index is *Normalized Mutual Information* (NMI). NMI provides an information that is impartial with respect to the number of clusters [10]. It measures how clustering results share the information with the true class assignment. NMI is computed as the average mutual information between every pair of clusters and classes:

$$\mathbf{NMI} = \frac{\sum_{i=1}^{I} \sum_{j=1}^{J} x_{ij} \log \frac{n x_{ij}}{x_i x_j}}{\sqrt{\sum_{i=1}^{I} x_i \log \frac{x_i}{n} \sum_{j=1}^{J} x_j \log \frac{x_j}{n}}}$$

where x_{ij} is the cardinality of the set of objects that occur both in cluster C_j and in class P_i; x_j is the number of objects in cluster C_j; x_i is the number of objects in class P_i. Its values range between 0 and 1.

The second measure is *purity.* In order to compute purity each cluster is assigned to the majority class of the objects in the cluster. Then, the accuracy of this assignment is measured by counting the number of correctly assigned objects divided by the total number of objects n.

$$\mathbf{Purity}(\mathbf{C}, \mathbf{P}) = \frac{1}{n} \sum_{j} max_i |C_j \cap P_i|$$

The third measure is the adjusted Rand index [11]. Let a be the number of object pairs belonging to the same cluster in $\mathbf{C}$ and to the same class in $\mathbf{P}$. This metric captures the deviation of a from its expected value corresponding to the hypothetic value of a obtained when $\mathbf{C}$ and $\mathbf{P}$ are two random, independent partitions. The expected value of a denoted by $E[a]$ is computed as follows:

$$E[a] = \frac{\pi(C) \cdot \pi(P)}{n(n-1)/2}$$

where $\pi(C)$ and $\pi(P)$ denote respectively the number of object pairs from the same clusters in $\mathbf{C}$ and from the same class in $\mathbf{P}$. The maximum value for a is defined as:

$$max(a) = \frac{1}{2} (\pi(C) + \pi(P))$$

The agreement between $\mathbf{C}$ and $\mathbf{P}$ can be estimated by the adjusted rand index as follows:

$$AR(\mathbf{C}, \mathbf{P}) = \frac{a - E[a]}{max(a) - E[a]}$$

Notice that this index can take negative values, and when $AR(\mathbf{C}, \mathbf{P}) = 1$, we have identical partitions.

5.3 Comparison Results

In this section we evaluate HiCC performance w.r.t ITCC [4]. HiCC is not deterministic like other well-known clustering algorithms such as K-means or ITCC itself. At each run we can obtain similar, but not equal, hierarchies. For this reason we run HiCC 30 times over each dataset (setting the number of iterations to 50,000). From these row/column co-hierarchies we choose the run that better optimizes an internal evaluation function.

To obtain a measure of quality of HiCC, for each level i of the hierarchy on the rows we select the corresponding level of the hierarchy on the columns. These levels define a pair of partitions: the first partition comes from the row cluster hierarchy, and the second one from the columns cluster hierarchy. On the pair of partitions from level i, an evaluation function EF_i is computed on the basis of Goodman-Kruskal τ_S [12], which is a symmetrical version of τ [7]. In order to compute an overall measure for the co-clustering we compute the following weighted mean:

$$Goodness = \frac{\sum_{i=1} \alpha_i * EF_i}{\sum_{i=1} \alpha_i} \quad (1)$$

Table 2. Average values of External Indices

Dataset	NMI		Purity		Adjusted Rand Index	
	Avg.	std. dev.	Avg.	std. dev.	Avg.	std. dev.
oh0	0.5176	±0.0134	0.7768	±0.0250	0.1150	±0.0129
oh15	0.4223	±0.0103	0.6653	±0.0207	0.0790	±0.0080
tr11	0.4606	±0.0087	0.7510	±0.0145	0.1091	±0.0119
tr21	0.2387	±0.0120	0.8103	±0.0122	0.0323	±0.0045
re0	0.3588	±0.0273	0.7317	±0.0235	0.0381	±0.0140
re1	0.5005	±0.0267	0.7616	±0.0397	0.0714	±0.017

where α_i is the weight associated to i-th level of the hierarchy, and allows to specify the significance assigned to the i-th level w.r.t. to the other levels of the hierarchy.

(1) is a general formula for the evaluation the goodness of a run of our method. In this work we set $\alpha_i = 1/i$. Indeed, we give a heavy weight to the top level and the lowest weight to the last level. This choice is motivated by observation that in a hierarchical solution the clusters on a level depend on the clusters at previous level. If we start with a good clustering, then next levels are more likely to produce good results too.

To compare each level of our hierarchies with ITCC results, we need to fix a number of row/column clusters to set ITCC parameters. We recall that ITCC is flat and does not produce hierarchies. For this reason we plan our experiments in the following way. Since HiCC is not deterministic, each run may produce partitions of different cardinality at each level. For this reason, we need to select one specific run of HiCC. Using *Goodness* function with τ_S as evaluation function, we choose the best hierarchies from HiCC for each dataset. From these hierarchies, we obtain a set of pairs (#numberRowCluster,#numberColCluster), where each pair specifies the number of clusters in a level of the hierarchies. For each of these combinations, we run ITCC 50 times with (#numberRowCluster,#numberColCluster) as parameters, and average for each level the obtained results.

In Table 3 we show the experimental results. To obtain a single index value for each dataset we compute the previously proposed *Goodness* function for each of the three external validation indices. We can see that our approach is competitive w.r.t. ITCC. Notice however that, for a given number of desired co-clusters, ITCC tries to optimize globally its objective function, and each time it starts from scratch. On the contrary, our algorithm is constrained by previous level which helps to speed-up its execution.

We can notice that ITCC is not more accurate than our algorithm. To clarify this point we report the complete behavior of the two algorithms in an example. In Table 4 we report the value of the three indices for each level of the hierarchy obtained by HiCC for **re1**. In the same table we show the values obtained by ITCC using the same number of clusters (#numberRowCluster,#numberColCluster) discovered by HiCC. We also report the standard deviation for ITCC, since for each pair of cluster numbers, we run it 50 times. We can see that HiCC outperforms ITCC, especially at the higher levels (first, second and third) of the row hierarchy.

Table 3. Comparison between ITCC and HiCC with the same number of cluster for each level

Dataset	ITCC			HiCC		
	NMI	Purity	Adj. Rand Index	NMI	Purity	Adj. Rand Index
oh0	0.4311	0.5705	0.1890	0.4607	0.5748	0.1558
oh15	0.3238	0.4628	0.1397	0.3337	0.4710	0.1297
tr11	0.3861	0.5959	0.1526	0.3949	0.6028	0.1325
tr21	0.1526	0.7291	0.0245	0.1277	0.7332	0.0426
re0	0.2820	0.5695	0.0913	0.2175	0.5388	0.0472
re1	0.3252	0.4957	0.0832	0.3849	0.5513	0.1261

Table 4. Complete view of performance parameters for **re1**

		ITCC			HiCC		
RowClust	ColClust	NMI	Purity	Adj. Rand	NMI	Purity	Adj. Rand
3	3	0.1807±0.0519	0.3282±0.0448	0.1028±0.0466	0.3290	0.4255	0.2055
6	6	0.2790±0.0462	0.3991±0.0393	0.1723±0.0530	0.2914	0.4255	0.1828
15	257	0.2909±0.0178	0.4357±0.0226	0.1358±0.0183	0.3098	0.4472	0.1555
330	1291	0.3499±0.0028	0.4013±0.0059	0.0021±0.0003	0.3950	0.51	0.0291
812	2455	0.4857±0.0019	0.5930±0.0056	0.0013±0.0002	0.4810	0.6530	0.0031
1293	3270	0.5525±0.0013	0.8284±0.0041	0.0008±0.0001	0.5517	0.8461	0.0009
1575	3629	0.5864±0.0006	0.9602±0.0020	0.0005±0	0.5854	0.9638	0.0001
1646	3745	0.5944±0.0002	0.9926±0.0008	0.0004±0	0.5940	0.9952	0
1657	3757	0.5951±0	1±0	0±0	0.5952	1	0
1657	3758	0.5951±0	1±0	0±0	0.5952	1	0

We notice also that *purity* index in HiCC always increases monotonically. This experiment shows that, when we explore deeper levels of the hierarchy, the confusion inside each cluster decreases. For sake of brevity we can only show one example, but in all the other experiments we observe the same trend.

In Table 5 we report the average Goodness of τ_S for each dataset, and the related standard deviation. We observe that standard deviation is very low w.r.t. the average Goodness. From this empirical evaluation we can conclude that the algorithm is quite stable.

In Table 2 we show the average values of the three external indices for each dataset. To perform this analysis we first compute the average of NMI, Purity and Adjusted Rand Index between all levels of the row hierarchy and for each run of the algorithm. Then we compute the average and the standard deviation over all 30 runs.

Finally, we report in Table 5 the average time needed to generate the two hierarchies for each dataset. The largest dataset is **re1**, and for this dataset HiCC takes about 4 hours to complete a run. In the same table we show the mean depth of the row hierarchy and of the column hierarchy for each dataset. We observe that the standard deviation is low. This points out that our algorithm is stable also from this point of view. Notice that HiCC has the ability to generate hierarchies which are not deep. Shorter hierarchies are preferable to hierarchies obtained only by binary splits, since they allow to identify the natural partitioning of the data, and improve the exploration of the results.

Table 5. Average τ_S, mean depth of hierarchies and mean time

Dataset	Goodness	Row Hier. Depth	Col. Hier. Depth	Avg. Time
oh0	0.2680±0.0063	9.37±0.71	9.83±0.64	6647.99 sec
oh15	0.2783±0.0012	10.63±0.71	11.23±0.62	5866.92 sec
tr11	0.2294±0.0032	10.27±0.63	15.63±1.99	5472.37 sec
tr21	0.3003±0.0012	10.87±0.67	14.4±0.99	5493.98 sec
re0	0.2350±0.0166	10.7±0.74	10.6±0.88	9359.04 sec
re1	0.1697±0.0112	8.8±0.65	9.2±0.65	14887.18 sec

Table 6. Row hierarchy of oh15

Enzyme-Activation	Enzyme-Activation	Enzyme-Activation
		Enzyme-Activation
	Cell-Movement	Cell-Movement
		Adenosine-Diphosphate
Staphylococcal-Infections	Uremia	Uremia
		Staphylococcal-Infections
	Staphylococcal-Infections	Staphylococcal-Infections
		Memory

Table 7. Column hierarchy of re1

oil, compani, opec, gold, ga, barrel, strike, mine, lt, explor		tonne, wheate, sugar, corn, mln, crop, grain, agricultur, usda, soybean		coffee, buffer, cocoa, deleg, consum, ico, stock, quota, icco, produc	
oil, opec, tax, price, dlr, crude, bank, industri, energi, saudi	compani, gold, mine, barrel, strike, ga, lt, ounce, ship, explor	tonne, wheate, sugar, corn, grain, crop, agricultur, usda, soybean, soviet	mln, export, farm, ec, import, market, total, sale, trader, trade	quota, stock, produc, meet, intern, talk, bag, agreem, negoti, brazil	coffee, deleg, buffer, cocoa, consum, ico, icco, pact, council, rubber

5.4 Inspecting Hierarchies

In this section we analyze in a qualitative way two hierarchies built by our algorithm. In particular we analyze the row hierarchy for **oh0** data and the column hierarchy from **re1** data. In Table 6 we report the first three levels of the row hierarchy produced by HiCC. To obtain this hierarchy we assigned at each cluster a label. We chose the label of the majority class in the cluster. Each column represents a level of the hierarchy and each cell of the table represents a single cluster. Cluster *Enzyme-Activation* at first level is split in two clusters: the first one has the same label, the second is *Cell-movement*, a topic more related to *Enzyme-Activation* than *Staphylococcal-Infections.* Then *Cell-movement* is split into two clusters. One of these is *Adenosine-Diphosphate*, which is known to be correlated with the topic *Cell-movement.* In the other branch of the hierarchy we notice that *Uremia* cluster is a child of *Staphylococcal-Infections*(a renal failure with bacterial infection as one of its causes). In Table 7 we report the first two levels of the column hierarchy produced by HiCC. For each cluster, we computed the mutual information for each word, and ranked the set of words, as described in [13]. Then we selected the 10-top words w.r.t. the mutual information for each cluster. At the first level each of the three clusters is about a well distinct topic: the first one is about *mineral*, the second one is on *agriculture* and the third one is on *coffee and cocoa*. From *mineral* cluster, two clusters are produced: *oil*, and *gold mineral*. Cluster on *agriculture* is split into two clusters: *agricultural products* and *buying and selling in agriculture*. Finally, we observe that *coffee and cocoa* cluster is split in a similar way.

6 Related Work

One of the earliest co-clustering formulations was introduced by Hartigan [2]. This algorithm begins with the entire data in a single block and then at each stage finds the row or column split of every block into two pieces, choosing the one that produces largest reduction in the total within block variance. The splitting is continued till the reduction of within block variance due to further splitting is less than a given threshold. This approach is clearly hierarchical, but it does not build any cluster hierarchy. Moreover, it does not optimize any global objective function. Kluger et al. [3] propose a spectral co-clustering method. First, they perform an adequate normalization of the data set to accentuate co-clusters if they exist. Then, they consider that the correlation between two columns is better estimated by the expression level mean of each column w.r.t. a partition of the rows. The bipartition is computed by the algebraic eigenvalue decomposition of the normalized matrix. Their algorithm critically depends on the normalization procedure. Dhillon et al. [4] and Robardet et al. [8] have considered the two searched partitions as discrete random variables whose association must be maximized. Different measures can be used. Whereas COCLUSTER [4] uses the loss in mutual information, BI-CLUST [8] uses Goodman-Kruskal's τ coefficient to evaluate the link strength between the two variables. In both algorithms, a local optimization method is used to optimize the measure by alternatively changing a partition when the other one is fixed. The main difference between these two approaches is that the τ measure is independent of the number of co-clusters and thus BI-CLUST can automatically determine the number of co-clusters. Another co-clustering formulation was presented in [14]. Authors propose two different residue measure, and introduce their co-clustering algorithm which optimizes the sum-squared residues function. Recently, Banerjee et al. have proposed in [15] a co-clustering setting based on matrix approximation. The approximation error is measured using a large class of loss functions called Bregman divergences. They introduce a meta-algorithm whose special cases include the algorithms from [4] and [14]. Another recent and significant theoretical result has been presented in [16]. The authors show that the co-clustering problem is NP-hard, and propose a constant-factor approximation algorithm for any norm-based objective functions.

To the best of our knowledge, our approach is the first one that performs a simultaneous hierarchical co-clustering on both dimensions, and that returns two coupled hierarchies. However, in recent literature, several approaches have been proposed that could be related to our work, even though they do not produce the same type of results. In [17] a hierarchical co-clustering for queries and URLs of a search engine log is introduced. This method first constructs a bipartite graph for queries and visited URLs, and then all queries and related URLs are projected in a reduced dimensional space by applying singular value decomposition. Finally, all connected components are iteratively clustered using *k-means* for constructing hierarchical categorization. In [18], the authors propose a hierarchical, model-based co-clustering framework that views a binary dataset as a joint probability distribution over row and column variables. Their approach starts by clustering tuples in a dataset, where each cluster is characterized by a different probability distribution. Then, the conditional distribution of attributes over tuples is exploited to discover natural co-clusters in the data. This method does not construct any coupled hierarchy, moreover, co-cluster are identified in a separate step, only after the set of tuple has been partitioned. In [19], a method is proposed that construct two hierarchies on gene expression data, but they are not generated simultaneously. In our approach, levels of the two hierarchies are alternately generated, so that each level of both hierarchies identifies a strongly related set of co-clusters of the matrix.

7 Conclusion

Quality of flat clustering solutions in high-dimensional data results often degraded. In this paper we have proposed a co-clustering approach. HiCC is a novel hierarchical algorithm, which builds two coupled hierarchies, one on the objects and one on features thus providing insights on both them. Hierarchies are high quality. We have validated them by objectives functions like NMI, Purity and Adjusted Rand Index on many high-dimensional datasets. In addition, HiCC has other benefits: it is parameter-less; it does not require a pre-specified number of clusters, produces compact hierarchies because it makes $n-$ary splits, with n automatically determined.

References

1. Han, J., Kamber, M.: Data Mining: Concepts and Techniques. The Morgan Kaufmann Series in Data Management Systems. Morgan Kaufmann, San Francisco (2000)
2. Hartigan, J.A.: Direct clustering of a data matrix. Journal of the American Statistical Association 67(337), 123–129 (1972)
3. Kluger, Y., Basri, R., Chang, J., Gerstein, M.: Spectral biclustering of microarray data: coclustering genes and conditions. Genome Research 13, 703–716 (2003)
4. Dhillon, I.S., Mallela, S., Modha, D.S.: Information-theoretic co-clustering. In: Proc. ACM SIGKDD 2003, Washington, USA, pp. 89–98. ACM, New York (2003)
5. Robardet, C., Feschet, F.: Comparison of three objective functions for conceptual clustering. In: Siebes, A., De Raedt, L. (eds.) PKDD 2001. LNCS (LNAI), vol. 2168, pp. 399–410. Springer, Heidelberg (2001)
6. Robardet, C.: Contribution à la classification non supervisée: proposition d'une methode de bi-partitionnement. PhD thesis, Université Claude Bernard - Lyon 1 (Juliet 2002)

7. Goodman, L.A., Kruskal, W.H.: Measures of association for cross classification. Journal of the American Statistical Association 49, 732–764 (1954)
8. Robardet, C., Feschet, F.: Efficient local search in conceptual clustering. In: Jantke, K.P., Shinohara, A. (eds.) DS 2001. LNCS (LNAI), vol. 2226, pp. 323–335. Springer, Heidelberg (2001)
9. Forman, G.: An extensive empirical study of feature selection metrics for text classification. Journal of Machine Learning Research 3, 1289–1305 (2003)
10. Strehl, A., Ghosh, J.: Cluster ensembles — a knowledge reuse framework for combining multiple partitions. Journal of Machine Learning Research 3, 583–617 (2002)
11. Hubert, L., Arabie, P.: Comparing partitions. Journal of Classification 2, 193–218 (1985)
12. Goodman, L.A., Kruskal, W.H.: Measure of association for cross classification ii: further discussion and references. Journal of the American Statistical Association 54, 123–163 (1959)
13. Slonim, N., Tishby, N.: Document clustering using word clusters via the information bottleneck method. In: Proc. SIGIR 2000, New York, NY, USA, pp. 208–215 (2000)
14. Cho, H., Dhillon, I.S., Guan, Y., Sra, S.: Minimum sum-squared residue co-clustering of gene expression data. In: Proc. SIAM SDM 2004, Lake Buena Vista, USA (2004)
15. Banerjee, A., Dhillon, I., Ghosh, J., Merugu, S., Modha, D.S.: A generalized maximum entropy approach to bregman co-clustering and matrix approximation. JMLR 8, 1919–1986 (2007)
16. Anagnostopoulos, A., Dasgupta, A., Kumar, R.: Approximation algorithms for co-clustering. In: Proc. PODS 2008, Vancouver, BC, Canada, pp. 201–210 (2008)
17. Hosseini, M., Abolhassani, H.: Hierarchical co-clustering for web queries and selected urls. In: Benatallah, B., Casati, F., Georgakopoulos, D., Bartolini, C., Sadiq, W., Godart, C. (eds.) WISE 2007. LNCS, vol. 4831, pp. 653–662. Springer, Heidelberg (2007)
18. Costa, G., Manco, G., Ortale, R.: A hierarchical model-based approach to co-clustering high-dimensional data. In: Proc. of ACM SAC 2008, Fortaleza, Ceara, Brazil, pp. 886–890 (2008)
19. Heard, N.A., Holmes, C.C., Stephens, D.A., Hand, D.J., Dimopoulos, G.: Bayesian coclustering of anopheles gene expression time series: Study of immune defense response to multiple experimental challenges. Proc. Natl. Acad. Sci. (102), 16939–16944

Mining Peculiar Compositions of Frequent Substrings from Sparse Text Data Using Background Texts

Daisuke Ikeda and Einoshin Suzuki

Department of Informatics, Kyushu University
Motooka 744, Fukuoka 819-0395, Japan
{daisuke,suzuki}@inf.kyushu-u.ac.jp

Abstract. We consider mining unusual patterns from text T. Unlike existing methods which assume probabilistic models and use simple estimation methods, we employ a set B of *background* text in addition to T and *compositions* $w = xy$ of x and y as patterns. A string w is *peculiar* if there exist x and y such that $w = xy$, each of x and y is more frequent in B than in T, and conversely $w = xy$ is more frequent in T. The frequency of xy in T is very small since x and y are infrequent in T, but xy is relatively abundant in T compared to xy in B. Despite these complex conditions for peculiar compositions, we develop a fast algorithm to find peculiar compositions using the suffix tree. Experiments using DNA sequences show scalability of our algorithm due to our pruning techniques and the superiority of the concept of the peculiar composition.

1 Introduction

As text data, such as Web documents and genome sequences, are becoming abundant in various areas, it is becoming more important to analyze text data and to extract useful knowledge from it. Hence, an increasing attention has been paid to text mining.

Exceptionality has attracted attention of researchers in various areas as an essential property of discovered knowledge, since a discovery is often inspired by unusual events which can not be explained by the current theory. In the field of data mining, the term "unexpected" has been included as a mandatory property of the target patterns [1] from the beginning of the field and various methods have been proposed for finding exceptions, such as outliers [2], rules [3], and other types of patterns [4]. Unusualness for time series data has been also studied [5,6]. For text data, finding unusual text patterns have been studied extensively, especially in bioinformatics [7,8,9,10].

For the definition of being *unusual*, it is natural to define *usual* states and to measure unusualness by deviation from the states. In the z-score, which is a popular measure for unusual patterns in text data [8,9,10], the popularity of a usual state for a pattern w is an estimated expectation for w based on a given probability model. More formally, the z-score $z(w)$ for w is defined

$$z(w) = \frac{f(w) - E(w)}{N(w)},$$

where $f(w) > 0$ denotes the observed frequency, $E(w)$ its expectation and $N(w)$ a normalization factor. Given a threshold α, w is unusual if $z(w) > \alpha$ or $z(w) < -\alpha$. In other

W. Buntine et al. (Eds.): ECML PKDD 2009, Part I, LNAI 5781, pp. 596–611, 2009.

words, an unusual pattern based on the z-score is a pattern whose frequency deviates far from its estimate under the given probability model.

However, scores to measure unusualness borrowed from statistical testing, such as the z-score, have the following problems: (1) they require an appropriate probabilistic model in advance, (2) discovered unusual patterns lack of clear interpretations and (3) an estimation based on the probabilistic model is inappropriate for sparse text data.

Firstly, such a score requires an appropriate probabilistic model. A simple model is easy to compute the corresponding score but it can not describe details of given data, while a sophisticated model well describes the given data but it is computationally costly to obtain parameters for the model. When we assume the Bernoulli model, which is a simple model, we obtain the ζ-score as a variant of the z-score [7,8], where the expectation $\hat{p}(w)$ for a pattern w is given by $\hat{p}(w) = \prod p(a)$, which is the product of all probabilities $p(a)$ for letters a in w. In this model, each letter is supposed to occur independently. The Markov model is also considered for the probabilistic model of the z-score [10]: the probability of the ith letter a_i depends on the probability of the previous k letters and therefore $\hat{p}(w)$ is estimated by a conditional probability $p(a_i|a_{i-1} \cdots a_{i-k+1})$. However, the value of k is fixed and must be given in advance, and it is difficult to decide an appropriate value for k automatically.

Secondly, it is difficult for us to understand the meaning of patterns obtained by a score based on statistical testing. Such a pattern is evaluated from only the view point of the statistics. Therefore, we only know that the frequency of the pattern is rare or abundant against its expectation but the frequency does not tell us about its meaning.

Finally, the estimation based on simple probabilistic models is inappropriate to find unusual patterns from sparse text data because the models use short sub-patterns to estimate longer patterns' probabilities. Estimating probabilities for long patterns is critical to find unusual patterns since unusual patterns must be long from the definition of being unusual. However, the simple probabilistic models provide inaccurate estimates to all long patterns as many long patterns do not appear in practical sparse data. Thus, we need a better estimation method for sparse text data.

To overcome these problems, first, we introduce a *background set* in addition to a target set of text data. From the background set, we find sub-patterns which compose unusual patterns. In this sense, the background set is used as a mother population. In practical usage, it is natural to compare a given data with other data, instead of comparing with the population. For example, to examine DNA sequences of a species, the same or similar subsequences of other, well-known species can be a good hint.

Next, we introduce a new form of a pattern, called a *composition*, which is defined as the concatenation of two frequent substrings x and y. The lengths of x and y are not restricted and hence they can be long. Thus, we can expect x and y help us to understand implications of xy because x and y are enough long and frequent. For example, three nucleotides in genome sequences compose a codon and thus we might try to understand a sequence of nucleotides using the codon table in a bottom up manner.

Finally, we define a composition to be *peculiar* using two ratios of frequencies for both target and background sets. That is, $w = xy$ is peculiar if both x and y are more frequent in B than in T and conversely their composition xy is more frequent in T than in B. Therefore, the frequency of w is small in the target set since both x and y are infrequent in T, and the frequency of w is larger than its expectation of $f(x)f(y)$ in B. In this sense, peculiar compositions are unusual. Hence, we only need to provide simple parameters to a data mining algorithm instead of a complicated probabilistic model.

Our problem definition raises a new challenging problem: we have to find frequent components as well as infrequent unusual patterns. Therefore, it is computationally hard to find peculiar compositions, compared to just finding substrings. We develop an algorithm, *FPCS* (Finding Peculiar Compositions), which exploits the generalized suffix tree and find peculiar compositions in $O(N^2)$ time, where N is the total length of input strings. The generalized suffix tree enables FPCS to find simultaneous discovery of a peculiar composition xy, its prefix x, and suffix y.

This paper is organized as follows: related work is surveyed in section 2. We define the problem in section 3 and then propose our main algorithm FPCS in section 4. Section 5 is devoted for experimental evaluation. Section 6 concludes.

2 Related Work

The z-score has attracted attention of bioinformatics researchers as a measure to find unusual patterns in strings. For a threshold α, if $z(w) > \alpha$ (resp. $z(w) < -\alpha$) then w is said to be *overrepresented* (resp. *underrepresented*). In [7,8], the Bernoulli model is assumed as a probabilistic model in estimating the expectations and the Markov model or the so called n-gram model is considered in [10]. The χ^2-score is also used to measure interestingness [9].

Many supervised learning algorithms, e.g., those for formal languages, and text mining algorithms [11,12,13,14] also use another set of text data in addition to a target set. However, they discover a pattern which satisfies many examples in a set but few in the other set. Therefore, patterns output by them are usual.

Indexing scores in information retrieval also consider a background set. For example, TF/IDF marks a high score to a word which appears frequently in a document but not in the background set. This criteria is similar to our notion of being peculiar. However, indexing scores are not for compositions of words but for single words, and the target set for TF/IDF is a single document not a set of documents like in our setting.

The concept of the peculiar composition borrows its essential idea from the simultaneous discovery of exceptional rules [15,16,17,18]. It tries to discover a set of rule pairs each of which corresponds to $Y \rightarrow x = v$ and $YZ \rightarrow x = v'$, where each of Y and Z represents a conjunction of "attribute = value"s, x is an attribute, v and v' are different values, and YZ is the conjunction of Y and Z. [15,16,18] assume that $Z \rightarrow x = v'$ does not hold and in this case a rule pair corresponds to a situation that an ensemble of two conditions result in an atypical result as shown in Fig. 1. Discovered rule pairs are shown to be valid, useful, novel, and unexpected in a medical application [18]. Our work can be considered as a

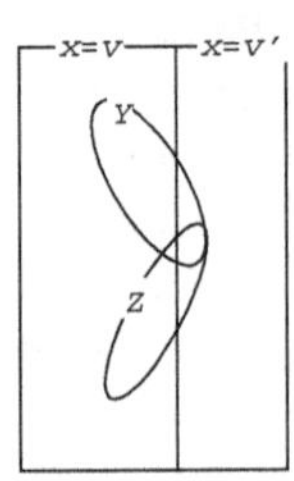

Fig. 1. Venn diagram of a discovered pattern in the simultaneous discovery of exception rules

text mining version of the simultaneous discovery of exception rules. It is difficult for algorithms in [15,16,17,18] to discover long patterns because the algorithms generate candidate patterns and then check their frequencies. One of our contributions lies in FPCS which exploits the generalized suffix tree in a sophisticated manner and hence the time complexity of FPCS does not depend on the lengths of x and y in discovered patterns.

3 Peculiar Composition Discovery Problem

We first define mandatory notions for dealing string data, introduce the peculiar composition, and its discovery problem.

Let Σ be a finite set of *characters*. We call Σ an *alphabet*. The set of all the finite sequences of zero or more characters is denoted by Σ^*. An element of Σ^* is called a *string*. We denote the length of a string x by $|x|$. The empty string, which is a string of zero character, is denoted by ε. The set of sequences each of which is composed of at least one character is denoted by Σ^+ i.e. $\Sigma^+ = \Sigma^* - \{\varepsilon\}$.

For strings $x, y \in \Sigma^+$, the concatenation of x and y is denoted by xy. We call xy the *composition* of x and y. For instance, if $x =$ "every" and $y =$ "thing" then $xy =$ "everything". Conversely, a pair of two strings (x, y) is called a *division* of w if $w = xy$.

For a string $x = a_1 \cdots a_n$ ($a_i \in \Sigma$), if there exist $u, v, w \in \Sigma^*$ such that $x = uvw$ then u (resp. v and w) is a *prefix* (resp. *substring* and *suffix*) of x.[1] For instance, if $uvw =$ "amazingly" then u="amaz" is a prefix, v="ing" is a substring, and w="ly" is a suffix. The notion of suffix will be crucial in introducing data structures in section 4.2. An *occurrence* of v in x is a positive integer i such that $a_i \cdots a_{i+|v|-1} = v$. The occurrence of "ing" in "amazingly" is 5, and "a" has two occurrences: 1 and 3. The *frequency* of v in x is the number of occurrences of v in x.

For $x, y \in \Sigma^*$, we consider that the frequency of x in y represents a kind of popularity of x in y. To treat this notion for a set D of strings, we use $\mathrm{f}(x|D)$ to denote the add-sum of the frequencies of x in all strings in D. Since the frequency is affected by the absolute size for D, we use relative frequencies or empirical probabilities $\mathrm{P}(x|D) = \mathrm{f}(x|D)/\#_D$, where $\#_D$ is the add-sum of frequencies of all substrings in D. For examplle, since AG appears 4 times in $D = \{CTAGAG, CTAGCTAG\}$ and $\#_D = 6 \cdot 7/2 + 8 \cdot 9/2 = 57$, $\mathrm{f}(AG|D) = 4$ and $\mathrm{P}(AG|D) = 4/57$.

We assume two sets T and B of strings, and we call T the *target* set and B the *background* set. Given a threshold $\theta > 1$, x is *contrastive* w.r.t. θ in target (resp. background) if $\mathrm{P}(x|T) > \theta\mathrm{P}(x|B)$ (resp. $\mathrm{P}(x|B) > \theta\mathrm{P}(x|T)$). θ represents the minimum ratio of the probability of x in one document set to that in the other set.[2] If it is clear from the context, we omit the threshold and simply say that x is contrastive in target or background. For instance, if T and B represent the sets of e-mails of A and B, respectively, and only A lives in Kyushu then "Kyushu" is likely to be contrastive w.r.t a relatively large θ.

Let $x, y \in \Sigma^+$. Given θ_T and θ_B, a composition xy is said to be *peculiar* in T against B if xy is contrastive in target w.r.t θ_T, and both x and y are contrastive in background

[1] A prefix or suffix is a substring because u or w can be the empty string.

[2] An appropriate value of θ in practice is decided by a trial and error process.

w.r.t θ_B. If it is clear from the context, we omit the string sets and simply call xy a peculiar composition.

In addition to the notion of being contrastive defined by the proportion of the relative frequencies, we also use the minimum support which is a minimum threshold for frequencies. For a set D of strings, we denote the minimum support by η_D, and we say that a string x is η_D-frequent in D if x appears more than η_D times in D. η_D represents the minimum support for the frequency of peculiar compositions.

If "Kyushu University" is a peculiar composition then its prefixes, such as "Kyushu Universi", are likely to be peculiar. These prefixes increase the number of patterns discovered by mining algorithms, but can be considered as irrelevant. Several pattern discovery algorithms discover only closed patterns which are maximal among equivalent patterns [19,20]. We define that two composition xy and $x'y'$ ($|xy| > |x'y'|$) are *equivalent* if $x = x'$, y' is a prefix of y, and $\mathrm{f}(y|D) = \mathrm{f}(y'|D)$ ($D = T, B$). We say that xy is *maximal* of equivalent compositions if xy is longest in the equivalent compositions.

The peculiar composition discovery problem is formally defined as follows:

Definition 1. *The peculiar composition discovery problem is, given two sets T and B of strings and threshold values θ_T, θ_B and η_T, to find all maximal, η_T-frequent, peculiar compositions in T against B.*

Example 1. Let $B = \{CTAGAG, CTAGCTAG\}$ and $T = \{AGCT, AGCT, AAAAAAA\}$. While AG and CT are popular but $AGCT$ is rare in B, $AGCT$ is popular in T. Thus $AGCT$ is a peculiar composition of AG and CT for $\theta_T = 1.2$, $\theta_B = 2.3$ and $\eta_T = 2$ because $\#_T = 48$, $\#_B = 57$, $\mathrm{f}(AGCT|T) = 2$ and $\mathrm{f}(AG|B) = \mathrm{f}(CT|B) = 3$.

It is easy to solve the problem if we neglect time-efficiency or if we limit the maximal length of the strings in the discovered patterns to a trivially-small number. A straightforward solution would be to count substrings and check if x, y and xy are contrastive. The number of possible substrings is $O(N^2)$, where N is the total length of input strings. For a substring of the form xy, we have to check $O(|xy|)$ compositions since there exist $O(|xy|)$ divisions. For each string w (or composition), $O(|w|)$ time is required to check if a string is contrastive. Thus the time complexity of this naive algorithm is $O(N^2 \times N \times N) = O(N^4)$. Certainly this solution is prohibitive for real problems.

The suffix tree [21,22], which is a popular data structure for text data, reduces this time complexity. Although there exist $O(N^2)$ possible substrings, we only need to check $O(N)$ substrings which correspond to nodes in a suffix tree, and it is done in constant time to check if a string is contrastive by storing frequencies on nodes of the tree. Thus, the time complexity is reduced to $O(N^2)$.

However, this is larger than the time complexity to compute z-score, which is calculated in $O(N)$ time [7]. The main difficulty arises from the fact that being contrastive does not satisfy anti-monotonicity because our problem is defined over two string sets instead of a single set. For a string w, there exist $O(|w|)$ divisions (x_i, y_i), where $w = x_i y_i$. Even if x_i, y_i and $x_i y_i$ are contrastive for some division (x_i, y_i), we do not conclude that, for another division (x'_i, y'_i), x'_i, y'_i and $x'_i y'_i$ are contrastive. Thus we need to check all divisions.

4 Our Algorithm FPCS

In this section, we present our main algorithm *FPCS* (Finding Peculiar Compositions).

4.1 Overview of FPCS

FPCS exploits the generalized suffix tree, which is a well-known data structure for text data. First FPCS constructs the generalized suffix tree for all input strings. A node of the tree corresponds to a substring, called a *branching string*, of the given strings. Then FPCS counts the frequencies for branching strings in both background and target sets simultaneously. As the result, the algorithm finds all nodes such that their branching strings are contrastive in target and, for such a node, there exists an ancestor node whose branching string is contrastive in background. Such a branching string xy is a candidate for a peculiar composition because xy is contrastive in target and x is contrastive in background. Finally FPCS checks if y is contrastive in background for all possible divisions (x, y) of xy efficiently using suffix links, which are pointers to nodes of the generalized suffix tree.

4.2 Generalized Suffix Tree

In this section, we briefly explain the generalized suffix tree used in FPCS. The generalized suffix tree is a suffix tree for a set of strings. The tree can be used to count all the frequencies of substrings in $O(N)$, where N is the total length of input strings. First we explain the suffix tree [21,22] for a single string and then extend it to the generalized suffix tree [23].

Let \$ be a special character such that $\$ \notin \Sigma$. For a string $x \in \Sigma^*$, $A = x\$$ and an integer p ($1 \leq p \leq |A|$), A_p denotes A's suffix starting at the pth character of A. The special character \$ is used to guarantee any suffix of x is not a prefix of another suffix, and hence A_p is uniquely identified.

A trie for strings is the tree in which there is one node for every common prefix. A suffix trie for x is the trie for $A_{p_1}, A_{p_2}, \ldots, A_{p_n}$, where $A_{p_1}, A_{p_2}, \ldots, A_{p_n}$ are all suffixes of A in lexicographic order. The *suffix tree* for x is the compact trie for $A_{p_1}, A_{p_2}, \ldots, A_{p_n}$, where all nodes with one child are merged with their parents. The number of nodes in a suffix tree is $O(N)$, since the number of internal nodes is $O(N)$ due to the compactness of the tree and the number of leaves is exactly N.

For each node v of the tree, $BS(v)$ denotes the string obtained by concatenating all strings labeled on the edges on the path from the root to v. We call $BS(v)$ a *branching string*.

Example 2. Fig. 2 is the suffix tree for "*mississippi\$*". For nodes u and v, we have $BS(u) = issi$ and $BS(v) = i$.

A suffix tree is often used with suffix links to speed up related procedures. A suffix link is a pointer from a node u to another node w, where $|BS(w)| + 1 = |BS(u)|$ and $BS(w)$ is a suffix of $BS(u)$. In other words, $BS(w)$ is obtained by deleting the first character of $BS(u)$. Suffix links are generated simultaneously during the construction of the suffix

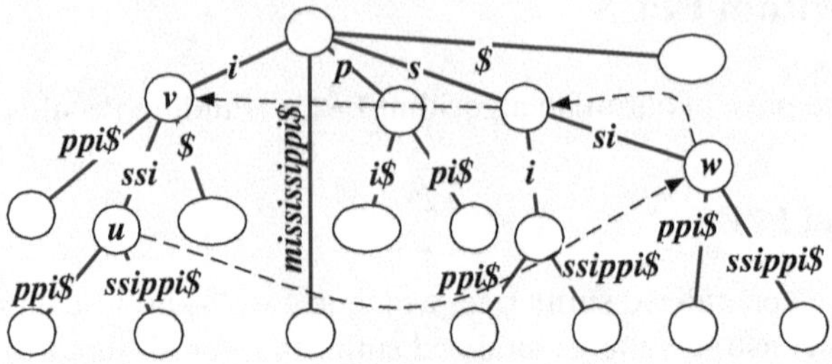

Fig. 2. The suffix tree for "*mississippi*$". A dotted line denotes a suffix link.

tree. In the above example, we have a suffix link from u to w because $BS(u) = issi$ and $BS(w) = ssi$.

For a node of the suffix tree of a string x, its branching string appears in x as many times as the number of leaves below the node. Node v of Fig. 2 has four leaves below, and so we find that $BS(v) = i$ appears four times in *mississippi*.

Let u be a child node of v. Then $BS(v)$ is a proper prefix of $BS(u)$ from the definition of the branching string. In Fig. 2, $BS(v) = i$ is a proper prefix of $BS(u) = issi$. Moreover, for a prefix s of $BS(u)$ which includes $BS(v)$ as a prefix, the frequency of s is the same as that of $BS(u)$. For example, $BS(u) = issi$ appears twice, and so do two of its prefixes *is* and *iss*. Therefore, when we count substring frequencies, all we have to do is to count only branching strings, i.e., to count leaves bellow nodes. Thus counting substring frequencies is completed in $O(N)$ time.

Now we introduce the generalized suffix tree to deal with a set D of strings instead of a single string. Let $D = \{x_1, x_2, \ldots, x_m\}$. The generalized suffix tree is the suffix tree for a string $x_1\$_1 x_2\$_2 \ldots x_m\$_m$.

For a generalized suffix tree, we assume that a node v of the tree stores a pair of the frequencies for $BS(v)$ in background and target sets. Formally, this is defined recursively as follows: a leaf node v has $(1, 0)$ (resp. $(0, 1)$) if $BS(v) \in B$ (resp. T), and an internal node v has

$$\left(\sum_{u \in C(v)} \mathrm{f}(BS(u)|B), \sum_{u \in C(v)} \mathrm{f}(BS(u)|T) \right),$$

where $C(v)$ is the set of the children of v. Note that $BS(v)$ appears only once in the input strings for a leaf node v due to \$.

4.3 Details of FPCS

Now we explain the details of FPCS using Algorithm 1 and Algorithm 2. Algorithm 1 is the main procedure which finds candidates for peculiar compositions, and Algorithm 2, which is called from the main procedure, checks the conditions to be peculiar for each of the candidates.

Given two string sets, first FPCS constructs the generalized suffix tree and stores a pair $(\mathrm{f}(BS(v)|T), \mathrm{f}(BS(v)|B))$ of the frequencies for $BS(v)$ in background and target sets on a node during a postorder traversal (see Algorithm 1).

Algorithm 1. FPCS for discovery of all maximal, frequent peculiar compositions.

Input: T, B, θ_T, θ_B and η_T
Output: all maximal peculiar compositions in T against B

```
construct the generalized suffix tree ST for all strings in T, B
for v in postorder traversal of ST do
  store (f(BS(v)|T), f(BS(v)|B)) to v
end for
for v in postorder traversal of ST do
  if BS(v) is contrastive in target and f(BS(v)|T) ≥ η_T then
    for an ancestor u of v do
      if BS(u) is contrastive in background then
        isContrast(ST, u, v, θ_B) {find appropriate divisions for xy}
      end if
    end for
  end if
end for
```

Next FPCS performs a postorder traversal again, and calls `isContrast()` (see Algorithm 2) for nodes whose branching strings are candidates for peculiar compositions. The branching string $BS(v)$ is a candidate for a peculiar composition $xy = BS(v)$. The subroutine `isContrast()` checks if y is contrastive in background or not for all divisions (x, y) of xy. If so, it outputs xy as a peculiar composition. To check these combinations, the suffix link plays an important role in `isContrast()`.

Theorem 1. *FPCS finds all maximal peculiar compositions in T against B.*

Proof. FPCS traverses all nodes of the suffix tree constructed from T and B in postorder (the outer `for` loop in Algorithm 1) and then checks if the branching string is peculiar or not. We do not need to check a substring w which is not a branching string of the suffix tree since there exists a branching string w' such that it is peculiar and its prefix is w and we have to find only maximal if the substring is peculiar. Let v be a node in the traversal. $BS(v)$ is a candidate for a peculiar composition if it is contrastive in target and $BS(v) \geq \eta_T$.

Next we check if there exists a division (x, y) of $BS(v)$ such that both x and y are contrastive in background (`isContrast()`). We do not have to check all $|BS(v)| - 1$ divisions. Let u be an ancestor of v such that $BS(u)$ is contrastive in background (see Fig. 3). In this case, we have to check for $(x, y) = (a, bcdefgh), (ab, cdefgh)$, because a longer prefix of $BS(v)$, such as *abc*, is not contrastive in background.[3] Now we know that $BS(u)$ is contrastive in background and so the next problem is *bcdefgh* or *cdefgh* is contrastive in background or not.

Consider that FPCS visits a node w in Fig. 3 after some suffix link traversal. If $BS(w)$ is contrastive in background then FPCS outputs $BS(u)BS(w)$ as a peculiar composition. If not, then FPCS visits nodes above w because branching strings of these nodes might be contrastive in background, and if FPCS finds such a node w' then outputs $BS(u)BS(w')$ as a peculiar composition.

[3] However, we have to check for $(x, y) = (abcde, fgh), (abcdef, gh)$ when $u = \text{parent}(v)$.

Algorithm 2. The subroutine `isContrast()`

Input: ST: generalized suffix tree, u, v: node, θ_B: ratio
Output: **true** or **false**

```
if v = root(T) then return end if {root(T) returns the root of ST}
x := BS(u) {we know that x (resp. xy) is contrastive in background (resp. target)}
y := edge(u, v) {edge(u, v) returns the string label between u and v}
w := v {w is the start node of suffix link traversal}
for l := 1 to |BS(u)| do
  w := node pointed by the suffix link of w;
  if BS(w) is contrastive in background then
    print BS(u)BS(w) {Found!}
  else
    w' := w {w' is the start node of upword traversal}
    for k := 1 to |edge(v, u)| do
      k := k + |edge(parent(w'), w')| {parent(u) returns the parent node of u}
      w' := parent(w')
      if k > |edge(v, u)| then break end if
      if w' is contrastive in background then
        print BS(u)BS(w') {Found!}
        break {since upword nodes are not maximal}
      end if
    end for
  end if
end for
```

This procedure corresponds to the check for $y = cdefgh$, and then FPCS visits the next node followed by the suffix link and check for $y = bcdefgh$. Thus all possible divisions are checked. □

The time complexity of the proposed algorithm is given in the following theorem.

Theorem 2. *The time complexity of FPCS is $O(N^2)$.*

Proof. Construction and traversal of the suffix tree are done in linear time with respect to the total length of input strings [21,22].

To find appropriate divisions, we need to check u which is an ancestor of v (see Fig. 3). u corresponds to x of a peculiar composition. In addition to u, FPCS visits other nodes w and w'. These three nodes are different. Therefore, for each node v, FPCS visits at most $O(N)$ nodes.

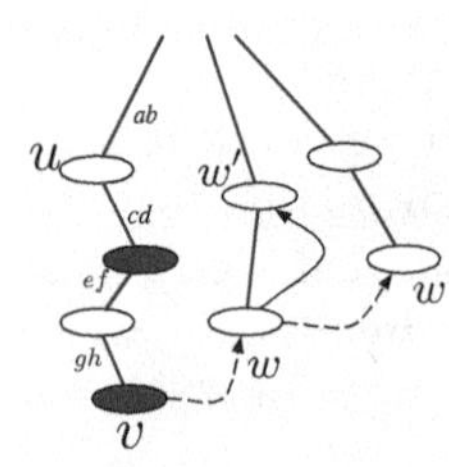

Fig. 3. Dotted arrows are suffix links and black nodes contain branching strings which are contrastive in target

At each node, FPCS tests if the corresponding branching string is contrastive or not if it is η_T-frequent. This tests can be done in constant time, using $(f(v|B), f(v|T))$ stored at all nodes. Thus the time complexity of the algorithm is $O(N^2)$. □

5 Experiments

We have implemented FPCS in C and conducted experiments on DNA sequences. All experiments were conducted on PowerMacG5 (OS: Mac OS X 10.5, CPU: 4×2.5GHz PowerPC G5, Memory: 8GB, and Compiler: gcc 4.0.1 with -O3, -arch ppc64 and -fast flags).

As input strings, we use DNA sequences. In particular, we often use the entire DNA nucleotide sequence of *Escherichia coli K-12* (RefSeq NC_00096, GI:50812173) and *Bacillus subtilis* (RefSeq NC_00096, GI:50812173). In addition to a sequence of GenBank, we add the complementary strand of the sequence into an input set for FPCS. For example, the input set of the entire DNA sequence of *E. coli K-12* consists of two strings, one is the original sequence and the other is its complementary strand. The sizes of these sets are 9279350bp (*E. coli K-12*) and 8429260bp (*B. subtilis*). Other sequences we used will be described at the corresponding experiments.

We first study the performance of FPCS with different data sizes or parameters in section 5.1 and then examine peculiar compositions output by FPCS from DNA sequences in section 5.2.

5.1 Performance Study

Firstly, we conduct two types of perfomance studies: one examines the relationship between the execution times and the number of peculiar compositions output by FPCS with different parameters, such as η_T or θ_B, and the other the execution times with different data sizes. For all experiments in this subsection, the background and target sets are fixed to the entire DNA sequence of *E. coli K-12* and one of *B. subtilis*, respectively.

Performance on Different Parameters. We examine the execution times and the number of peculiar compositions with different parameters, such as η_T or θ_B, for FPCS. Fig. 4 shows execution times of FPCS in second as η_T increase from 2 to 15, where $\theta_T = 5$ or 10 and $\theta_B = 1.5, 2.0, 2.5$ or 3.0. For any pair of parameters in Fig. 4, an execution time decreases as η_T increases. When θ_T or θ_B is large, the execution time seems not to decrease even if η_T increases. In such a case, however, the execution time does decrease drastically (see Fig. 5). Thus, pruning by η_T works effectively.

A decreasing rate becomes bigger as θ_T and θ_B decrease. These decreasing rates basically depend on θ_B but not on θ_T. In fact, we see every two lines for two values of θ_T with the same value of θ_B are close to each other.[4] Although θ_T and θ_B are not directly used to stop a traversal of FPCS, they can reduce the number of the candidates of xy, x or y. However, since there does not exist so many candidates of xy, θ_T does not decrease execution times effectively. On the other hand, we have many candidates for x and y, and thus θ_B can reduce the number of these candidates and execution times.

From these observations, we have the assumption that the execution time is closely related with the number of outputs. Two graphs in Fig. 5 show both the execution time (y-axis at the left-hand side) and the number of peculiar compositions output by FPCS (y-axis at the right-hand side), where $\theta_B = 1.5$ or 2.5.

[4] Exceptionally execution times are quite different when $\theta_B = 2.0$ and $1 \leq \eta_T \leq 5$.

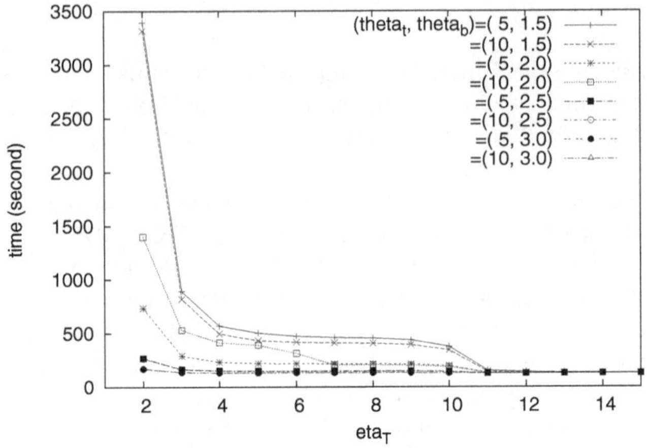

Fig. 4. Execution times of FPCS in second as η_T increases for various values of (θ_T, θ_B)

From these two graphs, we find that execution times are proportional to the number of peculiar compositions and the differences between $\theta_T = 5$ and 10 for a fixed η_T are negligible.

Scalability on Input Size. We examine the execution times with different data sizes. All execution times we show in this section are the average execution times of 5 trials on the same input.

As we increase the data size, we measure execution times of FPCS. To change the data size, we automatically generate *m* sequences with the same length *n* from sequences of *E. coli K-12* and *B. subtilis*, given *m* and *n*.

The graph on the left-hand side in Fig. 6 shows execution times of FPCS as *n* increases from 100 to 5000, where *m* is fixed to 1000, while execution times when *m* increases from 100 to 10000 in case $n = 500$ are given on the right-hand side.

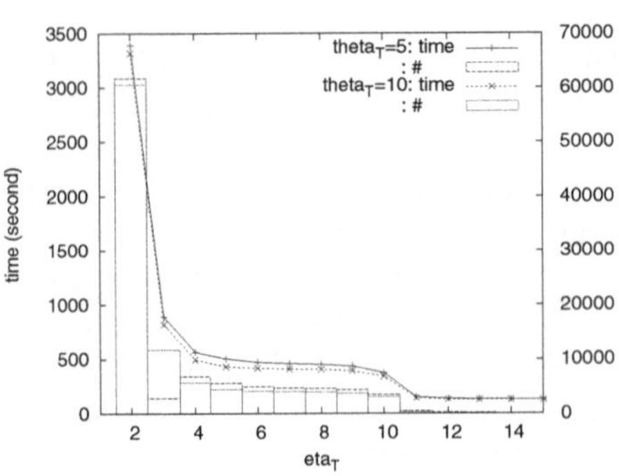

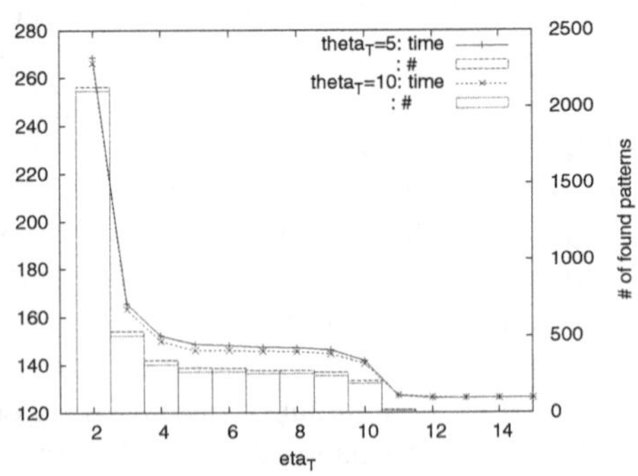

Fig. 5. Execution times of FPCS in second and the number of peculiar compositions output by FPCS as η_T increase in case of $\theta_B = 1.5$ (left-hand side) and $\theta_B = 2.5$ (right-hand side)

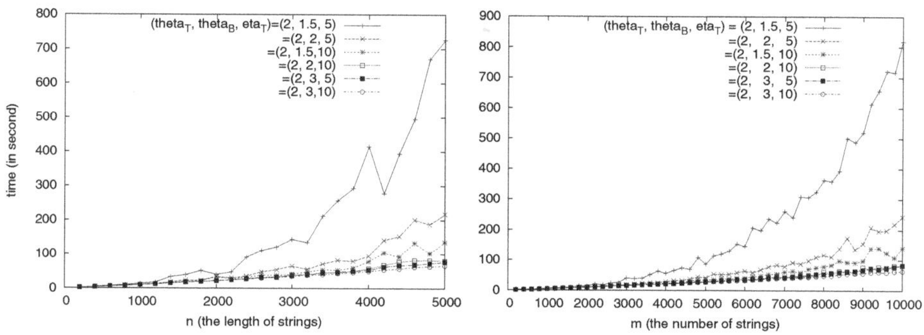

Fig. 6. Execution times for FPCS versus n (the length of strings) and m (the number of strings)

The execution times with 6 different sets of parameters, $(\theta_T, \theta_B, \eta_T) = (2, 1.5, 5)$, $(2, 1.5, 10)$, $(2, 2, 5)$, $(5, 2, 10)$, $(2, 3, 5)$, and $(2, 3, 10)$ are given. For any set of parameters, execution times in both graphs typically scale linearly with respect to the input size, except for $(\theta_T, \theta_B, \eta_T) = (2, 1.5, 5)$, although the time complexity of FPCS is $O(N^2)$ (see Section 4).

The increasing rate of the execution times in the right-hand side graph is larger than those in the left-hand side. This fact indicates that the execution time depends heavily on the depth of the suffix tree because the depth increases as the length n of the strings increases.

5.2 Properties of Peculiar Compositions

Secondly, we examine peculiar compositions from the entire DNA sequence of *E. coli K-12* or *B. subtilis*, as a target set using different background sets.

Peculiar Compositions in *B. Subtilis* against *E. Coli K-12*. We examine peculiar compositions in *B. subtilis* against *E. coli K-12* with different parameters. First, we set $(\theta_T, \theta_B, \eta_T) = (1.1, 5, 10)$ (see Table 1) and then $(10, 2, 15)$ (see Table 2), where one of θ_T or θ_B is small and the other is large.

Table 1 shows found peculiar compositions and related values, such as their frequencies, in case $(\theta_T, \theta_B, \eta_T) = (1.1, 5, 10)$. The z-score for each composition is also given.

Table 1. Peculiar compositions in *B. subtilis* against *E. coli K-12* with $(\theta_T, \theta_B, \eta_T) = (1.1, 5, 10)$. From left to right, prefix x and suffix y of compositions of the form xy, their lengths, the frequencies of xy, x, y in T and B, and z-scores are given. Probabilities of letters are computed over the whole sequences of the target set to compute the z-score.

(x, y)	length	$(\mathrm{f}(xy\|T), \mathrm{f}(xy\|B))$	$(\mathrm{f}(x\|T), \mathrm{f}(y\|B))$	$(\mathrm{f}(y\|T), \mathrm{f}(y\|B))$	z-score
$(CGGCGTGG, ACTACCAG)$	(8, 8)	(10, 7)	(66, 450)	(19, 154))	3.572e+02
$(CTGGTAGT, CCACGCCG)$	(8, 8)	(10, 7)	(19, 154)	(66, 450)	3.572e+02
$(GCGTGG, ACTACCAG)$	(6, 8)	(10, 7)	(529, 3845)	(19, 154)	7.759e+01
$(GGCGTGG, ACTACCAG)$	(7, 8)	(10, 7)	(161, 1407)	(19, 154)	1.666e+02

Table 2. Peculiar compositions in *B. subtilis* against *E. coli K-12* with $(\theta_T, \theta_B, \eta_T) = (10, 2, 15)$

(x, y)	length	$(f(xy\|T), f(xy\|B))$	$(f(x\|T), f(y\|B))$	$(f(y\|T), f(y\|B))$	z-score
$(CAGCG, GCGCC)$	(5, 5)	(17, 0))	(9816, 24161)	(6759, 17014)	8.924
$(GGCGC, CGCTG)$	(5, 5)	(17, 0)	(6759, 17014)	(9816, 24161)	8.924
$(CGCG, GCGCC)$	(4, 5)	(16, 1)	(16950, 56436)	(6759, 17014)	2.235
$(GGCGC, CGCG)$	(5, 4)	(16, 1)	(6759, 17014)	(16950, 56436)	2.235

To compute the score, we use the Bernoulli model where probabilities of individual letters, A, C, G, and T, are computed from the whole sequences of the target set and we have $p(A) = p(T) = 0.282$ and $p(C) = p(G) = 0.218$.

Four compositions are found as peculiar compositions and first two compositions are complementary each other. The lengths of the compositions are about 15. All of them appear 10 times in the target set while 7 times in the background set. These frequencies in the two sets are close since the given value for θ_T is nearly 1. However the peculiar compositions are unusual since frequencies of x and y in the background set are much larger than those in the target set.

The z-score is a normalized score whose average value is zero and variance is one [9]. Therefore, z-scores in Table 1 seem to be quite large and these composition are also unusual from the viewpoint of the z-score.

Next, we set $(\theta_T, \theta_B, \eta_T) = (10, 2, 15)$ for the same data sets. We have two pairs of two peculiar compositions which are complementary each other (see Table 2). All of found compositions appear 16 or 17 times in the target set while at most once in the background set.

Compared to those in Table 1, frequencies of x or y are quite large in Table 2 since we set a large value for θ_T and a small one for θ_B in Table 2 conversely. Compared to those in Table 1, the absolute values of the z-scores of found peculiar compositions are small. This means that, when we use the z-score to measure unusualness, it is difficult to find these compositions as unusual patterns because of the following reason. Since the z-score is a normalized score, we can calculate, for a positive number a, the number of substrings such that the absolute values of their z-scores are greater than $a = a\sigma$, where $\sigma = 1$. For example, in case $a = 3$, that number is 0.3% of the all substrings in given input data. Lengths of sequences used in the previous experiments, except for the performance study, are at least $N = 10^6$ and then we can estimate the number of all substrings by $O(N^2) = 10^{12}$. Hence, there exist a huge number of substrings whose z-scores are greater than $a = 2.235$.

Peculiar Compositions against *Saccharomyces Cerevisiae*. Now we examine compositions from the same data, *B. subtilis*, as a target set but against a set of sequences of *Saccharomyces Cerevisiae*, where we use sequences of Saccharomyces cerevisiae chromosome I–XVI (complete sequences) and Saccharomyces cerevisiae mitochondrion (complete genome) with their complementary strands. The total size of these sequences is about 24Mbp. The values of the parameters are set as $(\theta_T, \theta_B, \eta_T) = (20, 5, 20)$. 34 peculiar compositions are found from the target set and their lengths are about 14,

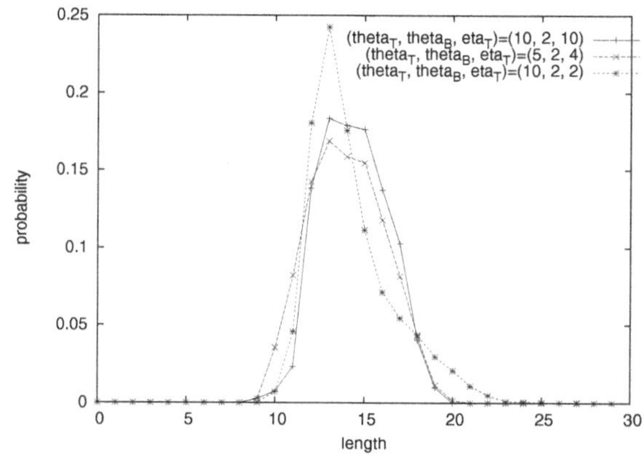

Fig. 7. The probability distributions for lengths of peculiar compositions for three parameter sets

ranging from 12 to 16. The z-scores for them are in the order of 10^1 or 10^2. We have many compositions with different divisions among 34 compositions. For example, we have three divisions, (x, y)=$(AAAACTAA, ACAAGACA)$, $(AAAACTAAA, CAAGACA)$ and $(AAAACTAAAC, AAGACA)$ for $xy = AAAACTAAACAAGACA$.

Next we try to find peculiar compositions from *E. coli K-12* against the same background set. 26 peculiar compositions are found, their lengths are about 13, the shortest one is 12 and the longest one is 16, and the z-scores are also in the order of 10^1 or 10^2. No composition of them is included in above 36 compositions.

Distribution of Lengths of Peculiar Compositions. In the previous experiments under many parameter settings, found peculiar compositions have similar lengths for some fixed parameter setting. Now we examine a distribution of their lengths.

Fig. 7 shows three probability distributions for lengths of found peculiar compositions under $(\theta_T, \theta_B, \eta_T)$=$(10, 2, 10)$, $(10, 2, 2)$ and $(5, 2, 4)$. Here, the probability means the relative frequency. We see that these distributions form bell shaped curves.

When we consider the z-score to measure unusualness, there exist many substrings whose lengths are quite small, such as 4 or 5. On the other hand, there are no short peculiar compositions of them. In fact, 9 is the shortest length of the peculiar compositions in Fig. 7. Moreover, this fact also holds when we found the small number of peculiar compositions while we have many short substrings according to the z-score even if we found the small number of substrings given a large threshold for the score. In fact, the number of found peculiar compositions is small and their lengths are not short in the previous experiments.

We think that FPCS collaterally discovers an appropriate model in which frequent substrings have similar lengths for a set of fixed parameters. This model is similar to the Markov model in which the estimation is calculated by substrings with length k. However, it is required to decide k in advance in the Markov model. On the other hand, FPCS simultaneously decides appropriate lengths in addition to finding peculiar compositions and their frequent components. Moreover, in our setting, substrings of peculiar compositions are not strictly restricted to have the same length k.

6 Conclusion

We have defined the peculiar composition discovery problem and then developed our algorithm, called FPCS, which runs in $O(N^2)$ time. Despite of its quadratic complexity, we have shown that FPCS typically runs linearly for many sets of parameters. We have also conducted experiments using DNA sequences and found, without assuming a probabilistic model, peculiar compositions which seem to be difficult to be found according to the z-score.

It is an important future work to apply our algorithm to other real data and to evaluate found peculiar compositions by domain experts. It is a challenging future work to consider more sophisticated patterns instead of compositions since a composition is defined by concatenation of only two strings without overlaps and hence we assume these strings occur independently. We think the MO method, introduced in [24], is promising for such a pattern. Using this method, we can estimate a probability of a given word w as follows: $p(w) = p(w_1)p(w_2|w'_1)\cdots p(w_k|w'_{k-1})$, where w'_i is a suffix of w_i. Note that the length of w'_i is variable. In [25], a linear time algorithm is presented to construct the model and to estimate probabilities for all given documents.

Acknowledgments

This research was partially supported by the Ministry of Education, Science, Sports and Culture, Grant-in-Aid for Scientific Research (B) 21300053 and Grant-in-Aid for Young Scientists (B) 19700150, and the Strategic International Cooperative Program funded by Japan Science and Technology Agency (JST).

References

1. Fayyad, U.M., Piatetsky-Shapiro, G., Smyth, P.: From Data Mining to Knowledge Discovery: An Overview. In: Advances in Knowledge Discovery and Data Mining, pp. 1–34. AAAI/MIT Press, Menlo Park, Calif (1996)
2. Knorr, E.M., Ng, R.T.: Finding Intensional Knowledge of Distance-Based Outliers. In: 25th International Conference on Very Large Data Bases, pp. 211–222 (1999)
3. Padmanabhan, B., Tuzhilin, A.: Small is Beautiful: Discovering the Minimal Set of Unexpected Patterns. In: Sixth ACM SIGKDD International Conference on Knowledge Discovery and Data Mining, pp. 54–63 (2000)
4. Sarawagi, S., Agrawal, R., Megiddo, N.: Discovery-Driven Exploration of OLAP Data Cubes. In: Schek, H.-J., Saltor, F., Ramos, I., Alonso, G. (eds.) EDBT 1998. LNCS, vol. 1377, pp. 168–182. Springer, Heidelberg (1998)
5. Ide, T., Inoue, K.: Knowledge Discovery from Heterogeneous Dynamic Systems using Change-Point Correlations. In: 2005 SIAM International Conference on Data Mining, pp. 571–576 (April 2005)
6. Keogh, E., Lonardi, S., Chiu, B.Y.: Finding Surprising Patterns in a Time Series Database in Linear Time and Space. In: Eighth International Conference on Knowledge Discovery and Data Mining, pp. 550–556 (2002)
7. Apostolico, A., Bock, M.E., Lonardi, S., Xu, X.: Efficient Detection of Unusual Words. Journal of Computational Biology 7(1/2), 71–94 (2000)

8. Leung, M.Y., Marsh, G.M., Speed, T.P.: Over- and Underrepresentation of Short DNA Words in Herpesvirus Genomes. Journal of Computational Biology 3(3), 345–360 (1996)
9. Parida, L.: Pattern Discovery in Bioinformatics: Theory & Algorithms. Chapman & Hall Crc Mathematical & Computational Biology Series. Chapman & Hall/CRC, Boca Raton (2007)
10. Schbath, S.: An Efficient Statistic to Detect Over- and Under-represented Words in DNA Sequences. Journal of Computational Biology 4(2), 189–192 (1997)
11. Arimura, H., Shimozono, S.: Maximizing agreement with a classification by bounded or unbounded number of associated words. In: Chwa, K.-Y., Ibarra, O.H. (eds.) ISAAC 1998. LNCS (LNAI), vol. 1533, pp. 39–48. Springer, Heidelberg (1998)
12. Ikeda, D.: Characteristic Sets of Strings Common to Semi-structured Documents. In: Arikawa, S., Furukawa, K. (eds.) DS 1999. LNCS (LNAI), vol. 1721, pp. 139–147. Springer, Heidelberg (1999)
13. Ji, X., Bailey, J., Dong, G.: Mining Minimal Distinguishing Subsequence Patterns with Gap Constraints. In: Fifth IEEE International Conference on Data Mining, pp. 194–201 (2005)
14. Nakanishi, M., Hashidume, M., Ito, M., Hashimoto, A.: A Linear-Time Algorithm for Computing Characteristic Strings. In: Du, D.-Z., Zhang, X.-S. (eds.) ISAAC 1994. LNCS, vol. 834, pp. 315–323. Springer, Heidelberg (1994)
15. Suzuki, E.: Autonomous Discovery of Reliable Exception Rules. In: Third International Conference on Knowledge Discovery and Data Mining, pp. 259–262 (1997)
16. Suzuki, E.: Undirected Discovery of Interesting Exception Rules. International Journal of Pattern Recognition and Artificial Intelligence 16(8), 1065–1086 (2002)
17. Suzuki, E., Shimura, M.: Exceptional Knowledge Discovery in Databases Based on Information Theory. In: Second International Conference Knowledge Discovery and Data Mining, pp. 275–278 (1996)
18. Suzuki, E., Tsumoto, S.: Evaluating Hypothesis-Driven Exception-Rule Discovery with Medical Data Sets. In: Terano, T., Chen, A.L.P. (eds.) PAKDD 2000. LNCS, vol. 1805, pp. 208–211. Springer, Heidelberg (2000)
19. Wang, J., Han, J., Pei, J.: CLOSET+: Searching for the Best Strategies for Mining Frequent Closed Itemsets. In: 9th ACM SIGKDD International Conference on Knowledge Discovery and Data Mining, pp. 236–245 (2003)
20. Yan, X., Han, J., Afshar, R.: CloSpan: Mining Closed Sequential Patterns in Large Databases. In: The 4th SIAM International Conference on Data Mining (May 2003)
21. McCreight, E.M.: A Space-Economical Suffix Tree Construction Algorithm. Journal of the ACM 23(2), 262–272 (1976)
22. Ukkonen, E.: On-line Construction of Suffix Trees. Algorithmica 14(3), 249–260 (1995)
23. Gusfield, D.: Algorithms on Strings, Trees and Sequence. Cambridge University Press, New York (1997)
24. Jagadish, H.V., Ng, R.T., Srivastava, D.: Substring Selectivity Estimation. In: Eighteenth Symposium on Principles of Database Systems, pp. 249–260 (1999)
25. Uemura, T., Ikeda, D., Arimura, H.: Unsupervised Spam Detection by Document Complexity Estimation. In: Boulicaut, J.-F., Berthold, M.R., Horváth, T. (eds.) DS 2008. LNCS (LNAI), vol. 5255, pp. 319–331. Springer, Heidelberg (2008)

Minimum Free Energy Principle for Constraint-Based Learning Bayesian Networks

Takashi Isozaki[1,2] and Maomi Ueno[1]

[1] Graduate School of Information Systems, The University of Electro-Communications, 1-5-1 Chofugaoka, Chofu-shi, Tokyo 182-8585, Japan
[2] Research and Technology Group, Fuji Xerox Co., Ltd. 3-1-1 Roppongi, Minato-ku, Tokyo 106-0032, Japan
{t-isozaki,ueno}@ai.is.uec.ac.jp

Abstract. Constraint-based search methods, which are a major approach to learning Bayesian networks, are expected to be effective in causal discovery tasks. However, such methods often suffer from impracticality of classical hypothesis testing for conditional independence when the sample size is insufficiently large. We propose a new conditional independence (CI) testing method that is effective for small samples. Our method uses the minimum free energy principle, which originates from thermodynamics, with the "Data Temperature" assumption recently proposed for relating probabilistic fluctuation to virtual thermal fluctuation. We define free energy using Kullback–Leibler divergence in a manner corresponding to an information-geometric perspective. This CI method incorporates the maximum entropy principle and converges to classical hypothesis tests in asymptotic regions. We provide a simulation study, the results of which show that our method improves the learning performance of the well known PC algorithm in some respects.

Keywords: Bayesian networks, Structure learning, Conditional independence test, Minimum free energy principle.

1 Introduction

Bayesian networks (BNs) [1] are graphical models and compact representations of joint probability distributions. This combination is suitable for modeling the uncertainty surrounding random variables. Actually, BNs are widely studied for various applications, such as expert systems, human modeling, autonomous agents, natural language processing, and computational biology. The network structure expresses conditional dependence–independence relations among random variables. A wide range of applications is considered. Therefore, structure discovery of BNs from observational data has become an attractive problem tackled by many researchers over the past decade.

The many proposed methods of structure learning can be categorized into two major approaches: score-and-search based methods (e.g. [2]) and constraint-based methods (e.g. [3]). Score-and-search based methods describe the fitness of

W. Buntine et al. (Eds.): ECML PKDD 2009, Part I, LNAI 5781, pp. 612–627, 2009.

each possible structure to the observed data while retaining appropriate complexity of models, for which scores such as AIC, BIC, MDL, and BDeu are typically used (e.g. [2]). Constraint-based methods are designed to estimate properties of conditional independence among the variables in the data. Both methods present distinct advantages and disadvantages. Constraint-based methods are computationally efficient and expected to find causal models and latent common causes under certain conditions [3].

However, in contrast to recent development of score-and-search based methods, disadvantages of constraint-based methods have increasingly achieved prominence [4]. Instances of conditional independence are often decided using classical hypothesis tests with χ^2 or G^2 test [3,4]. One disadvantage of the classical tests is their impracticality for use with small samples because of their use of asymptotic approximation of the statistic to the χ^2 distribution, which is justified for a sufficiently large sample size.

As described in this paper, we propose an alternative conditional independence testing method that presents an advantage of effectiveness for small samples and which has connectivity with the classical hypothesis tests. To realize this, we use the *minimum free energy principle*, which originated from thermodynamics (e.g. [5]). In fact, the minimum free energy (MFE) principle and similar ideas have recently attracted the attention of researchers in some domains of computer science such as clustering [6] and learning [7,8]. In thermal physics, free energy consists of an internal energy, an entropy, and temperature. All of these are expected to play important roles. Nevertheless, many studies that have applied free energy to statistical science have treated temperature as a fixed parameter or a free parameter, apparently because of its lack of clarity of the meaning in data science. Consequently, we consider that the potentials of free energies have not been well extracted.

We remain acutely aware that an advantage of using free energies in statistical science is their capability of expressing a tradeoff between the maximum likelihood (ML) [9] and the maximum entropy (ME) [10] principles, which are best used, respectively, for sufficiently large datasets and small datasets. As described in this paper, for solving this problem, we use a metaphor of the tradeoff between minimizing internal energies, which are dominant for low temperatures, and maximizing entropies, which are dominant for high temperatures in the MFE principle on thermodynamics. We can regard the temperature in free energies as a tradeoff parameter that determines the component fraction of the ML and ME concepts. Therefore, if we pursue the program, it is reasonable to relate temperature with the available data size. Consequently, we recently proposed the "Data Temperature" assumption by which the inverted temperature is a monotonic increasing function of the available data size. We demonstrated its effectiveness in parameter learning of BNs [11].

For this work, we adopt this assumption for learning structure BNs. However, a new manner of definition of free energies must be developed for constraint-based learning. These new definitions are related to a fact described in *information geometry* [12]. As a result, we obtain a new unified manner between the constraint-based structure and parameter learning of BNs using the MFE principle with the "Data Temperature" assumption, in which we use only one hyperparameter introduced in our previous work.

Advantages of the proposed method are the following.

- Improving accuracy in some respects for constraint-based learning methods on BNs, especially for small samples, and having connectivity with the classical hypothesis tests for asymptotic regions by introducing the minimum free energy principle, which can treat the tradeoff of the maximum likelihood and maximum entropy principles explicitly.
- Developing a novel theoretical framework of unifying structure and parameter learning methods of BNs, which corresponds to an information-geometrical perspective.

Furthermore, we demonstrated the performance of our methods incorporated with the PC algorithm [3], which is a typical benchmark constraint-based algorithm. Our robust independence testing method turns out to be more effective on one level than the standard hypothesis tests for small or medium data sizes.

2 Background

2.1 Bayesian Networks

A Bayesian network (BN) is a set $\mathbb{B} = \langle \mathbb{G}, P \rangle$. Actually, $\mathbb{G} = \langle \mathrm{V}, \mathrm{E} \rangle$ denotes a directed acyclic graph (DAG) with nodes representing random variables V. In addition, P is a joint probability distribution on V. Furthermore, $\mathbb{G}$ and P must satisfy the Markov condition: all variables $X \in \mathrm{V}$ are independent of any subset of its non-descendant variables conditioned on the set of its parents [1]. The set of the parents of a variable X_i is in the graph $\mathbb{G}$ as Π_i. For a distribution P of n variables $\mathrm{V} = \{X_1, \ldots, X_n\}$, a BN $\mathbb{B}$ can be factorized as conditional probability distributions

$$P(X_1, \ldots, X_n) = \prod_{i=1}^{n} P(X_i \mid \Pi_i) \ , \tag{1}$$

as proved easily using the Markov condition.

The graph of a BN presents some instances of entailed independence of the probability distribution. The d-separation [1] is a concept in DAGs, which characterizes entailed conditional independence in the graph. Two nodes X and Y are d-separated by $\boldsymbol{Z}$ in graph $\mathbb{G}$, denoted as $Dsep^{\mathbb{G}}(X;Y|\boldsymbol{Z})$. In addition, $Ind(X;Y|\boldsymbol{Z})$ is denoted as the conditional independence of X and Y given $\boldsymbol{Z}$ in P. It is known that $Dsep^{\mathbb{G}}(X;Y|\boldsymbol{Z}) \Rightarrow Ind(X;Y|\boldsymbol{Z})$ for a BN $\mathbb{B} = \langle \mathbb{G}, P \rangle$.

A BN $\mathbb{B} = \langle \mathbb{G}, P \rangle$ satisfies the *faithfulness condition* if the Markov condition entails all and only the instances of conditional independence in P [3]. In a faithful BN $\mathbb{B} = \langle \mathbb{G}, P \rangle$, $Dsep^{\mathbb{G}}(X;Y|\boldsymbol{Z}) \Leftrightarrow Ind(X;Y|\boldsymbol{Z})$ [1]. This equivalence relation enables us to infer structures of BNs from conditional independence relations. We assume for this discussion that BNs satisfy the faithfulness condition.

We describe some other assumptions used to conduct DAG structure inference in this paper, further to *i.i.d.*, as follows.

- Discrete Variables: each variable in V has a finite, discrete number of possible values.
- No Missing Values: let D be a database set with sample size N such that each sample has no missing values for all variables in set V.
- No Latent Variables: we consider DAGs without latent variables.

2.2 Constraint-Based Learning on BNs

Here we state a basic concept of the constraint-based structure learning methods using the preceding notation. For a BN that satisfies the faithfulness condition, the basic concepts are the following [13].

- Search for a set $\boldsymbol{Z}$ for each pair of variables X and Y in V such that $Ind(X;Y|\boldsymbol{Z})$ holds in P. Therefore, X and Y are conditionally independent given a set $\boldsymbol{Z}$ in P. Construct an undirected graph such that nodes X and Y are connected with an undirected edge if and only if no set $\boldsymbol{Z}$ can be found.
- For each pair of nonadjacent variables X and Y with a common neighbor W, check if $W \in \boldsymbol{Z}$. If not, then add arrowheads pointing at W (i.e., $X \rightarrow W \leftarrow Y$), the type of which is called a v-structure.
- Orient as many of the undirected edges as possible subject to two conditions: (i) the orientation should not create a new v-structure; and (ii) the orientation should not create a directed cycle graph.

For checking conditional independence, we describe classical hypothesis testing, which is frequently used within BN learning algorithm under a faithfulness assumption [3,4]; we also use it afterwards. Statistics such as χ^2 or G^2 are expressed here as S^2. If S^2 can be approximated to a χ^2 distribution with degrees of freedom df: χ^2_{df} and $S^2 < \chi^2_{\alpha,df}$, where $\chi^2_{\alpha,df}$ is a threshold value such that $P(\chi^2_{df} \geq \chi^2_{\alpha,df}) = \alpha$, in which α is a fixed confidence level, then we do not reject the null hypothesis of (conditional) independence between two selected variables given selected conditional sets; otherwise we reject it. The validity of approximation of statistics such as χ^2 or G^2 is proved in asymptotic regions [14]. However, for a small sample size, it is not justified. Spirtes et al. [3] have used, in their PC algorithm, a criterion for the validity: the algorithm does not perform an independence test if the sample size is less than 10 times the number of different possible joint patterns of the two variables and conditional sets, which means the variables are assumed to be conditionally dependent. This impracticality is a weak point of the constraint-based learning methods of BNs because learning BNs often process insufficient data.

3 Conditional Independence Testing Using the MFE Principle

In this section, we describe a new conditional independence testing method that is designed to be especially effective for small data size, and which is designed to be connected asymptotically with classical hypothesis testing. We take an approach from the metaphor of thermodynamics, where entropy, energy, and temperature play important roles.

3.1 Why Is the MFE Principle Needed?

Many studies of learning BNs have often used mutual information (e.g., [15]) for measuring dependence, which often means *minimizing entropy*, as described below. For asymptotic regions, for which sufficiently large samples are available, the guiding principle in statistics is the maximum likelihood (ML) principle [9]. Friedman et al. [15] derived that, for a BN $\mathbb{G}$, given a dataset D with N data size for n random variables, maximizing the log likelihood $LL(\mathbb{G}|D)$ is equivalent to maximizing empirical mutual information between a node and its parent nodes (represented as Π_i for a node X_i): $LL(\mathbb{G}|D) = N(\sum_{i=1}^{n} \hat{I}(X_i; \Pi_i) - \sum_{i=1}^{n} \hat{H}(X_i))$, where $\hat{I}$ and $\hat{H}$ denotes empirical mutual information and Shannon entropy [10], and the second term of the right-hand-side of the equation has nothing to do with the learning structure. Therefore, from the definition of mutual information, it is readily derived that

$$LL(\mathbb{G}|D) = -N \sum_{i=1}^{n} \hat{H}(X_i \,|\, \Pi_i) = -N\, \hat{H}(X_1, \ldots, X_n)\ . \tag{2}$$

The last equation is derived from the definition of BNs described in (1). This equation means that maximizing the log likelihood is equivalent to minimizing the entropy of BNs. This equation also implies that maximizing the log likelihood for constructing the DAG structures engenders *complete DAG* because the following inequality is justified: $-N\sum_i \hat{H}(X_i \,|\, \Pi_i) \geq -N\sum_i \hat{H}(X_i)$, because $0 \leq H(X|Y) \leq H(X)$ [10].

In contrast, when we obtain insufficient data, it is reasonable to use the maximum entropy (ME) principle [10], which states that the most preferable probabilistic model should maximize its entropies under some constraint related to available data. Consequently, with no constraint, maximizing entropies of BNs engenders the DAG with *no edges*, which means that a BN is a collection of complete independent distributions: $P(\boldsymbol{X}) = \prod_i P(X_i)$.

A tradeoff exists between maximum likelihood and maximum entropy for obtaining the valid structures. In the asymptotic region, the ML principle is expected to be dominant; in an insufficient sample region, the ME principle is expected to be dominant. Therefore, we can set a problem of how to decide the tradeoff between the ML and ME principles according to an arbitrarily given sample size. We regard the situation as a metaphor of thermodynamics. The tradeoff between minimizing internal energy and maximizing entropy in thermodynamics seems to correspond to the tradeoff between maximizing likelihood and entropy in statistics, and temperature can be regarded as a parameter that brings harmony of the two amounts. These amounts can be treated in a unified manner as a *free energy* that is well known in thermal physics. Furthermore, this tradeoff can be determined using the *minimum free energy* (MFE) principle.

During recent decades, many researchers investigated Bayesian methods, which can avoid overfitting derived from using the ML with insufficient data. This can be regarded as the same problem setting described in this paper. For example, Dash et al. proposed a robust conditional independence testing procedure using Bayesian Dirichlet smoothing [16]. However, the Bayesian method presents the difficulty of deciding optimal hyperparameters simultaneously in both theoretical (e.g. [17]) and practical [18] perspectives, when no prior knowledge exists.

Similarly, temperature is an unknown parameter in the MFE principle. However, we recently presented a model of inverted temperature that is a monotonic increasing function of available data size, which we call the "Data Temperature" assumption [11]. Using this approach, we give the meaning of temperature of free energy in data/statistical science by regarding the probability fluctuation as a virtual thermal fluctuation. Furthermore, we showed that the approach is effective for parameter learning of BNs, and that the effect is not sensitive for selecting a hyperparameter. We consider that the approach can also be useful in structure learning BNs for estimating optimal entropies of the network structure. To realize this, we regard that remaining problems are how to define amounts corresponding to energies, entropies, and temperature.

3.2 Minimum Free Energy Principle

The (Helmholtz) free energies were introduced originally into the field of thermodynamics. The energies are defined such that a maximum thermodynamical work is the difference between values of free energies in two distinct states, [5] where the maximum work is obtained using an isothermal quasistatic operation from a closed system under the condition of a constant temperature. Therefore, the free energy can be regarded as an amount, in a constant temperature, corresponding to a potential energy in dynamics (e.g., gravitational and electro-magnetic potential energy). In this meaning, the free energy is viewed as an amount that is extracted freely from a thermodynamical system.

We regard the free energy in information systems as an amount that has a similar effect in thermodynamical systems: it is extracted freely from a data system under a given data size (corresponding to inverted temperature). This property is apparently preferred for various tasks such as inference, learning, and estimation under a finite available data size because we wish to obtain maximum effective information from limited exploitable data.

We use a principle of minimum free energy (MFE) for statistical testing of conditional independence. A free energy F is defined by an internal energy U, an entropy H, and inverted temperature β_0 ($= 1/temperature$) as

$$F := U - \frac{H}{\beta_0}\ , \tag{3}$$

where (inverted) temperature β_0, which is a parameter, balances the respective contributions to F of U and H. According to the principle of MFE, given some temperature β_0, the stable state of the system is realized to minimize F [5], where minimizing U and maximizing H are balanced.

3.3 Representation of Free Energy in Probabilistic Models

Different from usual application of the MFE principle in data science, we start with description of the free energy definitely as a function of internal energy, entropy, and temperature to recognize clearly important properties of temperature and use effectively free energies. Fortunately, entropy was introduced into information theory by Shannon. It has since become a fundamental concept in computer science and statistical science [10]. Therefore, we define the entropy of a random variable X as Shannon entropy. We hope that the entropy serves to avoid overfitting for small samples. We adopt the Kullback–Leibler (KL) divergence between two probabilistic distributions, which are an empirical distribution and a true distribution. Here, we follow the "Data Temperature" assumption [11], which makes the MFE principle express a harmony between the ML and ME principle according to available data size: *temperature is defined as a monotonic function of the available data size such that temperature* $\beta_0 \to \infty$ *if data size* $N \to \infty$, *and* $\beta_0 \to 0$ *if* $N \to 0$.

3.4 An MFE Representation of Hypothesis Testing on BNs

We represent the conditional independence tests using the MFE principle. To do so, as in the usual manner [3,4], we represent the null hypothesis as conditional independent relations, and the opposite hypothesis as conditional dependent relations between two variables X and Y given conditional sets $\boldsymbol{Z}$.

We define the internal energies for each hypothesis. First, we represent the internal energy U such that the relative entropy (KL divergence) between the graphs expressing the null hypothesis (expressed as $\mathbb{H}_1$, corresponding distributions as $\hat{P}_1$) and the true graphs (as $\mathbb{H}_0$, corresponding as P_0), where $\hat{P}_1$ of the null hypothesis is defined as a maximum likelihood distribution. Therefore, we can define an internal energy U_1 which expresses the null hypothesis such as

$$\begin{aligned} U_1(X,Y,\boldsymbol{Z}) :&= -D(\hat{P}_1(X,Y,\boldsymbol{Z}) \,||\, P_0(X,Y,\boldsymbol{Z})) \\ &= \sum_{x,y,\boldsymbol{z}} \hat{P}(x,y,\boldsymbol{z}) \log \frac{P(x,y \mid \boldsymbol{z})}{\hat{P}(x \mid \boldsymbol{z})\hat{P}(y \mid \boldsymbol{z})} \,, \end{aligned} \tag{4}$$

where $\hat{P}$ is a maximum likelihood distribution and P is a distribution that will be estimated using the MFE principle with a "Data Temperature" model. In turn, the internal energy U_2 expresses the opposite hypothesis, which expresses a dependent relation as

$$\begin{aligned} U_2(X,Y,\boldsymbol{Z}) :&= -D(\hat{P}_2(X,Y,\boldsymbol{Z}) \,||\, P_0(X,Y,\boldsymbol{Z})) \\ &= \sum_{x,y,\boldsymbol{z}} \hat{P}(x,y,\boldsymbol{z}) \log \frac{P(x,y \mid \boldsymbol{z})}{\hat{P}(x,y \mid \boldsymbol{z})} \,. \end{aligned} \tag{5}$$

In the next step, we define the entropy term with respect to each hypothesis. Probability distributions which constitute the entropy are estimated under given available samples. We describe the entropy of the null hypothesis as

$$H_1(X, Y, \boldsymbol{Z}) := -\sum_{x,y,\boldsymbol{z}} P(x, y, \boldsymbol{z}) \log(P(x \mid \boldsymbol{z})P(y \mid \boldsymbol{z})P(\boldsymbol{z})) . \tag{6}$$

The other entropy, that of the opposite hypothesis, is

$$H_2(X, Y, \boldsymbol{Z}) := -\sum_{x,y,\boldsymbol{z}} P(x, y, \boldsymbol{z}) \log(P(x, y \mid \boldsymbol{z})P(\boldsymbol{z})) . \tag{7}$$

Now we are almost prepared to express the free energy of each hypothesis. We regard the temperature in each hypothesis (β_1 and β_2) as a *global temperature* over related variables. According to the "Data Temperature" assumption, we can consider that $\beta_1 = \beta_2 = \beta_0$, which means the same sample size. Therefore, the difference of the free energy of each hypothesis is

$$F_1(X, Y, \boldsymbol{Z}) - F_2(X, Y, \boldsymbol{Z}) = \hat{I}(X; Y|\boldsymbol{Z}) - \frac{1}{\beta_0} I(X; Y|\boldsymbol{Z}) , \tag{8}$$

where

$$\hat{I}(X; Y|\boldsymbol{Z}) = \sum_{x,y,\boldsymbol{z}} \hat{P}(x, y, \boldsymbol{z}) \log \frac{\hat{P}(x, y \mid \boldsymbol{z})}{\hat{P}(x \mid \boldsymbol{z})\hat{P}(y \mid \boldsymbol{z})} , \tag{9}$$

and

$$I(X; Y|\boldsymbol{Z}) = \sum_{x,y,\boldsymbol{z}} P(x, y, \boldsymbol{z}) \log \frac{P(x, y \mid \boldsymbol{z})}{P(x \mid \boldsymbol{z})P(y \mid \boldsymbol{z})} . \tag{10}$$

According to the notation used in our previous work, we define a new parameter β, which we call "Data Temperature" hereinafter as

$$\beta := \beta_0/(\beta_0 + 1) , \tag{11}$$

where if $\beta_0 \to 0$, then $\beta \to 0$ (high temperature limit); if $\beta_0 \to \infty$, then $\beta \to 1$ (low temperature limit).

For estimating the non-empirical conditional mutual information $I(X, Y|\boldsymbol{Z})$, as described above, we follow our previous work associated with parameter learning method, for which a different definition of internal energies U is needed [11]. Let $P(\boldsymbol{X})$ and $\hat{P}(\boldsymbol{X})$ respectively represent probability distributions of joint random variables $\boldsymbol{X}$ to be estimated from the MFE principle and ML principle. Internal energies $U(\boldsymbol{X})$ are defined for parameter learning as

$$U(\boldsymbol{X}) = D(\, P(\boldsymbol{X}) \,||\, \hat{P}(\boldsymbol{X})\,) = \sum_{\boldsymbol{x}} P(\boldsymbol{x}) \log \frac{P(\boldsymbol{x})}{\hat{P}(\boldsymbol{x})} . \tag{12}$$

Then, using Lagrange multipliers corresponding to minimizing the free energy with a constraint as $\sum_{\boldsymbol{X}=\boldsymbol{x}} P(\boldsymbol{X} = \boldsymbol{x}) = 1$, the estimated probability $P_\beta(\boldsymbol{x})$ is expressed in Boltzmann's formula, as shown below [11].

$$P_\beta(\boldsymbol{x}) = \frac{\exp(-\beta(-\log \hat{P}(\boldsymbol{x})))}{\sum_{\boldsymbol{x}'} \exp(-\beta(-\log \hat{P}(\boldsymbol{x}')))} = \frac{[\hat{P}(\boldsymbol{x})]^\beta}{\sum_{\boldsymbol{x}'} [\hat{P}(\boldsymbol{x}')]^\beta} \tag{13}$$

Therein, $\hat{P}$ is a relative frequency: the ML estimator.

Finally, we obtain the condition of conditional independence as

$$\hat{I}(X;Y|\boldsymbol{Z}) < \frac{1-\beta}{\beta} I_\beta(X;Y|\boldsymbol{Z}) \ , \tag{14}$$

where I_β is defined as

$$\begin{aligned} I_\beta(X;Y|\boldsymbol{Z}) &= \sum_{x,y,\boldsymbol{z}} P_\beta(x,y,\boldsymbol{z}) \log \frac{P_\beta(x,y \mid \boldsymbol{z})}{P_\beta(x \mid \boldsymbol{z}) P_\beta(y \mid \boldsymbol{z})} \\ &= \sum_{x,y,\boldsymbol{z}} P_\beta(x,y,\boldsymbol{z}) \log \frac{P_\beta(x,y,\boldsymbol{z}) P_\beta(\boldsymbol{z})}{P_\beta(x,\boldsymbol{z}) P_\beta(y,\boldsymbol{z})} \ . \end{aligned} \tag{15}$$

Therein, β plays only the role of a symbolic index; it does not represent a sole parameter. In each estimator, β should be calculated using the explicit model of "Data Temperature," as described in the next subsection. Therefore, β in (15) represents *local temperature*. In (14), the left-hand-side corresponds to the likelihood term, which is dominant for a large data size (large β), and the right-hand-side corresponds to the entropy term, which is dominant for a small data size (small β). We designate g^2_β and represent the conditional independence (CI) condition with it as

$$g^2_\beta = \hat{I}(X;Y|\boldsymbol{Z}) - \frac{1-\beta}{\beta} I_\beta(X;Y|\boldsymbol{Z}) < 0 \ . \tag{16}$$

This is useful for combination with the classical hypothesis tests.

In these formulations, it might seem strange that two distinct definitions of internal energies exist between hypothesis tests and parameter estimation. We express internal energies in the hypothesis tests as $D(\hat{P} \,||\, Q)$, where $\hat{P}$ is an ML distribution. That differs from our formula in parameter learning, where we expressed the internal energies as $D(P \,||\, \hat{Q})$, where $\hat{Q}$ is an ML distribution. This difference is pointed out from the perspective of *information geometry* [12]. From an information theoretical viewpoint, hypothesis testing is related to the large deviation theorem via Sanov's theorem [10], and then, from an information geometrical perspective, it can be interpreted as a $\nabla^{(\mathrm{e})}$-projection, whereas the ML estimation can be interpreted as $\nabla^{(\mathrm{m})}$-projection [12]. In other words, the hypothesis testing and ML estimation are then different concepts in view of information theory. Consequently, in the definitions for learning structures and parameters, the difference is reasonable from this perspective.

3.5 "Data Temperature" Model

In searching for the values of β, we use our simple model of temperature, which is proposed as a function of data size N [11]. The model function of β is defined as

$$\begin{aligned} \beta &:= 1 - \exp\left(-\frac{N}{\gamma N_c}\right) \ , \\ \gamma &:= |\boldsymbol{X}| - 1 \ , \end{aligned} \tag{17}$$

where γ is defined as the degrees of freedom of related random variables $\boldsymbol{X}$, and where N_c is a decoupling constant, which can be regarded as a hyperparameter for β. We use N_c as only one *common* hyperparameter in learning of both parameter and structure. This explicit model shows good performance and robustness against selected hyperparameters N_c in classification tasks using Bayesian network classifiers with structure learning [11].

3.6 Asymptotic Theoretical Analysis

We hope that the proposed method has consistency with the classical hypothesis test for an asymptotic region because it is theoretically justified. However, our conditional independence conditions using the inequality (14) cannot be used straightforwardly for large data sizes because $g_\beta^2 \geq 0$ always for sufficiently large data size. That is true because $\hat{I}(X;Y|\boldsymbol{Z}) \geq 0$ and $[(1-\beta)/\beta]\, I_\beta(X;Y|\boldsymbol{Z}) \to 0$ as β goes to 1 (as N becomes sufficiently large), which means that our method would produce an overly dense graph for sufficiently large data size. In such regions, the effect of enlarging the entropy term has vanished and the likelihood term has become dominant. However, different from parameter learning, hypothesis testing for BNs means that extra edges should be removed even for a large sample size, based on *Occam's razor* [13]. This connecting problem is solved as described below.

For a large sample size region, we wish to use the G^2 statistic for conditional independence testing, which is often used [3,4]. The G^2 test is used to identify $Ind(X, Y|\boldsymbol{Z})$, by which the null hypothesis of conditional independence is represented. Let $N_{xy\boldsymbol{z}}$ represent the number of times in the data where $X = x, Y = y$ and $\boldsymbol{Z} = \boldsymbol{z}$. We define $N_{x\boldsymbol{z}}$, $N_{y\boldsymbol{z}}$, and $N_{\boldsymbol{z}}$ similarly. Consequently, the G^2 statistic is defined as follows:

$$G^2 = 2 \sum_{x,y,\boldsymbol{z}} N_{xy\boldsymbol{z}} \log \frac{N_{xy\boldsymbol{z}} N_{\boldsymbol{z}}}{N_{x\boldsymbol{z}} N_{y\boldsymbol{z}}} \ . \tag{18}$$

The degrees of freedom *df* are defined as

$$df = (|X|-1)(|Y|-1) \prod_{Z \in \boldsymbol{Z}} |Z| \ , \tag{19}$$

where we designate $|X|$ as the number of states in X. It is noteworthy that the G^2 statistics have a relation with the empirical mutual information with data size N [14] as

$$G^2 = 2N\, \hat{I}(X;Y|\boldsymbol{Z}) \ . \tag{20}$$

The statistic is proven to be approximated asymptotically to a χ^2 distribution with degrees of freedom *df* [14]. Therefore, in a large sample size region, we should set the condition in which the null hypothesis (i.e., conditional independence) is not rejected, as

$$G^2 < \chi^2_{\alpha,df} \ , \tag{21}$$

where α is a significance level such as 0.05, and where df are the degrees of freedom, as defined in (19).

Here, we intend to connect the classical condition with the MFE condition. We define a formal correspondence amount G^2_β to G^2, using (16) and (20) as

$$G^2_\beta := 2Ng^2_\beta = G^2 - 2N\,\frac{1-\beta}{\beta}\,I_\beta(X;Y|\boldsymbol{Z})\,. \tag{22}$$

Using the explicit model of β expressed in (17), in an asymptotic region,

$$G^2_\beta = G^2 - 2N\,\frac{\exp(-N/\gamma N_c)}{1-\exp(-N/\gamma N_c)}\,I_\beta(X;Y|\boldsymbol{Z}) \to G^2\,. \tag{23}$$

Then, we can treat the MFE and the classical condition uniformly because a condition $G^2_\beta < \chi^2_{\alpha,df}$ can include the CI condition (16) and the classical condition (21). Even when the data size is small and $G^2_\beta \geq 0$, we conduct the classical hypothesis tests because our method was shown to generate pseudo-samples similarly to the Bayesian methods [19]. Consequently, we can conduct conditional independence tests on variables X and Y given $\boldsymbol{Z}$ using the MFE principle and G^2 tests as described below.

- If $G^2_\beta < 0$, because of MFE principle, then we set $X \perp\!\!\!\perp Y \mid \boldsymbol{Z}$ (conditional independence),
- else if $0 \leq G^2_\beta < \chi^2_{\alpha,df}$, because of the classical test, then we set $X \perp\!\!\!\perp Y \mid \boldsymbol{Z}$,
- else, we set $X \not\perp\!\!\!\perp Y \mid \boldsymbol{Z}$ (conditional dependence).

We designate this conditional independence method as MFE-CI.

4 Experiments

We next demonstrate the performances of our approach compared with traditional statistical testing methods. Some experiments of learning BNs can be done using the PC algorithm [3], which is a well known benchmark algorithm of constraint-based methods, embedding our conditional independence tests or classical independence tests using χ^2 distributions with fixed significant level $\alpha = 0.05$ for each hypothesis test of conditional independence. We implemented the PC algorithm for embedding the MFE-CI method using C++ programming language. The PC algorithm, which constructs partial DAGs (PDAGs), is the following [20]:

The PC algorithm

1. Assume a non-negative integer $m = 0$.
2. Let $\mathbb{G}$ *be a complete undirected graph.*
3. Repeat:
(a) For all pairs of variables (X, Y), *check* $Ind(X, Y|\boldsymbol{Z})$ *for all subsets* $\boldsymbol{Z}$ *such that* $|\boldsymbol{Z}| = m$ *and* $\boldsymbol{Z} \subset Adj(X)$ *or* $\boldsymbol{Z} \subset Adj(Y)$.
If there exists a $\boldsymbol{Z}$ *such as* $Ind(X, Y|\boldsymbol{Z})$,

then remove the edge $X - Y$ from $\mathbb{G}$, and add $\boldsymbol{Z}$ to SepSet(XY).
(b) Set $m = m + 1$.
Until no variable has more than m adjacencies,
or a stopping condition is satisfied.
4. Orientation rules are performed.
5. Return the partially directed acyclic graph $\mathbb{G}$.

Therein, $|\boldsymbol{X}|$ denotes the size of members in $\boldsymbol{X}$; $Adj(X)$ is a set of adjacent nodes to X. The orientation rules [21], described in step 4 of the algorithm, are as follows:

4-1. If $U \notin SepSet(XY)$, orient $X - U - Y$ as $X \rightarrow U \leftarrow Y$ (v-structure) for each uncoupled set of X and Y such as $X - U - Y$.
4-2. Repeat this step while more edges can be oriented.
4-2-1. Orient $U - Y$ as $U \rightarrow Y$ for each uncoupled set of X and Y such as $X \rightarrow U - Y$.
4-2-2. Orient $X - Y$ as $X \rightarrow Y$ for each set of X and Y such that a path exists from X to Y.
4-2-3. Orient $U - W$ as $U \rightarrow W$ for each uncoupled set of X and Y such as $X - U - Y$, $X \rightarrow W$, $Y \rightarrow W$, and $U - W$.

The completeness of the rules was proved by Meek [22].

The PC algorithm is performed under the *faithfulness assumption* described in section 2. Consequently, the algorithm can infer correct graph structures by finding conditional independence for probability distributions. However, if the assumption is violated, even though the true graph means $Ind(X, Y|\boldsymbol{Z})$ for X and Y and a conditional set $\boldsymbol{Z}$, the algorithm might find another false conditional set $\boldsymbol{Z'}$ for the test between X and Y, and then add $\boldsymbol{Z'}$ to *SepSet(XY)*. This false detection has *no influence* on removing the edge between X and Y correctly. However, the algorithm decides the wrong direction of edges using the orientation rules described above. In this situation, finding correctly conditional sets has a large influence on the directionality of edges in BNs.

We conducted the simulation study with various quantities of variables: {10, 20, 40, 80}, where each variable has all four possible states, and with networks of two types, i.e. the sparser and denser graphs, where sparser cases have the same number of edges as variables; the denser cases have twice. For each such graph, a random structure network was constructed with conditional probability tables (CPTs) of five types that were set by random numbers. The available sample size is varied in a range of {500, 1000, 2500, 5000, 10000}. When the number of conditional sets $|\boldsymbol{Z}|$ is large, the number of CI tests is intractably large because of a combinatorial explosion. Therefore, we did not perform CI tests and assume conditional dependence when $|\boldsymbol{Z}| \geq 5$. We selected a value of the hyperparameter N_c for β in (17) as 2.0, which shows good performance in preliminary experiments.

We set the performance criterion as counting *added edges*, *removed edges*, and *reversed edges*. Counting added edges expresses the consequence where two variables X and Y are not adjacent in original BNs but where an edge exists between them in reconstructed BNs. On the other hand, counting removed edges mean the opposite. Counting reversed edges means that if $X \rightarrow Y$ in the original, then $Y \rightarrow X$ in the output. The results are presented in Table 1 for sample sizes of 500, 1000, and 2500, and in Table 2 for sample sizes of 5000 and 10000. The values in the tables are averaged values of simulations for five different randomly set CPTs. We designate the PC algorithm with a standard G^2 test as *Std-PC* or *Std*, and the PC embedded with the MFE-IC as *MFE-PC*. These tables show that the counted quantities of extra added edges were very small, even for a small sample size such as 500 and even for denser structures. In contrast, quantities of removed edges are very large in both Std-PC and MFE-PC. The MFE-PC removed true edges more than Std-PC. In reversed checks, many errors were found in Std-PC. These characteristics were noticeable in large and denser networks. We discuss these results later. A key is apparently the *faithfulness condition* for understanding the results.

The MFE-PC seemed to underscore the effectiveness for deciding the direction of edges. It might be unfair, however, to conclude that because the MFE-PC removed more edges than Std-PC. Therefore, we defined *reversed ratio* as (number of reversed edges)/((true number of edges) – (number of removed)). Results of *reversed ratio* for denser networks are portrayed in Fig. 1, where the horizontal axis expresses the true number of edges, the vertical axis expresses the *reversed ratio*, and G2 and MFE respectively signify Std-PC and MFE-PC. These figures show that the MFE-PC outperforms Std-PC in deciding the direction of edges, especially for denser networks, even using samples such as 5000, which are not small.

We discuss these comparative results. As designed using the MFE principle, MFE-PC is expected to remove edges more than PC does for two reasons. The first is that MFE-PC performs CI tests in more cases than Std-PC, which does the test only for sufficiently large data size. For example, even when the data

Table 1. Results for the simulation using data sizes of 500, 1000, and 2500

		500				1000				2500			
		Sparser		Denser		Sparser		Denser		Sparser		Denser	
Type	Nodes	Std	MFE	Std	MFE	Std	MFE	Std	MFE	Std	MFE	Std	MFE
Added	10	0	0	0.6	0.2	0.4	0	1.8	0	0	0	0	0
	20	0	0	0.4	0.4	0.2	0	1.2	0	0	0	0.2	0
	40	0.2	0	0.4	0	0.6	0.4	0.8	0	0	0	0	0
	80	0.6	0.4	1.4	0.8	0.4	0.2	2.2	0.2	0.2	0.2	0	0
Removed	10	3.0	3.4	9.0	14.0	2.4	2.8	3.6	11.6	1.8	1.8	4.4	8.6
	20	4.2	7.6	19.0	26.2	3.0	5.8	11.4	22.0	2.0	2.8	12.8	18.6
	40	11.2	15.8	41.6	53.6	6.4	12.0	25.0	46.6	6.0	7.2	26.0	37.2
	80	21.4	32.0	81.2	109	11.0	21.2	48.6	93.8	9.0	12.0	54.4	73.8
Reversed	10	1.8	1.4	7.4	2.2	0.8	0.8	12.2	4.2	0.6	0.6	9.2	5.2
	20	5.4	2.0	14.6	6.4	5.2	2.6	22.2	7.6	2.8	2.4	13.2	7.6
	40	10.6	5.2	26.6	11.8	10.2	4.0	42.6	16.8	6.6	6.4	31.8	19.4
	80	18.8	10.6	54.2	26.6	21.0	11.2	86.4	26.0	11.6	10.8	56.6	38.8

Table 2. Results for simulations using data sizes 5000 and 10000

		5000				10000			
		Sparser		Denser		Sparser		Denser	
Type	Nodes	Std	MFE	Std	MFE	Std	MFE	Std	MFE
Added	10	0	0	1.2	0.2	0	0	0	0
	20	0	0	0.4	0.0	0	0	0	0
	40	0.4	0.4	0.0	0.0	0	0	0.2	0.2
	80	0.4	0.4	0.0	0.0	0.4	0.4	0	0
Removed	10	1.0	1.0	1.8	6.4	0.6	0.6	2.6	4.6
	20	0.4	1.4	6.2	14.6	0.6	0.6	8.6	10.4
	40	2.6	4.6	16.0	30.0	1.8	2.0	20.2	24.4
	80	4.6	7.2	24.2	59.2	4.4	4.6	40.0	48.0
Reversed	10	0.6	0.6	7.2	4.4	4.0	4.0	4.2	3.6
	20	2.6	2.4	20.8	9.6	6.2	6.2	12.0	11.4
	40	5.0	3.8	36.6	20.8	14.0	14.2	21.0	19.0
	80	8.4	7.8	78.0	38.0	24.0	24.2	48.5	44.3

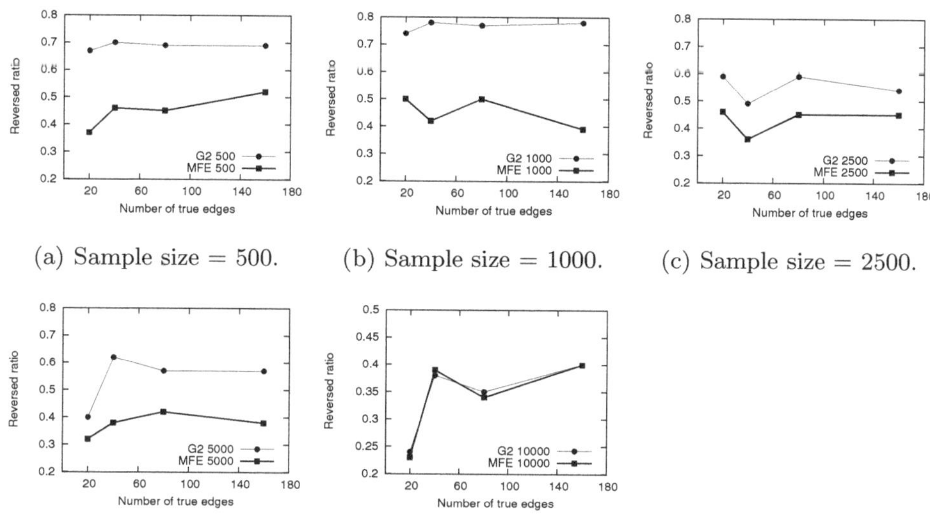

(a) Sample size = 500. (b) Sample size = 1000. (c) Sample size = 2500.

(d) Sample size = 5000. (e) Sample size = 10000.

Fig. 1. Ratio of reversed edges in the resultant graphs with denser BNs from use of a standard PC and PC embedded with the MFE-IC method

size $N = 5000$, Std-PC was unable to perform CI tests for $|\boldsymbol{Z}| \geq 3$ in this simulation, which implies that Std-PC might sometimes correctly happen to maintain some existing edges. The second is that, when confidence of existing edges is small because the data size is insufficiently large, the null hypothesis is not rejected because of the effect of maximum entropy. Therefore, MFE-IC method seemingly tends to prefer sparser graphs. These mean that MFE-IC method does draw edges only when high confidence for dependence is obtained. Additionally, we can comment on the fact that Std-PC incorrectly decided the direction of edges more than MFE-PC. This fact means that Std-PC detected conditional independence for invalid conditional sets $\boldsymbol{Z}$. In this case, wrong v-structures, divergence connections such as $X \leftarrow U \rightarrow Y$, and serial connections such as $X \rightarrow U \rightarrow Y$ were generated. This suggests that the simulation had difficulty realizing the faithfulness assumption. In other words, unfortunately, $Ind(X;Y|\boldsymbol{Z}) \Rightarrow Dsep^{\mathbb{G}}(X;Y|\boldsymbol{Z})$ was often violated in this simulation. This situation was also reported by Ramsey et al. for linear Gaussian models of DAGs [23]. However, we found the large difference of the reversed ratio, which shows that MFE-PC correctly found the conditional sets more than Std-PC. According to the discussion, we regard that MFE-IC method is especially preferable for causal discovery, where existing edges are expected to represent definite direct dependence between variables, and where direction has important meanings. In contrast, MFE-IC seems to be unsuitable for finding BNs in view of the predictive sense, for which the edges are allowed to be reversed for maintaining adequate parametric space size.

5 Conclusion

We proposed a method for improvement of conditional independence (CI) testing in small samples, which is a weak point of constraint-based learning Bayesian networks using the classical hypothesis tests. To do this, we introduced the minimum free energy principle with a "Data Temperature" assumption that relates probabilistic fluctuation to virtual thermal fluctuation. We defined a free energy using Kullback–Leibler divergence, which corresponds to an information-geometric view. This CI method incorporates the maximum entropy and maximum likelihood principles and converges to the classical hypothesis tests in asymptotic regions.

We also demonstrated the effectiveness of our method by embedding it in the well known PC algorithm. The results show that our method correctly identified the direction of the edges better than the standard tests did, which is expected to be effective for causal discovery where the orientation of edges is significant.

Acknowledgements

One author (T.I.) particularly thanks M. Kyojima, K. Shinozaki and T. Tanaka of Fuji Xerox Co., Ltd. for support and encouragement, and thanks T. Ogawa of the University of Electro-Communications for advice related to information geometry. The authors thank anonymous reviewers for fruitful comments, which help them improve the paper.

References

1. Pearl, J.: Probabilistic Reasoning in Intelligent Systems. Morgan Kaufmann, San Mateo (1988)
2. Heckerman, D., Geiger, D., Chickering, D.: Learning Bayesian networks: The combination of knowledge and statistical data. Machine Learning 20, 197–243 (1995)

3. Spirtes, P., Glymour, C., Scheines, R.: Causation, Prediction and Search, 2nd edn. MIT Press, Cambridge (2000)
4. Tsamardinos, I., Brown, L.E., Aliferis, C.F.: The max–min hill-climbing Bayesian network structure learning algorithm. Machine Learning 65(1), 31–78 (2006)
5. Callen, B.H.: Thermodynamics and An Introduction to Thermostatistics, 2nd edn. John Wiley & Sons, Hoboken (1985)
6. Pereira, F., Tishby, N., Lee, L.: Distributional clustering of English words. In: Proc. of Annual Meeting on Association for Computational Linguistics (ACL 1993), pp. 183–190 (1993)
7. Hofmann, T.: Probabilistic latent semantic analysis. In: Proc. of Conference on Uncertainty in Artificial Intelligence (UAI 1999), pp. 289–296 (1999)
8. LeCun, Y., Huang, F.J.: Loss functions for discriminative training of energy-based models. In: Proc. of International Workshop on Artificial Intelligence and Statistics (AISTATS 2005), pp. 206–213 (2005)
9. Lehmann, L.E.: Testing Statistical Hypotheses, 2nd edn. John Wiley & Sons, New York (1986)
10. Cover, T.M., Thomas, J.A.: Elements of Information Theory, 2nd edn. John Wiley & Sons, Hoboken (2006)
11. Isozaki, T., Kato, N., Ueno, M.: Minimum free energies with "data temperature" for parameter learning of Bayesian networks. In: Proc. of IEEE International Conference on Tools with Artificial Intelligence (ICTAI 2008), pp. 371–378 (2008)
12. Amari, S., Nagaoka, H.: Method of Information Geometry. Oxford University Press, New York (2000)
13. Pearl, J.: Causality, models, reasoning, and inference. Cambridge University Press, New York (2000)
14. Kullback, S.: Information Theory and Statistics. Dover Publications, Mineola (1968)
15. Friedman, N., Geiger, D., Goldszmidt, M.: Bayesian network classifiers. Machine Learning 29(2-3), 131–163 (1997)
16. Dash, D., Druzdzel, M.J.: Robust independence testing for constraint-based learning of causal structure. In: Proc. of Conference on Uncertainty in Artificial Intelligence (UAI 2003), pp. 167–174 (2003)
17. Clarke, B., Barron, A.: Jeffreys' prior is asymptotically least favorable under entropy risk. Journal of Statistical Planning and Inference 41, 37–60 (1994)
18. Silander, T., Kontkane, P., Myllymaki, P.: On sensitivity of the map Bayesian network structure to the equivalent sample size parameter. In: Proc. of Conference on Uncertainty in Artificial Intelligence (UAI 2007), pp. 360–367 (2007)
19. Isozaki, T., Kato, N., Ueno, M.: "Data temperature" in minimum free energies for parameter learning of Bayesian networks (to appear)
20. Neapolitan, R.E.: Learning Bayesian Networks. Prentice-Hall, Upper Saddle River (2004)
21. Verma, T., Pearl, J.: An algorithm for deciding if a set of observed independencies has a causal explanation. In: Proc. of Conference on Uncertainty in Artificial Intelligence (UAI 1992), pp. 323–330 (1992)
22. Meek, C.: Causal inference and causal explanation with background knowledge. In: Proc. of Conference on Uncertainty in Artificial Intelligence (UAI 1995), pp. 403–410 (1995)
23. Ramsey, J., Spirtes, P., Zhang, J.: Adjacency-faithfulness and conservative causal inference. In: Proc. of Conference on Uncertainty in Artificial Intelligence (UAI 2006), pp. 401–408 (2006)

Kernel-Based Copula Processes

Sebastian Jaimungal[1] and Eddie K.H. Ng[2]

[1]Department of Statistics, University of Toronto
sebastian.jaimungal@utoronto.ca
[2]The Edward S. Rogers Sr. Department of Electrical and Computer Engineering, University of Toronto
eddie@psi.toronto.edu

Abstract. Kernel-based Copula Processes (KCPs), a new versatile tool for analyzing multiple time-series, are proposed here as a unifying framework to model the interdependency across multiple time-series and the long-range dependency within an individual time-series. KCPs build on the celebrated theory of copula which allows for the modeling of complex interdependence structure, while leveraging the power of kernel methods for efficient learning and parsimonious model specification. Specifically, KCPs can be viewed as a generalization of the Gaussian processes enabling non-Gaussian predictions to be made. Such non-Gaussian features are extremely important in a variety of application areas. As one application, we consider temperature series from weather stations across the US. Not only are KCPs found to have modeled the heteroskedasticity of the individual temperature changes well, the KCPs also successfully discovered the interdependencies among different stations. Such results are beneficial for weather derivatives trading and risk management, for example.

Keywords: Copula, Kernel Methods, Gaussian Processes, Time-Series Analysis, Heteroskedasticity, Maximum Likelihood Estimation, Financial Derivatives, Risk Management.

1 Introduction

Gaussian processes (GPs) [3] have enjoyed wide acceptance in many machine learning applications as a non-linear regression and classification tool. GPs cleverly exploited the closure property of Gaussian random variables, under marginalization and conditioning, for efficient learning and inference. Model selection can be handled naturally under a Bayesian framework, and missing data can be treated effortlessly. The same closure property also limits the GPs to make predictions with Gaussian distributions. However, the world is an inherently non-Gaussian place. The Gaussian prior assumption is simply not valid in many real-world applications such as the modeling of financial data[12], wind-speeds and temperatures, and catastrophic damages (e.g. floods, earth-quakes and forest fires[6] [8]) where it has been shown that risk-management based on Gaussian assumptions severely under-estimates the real-risks and greatly inflates the social and economical costs. The ability to accommodate non-Gaussian features is therefore crucial.

W. Buntine et al. (Eds.): ECML PKDD 2009, Part I, LNAI 5781, pp. 628–643, 2009.

Copula theory [13] on the other hand has been proven to be an effective way to model complex dependencies and distributional behavior by decomposing the joint-distribution into the dependency structure and the individual marginal distributions. However, the applications of copula functions are mostly static while its dynamic extensions are either limited to first-order Markov processes [16] or simply using copula functions as a way to combine multiple time-series [5] [11] [14] [15].

This work proposes a novel time-series analysis model dubbed *Kernel-based Copula Processes* (KCPs), and offers a way to better address the non-Gaussian features of real-world data. KCPs are a powerful union of GP and copula functions. Unlike GPs where a time-series is modeled as a high-dimensional Gaussian distribtuion, KCPs use a copula function to separate the specification of dependency structure and the behavior in the marginal distributions; thus allowing for complex non-Gaussian behavior, while retaining most of the attractiveness of GPs such as efficient learning, inference, and natural missing data handling. For example, in the univariate setting, a high-dimensional elliptical copula function is used to capture any long-range temporal dependencies, while an arbitrary non-Gaussian distribution is used to match the non-Gaussian behavior in the marginal distribution. A kernel function is used along with the elliptical copula to keep the model parsimonious. As will be shown in this work, heteroskedastic processes can also be captured with KCPs through proper design of this kernel function.

Under this proposed framework, we further propose a natural extension for multi-variate time-series which allows multiple time-series with drastically different characteristics to be described under a single model. Further, the modular formulation of multivariate KCPs lends itself naturally to a multi-core/processor environment for parallel computation.

Section 2 presents the details of the KCPs framework in both univariate and multivariate settings. Section 3 presents a synthetic univariate example to illustrate the modeling power enjoyed by the non-Gaussian features of KCPs. Finally, section 4 presents a real-life example of modeling multiple temperature series from weather stations across the United States. A new kernel function is derived especially to accomodate the heteroskedastic nature of the temperature data. The performance of different flavors of KCPs are compared with GPs and GARCH [2].

2 Kernel-Based Copula Processes (KCPs)

This section introduces the basic formulation of copula processes for time-series analysis. Section 2.1 provides the formulation for the univariate time-series case which shows how kernel methods can bring a sense of time or parametric ordering to a high-dimensional copula, thus allowing the modeling of random processes instead of just a collection of random variables. This approach is reminiscent of the formulation of Gaussian processes (GP) as will be discussed in this section. Section 2.2 extends the basic formulation to the multiple time-series case. Section 2.3 presents the estimation methods for KCPs.

2.1 Univariate Time-Series

Consider a one-dimensional function $g(x)$ with a set of noisy observations $\mathbf{y} = \{y_1, ..., y_N\}^T = g(\mathbf{x}) + \varepsilon_n$ at $\mathbf{x} = \{x_1, ..., x_N\}^T$. In performing tasks such as regression or forecasting, the objective is to predict the values of the function, $\mathbf{y}^* = \{y_1^*, ..., y_N^*\}^T$, at $\mathbf{x}^* = \{x_1^*, ..., x_M^*\}^T$ where $\mathbf{x} \bigcap \mathbf{x}^* = \emptyset$. Under the KCP assumption, the joint distribution of the observed and predicted values is given by:

$$\mathbb{P}(\mathbf{Y} \leq \mathbf{y}, \mathbf{Y}^* \leq \mathbf{y}^* | \mathbf{x}, \mathbf{x}^*) = \mathbb{C}(\{F_i(y_i)\}_{i=1}^N, \{F_j(y_j^*)\}_{j=1}^M | \mathbf{x}, \mathbf{x}^*) \quad (1)$$

$$= \mathbb{C}(\{u_i\}_{i=1}^N, \{u_j^*\}_{j=1}^M | \mathbf{x}, \mathbf{x}^*) \quad (2)$$

where $\mathbb{C}(\cdot)$ is the copula function[1] of the process, $F_i(\cdot)$ are the marginal cumulative distribution functions (CDFs), and u_i is the transformed y_i via the CDFs, i.e. $u_i \doteq F_i(y_i)$.

The joint density of $(\mathbf{y}, \mathbf{y}^*)$ can be obtained by differentiating with respect to $\{\mathbf{y}, \mathbf{y}^*\}$:

$$\mathbb{P}(\mathbf{y}, \mathbf{y}^* | \mathbf{x}, \mathbf{x}^*) = \mathfrak{c}(\{F_i(y_i)\}_{i=1}^N, \{F_j(y_j^*)\}_{j=1}^M | \mathbf{x}, \mathbf{x}^*) \prod_{i=1}^N f_i(y_i) \prod_{j=1}^M f_j(y_j^*) \quad (3)$$

where $f_i(\cdot)$ are the marginal PDFs and $\mathfrak{c}(\cdot)$ is the so-called *copula density function* given by the derivative of the copula function with respect to its parameters:

$$\mathfrak{c}(\{u_i\}_{i=1}^N) = \frac{\partial^N \mathbb{C}}{\partial u_1 \partial u_2 \ldots \partial u_N} \quad (4)$$

In this formulation, all the marginal distributions are constrained to have the same parametric form. The role of the copula function $\mathbb{C}$ is to capture the entire dependency structure of $\{\mathbf{y}, \mathbf{y}^*\}$. There exists a vast array of parametric or empirical copula functions which allow the specification of complex and heavy-tail dependencies. Nelson [13] provides a wide collection of examples. However, when attention is focused on the elliptical class of copula functions, the dependency structure in the copula function can be specified by a kernel function to achieve parsimonious model specification.

For example, consider a copula process with a *Gaussian copula* function as follows:

$$\mathbb{C}_\Lambda(\{F_i(y_i)\}_{i=1}^N) = \Phi_\Lambda(\{\Phi^{-1}(F_i(y_i))\}_{i=1}^N) \quad (5)$$

[1] The copula function is a joint distribution function of uniform random variables. Sklar's theorem [13] states that given any joint distribution $H(y_1, \cdots, y_N)$ of random variables $\{y_i\}$ and margins $\{F_i(y_i)\}$. There exists a copula function $\mathbb{C}$ such that for all y_i in the respective domain, $H(y_1, \cdots, y_N) = \mathbb{C}(F_1(y_1), \cdots, F_N(y_N))$. If $\{F_i(y_i)\}$ are continuous, then $\mathbb{C}$ is unique. The converse is also true.

where the $\Phi(\cdot)$ and $\Phi_\Lambda(\cdot)$ are the standardized univariate normal CDF and the zero-mean multi-variate CDF (with covariance matrix Λ) respectively. The covariance matrix Λ can be defined with the Gram matrix:

$$\Lambda = \begin{bmatrix} k(x_1, x_1) & \dots & k(x_1, x_N) \\ \vdots & \ddots & \vdots \\ k(x_N, x_1) & \dots & k(x_N, x_N) \end{bmatrix} + \sigma_n^2 \mathbf{I} \tag{6}$$

where σ_n^2 is the noise variance and $k(x_i, x_j)$ is a kernel function describing the covariance[2] between the pairs of random variables y_i and y_j. The radial basis function (RBF),

$$k_{RBF}(x_i, x_j) = \exp\left(-\frac{1}{2h}(x_i - x_j)^2\right) \tag{7}$$

and the Ornstein-Uhlenbeck functions,

$$k_{OU}(x_i, x_j) = \exp\left|-\frac{1}{h}(x_i - x_j)\right| \tag{8}$$

are examples of commonly used and versatile kernel functions. More examples of kernel functions can be found in [3]. In the Appendix, we show how to construct a nonstationary kernel function from a stochastic differential equation (SDE) based on the understanding of the dynamics of the temperature data studied in Section 4.

Notice that if the copula is Gaussian and the marginal distributions are assumed to be normal, then the copula process reduces to the familiar GPs.

2.2 Multivariate Time-Series

In practical problems, random processes rarely exist in logical isolation but rather typically form an intricate web of dependencies. Thus practitioners are often faced with the challenge of analyzing not a univariate time-series, but multiple codependent time-series. To this end, we propose a multivariate KCP framework to capture such complex interdependencies.

In the cases where the daily values from each series are highly related and yet there exists no ordering among them, one natural parametrization of the joint distribution is as follows:

$$\begin{aligned} &\mathbb{P}(\mathbf{Y}^{(1)} \leq \mathbf{y}^{(1)}, \dots, \mathbf{Y}^{(M)} \leq \mathbf{y}^{(M)}) = \\ &\qquad \mathbb{C}_b\left[\mathbb{C}_1\left(\{F_i^{(1)}(y_i^{(1)})\}_{i=1}^{N_1}\right), \dots, \mathbb{C}_M\left(\{F_i^{(M)}(y_i^{(M)})\}_{i=1}^{N_M}\right)\right] \end{aligned} \tag{9}$$

where $\mathbb{C}_b(\cdot)$ is the *binding copula function* connecting M individual time-series together. The individual time-series are modeled as in the single KCP case

[2] For non-Gaussian random variables, the kernel function is not the covariance itself, but rather it dictates the behavior of the covariance.

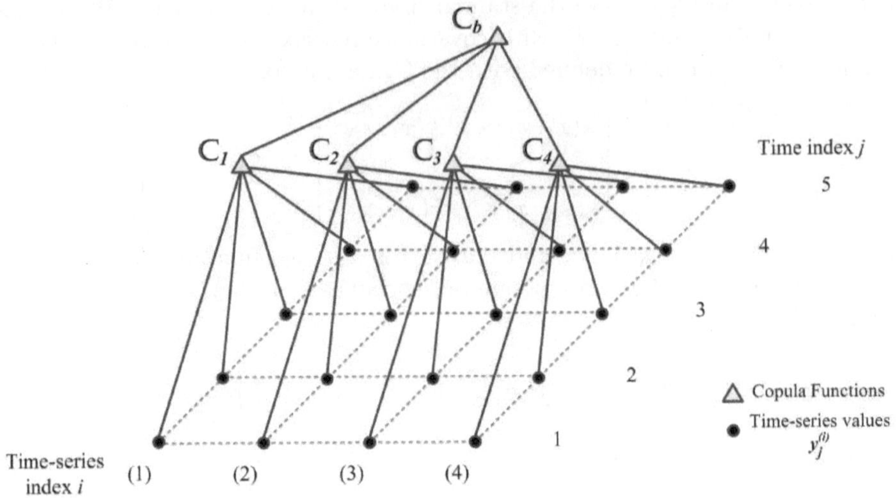

Fig. 1. A schematic illustration of a binding copula linking multiple KCPs. In this case, there are four time-series, each with five time samples. The temporal dependencies of each time-series are first described by the individual KCPs $\{\mathbb{C}_i\}_{i=1}^4$. The inter-time-series dependency is then described by a binding copula $\mathbb{C}_b$.

(Section 2.1), while the dependent structure resides with the binding copula. Figure 2.2 provides a schematic depiction of a binding copula joining multiple copula processes together. The advantage of this configuration is that time-series with greatly different individual dynamics can be captured by the individual copula processes, before they are joined together by the binding copula. For example, one time-series can be the changes in the LIBOR[3] rate while others could be the returns of any stocks that are sensitive to interest-rates, such as utilities companies or companies with high debt load. In such cases, the stock returns would have much higher volatility but shorter term correlation than the changes in the LIBOR rate.

2.3 Learning

The learning of KCPs can be performed by maximizing likelihood. The likelihood function of a generic KCP is given by:

$$\mathcal{L}(\theta) = \mathbb{P}(\mathbf{y}|\mathbf{x}, \theta) = \mathfrak{c}(\{F_i(y_i)\}_{i=1}^N) \prod_{i=1}^N f_i(y_i) \tag{10}$$

[3] The London Interbank Offered rate (LIBOR) is a daily reference rate based on the interest rates at which banks offer to lend to other banks [10].

where $\mathbf{y}$ is the set of training data and θ is the set of hyper-parameters of the copula process.

For instance, assuming the choice of Gaussian copula and margins are Student's t-distributed, then the negative log-likelihood function is given as

$$-\log(\mathcal{L}(\theta)) = \frac{1}{2}\log|\Lambda| + \frac{1}{2}\mathbf{z}^T\Lambda\mathbf{z} - \frac{1}{2}\mathbf{z}^T\mathbf{z} - \sum_{i=1}^{N}\log(t(y_i, \upsilon)) \tag{11}$$

where $\mathbf{z}^T = \{z_1 = \Phi^{-1}(T_\upsilon(y_1)), \ldots, z_N = \Phi^{-1}(T_\upsilon(y_N))\}$ is the transformed random vector $z_i \sim \mathcal{N}(0,1)$, and $T_\upsilon(y_i)$ and $t_i(y_i, \upsilon)$ are the univariate Student's t-distribution and density functions (with degree of freedom υ) respectively; and θ are the hyper-parameters of the kernel function. Here, without loss of generality the time-series samples are assumed to be standardized.

The likelihood function for multivariate KCPs can be derived by taking the $N \cdot M - th$ order derivative of Eq. (9) with respect to $y_i^{(j)}$:

$$\begin{aligned}
\mathcal{L}(\theta) &= \mathbb{P}(\mathbf{y}^{(1)}, \ldots, \mathbf{y}^{(M)}) \\
&= \frac{\partial^M \mathbb{C}_b}{\partial \mathbb{C}_1 \ldots \partial \mathbb{C}_M} \cdot \left[\frac{\partial^{N_1}\mathbb{C}_1}{\partial u_1^{(1)} \ldots \partial u_{N_1}^{(1)}} \prod_{j=1}^{N_1} f_j^{(1)}(y) \right] \cdot \ldots \\
&\qquad \cdot \left[\frac{\partial^{N_M}\mathbb{C}_M}{\partial u_1^{(M)} \ldots \partial u_{N_M}^{(M)}} \prod_{j=1}^{N_M} f_j^{(M)}(y) \right] \\
&\doteq \mathfrak{c}_b \cdot \mathcal{L}_1(\theta_1) \cdot \ldots \cdot \mathcal{L}_M(\theta_M)
\end{aligned} \tag{12}$$

The overall likelihood function factorizes into the product of the binding copula density, $\mathfrak{c}_b$, and the likelihood functions of the copula processes of the individual time-series, $\mathcal{L}_i(\theta_i)$. This modular property allows for two ways to learn the model parameters: (i) all the parameters are learnt at the same time; or (ii) the parameters for each time-series are learnt separately and are fixed while learning the parameters of the binding copula. The latter approach is particular convenient for equity portfolio analysis where individual names are often added and removed as portfolio compositions change.

2.4 Inference

One of the strengths for KCPs (which is also enjoyed by GPs) is that the entire predictive distribution is available during inference, thus predictions are not limited to point prediction, which allows for much more accurate risk management.

The predictive distribution of a univariate KCP is given by:

$$\mathbb{P}(\mathbf{y}^*|\mathbf{y}) = \frac{\mathbb{C}_{\tilde{\Lambda}}(\{F_i(y_i)\}_{i=1}^N, \{F_i(y_j^*)\}_{j=1}^M)}{\mathbb{C}_\Lambda(\{F_i(y_i)\}_{i=1}^N)} \tag{13}$$

where Λ and $\tilde{\Lambda}$ are the Gram matrices of the observations and the combined observations and targets respectively as defined in Eq. 6. The dependency on $\mathbf{x}$ and $\mathbf{x}^*$ is implicit and it is dropped from the notation for clarity of presentation.

The *maximum a-posterior* (MAP) estimator could be used to provide a point estimate of the target values:

$$\bar{\mathbf{y}^*} = \max_{\mathbf{y}^*} \mathbb{P}(\mathbf{y}^*|\mathbf{y}) \tag{14}$$

When the choice of copula functions and marginal distributions result in a symmetric predictive distribution, the MAP estimate will coincide with the mean-squared estimator $\mathbb{E}|_{\mathbb{P}(\mathbf{y}^*|\mathbf{y})}[\mathbf{y}]$.

2.5 Model Selection and Performance Metrics

A model selection method is required to select the best of parametric functions for both the copula and marginal component in KCPs. Given the modularity of multivariate KCPs, model selection can be performed in a modular fashion as well. That is, the best univariate KCP is selected for the individual time series and the associated results are subsequently used to learn the best binding copula. Furthermore, in regression and out-of-sample prediction of stochastic processes, attention must be focused on obtaining an accurate predictive distribution. Point-predictions are meaningless since the realized value from a stochastic process will almost always differ from the prediction. Knowing the entire predictive distribution on the other hand yields a much more complete picture. To this end, conventional performance metrics such as least-square errors between the point prediction and a set of realized values is of little use.

In this study, in addition to the maximized likelihood, the probability-integral-transform (PIT) [7] is employed as a complementary criteria for model selection. The maximized likelihood relates to the predicted density having minimum variance about the observations[4], while PIT measures how well the the predictive distribution corresponds to the true underlying distribution. Diebold *et al.* [7] evaluate PIT graphically, but it can also be evaluated quantitatively with the Anderson-Darling criterion[4] [1]. We will use log-likelihood and likelihood ratios to rank the competing models, while using the PIT with the Anderson-Darling criterion to evaluate the validity of the model on a standalone basis.

In evaluating the goodness-of-fit in the multiple time-series level, a visual comparison of the Gram and pair-wise Pearson's correlation matrices will also be given.

[4] The Anderson-Darling criterion has the added advantage that it produces a score that focuses goodness-of-fit at the tail regions of the distributions.

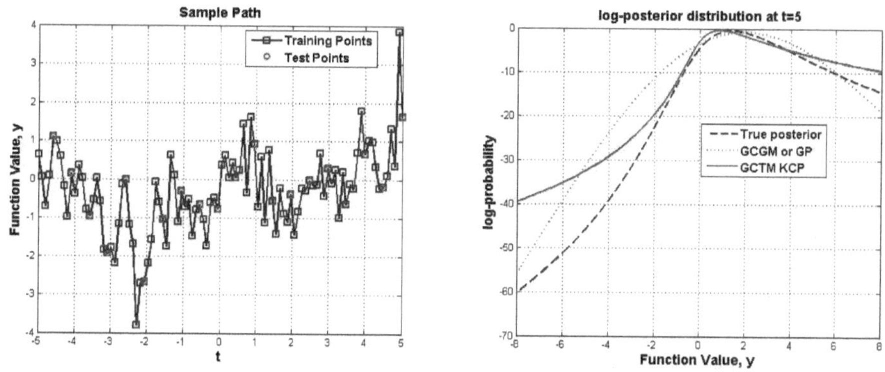

Fig. 2. A synthetic data set generated by a AR(2) process with skew-normal innovations is shown on the left panel. The posterior distribution of the process at $t = 5$ and the corresponding predictive distributions given by GP and GCTM KCP are shown on the right panel.

3 Synthetic Example

In this section a synthetic data set is used to illustrate the KCP in action in its simplest form. Here, a sample path is drawn from an AR(2) process with skew-normal[5] innovations:

$$y_t = 0.1y_{t-1} + 0.05y_{t-1} + 1 + \varepsilon_t \tag{16}$$

where ε_t is a standardized skew-normal with skewness of 0.5.

One hundred samples were taken at regular intervals within the range of $x = [-5, 5]$. Every fifth sample is taken as the test set (i.e. the regression / prediction targets), while the rest of the samples are used as the training set. Then two KCPs: Gaussian copula with Gaussian marginal distribution (GCGM) (or simply a Gaussian process (GP)) and Gaussian copula with t-marginal (GCTM), are learnt using MLE. The Ornstein-Uhlenbeck kernel (8) is used in both cases due to its auto-regressive nature.

[5] The skew-normal distribution can be obtained by taking the limit of the skew-t distribution by Hansen[9] as the degree of freedom η approaches infinity. The full formulation of the standardized skew-t distribution is reproduced here for reference:

$$g(z|\eta, \lambda) = \begin{cases} bc\left(1 + \frac{1}{\eta-2}\left(\frac{bz+a}{1-\lambda}\right)^2\right)^{-(\eta+1)/2} & z < -\frac{a}{b} \\ bc\left(1 + \frac{1}{\eta-2}\left(\frac{bz+a}{1+\lambda}\right)^2\right)^{-(\eta+1)/2} & z \geq -\frac{a}{b} \end{cases} \tag{15}$$

where $2 < \eta < \infty$, and $|\lambda| < 1$. The constants a, b, and c are given by $a = 4\lambda c\left(\frac{\eta-2}{\eta-1}\right)$, $b^2 = 1 + 3\lambda^2 - a^2$, and $c = \frac{\Gamma\left(\frac{\eta+1}{2}\right)}{\sqrt{\pi(\eta-2)}\Gamma\left(\frac{\eta}{2}\right)}$.

To illustrate the superior modeling power of KCPs over GPs, we consider the predictive distribution generated by a GP and a GCTM KCP in Fig. 2. This experiment is a simple example to illustrate the power and flexibility of copula processes simply by changing the marginal distributions from Gaussian to Student's t. Recall a GP is a special case of the GCTM KCP as the degree of freedom parameter $\upsilon \to \infty$. The two particular types of KCPs were chosen to illustrate what a slight departure from GPs framework can achieve.

As depicted, the predictive distribution produced by GCTM KCP is much closer to the true posterior distribution of the AR(2) process than that produced by GP. GP, restricted by its symmetric nature, must over shift its mean to the positive side to compensate for the excess probability mass in that region. This would introduce even greater prediction errors if point predictions are considered. GCTM KCP on the other hand is able to produce an asymmetric predictive distribution based on observed data to much better match the true distribution.

4 Real-Life Application

4.1 Data

To showcase the modeling power of KCPs with multiple time-series, we consider the daily maximum temperatures recorded by the United States Historical Climatology Network (HCN) [?]. Data from as far back as the early 1800s from a network of 1219 weather stations are freely downloadable from the Network's website. This study will consider the data from a few arbitrary three-year periods from 156 stations of various states. Missing data is commonplace in such data as equipment is moved and maintained; however, KCPs handle missing data naturally.

The latitude/longtitude and elevation coordinates for each weather station are also known. We will later use this information to help learn the dependency structure.

To focus on the stochastic nature of the data, we first subtract a sinusoidal seasonal trend from that data. A customized sinusoidal function is used for each weather station based on the least-square error criteria. An example of the temperature data with the fitted sinusoidal trend is given in Fig. 3a with the detrended version in Fig. 3b.

4.2 Model

Since there exists no natural ordering among the weather stations, we will take advantage of the modularity of the KCPs and first learn a univariate KCP for each weather station and connect them with a binding copula.

Upon closer examination of the detrended data, some notable features are observed which helped narrow down the modeling choices.

First, the second moment autocorrelation function (ACF), i.e. the ACF of the square of the data, experiences an annual cycle as illustrated in Fig. 3c. This implies the rate of fluctuation, or volatility, of maximum temperature goes through

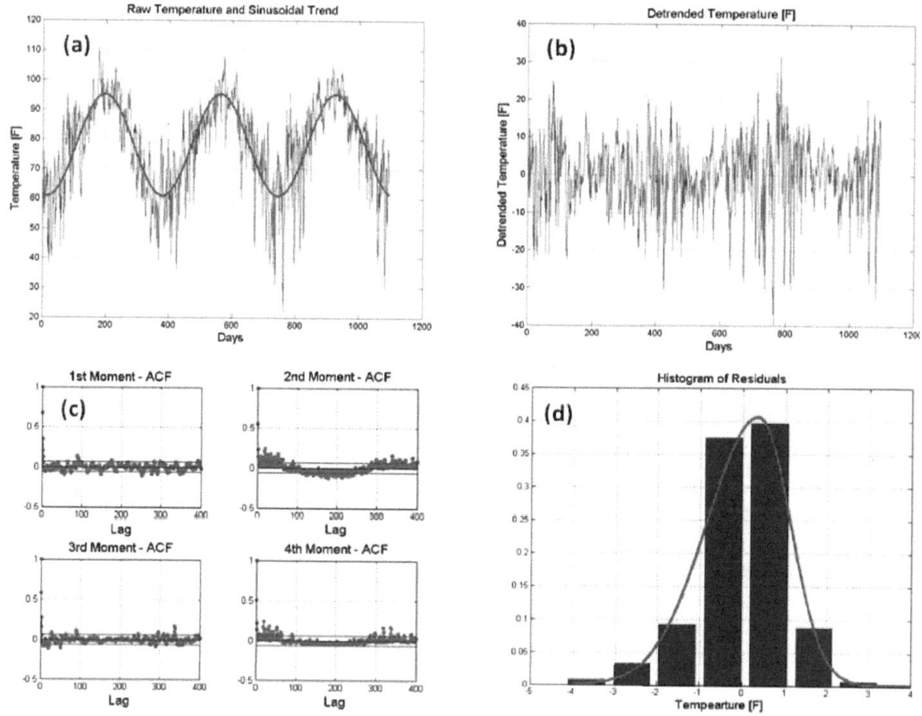

Fig. 3. Weather station TX410120 of Texas. Panel (a) depicts the raw maximum temperature data and the sinusoidal trend; Panel (b) shows the detrended data; Panel (c) shows the auto-correlation function of the first four moments of the detrended data; Panel (d) shows the empirical density and the estimated skew-normal distribution. The second moment ACF in (c) and the detrended temperature in (b) clearly shows the periodic volatility of the residuals.

an annual cycle. On the other hand, the first moment ACF drops off relatively quickly (in a few days time), but this also implies recent temperature trends tend to persist. This provides an important hint in designing or selecting the type of kernel function that would be appropriate for the analysis. In this case, we designed a heteroskedastic kernel function that satisfies the above features:

$$\begin{aligned} k(t_1, t_2) = \sigma^2 \cdot e^{-\kappa \Delta t} \Bigg[\frac{1}{2\kappa} + \frac{\alpha}{4\kappa^2 + T^{-2}} \Bigg(2\kappa \cdot \sin\left(\frac{\min(t_1, t_2)}{T} + \phi \right) \\ - \frac{1}{T} \cos\left(\frac{\min(t_1, t_2)}{T} + \phi \right) \Bigg) \Bigg] \end{aligned} \tag{17}$$

where σ is roughly the average volatility, κ is the inverse of length scale of dependency in days, T is the period (= 1 year) and ϕ is the phase of volatility. Please see the Appendix for the derivation. Note the kernel function is non-stationary as the volatility of maximum temperature depends on the day of the year.

Another helpful observed feature in the detrended data, is that a skewed-normal fits the empirical histogram well. Figure 3d shows a sample empirical histogram and the best-fit skew-normal distribution. Although the histograms are not strictly valid[6], it helps us narrow down the choice of marginal distributions.

It is important to note that the above analysis is not necessary for using KCPs. However, it is a simple way to incorporate additional domain knowledge by providing the initial values of the parameters[7] for MLE and as a basis for forming the prior distribution for Bayesian learning.

For this study, five competing models for the univariate time-series were investigated. These are (i) GP, (ii) Gaussian copula with skew-normal marginal (GCSM) KCP, (iii) t-copula with Gaussian marginal (TCGM) KCP, (iv) t-copula with skew-normal marginal (TCSM) KCP, and (v) GARCH(1,1) (with t-innovations) model. The GARCH(1,1) model is included here because it is one of the best-known models for heteroskedastic processes [2][10]. All KCPs above use the heteroskedastic kernel function (17).

The Gaussian- and t-copula are used as a binding copula to describe the interdependencies among the temperature series. In this case the Gaussian RBF kernel (7) is used, with the weighted distance metric:

$$d(i,j)^2 = d^2_{\texttt{lat}} + d^2_{\texttt{long}} + w \cdot d^2_{\texttt{elev}} \tag{18}$$

where $d(i,j)$ is the weighted distance between weather station i and j, d_x are distances along latitude, longitude, and elevation, and w is the weight on the distance along elevation.

4.3 Results

First we examine the results from the univariate KCPs. Figure 4 provides a sample comparison of the detrended data from weather station MD181750 of Maryland. The first column of panels from left to right show the time series, the corresponding ACF of the second moment, the histograms, and the q-q plots. The first row shows the plots of the detrended data while the subsequent rows show the samples generated from the corresponding models after learning on the same set of the detrended data. All KCP models work relatively well in the sense that the annual periodic change in volatilities was recovered as evident in both the time-domain plots and the ACF plots. The GARCH(1,1) model picked up some periodicity in

[6] Histograms are conventionally used to visually evaluate the probability density of a random variables. A random process on the other hand is a collection of random variables. Thus a sample path of a time-series (which is a random process), represents a series of draws from a potentially different distribution. Therefore, collapsing all the samples from a time-series to form a histogram is axiomatically incorrect.

[7] The skewness parameter can be obtained by the maximum likelihood fit of histograms of the residuals as depicted in 3d and the length scale parameter κ in the kernel function (17) can be estimated by $\hat{\kappa} = -\log(\rho|_{\tau=1})$ where $\rho(\tau)$ denotes the ACF for the first moment at lag τ.

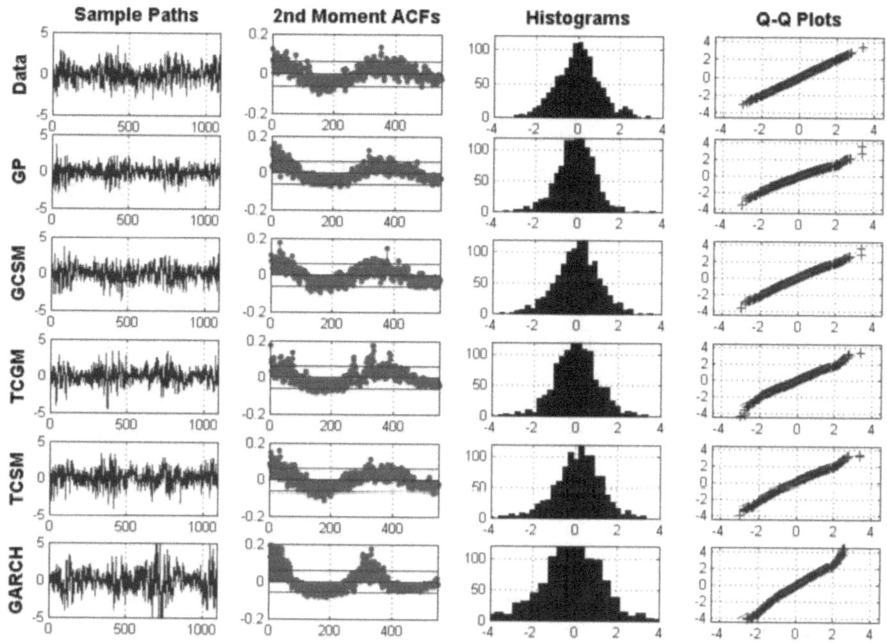

Fig. 4. Comparing stylized facts of the detrended data and sample paths generated by models with MLE parameters based on the weather station MD181750 of Maryland

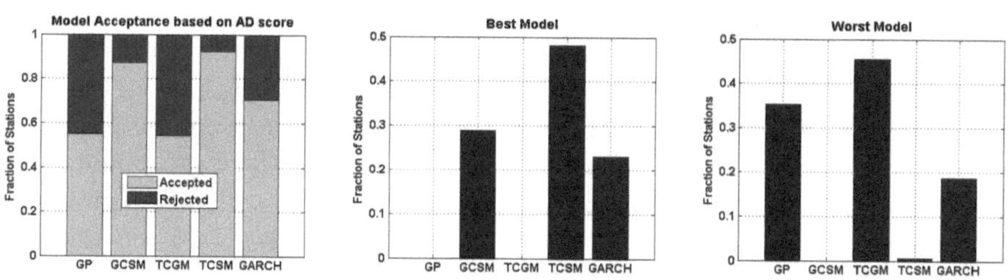

Fig. 5. Univariate model selection results. The left panel shows the fraction of model acceptance/rejection based on the Anderson-Darling criterion; the middle and right panels show the fractions of each models ranked to be the most and the least appropriate model for the univariate temperature series according to the achieved log-likelihood.

volatility as well, but the period was incorrect. Examining the histograms and the q-q plots visually, the TCSM KCP seems the best match with the data.

More generally, using the data from 156 weather stations across the US, the Anderson-Darling test rejects the hypothesis that GP and TCGM KCP are sufficiently good models (with 95% confidence level) in ~43% and ~42% of the cases; whereas, the TCSM KCP was shown to be the best model as it was rejected in only ~7.5% of the cases. The acceptance and rejection frequencies for

Table 1. Medians of pair-wise likelihood ratios among different models and the corresponding critical values for the likelihood ratio tests at 95% confidence level. Note that only the upper triangular part of this table is shown as the table is anti-symmetric.

	p_{GP}	p_{GCSM}	p_{TCGM}	p_{TCSM}	p_{GARCH}
q_{GP}	-	5.41 (3.84) dB	0.0 (3.84) dB	6.05 (5.99) dB	4.6 (5.99) dB
q_{GCSM}	-	-	-5.6 (0.00) dB	0.0 (3.84) dB	-4.6 (3.84) dB
q_{TCGM}	-	-	-	5.8 (3.84) dB	5.1 (3.84) dB
q_{TCSM}	-	-	-	-	-5.0 (0.00) dB
q_{GARCH}	-	-	-	-	-

all models are shown in the left panel of Fig. 5. The middle and the right panels of Fig. 5 show the frequency of each model being the *best* and the *worst* model for each data set as ranked by the maximized likelihoods. The GCSM and TCSM KCPs clearly dominate the GP and TCGM KCP, demonstrating that most of the modeling power (for this data set) comes from the skewness of the marginal distributions while the extra tail-dependency from the t-copula provides an incremental gain. The performance of the GARCH(1,1) model with t-innovations seems to be hit-and-miss with this data set, yet it is quite impressive that it ranks first in about 22% of the time, even though the KCPs use a kernel function that is custom designed for this data set. However, it is important to note that only the maximized likelihoods are used to rank models here. Model complexity (i.e. number of parameters in each model) is not considered. The likelihood ratio test that follows will address this issue by using a significance threshold which depends on the number of model parameters.

Table 1 lists the *median* pair-wise likelihood ratios computed across the data from 156 weather stations for each model-pair, where $\frac{p_x}{q_y}$ denotes the likelihood ratios of model x over model y. The corresponding critical values for the likelihood ratio tests at a 95% confidence level are also listed in parentheses. Note, if a particular pair-wise likelihood ratio $\frac{p_x}{q_y}$ is greater than the corresponding critical value, then model x is preferred over model y with statistical significance. The number of parameters used in each model is also taken into account when computing the critical values, thus balancing the desire for performance and the aversion to model complexity. Here, the likelihood ratios again show the superiority of KCPs over GPs and GARCH(1,1) by allowing different combinations of copulas and marginal distributions for the specific applications. In particular, both GCSM and TCSM are better than GP, TCGM, and GARCH(1,1) with statistical significance. Further, the likelihood ratios for $\frac{p_{TCGM}}{q_{GP}}$ and $\frac{p_{TCSM}}{q_{GCSM}}$ confirm that the t-copula provides statistically insignificant performance gain over the Gaussian copula for this data set. Thus, on average, the GCSM KCP is the most powerful and yet parsimonious model for this application.

Now we shift our attention to the effectiveness of the binding copula in capturing the dependencies of multiple time-series for weather stations located in the same state. Figure 6 shows the Gram matrices of binding copulas and pair-wise correlation matrices of the detrended data after transforming by the distribution induced by the univariate KCPs that was ranked to be the best earlier. Recall

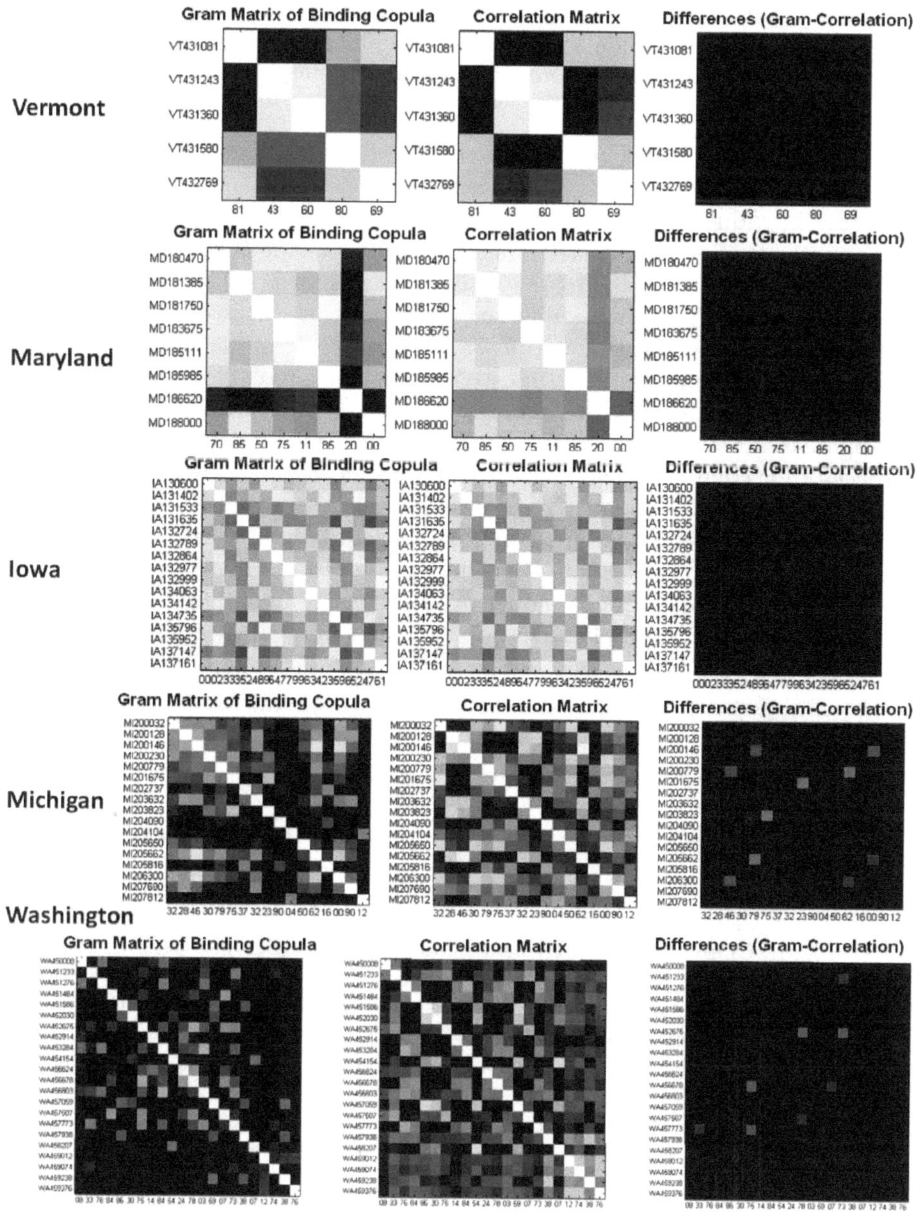

Fig. 6. Sample multivariate KCP results from a few states. The Gram matrix (left) of the binding copula, pairwise correlation (middle), and the difference of the two matrices (right). The IDs of the weather stations are labeled on the vertical axes while only the last two digits of the IDs are shown on the horizontal axes.

that for non-Gaussian random variables, the Gram matrix is not identical to the covariance matrix, but rather it dictates the behavior of the covariance. Nevertheless, the simple t-copula with only geographical distance information was able to recover a substantial amount of interdependency on a visual basis.

5 Conclusion and Future Work

This paper introduced the kernel-based copula processes (KCPs), a novel time-series analysis tool, which inherits features from both the theory of copula and Gaussian processes. The KCPs allows the modeling of non-Gaussian processes in a concise and parsimonious manner by separating the specification of the dependency structure and the margins.

An extension to multivariate time-series was also presented. A binding copula was used to encapsulate the interdependency among time series. This modular approach not only lends itself naturally to implementation on a multi-core / -processor for computational efficiency, it also allows time-series with different lengths and sampling frequencies to be described seamlessly under a single model.

Further research directions include possible sampling or variational methods to accommodate very large data set; the incorporation of a latent process for the kernel hyper-parameters to model processes with stochastic variances; and the application of KCPs in classification problems in similar capacity as GPs.

References

1. Anderson, T.W., Darling, D.A.: A test of goodness of fit. The Journal of American Statistical Association 49(268), 765–769 (1954)
2. Bollerslev, T.: Generalized autoregressive conditional heteroskedasticity. Journal of Econometrics, 31
3. Williams, C.K.I., Rasmussen, C.E.: Gaussian Processes for Machine Learning. The MIT Press, Cambridge (2006)
4. Carney, M., Cunningham, P.: Evaluating density forecast models. tcd-cs-2006- 21. Database, Trinity College Dublin, Department of Computer Science (2006)
5. Chen, X., Fan, Y.: Estimation of copula-based semiparameteric time series models. Journal of Econometrics 130(2), 307–335 (2006)
6. Sandberg, D.V., Alvarado, E., Stewart, G.: Modeling large forest fires as extreme events. Northwest Science, 72
7. Gunther, T.A., Diebold, F.X., Tay, A.S.: Evaluating density forecasts with applications to financial risk management. International Economic Review 39(4), 863–883 (1998)
8. Rosso, R., De Michele, C., Salvadori, G., Kottegoda, N.T.: Extremes in nature: an approach using Copulas. Springer, New York (2007)
9. Hansen, B.E.: Autoregressive conditional density estimation. International Economic Review 35(3), 705–730 (1994)
10. Hull, J.C.: Options, Futures, and Other Derivatives, 5th edn. Prentice-Hall, New Jersey (2003)
11. Lee, T.-H., Long, X.: Copula-based multivariate garch model with uncorrelated dependent errors. Database, UCR Working Paper 2005-16 (2005)

12. Mandelbrot, B.: The variation of certain speculative prices. Journal of Business, 26
13. Nelsen, R.B.: An Introduction to Copulas. Springer Series in Statistics. Springer-Verlag New York, Inc, New York (2006)
14. Fermanian, J.-D., Doukhan, P., Lang, G.: Copulas of a vector-valued stationary weakly dependent process. Database, Working paper CREST (2004)
15. Patton, A.J.: Modelling asymmetric exchange rate dependence. International Economic Review 47(2), 527–556 (2006)
16. Olsen, E.T., Darsow, W.F., Nguyen, B.: Copulas and markov processes. Illinois Journal of Mathematics, pp. 600–642 (1992)
17. Jr. M.J. Menne R.S. Vose Williams, C.N. and D.R. Easterling. United states historical climatology network daily temperature, precipitation, and snow data. Database, The Carbon Dioxide Information Analysis Center, Oak Ridge National Laboratory, U.S. Department of Energy (2006)

Appendix: Heteroskedastic Kernel Function

To derive the heteroskedastic kernel function, consider a random process (or time-series) X_t decomposed into a deterministic trend g_t and a stochastic component with heteroskedastic variance, i.e. $X_t = g_t + y_t$. Assume y_t follows a common Vasicek mean-reverting process:

$$dy_t = -\kappa \cdot (\theta - y_t) \cdot dt + \sigma_t \cdot dW_t \tag{19}$$

where κ is the rate of mean-reversion, θ is the mean-reversion level, σ_u is the variance process of y_t, and W_t is a Wiener process in a usual SDE setup. This generalized Vasicek process has solution:

$$y_t = y_0 \cdot e^{-\kappa t} + \theta \cdot (1 - e^{-\kappa t}) + \int_0^t \sigma_u e^{-\kappa(t-u)} dW_u \tag{20}$$

Thus the auto-covariance of X_t is given by

$$\begin{aligned} Cov[X_{t_1}, X_{t_2}] &= \mathbb{E}\left[\left(\int_0^{t_1} \sigma_u e^{-\kappa(t_1-u)} dW_u\right)\left(\int_0^{t_2} \sigma_u e^{-\kappa(t_2-u)} dW_u\right)\right] \\ &= e^{-\kappa(t_1+t_2)} \int_0^{\min(t_1,t_2)} \sigma_u^2 e^{-2\kappa u} du \end{aligned} \tag{21}$$

Notably, that this kernel function is not a function of the difference between t_1 and t_2 and is therefore non-stationary. Now we must specify the form of the variance process. For the detrended temperature data, the variance process is periodic, consequently, we assume it takes on the following form

$$\sigma_t^2 = \sigma^2 \cdot \left(\alpha \cdot \sin\left(\frac{t}{T} + \phi\right)\right) \tag{22}$$

where σ, α, T, and ϕ are all constants. Substituting (22) into (21), simplifying gives the final form of the kernel function found in (17).

Compositional Models for Reinforcement Learning

Nicholas K. Jong and Peter Stone

The University of Texas at Austin
1 University Station C0500
Austin, Texas 78712
United States
{nkj,pstone}@cs.utexas.edu

Abstract. Innovations such as optimistic exploration, function approximation, and hierarchical decomposition have helped scale reinforcement learning to more complex environments, but these three ideas have rarely been studied together. This paper develops a unified framework that formalizes these algorithmic contributions as operators on learned models of the environment. Our formalism reveals some synergies among these innovations, and it suggests a straightforward way to compose them. The resulting algorithm, Fitted R-MAXQ, is the first to combine the function approximation of fitted algorithms, the efficient model-based exploration of R-MAX, and the hierarchical decompostion of MAXQ.

1 Introduction

Research into *reinforcement learning* (RL) has yielded diverse techniques for more efficiently converging to rewarding behaviors in initially unknown environments, but too often these techniques are studied in isolation. Without assembling these ideas into a single algorithm, we cannot fully understand how they synergize or even conflict. In this paper, we develop a compositional framework that allows us easily to integrate three of the most important advances in RL.

The first of these advances is *model-based* RL. Early work in this direction demonstrated that summarizing an agent's experience into a model facilitates the efficient reuse of data [1]. Later work investigated how the uncertainty in the model can guide exploration, yielding the first (probabilistic) finite bounds on the amount of data required to learn near-optimal behaviors [2,3]. Still, these guarantees require that the agent exhaustively explore every state. Particularly in large domains, this exploration can be impractical.

Second, *function approximation* allows RL to cope with large or even infinite state spaces by introducing generalization. It allows an algorithm to approximate the long-term value of every action in every state using only a relatively small set of parameters. Many state-of-the-art approaches employ model-free algorithms and representations that attempt to estimate these values directly from data [4,5], but they often still rely on random exploration to acquire this data.

W. Buntine et al. (Eds.): ECML PKDD 2009, Part I, LNAI 5781, pp. 644–659, 2009.

Third, *hierarchical decomposition* is perhaps the most intuitively appealing extension of the standard approach, since we would like to imbue our learning algorithms with the same awareness of structure that seems to allow us to cope with the extraordinary complexity of the real world. Hierarchical RL has been explored via work on temporal abstraction, in which temporally extended abstract actions allow agents to reason above the level of primitive actions [6]. However, we still do not have a complete understanding of how hierarchy benefits learning and therefore how to design or discover hierarchies.

One main contribution of this work is a formulation of important algorithms from all of these branches of RL into a concise, unified notation that makes their synergies apparent. We introduce our notation in Sect. 2 and use it to review fitted function approximation, the model-based R-MAX algorithm, and the hierarchical MAXQ framework. In Sect. 3, this powerful reformulation allows us easily to define the first algorithm for model-based, continuous-state, hierarchical RL. This algorithm, which we call Fitted R-MAXQ, constitutes our second main contribution, and we evaluate it empirically in Sect. 4. Finally, we discuss future and related work in Sect. 5 before concluding in Sect. 6.

2 Model Components

A *Markov decision process* (MDP) $\langle S, A, R, P\rangle$ comprises a finite set of states S, a finite set of actions A, a $|S||A| \times 1$ reward vector R, and a $|S||A| \times |S|$ transition matrix P. Executing action a in state s earns reward $R[sa]$ on average and transitions to state s' with probability $P[sa, s']$. We define a *policy* π as a $|S| \times |S||A|$ matrix,[1] where $\pi[s, sa]$ gives the probability of executing a in s, and where $\pi[s_1, s_2a] = 0$ for $s_1 \neq s_2$. The $|S| \times 1$ *value function* V^π maps each state s to the expected discounted reward $V^\pi[s]$ of following π from that state. The value function satisfies the Bellman equation

$$V^\pi = \pi \left(R + \gamma P V^\pi\right), \tag{1}$$

where $\gamma \in [0, 1]$ is a discount factor.

Given R and P, which comprise a *model*, a planning algorithm computes a policy π that maximizes each entry of V^π. The standard policy iteration and value iteration algorithms [7] both simply alternate between updating V^π and improving π greedily. Policy iteration updates V^π by solving (1) for a fixed π, while value iteration updates V^π by evaluating the right-hand side of (1) using the fixed previous value of V^π. Both algorithms converge to the optimal value function if $\gamma < 1$.

In the RL setting, the model is not given, but model-based algorithms estimate R and P from data. We define an *instance* $i = \langle s_i, a_i, r_i, s'_i\rangle$ as a record containing a state s_i, the action a_i executed from s_i, the one-step reward r_i earned, and the successor state s'_i. Let D be a list of instances, and define $D_a = \{i \in D \mid a_i = a\}$.

[1] The policy may thus be interpreted as a matrix transitioning states to state-action pairs.

Then the maximum-likelihood reward model $\bar{R}$ and transition model $\bar{P}$ are matrices defined as follows:

$$\bar{R}[sa] = \frac{\sum_{i \in D_a} \delta(s, s_i) r_i}{n(s, D_a)} \tag{2}$$

$$\bar{P}[sa, s'] = \frac{\sum_{i \in D_a} \delta(s, s_i) \delta(s', s'_i)}{n(s, D_a)}, \tag{3}$$

where $n(s, D_a) = \sum_{i \in D_a} \delta(s, s_i)$, and δ is the Kronecker delta.

A complete RL algorithm must specify an exploration policy that guides an agent to acquire the data used to estimate the model. The R-MAX algorithm [3] maintains a set U of state-action pairs where insufficient data exists to estimate the model accurately. For state-action pairs in U, the algorithm uses an optimistic model in which the action terminates the trajectory after earning immediate reward $V^{\max}$, an upper bound on the value function. In this manner, the "unknown" state-action pairs are seen as optimal, encouraging the agent to visit them and gather data. Otherwise, the maximum-likelihood estimate of the state-action's effects is used. We can express this exploration mechanism in our matrix notation for V^π as follows:

$$V^\pi = \pi \left(U V^{\max} + (I - U)(\bar{R} + \gamma \bar{P} V^\pi) \right), \tag{4}$$

where U is represented as a $|S||A| \times |S||A|$ diagonal binary matrix where $U[sa, sa] = 1$ iff $sa \in U$. R-MAX has some appealing theoretical properties, such as a probabilistic polynomial bound on the number of times it departs from a near-optimal policy [8]. However, in practice, its thorough exploration behavior is impractical, and it only directly applies to finite MDPs where it is reasonable to gather ample data for every reachable state-action pair.

2.1 Function Approximation

Function approximation scales RL to environments where exhaustive exploration is infeasible or impossible (such as any domain with a continuous or otherwise infinite state space) by introducing the idea of generalization. A function approximator defines a family of value functions with some finite parameterization. In this paper, we focus on *averagers*, which approximate the value of a given state as a weighted average of the values of a finite subset $X \subset S$. In particular, it approximates $V^\pi[s]$ as a weighted average $\sum_{x \in X} \Phi[s, x] V^\pi[x]$, for a given $|S| \times |S|$ matrix Φ. Fitted Value Iteration [9] uses this approach by performing value iteration using the following approximation of V^π:

$$V^\pi = \pi \left(R + \gamma P \Phi V^\pi \right). \tag{5}$$

Some recent algorithms have also applied function approximation to the estimated reward and transition models. The instance-based Kernel-Based Reinforcement Learning algorithm [10] estimates a value function using the equations:

$$V^\pi = \pi \left(\hat{R} + \gamma \hat{P} V^\pi \right) \tag{6}$$

$$\hat{R}[sa] = \frac{\sum_{i \in D_a} \hat{\delta}(s, s_i) r_i}{\hat{n}(s, D_a)} \tag{7}$$

$$\hat{P}[sa, s'] = \frac{\sum_{i \in D_a} \hat{\delta}(s, s_i) \delta(s', s'_i)}{\hat{n}(s, D_a)}, \tag{8}$$

where $\hat{n}(s, D_a) = \sum_{i \in D_a} \hat{\delta}(s, s_i)$ tallies the weights given by the kernel function $\hat{\delta} : S \times S \to [0, 1]$. The kernel function determines the degree $\hat{\delta}(s, s_i)$ to which data at state s_i generalizes to the model at state s. The experiments in this paper use a Gaussian kernel, as specified in Sect. 4.1. Note that the kernel-based approximate model in (7) and (8) modify the maximum-likelihood model in (2) and (3) only by substituting weights $\hat{\delta}(s, s_i) \in [0, 1]$ in place of the exact binary indicator function $\delta(s, s_i) \in \{0, 1\}$.

2.2 Hierarchy

Given the structure we perceive in the real world, it seems natural to apply hierarchy to reinforcement learning. The MAXQ decomposition [11] and options [12] are the two most popular frameworks for hierarchical RL, which defines temporally abstract actions that represent sequences of primitive actions. MAXQ decomposes an overall learning problem using a given task hierarchy, where each abstract action is a task that induces its own individual learning problem. In contrast, the options framework formalizes an abstract action as a partial policy, which can be construed as a solution to a task. We will find it convenient to interpret an abstract action o in both ways, depending on context.

A *task* $o = \left\langle T^o, A^o, \tilde{R}^o \right\rangle$ comprises a set of terminal states $T^o \subset S$, a set of child actions or tasks A^o, and a "pseudoreward" (goal) function $\tilde{R}^o : T^o \to \mathbb{R}$. It imposes an objective onto the system defined by the state space S and the child actions A^o, which may include both other tasks and primitive actions such as those assumed by the preceding sections.[2] The task terminates upon reaching a state $s \in T^o$ and then awards itself an artificial goal value $\tilde{R}^o[s]$, where $\tilde{R}^o$ is a $|S| \times 1$ vector. We can also represent T^o as a $|S| \times |S|$ diagonal binary matrix such that $T^o[s, s] = 1$ iff o terminates upon entering s.

The optimal policy π^o for task o maximizes

$$\tilde{V}^o = T^o \tilde{R}^o + (I - T^o) \pi^o \left(R^o + \gamma P^o \tilde{V}^o \right), \tag{9}$$

where R^o and P^o are the (abstract) reward and transition matrices, respectively, for the actions in A^o. This policy π^o chooses children $c \in A^o$ in a way that maximizes a combination of one-step rewards during execution and goal values upon termination. The task value function $\tilde{V}^o$ captures this combined value, but it's also possible to compute the value function V^o that only includes the one-step rewards (and is not "contaminated" with the goal rewards). In particular, V^o is given by solving

[2] We will index the child actions A^o using c instead of a to emphasize that $c \in A^o$ may be either a task/option or a primitive action.

$$V^o = \pi^o \left(R^o + \gamma P^o (I - T^o) V^o\right). \tag{10}$$

A key insight of MAXQ is that V^o can be interpreted as the reward model for the *option*[3] $o = \langle T^o, A^o, \pi^o \rangle$ that, when initiated in a state $s \notin T^o$, simply selects actions according to π^o until reaching a state $s' \in T^o$. Suppose that $o \in A^p$ for some parent task p. Then we can capture the insight of MAXQ as $R^p[so] = V^o[s]$. In other words, we can use V^o to construct part of the reward vector for any MDP learning task p that includes o as an executable action.

The same recursive approach can also apply to the transition function. Just as the abstract reward function for an option specifies the expected (discounted) sum of one-step rewards earned before reaching a terminal state, the abstract transition function for an option should specify the expected (discounted) probability of terminating in each terminal state. To this end, we define the $|S| \times |S|$ *terminal-state matrix* Ω^o for an option o with the following Bellman-like equation:

$$\Omega^o = \pi^o \left(P^o T^o + \gamma P^o (I - T^o) \Omega^o\right). \tag{11}$$

Note intuitively that each column of Ω^o can be interpreted as a value function for a task which gives a reward of 1 upon terminating in the state corresponding to that column. As a result, Ω^o can be computed using standard MDP planning algorithms. Finally, we observe that if $o \in A^p$ for some parent task p, then $P^p[so, s'] = \Omega^o[s, s']$. Here, P^p is a multi-time model [12], so its rows may not sum to 1, reflecting the effect of the discount factor over time. This representation thus folds the duration of actions (typically represented explicitly in the standard SMDP formalism) into the discounted transition probabilities.

Our hierarchical decomposition thus specifies how to construct the reward and transition matrices for a task p recursively given the value functions and terminal-state matrices of the options $o \in A^p$. To complete this recursive specification, we need only give the base case. For a primitive action a, the value function V^a and terminal-state matrix Ω^a correspond exactly to the reward and transition models for that action.

3 Compositional Algorithms

The algorithms described in Sect. 2 generate exploration policies by solving modified forms of the standard Bellman equation (1). However, each of the modified equations (4), (5), and (9) share the same general form of (1): we can construe the right-hand side as π multiplied by the sum of a $|S| \times 1$ vector (that doesn't depend on V^π) and a $|S| \times |S|$ matrix multiplied by V^π. These equations are therefore equivalent to the standard Bellman equations for a modified version of the original Markov decision process.[4]

[3] It is straightforward to support the stochasic termination of the standard options framework by defining the diagonal entries of T as the termination probabilities.

[4] Equation (9) can be rewritten into this form by handling termination after applying the policy, but the representation of T^o and $\tilde{R}^o$ is then less intuitive.

We formalize such modifications as follows. For a given set of states S and primitive actions A, let $\mathcal{D}$ be the space of all possible lists of instances, $\mathcal{R}$ be the space of all possible $|S| \times 1$ reward vectors, and $\mathcal{P}$ be the space of all possible $|S| \times |S|$ transition matrices. The $|S||A| \times 1$ reward vector for a given learning task o is then obtained by composing the reward vectors for each child action $c \in A^o$ in the appropriate way. Similarly, the $|S||A| \times |S|$ task transition matrix is composed from the $|S| \times |S|$ action transition matrices. We now define a *model generator* $G : \mathcal{D} \to \mathcal{R} \times \mathcal{P}$ as a mapping from a list of instances to a reward and transition model for a given primitive action, and a *model operator* $M : \mathcal{R} \times \mathcal{P} \to \mathcal{R} \times \mathcal{P}$ as a mapping from one reward and transition model to another.

We can formalize the maximum-likelihood model as a family of model generators MLE_a for each primitive action a:

$$\text{MLE}_a(D) = \left(\bar{V}^a, \bar{\Omega}^a\right), \tag{12}$$

where $\bar{V}^a[s]$ and $\bar{\Omega}^a[s, s']$ are given by the right-hand sides of (2) and (3), respectively. The R-MAX algorithm then generates an exploration policy by using a standard planning algorithm on the learning task composed with the action models $\text{R-MAX}_a(\text{MLE}_a(D))$, where

$$\text{R-MAX}_a(R, P) = (U_a V^{\max} + (I - U_a)R, (I - U_a)P), \tag{13}$$

where U_a is the $|S| \times |S|$ submatrix of U such that $U_a[s, s] = 1$ iff sa is unknown. Note that the R-MAX operator is defined as a function of the data D, which are required to define U.

The Fitted R-MAX algorithm, which extends R-MAX using the model approximation of KBRL and fitted planning [13], can now be seen as planning with the action models $\text{FVI}(\text{R-MAX}_a(\text{KBRL}_a(D)))$, where

$$\text{FVI}(R, P) = (R, P\Phi) \tag{14}$$

encapsulates Fitted Value Iteration for a given Φ, and

$$\text{KBRL}_a(D) = \left(\hat{V}^a, \hat{\Omega}^a\right), \tag{15}$$

where $\hat{V}^a[s]$ and $\hat{\Omega}^a[s, s']$ are given by the right-hand sides of (7) and (8), respectively.

R-MAXQ, another recent extension to R-MAX, incorporates the MAXQ-based model decomposition described in Sect. 2.2 [14]. For a given task o, the computation of the option policy π^o requires planning with modified action models $\text{MAXQ}^o(V^c, \Omega^c)$ for each child $c \in A^o$, where

$$\text{MAXQ}^o(R, P) = \left(T^o \tilde{R}^o + (I - T^o)R, (I - T^o)P\right). \tag{16}$$

Given the option policy π^o, as well as policies for each descendent of o, the MAXQ decomposition also defines a model (V^o, Ω^o) of o. In this sense, each task o defines a model generator (since each option policy is a function of D).

Algorithm 1. MODEL(c)

if c is a (primitive) action **then**
  $a \leftarrow c$
  $(V^a, \Omega^a) \leftarrow G^a(D)$
  Return (V^a, Ω^a)
else {c is a (composite) task/option}
  $o \leftarrow c$
  $(\pi^o, R^o, P^o) \leftarrow$ PLAN(o)
  $V^o \leftarrow$ solution to $V^o = \pi^o(R^o + \gamma P^o(I - T^o)V^o)$
  $\Omega^o \leftarrow$ solution to $\Omega^o = \pi^o(P^o T^o + \gamma P^o(I - T^o)\Omega^o)$
  Return (V^o, Ω^o)
end if

Algorithm 1 precisely defines this model generator. Note that Algorithms 1 and 2 are mutually recursive, and they assume the following global parameters: a set of instances D, model generators G^a for each primitive action a, and model operators M^o for each task o. Algorithm 2 constrains the planning algorithm to only execute an option in states not in the option's termination set. In other words, for all options o and parent tasks p, $pi^p[s, so] > 0$ implies $s \notin T^o$. (The option's initiation set is the complement of its set of terminal states.)

Algorithm 2. PLAN(o)

for all children $c \in A^o$ **do**
  $(V^c, \Omega^c) \leftarrow M^o(\text{MODEL}(c))$
end for
$R^o \leftarrow$ choose so that $R^o[sc] = V^c[s]$
$P^o \leftarrow$ choose so that $P^o[sc, s'] = \Omega^c[s, s']$
$\pi^o \leftarrow$ optimize $\tilde{V}^o = \pi^o\left(R^o + \gamma P^o \tilde{V}^o\right)$ {subject to initiation constraints}
Return (π^o, R^o, P^o)

Given these subroutines, Algorithm 3 describes a broad family of model-based RL algorithms parameterized by a task hierarchy, model generators attached to each primitive action, and model operators attached to each task. For example, suppose that for a given task hierarchy we define $G^a = \text{R-MAX}_a \circ \text{MLE}_a$ for each primitive action a and $M^o = \text{MAXQ}$, for each task o. Then running EXECUTE on the root task of this hierarchy is exactly equivalent to R-MAXQ. Note that this algorithm uses the same model (R^o, P^o) to compute both the "contaminated" value function $\tilde{V}^o$ and the real value function V^o. It can use this model for both purposes, since the additional goal-reward values only affect the reward vector at terminal states. Equation (10) disregards values of terminal states, and Algorithm 2 prevents the execution of the option at terminal states.

These procedures can also implement non-hierarchical algorithms by using a "flat" hierarchy. Let A be the set of available primitive actions and define the task Root such that $A^{\text{Root}} = A$ and $T^{\text{Root}} = 0$. If we define $G^a = \text{R-MAX}_a \circ \text{MLE}_a$

and $M^{\mathsf{Root}} = \text{I}$, the identity operator, then EXECUTE(Root) is exactly equivalent to the original R-MAX algorithm. If we instead define $G^a = \text{R-MAX}_a \circ \text{KBRL}_a$ and $M^{\mathsf{Root}} = \text{FVI}$, then we obtain the Fitted R-MAX algorithm.

Algorithm 3. EXECUTE(c)

```
if c is a (primitive) action then
  a ← c
  Execute action a in the environment
  r ← reward
  s' ← successor state
  D ← D ∪ {⟨s, a, r, s'⟩}
else {c is a (composite) task/option}
  o ← c
  repeat
    s ← current state
    (π^o, V^o, Ω^o) ← PLAN(o)                          {V^o and Ω^o ignored}
    c ← choose child action/option according to π^o
    EXECUTE(c)
  until T^c[s', s'] = 1
end if
```

3.1 Fitted R-MAXQ

Our compositional approach to model-based RL immediately suggests a novel algorithm, obtained by applying all of our available model operators. We define $G^a = \text{R-MAX}_a \circ \text{KBRL}_a$, to obtain the optimistic exploration of R-MAX and the instance-based generalization of primitive action models of Kernel-Based Reinforcement Learning. We define $M^o = \text{MAXQ} \circ \text{FVI}$ to obtain the subtask decomposition of MAXQ and the value function approximation of Fitted Value Iteration. In keeping with prior algorithms extending R-MAX, we refer to this novel algorithm as Fitted R-MAXQ. For concreteness, we specify Fitted R-MAXQ in Algorithm 4, which also optimizes the computation of Algorithms 1–3 by using dynamic programming to unroll the mutual recursion. Note that Algorithm 4 assumes a continuing task, but the modifications for episodic tasks are straightforward.[5]

To our knowledge, Fitted R-MAXQ is the first model-based RL algorithm to combine function approximation and hierarchical decomposition. Interestingly, a close inspection reveals notable structural similarities among the operators that comprise Fitted R-MAXQ, which creates opportunities for synergies. For example, consider the R-MAX and MAXQ operators, (13) and (16). Both modify a model by changing the rewards at a subset of the states, which also become terminal. In the case of R-MAX, this subset is the set of unknown states U_a; for MAXQ, it is the set of terminal states T^o. Entering this set terminates the

[5] The T^{Root} matrix should remain zero even in episodic tasks, since it is the environment and not the agent terminating.

trajectory (by assigning zero probability to every successor) but earns a final reward given by the upper bound $V^{\max}$ or the pseudoreward function $\tilde{R}^o$. The net effect of these operators is to bias the exploration policy of the algorithm, towards leaving the known set and towards manually specified subgoal states, respectively. This characterization leads us to conjecture that one important role of hierarchy is to focus the otherwise too thorough exploration of R-MAX by requiring the optimistic value of exploration to overcome the perceived subgoal rewards.

Algorithm 4. FITTED-R-MAXQ(Root)

```
Initialize stack to [Root]
loop
  s ← current state
  c ← top of stack
  while c is a task, not a primitive action do
    o ← c
    c ← choose child in A^o using π^o
    Push c onto stack
  end while
  a ← c
  Execute action a
  r ← one-step reward
  s' ← successor state
  D ← D ∪ {⟨s, a, r, s'⟩}
  V^a ← U_a V^max + (I − U_a) V̂^a          {V̂^a[s] equals right-hand side of (7)}
  Ω^a ← (I − U_a) Ω̂^a                      {Ω̂^a[s, s'] equals right-hand side of (8)}
  for all tasks o do {bottom-up}
    R^o ← construct so that R^o[sc] = V^c[s]
    P^o ← construct so that P^o[sc, s'] = Ω^c[s, s']
    π^o ← optimize Ṽ^o = T^o R̃^o + (I − T^o) π^o (R^o + γ P^o Φ V^o)
    V^o ← solve V^o = π^o (R^o + γ P^o (I − T^o) V^o)
    Ω^o ← solve Ω^o = π^o (P^o T^o + γ P^o (I − T^o) Ω^o)
  end for
  repeat {stack begins with a primitive on top}
    Pop stack
    o ← top of stack
  until T^o[s', s'] = 0                      {s' not terminal}
end loop
```

We have already seen that some model operators modify the transition matrix by forcing some states to be terminal, but the FVI operator (14) instead "redirects" transitions into a relatively small subset $X \subset S$. By construction, at most $|X|$ of the columns of Φ are nonzero. Since this matrix is multiplied to the right of the original transition matrix, at most $|X|$ of the rows of V^{π} affect the Bellman equation. Even when S is infinite, this property permits our implementation to employ a sparse representation of the matrices involved. In general, in any one time step, the algorithm must store only $|X| + 1$ rows of each matrix, one for the current state and the rest for X.

Interestingly, the MAXQ decomposition can have a similar effect on planning efficiency. Note that (11) defines the abstract transition matrix for an option o in such a way that the nonzero columns correspond to the terminal states in T^o. Since many tasks achieve subgoal states that comprise a small fraction of the state space, planning at the abstract level of tasks may permit a compact representation of the value function which compounds with any reductions due to fitted function approximation.

4 Experiments

In order to exercise the full capabilities of Fitted R-MAXQ, we introduce a new domain modeled after the RL benchmark environment Puddle World. Puddle World already has a continuous state space, and no modifications are necessary to enable model-based reasoning (which is purely an algorithmic issue), but we introduce a task hierarchy that enables hierarchical reasoning. We intend our extensions to give the overall task more structure of the sort found in real-world tasks.

First, we describe the original Puddle World environment, depicted in Fig. 1. The agent must navigate the unit square to reach a goal state in the upper-right corner, which terminates each episode. Four primitive actions move the agent 0.05 in each of the four cardinal directions, with some Gaussian noise ($\sigma = 0.01$) added to each of the two state variables after every action. Each action incurs a -1 penalty until reaching the goal, but each time step spent in a puddle incurs an additional penalty between 0 and -40, depending on the proximity to the middle of the puddle.

We modify this environment by removing the goal state in the corner and instead giving the agent a set of four different resources it must harvest in each

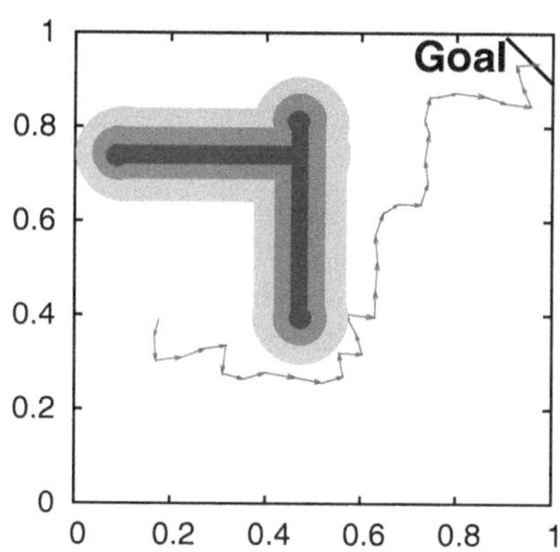

Fig. 1. A trajectory in the original Puddle World environment

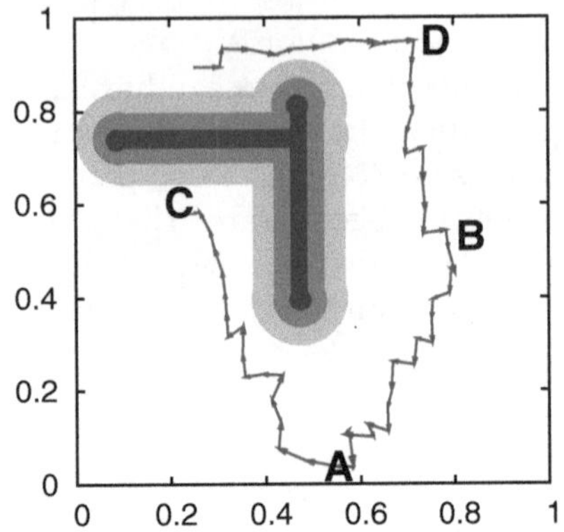

Fig. 2. A trajectory in the modified Puddle World environment. The agent gathers resources D, B, A, then C, but any order is permissible.

episode, as shown in Fig. 2. Each resource can only be collected in the neighborhood (within distance 0.1) of a specific spot in the unit square, which is initially unknown. For each resource, a binary state variable tracks whether the resource has been collected, and a distinct primitive action allows the agent to search the current location for the resource and harvest it if present. In each episode, the agent begins in a random location, so over time it must learn the locations of the resources, the locations of the costly puddles, and how to harvest all four resources as cheaply as possible. This environment has six state variables, two continuous and four binary, and eight primitive actions, four movement actions and the four collection actions.

We ran each algorithm tested for 50 independent trials, using for each algorithm the same set of 50 configurations of resource locations, generated uniformly at random but with no location inside of a puddle. For each configuration, we generated a fixed sequence of 500 start states, again uniformly at random. Each trial lasted for 500 episodes, and we limited each episode to 1000 time steps.

4.1 Algorithm Configurations

We compare several different instantiations of our compositional framework for model-based RL. In this section, we describe the precise configuration of each model generator and model operator used. We use value iteration with prioritized sweeping [1] both to compute the optimal policy π^o given a model (R^o, P^o) and also to evaluate π^o to obtain the abstract model (V^o, Ω^o). We used a discount factor of $\gamma = 1$, since the task is episodic.

The maximum-likelihood model generator MLE has no parameters, but to apply finite algorithms in PuddleWorld we used a discretization of the unit square into a 16×16 grid. This discretization seems quite coarse, but finer grids only lead to more excessive exploration without a concomitant increase in policy quality.

For the instance-based model approximation of KBRL, we adopted a Gaussian kernel function:

$$\hat{\delta}(s_1, s_2) = e^{-(d(s_1,s_2)/b)^2}, \tag{17}$$

where $d(s_1, s_2)$ is the Euclidean distance between s_1 and s_2, and $b = \frac{1}{16}$ is a bandwidth parameter that controls the breadth of generalization, chosen using coarse optimization. To compute $\hat{\delta}$ efficiently, we stored the instances in a cover tree [15] and rounded down to zero any value of $\hat{\delta}(s_1, s_2) < 0.01$. Finally, we adopt the "relative transition model" of [13], which modifies (8) by using the vector displacement observed at instance i instead of the absolute successor state observed:

$$\hat{P}[sa, s'] = \frac{\sum_{i \in D_a} \hat{\delta}(s, s_i)\delta(s', s + (s'_i - s_i))}{\hat{n}(s, D_a)}. \tag{18}$$

All of the algorithms we tested rely on the R-MAX approach to exploration. We set $V^{max} = 0$, since all the immediate rewards in Puddle World are negative. When used with MLE, we defined $U_a = \{s \in S \mid n(s, D_a) < 2\}$. Since the stochasticity in Puddle World is relatively benign, gathering more data for each state-action didn't improve the final policy quality but resulted in much more expensive exploration. When used with KBRL, we defined $U_a = \{s \in S \mid \hat{n}(s, D_a) < 1\}$. This low threshold seemed adequate since KBRL must typically generalize from several instances to reach a kernel weight of 1.

For FVI, we defined the averager Φ using linear interpolation over a uniformly spaced grid, with a resolution of $\frac{1}{16}$. This function approximation scheme therefore approximates the value of a point in the unit square (for a particular setting of the binary state variables) as an interpolation between the four surrounding points. Again, increasing the resolution did not improve the quality of the learned policy, but it did increase the computational burden of planning.

For the hierarchical algorithms, we defined a simple task hierarchy for our modified Puddle World that corresponds to the prior knowledge that the four resource collection actions are independent of one another.[6] For each resource, we define a task o such that the children actions A^o include the four movement primitives and the action that collects that resource. The terminal set T^o includes all states where the resource's boolean flag is set, and $\tilde{R}^o = 0$. The root of the hierarchy has these four tasks as children; it cannot execute any primitive actions directly.

Finally, all the algorithms benefitted from state abstraction. The four primitive movement actions neither depend on nor affect the boolean state variables for the four activities. Similarly, each activity only depends on the coordinates and only affects the corresponding boolean state variable. Due to limitations on space, we omit the details of these state abstractions, which were implemented in the obvious way.

---

[6] This hierarchy therefore imparts less domain knowledge than the hierarchy Dietterich provided for learning in the Taxi domain [11], where the possible passenger coordinates were all known a priori.

4.2 Results

Figure 3 shows learning curves for four algorithms: R-MAX, R-MAXQ, Fitted R-MAX, and our combination of these algorithms, Fitted R-MAXQ. All four algorithms converge to statistically the same policy quality after only 25-30 episodes, but they incur very different exploration costs before getting there.

Figure 4 integrates under the curves in Fig. 3 to show the total learning costs. Note that both figures only show the first several episodes, to focus on the period of learning when the algorithms' performance differs. Note that the benefit of adding both hierarchical decomposition and function approximation to R-MAX is greater than the sum of the benefits for adding each innovation by itself!

An inspection of the behavior of Fitted R-MAXQ reveals that it outperforms the other algorithms largely by avoiding excessive exploration in the puddles. Consider a state in the middle of a puddle that is in the set of unknown states U_a for some primitive action a. The R-MAX operator will assign this state the optimistic value $V^{\max}$, but this value does not guarantee that the agent will attempt to reach this state. If the predicted cost of completing the current task is smaller than the predicted cost of wading through the puddle to the unknown state, the agent will choose to ignore the unknown state and instead exploit a path through known states. In the non-hierarchical case, the current task is always to complete all the remaining activities, which may have a rather high cost. In the hierarchical case, the current task is to complete a particular one

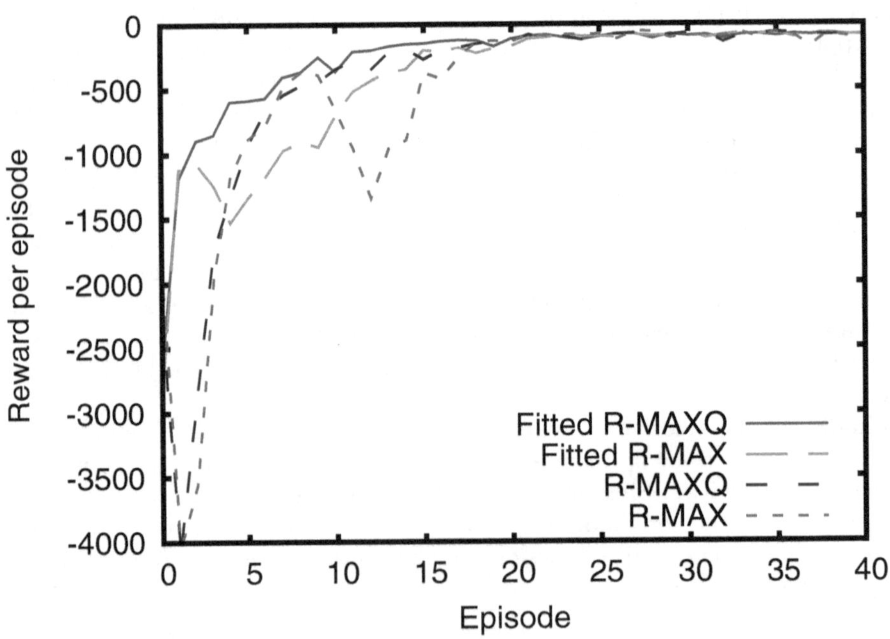

Fig. 3. Reward per episode for variations of R-MAX with and without function approximation and hierarchical decomposition

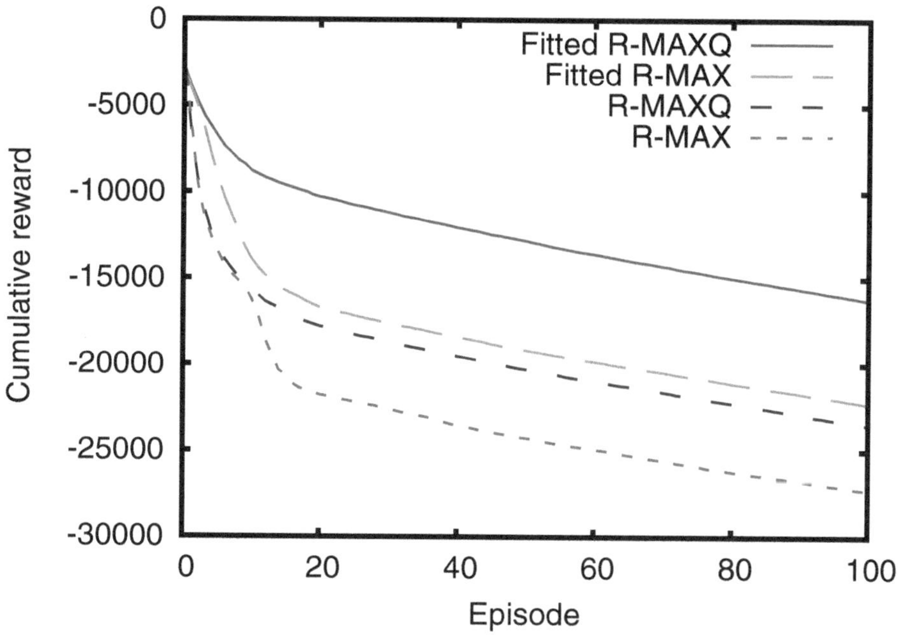

Fig. 4. Cumulative reward for variations of R-MAX with and without function approximation and hierarchical decomposition

of the activities, which is more likely to have a lower cost than wading into the puddle. In a sense, the hierarchical decomposition limits the optimism applied to unknown states, which R-MAXQ models as terminating only the current task, not the entire episode. Meanwhile, discretization interferes with the accurate prediction of the costs of exploring versus exploiting. The coarse discretization we used is very effective in most of the state space, where the dynamics are the same, but not near the puddles, where the immediate reward varies quickly as a function of the coordinates.

5 Discussion

In this paper, we cast certain existing algorithms into a unified framework with an eye towards defining a new algorithm that combined all the desired existing features. The generality of our framework leaves open the possibility that still more algorithms can be formalized in terms of model generators and model operators. Investigating combinations of these algorithms can only help us to develop deeper understandings of their individual contributions in context.

One important direction for future work is to derive general properties of the class of algorithms defined by our compositional framework. Each algorithm in this class behaves strictly according to the optimal policy for an MDP derived in some way from data. An understanding of the expressivity and limitations of such algorithms might either inspire the creation of new operators within this framework or of new algorithms that meaningfully break out of it. For example, hierarchical RL requires the ability to work with a derived MDP that has a different action space than the original MDP. We observe that the problem of optimal exploration in RL can be reduced to a planning problem in a derived MDP, where the state space is augmented with beliefs concerning the underlying MDP [16]. In what ways can model operations productively change the state and action spaces of the underlying MDP? Finally, integration with the ongoing work on MDP homomorphisms [17] may allow our framework to deal more explicitly with state abstraction.

6 Conclusion

This paper developed two main contributions to the literature on scaling reinforcement learning to increasingly complex environments. First, it introduced a novel, unifying notation and formulation of three previously disjoint ideas in RL: model-based exploration, function approximation, and hierarchy. This formulation construed existing algorithms as essentially the application of a standard planning algorithm to a transformed reward and transition model. Second, the paper leveraged this new notation to unify these three ideas into Fitted R-MAXQ, the first algorithm for hierarchical, model-based RL in continuous domains. Fitted R-MAXQ is fully implemented and evaluated in a hierarchical Puddle World, significantly outperforming algorithms that utilize only a subset of its components.

References

1. Moore, A.W., Atkeson, C.G.: Prioritized sweeping: Reinforcement learning with less data and less real time. Machine Learning 13, 103–130 (1993)
2. Kearns, M., Singh, S.: Near-optimal reinforcement learning in polynomial time. In: Proceedings of the Fifteenth International Conference on Machine Learning, pp. 260–268 (1998)
3. Brafman, R.I., Tennenholtz, M.: R-MAX – a general polynomial time algorithm for near-optimal reinforcement learning. Journal of Machine Learning Research 3, 213–231 (2002)
4. Lagoudakis, M.G., Parr, R.: Least-squares policy iteration. Journal of Machine Learning Research 4, 1107–1149 (2003)
5. Riedmiller, M.: Neural fitted Q iteration – first experiences with a data efficient neural reinforcement learning method. In: Proceedings of the European Conference on Machine Learning (2005)
6. Barto, A.G., Mahadevan, S.: Recent advances in hierarchical reinforcement learning. Discrete-Event Systems 13, 41–77 (2003); Special Issue on Reinforcement Learning
7. Puterman, M.L.: Markov Decision Processes: Discrete Stochastic Dynamic Programming. John Wiley & Sons, Inc., Chichester (1994)
8. Kakade, S.M.: On the Sample Complexity of Reinforcement Learning. PhD thesis, University College London (2003)

9. Gordon, G.J.: Stable function approximation in dynamic programming. In: Proceedings of the Twelfth International Conference on Machine Learning (1995)
10. Ormoneit, D., Sen, Ś.: Kernel-based reinforcement learning. Machine Learning 49(2), 161–178 (2002)
11. Dietterich, T.G.: Hierarchical reinforcement learning with the MAXQ value function decomposition. Journal of Artificial Intelligence Research 13, 227–303 (2000)
12. Sutton, R.S., Precup, D., Singh, S.: Between MDPs and semi-MDPs: A framework for temporal abstraction in reinforcement learning. Artificial Intelligence 112(1–2), 181–211 (1999)
13. Jong, N.K., Stone, P.: Model-based exploration in continuous state spaces. In: Proceedings of the Seventh Symposium on Abstraction, Reformulation and Approximation (2007)
14. Jong, N.K., Stone, P.: Hierarchical model-based reinforcement learning: R-MAX + MAXQ. In: Proceedings of the Twenty-Fifth International Conference on Machine Learning (2008)
15. Beygelzimer, A., Kakade, S., Langford, J.: Cover trees for nearest neighbor. In: Proceedings of the Twenty-Third International Conference on Machine Learning (2006)
16. Duff, M.: Design for an optimal probe. In: Proceedings of the Twentieth International Conference on Machine Learning, pp. 131–138 (2003)
17. Ravindran, B., Barto, A.G.: SMDP homomorphisms: An algebraic approach to abstraction in semi-Markov decision processes. In: Proceedings of the Eighteenth International Joint Conference on Artificial Intelligence (2003)

Feature Selection for Value Function Approximation Using Bayesian Model Selection

Tobias Jung and Peter Stone

Department of Computer Sciences, University of Texas at Austin, USA
{tjung,pstone}@cs.utexas.edu

Abstract. Feature selection in reinforcement learning (RL), i.e. choosing basis functions such that useful approximations of the unkown value function can be obtained, is one of the main challenges in scaling RL to real-world applications. Here we consider the Gaussian process based framework GPTD for approximate policy evaluation, and propose feature selection through marginal likelihood optimization of the associated hyperparameters. Our approach has two appealing benefits: (1) given just sample transitions, we can solve the policy evaluation problem fully automatically (without looking at the learning task, and, in theory, independent of the dimensionality of the state space), and (2) model selection allows us to consider more sophisticated kernels, which in turn enable us to identify relevant subspaces and eliminate irrelevant state variables such that we can achieve substantial computational savings and improved prediction performance.

1 Introduction

In this paper, we address the problem of approximating the value function under a stationary policy π for a continuous state space $\mathcal{X} \subset \mathbb{R}^D$,

$$V^\pi(\mathbf{x}) = \mathbb{E}_{\mathbf{x}'|\mathbf{x},\pi(\mathbf{x})} \; \{R(\mathbf{x},\pi(\mathbf{x}),\mathbf{x}') + \gamma V^\pi(\mathbf{x}')\} \tag{1}$$

using a linear approximation of the form $\tilde{V}(\cdot\,;\mathbf{w}) = \sum_{i=1}^m w_i \phi_i(\mathbf{x})$ to represent V^π. Here $\mathbf{x}$ denotes the state, R the scalar reward and γ the discount factor. Given a trajectory of states $\mathbf{x}_1,\ldots,\mathbf{x}_n$ and rewards $r_1,\ldots,r_{n-1}$ sampled under π, the goal is to determine weights w_i (and basis functions ϕ_i) such that $\tilde{V}$ is a good approximation of V^π. This is the fundamental problem arising in the policy iteration framework of infinite-horizon dynamic programming and reinforcement learning (RL), e.g. see [21,3]. Unfortunately, this problem is also a very difficult problem that, at present, has no completely satisfying solution. In particular, deciding which features (basis functions ϕ_i) to use is rather challenging, and in general, needs to be done manually: thus it is tedious, prone to errors, and most important of all, requires considerable insight into the domain. Hence, it would be far more desirable if a learning system could automatically choose its own representation. In particular, considering efficiency, we want to adapt to the actual difficulties faced, without wasting resources: often, there are many factors

W. Buntine et al. (Eds.): ECML PKDD 2009, Part I, LNAI 5781, pp. 660–675, 2009.

that can make a particular problem easier than it initially appears to be, for example, when only a few of the inputs are relevant, or when the input data lies on a low-dimensional submanifold of the input space.

Recent work in applying nonparametric function approximation to RL, such as Gaussian processes (GP) [6,16,18,5], or equivalently, regularization networks [8], is a very promising step in this direction. Instead of having to explicitly specify individual basis functions, we only have to specify a more general kernel that just depends on a very small number of hyperparameters. The key contribution of this paper is to demonstrate that feature selection in RL from sample transitions can be automated, using any of several possible model selection methods for these hyperparameters, such as marginal likelihood optimization in a Bayesian setting, or leave-one-out (LOO) error minimization in a frequentist setting. Here, we will focus on the Bayesian setting, and adapt marginal likelihood optimization for the GP-based approximate policy evaluation method GPTD, introduced without model selection in [6]. Overall, this will have the following benefits: First, only by automatic model selection (as opposed to a grid based search or manual tweaking of kernel parameters) will we be able to use more sophisticated kernels, which will allow us to uncover the "hidden" properties of given problem. For example, by choosing an RBF kernel with independent lengthscales for the individual dimensions of the state space, model selection will automatically drive those components to zero that correspond to state variables irrelevant (or redundant) to the task. This will allow us to concentrate our computational efforts on the parts of the input space that really matter and will improve computational efficiency. Second, because it is generally easier to learn in "smaller" spaces, it may also benefit generalization and thus help us to reduce sample complexity.

Despite its many promises, previous work with GPs in RL rarely explores the benefits of model selection: in [18], a variant of stochastic search was used to determine hyperparameters of the covariance for GPTD using as score function the online performance of an agent. In [16], standard GPs with marginal likelihood based model selection were employed; however, since their approach was based on fitted value iteration, the task of value function approximation was reduced to ordinary regression. The remaining paper is structured as follows: Section 2-3 contain background information and summarize the GPTD framework. As one of the benefits of model selection is the reduction of computational complexity, Section 4 describes how GPTD can be solved for large-scale problems using SR-approximation. Section 5 introduces model selection for GPTD and derives in detail the associated gradient computation. Finally, Section 6 illustrates our approach by providing experimental results.

2 Related Work

The overall goal of learning representations and feature selection for linearly parameterized $\tilde{V}$ is not new within the context of RL. Roughly, past methods can be categorized along two dimensions: how the basis functions are represented (e.g. either by parameterized and predefined basis functions such as RBF, or by nonparameterized basis functions directly derived from the data) and what quantity/target function is considered to guide their construction process (e.g. either supervised methods that consider the Bellman error and depend on the particular reward/goal, or unsupervised graph-based methods that consider connectivity properties of the state space). Conceptually closely related to our work is the approach described in [12], which adapts the hyperparameters of RBF-basis functions (both their location and lengthscales) using either gradient descent or the cross-entropy method on the Bellman error. However, because basis functions are adapted individually (and their number is chosen in advance), the method is prone to overfitting: e.g. by placing basis functions with very small width near discontinuities. The problem is compounded when only few data points are available. In contrast, using a Bayesian approach, we can automatically trade-off model fit and model complexity with the number of data points, choosing always the best complexity: e.g. for small data sets we will prefer larger lengthscales (less complex), for larger data sets we can afford smaller lengthscales (more complex).

Other alternative approaches do not rely on predefined basis functions: The method in [9] is an incremental approach that uses dimensionality reduction and state aggregation to create new basis functions such that for every step the remaining Bellman error for a trajectory of states is successively reduced. A related approach is given in [14] which incrementally constructs an orthogonal basis for the Bellman error. A graph-based unsupervised approach is presented in [11], which derives basis functions from the eigenvectors of the graph Laplacian induced from the underlying MDP.

3 Background: GPs for Policy Evaluation

In this section we briefly summarize how GPs [17] can be used for approximate policy evaluation; here we will follow the GPTD formulation of [6].

Suppose we have observed the sequence of states $\mathbf{x}_1, \mathbf{x}_2, \dots, \mathbf{x}_n$ and rewards $r_1, \dots, r_{n-1}$, where $\mathbf{x}_i \sim p(\cdot \,|\, \mathbf{x}_{i-1}, \pi(\mathbf{x}_{i-1}))$ and $r_i = R(\mathbf{x}_i, \pi(\mathbf{x}_i), \mathbf{x}_{i+1})$. In practice, MDPs considered in RL will often be of an episodic nature with absorbing terminal states. Therefore we have to transform the problem such that the resulting Markov chain is still ergodic: this is done by introducing a zero reward transition from the terminal state of one episode to the start state of the next episode. In addition to the sequence of states and rewards our training data thus also includes a sequence $\gamma_1, \dots, \gamma_{n-1}$, where $\gamma_i = \gamma$ (the discount factor in Eq. (1)) if $\mathbf{x}_{i+1}$ was a non-terminal state, and $\gamma_i = 0$ if $\mathbf{x}_i$ was a terminal state (in which case $\mathbf{x}_{i+1}$ is the start state of the next episode).

Assume that the function values $V(\mathbf{x})$ of the unknown value function $V : \mathcal{X} \subset \mathbb{R}^D \to R$ from Eq. (1) form a zero-mean Gaussian process with covariance function $k(\mathbf{x}, \mathbf{x}')$ for $\mathbf{x}, \mathbf{x}' \in \mathcal{X}$; in short $V \sim \mathcal{GP}(\mathbf{0}, k(\mathbf{x}, \mathbf{x}'))$. In consequence, the function values for the n observed states, $\mathbf{v} := \left(V(\mathbf{x}_1), \dots, V(\mathbf{x}_n)\right)^{\mathsf{T}}$, will have a Gaussian distribution

$$\mathbf{v} \,|\, \mathbf{X}, \boldsymbol{\theta} \sim \mathcal{N}(\mathbf{0}, \mathbf{K}), \tag{2}$$

where $\mathbf{X} := [\mathbf{x}_1, \ldots, \mathbf{x}_n]$ and $\mathbf{K}$ is the $n \times n$ covariance matrix with entries $[\mathbf{K}]_{ij} = k(\mathbf{x}_i, \mathbf{x}_j)$. Note that the covariance $k(\cdot, \cdot)$ alone fully specifies the GP; here we will assume that it is a simple (positive definite) function parameterized by a number of scalar parameters collected in vector $\boldsymbol{\theta}$ (see Section 4).

However, unlike in ordinary regression, in RL we cannot observe samples from the target function V directly. Instead, the values can only be observed indirectly: from Eq. (1) we have that the value of one state is recursively defined through the value of the successor state(s) and the immediate reward. To this end, Engel et al. propose the following generative model:[1]

$$R(\mathbf{x}_i, \mathbf{x}_{i+1}) = V(\mathbf{x}_i) - \gamma_i V(\mathbf{x}_{i+1}) + \eta_i, \tag{3}$$

where η_i is a noise term that may depend on the inputs.[2] Plugging in the observed training data, and defining $\mathbf{r} := (r_1, \ldots, r_{n-1})^\mathsf{T}$, we obtain

$$\mathbf{r} = \mathbf{H}\mathbf{v} + \boldsymbol{\eta}, \tag{4}$$

where the $(n-1) \times n$ matrix $\mathbf{H}$ is given by

$$\mathbf{H} := \begin{bmatrix} 1 & -\gamma_1 & & \\ & \ddots & \ddots & \\ & & 1 & -\gamma_{n-1} \end{bmatrix} \tag{5}$$

and noise $\boldsymbol{\eta} := (\eta_1, \ldots, \eta_{n-1})^\mathsf{T}$ has distribution $\boldsymbol{\eta} \sim \mathcal{N}(\mathbf{0}, \boldsymbol{\Sigma})$. One first choice for the noise covariance $\boldsymbol{\Sigma}$ would be $\boldsymbol{\Sigma} = \sigma_0^2 \mathbf{I}$, where σ_0^2 is an unknown hyperparameter (see Section 4). However, this model does not capture stochastic state transitions and hence would only be applicable for deterministic MDPs. If the environment is stochastic, the noise model $\boldsymbol{\Sigma} = \sigma_0^2 \mathbf{H}\mathbf{H}^\mathsf{T}$ is more appropriate, see [6] for more detailed explanations. For the remainder we will solely consider the latter choice, i.e. $\boldsymbol{\Sigma} = \sigma_0^2 \mathbf{H}\mathbf{H}^\mathsf{T}$.

Let $\mathcal{D} := \{\mathbf{X}, \gamma_1, \ldots, \gamma_{n-1}\}$ be an abbreviation for the training inputs. Using Eq. (4), it can be shown that the joint distribution of the observed rewards $\mathbf{r}$ given inputs $\mathcal{D}$ is again a Gaussian,

$$\mathbf{r} \,|\, \mathcal{D}, \boldsymbol{\theta} \sim \mathcal{N}(\mathbf{0}, \mathbf{Q}), \tag{6}$$

where the $(n-1) \times (n-1)$ covariance matrix $\mathbf{Q}$ is given by

$$\mathbf{Q} = (\mathbf{H}\mathbf{K}\mathbf{H}^\mathsf{T} + \sigma_0^2 \mathbf{H}\mathbf{H}^\mathsf{T}). \tag{7}$$

To predict the function value $V(\mathbf{x}^*)$ at a new state $\mathbf{x}^*$, we consider the joint distribution of $\mathbf{r}$ and $V(\mathbf{x}^*)$

$$\begin{bmatrix} \mathbf{r} \\ V(\mathbf{x}^*) \end{bmatrix} \,|\, \mathcal{D}, \mathbf{x}^*, \boldsymbol{\theta} \sim \mathcal{N}\left(\begin{bmatrix} \mathbf{0} \\ 0 \end{bmatrix}, \begin{bmatrix} \mathbf{Q} & \mathbf{H}\mathbf{k}(\mathbf{x}^*) \\ [\mathbf{H}\mathbf{k}(\mathbf{x}^*)]^\mathsf{T} & k^* \end{bmatrix} \right)$$

---

[1] Note that this model is just a linearly transformed version of the standard model in GP regression, i.e. $y_i = f(\mathbf{x}_i) + \varepsilon_i$.

[2] Formally, in GPTD noise is modeled by a second zero-mean GP that is independent from the value GP. See [6] for details.

where $n \times 1$ vector $\mathbf{k}(\mathbf{x}^*)$ is given by $\mathbf{k}(\mathbf{x}^*) := \big(k(\mathbf{x}^*, \mathbf{x}_1), \ldots, k(\mathbf{x}^*, \mathbf{x}_n)\big)^\mathsf{T}$ and scalar k^* by $k^* := k(\mathbf{x}^*, \mathbf{x}^*)$. Conditioning on $\mathbf{r}$, we then obtain

$$V(\mathbf{x}^*) \,|\, \mathcal{D}, \mathbf{r}, \mathbf{x}^*, \boldsymbol{\theta} \;\sim\; \mathcal{N}(\mu(\mathbf{x}^*), \sigma^2(\mathbf{x}^*)) \tag{8}$$

where

$$\mu(\mathbf{x}^*) := \mathbf{k}(\mathbf{x}^*)^\mathsf{T}\mathbf{H}^\mathsf{T}\mathbf{Q}^{-1}\mathbf{r} \tag{9}$$

$$\sigma^2(\mathbf{x}^*) := k^* - \mathbf{k}(\mathbf{x}^*)^\mathsf{T}\mathbf{H}^\mathsf{T}\mathbf{Q}^{-1}\mathbf{H}\mathbf{k}(\mathbf{x}^*). \tag{10}$$

Thus, for any given single state $\mathbf{x}^*$, GPTD produces the distribution $p(V(\mathbf{x}^*)| \mathcal{D}, \mathbf{r}, \mathbf{x}^*, \boldsymbol{\theta})$ in Eq. (8) over function values.

4 Computational Considerations

Regarding its implementation, GPTD for policy evaluation shares the same weakness that GPs have in traditional machine learning tasks: solving Eq. (8) requires the inversion[3] of a dense $(n-1) \times (n-1)$ matrix, which when done exactly would require $\mathcal{O}(n^3)$ operations and is hence infeasible for anything but small-scale problems (say, anything with $n < 5000$).

4.1 Subset of Regressors

In the subset of regressors (SR) approach initially proposed for regularization networks [15,10], one chooses a subset $\{\tilde{\mathbf{x}}\}_{i=1}^m$ of the data, with $m \ll n$, and approximates the covariance for arbitrary $\mathbf{x}, \mathbf{x}'$ by taking

$$\tilde{k}(\mathbf{x}, \mathbf{x}') = \mathbf{k}_m(\mathbf{x})^\mathsf{T}\mathbf{K}_{mm}^{-1}\mathbf{k}_m(\mathbf{x}'). \tag{11}$$

Here $\mathbf{k}_m(\cdot)$ denotes $\mathbf{k}_m(\cdot) := \big(k(\tilde{\mathbf{x}}_1, \cdot), \ldots, k(\tilde{\mathbf{x}}_m, \cdot)\big)^\mathsf{T}$, and $\mathbf{K}_{mm}$ is the submatrix $[\mathbf{K}_{mm}]_{ij} = k(\tilde{\mathbf{x}}_i, \tilde{\mathbf{x}}_j)$ of $\mathbf{K}$. The approximation in Eq. (11) can be motivated for example from the Nyström approximation [22]. Let $\mathbf{K}_{nm}$ denote the submatrix $[\mathbf{K}_{nm}]_{ij} = k(\mathbf{x}_i, \tilde{\mathbf{x}}_j)$ corresponding to the m columns of the data points in the subset. We then have the rank-reduced approximation $\mathbf{K} \approx \tilde{\mathbf{K}} = \mathbf{K}_{nm}\mathbf{K}_{mm}^{-1}\mathbf{K}_{nm}^\mathsf{T}$ and $\mathbf{k}(\mathbf{x}) \approx \tilde{\mathbf{k}}(\mathbf{x}) = \mathbf{K}_{nm}\mathbf{K}_{mm}^{-1}\mathbf{k}_m(\mathbf{x})$. Plugging these into Eq. (8), we obtain for the mean

$$\begin{aligned} \mu(\mathbf{x}^*) &\approx \tilde{\mathbf{k}}(\mathbf{x}^*)^\mathsf{T}\mathbf{H}^\mathsf{T}\big(\mathbf{H}\tilde{\mathbf{K}}\mathbf{H}^\mathsf{T} + \sigma_0^2\mathbf{H}\mathbf{H}^\mathsf{T}\big)^{-1}\mathbf{r} \\ &= \mathbf{k}_m(\mathbf{x}^*)^\mathsf{T}\big(\mathbf{G}^\mathsf{T}\mathbf{W}\mathbf{G} + \sigma_0^2\mathbf{K}_{mm}\big)^{-1}\mathbf{G}^\mathsf{T}\mathbf{W}\mathbf{r}, \end{aligned} \tag{12}$$

where we have defined $\mathbf{G} := \mathbf{H}\mathbf{K}_{nm}$, $\mathbf{W} := (\mathbf{H}\mathbf{H}^\mathsf{T})^{-1}$ and applied the SMW identity[4] to show that

$$\mathbf{K}_{mm}^{-1}\mathbf{G}^\mathsf{T}\big(\mathbf{G}\mathbf{K}_{mm}^{-1}\mathbf{G}^\mathsf{T} + \sigma_0^2\mathbf{W}^{-1}\big)^{-1} = \big(\mathbf{G}^\mathsf{T}\mathbf{W}\mathbf{G} + \sigma_0^2\mathbf{K}_{mm}\big)^{-1}\mathbf{G}^\mathsf{T}\mathbf{W}. \tag{13}$$

[3] For numerical reasons we implement this step using the Cholesky decomposition, which has the same computational complexity.

[4] $(\mathbf{A} + \mathbf{B}\mathbf{D}^{-1}\mathbf{C})^{-1}\mathbf{B}\mathbf{D}^{-1} = \mathbf{A}^{-1}\mathbf{B}(\mathbf{D} + \mathbf{C}\mathbf{A}^{-1}\mathbf{B})^{-1}$.

Similarly, we obtain for the predictive variance

$$\begin{aligned}\sigma(\mathbf{x}^*) &\approx \tilde{k}(\mathbf{x}^*, \mathbf{x}^*) - \tilde{\mathbf{k}}(\mathbf{x}^*)^{\mathsf{T}} \mathbf{H}^{\mathsf{T}} \left(\mathbf{H}\tilde{\mathbf{K}}\mathbf{H}^{\mathsf{T}} + \sigma_0^2 \mathbf{H}\mathbf{H}^{\mathsf{T}}\right)^{-1} \mathbf{H}\tilde{\mathbf{k}}(\mathbf{x}^*) \\ &= \sigma_0^2 \mathbf{k}_m(\mathbf{x}^*)^{\mathsf{T}} \left(\mathbf{G}^{\mathsf{T}}\mathbf{W}\mathbf{G} + \sigma_0^2 \mathbf{K}_{mm}\right)^{-1} \mathbf{k}_m(\mathbf{x}^*). \end{aligned} \tag{14}$$

Doing this means a huge gain in computational savings: solving the reduced problem in Eq. (12) costs $\mathcal{O}(m^2 n)$ for initialization, requires $\mathcal{O}(m^2)$ storage and every prediction costs $\mathcal{O}(m)$ (or $\mathcal{O}(m^2)$ if we additionally evaluate the variance). This has to be compared with the complexity of the full problem: $\mathcal{O}(n^3)$ initialization, $\mathcal{O}(n^2)$ storage, and $\mathcal{O}(n)$ prediction. Thus computational complexity now only depends linearly on n (for constant m).

Note that the SR-approximation produces a degenerate GP. As a consequence, the predictive variance in Eq. (14) will underestimate the true variance. In particular, it will be near zero when $\mathbf{x}$ is far from the subset $\{\tilde{\mathbf{x}}\}_{i=1}^m$ (which is exactly the opposite of what we want, as the predictive variance should be high for novel inputs). The situation can be remedied by considering the projected process approximation [4,19], which results in the same expression for the mean in Eq. (12), but adds the term

$$k(\mathbf{x}^*, \mathbf{x}^*) - \mathbf{k}_m(\mathbf{x}^*)^{\mathsf{T}} \mathbf{K}_{mm}^{-1} \mathbf{k}_m(\mathbf{x}^*) \tag{15}$$

to the variance in Eq. (14)

4.2 Selecting the Subset (Unsupervised)

Selecting the best subset is a combinatorial problem that cannot be solved effeciently. Instead, we try to find a compact subset that summarizes the relevant information by incremental forward selection. In every step of the procedure, we add that element from the set of remaining unselected elements to the active set that performs best with respect to a given specific criterion. In general, we distinguish between supervised and unsupervised approaches, i.e. those that consider the target variable we regress on, and those that do not. Here we focus on the incomplete Cholesky decomposition (ICD) as an unsupervised approach [7,1,2].

ICD aims at reducing at each step the error incurred from approximating the covariance matrix: $\left\|\mathbf{K} - \tilde{\mathbf{K}}\right\|_F$. Note that the ICD of $\mathbf{K}$ is the dual equivalent of performing partial Gram-Schmidt on the Mercer-induced feature representation: in every step, we add that element to the active set whose distance from the span of the currently selected elements is largest (in feature space). The procedure is stopped when the residual of remaining (unselected) elements falls below a given threshold, or a given maximum number of allowed elements is exceeded. In [4,8,6] online variants thereof are considered (where instead of repeatedly inspecting all remaining elements only one pass over the dataset is made and every element is examined only once). In general, the number of elements selected by ICD will depend on the effective rank of $\mathbf{K}$ (and thus its eigenspectrum).

5 Model Selection for GPTD

The major advantage of using GP-based function approximation (in contrast to, say, neural networks or tree-based approaches) is that both 'learning' of the weight vector and specification of the architecture/hyperparameters/basis functions can be handled in a principled and essentially automated way.

5.1 Optimizing the Marginal Likelihood

To determine hyperparameters for GPTD, we consider the marginal likelihood of the process, i.e. the probability of generating the rewards we have observed given the sequence of states and a particular setting of the hyperparameter $\boldsymbol{\theta}$. We then maximize this function (its logarithm) with respect to $\boldsymbol{\theta}$. From Eq. (6) we see that for GPTD we have $p(\mathbf{r}|\mathcal{D}, \boldsymbol{\theta}) = \mathcal{N}(\mathbf{0}, \mathbf{Q})$. Thus plugging in the definition for a multivariate Gaussian and taking the logarithm, we obtain

$$\mathcal{L}(\boldsymbol{\theta}) = -\frac{1}{2}\log\det\mathbf{Q} - \frac{1}{2}\mathbf{r}^{\mathsf{T}}\mathbf{Q}^{-1}\mathbf{r} - \frac{n}{2}\log 2\pi. \tag{16}$$

Optimizing this function with respect to $\boldsymbol{\theta}$ is a nonconvex problem and we have to resort to iterative gradient-based solvers (such as scaled conjugate gradients, e.g. see [13]). To do this we need to be able to evaluate the gradient of $\mathcal{L}$. The partial derivatives of $\mathcal{L}$ with respect to each individual hyperparameter θ_i can be obtained in closed form as

$$\frac{\partial\mathcal{L}}{\partial\theta_i} = -\frac{1}{2}\mathrm{tr}\left(\mathbf{Q}^{-1}\frac{\partial\mathbf{Q}}{\partial\theta_i}\right) + \frac{1}{2}\mathbf{r}^{\mathsf{T}}\mathbf{Q}^{-1}\frac{\partial\mathbf{Q}}{\partial\theta_i}\mathbf{Q}^{-1}\mathbf{r}. \tag{17}$$

Note that $\mathcal{L}$ automatically incorporates the trade-off between model fit (training error) and model complexity and can thus be regarded as an indicator for generalization capabilities, i.e. how well GPTD will predict the values of states not in its training set. The first term in Eq. (16) measures the complexity of the model, and will be large for 'flexible' and small for 'rigid' models.[5] The second term measures the model fit and can shown to be the value of the error function for a penalized least-squares that would (in a frequentist setting) correspond to GPTD.

[5] A property that manifests itself in the eigenvalues of $\mathbf{K}$ (since the determinant equals the sum of the eigenvalues). In general, flexible models are achieved by smaller bandwidths in the covariance, meaning that $\mathbf{K}$'s effective rank will be large and its eigenvalues will fall off more slowly. On the other hand, more rigid models are achieved by larger bandwidths, meaning that $\mathbf{K}$'s effective rank will be low and its eigenvalues will fall off more quickly. Note that the effective rank of $\mathbf{K}$ is also important for the SR-approximation (see Section 3), since the effectiveness of SR depends on building a low-rank approximation of $\mathbf{K}$ spending as few resources as possible.

5.2 Choosing the Covariance

A common choice for $k(\cdot,\cdot)$ is to consider a (positive definite) function parameterized by a small number of scalar parameters, such as the stationary isotropic Gaussian (or squared exponential), which is parameterized by the lengthscale (bandwidth h). In the following we will consider three variants of the form [13,17]:

$$k(\mathbf{x},\mathbf{x}') = v_0 \exp\left\{-\frac{1}{2}(\mathbf{x}-\mathbf{x}')^{\mathsf{T}}\boldsymbol{\Omega}(\mathbf{x}-\mathbf{x}')\right\} + b \tag{18}$$

where hyperparameter $v_0 > 0$ denotes the vertical lengthscale, $b > 0$ the bias, and symmetric positive semidefinite matrix $\boldsymbol{\Omega}$ is given by

- **Variant 1 (isotropic): $\boldsymbol{\Omega} = h\mathbf{I}$**
  with hyperparameter $h > 0$.
- **Variant 2 (axis-aligned ARD): $\boldsymbol{\Omega}$** $= \mathrm{diag}(a_1,\dots,a_D)$
  with hyperparameters $a_1,\dots,a_D > 0$.
- **Variant 3(factor analysis): $\boldsymbol{\Omega}$** $= \mathbf{M}_k\mathbf{M}_k^{\mathsf{T}} + \mathrm{diag}(a_1,\dots,a_D)$
  where $D \times k$ matrix $\mathbf{M}_k$ is given by $\mathbf{M}_k := [\mathbf{m}_1,\dots,\mathbf{m}_k]$, $k < D$, and both the entries of $\mathbf{M}_k$, i.e. $m_{11},\dots,m_{1D},\dots,m_{k1},\dots,m_{kD}$ and $a_1,\dots,a_D > 0$ are adjustable hyperparameters.[6]

The first variant (see Figure 1) assumes that every coordinate of the input (i.e. state-vector) is equally important for predicting its value. However, in particular for high-dimensional state vectors, this might be too simple: along some dimensions this will produce too much resolution where it will be wasted, along other dimensions this will produce too little resolution where it would otherwise be needed. The second variant is more powerful and includes a different parameter for every coordinate of the state vector, thus assigning a different scale to every state variable. This covariance implements automatic relevance determination (ARD): since the individual scaling factors are automatically adapted from the data via marginal likelihood optimization, they inform us about how relevant each state variable is for predicting the value. A large value of a_i means that the i-th state variable is important and even small variations along this coordinate are relevant. A small value of a_j means that the j-th state variable is less important and only large variations along this coordinate will impact the prediction (if at all). A value close to zero means that the corresponding coordinate is irrelevant and could be left out (i.e. the value function does not rely on that particular state variable). The benefit of removing irrelevant coordinates is that the complexity of the model will decrease while the fit of the model stays the same: thus likelihood will increase. The third variant first identifies relevant directions in the input space (linear combinations of state variables) and performs a rotation of the coordinate system (the number of relevant directions is specified in advance by k). As in the second variant, different scaling factors are then applied along the rotated axes.

[6] The number of directions k is also determined from model selection: we systematically try different values of k, find the corresponding remaining hyperparameters via scg-based likelihood optimization and compare the final scores (likelihood) of the resulting models.

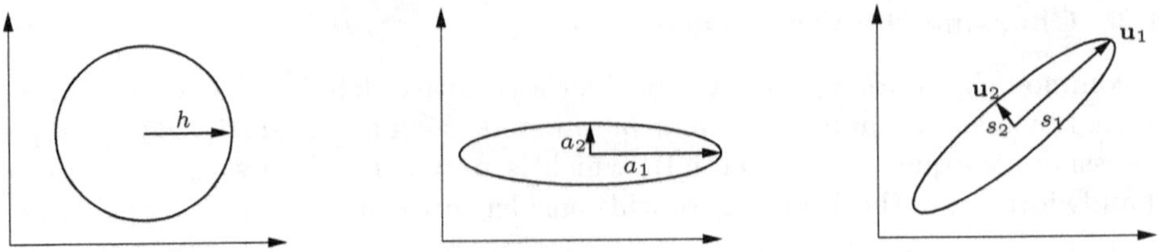

Fig. 1. Three variants of the stationary squared exponential covariance. The directions/scaling factors in the third case are derived from the eigendecomposition of $\boldsymbol{\Omega}$, i.e. $\mathbf{USU}^\mathsf{T} = \mathbf{M}_k\mathbf{M}_k^\mathsf{T} + \mathrm{diag}(a_1, \ldots, a_D)$.

5.3 Example: Gradient for ARD

As an example, we will now show how the gradient $\nabla_{\boldsymbol{\theta}}\mathcal{L}$ of Eq. (16) is calculated for the ARD covariance. Note that since all hyperparameters in this model, i.e. $\{v_0, b, \sigma_0^2, a_1, \ldots, a_D\}$, must be positive, it is more convenient to consider the hyperparameter vector $\boldsymbol{\theta}$ in log space: $\boldsymbol{\theta} = \left(\log v_0, \log b, \log \sigma_0^2, \log a_1, \ldots, \log a_D\right)$. We start by establishing some useful identities: for any $n \times n$ matrix $\mathbf{A}$ we have

$$[\mathbf{HAH}^\mathsf{T}]_{ij} = a_{ij} - \gamma_i a_{i+1,j} - \gamma_j a_{i,j+1} + \gamma_i\gamma_j a_{i+1,j+1}.$$

Furthermore, we have

$$[\mathbf{HH}^\mathsf{T}]_{ij} = \begin{cases} 1+\gamma_i^2 & , i=j \\ -\gamma_i & , i=j-1 \text{ or } i=j+1 \\ 0 & , \text{ otherwise} \end{cases}$$

Now write $\mathbf{K}$ as $\mathbf{K} = v_0\mathbf{C} + b\mathbb{1}_{n,n}$, where $[\mathbf{C}]_{ij} = \exp\left\{-0.5\sum_{d=1}^D a_d\left(x_d^{(i)} - x_d^{(j)}\right)^2\right\}$ and $\mathbb{1}_{n,n}$ is the $n \times n$ matrix of all ones. Computing the partial derivative of $\mathbf{K}$, we then obtain

$$\frac{\partial \mathbf{K}}{\partial v_0} = \mathbf{C}, \quad \frac{\partial \mathbf{K}}{\partial b} = \mathbb{1}_{n,n}$$

$$\left[\frac{\partial \mathbf{K}}{\partial a_\nu}\right]_{ij} = -\frac{1}{2}v_0 c_{ij}\left(x_\nu^{(i)} - x_\nu^{(j)}\right)^2, \quad \nu = 1 \ldots D$$

Next, we will compute the partial derivatives of $\mathbf{Q} = (\mathbf{HKH}^\mathsf{T} + \sigma_0^2\mathbf{HH}^\mathsf{T})$, giving for b:

$$\frac{\partial \mathbf{Q}}{\partial \log b} = b\frac{\partial \mathbf{Q}}{\partial b} = b\mathbf{H}\left[\frac{\partial \mathbf{K}}{\partial b}\right]\mathbf{H}^\mathsf{T} = b\mathbf{H}\mathbb{1}_{n,n}\mathbf{H}^\mathsf{T}$$

$$\Rightarrow \left[\frac{\partial \mathbf{Q}}{\partial \log b}\right]_{ij} = b(1 - \gamma_i - \gamma_j + \gamma_i\gamma_j).$$

For v_0 we have

$$\frac{\partial \mathbf{Q}}{\partial \log v_0} = v_0\frac{\partial \mathbf{Q}}{\partial v_0} = v_0\mathbf{H}\left[\frac{\partial \mathbf{K}}{\partial v_0}\right]\mathbf{H}^\mathsf{T} = v_0\mathbf{HCH}^\mathsf{T}$$

$$\Rightarrow \left[\frac{\partial \mathbf{Q}}{\partial \log v_0}\right]_{ij} = v_0\left(c_{ij} - \gamma_i c_{i+1,j} - \gamma_j c_{i,j+1} + \gamma_i\gamma_j c_{i+1,j+1}\right).$$

For σ_0^2 we have

$$\frac{\partial \mathbf{Q}}{\partial \log \sigma_0^2} = \sigma_0^2 \frac{\partial \mathbf{Q}}{\partial \sigma_0^2} = \sigma_0^2 \frac{\partial}{\partial \sigma_0^2}[\sigma_0^2 \mathbf{H}\mathbf{H}^\top] = \sigma_0^2 \mathbf{H}\mathbf{H}^\top$$

$$\Rightarrow \left[\frac{\partial \mathbf{Q}}{\partial \log \sigma_0^2}\right]_{ij} = \begin{cases} \sigma_0^2(1+\gamma_i^2) & , i = j \\ -\sigma_0^2 \gamma_i & , i = j-1 \text{ or } i = j+1 \\ 0 & , \text{ otherwise} \end{cases}$$

Finally, for each of the a_ν, $\nu = 1, \ldots, D$ we get

$$\frac{\partial \mathbf{Q}}{\partial \log a_\nu} = a_\nu \frac{\partial \mathbf{Q}}{\partial a_\nu} = a_\nu \mathbf{H} \left[\frac{\partial \mathbf{K}}{\partial a_\nu}\right] \mathbf{H}^\top$$

$$\Rightarrow \left[\frac{\partial \mathbf{Q}}{\partial \log a_\nu}\right]_{ij} = -\frac{1}{2} a_\nu v_0 (c_{ij} d_{ij}^\nu - \gamma_i c_{i+1,j} d_{i+1,j}^\nu - \gamma_j c_{i,j+1} d_{i,j+1}^\nu + \gamma_i \gamma_j c_{i+1,j+1} d_{i+1,j+1}^\nu)$$

where we have defined $d_{ij}^\nu := \left(x_\nu^{(i)} - x_\nu^{(j)}\right)^2$. Thus, with $\mathbf{w} := \mathbf{Q}^{-1}\mathbf{r}$ we have for Eq. (17)

$$\operatorname{tr}\left(\mathbf{Q}^{-1} \frac{\partial \mathbf{Q}}{\partial \theta_\nu}\right) = \sum_{i=1}^{n-1} \sum_{j=1}^{n-1} \left[\mathbf{Q}^{-1}\right]_{ij} \left[\frac{\partial \mathbf{Q}}{\partial \theta_\nu}\right]_{ji}$$

$$\mathbf{w}^\top \frac{\partial \mathbf{Q}}{\partial \theta_\nu} \mathbf{w} = \sum_{i=1}^{n-1} \sum_{j=1}^{n-1} [\mathbf{w}]_i [\mathbf{w}]_j \left[\frac{\partial \mathbf{Q}}{\partial \theta_\nu}\right]_{ij}$$

which can be used to calculate the partial derivates with computational complexity $\mathcal{O}(n^2)$ each (except for σ_0^2, where the matrix of derivatives is tridiagonal).

6 Experiments

This section demonstrates that our proposed model selection can be used to solve the approximate policy evaluation problem in a completely automated way – without any manual tweaking of hyperparameters. We will also show some of the additional benefits of model selection, which are improved accuracy and reduced complexity: because we automatically set the hyperparameters we can use more sophisticated covariance functions (see Section 5.2) that depend on a larger number[7] of hyperparameters, thus better fit the regularities of a particular dataset, and therefore do not waste unnecessary resources on irrelevant aspects of the state-vector. The latter aspect is particularly interesting for computational reasons (see Section 4) and becomes important in large-scale applications.

---

[7] Setting these hyperparameters by hand would require even more trial and error; therefore, these covariances are seldom employed without model selection.

6.1 Pendulum Swing-Up Task

First, we consider the pendulum swing-up task, a common benchmark in RL. The goal is to swing up an underpowered pendulum and balance it around the inverted upright position (here formulated as an episodic task). More details and the equations of motion can be found in e.g. [5]. Since GPTD only solves (approximate) policy evaluation, to test our model selection approach we chose to generate a sample trajectory under the optimal policy (obtained from fitted value iteration). We generated a sequence of 1000 state-transitions under this policy (which corresponds to about 25 completed episodes) and applied GPTD for the three choices of covariance: isotropic (I), axis-aligned ARD (II), and factor analyis (III). In each case, the best setting of hyperparameters was found from running[8] scaled conjugate gradients on Eq. (16), giving

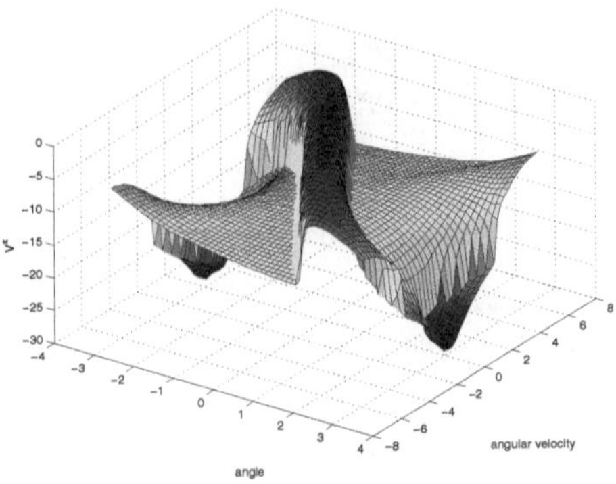

Fig. 2. Optimal value function for the pendulum domain, computed with fitted value iteration over a discretized state space (400 × 400 grid)

$$\begin{aligned}
&\text{I: } v_0 = 18.19 \;\; \sigma_0^2 = 0.05 \;\; b = 0.11 \;\; h = 7.48 \\
&\text{II: } v_0 = 15.95 \;\; \sigma_0^2 = 0.05 \;\; b = 0.10 \;\; a_1 = 3.62 \;\; a_2 = 6.63 \\
&\text{III: } v_0 = 10.82 \;\; \sigma_0^2 = 0.08 \;\; b = 0.10 \;\; s_1 = 13.91 \;\; s_2 = 0.36 \;\; \mathbf{u}_1 = [0.58\ 0.81] \;\; \mathbf{u}_2 = [-0.81\ 0.58]
\end{aligned}$$

(the last ones given in terms of the eigendecomposition of $\boldsymbol{\Omega}$). Figure 3 shows the results: all three produce an adequate representation of the true value function shown in Figure 2 in and near the states visited in the trajectory (MSE in states of the sample trajectory: (I) 0.27, (II) 0.24, and (III) 0.26), but differ once they start predicting values of states not in the training data (MSE for states on a 50 × 50 grid: (I) 46.36, (II) 48.89, and (III) 12.24). Despite having a slightly higher error on the known training data, (III) substantially outperforms the other models when it comes to predicting the values of new states. With respect to model selection, (III) also has the highest likelihood. Note that (III) chooses one dominant direction ($\mathbf{u}_1 = [0.58\ 0.81]$) to which it assigns high relevance ($s_1 = 13.91$); the remainder ($\mathbf{u}_2 = [-0.81\ 0.58]$) has only little impact ($s_2 = 0.36$). Taking a closer look at Figure 2, we see that indeed the value function varies more strongly along the diagonal direction lower left to upper right, whereas it varies only slowly along the opposite diagonal upper left to lower right. For (II), relevance can only be assigned along the φ and $\dot{\varphi}$ coordinates (state-variables), which in this case gives us no particular benefit; and (I) is not at all able to assign different importance to different state variables.

[8] We used the full data set for model selection, to avoid the complexities involved with subset-based likelihood approximation, e.g. see [20]. In our implementation, model selection for all 1000 data points took about 15-30 secs on a 1.5GHz PC.

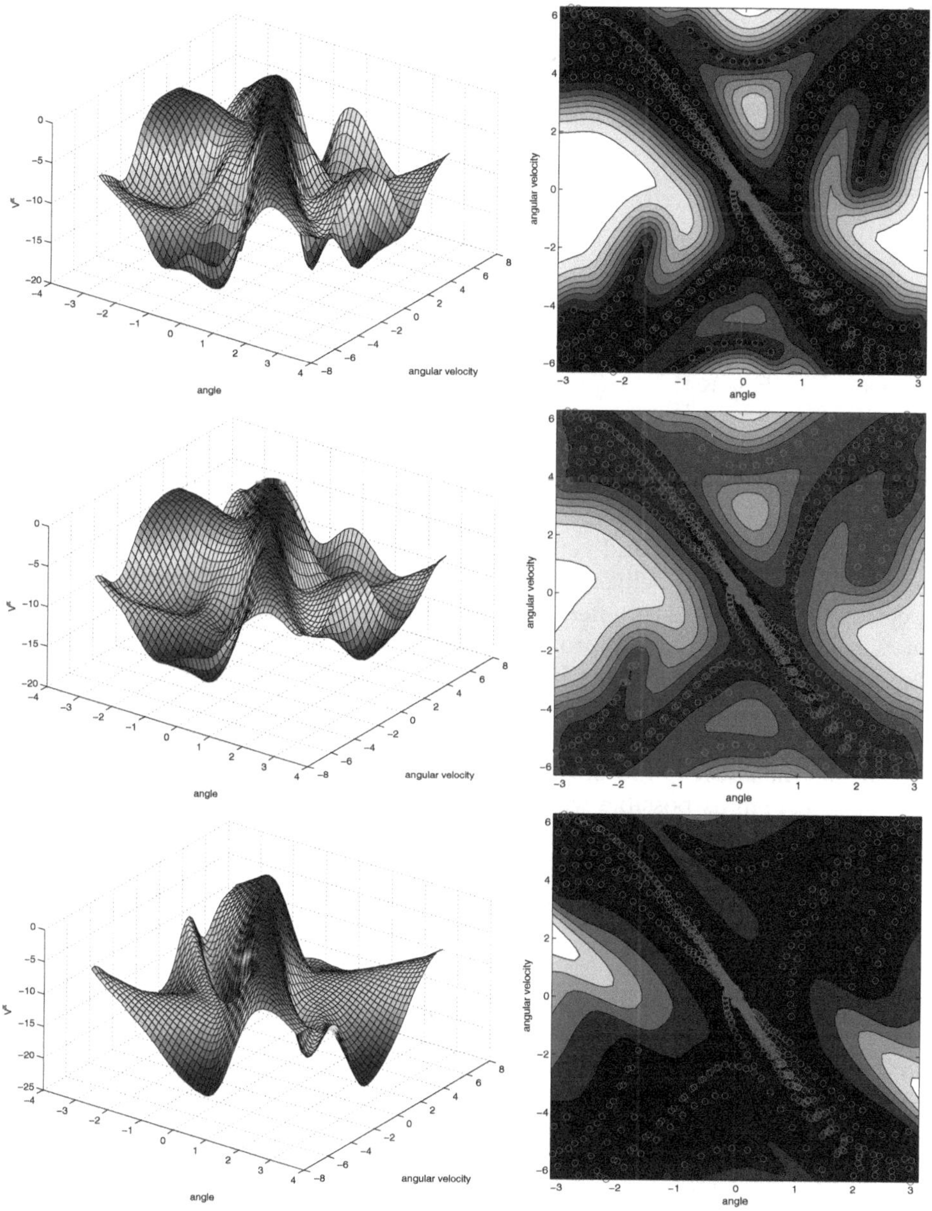

Fig. 3. From top to bottom: GPTD approximation of the value function from Figure 2 for the covariances (I),(II),(III), where in each case the hyperparameters were obtained from marginal likelihood optimization for the GPTD process in Eq. (16). Right: Associated predictive variance. Black indicates low variance, white indicates high variance and red circles indicate the location of the states in the training set (which was the same for all three experiments).

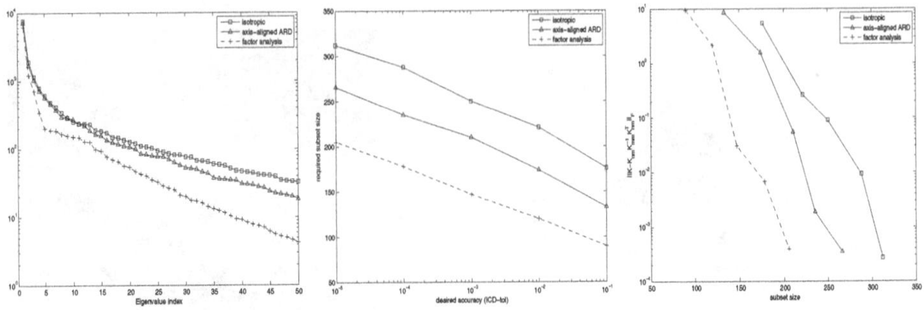

Fig. 4. Properties of **K** for different choices of $k(\cdot,\cdot)$. Left: Eigenspectrum. Center: Number of elements incomplete Cholesky selects for a given threshold. Right: Approximation error $\left\|\mathbf{K}-\tilde{\mathbf{K}}\right\|_F$, given the size of the subset.

Additional insight is gained by looking at the eigenspectrum of **K**. Figure 4 (left) shows that (I)'s eigenvalues decrease the slowest, whereas (III)'s decrease the fastest. This has two consequences. First, the eigenspectrum is intimately related with complexity and generalization capabilities (see Eq. (16)) and thus helps explain why (III) delivers better prediction performance. Second, the eigenspectrum also indicates the effective rank of **K** and strongly impacts our ability to build an efficient low-rank approximation of **K** using as small a subset as possible (see Section 4). A small subset in turn is important for computational efficiency because its size is the dominant factor when we employ the SR-approximation: both for batch and online learning the operation count depends quadratically on the size of the subset (and only linear on the number of datapoints). Keeping this size as small as possible without losing predictive performance is essential. Figure 4 (center and right) shows that in this regard (III) performs best and (I) worst: for example, if we were to approximate **K** using SR-approximation with ICD selection at a tolerance level of 10^{-1}, out of our 1000 samples (I) would choose ~ 175, (II) would chose ~ 140, and (III) would choose ~ 80 elements.

6.2 A 2D Gridworld with 1 Latent Dimension

To illustrate in more detail how our approach handles irrelevant state variables, we use a specifically designed 2D gridworld with 11×11 states. Every step entails a reward of -1 except when being in a state with $x=6$, which starts a new episode (teleports to a new random state with zero reward). We consider the policy that moves left when $x>6$ and right when $x<6$. In addition, every time we move left or right we will also move randomly up or down (with 50% each). The corresponding value function is shown in Figure 5 (left). We generated 500 transitions and applied GPTD with covariance (I) and (II) with automatic model selection resulting in[9]

[9] Here we do not include results for (III) which operates on linear combinations of states and in this scenario would have to find a direction that is perfectly aligned with the x-axis (which is more difficult).

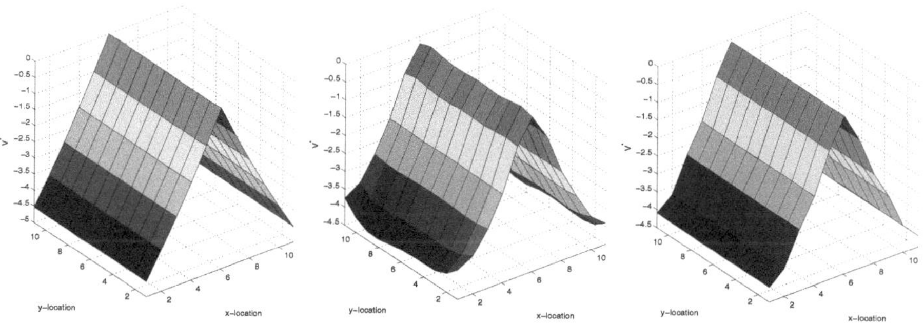

Fig. 5. Learned value functions for the 2D-gridworld domain. Left: true value function. Note how the y-coordinate is irrelevant for the value. Center: approximation with isotropic covariance. Right: approximation for axis-aligned ARD covariance (after removal of irrelevant input).

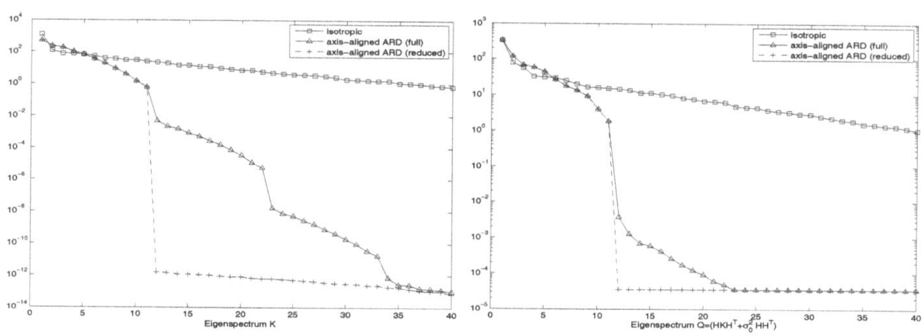

Fig. 6. Effect of dimensionality reduction on the complexity of the model. Left: Eigenvalues of **K**. Right: Eigenvalues of **Q**.

	Hyperparameters $\boldsymbol{\theta}$	Complexity	Data fit	$\mathcal{L}$ (smaller is better)
(I)	$h = 2.89$	-2378.2	54.78	-2323.4
(II)	$a_1 = 3.53 \quad a_2 = 10^{-5}$	-2772.7	13.84	-2758.8
(II) without y	$a_1 = 3.53 \quad a_2 = 0$	-2790.7	13.84	-2776.8

As can be seen from Figure 5 (center and right), both obtain a very reasonable approximation. However, (II) automatically detects that the y-coordinate of the state is irrelevant and thus assigns a very small weight to it ($a_2 < 10^{-5}$). With a uniform lengthscale, (I) is unable to do that and has to put equal weight on both state variables. As a consequence, its estimate is less exact and more wiggly (MSE: (I) 0.030, and (II) 0.019). Additional insight can be gained by looking at the likelihood $\mathcal{L}$ of the models (cf. Eq. (16)). Here we see that (II) has lower complexity (cf. eigenspectrum of $\mathbf{Q}$ in Figure 6), fits the data better and thus has a higher combined likelihood (note that the values in the table show the negative log likelihood which we minimize). Moreover, if we completely remove the y state variable (setting $a_2 := 0$), the eigenspectrum of $\mathbf{Q}$ decreases more rapidly; thus (II) without y has an even lower complexity while still having the same fit. This indicates that state component y can be safely ignored in this task/domain. In addition, as was mentioned before, the lower effective rank of $\mathbf{K}$ will also allow us to make more efficient use of SR-based approximations.

7 Future Work

It should be noted that the proposed framework for automatic feature generation and model selection should primarily be thought of as a practical tool: despite offering a principled solution to an important problem in RL, ultimately it does not come with any theoretical guarantees (due to some modeling assumptions from GPTD and the way the hyperparameters are obtained). For most practical applications this might be less of an issue, but in general care has to be taken.

The framework can be easily extended to perform policy evaluation over the joint state-action space to learn the model-free Q-function (instead of the V-function): we just have to choose a different covariance function, taking for example the product $k([\mathbf{x},a],[\mathbf{x}',a']) = k(\mathbf{x},\mathbf{x}')k(a,a')$ with $k(a,a') = \delta_{a,a'}$ for problems with a small number of discrete actions [8]. This opens the way for model-free policy improvement and thus optimal control via approximate policy iteration. Our next step then is to apply this approach to real-world high-dimensional control tasks, both in batch settings and hybrid batch/online settings; in the latter case exploiting the gain in computational efficiency obtained through model selection to improve [6].

Acknowledgments

This work has taken place in the Learning Agents Research Group (LARG) at the Artificial Intelligence Laboratory, The University of Texas at Austin. LARG research is supported in part by grants from the NSF (CNS-0615104), DARPA (FA8750-05-2-0283 and FA8650-08-C-7812), the Federal Highway Administration (DTFH61-07-H-00030), and General Motors.

References

1. Bach, F.R., Jordan, M.I.: Kernel independent component analysis. JMLR 3, 1–48 (2002)
2. Bach, F.R., Jordan, M.I.: Predictive low-rank decomposition for kernel methods. In: Proc. of ICML, vol. 22 (2005)
3. Bertsekas, D.: Dynamic programming and Optimal Control, vol. II. Athena Scientific (2007)
4. Csató, L., Opper, M.: Sparse online Gaussian processes. Neural Computation 14(3), 641–668 (2002)
5. Deisenroth, M.P., Rasmussen, C.E., Peters, J.: Gaussian process dynamic programming. Neurocomputing 72(7-9), 1508–1524 (2009)
6. Engel, Y., Mannor, S., Meir, R.: Reinforcement learning with Gaussian processes. In: Proc. of ICML, vol. 22 (2005)

7. Fine, S., Scheinberg, K.: Efficient SVM training using low-rank kernel representation. JMLR 2, 243–264 (2001)
8. Jung, T., Polani, D.: Learning robocup-keepaway with kernels. In: JMLR: Workshop and Conference Proceedings (Gaussian Processes in Practice), vol. 1, pp. 33–57 (2007)
9. Keller, P., Mannor, S., Precup, D.: Automatic basis function construction for approximate dynamic programming and reinforcement learning. In: Proc. of ICML, vol. 23 (2006)
10. Luo, Z., Wahba, G.: Hybrid adaptive splines. J. Amer. Statist. Assoc. 92, 107–116 (1997)
11. Mahadevan, S., Maggioni, M.: Proto-value functions: A Laplacian framework for learning representation and control in Markov decision processes. JMLR 8, 2169–2231 (2007)
12. Menache, N., Shimkin, N., Mannor, S.: Basis function adaptation in temporal difference reinforcement learning. Annals of Operations Research 134, 215–238 (2005)
13. Nabney, I.T.: Netlab: Algorithms for Pattern Recognition. Springer, Heidelberg (2002)
14. Parr, R., Painter-Wakefield, C., Li, L., Littman, M.: Analyzing feature generation for value-function approximation. In: Proc. of ICML, vol. 24 (2007)
15. Poggio, T., Girosi, F.: Networks for approximation and learning. Proceedings of the IEEE 78(9), 1481–1497 (1990)
16. Rasmussen, C.E., Kuss, M.: Gaussian processes in reinforcement learning. In: Advances in Neural Information Processing Systems, vol. 16, pp. 751–759. MIT Press, Cambridge (2004)
17. Rasmussen, C.E., Williams, C.K.I.: Gaussian Processes for Machine Learning. MIT Press, Cambridge (2006)
18. Reisinger, J., Stone, P., Miikkulainen, R.: Online kernel selection for Bayesian reinforcement learning. In: Proc. of ICML, vol. 25 (2008)
19. Seeger, M., Williams, C.K.I., Lawrence, N.: Fast forward selection to speed up sparse Gaussian process regression. In: Proc. of 9th Int'l Workshhop on AI and Statistics. Soc. for AI and Statistics (2003)
20. Snelson, E., Ghahramani, Z.: Sparse Gaussian processes using pseudo-inputs. In: NIPS, vol. 18 (2006)
21. Sutton, R., Barto, A.: Reinforcement Learning: An Introduction. MIT Press, Cambridge (1998)
22. Williams, C., Seeger, M.: Using the Nyström method to speed up kernel machines. In: NIPS, vol. 13, pp. 682–688 (2001)

Learning Preferences with Hidden Common Cause Relations

Kristian Kersting and Zhao Xu

Fraunhofer IAIS, Schloss Birlinghoven, 53754 Sankt Augustin, Germany
{kristian.kersting,zhao.xu}@iais.fraunhofer.de

Abstract. Gaussian processes have successfully been used to learn preferences among entities as they provide nonparametric Bayesian approaches for model selection and probabilistic inference. For many entities encountered in real-world applications, however, there are complex relations between them. In this paper, we present a preference model which incorporates information on relations among entities. Specifically, we propose a probabilistic relational kernel model for preference learning based on Silva et al.'s mixed graph Gaussian processes: a new prior distribution, enhanced with relational graph kernels, is proposed to capture the correlations between preferences. Empirical analysis on the LETOR datasets demonstrates that relational information can improve the performance of preference learning.

1 Introduction

Largely motivated by applications in search engines, information retrieval, and collaborative filtering, preference learning has recently received a lot of attention in the machine learning and information retrieval communities, see e.g. [8,2,1,3,12]. In a typical formulation, the goal of preference learning is to compare two entities such as documents, webpages, products, songs etc., and to decide which one is better or preferred e.g. by a costumer according to some application-specific criteria.

Consider a typical interaction between an information retrieval system and a user. When a user submits a query to the system, the search engine returns a list of document hyperlinks to the user, along with a title and query-related snippet extracted from the document. The user reads the list, and based on titles, snippet, and probably abstracts, decides whether a document in the list is more relevant to the query than another ones. Hence, in contrast to standard supervised learning problems such as regression and classification, preference learning is characterized by the fact that the training set consists of pairwise rankings between entities, instead of explicit entity-wise values. For example, we may only know that a webpage e_i is more relevant than another one e_j, denoted as $e_i \succ e_j$, or that a user prefers an item to another, but we do not know the exact degrees of relevance of webpages or preferences of users.

Because of their flexible nonparametric nature and good performance on regression/classification problems, Gaussian process (GP) models have recently

W. Buntine et al. (Eds.): ECML PKDD 2009, Part I, LNAI 5781, pp. 676–691, 2009.

been explored for learning to rank [4,5,10]. Basically, GP based preference learning models introduce for each entity a latent variable, which is a function value $f(x_i)$ (shortened as f_i in the rest of the paper) of entity attributes x_i. We can intuitively view the latent function values as preference/relevance degrees (called preference degrees later) of entities. Then entities are ranked according to the latent values. Namely if an entity e_i is ranked above another one e_j, i.e., $e_i \succ e_j$, then the latent function value f_i of the entity is larger than that f_j of another one, i.e., $f_i > f_j$.

Existing GP ranking models, however, only exploit the available information about entity attributes and typically ignore any relations among the entities. Intuitively, however, we would like to use our information about one entity to help us reach conclusions about other, related entities. Reconsider our information retrieval example. Here, we should be able to propagate our preference relation among two documents to documents that the two documents have links to and to documents that link to the two documents.

The main contribution of the present paper is the first nonparametric Bayesian approach to learn preferences from relational data based on Gaussian processes. Specifically, we employ the concept of hidden common causes to incorporate relational information. Hidden common causes were first introduced by Silva *et al.* [21] within the *mixed graph Gaussian process* framework (XPGs) and have been demonstrated to be quite successful for classification problems. The key insight for preference learning is that some hidden, but existing common causes lurk in relational graphs, and the hidden common causes are important factors to influence the preference degrees of entities. The overall preference degree of an entity is a comprehensive result of both the entity attributes and the hidden common causes. Technically, under the GP framework, we introduce for each entity an additional latent function value $g(r_i)$ (shortened as g_i) for the relations r_i the entity participate. This latent function value encodes the preference causes hidden in the relation. Then, we model the entity preference degree ξ_i as a linear combination of related function values, i.e., f_i and g_i. In turn, each preference $e_i \succ e_j$ is modeled as a random variable conditioned on an indicator that is a function of the preference degrees ξ_i and ξ_j of the involved entities. In other words, our relational GP framework for preference learning, which we call *mixed graph preference Gaussian process* (XPGP), ranks entities taking all available information into account, attributes and relations. As shown in our experimental results on real-world LETOR datasets, a significant improvement on preference prediction quality can be achieved when employing relational information.

Our second contribution is an active exploration scheme to relational preference learning. Providing preference labels is typically quite costly as the user has e.g. to read and understand abstracts of documents. Fortunately, the uncertainty model provided by the XPGP framework, offers predictive uncertainty estimates for preferences, and therefore *naturally* – in contrast to other kernel approaches such as SVMs – allows us to develop an active exploration scheme that guides the user by asking actively for her preferences among entities so as to provide more useful observations. As our experimental analysis on a real-world LETOR dataset showed, this improves the prediction quality faster than collecting preferences naively.

The rest of the paper is organized as follows. We start off by touching upon related work. Then, in Sec. 3, we will introduce XPGPs. Sec. 4 will develop approximate inference and learning methods, and Sec. 5 the active exploration scheme. Before concluding, we will present our experimental analysis and discuss extensions of XPGPs to domains with multiple types of relations.

2 Related Work

The present work joins two lines of research within the Gaussian process community, namely preference and relational learning.

In the first stream, Chu and Ghahramani [4] introduced a probabilistic kernel approach to ordinal regression based on Gaussian process models. In contrast to the setting discussed in this paper, the work focused on scenarios where labels of entities are ordered. Chu and Ghahramani presented a threshold model to encode the label-wise ordinal information. The work was later extended by Chu and Ghahramani to entity ranking problem (i.e. the setting discussed here) by introducing a novel likelihood function to express the entity-wise ordinal information [5]. Guiver and Snelson [10] recently presented a sparse Gaussian process model for soft ranking problem for large-scale datasets. All these models are reported to provide good performance on real-world datasets but they do not consider relational information.

The second line of research aims at incorporating relations into probabilistic kernel models. There are essentially two strategies to accomplish this. One is encoding relations in the covariance matrixes [25,21]. The other is encoding relations as random variables conditioned on the latent function values of entities involved in relations [6,24]. Recently Xu *et al.* [23] introduced a combination of both approach for multi-relational learning with Gaussian processes. The approach of representing relational information as hidden common causes developed by Silva *et al.* [21] is a straightforward way to encode relations, which was successfully applied on entity classification problems. So far, however, preference learning has not been considered for any of them.

Outside the Gaussian process community, several ranking and preference learning approaches have been proposed, see e.g. [12] for a nice classification of the existing approaches. Relational approaches have also been developed. For instance, Geerts *et al.* [9] provide a general ranking framework for relational databases. Agarwal [1] introduced a kernel-based approach with graph techniques (spectral relaxation), where entities and relations are respectively viewed as vertexes and edges in graphs. So the task of ranking entities is transformed to ranking vertices. It is not clear how to exploit attributes and relations simultaneously and how to distinguish different types of relations. The work closest to our is that of Qin *et al.* [18]. They proposed a kernel-based but *not probabilistic* method to rank relational entities. Their method enhanced attribute-based ranking function with the regularized Laplacian of the relational graph, then applied

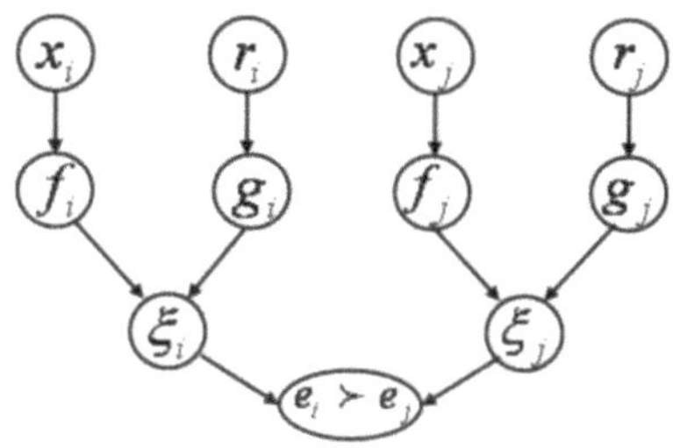

Fig. 1. Graphical representation of the XPGP model. Both entity attributes and relations are taken into account for predicting preferences $e_i \succ e_j$. Specifically, the f_is are latent function values of entity attributes following a Gaussian process (GP) prior. Respectively, the g_is are a latent function values of relations following another GP prior. The ξ_is are the overall preference degree of entities. They are weighted sums of the corresponding f_i and g_i.

SVM techniques to solve the learning, i.e., optimization task. All of these approaches do not provide natural probabilistic models so that active exploration is more complicated than in our model, see e.g. [19].

3 The Model

In this section, we will introduce the XPGP model for learning preferences. Assume that there are (1) a set of n entities $E = \{e_1, \ldots, e_n\}$ with attributes $X = \{x_i : x_i \in \mathbb{R}^D, i = 1, \ldots, n\}$, (2) relations $R = \{r_{i,j} : i, j \in 1, \ldots, n\}$ among the entities, and (3) a set of m observed pairwise preferences (a.k.a. ordinal relations/ranks) among entities, $O = \{e_{i_s} \succ e_{j_s} : s = 1, \ldots, m; i_s, j_s \in 1, \ldots, n\}$ (i_s and j_s are entities involved in s-th observed preference). With r_i, we will denote all relations in which entity e_i participates.

The XPGP model is graphically summarized in Fig. 1. Essentially, we introduce for each entity two latent function values $f(x_i)$ and $g(r_i)$ (shortened as f_i, g_i) such that $f(\cdot)$ and $g(\cdot)$ are functions of attributes and relations, respectively. Now, we form the linear combination of both values, i.e., $\xi_i = \omega_1 f_i + \omega_2 g_i$. The value ξ_i represents the preference degree of the entity e_i taking both attribute- and relation-wise factors into account. Finally, a preference $e_i \succ e_j$ is viewed as a random variable conditioned on the corresponding indicators of the involved entities with a likelihood distribution $P(e_i \succ e_j | \xi_i, \xi_j)$. In the following subsections, we will provide more details.

3.1 Prior Distributions

Let us start with defining the prior distributions. We essentially define priors for the attribute-wise and for the relation-wise latent function values separately and combine them using a linear model. Specifically, we assume an infinite number of latent function values $\{f_1, f_2, \ldots\}$ that follow a Gaussian process prior with mean function $m_a(x_i)$ and covariance function $k_a(x_i, x_j)$. Here we used the subscript a to emphasize that they are attribute-wise. In turn, any finite set of function values $\{f_i : i = 1, \ldots, n\}$ has a multivariate Gaussian distribution with mean and covariance matrix defined in terms of the mean and covariance functions of the GP [20]. Without loss of generality, we assume zero mean so that the GP is completely specified by the covariance function only. A typical choice is the squared exponential covariance function with isotropic distance measure:

$$k_a(x_i, x_j) = \kappa^2 \exp(-\frac{\rho^2}{2} \sum_d^D (x_{i,d} - x_{j,d})^2), \tag{1}$$

where κ and ρ are parameters of the covariance function, and $x_{i,d}$ denotes the d-th dimension of the attribute vector x_i.

Similarly, we place a zero-mean GP over $\{g_1, g_2, \ldots\}$. Again, $\{g_i : i = 1, \ldots, n\}$ follow a multivariate Gaussian distribution. In contrast to the attribute-wise GPs, however, the covariance function $k_r(r_i, r_j)$ should represent correlation of i and j on relations. There are essentially two strategies to define such kernel functions. The simplest way is to represent the known relations of entity i as a vector. The kernel function $k_r(r_i, r_j)$ can then be any Mercer kernel function, and the computations are essentially the same as for the attributes. Alternatively, we notice that entities and relations form a graph, and we can naturally employ graph-based kernels to obtain the covariances, see e.g. [22,25,21]. The simplest graph kernel might be the regularized Laplacian

$$K_r = [\beta(\Delta + I/\iota^2)]^{-1}, \tag{2}$$

where β and ι are two parameters of the graph kernel. Δ denotes the combinatorial Laplacian, which is computed as $\Delta = D - W$, where W denotes the adjacency matrix of a weighted, undirected graph, i.e., $W_{i,j}$ is taken to be the weight associated with the edge between i and j encoding for examples the extent of interactions between two genes or the communication frequency between two persons. D is a diagonal matrix with entries $d_{i,i} = \sum_j w_{i,j}$.

Finally, the prior distributions of f and g are combined as follows:

$$P(f, g|X, R) = \frac{1}{(2\pi)^n |K_a|^{\frac{1}{2}} |K_r|^{\frac{1}{2}}} \exp\left(-\frac{f^T K_a^{-1} f + g^T K_r^{-1} g}{2}\right),$$

where f and g denote $\{f_1, \ldots, f_n\}$ and $\{g_1, \ldots, g_n\}$. K_a (resp. K_r) denotes the $n \times n$ covariance matrix whose ij-th entry is computed with the corresponding covariance function.

3.2 Preference Likelihood

What is left is the definition of the preference likelihood. We essentially extend Chu and Ghahramani's likelihood function to the relational case [5]. Recall that, in the relational case, the preference degree of an entity consists of two components: the attribute-wise factor and the relation-wise factor, respectively represented as the latent function values f_i and g_i. To combine both we represent the overall preference degree of the entity as the weighted sum of both latent functions, i.e.,

$$\xi_i = \omega_1 f_i + \omega_2 g_i.$$

In the ideal, noise-free case it is natural to assume that *if e_i is preferred to e_j, then the preference degree of e_i is larger than that of e_j*. Denoting the non-contaminated preference degrees as $\tilde{\xi}_i$ and $\tilde{\xi}_j$, we have

$$P(\, e_i \succ e_j \mid \tilde{\xi}_i - \tilde{\xi}_j \,) = \begin{cases} 1 & \text{if } \tilde{\xi}_i - \tilde{\xi}_j \geq 0 \\ 0 & \text{otherwise.} \end{cases}$$

For real-world situations, however, it is more realistic to consider latent function values corrupted by Gaussian noise, i.e., $\xi_i = \tilde{\xi}_i + \epsilon$, $\epsilon \sim N(0, \sigma^2)$ and $P(\tilde{\xi}_i - \tilde{\xi}_j | \xi_i - \xi_j) = N(\cdot | \xi_i - \xi_j, 2\sigma^2)$. Now, we can define the preference likelihood function $P(e_i \succ e_j|\xi_i - \xi_j)$ as follows

$$\int P(e_i \succ e_j|\tilde{\xi}_i - \tilde{\xi}_j) P(\tilde{\xi}_i - \tilde{\xi}_j|\xi_i - \xi_j) d(\tilde{\xi}_i - \tilde{\xi}_j) = \int_{-\infty}^{\frac{\xi_i - \xi_j}{\sqrt{2}\sigma}} N(t|0,1)dt \equiv \Phi(\frac{\xi_i - \xi_j}{\sqrt{2}\sigma}).$$

This encodes the natural assumption:

> *The larger the difference between the preference degrees of e_i and e_j, the more likely is it that e_i is preferred to e_j.*

Finally, the marginal likelihood $P(O|\xi)$ of m observed ordinal relations given preference degrees $\xi = \{\xi_1, \ldots, \xi_n\}$ can be found to be

$$\prod\nolimits_s^m P(e_{i_s} \succ e_{j_s}|\xi_{i_s}, \xi_{j_s}) = \prod\nolimits_s^m \Phi(\frac{\xi_{i_s} - \xi_{j_s}}{\sqrt{2}\sigma}).$$

4 EMEP-Based Approximate Inference and Learning

So far we have described the XPGP model. In this section, we will present approximate algorithms for inferring posterior distributions, for making preference predictions, and for estimating the hyperparameters.

4.1 Posterior Inference

The key inference problem is computing the posterior distribution of the latent function values given attributes X, relations R, and preferences O, i.e.,

$$P(f, g|X, R, O) \propto P(f, g|X, R) \prod_s^m P(e_{i_s} \succ e_{j_s}|f_{i_s}, g_{i_s}, f_{j_s}, g_{j_s}). \quad (3)$$

Unfortunately, computing the posterior distribution is intractable for two reasons: (1) $P(e_{i_s} \succ e_{j_s}|f_{i_s}, g_{i_s}, f_{j_s}, g_{j_s})$ is not conjugated to the Gaussian prior distribution and (2) the attribute-wise GP and the relation-wise GP are coupled together according to the overall preference degrees that are weighted sums of attribute-wise and relation-wise latent function values. Therefore, we stick to the expectation propagation (EP) algorithm [17] to approximate the posterior distribution, which we will now derive.

First, to counterattack the computation complexity due to (2), i.e., due to the coupling of the different GPs, we introduce a new variable

$$\hat{\xi}_i = \frac{\omega_1}{\sqrt{2}\sigma} f_i + \frac{\omega_2}{\sqrt{2}\sigma} g_i. \tag{4}$$

Since $f = \{f_1, \ldots, f_n\}$ and $g = \{g_1, \ldots, g_n\}$ respectively follow two independent multivariate Gaussian distributions with mean zero and covariance matrixes K_a and K_r, we have that $\hat{\xi} = \{\hat{\xi}_1, \ldots, \hat{\xi}_n\}$ also follows a multivariate Gaussian distribution with mean zero and covariance matrix:

$$K = \frac{\omega_1^2}{2\sigma^2} K_a + \frac{\omega_2^2}{2\sigma^2} K_r. \tag{5}$$

Now, we convert the computation of the posterior distribution of f and g in Eq. (3) to the computation of the posterior distribution of $\hat{\xi}$:

$$P(\hat{\xi}|X, R, O) \propto P(\hat{\xi}|X, R) \prod_{s=1}^{m} P(e_{i_s} \succ e_{j_s} | \hat{\xi}_{i_s}, \hat{\xi}_{j_s}), \tag{6}$$

where the prior is a Gaussian distribution with covariance matrix as defined in Eq. (5). The likelihood function is $\Phi(\hat{\xi}_{i_s} - \hat{\xi}_{j_s})$, where $\Phi(\cdot)$ is the cumulative Gaussian distribution. Without loss of generality, we assume its mean and variance are zero and one respectively.

Now, we tackle the computational complexity due to (1), i.e., due to the non-conjugation of the likelihood and the prior distributions. In the EP framework, we use unnormalized Gaussian distributions

$$t_s(\hat{\xi}_{i_s}, \hat{\xi}_{j_s} | \tilde{\mu}_s, \tilde{\Sigma}_s, \tilde{Z}_s) \equiv \tilde{Z}_s \, \mathcal{N}(\hat{\xi}_{i_s}, \hat{\xi}_{j_s} | \tilde{\mu}_s, \tilde{\Sigma}_s) \tag{7}$$

to approximate the real likelihood distributions $\Phi(\hat{\xi}_{i_s} - \hat{\xi}_{j_s})$ where $\tilde{Z}_s$ is a real-valued scale, the unnormalized term. Since each t_s is a Gaussian distribution, the approximate posterior $q(\hat{\xi}) = \mathcal{N}(\hat{\xi}|\mu, \Sigma) \approx P(\hat{\xi}|X, R, O)$ is also a Gaussian distribution, whose mean μ and covariance matrix Σ can be found to be

$$\begin{aligned} \Sigma &= (K^{-1} + \tilde{\Sigma}_1^{-1} + \ldots + \tilde{\Sigma}_m^{-1})^{-1} \\ \text{and} \quad \mu &= \Sigma(\tilde{\Sigma}_1^{-1}\tilde{\mu}_1 + \ldots + \tilde{\Sigma}_m^{-1}\tilde{\mu}_m). \end{aligned} \tag{8}$$

Note that the t_s are two-dimensional, but Σ and μ are $n \times n$ and n dimensions, thus we need to extend $\tilde{\Sigma}_s^{-1}$ to a $n \times n$ matrix and $\tilde{\mu}_s$ to a n-dimensional column vector by adding zeros in the corresponding positions in the computation.

The approximate distributions are sequentially updated until convergence. Specifically, the approximate distribution t_s is updated at iteration $t+1$ so that it satisfies:

$$Q_{-s} \times P(e_{i_s} \succ e_{j_s} | \hat{\xi}_{i_s}, \hat{\xi}_{j_s}) \leftarrow Q_{-s} \times t_s(\hat{\xi}_{i_s}, \hat{\xi}_{j_s} | \tilde{\mu}_s^{(t+1)}, \tilde{\Sigma}_s^{(t+1)}, \tilde{Z}_s^{(t+1)}),$$
$$\text{where } Q_{-s} = \frac{\mathcal{N}(\hat{\xi}_{i_s}, \hat{\xi}_{j_s} | \mu_s^{(t)}, \Sigma_s^{(t)})}{t_s(\hat{\xi}_{i_s}, \hat{\xi}_{j_s} | \tilde{\mu}_s^{(t)}, \tilde{\Sigma}_s^{(t)}, \tilde{Z}_s^{(t)})}. \quad (9)$$

Both sides of Eq. (9) are the approximate marginal distributions of $\hat{\xi}_{i_s}$ and $\hat{\xi}_{j_s}$. They are the integrals of the approximations to Eq. (6): the left side replaces all actual likelihood distributions with approximations from the t-th iteration except for the likelihood of the preference pair currently being updated; the right side also replaces all actual likelihood distributions with approximations, but the approximation to the likelihood of the preference pair being updated is a new one, i.e., the mean, covariance matrix and unnormalized term need to be updated now. To satisfy Eq. (9), we only need to match their first and second moments [20]. This can be found to yield the following equations for computing $\tilde{\mu}_s^{(t+1)}$, $\tilde{\Sigma}_s^{(t+1)}$ and $\tilde{Z}_s^{(t+1)}$:

$$\Sigma_{-s} = (\Sigma_s^{-1} - \tilde{\Sigma}_s^{-1})^{-1}; \qquad \mu_{-s} = \Sigma_{-s}(\Sigma_s^{-1}\mu_s - \tilde{\Sigma}_s^{-1}\tilde{\mu}_s). \quad (10)$$

$$z = \frac{\varrho^T \mu_{-s}}{\sqrt{1 + \varrho^T \Sigma_{-s} \varrho}}; \quad Z = \Phi(z); \quad S = \frac{1}{\sqrt{2\pi}} \exp(-\frac{z^2}{2}); \quad \varrho^T = [1, -1];$$
$$\hat{\mu}_s = \mu_{-s} + \frac{y_s S \Sigma_{-s} \varrho}{Z\sqrt{1 + \varrho^T \Sigma_{-s} \varrho}}; \quad \hat{\Sigma}_s = \Sigma_{-s} - \frac{zSZ + S^2}{Z^2(1 + \varrho^T \Sigma_{-s} \varrho)} \Sigma_{-s} \varrho \varrho^T \Sigma_{-s} \quad (11)$$

$$\tilde{\Sigma}_s^{(t+1)} = (\hat{\Sigma}_s^{-1} - \Sigma_{-s}^{-1})^{-1}; \quad \tilde{\mu}_s^{(t+1)} = \tilde{\Sigma}_s^{(t+1)}(\hat{\Sigma}_s^{-1}\hat{\mu}_s - \Sigma_{-s}^{-1}\mu_{-s}); \quad \tilde{Z}_s^{(t+1)} = CZ;$$
$$C = (2\pi)^{-1}|\Sigma_{-s} + \tilde{\Sigma}_s|^{\frac{1}{2}} \exp(\frac{1}{2}(\mu_{-s} - \tilde{\mu}_s)^T(\Sigma_{-s} + \tilde{\Sigma}_s)^{-1}(\mu_{-s} - \tilde{\mu}_s)) \quad (12)$$

Eq. (10) is computing mean and covariance matrix of Q_{-s}. $\tilde{\mu}_s$ and $\tilde{\Sigma}_s$ are the mean and covariance matrix of the approximate distribution t_s, optimized in the last iteration t. To avoid a cluttering of notation, we do not use the superscript (t) to highlight the iteration. μ_s and Σ_s are the mean and the covariance matrix of the approximate posterior distribution of $\hat{\xi}_{i_s}$ and $\hat{\xi}_{j_s}$ that is the marginalized $q(\hat{\xi})$. Eq. (11) is computing the mean and the covariance matrix of the distribution at the left side of Eq. (9), which is derived by moment matching [20]. Finally, y_s is one if e_{i_s} is preferred to e_{j_s}, i.e. $e_{i_s} \succ e_{j_s}$; -1 otherwise.

At convergence, we obtain the optimized EP parameters that can be used to compute the approximate posterior distribution of $\hat{\xi}$ using Eq. (8).

4.2 Transductive Preference Prediction

The key prediction problem in preference learning is to predict preferences of new pairs of entities. Here, we consider the predictive inference in a transductive setting, i.e. there is no new entity introduced in prediction. Once the procedure of posterior inference reaches stationarity, we obtain the optimized distribution on $\hat{\xi} = \{\hat{\xi}_1, \ldots, \hat{\xi}_n\}$ with the EP parameters $\{\tilde{\mu}_s, \tilde{\Sigma}_s, \tilde{Z}_s\}_{s=1}^m$. It can be used to approximate the predictive distribution of the preference pair s' on entities i' and j' as follows

$$\begin{aligned} P(e_{i'} \succ e_{j'}|X, R, O) &= \int P(e_{i'} \succ e_{j'}|\hat{\xi}_{i'}, \hat{\xi}_{j'}) P(\hat{\xi}_{i'}, \hat{\xi}_{j'}|X, R, O)\, d\hat{\xi}_{i'}\, d\hat{\xi}_{j'} \\ &\approx \int \Phi(\hat{\xi}_{i'} - \hat{\xi}_{j'}) \mathcal{N}(\hat{\xi}_{i'}, \hat{\xi}_{j'}|\mu_{s'}, \Sigma_{s'}) d\hat{\xi}_{i'}\, d\hat{\xi}_{j'}, \\ &= \Phi(\frac{\varrho^T \mu_{s'}}{\sqrt{1 + \varrho^T \Sigma_{s'} \varrho}}), \end{aligned}$$

where $\mathcal{N}(\hat{\xi}_{i'}, \hat{\xi}_{j'}|\mu_{s'}, \Sigma_{s'})$ is the marginalized $q(\hat{\xi})$. It is the approximation to the real marginal posterior distribution $P(\hat{\xi}_{i'}, \hat{\xi}_{j'}|X, R, O)$. Since $q(\hat{\xi})$ is Gaussian, so the approximation is still Gaussian, and its mean $\mu_{s'}$ and covariance matrix $\Sigma_{s'}$ are the corresponding entries of μ and Σ (Eq. 8). The preference relation $e_{i'} \succ e_{j'}$ is conditioned on their difference on preference degree, i.e. $\hat{\xi}_{i'} - \hat{\xi}_{j'}$. Since both $\hat{\xi}_{i'}$ and $\hat{\xi}_{j'}$ are Gaussian random variables, their difference is Gaussian, too, with mean and variance $\varrho^T \mu_{s'}$ respectively $\varrho^T \Sigma_{s'} \varrho$, where ϱ denotes a column vector $[1, -1]^T$. Thus we have $\hat{\xi}_{i'} - \hat{\xi}_{j'} \sim \mathcal{N}(\cdot|\varrho^T \mu_{s'}, \varrho^T \Sigma_{s'} \varrho)$, where

$$\varrho^T \mu_{s'} \tag{13}$$

is just the difference of the means of the two preference degrees ($\hat{\xi}_{i'}$ and $\hat{\xi}_{j'}$). The variance $\varrho^T \Sigma_{s'} \varrho$ just equals

$$\mathrm{Var}(\hat{\xi}_{i'}) + \mathrm{Var}(\hat{\xi}_{j'}) - 2\mathrm{Cov}(\hat{\xi}_{i'}, \hat{\xi}_{j'})\ . \tag{14}$$

The larger the variance, the more uncertain we are about the preference relation.

4.3 Hyperparameter Estimation

Finally, we will describe how to estimate the hyperparameters under the empirical Bayesian framework. The hyperparameters of the XPGP model consists of the parameters of the kernel functions as well as the mixing weights of the attribute-wise and the relation-wise GPs. Note, however, that the mixing weights are just scaling the latent function values. Therefore, we can directly integrate them into the covariance functions. In other words, estimating the hyperparameters of the XPGP model can be reduced to estimating the hyperparameters of the covariance functions.

Let us denote all hyperparameters as θ. We now need to seek θ^* that maximizes the log-likelihood of the data, i.e.

$$\theta^* = \arg\max_\theta \log P(O|\theta, X, R) = \arg\max_\theta \log \int P(\hat{\xi}|\theta, X, R) P(O|\hat{\xi})\, d\hat{\xi},$$

where the prior $P(\hat{\xi}|\theta, X, R)$ is a Gaussian distribution. Unfortunately, the likelihood $P(O|\hat{\xi})$ is not Gaussian, thus the integral is analytically intractable.

To solve the problem, we follow an approximate Expectation Maximization (EM) approach [14]. The algorithm alternates the following steps until convergence. In the **E-step**, given the hyperparameters, the EP parameters ($\tilde{\mu}_s$, $\tilde{\Sigma}_s$, $\tilde{Z}_s$) are optimized to approximate the posterior distribution of the latent function variables with the current values of hyperparameters. In the **M-step**, given the density of the preference degrees, the hyperparameters are selected to maximize a lower bound of the marginal likelihood:

$$\int q(\hat{\xi}) \log \frac{P(O|\hat{\xi})P(\hat{\xi}|\theta, X, R)}{q(\hat{\xi})} d\hat{\xi} = \int q(\hat{\xi}) \log P(\hat{\xi}|\theta, X, R) d\hat{\xi} + \int q(\hat{\xi}) \log P(O|\hat{\xi}) d\hat{\xi} - \int q(\hat{\xi}) \log q(\hat{\xi}) d\hat{\xi}. \quad (15)$$

Note that the last two terms on the right-hand side are independent of the hyperparameters θ, thus we only need to optimize the first term, i.e., $\mathcal{L} :=$

$$\int q(\hat{\xi}) \log P(\hat{\xi}|\theta, X, R) d\hat{\xi} = -\frac{1}{2} \log |2\pi K| - \frac{1}{2} \mathbb{E}_q[\hat{\xi}]^T K^{-1} \mathbb{E}_q[\hat{\xi}] - \frac{1}{2} \mathrm{tr}(K^{-1} \Sigma),$$

where K is the covariance matrix of the prior of $\hat{\xi}$ as defined in Eq. (5), and $\mathbb{E}_q[\hat{\xi}]$ is the expectation of $\hat{\xi}$ on the approximate posterior $q(\hat{\xi})$, i.e. the mean μ of $q(\hat{\xi})$ as defined in Eq. (8). Differentiating $\mathcal{L}$ with respect to θ, the gradient can be found to be:

$$\frac{\partial \mathcal{L}}{\partial \theta} = -\frac{1}{2} \mathrm{tr}(K^{-1} \frac{\partial K}{\partial \theta}) + \frac{1}{2} \alpha^T \frac{\partial K}{\partial \theta} \alpha + \frac{1}{2} \mathrm{tr}(K^{-1} \frac{\partial K}{\partial \theta} K^{-1} \Sigma), \quad (16)$$

where α denotes $K^{-1}\mu$. It is possible to use any gradient-based optimizer to find the hyperparameters. In the experiments, we used a scaled conjugate gradient.

5 Active Preference Exploration

So far, we have assumed that observed preferences are provided in a batch, i.e., they are collected in a rather naive way. Preference prediction, however, could be made more efficient if we can actively select *uncertain* preferences. Consider a typical interaction between a search engine and a user. A user executes a query and considers the results presented. The system now requests the user to provide a set of preference relations in the result with the goal to improve the preference prediction being shown to the user.

Typically, however, the user pays attention to only a few entities in the result. Note that providing preference labels can be quite costly as the user has e.g. to read and understand abstracts of research papers. In turn, important entities may never be considered by the user and no preference feedback is provided. This is likely to lead to suboptimal preference predictions. To avoid this presentation effect [19], we guide the user by asking actively for her preferences among entities so as to provide more useful observations. Usually, we are limited to deploying a small number of preferences only, and thus must carefully choose them.

Being probabilistic models, XPGPs are extremely powerful for active exploration. If we observe a set of preferences corresponding to a finite subset $\mathcal{A} \subset \mathcal{R}$

of all possibly observable preferences, we can easily predict the uncertainty about any other preference $r \in \mathcal{R}\backslash\mathcal{A}$ conditioned on these observed ones using $P(r|\mathcal{A})$. An approximation of this marginal distribution can be found using the EP method discussed above. It is a Gaussian whose conditional mean $\mu_{r|\mathcal{A}}$ and variance $\sigma^2_{e_i \succ e_j|\mathcal{A}}$ are given by Eqs. (13) and (14) replacing i' with i and j' with j. The values $\hat{\xi}_i$ and $\hat{\xi}_j$ are the overall preference degrees as defined in Eq. (4) of the entities e_i and e_j. Their variances and covariance can be directly obtained from the covariance matrix in Eq. (8) of the approximate posterior distribution $q(\hat{\xi})$.

To solve the active exploration problem, we follow the commonly used greedy approach [15]. That is, we start from an empty set of preferences, $\mathcal{A} = \emptyset$, and greedily add preferences until $|\mathcal{A}| = k$. At each iteration, the greedy rule used is to add the preference $r \in \mathcal{R}\backslash\mathcal{A}$ that has the highest ratio of variance and squared difference in mean, i.e., *the preference with closest latent preference values that we are most uncertain about given the preferences observed so far.* Other scores such as mutual information could also be used [15] but they are out of scope of the present paper.

6 Experimental Analysis

We evaluated the XPGP model for relevance feedback on two real-world datasets, namely OHSUMED and TREC in the LETOR repository [16]. Relevance feedback is an important task in information retrieval: the original answer to user queries are refined based on user feedback (e.g. click or not, browsing time, etc). Thus, the "personalized" ranking list is presented to the user. We used XPGP models to predict preferences on articles (resp. webpages) based on some known preferences. This corresponds to a transductive preference learning. We compared the XPGP model with standard GP [5] and SVM models [13], denoted as PGP and SVM in the experiments. For the GP-based approaches, we used Gaussian kernels, Eq.(1), to compute the covariances on entity attributes and two different graph kernels to obtain correlations on relations: (1) the regularized Laplacian Eq.(2); and (2) Silva *et al.*'s kernel from [21]. For the SVM approach, the radial basis function (RBF) was chosen to compute the kernels. We report on the prediction error rate (ERR) and the area under ROC curve (AUC).

OHSUMED Dataset: The dataset was originally collected by Hersh *et al* [11], and was processed by Liu *et al.* [16] as a benchmark data for learning to rank in information retrieval. As a subset of MEDLINE medical database, the document collection contains 348,566 publications from 270 journals during 1937-1991. Each document consists of title, abstract, MeSH indexing terms, author, source, and publication type. In the dataset, there are 106 queries, each of which are associated with some relevant documents evaluated by humans. The relevance degree has three levels: definitely relevant, partially relevant and not relevant. In total, there are 16,140 query-document pairs. Due to practicality, Liu *et al.* sampled some "possible" relevant documents from the large scale document

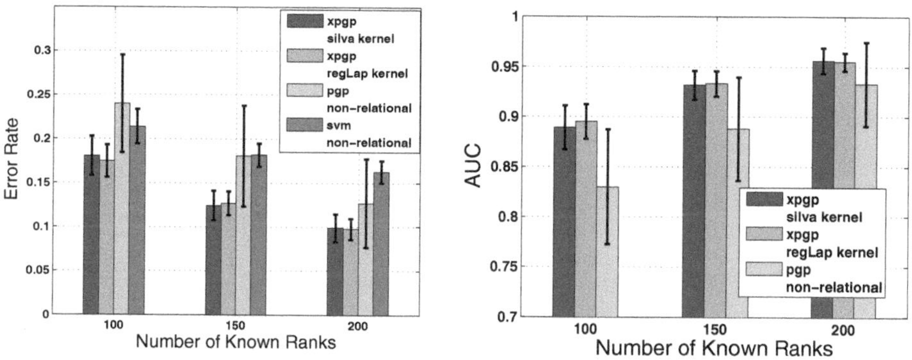

Fig. 2. OHSUMED: Experimental results averaged over 10 queries with 20 random reruns each on predicting preference pairs given different numbers of given preference pairs.(**Left**) Prediction error rate; the lower, the better. (**Right**) AUC values; the larger, the better.

collection and got on average about 152 documents per query. They extracted 25-dimensional feature vectors for each query-document pair. We refer to [16] for more details. The relations between documents are based on similarities, i.e. there is a weighted complete graph between documents; the weight of each edge is the cosine similarity between the contents (keywords) of two documents.

Given a query, each document was originally associated with a relevance degree. Based on this degree we obtained preference pairs of the form $e_i \succ e_j$. This is not only due to XPGPs modeling assumption but more importantly this information is also more realistic for real-world applications [13]. In the experiments, we randomly selected 100 (150, 200) preference pairs for each query as evidence. Then, the task was to predict the remaining ones. For each setting (100, 150, 200), the selection was repeated 20 times. Note that generating preference pairs was not costly, e.g. one evaluation on relevance provides $n-1$ preference pairs if there are n documents in a collection.

Fig. 2 shows the experimental results averaged over 10 randomly selected queries. Note that the output of [13] is not "soft" preference pairs, so we report the prediction error rates only. As one can see, in all cases (different number of known preferences), the XPGP model outperforms non-relational GP and SVM models. The significance is demonstrated with Wilcoxon rank sum test (p-value 0.01). The XPGP model performs well, especially when the number of known preference pairs is small. Overall, XPGPs reduced the mean error rates between 12% and 40%. To summarize, modeling relations among entities allows for information to be shared between entities and, in turn, to improve prediction quality. Which graph kernel to use seems to be less important: the two different graph kernels we used performed similarly well. To further verify both results, we compared XPGPs and PGPs on another dataset.

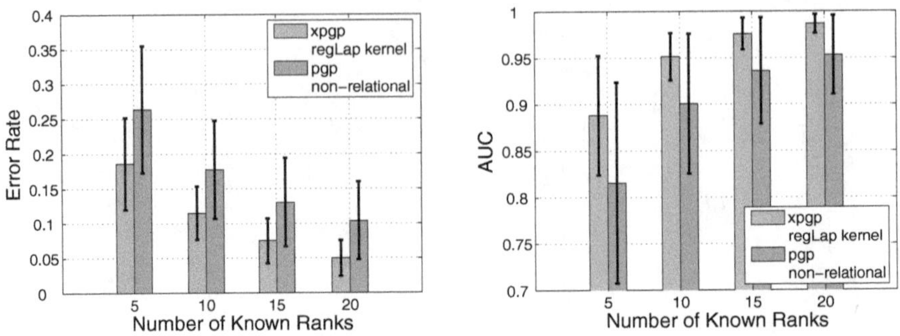

Fig. 3. TREC: Experimental results averaged over 10 queries with 20 reruns each on predicting preference pairs with different numbers of known preference pairs. (**Left**) Prediction error rate; the lower, the better. (**Right**) AUC values; the larger, the better.

TREC Dataset: The TREC dataset was originally collected for a special track on web information retrieval at TREC 2004. The goal of the track was to explore the performance of retrieval methods on large-scale data with hyperlinked structure such as the World Wide Web. The data was crawled from .gov domain in January, 2002. In total, there are 1,053,110 html documents with 11,164,829 hyperlinks. To each query, documents were assigned labels by human experts. Each document has two possible states: relevant/irrelevant. The unlabeled documents are viewed as irrelevant ones. Liu *et al.* [16] processed the TREC dataset and turned it into a benchmark for information retrieval. There were totally 75 queries left. For each query, they ranked the documents with the BM25 scores, and only kept (1) the first 1000 documents and (2) the documents labeled to be relevant. Since the original labels were entity-wise, we converted them into pairwise ones as follows: we set $e_i \succ e_j$ if e_i is relevant but e_j is irrelevant, $e_i \prec e_j$ otherwise. Again, the processing was not only because of the XPGP setting but because it is more realistic.

In the experiments, we randomly selected 10 queries. On average, there were about 1000 documents associated with each query, which are linked with about 2387 hyperlinks. About 16 of the documents were labeled as relevant. We notice that the hyperlinks are *directed*, i.e. the two webpages involved in a hyperlink play different roles: one is source, the other is target. We convert the directed relations to undirected ones by introducing a relation between two documents, which are linked by the same webpages [21]. For each query, 5 (10, 15, 20) preference pairs were chosen randomly as the known ones. The task was to predict the remaining ones. For each query, the random selection was repeated 20 times, and the mean and the standard deviation of AUC and error rate were computed to measure the prediction performance. Fig. 3 summarizes the experimental results averaged over the 10 queries. As one can see, in all settings (different number of known preferences), the XPGP provides better predictions than the non-relational GP method. A p-value threshold of 0.01 in Wilcoxon

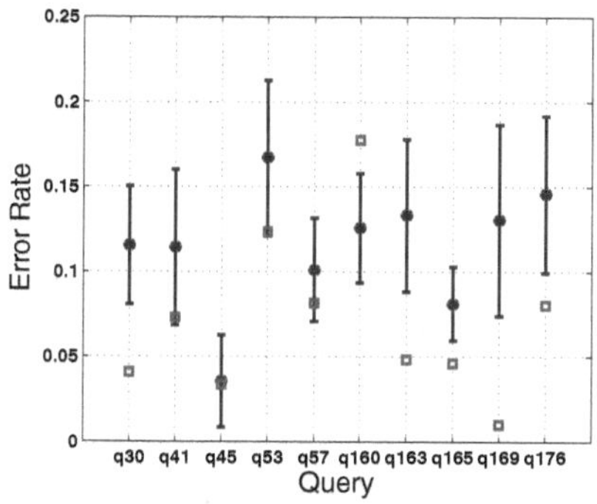

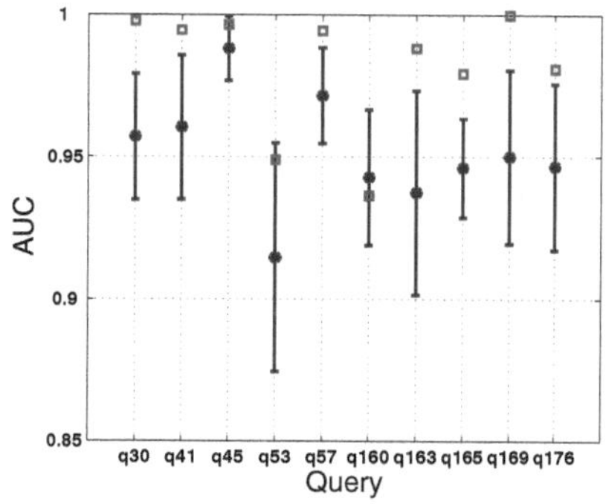

Fig. 4. Actively selecting 10 known preference pairs for 10 TREC queries. The prediction error rate (**left**) and AUC (**right**) values for the naive (means and std. taken from the previous experiment) and for the active selection (red squares) are shown. In 9 out of 10 cases, the active scheme is better than the (mean of the) naive one.

rank sum test showed that the difference is indeed significant. Overall, XPGPs reduced the mean error rates between 30% and 50%.

Active Preference Exploration: So far, the experiments were performed with randomly selected known preferences. In the final experiment, we evaluated the active exploration scheme by automatically selecting 10 known preferences on the TREC dataset. Fig. 4 summarizes the results. As one can see, active selection can indeed improve the prediction quality with higher efficiency than collecting preferences naively.

7 Extensions: Directed, Bipartite and Multiple Relations

Real-world domains typically show multiple relations of different types such as bipartite and directed relations. We will finally show that this situation can easily tackled using XPGPs.

We use distinct latent function values to represent preference factors driven by different types of relations, i.e., we introduce for each entity multiple relational function values, one for each type of relations: $\{g_i^{r1}, g_i^{r2}, \ldots\}$. The overall preference degree is now the weighted sum of all latent function values associated with the entity: $\xi_i = \omega_1 f_i + \omega_2 g_i^{r1} + \omega_3 g_i^{r2} + \ldots + \epsilon_i$. We now assume that the latent function values of the same type of relations, e.g. $\{g_1^{r1}, \cdots, g_n^{r1}\}$, are realizations of random variables in a Gaussian process, i.e., they follow a multivariate Gaussian distribution with mean zero and covariance matrix K_{r1}, which can be specified with graph kernel as discussed before.

In a directed relation, the two involved entities play different roles. Consider e.g. links of webpages `link`: webpage $\times$ webpage. The entities typically serve as the linking and linked webpages. It is reasonable to introduce for each webpage two latent function values representing preference factors for linking and linked "roles" of the entity. The overall preference degree of a webpage is again a weight sum: $\xi_i = \omega_1 f_i + \omega_2 g_i^{linking} + \omega_3 g_i^{linked} + \epsilon_i$. Alternatively, we can use random-walk-based methods to generate kernels on directed graphs [7].

There is only one difficulty when encoding bipartite relations: different types of entities are involved. In turn, graph kernels for univariate relations cannot be used. We address this problem by projecting bipartite relations to univariate ones. Specifically, we add a relation between entities i and j iff. both entities link to the same (heterogeneous) entity. All entities linking to the same (heterogeneous) entity form a clique. Then we can compute the graph kernels on the projected graphs. For example, we can convert a bipartite relation `Direct`: movie $\times$ person to a undirected one `Co-Directed`: movie $\times$ movie.

8 Conclusion

In this paper we have proposed the first nonparametric Bayesian approach to learn preferences from relational data using Gaussian processes. Modeling relations among entities allows for information to be shared between entities and, in turn, to improve prediction quality. The uncertainty model provided by the Gaussian process framework offers predictive uncertainty estimates for preferences, and naturally allowed us to develop an active exploration scheme in which preferences are optimally selected for interactive labeling. Our empirical results showed a significant improvement of preference prediction quality when employing relational information. Furthermore, active preference exploration improved the quality faster than collecting preferences naively.

A natural extension of this work is the adaption of sparse Gaussian processes to the relational case to tackle large-scale datasets. Furthermore, (non)myopic analysis of active preference exploration is interesting as it potentially yields provable bounds on the quality of the estimates. It is likely that other criteria than variance such as mutual information will show better performance.

We believe that this work is an interesting step towards increasing the quality of search engines and information retrieval systems as well as towards well-founded active learning algorithms for relational (preference) learning.

Acknowledgments. The authors would like to thank the anonymous reviewers for their comments. They would also like to thank John Guiver, Tie-Yan Liu and Tao Qin for sharing the relational LETOR datasets. Furthermore, they thank Thorsten Joachims for valuable discussions and for encouraging the active exploration scheme. The work was funded by the Fraunhofer ATTRACT Fellowship "Statistical Relational Activity Mining" (STREAM).

References

1. Agarwal, S.: Ranking on graph data. In: ICML, pp. 25–32 (2006)
2. Burges, C., Shaked, T., Renshaw, E., Lazier, A., Deeds, M., Hamilton, N., Hullender, G.: Learning to rank using gradient descent. In: ICML (2005)

3. Chakrabarti, S.: Learning to rank in vector spaces and social networks. In: WWW 2007 (2007)
4. Chu, W., Ghahramani, Z.: Gaussian processes for ordinal regression. Journal of Machine Learning Research 6, 1019–1041 (2005)
5. Chu, W., Ghahramani, Z.: Preference learning with gaussian processes. In: ICML (2005)
6. Chu, W., Sindhwani, V., Ghahramani, Z., Keerthi, S.: Relational learning with gaussian processes. In: Neural Information Processing Systems (2006)
7. Chung, F.R.K.: Laplacians and the cheeger inequality for directed graphs. Annals of Combinatorics 9, 1–19 (2005)
8. Freund, Y., Iyer, R., Schapire, R.E., Singer, Y.: An efficient boosting algorithm for combining preferences. JMLR 4, 933–969 (2003)
9. Geerts, F., Mannila, H., Terzi, E.: Relational link-based ranking. In: Proceedings of VLDB 2004, pp. 552–563 (2004)
10. Guiver, J., Snelson, E.: Learning to rank with softrank and gaussian processes. In: SIGIR (2008)
11. Hersh, W., Buckley, C., Leone, T., Hickam, D.: Ohsumed: an interactive retrieval evaluation and new large test collection for research. In: SIGIR (2007)
12. Hüllermeier, E., Fürnkranz, J., Cheng, W., Brinker, K.: Label ranking by learning pairwise preferences. Artificial Intelligence 172(16–17), 1897–1916 (2008)
13. Joachims, T.: Optimizing search engines using clickthrough data. In: SIGKDD (2002)
14. Kim, H.-C., Ghahramani, Z.: Bayesian gaussian process classification with the em-ep algorithm. IEEE Transactions on Pattern Analysis and Machine Intelligence 28(12), 1948–1959 (2006)
15. Krause, A., Singh, A., Guestrin, C.: Near-optimal sensor placements in gaussian processes: Theory, efficient algorithms and empirical studies. Journal of Machine Learning Research (JMLR) 9, 235–284 (2008)
16. Liu, T.-Y., Qin, T., Xu, J., Xiong, W.-Y., Li, H.: Letor: benchmark dataset for research on learning to rank for information retrieval. In: SIGIR 2007 Workshop on LR4IR (2007)
17. Minka, T.: A family of algorithms for approximate Bayesian inference. PhD thesis, MIT (2001)
18. Qin, T., Liu, T., Zhang, X., Wang, D., Xiong, W., Li, H.: Learning to rank relational objects and its application to web search. In: WWW (2008)
19. Radlinski, F., Joachims, T.: Active exploration for learning rankings from clickthrough data. In: Proceedings of KDD 2007, pp. 570–579 (2007)
20. Rasmussen, C.E., Williams, C.K.I.: Gaussian Processes for Machine Learning. MIT Press, Cambridge (2006)
21. Silva, R., Chu, W., Ghahramani, Z.: Hidden common cause relations in relational learning. In: Neural Information Processing Systems (2007)
22. Smola, A.J., Kondor, I.: Kernels and regularization on graphs. In: Annual Conference on Computational Learning Theory (2003)
23. Xu, Z., Kersting, K., Tresp, V.: Multi-relational learning with gaussian processes. In: Boutilier, C. (ed.) Proceedings of the International Joint Conference on Artificial Intelligence, IJCAI 2009 (to appear, 2009)
24. Yu, K., Chu, W., Yu, S., Tresp, V., Xu, Z.: Stochastic relational models for discriminative link prediction. In: Neural Information Processing Systems (2006)
25. Zhu, X., Kandola, J., Lafferty, J., Ghahramani, Z.: Graph kernels by spectral transforms. In: Chapelle, O., Schoelkopf, B., Zien, A. (eds.) Semi-Supervised Learning. MIT Press, Cambridge (2005)

Feature Selection for Density Level-Sets

Marius Kloft[1], Shinichi Nakajima[2], and Ulf Brefeld[1]

[1] Machine Learning Group, Technische Universität Berlin, Berlin, Germany
{mkloft,brefeld}@cs.tu-berlin.de
[2] Optical Research Laboratory, Nikon Corporation, Tokyo, Japan
nakajima.s@nikon.co.jp

Abstract. A frequent problem in density level-set estimation is the choice of the right features that give rise to compact and concise representations of the observed data. We present an efficient feature selection method for density level-set estimation where optimal kernel mixing coefficients *and* model parameters are determined simultaneously. Our approach generalizes one-class support vector machines and can be equivalently expressed as a semi-infinite linear program that can be solved with interleaved cutting plane algorithms. The experimental evaluation of the new method on network intrusion detection and object recognition tasks demonstrate that our approach not only attains competitive performance but also spares practitioners from *a priori* decisions on feature sets to be used.

1 Introduction

The set of points on which a function f exceeds a certain value ρ, e.g., $D_\rho = \{\boldsymbol{x} : f(\boldsymbol{x}) \geq \rho\}$, is called a level-set D_ρ. Boundaries of such sets typically constitute submanifolds in feature space whereas level-set approaches are frequently used for function estimation and denoising.

For anomaly and outlier detection tasks, level-set methods are often observed to outperform probability density estimators which have to be thresholded accordingly to act as detectors for unlikely and rare events. Statistical approaches frequently focus on *high density* regions to capture the underlying probability distribution. By contrast, density level-set estimators are specially tailored to work well in *low density* regions which is a crucial property for detecting anomalous events.

In this paper, we focus on level-set estimation for anomaly and outlier detection [9,4], where a model of normality is devised from available observations. Anomality of new objects is measured by their distance (in some metric space) from the learned model of normality. Apart from theoretical observations, in practice the effectiveness of density level-set estimation crucially depends on the representation of the observations and thus on the choice of features.

However, characteristic traits of particular learning problems are often spread across multiple features that capture various properties of data, giving rise to a set of kernel matrices $K_1, \ldots, K_m$ that have to be combined appropriately. As

W. Buntine et al. (Eds.): ECML PKDD 2009, Part I, LNAI 5781, pp. 692–704, 2009.

a motivating example, consider network intrusion detection where various sets of features have been deployed, including raw values of IP and TCP protocol headers [15,16], time and connection windows [13], byte histograms and n-grams [29,28], and "bag-of-tokens" language models [21,22]. While packet header based features have been shown to be effective against probes and scans, other kinds of attacks, e.g. remote buffer overflows, require more advanced payload processing techniques. The right kind of features for a particular application has always been considered as the matter of a judicious choice (or trial and error).

But what if this decision is really difficult to make? Given the choice of several kinds of features, a poor a priori decision would lead to an inappropriate model of normality being learned. A better strategy is to have a learning algorithm itself decide which set of features is the best. The reason for that is that learning algorithms find models with optimal generalization properties, i.e. the ones that are valid not only for observed data but also for the data to be dealt with in the future. The a priori choice of features may bias the learning process and lead to worse detection performance. By leaving this choice to the learning algorithm, the possibility of such bias is eliminated.

A natural way to address the kernel fusion problem is to learn a linear combination $K = \sum_{j=1}^{m} \theta_j K_j$ with mixing coefficients $\boldsymbol{\theta}$ together with model parameters, so as to maximize the generalization ability. To promote sparse solutions in terms of the linear kernel mixture, one frequently employs 1-norm simplex constraints on the mixing coefficients. This framework, known as multiple kernel learning (MKL), was first introduced for binary classification by [12]. Recently, efficient optimization strategies have been proposed for semi-infinite linear programming [25], second order approaches [3], and gradient-based optimization [20]. Other variants of two-class MKL have been proposed in subsequent work addressing practical algorithms for multi-class [19,32] and multi-label [8] problems.

We translate the multiple kernel learning framework to density level-set estimation to find a linear combination of features that realizes a minimal-volume description of the data. Furthermore, we generalize the MKL simplex constraint on the mixing coefficients to allow for arbitrary p-norms regularizations, where $p \geq 1$, hence leading to non-sparse kernel mixtures. Our approach also generalizes the one-class support vector machine [23] that is obtained as a special case for learning with only a single kernel. The optimization problem of our new method is efficiently solved by interleaved column generation and semi-infinite programming. Empirically, we evaluate our approach on network intrusion detection and object recognition tasks and compare its performance for different norms with unweighted-sum kernel mixtures. We observe our approach to attain higher predictive performances than baseline approaches.

The remainder of this paper is structured as follows. Section 2 briefly reviews the one-class support vector machine and presents our main contribution to density level-set estimation with multiple kernels. Section 3 reports on empirical results and Section 4 concludes.

2 Multiple Kernel Learning for Density Level-Sets

2.1 Density Level-Sets

In this paper, we focus on one-class classification problems. That is, we are given n data points $\boldsymbol{x}_1, \ldots, \boldsymbol{x}_n$, where $\boldsymbol{x}_i$ lies in some input space $\mathcal{X}$. The goal is to find a model $f : \mathcal{X} \rightarrow \mathbb{R}$ and a density level-set $D_\rho = \{\boldsymbol{x} : f(\boldsymbol{x}) \geq \rho\}$ that generalizes well on new and unseen data such that the level-set encloses the normal data, i.e., $\boldsymbol{x} \in D_\rho$, while for outliers $\boldsymbol{x}' \notin D_\rho$ holds. A common approach is to employ linear models of the form

$$f(\boldsymbol{x}) = \boldsymbol{w}'\psi(\boldsymbol{x}) \tag{1}$$

together with a (possibly non-linear) feature mapping $\psi : \mathcal{X} \rightarrow \mathcal{H}$. A max-margin approach leads to the (primal) one-class SVM optimization problem [23] for $\nu \in]0,1]$,

$$\begin{aligned} \min_{\boldsymbol{w},\rho,\boldsymbol{\xi}} \quad & \frac{1}{2}\boldsymbol{w}'\boldsymbol{w} + \frac{1}{\nu n}\|\boldsymbol{\xi}\|_1 - \rho \\ \text{s.t.} \quad & \forall i : \boldsymbol{w}'\psi(\boldsymbol{x}_i) \geq \rho - \xi_i, \quad \forall i : \xi_i \geq 0. \end{aligned} \tag{2}$$

Once optimal parameters $\boldsymbol{w}^*$ and ρ^* are found, these are plugged into Equation (1), and new instances $\tilde{\boldsymbol{x}}$ are classified according to $\text{sign}(f(\tilde{\boldsymbol{x}}) - \rho^*)$.

2.2 Density Level-Set Estimation with Multiple Kernels

When learning with multiple kernels, we are given m different feature mappings $\psi_1, \ldots, \psi_m$ in addition to the data points $\boldsymbol{x}_1, \ldots, \boldsymbol{x}_n$. Every mapping $\psi_j : \mathcal{X} \rightarrow \mathcal{H}_j$ gives rise to a reproducing kernel k_j of $\mathcal{H}_j$ such that

$$k_j(\boldsymbol{x}, \tilde{\boldsymbol{x}}) = \langle \psi_j(\boldsymbol{x}), \psi_j(\tilde{\boldsymbol{x}}) \rangle_{\mathcal{H}_j}.$$

The goal of one-class multiple kernel learning is to find a linear combination $\sum_{j=1}^m \theta_j K_j$ of kernels and parameters $\boldsymbol{w}$, $\boldsymbol{\xi}$, and ρ simultaneously, such that the resulting hypothesis f leads to a minimum-volume description of the normal data. We incorporate the kernel mixture into the model in Equation (1) and arrive at

$$f(\boldsymbol{x}) = \sum_{k=1}^m \theta_j \, \boldsymbol{w}_j'\psi_j(\boldsymbol{x}) = \boldsymbol{w}_{\boldsymbol{\theta}}'\psi_{\boldsymbol{\theta}}(\boldsymbol{x}),$$

where the weight vector and the feature mapping have a block structure

$$\boldsymbol{w}_{\boldsymbol{\theta}} = (\sqrt{\theta_j}\boldsymbol{w}_j)_{j=1,\ldots,m}, \quad \psi_{\boldsymbol{\theta}}(\boldsymbol{x}_i) = (\sqrt{\theta_j}\psi_j(\boldsymbol{x}_i))_{j=1,\ldots,m}, \tag{3}$$

with mixing coefficients $\theta_j \geq 0$.

Incorporating (3) into (2) and imposing a general p-norm constraint $\|\boldsymbol{\theta}\|_p = 1$ for $p \geq 1$ on the mixing coefficients leads to the following primal optimization problem for $\nu \in]0,1]$, and $p \geq 1$.

$$\min_{\boldsymbol{\theta},\boldsymbol{w},\rho,\boldsymbol{\xi}} \quad \frac{1}{2}\boldsymbol{w}'_{\boldsymbol{\theta}}\boldsymbol{w}_{\boldsymbol{\theta}} + \frac{1}{\nu n}\|\boldsymbol{\xi}\|_1 - \rho \tag{3a}$$

$$\text{s.t.} \quad \forall i: \ \boldsymbol{w}'_{\boldsymbol{\theta}}\psi_{\boldsymbol{\theta}}(\boldsymbol{x}_i) \geq \rho - \xi_i; \quad \boldsymbol{\xi} \geq \mathbf{0}\ ; \quad \boldsymbol{\theta} \geq \mathbf{0}\ ; \quad \|\boldsymbol{\theta}\|_p = 1. \tag{3b}$$

The above optimization problem is non-convex because (i) the products $\theta_j \boldsymbol{w}_j$ are non-convex which, however, can be easily removed by a change of variables $\boldsymbol{v}_j := \theta_j \boldsymbol{w}_j$ (e.g. see [2]), and (ii) the set $\{\boldsymbol{\theta} : \|\boldsymbol{\theta}\|_p = 1\}$ is not convex. As a remedy to (ii), we relax the constraint on $\boldsymbol{\theta}$ to become an inequality constraint, i.e., $\|\boldsymbol{\theta}\|_p \leq 1$. Treating the above optimization problem as interleaved minimization – over $\boldsymbol{\theta}$ *and* $\boldsymbol{w}$, $\boldsymbol{\xi}$, and ρ – it is easily verified that the optimal $\boldsymbol{\theta}^*$ in the $\boldsymbol{\theta}$-step always fulfills $\|\boldsymbol{\theta}^*\|_p = 1$ for all $p \geq 1$; essentially, we solve $\min_{\boldsymbol{\theta}} \sum_j c_j/\theta_j$ *s.t.* $\|\boldsymbol{\theta}\|_p \leq 1$ which induces solutions $\boldsymbol{\theta}^*$ at the border $\|\boldsymbol{\theta}^*\|_p = 1$. We thus arrive at the following equivalent optimization problem, which now is convex.

$$\min_{\boldsymbol{\theta},\boldsymbol{v},\boldsymbol{\xi},\rho} \quad \frac{1}{2}\sum_{j=1}^{m}\frac{\boldsymbol{v}'_j\boldsymbol{v}_j}{\theta_j} + \frac{1}{\nu n}\|\boldsymbol{\xi}\|_1 - \rho \tag{4a}$$

$$\text{s.t.} \quad \forall i: \ \sum_{j=1}^{m}\boldsymbol{v}'_j\psi_j(\boldsymbol{x}_i) \geq \rho - \xi_i; \quad \boldsymbol{\xi} \geq \mathbf{0}\ ; \quad \boldsymbol{\theta} \geq \mathbf{0}\ ; \quad \|\boldsymbol{\theta}\|_p \leq 1. \tag{4b}$$

Several previous algorithms for two-class multiple kernel learning utilized a two-step structure by alternating full SVM steps with $\boldsymbol{\theta}$ steps of different flavor [32,20,30]. In contrast, we follow [25] and propose to alternate $\boldsymbol{\theta}$ steps with *minor iterations* of SVM optimizers *without running them to completion.* We chose SVM^{light} [10] as a basic solver, since its underlying chunking idea employs efficient $\boldsymbol{\alpha}$ minimization steps, making it well-suited for an interleaved $\boldsymbol{\alpha}, \boldsymbol{\theta}$ minimization. To solve the p-norm one-class MKL problem, we now devise a semi-infinite programming (SIP) approach similar to [25].

The underlying idea is to interleave the optimization of the upper bound on the objective of the SVM step and the $\boldsymbol{\theta}$ step. Fixing $\boldsymbol{\theta} \in \Theta$, where $\Theta = \{\boldsymbol{\theta} \in \mathbb{R}^n \mid \boldsymbol{\theta} \geq \mathbf{0}\ , \ \ \|\boldsymbol{\theta}\|_p \leq 1\}$, we build the partial Lagrangian with respect to $\boldsymbol{v}$, $\boldsymbol{\xi}$, and ρ by introducing componentwise non-negative Lagrange multipliers $\boldsymbol{\alpha}, \boldsymbol{\gamma} \in \mathbb{R}^n, \delta \in \mathbb{R}$. The partial Lagrangian is given by

$$L = \frac{1}{2}\sum_{j=1}^{m}\frac{\boldsymbol{v}'_j\boldsymbol{v}_j}{\theta_j} + \frac{1}{\nu n}\sum_{i=1}^{n}\xi_i - \sum_{i=1}^{n}\gamma_i\xi_i - \sum_{i=1}^{n}\alpha_i\Big(\sum_{j=1}^{m}\boldsymbol{v}'_j\psi_j(\boldsymbol{x}_i) - \rho + \xi_i\Big) - \delta\rho.$$

Setting the partial derivatives with respect to the primal variables to zero yields the relations $0 \leq \alpha_i \leq \frac{1}{\nu n}$, $\sum_i \alpha_i = 1$, and $\boldsymbol{v}_j = \sum_i \alpha_i\theta_j\psi_j(\boldsymbol{x}_i)$ for $1 \leq i \leq n$ and $1 \leq j \leq p$. The KKT conditions trivially hold and re-substitution into the Lagrangian gives rise to the min-max formulation for $\nu \in]0,1]$ and $p \geq 1$,

$$\min_{\boldsymbol{\theta}} \max_{\boldsymbol{\alpha}} \quad -\frac{1}{2}\sum_{i,l=1}^{n} \alpha_i \alpha_l \sum_{j=1}^{m} \theta_j k_j(\boldsymbol{x}_i, \boldsymbol{x}_l) \tag{5a}$$

$$\text{s.t.} \quad \mathbf{0} \leq \boldsymbol{\alpha} \leq \frac{1}{\nu n}\mathbf{1}; \quad \mathbf{1}'\boldsymbol{\alpha} = 1; \quad \boldsymbol{\theta} \geq \mathbf{0}; \quad \|\boldsymbol{\theta}\|_p \leq 1. \tag{5b}$$

The above optimization problem can be solved directly by gradient-based techniques exploiting the smoothness of the objective [1]. Alternatively, we can translate it into an equivalent semi-infinite program (SIP) as follows. Suppose $\boldsymbol{\alpha}^*$ is optimal, then denoting the value of the target function by $t(\boldsymbol{\alpha}, \boldsymbol{\theta})$, we have $t(\boldsymbol{\alpha}^*, \boldsymbol{\theta}) \geq t(\boldsymbol{\alpha}, \boldsymbol{\theta})$ for all $\boldsymbol{\alpha}$ and $\boldsymbol{\theta}$. Hence we can equivalently minimize an upper bound λ on the optimal value. We thus arrive at the following optimization problem,

$$\min_{\lambda, \boldsymbol{\theta}} \quad \lambda \quad \text{s.t.} \quad \lambda \geq -\frac{1}{2}\boldsymbol{\alpha}' \sum_{j=1}^{m} \theta_j K_j \boldsymbol{\alpha} \tag{6}$$

for all $\boldsymbol{\alpha} \in \mathbb{R}^n$ with $\mathbf{0} \leq \boldsymbol{\alpha} \leq \frac{1}{\nu n}\mathbf{1}$, $\mathbf{1}'\boldsymbol{\alpha} = 1$, and $\boldsymbol{\alpha} \geq \mathbf{0}$, as well as $\|\boldsymbol{\theta}\|_p \leq 1$ and $\boldsymbol{\theta} \geq \mathbf{0}$. The optimization problem in Equation (6) generalizes the idea of [25] to the case $p \geq 1$. Analogously, it can be optimized with interleaving cutting plane algorithms, that is, the solution of a quadratic program (here a one-class SVM) generates the most strongly violated constraint for the actual mixture $\boldsymbol{\theta}$. The optimal $(\boldsymbol{\theta}^*, \lambda)$ however depends on the value of p. We differentiate between two cases, $p = 1$ and $p > 1$.

Optimizing $\boldsymbol{\theta}$ for $p = 1$: for $p = 1$ is then identified by solving a linear program with respect to set of active constraints.

Optimizing $\boldsymbol{\theta}$ for $p > 1$: For the general case $p > 1$, a non-linearity is introduced by requiring $\|\boldsymbol{\theta}\|_p \leq 1$. Such constraint is rather uncommon in standard optimization toolboxes that often handle only linear and quadratic constraints. As a remedy we propose to solve a sequence of quadratically constrained subproblems. To this end, we substitute the p-norm constraint by sequential second-order Taylor approximations of the form

$$\begin{aligned}
||\boldsymbol{\theta}||_p^p &\approx 1 + p(\boldsymbol{\theta}_k^{p-1})'(\boldsymbol{\theta} - \boldsymbol{\theta}^{old}) \\
&\quad + \frac{p(p-1)}{2}(\boldsymbol{\theta} - \boldsymbol{\theta}^{old})'\text{diag}((\boldsymbol{\theta}^{old})^{p-2})(\boldsymbol{\theta} - \boldsymbol{\theta}^{old}) \\
&= 1 - \frac{p(3-p)}{2} - \sum_j p(p-2)(\theta_j^{old})^{p-1}\,\theta_j \\
&\quad + \frac{p(p-1)}{2}\sum_j (\theta_j^{old})^{p-2}\,\theta_j^2,
\end{aligned}$$

where $\boldsymbol{\theta}^p$ is defined element-wise, that is $\boldsymbol{\theta}^p := (\theta_1^p, ..., \theta_m^p)$. We use $\boldsymbol{\theta}^{old} = \sqrt[p]{\frac{1}{m}}\mathbf{1}$ as a starting point. Note that the quadratic term in the approximation is diagonal. As a result the quadratically constrained problem can be solved very efficiently. For

Algorithm 1. p-Norm MKL chunking-based training algorithm. It simultaneously optimizes $\boldsymbol{\alpha}$ and the kernel weighting $\boldsymbol{\theta}$. The accuracy parameter ϵ and the subproblem size Q are assumed to be given to the algorithm. For simplicity, a few speed-up tricks are not shown: the removal of inactive constraints and hot-starts.

1: $g_{j,i} = 0,\ \hat{g}_i = 0,\ \alpha_i = 0,\ \theta_j = \sqrt[p]{1/m}$ for $j = 1, \ldots, m$ and $i = 1, \ldots, n$
2: **for** $t = 1, 2, \ldots$ and while SVM and MKL optimality conditions are not satisfied **do**
3: Select Q suboptimal variables $\alpha_{i_1}, \ldots, \alpha_{i_Q}$ based on the gradient $\hat{\mathbf{g}}$ and $\boldsymbol{\alpha}$; store $\boldsymbol{\alpha}^{old} = \boldsymbol{\alpha}$
4: Solve SVM dual with respect to the selected variables and update $\boldsymbol{\alpha}$
5: Update gradient $g_{j,i} \leftarrow g_{j,i} + \sum_{q=1}^{Q} (\alpha_{i_q} - \alpha_{i_q}^{old}) k_j(\boldsymbol{x}_{i_q}, \boldsymbol{x}_i)$ for all $j = 1, \ldots, m$ and $i = 1, \ldots, n$
6: **for** $j = 1, \ldots, m$ **do**
7: $S_j^t = \frac{1}{2} \sum_i g_{j,i} \alpha_i$
8: **end for**
9: $S^t = \sum_j \theta_j S_j^t$
10: **if** $|1 - \frac{S^t}{\lambda}| \geq \epsilon$
11: **for** $k = 1, 2, \ldots$ and while MKL optimality conditions are not satisfied **do**
12: $\boldsymbol{\theta}^{old} = \boldsymbol{\theta}$
13: $(\boldsymbol{\theta}, \lambda) \leftarrow \text{argmax } \lambda$
14: w.r.t. $\boldsymbol{\theta} \in \mathbb{R}^m, \lambda \in \mathbb{R}$
15: s.t. $\mathbf{0} \leq \boldsymbol{\theta} \leq \mathbf{1}, \quad \sum_j \theta_j S_j^r \geq \lambda$ for $r = 1, \ldots, t$
16: $\frac{p(p-1)}{2} \sum_j (\theta_j^{old})^{p-2}\, \theta_j^2 - \sum_j p(p-2)(\theta_j^{old})^{p-1}\, \theta_j \leq \frac{p(3-p)}{2}$
17: $\boldsymbol{\theta} \leftarrow \boldsymbol{\theta} / ||\boldsymbol{\theta}||_p$
18: **end for**
19: **end if**
20: $\hat{g}_i = \sum_j \theta_j g_{j,i}$ for all $i = 1, \ldots, n$
21: **end for**

the special case $p = 2$, the Taylor approximation is tight and hence the sequence of quadratically constrained sub-problems converges after one iteration.

Optimization Algorithm. Algorithm 1 outlines the interleaved $\boldsymbol{\alpha}, \boldsymbol{\theta}$ MKL training algorithm. Lines 3-5 are standard in chunking based SVM solvers and carried out by SVM^{light}. Lines 6-9 compute (parts of) SVM-objective values for each kernel independently. Finally lines 11 to 18 solve a sequence of semi-infinite programs with the p-norm constraint being approximated as a sequence of second-order constraints. The algorithm terminates if the maximum KKT violation (see [10]) falls below a predetermined precision ε_{svm} and for MKL if the normalized maximal constraint violation $|1 - \frac{S^t}{\lambda}| < \varepsilon_{mkl}$.

3 Empirical Results

In this section we study p-norm multiple kernel learning for density level-sets in terms of efficiency and accuracy. We experiment on network intrusion detection and object recognition tasks and compare our approach to baseline one-class SVMs with unweighted-sum kernels $K = \sum_{j=1}^{m} K_j$ wich we refer to as ∞-norm MKL. We choose this baseline because for two-class multiple kernel learning approaches, unweighted-sum kernel mixtures have frequently been observed to outperform sparse kernel mixtures in practical applications.

3.1 Network Intrusion Detection

For the intrusion detection experiments we use HTTP traffic recorded at Fraunhofer Institute FIRST Berlin. The unsanitized data contains 2500 normal HTTP requests drawn randomly from incoming traffic recorded over two months. Malicious traffic is generated using the Metasploit framework [18]. We generate 30 instances of 10 real attack classes from recent exploits, including buffer overflows and PHP vulnerabilities. Every attack is recorded in different variants using virtual network environments and decoy HTTP servers.

The malicious data are normalized to match frequent attributes of the normal HTTP requests such that the payload provides the only indicator for separating normal from attack data. We deploy 10 spectrum kernels [14,24] for $1, 2, \ldots, 10$-gram feature representations. All kernels are normalized according to Equation (7) to avoid dependencies on the HTTP request length.

$$K(\boldsymbol{x}, \tilde{\boldsymbol{x}}) \longmapsto \frac{K(\boldsymbol{x}, \tilde{\boldsymbol{x}})}{\sqrt{K(\boldsymbol{x}, \boldsymbol{x})K(\tilde{\boldsymbol{x}}, \tilde{\boldsymbol{x}})}}, \tag{7}$$

We randomly split the normal data into 1000 training, 500 validation and 1000 test examples. The training partition is used as it is since centroid-based learners assume uncorrupted training data. The validation and test partitions are mixed with 15 attack instances that are randomly chosen from the malicious pool. We make sure that attacks of the same class occur either in the holdout or in the test data but not in both, hence reflecting the goal of anomaly detection to recognize *previously unknown* attacks. We report on average areas under the ROC curve in the false-positive interval $[0, 0.01]$ ($\text{AUC}_{[0,0.01]}$) over 100 repetitions with distinct training, holdout, and test sets.

Table 1 shows the results for one-class multiple kernel learning with $p \in \{\infty, 1, \frac{4}{3}, 2, 4\}$. Depending on the actual value of p, the performances are quite different. The unweighted-sum kernel (∞-norm MKL) outperforms most of the one-class MKL approaches. However, employing a 2-norm constraint on the mixing coefficients leads to better results than the ∞-norm mixture. Notice that the 2-norm mixture is about 10% better than its sparse 1-norm counterpart.

Figure 1 reports on the optimal kernel mixture coefficients $\boldsymbol{\theta}$ for $p \in \{1, \frac{4}{3}, 2, 4\}$-norm MKL and the unweighted-sum kernel. The sparse 1-norm solution places all the weight into 1-grams that – although leading to concise representations because of the low dimensional feature space – result in inappropriate performances (see Table 1). The higher the value of p, the less weight is placed on the 1-gram kernel but spread across higher n-gram kernels. The 4-norm mixture is similar to the trivial ∞-norm solution. The best solution (2-norm) still places weight to 1-grams but incorporates all other n-gram kernels to some extend.

Table 1. Results for intrusion detection

MKL	$AUC_{0.01}$
∞-norm	89.4 ± 0.7
1-norm	79.4 ± 0.9
$\frac{4}{3}$-norm	85.7 ± 0.8
2-norm	90.7 ± 0.8
4-norm	88.9 ± 0.9

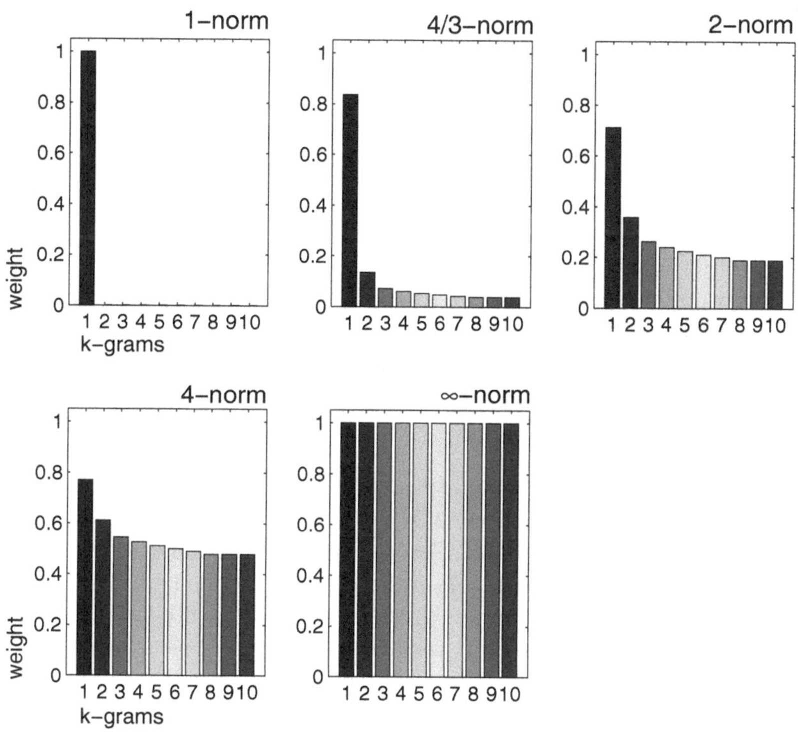

Fig. 1. Mixing coefficients for the intrusion detection task

3.2 Multi-label Image Categorization

Besides anomaly and outlier detection, one-class learning techniques are frequently applied to multi-class classification problems with temporally varying numbers of categories such as event detection and object recognition tasks. Their advantage lies in training a single model for every (new) category in contrast to maintaining expensive multi-class classifiers that have to be re-trained once a new category is included in the task.

To study one-class multiple kernel learning in this alternative scenario, we apply our approach to the multi-label classification task of the VOC 2008 challenge [7]. The data set contains 8780 images, divided into 2113 training, 2227 validation, and 4340 test images. Images are annotated with a subset of 20 class labels such as *aeroplane*, *bicycle*, and *bird*. Since the ground-truth of the test set is not yet disclosed by the challenge organizers, we focus on the training and validation splits. From these two original sets, we draw 2111 training, 1111 validation, and 1110 test images at random and report on average precisions (AP) for all recall values over 10 runs with distinct training, holdout, and test sets.

We employ two sets of kernels inspired from the VOC 2007 winner (K12) [17] and the VOC 2008 winner (K30) [26]. For both approaches, all basic features are combined with the respective pyramid levels and translated into a χ^2 kernel [31], where the widths of the χ^2 kernels are chosen according to a heuristic [11]. The sets of kernels are obtained as follows.

K12. We extract 12 kernels based on four basic features: histograms of visual words [5] in the grey (HOW-G) and in the hue color channel (HOW-H), histogram of oriented gradient (HOG) [6], and histograms of the hue color channel (HOCOL) [17]. These representations are combined with a pyramidal representation of level 2 to capture spatial dependencies, i.e., each image is tiled into 1, 4, and 16 parts.

K30. We extract 30 kernels based on histograms of visual words with 2 different sampling methods (dense and interest points), 5 different sets of colors (grey, opponent color, normalized opponent color, normalized RG, and RGB) [27] and 3 different tilings (level-0 and level-1 of the pyramid, and 1×3 tiling) [26].

We compare the performance of the unweighted-sum kernel ∞, and 1- and 2-norm MKL with the optimal p-norm MKL that maximizes the average precision on the validation set for each class. For the latter approach, model selection is not only performed for trade-off parameter ν but extended to the MKL norm p. Table 2 shows the mean average precisions over 20 categories for the test data. Bold faces indicate significant results, that is, the best method and ones that are not comparably different from the best result according to a Wilcoxon signed-ranks test using a 5% confidence-level.

For the K12 set of kernels, 1-norm MKL outperforms both, the unweighted-sum kernel ∞-norm and a non-sparse 2-norm MKL, which perform equally well. However, model selection over p for each class leads to comparable results as 1-norm MKL. We do not display the optimal p^* values for all 20 classes, however, the respective mixtures are non-sparse (see also Figure 2) so that the sparse 1-norm approach denotes the best solution for K12 in terms of accuracy and interpretability.

For the K30 set of kernels, the outcome is different. Here, the 1-norm MKL performs significantly worse compared to its non-sparse counterparts. Although model selection over p leads to the highest average precisions, the results are not significantly different to 2-norm MKL and unweighted-sum kernel mixtures. Our experiments show that the right choice of the value p depends highly on the employed kernels. Vice versa, once a set of kernels is fixed, it is necessary to include the norm parameter p in the model selection to find the best kernel mixture.

Table 2. Results for the VOC 2008 data set

	1-norm	p^*-norm	2-norm	∞-norm
mean AP (K12)	**17.6±0.8**	**17.8±1.0**	17.1±0.8	17.0±0.6
mean AP (K30)	16.3±0.5	**17.1±0.9**	**17.1±0.6**	**17.0±0.7**

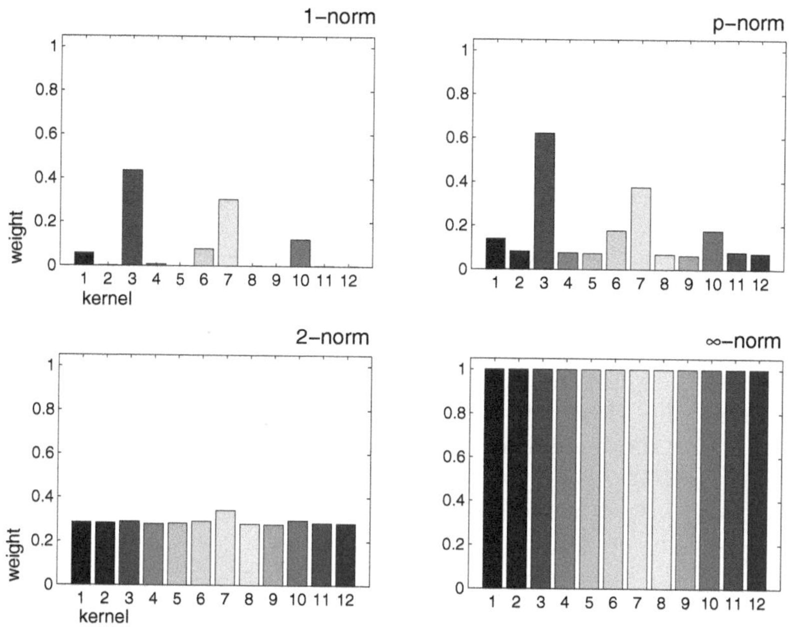

Fig. 2. Mixing coefficients for the multi-label image categorization experiment

Figure 2 shows the optimal mixing coefficients for the K12 task, averaged over 10 repetitions. The 1-norm solution picks a sparse combination resulting in a minimum volume description of the data. While a 2-norm solution distributes the weights almost uniformly on the 12 kernels, the p-norm solution lies in between and considers all kernels with non-zero mixing coefficients in the solution.

3.3 Execution Time

We show the efficiency of one-class MKL and compare the execution times for our approach with $p \in \{1, 1.333, 2, 3, 4, \infty\}$ to one-class SVMs using the unweighted sum-kernel as implemented in [10]. To show different aspects of our approach, we draw a sample of size n from a 10-dimensional Gaussian distribution for various values of n. Kernel matrices are computed using RBF-kernels with different bandwidth parameters. We optimize the duality gap for all methods up to a precision of 10^{-3}.

Figure 3 (left) displays the results for varying sample sizes in a log-log plot; errorbars indicate standard error over 5 repetitions. Unsurprisingly, the baseline one-class SVM using the sum-kernel is the fastest method. The execution time of

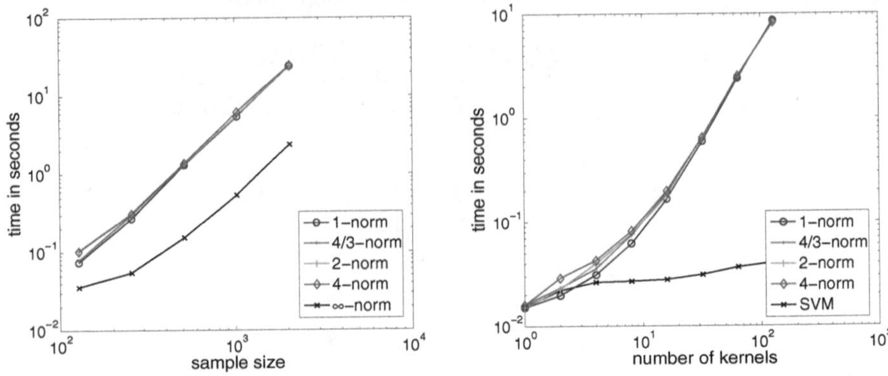

Fig. 3. Execution times for one-class MKL. Left: results for varying sample sizes. Right: execution times for varying numbers of kernels.

non-sparse MKL depends on the value p. We observe longer computation times for large values of p. However, all approaches scale similarly.

Figure 3 (right) shows execution times for varying numbers of kernels and fixed sample size $n = 100$. Again, the baseline one-class SVM with the unweighted-sum kernel is the fastest method. All one-class MKL approaches show reasonable run-times and converge quickly for 128 kernels.

4 Conclusion

We presented an efficient and accurate approach to multiple kernel learning for density level-set estimation. Our approach generalizes the standard setting of multiple kernel learning by allowing for arbitrary norms for the kernel mixture. This enabled us to study sparse and non-sparse kernel mixtures. Our method contains the one-class SVM as a special case for training with only a single kernel. Our optimization strategy is based on interleaved semi-infinite programming and chunking based SVM training. Empirical results proved the efficiency and accuracy of our methods compared to baseline approaches. We observed one-class MKL to be robust in situations where unweighted-sum kernels are prone to fail.

Acknowledgments

The authors wish to thank Sören Sonnenburg, Alexander Zien, and Pavel Laskov for fruitful discussions and helpful comments. Furthermore we thank Patrick Düssel and Christian Gehl for providing the network traffic and Alexander Binder, Christina Müller, Motoaki Kawanabe, and Wojciech Wojcikiewicz for sharing kernel matrices for the VOC data with us. This work was supported in part by the German Bundesministerium für Bildung und Forschung (BMBF) under the project REMIND (FKZ 01-IS07007A) and by the FP7-ICT Programme of the European Community, under the PASCAL2 Network of Excellence, ICT-216886.

References

1. Bach, F.R., Lanckriet, G.R.G., Jordan, M.I.: Multiple kernel learning, conic duality, and the SMO algorithm. In: Proceedings of the Twenty-first International Conference on Machine Learning (2004)
2. Boyd, S., Vandenberghe, L.: Convex Optimization. Cambrigde University Press, Cambridge (2004)
3. Chapelle, O., Rakotomamonjy, A.: Second order optimization of kernel parameters. In: Proceedings of the NIPS Workshop on Kernel Learning: Automatic Selection of Optimal Kernels (2008)
4. Chhabra, P., Scott, C., Kolaczyk, E.D., Crovella, M.: Distributed spatial anomaly detection. In: Proceedings of the IEEE Infocom 2008 (2008)
5. Csurka, G., Bray, C., Dance, C., Fan, L.: Visual categorization with bags of keypoints. In: Workshop on Statistical Learning in Computer Vision, ECCV, Prague, Czech Republic, May 2004, pp. 1–22 (2004)
6. Dalal, N., Triggs, B.: Histograms of oriented gradientsfor human detection. In: IEEE Computer Society Conference on Computer Vision and Pattern Recognition, San Diego, USA, June 2005, vol. 1, pp. 886–893 (2005)
7. Everingham, M., Van Gool, L., Williams, C.K.I., Winn, J., Zisserman, A.: Proceedings of the the PASCAL Visual Object Classes Challenge 2008, VOC 2008 (2008)
8. Ji, S., Sun, L., Jin, R., Ye, J.: Multi-label multiple kernel learning. In: Advances in Neural Information Processing Systems (2009)
9. Jiang, Z., Luosheng, W., Yong, F., Xiao, Y.C.: Intrusion detection based on density level sets estimation. In: NAS 2008: Proceedings of the 2008 International Conference on Networking, Architecture, and Storage (2008)
10. Joachims, T.: Making large–scale SVM learning practical. In: Schölkopf, B., Burges, C.J.C., Smola, A.J. (eds.) Advances in Kernel Methods — Support Vector Learning, pp. 169–184. MIT Press, Cambridge (1999)
11. Lampert, C.H., Blaschko, M.B.: A multiple kernel learning approach to joint multi-class object detection. In: Rigoll, G. (ed.) DAGM 2008. LNCS, vol. 5096, pp. 31–40. Springer, Heidelberg (2008)
12. Lanckriet, G., Cristianini, N., Ghaoui, L.E., Bartlett, P., Jordan, M.I.: Learning the kernel matrix with semi-definite programming. Journal of Machine Learning Research 5, 27–72 (2004)
13. Lee, W., Stolfo, S.J.: A framework for constructing features and models for intrusion detection systems. ACM Transactions on Information Systems Security 3, 227–261 (2000)
14. Leslie, C., Eskin, E., Noble, W.S.: The spectrum kernel: A string kernel for SVM protein classification. In: Proc. Pacific Symp. Biocomputing, pp. 564–575 (2002)
15. Mahoney, M.V., Chan, P.K.: Learning nonstationary models of normal network traffic for detecting novel attacks. In: Proc. of ACM SIGKDD International Conference on Knowledge Discovery and Data Mining (KDD), pp. 376–385 (2002)

16. Mahoney, M.V., Chan, P.K.: Learning rules for anomaly detection of hostile network traffic. In: Proc. of International Conference on Data Mining (ICDM) (2003)
17. Marszalek, M., Schmid, C.: Learning representations for visual object class recognition. In: Proceedings of the PASCAL Visual Object Classes Challenge 2007, VOC 2007 (2007)
18. Maynor, K., Mookhey, K., Cervini, J.F.R., Beaver, K.: Metasploit toolkit. Syngress (2007)
19. Rakotomamonjy, A., Bach, F., Canu, S., Grandvalet, Y.: More efficiency in multiple kernel learning. In: ICML, pp. 775–782 (2007)
20. Rakotomamonjy, A., Bach, F., Canu, S., Grandvalet, Y.: SimpleMKL. Journal of Machine Learning Research 9, 2491–2521 (2008)
21. Rieck, K., Laskov, P.: Detecting unknown network attacks using language models. In: Büschkes, R., Laskov, P. (eds.) DIMVA 2006. LNCS, vol. 4064, pp. 74–90. Springer, Heidelberg (2006)
22. Rieck, K., Laskov, P.: Language models for detection of unknown attacks in network traffic. Journal in Computer Virology 2(4), 243–256 (2007)
23. Schölkopf, B., Platt, J., Shawe-Taylor, J., Smola, A.J., Williamson, R.C.: Estimating the support of a high-dimensional distribution. Neural Computation 13(7), 1443–1471 (2001)
24. Shawe-Taylor, J., Cristianini, N.: Kernel methods for pattern analysis. Cambridge University Press, Cambridge (2004)
25. Sonnenburg, S., Rätsch, G., Schäfer, C., Schölkopf, B.: Large Scale Multiple Kernel Learning. Journal of Machine Learning Research 7, 1531–1565 (2006)
26. Tahir, M., van de Sande, K., Uijlings, J., Yan, F., Li, X., Mikolajczyk, K., Kittler, J., Gevers, T., Smeulders, A.: Surreyuva srkda method. In: Proceedings of the PASCAL Visual Object Classes Challenge 2008, VOC 2008 (2008)
27. van de Sande, K.E.A., Gevers, T., Snoek, C.G.M.: Evaluation of color descriptors for object and scene recognition. In: IEEE Conference on Computer Vision and Pattern Recognition (2008)
28. Wang, K., Parekh, J.J., Stolfo, S.J.: Anagram: A content anomaly detector resistant to mimicry attack. In: Zamboni, D., Krügel, C. (eds.) RAID 2006. LNCS, vol. 4219, pp. 226–248. Springer, Heidelberg (2006)
29. Wang, K., Stolfo, S.J.: Anomalous payload-based network intrusion detection. In: Jonsson, E., Valdes, A., Almgren, M. (eds.) RAID 2004. LNCS, vol. 3224, pp. 203–222. Springer, Heidelberg (2004)
30. Xu, Z., Jin, R., King, I., Lyu, M.R.: An extended level method for efficient multiple kernel learning. In: Advances in Neural Information Processing Systems (2009)
31. Zhang, J., Marszalek, M., Lazebnik, S., Schmid, C.: Local features and kernels for classification of texture and object categories: A comprehensive study. International Journal of Computer Vision 73(2), 213–238 (2007)
32. Zien, A., Ong, C.S.: Multiclass multiple kernel learning. In: Ghahramani, Z. (ed.) ICML. ACM International Conference Proceeding Series, vol. 227, pp. 1191–1198. ACM, New York (2007)

Efficient Multi-start Strategies for Local Search Algorithms*

Levente Kocsis and András György

Machine Learning Research Group
Computer and Automation Research Institute
of the Hungarian Academy of Sciences
Kende u. 13-17, 1111 Budapest, Hungary
kocsis@sztaki.hu, gya@szit.bme.hu

Abstract. Local search algorithms for global optimization often suffer from getting trapped in a local optimum. The common solution for this problem is to restart the algorithm when no progress is observed. Alternatively, one can start multiple instances of a local search algorithm, and allocate computational resources (in particular, processing time) to the instances depending on their behavior. Hence, a multi-start strategy has to decide (dynamically) when to allocate additional resources to a particular instance and when to start new instances. In this paper we propose a consistent multi-start strategy that assumes a convergence rate of the local search algorithm up to an unknown constant, and in every phase gives preference to those instances that could converge to the best value for a particular range of the constant. Combined with the local search algorithm SPSA (Simultaneous Perturbation Stochastic Approximation), the strategy performs remarkably well in practice, both on synthetic tasks and on tuning the parameters of learning algorithms.

1 Introduction

Local search algorithms for global optimization often suffer from getting trapped in a local optimum. Moreover, local search algorithms that are guaranteed to converge to a global optimum under some conditions (such as Simulated Annealing or SPSA) usually converge at a very slow pace in order to provide consistency. On the other hand, if the algorithms are employed with more aggressive settings, much faster convergence to local optima is achievable, but with no guarantee to find the global optimum. The common solution to escape from a local optimum is to restart the algorithm when no progress is observed. Alternatively, one can start multiple instances of the local search algorithm, and allocate computational resources, in particular, processing time, to the instances depending on their behavior. The moment of starting an instance can vary in time, and so the number of instances can also grow over time depending on the allocation strategy.

* This research was supported in part by the Mobile Innovation Center of Hungary, the Hungarian Scientific Research Fund and the Hungarian National Office for Research and Technology (OTKA-NKTH CNK 77782), and by the PASCAL2 Network of Excellence (EC grant no. 216886).

W. Buntine et al. (Eds.): ECML PKDD 2009, Part I, LNAI 5781, pp. 705–720, 2009.

Running several instances of an algorithm or several algorithms in parallel and selecting among the algorithms have been intensively studied in the area of meta-learning [22]. The main problem here is to allocate time slices to particular algorithms with the aim of maximizing the best result returned. This allocation may depend on the intermediate performance of the algorithms.

A simplified version of this problem is the so called Maximum K-Armed Bandit [6], where at each time instance exactly one algorithm (called arm in this framework) can be run, and it is assumed that the intermediate results returned by an instance of the algorithm are drawn independently from a particular distribution. Note, however, that this assumption is quite unrealistic, and usually leads to oversimplification, as local search algorithms steadily improve their performance, and thus the distribution from which the next value is drawn differs significantly from the ones from which the values in the previous several iterations were drawn.

Nevertheless, the Maximum K-Armed Bandit problem provides a convenient framework, and is often used for the analysis of multi-start search strategies [6,7,20]. A generic algorithm for this problem is provided in [1], where the, so called, reservation price of an instance is introduced. If an instance achieves its reservation price, it is useless to select it again. The computation of the reservation price depends on a model of the algorithm that can be learned under some specific constraints. Several algorithms [6,7,20] are based on the (generalized) extreme value theory: these algorithms are similar to standard multi-armed bandit algorithms (see, e.g., [3]) except that the upper confidence bounds are derived from different distributions (e.g., from Gumbel distribution). In [21] a distribution free approach is proposed that combines a multi-armed bandit exploration strategy with a heuristic selection among the available arms. The resulting algorithm (called ThresholdAscent) appeared to have attractive practical performance compared to the ones based on extreme value theory. Finally, a natural strategy is to probe the algorithm instances for a while, estimate their future performance based on the results of this trial phase, and then use the most promising algorithm for the time remaining. Simple rules were suggested in [4] to predict the future performance of each algorithm, while [5] employs Bayesian prediction.

Another related problem is the following. Several algorithm instances are available that all produce the correct answer to a certain question if run for a sufficiently long time. The time needed for an algorithm instance to find the answer is a random quantity, and the goal is to combine the given algorithms to minimize the expected running time until the answer is found. When the distribution of the running time is known, an optimal time allocation strategy to minimize the expected running time is to perform a sequence of runs with a certain cut-off time that depends on the distribution [17]. If the distribution is unknown, a particular running time sequence can be chosen that results in an expected total running time that is only a logarithmic factor larger than the optimum achievable if the distribution is known. We note that this strategy is among the few that provide a schedule that increases the number of algorithm instances. For the same set-up, an allocation strategy is proposed in [16] based on updating dynamically the belief over the run-time distribution. Finally, when a set of time allocation strategies are available and the optimization problem is to be solved several times, one can use the standard multi-armed bandit framework as in [9,10,11].

In this paper, we propose a new multi-start strategy, MetaMax, that assumes a convergence rate of the local search known up to an unknown constant. In every phase those instances are selected that may converge to the best value for a particular range of the constant. The selection mechanism is analogous to the DIRECT algorithm [15,8,14] for optimizing Lipschitz-functions with an unknown constant, where preference is given to rectangles that may contain the global optimum. The optimum within each rectangle is estimated in an optimistic way, and the estimate depends on the size of the rectangle. In our algorithm we use the function describing the convergence rate of the local search algorithms in a similar way as the size of the rectangles are used in the DIRECT algorithm.

The rest of the paper is organized as follows. Basic definitions and assumptions are given in Section 2. The new multi-start local search strategies of this paper, MetaMax(K) and MetaMax, are described and analyzed in Section 3: in Section 3.1 we deal with a selection mechanism among a fixed number of instances of the local search algorithm, while, in addition, a simple schedule for starting new instances is also considered in Section 3.2. Simulation results on real and synthetic data are provided in Section 4. Conclusions and possible further work are described in Section 5.

2 Preliminaries

Assume we wish to maximize a nonnegative real valued function f on the d-dimensional unit hypercube $[0,1]^d$, that is, the goal is to find a maximizer $x^* \in [0,1]^d$ such that $f(x^*) = \max_{x\in[0,1]^d} f(x)$. Without loss of generality, we assume that f is non-negative, and suppose, for simplicity, that f is continuous on $[0,1]^d$.[1] The continuity of f implies the existence of x^*, and, in particular, that f is bounded.

If the form of f is not known explicitly, search algorithms usually sample f at several locations and return an estimate of x^* and $f(x^*)$ based on these observations. There is an obvious trade-off between the number of samples used and the quality of the estimate, and the performance of any search strategy may be measured by the accuracy it achieves in estimating $f(x^*)$ under a constraint on the cost measured by the number of samples used.

Given a relatively good local search algorithm A, a general strategy for finding a good approximation of the optimum x^* is to run several instances of A initialized at different starting points and approximate $f(x^*)$ with the maximum f value observed. We concentrate on local search algorithms A defined formally by a sequence of possibly randomized sampling functions $s_n : [0,1]^{d\cdot n} \to [0,1]^d$,

[1] The results can easily be extended to (arbitrary valued) bounded piecewise continuous functions with finitely many continuous components.

$n = 1, 2, \ldots$: A samples f at locations $X_1, X_2, \ldots$ where $X_{i+1} = s_i(X_1, \ldots, X_i)$ for $i \geq 1$, and the starting point $X_1 = s_0$ is chosen uniformly at random from $[0,1]^d$; after n observations A returns the estimate of x^* and the maximum $f(x^*)$, respectively, by

$$\widehat{X}_n = \operatorname*{argmax}_{1 \leq k \leq n-1} f(X_k) \qquad \text{and} \qquad f(\widehat{X}_n).$$

It is clear that if the starting points are sampled uniformly from $[0,1]^d$ and each algorithm is evaluated at its starting point then this strategy is asymptotically consistent as the number of instances tend to infinity (as in the worst case we perform a random search that is known to converge to the maximum almost surely). On the other hand, if algorithm A has some favorable properties then it is possible to design multi-start strategies that still keep the random search based consistency, but provide much faster convergence to the optimum in terms of the number of evaluations of f.

In particular, we assume the following on the local behavior of the search algorithms:

Assumption 1. *The search algorithm A converges to a local maximum at a guaranteed rate, that is,*

(i) the limit $\lim_{t\to\infty} \widehat{X}_t$ exists and it is a deterministic function of the starting point X_1 for almost all values of X_1 with respect to the Lebesgue measure;
(ii) for any $0 < \delta < 1$, there exists a function $g_\delta(n)$ such that

$$\mathbb{P}\left(\lim_{t\to\infty} f(\widehat{X}_t) - f(\widehat{X}_n) \leq c g_\delta(n) \big| X_1\right) \geq 1 - \delta \tag{1}$$

where $g_\delta(n)$ is a positive, monotone decreasing function of n that converges to 0 for any $0 < \delta < 1$, that is, $\lim_{n\to\infty} g_\delta(n) = 0$, and c is a positive constant that depends on f and g_δ.

In certain cases $g_\delta(n) = O(e^{-\alpha n})$ or $g_\delta(n) = O(n^{-\alpha})$ for some $\alpha > 0$, see, e.g., [12]. The constant c depends on certain characteristics of f, e.g., on the maximum local steepness, or, if f is a random function, on the variance of f (note, however, that in this paper we consider only deterministic functions). The above assumed property of the local search algorithms will be utilized in the next section to derive efficient multi-start search strategies.

3 Multi-start Search Strategies

Standard multi-start search strategies run an instance of A until it seems to converge to a location where there is no hope to beat the current best approximation of $f(x^*)$. An alternative way of using multiple instances of local search algorithms is to run all algorithms parallel, and in each round we decide which algorithms can take an extra step. This approach may be based on estimating the potential performance of a local search algorithm A based on Assumption 1.

Note that if c were known, an obvious way would be to run each instance as long as their possible performances become separated with high probability. Then we could just pick the best instance and run it until the end of our computational budget (this would be a simple adaptation of the idea of choosing the best algorithm based on a trial phase as in [4,5]). However, if c is unknown, we cannot make such separation. On the other hand, using ideas from the general methodology for Lipschitz optimization with unknown constant [15], we can get around this problem and estimate, in a certain optimistic way, the potential performance of each algorithm instance, and in each round we can step the most promising ones.

The main idea of the resulting strategy can be summarized as follows. Assume we have K instances of an algorithm A, denoted by $A_1, \ldots, A_K$ satisfying Assumption 1. Let $X_{i,n}, i = 1, \ldots, K$ denote the location at which f is sampled by A_i at the nth time it can take a sample, where $X_{i,1}$ is the starting point of A_i. The estimate of the location of the maximum by algorithm A_i after n samples is

$$\widehat{X}_{i,n} = \operatorname*{argmax}_{1 \le t \le n} f(X_{i,t})$$

and the maximum value of the function is estimated by $\hat{f}_{i,n} = f(\widehat{X}_{i,n})$. Now if A_i observes $n_{i,r}$ samples by the end of the rth round, by Assumption 1 we have, with probability at least $1 - \delta$,

$$f(\bar{X}_i) - \hat{f}_{i,n_{i,r}} \le c g_\delta(n_{i,r})$$

where $\bar{X}_i = \lim_{n \to \infty} \widehat{X}_{i,n}$ is the point where A_i converges. Thus, an optimistic estimate of the maximum of f for a constant $\hat{c}$, based on A_i, is

$$\hat{f}_{i,n_{i,r}} + \hat{c} g_\delta(n_{i,r}).$$

Then it is reasonable to choose, in each round, those algorithms to take another step that provide the largest estimate for some values of $\hat{c}$ (in this way we can get around the fact that we do not know c).

3.1 Constant Number of Instances

The above idea can be translated to the algorithm MetaMax(K) shown in Figure 1. Here we consider the case when we have a fixed number of instances, and our goal is to perform (almost) as well as the best of them (in hindsight), while using the minimum number of evaluations of f. Note the slight abuse of notation that in the MetaMax(K) algorithm $\widehat{X}_r$ and $\hat{f}_r$ denote the estimates of the algorithm after r rounds (and not r samples).

In each round, the strategy MetaMax(K) selects the local search algorithms A_i for which the point $(g_\delta(n_{i,r-1}), \hat{f}_{i,n_{i,r-1}})$ lies on the upper convex hull of the set

$$\mathcal{P}_r = \{(g_\delta(n_{j,r-1}), \hat{f}_{j,n_{j,r-1}}) : j = 1, \ldots, K\} \cup \{(0, \max_{1 \le j \le K} \hat{f}_{j,n_{j,r-1}})\}. \quad (3)$$

METAMAX(K): A MULTI-START STRATEGY WITH K ALGORITHM INSTANCES.

Parameters: $K > 0$, $0 < \delta < 1$, $g_\delta(n)$.
Initialization: For each $i = 1, \ldots, K$, let $n_{i,0} = 0$, $\hat{f}_{i,0} = 0$.

For each round $r = 1, 2, \ldots$

(a) For $i = 1, \ldots, K$ select algorithm A_i if there exists a $\hat{c} > 0$ such that

$$\hat{f}_{i,n_{i,r-1}} + \hat{c} g_\delta(n_{i,r-1}) > \hat{f}_{j,n_{j,r-1}} + \hat{c} g_\delta(n_{j,r-1}) \qquad (2)$$

for all $j \neq i$ such that $n_i \neq n_j$. If there are several values of i selected that have the same sample number $n_{i,r-1}$ then keep only one of these selected uniformly at random.

(b) Step each selected A_i, and update variables. That is, set $n_{i,r} = n_{i,r-1} + 1$ if A_i is selected, and $n_{i,r} = n_{i,r-1}$ otherwise. For each selected A_i sample $f(X_{i,n_{i,r}})$ and compute the new estimates $\widehat{X}_{i,n_{i,r}}$ and $\hat{f}_{i,n_{i,r}}$.

(c) Let $I_r = \operatorname{argmax}_{i=1,\ldots,K} \hat{f}_{i,n_{i,r}}$ denote the index of the algorithm with the currently largest estimate of $f(x^*)$, and estimate the location of the maximum with $\widehat{X}_r \widehat{X}_{I_r,n_{I_r,r}}$ and its value with $\hat{f}_r = \hat{f}_{I_r,n_{I_r,r}}$.

Fig. 1. The METAMAX(K) algorithm

Note that it is guaranteed that in each round we sample at least two algorithms, one with the largest estimate $\hat{f}_{n_{i,r-1}}$, and one with the smallest sample size $n_{i,r-1}$. Thus, at most half of the total number of samples can be used by any optimal local search algorithm.

The randomization in step (a) that precludes sampling multiple instances with the same sample size is motivated by the following example. Assume that A_1 converges to the correct estimate, while all the other algorithms $A_2, \ldots, A_K$ produce the same estimate in each round, independently of their samples, that is inferior to the estimates of A_1. If we use the randomization, half of the calls to compute f will be made by A_1, but without the randomization this would drop down to $1/K$ as in each round we would sample each algorithm. The randomization is used to exclude such pathological cases.

The following proposition shows that the algorithm is consistent.

Proposition 1. *Assume that $g_\delta(n)$ is positive and monotone decreases to 0 as $n \to \infty$. Then the* METAMAX(K) *algorithm is consistent in the sense that $\hat{f}_r \leq f(x^*)$ for all r, and*

$$f(x^*) - \lim_{r\to\infty} \hat{f}_r = \min_{i=1,\ldots,K} \left\{ f(x^*) - \lim_{n\to\infty} \hat{f}_{i,n} \right\}.$$

Proof. The proof follows trivially from the fact that each algorithm is selected infinitely often, that is, $\lim_{r\to\infty} n_{i,r} = \infty$. To see the latter, we show that in every K rounds the number of samples received by the least used algorithm, that is, $\min_{i=1,\dots,K} n_{i,r}$, is guaranteed to increase by one. That is, for all $k \geq 0$,

$$\min_{i=1,\dots,K} n_{i,kK} \geq k. \tag{4}$$

As described above, in each round we select exactly one of the algorithms with minimal sample size. Thus, if there are k such algorithms, the minimal sample size will increase in k steps, which completes the proof. □

Let $\hat{f}_i^* = \lim_{r\to\infty} \hat{f}_{i,n_{i,r}}$ be the asymptotic estimate of algorithm A_i for $f(x^*)$, and let $\hat{f}^* = \max_{1\leq i\leq K} \hat{f}_i^*$ denote the best estimate achievable using algorithms $A_1, \dots, A_K$. Let $O \subset \{1, \dots, K\}$ be the set of optimal algorithms that converge to the best estimate $\hat{f}^*$ (for these algorithms), and let $|O|$ denote the cardinality of O (i.e., the number of optimal algorithm instances). The next lemma shows that if $i \notin O$, then A_i is not sampled at a round r if it has been sampled too often so far.

Lemma 1. *Suppose Assumption 1 holds, and let*

$$\Delta = \hat{f}^* - \max_{i\notin O} \hat{f}_i^*$$

denote the margin between the estimates of the best and the second best algorithms. Then there is an $R^{(\delta)} > 0$ such that A_i is not sampled by MetaMax(K) *at a round $r + 1 > R^{(\delta)}$ with probability at least $1 - K\delta$ given the starting points $X_{i,1}, i = 1, \dots, K$, if*

$$n_{i,r} > g_\delta^{-1}\left(g_\delta(\min_j n_{j,r})\left(1 - \frac{\hat{f}_i^* + \Delta/2}{\hat{f}^*}\right)\right) \tag{5}$$

simultaneously for all $i \notin O$.

Proof. For each $i = 1, \dots, K$, the conditions of Assumption 1 and (4) imply that there exists an $R_i^{(\delta)} > 0$ such that for all $r > R_i^{(\delta)}$

$$\mathbb{P}\left(\hat{f}_i^* - \hat{f}_{i,n_{i,r}} \leq \Delta/2 \mid X_{i,1}\right) \geq 1 - \delta.$$

Note that the above inequality holds simultaneously for all $r > R_i^{(\delta)}$ since the difference $\hat{f}_i^* - \hat{f}_{i,n_{i,r}}$ is a non-increasing sequence in r. By the union bound, for any $r > R^{(\delta)} = \max\{R_1^{(\delta)}, \dots, R_K^{(\delta)}\}$

$$\mathbb{P}\left(\hat{f}_i^* - \hat{f}_{i,n_{i,r}} \leq \Delta/2 \text{ for all } i = 1, \dots, K | X_{1,1}, \dots, X_{K,1}\right) \geq 1 - K\delta. \tag{6}$$

This implies that after $R^{(\delta)}$ rounds the algorithm will pick, with high probability, the estimate of one of the best algorithms. It is easy to see that, given the starting points $X_{i,1}, i = 1, \dots, K$, if (6) holds then an algorithm A_i with $i \notin O$ cannot be sampled at round $r+1$ if $n_{i,r} \geq n_{I_r,r}$ (that is, if it has been sampled more than an algorithm with a better maximum estimate). Furthermore, if $n_{i,r} < n_{I_r,r}$, it is easy to see that A_i is not sampled at a round $r+1$ if

$$\frac{\hat{f}_r - \hat{f}_{i,r}}{g_\delta(n_{i,r}) - g_\delta(n_{I_r,r})} > \frac{\hat{f}_r - 0}{g_\delta(\min_j n_{j,r}) - g_\delta(n_{I_r,r})}$$

since the line connecting $(g_\delta(\min_j n_{j,r}), 0)$ and $(g_\delta(n_{I_r,r}), \hat{f}_{I_r, n_{I_r,r}})$ is a lower bound to the upper convex hull of $\mathcal{P}_{r+1}$ (defined by (3)). Since $\hat{f}_r - \hat{f}_{i,r} \geq \hat{f}^* - \hat{f}_i^* - \Delta/2$ and $\hat{f}^* \geq \hat{f}_r$, given the starting points $X_{i,1}$, A_i is not selected at round $r+1$ simultaneously for all i, with probability at least $1 - K\delta$, if

$$\frac{\hat{f}^* - \hat{f}_i^* - \Delta/2}{g_\delta(n_{i,r}) - g_\delta(n_{I_r,r})} > \frac{\hat{f}^*}{g_\delta(\min_j n_{j,r}) - g_\delta(n_{I_r,r})}. \tag{7}$$

The latter is equivalent to the upper bound

$$g_\delta(n_{i,r}) < g_\delta(\min_j n_{j,r}) \left(1 - \frac{\hat{f}_i^* + \Delta/2}{\hat{f}^*}\right) + g_\delta(n_{I_r,r}) \frac{\hat{f}_i^* + \Delta/2}{\hat{f}^*}.$$

Taking into account that g_δ is non-increasing and positive, this is surely satisfied if (5) holds. □

A simple corollary of the above lemma is that if the local search algorithm converges fast enough (exponentially with a problem dependent rate, or faster than exponential) then half of the samples taken from f correspond to optimal algorithm instances.

Corollary 1. *Assume that the performance of the algorithms $A_i, i = 1, \ldots, K$ are not all the same, that is, $|O| < K$, and that the conditions of Assumption 1 hold. Furthermore, suppose that*

$$\limsup_{n \to \infty} \frac{g_\delta(n+1)}{g_\delta(n)} < \min_{i \notin O} \left\{1 - \frac{\hat{f}_i^* + \Delta/2}{\hat{f}^*}\right\}. \tag{8}$$

Then asymptotically at least half of the sampling of f in MetaMax(K) *corresponds to an optimal algorithm. That is,*

$$\mathbb{P}\left(\liminf_{r \to \infty} \frac{\sum_{i \in O} n_{i,r}}{\sum_{i=1}^K n_{i,r}} \geq \frac{1}{2} \,\middle|\, X_{1,1}, \ldots, X_{K,1}\right) \geq 1 - K\delta.$$

Proof. We show that A_i is not chosen for large enough r if $n_{i,r} > \min_j n_{j,r}$. By Lemma 1, it is sufficient to prove that, for large enough r,

$$\min_j n_{j,r} + 1 > g_\delta^{-1}\left(g_\delta(\min_j n_{j,r})\left(1 - \frac{\hat{f}_i^* + \Delta/2}{\hat{f}^*}\right)\right).$$

The latter is clearly satisfied by (8) as $\lim_{r \to \infty} \min_j n_{j,r} = \infty$ by (4). This fact finishes the proof. □

METAMAX: A MULTI-START STRATEGY WITH INFINITELY MANY ALGORITHM INSTANCES.

Parameters: $0 < \delta < 1$, $g_\delta(n)$.
Initialization: Set $K = 0$. For each round $r = 1, 2, \ldots$

(a) Increase the value of K by 1. Initialize algorithm A_K by setting $n_{K,r-1} = 0$, $\hat{f}_{K,0} = 0$.
(b) For $i = 1, \ldots, K$ select algorithm A_i if there exists a $\hat{c} > 0$ such that

$$f_{i,n_{i,r-1}} + \hat{c} g_\delta(n_{i,r-1}) > f_{j,n_{j,r-1}} + \hat{c} g_\delta(n_{j,r-1}) \tag{9}$$

for all $j \neq i$ such that $n_{i,r-1} \neq n_{j,r-1}$. If there are several values of i selected that have the same sample number $n_{i,r-1}$ then keep only one of these selected uniformly at random.
(c) Step each selected A_i, and update variables. That is, set $n_{i,r} = n_{i,r-1} + 1$ if A_i is selected, and $n_{i,r} = n_{i,r-1}$ otherwise. For each selected A_i sample $f(X_{i,n_{i,r}})$ and compute the new estimates $\widehat{X}_{i,n_{i,r}}$ and $\hat{f}_{i,n_{i,r}}$.
(d) Let $I_r = \operatorname{argmax}_{i=1,\ldots,K} \hat{f}_{i,n_{i,r}}$ denote the index of the algorithm with the currently largest estimate of $f(x^*)$, and estimate the location of the maximum with $\widehat{X}_r = \widehat{X}_{I_r,n_{I_r,r}}$ and its value with $\hat{f}_r = \hat{f}_{I_r,n_{I_r,r}}$.

Fig. 2. The METAMAX algorithm

The above corollary immediately yields the following result on the rate of convergence of the METAMAX(K) algorithm.

Corollary 2. *Assume the conditions of Corollary 1 hold. Then for any $\delta > 0$ and $\epsilon > 0$ there exists a threshold $T_{\delta,\epsilon} > 0$ depending on the starting points $X_{1,1}, \ldots, X_{1,K}$ such that if the METAMAX(K) algorithm is run for r rounds such that the total number of samples $T = \sum_{i=1}^{K} n_{i,r}$ satisfies $T > T_{\delta,\epsilon}$, then, with probability at least $(1 - (K + |O|)\delta)$,*

$$\hat{f}^* - \hat{f}_r \leq c g_\delta \left(\frac{T}{(2+\epsilon)|O|} \right).$$

3.2 Unbounded Number of Instances

It is clear that if the local search algorithms are not consistent, then, despite its favorable properties, the METAMAX(K) strategy is inconsistent, too. However, if we increase the number of algorithms in each round, then we get the consistency from random search, while still keeping the reasonably fast convergence rate from METAMAX(K). This results in the multi-start strategy METAMAX, shown in Figure 2.

Since the MetaMax strategy includes a random search (the number of algorithms tends to infinity as the length of each round is finite), the algorithm is consistent:

Proposition 2. *The strategy* MetaMax *is consistent. That is,*

$$\lim_{r\to\infty} \hat{f}_r = f(x^*).$$

Under Assumption 1 we can prove, similarly to Lemma 1, that any suboptimal algorithm is selected only finitely many times.

Lemma 2. *Suppose Assumptions 1 holds, and assume that an instance A_i is suboptimal, that is, $\hat{f}_i^* = \lim_{r\to\infty} \hat{f}_{i,n_{i,r}} < f(x^*)$. Let*

$$\Delta_i = f(x^*) - \hat{f}_i^*.$$

Then, for sufficiently large r, A_i is not sampled at a round $r+1$, with probability at least $1-2\delta$, if

$$n_{i,r} > \max\left\{ g_\delta^{-1}\left(g_\delta(0)\left(1 - \frac{\hat{f}_i^* + \Delta_i/2}{f(x^*)}\right)\right),\ g_\delta^{-1}\left(\frac{\Delta_i}{2c}\right)\right\}.$$

Proof. By the assumptions of the lemma there exists a positive integer $R > 0$ such that $f(x^*) - \hat{f}_R < \Delta_i/2$ with probability at least $1-\delta$. (A crude estimate of R can be obtained by considering the probabilities that $X_{r,1}$ falls in a sufficiently small neighborhood of x^*. Better values of R can be calculated by considering the probability of choosing a starting point such that the corresponding algorithm converges to a global optimum, and then considering the rate of convergence of this local search algorithm.) By Assumption 1, $\hat{f}_i^* - \hat{f}_{i,n} < \Delta_i/2$ with probability at least $1-\delta$ for all $n > n_i = g_\delta^{-1}\left(\frac{\Delta_i}{2c}\right)$. Thus, by the union bound, for any round $r+1 > R$ such that $n_{i,r} > g_\delta^{-1}\left(\frac{\Delta_i}{2c}\right)$, we have, with probability at least $1-2\delta$, simultaneously

$$f(x^*) - \hat{f}_r < \Delta_i/2 \qquad \text{and} \qquad \hat{f}_i^* - \hat{f}_{i,n_{i,r}} < \Delta_i/2. \tag{10}$$

Now a similar argument as in Lemma 1 finishes the proof. Note that compared to (5) we replaced $\min_i n_{i,r}$ with 0 since a new algorithm is started in every round with $n_{r+1,r} = 0$. □

Experimental results in Section 4 show that K, the number of algorithm instances, increases sublinearly: in all simulations K was of the order $T/\ln(T)$ with $K < 2T/\ln T$. It is interesting to note that this is roughly the same as the number of algorithm instances introduced by the algorithm Luby [17] (see Section 4 for more details). If Assumption 1 is satisfied with $\delta = 0$ and there are only finitely many local maxima of f, this fact gives rise to the following heuristic derivation for the convergence rate of MetaMax. Let p be the probability that a random starting point is chosen such that the resulting algorithm converges to a global optimum. Then, for T large enough, there are approximately

$(1-p)T/\ln T$ suboptimal algorithms, and, after some initial phase with S samples, each of them can be sampled at most N times for some integer N. Then the $pT/\ln T$ optimal algorithms are sampled at least $T-N(1-p)T/\ln T-S$ times, so one of them is sampled at least $(T-N(1-p)T/\ln T-S)/(pT/\ln T)=O(\ln T)$ times, resulting in a convergence rate of $cg_0(\kappa \ln T)$ for some $\kappa>0$.

4 Experiments

In the experiments we compared the two versions of the MetaMax algorithm with four algorithms including two with fixed (K), and two with variable number of algorithm instances (arms). We used the SPSA ([18], Simultaneous Perturbation Stochastic Approximation) as the base local search algorithm in all cases. SPSA is a local search algorithm with a sampling function that uses gradient descent with a stochastic approximation of the derivative: at the actual location $X_t=(X_{t,1},\ldots,X_{t,d})$, SPSA estimates the derivative of f by

$$\tilde{f}_{t,l}(X_{t,l})=\frac{f(X_t+\phi_t B_t)-f(X_t-\phi_t B_t)}{2\phi_t B_{t,l}},$$

where the $B_{t,l}$ are i.i.d. Bernoulli random variables that are the components of the vector B_t, and then uses the sampling function $s_t(X_t)=X_t+a_t\tilde{f}_t(X_t)$ to choose the next point to be sampled, that is,

$$X_{t+1,l}=X_{t,l}+a_t\tilde{f}_{t,l}(X_{t,l})$$

for $l=1,\ldots,d$.

In the implementation of the algorithm we have followed the guidelines provided in [19], with the gain sequence $a_t=a/(A+t+1)^\alpha$, and perturbation size $\phi_t=\phi/(t+1)^\gamma$, where $A=60$, $\alpha=0.602$ and $\gamma=0.101$. The values of a and ϕ vary in the different experiments. Next to the two evaluations required at the perturbed points, we also evaluate the function at the current point X_t. The starting point is chosen randomly, and the function is evaluated first at this point.

The four reference algorithms the MetaMax(K) and MetaMax algorithms are compared to are the following:

Unif: This algorithm selects from a constant number of instances of SPSA uniformly. In our implementation the instance $I_t=t \bmod K$ is selected at time t, where K denotes the number of instances.

ThrAsc: The ThresholdAscent algorithm of [21]. The algorithm begins with selecting each of a fixed number of instances once. After this phase at each time step t ThrAsc selects the best s estimates produced so far by all algorithm instances $A_i, i=1,\ldots,K$ in all the previous time steps, and for each A_i it counts how many of these estimates were produced by A_i. Denoting the latter value by $S_{i,t}$, at time t the algorithm selects the instance with index

$I_t = \text{argmax}_i\, U(S_{i,t}/n_{i,t}, n_{i,t})$, where $n_{i,t}$ is the number of times the ith instance has been selected up to time t,

$$U(\mu, n) = \mu + \frac{\alpha + \sqrt{2n\mu\alpha + \alpha^2}}{n}$$

and $\alpha = \ln(2TK/\delta)$. s and δ are the parameters of the algorithm, and in the experiments the best value for s appeared to be 100, while δ was set to 0.01.

RAND: The random search algorithm. It can be seen as running a sequence of SPSA algorithms such that each instance is used for exactly one step, which is the evaluation of the random starting point of the SPSA algorithm.

LUBY: The algorithm based on [17]. This method runs several instances of SPSA sequentially after each other, where the ith instance is run for t_i steps, with t_i defined by

$$t_i = \begin{cases} 2^{k-1}, & \text{if } i = 2^k - 1 \\ t_{i-2^{k-1}+1}, & \text{if } 2^{k-1} \le i < 2^k - 1 \end{cases}$$

The above definition produces a scheduling such that from the first $2^k - 1$ algorithm instances one is run for 2^{k-1} steps, two for 2^{k-2} steps, four for 2^{k-3} steps, and so on.

Both versions of the METAMAX algorithm were tested. Motivated by the fact that SPSA is known to converge to a global optimum exponentially fast if f satisfies some restrictive conditions [12], we chose a $g_\delta(n)$ that decays exponentially fast (recall that this is the rate of convergence to a local minimum). Furthermore, to control the exploration of the so far suboptimal algorithm instances heuristically, we allowed $g_\delta(n)$ to be a time-varying function, that is, it changes with t, the total number of samples seen so far. Thus, at round r we used $g_\delta(n) = e^{-n/\sqrt{t_r}}$, where $t_r = \sum_i n_{i,r-1}$.

For the algorithms with a fixed number of arms (METAMAX(K), UNIF, and THRASC), the number of arms K was set to 100 in the simulations, which turned out to be a good choice overall.

The multi-start algorithms were tested using two versions of a synthetic function, and by tuning the parameters of a learning algorithm on two standard data sets.

The synthetic function was a slightly modified[2] version of the Griewank function [13]:

$$f(x) = \sum_{l=1}^{d} \frac{4\pi^2 x_l^2}{100} - \prod_{l=1}^{d} \cos \frac{2\pi x_l}{\sqrt{l}} + 1$$

where $x = (x_1, \ldots, x_l)$ and the x_i were constrained to the interval $[-1, 1]$. We show the results for the two-dimensional and the ten-dimensional cases.

The parameters of SPSA were $a = 0.05$ and $\phi = 0.1$ for the two-dimensional case, and $a = 0.5$ and $\phi = 0.1$ for the ten-dimensional case. The performance

[2] The modification was made in order to have more significant differences between the values of the function at the global maximum and at other local maxima.

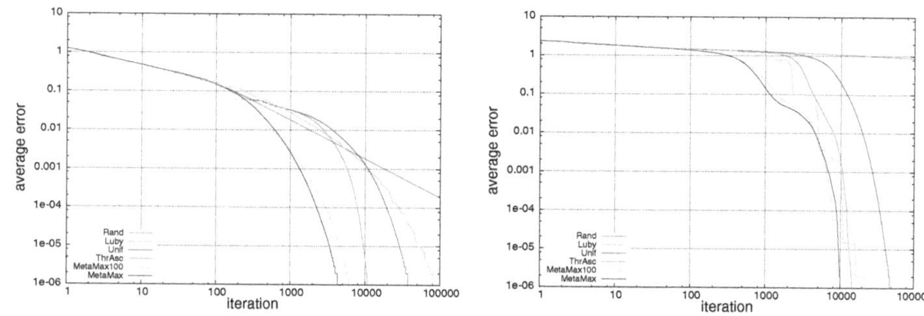

Fig. 3. The average error for the multi-start strategies on the two-dimensional (left) and ten-dimensional (right) modified Griewank function

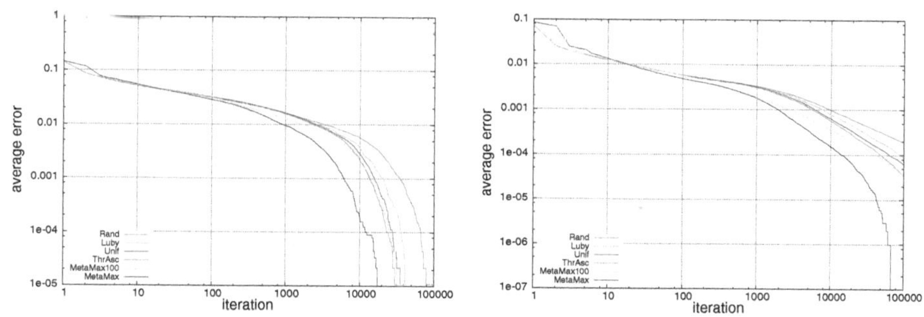

Fig. 4. The average error for the multi-start strategies on tuning the parameters of MultilayerPerceptron for the *vehicle* data set (left) and the *letter* data set (right)

of the search algorithms were measured by the error defined as the difference between the maximum value of the function (in this case 1) and the best result obtained by the search algorithm in a given number of steps. The results for the above multi-start strategies for the two- and the ten-dimensional test functions are shown in Figure 3. Each error curve is averaged over 10,000 runs, and each strategy was run for 100,000 steps (or iterations). One may observe that in both cases the two versions of the METAMAX algorithm converge the fastest. THRASC is better then UNIF, while LUBY seems fairly competitive with these two. The random search seems an option only for the low-dimensional function. Similar results were obtained for dimensions between 2 and 10.

For tuning the parameters of a learning algorithm, we have used two standard data sets from the UCI Machine Learning Repository [2]: *vehicle* and *letter*, and the MultilayerPerceptron learning algorithm of Weka [23]. Two parameters were tuned: the learning rate and the momentum, both in the range of $[0, 1]$. The size of the hidden layer for the MultilayerPerceptron was set to 8, while the number of epochs to 100. The parameters of the SPSA algorithm were $a = 0.5$ and $\phi = 0.1$. The rate of correctly classified items on the test set for *vehicle* using MultilayerPerceptron with varying values of the two parameters is shown

Fig. 5. Classification rate on the *vehicle* data set. The rates are plotted by subtracting them from 0.911112 and thus global optima are at the scattered black spots corresponding to a value equal to 0.001.

Fig. 6. Classification rate on the *letter* data set. The rates are plotted by subtracting them from .7515 and thus global optima are at the scattered black spots corresponding to a value equal to 0.001.

in Figure 5, with the highest rate being 0.910112. Similarly, the classification rate for *letter* is shown in Figure 6, with the highest rate being 0.7505.

The error rates of the optimized MultilayerPerceptron on the data sets *vehicle* and *letter* are shown in Figure 4, when the parameters of the learning algorithm were tuned by the multi-start strategies above. The error in these cases is the difference between the best classification rate that can be obtained (0.910112 and

0.7505, respectively) and the best classification rate obtained by the multi-start strategies in a given number of steps. The results shown are averages over 1,000 runs. We observe that the MetaMax algorithm (with an increasing number of arms) converged the fastest in average, the three strategies with a fixed number of arms had nearly identical results, Luby was slightly worse than these, and the random search was the slowest, although it did not perform as bad as for the synthetic functions.

In summary, the MetaMax algorithm (with an increasing number of arms) provided far the best performance in all tests, usually requiring an order of magnitude less steps to find the optimum than the other algorithms. E.g., for the *letter* data set only the MetaMax algorithm found the global optimum in all runs in 100,000 time steps. We can conclude that MetaMax converged faster than the other multi-start strategies investigated in all four test cases, with a notable advantage on the difficult surfaces induced by the classification tasks.

5 Conclusions

We provided a multi-start strategy for local search algorithms. The method assumes a convergence rate of the local search algorithm up to an unknown constant, and in every phase it gives preference to those local search algorithm instances that could converge to the best value for a particular range of the constant. Two versions of the algorithm were presented, one that is able to follow the performance of the best of a fixed number of local search algorithm instances, and one that, with gradually increasing the number of the local search algorithms, achieves global consistency. Some theoretical properties of the algorithms were explored, although the methods used do not seem to fully explain the superior behavior of the strategy in experiments, requiring further study of the problem.

References

1. Adam, K.: Learning while searching for the best alternative. Journal of Economic Theory 101, 252–280 (2001)
2. Asuncion, A., Newman, D.J.: UCI machine learning repository (2007)
3. Auer, P., Cesa-Bianchi, N., Fischer, P.: Finite time analysis of the multiarmed bandit problem. Machine Learning 47(2-3), 235–256 (2002)
4. Beck, C.J., Freuder, E.C.: Simple rules for low-knowledge algorithm selection. In: Régin, J.-C., Rueher, M. (eds.) CPAIOR 2004. LNCS, vol. 3011, pp. 50–64. Springer, Heidelberg (2004)
5. Carchrae, T., Beck, J.C.: Low-knowledge algorithm control. In: Proceedings of the 19th National Conference on Artificial Intelligence (AAAI), pp. 49–54 (2004)
6. Cicirello, V.A.: The max k-armed bandit: A new model of exploration applied to search heuristic selection. In: Proceedings of the Twentieth National Conference on Artificial Intelligence, pp. 1355–1361 (2005)
7. Cicirello, V.A., Smith, S.F.: Heuristic selection for stochastic search optimization: Modeling solution quality by extreme value theory. In: Wallace, M. (ed.) CP 2004. LNCS, vol. 3258, pp. 197–211. Springer, Heidelberg (2004)

8. Finkel, D.E., Kelley, C.T.: Convergence analysis of the direct algorithm. Technical Report CRSC-TR04-28, NCSU Mathematics Department (2004)
9. Gagliolo, M., Schmidhuber, J.: Learning dynamic algorithm portfolios. Annals of Mathematics and Artificial Intelligence 47(3–4), 295–328 (2006); AI&MATH 2006 Special Issue
10. Gagliolo, M., Schmidhuber, J.: Learning restart strategies. In: Veloso, M.M. (ed.) IJCAI 2007 — Twentieth International Joint Conference on Artificial Intelligence, vol. 1, pp. 792–797. AAAI Press, Menlo Park (2007)
11. Gagliolo, M., Schmidhuber, J.: Algorithm selection as a bandit problem with unbounded losses. Technical Report IDSIA - 07 - 08, IDSIA (July 2008)
12. Gerencsér, L., Vágó, Z.: The mathematics of noise-free SPSA. In: Proceedings of the IEEE Conference on Decision and Control, pp. 4400–4405 (2001)
13. Griewank, A.O.: Generalized descent for global optimization. Journal of Optimization Theory and Applications 34, 11–39 (1981)
14. Horn, M.: Optimal algorithms for global optimization in case of unknown Lipschitz constant. Journal of Complexity 22(1), 50–70 (2006)
15. Jones, D.R., Perttunen, C.D., Stuckman, B.E.: Lipschitzian optimization without the lipschitz constant. Journal of Optimization Theory and Applications 79(1), 157–181 (1993)
16. Kautz, H., Horvitz, E., Ruan, Y., Gomes, C., Selman, B.: Dynamic restart policies. In: Proceedings of the Eighteenth National Conference on Artificial Intelligence (AAAI), pp. 674–681 (2002)
17. Luby, M., Sinclair, A., Zuckerman, D.: Optimal speedup of Las Vegas algorithms. Information Processing Letters 47, 173–180 (1993)
18. Spall, J.C.: Multivariate stochastic approximation using a simultaneous perturbation gradient approximation. IEEE Transactions on Automatic Control 37, 332–341 (1992)
19. Spall, J.C.: Implementation of the simultaneous perturbation algorithm for stochastic optimization. IEEE Transactions on Aerospace Electronic Systems 34, 817–823 (1998)
20. Streeter, M.J., Smith, S.F.: An asymptotically optimal algorithm for the max k-armed bandit problem. In: Proceedings, The 21st National Conference on Artificial Intelligence and the 18th Innovative Applications of Artificial Intelligence Conference, pp. 135–142 (2006)
21. Streeter, M.J., Smith, S.F.: A simple distribution-free approach to the max k-armed bandit problem. In: Benhamou, F. (ed.) CP 2006. LNCS, vol. 4204, pp. 560–574. Springer, Heidelberg (2006)
22. Vilalta, R., Drissi, Y.: A perspective view and survey of meta-learning. Artificial Intelligence Review 18(2), 77–95 (2002)
23. Witten, I.H., Frank, E.: Data Mining: Practical Machine Learning Tools and Techniques, 2nd edn. Morgan Kaufmann, San Francisco (2005)

Considering Unseen States as Impossible in Factored Reinforcement Learning

Olga Kozlova[1,2], Olivier Sigaud[1], Pierre-Henri Wuillemin[3], and Christophe Meyer[4]

[1] Institut des Systèmes Intelligents et de Robotique
Université Pierre et Marie Curie - Paris 6, CNRS UMR 7222
4 place Jussieu, F-75005 Paris, France
Olivier.Sigaud@upmc.fr
[2] Thales Security Solutions & Services, Simulation
1 rue du Général de Gaulle, Osny BP 226
F95523 Cergy Pontoise, France
Olga.Kozlova@thalesgroup.com
[3] Laboratoire d'Informatique de Paris 6
Université Pierre et Marie Curie - Paris 6, CNRS UMR 7606
4 place Jussieu, F-75005 Paris, France
Pierre-Henri.Wuillemin@lip6.fr
[4] Thales Security Solutions & Services, ThereSIS Research and Innovation Office
Campus Polytechnique
1, avenue Augustin Fresnel
91767 Palaiseau, France
Christophe.Meyer@thalesgroup.com

Abstract. The Factored Markov Decision Process (FMDP) framework is a standard representation for sequential decision problems under uncertainty where the state is represented as a collection of random variables. Factored Reinforcement Learning (FRL) is an Model-based Reinforcement Learning approach to FMDPs where the transition and reward functions of the problem are learned. In this paper, we show how to model in a theoretically well-founded way the problems where some combinations of state variable values may not occur, giving rise to impossible states. Furthermore, we propose a new heuristics that considers as impossible the states that have not been seen so far. We derive an algorithm whose improvement in performance with respect to the standard approach is illustrated through benchmark experiments.

1 Introduction

Based on the Markov Decision Process (MDP) framework (Sutton & Barto, 1998), the Factored Markov Decision Process (FMDP) framework (Boutilier et al., 1995) is the standard representation of sequential decision problems under uncertainty when the state of the problem can be decomposed as a set of random variables. In this framework, the state space of a sequential decision problem is represented as a collection of random variables $X = \{X_1, \ldots, X_n\}$. A state is then defined by a vector $x = (x_1, \ldots, x_n)$ with $\forall i, x_i \in \text{Dom}(X_i)$. FMDPs exploit the structure of the dependencies between variables to represent large MDPs compactly. For each action a, the transition model is defined by a separate Dynamic Bayesian Network (DBN) model (Dean &

W. Buntine et al. (Eds.): ECML PKDD 2009, Part I, LNAI 5781, pp. 721–735, 2009.

Kanazawa, 1989). The model G_a is a two-layer directed acyclic graph whose nodes are $\{X_1, \ldots, X_n, X'_1, \ldots, X'_n\}$ with X_i a variable at time t and X'_i the same variable at time $t+1$. The parents of X'_i are noted $\text{Parents}_a(X'_i)$. The transition model is quantified by *Conditional Probability Distributions* (CPDs), noted $P^a(X'_i|\text{Parents}_a(X'_i))$, associated to each node $X'_i \in G_a$.

The authors of (Boutilier et al., 2000) propose two structured Dynamic Programming (SDP) algorithms, namely Structured Value Iteration (SVI) and Structured Policy Iteration (SPI). These algorithms and later extensions like SPUDD (Hoey et al., 1999) deal with the case where there are no synchronic arcs in the DBNs, i.e. $\forall i, \text{Parents}_a(X'_i) \subseteq \{X_1, \ldots, X_n\}$ and $\text{Parents}_a(X_i) = \emptyset$. Thus, in such a model, the X'_i are independent of each other conditionally to $\{X_1, \ldots, X_n\}$. The independence assumption results in the opportunity to compute the joint probability of a collection of variables as a product: $P(X'|X) = \Pi_i P(X'_i|\text{Parents}_a(X'_i))$.

In this paper, we focus on the case where the structure and parameters of the FMDP are learnt from experience, and we address a wider class of problems where particular combinations of values of variables may not occur. We argue that this situation often happens in practice and we show through benchmark experiments that modifying standard algorithms to deal with such "impossible states" is more efficient than just ignoring this phenomenon. Finally, we show that considering as impossible a state that has never been seen so far is an efficient heuristic to learn quickly in FMDPs.

The paper is organized as follows. In the next section, we give a brief overview of SDP algorithms, insisting on the steps where the independence assumption is used and we give a brief overview of an algorithm that learns the structure of the FMDP while solving it. Then we show in section 3 how the presence of impossible states can be modeled and how the algorithms must be modified as a consequence. In section 4, we examine through benchmark experiments the benefits that can result from our method when impossible states are present. In section 5, we discuss these benefits depending on the rate of impossible states, before concluding on the possibility to extend this work in different directions.

2 Factored Reinforcement Learning

In this section we briefly present structured dynamic programming algorithms and then SPITI, a model-based reinforcement learning approach dedicated to FMDPs.

2.1 Structured Dynamic Programming

Standard SDP algorithms such as SVI and SPI use decision trees as factored representation. In the rest of the paper, a function F represented as a decision tree is noted $Tree(F)$. SVI and SPI can be seen as an efficient way to perform the Bellman-backup operation on trees, expressed as follows:

$$Tree(Q_a^V)(x) = Tree(R(x,a)) + \gamma Tree(\sum_{x'} P(x'|x,a)V(x')). \quad (1)$$

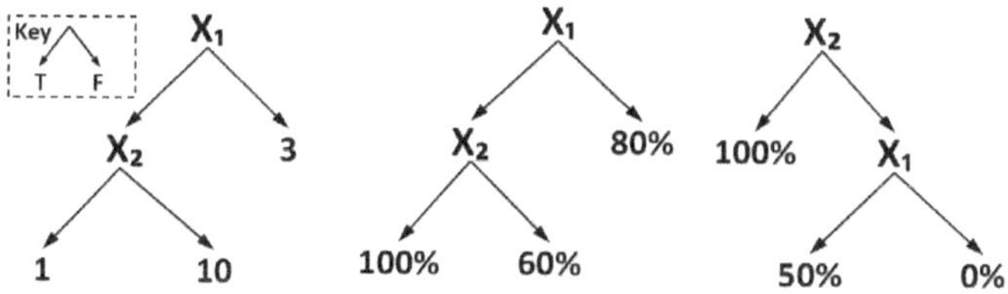

(a) *Reward* (b) $P(x_1'|X_1, X_2, a_0)$ (c) $P(x_2'|X_1, X_2, a_0)$

Fig. 1. A toy example: the tree representation of a reward function and of the transition functions of binary variables X_1 and X_2 given an unique action a_0. Notations are similar to the ones in (Boutilier et al., 2000), but in the boolean case we note x_i for $X_i = true$ and $\bar{x}_i$ for $X_i = false$.

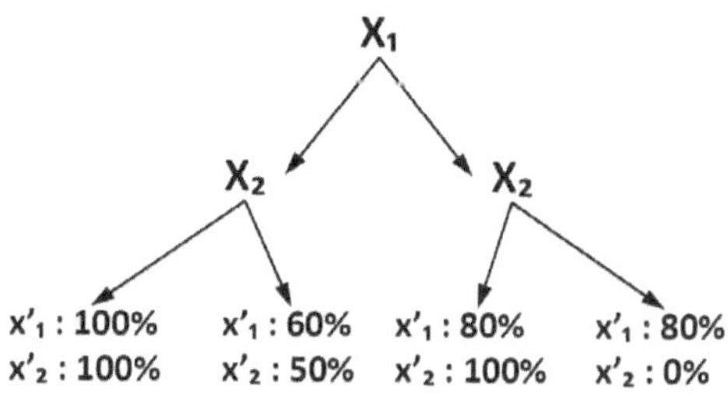

Fig. 2. $P(X'|X, a_0)$ computed by the $PRegress$ operator from the example of Fig. 1

In SVI and SPI, the $Regress(Tree(V), a)$ algorithm performs Bellman-backup operations. Inside $Regress$, $PRegress$ computes $Tree(\sum_{x'} P(x'|x, a)V(x'))$ using the structure of $Tree(P)$ and $Tree(V)$.

Consider the toy example whose reward and transition functions for an unique action a_0 are given in Fig. 1.

From the transition trees, $PRegress$ first computes the $Tree(P(X'|X, a_0))$ shown in Fig. 2. This tree represents the individual probabilities of each variable value at $t+1$ given the variable values at t. For instance, the rightmost branch of the tree reads as follows: if X_1 and X_2 were false, the probability that X_1' and X_2' are true are 80% and 0% respectively. Note that the probability of one of the values can be omitted in the representation and inferred from the other probabilities. Furthermore, $PRegress$ only computes the combinations of values that are necessary to perform regression from the current value function (see (Boutilier et al., 2000) for details).

Given the tree represented in Fig. 2 and considering that the variables at $t + 1$ are independent conditionally to those at t, $PRegress$ computes the joint probabilities as a product, as shown in Table 1. Note that the table representation is not computed explicitly in the algorithm: in the more general case with any number of variables and enumerated values, this calculation is implemented by expanding a tree of all variable values combinations and computing probabilities at the leaves as a product on individual probabilities.

The last step of $PRegress$ computes $\sum_{x'} P(x'|x, a)V(x')$ using the structure shown in Fig. 3. Then $Regress$ computes $Tree(Q_a^V)$ according to equation (1) by performing the product with γ and the sum with $Tree(R(X, a))$.

Table 1. Probabilities for joint variables, resulting from Fig. 2

$P(X'\|X, a_0)$	$x_1'x_2'$	$x_1'\bar{x_2}'$	$\bar{x_1}'x_2'$	$\bar{x_1}'\bar{x_2}'$
x_1x_2	100%	0%	0%	0%
$x_1\bar{x_2}$	30%	30%	20%	20%
$\bar{x_1}x_2$	80%	0%	20%	0%
$\bar{x_1}\bar{x_2}$	0%	80%	0%	20%

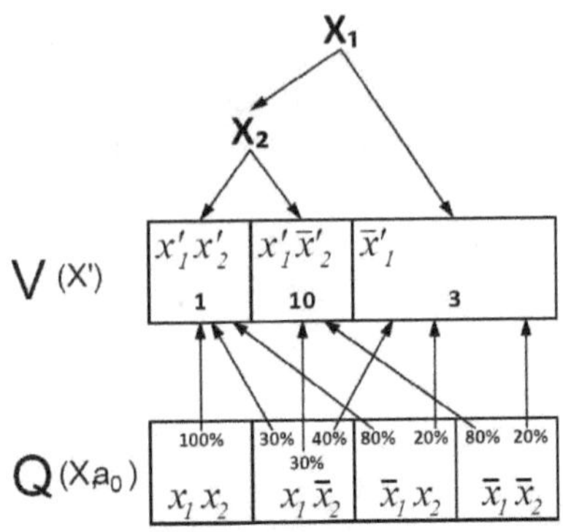

Fig. 3. Regression with structured representations using the values computed in Table 1. We consider the first iteration where $Tree(V) = Tree(R)$. Notice that transitions to $\bar{x_1}x_2$ and $\bar{x_1}\bar{x_2}$ are summed to $\bar{x_1}$.

On top of *Regress*, SVI and SPI behave differently. In SVI, the value function $Tree(V)$ is computed by merging the set of action-value functions $Tree(Q_a^V)$ using maximization as combination function. It is shown in (Boutilier et al., 2000) that, given a perfect knowledge of the transition and reward functions and starting with $Tree(V_0) = Tree(R)$, $Tree(V)$ converges in a finite number of time steps to the optimal value function $Tree(V^*)$. Then one can extract $Tree(\pi^*)$ from $Tree(V^*)$ using a simple tree-based greedy operator.

In SPI, the process is slightly more complex. Policies $Tree(\pi)$ and value functions $Tree(V^\pi)$ are computed iteratively until convergence, making profit of the structure of $Tree(\pi)$ to optimize the computation of $Tree(V^\pi)$. We do not detail the algorithm since ours is based on SVI, but transferring the approach described in section 3 to SPI is straightforward.

2.2 SPITI

Reinforcement Learning in FMDPs is generally about the case where the structure of the DBNs are given, but the parameters of the CPDs are learnt from experience. By contrast, we call *Factored Reinforcement Learning* (FRL) the case where the structure of the DBNs itself is learnt.

An implementation of FRL is expressed in the SDYNA framework (Degris et al., 2006a; Degris et al., 2006b) as a structured version of the DYNA architecture (Sutton,

1991). In SDYNA, the model of transitions and of the reward are learned from experience under a compact form. The inner loop of SDYNA is decomposed into three phases:

- *Acting*: choosing an action according to the current policy, including some exploration;
- *Learning*: updating the model of the transition and reward functions of the FMDP from $\langle X, a, X', R\rangle$ observations;
- *Planning*: updating the value function $Tree(V)$ and policy $Tree(\pi)$ using one sweep of SDP algorithms.

SPITI is a particular instance of SDYNA using ϵ-$greedy$ as exploration method, the *Incremental Tree Induction* (ITI) algorithm (Utgoff, 1989) to learn the model of transitions and reward functions as a collection of decision trees, and the inner loop of SVI as planning method. An algorithmic description is given in (Degris et al., 2006b).

3 Dealing with Impossible States

The techniques presented so far address the case where there are no synchronic arcs in the DBNs that represent the structure of FMDPs. On the other extreme, the authors of (Boutilier et al., 2000) propose to model the case where the variables are not independent by adding synchronic arcs between variables at $t + 1$ and recording the joint probabilities of all groups of variables that are connected by such synchronic arcs. This results in more complex, slower and more memory-intensive algorithms, but that can deal with a much wider class of problems. A more detailed study of that case is presented in (Boutilier, 1997).

In this paper, we address a class of problems that is intermediate between the "no synchronic arcs" and the "any synchronic arcs" classes. It corresponds to problems where the variables at $t+1$ behave as if they were independent, but some combinations of values for some variables do not occur in practice, which contradicts the independence assumption. We show below how such a situation can be modeled without using the general class of problems with synchronic arcs in the DBNs.

3.1 Modeling Impossible States

The class of problems we want to address can be modeled with the kind of DBNs shown in Fig. 4, given that there is one such DBN for each action. In this representation, there are no synchronic arcs between variables at $t + 1$, but there are some constraints K on whether some combinations of variable values are possible or not. K (resp. K') stands for the knowledge of impossible states x (resp. x'). The values of K and K' are either $true$ or $false$ depending on the possibility of the corresponding states.

As we illustrate in the experimental section, without using such constraints, the $Tree(V)$ and $Tree(Q_a^V)$ structures may represent many states that do not occur in practice. Dealing with the constraints explicitly is a way to avoid the computational and memory overhead resulting from this useless information by filtering out all impossible states in the data structures.

We show below that this filtering can be performed safely just by discarding the impossible states and normalizing again the probabilities when it is necessary.

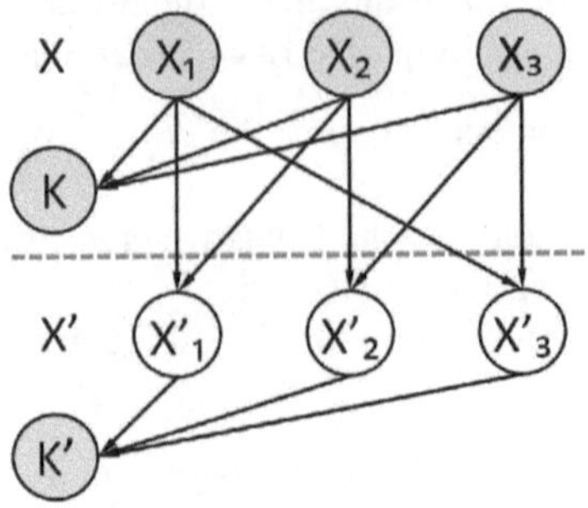

Fig. 4. DBN representation for problems with independent variables and impossible states. K and K' stand for constraints stating if a particular combination of values is possible. Variables in gray are observed. K' and relevant X_j are used to predict X_i' for all i. K' does not necessarily depend on all X_i'.

3.2 Impact on Regression

Let us show that the values computed by $PRegress$ taking the constraints into account, i.e. $\Sigma_{x'} P(X'|x, a, k').V(x')$, are equal to the values one would obtain without taking these constraints into account, just leaving the impossible states away and normalizing again the probabilities.

First, with the representation above, the probability distribution of X' for a particular action a can be computed given the values of the variables x_i and the value k' of the common constraint K'. Indeed, the distribution of X' can be expressed as $P(X'|x, k') \propto P(k'|X', x)P(X'|x)$. Furthermore we have $P(K'|X', X) = P(K'|X')$, since $K' \perp\!\!\!\perp X | X'$. Thus, we have:

$$P(X'|x, k') \propto P(k'|X')P(X'|x) \tag{2}$$

and $P(k'|X') = 0$ or 1 depending on the constraint.

If the variables are independent, we have $P(X'|x, a, k') = \prod_i P(X_i'|x, a) = \prod_i P(X_i'|parent(X_i'), a)$, thus we are in the standard context where the proof of convergence given in (Boutilier et al., 2000) applies.

Now, if we consider impossible states, from (2) we have

$$\sum_{x'} P(X'|x, a, k') = \sum_{x'} N_{x,a} P(k'|X')P(X'|X, a) \tag{3}$$

where $N_{x,a}$ is a normalization factor such that

$$\forall x, \forall a, \sum_{x'} N_{x,a} P(k'|X')P(X'|X, a) = 1.$$

In (3), there are two categories of terms. If X' corresponds to impossible states, we have $P(k'|X') = 0$ and the corresponding term is removed. Otherwise, the state variables in X' are independent thus $P(X'|x, a, k') = \prod_i P(X_i'|parent(X_i'), a)$ and $P(k'|X') = 1$. Thus (3) can be simplified as

Table 2. Transition probabilities for joint variables resulting from Fig. 2, given that $\bar{x_1}'x_2'$ is impossible

$P(X'\|X, a_0, k')$	$x_1'x_2'$	$x_1'\bar{x_2}'$	$\bar{x_1}'\bar{x_2}'$
x_1x_2	100%	0%	0%
$x_1\bar{x_2}$	37.5%	37.5%	25%
$\bar{x_1}\bar{x_2}$	0%	0%	100%

$$\sum_{x'} N_{x,a} P(k'|X') \prod_i P(X_i'|parent(X_i'), a))$$

where only the existing states remain. We are back to the situation where we consider only possible states with independent variables and the proof from (Boutilier et al., 2000) applies again.

From the result above, it turns out that, if we take the constraints into account, the values can be computed in $PRegress$ as in the case without dependencies, just discarding the impossible states and normalizing again so that the sum of probabilities over all remaining states is 1.

Note that this way to remove impossible states in the computation of $PRegress$ is the only one which results in the possibility to renormalize. Otherwise, if, for instance, we remove the leaves corresponding to impossible states in $Tree(Q_a^V)$ or $Tree(V)$, we cannot perform the normalization since the probability information is lost in these trees. Furthermore, in addition to being theoretically well-founded, this way of filtering out impossible states at the heart of the $PRegress$ operator is much more efficient than filtering later on, since this other solution would result in expanding the number of leaves in the value tree before reducing it, which is exactly what we want to prevent.

To illustrate our approach, let us consider again the example given in the previous section. Now, assume that we have some information that the state $\bar{x_1}x_2$ is impossible. As a consequence, $P(\bar{x_1}'x_2')$ is null and the probability of any state at $t+1$ given $\bar{x_1}x_2$ is pointless. Thus, the corresponding probabilities in Table 1 must be filtered out and the remaining values must be renormalized as shown in Table 2. Here again, Table 2 is not computed explicitly, it is represented as a tree as in the standard case, adding in the algorithm the filtering out of the branches corresponding to impossible states and finally normalizing again the values at the leaves.

3.3 Impact on Other Tree Operations

After modifying $PRegress$ as presented above, the trees corresponding to (3) are free from impossible states for all actions.

But then, performing a Bellman-backup according to equation (1) and the other operations required to run SVI or SPI implies some operations over the resulting trees. As exemplified in Fig. 5, despite the filtering performed in $PRegress$, simple operations on several value-related trees ($Tree(R)$, $Tree(V)$, $Tree(Q_a^V)$) can generate leaves representing impossible states if these trees do not share the same structure. Indeed, whereas generalization over identical values can "hide" the expression of impossible states in

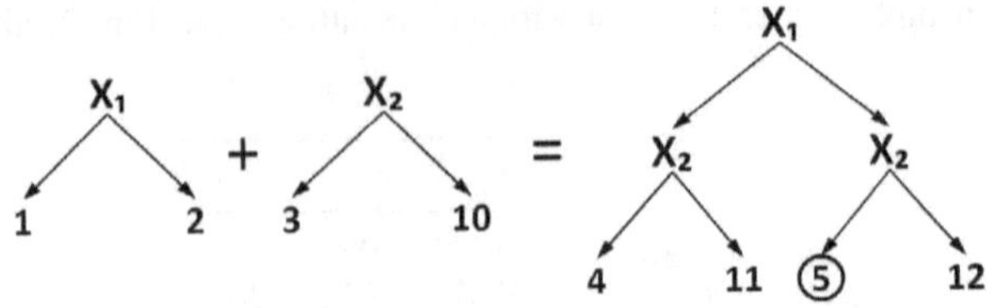

Fig. 5. Combining two trees can generate leaves about impossible states (here, $\bar{x_1}x_2$ is impossible)

the source trees, the operations can make these states appear in the resulting trees. As a result, we must filter impossible states out in such operations.

3.4 IMPSPITI

So far, we have described modifications that apply to SDP algorithms. To evaluate the impact of these considerations in the context of FRL, we design a new system inspired from SPITI, called IMPSPITI, that incorporates the following modifications in its *Planning* phase:

- in $PRegress$, when computing the joint probabilities over variable values, the branches in the tree corresponding to impossible states are discarded and the probabilities are normalized again;
- in the sum $Tree(R(x,a)) + \gamma \sum_{x'} Tree(P(x'|X,a))Tree(V(x'))$ and the maximization $Tree(V) = \text{argmax}_a\, Tree(Q_a^V)$, the branches of the resulting tree corresponding to impossible states are discarded;

The model of transition and reward functions being learnt from experience, they cannot contain impossible states, thus the *Learning* phase does not need to be modified. Algorithm 1 schematically describes the planning phase if IMPSPITI.

Finally, we need a function to decide that a state is impossible. Since this function is called at three places in each step of the central iteration, it may result in a significant time overhead.

Algorithm 1. IMPSPITI- *Planning* phase

Input : FMDP $\mathcal{F}[Tree(P(x'|x,a))], Tree(V_{t-1})$;
1. $Tree(Q_a^t) = addTrees[Tree(R(x,a), \gamma.PRegress[Tree(V_t), a, N_{x,a}]]$ for each action $a \in A$ discarding impossible states in addTrees()
2. $Tree(V_t) = MaxMerge_a[Tree(Q_a^t)]$ discarding impossible states in MaxMerge().
3. Return : $Tree(V_t)$ and $\{Tree(Q_a^t), \forall a \in A\}$.

In the experimental section, we will compare two approaches. One consists in using problem specific expert rules. The second is more general, it consists in building a tree where the states already visited by the agent are stored as possible. Thus we consider all states that have not been visited yet as impossible. In the worst case, this representation would boil down to the complete enumeration of states, but this is also true for the trees manipulated in standard SDP algorithms (see (Boutilier et al., 2000) for a discussion). In most cases of interest, however, this representation will benefit from factorization over states.

4 Experimental Study

To illustrate the benefits of our approach, we perform experiments on two benchmarks: MAZE6 and BLOCKS WORLD. The algorithms are coded in C# and run on Intel Core2Duo 1.80GHz processor with 2Go RAM. All results presented below are averaged over 50 runs where each run performs 50 episodes limited to 50 steps. The ϵ-*greedy* exploration policy uses $\epsilon = 0.1$.

4.1 Maze6

Maze environments are standard benchmark problems in the Learning Classifier Systems (LCSs) literature, LCSs being a heuristic approach to FRL (see (Sigaud & Wilson, 2007)). Mazes are represented by a two-dimensional grid. Each cell can be occupied by an obstacle, denoted as variable value by a `'1'`, a reward, denoted by a `'R'`, or can be empty, denoted by a `'0'`. The agent perceives the eight adjacent cells starting with the cell to the north and coding clockwise. Fig. 6 shows MAZE6, one of such mazes, designed so that Markov property holds.

For example, an agent located in the cell below the reward perceives `'R1110011'` whereas an agent located as shown in Fig. 6 perceives `'00110101'`. Although there are only 37 actual states within the problem, the combinatorial representation results in $3^8 = 6561$ states. The agent can perform eight actions, the movements to adjacent cells. If a movement leads to a cell containing an obstacle, the action has no effect and there is no penalty. In the stochastic case, the chosen action may result in a move corresponding to an immediately adjacent action, with probability 10%. Once the reward position is reached, the environment provides a reward of 1000 and the episode ends. In that case, the agent starts again in a randomly chosen empty cell.

Fig. 6. Maze6

4.2 Blocks World

In the BLOCKS WORLD problem, introduced in (Butz et al., 2002) (see Fig. 7), b blocks are distributed over a given number s of stacks. At the beginning of each episode, blocks are distributed randomly. The agent can manipulate the stacks by the means of a gripper that can either grip or release a block on a certain stack. Additionally, the problem contains a goal state, which consists in putting a particular number $y \leq b$ of blocks on the left hand stack. The stacks are not limited in height. Fig. 7(d) shows the goal in the problem with $b = 4, s = 3, y = 3$.

The complexity of BLOCKS WORLD highly depends on the representation chosen to encode the states. We developed three such representations.

The first is the one used in (Butz et al., 2002). We call it *Binary* representation. The agent perceives the current blocks distribution coding each stack with b variables. One additional variable indicates if the gripper is currently holding a block. In this representation, many arbitrary combinations of variable values correspond to states that

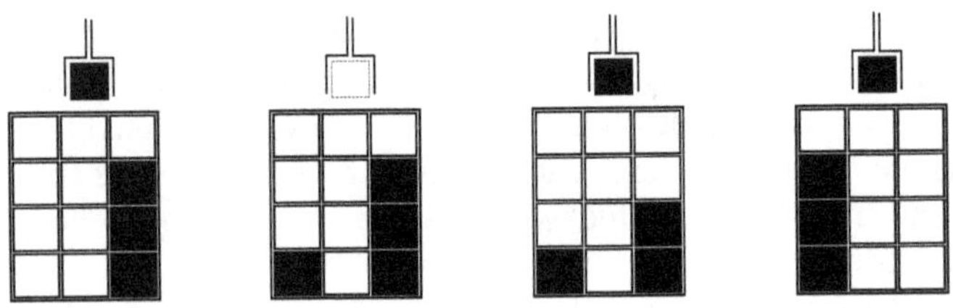

(a) (b) (c) (d)

Fig. 7. A BLOCKS WORLD scenario, from a random initial position (a) to the goal position (d)

Table 3. BLOCKS WORLD representations of the situation given in Fig. 7 (G stands for gripper)

(a)	(b)	(c)	(d)
	BINARY		
0000,0000,1110,1	1000,0000,1110,0	1000,0000,1100,1	1110,0000,0000,1
	STACKS		
STACK1=0	STACK1=1	STACK1=1	STACK1=3
STACK2=0	STACK2=0	STACK2=0	STACK2=0
STACK3=3	STACK3=3	STACK3=2	STACK3=0
G =TRUE	G =FALSE	G =TRUE	G =TRUE
	BLOCKS		
BLOCK1=S3	BLOCK1=S3	BLOCK1=S3	BLOCK1=G
BLOCK2=S3	BLOCK2=S3	BLOCK2=S3	BLOCK2=S1
BLOCK3=S3	BLOCK3=S3	BLOCK3=G	BLOCK3=S1
BLOCK4=G	BLOCK4=S1	BLOCK4=S1	BLOCK4=S1

do not occur in practice. Indeed, all states where a block is lying neither on top of another block nor on the table are impossible. The more empty cells in the problem, the more such impossible states.

In the second representation (called *Stacks*), there is one variable per stack giving the number of blocks it contains, and one additional variable indicating if the gripper is holding a block. The impossible states are the states where the total number of blocks is not b. Thus, in that case, an *ad hoc* way to decide if a state is possible consists in simply summing the represented blocks and comparing to b.

Finally, the third representation (called *Blocks*), there are b variables whose values are given by the gripper or stack where the corresponding block is currently placed. The only impossible states are the ones where several blocks are in the gripper, which gives a straightforward *ad hoc* rule to decide if a state is possible. The major drawback of this representation is that, blocks being identical, there are many ways to represent the same state of the problem. For instance, $\{block_1{=}s_1, block_2{=}s_2, \ldots\}$ and $\{block_1{=}s_2, block_2{=}s_1, \ldots\}$ represent the same configuration and there are 12 different ways to represent Fig. 7(c). This results in a greater number of possible states than necessary, thus in more complex value and policy trees structures.

Table 3 shows an example of these three representations with $b = 4, s = 3, y = 3$.

4.3 Empirical Results

Table 4 recaps the performance on MAZE6 of SPITI and IMPSPITI and Fig. 8 shows their convergence speed in number of episodes considering the number of steps needed to perform each episode.

IMPSPITI clearly outperforms SPITI both in the deterministic and in the stochastic case: it converges faster in time and in number of learning episodes, but also requires less memory to represent value and policy functions. More precisely, IMPSPITI only considers the 37 states that are actually possible (the variance in value and policy size comes from cases where the run ended before all states were explored).

Fig. 9 shows on different BLOCKS WORLD problems the value function size, computed as the number of leaves in $Tree(V)$ after 50 episodes.

Table 5 gives the rate of impossible states derived from the number of states shown in labels in Fig. 9 and the time required to perform one learning step by SPITI and IMPSPITI respectively. IMPSPITI uses trees to represent impossible states (the time with the *ad hoc* rules described in section 4.2 is indicated between parentheses).

Fig. 10 shows the convergence speed in number of episodes for the BLOCKS WORLD of size 4-3-4, that is a representative middle size problem. IMPSPITI takes less episodes

Table 4. MAZE6 performance (cost in memory and time)

	Value size	Policy size	Time/step(sec)
SPITI det	214 ± 10	241 ± 4	1.7 ± 0.8
SPITI stoc	252 ± 14	240 ± 12	3.7 ± 1.2
IMPSPITI det	35 ± 2	35 ± 2	0.1 ± 0.08
IMPSPITI stoc	35 ± 2	35 ± 2	0.24 ± 0.1

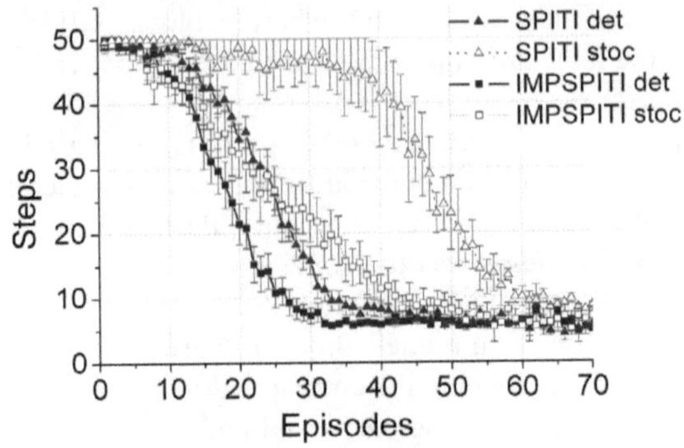

Fig. 8. Convergence over episodes on MAZE6

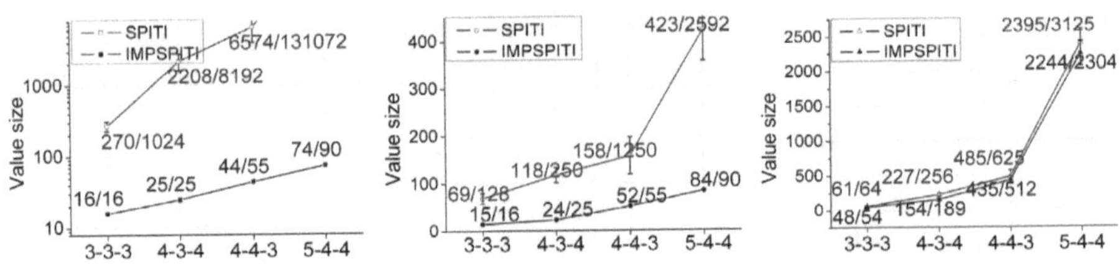

(A) BINARY (B) STACKS (C) BLOCKS

Fig. 9. BLOCKS WORLD: value function size as a function of the size of the problem. Labels on points indicate this value / the total number of states considered. Note the log scale for *Binary*.

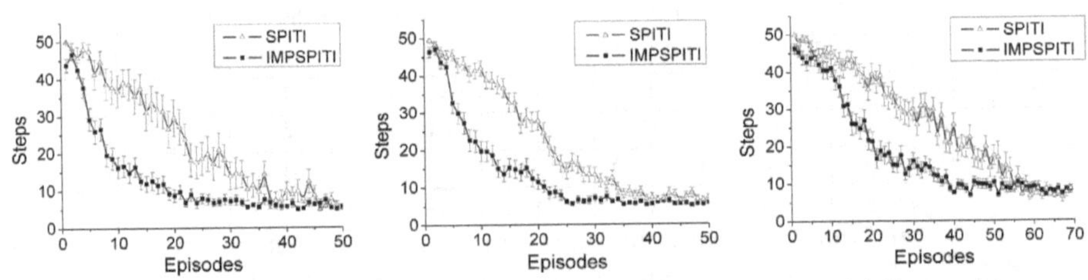

(A) BINARY (B) STACKS (C) BLOCKS

Fig. 10. BLOCKS WORLD performance along episodes (size 4-3-4). We run 70 episodes with *Blocks* to wait for convergence.

than SPITI to reach the optimal policy with all representations. This result is explained by the fact that IMPSPITI uses smaller trees with a simpler structure, therefore it takes less steps to propagate values over the trees. But note that, as expected, the difference is smaller when the rate of impossible states is smaller. Note also that, even when there are very few impossible states as is the case with the *Blocks* representation, the policy improves faster with IMPSPITI.

Now, comparing the performance in time on a single step from Table 5, in *Binary* representation, where the rate of impossible states grows very fast, IMPSPITI

Table 5. BLOCKS WORLD: rate of impossible states and time to perform one step (in seconds). A "-" indicates that the value could not be obtained after three days of computation.

		BINARY		STACKS		BLOCKS	
B-S-Y		SPITI	IMPSPITI	SPITI	IMPSPITI	SPITI	IMPSPITI
3-3-3	% IMP.	98.5%		87.5%		15.6%	
	TIME	0.28 ±0.03	0.04 ± 0.01	0.05 ±0.02	0.01 (0.01)	0.04	0.03 (0.03)
4-3-4	% IMP.	99.7%		90%		26.1%	
	TIME	2.9 ± 0.4	0.08 ± 0.01	0.06 ±0.02	0.01 (0.01)	0.17 ± 0.02	0.18 ± 0.01 (0.13 ± 0.01)
4-4-3	% IMP.	>99.99%		95.6%		18%	
	TIME	34 ± 7	1.2 ± 0.2	0.13 ± 0.06	0.02 (0.02)	0.34 ± 0.2	0.9± 0.07 (0.5 ± 0.03)
5-4-4	% IMP.	>99.99%		96.5%		26.3%	
	TIME	-	2.6 ± 0.3	0.2 ± 0.09	0.05 (0.04)	2.6 ± 0.3	6.5 ± 0.9 (3.1 ± 0.2)

unquestionably outperforms SPITI in time and memory use. With the *Stacks* representation, both the size of the problem and the rate of impossible states grows slower, thus the time difference between SPITI and IMPSPITI keeps small. Finally, the rate of impossible states is much smaller in the *Blocks* representation, therefore both algorithms perform similarly for small BLOCKS WORLDs and SPITI takes less time per step than IMPSPITI as the problem size grows, due to the time overhead required to check for the existence of states. This is also true using *ad hoc* rules to detect impossible states, even if the overhead is smaller in that case.

5 Discussion

The first message of this paper is that, although using a factored representation in reinforcement learning results in the possibility to address larger problems, designing a factored representation so that it does not artificially increase the number of considered states may be very difficult. Consider the MAZE6 problem, where there are only 37 states, but a somewhat standard factored representation may result in 6561 potential states. This problem is an idealization of a standard robot navigation problem where, given usual robot sensors, one could not say in advance which sensory values cannot occur simultaneously. Similarly, if we take the BLOCKS WORLD problem, it proved difficult to design a representation that would fit the number of actual states. The *Blocks* representation results in fewer impossible states, but at the price of some redundancy that makes it very inefficient for larger problems (for instance, with a size 6-5-5 problem, we have 34375 possible states represented whereas there are only 334 actual states). We have shown that IMPSPITI was able to stick to the number of possible states of the problem given a representation, resulting in the possibility to address a much wider class of FMDPs than standard SDP methods.

Our second message is about the time overhead resulting from the necessity to check whether a state is possible. Considering the computation time per step and the number of steps required to converge, IMPSPITI always performs faster if there are enough impossible states. More precisely, as illustrated with the *Blocks* representation, the larger the problem, the larger the necessary rate of impossible states, since the tree of possible states will grow larger, resulting in a large overhead. However, using domain specific rules to detect impossible states generally results in a further gain in speed, but it is at the price of generality.

Finally, the fact that the policy improves faster with IMPSPITI than with SPITI even when there are very few impossible states, as is the case with the *Blocks* representation, tends to indicate that considering states as impossible until they are seen is an efficient heuristics even in the absence of actually impossible states. This heuristics itself will deserve further analyses. In particular, it would be of much interest to study the potential interactions between this *pessimistic* heuristics and using efficient optimistic exploration strategies such as "Optimism in the Face of Uncertainty" (Szita & Lőrincz, 2008) that drives the agent towards unseen states. At first glance, these heuristics are contradictory, since in the former we do not want to represent impossible states which we do not distinguish from unseen states, whereas in the latter we want to attribute a large value to unseen states, thus we need to represent them.

6 Conclusion

We have shown that there exists a practically relevant class of FMDPs between the "no synchronic arcs" and the "any synchronic arcs" classes that corresponds to problems where some combinations of state variable values do not occur. Though standard SDP algorithms such as SVI and their derivatives such as SPITI can be applied to this class of problems, we have shown that modifying the algorithm to take the presence of impossible states into account can result in significant performance improvements. Moreover, we have shown that, in the context where an agent has to explore its environment to learn its structure, considering as impossible the states that have not yet been encountered is also beneficial to the performance.

Note that the approach described in section 3 can be applied to other standard SDP algorithms. Here, we focused on one particular FRL method, namely SPITI, but the impact on other instances of SDYNA using SPUDD or Guestrin's linear programming approach (Guestrin et al., 2003) remains to be studied. Furthermore, we want to compare IMPSPITI with XACS (Butz et al., 2002), an Anticipatory Learning Classifier System endowed with most FRL systems properties, but which uses genetic algorithm heuristics instead of SDP and incremental tree induction methods. A previous comparison between XACS and SPITI (Sigaud et al., 2009) has shown that XACS can deal efficiently with impossible states, but a closer comparison between the mechanisms of IMPSPITI presented here and those of XACS remains to be performed.

References

Boutilier, C.: Correlated action effects in decision theoretic regression. In: Proceedings of the 13th International Conference on Uncertainty in Artificial Intelligence, pp. 30–37. AUAI Press (1997)

Boutilier, C., Dearden, R., Goldszmidt, M.: Exploiting structure in policy construction. In: Proceedings of the 14th International Joint Conference on Artificial Intelligence, Montreal, pp. 1104–1111 (1995)

Boutilier, C., Dearden, R., Goldszmidt, M.: Stochastic dynamic programming with factored representations. Artificial Intelligence 121, 49–100 (2000)

Butz, M.V., Goldberg, D.E., Stolzmann, W.: The Anticipatory Classifier System and Genetic Generalization. Natural Computing 1, 427–467 (2002)

Dean, T., Kanazawa, K.: A model for reasoning about persistence and causation. Computational Intelligence 5, 142–150 (1989)

Degris, T., Sigaud, O., Wuillemin, P.-H.: Chi-square tests driven method for learning the structure of factored MDPs. In: Proceedings of the 22nd International Conference on Uncertainty in Artificial Intelligence, Massachusetts Institute of Technology, pp. 122–129. AUAI Press, Cambridge (2006a)

Degris, T., Sigaud, O., Wuillemin, P.-H.: Learning the structure of factored markov decision processes in reinforcement learning problems. In: Proceedings of the 23rd International Conference on Machine Learning, pp. 257–264. ACM Press, Pittsburgh (2006b)

Guestrin, C., Koller, D., Parr, R., Venkataraman, S.: Efficient Solution Algorithms for Factored MDPs. Journal of Artificial Intelligence Research 19, 399–468 (2003)

Hoey, J., St-Aubin, R., Hu, A., Boutilier, C.: SPUDD: Stochastic planning using decision diagrams. In: Proceedings of the 15th Conference on International Conference on Uncertainty in Artificial Intelligence, Stockholm, pp. 279–288 (1999)

Sigaud, O., Butz, M.V., Kozlova, O., Meyer, C.: Anticipatory Learning Classifier Systems and Factored Reinforcement Learning. In: ABiALS 2008. LNCS (LNAI), vol. 5499. Springer. Heidelberg (2009)

Sigaud, O., Wilson, S.W.: Learning Classifier Systems: a survey. Journal of Soft Computing 11, 1065–1078 (2007)

Sutton, R.S.: DYNA, an integrated architecture for learning, planning and reacting. In: Working Notes of the AAAI Spring Symposium on Integrated Intelligent Architectures (1991)

Sutton, R.S., Barto, A.G.: Reinforcement learning: An introduction. MIT Press, Cambridge (1998)

Szita, I., Lőrincz, A.: The many faces of optimism: a unifying approach. In: ICML 2008: Proceedings of the 25th international conference on Machine learning, pp. 1048–1055. ACM Press, New York (2008)

Utgoff, P.E.: Incremental induction of decision trees. Machine Learning 4, 161–186 (1989)

Relevance Grounding for Planning in Relational Domains

Tobias Lang and Marc Toussaint

TU Berlin, Franklinstrasse 28/29, 10587 Berlin, Germany
{lang,mtoussai}@cs.tu-berlin.de

Abstract. Probabilistic relational models are an efficient way to learn and represent the dynamics in realistic environments consisting of many objects. Autonomous intelligent agents that ground this representation for all objects need to plan in exponentially large state spaces and large sets of stochastic actions. A key insight for computational efficiency is that successful planning typically involves only a small subset of relevant objects. In this paper, we introduce a probabilistic model to represent planning with subsets of objects and provide a definition of object relevance. Our definition is sufficient to prove consistency between repeated planning in partially grounded models restricted to relevant objects and planning in the fully grounded model. We propose an algorithm that exploits object relevance to plan efficiently in complex domains. Empirical results in a simulated 3D blocksworld with an articulated manipulator and realistic physics prove the effectiveness of our approach.

1 Introduction

Artificial Intelligence investigates systems that act autonomously in complex environments. Such systems need to be able to reason under time pressure about their world in order to derive plans of actions and achieve their goals. This requires probabilistic relational knowledge representations that can deal with the stochasticity of actions, cope with noise and generalize over object instances. Complex environments typically contain very many objects. Consider for example a household robot, that has to represent all kinds of furnitures, dishes, house inventary etc. together with their properties and relationships. Such realistic domains comprise state spaces that are exponential in the number of represented objects and large sets of stochastic actions. Research in A.I. over the last years has led to world models that describe the action effects and state transitions compactly in terms of abstract logical formulae and can be learned from experience. However, how to exploit this model compactness for planning remains a major challenge. Planning in the fully grounded representation is often a hopeless undertaking as the state space quickly grows for all but the smallest problems. This problem is often simply ignored by designing the domain carefully to only contain those domain aspects which are relevant for successful planning. For truly autonomous agents operating continuously with changing tasks, however, we require principled ways to make planning in complex environments tractable.

W. Buntine et al. (Eds.): ECML PKDD 2009, Part I, LNAI 5781, pp. 736–751, 2009.

Models of human cognition provide an inspiring idea of how one may plan in a highly complex world. Humans are often assumed to possess declarative world knowledge about the types of objects they encounter in daily life [1] which is presumably stored in the long-term memory. For instance, they know that piling dishes succeeds the better the more exactly aligned these dishes are. This abstract knowledge is independent of any concrete dish instances or other unrelated objects (such as lamps and cars) and is akin to the abstract probabilistic relational models in A.I.. When planning, human beings may reason about objects according to their abstract world knowledge [3], i.e., they ground their abstract world model with respect to these objects. Such reasoning is often assumed to take place in the working memory, a cognitive system functioning as a work-space in which recently acquired sensory information and information from long-term memory are processed for further action such as decision-making [2] [16]. This system has *limited capacity* and humans can only take some selected objects into account – those they deem relevant for the problem at hand. For example, when planning to prepare a cup of tea, they do not consider the frying pan in the shelf or a soccer ball in the garage. One can view this as grounding the abstract world knowledge only with respect to these relevant objects, thereby enabling tractable planning.

In this paper, we take up this idea and exploit the great advantage of abstract relational world models to be applicable to arbitrary subsets of objects. First, we define object relevance in terms of a graphical model. This allows us then to prove consistency between repeated planning in partially grounded models restricted to relevant objects and planning in the fully grounded model. Thereby, we reformulate the original intractable problem into tractable versions where we can apply any efficient planning method to solve our problem at hand, enabling real-time planning and planning with quickly changing goals. Empirical results in an extended simulated 3D blocksworld with realistic physics and an articulated manipulator using a learned world model show the effectiveness of our approach.

The remainder of this paper is organized as follows. In the next section, we present relational world models and discuss the difficulties of planning in the fully grounded representation. In Section 3, we introduce our approach of *Relevance Grounding*. In Section 4, we present our empirical results. In Section 5, we discuss related work before we conclude.

2 Background

2.1 Compact World Models

A relational domain is represented by a relational logic language $\mathcal{L}$: the set of logical predicates $\mathcal{P}$ and the set of logical functions $\mathcal{F}$ contain the relationships and properties that can hold for domain objects. The set of logical predicates $\mathcal{A}$ comprises the possible actions in the domain.

A concrete instantiation of a relational domain is made up of a finite set of objects $\mathcal{O}$. If the arguments of a predicate or function are all concrete, i.e. taken from $\mathcal{O}$, we call it *grounded*. A concrete world state s is fully described by all grounded predicates and functions. Concrete actions a are described by positive grounded predicates from $\mathcal{A}$.

The arguments of predicates and functions can also be abstract logical variables which can represent any object. If a predicate or function has only abstract arguments, we call it *abstract*. We will speak of grounding an abstract formula ψ if we apply a substitution σ that maps all of the variables appearing in ψ to objects in $\mathcal{O}$.

A relational model $\mathcal{T}$ of the transition dynamics specifies $P(s'|a,s)$, the probability of a successor state s' if action a is performed in state s. $\mathcal{T}$ is usually defined compactly in terms of abstract predicates and functions. This enables abstraction from object identities and concrete domain instantiations. For instance, the effects of trying to grab a cup may be defined by a single abstract model for all concrete cups. To apply $\mathcal{T}$ in a given world state, one needs to ground $\mathcal{T}$ with respect to some of the objects in the domain.

Examples of abstract transition models include relational probability trees for predicates and functions based on abstract logical formulae [6] and probabilistic relational rules, e.g. in the form of STRIPS-operators. An example of the latter are the *noisy indeterministic deictic (NID) rules* [15] which will be our running example in this paper and which we briefly review here. A NID rule r is given as follows

$$a_r(\mathcal{X}):\ \Phi_r(\mathcal{X}) \quad \rightarrow \begin{cases} p_{r,1} & : \Omega_{r,1}(\mathcal{X}) \\ & \vdots \\ p_{r,m_r} & : \Omega_{r,m_r}(\mathcal{X}) \\ p_{r,0} & : \Omega_{r,0} \end{cases}, \tag{1}$$

where $\mathcal{X}$ is a set of logic variables in the rule (which represent a (sub-)set of abstract objects). The rule r consists of preconditions, namely that action a_r is applied on $\mathcal{X}$ and that the state context Φ_r is fulfilled, and m_r+1 different outcomes with associated probabilities $p_{r,i} > 0$, $\sum_{i=0} p_{r,i} = 1$. Each outcome $\Omega_{r,i}(\mathcal{X})$ describes which predicates and functions change when the rule is applied. The context $\Phi_r(\mathcal{X})$ and outcomes $\Omega_{r,i}(\mathcal{X})$ are conjunctions of literals constructed from the predicates in $\mathcal{P}$ as well as equality statements comparing functions from $\mathcal{F}$ to constant values. The so-called *noise outcome* $\Omega_{r,0}$ subsumes all possible action outcomes which are not explicitly specified by one of the other $\Omega_{r,i}$. The arguments of the action $a(\mathcal{X}_a)$ may be a true subset $\mathcal{X}_a \subset \mathcal{X}$ of the variables $\mathcal{X}$ of the rule. The remaining variables are called deictic references $DR = \mathcal{X} \setminus \mathcal{X}_a$ and denote objects relative to the agent or action being performed.

As above, let σ denote a substitution that maps variables to constant objects, $\sigma : \mathcal{X} \rightarrow \mathcal{O}$. Applying σ to an abstract rule $r(\mathcal{X})$ yields a *grounded rule* $r(\sigma(\mathcal{X}))$. We say a grounded rule r *covers* a state s and a ground action a if $s \models \Phi_r$ and $a = a_r$. By grounding NID rules, we can predict successor states for a given state. NID rules can be learned from experience triples (s, a, s') using a batch algorithm that trades off the likelihood of these triples with the complexity of the learned rule-set. For more details, we refer the reader to Pasula et al. [15].

2.2 Planning in Ground Representations

Our goal is to plan in the ground relational domain: find a "satisficing" action sequence that will lead with high probability to states with large rewards. While a relational transition model $\mathcal{T}$ provides a very compact description of the dynamics of the world, this compactness does not carry over to planning. Grounding the relational representation language $\mathcal{L}$ w.r.t. all domain objects $\mathcal{O}$ results in a state space that is exponential in $|\mathcal{O}|$. Thus, evaluating an action sequence is exponential in $|\mathcal{O}|$. Furthermore, the set of ground actions and thus the search space of plans scales with the number of objects. (Planning is even further complicated due to the stochasticity of actions.) Planning is only tractable in case $|\mathcal{O}|$ is very small. In most realistic scenarios, $|\mathcal{O}|$ is rather large, however. Fortunately, it often suffices to take only the objects that are relevant for the planning problem into account. In the next section, we will introduce *Relevance Grounding* which formalizes this idea in a systematic way. To plan in grounded models, we use the PRADA algorithm [14] in this paper. PRADA converts NID rules into dynamic Bayesian networks, predicts the effects of action sequences on states and rewards by means of approximate inference and samples action sequences in an informed way. PRADA has the crucial advantage to evaluate an action sequence very efficiently, in particular in time linear in its length.

3 Relevance Grounding

In the following, we introduce a probabilistic model which expresses the coupling between state sequences, action sequences, objects and rewards. This model will help to formalize what planning with subsets of objects implies. In particular, we will be able to derive results on planning with subsets of objects – which corresponds to conditioning on object-sets $\mathbf{o}$.

Assume our domain contains objects $\mathcal{O}$. By M we denote the model grounded for all objects, including the complete state and action space (all ground predicates and functions w.r.t. $\mathcal{O}$), which defines the state transition dynamics according to some given relational transition model $\mathcal{T}$. Let $\mathbf{a} = (a_1, \dots, a_T)$ denote a *plan*, i.e., a sequence of actions. Let $\mathbf{s} = (s_1, \dots, s_T)$ denote a sequence of encountered states. We assume that in a given trial $(\mathbf{s}, \mathbf{a})$ certain objects are relevant while others are not. For example, an object o is relevant if it is an argument of one of the actions in $\mathbf{a}$. We give a concrete definition of object relevance in Sec. 3.1. For now, we only assume that, in general, object relevance can be expressed by a conditional probability $P(\mathbf{o}|\mathbf{s}, \mathbf{a})$ where $\mathbf{o}$ is a random variable referring to a subset of $\mathcal{O}$. Let R denote the event of achieving a reward at the end of a trial[1]. We assume the following joint distribution over these random variables:

$$P(R, \mathbf{o}, \mathbf{s}, \mathbf{a}; M) = P(R \mid \mathbf{s}, \mathbf{a}; M)\, P(\mathbf{o} \mid \mathbf{s}, \mathbf{a}; M)\, P(\mathbf{s} \mid \mathbf{a}; M)\, P(\mathbf{a}; M) \qquad (2)$$

[1] When we assume a geometric prior on the trial length, the expected reward is equivalent to the sum of discounted rewards when rewards are given in each time-step – see Toussaint et al. [18] for details.

Note that all conditional distributions depend on the model M. In absence of goals or rewards, we assume a uniform prior over plans $P(\mathbf{a}; M)$, discounted by their length T by a discount factor $0<\gamma<1$ (see footnote 1). In the following, we will eliminate $\mathbf{s}$, i.e., we consider

$$P(R, \mathbf{o}, \mathbf{a}; M) = \sum_{\mathbf{s}} P(R, \mathbf{o}, \mathbf{s}, \mathbf{a}; M) = P(R \mid \mathbf{o}, \mathbf{a}; M) P(\mathbf{o} \mid \mathbf{a}; M) P(\mathbf{a}; M) , \tag{3}$$

where R now conditionally depends on $\mathbf{o}$.

Generally, planning requires finding the maximizing argument $\mathbf{a}^*$ of the following distribution:

$$P(\mathbf{a} \mid R; M) \propto P(R \mid \mathbf{a}; M) P(\mathbf{a}; M) . \tag{4}$$

Finding plans with high $P(\mathbf{a} \mid R; M)$ is a difficult task for the following reasons: *(i)* the search space of $\mathbf{a}$ scales with the number of objects; *(ii)* evaluating $P(R \mid \mathbf{a}; M)$ is difficult as M's state space is exponential in the number of objects $\mathcal{O}$. To overcome this problem, we observe that we can decompose

$$P(\mathbf{a} \mid R; M) = \textstyle\sum_{\mathbf{o}} P(\mathbf{a} \mid \mathbf{o}, R; M) P(\mathbf{o} \mid R; M) \tag{5}$$

where $P(\mathbf{o} \mid R; M)$ is defined as

$$P(\mathbf{o} \mid R; M) \propto \sum_{\mathbf{a}} P(R \mid \mathbf{o}, \mathbf{a}; M) P(\mathbf{o} \mid \mathbf{a}; M) P(\mathbf{a}; M) . \tag{6}$$

$P(\mathbf{o} \mid R; M)$ is a measure for the relevance of object-sets with respect to the reward. Note that this posterior favors small object-sets due to the prior over plan lengths in $P(\mathbf{a}; M)$. If every successful plan makes use of object o, then for each $\mathbf{o}$ with $P(\mathbf{o} \mid R; M) > 0$ we have $o \in \mathbf{o}$. In this case, we call o necessary for R. Using this formalization of the relevance of object-sets, Eq. (5) provides us a way to decompose the above planning problem into two stages: *(i)* sampling of object-sets using $P(\mathbf{o} \mid R; M)$; *(ii)* finding plans with high $P(\mathbf{a} \mid \mathbf{o}, R; M)$ corresponding to planning conditioned on a set of relevant objects. The key idea is that the conditioning on $\mathbf{o}$ in stage *(ii)* may significantly reduce the cost of planning, as we will discuss below.

3.1 A Sufficient Definition of Relevance

We will now provide a definition of $P(\mathbf{o}|\mathbf{s}, \mathbf{a})$ which we have neglected thus far. For a given pair $(\mathbf{s}, \mathbf{a})$, we define the set Ω of relevant object-sets as

$$\begin{aligned} \Omega(\mathbf{s}, \mathbf{a}) = \{\mathbf{o} \subseteq \mathcal{O} \mid \forall t, 0 \le t < T : P(s_{t+1} \mid s_t, a_t; M) = P(s_{t+1} \mid s_t, a_t; M_{\mathbf{o}}) \\ \wedge \ \forall \mathbf{o}' \subset \mathbf{o}, \mathbf{o}' \neq \mathbf{o} \ \exists t, 0 \le t < T : P(s_{t+1} \mid s_t, a_t; M) \neq P(s_{t+1} \mid s_t, a_t; M_{\mathbf{o}'})\} \end{aligned} \tag{7}$$

where $M_{\mathbf{o}}$ is the *reduced model* including only the objects $\mathbf{o}$ with their groundings of predicates and functions. $P(\mathbf{s}_{t+1}|\mathbf{s}_t, \mathbf{a}_t; M_{\mathbf{o}})$ is defined such that all predicates and functions with at least one argument $o \notin \mathbf{o}$ persist from $\mathbf{s}_t$ and are ignored while calculating transition probabilities, and we define the action prior in the reduced model as $P(\mathbf{a}; M_\mathbf{o}) = P(\mathbf{a}|\mathbf{o}; M)$. $\Omega(\mathbf{s}, \mathbf{a})$ comprises minimal object-sets that are required to predict the state transitions correctly for a specific trial $(\mathbf{s}, \mathbf{a})$. Note that $|\Omega(\mathbf{s}, \mathbf{a})| \geq 1$ for all $(\mathbf{s}, \mathbf{a})$. We define $P(\mathbf{o}|\mathbf{s}, \mathbf{a})$ as

$$P(\mathbf{o} \mid \mathbf{s}, \mathbf{a}) = \frac{I(\mathbf{o} \in \Omega(\mathbf{s}, \mathbf{a}))}{|\Omega(\mathbf{s}, \mathbf{a})|} . \tag{8}$$

Intuitively, relevant object-sets are those that are taken into account to calculate the transition probabilities in $\mathbf{s}$ for a given $\mathbf{a}$. Clearly, they include the objects which are manipulated, i.e., whose properties or relationships change. We call these *actively relevant*. There are also *passively relevant* objects which are taken into account by the world dynamics model $\mathcal{T}$. For instance, imagine the task to go to the kitchen and prepare a cup of tea. The tea bag, the cup and the water heater are actively relevant objects. If the kitchen has two doors and one of them is locked, then the latter is passively relevant: we cannot manipulate, i.e. open, it, but it plays a role in planning as its being locked determines the other door to be necessary. A more technical example of object relevance is given below.

In general, there might be alternative interesting definitions of object relevance, e.g. where the transition probabilities in Eq.(7) only hold approximately. We chose the above definition because it is sufficient to a certain consistency for planning in reduced models:

Lemma 1. *When conditioning on a subset* $\mathbf{o}$ *of relevant objects, the following probabilities in the reduced model* $M_\mathbf{o}$ *are the same as in the full model* M*:*
(a) State sequences: $P(\mathbf{s} \mid \mathbf{o}, \mathbf{a}; M) = P(\mathbf{s} \mid \mathbf{a}; M_\mathbf{o})$
(b) Rewards: $P(R \mid \mathbf{o}, \mathbf{a}; M) = P(R \mid \mathbf{a}; M_\mathbf{o})$
(c) Action sequences: $P(\mathbf{a} \mid \mathbf{o}, R; M) = P(\mathbf{a} \mid R; M_\mathbf{o})$

Proof. If $\mathbf{o} \in \Omega(\mathbf{s}, \mathbf{a})$, we have:

$$P(\mathbf{s}|\mathbf{o}, \mathbf{a}; M) = \prod_{t=0}^{T-1} P(s_{t+1}|\mathbf{o}, s_t, a_t; M) = \prod_{t=0}^{T-1} P(s_{t+1}|s_t, a_t; M_\mathbf{o}) = P(\mathbf{s}|\mathbf{a}; M_\mathbf{o}) . \tag{9}$$

If $\mathbf{o} \notin \Omega(\mathbf{s}, \mathbf{a})$, we have $P(\mathbf{s} \mid \mathbf{o}, \mathbf{a}; M) = 0$. Similarly, $\mathbf{s}$ cannot be predicted in $M_\mathbf{o}$ as only the irrelevant object-set $\mathbf{o}$ is available, so we get $P(\mathbf{s} \mid \mathbf{a}; M_\mathbf{o}) = 0$. Furthermore, we have:

$$P(R \mid \mathbf{o}, \mathbf{a}; M) = \sum_{\mathbf{s}} P(R, \mathbf{s} \mid \mathbf{o}, \mathbf{a}; M) = \sum_{\mathbf{s}} P(R|\mathbf{s}, \mathbf{o}, \mathbf{a}; M) P(\mathbf{s}|\mathbf{o}, \mathbf{a}; M) \tag{10}$$

$$= \sum_{\mathbf{s}} P(R \mid \mathbf{s}, \mathbf{a}; M)\, P(\mathbf{s} \mid \mathbf{a}; M_\mathbf{o})\, I(\mathbf{o} \in \Omega(\mathbf{s}, \mathbf{a})) \tag{11}$$

$$= \sum_{\mathbf{s}:\mathbf{o} \in \Omega(\mathbf{s}, \mathbf{a})} P(R \mid \mathbf{s}, \mathbf{a}; M) P(\mathbf{s} \mid \mathbf{a}; M_\mathbf{o}) = P(R \mid \mathbf{a}; M_\mathbf{o}) \tag{12}$$

Finally, we have:

$$P(\mathbf{a} \mid \mathbf{o}, R; M) \propto P(R \mid \mathbf{o}, \mathbf{a}; M)\, P(\mathbf{a} \mid \mathbf{o}; M) \tag{13}$$

$$= P(R \mid \mathbf{a}; M_\mathbf{o})\, P(\mathbf{a}; M_\mathbf{o}) \;\; \propto \;\; P(\mathbf{a} \mid R; M_\mathbf{o}) \;\; \square \tag{14}$$

Table 1. Example of active and passive object relevance. Objects B and C are actively relevant as their properties are changed. A is passively relevant as it determines rule 2 to model the transition dynamics. If it was ignored, rule 1 would be used instead yielding wrong state transition probabilities.

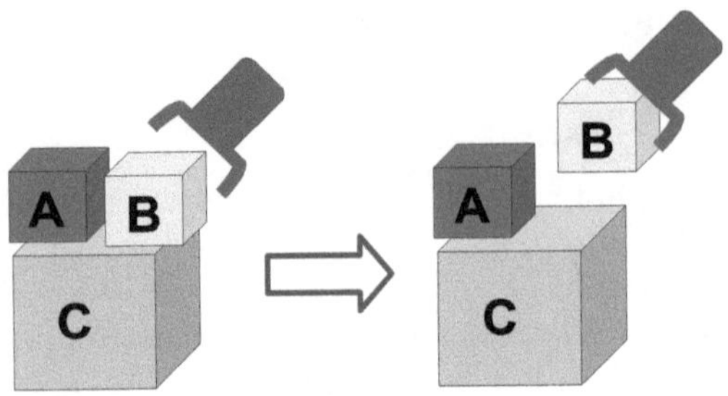

Action: *grab*(B)

Rule 1:
$grab(X): \quad on(X,Y), \neg on(Z,Y)$
$\rightarrow 1.0: inhand(X),\ \neg on(X,Y)$

Rule 2:
$grab(X): \quad on(X,Y), on(Z,Y)$
$\rightarrow 0.8: inhand(X),\ \neg on(X,Y)$
$\quad 0.2: \neg on(X,Y)$

From Lemma 1 and Eq. (5) the following proposition follows directly:

Proposition 1. *Given the joint in Eq. (2) and the definition of* $P(\mathbf{o} \mid \mathbf{s}, \mathbf{a})$ *in Eq. (8), it holds:*

$$P(\mathbf{a} \mid R; M) = \sum_{\mathbf{o}} P(\mathbf{a} \mid R; M_{\mathbf{o}})\ P(\mathbf{o} \mid R; M) \tag{15}$$

An illustrative example of object relevance. The scenario in Table 1 illustrates active and passive object relevance. Two small blocks A and B are on top of a big block C. Our goal is to hold B inhand. This can be achieved by means of a plan consisting of a single action *grab*(B). Our transition dynamics model contains two NID rules to model the *grab*-action. Rule 1 applies if the target block is the only block on top, in which case *grab* always succeeds. Rule 2 applies if the target block is not the only block on top. In this case, *grab* only succeeds with probability 0.8; otherwise with probability 0.2, grabbing fails due to lack of space and the target block is pushed off the big block instead. Clearly, in our situation we have to use rule 2. Blocks B and C are manipulated and thus are actively relevant, whereas A is passively relevant as it determines rule 2 to apply. If A was ignored, we would use rule 1 – yielding an erroneous higher success probability. Similar scenarios are typical in physical worlds: the probabilities of successful planning change when objects (e.g. potential obstacles) are added to or removed from the scene even when they are not actively manipulated.

3.2 Planning with Relevant Objects

Our definition of object relevance and the subsequent discussion led to a crucial observation: to find plans with high $P(\mathbf{a} \mid R; M)$, it is not necessary to use the full model M including all objects $\mathcal{O}$. As Proposition 1 shows, an alternative is to find plans in the reduced models $P(\mathbf{a} \mid R; M_{\mathbf{o}})$ for object-sets $\mathbf{o}$ with high

Algorithm 1. Relevance Grounding

Input: objects $\mathcal{O}$, goal τ, relational transition model $\mathcal{T}$, relevance distribution $q(\cdot)$, number of relevance groundings N_{rel}, number of verifications N_{ver}
Output: action sequence $\mathbf{a}$

```
for i = 1 to N_rel do                                   ▷ Relevance grounding
    Sample object-set o_i ⊂ O according to q
    Build reduced model M_{o_i}
    a_i = plan(τ; M_{o_i}, T)                           ▷ Plan in reduced model
    ψ(a_i) = P(R | a_i; M_{o_i}, T)                     ▷ Value in reduced model
end for
for i = 1 to N_ver do                                   ▷ Verifying in original model
    Let a denote plan with i-th largest ψ
    Calculate Ψ(a) = P(R | a; M, T)                     ▷ Value in original model
end for
return argmax_a Ψ(a)
```

relevance $P(\mathbf{o} \mid R; M)$. This makes planning more efficient due to the reduced state and action spaces in $M_{\mathbf{o}}$.

Obviously, we do not know $P(\mathbf{o} \mid R; M)$. If we knew the reward likelihoods for all plans, i.e., if we had already planned, we could calculate this quantity according to Eq. (6). However, planning is just the problem we are trying to solve. Thus, we have to estimate this quantity by some distribution $q(\cdot)$ over object-sets resulting in the approximate distribution $Q(\cdot)$ over plans defined as

$$Q(\mathbf{a}; R, M) = \sum_{\mathbf{o}} P(\mathbf{a} \mid R; M_{\mathbf{o}})\, q(\mathbf{o}; R, M) \approx P(\mathbf{a} \mid R; M)\,. \tag{16}$$

The quality of $Q(\cdot)$ depends on the quality of the approximate relevance distribution $q(\cdot)$. If $q(\cdot)$ is not exact, then a plan found in $M_{\mathbf{o}}$ may have lower success probability when planning in M instead (cf. Table 1). Therefore, it is a good idea to verify the quality of the proposed plan in the original model M or in a less reduced model $M_{\mathbf{o}'}$ with $\mathbf{o} \subset \mathbf{o}'$. This requires algorithms that can exploit the transition dynamics $\mathcal{T}$ to efficiently calculate $P(R \mid \mathbf{a})$ also in large models.

Algorithm 1 presents our complete *Relevance Grounding* method. Given an estimator for object-set relevance $q(\cdot)$, we can find plans with approximately high $P(\mathbf{a} \mid R; M)$ as follows: *(i)* we take samples $\mathbf{o}$ from $q(\cdot)$; *(ii)* we plan in the reduced models $M_{\mathbf{o}}$; *(iii)* we verify the resulting plans in the original or a less reduced model; *(iv)* we return the plan with the best verified value. In this paper, we employ NID rules as transition dynamics model $\mathcal{T}$ and use the PRADA algorithm for planning which is in particular appropriate for verification as it evaluates an action sequence in time linear in its length.

3.3 Learning Object Relevance

A crucial part in our proposed method is the relevance estimator of object-sets. Learning such an estimator is a novel and interesting machine learning problem.

As we are using relational representations, we can generalize over object identities in our planning goals and transfer the knowledge gained in previous planning trials to new, but similar problems. For a given goal τ, we can use our world dynamics model to create training instances $(\sigma_0, \tau, \mathbf{o}, P(\mathbf{o} \mid R))$. This enables us to learn object relevance based on nothing more than internal simulation (in contrast to "real" experiences) – akin to human reflection about a problem. σ_0 is a description of the start state s_0 and may involve all types of information, such as discrete, relational and continuous features. We can employ any regressor that can make use of the chosen features to learn a function $q(\mathbf{o}; s_0, \tau) \rightarrow \mathbb{R}$. A full approach to learning object relevance is beyond the scope of this paper, but in our first experiment (cf. Section 4.1) we will present an example of how to learn object relevance in a straightforward way, based purely on internal simulation.

4 Experiments

We test our *Relevance Grounding* approach in an extended simulated blocks world where a robot manipulates blocks and also balls scattered on a table. We use a 3D rigid-body dynamics simulator (ODE) that enables a realistic behavior of the objects. For instance, piles of objects may topple over or objects may even fall off the table (in which case they become out of reach for the robot). Object classes show different characteristics. For example, it is almost impossible to successfully put an object on top of a ball, and building piles with small objects is more difficult. The robot can grab objects and try to put them on top of other objects or on the table. Its actions are affected by noise so that resulting object piles are not straight-aligned. We assume full observability of triples (s, a, s') that specify how the world changed when an action was executed in a certain state. We represent the data with predicates $block(X)$, $ball(X)$, $table(X)$, $on(X, Y)$, $out(X)$, $inhand(X)$, $upright(X)$, $clear(X) \equiv \forall Y. \neg on(Y, X)$ and functions $size(X), color(X)$ for state descriptions and $puton(X)$, $grab(X)$ and $doNothing()$ for actions. If there are o objects and f different object sizes and colors, the action space contains $2o+1$ actions while the state space is huge with $f^{2o} 2^{o^2+6o}$ different states (not excluding states one would classify as impossible given some intuition about real world physics).

We use NID rules described in Sec. 2.1 to model the state transitions. We employ the rule learning algorithm of Pasula et al. [15] with the same parameter settings to learn three different sets of fully abstract NID rules from independent training sets of 500 experience triples each. Training data to learn rules are generated in a world of ten objects (six blocks, four balls) of two different sizes by performing random actions with a slight bias to build high piles. The resulting rule-sets contain 11, 12 and 12 rules respectively. We use the PRADA algorithm [14] for planning. We test our approach in worlds with varying numbers of blocks and balls of two different sizes. Thus, we transfer the knowledge gained in the training world to different, but similar worlds by using abstract NID rules. In each experiment, for each object number we create five start situations with different objects. Per rule-set and start situation, we perform three independent runs with different random seeds. In all scenarios, we make the assumption that relevant object-sets contain 5 objects. We investigate different estimators to determine these 5 objects. To evaluate each approach, we compute the planning times and the mean performance over the fixed (but randomly generated) set of 45 test scenarios (3 *learned* rule-sets, 5 situations, 3 seeds).

4.1 Building High Piles

In our first experiment, we repeat the experiment of Pasula et al. which investigates building high piles. Our starting situations are chosen such that all objects have height 0 (are on the table) and our reward is the total change in object heights. We let the algorithm run for 10 time-steps. We set PRADA's planning horizon to $d = 6$ and use a discount factor of $\gamma = 0.95$. If the world was deterministic and objects could be stacked perfectly (such that objects could also be stacked on balls), the optimal discounted total reward would be 37.04.

We investigate three different relevance estimators to determine the sets of relevant objects. The *random* estimator samples objects randomly and independently. The *hand-made heuristic* assigns high probability to big blocks (since these are best to build with) and to objects that are either part of a high pile or on the table (in order to build higher piles). Once it has sampled an object, it assigns high probability to objects within the same pile as these might be required for deictic referencing in the NID rules (passive object relevance).

Furthermore, we investigate a simple *learned* estimator of object relevance from which we sample objects independently. We use linear regression to learn from discrete and logical object features, namely object size, type, color, height and clearedness. Training data are generated solely by internal simulation with the PRADA algorithm (in contrast to using the "real" ODE simulator) as follows: for a given situation, we randomly sample 5 objects and derive a plan in the partially grounded network; this plan is then evaluated in the full network and the resulting value is used as relevance estimate for these 5 objects. Note that this procedure does not require real experiences as it is fully based on internal reasoning about which features make an object relevant according to the learned world model (the NID rules in our case). The resulting learned estimator ignores object color as expected, but takes all other features into account, favoring clear big blocks at high heights. We compare these three relevance estimators to the *full-grounding* baseline which plans in the fully grounded model.

Table 2 presents our results. The mean performance of the heuristic is comparable to the full-grounding baseline. The performance of the learned estimator is comparable or only slightly worse than the heuristic, depending on the number of objects and the number N_{rel} of partially grounded models, but always significantly better than the random estimator. The performance of all estimators improves with increasing N_{rel}, but this effect diminishes if N_{rel} is large. Planning in the fully grounded model is hopelessly inefficient, in particular for large worlds. The same performance levels can be achieved by means of Relevance Grounding in only tiny fractions of this planning time, which is independent of the object number and thus constant over all investigated domain sizes.

Table 2. *Building high piles problem*: (a) Mean rewards (changes in tower heights), (b) planning times, (c) details over 45 runs (3 rule-sets, 5 start situations, 3 seeds). Error bars for the rewards give the std. dev. of the mean estimator. N_{rel} denotes the number of relevant reduced models (cf. Algorithm 1). Performing no actions gives a reward of 0.

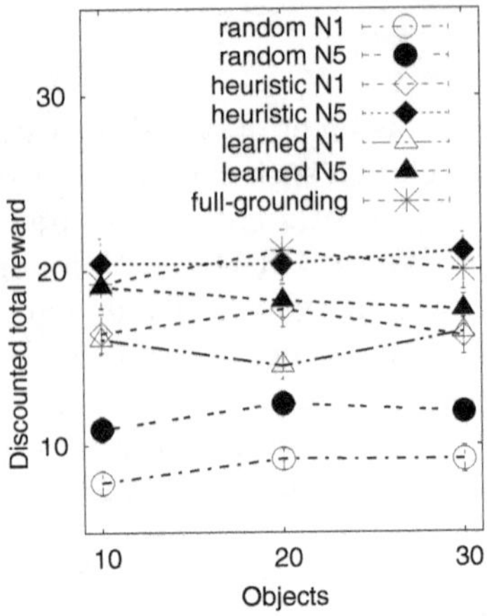

(a)

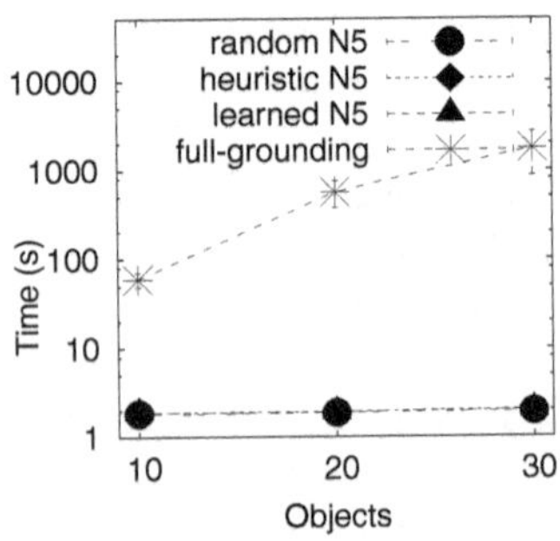

(b)

Obj.	Config	Reward	Time
10	random $N_{rel}=1$	7.85 ± 0.70	**0.37 ± 0.03**
	random $N_{rel}=5$	10.94 ± 0.73	1.84 ± 0.18
	random $N_{rel}=10$	14.59 ± 1.20	3.63 ± 0.26
	heuristic $N_{rel}=1$	16.42 ± 1.10	0.38 ± 0.02
	heuristic $N_{rel}=5$	20.43 ± 1.47	1.89 ± 0.11
	heuristic $N_{rel}=10$	**20.83 ± 1.32**	3.78 ± 0.23
	learned $N_{rel}=1$	16.07 ± 0.85	0.39 ± 0.02
	learned $N_{rel}=5$	19.07 ± 1.23	1.91 ± 0.11
	learned $N_{rel}=10$	18.63 ± 1.12	3.76 ± 0.29
	full-grounding	19.26 ± 1.38	59.51 ± 10.72
20	random $N_{rel}=1$	9.20 ± 0.62	0.39 ± 0.03
	random $N_{rel}=5$	12.38 ± 0.69	1.87 ± 0.15
	random $N_{rel}=10$	14.93 ± 0.74	3.78 ± 0.21
	heuristic $N_{rel}=1$	17.77 ± 0.99	**0.39 ± 0.02**
	heuristic $N_{rel}=5$	20.34 ± 0.91	1.93 ± 0.11
	heuristic $N_{rel}=10$	20.69 ± 1.16	3.85 ± 0.15
	learned $N_{rel}=1$	14.53 ± 0.77	0.42 ± 0.03
	learned $N_{rel}=5$	18.26 ± 0.94	1.90 ± 0.13
	learned $N_{rel}=10$	18.34 ± 0.82	3.81 ± 0.11
	full-grounding	**21.12 ± 1.21**	561.78 ± 186.76
30	random $N_{rel}=1$	9.16 ± 0.76	**0.38 ± 0.03**
	random $N_{rel}=5$	11.90 ± 0.64	1.93 ± 0.12
	random $N_{rel}=10$	14.00 ± 0.69	3.84 ± 0.25
	heuristic $N_{rel}=1$	16.23 ± 1.05	0.39 ± 0.02
	heuristic $N_{rel}=5$	**21.08 ± 1.03**	2.01 ± 0.13
	heuristic $N_{rel}=10$	20.21 ± 1.10	3.84 ± 0.16
	learned $N_{rel}=1$	16.45 ± 0.77	0.42 ± 0.04
	learned $N_{rel}=5$	17.72 ± 0.88	1.99 ± 0.04
	learned $N_{rel}=10$	18.44 ± 0.75	3.78 ± 0.24
	full-grounding	19.99 ± 1.11	1770.55 ± 916.44

(c)

4.2 Desktop Clearance

The goal in our second experiment is to clear up the desktop (see Fig. 1). Objects are lying splattered all over the desktop. An object is cleared if it is part of a pile containing all other objects of the same class, which can be defined as

$$cleared(X) \equiv \forall Y : sameClass(X, Y) \rightarrow samePile(X, Y) \,. \tag{17}$$

A class is defined in terms of color and size, but not type so that a class contains both blocks and balls. In our experiments, classes are made up of 2-4 objects with at most 1 ball (in order to enable successful piling). Our starting situations contain some piles, but only with objects of different classes. We let the algorithm run for 30 time-steps. For planning, we set PRADA's planning horizon

Fig. 1. *Clearance task.* The robot has to clear up the desktop by piling objects of the same size and color.

to $d = 20$ and use a discount factor of $\gamma = 0.95$. If the world was deterministic and objects could be stacked perfectly, the optimal values would be 86.91 for worlds with 10 objects and 110.85 for worlds with 20 and 30 objects. We investigate two relevance estimators. The random estimator samples randomly and independently among all objects. The heuristic estimator chooses randomly among the objects which are not cleared yet and then takes all other objects of the same class into account. Nearest neighbors are used to fill up the object-set. While this heuristic is hand-made, its idea can be derived from the logical reward description in Eq. (17) which states the importance of classes in relevant object-sets on the left side of the implication. How this can be done in principled ways is a major direction of future work.

Table 3 presents our results. The random estimator performs poorly since its reduced models contain mostly only single instances of a class. This is disadvantageous as planning requires at least a second object of the same class and singleton instances are always cleared within reduced models which are thus a bad approximation of the full model. The mean performance of the heuristic estimator is significantly better than the full-grounding baseline, in particular in worlds with many objects. Note that the full-grounding baseline cannot find an optimal solution due to the huge search space. In contrast to planning in the fully ground model, the relevance grounding planning approaches are independent of the number of objects and thus several orders of magnitude faster.

We also investigate the use of verification of the plans found in the reduced models (cf. Algorithm 1). We evaluate the $N_{ver} = 3$ best reduced-model plans in a less reduced model containing 10 objects where the missing 5 slots are filled in by nearest neighbors. Thereby, information about objects within the same piles may be taken into account. Our results show that this improves the mean performance for both relevance estimators significantly at only a small increase in computational cost. In particular, this greatly increases the performance of the random estimator in worlds with 10 objects. In larger worlds, the random estimator almost never finds good plans in which case verification cannot help.

Table 3. *Clearance problem*: (a) Mean rewards, (b) planning times, (c) details over 45 runs (3 rule-sets, 5 start situations, 3 seeds). Error bars for the rewards give the std. dev. of the mean estimator. N_r denotes the number of relevant reduced models N_{rel}, N_v the number of partial plans N_{ver} that are verified in a less reduced model (cf. Algorithm 1). Performing no actions gives a reward of 0.

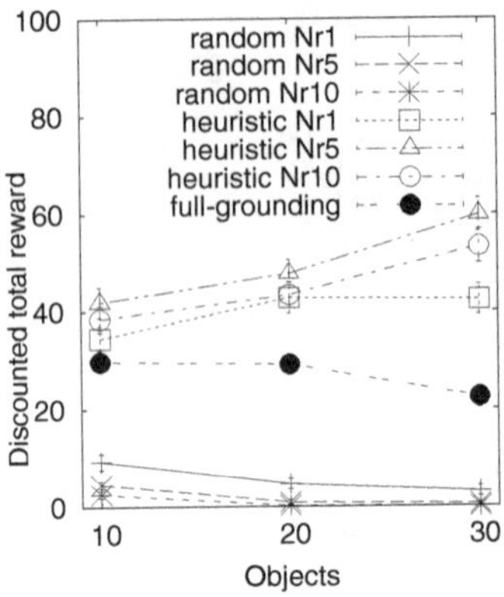

(a)

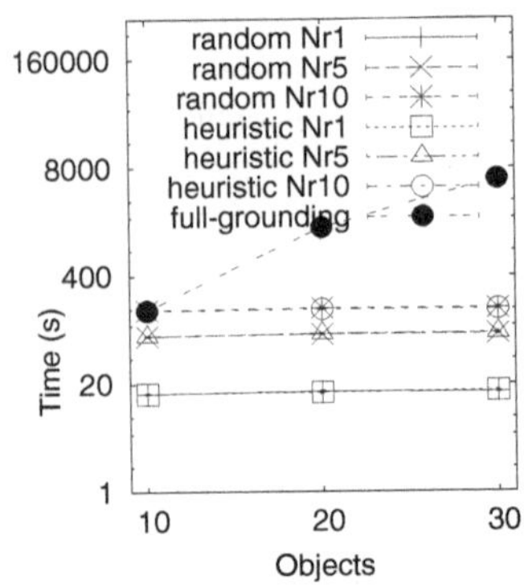

(b)

Obj.	Config	Reward	Time
10	random N_r=1	9.33 ± 1.67	**15.26 ± 0.21**
	rand. N_r=5	4.59 ± 0.78	75.73 ± 1.53
	rand. N_r=10	2.58 ± 0.73	153.86 ± 3.37
	rand. N_r=5, N_v=3	17.54 ± 1.92	84.63 ± 2.42
	rand. N_r=10, N_v=3	14.76 ± 1.85	162.25 ± 7.60
	heuristic N_r=1	34.50 ± 2.44	15.31 ± 0.27
	heur. N_r=5	41.88 ± 3.08	75.02 ± 1.64
	heur. N_r=10	38.48 ± 2.67	151.92 ± 2.94
	heur. N_r=5, N_v=3	46.46 ± 3.12	84.98 ± 1.83
	heur. N_r=10, N_{vr}=3	**49.15 ± 2.81**	161.79 ± 3.63
	full-grounding	29.77 ± 2.02	153.93 ± 13.51
20	random N_r=1	4.68 ± 1.07	16.24 ± 0.44
	rand. N_r=5	0.88 ± 0.45	78.92 ± 1.08
	rand. N_r=10	0.16 ± 0.11	163.21 ± 4.93
	rand. N_r=5, N_v=3	2.62 ± 0.90	90.25 ± 1.74
	rand. N_r=10, N_v=3	0.72 ± 0.58	168.91 ± 3.31
	heuristic N_r=1	42.77 ± 3.19	**15.89 ± 0.48**
	heur. N_r=5	47.72 ± 2.96	80.24 ± 1.38
	heur. N_r=10	43.34 ± 2.60	158.53 ± 2.06
	heur. N_r=5, N_v=3	51.03 ± 3.12	88.94 ± 1.99
	heur. N_r=10, N_v=3	**56.76 ± 2.68**	172.27 ± 5.01
	full-grounding	29.19 ± 1.98	1537.37 ± 225.96
30	random N_r=1	3.09 ± 1.09	**16.06 ± 0.41**
	rand. N_r=5	0.52 ± 0.26	80.11 ± 1.47
	rand. N_r=10	0.02 ± 0.02	162.17 ± 6.34
	rand. N_r=5, N_v=3	1.16 ± 0.51	92.17 ± 2.52
	rand. N_r=10, N_v=3	0.16 ± 0.10	178.96 ± 3.56
	heuristic N_r=1	42.31 ± 3.06	16.47 ± 0.43
	heur. N_r=5	59.79 ± 3.33	81.34 ± 3.87
	heur. N_r=10	53.29 ± 3.50	159.01 ± 3.39
	heur. N_r=5, N_v=3	55.00 ± 3.54	90.26 ± 3.46
	heur. N_r=10, N_v=3	**58.01 ± 3.42**	168.51 ± 5.16
	full-grounding	22.42 ± 2.14	5893.81 ± 1006.56

(c)

5 Related Work

The problem of planning in stochastic relational domains has been approached in quite different ways. The field of Relational Reinforcement Learning (RRL) [19] investigates value functions and Q-functions that are defined over all possible ground states and actions of a relational domain. The idea is to describe important world features in terms of abstract logical formulas enabling generalization over objects and situations. Examples of model-free approaches employ relational regression trees [8] or instance-based regression using distance metrices between relational states such as graph kernels [7] to learn Q-functions. Model-free approaches have the disadvantage to be inflexible as they enable planning only for the specific problem type used in the training examples. In contrast, model-based RRL approaches first learn a relational world model from the state transition experiences, for example in form of relational probability trees for individual state properties [6] or SVMs using graph kernels [12]. One way to make use of the resulting model is to sample look-ahead trees of state transitions in Q-learning, i.e., to work with ground states. All approaches discussed thus far make use of ground states and actions and may well profit from our relevance grounding approach. Consider for example the instance-based approaches where relevance grounding will lead to tractable instance representations.

A promising alternative is to compute abstract value functions by working in the "lifted" abstract representation without grounding or referring to particular problem instances. This requires the learned (or prespecified) model to be complete. Symbolic Dynamic Programming [5] investigates exact solution methods for relational MDPs. The idea is to construct minimal logical partitions of the state space required to make all necessary value function distinctions. For example, Kersting et al. [13] present an exact value iteration for relational MDPs. Sanner et al. [17] exploit factored transition models of first-order MPDs to approximate the value function based on linear combinations of abstract first-order value functions. Their work shows that under certain assumptions (such as additive rewards), it is possible to derive efficient solution techniques. Nonetheless, this promising line of research is only in its beginnings and is confronted with serious technical challenges. It requires complex theorem proving to keep the logical formulas that represent sets of underlying states manageable.

All of the above approaches compute full policies over complete state and action spaces. Instead, one may restrict oneself to deriving plans for a given start state. When grounding the full model, one might in principle use any of the traditional A.I. planning methods used for propositional representations, see [20] and [4]. An interesting strategy to work in a grounded model in a principled way is to consider only a small relevant subset of the state space which is derived from the start state and the planning goal. In contrast to our approach, the resulting subspace still represents all objects, thus the action space size is not decreased. A straight-forward way to create such a subspace are look-ahead trees for the start state that estimate the value of an action by taking samples of the corresponding successor state distribution [15]. Another idea is to maintain an *envelope* of states, a high-utility subset of the state space [9] which can be used to define a relational MDP. This envelope can be further refined by incorporating nearby states in order to improve planning quality. A crucial part of this approach is the initialization of the envelope which is based on an initial straight-line path from the start state to a goal state using a heuristic forward planner (e.g., by making this planning problem deterministic by only considering the most-probable successor state of an action). The envelope-based approach depends strongly on the efficiency and quality of this initial planner which is still faced with the complexity of the action space and its dependence on the number of objects, thus being applicable only for rather small planning horizons.

Action space complexity can be decreased by noting that if the identities of objects do not matter but only their relationships, then different equivalent actions may lead to equivalent successor states [10]. These are states where the same relationships hold, but not necessarily with the same objects. Relevance grounding accounts for this idea by defining different object subsets to be relevant for the planning problem at hand. Action equivalence can be exploited during planning by only considering one sampled action per action equivalence class which significantly reduces the search space. If identity matters for a large number of objects, however, then this approach does not yield significant improvements. Another way to reduce the state space complexity is to look only at a subset of the logical vocabulary, i.e., ignore certain predicates and functions [11]. This helps when combined with the action equivalence approach as state descriptions become shorter and more approximate and the number of state equivalences increases. All these methods just discussed are complementary to our approach and when applied in a reduced grounded model within the relevance grounding framework might yield a strong way to plan efficiently in highly complex domains.

6 Discussion and Conclusions

In this paper, we have presented an approach for efficient planning in stochastic relational worlds based on exploiting object relevance. We define object relevance in terms of a graphical model. We have derived a systematic framework to plan in partially grounded models which we have proven to be consistent with planning in the fully grounded model. Empirical results show our approach to be effective in complex relational environments. Also, we have argued that our approach has interesting analogies to human cognition. Our framework is independent of the concrete planning algorithm used within the reduced models. In particular, it can well be combined with other approaches to increase planning efficiency in stochastic relational domains that have recently been introduced, such as envelope-based methods [9].

A key part for our framework and our major direction of future research is the estimator of object relevance. We have provided a successful example of how relevance can be learned from object features by means of nothing more than internal simulation, but this is clearly only preliminary. In our point of view, learning to estimate the relevance of objects is a formidable problem for machine learning, as a huge variety of methods using discrete, continuous, and logical features can be applied. Clearly, this is a difficult problem, bearing in mind that human beings often take a long time until they master certain types of planning problems. However, estimating object relevance appears to us to be a crucial prerequisite to be able to plan in the highly complex real world.

Acknowledgments

We would like to thank the anonymous reviewers for their helpful comments. This work was supported by the German Research Foundation (DFG), Emmy Noether fellowship TO 409/1-3.

References

1. Anderson, J.: Rules of the mind. Lawrence Erlbaum, Hillsdale (1993)
2. Baddeley, A.: The episodic buffer: a new component of working memory? Trends in Cognitive Sciences 4(11), 417–423 (1999)
3. Botvinick, M.M., An, J.: Goal-directed decision making in prefrontal cortex: a computational framework. In: Advances in Neural Information Processing Systems, NIPS (2009)
4. Boutilier, C., Dean, T., Hanks, S.: Decision-theoretic planning: Structural assumptions and computational leverage. Artificial Intelligence Research 11, 1–94 (1999)
5. Boutilier, C., Reiter, R., Price, B.: Symbolic dynamic programming for first-order MDPs. In: IJCAI, pp. 690–700. Morgan Kaufmann, San Francisco (2001)
6. Croonenborghs, T., Ramon, J., Blockeel, H., Bruynooghe, M.: Online learning and exploiting relational models in reinforcement learning. In: Proc. of the Int. Conf. on Artificial Intelligence, IJCAI (2007)
7. Driessens, K., Ramon, J., Gärtner, T.: Graph kernels and Gaussian processes for relational reinforcement learning. Machine Learning (2006)
8. Dzeroski, S., de Raedt, L., Driessens, K.: Relational reinforcement learning. Machine Learning 43, 7–52 (2001)
9. Gardiol, N.H., Kaelbling, L.P.: Envelope-based planning in relational MDPs. In: Proc. of the Conf. on Neural Information Processing Systems, NIPS (2003)
10. Gardiol, N.H., Kaelbling, L.P.: Action-space partitioning for planning. In: Proc. of the National Conference on Artificial Intelligence (AAAI), Vancouver, Canada (2007)
11. Gardiol, N.H., Kaelbling, L.P.: Adaptive envelope MDPs for relational equivalence-based planning. Technical Report MIT-CSAIL-TR-2008-050, MIT CS & AI Lab, Cambridge, MA (2008)
12. Halbritter, F., Geibel, P.: Learning models of relational MDPs using graph kernels. In: Proc. of the Mexican Conference on Artificial Intelligence (2007)
13. Kersting, K., Otterlo, M.V., de Raedt, L.: Bellman goes relational. In: Proc. of the Int. Conf. on Machine Learning, ICML (2004)
14. Lang, T., Toussaint, M.: Approximate inference for planning in stochastic relational worlds. In: Proc. of the Int. Conf. on Machine Learning, ICML (2009)
15. Pasula, H.M., Zettlemoyer, L.S., Kaelbling, L.P.: Learning symbolic models of stochastic domains. Artificial Intelligence Research 29 (2007)
16. Ruchkin, D.S., Grafman, J., Cameron, K., Berndt, R.S.: Working memory retention systems: a state of activated long-term memory. Behavioral and Brain Sciences 26, 709–777 (2003)
17. Sanner, S., Boutilier, C.: Approximate solution techniques for factored first-order MDPs. In: Proc. of the Int. Conf. on Automated Planning and Scheduling, ICAPS (2007)
18. Toussaint, M., Storkey, A.: Probabilistic inference for solving discrete and continuous state Markov decision processes. In: Proc. of the Int. Conf. on Machine Learning, ICML (2006)
19. van Otterlo, M.: The Logic of Adaptive Behavior. IOS Press, Amsterdam (2009)
20. Weld, D.S.: Recent advances in AI planning. AI Magazine 20(2), 93–123 (1999)

Author Index

Kolibri 360 GmbH
Ferdinand-Jühlke-Straße 7
99095 Erfurt, Deutschland
produktsicherheit@kolibri360.de